Clock Distribution Networks in VLSI Circuits and Systems

IEEE PRESS
445 Hoes Lane, PO Box 1331
Piscataway, NJ 08855-1331

Clock Distribution Networks in VLSI Circuits and Systems

Edited by

Eby G. Friedman
Department of Electrical Engineering
University of Rochester

A Selected Reprint Volume
IEEE Circuits and Systems Society, *Sponsor*

The Institute of Electrical and Electronics Engineers, Inc., New York

This book may be purchased at a discount from the publisher
when ordered in bulk quantities. For more information, contact

IEEE PRESS Marketing
Attn: Special Sales
PO Box 1331
445 Hoes Lane
Piscataway, NJ 08855-1331

Photo credit: Front cover photo courtesy of Digital Semiconductor,
a Digital Equipment Corporation business

Printed in the United States of America
10 9 8 7 6 5 4 3 2 1

ISBN 0-7803-1058-6
IEEE Order Number: PC4127

Library of Congress Cataloging-in-Publication Data

Clock distribution networks in VLSI circuits and systems / edited by
Eby G. Friedman.
p. cm.
"A selected reprint volume, IEEE Circuits and Systems Society,
sponsor."
Includes bibliographical references and index.
ISBN 0-7803-1058-6 (hardcover)
1. Integrated circuits--Very large scale integration--Design and
construction. 2. Timing circuits--Design and construction.
I. Friedman, Eby G.
TK7874.75.C58 1995
621.39'5--dc20 95-14764
CIP

Dedication

To my wife, Laurie Ann, and my children, Joseph Shimon and Samuel Caleb, who provide my life with energy, inspiration, and focus.

Contents

Preface

The purpose of this book is to unify the interdisciplinary concepts of clock distribution networks and to integrate related material from divergent sources. With this book, a detailed reference on the primary issues that relate to the synchronization of high speed digital systems is now available. To the best of my knowledge, this book is the first self-contained volume on clock distribution networks in VLSI circuits and systems.

As semiconductor technologies operate at increasingly higher speeds, system performance has become limited not by the delays of the individual logic elements and interconnect but by the ability to synchronize the flow of the data signals. Different synchronization strategies have been considered, ranging from completely asynchronous to fully synchronous. Throughout industry over the past several decades, however, the dominant synchronization strategy has been, is presently, and will continue for a long time to be that of fully synchronous clocked systems. System applications ranging from medium scale circuits to large megatransistor microprocessors and ultra-high speed supercomputers utilize fully synchronous operations that require high speed and highly reliable clock distribution networks. Distributing the clock signals is one of the primary limitations to building high performance synchronous digital systems. Greater attention is therefore being placed on the design of clock distribution networks for large VLSI systems.

The focus of this book is on the design and analysis of clock distribution networks for application to VLSI-based computing systems. These networks provide the synchronizing signals that define the temporal reference point for the data flow within the entire system. Issues such as clock skew, optimal clock schedules, distributed buffers, interconnect delay, and the automated layout of clock nets are considered in terms of how the maximum performance and reliability of a synchronous system is effected.

The literature on clocking and clock distribution is distributed among a host of journals and conference proceedings, and is discussed among a variety of IEEE societies and technical groups. These disparate publications have been synthesized in this book into a collective whole, permitting the presentation of a unified theory of synchronous clocking and the design and analysis of clock distribution networks. An original tutorial introduction is included along with previously published articles to provide background and perspective to the field. Furthermore, an extensive bibliography on clocking and clock distribution networks follows the introduction. The great variety of journals and conferences in which these papers were originally published is worth noting.

This book is intended to provide a theoretical grounding for building high performance synchronous digital systems. It is hoped that this book will help standardize the terminology related to synchronous clocking and clock distribution networks. In the selection of papers to publish in this book, special effort and care were taken to provide very recent papers as well as classic ones, since it is the editor's opinion that these early papers provide tremendous insight into the evolution and development of existing solutions and approaches to high speed clock distribution. It is also hoped that by considering both early work and recent material, a path for further ingenuity and innovation untampered by acquiescence to existing and accepted practices will be provided.

The area of high speed VLSI-based system design is a growing and active field in which many new papers are published every month. Significant effort was placed in the introduction on dividing the topic of clock distribution networks into individual subtopics. As the topic of clock distribution networks increases in importance, these subtopics will emerge as fields of study in and of themselves. To a significant degree this trend has appeared already, as evidenced by recent conferences in which individual sessions are focused solely on subtopics such as the automated layout of clock nets, optimal clock scheduling, and the timing and clock distribution of structured custom VLSI circuits.

This book contains an introduction followed by a bibliography and 60 reprinted articles grouped by subtopic. Careful attention was placed on relating specific sections within the introduction to particular subtopics of the reprinted articles. Each subtopic is therefore reviewed and summarized in the introduction, providing some insight to the reprinted papers. The twelve papers in Part 1 are more general and tutorial in nature and provide a theoretical background to the basic concepts of clocking and clock distribution networks. In Part 2, various strategies for distributing clock signals in structured custom VLSI circuits are described. Eleven papers are included in this section to provide a broad background to the types of design and test approaches used throughout industry and academe. A burgeoning field of interest, particularly with the importance of the semi-custom application-specific marketplace, is the automated layout and synthesis of clock distribution networks. Fifteen recent papers are presented in Part 3 that discuss this area. The important topic of characterizing and modeling the timing characteristics of clock distribution networks is addressed by seven papers in Part 4. Of growing importance is the scheduling of the clock delays within a synchronous system. Five recent papers describing this new field are included in Part 5. Six papers describing clock distribution networks tailored for specialized VLSI/WSI architectures are presented in Part 6. Finally, four papers presenting example implementations of high speed clock distribution networks are provided in Part 7.

This reprint book has been developed for the benefit of integrated circuit design engineers and graduate students. The book can be used either as a reference or as a supplement to a graduate level course on integrated circuits/VLSI design and analysis. The minimum academic preparation required to fully utilize the material in this book is an undergraduate background in electrical engineering.

I would like to thank my mentor and friend, Dr. J. H. Mulligan, Jr., for providing the spark and direction for the development of this book. I would also like to thank my students, B. Cherkauer, J. L. Neves, V. Adler, and T. Soyata, research collaborator, K. Gaj, and wife, L. Friedman, for their help in reviewing this manuscript. Also, I would like to thank the anonymous reviewers for their very detailed and useful comments. These knowledgeable individuals significantly enhanced the quality of this book. All errors including those of omission are solely my own.

The lack of cohesiveness that has existed within the topic of clock distribution networks has limited its ability to develop into an area of research commensurate with the importance that it has on the entire field of microelectronics. It is hoped that this book will help consolidate this field and provide a base for further advances in system performance.

Eby G. Friedman
Department of Electrical Engineering
University of Rochester

Clock Distribution Networks in VLSI Circuits and Systems

Introduction

Clock Distribution Networks in VLSI Circuits and Systems

EBY G. FRIEDMAN
DEPARTMENT OF ELECTRICAL ENGINEERING
UNIVERSITY OF ROCHESTER

Abstract—**Clock distribution networks synchronize the flow of data signals among data paths. The design of these networks can dramatically affect system wide performance and reliability. In order to better understand how clock distribution networks interact with data paths, a theoretical background of clock skew is provided. Minimum and maximum timing constraints are developed from the relative timing between the localized clock skew and the data paths. These constraint relationships are reviewed and compensating design techniques are discussed.**

Significant interest in clock distribution networks exists within both the industrial and academic communities, and a diverse spectrum of results has been developed. These results, representing the field of clock distribution network design and analysis, can be grouped into a number of subtopics: 1) circuit and layout techniques for structured custom VLSI systems; 2) the automated synthesis of clock distribution networks with application to automated placement and routing of gate arrays, standard cells, and larger block-oriented circuits; 3) the analysis and modeling of the timing and power dissipation characteristics of clock distribution networks; 4) the specification of the optimal timing characteristics of clock distribution networks based on architectural and functional performance requirements; and 5) the design of clock distribution networks for specific architectures, such as systolic arrays and wafer scale integration. Each of these areas is described, the clock distribution networks of specific example circuits are surveyed, and future trends discussed.

1. INTRODUCTION

In a synchronous digital system, the clock signal is used to define a time reference for the movement of data within that system. Since this function is vital to the operation of synchronous systems, much attention has been given to the characteristics of these clock signals and the networks used in their distribution. Clock signals are often regarded as simple control signals; however, these signals have some very special characteristics and attributes. Clock signals are typically loaded with the greatest fanout, travel over the greatest distances, and operate at the highest speeds of any signal (either control or data), within the entire system. Since the data signals are provided with a temporal reference by the clock signals, the clock waveforms must be particularly clean and sharp. Furthermore, these clock signals are particularly affected by technology scaling, in that long global interconnect lines become much more highly resistive as line dimensions are decreased. This increased line resistance is one of the primary reasons for the increasing significance of clock distribution networks on synchronous performance. Finally, the control of any differences in the delay of the clock signals can severely limit the maximum performance of the entire system and create catastrophic race conditions in which an incorrect data signal may latch within a register.

Most synchronous digital systems consist of cascaded banks of sequential registers with combinatorial logic between each set of registers. The functional requirements of the digital system are satisfied by the logic stages. The global performance and local timing requirements are satisfied by the careful insertion of pipeline registers into equally spaced time windows to satisfy critical worst case timing constraints. The proper design of the clock distribution network further ensures that these critical timing requirements are satisfied and that no race conditions exist [B1–B25, 1, 2]. With the careful design of the clock distribution network, system-level synchronous performance can actually increase, surpassing the performance advantages of asynchronous systems by permitting synchronous performance to be based on average path delays rather than worst case path delays, without incurring the handshaking protocol delay penalties required in most asynchronous systems. Thus, the upper limit to system clock frequency is strongly tied to the synchronization and clock distribution strategies used in the design and implementation of the particular system.

In a synchronous system, each data signal is typically stored in a latched state within a bistable register [3] awaiting the incoming clock signal, which determines when the data signal leaves the register. Once the enabling clock signal reaches the register, the data signal leaves the bistable register, propagates through the combinatorial network and, for a properly working system, enters the next register, and is fully latched into that register before the next clock signal appears. Thus, the delay components that make up a general synchronous system are composed of the following three individual subsystems [B128, B129, 4]:

1) the memory storage elements,
2) the logic elements, and
3) the clocking circuitry and distribution network.

Interrelationships among these three subsystems of a synchronous digital system are critical to achieving maximum levels of performance and reliability, and represent the primary and most significant material within this introductory chapter. The important area of clock generation, as compared to clock distribution (which is the primary topic of this book), is only briefly mentioned and bears separate focus.

The introductory chapter is organized as follows. In Section 2, fundamental definitions and the timing characteristics of clock skew are discussed. The timing relationships between a local data path and the clock skew of that path are described in Section 3, which also contains a review of globally asynchronous, locally synchronous systems. The interplay among the aforementioned three subsystems making up a synchronous digital system is described in Section 4, particularly, how the timing characteristics of the memory and logic elements constrain the design and synthesis of clock distribution networks. Different forms of clock distribution networks, such as buffered trees and H-trees, are discussed along with compensation techniques, microwave frequency clock distribution networks, clock distribution networks optimized for low power, and the testing and evaluation of clock distribution networks. The automated layout and synthesis of clock distribution networks are described in Section 5. Various models exist to calculate the timing characteristics of clock distribution networks. Some examples of these models, both deterministic and probabilistic, are presented in Section 6, as well as techniques for designing process insensitive clock distribution networks. The information describing the localized scheduling of the clock delays is useful in optimizing the performance of high speed synchronous circuits. The specification of the optimal timing characteristics of clock distribution networks is reviewed in Section 7. In VLSI systems, certain architectures exist that place unusual constraints on the design of the clock distribution network. The design of clock distribution networks for these specific VLSI-based architectures are discussed in Section 8. The application of clock distribution networks to high speed circuits has existed for many years. The design of the clock distribution networks of certain important VLSI-based systems has been described in the literature, and some examples of these circuits are described in Section 9. In an effort to provide some insight into future and evolving areas of research relevant to high performance clock distribution networks, some potentially important topics for future research are discussed in Section 10. Finally, a summary with some concluding remarks is provided in Section 11. Note that each section is independent and referenced within other sections as appropriate.

2. THEORETICAL BACKGROUND OF CLOCK SKEW

A schematic of a generalized synchronous data path is presented in Figure 1, where C_i and C_f represent the clock signals driving a sequentially-adjacent pair of registers, specifically the initial register R_i and the final register R_f of a data path, respectively. Both clock signals originate from the same clock signal source and a pair of registers are sequentially-adjacent if only combinatorial logic (no sequential elements) exists between the two registers. The propagation delay from the clock source to the jth clocked register is the **clock delay**, T_{Cj}. The clock delays to the initial clock signal T_{Ci} and the final clock signal T_{Cf} define the time reference when the data signals begin to leave their respective registers, R_i and R_f. These clock signals originate from a clock distribution network designed to generate a specific clock signal waveform, used to synchronize each register. This standard clock distribution network structure is based on *equipotential clocking*, where the entire network is considered a surface that must be brought to a specific voltage (clock signal polarity) at each half of the clock cycle. Ideally, clocking events occur at all registers simultaneously. Given this global clocking strategy, the clock signal arrival times (at each register) are defined with respect to a universal time reference.

Definition of Clock Skew

The difference in clock signal arrival times between *two sequentially-adjacent registers*, as shown in (1), is the **clock skew**, T_{Skew}. A sequentially-adjacent pair of registers is a path with only combinational logic and/or interconnect between two registers. If the clock signals C_i and C_f are in complete synchronism (i.e., the clock signals arrive at their respective registers at exactly the same time), the clock skew is zero. A definition of clock skew is provided below:

DEFINITION 1: *Given two sequentially-adjacent registers, R_i and R_j, and an equipotential clock distribution network, the clock skew between these two registers is defined as*

$$T_{Skewi,j} = T_{Ci} - T_{Cj}, \tag{1}$$

where T_{Ci} and T_{Cj} are the clock delays from the clock source to the registers R_i and R_j, respectively.

It is important to observe that the temporal skew between the arrival time of different clock signals is only relevant in sequentially-adjacent registers making up a single data path, as shown in Figure 1. Thus, system-wide (or chip-wide) clock

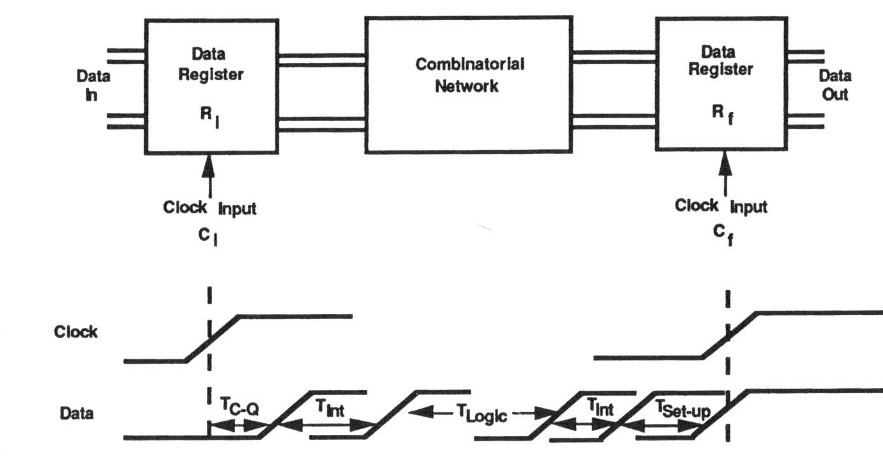

Fig. 1. Timing Diagram of Clocked Data Path.

skew between nonsequentially connected registers, from an analysis viewpoint, has no effect on the performance and reliability of the synchronous system and is essentially meaningless. However, from a design perspective, system-wide global clock skew places constraints on the permissible local clock skew. It should be noted that in [B11, 5], Hatamian designates the lead/lag clock skew polarity (positive/negative clock skew) notation as the opposite of that used here.

Different clock signal paths can have different delays for a variety of reasons. Wann and Franklin [B3] present the following causes of clock skew:

1) differences in line lengths from the clock source to the clocked register;
2) differences in delays of any active buffers (e.g., distributed buffers) within the clock distribution network (due to 3 and 4 below);
3) differences in passive interconnect parameters, such as line resistivity, dielectric constant and thickness, via/contact resistance, line and fringing capacitance, and line dimensions; and
4) differences in active device parameters, such as MOS threshold voltages and channel mobilities, which affect the delay of the active buffers.

It should be noted that for a well designed and balanced clock distribution network, the distributed clock buffers are the principal source of clock skew.

Delay Components of Data Path

The minimum allowable clock period $T_{CP}(min)$ between any two registers in a sequential data path is given by

$$\frac{1}{f_{Clkmax}} = T_{CP}(min) \geq T_{PD} + T_{Skew} , \qquad (2)$$

where

$$T_{PD} = T_{C-Q} + T_{Logic} + T_{Int} + T_{Set\text{-}up} , \qquad (3)$$

and the total path delay of a data path, T_{PD}, is the sum of the time required for the data to leave the initial register once the clock signal C_i arrives, T_{C-Q}, the time necessary to propagate through the logic and interconnect, $T_{Logic} + T_{Int}$, and the time required to successfully propagate to and latch within the final register of the data path, $T_{Set\text{-}up}$. Observe that the latest arrival time is given by $T_{Logic(max)}$ and the earliest arrival time is given by $T_{Logic(min)}$, since data is latched into each register within the same clock period.

The sum of the delay components in (3) must satisfy the timing constraint of (2) in order to attain the clock period $T_{CP}(min)$, which is the inverse of the maximum possible clock frequency, f_{Clkmax}. Note that in (1), the clock skew $T_{Skewi,j}$ can be positive or negative depending on whether C_j leads or lags C_i, respectively. The clock period is chosen such that the latest data signal generated by the initial register is latched by the final register with the next clock edge after the clock edge that activated the initial register. Furthermore, in order to avoid race conditions, the local path delay must be chosen such that for any two sequentially-adjacent registers in a multistage data path, the latest data signal must arrive and be latched within the final register before the earliest data signal generated with the next clock pulse at the output of the initial register arrives. The waveforms in Figure 1 show the timing requirement of (2) being barely satisfied (i.e., the data signal arrives at R_f just before the clock signal arrives at R_f).

To determine the clock delay from the clock source to each register, it is important to investigate the relationships among the clock skews of the sequentially-adjacent registers occurring within a global data path. Furthermore, it is necessary to consider the effects of feedback within global data paths on the clock skew.

The path between two sequentially-adjacent registers is described in this introductory chapter as a local data path, in contrast to a global data path, where a local data path is composed of two sequentially-adjacent registers and a global data path can consist of one or more local data paths. The relationship between the clock skew of sequentially-adjacent registers in a global data path is called **conservation of clock skew** and is formalized below:

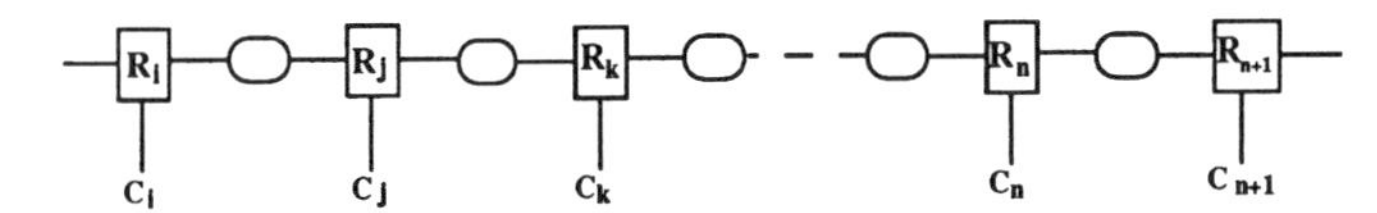

Fig. 2. Global Data Path Composed of Multiple Local Data Paths.

THEOREM 1: *For any given global data path, clock skew is conserved. Alternatively, the clock skew between any two registers in a global data path that are not necessarily sequentially-adjacent is the sum of the clock skews between each pair of registers along the global data path between those same two registers.*

PROOF: For a global data path with only two registers, the clock skew is defined by Definition 1. Now add an extra register to the global data path as illustrated in Figure 2. The clock skew between registers R_i and R_j is $T_{Skewi,j} = T_{CDi} - T_{CDj}$, and the clock skew between registers R_j and R_k is $T_{Skewj,k} = T_{CDj} - T_{CDk}$. Adding $T_{Skewi,j}$ to $T_{Skewj,k}$, $T_{Skewi,j} + T_{Skewj,k} = T_{CDi} - T_{CDk}$ is obtained, which is the clock skew between registers R_i and R_k. By adding n registers to the global data path, the clock skew between the first and the last register is $T_{Skewi,n} = T_{Skewi,j} + \ldots + T_{Skewn-1,n} = T_{CDi} - T_{CDn}$. If one register is added to the global data path, the clock skew between registers R_n and R_{n+1} is $T_{Skewn,n+1} = T_{CDn} - T_{CDn+1}$. Substituting $T_{Skewn,n+1}$ into$T_{Skewi,n}$, $T_{Skewi,n+1} = T_{Skewi,j} + \ldots + T_{Skewn,n+1}$ is obtained. Since n can be any value greater than two, this theorem is proved. □

Although clock skew is defined between two sequentially-adjacent registers, Theorem 1 shows that clock skew can exist between any two registers in a global data path. Therefore, it extends the definition of clock skew introduced by Definition 1 to any two nonsequentially-adjacent registers belonging to the same global data path. It also illustrates that the clock skew between any two nonsequentially-adjacent registers that do not belong to the same global data path has no physical meaning, since no functional data transfer between these registers occurs.

A typical sequential circuit may contain sequential feedback paths, as illustrated in Figure 3. It is possible to establish a relationship between the clock skew in the forward path and the clock skew in the feedback path because the initial and final registers in the feedback path are also registers in the forward path. As shown in Figure 3, the initial and final registers in the feedback path R_l–R_j are the final and initial registers of the forward path R_j–R_k–R_l. This relationship is formalized in the following:

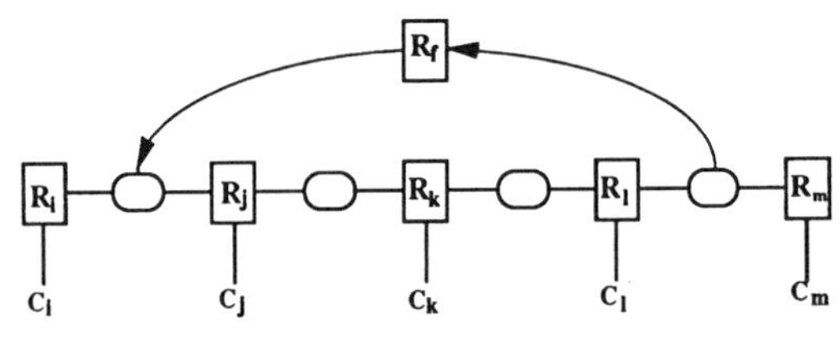

Fig. 3. Data Path with Feedback Paths.

THEOREM 2: *For any given global data path containing feedback paths, the clock skew in a feedback path between any two registers, say R_l and R_j, is related to the clock skew of the forward path by the following relationship,*

$$T_{Skewfeedbackl,j} = -T_{Skewforwardj,l}\,. \tag{4}$$

PROOF: Applying Theorem 1 to the forward path between registers R_l and R_j, $T_{Skewl,j} = T_{CDl} - T_{CDj}$ is obtained. Likewise, applying Theorem 1 to the feedback path between registers R_j and R_l, $T_{Skewj,l} = T_{CDj} - T_{CDl}$ is obtained, which is the negative of $T_{Skewl,j}$, validating the theorem. □

Both Theorems 1 and 2 are useful for determining the optimal clock skew schedule within a synchronous digital system, specifically, the set of local clock skew values that maximizes system performance and reliability. The process for determining these clock skew values is discussed in subsection 7.1.

3. TIMING CONSTRAINTS DUE TO CLOCK SKEW

The magnitude and polarity of the clock skew have a two-sided effect on system performance and reliability. Depending upon whether C_i leads or lags C_f and upon the magnitude of T_{Skew} with respect to T_{PD}, system performance and reliability can either be degraded or enhanced. These cases are discussed below:

3.1 MAXIMUM DATA PATH/CLOCK SKEW CONSTRAINT RELATIONSHIP

For a design to meet its specified timing requirements, the greatest propagation delay of any data path between a pair of data registers, R_i and R_f, being synchronized by a clock distribution network must be less than the minimum clock period (the inverse of the maximum clock frequency) of the circuit as shown in (2) [B7, B8, B11, B16, B28, B126, 4–7]. If the time of arrival of the clock signal at the final register of a data path T_{Cf} *leads* that of the time of arrival of the clock signal at the initial register of the same sequential data path T_{Ci} (see Figure 4A), the clock skew is referred to as **positive clock skew** and, under this condition, the maximum attainable operating frequency is decreased. Positive clock skew is the additional amount of time that must be added to the minimum clock period to reliably apply a new clock signal at the final register. Reliable operation implies that the system will function correctly at both low and high frequencies (assuming fully static logic). It should be noted that positive clock skew only affects the maximum frequency of a system and cannot create race conditions.

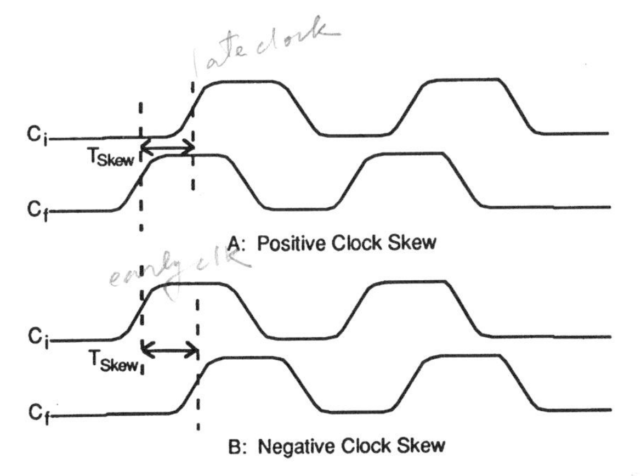

Fig. 4. Clock Timing Diagrams.

In the positive clock skew case, the clock signal arrives at R_f before it reaches R_i. From (2) and (3), the maximum permissible positive clock skew can be expressed as [B7, B8, B11, B16, B28, B126, 4–7]

$$T_{Skewi,f} \leq T_{CP} - T_{PD(max)} = T_{CP} - (T_{C\text{-}Qi} + T_{Logic(max)} + T_{Int} + T_{Set\text{-}upf}) \quad for\ T_{Ci} > T_{Cf}. \tag{5}$$

where $T_{PD(max)}$ is the maximum path delay between two sequentially-adjacent registers. This situation is the typical critical path timing analysis requirement commonly seen in most high performance synchronous digital systems. If (5) is not satisfied, the system will not operate correctly at that specific clock period (or clock frequency). Therefore, T_{CP} must be increased for the circuit to operate correctly, thereby decreasing the system performance. In circuits where the tolerance for positive clock skew is small [T_{Skew} in (5) is small], the clock and data signals should be run in the same direction, thereby forcing C_f to lag C_i and making the clock skew negative.

3.2 Minimum Data Path/Clock Skew Constraint Relationship

If the clock signal arrives at R_i before it reaches R_f (see Figure 4B), the clock skew is defined as being negative. **Negative clock skew** can be used to improve the maximum performance of a synchronous system by decreasing the delay of a critical path; however, a potential minimum constraint can occur, creating a race condition [B11, B12, B31, B118, B119, B126, 4, 5, 7]. In this case, when C_f *lags* C_i, the clock skew must be less than the time required for the data to leave the initial register, propagate through the interconnect and combinatorial logic, and latch in the final register (see Figure 1). If this condition is not met, the data signal stored in register R_f is overwritten by the data signal that had been stored in register R_i and has propagated through the combinatorial logic. Furthermore, a circuit operating close to this condition might pass system diagnostics but malfunction at unpredictable times due to fluctuations in ambient temperature or power supply voltage [B126]. Correct operation requires that R_f latches the data signal corresponding to the data signal R_i latched during the previous clock period. This constraint on clock skew is

$$|T_{Skewi,f}| \leq T_{PD(min)} = T_{C\text{-}Qi} + T_{Logic(min)} + T_{Int} - T_{Holdf} \quad for\ T_{Cf} > T_{Ci}, \tag{6}$$

where $T_{PD(min)}$ is the minimum path delay between two sequentially-adjacent registers and T_{Hold} is the amount of time the input data signal must be stable once the clock signal changes state.

An important example in which this minimum constraint can occur is in those designs using cascaded registers, such as a serial shift register or a k-bit counter, as shown in Figure 5. Note that a distributed RC impedance is between C_i and C_f. In cascaded register circuits, $T_{Logic(min)}$ is zero and T_{Int} approaches zero (since cascaded registers are typically designed, at the geometric level, to abut). If $T_{Cf} < T_{Ci}$ (i.e., negative clock skew), then the minimum constraint becomes

$$|T_{Skewi,f}| \leq T_{C\text{-}Qi} - T_{Holdf} \quad for\ T_{Cf} > T_{Ci}, \tag{7}$$

and all that is necessary for the system to malfunction is a poor relative placement of the flip flops creating a highly resistive connection between C_i and C_f. In a circuit configuration such as a shift register or counter, where negative clock skew is a more serious problem than positive clock skew, provisions should be made to force C_f to lead C_i, as shown in Figure 5.

As higher levels of integration are achieved in high complexity VLSI circuits, on-chip testability [8] becomes necessary. Data registers, configured in the form of serial set/scan chains when operating in the test mode, are a common example of a design for testability (DFT) technique. The placement of these circuits is typically optimized around the functional flow of the data. When the system is reconfigured to use the registers in the role of the set/scan function, different local data path delays are possible. In particular, the clock skew of the reconfigured local data path can be negative and greater in magnitude than the local register delays. Therefore, with increased negative clock skew, (7) may no longer be satisfied and incorrect data may latch into the final register of the reconfigured local data path. Therefore, it is imperative that attention be placed on the clock distribution of those paths that have nonstandard modes of operation.

In ideal scaling of MOS devices, all linear dimensions and voltages are multiplied by the factor $1/S$, where $S > 1$ [B16,

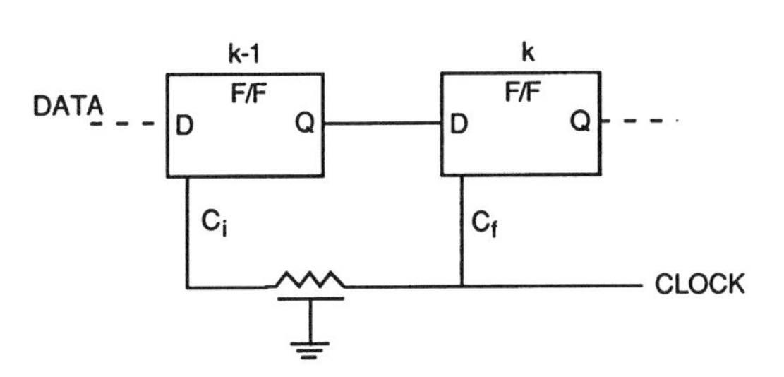

Fig. 5. K-Bit Shift Register with Positive Clock Skew.

9–11]. Device dependent delays, such as T_{C-Q}, $T_{Set\text{-}up}$, and T_{Logic}, scale as $1/S$ while interconnect dominated delays such as T_{Skew} remain constant to first order, and if fringing capacitance and electromigration are considered, actually increase with decreasing dimensions. Therefore, when examining the effects of dimensional scaling on system reliability, (6) and (7) should be considered carefully [12]. One straightforward method to avoid the effect of technology scaling on those data paths particularly susceptible to negative clock skew is to not scale the clock distribution lines. Svensson and Afghahi [B116] show that, by using coarser than ordinary lines for global clock distribution, 20-millimeter-wide chip sizes with CMOS circuits scaled to 0.3-μm polysilicon lines would have comparable logic and cross-chip interconnect delays (on the order of 0.5 ns), making possible synchronous clock frequencies of up to 1 GHz. Therefore, the scaling of device technologies can severely affect the design and operation of clock distribution networks, necessitating specialized strategies and compensation techniques.

3.3 Enhancing Synchronous Performance by Applying Localized Clock Skew

Localized clock skew can be used to improve synchronous performance by providing more time for the critical worst case data paths [B122, B126, B128, B129, 4, 7]. By forcing C_i to *lead* C_f at each critical local data path, excess time is shifted from the neighboring less critical local data paths to the critical local data paths. This negative clock skew represents the additional amount of time that the data signal at R_i has to propagate through the logic stages and interconnect sections, and into the final register. Negative clock skew subtracts from the logic path delay, thereby decreasing the minimum clock period. Thus, applying negative clock skew, in effect, increases the total time that a given critical data path has to accomplish its functional requirements by giving the data signal released from R_i more time to propagate through the logic and interconnect stages and latch into R_f. Thus, the differences in delay between each local data path are minimized, thereby compensating for any inefficient partitioning of the global data path into local data paths that may have occurred, a common situation in many practical systems. Different terms have been used in the literature to describe negative clock skew, such as "double-clocking" [B126], "deskewing data pulses" [B31], "cycle stealing" [B121, B122], "useful clock skew" [11], and "prescribed skew" [B72].

The maximum permissible negative clock skew of a data path, however, is dependent upon the clock period itself as well as the time delay of the previous data paths. This dependence results from the temporal nature of the serially cascaded local data paths making up the global data path. Since a particular clock signal synchronizes a register that functions in a dual role, as the initial register of the next local data path and as the final register of the previous data path, the earlier C_i is for a given data path, the earlier that same clock signal (now C_f) is for the previous data path. Thus, the use of negative clock skew in the ith path results in a more positive clock skew for the preceding path, which may then establish the new upper limit for the system clock frequency.

An Example of Applying Localized Negative Clock Skew to Synchronous Circuits

Consider the nonrecursive synchronous circuit shown in Figure 6, where the horizontal ovals represent logic elements with logic delays and the vertical ovals represent clock delays. Since the local data path from R_2 to R_3 represents the worst case path (assuming the register delays are equal), by delaying C_3 with respect to C_2, negative clock skew is added to the R_2–R_3 local data path. If C_1 is synchronized with C_3, then the R_1–R_2 local data path receives some positive clock skew. Thus, assuming the register delays are both 2 ns, C_2 should be designed to lead C_3 by 1.5 ns, forcing both paths to have the same total local data path delay, $T_{PD} + T_{Skew} = 7.5$ ns. The delay of the critical path of the synchronous circuit is temporally refined to the precision of the clock distribution network, and the entire system (for this simple example) could operate at a clock frequency of 133.3 MHz, rather than 111.1 MHz if no localized clock skew is applied. The performance characteristics of the system, both with and without the application of localized clock skew, are summarized in Table 1.

Note that $|T_{Skew}| < T_{PD}$ ($|-1.5 \text{ ns}| < 9$ ns) for the R_2–R_3 local data path; ensuring that the correct data signal is successfully latched into R_3 and no minimum data path/clock skew constraint relationship exists. This design technique of applying localized clock skew is particularly effective in sequentially-adjacent, temporally irregular local data paths; however, it is applicable to any type of synchronous sequential system. For certain architectures, a significant improvement in performance is both possible and likely.

The limiting condition for applying localized negative clock skew is determined by the control of the clock skew variations

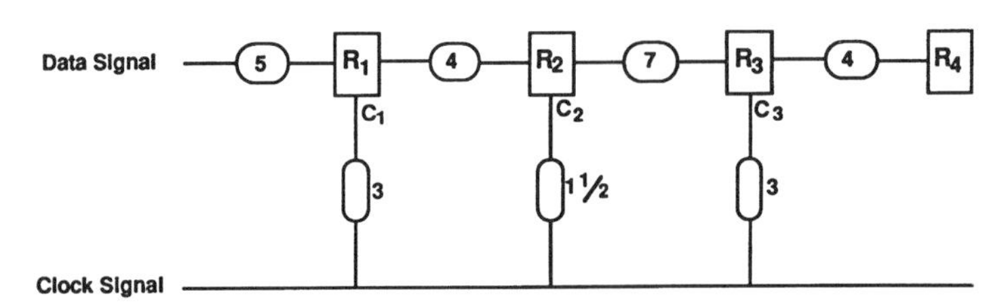

Fig. 6. Example of Applying Localized Negative Clock Skew to Synchronous Circuit.

TABLE 1. PERFORMANCE CHARACTERISTICS OF CIRCUIT OF FIGURE 6 WITHOUT AND WITH LOCALIZED CLOCK SKEW.

Local Data Path	$T_{PD(min)}$ − zero skew	T_{Ci}	T_{Cf}	T_{Skew}	$T_{PD(min)}$ − with non-zero skew
R_1 to R_2	$4 + 2 + 0 = 6$	3	1.5	1.5	$4 + 2 + 1.5 = 7.5$
R_2 to R_3	$7 + 2 + 0 = 9$	1.5	3	−1.5	$7 + 2 - 1.5 = 7.5$
f_{Max}	111.1 MHz				133.3 MHz

(All time units are in nanoseconds.)

and by the differences in path delays among neighboring local data paths. These clock skew variations are due to power supply variations, to process tolerances (where process parameters may vary over a specified range), and to environmental effects, such as temperature or radiation, which, for example, can affect both MOS threshold voltages and channel mobilities.

3.4 Globally Asynchronous and Locally Synchronous Systems

In order to minimize the difficulty in distributing global clock signals across large systems, specifically, the creation of large amounts of localized clock skew, asynchronous approaches have been developed to constrain intermodule timing [B16]. These different asynchronous approaches may be divided into two general classes: fully asynchronous systems (including self-timed systems requiring handshaking protocols and wavefront processors not requiring specialized intermodule communication [13]) and hybrid globally asynchronous, locally synchronous systems [14].

The concept of globally synchronous, locally asynchronous timing does not exist, since it is impossible for a globally synchronous system to synchronously control the timing of a locally asynchronous system. Moving data with no information describing the local temporal data flow to another module requiring some information about the local temporal data flow is not possible without providing some form of temporal reference point. If a temporal reference point is provided, the initial module is no longer fully asynchronous and contains some level of synchronous information. Therefore, a globally synchronous, locally asynchronous system can not effectively exist.

However, a *globally asynchronous, locally synchronous* system is possible, since a system with some temporal information (the locally synchronous system) can interact globally at the asynchronous level since no additional timing information is necessary. Thus, this approach is both possible and practical and is briefly reviewed here. The topic of fully asynchronous systems represents a field unto itself and therefore is not discussed in any great detail here. However, a useful and highly interesting comparison between the performance characteristics of synchronous and asynchronous systems by Afghahi and Svensson [B16] is included in this book.

In a globally asynchronous, locally synchronous system, individual modules operate completely synchronously, using self-contained local clocks within the module and distributing these clock signals via clock distribution networks. These synchronous modules, for example, can occur at the IC level or at the board level, and communicate asynchronously at the system level. The global asynchronous signals are converted to local fully synchronous signals by using local arbiters or synchronizers [15, 16]. Arbiters accept asynchronous signals and synchronize them to a temporal reference, while minimizing the probability of a synchronous register becoming metastable [3]. **Metastability** is the state of a register in which neither binary state (a "0" or a "1") appears at the output within a time period consistent with the normal operation of the register. The register will remain in this tenuous state of equilibrium until some circuit parameter varies sufficiently so as to drive the state of the register into one of the two binary states. This state of metastability is categorized as a type of reliability problem. The register could remain in this state for milliseconds, thereby forcing the overall system to function incorrectly. Therefore, it is important to ensure that the local synchronous clock signals are correctly received at the global asynchronous level.

An approach to minimizing the occurrence of metastability is to increase the clock period so as to allow sufficient time for the synchronous register to return to its proper state. This method, however, assumes that the register will return to its correct digital state (instead of the reverse polarity of the proper state) and will return in a sufficiently short amount of time. This concept can be applied to pausible or extendible clock schemes [B16], where the local synchronous clock is stopped until the output state of the register is properly defined, whereupon a new clock signal is generated once a proper register output is achieved. This strategy requires that the arbiters are monitored to guarantee that the register output is both stable and logically well defined. Since there is no absolute bound on the register resolution time, the clock period can become unacceptably long. If a specified maximum delay is exceeded, the system fails. Therefore, the primary performance penalty of globally asynchronous, locally synchronous systems is the delay due to signal arbitration and any required metastable resolving time. Note how this disadvantage compares with fully synchronous systems, in which the primary performance penalty is due to positive clock skew.

An important attribute of globally asynchronous, locally synchronous timing is that different modules can operate at different locally optimal clock frequencies. The system communication is performed using some form of handshaking protocol among the independent synchronous modules. The delay of the handshaking protocol can significantly affect the overall data rate of the system. In general, it has been shown that local data paths perform faster when operated fully synchronously, while larger modules tend to perform faster when operated asynchronously [B16], where the speed of an asynchronous system is approximately the average speed of the modules plus some handshaking delay.

4. CLOCK DISTRIBUTION DESIGN OF STRUCTURED CUSTOM VLSI CIRCUITS

Many different approaches, from ad hoc to algorithmic, have been developed for designing clock distribution networks in VLSI circuits. The requirement of distributing a tightly controlled clock signal to each synchronous register on a large nonredundant hierarchically structured VLSI circuit (an example floorplan is shown in Figure 7) within specific temporal bounds is difficult and problematic. Furthermore, the tradeoffs that exist among system speed, physical die area, and power dissipation are greatly affected by the clock distribution network. The design methodology and structural topology of the clock distribution network should be considered in the development of a system for distributing the clock signals. Therefore, various clock distribution strategies have been developed. The most common and general approach to equipotential clock distribution is the use of buffered trees, discussed in subsection 4.1. In contrast to these asymmetric structures, symmetric trees such as H-trees are used to distribute high speed clock signals. This topic is described in subsection 4.2. In developing structured custom VLSI circuits, such as exemplified by the floorplan pictured in Figure 7, specific circuit design techniques are used to control the delays within the clock distribution network. One important compensation technique is described in subsection 4.3. Furthermore, additional design issues exist if the VLSI circuit operates at extremely high frequencies. Some discussion on distributing microwave clock frequencies is provided in subsection 4.4. Low power design is an area of significant currency and importance. Some very recent efforts to reduce the power dissipated within the clock distribution network are reviewed in subsection 4.5. Finally, only minimal material has been published on the testing and evaluation of high performance clock distribution networks. This topic is briefly discussed in subsection 4.6.

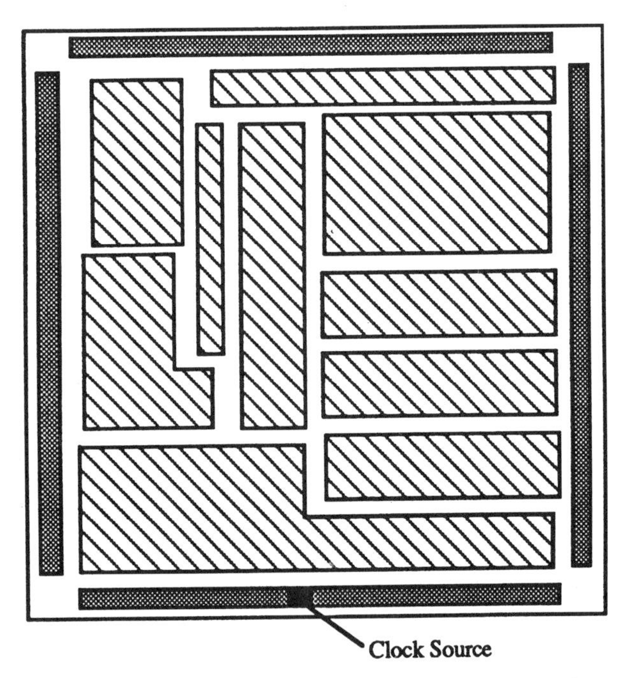

Fig. 7. Floorplan of Structured Custom VLSI Circuit Requiring Synchronous Clock Distribution.

4.1 Buffered Clock Distribution Trees

The most common strategy for distributing clock signals in VLSI-based systems is to insert buffers at the clock source and/or along a clock path, forming a tree structure. Thus, the clock source can be described as the root of the tree, the initial portion of the tree as the trunk, individual paths driving each register as the branches, and the driven registers as the leaves. This metaphor for describing a clock distribution network is commonly accepted and used throughout the literature, and is illustrated in Figure 8. Occasionally a mesh version of the clock tree structure is used, in which shunt paths farther down the clock distribution network are placed to minimize the interconnect resistance within the clock tree. This mesh structure effectively places the branch resistances in parallel, minimizing both the clock delay and the clock skew. An example of this mesh structure is described and illustrated in subsection 9.2. The mesh version of the clock tree is considered in this introductory chapter as an extension of the standard, more commonly used, clock tree depicted in Figure 8.

If the interconnect resistance of the buffer at the clock source is small compared to the buffer output resistance, often a single buffer is used to drive the entire clock distribution network. This strategy may be appropriate if the clock is distributed entirely on metal, making load balancing of the network less critical. The primary requirement of a single buffer system is that the buffer provide enough current to drive the network capacitance (both interconnect and fanout) while maintaining high quality waveform shapes (i.e., short transition times) and minimizing the effects of the interconnect resistance. The output resistance of the buffer must be designed to

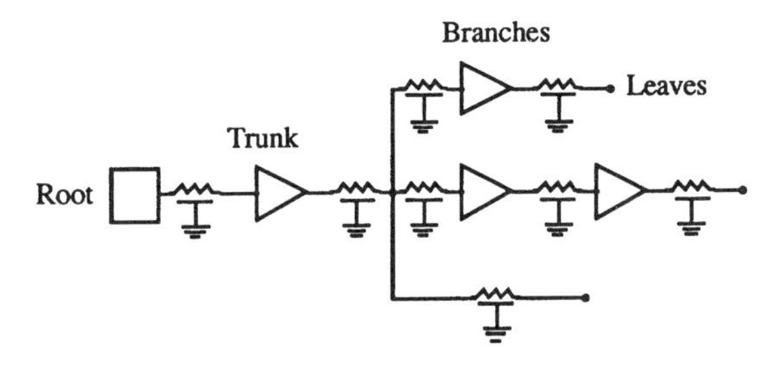

Fig. 8. Tree Structure of Clock Distribution Network.

be much greater than the resistance of the interconnect section being driven.

An alternative approach to using only a single buffer at the clock source is to distribute buffers throughout the clock distribution network, as shown in Figure 8. This approach requires additional area but greatly improves the precision and control of the clock signal waveforms and is necessary if the resistance of the interconnect lines is not negligible. The distributed buffers serve the double function of amplifying the clock signals degraded by the distributed interconnect impedances and isolating the local clock nets from upstream load impedances [B28]. A three-level buffer clock distribution network utilizing this strategy is shown in Figure 9. In this approach a single buffer drives multiple clock paths (and buffers). The number of buffer stages between the clock source and each clocked register depends on the total load capacitance, in the form of registers and interconnect, and the permissible clock skew [B39]. It is worth noting that the buffers are a primary source of the total clock skew within a well-balanced clock distribution network since the active device characteristics vary much more greatly than the passive device characteristics. The maximum number of buffers driven by a single buffer is determined by the current drive of the source buffer and the capacitive load (assuming an MOS technology) of the destination buffers. The final buffer along each clock path provides the control signal driving the register.

Historically, the primary goal in designing clock distribution networks has been to ensure that a clock signal arrives at every register within the entire synchronous system at precisely the same time. This concept of zero clock skew design has been extended, as is explained in subsection 3.3, to provide either a positive or a negative clock skew at a magnitude depending upon the temporal characteristics of each local data path to improve system performance and enhance system reliability.

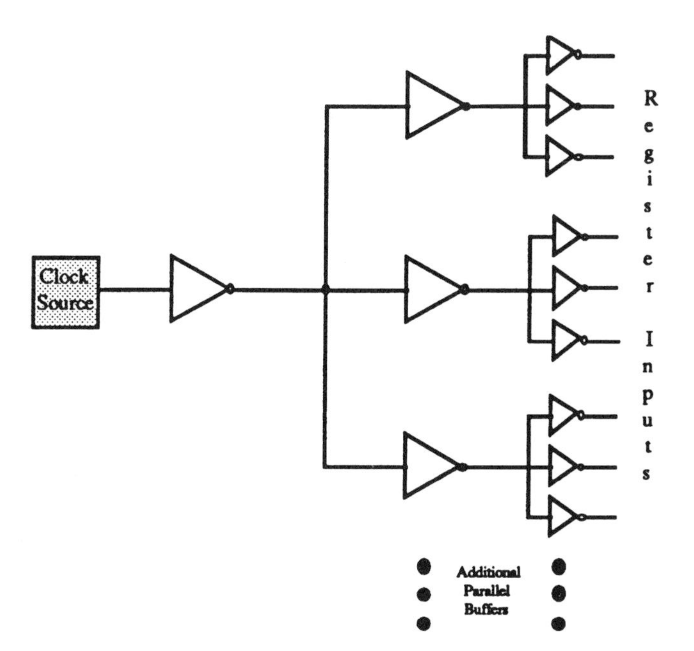

Fig. 9. Three-Level Buffer Clock Distribution Network.

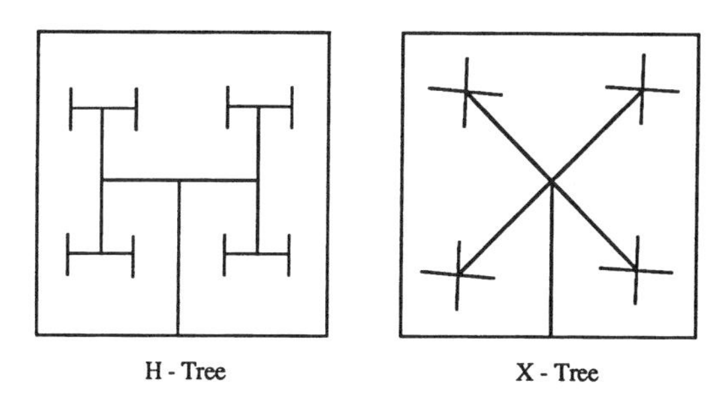

Fig. 10. Symmetric H-Tree and X-Tree Clock Distribution Networks.

4.2 Symmetric H-Tree Clock Distribution Networks

Another approach for distributing clock signals, a subset of the distributed buffer approach depicted in Figure 8, utilizes a hierarchy of planar symmetric H-tree or X-tree structures (see Figure 10) [B30, B44, 11]. Zero clock skew is ensured by maintaining identical paths from the clock signal source to the clocked register of each clock path. In the H-tree network, the primary clock driver is connected to the center of the main "H" structure. The clock signal is transmitted to the four corners of the main "H." These four close-to-identical clock signals provide the inputs to the next level of the H-tree hierarchy, represented by the four smaller "H" structures. The distribution process then continues through several levels of progressively smaller "H" structures. The final destination points of the H-tree are used to drive the local registers or are amplified by local buffers that drive the local registers. Thus, each clock path from the clock source to a clocked register has practically the same delay. The primary delay difference among the clock signal paths is due to variations in process parameters that affect the interconnect impedance and, in particular, any active distributed buffer amplifiers. As described in Section 8, the amount of clock skew within an H-tree structured clock distribution network is strongly dependent upon the physical size, the control of the semiconductor process, and the degree to which active buffers and clocked latches are distributed within the H-tree structure.

The conductor widths in H-tree structures are designed to decrease progressively as the signal propagates to lower levels of the hierarchy. This strategy minimizes reflections of the high speed clock signals at the branching points. Specifically, the impedance of the conductor leaving each branch point Z_{K+1} must be twice the impedance of the conductor providing the signal to the branch point Z_K for an H-tree structure [B30, B44, B142, 11] and four times the impedance for an X-tree structure. This tapered H-tree structure is illustrated in Figure 11.

$$Z_k = \frac{Z_{k+1}}{2} \quad \textit{for an H-tree structure} \qquad (8)$$

The planar H-tree structure places constraints on the physical layout of the clock distribution network as well as on the design methodology used in the development of the VLSI system. For example, in an H-tree network, clock lines must be routed in

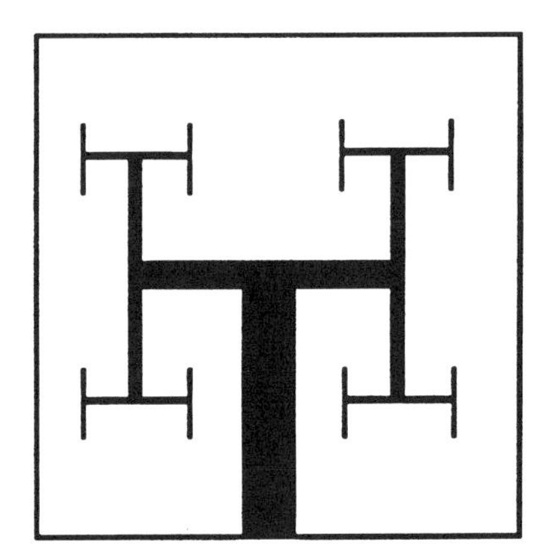

Fig. 11. Tapered H-Tree Clock Distribution Network.

both the vertical and horizontal directions. For a standard two-level metal CMOS process, this requirement creates added difficulty in routing the clock lines without using either resistive interconnect or multiple high resistance vias between the two metal lines. This difficulty is a primary reason for the development of three or more layers of metal in logic-based CMOS processes. Furthermore, the interconnect capacitance (and therefore the power dissipation) is much greater for the H-tree as compared with the standard clock tree, since the total wire length tends to be much greater [B170]. This increased capacitance of the H-tree structure exemplifies an important tradeoff between clock delay and clock skew in the design of high speed clock distribution networks. Symmetric structures are used to minimize clock skew; however, an increase in clock signal delay is incurred. Therefore, the increased clock delay must be considered when choosing between buffered tree and H-tree clock distribution networks. Also, since clock skew only affects sequentially-adjacent registers, the obvious advantages to using highly symmetric structures to distribute clock signals are significantly degraded. There may be, however, certain sequentially-adjacent registers distributed across the integrated circuit. For this situation, a symmetric H-tree structure may be appropriate.

Another consideration in choosing a clock distribution topology is that the H-tree and X-tree clock distribution networks are difficult to implement in those VLSI-based systems that are irregular in nature, such as pictured in Figures 7 and 12. In these types of systems, buffered tree topologies integrated with structured custom design methodologies [17] should be used in the design of the clock distribution networks in order to maximize system clock frequency, minimize clock delay, and control any deleterious effects of local (particularly negative) clock skew.

4.3 Compensation Techniques for Controlling Clock Skew

One structured custom approach, oriented to hierarchical VLSI-based circuits, utilizes compensation techniques to minimize the variation of interconnect impedances and capacitive loads between clock signal paths [B28, B31, B34, B36, 18, 19]. A general schematic of a clock distribution network is shown in Figure 13, in which the nodes i, j, and k represent different clock signal destinations (i.e., clocked registers). Different clock paths could conceivably have different levels of buffering, where each buffer drives a localized distributed *RC* impedance. The location of these buffers is often chosen so that the active buffer output impedance is comparable to or greater than the interconnect resistance seen at the buffer output. The buffer location ensures that the locally distributed *RC* interconnect section can be modeled accurately as being primarily capacitive. The use of distributed buffers in this manner is described as buffer repeaters [11, 20]. However, in general the interconnect impedance should be modeled as a distributed resistive-capacitive section of interconnect.

The difficulty with applying symmetric clock distribution strategies is that they do not easily support the ability to partition large VLSI systems into hierarchically structured functional blocks. Preferably, each large functional block would contain its own locally optimized clock distribution network to satisfy the local timing and loading of that particular functional block. For a globally synchronous system, however, local optimization within a functional element does not necessarily lead to global optimization of the overall on-chip clock distribution system.

If the interconnect resistance of the global clock distribution network is relatively small, a chip-level centralized clock buffer circuit can be used to satisfy the synchronization requirements of a VLSI circuit. However, in most large VLSI circuits the physical distances are such that line resistances, coupled with any via/contact resistances and the typically significant line and coupling capacitances, will create large interconnect impedances. Therefore, even with a centrally located clock generation and distribution circuit, additional techniques are

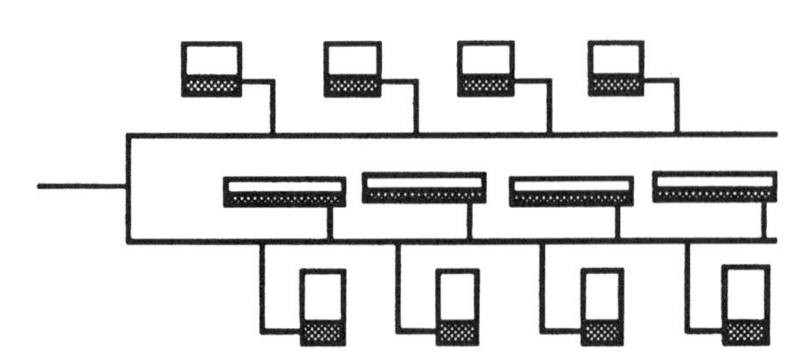

Fig. 12. Clock Distribution Network for Structured Custom VLSI Circuit.

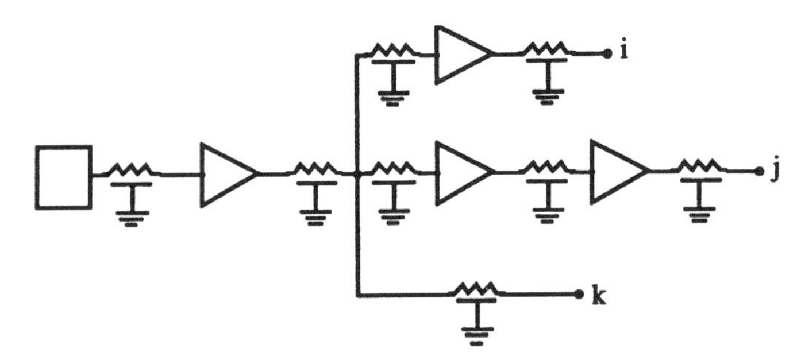

Fig. 13. Clock Distribution Network.

required to compensate for variations in interconnect and register loading.

In order to control the delay of each clock signal path and to minimize the skew between these paths, passive *RC* delay elements [B31] or geometrically sized transistor widths [B28] are used to compensate for the variation of the delay of each clock signal path. These delay variations are caused by different on-chip locations (i.e., different path-dependent interconnect impedances) and capacitive loading of the clock destinations (i.e., the number and load of the clocked registers per clock signal path). Clock buffers are placed along the clock path such that the highly resistive interconnect lines (typically long lines) drive loads with low capacitance, while the low resistance interconnect lines (typically short lines) drive loads with high capacitance. Thus, either a centralized module of clock buffer drivers can be used or those clock buffers driving large capacitive loads can be placed close to the registers, thereby decreasing the interconnect resistance. This design strategy of using compensation techniques to control the local clock skew is depicted graphically in Figure 14. The variation of clock delay between each of the functional ele-ments is compensated for by parameterizing the current drive of each of the functional block clock buffers resident in the centrally located clock buffering circuit (see Figure 14). If feedback circuitry is being used to further control the delays and skews within the clock distribution network, as in on-chip phase lock loops (PLLs), taps are placed close to the register, which are fed back to maintain lock.

In order to ensure that the clock distribution network is successfully designed, the following practices should be followed: 1) the number of stages of clock buffering within each of the functional blocks should be the same to maintain equal polarity, 2) the maximum clock signal rise and fall times within each functional block should be specified and controlled, and 3) the internal functional block clock skew should be specified and controlled using the same hierarchical clock distribution strategy as is used at the global VLSI system level [B28].

Advantages and Disadvantages of Compensation Technique

The primary advantage of using a compensation technique is controlling (and reducing) on-chip clock skew. Also, the clock delay from the clock source to the clocked registers is reduced, due to improved partitioning of the *RC* loads. Since the inverters, located within each functional block, drive large capacitive loads, the interconnect impedance, and in particular the interconnect resistance driven by any specific clock buffer, is small in comparison to the buffer output impedance. The fairly long distances of the intra-block clock signal paths are fairly resistive. These paths, however, are isolated from the highly capacitive loads. Thus, the *RC* time constants are reduced, reducing the overall clock delay. Another important advantage of this

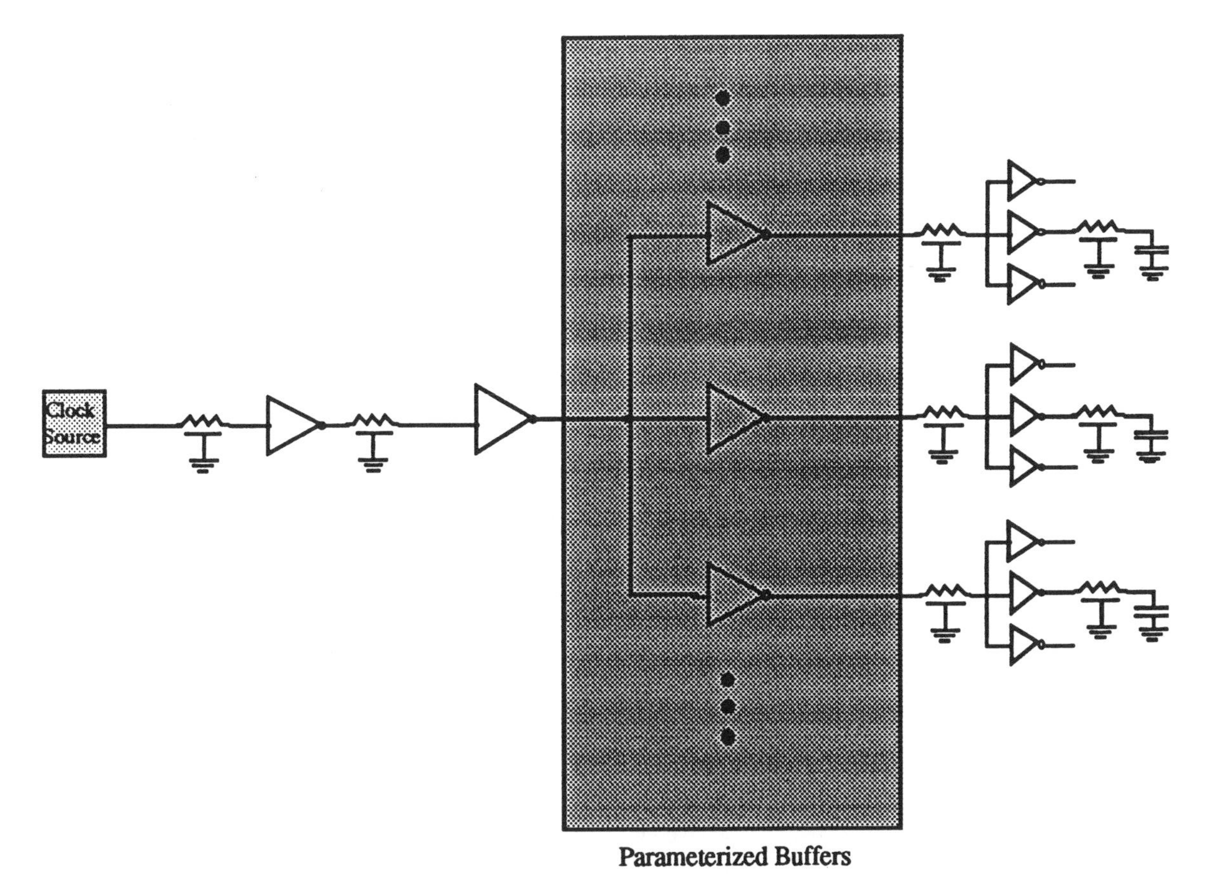

Fig. 14. Parameterized Buffers Integrated into a Clock Distribution Network to Control Local Clock Skew [B28].

design technique is the ease of partitioning the clock distribution problems among a team of VLSI circuit designers. The overall VLSI system design can be partitioned hierarchically into a manageable domain of information while still providing a strategy for implementing optimal clock distribution networks. The usefulness of this compensation technique is dependent upon the ability to characterize the device and interconnect impedances within the VLSI circuit. With an accurate estimate of these impedances, the parameterized buffers can be designed to satisfy a specific clock skew schedule.

It is important to note a disadvantage of this compensation technique. Unlike interconnect impedances, transistor conductances tend to be very sensitive to variations in supply voltage, and to process and environmental conditions (e.g., temperature, radiation). A clock signal path whose delay is dominated by interconnect impedances may vary differently from a clock signal path whose delay is dominated by device impedances [B28].

Several specific examples of clock distribution networks are discussed in the literature [e.g., B1, B11, B28, B35, B36, B39, B161]. Some examples of clock distribution networks applied to high speed circuits are described in Section 9. In each of these clock distribution networks, significant effort has been placed on accurately estimating the magnitude of the resistive and capacitive interconnect impedances to determine the effect of these *RC* loads on the shape of the clock signal waveform. This information is typically back-annotated into a SPICE-like circuit simulator to adjust the clock delays for minimum clock skew [21]. Minimal work exists, however, in developing circuit procedures and algorithms for automating the circuit design of clock distribution networks in structured custom VLSI circuits [B95–B98, B174]. One primary requirement for developing these algorithms is a physical model for estimating the delay of the clock signal path. A distributed *RC* interconnect delay model must be integrated with a model of the delay of a distributed buffer in order to estimate the local clock skews. This topic is further discussed in Section 6. An important observation is that the accuracy required to calculate delay differences (as in clock skew) is much greater than that required when calculating absolute delay values (as in the delay of a clock path).

4.4 Clock Distribution Techniques for Microwave Clock Frequencies

Those systems with signals propagating at very high speeds over long lengths may induce reflections into the propagating signal that could seriously degrade the signal waveform shape. Therefore, the design of clock distribution networks where clock signals are distributed over a large area, as is common in VLSI systems, may become increasingly more difficult as clock frequencies approach or exceed microwave frequencies ($>10^9$ Hz or 1 GHz). A simple convention for estimating whether a transmission line model is necessary to analyze the waveform characteristics rather than a distributed *RC* impedance is to compare the length of the line over which the signal is propagating $l_{Clockpath}$ to the speed of light in the dielectric material c_{Di} (where $c = 3 \times 10^8$ meters per second in a vacuum) multiplied by the waveform transition time τ_t, as shown below [11].

$$\tau_t c_{Di} = l_{Clockpath} \tag{9}$$

If $l_{Clockpath}$ is greater than $\tau_t c_{Di}$, a transmission model for the clock signal path becomes necessary; otherwise the distributed *RC* impedance model is satisfactory.

A transmission line model to analyze clock distribution networks is described in [B46] and is based on a finite Fourier Series expansion of an approximate square wave clock signal assuming finite rise and fall times. Complex transmission and reflection coefficients are determined for each harmonic at each termination (clocked register) and branch point within the clock distribution network. Assuming TEM mode propagation, accurate results up to 2 GHz can be achieved. If corrections are made to the dielectric constant, higher frequency results could accurately be attained. Load and source impedances lead to reflection coefficients at all net terminations and tapping points within the clock tree, while the interconnect is treated as sections of linear transmission lines with a fixed characteristic impedance. A time-domain waveform is reconstructed from steady-state solutions of the reflections at each harmonic frequency for any desired point within the clock distribution network.

With this high frequency model, important characteristics of the clock distribution network can be determined, such as clock skew, waveshape, and ringing. Specifically, the magnitude and location of the terminating resistors can be determined and any ringing or false clock edges can be anticipated [B46].

Bußmann and Langmann [B47] describe an alternative approach to standard tree-structured equipotential clock distribution networks. Recognizing the difficulty in accurately and efficiently transmitting microwave clock signals, sine waves rather than square waves typically are used. The authors apply damping compensation techniques to the design of high frequency clock distribution networks. Active compensation with negative impedance converters is used to improve the performance of a clock distribution network without significantly increasing power dissipation or clock signal delay. The idea behind applying active compensation to clock distribution networks is to change the propagation characteristics of the microstrip, reducing the attenuation constant of the transmission line. An on-chip differential microstrip is used to distribute the clock signals, thereby minimizing adjacent line coupling, resulting in well defined signal propagation characteristics. Differential current switch circuits operating into the Gigabit-per-second range are used to complement the high speed differential microstrip.

The capacitive loading of the registers is considered to be part of the transmission line structure itself. This loading dominates the transmission line characteristics in integrated circuits, affecting the characteristic impedance and propagation constant of the transmission line. The distributed load model is applied to the periodically loaded transmission lines since the distance between loads is chosen not to exceed the limit set by the operating frequency [B47]. This frequency-domain approach provides insight into the basic performance characteristics of clock distribution networks operating at microwave frequencies.

In distributing sinusoidal clock signals across the backplane of a large massively parallel computer operating at a frequency of 160 MHz, Chi [B17, B18] presents a novel strategy for distributing clock signals over long distances (up to several meters). This strategy is denoted by Chi as "salphasic" clock distribution, where the word **salphasic** is intended to connote the concept of sudden transitions in phase between regions of relative constancy. Clock skew is minimized with this approach by exploiting the spatial properties of standing waves, particularly the property that the amplitude and frequency of the forward and reverse traveling waves are identical. This strategy has been demonstrated on a computing system, producing a clock skew of only 175 ps, effectively an order-of-magnitude improvement over conventional approaches (~2 ns). The computing system being synchronized by the salphasic clock distribution network, after an initial tuning procedure, "has performed continuously without attention since August 1990" [B17, B18]. Optimizing salphasic networks, however, does require selecting specific reactive terminations to control the position of the standing waves and, if the system frequency is changed, the values of the termination impedances must also be changed. In general, salphasic networks must operate in relatively lossless transmission lines; therefore, the applicability of this strategy to VLSI systems is most appropriate for synchronizing systems at the board level rather than on the VLSI circuit itself, due to the highly resistive, lossy nature of on-chip interconnect lines.

4.5 Design of Low Power Clock Distribution Networks [B52–B54]

In a modern VLSI system, the clock distribution network may drive thousands of registers, creating a large capacitive load that must be sourced efficiently. Furthermore, each transition of the clock signal changes the state of each capacitive node within the clock distribution network, in contrast to the switching activity in combinational logic blocks, where the change of logic state is dependent on the logic function. The combination of large capacitive loads and a continuous demand for higher clock frequencies has led to an increasingly larger proportion of the total power of a system being dissipated within the clock distribution network, in some applications much greater than 25% of the total power [B53, B166].

The primary component of power dissipation in most CMOS-based digital circuits is dynamic power. It is possible to reduce CV^2f dynamic power by lowering the clock frequency, the power supply, and/or the capacitive load of the clock distribution network. Lowering the clock frequency, however, conflicts with the primary goal of developing high speed VLSI systems. Therefore, for a given circuit implementation, low dynamic power dissipation is best achieved by employing certain design techniques that minimize the power supply and/or the capacitive load.

Recently, De Man [B52] introduced a technique for designing clock buffers and pipeline registers such that the clock distribution network operates at half the power supply swing, reducing the power dissipated in the clock tree by 60% without compromising the clock frequency of the circuit. Kojima, Tanaka, and Sasaki [B53] describe a similar strategy in which the clock signals operate also only over half of the power supply rail, reducing the power dissipated in the clock tree by ideally 75%. The degradation in system speed is very small since, unlike the clock signals, the data signals operate over the full power supply rail. Thus the voltage is reduced only in the clocking circuitry, resulting in significantly reduced power with a minimal degradation in system speed. Experimentally derived savings of 67% were demonstrated on a test circuit (a 16-stage shift register) fabricated in a 0.5-μm CMOS technology with only a 0.5 ns degradation in speed using this half-swing clocking scheme.

Other approaches exist for reducing the power dissipated within a clock distribution network. These approaches reduce power by decreasing the total effective capacitance required to implement a clock tree. Reductions of 10% to 25% in power dissipated within the clock tree are reported with no degradation in clock frequency [B54]. As described in Section 10, the development of design strategies for minimizing the power dissipated both internal to the clock tree as well as the overall system being synchronized is a research topic of great relevance to a variety of important applications.

4.6 Test and Evaluation of High Speed Clock Distribution Networks

Following the design of clock distribution networks, these circuits must also be tested. Deol [B41] and Keezer [B42] describe different functional test systems for performing prototype testing in which the time differences within the clock distribution networks are evaluated. In [B41] an analysis system is described specifically tailored for evaluating tester skew and clock distribution networks, permitting both enhanced test programs and improved circuit implementations. Non-invasive electron beam testing is demonstrated in [B42] for both characterizing and optimizing clock distribution networks within VLSI circuits. Experimental results are described in which race conditions exist due to negative clock skew [see (6)]. These nonfunctional clock distribution networks are detected using electron beam probing. Both of these papers exemplify the importance of verifying and evaluating the design and implementation of the clock distribution networks when building high speed and highly reliable VLSI systems.

5. AUTOMATED SYNTHESIS AND LAYOUT OF CLOCK DISTRIBUTION NETWORKS

Different approaches have been taken in the automated synthesis and layout of clock distribution networks, ranging from procedural behavioral synthesis of pipelined registers [B100–B106] to the automated layout of clock distribution nets for application to gate array and standard cell-based integrated circuits [B35, B55–B85, B91, 11]. In the area of automated layout, two paths have been taken, although with time these approaches should converge. These two paths, commercial ad hoc and algorithmic, are described below in subsections 5.1 and 5.2, respectively. The integration of the effects of clock distribution into behavioral synthesis methodologies is described in subsection 5.3.

5.1 Commercial Strategies for Implementing Clock Distribution Networks in In-House Design Systems

This research path is oriented to commercial semiconductor foundries and supporting design tools [B35, B55–B58, B61, B68, B69] in which a variety of strategies are in use. Each of these strategies is aimed at increasing the routing prioritization of the clock signal nets over the data signal nets and connecting these clock nets to previously placed distributed local buffers. These buffers are used for amplifying the clock signals, since the clock signals traverse long interconnect sections. Empirical delay models coupled with back annotation are typically used to calculate the clock path delays. Either the clock skews are estimated for inclusion in timing analysis or the clock paths are compensated for, thereby forcing the clock skew to be of negligible magnitude.

An interesting example of this type of strategy, developed for gate arrays, is described in [B58]. The clock distribution system consists of a single chip-wide clock conditioner and, for each clock net, a clock ring, a primary clock buffer, and a set of secondary clock buffers. The clock ring is a global clock net that surrounds the entire integrated circuit just inside the I/O buffers. The clock conditioner is located physically inside the ring; it is sourced by the off-chip input clock signal and drives a primary clock buffer. The primary clock buffer drives the clock ring, which in turn drives the secondary clock buffers and any registers within the I/O buffers. Each secondary buffer drives a portion of the on-chip registers. For reasons of symmetry (i.e., to minimize the clock skew), the secondary buffers are located along the vertical sides of the integrated circuit, and the primary buffers are centered on one horizontal row. All connections between buffers are in metal for low resistance and the number of registers clocked by a particular buffer is limited by load balancing requirements. A primary source of delay within the clock path is the *RC* delay within the clock ring; however, internal clock skews of less than 500 ps are maintained. Due to area concerns, internal strapping of the clock lines is not done. A design methodology is described that extends commercial software by performing register pin distribution (physical location, capacitive load, and number), partitioning of registers by chip region for load balancing, additional balancing of capacitive loads, clock net routing, and back-annotation for precise timing analysis.

5.2 Automated Layout of Clock Distribution Networks

A second research path in the area of automated layout has been the development of algorithms that carefully control the variations in delay between clock signal net length so as to minimize clock skew [e.g., B59, B62–B67, B72, B78]. The strategy used is to construct binary tree-like structures with the clock pins at the leaf nodes. Minimal skew clock distribution networks are created using a recursive bottom-up approach. At each of the clock pins of the registers, which represent the leaves of the clock distribution tree, a clock net is defined. The point where two zero skew clock nets connect is chosen such that the effective delay from that point to each clocked register is identical (within the accuracy of the delay model). This process continues up the clock distribution tree, the point of connection of each new branch being chosen to satisfy the zero skew design goal. The layout process terminates when the root (or source) of the clock tree is reached. The schematic diagram of this geometric matching process is illustrated in Figure 15. Thus, the automated layout algorithm attempts to balance the delay of each clock branch in a recursive manner, moving from the leaves to the root of the tree. The appropriate branching points of the zero skew subtree are chosen to maintain equal delay. If the zero skew tapping point falls directly on an unroutable location, such as an existing macrocell, a non-zero clock skew would be realized [B91–B93].

Some early clock routing algorithms [B59, B65, B66] define the delay as a measure of the total wire length along a path. These algorithms attempt to equalize the lengths of each net from the root of the clock tree to each of the leaf nodes. Thus, the clock skew is minimized during the routing phase of the layout process. No attempt is made to postprocess the layout database to further improve the skew characteristics.

In [B91–B93] the automated layout of the clock tree is composed of a two-phase process. The clock net is initially routed in a binary-tree manner with the clock pins as leaf nodes and the clock buffer as the root. This layout phase is followed by a post-

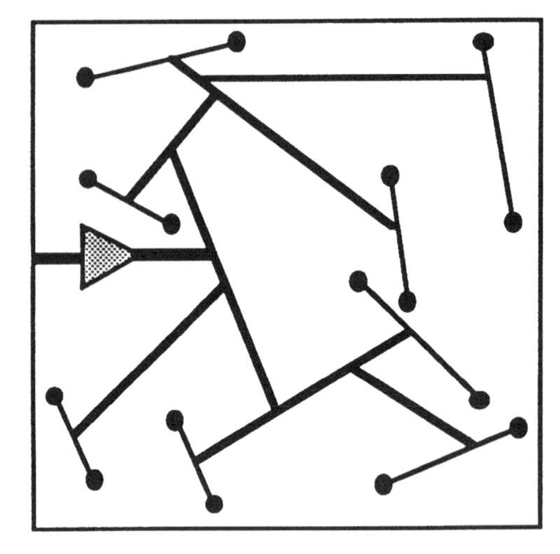

Fig. 15. Geometric Matching for Zero Clock Skew.

layout phase in which the clock nets are widened according to the zero skew specifications, thereby giving the clock layout system additional flexibility in routing around possible blockages. The choice of which clock net to widen is determined by analyzing the sensitivity of the clock net impedance. These sensitivities provide a means of choosing those nets that will decrease the average delay of the *RC* trees as near as possible to a specified target delay. The nets that are widened are less sensitive to increasing capacitance and whose clock path delay must be increased. However, if all the clock nets are relatively thin, statistical variations in the widths of those wires closest to the clock driver may affect the actual clock skew the greatest [B91]. This effect occurs since the section of interconnect closest to the driver sees the greatest portion of the distributed *RC* impedance of the interconnect line. Therefore, the greatest change in delay will occur in those clock paths in which the interconnect impedance closest to the clock source varies significantly.

These automated clock layout algorithms tend to use simplified delay models, such as linear delay, where the delay is linearly related to the path length, or the Elmore delay [22–24], where the delay along a path is the summation of the products of the branch resistance and the downstream capacitance of every branch on the path from the root to the downstream node i (the clock pin of the register). The Elmore delay is

$$T_{Di} = \sum_{k} R_{ki} C_k \, , \tag{10}$$

where C_k is the capacitance at node k, and R_{ki} is the resistance of the portion of the (unique) path between the input and the output node i that is common with the (unique) path between the input and node k. The Elmore delay is a first-order step response approximation of the delay through a distributed resistive-capacitive interconnect section. For slow input waveforms, the Elmore delay approximation can become highly inaccurate, since the shape and magnitude of the clock waveforms are not considered.

The fundamental difficulty with these delay models, however, is the inability to consider accurately the effects of active devices, such as distributed buffers, when estimating delay as well as more subtle considerations, such as bias-dependent loading and varying waveform shapes. The primary focus of the existing research into the automatic layout of clock distribution networks has been placed on minimizing total wirelength, metal-to-metal contacts and crossovers, as well as attaining zero system-wide clock skew (assuming nominal process and environmental conditions).

5.3 Integration of Clock Distribution into Behavioral Synthesis

Localized clock distribution has been considered only minimally in automated layout or physical synthesis. However, early work in applying local clock skew to behavioral synthesis is described in [B95–B106, B174]. These papers represent early efforts to develop strategies that consider the effects of clock distribution networks in the behavioral synthesis process, rather than after the circuit has been partitioned into logic (or register transfer level) blocks. This capability will improve high level exploratory design techniques as well as optimize the performance of circuits implemented with high level synthesis tools.

As described in [B95–B97], the automated synthesis of clock distribution networks can be broken up into four phases: 1) optimal clock scheduling, 2) topological design, 3) circuit design, and 4) physical layout. Optimal scheduling represents a primary research activity in clock distribution networks and is discussed further in subsection 7.1. The area of topological design, in which the structure of the clock distribution network is derived from the scheduling of the local clock skew schedule, is discussed in [B95]. Algorithms are developed for converting the clock skew information into path-specific clock delays. With this information and some information describing the hierarchy of the circuit function, a clock distribution tree is developed with delay values assigned to each branch of the tree. With the topological structure and delay information determined, circuit delay elements are synthesized that satisfy the individual branch delays. Circuit techniques to implement the clock tree delay elements are discussed further in [B96, B97]. Finally, a variety of techniques exist to lay out the clock distribution trees. Some of these layout techniques are discussed in subsections 5.1 and 5.2. This work represents early research in the development of a systematic methodology for synthesizing tree-structured clock distribution networks that contain distributed cascaded buffers and, furthermore, exploit non-zero localized clock skew.

In [B101–B103], a delay model characterizing the timing components of a local data path, similar to (2) and (3), is used to incorporate the effects of local clock distribution delays into the retiming process. This improvement in model accuracy is accomplished by assuming that physical regions of similar clock delay exist throughout an integrated circuit. Retiming is an automated synthesis process for relocating pipeline registers such that the critical worst case path delay is minimized, creating a synchronous system with the highest possible clock frequency while maintaining the function and latency of the original system. Previous work in the area of retiming ignored clock skew in the calculation of the minimum clock period. In the algorithm presented in [B102, B103], clock delays are attached to individual paths between logic gates. As a register is placed on a new path, it assumes the clock delay of that path. Since each sequentially-adjacent pair of registers defines a local data path, as registers are moved from one local data path to another local data path during the retiming process, the displaced registers assume the clock delay of the new physical region. Thus, the local clock skews of each data path are determined at each iteration of the retiming process, permitting both increased accuracy in estimating the maximum clock frequency and detection and elimination of any catastrophic race conditions. If a choice of register locations does not satisfy a particular clock period or if a race condition is created, that specific register instantiation is disallowed. This algorithm, therefore, integrates the effects of clock skew (and variable register and interconnect delays) directly into the synchronous retiming process.

It is interesting to note that adding clock delay to a clock path (applying localized clock skew) has an effect similar to retiming, where the register crosses logic boundaries. Thus, time can be shifted by moving the registers or changing the local clock delays, where retiming is discrete in time and localized clock skew is continuous in time. In general, the two methods complement each other [B126]. As Fishburn mentions in [B126], since both methods are linear, "it is likely that efficient procedures could be given for optimizing systems by jointly considering both sets of variables."

6. ANALYSIS AND MODELING OF THE TIMING CHARACTERISTICS OF CLOCK DISTRIBUTION NETWORKS

This research area is composed of a number of disparate topics, all of which have in common the attributes of modeling the general performance characteristics of clock distribution networks. An important and active area of research in clock distribution networks is the design of circuits less sensitive to variations in process parameters. This topic is discussed in subsection 6.1. Deterministic approaches for calculating clock skew are summarized in subsection 6.2. In contrast to the use of deterministic approaches, probabilistic and statistical estimates of clock skew are possible, particularly for highly regular circuit structures. Some probabilistic and statistical models of clock skew are described in subsection 6.3.

6.1 Design of Process Insensitive Clock Distribution Networks

A primary disadvantage of clock distribution networks is that the delay of each of the elements of a clock path (the distributed buffers and the interconnect impedances) is highly sensitive to geometric, material, and environmental variations that exist in an implementing technology. Thus, as device and interconnect parameters vary from process lot to process lot, the specific performance characteristics of the clock distribution network may change. This phenomenon can have a disastrous effect on both the performance and the reliability of a synchronous system, thereby limiting the precision and the design methodology of the clock distribution network. It is essential for a robust clock distribution network to exhibit a certain degree of tolerance to variations in process parameters and environmental conditions. In an effort to overcome this problem, various approaches have been developed that mitigate the effects of process tolerances on the design of clock distribution networks while maintaining an effective methodology for designing these networks.

Threshold Tracking to Control Clock Skew

An important circuit design technique for making clock distribution networks less process sensitive is described by Shoji [B107]. The technique uses the MOS circuit characteristic that N-channel and P-channel parameters tend not to track each other as a process varies. Interestingly, the response times of these devices tend to move in opposite directions, since the cause of a positive threshold voltage shift in one type of MOS transistor (e.g., an N-channel device) will typically cause the P-channel threshold voltage to shift in the opposite direction (due to charge build-up in the thin oxide between the polysilicon gate and the conducting channel). Shoji quantitatively describes how the delays of the P-channel and N-channel transistors within the distributed buffers of a clock distribution network should be individually matched to ensure that, as the process varies, the path delays between different clock paths will continue to track each other.

The primary objective of this process insensitive circuit design technique is to match the two clock edges (of either a P-channel or an N-channel transistor) as the process parameters vary. Shoji presents two rules to minimize the effects of process variations on clock skew. The rules are:

1) match the sum of the pull-up delays of the P-channel MOSFET with the pull-up delays of any related clock signal paths, and
2) match the sum of the pull-down delays of the N-channel MOSFET with the pull-down delays of any related clock signal paths.

Although process variations may change the total clock delay along a given path, the difference in delay between paths will track each other, keeping the skew small.

A circuit utilizing this technique is shown in Figure 16. Delay times T_1, T_3, and T_A are directly related to the conductances of the N-channel devices, N_1, N_3, and N_A, respectively. Delay times T_2 and T_B are directly related to the conductances of the P-channel devices, P_2 and P_B, respectively. The conductance of

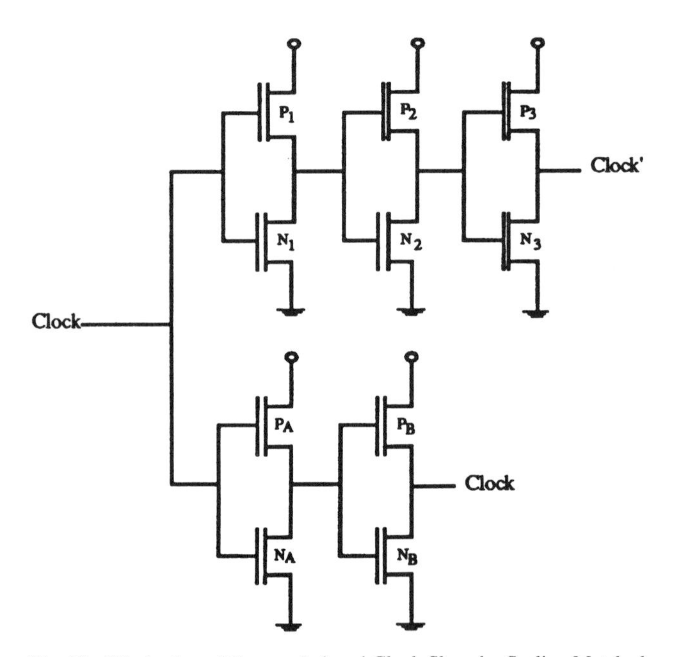

Fig. 16. Elimination of Process Induced Clock Skew by Scaling Matched Transistor Types.

single → Differential

each of these devices is proportional to the size of the MOSFET geometric width. In conventional CMOS circuit design, the transistor widths are adjusted to satisfy

$$T_A + T_B = T_1 + T_2 + T_3 \,, \tag{11}$$

thereby ensuring that the skew at the output of each clock branch is close to zero. If, for example, either the N-channel or the P-channel threshold voltage varies significantly from its assumed value, (11) would no longer be satisfied and a large skew would develop between the outputs, Clock and Clock′, shown in Figure 16. Instead of designing the circuit in Figure 16 to satisfy (11), the circuit is designed to satisfy the two equations, (12) and (13). Thus, the primary objective of (11) (zero clock skew) is maintained, while the added more stringent design constraint makes the entire clock distribution network more tolerant to process variations. This increased tolerance occurs, since by satisfying both (12) and (13), the N-channel and P-channel transistors of both branches individually track each other, making the system more tolerant to variations in the N-channel and P-channel transistor characteristics.

$$T_A = T_1 + T_3 \,, \tag{12}$$

$$T_B = T_2 \,. \tag{13}$$

This design technique can be used to make circuits less sensitive to process variations and environmental conditions even if the circuits are not inverters but are more general forms of logic gates. The technique also ensures similar behavior when interconnect impedances are included within the circuit. Simulated worst case clock skews of circuits using this technique exhibit skews that are 10% less than that of conventionally designed circuits [B107].

Interconnect Widening to Minimize Clock Skew Sensitivity

As described in subsection 5.2, one approach for the automated layout of clock nets is to lengthen specific clock nets to equalize the length of every clock line, thereby keeping the clock skew close to zero. A disadvantage of this approach is that these minimum width lines are very susceptible to variations in the etch rate of the metal lines, as well as to mask misalignment or local spot defects. Therefore, the effective interconnect impedance (and the delay) of these long thin clock nets can vary greatly from wafer to wafer as these line widths vary. In order to design these clock nets to be less sensitive to process variations, Pullela, Menezes, and Pillage [B91–B93] have developed an automated layout algorithm that widens rather than lengthens the clock nets while equalizing the line delays. These nets are therefore less sensitive to both under- and over-etching during the metal patterning process. By widening the clock lines, the interconnect resistance is decreased; however, the interconnect capacitance increases. It is interesting to note that increasing the line width of those branches closer to the root of the *RC* tree has a greater effect on the clock path delay than increasing the widths closer to the leaf nodes (the clocked registers). Thus, decreasing the resistance at the source by increasing the line width affects the total path delay more significantly than decreasing the resistance at the leaf node, since more capacitance is seen by the large source resistance than if the resistance is greater near the leaf. Therefore, the clock skew is particularly sensitive to changes in line width close to the clock source. One approach to making the clock lines more tolerant of process variations is to make the width of the clock interconnect lines widest near the clock source and thinner as the leaf nodes are approached. This strategy would provide a reasonable tradeoff between controlling the effect of process variations (particularly, metal etch rates) on the clock skew and minimizing line dimensions for process yield and circuit layout efficiency. The relative sensitivities of each net to changes in the capacitive and resistive interconnect impedances are analyzed and integrated into the Elmore delay model [22–24]. One of the primary advantages of this approach is that the process of automatically laying out the clock nets is separate from the clock skew reduction process. Thus, local layout techniques such as widening the clock nets can be used to make the overall circuit less sensitive to variations in process parameters.

6.2 Deterministic Models for Estimating Clock Skew

A clock signal path within a clock distribution network has a single input and, although paths branch off from the trunk of the tree, a single path (or branch) exists from the clock source to the clock input of a register. This branch is typically composed of distributed buffers and interconnect sections, as shown in Figure 13. In order to simplify the calculation of the path delay and to provide simple closed form delay models of the path, it is typically assumed that the buffer on-resistance is much greater than the interconnect resistance that the buffer is driving. This assumption permits the distributed *RC* interconnect section to be modeled as a simple lumped capacitor. Percent errors reflecting this assumption are provided in [25], where the errors are dependent upon the ratio of the load resistance to the output buffer on-resistance. However, if the line resistance is not significantly smaller than the buffer output resistance, repeaters [11, 20] are often inserted at a point within the clock line to ensure that the output resistance of the repeater buffer is much larger than the local line resistance of the interconnect section between the repeaters [B16].

In order to calculate the clock path delay and skew, a simple model of a CMOS inverter driving another inverter with line resistance and capacitance between the two inverters is often used. A well-known empirical estimate of the rise (or fall) time of a single CMOS inverter driving an *RC* interconnect section with a capacitive load (representing the following CMOS inverter) is [10, 11, 25]

$$T_{R/F} = 1.02 R_{Int} C_{Int} + 2.21 (R_{Tr} C_{Int} + R_{Tr} C_{Tr} + R_{Int} C_{Tr}) \,, \tag{14}$$

where R_{Int} and C_{Int} are the resistance and capacitance of the interconnect section, respectively, and R_{Tr} and C_{Tr} are the output on-resistance of the driving buffer and the input load capacitance ($= C_{OX}\, W\, L$) of the following buffer, respectively. Note that C_{OX} is the oxide capacitance per unit area, and W and L are the width and length, respectively, of the following buffer. An approximate estimate of the output resistance of the buffer may be obtained from [10, 11, 25]

$$R_O \approx \frac{L/W}{\mu C_{ox}(V_{DD} - V_T)}, \tag{15}$$

where μ is the channel mobility, V_{DD} is the power supply voltage, and V_T is the device threshold voltage. Equation (15) is derived from the large signal I-V equation of a MOSFET operating in the saturation region close to the linear region and is accurate for small channel geometries, since velocity saturation decreases the quadratic behavior of the MOS device operating in the saturation region. The physical delay model represented by (14) and (15) is a fairly simple approximation of the delay of a CMOS inverter driving a distributed *RC* impedance. More complex and accurate delay models exist. This area of inquiry represents an important topic of intensive research unto itself and is discussed in great detail throughout the literature.

An important research area in VLSI circuits is timing analysis, where simplified *RC* models are used to estimate the delay through a CMOS circuit. Clock characteristics are provided to a timing analyzer to define application-specific temporal constraints, such as the minimum clock period or hold time, on the functional timing of a specific synchronous system [B113]. In [B121, B122], Tsay and Lin continue this approach by describing an innovative timing analyzer which exploits negative clock skew (i.e., time is "stolen" from adjacent data paths to increase system performance). Therefore, the descriptive term "cycle stealing" is used to describe this process. In [B114, B115], Dagenais and Rumin present a timing analysis system that determines important clocking parameters from a circuit specification of the system, such as the minimum clock period and hold time. This approach is useful for top-down design when performing exploratory estimation of system performance.

6.3 Probabilistic/Statistical Models for Estimating Clock Skew

Clock skew is caused by variations in clock signal delay between sequentially-adjacent registers. Differences in process parameters across an integrated circuit die are a primary source of this delay variation. A strategy for analyzing these effects is to use a statistical approach to model the process parameter variations in order to estimate changes in the clock path delay. This method of estimating clock skew is quite different from classical deterministic techniques used within industry and described throughout the literature. The primary application of statistical models of clock skew is highly regular architectures, such as parallel processors.

In developing a statistical model of clock skew for parallel processors, three basic assumptions regarding the clock distribution network are made [B111]: 1) the clock paths from the clock source to the processing elements are identical; 2) the clock arrival times are considered to be random variables and are the sums of uncertain independent delays due to the distributed interconnect and buffers along the clock path; and 3) to make the results independent of topology, any processor can communicate with any other. Since the delay of each cascaded buffer is modeled as a real random variable, the entire clock path delay is modeled as the sum of these random variables. As described by Afghahi and Svensson in [B16, B117] and Kugelmass and Steiglitz in [B110, B111], the variance of the clock path is equal to the sum of the variances of each buffer along the path (see Figure 13). Therefore, according to the Central Limit Theorem [26], the distribution of the clock path delay tends to a normal Gaussian distribution, regardless of the distribution of the individual buffer delays. Kugelmass and Steiglitz provide upper bounds on clock skew, assuming a Gaussian distributed clock delay with a variance proportional to the wire length. Afghahi and Svensson analyze the individual device parameters characterizing the on-resistance of a transistor operating in the saturation region. The authors discern an interesting aspect: long highly resistive interconnects are often broken up into equally spaced interconnect sections by inserting individual buffers (i.e., repeaters [11, 20]), thereby making the line resistance negligible as compared with the buffer resistance (since each interconnect section is small in comparison to the output on-resistance of the buffer). This strategy effectively makes the interconnect delay linear with line length. Furthermore, the use of repeaters actually reduces the standard deviation of the interconnect delay, and the line delay becomes proportional to the square root of the interconnect length. Thus, the use of repeaters minimizes the effect of scaling the interconnect lines on the standard deviation of the clock signal paths.

Kugelmass and Steiglitz show that the expected range of variation of a set of related random variables is no larger than the expected range of the corresponding set of independent random variables. As shown in [B16, B110], for the set of normally distributed independent paths, the expected value and variance of the difference between the maximum and minimum delay (clock skew) is

$$T_{Skew} = \sigma \frac{4\, ln\, \mathrm{R} - ln\, ln\, \mathrm{R} - ln\, 4\, \pi + 2\mathrm{C}}{\sqrt{2\, ln\, \mathrm{R}}}, \tag{16}$$

$$Variance\ (T_{Skew}) \approx \frac{\pi^2 \sigma^2}{6\, ln\, \mathrm{R}}, \tag{17}$$

where C $\approx$ 0.5772 (Euler's constant), R is the number of clock signal paths, and σ is the standard deviation of a single path delay. Equation (16) is an asymptotic upper bound on the expected clock skew for a clock distribution network with R signal paths. Note that the clock skew grows with the square root of the logarithm of the size of the system (i.e., the number of clock paths). This statistical approach assumes a highly regular structure, thereby assuring that a balanced clock distribution network is possible. The asymptotic upper bound of the expected clock skew, using this model and assuming an H-tree clock distribution network, is $\Theta(N^{0.25}\sqrt{\log N})$, where N is the number of leaves (or destination points) in the H-tree.

A statistical approach has also been used for estimating the distribution of the delays of the leaf nodes of a clock tree with application to the automated layout problem [B88–B90]. The extremal points of the distribution determine the clock skew while the average of the distribution determines the delay. Attention is placed on reducing the variance of the nominal delay distribution as well as the variances due to any process variations [B91].

One major weakness in applying a statistical approach to the design and analysis of clock skew in VLSI systems is that a statistical methodology ignores the localized deterministic nature of clock skew in affecting system performance and reliability, and instead considers clock skew as a system-wide global parameter. The local data path dependent clock skew directly determines the reliability and performance of the entire system, rather than an average system-wide clock skew. Any individual local data path can limit the maximum speed or create a catastrophic race condition within the system. Therefore, the probabilistic models described in this subsection are appropriate only if applied to highly regular architectures, such as systolic parallel processors [B110, B111]. The primary reason that a probabilistic clock skew model is applicable to a parallel processor architecture is that the local data paths within these types of circuit structures tend to have similar path delays; therefore, any local clock skew is close in magnitude to an average global clock skew. The distribution of clock signals within parallel processors is presented in much greater detail in subsection 8.1.

7. SPECIFICATION OF THE OPTIMAL TIMING CHARACTERISTICS OF CLOCK DISTRIBUTION NETWORKS

An important element in the design of clock distribution networks is choosing the minimum local clock skews that increase circuit performance by reducing the maximum clock period while ensuring that no race conditions exist. This design process is called *optimal clock skew scheduling;* it has been extensively studied in [B126–B138] and is described in subsection 7.1. Starting with the timing characteristics of the circuit, such as the minimum and maximum delay of each combinational logic block and register, it is possible to obtain the localized clock skews and the minimum clock period. This process is accomplished by formulating the optimal clock scheduling problem as a linear programming problem and solving with linear programming techniques [B126, B133]. In subsection 7.2 a graphical technique for investigating architectural effects on pipelining is described. Specifically, the performance tradeoff between latency and clock frequency for increasing pipelining is presented.

7.1 Optimal Clock Skew Scheduling

The concept of scheduling the system-wide clock skews for improved performance while minimizing the likelihood of race conditions was first presented by Fishburn in 1990 [B126], although the application of localized clock skew to increase the clock frequency and to eliminate race conditions was known previously [7]. Fishburn presents a methodology in which a set of linear inequalities are solved using standard linear programming techniques in order to determine each clock signal path delay from the clock source to every clocked register. Two clocking hazards, identical to the constraint relationships described in Section 3, are eliminated by solving the set of linear inequalities derived from (5) and (6) for each local data path. The deleterious effect of positive clock skew (the maximum data path/clock skew constraint relationship) is described as *zero-clocking,* while the deleterious effect of negative clock skew (the minimum data path/clock skew constraint relationship) is described as *double-clocking*. This approach is demonstrated on a 4-bit ripple-carry adder with accumulation and input register in a 1.25-μm CMOS technology. The minimum clock period is decreased from 9.5 ns with zero clock skew to 7.5 ns with localized clock skew [B126]. Szymanski improves this methodology for determining an optimal clock skew schedule by selectively generating the short path constraints, permitting the inequalities describing the timing characteristics of each local data path to be solved more efficiently [B133].

In order to describe the process for determining an optimal clock skew schedule, a system timing model is presented. A block diagram of a multistage synchronous digital system is depicted in Figure 17. Between each stage of registers there is, typically, a block of combinational logic with possible feedback paths between the registers. Each register is either a multi- or single-bit register, and all the inputs are assumed to change at the same time point of the transition of the clock signal. Only one single-phase clock signal source in the circuit is assumed. The registers are composed of edge-triggered flip flops and assume a single value for each clock cycle. The combinational logic block is described in terms of the maximum and minimum delay values, $T_{Logic(max)}$ and $T_{Logic(min)}$. These logic delay values are obtained by considering the delay of all possible input to output paths within each combinational logic block. For simplicity and without loss of generality, the block diagram in Figure 17 only considers a single-input, single-output circuit. In this figure, Δ_1 and Δ_2 are off-chip clock delays outside the VLSI-based system and δ is that portion of the on-chip clock delay that is shared by each of the clock paths (the initial trunk of the clock tree). The registers R_{in} and R_{out} make up one set of registers placed at the input and output, respectively, of the VLSI-based

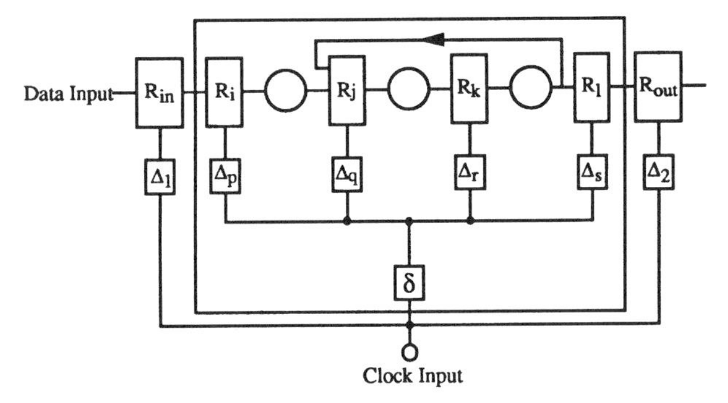

Fig. 17. Multistage Block Diagram of a Synchronous Digital System.

system. The circuit is composed of the registers, R_i and R_l, where the logic blocks are between the registers and the on-chip signal clock delays are Δ_p to Δ_s $(+\delta)$.

Off-Chip Clock Skew

This circuit model, pictured in Figure 17, also considers the relationship between off-chip and on-chip clock skew. The registers R_{in} and R_{out} symbolize off-chip registers and are controlled by the off-chip clock source, which also provides the on-chip clock signals, since the circuit is assumed to be a fully synchronous system. This relationship is represented by

$$T_{Skewin,out} = T_{Skewin,i} + T_{Skewi,j} + \ldots + T_{Skewl,out} = 0 \,. \tag{18}$$

Therefore, to satisfy (18), in Figure 17, $\Delta_1 = \Delta_2$.

Although it is possible to have off-chip non-zero clock skew, it is desirable to ensure that the clock skew between VLSI I/O approaches zero, in order to avoid complicating the design of a circuit board or the specification of the interface of the circuit with other components also controlled by the same clock source. For example, a circuit with intentional non-zero clock skew requires that the clock distribution network of any other synchronous circuit sharing the same global clock be offset by the same amount of temporal skew; otherwise, race conditions such as described in subsection 3.2 may occur at the board level. This strategy of minimizing clock skew as the level of design complexity shifts should be applied at each higher level of design, such as from the board level to the multiboard level.

A fully synchronous circuit must generate data at a rate defined by the clock period of the clock source; otherwise, race conditions may occur at the interface of the circuit with other parts of the system. These race conditions occur when there is a negative clock skew, intentionally or unintentionally introduced into the circuit. Observe that every circuit has unintentional clock skew caused by several factors, one common cause being variations in process parameters. This type of clock skew must be considered during the design of the circuit, and should be less than the intentional on-chip clock skew introduced to increase the performance of the circuit. Furthermore, the magnitude of the intentional clock skew at each register I/O may vary substantially, according to the optimization applied to the data path. Therefore, clock skew at the system level of a VLSI circuit should be constrained to approach zero, in order to allow the circuit to communicate correctly with other board-level circuits. For example, a symmetric zero clock skew distribution system should be used for the external registers (symmetric networks are discussed in subsection 4.2). Observe that restricting the off-chip clock skew to zero does not preclude the circuit from being optimized with localized clock skew. The primary effect is that the performance improvement is less than that obtained without this constraint.

Observe that (18) is valid only if the interface circuitry is controlled by the same clock source. The restriction does not apply to asynchronous circuits or synchronous circuits that communicate asynchronously (i.e., globally asynchronous, locally synchronous systems) (see subsection 3.4).

Global and Local Timing Constraints

As described in Section 3, in order to avoid either type of clock hazard (either a maximum or minimum data path/clock skew constraint relationship), a set of inequalities must be satisfied for each local data path in terms of the system clock period T_{CP} and the individual delay components within the data path. To avoid limiting the maximum clock rate between two sequentially-adjacent registers, R_i and R_j, (5) must be satisfied. To avoid race conditions between two sequentially-adjacent registers, R_i and R_j, (6) must be satisfied.

The system-wide clock period is minimized by finding a set of clock skew values that satisfy (5) and (6) for each local data path and (18) for each global data path. These relationships are sufficient conditions to determine the optimal clock skew schedule such that the overall circuit performance is maximized while eliminating any race conditions.

The timing characteristics of each local data path are asumed to be known. The minimum clock period is obtained when the problem is formalized as a linear programming problem, such as

Minimize

$$T_{CP}$$

subject to the local and global timing constraints:

$$T_{Skewi,j} \leq T_{CP} - T_{PD(max)} = T_{CP} - (T_{C\text{-}Qi} + T_{Logic(max)} + T_{Int} + T_{Set\text{-}upj}) \quad \textit{for } T_{Ci} > T_{Cj}$$

$$|T_{Skewi,j}| \leq T_{PD(min)} = T_{C\text{-}Qi} + T_{Logic(min)} + T_{Int} - T_{Holdj} \quad \textit{for } T_{Cj} > T_{Ci}$$

$$T_{CP} \geq T_{PD(max)} + T_{Skewi,j}$$

$$T_{PD(max)} = T_{C\text{-}Qi} + T_{Logic(max)} + T_{Int} + T_{Set\text{-}upj}$$

$$T_{Skewin,i} + T_{Skewi,j} + T_{Skewj,k} + \ldots + T_{Skewn,out} = 0.$$

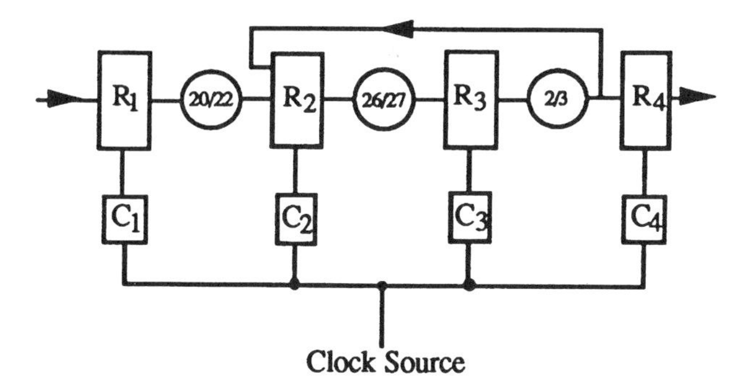

Fig. 18. Circuit Example with Feedback.

An Example of Determining the Optimal Clock Skew Schedule of a Pipelined System

An example of determining the minimum clock period of a multistage system with feedback paths is the circuit illustrated in Figure 18. This example is similar to that used in [27], adapted to consider zero clock skew between off-chip registers. The numbers inside the logic blocks are the minimum and maximum delays of each block, respectively. Similar to the approach taken in [27], for simplicity all the register and interconnect timing parameters are assumed to be zero.

The linear program that gives the minimum clock period and the optimal clock skew schedule for the circuit shown in Figure 18 is

Minimize

$$T_{CP}$$

subject to:

$$R_1 - R_2: \quad C_1 - C_2 = T_{Skew1,2} \geq -20\ ns$$

$$T_{Skew1,2} - T_{CP} \leq -22\ ns$$

$$R_2 - R_3: \quad C_2 - C_3 = T_{Skew2,3} \geq -26\ ns$$

$$T_{Skew2,3} - T_{CP} \leq -27\ ns$$

$$R_3 - R_4: \quad C_3 - C_4 = T_{Skew3,4} \geq -2\ ns$$

$$T_{Skew3,4} - T_{CP} \leq -3\ ns$$

$$R_3 - R_2: \quad C_3 - C_2 = T_{Skew3,2} \geq -2\ ns$$

$$T_{Skew3,2} - T_{CP} \leq -3\ ns$$

$$R_1 - R_4: \quad T_{Skew1,2} + T_{Skew2,3} + T_{Skew3,4} = 0$$

where the optimal clock schedule and minimum clock period are

$$T_{Skew1,2} = -3\ ns$$

$$T_{Skew2,3} = -12\ ns$$

$$T_{Skew3,4} = 15\ ns$$

$$T_{CP} = 19\ ns$$

If zero clock skew between off-chip registers is not considered, the minimum clock period is $T_{CP} = 15$ ns. Although the restriction of zero clock skew increases the clock period, there is still an improvement in performance by applying intentional localized non-zero clock skew to the circuit. With zero clock skew, the minimum period is $T_{CP} = 27$ ns due to the worst case path delay of the local data path between registers R_2 and R_3.

7.2 Effects of Clock Distribution Design on Architectural Tradeoffs

Before the design effort of a clock distribution network can commence, certain timing constraints and goals must be specified. These timing characteristics are typically application-specific and depend greatly on the architectural and circuit tradeoffs of a given system implementation. A few papers exist that consider the effects of clock distribution on different aspects of these architectural tradeoffs. For example, Friedman and Mulligan [B128, B129, 4] describe the tradeoff between latency and clock frequency when pipelining a synchronous digital system. The authors provide equations and a graphical technique for determining the optimal level of pipelining.

As described earlier, registers are inserted into global data paths in order to increase the clock frequency of a digital system, albeit with an increase in latency. This tradeoff between clock frequency and latency is illustrated graphically in Figure 19. In this figure, both the latency and the clock period are shown as a function of the number of pipeline registers M inserted into a global data path. Thus, as M increases, the latency increases by the delay of each inserted register and any local clock skew while the maximum possible clock frequency increases, because the critical path is shortened (since there is less logic and interconnect delay per local data path).

If no registers are inserted into a data path, the minimum latency L_{Min} is the summation of the individual logic delays, NT_{fN}, where N is the number of logic stages per global data path and T_{fN} is the average delay per logic stage. For each register inserted into the global data path, L increases by $T_{Reg} + T_{Skew}$, assuming a constant register delay and clock skew per local data path, where $T_{Reg} = T_{C\text{-}Q} + T_{Set\text{-}up}$. Thus, L increases linearly with M, as shown in Figure 19.

The local data path having the largest value of T_{PD}, defined as the critical data path, establishes the maximum clock frequency f_{Clkmax} for the system. This argument assumes that the circuit is not a simple shift register and logical operations are being performed between the registers of each local data path. If the global data path is a simple shift register, the maximum clock frequency is limited by the maximum path delay T_{PD} of a local data path. The MAX subscript in Figure 19 is used to emphasize that the critical local data path limits the minimum

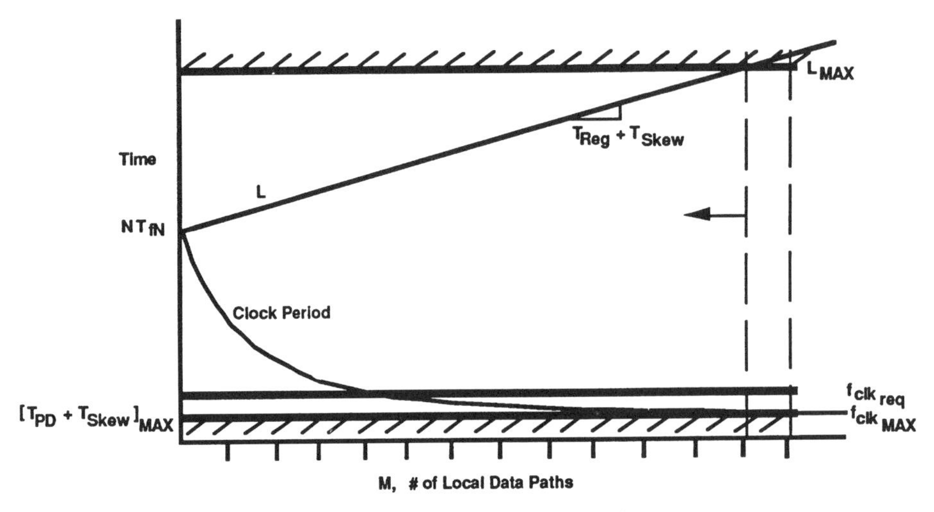

Fig. 19. Design Paradigm for Pipelined Synchronous Systems [B128, B129, 4].

clock period (maximum clock frequency) of the total global data path.

Most design requirements must satisfy some specified maximum time for latency while satisfying or surpassing a required clock frequency. The design constraints due to an application-specific limitation on the maximum permissible latency L_{Max} and the maximum possible clock frequency f_{Clkmax} are shown in Figure 19 by the vertical dashed lines. Thus, for a given L_{Max}, an appropriate maximum clock frequency and level of pipelining M are defined by the intersection of the L curve and the L_{Max} line. If L_{Max} is not specified or is very large and the desire is to make the clock frequency as high as possible, then an appropriate f_{Clk} is defined by the intersection of the clock period curve and the f_{Clkmax} line. Thus, for a particular L and f_{Clk}, the extent of the possible design space is indicated by the horizontal arrow. If L and f_{Clk} are both of importance and no L_{Max} or f_{Clkmax} is specified or constrains the design space, then some optimal level of pipelining is required to provide a "reasonably high" clock frequency while maintaining a "reasonable" latency. This design choice is represented by a particular value of M, defining an application-specific f_{Clk} and L.

Effect of Positive Clock Skew on Pipelined Architecture

The effects of clock skew, technology, and logic architecture on clock period and latency are demonstrated graphically in Figures 20 and 21. If the clock skew is positive or if a poorer (i.e., slower) technology is used, then, as shown in Figure 20, L reaches L_{Max} at an increasingly smaller value of M. Also, the minimum clock period increases, which decreases the maximum clock frequency. Furthermore, large positive clock skew or a very poor technology eliminates any possibility of satisfying a specified clock frequency f_{Clkreq} and limits the entire design space as defined by the intersection of L and L_{Max}. For a poorer technology or logic architecture, the intersection between either the clock period or the latency curve and the ordinate shifts up-

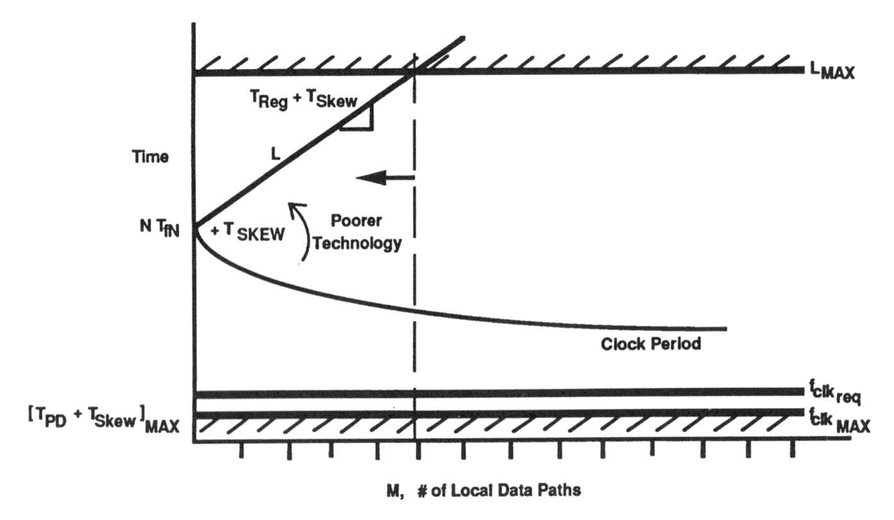

Fig. 20. Effect of Positive Clock Skew and Technology on Design Paradigm.

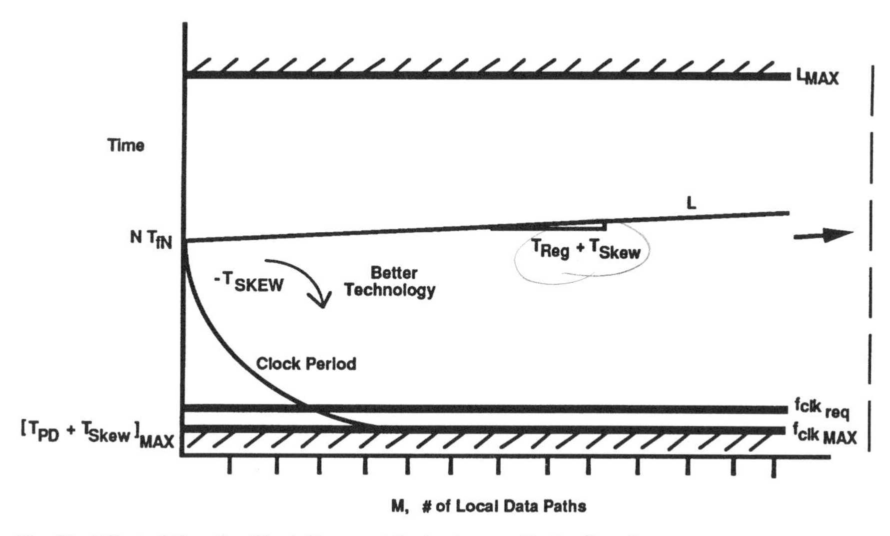

Fig. 21. Effect of Negative Clock Skew and Technology on Design Paradigm.

ward, since T_{fN} increases due to the slower technology and N increases for the less optimal architecture.

Effect of Negative Clock Skew on Pipelined Architecture

If the clock skew is negative or if a better (i.e., faster) technology is used, as shown in Figure 21, L reaches L_{Max} at a larger value of M. Also, the minimum clock period decreases, satisfying f_{Clkreq} and f_{Clkmax} with less pipelining. The possible design space, represented by the intersection of L and L_{Max}, is greatly increased, permitting higher levels of pipelining if very high clock rates are desired. In addition, for a better technology or logic architecture, the intersection between either the clock period or the latency curve and the ordinate shifts downward, since T_{fN} decreases due to the faster technology and N decreases for a more optimal architecture.

Thus, this graphical paradigm can be used to explore the tradeoff between latency and clock frequency in synchronous digital systems. Furthermore, it provides insight into how the design of the clock distribution network affects system-wide performance and pipelining decisions.

8. CLOCK DISTRIBUTION NETWORKS FOR TARGETED VLSI/WSI ARCHITECTURES

Unlike the general problem of designing clock distribution networks for structured custom VLSI circuits, certain circuit structures and architectures have been developed that lend themselves to specific synchronization strategies and can create special problems and constraints. Some examples of these architectures are parallel processors, neural networks, multi-chip modules (MCMs) or hybrid circuits [B171], and wafer scale integrated (WSI) circuits. Interestingly, there exist several examples in which the research areas of parallel processing and wafer scale integration have been merged in order to build highly dense, fine-grained parallel processors (in certain cases, performing neural processing) at the wafer level. The main reasons for this symbiotic relationship are that parallel processors become much more effective as the number of individual processors increase, and wafer scale integration is most appropriate when implemented with those circuit structures that are naturally redundant, thereby permitting localized wafer yield loss. This strategy permits those WSI processors that are faulty to be replaced via software or hardware techniques. In this section, clock distribution networks targeted to these architectures are described. Specifically, clock distribution in parallel processors and in wafer scale integrated circuits are discussed in subsections 8.1 and 8.2, respectively.

8.1 Clock Distribution in Parallel Processors

Many different parallel processing architectures exist. The clock distribution networks utilize a global clock to synchronize the data flow from one processor to the next, such as in a systolic array [13].

Much of the existing literature analyzing clock distribution networks in processor arrays uses graph theory to investigate the connectivity characteristics of these array structures. Fisher and Kung investigated clock distribution issues in large two-dimensional arrays clocked by a planar H-tree clock structure [B145, B146]. Two deterministic clock skew models are considered, one in which clock skew is proportional to the difference in path length from two nodes to the nearest common upstream tapping point of the clock distribution tree (*difference model*), and the other in which the clock skew is proportional to the sum of those lengths (*summation model*). Note that because the summation model is weaker and less constraining than the difference model, any clocking strategy that operates correctly with the summation model will operate correctly with the difference model. Since all sources of clock skew are treated collectively as a distance metric, the authors show that if small variations in delay are considered, clock

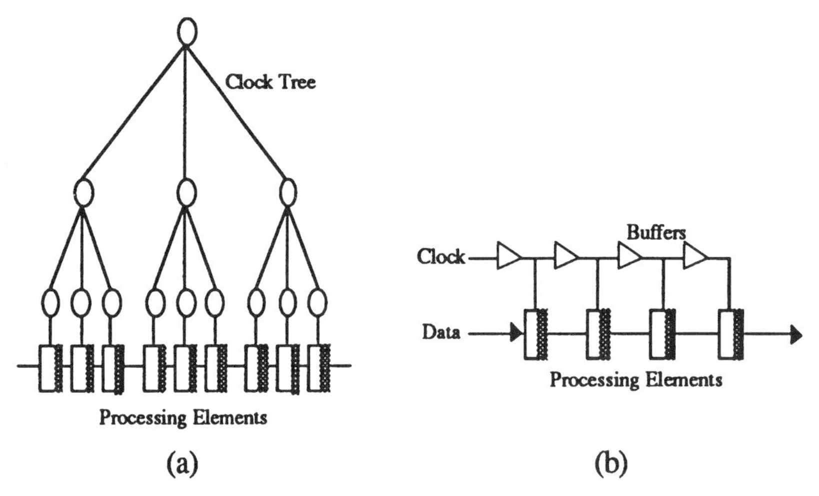

Fig. 22. Clocking Schemes Used in Systolic Arrays. a) Tree Structure and b) Straight-Line Structure [B150].

skew tends to grow with system size (the summation model). The summation model of Fisher and Kung [B146] predicts both an upper and lower bound of Θ ($\sqrt{N}$) for the clock skew of an H-tree clock distribution network, where N is the number of processing elements.

One of the limiting factors in implementing long systolic arrays is the difficulty in providing high speed and reliably synchronized signals. The simplest form of a systolic array is a linear array, composed of a one-dimensional pipeline of processing elements (PEs). These one-dimensional arrays are particularly important in practical systems due to the large number of applications, and the performance is typically bounded by the speed of the I/O circuitry [B146]. In the synchronization of a linear systolic array, two types of clock distribution strategies are possible.

Tree-Structured Clocking Scheme

One equipotential clocking approach, similar to that shown in Figure 9, is the use of a buffered tree structure, as depicted in Figure 22a [B149, B150, B153]. For these networks, the clock delay becomes more severe as the size of the network increases. Note that in Figure 22a the circles represent distributed buffers and the edges (i.e., the connecting wires) represent the wires of the clock tree, while the rectangles are individual processing elements. Thus, the larger the systolic array, the greater the number of branches in the clock distribution network. Assuming that the interconnect resistance and capacitance of the clock distribution network is significantly greater than the driver source resistance and register input capacitance, the clock signal waveform will degrade as the array size increases [B148]. This degradation of the clock signal for increasing array size does not necessarily introduce clock skew, but may limit the maximum clock frequency of the system [28].

As additional PEs are added to the string (or the processor is scaled), the clock distribution network must be completely redesigned, since different loading appears at different points within the clock tree. If the PEs are on a single integrated circuit, then adding PEs to the string requires redesigning the clock distribution network. If, however, the PEs are on different integrated circuits, then the difference in clock delays driving different PEs will be more pronounced.

These issues of scalability represent some of the primary problems in applying global clocking strategies to ever expanding parallel processors. Dikaiakos and Steiglitz [B150], however, show that as the length of a one-dimensional systolic array synchronized by a tree-structured clock distribution network increases by N (the number of processing elements or the pipeline length), the clock period must be increased in proportion to $\log N$, to guarantee that the probability of a negative clock skew constraint relationship not being satisfied, as described by (6), remains negligible.

Buffered Straight-Line Clocking Scheme

A second strategy for distributing clock signals in systolic arrays is shown in Figure 22b. This approach is called *buffered clocking* (also straight-line or pipelined clocking [B146]) and uses a straight-line data-driven structure. The clock distribution network is composed of a series of distributed buffers generating successive clock pulses. These pulses originate from the same global clock signal, allowing clocking events to be pipelined through the network, so that several clock signals are simultaneously active within the system [B150, B153, 28]. Consecutive events are separated by the buffer delay as well as any local interconnect delay, which is the effective clock delay. Note that with respect to clocking large networks, buffered clocking is less limiting. Note that in Figure 22b each PE is synchronized by the clock signal from its left and a new buffered clock signal is used in communication with the PE to its right. Thus, data transfers are resynchronized with the buffered clock at each PE. Each buffered clock delay is chosen to be equal to the logic delay T_{PD}. This strategy, however, ignores the data dependency of the path delays. The output from one logic function is synchronized and passed to the successive logic function with minimum delay [28]. Buffered clocking, however, is only practical if the clock transmission between processors is sufficiently uniform to gain a performance advantage over the buffered tree structure [B146].

Dikaiakos and Steiglitz [B150] show that differences in delay between the rising and falling clock signal edges may create synchronization failure. The failure mode of the buffered clocking strategy is different from that commonly seen in tree-structured clock distribution networks as described in (5) and (6). For buffered clocking, clock skew should be redefined to be the deviation of the actual clock buffer delay from the desired clock buffer delay [28]. Careful control of the buffer delay characteristics is therefore required to ensure that each PE is synchronized properly. Also, as N increases, the clock period increases proportionally in order to ensure that the systolic pipeline will function properly. Thus, as systolic arrays increase in length, tree-structured clock distribution networks become preferable to straight-line clock distribution networks [B150], despite the added requirement of redesigning the clock distribution network due to the increased asymmetric loading.

The straight-line clocking structure is also difficult to generalize to two-dimensional arrays. Therefore, a hybrid globally asynchronous, locally synchronous clocking approach (see subsection 3.4) is recommended for two-dimensional processor arrays [B146]. In noting the self-timed nature of the synchronization pulses when using straight-line or buffered clocking, one readily sees why asynchronous strategies are commonly used in parallel processors, probably best exemplified by wavefront processors [13].

Both deterministic and probabilistic clock skew models have been developed to analyze clock distribution networks in parallel processors. One of the primary applications of a probabilistic clock skew model is highly regular architectures, such as parallel processors. Probabilistic clock skew models are further discussed in subsection 6.3.

8.2 Clock Distribution Networks in Wafer Scale Integrated Circuits

As system complexity, circuit densities, and clock frequencies evolve to more challenging levels, specialized circuit architectures become more enticing. One such architecture is wafer scale integration (WSI) [28]. In WSI, an entire multi-inch wafer is kept undiced and is packaged as a single wafer, utilizing only those circuits that function correctly. Redundancy techniques, either in hardware or software, are used to insulate the system from any non-yielding circuits on the wafer by functionally removing those circuits that do not behave correctly. Since wafer scale distances are measured in terms of several centimeters rather than VLSI signal lengths, typically on the order of a few millimeters, speed-of-light limitations may become significant when clocking high speed WSI circuits. Furthermore, as circuits are removed from the WSI system, loading within the clock distribution network changes. Therefore, it is necessary to design a clock distribution network that is robust enough to handle this varying load environment. These issues can severely affect the performance of high speed clock distribution networks in WSI systems. Therefore, this type of high density architecture requires a synchronization strategy that can satisfy these difficult and unusual requirements.

Application of H-Trees to WSI

In developing strategies for distributing clock signals in constrained architectures, specific solutions have arisen that are similar to the more general problem of distributing clock signals in largely unstructured VLSI systems, but that incorporate aspects peculiar to the WSI architecture. For example, Keezer and Jain [B141–B143] apply both passive and active (with distributed buffers) H-tree structures, as shown in Figures 10 and 11, to the problem of distributing synchronous clock signals across a large multi-inch wafer, implementing a large neural network processor. The authors integrate distributed active buffers into an H-tree structure (see Figure 23) in order to offset the significant waveform degradation that would arise due to the great distances over which the global clock signals must propagate across the wafer (a 7-cm line for a 4-in wafer [B143]). The final buffer at the end of each H-tree path drives a specific VLSI circuit die located on the wafer. In [B141, B143], the authors note that the distributed active buffers provide the greatest portion of the total variation in delay caused by any process variations across the wafer. In the example cited in [B141, B143], assuming each of the process parameters vary by ±20% of nominal value, approximately 60% of the clock skew induced by process tolerances are due to the variation of the delay of the internal buffers placed within the H-tree clock distribution network. Clock skews of over 5 ns across a poorly processed wafer are not uncommon. Therefore, process variations across a wafer represent one of the primary limitations to wafer scale circuits operating at high frequencies.

In contrast, Bakoglu [11] describes a completely passive (no active buffers are distributed within the H-tree structure) 3-in wafer scale implementation of an H-tree exhibiting clock skews less than 30 ps. Thus, *passive* H-tree structures provide significantly less variation in clock skew than *active* H-tree structures; however, this advantage must be weighed against the increased signal degradation that occurs in completely passive networks. A summary of some of the salient features of the H-tree structure used by Bakoglu are 1) a consistent fanout of two at each breakpoint, 2) minimal induced crosstalk from differently phased clock signals, and 3) controlled impedance of tapered transmission lines to minimize the reflections in the high frequency clock signals.

Clocking Issues in WSI

In [29], Johnstone and Butcher describe the design of power and clock distribution networks for a wafer scale-based systolic processor. The authors note the interrelationship between the two distribution systems and describe how to minimize power transients within the power distribution system by exploiting localized clock skew to shift the delays of each of the local data paths within the system. Thus, the current being drawn from the

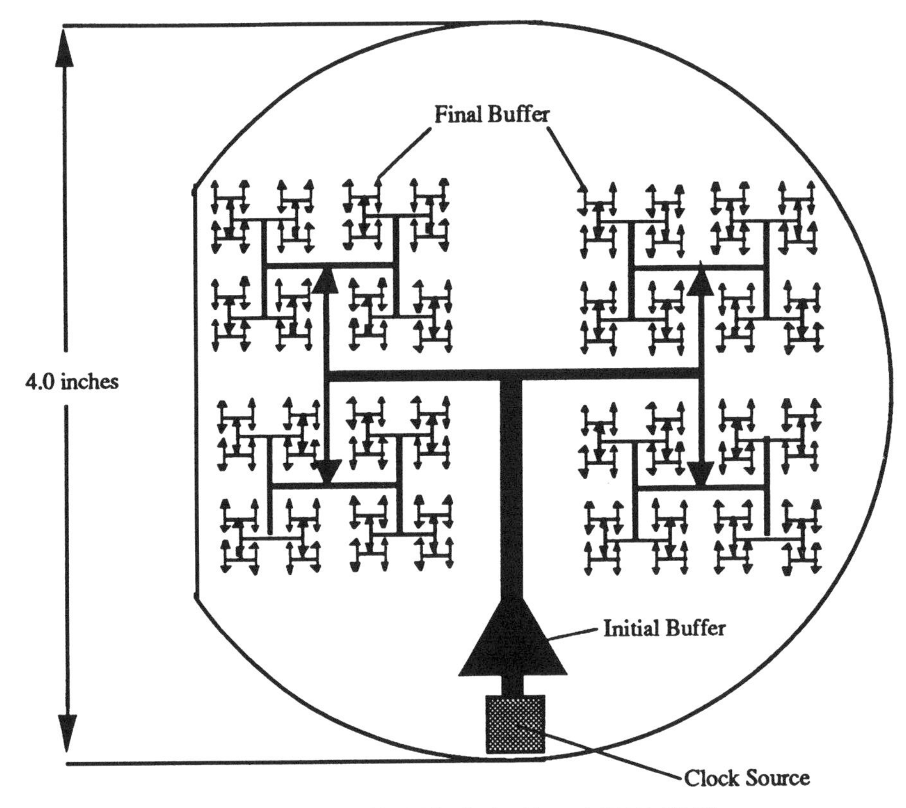

Fig. 23. Wafer Scale Four-Level H-Tree Clock Distribution Network [B141, B143].

power distribution circuitry is offset in time by applying localized clock skew, effectively "smoothing out" the transient current requirements of the power distribution over time.

In addition to the challenging requirements imposed on the design of a clock distribution network of a standard VLSI circuit, wafer scale integration places completely different and new requirements for synchronizing these large physical areas. Specifically, there is additional complexity due to the load impedance being variable; therefore, the buffers must be restructured. This variability in loading occurs since wafer yield is not expected to reach 100%. Redundancy techniques, both through programmable switches (soft restructuring) or through selective metal connections or fuses (hard restructuring), are therefore necessary [B143, 30]. In either case, the effects of these programmable circuit paths can significantly degrade or at best change the characteristics of the on-wafer clock distribution circuitry. High resistance elements placed near the driving source contribute more to the path delay than if placed near the receiving end (since the load resistance sees most of the load capacitance). One interesting result, described in [B143], is that high resistance fuses, when used in clock distribution networks, should be placed close to the clocked register element. Since it is preferable not to place high impedance links directly within the clock path, another strategy is to disable the clock drivers by building controllable tristate drivers within the clock buffers [30]. This procedure maintains the low impedance nature of the high speed clock lines while supporting the use of redundant drivers, important to wafer scale circuits.

Another option in synchronizing wafer scale circuits is to use some degree of asynchronous self-timing. Many synchronization possibilities exist, ranging from data flow architectures such as wavefront processors to the aforementioned globally clocked systolic arrays [13], including globally asynchronous, locally synchronous systems. One type of approach targeted for a fine-grained parallel processor utilizing wafer scale technology is summarized in [B140]. Two strategies are presented: either distribute a fast global clock signal and generate a slower clock locally, or distribute slower global out-of-phase clock signals and multiply them together to generate a high speed clock signal locally. Distributing a slow clock signal has the advantage of being less dependent on power supply variations across a wafer, although inaccuracies may develop during the out-of-phase multiplying process. The preferred synchronization strategy presented in [B140] is to distribute a slow clock signal globally and to multiply the clock signal further down the clock tree as the lower levels are reached. It should be noted, however, that the skew of the slow clock should be much less than the clock period of the fast clock.

In closing this topic on synchronizing WSI circuits, a highly interesting and provocative clock distribution strategy should be mentioned, specifically, distributing the clock signals across a large wafer (or printed circuit board) using optical transmitters to generate the clock signal and local electro-optical receivers (or detectors) to receive the clock signals. The clock signals are then distributed locally via standard electronic means. This strategy would have negligible clock skew and signal latency (i.e., clock delay). Chou and Franklin [B144] discuss using silicon detectors placed advantageously across a wafer to provide system-wide clock distribution. These detectors would receive the transmitted global clock signal, convert it to a current (or voltage), and distribute locally the clock signal to neighboring local data paths (or circuit modules), thereby ameliorating the clock distribution problem across the entire system. Thus, as the requisite technologies mature, electro-optical techniques could potentially be used to distribute global clock signals across a wafer or system.

9. EXAMPLE IMPLEMENTATIONS OF CLOCK DISTRIBUTION NETWORKS

A number of interesting and innovative examples of high performance fully synchronous clock distribution networks have been developed for highly specialized and high performance commercial processors and have been described in the literature. These VLSI-based systems required an unusual combination of methodologies and practices commensurate with large design teams while maintaining the localized circuit optimization requirements important to high speed VLSI circuit design. The design of the clock distribution network used to synchronize some well known industrial circuit examples are discussed in this section.

9.1 The Bell Telephone BELLMAC-32A and WE32100 32-bit Microprocessors [B157, B158]

In the early 1980s, a family of 32-bit microprocessors was developed at Bell Laboratories using a variety of advanced CMOS circuit design techniques and methodologies. Since performance was of fundamental importance, significant attention was placed on synchronization, particularly the design of the global clock distribution network. In 1982, Shoji [B157] described the clock distribution of the BELLMAC-32A, a 146,000-transistor CPU operating at 8 MHz and built using a 2.5-μm single-level metal silicide CMOS technology. The clock delay and maximum tolerance of the clock distribution network were specified at 15 ns ± 3.5 ns, defining the maximum permissible clock skew. A four-phase clocking strategy was utilized; each phase synchronizes in order: the slave latches, the slave latch logic, the master latches, and the master latch logic. Each time a clock signal crosses a power buss, a silicide crossunder is used to route the clock signal. In order to equalize the series resistance, each clock path is routed with an identical number of three power buss crossunders from either of the two primary clock lines around the chip periphery. Buffers are placed strategically after each crossunder to amplify the degraded clock signal. With this clock distribution strategy, the circuit satisfies the clock frequency specification of 8 MHz at 70°C with the clock skew not exceeding ±3.5 ns. It is worth noting that, due to the significantly increased complexity encountered when distributing four separate clock signals, a four-phase clocking strategy is not particularly appropriate for higher density, higher speed VLSI-based systems. This perspective is consistent with the processor described next.

In 1986, Shoji [B158] reported on the electrical design of the WE32100 CPU built using a 1.75-μm CMOS technology. The approach used in designing the WE32100 synchronizing clock system is described; local clock skew was optimized for a small number of the critical paths by applying negative clock skew. This strategy is consistent with the customized design methodology used in the design of the CPU. The clock distribution network utilizes a standard tree structure where the input clock signal is buffered by the clock driver and distributed over the entire circuit. Buffers are again placed after each crossunder. A strategy very similar to the approach presented in subsection 4.3 [B28] and depicted in Figure 14 is used to compensate for the variation in interconnect impedance and register load of each clock line. Clock edges at each register are further synchronized by adjusting the MOSFET transistor geometries of the distributed buffers within the clock distribution network.

Another circuit technique used in the WE32100 to minimize the dependence of the clock skew on any process variations is discussed in subsection 6.1. This technique minimizes process-induced clock skew caused by asymmetric variations of the device parameters of the N-channel and P-channel MOSFETs. Clock distribution networks with skews an order of magnitude less than conventionally designed circuits have been simulated with this technique [B107].

The issue of chip-to-chip synchronization is important in the design of integrated circuits because these circuits make up the components of a larger computing system. Since individual integrated circuits are processed in different wafer lots, the device parameters may differ, and therefore any internal delays, such as the clock line delays, will also differ. If the clock delay is T_{CD} for the slowest integrated circuit, as much as $T_{CD}/2$ clock skew may be developed when the fastest and slowest chips communicate synchronously. Therefore, a preferable strategy is to reduce the on-chip clock delay T_{CD} to a minimum and to preselect chips for similar clock delays [B158].

9.2 The DEC 64-bit Alpha Microprocessor [B166]

An important application area for high speed clock distribution networks is the development of high speed microprocessors. The performance of these circuits is often limited by the clocking strategy used in their implementation. The DEC Alpha chip currently represents a significant milestone in microprocessor technology. The VLSI circuit operates above 200 MHz with a 3.3-V power supply implemented in 0.75-μm CMOS three-level metal technology. A clock period of 5 ns must be satisfied for each local data path. Therefore, the clock skew should be of very small magnitude (e.g., less than 0.5 ns for a 10% positive clock skew requirement). This strategy assumes the clock skew to be a global effect rather than a local effect. Thus, careful attention to modeling the circuits and interconnects is required in order to design and analyze this type of high speed system. The Alpha microprocessor contains 1.68 million transistors and supports a fully pipelined 64-bit data structure. The functional attributes of the microprocessor are described in greater detail in [B166], since the focus herein is on the clocking strategy used within the circuit.

In designing this high speed microprocessor, significant attention has been placed on the circuit implementation. The single-phase clock signal is distributed globally on the topmost level of the trilevel metal process, as is the power distribution, since the third layer of metal is thicker and wider (7.5-μm pitch with contacts as compared to 2.625 μm and 2.25 μm for the second and first layers of metal, respectively). Therefore, the resistivity per unit length of the third layer of metal and the metal-to-substrate capacitance is less. A number of inherent difficulties exist within the clock distribution requirements of the Alpha chip. For example, a substantial capacitive load must be driven at high speed by the clock distribution network, 3250 pF (3.25 nF). Also, for both latch design and power dissipation reasons, so as to minimize short-circuit current [31], a fast clock edge rate (<0.5 ns) must be maintained throughout the clock distribution network. The huge capacitative load is due to the 63,000 transistor gates being driven by the clock distribution system. The distribution of the loads is non-symmetric, necessitating a specialized strategy for distributing the clock.

The single 200-MHz clock signal is distributed through five levels of buffering, where the total network consists of 145 separate elements. Each of the elements contains four levels of buffering with a final output stage locally driving the clocked registers. These distributed buffers are configured as a mesh as shown in Figure 24 [B43]. Vertical straps are placed on the second level of metal (M2) to minimize any skew that may develop within the initial four-stage portion of the buffer tree. The primary signal-wide distribution is on the third level of metal (M3), designed to be particularly thick to minimize any line resistance as well as to improve process yield.

The approach used in designing the clock distribution of the Alpha chip is to permit only positive clock skew, thereby assuring that no catastrophic race conditions induced by negative clock skew can occur. Thus, only a graceful degradation in maximum clock rate caused by localized positive clock skew is possible. This strategy is accomplished by locating the clock generation circuitry centrally within the integrated circuit. The clock signal is then distributed radially from the center of the chip die to its periphery. By carefully monitoring the design of this clock distribution methodology, the likelihood of developing a catastrophic amount of negative clock skew (i.e., $|T_{Skew}| > T_{PD(min)}$) is minimized.

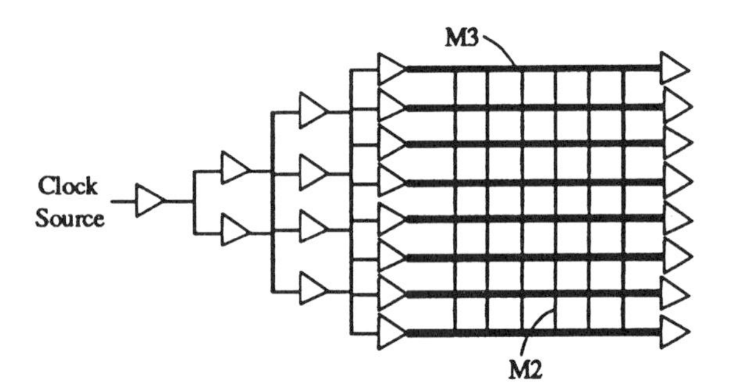

Fig. 24. Clock Distribution Network of DEC Alpha Microprocessor.

9.3 Pipelined Multiplier [B11, B139, B159, B160, B170, 32]

Another application area that requires sophisticated clock distribution is heavily pipelined digital signal processors (DSP), such as FIR/IIR digital filters, multiply-adders, multiply-accumulators, and frequency synthesizers. These types of circuits repeatedly use similar circuit elements, such as multipliers, adders, and registers. Careful attention is placed on developing high performance customized versions of these circuit elements, which are used repeatedly within larger DSP systems. The primary difference between different versions of these DSP circuit components is typically the size of the bit slice (e.g., 8-bit versus 64-bit) and the degree of pipelining. Substantial pipelining is applied to these circuits to increase system clock frequency (see [B128, B129, 4] and subsection 7.2).

The multiplier function is a good example of a complex circuit element capable of significant improvements in clock rate with high levels of pipelining. Since the data flow is nonrecursive, fewer pipeline registers are required as compared to those structures that contain substantial feedback. Furthermore, the multiplier tends to be the critical element (in terms of speed, area, and power) in most DSP circuits. Heavily pipelined multipliers requiring sophisticated clock distribution networks are the focus of considerable research. In this subsection, specific circuit examples of clock distribution networks in highly pipelined DSP-based multipliers implemented in VLSI technologies are described.

A common feature of these VLSI-based multipliers (and many VLSI-based systems) is the repetitive organization of the physical layout. Repetitive parallel arrays of abutted adder cells, pipelined at each bit (a register placed between each adder cell), provide worst case path delays of only a single adder and a register delay (T_{C-Q} and $T_{Set\text{-}up}$), permitting very high multiplication throughput. Specialized architectures beyond the scope of this book, such as carry-save addition, are used to further improve the throughput of these VLSI-based multipliers [see e.g., B11, B139, B159, B160, B170, 32].

In highly arrayed structures, clock skew can appear both horizontally (serial skew), in the direction of the data flow, and vertically (parallel skew), orthogonal to the data flow. As described by Hatamian and Cash in [B139, B159] and pictured in Figure 25, assuming the clock source originates from point A, the clock skew between points D and E at the cell inputs is quite close to the clock skew between points B and C. As long as this horizontal clock skew is less than the local data path delay between cells, no negative clock skew condition will occur [see subsection 3.2 and (6)], and the multiplier array will operate properly. Furthermore, additional cells can be added to the array without

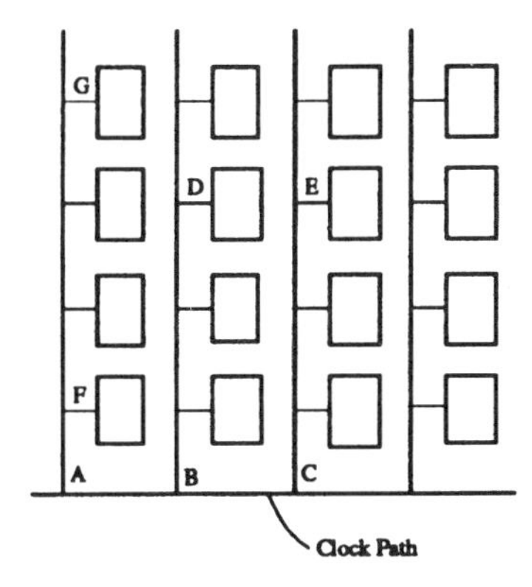

Fig. 25. Clock Distribution Network of Multiplier Array [B139, B159].

creating any race conditions as long as the same constraint is maintained. Unlike the horizontal skew, however, the vertical skew is cumulative. The clock skew increases as the signal propagates vertically from point A. Thus, the cumulative skew between points F and G dominates over the horizontal skew, again assuming the clock source originates from point A. It is worth noting that the highly arrayed structure of the multiplier, which makes it particularly amenable to a VLSI implementation, also constrains and limits the topology and layout of the clock distribution network used to synchronize the circuit.

In the 8-bit × 8-bit multiplier described in [B139, B159], which is implemented in a 2.5-μm CMOS technology and operates up to 70 MHz, the clock signals are distributed entirely on metal, except where short polysilicon crossunders are used to transverse the power lines. A two-level buffer clock distribution network is used, where the clock load is balanced at the output of the second buffer, and the path before the second buffer is kept symmetric with the other clock paths. This clock distribution strategy is similar to that described in [B160]. Also an 8-bit × 8-bit multiplier, the multiplier described in [B160] is implemented in a 1.0-μm NMOS technology and operates up to 330 MHz at room temperature and up to 600 MHz with liquid nitrogen cooling. Pipelining after each 1-bit addition, the multiplier architecture is very similar to that described in [B139]. The clock signals are distributed in metal outside the multiplier array and in polycide (strapped polysilicon with silicide to minimize the line resistance) inside the multiplier array. Two-phase clocking is used with a total master and slave register fanout of 8 pF. No special circuitry to prevent overlap of the two-phase clock is used, since this practice degrades the active-high portion of the clock signal.

Maximal pipelining of multiplying architectures is taken one step further by pipelining each half-bit of an 8-bit × 8-bit multiplier [B170]. Operating at 230 MHz and implemented in a 1.6-μm double-level metal CMOS process, the multiplier architecture is similar to that used in [B139, B159, B160]. The circuit uses a single-phase clocking scheme [33]. A standard three-level buffer network is used to distribute the clock signal. Each row of the multiplier provides a capacitive load of 5.3 pF. A common clock line runs horizontally in metal one (M1) and is driven by a large sized buffer. The complete buffer tree is composed of fourteen buffers, where ten of these buffers drive the registers (i.e., the leaves of the tree). The maximum skew measured between two clock lines is 150 ps [B170].

These three multiplier examples are intended to provide some insight into distributing clock signals within highly arrayed VLSI-based DSP systems. The primary attributes of the multiplier—repetitive circuit elements, abutted or closely spaced layouts, and extremely high throughput (i.e., many hundreds of MHz) due to heavy pipelining—make the multiplier an important class of VLSI system that requires highly specialized clock distribution networks.

10. DIRECTIONS FOR FUTURE RESEARCH IN THE DESIGN OF CLOCK DISTRIBUTION NETWORKS

Significant research opportunities remain in the design of clock distribution networks. Some examples of these research areas are briefly described in this section.

Automated Synthesis of Clock Distribution Networks

Much of the current research focuses on automating the synthesis of clock distribution networks to support higher performance requirements. The optimal placement of locally distributed buffers, improved delay models that account for nonlinear active transistor behavior, the use of localized clock skew to increase circuit speed, and the integration of RC interconnect and buffer physical delay models for more accurate delay analysis must be considered in the automated design and layout of clock distribution networks. The effects of clock skew, both positive and negative, must also be integrated into behavioral synthesis and RC timing analyzers to detect race conditions as well as to satisfy local and global performance constraints. Synchronous timing constraints must be integrated into high level behavioral synthesis algorithms, thereby improving the accuracy and generality of these synthesis (and simulation) tools.

Most clock distribution networks are tree-structured; however, in many customized VLSI circuits, certain portions of the network are strapped in a mesh configuration to minimize interconnect resistance. These mesh networks decrease clock skew as well as improve circuit reliability. Therefore, one area of future research is the automated layout of clock meshes [B91]. Both timing models and physical layout must be integrated to handle this important extension of tree-structured clock distribution networks.

Design of Process Insensitive Clock Distribution Networks

A research area of primary importance to the practical design of high speed clock distribution networks is improving the tolerance of these networks to both process and environmental variations. As functional clock periods approach 1 ns (i.e., 1 GHz clock frequency), variations in delay of tens of picoseconds could degrade the performance and reliability of these high speed synchronous systems. Variations of this magnitude are quite common in modern semiconductor technologies. This topic is of immediate importance to the existing high speed processor community.

Design of Microwave Frequency Clock Distribution Networks

As system-wide clock frequencies increase beyond Gigahertz levels, transmission line effects will begin to influence the performance characteristics of clock distribution networks. Models of interconnect sections will require the inclusion of accurate inductive elements. These and related microwave effects will become more apparent as ultra-high speed digital technologies become more readily available. For example, the superconductive digital electronic technology, Single Flux Quantum (SFQ) logic [B51, 34], can operate well into the 10- to 100-GHz frequency range.

System Issues Affecting the Design of Clock Distribution Networks

System issues in the design of clock distribution networks also necessitate focused research. Important and developing architectural structures, such as parallel processors, neural networks, supercomputers, hybrid circuits or monolithic multichip modules [B171], and wafer scale integration all require specialized synchronization strategies. Some approaches, such as WSI-based optical clock distribution, have been discussed briefly within this introductory chapter and represent important and challenging areas of research. Related systems issues, such as ultra-low power circuits requiring ultra-low power clock distribution networks [B52–B54], are becoming increasingly important. Improving the power dissipation characteristics of clock distribution networks is particularly important, since these networks dissipate a large portion of the total system-wide power budget. Topics such as distributing small differential signals and task monitoring (or power management) strategies are important areas of research in low power and low voltage clock distribution network design.

Debug and Production Test of Clock Distribution Networks

A necessary requirement in developing a product is evaluating the quality of that product. Both debug and production test of high performance clock distribution networks are of fundamental importance. As exemplified in subsection 4.6, minimal research exists that describes how to debug high speed clock distribution networks as well as how to test these networks in a production environment.

A testimony of the importance of the design and analysis of clock distribution networks to future VLSI systems is the large number of recent PhD dissertations on this topic. A few are referenced here [7, 14, 27, 32, 35–39].

11. SUMMARY AND CONCLUSIONS

All electronic systems are fundamentally asynchronous in nature; by the careful insertion of precise localized timing relationships and storage elements, an asynchronous system can be adapted to appear to behave synchronously. As long as the specific local timing and functional relationships are satisfied, synchronous systems can be used, easing the timing constraints on data flow, albeit requiring a clock distribution network to provide the synchronizing reference signal. By synchronizing with distributed clock signals, clock frequency (a measure of how often new data appear at the output of a system) will remain the primary performance metric in synchronous systems. Furthermore, systems can be developed by careful design of the clock distribution network that operate at performance levels otherwise unattainable without significant architectural or technological improvements.

It is often noted that the design of the clock distribution network represents the fundamental circuit limitation to performance in high speed synchronous digital systems. The local data path dependent nature of clock skew, rather than its global characteristics, requires extreme care in the design, analysis, and evaluation of high speed clock distribution networks. The design complexity and difficulty in scaling these networks to finer geometries are the primary reasons for the recent emphasis placed on asynchronous systems. Clearly, however, synchronous systems will be commonplace for a long time to come, necessitating improved techniques for designing and implementing high speed, highly reliable clock distribution networks. Furthermore, as tighter control of the clocking parameters improves, approaches such as localized clock skew will be applied more generally to the design of clock distribution networks to further enhance system performance.

A singular commentary on the current immaturity of the research area of clock distribution design is the complete lack of an agreed upon terminology and notation defining the primary concepts and terms. This problem is evidenced by the large variety of terms used to describe such issues as race conditions, negative clock skew, and T_{Skew}.

In this book on synchronous VLSI-based clock distribution networks, timing relationships are examined and are used to constrain the timing characteristics of the overall system. Various architectures and applications are considered, and circuit strategies for distributing the clock signals are offered. It is the intention of this introductory chapter to integrate the various subtopics and to provide some sense of cohesiveness to the field of clocking and, specifically, clock distribution networks.

ACKNOWLEDGMENT

I would like to acknowledge J.L.C. Neves for his significant assistance with Sections 2 and 7.

REFERENCES

Note: references beginning with a "B" are placed in the bibliography following this immediate list of references. All other references are listed below in the order of their citation.

[1] K. Wagner and E. McCluskey, "Tuning, Clock Distribution, and Communication in VLSI High-Speed Chips," Stanford University, Palo Alto, California, CRC Technical Report 84-5, June 1984.

[2] K. D. Wagner, "A Survey of Clock Distribution Techniques in High-Speed Computer Systems," Stanford University, Palo Alto, California, CRC Technical Report No. 86-20, December 1986.

[3] E. G. Friedman, "Latching Characteristics of a CMOS Bistable Register," *IEEE Transactions on Circuits and Systems I: Fundamental Theory and Applications,* Vol. CAS-40, No. 12, pp. 902–908, December 1993.

[4] E. G. Friedman and J. H. Mulligan, Jr., "Pipelining and Clocking of High Performance Synchronous Digital Systems," *VLSI Signal Processing Technology,* M. A. Bayoumi and E. E. Swartzlander, Jr., eds., pp. 97–133, Kluwer Academic Publishers, 1994.

[5] M. Hatamian, "Understanding Clock Skew in Synchronous Systems," *Concurrent Computations (Algorithms, Architecture and Technology),* S. K. Tewksbury, B. W. Dickinson, and S. C. Schwartz, eds., pp. 87–96, Plenum Publishing, 1988.

[6] J. Alves Marques and A. Cuhna, "Clocking of VLSI Circuits," *VLSI Architecture,* Randell and Treleaven, eds., pp. 165–178, Prentice-Hall, 1983.

[7] E. G. Friedman, *Performance Limitations in Synchronous Digital Systems,* PhD dissertation, University of California, Irvine, California, June 1989.

[8] T. W. Williams and K. P. Parker, "Design for Testability—A Survey," *Proceedings of the IEEE,* Vol. 71, No. 1, pp. 98–112, January 1983.

[9] R. H. Dennard, F. H. Gaensslen, H-N. Yu, V. L. Rideout, E. Bassous, and A. R. LeBlanc, "Design of Ion-Implanted MOSFET's with Very Small Physical Dimensions," *IEEE Journal of Solid-State Circuits,* Vol. SC-9, No. 5, pp. 256–268, October 1974.

[10] H. B. Bakoglu and J. D. Meindl, "Optimal Interconnection Circuits for VLSI," *IEEE Transactions on Electron Devices,* Vol. ED-32, No. 5, pp. 903–909, May 1985.

[11] H. B. Bakoglu, *Circuits, Interconnections, and Packaging for VLSI,* Addison Wesley, 1990.

[12] C. V. Gura, "Analysis of Clock Skew in Distributed Resistive-Capacitive Interconnects," University of Illinois, Urbana, Illinois, SRC Technical Report No. T87053, June 1987.

[13] S. Y. Kung, *VLSI Array Processors,* Prentice-Hall, 1988.

[14] D. M. Chapiro, *Globally-Asynchronous Locally-Synchronous Systems,* PhD dissertation, Stanford University, Palo Alto, California, October, 1984.

[15] D. J. Kinniment and J. V. Woods, "Synchronization and Arbitration Circuits in Digital Systems," *Proceedings of the IEE,* pp. 10–14, 1st Quarter 1980.

[16] T. Sakurai, "Optimization of CMOS Arbiter and Synchronization Circuits with Submicrometer MOSFET's," *IEEE Journal of Solid-State Circuits,* Vol. SC-23, No. 4, pp. 901–906, August 1988.

[17] D. D. Gajski, *Silicon Compilation,* Addison Wesley, 1988.

[18] R. H. Krambeck and M. Shoji, "Skew-Free Clock Circuit for Integrated Circuit Chip," Patent #4,479,216, AT&T Bell Laboratories, October 23, 1984.

[19] C. M. Lee and B. T. Murphy, "Trimmable Loading Elements to Control Clock Skew," Patent #4,639,615, AT&T Bell Laboratories, January 27, 1987; *IEEE Journal of Solid-State Circuits,* Vol. SC-22, No. 6, pp. 1220, December 1987.

[20] S. Dhar and M. A. Franklin, "Optimum Buffer Circuits for Driving Long Uniform Lines," *IEEE Journal of Solid-State Circuits,* Vol. SC-26, No. 1, pp. 32–40, January 1991.

[21] G. Yacoub, H. Pham, M. Ma, and E. G. Friedman, "A System for Critical Path Analysis Based on Back Annotation and Distributed Interconnect Impedance Models," *Microelectronics Journal,* Vol. 19, No. 3, pp. 21–30, May/June 1988.

[22] W. C. Elmore, "The Transient Response of Damped Linear Networks with Particular Regard to Wideband Amplifiers," *Journal of Applied Physics,* Volume 19, pp. 55–63, January 1948.

[23] J. Rubinstein, P. Penfield, Jr., and M. A. Horowitz, "Signal Delay in RC Tree Networks," *IEEE Transactions on Computer-Aided Design,* Vol. CAD-2, No. 3, pp. 202–211, July 1983.

[24] J. Vlach, J. A. Barby, A. Vannelli, T. Talkhan, and C. J. Shin, "Group Delay as an Estimate of Delay on Logic," *IEEE Transactions on Computer-Aided Design,* Vol. CAD-10, No. 7, pp. 949–953, July 1991.

[25] T. Sakurai, "Approximation of Wiring Delay in MOSFET LSI," *IEEE Journal of Solid-State Circuits,* Vol. SC-18, No. 4, pp. 418–426, August 1983.

[26] A. Papoulis, *Probability, Random Variables, and Stochastic Processes,* McGraw-Hill, 1965.

[27] C. T. Gray, *Optimal Clocking of Wave Pipelined Systems and CMOS Applications,* PhD dissertation, North Carolina State University, Raleigh, North Carolina, July 1993.

[28] S. K. Tewksbury, *Wafer-Level Integrated Systems: Implementation Issues,* Kluwer Academic Publishers, 1989.

[29] K. K. Johnstone and J. B. Butcher, "Power Distribution for Highly Parallel WSI Architectures," *Proceedings of the IEEE International Conference on Wafer Scale Integration,* pp. 203–214, January 1989.

[30] J. Fried, "An Analysis of Power and Clock Distribution for WSI Systems," *Wafer Scale Integration*, G. Saucier and J. Trilhe, eds., pp. 127–142, Elsevier Science Publishers B.V. (North-Holland), 1986.

[31] H.J.M. Veendrick, "Short-Circuit Dissipation of Static CMOS Circuitry and its Impact on the Design of Buffer Circuits," *IEEE Journal of Solid-State Circuits,* Vol. SC-19, No. 4, pp. 468–473, August 1984.

[32] M. R. Santoro, *Design and Clocking of VLSI Multipliers,* PhD dissertation, Stanford University, Palo Alto, California, October 1989.

[33] Y. Jiren, I. Karlsson, and G. Svensson, "A True Single-Phase-Clock Dynamic CMOS Circuit Technique," *IEEE Journal of Solid-State Circuits,* Vol. SC-22, No. 4, pp. 899–901, October 1987.

[34] K. K. Likharev and V. K. Semenov, "RSFQ Logic/Memory Family: A New Josephson-Junction Technology for Sub-Terahertz Clock-Frequency Digital Systems," *IEEE Transactions on Applied Superconductivity,* Vol. 1, No. 1, pp. 3–28, March 1991.

[35] D. C. Noice, *A Clocking Discipline for Two-Phase Digital Integrated Circuits,* PhD dissertation, Stanford University, Palo Alto, California, January 1983.

[36] Y. H. Kim, *Accurate Timing Verification for Digital VLSI Designs,* PhD dissertation, University of California, Berkeley, California, January 1989.

[37] P. R. Mukund, *Optimal Clock Distribution in VLSI Systems,* PhD dissertation, University of Tennessee, Knoxville, Tennessee, December 1990.

[38] D. Wong, *Techniques for Designing High-Performance Digital Circuits using Wave-Pipelining,* PhD dissertation, Stanford University, Palo Alto, California, September 1991.

[39] D. A. Joy, *Clock Period Minimization with Wave Pipelining,* PhD dissertation, University of Massachusetts, Amherst, Massachusetts 1991.

BIBLIOGRAPHY

Reference numbers with asterisks refer to papers included in the book. Citations beginning with a "B" refer to this list of publications.

1. Basic Concepts and Analysis

[1]* F. Anceau, "A Synchronous Approach for Clocking VLSI Systems," *IEEE Journal of Solid-State Circuits,* SC-17, No. 1, pp. 51–56, February 1982.

[2] M. A. Franklin and D. F. Wann, "Asynchronous and Clocked Control Structures for VLSI Based Interconnection Networks," *Proceedings of 9th Annual Symposium on Computer Architecture,* pp. 50–59, April 1982.

[3]* D. Wann and M. Franklin, "Asynchronous and Clocked Control Structures for VLSI Based Interconnection Networks," *IEEE Transactions on Computers,* Vol. C-32, No. 3, pp. 284–293, March 1983.

[4]* S. Dhar, M. Franklin, and D. Wann, "Reduction of Clock Delays in VLSI Structures," *Proceedings of IEEE International Conference on Computer Design,* pp. 778–783, October 1984.

[5] S. Unger and C-J. Tan, "Optimal Clocking Schemes for High Speed Digital Systems," *Proceedings of IEEE International Conference on Computer Design,* pp. 366–369, October 1983.

[6] J. Beausang and A. Albicki, "A Method to Obtain an Optimal Clocking Scheme for a Digital System," *Proceedings of IEEE International Conference on Computer Design,* pp. 68–72, October 1983.

[7]* S. H. Unger and C-J. Tan, "Clocking Schemes for High-Speed Digital Systems," *IEEE Transactions on Computers,* Vol. C-35, No. 10, pp. 880–895, October 1986.

[8] D. Noice, R. Mathews, and J. Newkirk, "A Clocking Discipline for Two-Phase Digital Systems," *Proceedings of IEEE International Conference on Circuits and Computers,* pp. 108–111, September 1982.

[9] C. Svensson, "Signal Resynchronization in VLSI System," *Integration, the VLSI Journal,* Vol. 4, No. 1, pp. 75–80, March 1986.

[10] M. S. McGregor, P. B. Denyer, and A. F. Murray, "A Single-Phase Clocking Scheme for CMOS VLSI," *Proceedings of the Stanford Conference on Advanced Research in VLSI,* pp. 257–271, March 1987.

[11]* M. Hatamian and G. L. Cash, "Parallel Bit-Level Pipelined VLSI Designs for High-Speed Signal Processing," *Proceedings of the IEEE,* Vol. 75, No. 9, pp. 1192–1202, September 1987.

[12] M. Hatamian, L. A. Hornak, T. E. Little, S. T. Tewksbury, and P. Franzon, "Fundamental Interconnection Issues," *AT&T Technical Journal,* Volume 66, Issue 4, pp. 13–30, July/August 1987.

[13]* K. D. Wagner, "Clock System Design," *IEEE Design & Test of Computers,* pp. 9–27, October 1988.

[14]* A. F. Champernowne, L. B. Bushard, J. T. Rusterholz, and J. R. Schomburg, "Latch-to-Latch Timing Rules," *IEEE Transactions on Computers,* Vol. C-39, No. 6, pp. 798–808, June 1990.

[15]* D. G. Messerschmitt, "Synchronization in Digital System Design," *IEEE Journal on Selected Areas in Communications,* Vol. 8, No. 8, pp. 1404–1419, October 1990.

[16]* M. Afghahi and C. Svensson, "Performance of Synchronous and Asynchronous Schemes for VLSI Systems," *IEEE Transactions on Computers,* Vol. C-41, No. 7, pp. 858–872, July 1992.

[17] V. Chi, "Designing Salphasic Clock Distribution Systems," *Proceedings of Symposium on Integrated Systems,* pp. 219–233, March 1993.

[18]* V. L. Chi, "Salphasic Distribution of Clock Signals for Synchronous Systems," *IEEE Transactions on Computers,* Vol. C-43, No. 5, pp. 597–602, May 1994.

[19] M. C. Papaefthymiou and K. H. Randall, "Edge-Triggering vs. Two-Phase Level-Clocking," *Proceedings of Symposium on Integrated Systems,* pp. 201–218, March 1993.

[20] W.K.C. Lam, R. K. Brayton, and A. L. Sangiovanni-Vincentelli, "Valid Clocking in Wavepipelined Circuits," *Proceedings of IEEE International Conference on Computer-Aided Design,* pp. 124–131, November 1992.

[21] D. A. Joy and M. J. Ciesielski, "Clock Period Minimization with Wave Pipelining," *IEEE Transactions on Computer-Aided Design,* Vol. CAD-12, No. 4, pp. 461–472, April 1993.

[22]* C. T. Gray, W. Liu, and R. K. Cavin, III, "Timing Constraints for Wave-Pipelined Systems," *IEEE Transactions on Computer-Aided Design,* Vol. CAD-13, No. 8, pp. 987–1004, August 1994.

[23] X. Zhang and R. Sridhar, "Synchronization of Wave-Pipelined Circuits," *Proceedings of IEEE International Conference on Computer Design,* pp. 164–167, October 1994.

[24] E. G. Friedman, "Clock Distribution Design in VLSI Circuits—an Overview," *Proceedings of IEEE International Symposium on Circuits and Systems,* pp. 1475–1478, May 1993.

[25]* K. A. Sakallah, T. N. Mudge, T. M. Burks, and E. S. Davidson, "Synchronization of Pipelines," *IEEE Transactions on Computer-Aided Design,* Vol. CAD-12, No. 8, pp. 1132–1146, August 1993.

[26] R. Peset Llopis, L. Ribas Xirgo, and J. Carrabina Bordoll, "Short Destabilizing Paths in Timing Verification," *Proceedings of IEEE International Conference on Computer Design,* pp. 160–163, October 1994.

[27] S.-Z. Sun, D.H.C. Du, Y.-C. Hsu, and H.-C. Chen, "On Valid Clocking for Combinational Circuits," *Proceedings of IEEE International Conference on Computer Design,* pp. 381–384, October 1994.

2. Clock Distribution Design of Structured Custom VLSI Circuits

[28]* E. G. Friedman and S. Powell, "Design and Analysis of a Hierarchical Clock Distribution System for Synchronous Standard Cell/Macrocell VLSI," *IEEE Journal of Solid-State Circuits,* Vol. SC-21, No. 2, pp. 240–246, April 1986.

[29] E. Friedman, "A Partitionable Clock Distribution System for Sequential VLSI Circuits," *Proceedings of IEEE International Symposium on Circuits and Systems,* pp. 743–746, May 1986.

[30]* H. B. Bakoglu, J. T. Walker, and J. D. Meindl, "A Symmetric Clock-Distribution Tree and Optimized High-Speed Interconnections for Reduced Clock Skew in ULSI and WSI Circuits," *Proceedings of IEEE International Conference on Computer Design,* pp. 118–122, October 1986.

[31] "Method of Deskewing Data Pulses," *IBM Technical Disclosure Bulletin,* Vol. 28, No. 6, pp. 2658–2659, November 1985.

[32] M. Afghahi and C. Svensson, "A Scalable Synchronous System," *Proceedings of IEEE International Symposium on Circuits and Systems,* pp. 471–474, May 1988.

[33]* M. Afghahi and C. Svensson, "A Unified Single-Phase Clocking Scheme for VLSI Systems," *IEEE Journal of Solid-State Circuits,* Vol. SC-25, No. 1, pp. 225–233, February 1990.

[34] B. Wu and N. A. Sherwani, "Effective Buffer Insertion of Clock Tree for High-Speed VLSI Circuits," *Microelectronics Journal,* Vol. 23, No. 4, pp. 291–300, July 1992.

[35]* G. M. Blair, "Skew-Free Clock Distribution for Standard-Cell VLSI Designs," *IEE Proceedings-G,* Vol. 139, No. 2, pp. 265–267, April 1992.

[36]* S. Padin, "Scheme for Distributing High-Speed Clock Signals in a Large Digital System," *Electronics Letters,* Vol. 25, No. 2, pp. 92–93, January 1989.

[37] M. A. Cirit, "Clock Skew Elimination in CMOS VLSI," *Proceedings of IEEE International Symposium on Circuits and Systems,* pp. 861–864, May 1990.

[38]* F. Minami and M. Takano, "Clock Tree Synthesis Based on RC Delay Balancing," *Proceedings of IEEE Custom Integrated Circuits Conference,* pp. 28.3.1–28.3.4, May 1992.

[39] D. Mijuskovic, "Clock Distribution in Application Specific Integrated Circuits," *Microelectronics Journal,* Vol. 18, pp. 15–27, July/August 1987.

[40] D. Renshaw and C. H. Lau, "Race-Free Clocking of CMOS Pipelines Using a Single Global Clock," *IEEE Journal of Solid-State Circuits,* Vol. SC-25, No. 3, pp. 766–769, June 1990.

[41]* I. Deol, "Automatic Analysis of Circuits for Tester Skew and Clock Distribution for VLSI Circuits," *Proceedings of the IEEE International Conference on Computer-Aided Design,* pp. 350–353, October 1987.

[42]* D. C. Keezer, "Design and Verification of Clock Distribution in VLSI," *Proceedings of IEEE International Conference on Communications,* pp. 811–816, April 1990.

[43] M. Horowitz, "Clocking Strategies in High Performance Processors," *Proceedings of the IEEE Symposium on VLSI Circuits,* pp. 50–53, June 1992.

[44] M. Nekili, Y. Savaria, G. Bois, and M. Bennani, "Logic-Based H-Trees for Large VLSI Processor Arrays: A Novel Skew Modeling and High-Speed Clocking Method," *Proceedings of the 5th International Conference on Microelectronics,* pp. 144–147, December 1993.

[45] W. Chuang, S. S. Sapatnekar, and I. N. Hajj, "A Unified Algorithm for Gate Sizing and Clock Skew Optimization to Minimize Sequential Circuit Area," *Proceedings of IEEE International Conference on Computer-Aided Design,* pp. 220–223, November 1993.

[46]* C. Kraft, "Harmonic Series Analysis of Digital Clock Distribution Circuits," *Proceedings of 32nd IEEE Midwest Symposium on Circuits and Systems,* pp. 206–211, August 1989.

[47]* M. Bußmann and U. Langmann, "Active Compensation of Interconnect Losses for Multi-GHz Clock Distribution Networks," *IEEE Transactions on Circuits and Systems-II: Analog and Digital Signal Processing,* Vol. CAS-39, No. 11, pp. 790–798, November 1992.

[48] Q. Zhu, W.W.-M. Dai, and J. G. Xi, "Optimal Sizing of High Speed Clock Networks Based on Distributed RC and Lossy Transmission Line Models," *Proceedings of IEEE International Conference on Computer-Aided Design,* pp. 628–633, November 1993.

[49] P. Ramanathan, A. J. Dupont, and K. G. Shin, "Clock Distribution in General VLSI Circuits," *IEEE Transactions on Circuits and Systems-I: Fundamental Theory and Applications,* Vol. CASI-41, No. 5, pp. 395–404, May 1994.

[50] W. D. Grover, "A New Method for Clock Distribution," *IEEE Transactions on Circuits and Systems-I: Fundamental Theory and Applications,* Vol. CASI-41, No. 2, pp. 149–160, February 1994.

[51] K. Gaj, E. G. Friedman, M. J. Feldman, and A. Krasniewski, "A Clock Distribution Scheme for Large RSFQ Circuits," *IEEE Transactions on Applied Superconductivity,* 1995.

[52]* E. De Man and M. Schobinger, "Power Dissipation in the Clock System of Highly Pipelined ULSI CMOS Circuits," *Proceedings of the International Workshop on Low Power Design,* pp. 133–138, April 1994.

[53] H. Kojima, S. Tanaka, and K. Sasaki, "Half-Swing Clocking Scheme for 75% Power Saving in Clocking Circuitry," *Proceedings of the IEEE Symposium on VLSI Circuits,* pp. 23–24, June 1994.

[54] J. L. Neves and E. G. Friedman, "Minimizing Power Dissipation in Non-Zero Skew-based Clock Distribution Networks," *Proceedings of the IEEE International Symposium on Circuits and Systems,* May 1995.

3. Automated Layout and Synthesis of Clock Distribution Networks

[55]* Y. Ogawa, T. Ishii, Y. Shiraishi, H. Terai, T. Kozawa, K. Yuyama, and K. Chiba, "Efficient Placement Algorithms Optimizing Delay for High-Speed ECL Masterslice LSI's," *Proceedings of ACM/IEEE 23rd Design Automation Conference,* pp. 404–410, June 1986.

[56]* S. Boon, S. Butler, R. Byrne, B. Setering, M. Casalanda, and A. Scherf, "High Performance Clock Distribution for CMOS ASICs," *Proceedings of IEEE Custom Integrated Circuits Conference,* pp. 15.4.1–15.4.5, May 1989.

[57] A. Chao, "Clock Tree Synthesis for Large Gate Arrays," *High Performance Systems,* p. 32, 1989.

[58]* D. Y. Montuno and R.C.S. Ma, "A Layout Methodology for the Synthesis of High Speed Global Clock Nets," *Proceedings of IEEE Custom Integrated Circuits Conference,* pp. 28.4.1–28.4.4, May 1992.

[59]* P. Ramanathan and K. G. Shin, "A Clock Distribution Scheme for Non-Symmetric VLSI Circuits," *Proceedings of the IEEE International Conference on Computer-Aided Design,* pp. 398–401, November 1989.

[60] K. D. Boese and A. B. Kahng, "Zero-Skew Clock Routing Trees with Minimum Wirelength," *Proceedings of IEEE International Conference on ASICs,* pp. 17–21, September 1992.

[61] J. Burkis, "Clock Tree Synthesis for High Performance ASICs," *Proceedings of IEEE International Conference on ASICs,* pp. 9.8.1–9.8.4, September 1991.

[62] T.-H. Chao, Y.-C. Hsu, and J.-M. Ho, "Zero Skew Clock Net Routing," *Proceedings of ACM/IEEE Design Automation Conference,* pp. 518–523, June 1992.

[63]* J. Cong, A. B. Kahng, and G. Robins, "Matching-Based Methods for High-Performance Clock Routing," *IEEE Transactions on Computer-Aided Design,* Vol. CAD-12, No. 8, pp. 1157–1169, August 1993.

[64] M. Edahiro, "A Clock Net Reassignment Algorithm Using Voronoi Diagrams," *Proceedings of IEEE International Conference on Computer-Aided Design,* pp. 420–423, November 1990.

[65]* M.A.B. Jackson, A. Srinivasan, and E. S. Kuh, "Clock Routing for High Performance ICs," *Proceedings of ACM/IEEE Design Automation Conference,* pp. 573–579, June 1990.

[66] A. B. Kahng, J. Cong, and G. Robins, "High-Performance Clock Routing Based on Recursive Geometric Matching," *Proceedings of ACM/IEEE Design Automation Conference,* pp. 322–327, June 1991.

[67] R. S. Tsay, "Exact Zero Skew," *Proceedings of IEEE International Conference on Computer-Aided Design,* pp. 336–339, November 1991.

[68] T. Saigo, S. Watanabe, Y. Ichikawa, S. Takayama, T. Umetsu, K. Mima, T. Yamamoto, J. Santos, and J. Buurma, "Clock Skew Reduction Approach for Standard Cell," *Proceedings of IEEE Custom Integrated Circuits Conference,* pp. 16.4.1–16.4.4, May 1990.

[69] P. D. Ta and K. Do, "A Low Power Clock Distribution Scheme for Complex IC System," *Proceedings of IEEE International Conference on ASICs,* pp. 1-5.1–1-5.4, September 1991.

[70] P. R. Mukund and D. W. Bouldin, "A Graph Theoretic Approach to the Clock Distribution Problem," *Proceedings of IEEE International Conference on ASICs,* pp. 7-5.1–7-5.4, September 1991.

[71] J. Cong, A. Kahng, and G. Robins, "On Clock Routing for General Cell Layouts," *Proceedings of IEEE International Conference on ASICs,* pp. 14.5.1–14.5.4, September 1991.

[72]* T.-H. Chao, Y.-C. Hsu, J.-M. Ho, K. D. Boese, and A. B. Kahng, "Zero Skew Clock Routing with Minimum Wirelength," *IEEE Transactions on Circuits and Systems-II: Analog and Digital Signal Processing,* Vol. CAS-39, No. 11, pp. 799–814, November 1992.

[73] N. A. Sherwani and B. Wu, "Clock Layout for High-Performance ASIC Based on Weighted Center Algorithm," *Proceedings of IEEE International Conference on ASICs,* pp. 15.5.1–15.5.4, September 1991.

[74] D. A. Joy and M. J. Ciesielski, "Placement for Clock Period Minimization with Multiple Wave Propagation," *Proceedings of ACM/IEEE 28th Design Automation Conference,* pp. 640–643, June 1991.

[75] Y-M. Li and M. A. Jabri, "A Zero-Skew Clock Routing Scheme for VLSI Circuits," *Proceedings of the IEEE International Conference on Computer-Aided Design,* pp. 458–463, November 1992.

[76] W. Khan, M. Hossain, and N. Sherwani, "Zero Skew Clock Routing in Multiple-Clock Synchronous Systems," *Proceedings of IEEE International Conference on Computer-Aided Design,* pp. 464–467, November 1992.

[77] Q. Zhu and W.W-M. Dai, "Perfect-Balance Planar Clock Routing with Minimal Path-Length," *Proceedings of IEEE International Conference on Computer-Aided Design,* pp. 473–476, November 1992.

[78]* R-S. Tsay, "An Exact Zero-Skew Clock Routing Algorithm," *IEEE Transactions on Computer-Aided Design,* Vol. CAD-12, No. 2, pp. 242–249, February 1993.

[79] W. Khan and N. Sherwani, "Zero Skew Clock Routing Algorithm for High Performance ASIC Systems," *Proceedings of IEEE International Conference on ASICs,* pp. 79–82, September 1993.

[80] W. Khan, S. Madhwapathy, and N. Sherwani, "An Hierarchical Approach to Clock Routing in High Performance Systems," *Proceedings of IEEE International Symposium on Circuits and Systems,* pp. 1.467–1.470, May/June 1994.

[81] S. Lin and C. K. Wong, "Process-Variation-Tolerant Zero Skew Clock Routing," *Proceedings of IEEE International Conference on ASICs,* pp. 83–86, September 1993.

[82] M. Edahiro, "A Clustering-Based Optimization Algorithm in Zero-Skew Routings," *Proceedings of ACM/IEEE Design Automation Conference,* pp. 612–616, June 1993.

[83] N.-C. Chou and C.-K. Cheng, "Wire Length and Delay Minimization in General Clock Net Routings," *Proceedings of the IEEE International Conference on Computer-Aided Design,* pp. 552–555, November 1993.

[84]* M. Edahiro, "Delay Minimization for Zero-Skew Routing," *Proceedings of the IEEE International Conference on Computer-Aided Design,* pp. 563–566, November 1993.

[85] M. Edahiro, "An Efficient Zero-Skew Routing Algorithm," *Proceedings of ACM/IEEE 31st Design Automation Conference,* pp. 375–380, June 1994.

[86] J. D. Cho and M. Sarrafzadeh, "A Buffer Distribution Algorithm for High-Speed Clock Routing," *Proceedings of ACM/IEEE Design Automation Conference,* pp. 537–543, June 1993.

[87] G. E. Tellez and M. Sarrafzadeh, "Clock-Period Constrained Minimal Buffer Insertion in Clock Trees," *Proceedings of the IEEE International Conference on Computer-Aided Design,* pp. 219–223, November 1994.

[88] C.W.A. Tsao and A. B. Kahng, "Planar-DME: Improved Planar Zero-Skew Clock Routing with Minimum Pathlength Delay," *Proceedings of the European Design Automation Conference,* pp. 440–445, September 1994.

[89] A. B. Kahng and C.-W.A. Tsao, "Low-Cost Single Layer Clock Trees with Exact Zero Elmore Delay Skew," *Proceedings of the IEEE International Conference on Computer-Aided Design,* pp. 213–218, November 1994.

[90] M. Seki, T. Inoue, K. Kato, K. Tsurusaki, S. Fukasawa, H. Sasaki, and M. Aizawa, "A Specified Delay Accomplishing Clock Router Using Multiple Layers," *Proceedings of the IEEE International Conference on Computer-Aided Design,* pp. 289–292, November 1994.

[91]* S. Pullela, N. Menezes, and L. T. Pillage, "Reliable Non-Zero Clock Trees Using Wire Width Optimization," *Proceedings of ACM/IEEE Design Automation Conference,* pp. 165–170, June 1993.

[92] N. Menezes, A. Balivada, S. Pullela, and L. T. Pillage, "Skew Reduction in Clock Trees Using Wire Width Optimization," *Proceedings of IEEE Custom Integrated Circuits Conference,* pp. 9.6.1–9.6.4, May 1993.

[93] S. Pullela, N. Menezes, J. Omar, and L. T. Pillage, "Skew and Delay Optimization for Reliable Buffered Clock Trees," *Proceedings of IEEE International Conference on Computer-Aided Design,* pp. 556–562, November 1993.

[94] K. Zhu and D. F. Wong, "Clock Skew Minimizing During FPGA Placement," *Proceedings of ACM/IEEE 31st Design Automation Conference,* pp. 232–237, June 1994.

[95]* J. L. Neves and E. G. Friedman, "Topological Design of Clock Distribution Networks Based on Non-Zero Clock Skew Specifications," *Proceedings of 36th IEEE Midwest Symposium on Circuits and Systems,* pp. 468–471, August 1993.

[96]* J. L. Neves and E. G. Friedman, "Circuit Synthesis of Clock Distribution Networks Based on Non-Zero Clock Skew," *Proceedings of IEEE International Symposium on Circuits and Systems,* pp. 4.175–4.178, May/June 1994.

[97] J. L. Neves and E. G. Friedman, "Synthesizing Distributed Buffer Clock Trees for High Performance ASICs," *Proceedings of IEEE International Conference on ASICs,* pp. 126–129, September 1994.

[98] J. Chung and C.-K. Cheng, "Optimal Buffered Clock Tree Synthesis," *Proceedings of IEEE International Conference on ASICs,* pp. 130–133, September 1994.

[99] J. Chung and C.-K. Cheng, "Skew Sensitivity Minimization of Buffered Clock Tree," *Proceedings of the IEEE International Conference on Computer-Aided Design,* pp. 280–283, November 1994.

[100] N. Park and A. Parker, "Synthesis of Optimal Clocking Schemes," *Proceedings of ACM/IEEE 22nd Design Automation Conference,* pp. 489–495, June 1985.

[101]* E. G. Friedman, "The Application of Localized Clock Distribution Design to Improving the Performance of Retimed Sequential Circuits," *Proceedings of IEEE Asia-Pacific Conference on Circuits and Systems,* pp. 12–17, December 1992.

[102] T. Soyata, E. G. Friedman, and J. H. Mulligan, Jr., "Integration of Clock Skew and Register Delays into a Retiming Algorithm," *Proceedings of IEEE International Symposium on Circuits and Systems,* pp. 1483–1486, May 1993.

[103]* T. Soyata and E. G. Friedman, "Retiming with Non-Zero Clock Skew, Variable Register, and Interconnect Delay," *Proceedings of the IEEE International Conference on Computer-Aided Design,* pp. 234–241, November 1994.

[104]* N. V. Shenoy, R. K. Brayton, and A. L. Sangiovanni-Vincentelli, "Resynthesis of Multi-Phase Pipelines," *Proceedings of ACM/IEEE Design Automation Conference,* pp. 490–496, June 1993.

[105] L.-F. Chao and E.H.-M. Sha, "Retiming and Clock Skew for Synchronous Systems," *Proceedings of IEEE International Symposium on Circuits and Systems,* pp. 1.283–1.286, May/June 1994.

[106] B. Lockyear and C. Ebeling, "The Practical Application of Retiming to the Design of High-Performance Systems," *Proceedings of IEEE International Conference on Computer-Aided Design,* pp. 288– 295, November 1993.

4. Analysis and Modeling of the Timing and Power Characteristics of Clock Distribution Networks

[107]* M. Shoji, "Elimination of Process-Dependent Clock Skew in CMOS VLSI," *IEEE Journal of Solid-State Circuits,* Vol. SC-21, No. 5, pp. 875–880, October 1986.

[108] S. Lin and C. K. Wong, "Process-Variation-Tolerant Clock Skew Minimization," *Proceedings of the IEEE International Conference on Computer-Aided Design,* pp. 284–288, November 1994.

[109] S. R. Kunkel and J. E. Smith, "Optimal Pipelining in Supercomputers," *Proceedings of IEEE International Symposium on Computer Architecture,* pp. 404–411, June 1986.

[110] S. D. Kugelmass and K. Steiglitz, "A Probabilistic Model for Clock Skew," *Proceedings of IEEE International Conference on Systolic Arrays,* pp. 545–554, May 1988.

[111]* S. D. Kugelmass and K. Steiglitz, "An Upper Bound on Expected Clock Skew in Synchronous Systems," *IEEE Transactions on Computers,* Vol. C-39, No. 12, pp. 1475–1477, December 1990.

[112] C.-S. Li and D. G. Messerschmitt, "Statistical Analysis of Timing Rules for High-Speed Synchronous Interconnects," *Proceedings of IEEE International Symposium on Circuits and Systems,* pp. 37–40, May 1992.

[113] E. Vanden Meersch, L. Claesen, and H. De Man, "Automated Analysis of Timing Faults in Synchronous MOS Circuits," *Proceedings of IEEE International Symposium on Circuits and Systems,* pp. 487–490, May 1988.

[114] M. R. Dagenais and N. C. Rumin, "Automatic Determination of Optimal Clocking Parameters in Synchronous MOS VLSI Circuits," *Proceedings of the 1988 Stanford Conference on Advanced Research in VLSI,* pp. 19–33, March 1988.

[115]* M. R. Dagenais and N. C. Rumin, "On the Calculation of Optimal Clocking Parameters in Synchronous Circuits with Level-Sensitive Latches," *IEEE Transactions on Computer-Aided Design,* Vol. CAD-8, No. 3, pp. 268–278, March 1989.

[116]* C. Svensson and M. Afghahi, "On *RC* Line Delays and Scaling in VLSI Systems," *Electronics Letters,* Vol. 24, No. 9, pp. 562–563, April 1988.

[117] M. Afghahi and C. Svensson, "Calculation of Clock Path Delay and Skew in VLSI Synchronous Systems," *Proceedings of IEEE European Conference on Circuit Theory and Design,* pp. 265–269, September 1989.

[118] S. C. Menon and K. A. Sakallah, "Clock Qualification Algorithm for Timing Analysis of Custom CMOS VLSI Circuits with Overlapping Clocking Disciplines and On-section Clock Derivation," *Proceedings of the First International Conference on Systems Integration,* pp. 550–558, April 1990.

[119] K. A. Sakallah, T. N. Mudge, and O. A. Olukoton, "Analysis and Design of Latch-Controlled Synchronous Digital Circuits," *Proceedings of ACM/IEEE Design Automation Conference,* pp. 111–117, June 1990.

[120]* K. A. Sakallah, T. N. Mudge, and O. A. Olukoton, "Analysis and Design of Latch-Controlled Synchronous Digital Circuits," *IEEE Transactions on Computer-Aided Design,* Vol. CAD-11, No. 3, pp. 322–333, March 1992.

[121]* R.-S. Tsay and I. Lin, "Robin Hood: A System Timing Verifier for Multi-Phase Level-Sensitive Clock Designs," *Proceedings of IEEE International Conference on ASICs,* pp. 516–519, September 1992.

[122]* I. Lin, J. A. Ludwig, and K. Eng, "Analyzing Cycle Stealing on Synchronous Circuits with Level-Sensitive Latches," *Proceedings of ACM/IEEE Design Automation Conference,* pp. 393–398, June 1992.

[123] S. Narayan and D. D. Gajski, "System Clock Estimation based on Clock Wastage Minimization," *Proceedings of the European Design Automation Conference,* pp. 66–71, September 1992.

[124] M. C. Papaefthymiou and K. H. Randall, "TIM: A Timing Package for Two-Phase, Level-Clocked Circuitry," *Proceedings of ACM/IEEE Design Automation Conference,* pp. 497–502, June 1993.

[125] P. V. Argade, "Sizing an Inverter with a Precise Delay: Generation of Complementary Signals with Minimal Skew and Pulsewidth Distortion in CMOS," *IEEE Transactions on Computer-Aided Design,* Vol. CAD-8, No. 1, pp. 33–40, January 1989.

5. Specification of the Optimal Timing Characteristics of Clock Distribution Networks

[126]* J. P. Fishburn, "Clock Skew Optimization," *IEEE Transactions on Computers,* Vol. C-39, No. 7, pp. 945–951, July 1990.

[127] K. A. Sakallah, T. N. Mudge, T. M. Burks, and E. S. Davidson, "Optimal Clocking of Circular Pipelines," *Proceedings of IEEE International Conference on Computer Design,* pp. 642–646, October 1991.

[128]* E. G. Friedman and J. H. Mulligan, Jr., "Clock Frequency and Latency in Synchronous Digital Systems," *IEEE Transactions on Signal Processing,* Vol. SP-39, No. 4, pp. 930–934, April 1991.

[129] E. G. Friedman and J. H. Mulligan, Jr., "Pipelining of High Performance Synchronous Digital Systems," *International Journal of Electronics,* Vol. 70, No. 5, pp. 917–935, May 1991.

[130]* K. A. Sakallah, T. N. Mudge, and O. A. Olukoton, "*checkTc* and *minTc:* Timing Verification and Optimal Clocking of Synchronous Digital Circuits," *Proceedings of IEEE International Conference on Computer-Aided Design,* pp. 552–555, November 1990.

[131] K. A. Sakallah, T. N. Mudge, and O. A. Olukoton, "Optimal Clocking of Synchronous Systems," *Proceedings of TAU90 ACM International Workshop on Timing Issues in the Specification and Synthesis of Digital Systems,* August 1990.

[132] T. M. Burks, K. A. Sakallah, K. Bartlett, and G. Borriello, "Performance Improvement through Optimal Clocking and Retiming," *Proceedings of International Workshop on Logic Synthesis,* May 1991.

[133]* T. G. Szymanski, "Computing Optimal Clock Schedules," *Proceedings of ACM/IEEE Design Automation Conference,* pp. 399–404, June 1992.

[134]* T. G. Szymanski and N. Shenoy, "Verifying Clock Schedules," *Proceedings of IEEE International Conference on Computer-Aided Design,* pp. 124–131, November 1992.

[135] N. Shenoy and R. K. Brayton, "Graph Algorithms for Clock Schedule Optimization," *Proceedings of IEEE International Conference on Computer-Aided Design,* pp. 132–136, November 1992.

[136] R. B. Deokar and S. S. Sapatnekar, "A Graph-theoretic Approach to Clock Skew Optimization," *Proceedings of IEEE International Symposium on Circuits and Systems,* pp. 1.407–1.410, May/June 1994.

[137] W. Chuang, S. S. Sapatnekar, and I. N. Hajj, "A Unified Algorithm for Gate Sizing and Clock Skew Optimization to Minimize Sequential Circuit Areas," *Proceedings of IEEE International Conference on Computer-Aided Design,* pp. 220–223, November 1993.

[138] T. M. Burks, K. A. Sakallah, and T. N. Mudge, "Identification of Critical Paths in Circuits with Level-Sensitive Latches," *Proceedings of IEEE International Conference on Computer-Aided Design,* pp. 137–141, November 1992.

6. Clock Distribution Networks for Targeted VLSI/WSI Architectures

[139] M. Hatamian and G. Cash, "A 70-MHz 8-bit × 8-bit Parallel Pipelined Multiplier in 2.5μm CMOS," *IEEE Journal of Solid-State Circuits,* Vol. SC-21, No. 4, pp. 505–513, August 1986.

[140] J. N. Coleman and R. M. Lea, "Clock Distribution Techniques for Wafer-Scale Integration," *Proceedings of Southhampton Workshop on Wafer Scale Integration,* pp. 1–8, 1985.

[141] D. C. Keezer and V. K. Jain, "Neural Network Clock Distribution," *Proceedings of IFIP Workshop on Silicon Architecture for Neural Nets,* pp. 101–111, November 1990.

[142] D. C. Keezer and V. K. Jain, "Clock Distribution Strategies for WSI: A Critical Survey," *Proceedings of IEEE International Conference on Wafer Scale Integration,* pp. 277–283, January 1991.

[143]* D. C. Keezer and V. K. Jain, "Design and Evaluation of Wafer Scale Clock Distribution," *Proceedings of IEEE International Conference on Wafer Scale Integration,* pp. 168–175, January 1992.

[144]* H. U. Chou and M. A. Franklin, "Optical Distribution of Clock Signals in Wafer Scale Digital Circuits," *Proceedings of IEEE International Conference on Computer Design,* pp. 117–121, October 1987.

[145] A. L. Fisher and H. T. Kung, "Synchronizing Large VLSI Processor Arrays," *Proceedings of 10th Annual International Symposium on Computer Architecture,* pp. 54–58, June 1983.

[146]* A. L. Fisher and H. T. Kung, "Synchronizing Large VLSI Processor Arrays," *IEEE Transactions on Computers,* Vol. C-34, No. 8, pp. 734–740, August 1985.

[147] M. Hatamian and G. Cash, "High Speed Signal Processing, Pipelining, and VLSI," *Proceedings of IEEE Conference on Acoustics, Speech, and Signal Processing,* pp. 1173–1176, April 1986.

[148]* S. Y. Kung and R. J. Gal-Ezer, "Synchronous Versus Asynchronous Computation in Very Large Scale Integrated (VLSI) Array Processors," *Proceedings of SPIE,* Vol. 341, Real Time Signal Processing V, pp. 53–65, May 1982.

[149] M. D. Dikaiakos and K. Steiglitz, "Comparison of Tree and Straight-Line Clocking for Long Systolic Arrays," *Proceedings of IEEE Conference on Acoustics, Speech, and Signal Processing,* pp. 1177–1180, April 1991.

[150]* M. D. Dikaiakos and K. Steiglitz, "Comparison of Tree and Straight-Line Clocking for Long Systolic Arrays," *Journal of VLSI Signal Processing,* Vol. 2, pp. 287–299, 1991.

[151] M. Roumeliotis, J. McKeeman, and G. Gray, "A Distributed Fault Tolerant Clocking Scheme for Systolic Array Architectures," *Proceedings of 6th Annual International Phoenix Conference on Computers and Communications,* pp. 105–109, February 1987.

[152] A. El-Amawy, "Branch-and-Combine Clocking of Arbitrarily Large Computer Networks," *Proceedings of IEEE International Conference on Parallel Processing,* pp. 409–417, August 1991.

[153]* A. El-Amawy, "Clocking Arbitrarily Large Computing Structures under Constant Skew Bound," *IEEE Transactions on Parallel and Distributed Systems,* Vol. PDS-4, No. 3, pp. 241–255, March 1993.

[154] N. Nigam and D. C. Keezer, "A Comparative Study of Clock Distribution Approaches for WSI," *Proceedings of IEEE International Conference on Wafer Scale Integration,* pp. 243–251, January 1993.

[155] N. G. Sheridan, C. M. Habiger, and R. M. Lea, "WSI Clock & Signal Distribution: a Novel Approach," *Proceedings of IEEE International Conference on Wafer Scale Integration,* pp. 252–261, January 1993.

[156] A. El-Amawy and U. Maheshwar, "Synchronous Clocking Schemes for Large VLSI Systems," *Proceedings of 27th Asilomar Conference on Signals, Systems, & Computers,* pp. 761–765, November 1993.

7. Example Implementations of Clock Distribution Networks

[157]* M. Shoji, "Electrical Design of BELLMAC-32A Microprocessor," *Proceedings of IEEE International Conference on Circuits and Computers,* pp. 112–115, September 1982.

[158]* M. Shoji, "Reliable Chip Design Method in High Performance CMOS VLSI," *Proceedings of IEEE International Conference on Computer-Aided Design*, pp. 389–392, October 1986.

[159] M. Hatamian and G. L. Cash, "High Speed Signal Processing, Pipelining, and VLSI," *Proceedings of IEEE International Conference on Acoustics, Speech, and Signal Processing*, pp. 1173–1176, April 1986.

[160] T. G. Noll, D. Schmitt-Landsiedel, H. Klar, and G. Enders, "A Pipelined 330-MHz Multiplier," *IEEE Journal of Solid-State Circuits*, Vol. SC-21, No. 3, pp. 411–416, June 1986.

[161] R. Maini, J. McDonald, and L. Spangler, "A Clock Distribution Circuit with a 100 PS Skew Window," *Proceedings of IEEE Bipolar Circuits and Technology Meeting*, pp. 41–43, September 1987.

[162] D. Chengson, L. Costantino, A. Khan, D. Le, and L. Yue, "A Dynamically Tracking Clock Distribution Chip with Skew Control," *Proceedings of IEEE Custom Integrated Circuits Conference*, pp. 15.6.1–15.6.4, May 1990.

[163] H. Itoh, N. Masuda, S. Kawashima, B. Fujita, S. Ishii, and M. Usami, "A Novel Design Concept for Small-Skew Clock LSIs with the Self-Delay-Adjustment," *Proceedings of IEEE Bipolar Circuits and Technology Meeting*, pp. 130–133, September 1991.

[164] M. Usami, S. Ishii, S. Kawashima, B. Fujita, N. Masuda, and H. Itoh, "An Automatic 5ps Skew-time Control Clock-pulse Adjustment LSI for High-speed Computers," *Proceedings of IEEE Symposium on VLSI Circuits*, pp. 53–54, May/June 1991.

[165] G. M. Blair, "Bit-Serial Correlator with Novel Clocking Scheme," *Proceedings of European Solid-State Circuits Conference*, pp. 157–160, 1991.

[166]* D. W. Dobberpuhl, *et al.*, "A 200-MHz 64-b Dual Issue CMOS Microprocessor," *IEEE Journal of Solid-State Circuits*, Vol. SC-27, No. 11, pp. 1555–1565, November 1992.

[167] R. B. Watson, Jr., H. A. Collins, and R. Iknaian, "Clock Buffer Chip with Absolute Delay Regulation Over Process and Environmental Variations," *Proceedings of IEEE Custom Integrated Circuits Conference*, pp. 25.2.1–25.2.5, May 1992.

[168] A. Ishibashi, A. Maeda, T. Arakawa, K. Higashitani, and M. Tatsuki, "High-Speed Clock Distribution Architecture Employing PLL for 0.6μm CMOS SOG," *Proceedings of IEEE Custom Integrated Circuits Conference*, pp. 27.6.1–27.6.4, May 1992.

[169] K. Ishibashi, T. Hayashi, T. Doi, N. Masuda, A. Yamagiwa, and T. Okabe, "A Novel Clock Distribution System for CMOS VLSI," *Proceedings of IEEE International Conference on Computer-Aided Design*, pp. 289–292, October 1993.

[170] D. Somasekhar and V. Visvanathan, "A 230-MHz Half-Bit Level Pipelined Multiplier Using True Single-Phase Clocking," *IEEE Transactions on VLSI Systems*, Vol. VLSI-1, No. 4, pp. 415–422, December 1993.

[171]* R. Reinschmidt and D. Leuthold, "Clocking Considerations for a Pentium Based CPU Module with a 512K Byte Secondary Cache," *Proceedings of IEEE Multi-Chip Module Conference*, pp. 26–31, March 1994.

8. Recent Papers

[172] Y.-M. Li, S. Ashtaputre, J. Greidinger, M. Hartoog, M. Hossain, and S.-T. Hui, "Skew Controllable Buffered Clock Tree," *Proceedings of 1st International Conference on ASIC*, pp. 29–32, October 1994.

[173] P. Popescu, C. Kurowski, G. Thomsen, and W. Jager, "High Speed Clocking Methodology for ASIC Design in Fault Tolerant Systems," *Proceedings of 1st International Conference on ASIC*, pp. 271–274, October 1994.

[174] A. Balboni, C. Costi, A. Pellencin, M. Quadrini, and D. Sciuto, "Automatic Clock Tree Generation in ASIC Designs," *Proceedings of European Design and Test Conference*, March 1995.

[175] A. Erdal, M. Yue, L. Hiramoto, and J. Stahler, "An Implementation of a Clock-Tree Distribution Scheme for High-Performance ASICs," *Proceedings of IEEE International ASIC Conference*, pp. 26–29, September 1992.

[176] W. Khan, M. Hossain, and N. Sherwani, "Minimum Skew Multiple Clock Routing in Synchronous ASIC Systems," *Proceedings of IEEE International ASIC Conference*, pp. 22–25, September 1992.

[177] W. Chuang, S. S. Sapatnekar, and I. N. Hajj, "Timing and Area Optimization for Standard-Cell VLSI Circuit Design," *IEEE Transactions on Computer-Aided Design*, vol. CAD-14, no. 3, pp. 308–320, March 1995.

[178] J.-D. Cho and M. Sarrafzadeh, "A Buffer Distribution Algorithm for High-Performance Clock Net Optimization," *IEEE Transactions on VLSI Systems*, vol. VLSI-3, no. 1, pp. 84–98, March 1995.

[179] N.-C. Chou and C.-K. Cheng, "On General Zero-Skew Clock Net Construction," *IEEE Transactions on VLSI Systems*, vol. VLSI-3, no. 1, pp. 141–146, March 1995.

[180] T. Soyata, E. G. Friedman, and J. H. Mulligan, Jr., "Monotonicity Constraints on Path Delays for Efficient Retiming with Localized Clock Skew and Variable Register Delay," *Proceedings of the International Symposium on Circuits and Systems*, May 1995.

Part 1

Basic Concepts and Analysis

A Synchronous Approach for Clocking VLSI Systems

FRANÇOIS ANCEAU

***Abstract*—This paper presents a synchronous solution for clocking VLSI systems organized as distributed systems. This solution avoids the drawbacks of the self-timed approach. These VLSI systems are constituted of modules which represent synchronous areas driven by their own fast clock, interconnected by a synchronous communication mechanism driven by a slow clock. In order to avoid the risk of metastability in flip-flop between the modules and the communication mechanism, we suggest to resynchronize the phase of each module clock on the transitions of the communication clock by a phase locked loop circuitry added to each module.**

I. Introduction

The clock mechanisms of VLSI systems may not be designed as a simple extension of those used in a synchronous approach in the existing LSI circuits. The propagation delays in the wires which distribute clock signals become so large when compared to gate delays, that they introduce unacceptable skew between the clock signals which drive the different parts of a VLSI.

Some authors [1], [2] suggest using a self-timed approach in order to avoid this problem; such an approach has several drawbacks, mainly, the loss of a unique time reference for the system, and the need of changing design philosophy.

This paper presents a solution for clocking a family of VLSI systems using a single accurate clock (like an X-Tal clock) without introducing metastability problems.

II. Isochronic Regions

Let us call isochronic regions (or equipotential regions [1]) those parts of a VLSI system which can be considered as synchronous and may be driven by their own single clock. In order to keep the integrity of the clocking scheme, the clock skew in every point of an isochronic region must be kept below a given fraction s of the clock period τ (e.g., 5 to 10 percent). The evaluation of the clock skew permits us to define a relationship between the diameter D of an isochronic region and the period τ of its clock.

The maximum length of the wires used to distribute the clock inside an isochronic region may be evaluated as being KD (with $1 \leqslant K \leqslant 3$).

The maximum resistance of these wires is

$$R = R_{\square} \frac{KD}{W}$$

where W is the width of wires and $R_{\square}$ the resistance per square of this layer.

The maximum capacitance of these wires is

$$C = C_{\square} WKD$$

where $C_{\square}$ is the capacitance per square of this layer.

The propagation through an RC distributed wire is given by a diffusion equation when the input is a step function

$$v(t) = V_0 \operatorname{erfc}\left(\frac{1}{2\sqrt{\frac{t}{RC}}}\right)$$

$$\text{where } \operatorname{erfc}(x) = 1 - \frac{2}{\sqrt{\pi}} \int_0^x e^{-\lambda^2} d\lambda$$

R is the total resistance of the line
C is the total capacitance of the line.

The practical propagation delay (skew) t_s is given by

$$t_s = RC = R_{\square} C_{\square} K^2 D^2.$$

Hence, the clock period must satisfy the relation

$$\tau \geqslant \frac{1}{s} t_s = \underbrace{\frac{1}{s} R_{\square} C_{\square} K^2}_{S} D^2 = SD^2. \tag{1}$$

From another point of view, the clock period of a module must be wide enough to allow combinatorial circuits to give stable outputs.

$$\tau \geqslant 2n\, \tau g \tag{2}$$

where τg is the typical gate delay and n the length of critical path measured in terms of logical layers (typically, $20 \geqslant 2n \geqslant 100$).

From (1) and (2)

$$\tau \geqslant \max (2n\, \tau g, SD^2).$$

When scaling down technology by a factor α, the diameter of a module is divided by α and the length of its clock wires is also divided by α. The oxide and metal thicknesses are also assumed to be divided by α (which is a worse case for metal layers); hence, $C_{\square}$ and $R_{\square}$ are multiplied by α during scaling. The propagation delay into the clock wires of this module thus remains constant during scaling. The gate delay τ_g and capacitive effects are divided by α. The relative importance of gate and propagation delays may change during scaling. When the propagation delays become predominant over gate delays, the clock period of a module becomes constant during scaling.

Reprinted from *IEEE J. Solid-State Circuits,* vol. SC-17, no. 1, pp. 51–56, Feb. 1982.

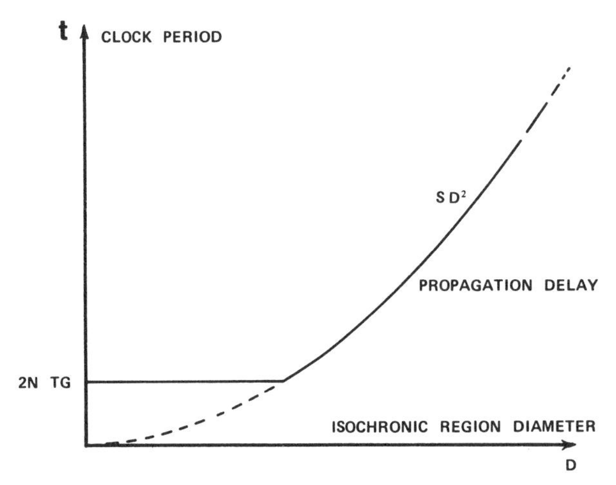

Fig. 1. Clock period for an isochronic region.

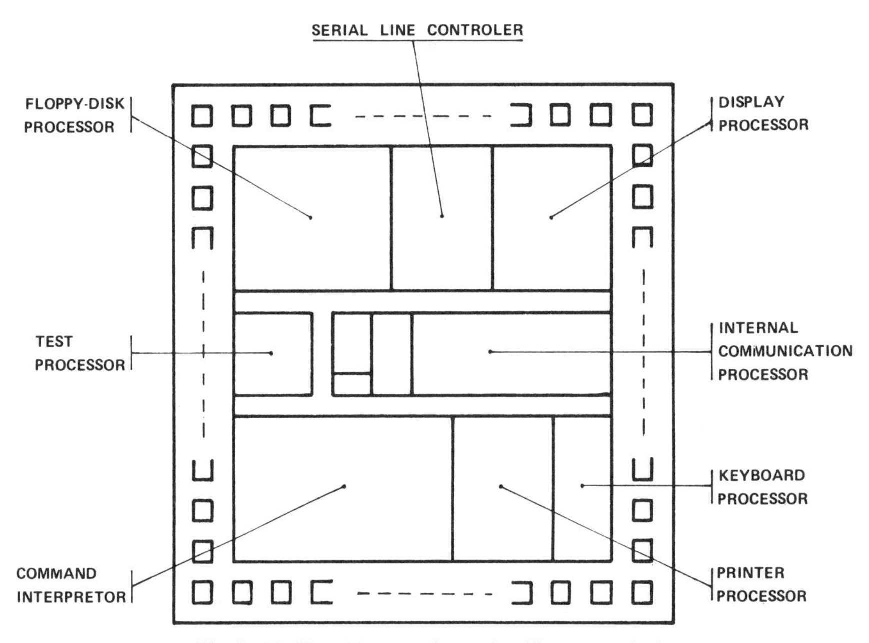

Fig. 2. VLSI architecture for an intelligent terminal.

III. Internal Architecture of a Family of VLSI Systems

A large application area of VLSI systems will be their use instead of microprocessors systems to control electromechanical devices, smart terminals, etc.

A possible internal architecture for these VLSI systems [3], [4] could be the assembly of predesigned intelligent functional modules (processors, I/O controllers, etc.) interconnected as distributed systems by a communication mechanism. The intelligence of functional modules reduces the amount of data they exchange to perform a given job.

Such an architecture is well adapted to the use of a communication clock slower than the module clocks. This property justifies this proposal of architecture for VLSI systems. Each of these functional modules may be small enough to define an isochronic region driven by a single clock (called module clock) with a period defined from gate delays (the propagation delays are negligible inside modules).

A synchronous communication mechanism may be used to interconnect functional modules. It spans the complete chip and defines an isochronic area as large as the complete chip.

TABLE I
Functional Module Scaling

NMOS technology	6 μ	1 μ	0.3 μ
Scaling factor	1	6	20
Typical gate delay	5 ns	0.8 ns	0.25 ns
Functional module diameter D	7 mm	1.16 mm	0.35 mm
Propagation delay (1.5 D metal wire)	0.1 ns	0.1 ns	0.1 ns
Clock period	500 ns	80 ns	25 ns
	from gate delay ($2n$ = 100)		

Its clock, called the communication clock, is slower than the clock used by the functional modules.

If we assume that the overall size of the chips remains constant during technology scaling, the propagation delays into the communication clock wires is multiplied by α^2. The driven delays (gate) will be divided by α.

For submicron technologies, the communication clock period will be defined by propagation delays rather than driver delays.

TABLE II
COMMUNICATION MECHANISM SCALING

NMOS technology	6 μ	1 μ	0.3 μ
Scaling factor	1	6	20
Chip size	7 mm	7mm	7 mm
Propagation delays (10 mm metal wires)	0.1 ns	3.6 ns	40 ns
Communication clock period	500 ns	80 ns	400 ns
	from driver delays ($2n = 100$)		from propagation delays ($1/s = 10$)

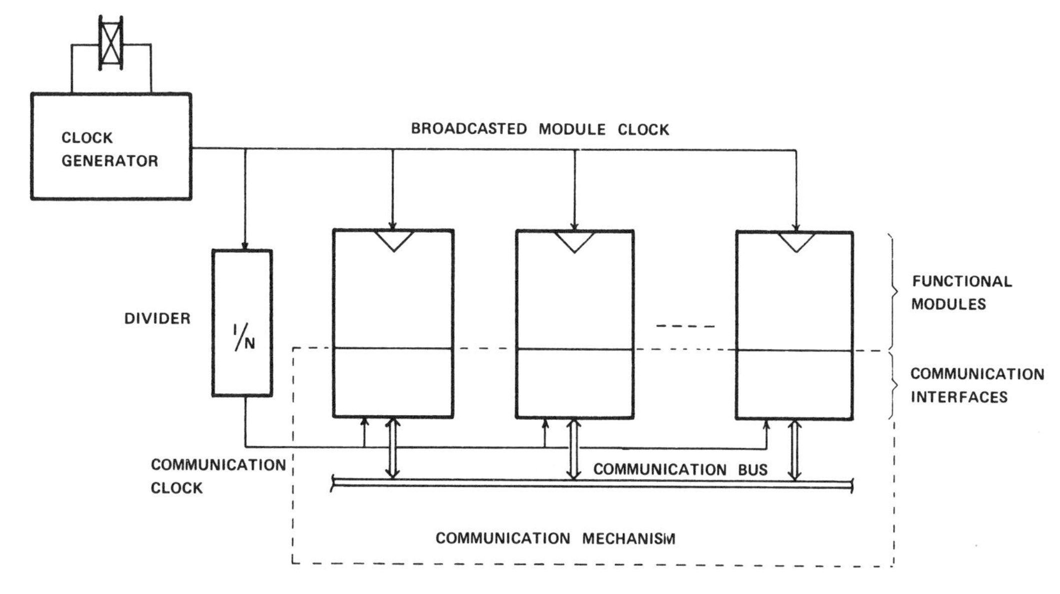

Fig. 3. Clock distribution.

IV. Clock Mechanism

Clock signals which drive each functional module may be obtained from a common source (external or on-chip oscillator) and broadcasted inside the chip to the functional modules. The slower communication clock may be obtained by dividing the oscillator output.

V. Communication Mechanism

The communication mechanism consists of the communication interfaces of each functional module and of the communication bus. Each communication interface is driven by the communication clock.

The propagation delays, both in the wires which distribute the broadcasted module clock and in the wires which distribute the communication clock, lead us to consider a functional module and its communication interface as asynchronous.

VI. Metastability Avoidance

The control of data exchange between a functional module and its communication interface introduces a risk of setting synchronizer flip-flops ff_1 and ff_2 in a intermediate state where these flip-flops may remain a longtime out of a logical state [5], [6]. This phenomenon is called metastability.

Metastability phenomena may occur only for very specific time relationships between the active transitions of internal module clocks and of the communication clock which corresponds to changing the input of a flip-flop when it latches.

Metastability problems may be avoided by maintaining a correct phase relationship between module clocks and communication clock.

This phase relationship may be maintained by the use of a phase-locked loop circuitry in each module. That circuitry adjusts the phase of each internal module clock to communication clock in order to avoid metastability in the flip-flops which control dialog between functional modules and their communication interfaces.

Such a solution is made possible because the complete VLSI system uses a single clock source. Every internal module clock has exactly the same frequency as the common clock generator, and the frequency of the communication clock is an exact fraction of the module clock.

The technological variation and the thermal effects may introduce such a large skew between module and communication clocks, that metastability problems may also occur during the life of a VLSI system. A solution to avoid this is the use, for each functional module, of a phase-locked loop circuitry to dynamically adjust the relationship between the phase of the internal module clock and the phase of the communication clock which drives its communication interface.

The amount of circuitry required to build the phase locked loop is small when compared to the expected size of a functional module and its communication interface.

An example of a possible NMOS realization is given below.

The accuracy and speed needed for such a phase locked loop

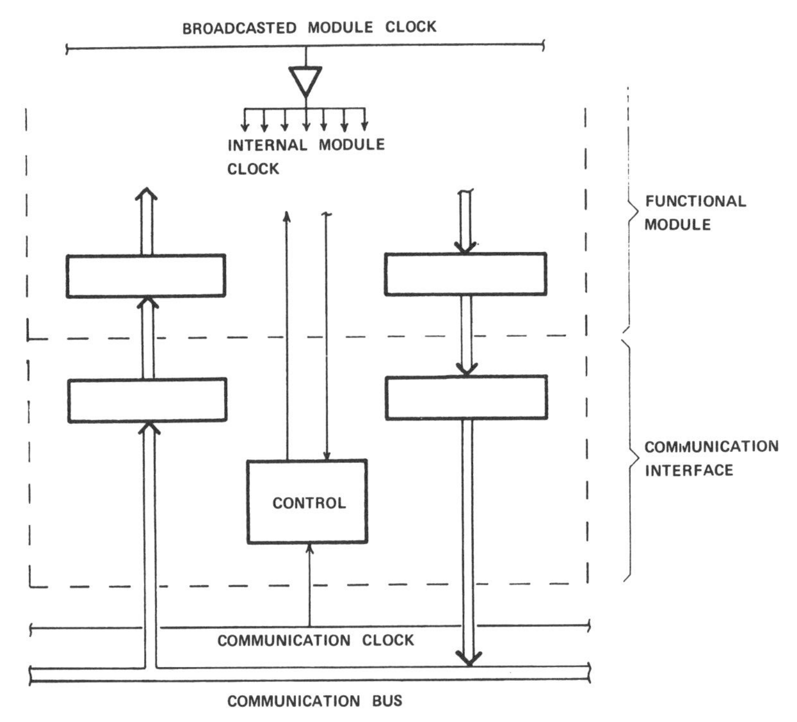

Fig. 4. Communication interface.

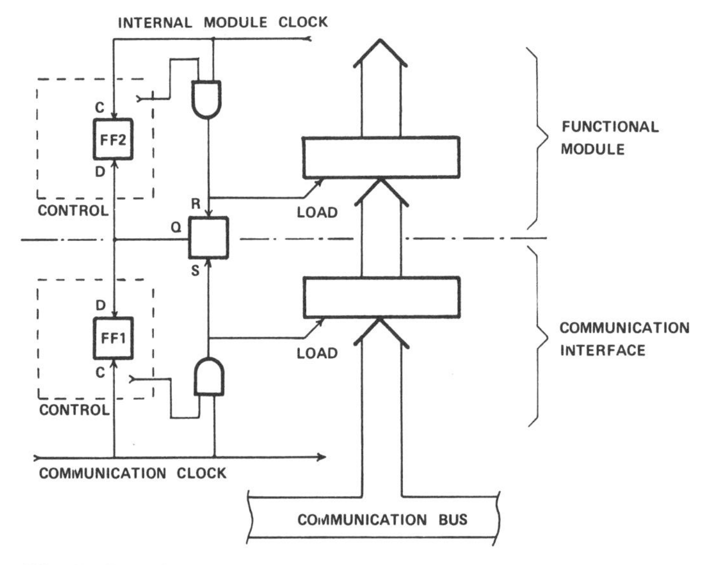

Fig. 5. Control of data exchanges between functional module and its communication interface (reception part).

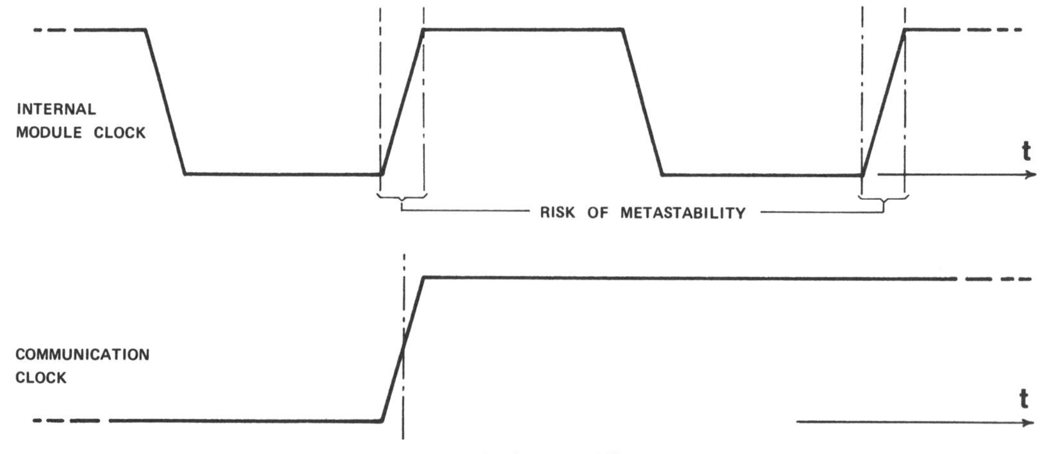

Fig. 6. Risk of metastability.

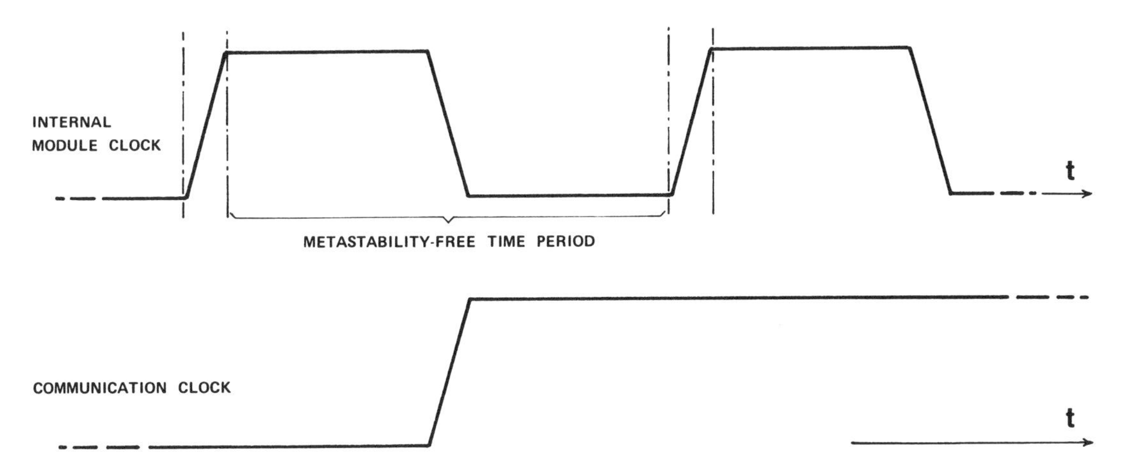

Fig. 7. Metastability avoidance.

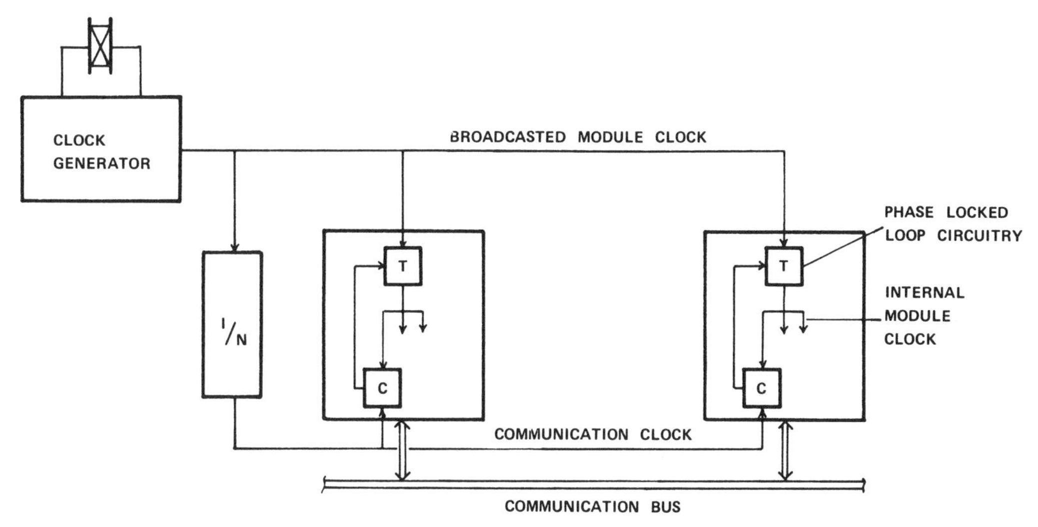

Fig. 8. Use of phase-locked loop circuitries.

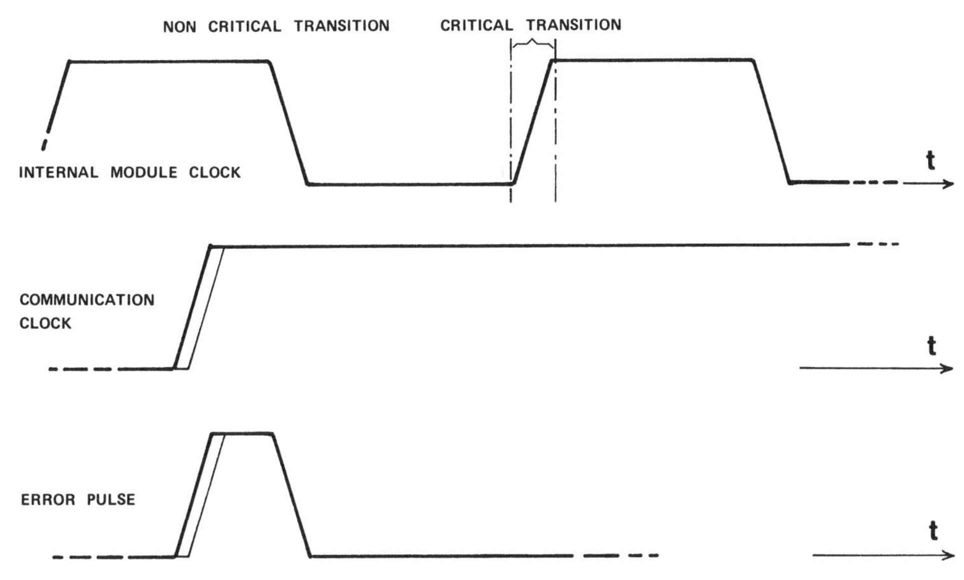

Fig. 9. Communication clock transition must occur before module clock transition to allow phase locking.

need not be very high. Its purpose is only to compensate technological variation and slow thermal influence on propagation delays in the wires which distribute the clock.

VII. Conclusion

VLSI systems, organized as distributed systems, may be built using a single and accurate clock generator. Their design ap-

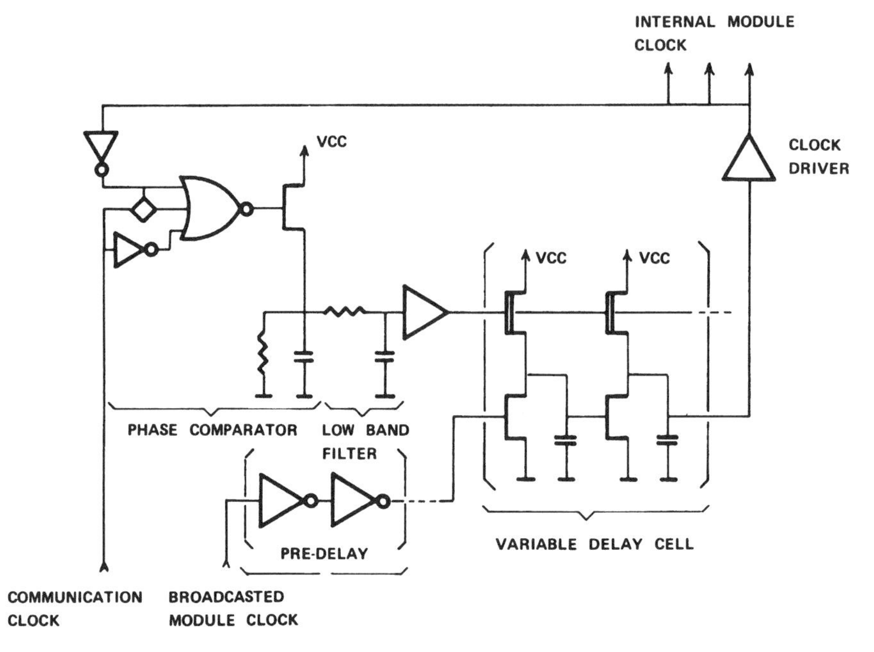

Fig. 10. Possible NMOS realization of the phase-locked loop circuitry.

proaches are very close to that used for designing existing distributed systems using synchronous modules and asynchronous communication. The dynamic phase adjustment mechanism into each internal module allows the avoidance of the risks of metastability introduced by propagation delays into clock distribution. This solution does not violate the theoretical conjectural unsolvability of metastability because such a VLSI system is not really a multiclock device since all its internal clocks are built from a single clock generator. This solution of local resynchronization is very similar to the one used in mainframe computers to resynchronize the clock in order to maintain a synchronous behavior between frames. It may be reused to avoid metastability in other kinds of distributed systems using synchronous communications, such as some local computer networks.

References

[1] C. Mead and L. Conway, *Introduction to VLSI Systems.* Reading, MA: Addison-Wesley, 1980.

[2] C. L. Seitz, "Self-timed VLSI systems," in *Proc. Caltech Conf. VLSI*, Pasadena, CA, Jan. 22-24, 1979, pp. 345-355.

[3] J. M. C. A. Marques, "A multiprocessor architecture adapted to VLSI custom design," in *Proc. Euromicro Conf.*, London, England, Sept. 16-18, 1980, pp. 321-327.

[4] —, "MOSAIC: A design methodology for VLSI custom circuits," in *Proc. ESSIRC '80*, Grenoble, Sept. 22-25, 1980, pp. 26-28.

[5] I. Catt, "Time loss through gating of asynchronous logic signal pulses," *IEEE Trans. Electron. Comput.*, vol. EC-12, pp. 108-111, Feb. 1966.

[6] T. J. Chaney and C. E. Molnar, "Anomalous behavior of synchronizers and arbiter circuits," *IEEE Trans. Comput.*, vol. C-22, pp. 421-422, Apr. 1973.

[7] M. Pechoucek, "Anomalous response times of input synchronizers," *IEEE Trans. Comput.*, vol. C-25, pp. 133-139, Feb. 1976.

Asynchronous and Clocked Control Structures for VSLI Based Interconnection Networks

DONALD F. WANN, MEMBER, IEEE, AND MARK A. FRANKLIN, SENIOR MEMBER, IEEE

***Abstract*—A central issue in the design of multiprocessor systems is the interconnection network which provides communication paths between the processors. For large systems, high bandwidth interconnection networks will require numerous "network chips" with each chip implementing some subnetwork of the original larger network. Modularity and growth are important properties for such networks since multiprocessor systems may vary in size. This paper is concerned with the question of timing control of such networks. Two approaches, asynchronous and clocked, are used in the design of a basic network switching module. The modules and the approaches are then modeled and equations for network time delay are developed. These equations form the basis for a comparison between the two approaches. The importance of clock distribution strategies and clock skew is quantified, and a network clock distribution scheme which guarantees equal length clock paths is presented.**

***Index Terms*—Asynchronous, clocked, clock skew, crossbar, interconnection, multiprocessors, networks, self-timed, switches, switching modules.**

I. INTRODUCTION

THE principal issues in computer architecture have continually evolved in response to changes in basic computer technology. In recent years the advent of VLSI technology has once again shifted the design space. While the implications of this shift are not yet fully understood, certain changes in direction are becoming apparent. This paper is concerned with a problem of central importance in the design of large multiprocessor computer systems. Such systems are typically based on the use of inexpensive yet powerful VLSI microprocessor chips and chip sets.

Over the past few years the availability of such microprocessors has led to numerous proposals for the design of a variety of physically local, closely coupled multiprocessor systems [27], [4], [23], [25]. The communication network utilized by such closely coupled processor configurations is of central importance to the performance of these systems. Various studies have focused on functional properties of such networks [2], [11], [13], [19], on their complexity and performance [18], [24], [15], and to a certain extent, on their actual design [7], [20], [8]. Some of these studies have emphasized the VLSI implementation of such connection networks [12], [28], [9], [17] and, in particular, just how the topology of various networks affects their chip area and time delay properties.

This paper considers the problem of timing control structures for large interconnection networks where numerous VLSI chips are required for full network implementation. A general interconnection network is shown in Fig. 1. The network is taken to have N' input ports and N' output ports where each port has B' data lines. Other lines not shown in the figure will also be required for control and synchronization of data transfer through the network. Clearly for N' and B' beyond a certain size, the network will require partitioning into a number of subnetworks where a single VLSI chip can be associated with each subnetwork. A principal motivation for partitioning the network relates to chip pin limitations and, as discussed in [10], there are several partitioning strategies available. One simple approach is shown in Fig. 2.

Related to the partitioning problem is the broader question of timing control; that is, how is data movement synchronized in the network? Consider, for instance, the mesh connected crossbar shown in Fig. 2 and assume that the network is to pass messages from input to output ports. Assume that local routing capabilities are present at each crosspoint in the network (i.e., any message proceeding through the network contains header information that is successively examined by each crosspoint switch to determine message routing through the switch), and that once a path through the network has been established, that path is held for the duration of the message. Assume furthermore that a message can be pipelined along a captured input/output path.

There are two principal methods which can be used in controlling data movement along such a path. These methods are referred to as asynchronous and synchronous (or clocked) control schemes. While they are well known as general and often opposing methodologies, there have been few cases where a quantitative comparison has been made between them on the basis of a common system design problem. This paper presents such a comparison, where the system in question is a crossbar interconnection network of the general type described above. Although not pursued here, the analysis presented extends to other network topologies as well.

In practice, clocked designs have usually been preferred due to their relative simplicity and generally lower hardware costs. When systems become physically large, when their size cannot be predicted in advance, or when there are numerous system inputs which operate independently (and on separate clocks),

Reprinted from *IEEE Trans. Computers,* vol. C-32, no. 3, pp. 284–293, March 1983.

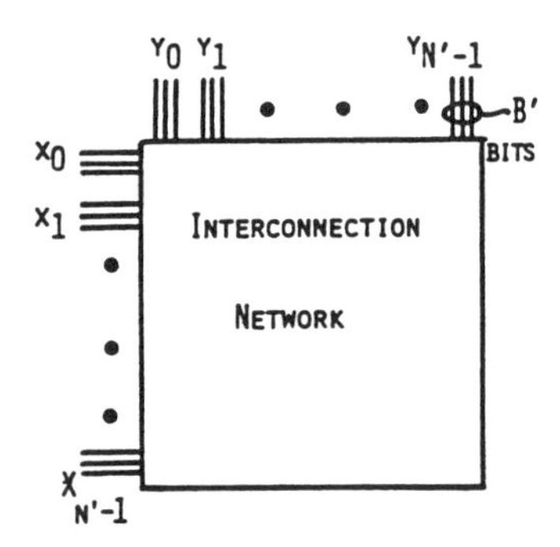

Fig. 1. A general $N' \times N'$ interconnection network.

then the advantages of asynchronous design begin to mount. One example of this is in processor bus design where asynchronous control schemes are common [5], [26]. Another example is in modular computer design where expandability and arbitrary system restructuring are key system features [3]. In such modular systems determining the appropriate clock period is difficult, if not impossible, since the final size and configuration of the system is not known in advance. Not knowing this final size makes estimation of clock skew and the design of clock distribution schemes problematic. Designing for a maximum size system, on the other hand, would require such large clock periods that system speed would be inordinately slow. A similar type of problem arises when designing timing control structures in the VLSI domain which are robust over a range of feature sizes, that is, which operate properly as the physical dimensions of the components are scaled [21]. The use of asynchronous control schemes in such environments is a natural solution to designing for system growth (or shrinkage). In this context it has been pointed out that use of asynchronous techniques can also be viewed as a structured design discipline in the time (or actually sequence) domain akin in spirit to structured programming in the program domain [22]. Design of such control schemes are, however, difficult, and in general engineers have shunned this approach focusing instead on extending conventional clocked schemes. An interesting tradeoff can be seen here. On the one hand, with clocked schemes the design of the control logic is simple while the clock distribution problem is difficult. On the other hand, with asynchronous schemes the design of the control logic is difficult while there is no clock distribution problem with which to contend.

All of this relates directly to the control problem encountered with interconnection networks used in multiprocessor systems. The ideal network should be modular and expandable so that it can be readily used to support a wide range of multiprocessor system sizes. This points to the use of an asynchronous scheme. Furthermore, since network chips will tend to be pin limited rather than component limited, any extra logic components needed for implementation of the asynchronous control could easily be absorbed on the chip. In contrast, any asynchronous control scheme would have to be implemented on a per port basis. That is, each input and output port of the network or subnetwork would require extra control lines (e.g., request, acknowledge). An asynchronous scheme would thus tend to have heavy pin requirements in a situation which al-

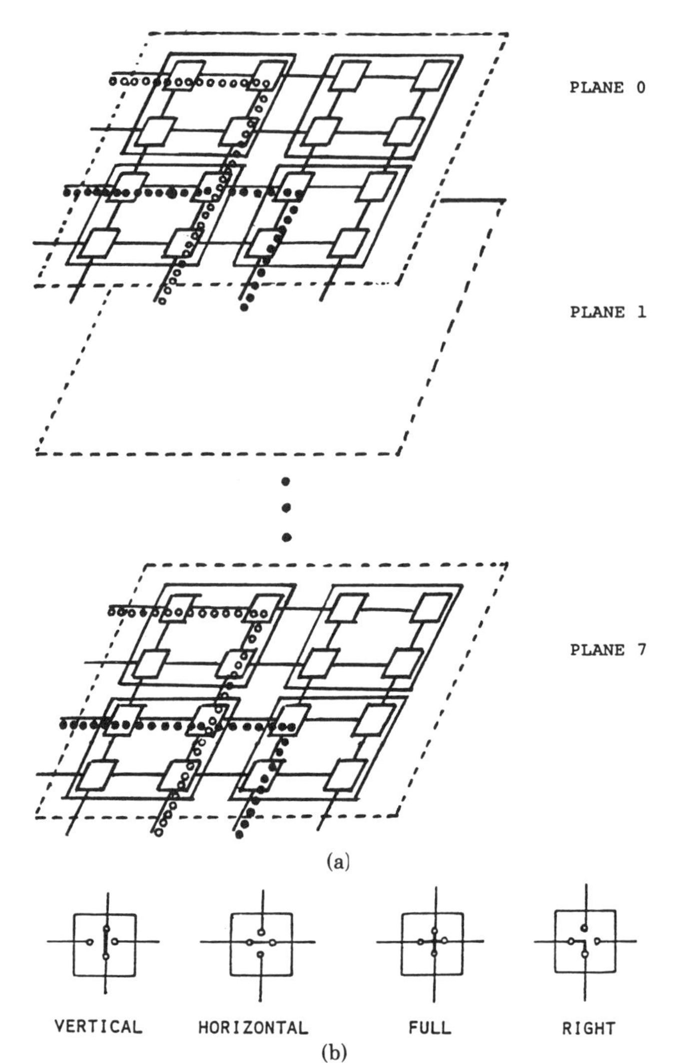

Fig. 2. Bit slice partitioning of a mesh connected crossbar ($N' = 4$, $B' = 8$). (a) Two sample input/output paths are shown using 2×2 subnetworks per chip. (b) Crosspoint positions for the mesh connected crossbar.

ready has pin constraints. The full range of considerations here is currently being studied.

This paper concentrates on a principal component of this study, namely, what are the relative speeds that can be achieved when implementing an interconnection network using asynchronous versus clocked control schemes. The section that follows defines asynchronous and clocked protocols for a simple but reasonable network switch module. Time delay models are then developed and general expressions for the delay presented and discussed. These models show that clock distribution delays, and associated clocking skew are key factors in determining which of the two approaches leads to a faster switch module. The role of clock skew is examined in the succeeding section and its physical origins are investigated under the assumption that the switch modules are fabricated using standard NMOS technology. Clock path layout is important in determining skew and an interesting tree structured layout is presented which provides for equal length paths to each switching module. The paper concludes with a detailed example demonstrating the use of the time delay models in a particular switch design situation.

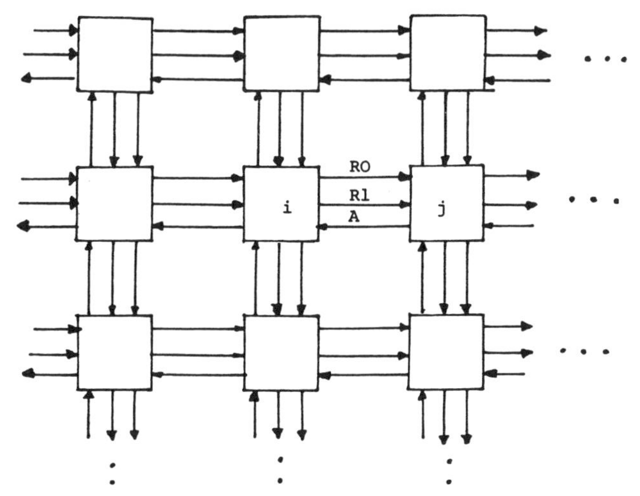

Fig. 3. Crossbar interconnection network constructed from asynchronous modules.

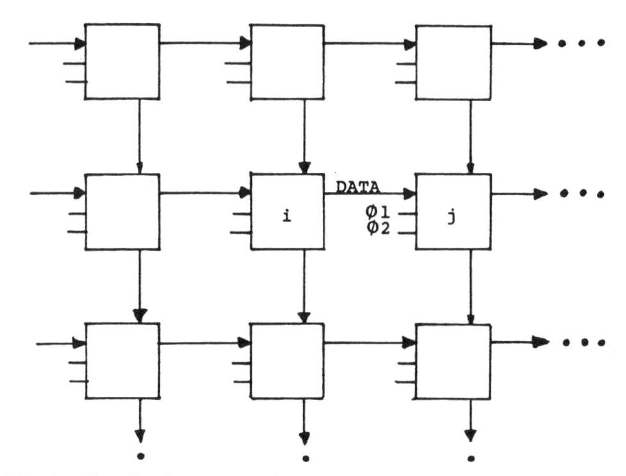

Fig. 4. Crossbar interconnection network constructed from synchronous modules

II. Protocols

A brief description of suggested protocols for the synchronous and asynchronous realizations of the crossbar interconnection network is given in this section. A complete interconnection network would require control provision for

- path establishment
- transfer of data from source to destination
- detection of a blocked path
- indication of end of transmission
- path clearing.

We have described how all of these requirements could be satisfied [10]. We have also shown that for many cases of practical importance, a bit slice architecture in which the network is partitioned into planes, each plane switching one bit of the incoming data words, is optimal from a chip count viewpoint [10]. For this reason the analysis here is restricted to a one bit plane network as shown in Fig. 2(a). Fig. 2(b) illustrates what positions may be established in an individual switch. Further, since data rate is the performance measure, only the protocol necessary to transfer data from module to module (assuming that the path from source to destination has already been established) is described.

A. Asynchronous Protocol

A delay insensitive protocol is adopted for the asynchronous modules, that is, insertion of any fixed or time-varying delay in any of the paths between modules will not cause the network to fail—although its speed may be modified. One such delay insensitive protocol that has the minimum number of signal changes (and thus probably will have the maximum data rate) uses transition sensitive logic. This protocol can be illustrated in Fig. 3 by considering a single switch pair (i, j) in which data are to be transmitted from a source at the left to a destination at the right. The protocol is as follows. If module i wants to send a logic zero to module j, it makes a change in the $R0$ line; if it wants to send a logic one to module j, it makes a change in the $R1$ line. Upon receipt of a change in $R0$ or $R1$, module j accepts the data and returns an acknowledge to module i by changing the A line. If module i has some new data (e.g., received from the module to its left), it may now transmit it to module j by again changing the appropriate line, $R0$ or $R1$. Note that this transaction technique is independent of any delay that is inserted in the lines between modules i and j since the change will eventually reach its intended module and the module cannot proceed with its next exchange until the previous exchange has been completed. This protocol will be assumed in the further discussions on asynchronous interconnection networks.

B. Synchronous Protocol

For the synchronous system the standard clocked-data protocol is adapted in which a level sensitive two-phase clock is employed. Two such communicating modules (i, j) are shown in Fig. 4 along with an entire network. This arrangement operates in the following manner. Let data be available at the input to module i. These data are captured by module i (i.e., stored) upon the assertion of the phase-one clock. On the assertion of the phase-two clock the data are transferred to the output of module i and propagate over the communication pathway to the input of module j. These data at the input of module j are then captured on the assertion of the phase-one clock and the procedure is repeated. Note that this protocol is not delay insensitive since if the period of the clock (e.g., the time between successive assertions of the phase-one clock) is too short, then the data captured at the input to module i at an assertion of the clock will not have had adequate time to propagate through module i, across the interconnecting pathway, and be available at the input to module j on the next clock assertion.

III. Delay Models

A. Asynchronous Delay Model

For the asynchronous switching elements shown in Fig. 3, consider only a single request from left to right with the corresponding acknowledge from right to left. The Huffman model (that satisfies the protocol discussed in the previous section) for a pair of modules along a path from source to destination is represented in Fig. 5. Consider module i in this

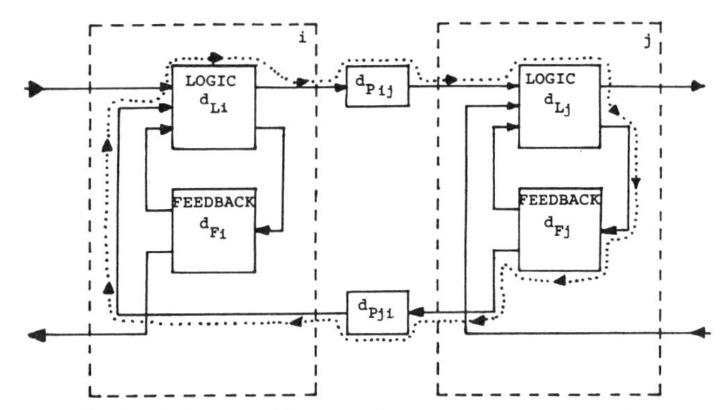

Fig. 5. Delay model for two adjacent asynchronous modules.

figure. Let the propagation delay of the combinational logic be d_{Li}, the propagation delay of the feedback path be d_{Fi}, and the propagation delay along the request path from module i to module j be d_{Pij}. Similarly for module j the logic and feedback delays are d_{Lj}, and d_{Fj}, while for the acknowledge path from module j to module i, the delay is d_{Pji}. There are certain constraints on the delays in the network that must be satisfied to ensure race free operation. Consider a change at the input to module i. A sufficient condition to guarantee deterministic behavior is that every part of the combinational logic circuit must sense this input change before the combinational logic circuit senses any response due to this input change via either the feedback path or the path from the module j output back to the module i input [6]. The delays, therefore, must satisfy the relation

$$d_F \geq d_L. \quad (1)$$

It is now possible to compute the minimum interarrival time between successive requests to module i. Assume that there is an acknowledge present at module i, and a first request to this module arrives from the module to its left. This request propagates along the dotted path shown in Fig. 5 and arrives back at the combinational logic input to module i. If the module to the left of module i has produced a new request in response to the acknowledge generated by module i, this second request could be processed immediately. Thus the time between servicing successive data bits (e.g., requests) for the asynchronous architecture is given by the loop delay, d_A, where

$$d_A = d_{Li} + d_{Pij} + d_{Lj} + d_{Fj} + d_{Pji}. \quad (2)$$

The values for the individual delays will vary from module to module due to processing and fabrication nonuniformities so the next question is just what values for the delays to adopt. Note that the network operates in a pipeline manner, and thus the data rate of the network is determined by the maximum value of this loop time (i.e., the maximum value of one of the pipe delays). For a network with N sources and N destinations and uniformly distributed addressing of sources and destinations, the average path through the network contains N modules. For large N there is a high probability that a loop between a pair of adjacent modules will be encountered that contains propagation delays that all have their largest values. This probability approaches one as N grows and to ensure worst case conditions it will be assumed that such a maximum loop delay, $\bar{d}_A$, is encountered. (It is interesting to observe here that because of the pipeline nature of this architecture the *maximum* delay determines its performance. This is in contrast to many applications of asynchronous systems in which a good estimate of the performance is given by the *average* delay.) These maximum delays will be identified by removing the subscripts i and j in (2). Hence,

$$\bar{d}_A = 2d_L + d_F + 2d_p. \quad (3)$$

B. Synchronous Delay Model

The model for two modules in the synchronous system of Fig. 4 can be constructed using a finite state machine representation of each module. This machine is designed to implement the classical two-phase level sensitive clocking scheme protocol described in Section II, and can be depicted as shown in Fig. 6. This model has combinational logic delay d_L, memory delay d_M, interconnection data path delay d_p, and delay along the clock line d_C. The delays d_L and d_p are assumed to have a distribution of values identical to those used for the asynchronous model; thus no distinguishing subscript is needed. The delay d_C is used to represent propagation delay along the path over which the clock is distributed. Each module contains two memory elements (i.e., a master–slave configuration) and these are identified by an additional subscript (e.g., d_{Mi1}, d_{Mi2}) corresponding to whether they receive a phase-one, or a phase-two clock.

Consider data stored in a memory element on a particular clock phase. These data can propagate along a path to a memory element and, to guarantee deterministic behavior, these data must arrive and be stable prior to the next occurrence of this clock phase. This imposes constraints on the minimum clock period. The module pair shown in Fig. 6 has four such paths along which data is transmitted: a path from memory $i1$ to memory $j1$, a path from memory $i2$ to memory $j2$, an internal feedback path from $i1$ to $i1$ and an internal feedback path from $j1$ to $j1$. Each of these path delays places a constraint on the clock period and the maximum delay determines the acceptable period. The constraint for Path 1 is as follows. Data at the input to memory $i1$ at an occurrence of the phase-one clock at memory $i1$ must propagate via d_{Mi1}, d_{Mi2}, d_{Pij}, and d_{Lj} and be stable at the input to memory $j1$ at the next occurrence of the phase-one clock at memory $j1$. A timing diagram for this constraint is shown in Fig. 7 where the period of the clock is specified as T. The occurrence of the phase-one clock at the two memory elements $i1$ and $j1$ is influenced by the value of the two clock line delays d_{Ci1} and d_{Cj1}. From the timing diagram the clock period can be expressed as

$$T > d_{Mi1} + d_{Mi2} + d_{Pij} + d_{Lj} + (d_{Ci1} - d_{Cj1}). \quad (4)$$

In a similar manner the other three paths consist of data propagating through the following elements:

Path 2: d_{Mi2}, d_{Pij}, d_{Lj}, and d_{Mj1}

Path 3: d_{Mi1}, d_{Mi2}, and d_{Li}

Path 4: d_{Mj1}, d_{Mj2}, and d_{Lj}.

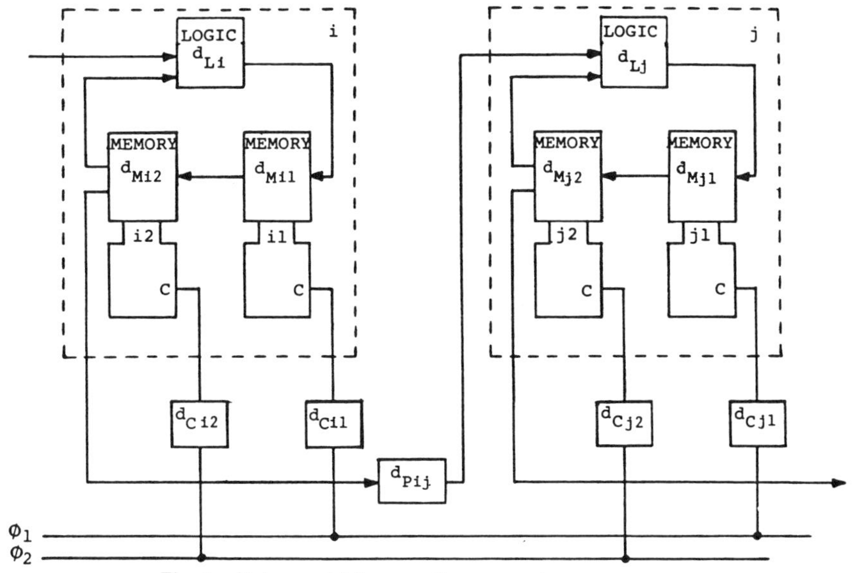

Fig. 6. Delay model for two adjacent synchronous modules.

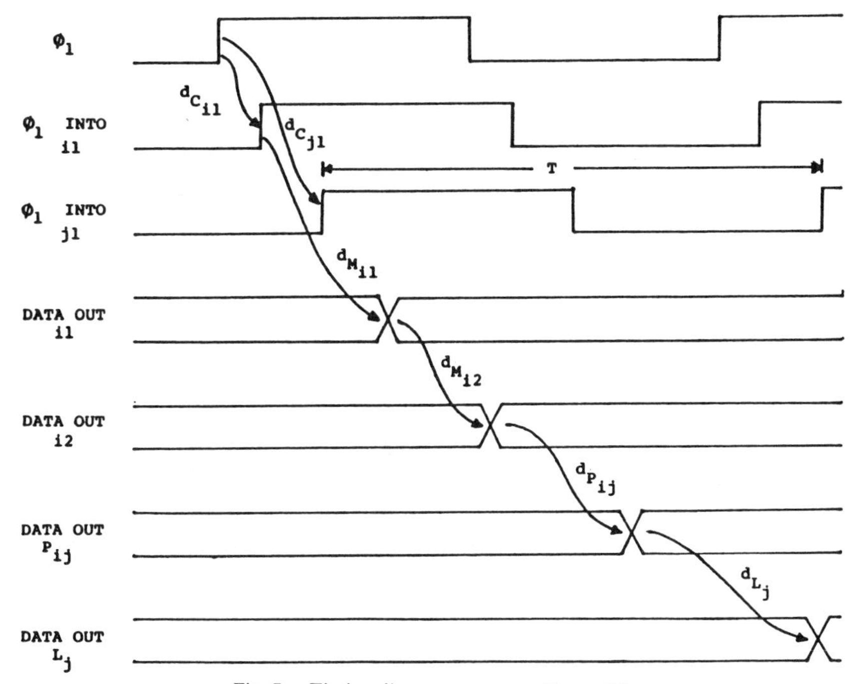

Fig. 7. Timing diagram corresponding to (4).

The constraint relations for these three paths can be found in a manner similar to that for Path 1 and are

$$T > d_{Mi2} + d_{Pij} + d_{Lj} + d_{Mj1} + (d_{Ci2} - d_{Cj2}) \tag{5}$$

$$T > d_{Mi1} + d_{Mi2} + d_{Li} \tag{6}$$

$$T > d_{Mj1} + d_{Mj2} + d_{Lj}. \tag{7}$$

Since in most designs the last two constraints imposed on T are smaller than either of the first two, they will not be considered further. Note that the bracketed quantity in (4) is the difference in arrival of the phase-one clock at the corresponding memory elements of the two modules. Likewise for the bracketed quantity in (5) and the phase-two clock. These differences are called *clock skew* and will be defined as

$$\delta_{C1} = d_{Ci1} - d_{Cj1}; \quad \delta_{C2} = d_{Ci2} - d_{Cj2}. \tag{8}$$

The delay values are statistically distributed and, due to the pipeline argument, the delays for the module pair under consideration are assumed to represent the largest delay encountered in the distribution. Hence, (4) and (5) are identical and the subscripts can be removed with the resulting equation indicating a worst case condition.

In addition to the clock period being constrained by the loop delays discussed above, the period must also be sufficiently large so that the entire clock distribution path (clock tree) has enough time to charge and discharge appropriately. That is, when charging the clock tree, a voltage in excess of the input high threshold voltage of devices driven by the clock must be reached, while when discharging the clock tree voltages below

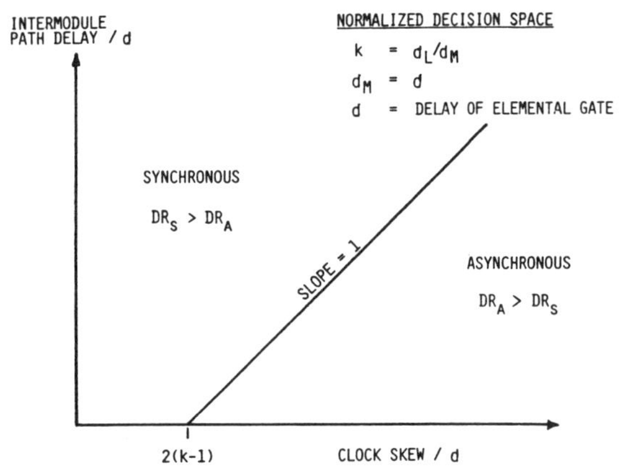

Fig. 8. Design space as a function of normalized clock skew and normalized path delay illustrating regions in which asynchronous and synchronous architectures have superior performance.

the input low threshold must be achieved. Assuming both clock, and charge and discharge time symmetry, then if τ is the time required to charge or discharge the clock tree, a second constraint on the clock period is

$$T > 2\tau. \tag{9}$$

The maximum delay for the synchronous architecture can now be written as

$$\overline{d}_S = \max\,[d_L + 2d_M + d_p + \delta, 2\tau] \tag{10}$$

where the clock skew is $\delta = d_{Ci} - d_{Cj}$.

IV. Data Rate Comparison

The data rate for the asynchronous system, DR_A, and the data rate for the synchronous system, DR_S, are the inverses of the corresponding delay times given in (3) and (10), respectively. Therefore

$$DR_A = 1/(2d_L + d_F + 2d_p) \tag{11}$$

$$DR_S = 1/\max\,[d_L + 2d_M + d_P + \delta, 2\tau]. \tag{12}$$

For many high performance interconnection networks of interest, $d_L + 2d_M + d_P + \delta > 2\tau$. Given this situation, the condition under which the asynchronous data rate is larger than the synchronous data rate can be written as

$$\delta > d_L + d_P + d_F - 2d_M. \tag{13}$$

Equations (11) and (12), which provide the data rates for the asynchronous and synchronous systems, and (13), which indicates how clock skew affects the data rate comparison between the two systems, are the major developments of this work. Once a design team has information about the specific implementation parameters and can determine their numerical delay values, these design equations can be used to compare system performance and make a selection of an appropriate system architecture.

The interpretation of this rate comparison can be simplified further by observing that the delay of the memory will normally be equal to the delay of an elemental transistor d. The combinational logic delay can be expressed as a multiple of this delay, that is, $d_L = kd$. The minimum delay constraint for d_F

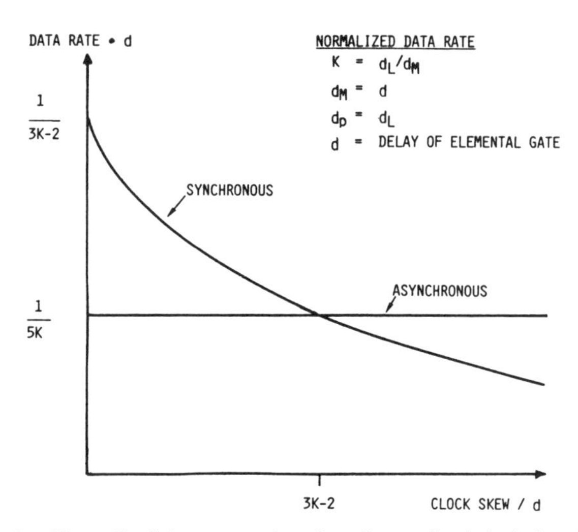

Fig. 9. Normalized data rate as function of normalized clock skew for the two architectures.

from (1) will be used so that $d_F = d_L = kd$. Using this nomenclature the condition of (13) then becomes

$$\delta/d > 2(k - 1) + d_p/d. \tag{14}$$

This relation is shown in Fig. 8 and illustrates the regions in which $DR_A > DR_S$ and in which $DR_A < DR_S$. For large d_p the synchronous system data rate is higher than that of the asynchronous system. This is because d_p appears only *once* in the path between two synchronous switch modules [see Fig. 6 and (12)] while, in the asynchronous system, the handshake protocol requires a round trip, thus d_P occurs *twice* [see Fig. 5 and (11)]. Likewise as δ is increased, the performance of the synchronous system is degraded [see (12)]. The interaction between these two variables can be shown graphically (for a specific value of d_P) by plotting a normalized data rate, $d \cdot DR$ versus δ/d. An example of this is illustrated in Fig. 9 for the case where $d_P = d_L$. The role of clock skew becomes clear from this figure. As the skew increases, there is a distinct crossover point beyond which an asynchronous design is superior from a data rate performance viewpoint. The effect of the intermodule propagation delay is also apparent. As this delay increases the synchronous design becomes superior.

V. Determination of System Delay Parameters

Let the crossbar network be implemented via NMOS technology with a collection of N^2 switching modules on a chip, and let a number of such chips be placed on a printed circuit board and interconnected via printed circuit wiring. In this section the factors that affect the values of the various system delays are examined and simple expressions for each of them are developed.

A. Combinational Logic, Memory and Feedback Delays

To obtain realistic values for the various delays, the synchronous module was implemented using a PLA for the combinational logic and pass transistor controlled gates for the dynamic memory. This configuration is depicted in Fig. 10. A functional specification of the synchronous crossbar that

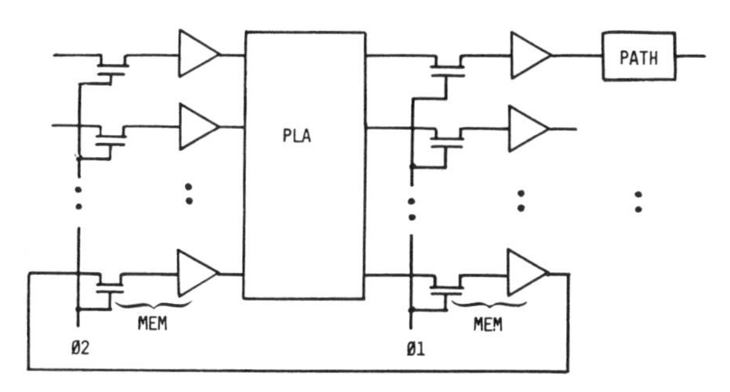

Fig. 10. Implementation of synchronous module using PLA and dynamic memory.

included all the features described in Section II (e.g., path establishment, path clearing, etc.) was formulated, and a state table having seven variables and 40 rows was constructed that satisfied this description. A design methodology for determining a PLA with the minimum delay was applied [1] and yielded a combinational delay, $d_L = 37.5$ ns and memory delay $d_M = 2$ ns. Although a complete state table for the asynchronous case was not developed, the preliminary analysis indicates that its complexity would be comparable so this same value of d_L is used for both architectures. Using the minimum delay constraint of (1), the feedback delay, d_F, for the asynchronous case is equal to the combinational logic delay.

B. Path Delay

There are two intermodule paths that must now be examined: paths between two adjacent modules on the same integrated circuit chip, and paths between two adjacent modules on different chips. Because of the planar topology of the crossbar network, modules that communicate can be implemented in close physical proximity. Compared to the other circuit delays, the intermodule path delay between modules on the same chip is then very small and can be ignored. This, however, is not true when a module on one chip must communicate with a module on another chip. The delay here is related to the resistance and capacitance of the path. Since the interconnection path will be short and will use metal conductors, its resistance is small and can be neglected. The propagation delay is then determined by the ratio of the capacitance of an elemental gate, C_g, and the capacitance of the load being driven, C_L. It has been shown [16] that if one uses an exponential buffer, this delay is given by the expression

$$d_p = d * e * [\ln (C_L/C_g)]. \tag{15}$$

where e is the base of the natural logarithm.

C. Clock Skew

The clock skew is the maximum delay between the clock signals that control a horizontally or vertically adjacent module pair in Fig. 4. Consider the simple model for this situation in which two modules $M1$ and $M2$ are driven from a common clock point P, with clock lines from this common point of lengths $L1$ and $L2$. Assume that the clock is asserted at $t = 0$. The clock skew is then the difference in time between when $M1$ and when $M2$ respond to the clock. This difference in response is determined by four factors.

1) Differences in the line lengths $L1$ and $L2$.

2) Differences in delays through any active elements inserted in the lines (e.g., clock buffers).

3) Differences in the line parameters (e.g., resistivity, dielectric constant) that determine the line time constant.

4) Differences in the threshold voltages of the two modules $M1$ and $M2$.

In the next section a clock distribution scheme that guarantees equal clock line lengths for all paths is developed. Thus, the first factor above can be eliminated.

The second factor can be significant if clock buffers are present. There are three principal reasons for the use of such buffers. First, in complex circuits such as microprocessors, various clock subcycles and pulses derived from the main two clock phases may be needed. In this situation, logic and buffers are needed to obtain these signals and distribute them within the chip. The synchronous module, however, only requires the unmodified phase-one and phase-two signals, and thus such clock buffers within the chip are unnecessary. Second, clock buffers may be required if the clock signal is not strong enough to drive all the chips in the interconnection network. Such buffers would increase the clock skew, and while the analysis presented can be expanded to deal with this situation, for simplicity, we assume here that sufficient drive is available from the central clock so that these buffers are not needed. Third, it is possible to include buffers to facilitate clock pipelining. In such a situation several clock pulses would be present in the clock tree simultaneously. Although this type of scheme would reduce the clock period constraint given in (9), such systems are difficult to design, and are not considered in this paper. This eliminates factor 2 above.

A model for the clockpath skew that includes the last two factors is developed in the Appendix. The clock skew is found in terms of the maximum and minimum time constants of a clock tree ($\overline{\tau}$ and $\underline{\tau}$), the maximum and minimum values of the threshold voltage of a typical logic gate ($\overline{V}_T$ and $\underline{V}_T$), and the supply voltage V_{DD}. The result of that derivation gives the clock skew as

$$\delta = \underline{\tau} \ln \left(1 - \frac{\underline{V}_T}{V_{DD}}\right) - \overline{\tau} \ln \left(1 - \frac{\overline{V}_T}{V_{DD}}\right) \tag{16}$$

VI. Clock Distribution

As shown in the Appendix, it is important to maintain the same on-chip clock line length to each of the N^2 modules. One technique that accomplishes this utilizes a binary tree layout for the clock path. An example of this distribution for a single-phase clock supplied to an 8 × 8 network is shown in Fig. 11. Note that as this tree is traversed from its root (clock input) to any leaf (module), the length to each switch is constant.

In order to estimate the clock skew, the delay associated with the clock tree must be computed. This has been considered by Kung [30] and Penfield [31] and, as indicated in the Appendix, can also be determined using simulation techniques. All such computations require a knowledge of R_o and C_o, the resistance and capacitance associated with a leaf of the tree. Their values depend on the length of the leaf and the material used in its fabrication. Although the clock could be distributed using metal for the horizontal clock paths and diffusion for the

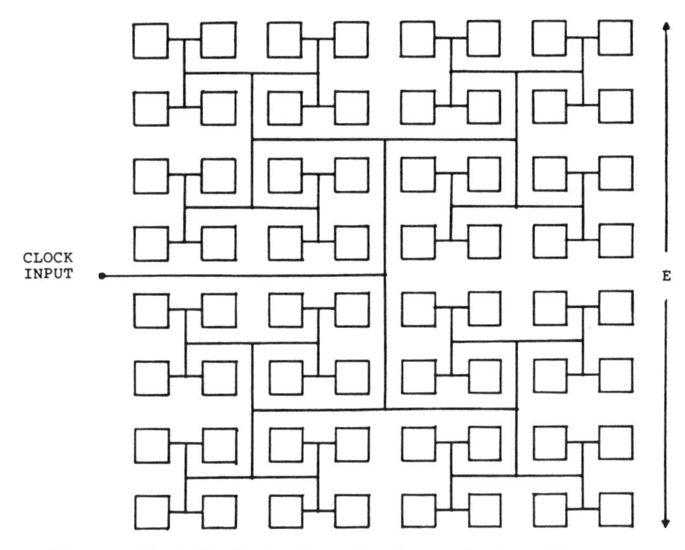

Fig. 11. Clock distribution for an 8 × 8 network using a binary tree.

vertical paths, metal should be used wherever possible. Some short sections of diffusion will be necessary, however, in order to bridge intermodule communication, power, ground and reset lines (assuming only a single layer of metal is available). An actual layout of a synchronous module indicates that these sections can be shortened so that only about 5 percent of the clock line is in diffusion. The distribution of the two-phase clock requires two such binary trees—this second tree can be constructed by merely displacing the first tree in the vertical and horizontal directions by the minimum line separation.

The length of the leaf depends in part on the relative areas taken up by switch modules versus the clock distribution lines. This is considered in the context of a particular example in the Appendix.

VII. Example

Consider a 1 cm × 1 cm chip packaged in a 2.5 cm × 2.5 cm carrier having 132 pins [14]. With 132 pins, one bit slice of an 8 × 8 asynchronous network could be constructed per chip. If a 25 cm × 25 cm printed circuit board is used, 64 of these chips could be assembled per board, thereby achieving one bit slice of a 64 × 64 network on a single board. Assume copper printed circuit connections between chips. The pin capacitance for this type of construction is about 4 pF and the capacitance of an elemental gate is about 0.02 pF. Then, the maximum delays for this type of circuit are approximately $d_L = 37.5$ ns, $d_M = 2$ ns, $d_F = 37.5$ ns, and $d_p = 43$ ns.

Since there is negligible clock skew due to the printed circuit board paths, the clock skew is dominated by intrachip parameters. From the Appendix, the τ value for the chip clock tree is found as 37 ns. Discussion with commercial fabricators has indicated a variation of ±20 percent in τ. Thus, $\overline{\tau} = 44$ ns and $\underline{\tau} = 30$ ns. For a supply voltage of 5 V, equal noise margins can be obtained with $V_T = 2.5$ V. The range of the threshold variation is also about ±20 percent, so $\overline{V}_T = 3.0$ and $\underline{V}_T = 2.0$ V. Using these values in (16) gives the clock skew as 26 ns. Equations (11) and (12) can now be used to obtain the asynchronous and synchronous data rates. The result is that the synchronous system has a higher data rate ($DR_S = 9$ MHz; $DR_A = 5$ MHz). These equations can also be used to compare asynchronous and synchronous data rates over a range of clock skew values as shown in Fig. 12. The dot indicates the operating point for the example synchronous switch network. The figure indicates that for a clock skew greater than 68.5 ns the asynchronous design has a higher data rate. For small values of skew, the synchronous design has a higher data rate. Notice that at $\delta = 45$ ns there is a discontinuity in the synchronous curve. For smaller values of skew the switch parameters dominate the delay (i.e., $d_L + 2d_M + d_P + \delta > 2\tau$) while above this point the clock tree charge and discharge times dominate (i.e., $2\tau > d_L + 2d_M + d_P + \delta$).

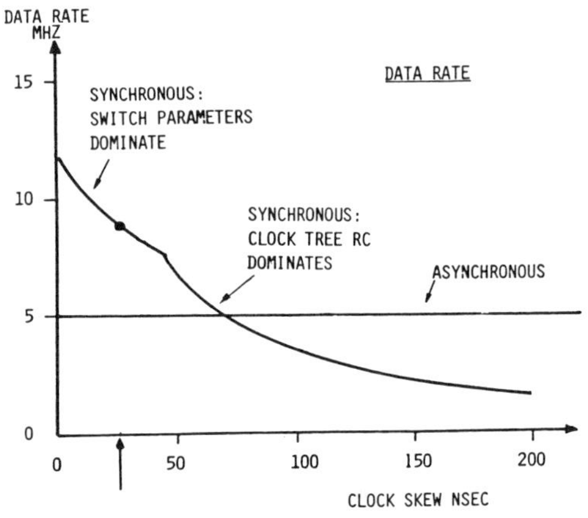

Fig. 12. Data rates for asynchronous and synchronous architectures for example system. Arrow indicates actual clock skew.

VIII. Conclusions

This paper has presented a comparison of asynchronous and clocked timing control structures in the context of the design of an interconnection network. The network has local routing control, is pipelined, and has a mesh connected crossbar topology. It is intended for use in a message based multiprocessor environment. The selection of the appropriate control structure is critical if a high bandwidth, modular interconnection network is to be achieved. Designing for growth and size uncertainty both in the large (i.e., the number of processors in the final system may change), and in the small (i.e., the VLSI feature size may change) appears to make an asynchronous approach attractive. Such an approach, however, will tend to have heavy pin requirements. Before a complete comparison of the methodologies can be achieved, it is necessary that fundamental models which allow one to compare their relative speeds are developed. This is the principal contribution of this paper.

First, two switch modules, one asynchronous and one clocked, and their respective data synchronization protocols were defined. Based on this a model of each module's operation was developed. These models allow one to determine the data transmission speed associated with the modules (and therefore for an entire network) and to compare the timing control methodologies in question. The model equations yield decision curves which can be used to compare these two control structures under a variety of design parameters. The equations indicate the key role played by clock path delay and clock skew, and clearly show how the speed of clocked systems must be lowered as the clock skew increases. Key equations are derived for path delay and clock skew, and a tree structured clock layout scheme is presented which results in equal length clock paths to each switching module. An example is developed where the switching modules are assumed to be fabricated in the NMOS technology, and for this example, the synchronous control is shown to yield a faster system.

APPENDIX

Leaf Characteristics

The values of R_o and C_o for a leaf are determined as follows. The physical geometry of a conductor can be represented as shown in Fig. 13 where the resistance and the capacitance are given by

$$R_o = \rho L/(Wh_c) \qquad C_o = \epsilon LW/(h_o).$$

In these relations ρ is the resistivity of the conductor, ϵ is the dielectric constant of the oxide, h_o is the thicknesss of the oxide, and h_c is the thickness, W is the width and L is the length of the conductor. Hence the time constant of the leaf can be expressed as

$$R_oC_o = \rho\epsilon L^2/(h_ch_o).$$

Note that the time constant is independent of the width of the line. If two clock lines have different lengths, the clock skew is related to the square of the differences in line lengths. It is for this reason that the binary tree distribution illustrated in Fig. 11, which provides equal lengths, has been chosen.

Consider one of the leaves in the binary tree, and let the diffusion length and width be L_D and W_D, and the metal length and width be L_M and W_M. The resistance will be expressed in terms of ohms per square (i.e., $L = W$) and capacitance in terms of pF per unit area (in this case square microns). These relations can be related to the basic parameters shown in Fig. 13 as

$$R/sq = \rho/h_c \text{ and } C/ar = \epsilon/h_o.$$

Since the resistivity of diffusion is very large compared to the resistivity of metal, the contribution of the resistance of the metal segment to the total leaf resistance will be negligible. However, the capacitance of both the diffusion and metal segments must be included in the computation. Therefore the leaf resistance and capacitance are approximately

$$R_o = (R_D/sq)(L_D/W_D)$$

$$C_o = C_M + C_D = (C_M/ar)W_ML_M + (C_D/ar)W_DL_D.$$

Leaf Time Constant

For a chip with N^2 modules that has an edge length of E, the length of the clock leaf is approximately $E/2N$. Suppose the diffusion length of this is $L_D = qE/2N$ and the metal length is $L_M = (1 - q)E/2N$ where $0 \leq q \leq 1$. If the minimum feature size is λ, then using the design rules of [16], the minimum width of a diffusion conductor is 2λ and the minimum width of a metal conductor is 3λ. Hence the resistance and capacitance can be written as

$$R_o = (R_D/sq)qE/4\lambda N$$

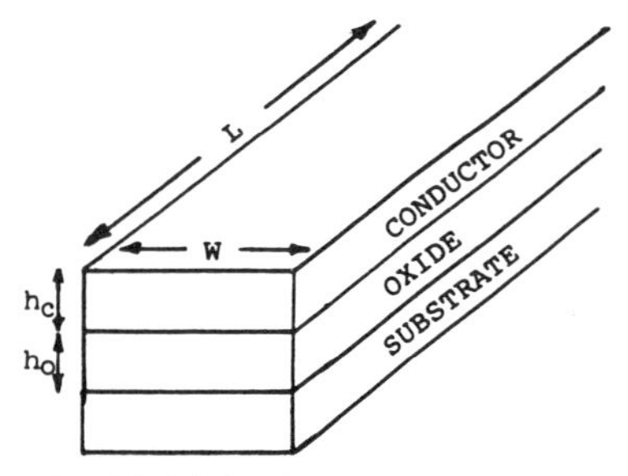

Fig. 13. Model of an integrated circuit conductor.

$$C_o = (C_M/ar)3E\lambda(1 - q)/2N + (C_D/ar)E\lambda q/N.$$

The leaf time constant is then

$$R_oC_o = \frac{(R_D/sq)qE^2}{8N^2}[3(1 - q)(C_M/ar) + 2q(C_D/ar)].$$

For the example discussed in Section VII in which $N = 8$ the following representative values for the parameters are used:

$$R_D/sq = 20\ \Omega/sq;\ E = 1 \text{ cm} = 10^4\ \mu\text{m}$$

$$C_M/ar = 0.3 \times 10^{-4}\ \text{pF}/\mu\text{m}^2;\ C_D/ar = 10^{-4}\ \text{pF}/\mu\text{m}^2$$

$$q = 0.05.$$

Substituting these values into the equation for R_oC_o gives $R_oC_o = 18.6$ ps.

Tree Time Constant

Obtaining the tree time constant is a problem that has been investigated by Kung [30] and Penfield [31], and both analytic expressions and bounds have been developed. For the example considered in this paper a SPICE [32] simulation of the tree was performed. This yields a tree time constant $\tau = 37$ ns, close to the value obtained with Kung's expression $\tau \simeq 1.43N^3(3 - 2/N)(R_oC_o)$.

Clock Skew

The waveform of the clock at the input to one of the modules in Fig. 11 due to a step change at the clock input may be approximated by an exponential of the form

$$v(t) = V_{DD}[1 - \exp(-t/\tau)]$$

where τ is the clock tree time constant and V_{DD} is the supply voltage. Observe that the response of two individual clock lines will not be the same because of the variability in the line parameters. Let $\overline{\tau}$ and $\underline{\tau}$ be the maximum and minimum values of this variability. The clock skew is also dependent on the differences in the threshold voltages of the two adjacent modules. Let the mean threshold voltage of a module be V_T with a range of $\overline{V}_T$ to $\underline{V}_T$. Then if one module has the maximum threshold and the maximum time constant from root to leaf, and the adjacent module has the minimum threshold and the minimum time constant, this causes the maximum time difference in the response of the two modules to the clock. This is illustrated in Fig. 14 where the delay difference is $\overline{t} - \underline{t}$. Substituting these worst case conditions into the above equa-

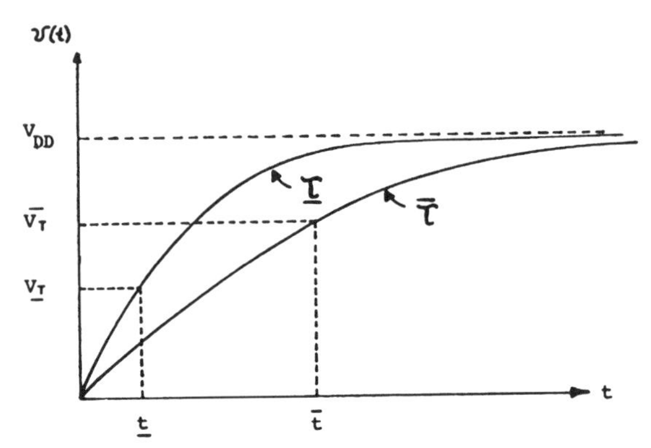

Fig. 14. Clock waveform at module input due to step change of clock at chip input. Shown as function of τ, the time constant of the on-chip clock tree.

tion allows the values of $\overline{V}_T$ and $\underline{V}_T$ to be expressed in terms of $\bar{t}$ and $\underline{t}$ as

$$\overline{V}_T = V_{DD}[1 - \exp(-\bar{t}/\bar{\tau})]$$

$$\underline{V}_T = V_{DD}[1 - \exp(-\underline{t}/\underline{\tau})].$$

Combining these two equations yields the clock skew due to clock tree time constant and threshold variations as

$$\delta = \bar{t} - \underline{t} = \bar{\tau}\ln\left(1 - \frac{\underline{V}_T}{V_{DD}}\right) - \bar{\tau}\ln\left(1 - \frac{\overline{V}_T}{V_{DD}}\right).$$

References

[1] S. Anantharaman, "Delays in PLA's: Analysis, reduction and computer aided optimization," M.S. thesis, Dep. Elec. Eng., Washington Univ., St. Louis, MO, 1982.

[2] V. E. Benes, *Mathematical Theory of Connecting Networks and Telephone Traffic.* New York: Academic, 1965.

[3] W. A. Clark, "Macromodular computer systems," in *Proc. AFIPS 1967 SJCC*, vol. 30, pp. 335–336.

[4] J. B. Dennis and D. P. Minsunas, "A preliminary architecture for a basic data-flow processor," in *Proc. 2nd Annu. Symp. Comput. Arch.*, Jan. 1975, pp. 126–132.

[5] Digital Equipment Corp., *Microcomputers and Memories.* Maynard, MA: DEC Press, 1981.

[6] T. P. Fang, "On the design of hazard free circuits," Comput. Syst. Lab., Washington Univ., St. Louis, MO, Tech. Memo. 285, Nov. 1981.

[7] M. A. Franklin, S. A. Kahn, and M. J. Stucki, "Design issues in the development of a modular multiprocessor communications network," in *Proc. 6th Annu. Symp. Comp. Arch.*, Apr. 1979, pp. 182–187.

[8] M. A. Franklin and D. F. Wann, "Pin limitations and VLSI interconnection networks," in *Proc. 1981 Int. Conf. Parallel Process.*, 1981, pp. 253–258.

[9] M. A. Franklin, "VLSI performance comparison of banyan and crossbar switching networks," *IEEE Trans. Comput.*, vol. C-30, pp. 283–291, Apr. 1981.

[10] M.A. Franklin, D. F. Wann, and W. J. Thomas, "Pin limitations and partitioning of VLSI interconnection networks," *IEEE Trans. Comput.* vol. C-31, pp. 1109–1116, Nov. 1982.

[11] L. R. Goke and G. J. Lipovski, "Banyan networks for partitioning multiprocessor systems," in *Proc. 1st Annu. Symp. Comput. Arch.*, 1973, pp. 21–28.

[12] D. Hoey and C. E. Leiserson, "A layout for the shuffle-exchange network," in *Proc. 1980 Int. Conf. Parallel Process.*, Aug. 1980, pp. 329–336.

[13] D. H. Lawrie, "Access and alignment of data in an array processor," *IEEE Trans. Comput.*, vol. C-24, pp. 1145–1155, Dec. 1975.

[14] J. Lyman, "Chip carriers, pin-grid arrays change the PC board landscape," *Electronics*, pp. 66–75, Dec. 29, 1981.

[15] M. Malek and W. W. Myre, "Figures of merit for interconnection networks," in *Proc. Workshop* on *Interconnection Networks Parallel Distribut. Process.*, Apr. 1980, pp. 74–83.

[16] C. Mead and L. Conway, *Introduction to VLSI Systems.* Reading, MA: Addison-Wesley, 1980.

[17] K. Padmanabhan, "Multiprocessor interconnection networks in a VLSI environment," M.S. thesis, Dep. Elec. Eng., Washington Univ., St. Louis, MO, 1981.

[18] J. H. Patel, "Performance of processor-memory interconnection networks for multiprocessors," *IEEE Trans. Comput.*, vol. C-30, pp. 771–780, Oct. 1981.

[19] M. C. Pease, "The indirect binary *n*-cube microprocessor array," *IEEE Trans. Comput.*, vol. C-26, pp. 458–473, May 1977.

[20] B. Quatember, "Modular crossbar switch for large-scale multiprocessor systems—Structure and implementation," in *Proc. AFIPS 1981 Nat. Comput. Conf.*, vol. 50, pp. 125–135.

[21] C. L. Seitz, "Self-timed VLSI systems," in *Proc. Caltech Conf. VLSI*, Jan. 1979, pp. 345–355.

[22] ——, "System timing," in *Introduction to VLSI Systems*, C. Mead and L. Conway, Eds. Reading, MA: Addison-Wesley, 1980, pp. 218–262.

[23] M. C. Sejnowski *et al.*, "An overview of the Texas reconfigurable array computer," in *Proc. AFIPS 1980 Nat. Comput. Conf.*, vol. 49, pp. 631–641.

[24] H. J. Siegel, R. J. McMillen, and P. T. Mueller, Jr., "A survey of interconnection methods for reconfigurable parallel processing systems," in *Proc. AFIPS 1979 Nat. Comput Conf.*, vol. 48, pp. 529–548.

[25] H. Sullivan, T. R. Bashkow and K. Klappholz, "A large scale, homogeneous, fully distributed parallel machine: I," in *Proc. 4th Annu. Symp. Comput. Arch.*, Mar. 1977, pp. 105–117.

[26] I. E. Sutherland et al., "The TRIMOSBUS," in *Proc. Caltech Conf. VLSI*, Jan. 1979, pp. 395–427.

[27] R. J. Swan, S. H. Fuller, and D. P. Siewiorek, "Cm*—A modular multi-microprocessor," in *Proc. AFIPS 1977 Nat. Comput. Conf.*, pp. 637–644.

[28] C. D. Thompson, "Area-time complexity for VLSI," in *Proc. 11th Annu. Ass. Comput. Mach. Symp. Theory Comput.*, Apr. 1979, pp. 81–88.

[29] M. A. Franklin and D. F. Wann, "Asynchronous and clocked control structures for VLSI based interconnection networks," in *Proc. 9th Annu. Symp. Comput. Arch.*, Apr. 1982, pp. 50–59.

[30] S. Y. Kung and R. J. Gal-Ezer, "Synchronous vs. asynchronous computation in VLSI array processors," *Proc. SPIE*, vol. 341, May 1982.

[31] P. Penfield, Jr. and J. Rubinstein, "Signal delay in *RC* tree networks," in *Proc. 18th Design Auto. Conf.*, June 1981, pp. 613–617.

[32] L. Nagel, "SPIECE2: A computer program to simulate semiconductor circuits," Elec. Res. Lab., Univ. of California, Berkeley, Tech. Rep. UCB ERL-M250, May 1975.

Reduction of Clock Delays in VLSI Structures

SANJAY DHAR, MARK A. FRANKLIN and DONALD F. WANN.

Center for Computer Systems Design
Washington University,
St. Louis, MO. 63130

ABSTRACT

With the growth in chip size and reduction in line width, delays in driving long lines have become increasingly important in determining overall chip level performance. In synchronous systems the proper distribution of the clock signal is critical in determining system throughput. This paper considers the problem of optimally driving clock lines. A general delay model is developed and applied to a clock tree where the path distances from the root node to each of the leaf nodes are all equal. This strategy reduces clock skew and increases clock rates. A tree delay model is developed and is used to determine the optimal number and placement of buffers within the tree so that the clock delay is minimized. An example of a clock tree driving a synchronous crossbar network is provided, and minimum delay and corresponding number of buffers are indicated as a function of the minimum line width and network size. For a 64*64 network this minimization technique yielded an order of magnitude delay reduction over standard single exponential buffer usage.

1.0 Introduction:

Developments in VLSI fabrication technology have rapidly increased the size of chips as well as reduced minimum line width. As this continues, delay in control and data lines is becoming critical in designing high performance systems [SINH82, CART84]. Driving long lines can be especially acute in globally synchronized clocked systems. Two such systems whose clock distribution properties have been analyzed are discussed in FISH83 and WANN83. These studies indicate a close relationship between delays in clock lines and system performance (data rate). To reduce clock skew the clock distribution often takes the form of a H-tree [WANN83]; Kung [KUNG82] has shown that the delay in a full H-tree increases as $O(N^3)$ where N is the network size. This paper presents a technique for minimizing delays in clock trees. The procedure is based on the sectioning and inserting of buffers in the clock line to minimize the propagation delay, and yields the number of buffers, their position and size. In a 64*64 crossbar network example an order of magnitude reduction in the clock distribution delay is obtained over the single optimum exponential buffer case.

The propagation delay along a single section of a line is first modeled by a buffer driving a line and a load. The section delay is minimized, and the total line is then divided into a number of sections such that the total line delay is minimized. This technique is then applied to a clock tree, which is first reduced to a single electrically equivalent line by repeatedly folding the tree. The number of buffers, their position and size that minimize the delay are obtained for the folded clock tree. These folded line parameters are mapped back into equivalent parameters for the unfolded clock tree preserving the minimum delay.

2.0 Delay Model of a Section:

The basic circuit configuration is shown in Figure 1a and consists of (a) a source which is an NMOS transistor with pullup to pulldown ratio of k, a gate capacitance of C_{gs}, and a pulldown resistance of R_{gs}; (b) a line of total length L with distributed resistance and capacitance lumped into values R_L and C_L; and (c) an NMOS transistor load with a gate capacitance of C_{gl}. The inductance of the line is neglected and it is assumed that the width and material of the line are uniform and fixed.

In general k is greater than 1 and hence the largest propagation delay occurs when point P_3 is charged to the power supply voltage. In this analysis the worst case pullup delay will be used, although if superbuffers were used the smaller pulldown delay would be appropriate. (It is also possible to employ the pair delay). The line is modeled by a series resistor and shunt capacitor. The load (gate capacitance of the next transistor) is in parallel with the line capacitance.

The complete delay model for a section is shown in Figure 1b. The signal propagation time from point P_1 to point P_3, defined as the product of the equivalent resistance and capacitance of the section, is given by

$$d_s = (k*R_{gs} + R_L)*(C_{gl} + C_L)$$
$$= k*R_{gs}*(C_{gl} + C_L) + R_L*C_L + R_L*C_{gl} \quad (1)$$

The first term is the delay in driving the line and load transistor capacitances by the source

* This research has been sponsored in part by funding from ONR Contract N00014-8D-C-0761 and NSI Grant MCS-78-20731.

Reprinted from *IEEE Int'l Conf. Computer Design*, pp. 778–783, Oct. 1984.

transistor. The second and third terms represent the delays involved in driving the line and load capacitances, (C_L and C_{gl}) by the line resistance (R_L). For a given L, since the width and material of the line are fixed, these two terms cannot be reduced any further. For a long line, C_L will be large and the delay associated with the first term can be minimized by using a technique proposed by [JAEG75]. A series of buffers starting with a minimum size buffer and exponentially increasing in size is used (Figure 2). The delay d_C in driving a capacitance C is then given by

$$d_C = k*tau*e*ln(C/C_{gs}) \quad (2)$$

where tau is the product of the resistance R_m and capacitance C_m of a minimum size transistor.

This technique is applied to the section in Figure 1a for driving the capacitance ($C_L + C_{gl}$) giving the minimized section delay as

$$d_s = k*tau*e*ln\{(C_L+C_{gl})/C_{gs}\} + R_L*C_L+R_L*C_{gl} \quad (3)$$

Consider a line of total length L′ with x indicating distance along the line, 0 <= x <= L′. Let R(x) and C(x) describe the variations of the resistance and capacitance. The line is divided into n sections with the position of the end of section i given by L_i, i=1,..,n (Figure 3). Notice that the load transistor of section i is also the source transistor for section (i+1) and L_0 = 0. The delay d_i in section i is given by

$$d_i = k*tau*e*ln[\{C(L_i)-C(L_{i-1})+C_{gs(i+1)}\}/C_{gsi}] + [R(L_i)-R(L_{i-1})]*[C(L_i)-C(L_{i-1})] + [R(L_i)-R(L_{i-1})]*C_{gs(i+1)}, \quad i=1,..,n \quad (4)$$

The total line delay, defined as the sum of the individual delays in all the sections, is given by:

$$d = \sum_{i=1}^{n} (k*tau*e*ln[\{C(L_i)-C(L_{i-1})+C_{gs(i+1)}\}/C_{gs(i+1)}]) + \sum_{i=1}^{n} (\{R(L_i)-R(L_{i-1})\}*\{C(L_i)-C(L_{i-1})\}) + \sum_{i=1}^{n} (\{R(L_i)-R(L_{i-1})\}*C_{gs(i+1)}) \quad (5)$$

The total delay is seen to be a function of the lengths of the individual sections as well as the number of sections. Assume that the number of sections into which the line is divided is known, then the partial derivative of d with respect to L_i, i=1,...,n can be determined, set equal to zero, and the resulting equations solved to obtain L_i, i=1,...,n. Thus, the lengths of the individual sections that give the minimum delay can be obtained for a given value of n. A global minimum is determined by repeated application of this process with varying n.

3.0 Clock Delay in Interconnection Networks:

Figure 4 shows the structure of an N*N synchronously controlled crossbar network built from 2*2 modules [DHAR83]. The analysis presented here focuses on minimizing the delay in the clock distribution. As shown in WANN83, unequal delays in the clock line give rise to clock skew, which reduces the throughput of the network. Using a tree structure ensures that all clock line lengths are equal, thus reducing clock skew and increasing throughput.

In general, large interconnection networks are built from many chips and the clock distribution tree consists of a part that resides on the board containing the chips and distributes the clock to each individual chip, and a part which distributes the clock to submodules within a chip. The part of the tree residing on the board consists of metal lines having negligible resistance; preliminary analysis indicates that the delay in this part of the tree is small compared to the internal chip delay and hence can be neglected. The analysis presented here is restricted to the clock tree internal to a chip and to network chips of sizes N*N, $N = 2^m$, m an integer.

Figure 4 shows the clock distribution tree for an 8*8 network. Due to the horizontal and vertical symmetry, the voltage at all points in the tree equidistant from the root must be equal at all times and all such points can be connected without altering electrical behaviour. Thus, two or more identical parts of the tree can be folded into one single structure. In doing so, the resistance of the folded tree decreases whereas the capacitance increases by the same factor. The tree for an N*N network can be reduced to a single line by repeated folding (Figure 5). The resulting structure has 2m+1 sections, with the resistance per unit length decreasing by a factor of 2 from one section to the next and the capacitance per unit length increasing by a factor of 2. Figure 6 shows the variation of capacitance, resistance and depth of fold with the length of the line. The depth of fold of a section of the folded line is defined as the number of sections of the original tree that were folded to form the equivalent section of the folded line. Notice that the depth of fold is constant over a given section of the folded tree, and increases by a factor of two from one section to the next.

While the equations previously developed for minimizing the delay are easily applied to continuous functions, the minimization procedure becomes complex and computationally intensive if the discontinuous functions of Figure 6 are used. Hence, these functions are approximated by the continous functions given below

$$C_a(x)=a_1(e^{b_1*x}-1)+a_2(e^{b_2*x^2}-1)+...+a_5(e^{b_5*x^5}-1) \quad (8)$$

$$R_a(x)=c_1(1-e^{d_1*x})+c_2(1-e^{d_2*x^2})+...+c_4(1-e^{d_4*x^4}) \quad (9)$$

$$N_a(x)=g_1*e^{h_1*x}+g_2*x*e^{h_2*x^2}+...+g_5*x^4*e^{h_5*x^5} \quad (10)$$

Once the fabrication parameters are known (e.g. resistance and capacitance of the line material), the coefficients (i.e., $a_i,b_i,...,h_i$) above can be found via a computer program.

3.1 Minimization of Clock Line Delay:

In order to apply equation (5), the source capacitances C_{gsi}, i=1,....,n have to be obtained. Consider the effect of increasing the value of C_{gsi} on the delay of the $(i-1)^{th}$ and i^{th} sections. The delay in the $(i-1)^{th}$ section increases because the capacitance in this section increases, while the delay in the ith section decreases since the buffer is larger. Evidently, there is an optimum value that results in minimum delay. To determine this optimum value of C_{gsi} requires the minimization process to treat C_{gsi} as an independent variable. While this could be done in a more general analysis, in this analysis we assume that all buffers in the actual (not folded) clock tree have a minimum size transistor at their input. Let the depth of fold at point x be N(x). Then in the folded clock line the input capacitance of a buffer located at distance x from the root will be $N(x)*C_m$.

Next consider the effect of dividing the folded clock line into n sections in the manner described in section 2.0. The source capacitance of the ith section, C_{gsi}, is then given by

$$C_{gsi} = N(L_1+...+L_{i-1})*C_m, \quad i=1,...,n \tag{11}$$

This expression for C_{gsi} can now be subsituted into (5); the expressions for capacitance, resistance and depth of fold functions also will be replaced by their approximate functions, $C_a(x)$, $R_a(x)$ and $N_a(x)$ respectively yielding the final approximation to the delay as

$$d = \sum_{i=1}^{n}(k*tau*e*\ln[\{C_a(L_i) - C_a(L_{i-1}) + N_a(L_{i-1})*C_m\}/N_a(L_i)*C_m]) + \sum_{i=1}^{n}(\{R_a(L_i) - R_a(L_{i-1})\}*\{C_a(L_i) - C_a(L_{i-1})\}) + \sum_{i=1}^{n}(\{R_a(L_i) - R_a(L_{i-1})\}*N_a(L_i)*C_m) \tag{12}$$

$$\text{where} \quad L' = L_i \tag{13}$$

From (12) and (13) the positions of the individual sections L_i, i=1,...,n that give the minimum delay can be determined using standard continuous function computer optimization schemes.

Let the resistance per square be denoted by R_{sq}, and the capacitance per square micron by C_{sq}. The typical values of R_{sq} and C_{sq} for a 5 micron linewidth NMOS technology are given in Table 1 [MEAD80]. Let lambda denote the minimum resolution of the process (minimum line width is 2*lambda). Let the parameters with a superscript (′) indicate the value of the scaled parameters. Then the scaling of the key parameters as lambda scales from lambda to k_s*lambda (k_s < 1) are as follows [MEAD80]:

$$R_{sq}' = R_{sq}/k_s \qquad C_{sq}' = C_{sq}/k_s$$

$$R_m' = R_m \qquad C_m' = k_s*C_m$$

Material	R_{sq} ohms/square	C_{sq} pF/sq. micron
Metal	0.03	0.00003
Polysilicon	15-100	0.00004
Diffusion	10	0.0001

Table 1

Furthermore, assume that the largest square chip that can fabricated be with adequate yield, has a side dimension less than or equal to 1.5 cm. Also assume that this restriction applies as the linewidth decreases. Experience obtained from designing a 2*2 network module [DHAR82] indicates that for a technology with lambda of 2.5 microns, a 32*32 network will occupy a chip area of about 2.25 sq. cm. If L_{ch} is the length of a square chip occupied by an N*N network with a line resolution of lambda, then the length of a square chip L_{ch}' occupied by an N′*N′ network with a line resolution of k_s*lambda is given by $L_{ch}' = L_{ch}*N'*k_s/N$.

Next the clock line parameters will be selected assuming that the fabrication technology provides only one layer of metal. Clock line routing will likely be laid out in part on a material other than metal. In the example presented here, 90% of the clock line is metal and 10% is diffusion.

Consider equation (12) and (13). The functions $C_a(x)$, $R_a(x)$ and $N_a(x)$ can be determined. Also, R_m, C_m and L' are known for a particular network size and lambda. Hence (12) can be minimized to obtain the minimum delay in the folded clock tree. The actual minimization of equation (12) was performed using a program that finds the minimum of a non-linear continuous function constrained with a set of non-linear but continuous equality and non-equality relations.

3.2 Minimization Results:

Figures 7 show the variation of capacitance with length, for various network sizes with lambda varying from 2.5 to 0.5 microns. The broken lines are the approximate curves and have been fitted with an error of less than 1%. Similar curves can be obtained for the variation of resistance with length. Figures 8 shows the variation of the depth of fold with line length.

Figure 9 shows the variation of the line delay with network size and lambda for the folded clock tree. Both minimized and unminimized (i.e., a single exponential buffer drives the entire folded line) delays are illustrated. In section 4.0 it is shown that these delays hold for the unfolded clock tree also. Comparison of the minimized and unminimized delays show that there is a substantial reduction in the line delay due to the minimization technique. For a network size of 64*64 there is more than an order of magnitude reduction in the delay. Observe that as the network size decreases and lambda increases, the difference in the minimized and unminimized delay decreases. This can be explained as follows: The delay in a line given by (3) is the sum of (a) delay due to driving the line capacitance by a buffer and (b) delay due to driving the line capacitance by the line resistance.

The first part of the delay is minimum when the total line capacitance is driven by one buffer whereas the second part is minimum when the line is divided into numerous small segments and each one is driven separately. When the network size decreases and lambda increases, the contribution to the unminimized delay due to the second part is small and hence there is little or no reduction in the delay due to minimization. Hence "small" lines (definition of a small line depends on the length of the line as well as the resistance and capacitance per unit length of the line) can be driven as a whole without incurring any significant delay penalty.

Figure 10 shows the variation of the number of buffers required to obtain the minimum delay with lambda and network size. As the size of the network increases and lambda decreases, more buffers are needed to obtain the minimum delay. This is because the contribution of the product of the resistance and capacitance to the total unminimized line delay increases as the network size increases and lambda decreases. Since this delay term decreases as the number of buffers increases, the variations shown in Figure 9 are produced. For the current state of VLSI technology, only a few buffers would be required to drive the longest lines in minimum time. Of course, more buffers would be needed for long diffusion or polysilicon lines whose resistance per square is much larger than metal.

4.0 Placement of Buffers:

Starting from the root of the tree of Figure 4, traverse the tree to each leaf. In doing so, mark off lengths of L_1, L_2, ... and L_n from the root. These are the positions of the buffers. Number the buffers so that all buffers in a path from the root to a leaf of the tree that are at a distance of L_i from the root are numbered i. Notice that the number of buffers as predicted for the folded clock tree has increased due to the process of unfolding the clock tree. The ith buffer in the folded clock tree has now been distributed in the actual clock tree into all buffers that are numbered i. This distribution of a buffer in the folded tree to that in the unfolded tree takes place according to the depth of fold function. The input capacitance of each of these distributed buffers is N times less than the input capacitance of the buffer in the unfolded tree, where N is the depth of fold at that point. Consider the ith buffer in the folded tree being distributed in the above manner in the unfolded tree. The delay in the ith section of the folded tree is given by equation 4. Consider the effect of unfolding the tree on each term of (4). The capacitance being driven by each buffer in the unfolded tree decreases by a factor of $N(L_{i-1})$ while the resistance increases by the same factor. The first term of (4) consists of a ratio of capacitances and since the numerator and denominator decrease by the same factor, it remains unchanged in the unfolded tree. The second and third terms are products of capacitance and resistance which also remain the same with unfolding. Hence, (4) also gives the delay in the ith section of the unfolded tree. Since no delay term is affected by the unfolding, the delay in the unfolded tree is equal to the delay in the folded tree and is also the minimum. Figure 11 shows a folded clock tree and the positions of the buffers in the actual clock tree.

5.0 Conclusion:

The delay minimization equations developed in this paper were applied to the case of a binary tree distributing clock to all sub-modules of an N*N interconnection network chip. For N = 64, the minimization resulted in more than an order of magnitude reduction in the delay. In order to apply the analysis of a line to the clock tree structure, the clock tree was first collapsed to the form of a single line by making use of its symmetric structure. The minimization of delay is achieved by recognizing that the contribution of the product of the resistance and capacitance to the total line delay increases faster than the line length. Since the resistance and capacitance of a line are typically proportional to the line length, less delay is encountered if a number of smaller length lines are driven rather than the whole line. For long lines, the reduction in delay is appreciable. Such reductions are possible only if the the contribution of the resistance-capacitance product to the delay is significant. The minimization technique gives the number, position and size of the buffers required to obtain the calculated delay.

REFERENCES

CART84 Carter, D.L. and Guise, D.F.: "Effects of Interconnections on Submicron Chip Performance", VLSI Design, Jan. 1984.

DHAR83 Dhar, S., Franklin, M.A. and Wann, D.F.: "Timing Control in VLSI Based NlogN and Crossbar Networks", Proc. 1983 Inter. Conf. on Parallel Processing, Aug. 1983.

DHAR82 Dhar, S. and Bhatia, P.: "Design and VLSI Implementation of a Synchronous Crosspoint Switch", CCSD Tech. Rpt. 101, Dept. of Elec. Engr., Washington Univ., St. Louis, MO., Sept. 1982.

FISH83 Fisher, A.L. and Kung, H.T.: "Synchronizing Large Processor Networks", 1983 Symp. on Comp. Arch., Jan. 1983.

JAEG75 Jaeger, R.C.: "Comments on "An optimized Output Stage for MOS Integrated Circuits"", IEEE J. Solid-State Circuits, June 1975.

KUNG82 Kung, S.Y. and Gal-Ezer, R.J.: "Synchronous vs Asynchronous Computation in VLSI Array Processor", Proc. SPIE, Vol. 341, May 1982.

MEAD80 Mead, C. and Conway, L.: "Introduction to VLSI Systems", Addison-Wesley Pub. Co., 1980.

PENF81 Penfield, P. and Rubinstein, J.: "Signal Delay in RC Tree Networks", Proc. 18th Design Auto. Conf., June 1981.

SINH82 Sinha, A.K., Cooper, J.A. and Levinstein, H.J.: "Speed Limitations Due to Interconnect Time Constants in VLSI Integrated Circuits", IEEE Electron Device Letters, EDL-3, No. 4, April 1982.

WANN83 Wann, D.F. and Franklin, M.A.: "Asynchronous and Clocked Control Structures for VLSI Based Interconnection Networks", IEEE Trans. on Comp., March 1983.

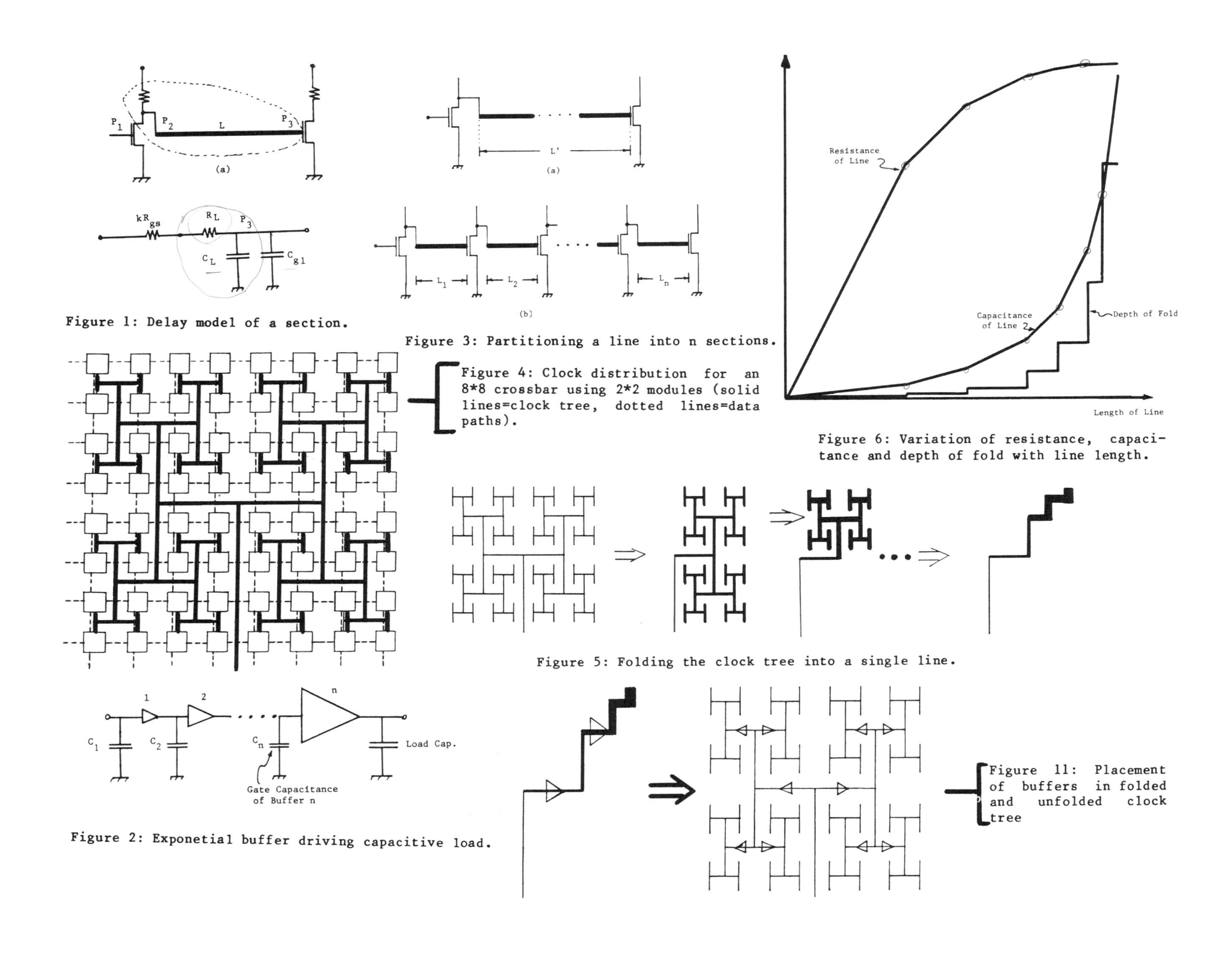

Figure 1: Delay model of a section.

Figure 3: Partitioning a line into n sections.

Figure 4: Clock distribution for an 8*8 crossbar using 2*2 modules (solid lines=clock tree, dotted lines=data paths).

Figure 6: Variation of resistance, capacitance and depth of fold with line length.

Figure 5: Folding the clock tree into a single line.

Figure 2: Exponetial buffer driving capacitive load.

Figure 11: Placement of buffers in folded and unfolded clock tree

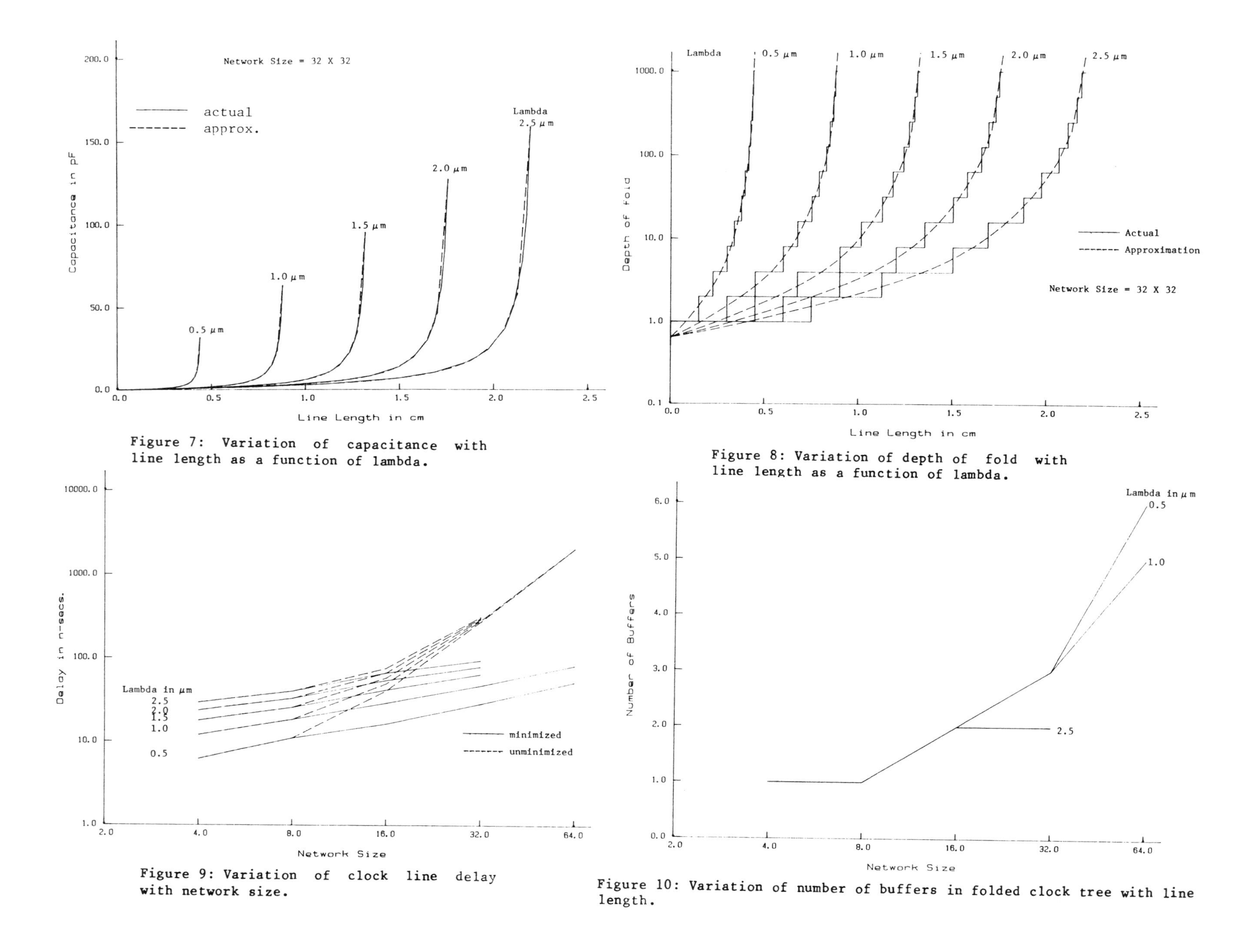

Figure 7: Variation of capacitance with line length as a function of lambda.

Figure 8: Variation of depth of fold with line length as a function of lambda.

Figure 9: Variation of clock line delay with network size.

Figure 10: Variation of number of buffers in folded clock tree with line length.

Clocking Schemes for High-Speed Digital Systems

STEPHEN H. UNGER, FELLOW, IEEE, AND CHUNG-JEN TAN, MEMBER, IEEE

Abstract—**A key element (one is tempted to say the *heart*) of most digital systems is the clock. Its period determines the rate at which data are processed, and so should be made as small as possible, consistent with reliable operation.**

Based on a worst case analysis, clocking schemes for high-performance systems are analyzed. These are 1- and 2-phase systems using simple clocked latches, and 1-phase systems using edge-triggered *D*-flip-flops. Within these categories (any of which may be preferable in a given situation), it is shown how optimal tradeoffs can be made by appropriately choosing the parameters of the clocking system as a function of the technology parameters. The tradeoffs involve the clock period (which of course determines the data rate) and the tolerances that must be enforced on the propagation delays through the logic. Clock-pulse edge tolerances are shown to be an important factor. It is shown that, for systems using latches, their detrimental effects on the clock period can be converted to tighter bounds on the short-path delays by allowing *D* changes to lag behind the leading edges of the clock pulses and by using wider clock pulses or, in the case of 2-phase systems, by overlapping the clock pulses.

Index Terms—**Clocking, clock pulses, delays, digital systems, edge-triggered flip-flops, edge tolerances, latches, one-phase clocking, skew, synchronous circuits, timing.**

I. Introduction

VIRTUALLY all contemporary computers and other digital systems rely on clock pulses to control the execution of sequential functions. A number of different general schemes are used, along with several different types of flip-flops or similar storage elements. Despite the deceptively simple outward appearance of the clocking system, it is often a source of considerable trouble in actual systems. The number of parameters involved, particularly in 2-phase systems, is large, and a close analysis reveals a surprising degree of conceptual complexity.

If one is not particularly interested in maximizing performance, then a 2-phase system with nonoverlapping clocks, or a 1-phase system with edge-triggered FF's is not difficult to design. However, if minimizing the clock period is a prime issue, then the problem becomes far more complex. However, significant performance gains are possible by carefully choosing the clocking parameters (period, pulse-widths, overlap), and further gains may be achieved by using well-designed latches.

In this study we develop sets of relations for three basic types of systems that make possible intelligent tradeoffs between speed maximization (period minimization) and the difficulty of satisfying constraints on the logic path delays. We begin with discussions of the state devices considered, the nature of imprecision in clock-pulse generation and distribution systems, logic block delays, and the design goals. We then analyze the simple case of the 1-phase system using edge-triggered FF's. After this warm-up, we proceed to treat the 1-phase system using latches, a considerably more complicated case. An extension of the methodology used in that section is then applied to the case of 2-phase systems using latches. Some overall conclusions are then presented in the final section.

A. State Devices and Their Parameters

The state devices (or storage elements) treated here are:

The latch [2], [6], [1] (sometimes referred to as the *polarity hold* latch. This is a device with inputs C and D, and output Q (often Q', the complement of Q is also generated), such that, ideally, while $C = 0$, Q remains constant (regardless of the value of D), and while $C = 1$, $Q = D$, changing whenever D changes (see Fig. 1). (For real latches, as is explained below, there are nonzero delays in the response times, and there must be constraints on the behavior of the inputs.) The C and D inputs are usually referred to as the clock and data inputs, respectively. Although it is not, in general, necessary to do so, in the applications treated here, the system clock signals are indeed fed to the C inputs of the latches. A variety of implementations of latches are known, differing in such factors as suitability for various technologies, load driving ability, and relative values of the parameters to be discussed subsequently. Latches with logic hazards have been used in some systems. In order to eliminate the possibility of malfunction due to those hazards, the *complement* of the C signal is distributed independently to the latches with its edges carefully controlled relative to the corresponding edges of the C signals. We do not discuss such systems here, where it is assumed that the latches are free of hazards.

The *edge-triggered D-flip-flop* (ETDFF) [2], [6] has the same inputs and outputs as the latch, but Q responds to changes in D only on one edge of the C pulse (see Fig. 2). That is, Q can change only at the time that C changes from 0 to 1 (the rising edge of the C signal), and then only if necessary to assume the same value that D has at that time. (There are also ETDFF's that change state on the negative-going edge of the C signal. Furthermore, it is possible to build a double-edge-triggered D-FF [9] that will respond on *both* edges of the C pulse.)

1) Latch Parameters: The significant parameters for a latch

Reprinted from *IEEE Trans. Comput.*, vol. C-35, no. 10, pp. 880–895, Oct. 1986.

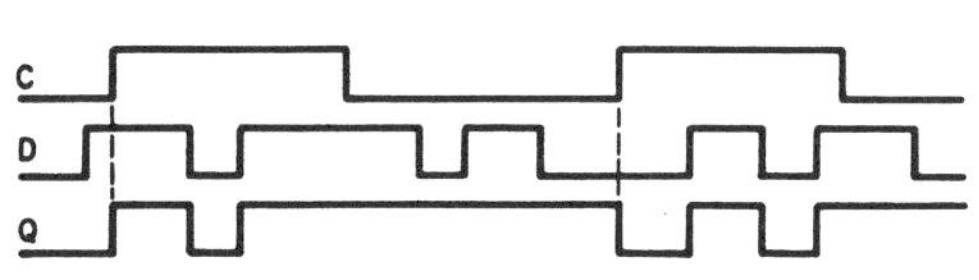

Fig. 1. Behavior of an ideal latch.

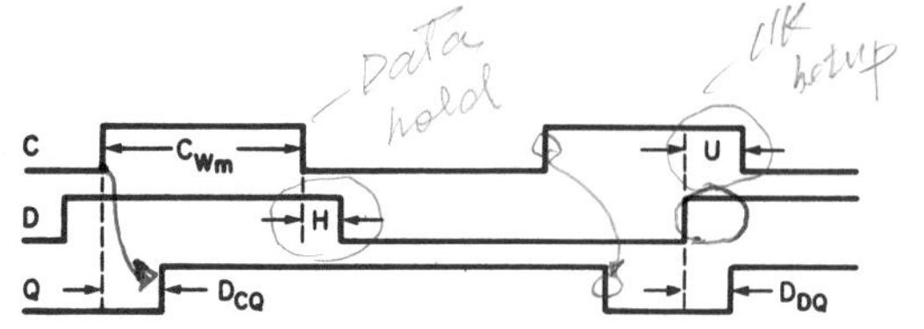

Fig. 3. Latch parameters.

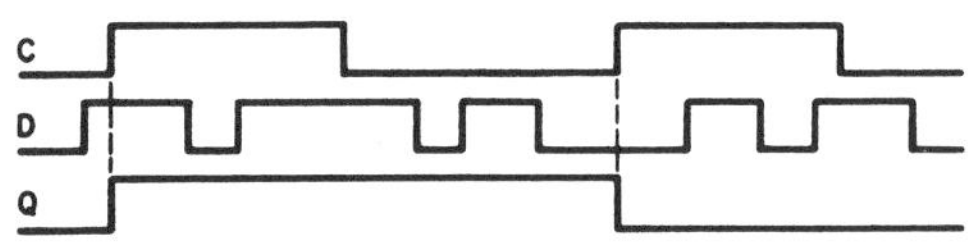

Fig. 2. Behavior of ideal positive triggered ETDFF.

are listed below, with rough definitions (illustrated in Fig. 3). These definitions are then refined to take into account dependencies that exist among the parameters.

C_{Wm}: Minimum clock-pulse width, the minimum width of the clock pulse such that the latch will operate properly even under worst case conditions, and such that widening the C pulse further by making its leading edge occur earlier will not affect the values of D_{DQ}, U, or H, as defined below.

D_{CQ}: Propagation delay from the C terminal to the Q terminal, assuming that the D signal has been set early enough relative to the leading edge of the C pulse.

D_{DQ}: Propagation delay from the D terminal to the Q terminal, assuming that the C signal has been turned on early enough relative to the D change.

U: The *setup time*, the minimum time between a D change and the trailing edge of the C pulse such that, even under worst case conditions, the Q output will be guaranteed to change so as to become equal to the new D value, assuming that the C pulse is sufficiently wide.

H: The *hold time*, the minimum time that the D signal must be held constant *after* the trailing edge of the C signal so that, even under worst case conditions, and assuming that the most recent D change occurred no later than U prior to the trailing edge of C, the Q output will remain stable after the end of the clock pulse. (It is not unusual for the value of this parameter to be negative.)

Note that D_{DQ}, for example, may vary significantly depending on whether the latch output is being changed from 0 to 1 or vice versa. A similar situation exists for D_{CQ}. Where appropriate it is useful to add subscripts R or F to these parameters to distinguish between the rising and falling output cases. This will not be done here. Instead, we shall confine ourselves to using overall maximum and minimum values, as indicated below.

The addition to the subscripts of D_{DQ} or D_{CQ} of an M or m make these parameters the maximum or minimum values, respectively. These are the extremes with respect to variations in the parameters of the components from which the latches are constructed, the directions of signal changes, and the destinations (Q or Q') of the signals.

In the definition of D_{CQ}, it is assumed that D has assumed its proper value early enough. We can make this concept more precise by requiring that the change in D occurs sufficiently early so that making it appear any earlier would have no effect on when Q changes. For any real latch it is always possible to define such an interval. Similarly, when defining D_{DQ}, it is assumed that the leading edge of C appears sufficiently early so that turning C on any earlier would not make Q change any sooner. Again this is possible for any real latch.

Now we state an important postulate regarding propagation delays:

Suppose that C goes on at time t_C, and that D changes, making D different from Q, at time t_D. Then we postulate that the time t_Q at which Q changes is, at the latest:

$$t_Q = \max\ [t_C + D_{CQM},\ t_D + D_{DQM}]. \quad (1)$$

Although for some latches there are higher order effects, depending on the technology, that may cause t_Q to be larger when the difference between the arguments of the max is small, the error is small enough to justify our postulate for most practical purposes. Refining the model to take such effects into account is left for further research.

A related assumption about latch behavior is that, provided that the setup, hold-time, and minimum pulse-width constraints are observed, the propagation delay will not be affected by the clock-pulse going off before the output changes in response to a D change. An examination of a variety of latch designs appears to justify this assumption.

There are other possibilities for refining our results, by using more complex definitions of latch parameters. If we define the actual interval between the occurrence of a D change and the trailing edge of C as u (note that proper operation requires that $u \geq U$), then, for many latch designs it will be found that the hold time H is, over some range of values of u, a decreasing function of u. There are also possibilities for reducing the clock-pulse width below C_{Wm} (within limits), usually at a cost of increasing propagation delays and/or setup and hold times. For the sake of making the analysis more tractable, we shall not consider these alternatives, but instead shall assume that there is a fixed, consistent, set of latch parameters, as described above.

In summary, we assume that the minimum clock-pulse width is large enough so that further increases cannot reduce any of the other latch parameters, that U is minimal, that H is minimal given U, and that the postulate stated above regarding propagation delays is valid.

2) Edge-Triggered-D-FF Parameters: The significant parameters for an ETDFF are defined below (see also Fig. 4).

U: The *setup time*, the minimum time that the D signal must be stable prior to the triggering edge of the C pulse.

H: The *hold time*, the minimum time that the D signal must be held constant *after* the triggering edge of the C pulse. (The value of H may be 0 or even negative for some ETDFF's.)

C_{Wm}: Minimum clock-pulse width, the minimum width

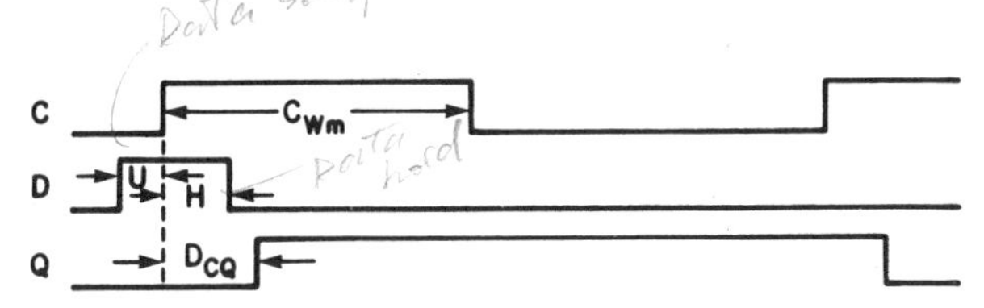

Fig. 4. Parameters of a positive-edge-triggered-D-FF.

of the clock pulse such that the ETDFF will operate properly even under worst case conditions.

D_{CQ}: *Propagation delay* from the C terminal to the Q terminal, assuming that the D signal has been set up sufficiently far in advance as specified by the setup time constraint.

B. Clock-Pulse Edge Deviation

In any real-world system there are limits to the precision with which events can be timed. Our concern here is with synchronous systems with clock-pulses distributed to a multitude of devices for the purpose of coordinating events. The intent is to have certain clock-pulse edges occur simultaneously at all devices (in some cases fixed displacements may be specified for corresponding signals at different devices). In designing clocking schemes, it is necessary to take into account the extent to which this goal cannot be fully attained.

The approach taken here is to assume that, at each significant clock-pulse edge, there is a specified tolerance range, within which we can assume the errors will be confined. This is, essentially, a "worst case" approach. No attempt will be made to exploit statistical information that could make possible more precise estimates of errors, nor will any effort be made to consider the effects of correlations between errors or between delays.

The most elaborate situation that we deal with is that of 2-phase systems using latches as storage elements. Here both the leading and trailing edges of both clock-pulses are of interest (although the analysis makes it clear that certain edges are more significant than others). We define tolerances for all 4 edges, designating them as T_{1L}, T_{1T}, T_{2L}, and T_{2T}, corresponding to the leading and trailing edges of $C1$ and $C2$, respectively. Assume that, for example (see Fig. 13), the leading edge of the $C1$ pulse for some period would have arrived at every latch at time t (which we refer to as its *nominal* arrival time) if there were no inaccuracies in timing. Then, in the actual system, this edge is received at every latch somewhere in the time interval, $(t - T_{1L}, t + T_{1L})$. Corresponding assumptions of course apply for the other three edges. Our goal is to design our systems so that if this assumption, and corresponding assumptions about the precision of the components used, are valid, then there will be no failures due to timing, even if some malicious demon is, in each case, permitted to choose the extreme deviations most likely to cause trouble. Of course in 1-phase systems we need only define two edge tolerances, T_L and T_T.

We are lumping together in these edge tolerances all sources of imprecision in clock timing and distribution. These are principally in the circuits used to determine the clock-pulse widths, often called "shapers," and in the networks used to distribute the pulses to the individual latches (or other similar devices). This latter factor is generally referred to as clock-pulse skew. In the case of 2-phase systems, it is also necessary to consider the circuits that determine the phase relationship between the $C1$- and $C2$-clocks.

Relative to other sources of error, the precision with which the clock frequency can be maintained, at least in high-performance systems, is so great (due to the use of crystal controlled oscillators) that we can safely neglect this factor. (If this assumption is not justified in any particular case, it is not difficult to introduce a tolerance factor on the clock period, which can be superposed on our basic results.)

By representing all of the timing deviations in terms of the edge tolerances, we simplify our analysis, making it easier to treat, as a separate issue, the mechanisms whereby precision is lost.

The precision with which clock-pulse widths can be controlled is generally a function of how precisely delay elements can be specified. The same factor usually is involved in controlling the phase between the $C1$ and $C2$ pulses of a 2-phase system. The ratio of 2 delays on the same chip can be specified with much greater precision than is the case for delays on different chips. Usually one edge of the output of a shaper can be controlled more precisely than the other. In the 2-phase case, there are techniques for minimizing the edge tolerances for particular pairs of edges. As is shown in the sequel, T_{2L} and T_{1T} are usually more significant. They should therefore be kept smaller, relative to the other two-edge tolerances.

Several factors contribute to clock-pulse skew. Despite all efforts to equalize conduction path lengths between the clock source and each clock-pulse "consumer," differences inevitably occur in both off-chip wiring and in paths on chips. Since it is usually necessary to provide amplifiers in the distribution paths, variations in the delays encountered in such devices along different paths produce significant amounts of skew.

Another contribution to skew results from the fact that pulse edges are never vertical as shown in our idealized diagrams, and that there is variability among individual latches, even on the same chip, with respect to the voltage thresholds that effectively distinguish 1's from 0's. Thus even if a pulse edge should arrive simultaneously at the inputs to two different latches, its effect might be felt at different times due to a difference in thresholds. The result is the same as if the delays in the paths leading to the two latches differed. Hence, such effects are considered as part of the skew. Note that, unlike the factor due to varying length conduction paths, this effect could result in the delayed sensing of a *positive*-going edge at a latch that is relatively quick in sensing a *negative*-going edge. (This would occur if the device involved had a relatively high threshold.)

C. Logic Block Delays

In addition to the various parameters associated with the clocking system and with the latches or FF's, a very important pair of parameters is that associated with the logic circuitry: the *maximum* and *minimum* delays in any path through the logic block, designated as D_{LM} and D_{Lm}, respectively. As is made evident in our analysis, large variations among logic path delays are clearly detrimental. That is, for a given value of D_{LM}, it is desirable to keep the *smallest* path delay as close to D_{LM} as possible.

It is frequently the case, when choosing the clocking parameters, that the value of D_{LM}, the *long-path* delay is given; it is a function of the maximum number of stages of logic, the amount of fan-in and fan-out associated with gates in the longest paths, and of the technology, which determines propagation delay through individual gates. The lower bound on the *short-path* delay D_{Lm}, on the other hand, can often be dictated, within limits, by the clock system designer, using such means as adding delay pads to increase the delays in the shortest paths, or adjusting the power levels of certain key gates.

The ultimate limits on how tightly the short-path delays can be controlled, that is, on how high a lower bound D_{LmB} on them is feasible, depends on the tolerances with which gate delays can be specified, as well as on how well wire lengths, both on and off chip can be predicted at design time. It is these factors that determine, for a given value of D_{LM}, what the largest feasible value of D_{LmB} is.

D. Goals for Design of Clocking Schemes

It is assumed here that a principal goal in the specification of a clocking scheme is to make the period as small as possible, which is tantamount to maximizing the speed of the system. But of course this must be done within the confines of a design that results in a system that can be made to operate reliably.

It is obvious that minimizing D_{LM} is basic to minimizing the clock period. But, as pointed out above, it is also important to keep the *smallest* path delay as large as possible. But it is by no means easy to make the logic path delays uniform in value. For this reason, we have developed procedures for finding the minimum possible value of P given the maximum achievable lower bound D_{LmB} on the short-path delays.

II. Optimum Parameters for 1-Phase Clocking with ETDFF's

For 1-phase systems using ETDFF's, the clocking parameters to be determined, (see Fig. 4) are the period P and the clock-pulse width W. A block diagram of the systems under consideration is shown as Fig. 5.

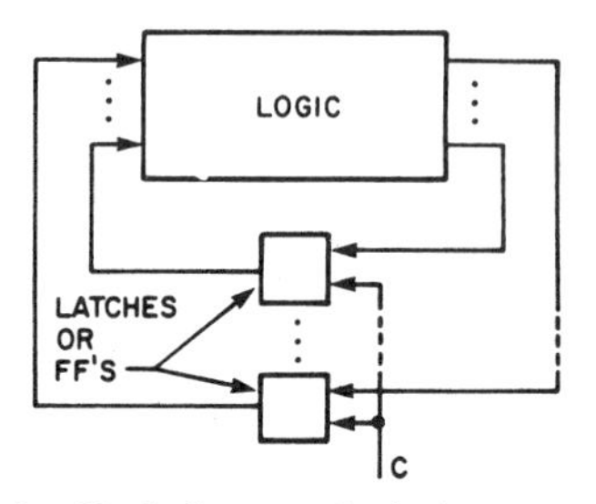

Fig. 5. Block diagram of a 1-phase system.

We develop a set of constraints, such that if all are satisfied, and if the D signals arrive on time for the first cycle, then they will also arrive on time for the next cycle and will remain stable long enough to ensure that the FF's react properly. By induction, it follows that, for all succeeding cycles, the FF inputs are also stable over the appropriate intervals, so that the system will behave according to specifications.

For any clock-pulse period, proper operation requires that the D signals become stable at least U prior to the earliest possible occurrence of the triggering edge. (It is assumed here that this is the positive-going edge. Precisely the same arguments apply where the triggering edge is negative going—or even if the FF's trigger on both edges.) If we assume that $t = 0$ coincides with the nominal time of the leading edge of the current clock pulse, then the earliest possible occurrence time of that edge is $-T_L$. (See Fig. 6(a).) Hence, the latest possible arrival time, under worst case conditions, of the D signals for the current clock pulse t_{DLArr} must meet the constraint

$$t_{DLArr} \leq -T_L - U.$$

Defining the latest possible arrival time, under worst case conditions, of the D signals for the *next* clock-pulse as t_{DLArrN}, it follows that "on time arrival" of D for the next cycle means

$$t_{DLArrN} \leq P - T_L - U. \tag{2}$$

Since the *latest* possible occurrence of the leading edge of the current clock-pulse is at T_L, it follows that the latest arrival time of the D signals for the next cycle is

$$t_{DLArrN} = T_L + D_{CQM} + D_{LM} \tag{3}$$

(see Fig. 6(b)).

Replacing t_{DLArrN} in (2) by its value from (3), we have the required constraint to ensure that D signals are not late

$$P - T_L - U \geq T_L + D_{CQM} + D_{LM}.$$

Solving for P converts it to the following more meaningful form:

$$P \geq 2T_L + U + D_{CQM} + D_{LM}. \tag{4}$$

Next it is necessary to constrain the system so as to ensure that the *earliest* arrival time of a D signal for the *next* cycle does not arrive so early as to violate the hold-time constraint for the current cycle. (See Fig. 7.)

Given that the latest occurrence time of the leading edge of a clock pulse is T_L, the hold-time constraint mandates that the earliest occurrence time of a D signal for the next cycle, t_{DEArrN}, satisfy

$$t_{DEArrN} > T_L + H. \tag{5}$$

Since the *earliest* occurrence of a leading edge of a clock pulse is at $-T_L$, we can express t_{DEArrN} in terms of the FF propagation delay and the logic delay as

$$t_{DEArrN} = -T_L + D_{CQm} + D_{Lm}.$$

Inserting the value of t_{DEArrN} from the above equation into (5) gives us a relation, the satisfaction of which is a necessary and sufficient condition for preventing, under worst case assumptions, premature changes in D signals:

$$-T_L + D_{CQm} + D_{Lm} > T_L + H.$$

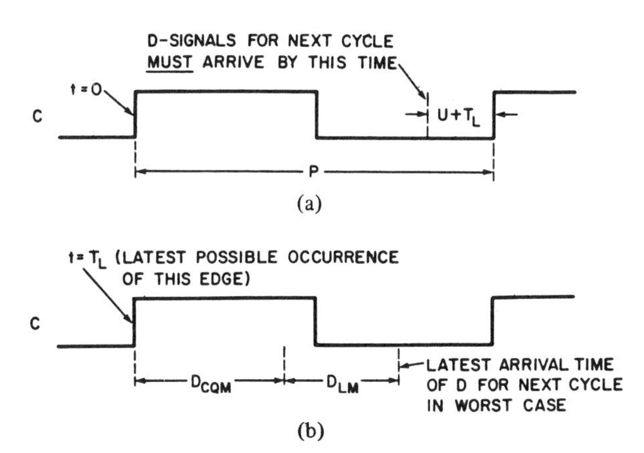

Fig. 6. Ensuring that D signals do not arrive too late in ETDFF systems. (a) Specification of latest permissible arrival time of D signal for next cycle. (b) Arrival time (worst case) for D signal for next cycle.

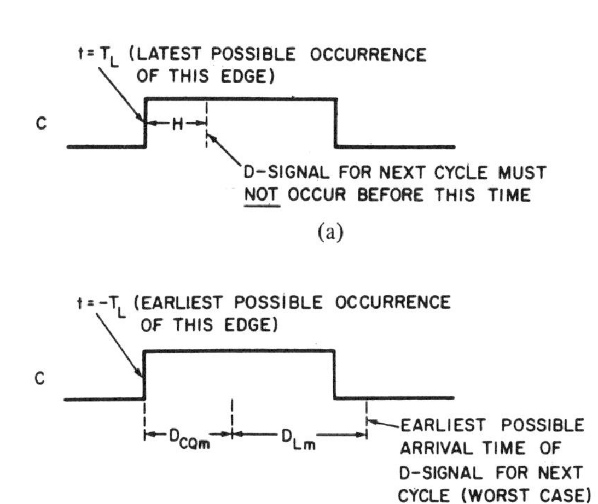

Fig. 7. Ensuring that D signals in ETDFF systems do not arrive too early. (a) Specification of earliest permissible arrival time of D signal for next cycle. (b) Computation of earliest possible arrival time of D signal for next cycle.

Simplifying and rearranging terms yields the basic constraint that defines D_{LmB}, the lower bound on the short-path delays:

$$D_{Lm} > D_{LmB} = 2T_L + H - D_{CQm}. \tag{6}$$

In addition to constraints (4) and (6) on the period and short-path delays, it is necessary to add a third constraint to ensure that the minimum pulse-width specification for the FF's is satisfied. Since, under worst case assumptions skew might make the leading edge late and the trailing edge early, the minimum width specification for the clock pulses is

$$W \geq T_L + T_T + C_{Wm}. \tag{7}$$

The procedure for choosing optimum clocking parameters for 1-phase systems using ETDFF's is usually very straightforward. We simply set W at any convenient value satisfying constraint (7) and set P to satisfy constraint (4) with equality. In most cases, it will be found that the constraint on the short-path bound given by (6) is not difficult to meet. In the unlikely event that this is not the case, it may be necessary to insert delay pads at the outputs of the FF's. The procedure for doing this is the same as that for the 1-phase case with latches, treated in Section III-D.

III. Optimum Parameters for 1-Phase Clocking with Latches

Fig. 5 is a block diagram of the 1-phase systems treated here. Clock signals with parameters noted are shown in Fig. 8. We shall develop a set of constraints, involving the various parameters we have discussed, such that if and only if they are all respected, the system will operate properly in the sense that the D inputs to all the latches will arrive on time for each clock cycle (as specified by the setup time parameter), and will remain stable for a sufficient interval (as specified by the hold-time parameter).

The argument is in the form of induction on the clock periods. It is assumed at the outset that the D signals arrive on time for the first clock cycle. Constraints are developed to ensure that, given this assumption, the D signals will arrive on time for the next cycle. Additional constraints are then found to ensure that the D signals remain stable for an adequate interval during the first cycle. It is then obvious by induction that the same will be true for all subsequent clock cycles.

More specifically, our initial assumption is that, under worst case conditions (of delay values, edge tolerances, etc.), every D signal must arrive (at a latch input terminal) no later than U prior to the trailing edge of the clock pulse. Taking $t = 0$ as the *nominal* time of occurrence of the *leading* edge of the clock pulse for the current cycle (i.e., the time this edge would arrive if the tolerance on this edge T_L were 0), the earliest possible occurrence time of the *trailing* edge would be $W - T_T$.

Since the D signal must arrive at least U prior to this edge, we have for the latest permissible arrival time for D, t_{DLArr}:

$$t_{DLArr} \leq W - T_T - U. \tag{8}$$

Assume now that the above constraint is satisfied for the first clock cycle.

A. Preventing Late Arrivals of D Signals

The latest (under worst case conditions) arrival time of D signals for the next cycle is designated as t_{DLArrN}. The maximum permitted value of t_{DLArrN} is found by simply adding P to the right side of (8)

$$t_{DLArrN} \leq W - T_T - U + P \tag{9}$$

(see Fig. 9(a)).

The worst case value of t_{DLArrN} is the latest time at which the output of a latch could respond to a D signal, plus the maximum delay through the logic. Designating the latest occurrence time of a leading edge of a clock pulse as t_{CLL}, and using postulate (1) for determining the latest time at which the output of a latch could change, we obtain

$$t_{DLArrN} = \max\ [t_{CLL} + D_{CQM},\ t_{DLArr} + D_{DQM}] + D_{LM}.$$

(The discussions pertaining to the left and right parts, respectively, of the max expression are illustrated by Fig. 9(b)

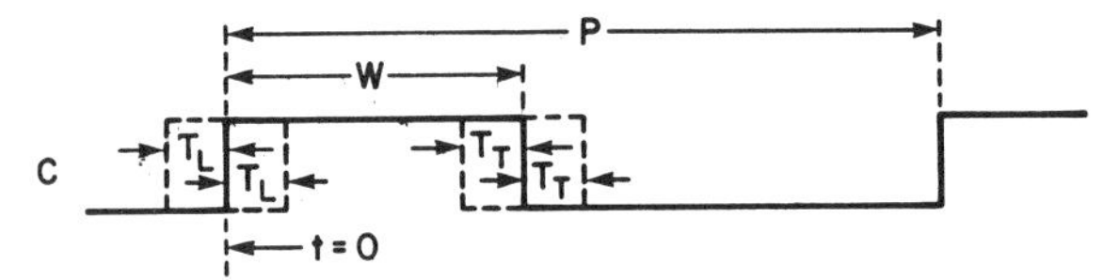

Fig. 8. Parameters for 1-phase systems.

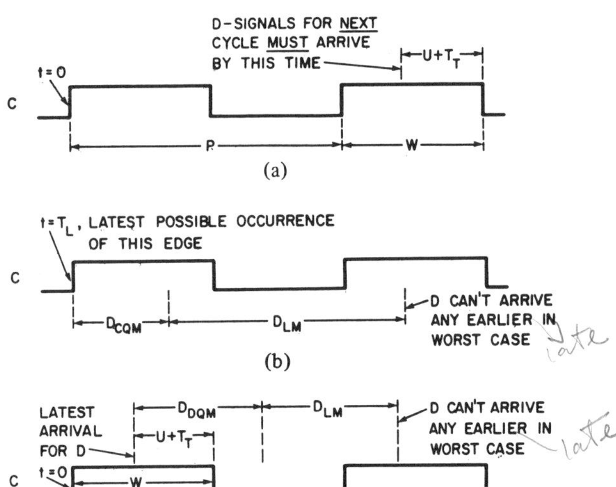

Fig. 9. Ensuring that D arrives sufficiently early. (a) Deadline for arrival of D signals. (b) Bound on D arrival due to leading edge of C. (c) Bound on next-cycle D arrival due to arrival time of D for current cycle.

and (c).) The value of t_{CLL} is clearly T_L, and the value of t_{DLArr} is given by (8), so replacing those variables in the above relation gives us

$$t_{DLArrN} = \max\ [T_L + D_{CQM},\ W - T_T - U + D_{DQM}] + D_{LM}. \tag{10}$$

Combining (9) with (10) produces

$$\max\ [T_L + D_{CQM},\ W - T_T - U + D_{DQM}] + D_{LM} \leq W - T_T - U + P.$$

Solving for P yields

$$P \geq \max\ [T_L + T_T + U + D_{CQM} - W,\ D_{DQM}] + D_{LM}.$$

This expression can be decomposed into 2 constraints that, in combination, are equivalent to it:

$$P \geq D_{CQM} + D_{LM} + U + T_L + T_T - W \tag{11}$$

and

$$P \geq D_{DQM} + D_{LM}. \tag{12}$$

The constraint (12) can be intuitively justified by noting that it represents the total time for a signal to traverse a complete loop, under worst case conditions. If the period were any less, then, if the worst case conditions were actually realized, a signal following a sequence of such maximum delay paths would fall increasingly far behind the clock pulses until it eventually violated a setup time constraint.

Constraint (11) can also be justified intuitively. (Transposing the W term makes this clearer.) It can be interpreted as stating that, starting at the leading edge of a clock pulse, there must be time, under even worst case conditions, before the trailing edge of the *next* clock pulse, for a signal to get through a latch, and the logic block in time to meet the setup time constraint at the input to some latch.

The D signals for the next cycle will arrive on time if, and only if, both (11) and (12) are satisfied, and if (8) is satisfied for the current cycle.

B. Preventing Premature Arrivals of D Signals

If the D signal for the next clock cycle is generated too soon, then the hold-time constraint for a latch might be violated. This is where the short-path delays become important. In order to prevent the possibility of a hold-time violation, it is necessary that, in the worst case, a D change for the next cycle not occur until at least H after the latest possible occurrence of the trailing edge of the clock-pulse defining the current cycle. With t_{CLT} as the latest occurrence of a clock-pulse trailing edge, and t_{DEArrN} as the earliest possible arrival of a D signal for the next cycle, this constraint is expressed as

$$t_{DEArrN} > t_{CLT} + H.$$

(This discussion is illustrated by Fig. 10(a).) Replacing t_{CLT} by its value, $W + T_T$, we obtain

$$t_{DEArrN} > W + T_T + H. \tag{13}$$

Letting t_{CEL} represent the earliest possible arrival time of a clock-pulse leading edge, and t_{DEArr} represent the earliest arrival time of a D signal for the *current* cycle, we again utilize (1) to obtain the following:

$$t_{DEArrN} = \max\ [t_{CEL} + D_{CQm},\ t_{DEArr} + D_{DQm}] + D_{Lm}.$$

(The discussion involving the left part of the max is illustrated in Fig. 10(b).) Replacing t_{CEL} by its value $-T_L$ and bringing D_{Lm} inside the max, yields

$$t_{DEArrN} = \max\ [-T_L + D_{CQm} + D_{Lm},\ t_{DEArr} + D_{DQm} + D_{Lm}]. \tag{14}$$

Inserting the above value of t_{DEArrN} in (13) yields

$$\max\ [-T_L + D_{CQm} + D_{Lm},\ t_{DEArr} + D_{DQm} + D_{Lm}] > W + T_T + H. \tag{15}$$

Now we show that, for a system that operates properly even under worst case conditions, (15) is satisfied if, and only if, the left part of the max in (15) exceeds the right side of the inequality. The "if" part of this assertion is obviously true.

To prove necessity (the "only if" part) let us assume that (15) is valid but that the left part of the max does *not* exceed the right part of the inequality. Then it follows that the *right* part of the max must satisfy the inequality, and hence must exceed the left part of the max. In that case, (14) is reduced to

$$t_{DEArrN} = t_{DEArr} + D_{DQm} + D_{Lm}. \tag{16}$$

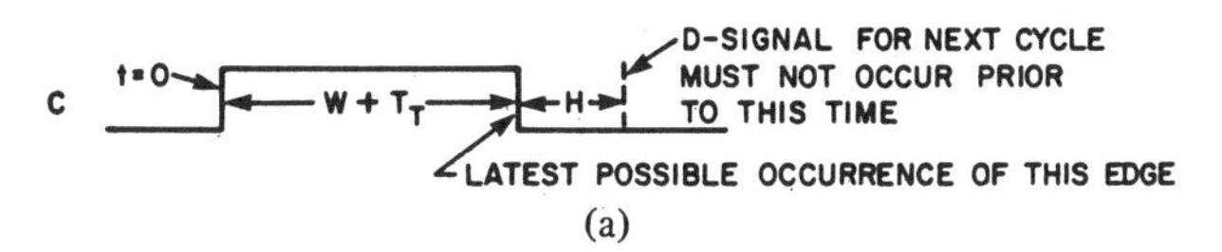

(a)

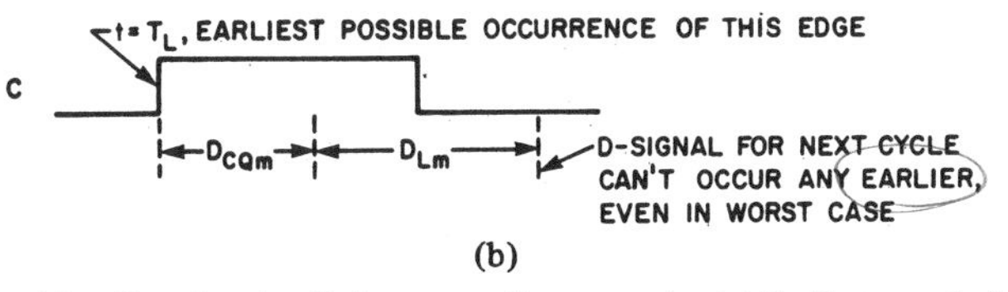

(b)

Fig. 10. Ensuring that D does not arrive too early. (a) Earliest permissible arrival time of D. (b) Lower bound on D arrival for next cycle due to leading edge of C.

But from (12) it is clear that

$$P > D_{DQm} + D_{Lm}.$$

Adding t_{DEArr} to both sides yields

$$t_{DEArr} + P > t_{DEArr} + D_{DQm} + D_{Lm}.$$

From the above and from (16) we then obtain

$$t_{DEArr} + P > t_{DEArrN}.$$

But this means that, for each cycle (in the worst case), D arrives earlier and earlier relative to the trailing edge of C. Therefore, even if t_{DEArr} is comfortably above the minimum for the first cycle, it will eventually violate the hold-time constraint, and hence the system would not operate properly under worst case conditions. Hence, by contradiction, we have completed the argument that (15) is equivalent to:

$$-T_L + D_{CQm} + D_{Lm} > W + T_T + H$$

or, solving for D_{Lm}:

$$D_{Lm} > D_{LmB} = T_L + T_T + H + W - D_{CQm}. \tag{17}$$

The above expression gives us the lower bound D_{LmB} on the short-path delay. Satisfying this bound is necessary and sufficient to ensure against the premature arrival of a D signal.

C. Consequences of the Constraints

The basic constraints derived in the previous subsections are reproduced below.

$$P \geq D_{CQM} + D_{LM} + U + T_L + T_T - W \tag{11}$$

$$P \geq D_{DQM} + D_{LM} \tag{12}$$

$$D_{Lm} > D_{LmB} = T_L + T_T + H + W - D_{CQm}. \tag{17}$$

To these we must add one more to ensure that, even under worst case conditions, the clock-pulse width at any latch input meets the minimum clock pulse width specifications of the latches. This is:

$$W \geq C_{Wm} + T_L + T_T. \tag{18}$$

W in (11) cannot usefully be increased beyond the point where the right side of (11) would, if equality held, violate (12), which of course also represents a lower bound on P. Note that it is undesirable to increase W gratuitously, since this would, as indicated by (17), raise the lower bound on the short-path delays. To find the maximum useful value of W, treat (11) and (12) as equalities and solve them simultaneously (eliminating P) to obtain

$$W = U + T_L + T_T + D_{CQM} - D_{DQM}. \tag{19}$$

When W is less than the above value, (11), with equality, specifies the minimum value of P. When W equals that value, the minimum value of P is given by (12). The maximum useful value of D_{LmB} is found by substituting into (17) the maximum useful value of W. This gives us

$$D_{LmB} = H + U + 2(T_L + T_T) + D_{CQM} - D_{CQm} - D_{DQM}. \tag{20}$$

If the value of the lower bound on the short-path delays given by the above relation is attainable, then the minimum P value of (12) is attainable. If not, then, to find the minimum P value as a function of an achievable value of D_{LmB}, solve (17) and (11) (as equations) simultaneously for P, eliminating W. This results in

$$P = H + U + 2(T_L + T_T) + D_{CQM} - D_{CQm} + D_{LM} - D_{LmB}. \tag{21}$$

Since W must also satisfy constraint (18), there is a corresponding lower bound on D_{LmB}, which is found by substituting into (17) the right side of (18) for W to obtain

$$D_{LmB} = 2(T_L + T_T) + H - D_{CQm} + C_{Wm}. \tag{22}$$

The relations developed here are the basis for the optimization procedure of the next subsection. First, however we must consider a possible variation of the development thus far.[1] The initial assumption in the discussion of 1-phase systems was that the D signals must appear at latch inputs no later than U prior to the trailing edges of the clock pulses. In what followed, this constraint was consistently observed. But what if we had made a stronger assumption, i.e., that the D changes must appear even earlier, say at $U + r$ ($r > 0$) prior to the trailing edges of the clock pulses? Is it possible that there might be some advantages to this?

The key to analyzing this question is to observe that the proposal is exactly equivalent to assuming a larger value of the setup time U. The effect of this can be determined by looking at those constraints and derived relations that involve U, namely (11), (19), (20), and (21). The value of D_{LmB} necessary to achieve the minimum P increases with U. So does the minimum value of P for any value of D_{LmB} in the range for which (21) is valid. Thus there are clear disadvantages to this alternative of effectively increasing U, and no apparent advantages to compensate for them. It follows then that any 1-phase clocking scheme that violates any of our constraints will, under the worst case assumption, either be vulnerable to failure, or will be suboptimum in that either P or D_{LmB} would be reducible without increasing the other.

[1] The necessity for considering this possibility was pointed out by V. Pitchimani and G. Smith.

D. When the Short-Path Bounds Cannot be Met

Now observe that neither the basic constraint (17) on D_{LmB}, nor either of the derived extremes of D_{LmB} given by (20) and (22) involve D_{LM}. Thus, there is no inherent reason why the range found for D_{LmB} (in terms of the afore-noted extremes) should be much *below* or indeed not *above* D_{LM}. If, despite all efforts, including the use of delay pads in critical paths, it is still not possible to satisfy the lower bound on the short-path delays represented by (22), then (assuming that the relevant latch or other parameters cannot be favorably altered so as to remedy this situation), it is necessary to resort to more drastic measures.

The most practical technique appears to be to introduce uniform delay elements into *all* logic paths so as to increase the minimum path delays by an amount sufficient to get us into the desired range. Suppose, for example that the largest value of D_{LmB} that can be reliably guaranteed, is less than the bound of (22) by the amount d_x. Then we could add delay pads with *minimum* values d_x to the outputs of all latches. The effect would be to increase the attainable D_{LmB} to the desired minimum, and to increase D_{LM} by the amount corresponding to the *maximum* value of delay elements with minimum values d_x. If we define T_d as the delay element tolerance ratio d_M/d_m then the addition to D_{LM} is $T_d d_x$. Note that P increases by $T_d d_x$ over the value obtained for it if the D_{LmB} from (22) is used in (21). The graph of Fig. 11 illustrates how P varies with the maximum attainable value of D_{LmB}. It is piece-wise linear, with the left part corresponding to the region where uniform pads must be added as just indicated, and with the right part generated directly from (21). The value P_1 corresponds to the value given by (12).

E. Procedure for Optimizing the Clocking Parameters

We are now in position to describe a procedure for finding the minimum clock period, given D_{MLmB}, the maximum lower bound we can enforce on the short-path delays. The corresponding value of W is also determined.

A complicating factor is the possibility that the lower bound on W given by (18), might exceed what we have called the "maximum useful value of W," given by (19). In that event, the W value is given by (18), and D_{LmB} is given by (22). Note that, when D_{MLmB} is less than the required value of D_{LmB}, it is necessary to pad the outputs of all latches with delay elements whose *minimum* values make up the difference. This adds to the period an amount T_d times this minimum value.

The procedure is as follows.

IF the right side of (18) $\leq$ the right side of (19)
THEN
  IF $D_{MLmB} \geq$ right side of (20)
  THEN
    $D_{LmB} \Leftarrow$ right side of (20)
    $W \Leftarrow$ right side of (19)
    $P \Leftarrow$ right side of (12)
  ELSE
    $D_{LmB} \Leftarrow D_{MLmB}$
    IF $D_{MLmB} \geq$ right side of (22)
    THEN
      $P \Leftarrow$ right side of (21)
      Solve (17) to determine W
    ELSE
      $d \Leftarrow$ right side of (22) $- D_{MLmB}$
      In all latch outputs put delay pads with minimum value d
      $W \Leftarrow$ right side of (18)
      $P \Leftarrow$ right side of (11) $+ T_d d$
ELSE
  $W \Leftarrow$ right side of (18)
  $D_{LmB} \Leftarrow$ right side of (22)
  IF $D_{MLmB} \geq$ right side of (22)
  THEN
    $P \Leftarrow$ right side of (12)
  ELSE
    $d \Leftarrow$ right side of (22) $- D_{MLmB}$
    In all latch outputs put delay pads with minimum value d
    $P \Leftarrow$ right side of (12) $+ T_d d$.

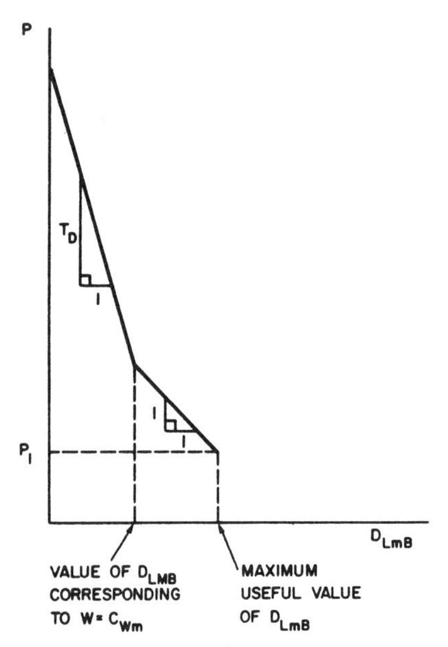

Fig. 11. P as a function of the largest achievable lower bound on short-path delay.

Other procedures based on the constraints developed here may be useful under special circumstances.

IV. Optimum Parameters for 2-Phase Clocking with Latches

Fig. 12 is a general block diagram of the 2-phase clocked systems treated here. Clock signals (shown in Fig. 13) go directly to the C inputs of the latches. Facilities for scan-in and scan-out are not included as they do not affect the basic arguments.

The strategy to be followed is based on the assumption that if the D inputs to all of the latches are valid in the intervals specified by the U and H parameters, then the system will operate as specified. A set of constraints will be derived, such that if the D inputs to all of the $L1$ latches arrive early enough for the first clock cycle, then if, and only if, all of the constraints are satisfied, the inputs to the $L2$ latches will arrive on time for the first $C2$ clock interval, and the D inputs to the

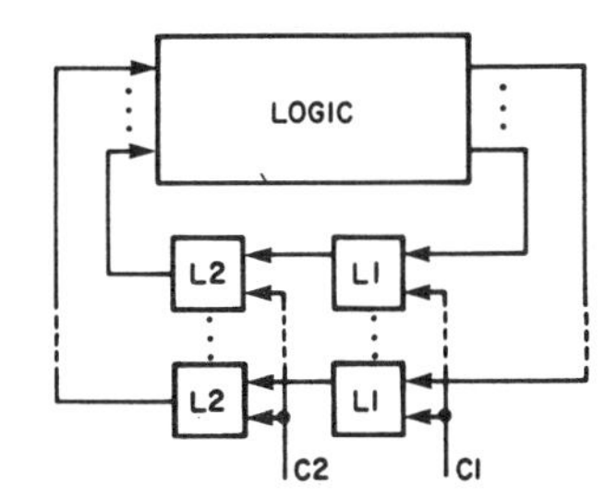

Fig. 12. Block diagram of a 2-phase clocked system.

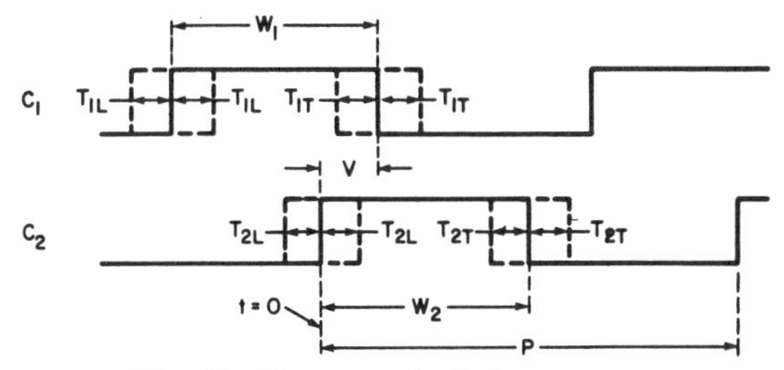

Fig. 13. Parameters for 2-phase systems.

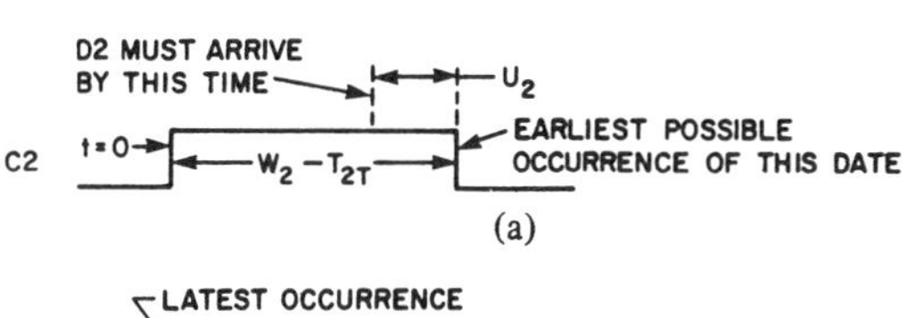

(a)

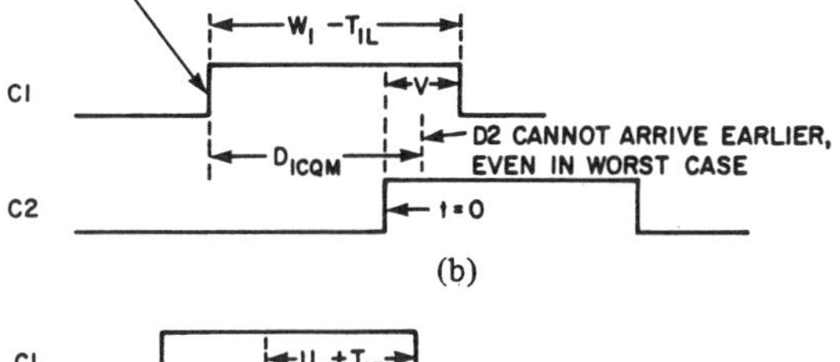

(b)

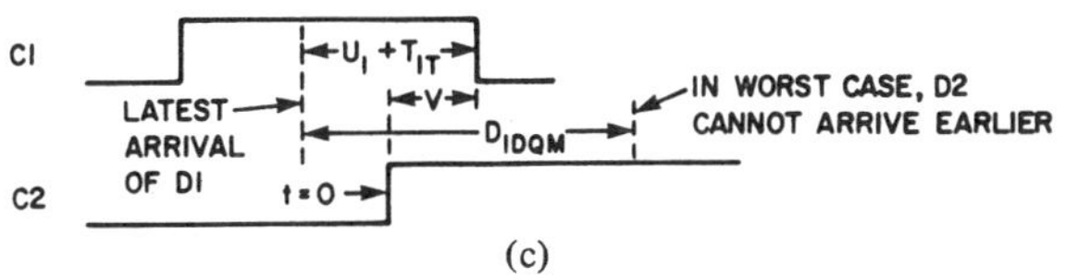

(c)

Fig. 14. $D2$ arrival time. (a) Deadline for arrival of $D2$ signals. (b) Lower bound on $D2$ arrival due to $C1$ leading edge. (c) Lower bound on $D2$ arrival due to arrival of $D1$ signal.

L1-latches will arrive early enough for the next C1-clock interval. Also, the D inputs to the $L1$ latches will remain valid long enough during the first $C1$ interval, and the D inputs to the $L2$ latches will remain valid sufficiently long during the first $C2$ interval. By induction, it then follows that, for all subsequent clock periods, the latches will all have valid inputs during the prescribed intervals.

Throughout the following discussion it is assumed that $t = 0$ at the *nominal* time (by "nominal time" we mean what the time would be if the edge tolerances were 0) of the leading edge of the $C2$ clock. (The *actual* arrival time of this edge at any $L2$ latch may be anywhere between $-T_{2L}$ and $+T_{2L}$). It follows then that the earliest arrival time of the trailing edge of the $C1$ pulse is $V - T_{1T}$. To ensure that the $L1$ latch setup time constraints are met, even under worst case conditions, t_{D1LArr}, the latest arrival time for $D1$ signals during the current clock cycle, must satisfy

$$t_{D1LArr} \leq V - T_{1T} - U_1. \tag{23}$$

In all that follows, it is assumed that, for the first clock period, all $D1$ signal arrival times satisfy (23).

The argument that the constraints developed here are necessary as well as sufficient is dependent on the assumption that, in the worst case, (23) is satisfied with equality. Since this is not actually necessary, it follows that the constraints are not strictly necessary. However, enforcing a more stringent constraint on arrival times of $D1$ signals, namely that they be required to be earlier by some additional amount, is equivalent to assuming that U_1 has increased by this same amount. The effect of this is considered at the end of this section, where it is shown that, as compared to the disadvantages, there is very little to be gained by increasing U_1 (or U_2, which is equivalent to insisting that the $D2$ signals arrive at a time earlier than required by the setup time requirements).

A. Latest Arrival Times of D2-Signals for First Clock Interval

First we develop constraints to ensure that, if the $D1$ signals arrive on time, the $D2$ signals will also arrive on time. (Refer here to Fig. 14(a).) In this case, "on time" means that in order to respect the setup time constraint for the $L2$ latches, the $D2$ signals must arrive no later than U_2 prior to the trailing edge of the $C2$ pulses. *At the earliest*, the trailing edge of a $C2$ pulse might occur at $W_2 - T_{2T}$.

So, the latest arrival time t_{D2LArr} of the $D2$ signals must satisfy

$$t_{D2Larr} \leq W_2 - T_{2T} - U_2. \tag{24}$$

Let t_{C1LL} be the latest arrival time of the leading edge of a $C1$ pulse. Then, recalling (1) about latch propagation delays, the latest time when the output of an $L1$ latch changes (an alternate description of t_{D2LArr}) is as follows (the left side of the max is illustrated by Fig. 14(c) and the right side by part Fig. 14(b)):

$$t_{D2LArr} = \max\ [t_{D1LArr} + D_{1DQM},\ t_{C1LL} + D_{1CQM}].$$

Replacing t_{C1LL} by its value $V - W_1 + T_{1L}$ and t_{D1LArr} by the value given in (23) (assuming that (23) is satisfied with equality) gives us

$$t_{D2LArr} = \max\ [V - U_1 - T_{1T} + D_{1DQM},\ V - W_1 + T_{1L} + D_{1CQM}]. \tag{25}$$

Combining (24) with (25) we obtain

$$\max\ [V - U_1 - T_{1T} + D_{1DQM},\ V - W_1 + T_{1L} + D_{1CQM}] \leq W_2 - T_{2T} - U_2.$$

This can be expressed as two separate constraints:

$$V - U_1 - T_{1T} + D_{1DQM} \leq W_2 - T_{2T} - U_2$$

and

$$V - W_1 + T_{1L} + D_{1CQM} \leq W_2 - T_{2T} - U_2$$

which can be rewritten, respectively, as

$$W_2 \geq V + U_2 - U_1 + D_{1DQM} + T_{2T} - T_{1T} \tag{26}$$

and

$$W_1 + W_2 \geq V + U_2 + D_{1CQM} + T_{1L} + T_{2T}. \quad (27)$$

If (23) is satisfied, then (26) and (27) are sufficient conditions for ensuring that even under worst case conditions, the $D2$ signals arrive on time. If (23) is satisfied with equality, then they are also sufficient for this purpose.

A. Latest Arrival Times of D1 Signals During the Next Cycle

Now consider what is required to ensure that the $D1$ signals arrive on time for the *next* clock cycle, assuming that the $D1$ and $D2$ signals are on time for the present cycle. (Refer here to Fig. 15(a).) The upper bound on the latest arrival time $t_{D1LArrN}$ of a $D1$ signal during the next cycle is obtained from (23), which gives the latest permissible arrival time for the first cycle by simply adding the period P to the right side. This gives us

$$t_{D1LArrN} \leq P + V - U_1 - T_{1T}. \quad (28)$$

Now consider how long it might take a signal to get through an $L1$ latch, through the following $L2$ latch, and through the logic to reach an $L1$ latch input in time for the next $C1$ pulse. (See Fig. 12). In terms of the latest arrival time at an $L2$ input t_{D2LArr} and the latest possible occurrence of a $C2$ leading edge t_{C2LL}, (1) gives us for the latest arrival time t_{Q2LArr} for a signal at an $L2$ output

$$t_{Q2LArr} = \max\ [t_{D2LArr} + D_{2DQM},\ t_{C2LL} + D_{2CQM}].$$

Adding the maximum delay through the logic D_{LM} gives us the latest arrival time, $t_{D1LArrN}$ for a signal at an $L1$ input during the *next* cycle

$$t_{D1LArrN} = \max\ [t_{D2LArr} + D_{2DQM},\ t_{C2LL} + D_{2CQM}] + D_{LM}.$$

Equation (25) gives us t_{D2LArr}, and t_{C2LL} is simply T_{2L}. Substituting in the above relation yields

$$\begin{aligned} t_{D1LArrN} = \max\ [&\max\ [V - U_1 - T_{1T} + D_{1DQM}, \\ &V - W_1 + T_{1L} + D_{1CQM}] + D_{2DQM}, \\ &T_{2L} + D_{2CQM}] + D_{LM}. \end{aligned}$$

Expanding the inner max yields

$$\begin{aligned} t_{D1LArrN} = \max\ [&V - U_1 - T_{1T} + D_{1DQM} + D_{2DQM}, \\ &V - W_1 + T_{1L} + D_{1CQM} + D_{2DQM}, \\ &T_{2L} + D_{2CQM}] + D_{LM}. \end{aligned} \quad (29)$$

There are three factors restricting the propagation of signals thru the two latches: propagation thru the D inputs of both $L1$ and $L2$ latches, propagation from the C inputs of the $L1$ latches (involving the location of the $C1$ leading edge) through the D inputs of $L2$ latches, and propagation from the C inputs of the $L2$ latches (involving the location of the $C2$ leading edge). These are all accounted for in the above expression. They are illustrated in Fig. 15(b), (c), and (d), respectively.

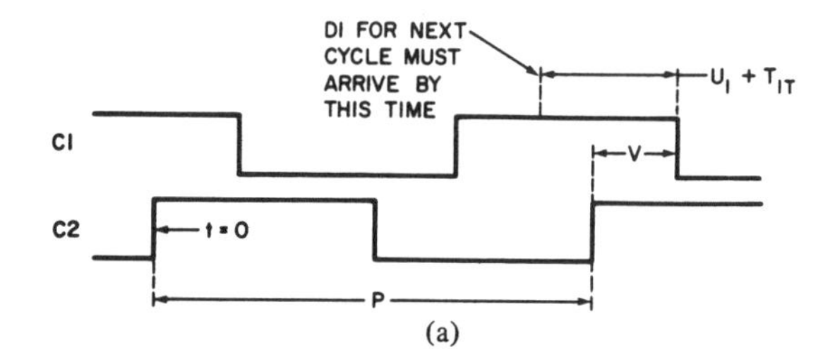

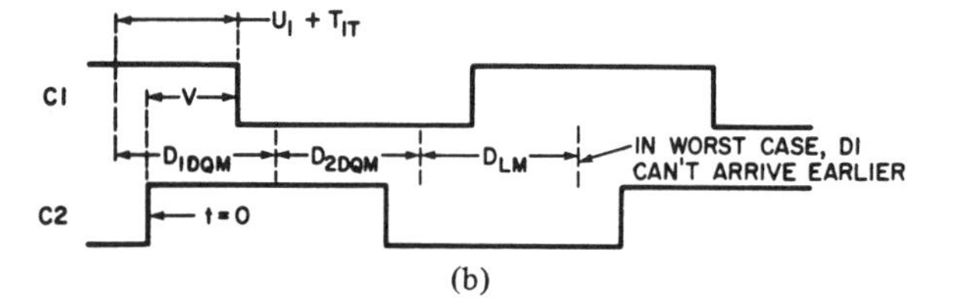

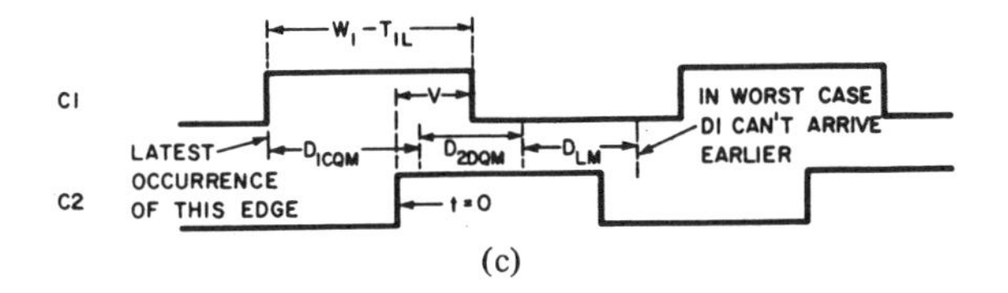

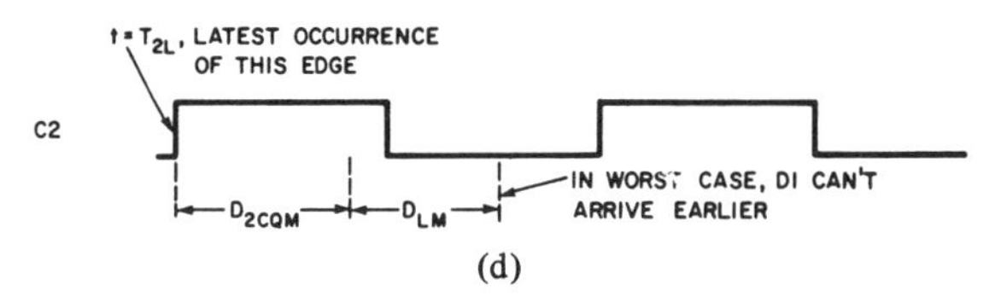

Fig. 15. $D1$ arrival time. (a) Deadline for $D1$ arrival during next cycle. (b) Lower bound on next-cycle $D1$ arrival due to propagation delays through D inputs of latches. (c) Lower bound on next cycle $D1$ arrival due to $C1$ leading edge. (d) Lower bound on $D1$ arrival for next cycle due to $C2$ leading edge.

Replacing $t_{D1LArrN}$ in (28) by the value found in (29) gives us

$$\begin{aligned} \max\ [&V - U_1 - T_{1T} + D_{1DQM} + D_{2DQM}, \\ &V - W_1 + T_{1L} + D_{1CQM} + D_{2DQM}, \\ &T_{2L} + D_{2CQM}] + D_{LM} \leq P + V - U_1 - T_{1T}. \end{aligned}$$

Solving for P and simplifying yields

$$\begin{aligned} P \geq \max\ [&D_{1DQM} + D_{2DQM}, \\ &-W_1 + D_{1CQM} + D_{2DQM} + U_1 + T_{1L} + T_{1T}, \\ &T_{1T} + T_{2L} - V + D_{2CQM} + U_1] + D_{LM}. \end{aligned} \quad (30)$$

Relation (30) can be decomposed into the following three equivalent constraints which, taken together, are equivalent to it.

$$P \geq D_{1DQM} + D_{2DQM} + D_{LM} \quad (31)$$

$$P \geq -W_1 + D_{1CQM} + D_{2DQM} + U_1 + D_{LM} + T_{1L} + T_{1T}$$

or, solving for W_1

$$W_1 \geq -P + D_{1CQM} + D_{2DQM} + U_1 + D_{LM} + T_{1L} + T_{1T} \quad (32)$$

$$P \geq -V + D_{2CQM} + U_1 + D_{LM} + T_{1T} + T_{2L}. \quad (33)$$

Each of the above constraints can be justified intuitively.

- Constraint (31) indicates that the period cannot be less than the total time it would take a signal, under worst case conditions, to propagate around a loop (i.e., thru an $L1 - L2$ latch pair and the logic).
- Constraint (33) (when the $-V$ is transposed) states that, starting at the leading edge of a $C2$ pulse, there must be time, prior to the end of the *next* $C1$ pulse, for signals to get through $L2$ latches and the logic to the inputs of $L1$ latches prior to the setup times for those latches, under worst case conditions of logic delay, latch delay and edge tolerances.
- Similarly, (32) states (transposing the $-P$ term helps make this clearer) that a similar relation holds with respect to starts made at the leading edge of $C1$ pulses and ending at the trailing edges of $C1$ pulses during the next cycle.

Note that if (26) is satisfied with equality, and if (27) is satisfied, then, it is not difficult to show, with the aid of (31), that (32) is implied. Alternatively, satisfying both (32) with equality and (27) ensures that (26) is satisfied.

C. Premature Changes of D1 Signals

Next we ensure that changes in $D1$ signals do not propagate through the $L1$ and $L2$ latches and the logic so fast that they cause some $D1$ inputs to change to their values for the *next* cycle prematurely, i.e., before the hold times for the current cycle have expired. (Refer here to Fig. 16(a).) The earliest arrival time $t_{D1EArrN}$ of such "short-path" signals for the next cycle must be later than H_1 after the latest possible occurrence of a $C1$ trailing-edge; that is

$$t_{D1EArrN} > V + T_{1T} + H_1. \quad (34)$$

The earliest time that a $D1$ signal can change as a result of signal changes generated during the same clock period getting around the loop is arrived at analogously to the way (29) was produced; the same three categories of constraints must be considered. Now, however, since we seek the *minimum* delays, we use *minimum* values for the delays within the max expressions, and the *earliest* times for the critical clock-pulse edges.

With t_{C2EL} as the earliest occurrence time of a $C2$ pulse leading edge, and with t_{D2EArr} as the earliest arrival time of a $D2$ input change, postulate (1) indicates that the earliest output from an $L2$ latch can occur at t_{Q2E}, given by

$$t_{Q2E} = \max\ [t_{C2EL} + D_{2CQm},\ t_{D2EArr} + D_{2DQm}]$$

Adding D_{Lm} to each component of the max of the right side of the above relation, and replacing t_{C2EL} by its value $-T_{2L}$ gives us $t_{D1EArrN}$, the earliest arrival time of a $D1$-change for the *next* clock cycle

$$t_{D1EArrN} = \max\ [-T_{2L} + D_{2CQm} + D_{Lm},\ t_{D2EArr} + D_{2DQm} + D_{Lm}]. \quad (35)$$

To find t_{D2EArr} is the same as finding the earliest output of an $L1$ latch. If we represent the earliest occurrence time of a $C1$ pulse leading edge by t_{C1EL}, and the earliest arrival of a $D1$ input for the current cycle as t_{D1EArr}, then we have

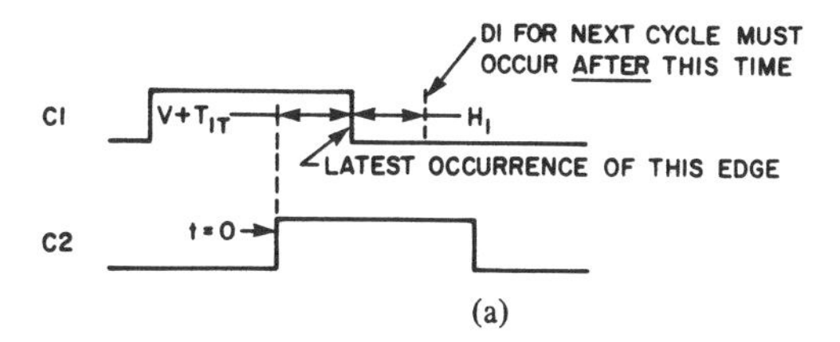

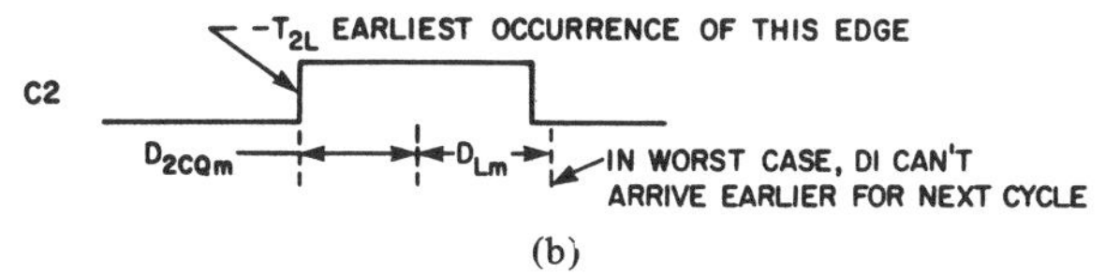

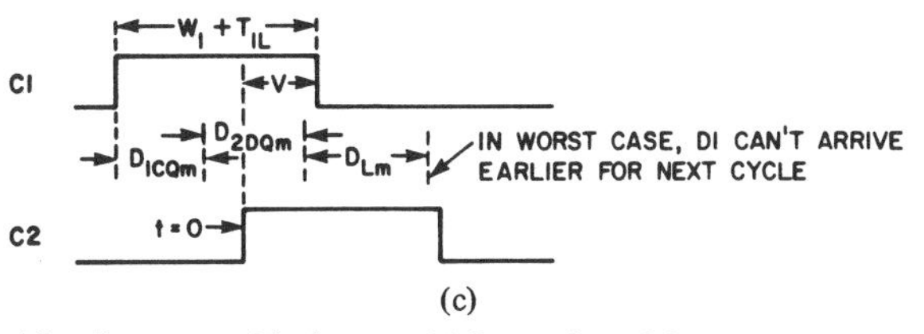

Fig. 16. Premature $D1$ changes. (a) Lower bound for occurrence time of $D1$ for next cycle. (b) Lower bound on next cycle $D1$ arrival due to leading edge of $C2$. (c) Lower bound on $D1$ arrival for next cycle due to leading edge of $C1$.

$$t_{D2EArr} = \max\ [t_{C1EL} + D_{1CQm},\ t_{D1EArr} + D_{1DQm}]. \quad (36)$$

Replacing t_{C1EL} in the above equation by $V - W_1 - T_{1L}$, and inserting the resulting expression for t_{D2EArr} in (35), yields

$$t_{D1EArrN} = \max\ [-T_{2L} + D_{2CQm} + D_{Lm},\ \max\ [V - W_1 - T_{1L} + D_{1CQm},\ t_{D1EArr} + D_{1DQm}] + D_{2DQm} + D_{Lm}].$$

Expanding the inside max in the above equation gives us

$$t_{D1EArrN} = \max\ [-T_{2L} + D_{2CQm} + D_{Lm},\ V - W_1 - T_{1L} + D_{1CQm} + D_{2DQm} + D_{Lm},\ t_{D1EArr} + D_{1DQm} + D_{2DQm} + D_{Lm}]. \quad (37)$$

(The first 2 parts of the max are illustrated in Fig. 16(b) and (c), respectively.)

Now we show that, for a system that operates properly even under worst case conditions, (34) is valid if, and only if, it is valid when the value used for $t_{D1EArrN}$ is that of (37) with the third part of the max deleted. The "if" part of this assertion is obviously true.

To prove necessity (the "only if" part), let us assume the contrary, namely that (34) is valid and that neither of the first 2 parts of the max of (37) exceeds the right side of (34).

Then, since $t_{D1EArrN}$ *must* satisfy (34), it follows that the *third* part of the max must do so. Therefore, it must exceed each of the first two parts, both of which can therefore be deleted from (37), reducing it to

$$t_{D1EArrN} = t_{D1EArr} + D_{1DQm} + D_{2DQm} + D_{Lm}. \quad (38)$$

But, from (31) it is clear that

$$P > D_{1DQm} + D_{2DQm} + D_{Lm}.$$

Adding t_{D1EArr} to both sides gives us

$$t_{D1EArr} + P > t_{D1EArr} + D_{1DQm} + D_{2DQm} + D_{Lm}.$$

From the above and from (38) we have

$$t_{D1EArrN} < t_{D1EArr} + P.$$

But this means that, for each cycle (in the worst case), $D1$ arrives earlier and earlier relative to the trailing edge of $C1$. Therefore, even if t_{D1EArr} is comfortably above the minimum for the first cycle it will eventually violate the hold-time constraint, so that the system would not operate properly. Hence, by contradiction, we have completed our argument.

Thus, we can replace $t_{D1EArrN}$ in (34) with the right side of (37), omitting the third part of the max (and factoring out D_{Lm}), which gives us

$$\max\left[-T_{2L} + D_{2CQm},\ V - W_1 - T_{1L} + D_{1CQm} + D_{2DQm}\right] + D_{Lm} > V + T_{1T} + H_1.$$

Solving for D_{Lm} produces

$$D_{Lm} > D_{LmB} = \min\left[V + H_1 + T_{1T} + T_{2L} - D_{2CQm},\ W_1 + H_1 + T_{1L} + T_{1T} - D_{1CQm} - D_{2DQm}\right].$$

The above expression can be partitioned into two relations, at least one of which must be satisfied.

$$D_{Lm} > D_{LmB} = V + H_1 + T_{1T} + T_{2L} - D_{2CQm} \quad (39)$$

$$D_{Lm} > D_{LmB} = W_1 + H_1 + T_{1T} + T_{1L} - D_{1CQm} - D_{2DQm}. \quad (40)$$

While it is conceivable that a system might exist for which the right side of (40) is less than the right side of (39), an examination of the 2 expressions suggests that this is very unlikely. Hence, in most cases it is constraint (39) that should be relied upon.

D. Premature Changes of D2 Signals

Now consider how to ensure that the $D2$ signals, once on, remain stable long enough for proper operation, i.e., that the hold-time constraints for the $L2$ latches are satisfied. It is necessary to ensure that $t_{D2EArrN}$ the time of the earliest change in a $D2$ signal resulting from a signal passed by the *next* $C1$ pulse satisfies the following relation where t_{C2LT} is the latest occurrence time of the trailing edge of $C2$.

$$t_{D2EArrN} > t_{C2LT} + H_2. \quad (41)$$

The latest appearance of the trailing edge of $C2$, C_{2LT}, occurs at $W_2 + T_{2T}$. (Refer now to Fig. 17(a).) Replacing t_{C2LT} in (41) by this value, we obtain

$$t_{D2EArrN} > W_2 + T_{2T} + H_2. \quad (42)$$

Noting that the earliest time that any $D1$ signal is permitted to change as a result of a previous $D1$ change during the same cycle is $V + H_1 + T_{1T}$ [see (34)], and that the leading edge of the next $C1$ pulse occurs no earlier than $P + V - W_1 - T_{1L}$, we can compute $t_{D2EArrN}$ as follows:

$$t_{D2EArrN} = \max\left[V + H_1 + T_{1T} + D_{1DQm},\ P + V - W_1 - T_{1L} + D_{1CQm}\right]. \quad (43)$$

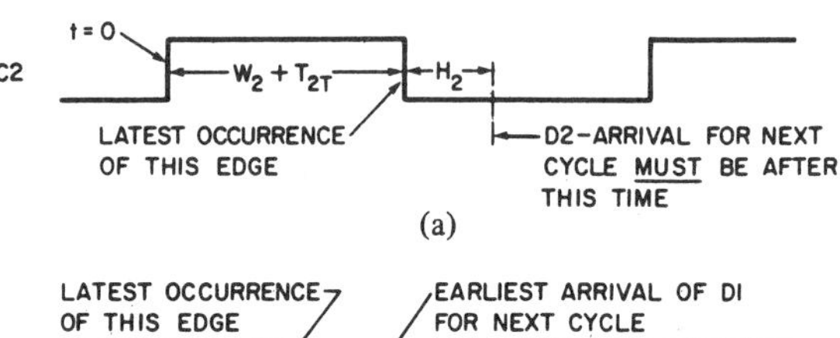

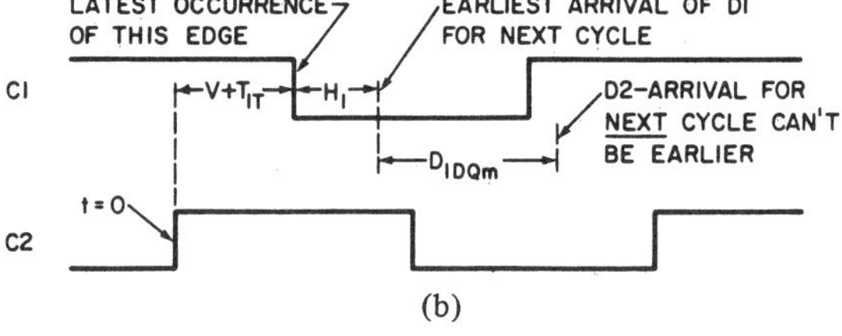

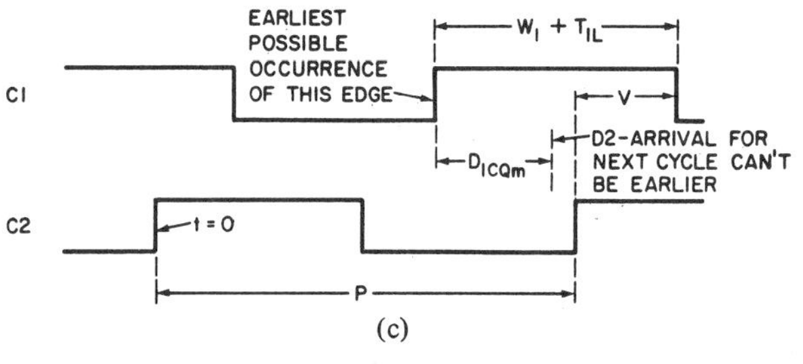

Fig. 17. Premature $D2$ changes. (a) Earliest permissible arrival time of $D2$. (b) Lower bound on $D2$ arrival time for next cycle due to $D1$ arrival time. (c) Lower bound on $D2$ arrival for next cycle due to the leading edge of $C1$ for the next cycle.

Combining (42) and (43) yields

$$\max\left[V + H_1 + T_{1T} + D_{1DQm},\ P + V - W_1 - T_{1L} + D_{1CQm}\right] > W_2 + H_2 + T_{2T}. \quad (44)$$

The left and right parts of the max of (44) are illustrated in Fig. 17(b) and (c), respectively.

Relation (44) can be expressed as the following pair of relations, *at least one of which* must be satisfied:

$$V + H_1 + T_{1T} + D_{1DQm} > W_2 + H_2 + T_{2T}$$

$$P + V - W_1 - T_{1L} + D_{1CQm} > W_2 + H_2 + T_{2T}.$$

These may be more conveniently expressed, respectively, as

$$W_2 < H_1 - H_2 + D_{1DQm} + V + T_{1T} - T_{2T} \quad (45)$$

and

$$W_1 + W_2 < D_{1CQm} + V + P - H_2 - T_{1L} - T_{2T}. \quad (46)$$

They constitute necessary and (along with the other constraints developed above) sufficient conditions for ensuring that the inputs to the $L2$ latches will remain on for a sufficiently long time relative to the trailing edges of the $C2$ pulses. Under most circumstances, it would appear that (46) is much more likely to be satisfied than is (45).

E. Intervals During Which Output Signals are Valid

(Since the material in this subsection is not essential to what follows, it may be skipped at first reading.)

If outputs are taken from the logic block, and are thereafter sent to external receivers instead of to $L1$ latches, then it is clear that those signals will be stable and valid at least over the interval during which we have ensured that the $D1$ signals are valid, namely

$$(V-U_1-T_{1T},\ V+H_1+T_{1T}).$$

If the outputs are taken directly from $L2$ latches, then we can compute the stable output interval as follows.

The *unstable* interval begins at the earliest time at which a $Q2$ signal can change (i.e., the earliest time an $L2$ latch output can change). This time t_{StUn} can be found in terms of the time of occurrence of the earliest leading edge of a $C2$ pulse, which is $-T_{2L}$, and t_{D2EArr}, the earliest time at which a $D2$ input can change.

$$t_{StUn}=\max\ [-T_{2L}+D_{2CQm},\ t_{D2EArr}+D_{2DQm}].$$

We have already found an expression for t_{D2EArr} in (36), which we can insert in the above expression. Let us do so, also replacing the t_{C1EL} term by its value as indicated by $V-W_1-T_{1L}$. This gives us

$$t_{StUn}=\max\ [-T_{2L}+D_{2CQm},$$
$$\max\ [V-W_1-T_{1L}+D_{1CQm},\ t_{D1EArr}+D_{1DQm}]+D_{2DQm}].$$

Expanding the inner max yields

$$t_{StUn}=\max\ [-T_{2L}+D_{2CQm},$$
$$V-W_1-T_{1L}+D_{1CQm}+D_{2DQm},$$
$$t_{D1EArr}+D_{1DQm}+D_{2DQm}]. \tag{47}$$

As was shown earlier [see (34)] the earliest change of $D1$ permitted for the next cycle is at time

$$H_1+V+T_{2T}.$$

Therefore, the earliest time we can expect $D1$ to change for the *current* cycle, i.e., the value of t_{D1EArr} is P less than that amount, or

$$t_{D1EArr}=-P+H_1+V+T_{1T}.$$

Substituting this value into (47) gives us

$$t_{StUn}=\max\ [-T_{2L}+D_{2CQm},$$
$$V-W_1-T_{1L}+D_{1CQm}+D_{2DQm},$$
$$-P+H_1+V+T_{1T}+D_{1DQm}+D_{2DQm}]. \tag{48}$$

The Q_2 signals become stable again after the latest $D2$ change prior to the setup time propagates to the latch outputs. Using the value for the latest $D2$ change given in (24) we get for $t_{\mathrm{End}\ Un}$, the latest time that the unstable period can end

$$t_{\mathrm{End}\ Un}=\max\ [W_2-T_{2T}-U_2+D_{2DQM},\ T_{2L}+D_{2CQM}]. \tag{49}$$

At all other times, the Q_2 signals are guaranteed to be stable and valid.

F. Consequences of the Constraints

The necessary and sufficient constraints derived above are reproduced below.

$$W_2\geq V+U_2-U_1+D_{1DQM}+T_{2T}-T_{1T} \tag{26}$$

$$W_1+W_2\geq V+U_2+D_{1CQM}+T_{1L}+T_{2T} \tag{27}$$

$$P\geq D_{1DQM}+D_{2DQM}+D_{LM} \tag{31}$$

$$W_1\geq -P+D_{1CQM}+D_{2DQM}+U_1+D_{LM}+T_{1L}+T_{1T} \tag{32}$$

$$P\geq -V+D_{2CQM}+U_1+D_{LM}+T_{1T}+T_{2L}. \tag{33}$$

At least one of the following 2 constraints on D_{Lm} must be satisfied. In most cases, (39) is less stringent (its right side is smaller) and so determines D_{LmB}, the lower bound on D_{Lm}.

$$D_{Lm}>D_{LmB}=V+H_1+T_{1T}+T_{2L}-D_{2CQm} \tag{39}$$

$$D_{Lm}>D_{LmB}=W_1+H_1+T_{1T}+T_{1L}-D_{1CQm}-D_{2DQm}. \tag{40}$$

At least one of the following 2 constraints must be satisfied. In most cases, this will be (46).

$$W_2<H_1-H_2+D_{1DQm}+V+T_{1T}-T_{2T} \tag{45}$$

$$W_1+W_2<D_{1CQm}+V+P-H_2-T_{1L}-T_{2T}. \tag{46}$$

In addition to the above constraints, two more are necessary to ensure that the clock-pulse widths satisfy the minimum requirements of the latches themselves. These are:

$$W_1\geq C_{W1m}+T_{1L}+T_{1T} \tag{50}$$

and

$$W_2\geq C_{W2m}+T_{2L}+T_{2T}. \tag{51}$$

Our objective is to choose the clock parameters (widths, period, and overlap) so as to maximize the speed of the system (clearly this is achieved when the period P is minimized), while making it as insensitive as possible to parameter variations. That is, we would like to make the tolerances as large as possible. We often start out with a desired value for the maximum logic delay D_{LM} in a logic path (the *long-path delay*) as this is largely determined by the given technology and the desired maximum number of stages of logic. The crucial factor determining feasibility with known tolerances for delay per logic stage is then the *minimum* delay in a logic path D_{Lm} or *short-path delay*. If the required lower bound on the short-path delay is too large compared to the long-path delay, then the system may be difficult or impossible to realize reliably.

We therefore define the problem as that of finding the minimum value of P such that the lower bound on the short-path delay (D_{LmB}) is acceptable (not too large). It is assumed that we are given all of the latch parameters, the clock-pulse edge tolerances, and the long-path delay D_{LM}.

The key constraint on D_{Lm} is almost always (39). Hence, we set D_{LmB} equal to the right side of that constraint and solve for

V:

$$V = D_{LmB} - H_1 - T_{1T} - T_{2L} + D_{2CQm}. \qquad (52)$$

Now substitute the above right side for V in (33), which is the key constraint on P, to obtain an expression for the minimum value of P as a function of the short-path delay:

$$P = H_1 + U_1 + D_{2CQM} - D_{2CQm} + D_{LM} - D_{LmB} + 2(T_{1T} + T_{2L}). \qquad (53)$$

This expression is valid provided that the value of P obtained does not violate (31). Thus, to find the maximum value of D_{LmB} beyond which no further reductions in P are possible, we must first find the maximum value of V for which (33) is valid (i.e., the value for which (31) is not violated). We do this by substituting the right side of (31) for P in (33) and, treating the resulting expression as an equality, solving for V:

$$V = T_{1T} + T_{2L} + D_{2CQM} + U_1 - D_{1DQM} - D_{2DQM}. \qquad (54)$$

There is clearly nothing to be gained by making the overlap any larger than the value given in (54), since the effect would be to increase the lower bound on the short-path delay without reducing P beyond the absolute minimum given by (31).

Now we can compute the maximum useful value of D_{LmB} by substituting into (39) the above value of V:

$$D_{LmB} = 2(T_{1T} + T_{2L}) + H_1 + U_1 - D_{1DQM} - D_{2DQM} + D_{2CQM} - D_{2CQm}. \qquad (55)$$

Now we are in position to discuss the question mentioned at the beginning of this section as to the consequences of forcing the $D1$ and/or the $D2$ signals to appear earlier than the minimum bounds dictated by the setup times for the latches. The effect of doing this is the same as if the values of the setup times (the U_i's) were increased. Let us examine the relations derived here to see what effects such increases would have.

First observe that U_1 appears in (26), (32), and (33), as well as in (54) for the maximum useful overlap, in (55) for the value of D_{LmB} corresponding to the absolute minimum bound on P, and in (53) for the minimum value of P as a function of the lower bound on the short-path delay. The direct effects of increasing U_1 are detrimental in all cases except that corresponding to (26). That is, the period would have to be increased and/or D_{LmB} would have to be increased (various tradeoffs are possible), both of which are bad, but the lower bound on the width of the $C2$ pulse would be relaxed, a benefit, but seldom one that is needed.

The U_2 term appears only in (26) and (27), and in (49) for the end of the unstable period for the outputs of $L2$. In the first two cases it tightens (by increasing) the lower bounds on the pulse widths, which is mildly bad, and in the last case it increases the interval during which the $Q2$ signals are stable, which might conceivably be advantageous in some situation.

It therefore does not seem useful to consider requiring the D inputs to the latches to arrive earlier than necessary, unless a very special circumstance should make important one of the factors discussed above. An interesting and perhaps useful added conclusion from the above discussion is that the setup time for the $L2$ latches is of less importance with respect to speed and tolerances than is the set-up time for the $L1$ latches.

G. Computing Optimum Clock Parameters

Let $D_{\max LmB}$ be the largest lower bound that we can enforce on the short-path delays. To compute optimum clock parameters, proceed as follows.

IF $D_{\max LmB} \geq$ right side of (55)
THEN
  $D_{LmB} \Leftarrow$ right side of (55)
  $P \Leftarrow$ right side of (31)
  $V \Leftarrow$ right side of (54)
ELSE
  $D_{LmB} \Leftarrow D_{\max LmB}$
  $P \Leftarrow$ right side of (53)
  Compute V from relation (39)
$W_2 \Leftarrow$ max[right side of (26), right side of (51)]
Compute W_1 from (27) (use equality)
Increase W_1 if necessary to satisfy (50)
IF $W_1 + W_2 >$ right side of (46) (Not likely.)
THEN
  IF W_2 violates (45) (It probably will.)
  THEN increase P to satisfy (46)
IF $D_{LmB} >$ right side of (40) (Not likely.)
THEN decrease D_{LmB} until (40) is satisfied with equality.

The procedure given above is intended as a general guide to the use of the constraints developed here. In particular cases alternative procedures may be more appropriate.

V. Conclusions

As is evident from the length of the corresponding section, the task of determining optimum clocking parameters for systems using ETDFF's is relatively simple. The clock-pulse width is not critical, and the constraint on the short-path delays is seldom stringent. The price paid for this is that the minimum clock period is the sum, not only of the maximum delays through the logic and the FF's, but also of the setup time and twice the edge tolerance. No tradeoffs are possible to reduce this quantity.

For 1-phase systems using latches, it may be possible to make the period as small as the sum of the maximum delays through a latch (from the D input) and the logic. In order to do this, the clock-pulse width must be made sufficiently wide (usually past the point where the leading edge of the clock-pulse precedes the appearance of the D signals). Wider clock pulses imply increased values of D_{LmB}, the lower bound on the short-path delays. If this bound is not to become unreasonably high, it is necessary to keep the edge tolerances small. It is also helpful if the difference between the maximum and minimum values of the propagation delays from the C inputs of the latches are small.

The 2-phase system with latches is inherently more complex in that more variables are involved. As in the previous case, tradeoffs are possible between P and D_{LmB}. Here the intermediate variable is V, the amount of overlap between the

$C1$ and $C2$ pulses. In very conservative designs there is a negative overlap and D_{LmB} is zero. If positive overlaps are permitted, P can be decreased, but at the cost of making D_{LmB} nonzero. A continuum of tradeoffs exists to the point where P is reduced to the sum of the maximum propagation delays through the $L1$ and $L2$ latches (from the D inputs) and the logic. Again it is possible to absorb the effect of edge tolerances in terms of short-path rather than long-path problems.

An important advantage of 2-phase over 1-phase systems is that, for every 2-phase system, simply by varying the overlap (i.e., the phasing between the $C1$ and $C2$ clock pulses), D_{LmB} can be varied continuously from zero to the highest useful value (with the minimum P of course changing in the opposite direction). On the other hand, for 1-phase systems, the range of variation of D_{LmB} possible by varying the clock-pulse width is often much smaller, particularly at the low end. As illustrated in the graph of Fig. 11, there may be a significant range of values of D_{LmB} that is attainable only by adding delay pads at the outputs of all latches.

In 1-phase systems, if the designer is overly aggressive and it becomes apparent during the test phase that the short-path bound cannot be met, then it is usually necessary to add delay pads at the latch outputs as well as to increase the clock period. This usually means very extensive changes, affecting many chips. Should the same situation arise in connection with a 2-phase system, in addition to increasing the clock period, all that need be done is to reduce the amount of overlap, adjustments that affect only the clocking system, usually a much simpler process affecting far fewer chips. Hence designers of 2-phase systems can afford to be bolder in choosing the clock period since the penalty for over-reaching is less severe.

With only one latch in each feedback path, the lower limit on the clock period is lower for 1-phase systems, although this factor is somewhat attenuated by the fact that some latches in 1-phase systems will have both inputs from sources that fan out to other latches, and outputs that fan-out to many gate inputs. Both of these are factors that reduce speed. But in 2-phase systems each $L1$ latch feeds only one other device (an $L2$ latch), and each $L2$ latch receives its D input from a source (an $L1$ latch) feeding no other device. Hence, all other things being equal, we would expect the delays through the two latches in the feedback paths of 2-phase systems to have less than twice the delays of the one latch in the feedback path of a 1-phase system.

An advantage of 2-phase systems over both of the other types considered here is that they are somewhat more compatible with the LSSD concept for system testing [1], [2].

It appears that all three types of systems have their places. Where there is a willingness to exert great efforts to suppress skew (e.g., by hand-tuning the delays in clock distribution paths), and to control other related factors very precisely, the 1-phase system may be the best choice, as in the case of the CRAY 1 machine. In other cases of high-performance machines, 2-phase clocking may be more suitable. Use of ETDFF's seems to have advantages for less aggressive designs.

The results presented here in such precise looking relations obviously depend heavily on the precision with which the parameters of those relations can be determined. Realistic figures must be obtained that take into account such matters as power supply and temperature variations, as well as data sensitive loading considerations.

The relations developed here may be useful in determining what latches to use in certain situations and to determine how to modify latch designs so as to improve system performance. For example, an examination of the constraints developed in Section III-C for 1-phase systems with latches suggests that the *minimum* value of D_{DQ} is of no importance, whereas the minimum value of D_{CQ} *is* important in that the larger it is, the less stringent is the constraint on short-path delays.

In the 2-phase case, minimizing $(D_{2CQM} - D_{2CQm})$ is clearly helpful. It relaxes the requirement on D_{LmB} imposed by (55), which, if it can be satisfied, allows P to be set to the minimum value given by (31). If (55) cannot be satisfied, then P is given by (53), and will therefore vary directly with $(D_{2CQM} - D_{2CQm})$.

On the other hand, neither D_{1CQM}, D_{1CQm}, D_{1DQm}, nor D_{2DQm} seem to be of primary importance. As was pointed out on page 893, the setup and hold-time requirements for the $L1$-latches are much more important than are the corresponding parameters for the $L2$ latches. It is clear that there are different optimum requirements for $L1$ and $L2$ latches. Furthermore, different choices may be appropriate depending upon whether or not an effort is being made to attain the minimum period corresponding to the maximum loop delay.

It is clear from the results developed here that minimizing clock edge tolerances is of considerable importance in high-performance digital systems. In 2-phase systems, a special effort is warranted to minimize T_{1T} and T_{2L}, which appear in key several constraints. Unfortunately, technology trends are such as to emphasize factors that cause skew. For example, as the dimensions of logic elements on chips shrink, the ratio of wiring delays to gate delays grows. A high priority must therefore be given in wiring algorithms to the clock distribution system. Off-chip wiring forming part of the clock distribution network must be carefully controlled. In some cases, the insertion of adjustable delays in these paths may be warranted. It is quite likely that the continuation of the trends that exacerbate the skew problem will soon make it worthwhile to consider systems that do not use clock pulses or that use clock pulses only locally. Discussions of such asynchronous, self-timed, or speed-independent systems are in [4] and [8].

Logic designers and those developing computer aids for logic design customarily pay a great deal of attention to minimizing long-path delays. It is also important to consider techniques for increasing short-path delays. In line with this there is a need for circuit designers to develop techniques for introducing *precisely controlled* delay elements where needed. At present, in many technologies, logic designers are forced to cascade inverters to produce delays. This is wasteful in terms of both chip area and power. In general, the idea that greater speed may result from better delay elements should be conveyed to those developing digital technology.

Further developments along the lines developed here would

include the use of statistical rather than worst-case analyses, which would allow us to choose clocking parameters such that the likelihood of a timing failure is very small, but not zero. This usually implies shorter clocking periods. In using this approach it is important to be able to take into account correlations among delay values, skew etc., in various parts of the system [5], [7].

It is also possible to speed up systems by exploiting detailed knowledge of the logic paths. There may be, for example, constraints on the sequencing of signals through certain combinations of paths that allow us to consider consecutive pairs, triples, etc., of cycles together and thereby realize that shorter periods are feasible than would be the case if each period were considered separately. Research along this line is being conducted by K. Maling [3].

An earlier presentation of the work discussed here, in a different form with different notation was issued by the authors several years ago [10], [11]. The idea that clocked systems could be speeded up by permitting the D-inputs to latches to lag behing the leading edges of the clock-pulses and by allowing the C1- and C2-clock pulses to overlap is not new. These ideas are included in the very interesting book on digital systems design by Langdon [2], and have been pointed out by D. Chang of IBM's Poughkeepsie Laboratories a number of years ago in at least one internal memorandum. Other pertinent work, in connection with pipelining, is by Kogge [12] and Cotten [13].

Acknowledgments

Substantial contributions to the form and substance of this work were made by R. Risch. The comments and criticisms of K. Maling were very helpful, and we also profited from discussions with V. Kumar, W. Donath, G. Buchanan, D. Kurshner, J. Shelly, G. Maley, P. Meehan, and D. Lee.

References

[1] E. B. Eichelberger and T. W. Williams, "A logic design structure for LSI testability," in *Proc. 14th Design Automat. Conf.*, June, 1977, pp. 462–468.
[2] G. Langdon, Jr., *Computer Design.* San Jose, CA: Computeach, 1982.
[3] K. Maling, "Automatic clock optimization," IBM internal memorandum, 1982.
[4] C. Mead and L. Conway, *Introduction to VLSI Systems.* Reading, MA: Addison-Wesley, 1980.
[5] J. Shelly and D. Tryon, "Statistical techniques of timing verification," in *Proc. 20th Design Automation Conf.*, 1983, pp. 396–402.
[6] *The TTL Data Book for Design Engineers*, Texas Instruments, Inc., Dallas, TX, 1981.
[7] D. Tryon, F. M. Armstrong, and M. R. Reiter, "Statistical failure analysis of system timing," *IBM J. Res. Develop.*, vol. 28, pp. 340–355, July, 1984.
[8] S. H. Unger, *Asynchronous Sequential Switching Circuits.* New York: Wiley, 1969; 3rd printing issued by R. E. Krieger, Malabar, FL, 1983.
[9] ——, "Double-edge-triggered flip-flops," *IEEE Trans. Comput.*, vol. C-30, pp. 447–451, June, 1981.
[10] S. H. Unger and C.-J. Tan, "Clocking schemes for high-speed digital systems," IBM T. J. Watson Res. Center, Yorktown Heights, NY, Tech. Rep. RC9754, Dec. 1982.
[11] ——, "Clocking schemes for high-speed digital systems," in *Proc. IEEE Int. Conf. Comput. Des.: VLSI Comput.*, Oct.-Nov., 1983, pp. 366–369.
[12] P. Kogge, *The Architecture of Pipelined Computers.* New York: McGraw-Hill, 1981.
[13] L. W. Cotten, "Circuit implementation of high-speed pipelined systems," in *Proc. AFIPS*, vol. 27, pt. 1, pp. 489–504, 1965.

Parallel Bit-Level Pipelined VLSI Designs for High-Speed Signal Processing

MEHDI HATAMIAN, MEMBER, IEEE, AND GLENN L. CASH

This paper explores the potential of bit-level pipelined VLSI for high-speed signal processing. We discuss issues involved in designing such fully pipelined architectures. These include clock skew, clock distribution networks, buffering, timing simulation, area overhead due to pipelining, and testing. A total of six bit-level pipelined designs, including a multiplier, an FIR filter block, and a multichannel multiply-accumulate/add chip, have now been fabricated in CMOS technology. These chips have been tested both for functionality and speed. The results of these tests and the applications of these chips are presented and discussed.

I. Introduction

Pipelining is a very well-known architectural technique for increasing the throughput of a given system implementing a computational task composed of several independent subtasks to be sequentially performed on each element of input data. Depending on the application, the individual subtasks (i.e., the stages of the pipeline) can range anywhere from a large set of operations to a single operation, or even the bit-level activities involved in a single operation. Obviously, the throughput (and the latency) of the pipelined system increases as the degree of pipelining is increased.

Due to their nature of dealing with continuous streams of data, most signal processing tasks are well-suited for pipelining. There are numerous examples of signal processing modules that employ pipelining anywhere from the board and system level down to the component or chip level. However, the potential of finer grain pipelining at the on-chip level has not been fully exploited in practice. In recent years, there has been some attention paid to bit-level pipelined VLSI architectures for digital signal processing [1]-[13]. Most of the published work, except for [2], [3], [11], and [12], although perfectly valid from an architectural point of view, do not include examples of practical VLSI implementation of such architectures. In this paper, we discuss issues involved in designing highly pipelined synchronous architectures in VLSI. These issues include clock skew, clock distribution networks, the question of suitable technologies, the problem of timing simulation, buffer design, area overhead due to pipelining, and testing. We use examples of bit-level pipelined multiply, multiply-accumulate, and multiply-add operations since they can be considered as the basic building blocks of many digital signal processing algorithms. Our first working example reported in [12] was an 8 × 8 bit-level pipelined multiplier fabricated in 2.5-μm CMOS technology. This chip was tested at multiplication rates up to 70 MHz with a power dissipation of 250 mW at that frequency. Achieving such a throughput from a mature technology like 2.5-μm CMOS indicates that, at least as far as pipelined signal processing applications are concerned, the speed capabilities of these well-developed technologies are not yet exhausted.

Following the multiplier chip, several other pipelined architectures have been designed and fabricated. These include a multiply-add chip, a reduced-stage multiplier chip, an FIR filter block, a multichannel multiply-accumulate/add chip, all in 2.5-μm CMOS technology, and a 14 × 14 multiplier-accumulator chip in 1.25-μm CMOS (presently being fabricated). These designs are all in two's complement format. The fabricated chips have all been tested both for functionality and speed. The results of these tests as well as the architecture of the chips will be reported and discussed in this paper. A high-speed programmable FIR filter based on chips presented here has been designed. We discuss this design and report its status.

A layout generator has been written in our CAD tool environment to generate most of the designs described in this paper for even word lengths. The generator is technology independent and is being improved to allow programmability in selecting the number of pipeline stages. The 14 × 14 multiplier-accumulator chip mentioned above was partially designed with the aid of this generator.

II. The Issues

Among the major issues to be considered in designing highly pipelined VLSI circuits are clock skew, clock distribution and buffering, ease of timing simulation, area overhead due to pipelining, and the degree of pipelining. These issues are discussed in this section.

Clock Skew in Synchronous Systems

Clock skew has often been termed the chief difficulty in designing high-speed synchronous systems [14]-[16]; a

Reprinted from *Proc. IEEE*, vol. 75, no. 9, pp. 1192–1202, Sept. 1987.

problem that grows with the speed and size of the system. With high throughput being a characteristic—and the only advantage—of pipelined circuits, the skew problem would appear to be a major stumbling block. However, we believe that clock skew is often blamed for impairing the high-speed operation of a synchronous circuit while it can be handled by first understanding how it affects the operation of the system, and then by using proper clock buffering and distribution networks to overcome the difficulties. The way clock skew affects the operation of a system generally depends on the type of clocking used as well as the direction of data transfer with respect to the clock. We use the synchronous pipelined array shown in Fig. 1 for clock skew

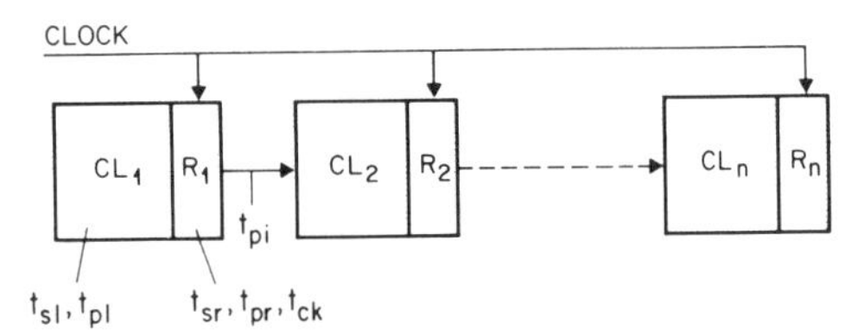

Fig. 1. A linear pipelined array. $t_{p\ell}$: cell propagation delay, $t_{s\ell}$: cell computation delay or settling time, t_{sr}: register settling time, t_{pr}: register propagation time, t_{ck}: clock skew with respect to a global reference, t_{pi}: interconnection propagation time.

discussion. In this figure, the computation cell "*CL*" and the register "*R*" form one module of the synchronous system, or one stage of the pipeline. Depending on the degree of pipelining, the computation cell can be a single gate, a bit-level operation (e.g., a full-adder), a word-level operation (e.g., addition, multiplication), etc. Each module has five time-parameters of interest associated with it: cell computation delay or settling time $t_{s\ell}$, cell propagation delay $t_{p\ell}$, register propagation delay t_{pr}, register settling time t_{sr}, and register clock time t_{ck} which is equal to the clock skew between the cell and a global reference (e.g., the clock source). t_{pi} is the propagation delay time corresponding to the interconnection between communicating modules (adjacent stages of pipeline in Fig. 1). We would like to emphasize that the cell propagation delay $t_{p\ell}$ is defined as the time it takes before at least one of the outputs of the computation cell starts a transition in response to a change at any one of its inputs. This time is generally much shorter than the settling time (computation delay) $t_{s\ell}$ of the cell, and plays an important role in studying the effects of clock skew. The propagation delay and settling times of the register (t_{pr} and t_{sr}) are also defined in the same way. For an array similar to Fig. 1 it can be easily shown that, as far as clock skew is concerned, only the skew between the adjacent stages of the pipeline is of importance and the cumulative skew (e.g., between the first and last cell of the array) imposes no difficulty.

We now focus our attention on the data transfer between two cells in the presence of clock skew. We carry out this discussion for three different clocking schemes: edge-triggered, single-phase level-sensitive, and two-phase clocking. For the purpose of simplicity in notation we assume that each time-parameter of interest has the same value for all modules.

Edge-Triggered Clocking: Consider the transfer of data between CL_1 and CL_2 in Fig. 1. At time $t = t_0$ the output of CL_1 begins to be loaded into R_1, and at time $t = t_0 + \delta$ the output of CL_2 begins to be loaded into R_2, where δ is the clock skew between the two modules defined as

$$\delta = t_{ck_2} - t_{ck_1}. \tag{1}$$

Also, at time $t = t_0 + t_{pr} + t_{pi} + t_{p\ell}$ the response to the change in R_1 has propagated all the way to the input of R_2. For the data transfer to take place properly, this event at $t = t_0 + t_{pr} + t_{pi} + t_{p\ell}$ must occur after the loading time of R_2 at $t_0 + \delta$. So the following condition must be satisfied:

$$\delta < t_{pr} + t_{pi} + t_{p\ell}. \tag{2}$$

In other words, the clock skew must be less than the sum of the propagation delay times of the register, interconnection, and logic. Notice that according to the definition of δ in (1), the clock skew can be either positive or negative depending on the direction of the clock with respect to data transfer. In the case of negative δ, since the propagation delays are always positive, condition (2) is always satisfied. This might sound very surprising because no matter how large the skew is, the system still operates. However, the discussion will not be complete without considering the effects on the throughput of the system. The clock period T must be large enough to allow for the computation to take place. The total computation (or settling) time is $t_c = t_{sr} + t_{pi} + t_{s\ell}$. If the skew is negative ($t_{ck_2} < t_{ck_1}$) then we must have

$$T > t_c + |\delta| = t_{sr} + t_{pi} + t_{s\ell} + |\delta| \tag{3}$$

and if the skew is positive ($t_{ck_2} > t_{ck_1}$) the relation becomes

$$T > t_c - \delta = t_{sr} + t_{pi} + t_{s\ell} - \delta. \tag{4}$$

Combining (2)-(4) the conditions to be satisfied are

$$\begin{cases} T > t_{sr} + t_{pi} + t_{s\ell} + |\delta|, & \delta \le 0 \\ \delta < t_{pr} + t_{pi} + t_{p\ell}, & \\ T > t_{sr} + t_{pi} + t_{s\ell} - \delta, & \delta > 0. \end{cases} \tag{5}$$

These relations can simply be interpreted as follows. If clock skew is negative, then condition (2) is always satisfied and the propagation delays of register, logic, and interconnection cannot cause disasters; the system will always operate properly provided the period is large enough. The price paid is loss of throughput because the clock period must be increased by $|\delta|$. On the other hand, if clock skew is positive, then condition (2) must be satisfied to prevent system failure. However, throughput is improved because the period can be shortened by δ (a case where skew helps the throughput!). Obviously the improvement in throughput as δ is increased is very limited because soon δ violates condition (2). Also notice that in the case of positive skew, if condition (2) is violated, no matter how large T is, the system will fail and cannot be made operational by just reducing the clock frequency of the synchronous system unless the dependency of skew on the clock frequency is strong enough such that lowering the frequency can reduce the skew to where it satisfies the condition.

If the goal is maximizing the throughput, it seems that the case of positive clock skew should be preferred over the negative skew case (at the cost of increasing the possibility of system failure). The objection one might raise though, is that for a general-purpose synchronous system, the clock skew parameter δ could take on both positive and negative values throughout the system, depending on the direction of data transfers between modules. Hence, in this case, the worst possible situation must be considered. However, if there are no feedbacks involved, the clock distribution network can be designed with the data transfer directions in mind such that the clock skew can be made positive for all cases of data transfer.

In their discussion of clock skew in synchronous systems, Franklin and Wann, in [17], derive a single condition for the clock period similar to relation (3) (without using absolute value function on δ) and do not consider the case of negative clock skew or derive any condition similar to (2). As a result, the effect of clock skew is always that of reducing the throughput without any failure modes, while such is not always the case as we discussed above.

Parameters defining the upper bound on clock skew in (2) are defined by the technology used, the speed of the computing cells, and the interconnection between modules. For the bit-level pipelined architectures, which are the subject of this paper, the interconnection propagation delay t_{pi} is very small because data transfer takes place between modules that are close to each other on the chip. $t_{p\ell}$ is also very small because the largest computing cell is a one-bit full-adder. For the 2.5-μm CMOS technology used in most of the designs discussed in this paper $t_{pr} + t_{pi} + t_{p\ell}$ amounts to about 1.2 ns.

Single-Phase Clocking: The case of a single-phase level-sensitive clocking scheme can also be discussed in exactly the same way as above. The discussion and conditions on the clock period are the same. However, another parameter enters into condition (2), and that is the clock pulsewidth t_ϕ. Following the same approach for the edge-triggered clocking, we arrive at the following condition for the system to operate properly:

$$\delta + t_\phi < t_{pr} + t_{pi} + t_{p\ell} \qquad (6)$$

or

$$\delta < t_{pr} + t_{pi} + t_{p\ell} - t_\phi.$$

Again δ can be either negative or positive, and whether the system fails or not is determined by the above relation. For example, if δ is positive and greater than $t_{pr} + t_{pi} + t_{p\ell}$ then we must have $t_\phi < 0$ which is impossible, resulting in system failure. Even if δ is smaller than $t_{pr} + t_{pi} + t_{p\ell}$, for a high-speed system, t_ϕ might become so small that it makes the implementation impractical. On the other hand, if δ is negative the upper bound on t_ϕ becomes $t_{pr} + t_{pi} + t_{p\ell} + |\delta|$ and makes the system more practical, but with a loss of throughput in the same manner as we discussed in the edge-triggered clocking section.

Notice that parameter t_ϕ can also be entered into the discussion of edge-trigger clocking as the hold time of the registers, in which case, it should be treated as a module property parameter rather than one that is globally controlled by the user, like clock pulsewidth.

Two-Phase Clocking: In this case, register "R" in Fig. 1 is considered as a two-phase-clock register. We use a dynamic two-phase-clock (level-sensitive) register widely used in MOS designs. t_{ϕ_1}, t_{ϕ_2}, and $t_{\phi_{12}}$, are defined as the phase-1 pulsewidth, phase-2 pulsewidth, and the delay time between phase-1 and phase-2, respectively. The clock period is T as before. The rest of the parameters stay the same except that we use t_{pr_1} and t_{sr_1} for the phase-1 register, and t_{pr_2} and t_{sr_2} for the phase-2 register. We can also assume that clock skew has the same pattern for both phase-1 and phase-2, which is not an impractical assumption. At time $t = t_0$ the output of CL_1 begins a transfer into the phase-1 register of R_1 and its effect appears at the phase-1 input of R_2 at time

$$t = t_0 + t_{\phi_1} + t_{\phi_{12}} + t_{pr_2} + t_{pi} + t_{p\ell}.$$

For the system to operate properly, this time should be greater than $t_0 + \delta + t_{\phi_1}$. So, the condition becomes

$$\delta < t_{pr_2} + t_{pi} + t_{p\ell} + t_{\phi_{12}}. \qquad (7)$$

It can also be easily shown that the conditions on the clock period are

$$T > t_{\phi_1} + t_{\phi_{12}} + t_{sr_2} + t_{pi} + t_{s\ell} + |\delta|, \qquad \delta \le 0$$

$$T > t_{\phi_1} + t_{\phi_{12}} + t_{sr_2} + t_{pi} + t_{s\ell} - \delta, \qquad \delta > 0. \qquad (8)$$

We also must have $t_{\phi_1} > t_{sr_1}$ and $t_{\phi_2} > t_{sr_2}$. The discussion for positive and negative clock skew is very similar to the edge-triggered clocking scheme except that in the case of positive δ, one can always increase $t_{\phi_{12}}$ to satisfy condition (7) at the cost of reducing the throughput, as can be seen from condition (8). In other words, unlike the edge-triggered clocking scheme, in a two-phase clock approach, it is always possible to prevent failure mode by slowing down the clock frequency (increasing $t_{\phi_{12}}$). $t_{\phi_{12}}$ acts as a global parameter that can be used to control the clock skew.

The approaches taken above can be followed for any other clocking scheme or register arrangement to study and understand the effect of clock skew on the operation of a synchronous system. The important point to be made here is that, contrary to what most designers believe, clock skew does not always result in loss of throughput; at the same time it can also create failure modes that are independent of the clock frequency no matter how low it is. Another point to remember is that of differentiating between the propagation delay and the settling time (computation delay). At the beginning of this section we made the assumption that each time-parameter of interest takes on the same value for all modules in Fig. 1. In a general-purpose synchronous system this may not be true, in which case we have to consider the worst possible situation. The relations can be generalized by using the appropriate min(·) and max(·) functions on the parameters.

Clock Distribution

Parameters that determine the clock skew picture in a given synchronous system are: a) shape of the clock distribution network, b) resistance, capacitance, and inductance of the interconnection material used in distributing the clock, c) number of modules (e.g., processing elements) in the system and the load presented by each module to the clock network, d) clock buffers and the buffering scheme used (global versus distributed), e) fabrication process variation over the area of the chip or wafer (e.g., variation of the threshold voltage of a MOS transistor), and f) rise and fall times and the frequency of the clock signal. The last parameter makes a difference in how clock propagation is studied. If the rise time of the clock pulse times the speed of light is less than the length of the clock path, then the clock propagation should be studied using a transmission line model of the network. However, if this figure is considerably greater than the length of the clock path, a distributed RC model of the clock distribution network is sufficient. For instance, for a clock signal having a rise time of 0.5 ns the latter will be the case for distributing clocks at the on-chip level, while at the wafer or printed circuit board level a transmission line model is more appropriate.

Probably the most widely studied clock distribution network that minimizes skew is the H-tree network [15]–[19]. Fig. 2 shows an example of such a network for an 8×8 array

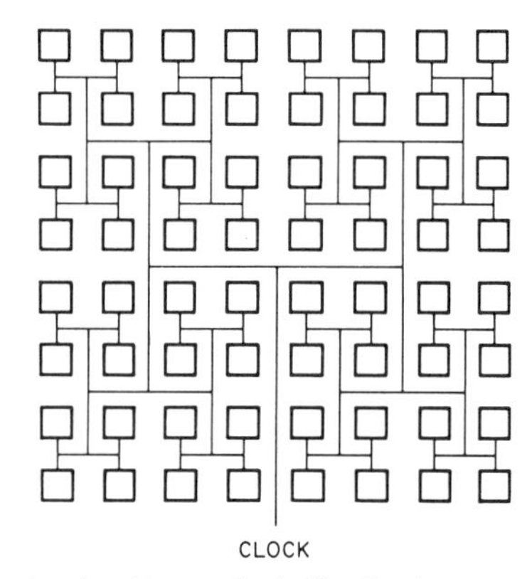

Fig. 2. Example of an H-tree clock distribution network for an 8×8 array of processing elements.

of cells. It is obvious from this figure that if the cells are all identical in size and present the same load, then the paths from the clock source to all cells are similar and clock skew is minimized. In [15], Kung and Gal-Ezer derive a transfer function for the clock path from the clock source to each individual cell for an $N \times N$ array of processing elements. They show that the equivalent time constant of the distribution network is $O(N^3)$ and conclude that the clock pulse rise time and the clock skew associated with it are $O(N^3)$ as well. Therefore, as N (i.e., the size of the system) increases, the clock skew should rapidly become a stumbling block. We find this conclusion about the clock skew rather surprising because it is the skew between different modules that is the determining factor, not that between the clock source and each processing element. As for the problem with the clock pulse rise time, it can be alleviated by using a distributed buffering scheme (Fig. 3) as is also suggested by Bakoglu and Meindl [18], [20]. The disadvantage of introducing the intermediate buffers in the H-tree network, other

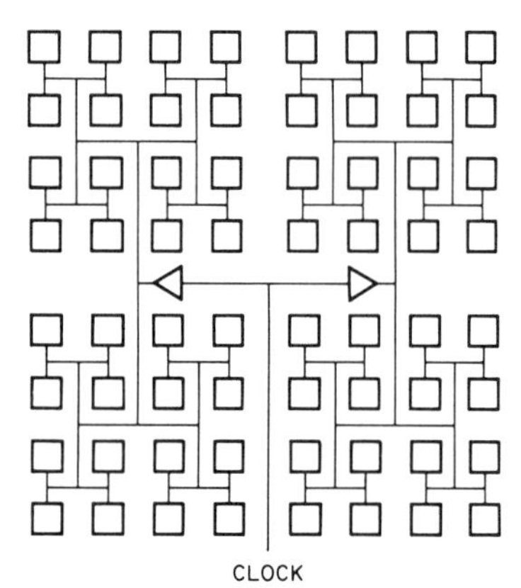

Fig. 3. H-tree clock distribution network with intermediate clock buffering.

than possible area overhead, is increasing the sensitivity of the design to process variations. This undesirable effect is, however, very well offset by the improvement gained in speed. An example of process variations in an actual chip is given at the end of this section. Another point to be noted about the analysis carried out in [15] is that the equivalent time constant τ_{eq} that is shown to be $O(N^3)$ is approximately $5N^3R_0C_0$ where R_0 and C_0 are the resistance and capacitance of each of the final branches of the H-tree network. For an array whose cells are one-bit full-adders as in the bit-level pipelined designs discussed in this paper, R_0 and C_0 are small enough that it takes a value of $N > 100$ (over 10 000 cells) to result in $\tau_{eq} = 1$ ns (for our 2.5-μm CMOS process).

Despite their attractiveness in providing minimum clock skew, H-tree networks are most useful when all processing cells in a synchronous system are identical and the clock can truly be distributed in a binary tree fashion. Friedman and Powell discuss cases where H-tree networks lose their attractiveness for hierarchical designs [19]. An alternative to the H-tree network, which is suitable for on-chip and on-wafer clock distribution, and is simple to design, is the network shown in Fig. 4. This figure shows a two-level dis-

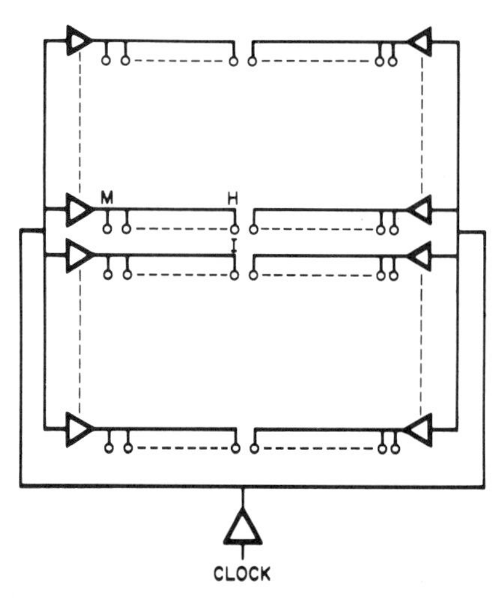

Fig. 4. A clock distribution network using two-level distributed buffering. Circles represent computation cells. Each row represents one stage of the bit-level pipeline architecture.

tributed buffering scheme. Of course, the number of intermediate buffering levels depends on the size of the system and the interconnection material used. Simulation examples of two- and three-level distributions are discussed in [13]. This network, in theory, does not provide the minimum clock skew feature offered by an *ideal* H-tree network. However, it can prove to be equally useful because the extent of skew minimization depends on the amount of clock skew tolerated by the system, which is often not as strict as it is thought to be (as we discussed before). A network similar to the one shown in Fig. 4 was used in the bit-level pipelined designs discussed in this paper. Simulation results of this network are presented in the following section on timing simulation.

As in any other distributed buffering scheme, introducing the intermediate buffers in the network of Fig. 4 reduces the tolerable clock skew margin because of process variation, but the effect is usually minimal. Studying the effects of the fabrication process variation via simulation is a tedious and time-consuming task and often the results are not very reliable. For that reason, one should consider the variation that can result in a worst case situation and subtract that from the tolerable clock skew margin of the system at design time. As an example of the degree of process variation, we show the results obtained from an actual design fabricated in 2.5-μm CMOS technology. As part of a test study in this design, an input pad buffer was connected to an output pad buffer adjacent to it through a piece of polysilicon wire. The same structure was repeated at the four corners of the chip. Fig. 5 shows the response of the fabricated and packaged design for two structures that are located about 1 cm apart. A skew of about 0.15 ns is measured between the two output waveforms (assuming that no skew is introduced by the measurement setup). Experiments with other fabricated chips show the same behavior across the wafer area.

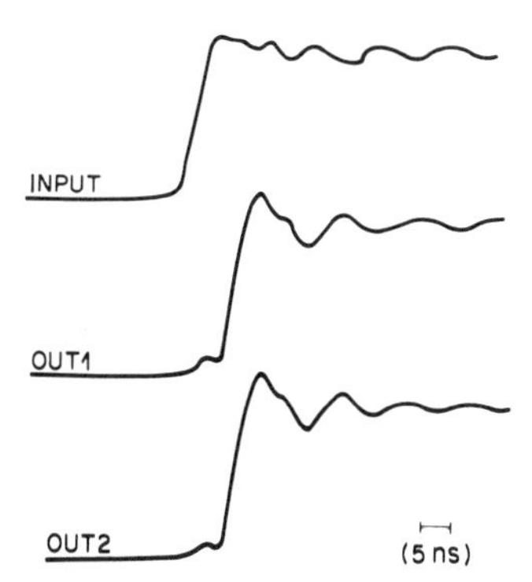

Fig. 5. Input/output waveforms for two buffers located about 1 cm apart on a die fabricated in 2.5-μm CMOS technology. OUT1 and OUT2 represent outputs of the two buffers. Process variation causes a 0.15-ns skew between the two outputs assuming that no skew is introduced by the measurement setup.

Timing Simulation

An important issue that requires special attention in designing highly pipelined VLSI circuits, or any other synchronous system for that matter, is timing simulation. In order to make sure that the clock skew criteria are satisfied at all cells, a timing simulation of the clock distribution network must be performed. This can prove to be extremely time-consuming if a SPICE-type simulation of the whole circuit is undertaken. However, it should be noted that, as far as clock timing and skew are concerned, only the load presented by each cell to the clock network is of importance; details of the internal timing of the cell need not be known for this purpose. Therefore, the cell loads can be modeled by only considering the input circuit of their clocked section in the simulation. This leads to one to three orders of magnitude reduction in the simulation time depending on the size of the system. In our studies, having access to accurate information about the fabrication process parameters, the clock interconnection lines are modeled as distributed *RC* networks including the effect of parasitics. Long interconnection lines are broken into smaller segments and each segment is represented by an *RC* model. For example, for the process used in fabricating the designs discussed in this paper, polysilicon lines of 2.5-μm width are broken into 20-μm segments with each segment having a resistance of 32 Ω and a capacitance of 5.2 fF. Fig. 6 shows the waveform obtained from a SPICE simulation of the clock distribution network of Fig. 4 used in the first design described in Section III. Waveforms show the clock signal at points *M*, *H*, and *I* in Fig. 4. As discussed in [13], these points represent the critical nodes as far as clock skew is concerned. For example, in the vertical direction, data transfer takes place between cells *H* and *I* (adjacent stages of the pipeline) and the situation is similar to what was discussed at the beginning of the section on clock skew. With the fastest computing cell at these nodes being a single register (very small t_{pr}), the worst case tolerable skew margin for our two-phase clocking scheme and fabrication process is about 1 ns. A skew of about 0.12 ns can be measured between waveforms *H* and *I* in Fig. 6. Even adding a maximum of 0.2 ns for the process variation, the measured skew is well within the acceptable margin. In [13] we show how replacing some metal clock lines with polysilicon lines can increase the clock skew between points *H* and *I* to 3.5 ns, and how this can be reduced to 0.4 ns using a three-level distributed buffering network. In this situation, the worst case process variation can reduce the 1-ns skew margin to 0.6 ns because now there are two intermediate buffering levels, each contributing a worst case value of 0.2 ns. Even in this case, the 0.4-ns skew between points *H* and *I* does not violate the skew criterion.

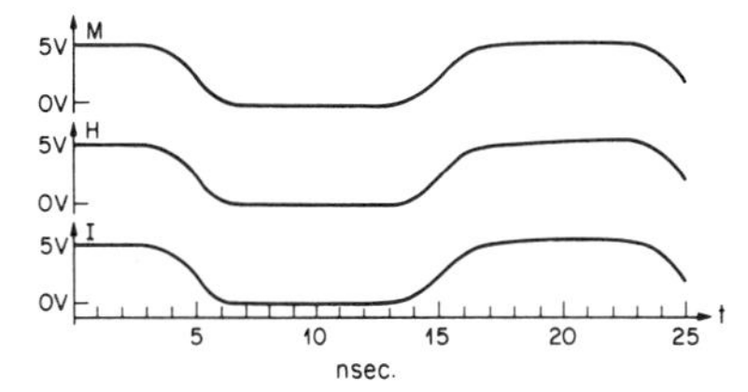

Fig. 6. Clock waveforms obtained from SPICE simulation of the clock distribution network of Fig. 4. A 0.12-ns skew is measured between points *H* and *I*, points *M* and *H* have a skew of about 0.1 ns.

Buffer Design

Knowing the characteristics of the load and the required rise and fall times of the clock pulse, one can design a buffer for the technology at hand. It is beyond the scope of this paper to cover the details and possible design equations for various technologies. However, in this section, we point out an important issue often neglected at the design time, namely, the effect of parasitics introduced by the buffer itself. This effect becomes more pronounced when the size of the buffer gets large, which is often the case for clock buffers driving large loads and required to have fast rise and fall times. We illustrate the situation for the CMOS technology. Consider a CMOS buffer driving a capacitive load C_L. If the parasitics are ignored, the size of the *n* and *p* transistors (gate width-to-length ratio) of the output stage of the buffer are determined from

$$\frac{W_n}{L_n} = K_n \frac{C_L}{t_f} \tag{9}$$

$$\frac{W_p}{L_p} = K_p \frac{C_L}{t_r} \tag{10}$$

where t_r and t_f are the rise and fall times of the output, respectively, and K_n and K_p are two factors determined by the power supply voltage, transistor's threshold voltage, oxide thickness, electron mobility (hole mobility for K_p), oxide's dielectric constant, and the voltage limits defining the rise and fall time regions (e.g., from 10 to 90 percent of the power supply voltage). According to the above equations, the sizes of n and p transistors can be arbitrarily increased to get the desired rise and fall times for a given load. However, if parasitics are considered, the above equations become more involved and can be shown to be

$$\frac{W_n}{L_n} = \frac{K_n t_r (C_L + K)}{t_r(t_f - K_n L_n P_n) - K_p L_p P_p t_f} \quad (11)$$

$$\frac{W_p}{L_p} = \frac{K_p t_f (C_L + K)}{t_r(t_f - K_n L_n P_n) - K_p L_p P_p t_f} \quad (12)$$

where P_n, P_p, and K are parameters defined by the fabrication process and the width of the source and drain of the transistors. If these parameters are set to zero, (11) and (12) reduce to (9) and (10), respectively. Expressing t_r and t_f in terms of these parameters and the transistor sizes we get

$$t_r = \frac{K_p L_p}{W_p}(C_L + P_n W_n + P_p W_p + K) \quad (13)$$

$$t_f = \frac{K_n L_n}{W_n}(C_L + P_n W_n + P_p W_p + K). \quad (14)$$

Equations (11)–(14) indicate that the size of the output transistors cannot be arbitrarily increased to achieve the desired rise and fall times. One cannot minimize the rise and fall times independent of each other. Minimizing the rise (fall) time can be achieved at the cost of fall (rise) time approaching infinity. It can easily be shown that if $t_r + t_f$ is minimized, then the resulting minimum is $t_r = t_f$. At this minimum point if we choose

$$\frac{K_p L_p}{W_p} = \frac{K_n L_n}{W_n}$$

then

$$t_r = t_f = \frac{K_n L_n}{W_n}\left(C_L + P_n W_n + P_p \frac{K_p L_p}{K_n L_n} W_n + K\right).$$

As an example, a CMOS buffer with $L_n = L_p = 3$ μm, $W_n =$ 500 μm, $W_p = 1500$ μm driving a 22-pF capacitive load should result in a rise time of 2.21 ns using the fabrication process parameters for our 2.5-μm CMOS technology. Fig. 7 shows the output waveform of such a buffer obtained by laying out the circuit, extracting the circuit description from the layout, and performing a SPICE simulation with a 22-pF load. The measured rise time is 2.1 ns which is in close agreement with the value predicted by (13). Neglecting the effect of parasitics and using (10) would result in $t_r = 1.63$ ns (an error of 26 percent compared to the simulation results).

Area Overhead

The major factors contributing to area increase due to pipelining are: a) area taken by the registers on the output of the computation cells as well as the skewing and deskewing registers often required (depending on the particular algorithm being pipelined), b) area taken by the clock interconnection lines, c) overhead due to clock buffers, and d) area overhead due to extra routing for the skewing registers. Reference [21] gives estimates of the area increase as a function of the degree of pipelining, however, these estimates do not take into account all factors mentioned above and are not supported by fabricated designs. It is our experience that the area overhead, aside from the degree of pipelining, is also a function of the technology used, the design tools, and the floorplanning of the chip (usually a function of the designer). For example, MOS technology offers dynamic registers in a relatively small area while such is not the case for bipolar technology. As for the design tools, usually tools that require more interaction and manual input on the side of the user result in a more area-efficient design than highly automated tools at the cost of a much lengthier design time.

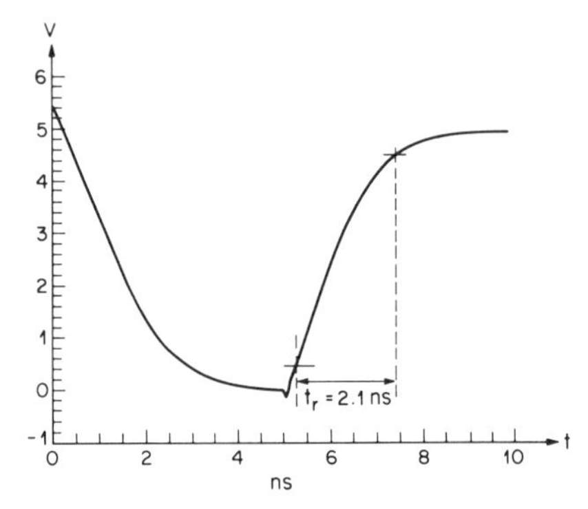

Fig. 7. Output response of a CMOS buffer driving a 22-pF load obtained via SPICE simulation using the fabrication parameters of a 2.5-μm CMOS process. $L_p = L_n = 3$ μm, $W_n = 500$ μm, $W_p = 1500$ μm. Measured rise time is 2.1 ns, predicted rise time using equations that consider parasitics is 2.21 ns. Predicted rise time using equations that ignore the effect of parasitics is 1.63 ns.

Table 1 shows the area overhead figures for the bit-level pipelined designs discussed in this paper. The first and second factors mentioned above are combined in one column in Table 1 because the clock distribution and routing is mostly designed as part of the registers. In all cases, a large percentage of the area overhead is due to pipeline registers and clock distribution. The total area overhead due to bit-level pipelining is in the range of 63 to 75 percent as can be seen from Table 1. This represents a 2.5- to 4-fold increase in area compared to the nonpipelined version. The gain in throughput is about 6- to 8-fold at present, however, as we will discuss in Section IV, reliable testing at high clock rates is the limiting factor in determining the actual gain in throughput. We expect to see higher gains once our improved test setup is operational.

The limit to the degree of pipelining is mostly determined by the underlying technology, and the rise and fall times of the operating clock. As the cell delay time approaches the rise and fall times of the clock pulse, pipelining at lower levels becomes less efficient. Also, smaller computing cells mean smaller propagation delays ($t_{p\ell}$) which makes the clock skew criterion more difficult to satisfy. As technologies are scaled down, or in the case of GaAs technology, the propagation delay and settling times of small computing cells and registers are reduced, while the interconnection delays are not reduced at the same pace. Therefore, these technologies place a more strict limit on the degree of pipelining.

Table 1 Area Overhead Due to Pipelining for Bit-Level Pipelined Designs Fabricated in CMOS Technology. The Second entry corresponds to a design not discussed in this paper. Last entry corresponds to a chip currently under fabrication.

Design	Area Overhead				
	Active Area (mm^2)	Registers and Clock Distribution (mm^2)	Clock Buffers (mm^2)	Skew Routing (mm^2)	Total
8 × 8 multiplier 2.5-μm CMOS	12.53	4.98 (39.74%)	2.04 (16.28%)	2.4 (19.15%)	75.18%
8 × 8 multiplier Single-phase clock 2.5-μm CMOS	8.91	3.49 (39.17%)	0.61 (6.85%)	2.11 (23.68%)	69.7%
8 × 8 multiplier 2's complement 2.5-μm CMOS	14.52	5.62 (38.71%)	2.71 (18.66%)	2.20 (15.15%)	72.52%
Multiply-add 2.5-μm CMOS	16.07	6.74 (41.94%)	2.76 (17.17%)	1.86 (11.57%)	70.69%
Multichannel Multiply-accumulate/add 2.5-μm CMOS	25.34	9.98 (39.38%)	4.27 (16.85%)	1.69 (6.67%)	62.9%
10-bit FIR filter block 2.5-μm CMOS	26.02	12.76 (49.04%)	3.69 (14.18%)	2.98 (11.45%)	74.67%
14 × 14 Multiply-accumulate (36 bits) 1.25-μm CMOS	15.68	6.92 (44.13%)	2.56 (16.33%)	2.01 (12.82%)	73.28%

III. Fabricated Pipelined Chips

With the issues discussed in the previous section in mind, a number of experimental bit-level parallel pipelined circuits have been designed and fabricated in CMOS technology. These chips implement variations of 2's complement multiplication, multiplication-addition, and multiplication-accumulation operations found in almost every signal processing application. The architecture and design of these chips are described in this section. Test results are presented in Section IV.

8-Bit Multiplier

This chip was the first in our series of pipelined designs to be fabricated. It has been reported in [12]. It performs 8 × 8 multiplication and has 16 pipeline stages. The chip has been fabricated in 2.5-μm CMOS technology and has been tested at multiplication rates up to 70 MHz with a power dissipation of 250 mW at that frequency. The complete architecture and floorplan of the chip is shown in Fig. 8. It is a straightforward pipelining of the carry save adder (CSA) array for multiplication, except that the last section of the CSA array which has a ripple carry path is implemented by a triangular array of half-adders as shown in the lower left-hand corner of Fig. 8. For a discussion of the advantages of this half-adder array compared to its full-adder type implementation see [12]. The registers (empty rectangles) in Fig. 8 are for the purpose of I/O data skewing and deskewing.

Bit-level pipelining of an array multiplier in a manner shown in Fig. 8 results in $N + M$ pipeline stages for an N bit × M bit multiplier. However, the half-adder array (lower left-hand corner of Fig. 8) can be implemented with half as many pipeline stages by combining two stages into one. The delay through two half-adders can be made almost equal to the delay through one full-adder, hence there will be no significant loss of throughput. The advantages are reduction in the area, the number of pipeline stages, and the load on the clock lines. This was implemented in a version of the 8 × 8 multiplier mentioned above resulting in 12 stages of pipeline instead of 16. The chip has been fabricated in 2.5-μm CMOS technology and is fully functional.

Multiply-Add Chip

The architecture of Fig. 8 can be easily modified to perform a multiply-add operation ($xy + z$) without the need to design any new cells. This can be accomplished by replacing the registers immediately adjacent to the main diagonal of the array with full-adder cells; the second inputs of these full-adders correspond to the z input. Usually, it is desired to have a larger word-length for the z input and the output, in which case the size of the half-adder array has to be increased accordingly and some extra skewing registers will be required. A chip with 8-bit x and y inputs, a 16-bit z input, and a 16-bit output was designed and fabricated in 2.5-μm CMOS technology. The chip has been tested and is fully functional.

FIR Filter Block Chip

This chip is essentially an extension of the multiply-add design. Its block diagram is shown in Fig. 9. The x and y inputs each have a 10-bit word-length with y representing the filter coefficient. The z input and the output have a

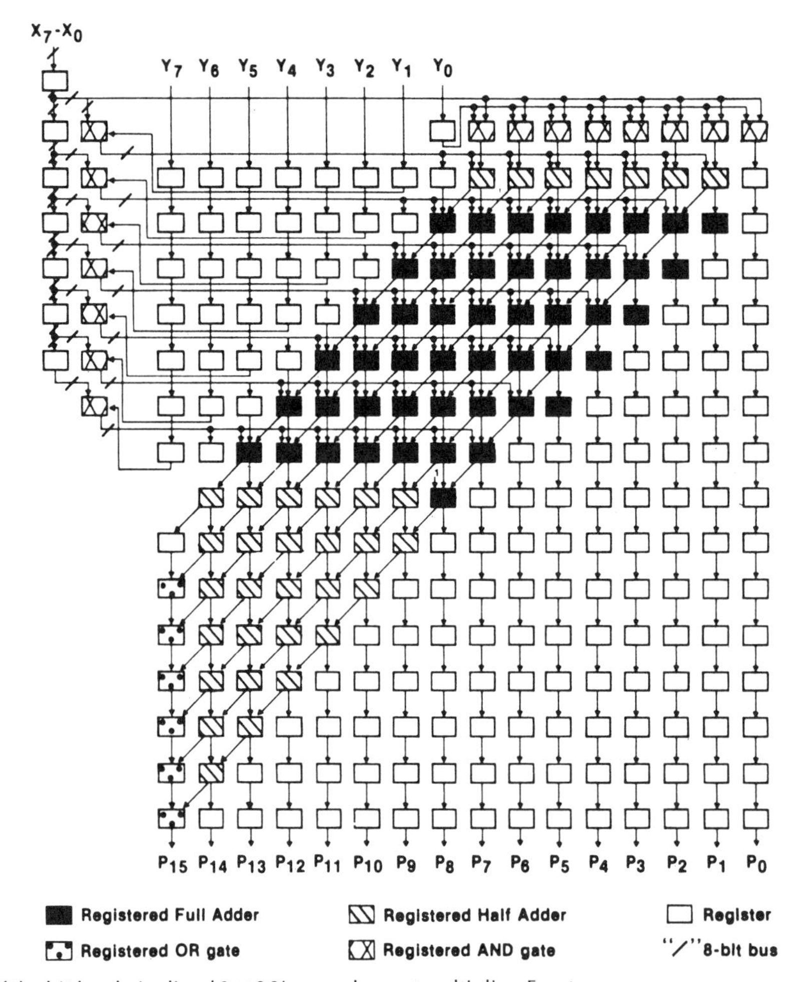

Fig. 8. The architecture of the bit-level pipelined 8 × 8 2's complement multiplier. Empty rectangles represent the registers required for I/O data skewing and deskewing.

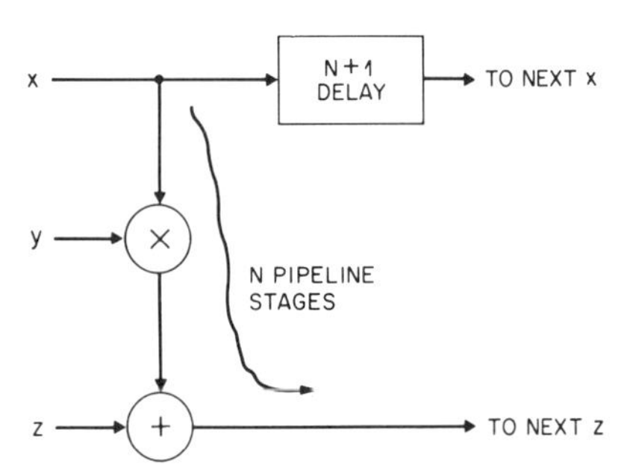

Fig. 9. Block diagram of the FIR-filter-block chip implemented in a bit-level pipelined architecture using an array similar to that shown in Fig. 8. The number of pipeline stages N is determined by the word-length of the x, y, and z inputs and the desired word-length of the summed output.

20-bit word-length. The y input is connected to an on-chip coefficient register. Due to a limitation on the number of pins, this register is programmed serially. The delay unit in Fig. 9 is an $(N + 1)$-bit shift register where N is the number of pipeline stages between the x input and the $xy + z$ output. N itself is determined by the word-lengths of the x, y, and z inputs, and the desired word-length of the summed output, which is a function of the maximum number of filter taps to be implemented with this FIR block. The FIR filter chip has been fabricated in 2.5-μm CMOS technology and is fully functional.

Multichannel Multiply-Accumulate/Add Chip

By replacing the registers immediately below the diagonal of the array of Fig. 8 with full-adder cells, as in the multiply-add function above, and feeding the outputs of the array to the inputs of these full-adders through a second set of skewing registers, a multichannel multiplier-accumulator is formed. The operation performed by this structure is schematically shown in Fig. 10 (P is the number of pipeline stages). By taking advantage of the time multiplexing inherent in the pipelined array with feedback, this architecture allows a fast multiply-accumulator to be used on a signal sampled at slower rates. For example, if the input signal x is sampled at a rate P times slower than the clock rate of the pipelined array, then, depending on the ordering of the coefficients at the y input, a variety of functions such as running DFT and multichannel FIR filtering can be performed. Fig. 11 shows the block diagram of an experimental chip designed based on this architecture. A pin on the chip selects between a multichannel operation as described above or a multiply-add operation ($xy + z$) discussed earlier. This chip has been fabricated in 2.5-μm CMOS technology; the x and y inputs are 8 bits each, z is a 16-bit input, and the output has a 24-bit word-length. The number of channels in the multichannel mode is equal to the number of pipeline stages (24 for this chip). The chip has been tested and is functional.

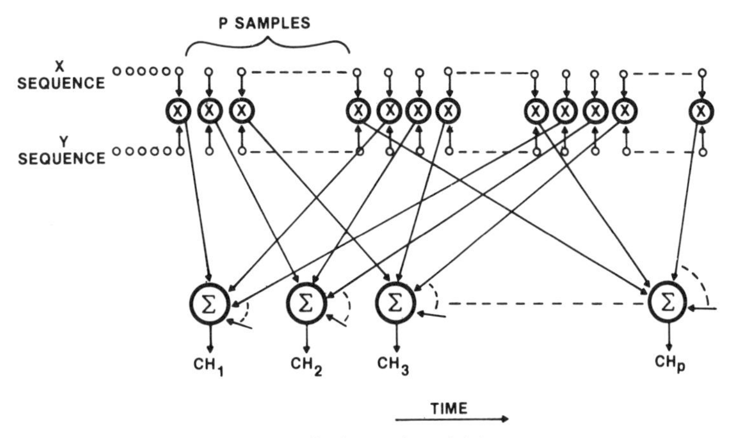

Fig. 10. Illustration of the operation performed by the multichannel multiply-accumulate chip implemented with a bit-level pipelined array.

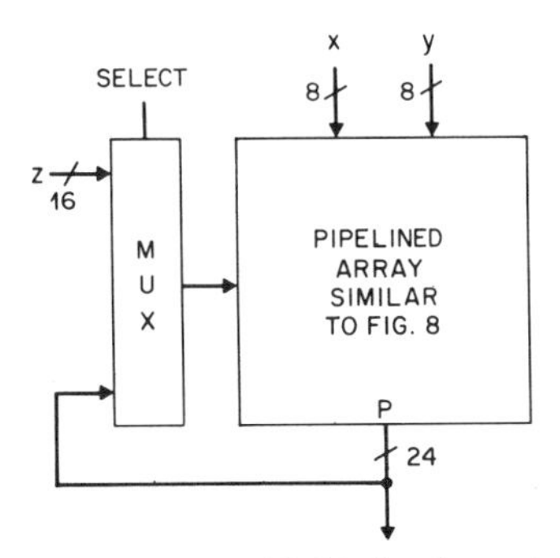

Fig. 11. Block diagram of the multichannel multiply-accumulate/add chip implemented based on the bit-level pipelined array of Fig. 8.

14 × 14 Multiplier-Accumulator

The array of Fig. 8 can be modified to implement a bit-level pipelined multiply-accumulate operation by feeding the outputs of the full-adders introduced in the multiply-add chip above (below the diagonal) back to their inputs through a single register. A 14 × 14 multiplier-accumulator chip (36-bits accumulation) based on this architecture is the latest design in our bit-level pipelined examples. A new feature has been included in this design to allow for programming the number of pipeline stages anywhere from zero (nonpipelined) to 38 (fully pipelined). This feature can also be used to obtain *experimental* data from a fabricated design for studying performance as a function of the degree of pipelining. This chip was designed with the aid of a layout generator program. We are presently in the process of completing the program to generate any of the designs described in this paper for arbitrary word-lengths. The chip is presently being fabricated in 1.25-μm CMOS technology.

IV. Speed Performance Tests

Testing of the bit-level pipelined chips described in this paper has been performed in three phases. In the first phase, the chip is tested in a nonpipelined mode by connecting the clock inputs to the supply voltage. Since all pipeline registers are dynamic, this makes the array look like a conventional nonpipelined circuit. In this mode, the chip is rather simply tested on a general-purpose microprocessor-based test system (pattern generation and capture) and nonworking chips are rejected. In the second phase, the chip is tested in a clocked pipeline mode at very low speeds. The clock signal in this mode is generated as a pattern by the general-purpose tester. This test assures that the clock paths are all functional. Usually chips that pass the phase-one test pass the second phase as well.

The most important, and indeed challenging, part of the testing is the third phase where the maximum speed of the chip in the clocked mode must be determined. This is especially more difficult for high-speed pipelined circuits than their nonpipelined counterparts. For example, the speed of a nonpipelined GaAs multiplier can be estimated by feeding (for an 8-bit multiplier) 1 1 1 1 1 1 1 1 to one input and 1 0 0 0 0 0 0 x_0 to the other input where x_0 is the inverse of the most significant bit of the output [22]. This causes the multiplier output to oscillate between two numbers while exercising the ripple carry path through the array. The frequency of this oscillation is an estimate of the maximum speed of the multiplier. Such a test, however, cannot be used for a bit-level pipelined multiplier because there is no longest ripple carry path and the chip must be tested in clocked mode. The ultimate test would be to apply test vectors to the chip at high clock rates and verify that *all* output bits are changing properly. To the best of our knowledge, such a test has never been reported on any high-speed multiplier implemented in GaAs or any other technology for that matter. The procedure reported for testing the 8 × 8 bit-level pipelined multiplier chip described in [11] cannot, in our opinion, be considered a complete test because only two input bits are clocked at high speed. It should also be noted that only the eight most significant bits of the 16-bit multiplication results are implemented in [11]. Our 8-bit pipelined multiplier was tested using a special-purpose test circuit implemented in ECL for generating test vectors at high rates. The outputs were monitored on a 400-MHz logic analyzer. As reported in [12], the multiplier has been tested at multiplication rates up to 70 MHz where we believe that we are limited by the test setup. The other pipelined chips reported in this paper have been tested in the following way. Two input vectors are fed into the first stage of the pipeline synchronous with a high-speed clock, and outputs are monitored on a 400-MHz logic analyzer. If the entire pipeline operates properly, the correct results should appear on the output pins exactly N clock cycles after the inputs are presented, where N is the number of pipeline stages of the chip under test. The test is then repeated for a different set of input vectors. Our preliminary test results indicate that all the pipeline chips are operating up to 60–65 MHz. However, we expect them to operate at faster rates. We are presently designing a new test setup to carry out reliable tests at higher frequencies.

V. Applications

Although the bit-level pipelined chips presented in this paper have all been successfully tested at high speeds, we believe that their true potential for high-speed digital signal processing can only be determined by using them in an actual working system. There are numerous examples of signal processing algorithms such as matrix–vector and matrix–matrix multiplication, fast transforms, FIR filtering, etc., that can be implemented using these bit-level pipelined designs. To demonstrate this potential and study the problems that might surface in a large scale (beyond the chip) bit-level pipelined system, an example has been planned: a very high sampling rate programmable 8-tap FIR filter for wide-bandwidth applications using the pipelined FIR block. A printed circuit board is presently being laid out for this system. Very-high-speed A/D and D/A chips are also included on the board for analog I/O. A wire-wrap prototype of this system has been built and debugged and is operational at close to 40-MHz sampling rate. The speed of this prototype is limited by clock and data interference on the wire-wrap board and the fact that chips are placed in rather large adapter sockets for the purpose of placing them on the board. We expect to achieve considerably higher rates with the printed circuit version of this system, but even at the present 40-MHz sample rate (320 million operations/s for the filter) and considering that the underlying technology is 2.5-μm CMOS, the performance is exceptional.

VI. Conclusion

As stated in our Abstract, the intent of this work has been that of exploring the potential of bit-level pipelining for high-speed digital signal processing, and understanding the issues and problems involved in implementing such architectures in VLSI. Most of the issues are, one way or another, related to the problem of clock skew in synchronous systems. However, one major conclusion to be drawn here is that clock skew does not seem to be a major limitation provided that appropriate clock buffers are used and the clock distribution network is carefully laid out and simulated. This is evidenced by examples of fabricated and tested designs presented in this paper. The real difficulties caused by clock skew are longer chip design time and area overhead due to clock buffers and clock routing.

The chips presented here are examples of building blocks for implementing those signal processing algorithms where throughput and real-time operation are the major concerns, and latency is not a critical factor. It is noted that these chips have mainly been fabricated in 2.5-μm CMOS technology, a technology that is low-cost, very mature, and far from being called the state of the art. This technology is not known to be a very-high-speed one, yet by employing bit-level pipelining and understanding the circuit and layout design issues involved, a multiplication throughput of 70 MHz has been demonstrated. Speed testing of such high-throughput pipelined chips presents a major challenge that should not be taken lightly. In Section IV we touched upon some differences between testing pipelined versus non-pipelined high-speed chips. Finally, we do not wish to leave the readers with the impression that a number of working bit-level pipelined chips necessarily mean a working system. To that end, and to satisfy system designers, we have chosen to implement a high sampling rate 8-tap programmable FIR filter using one of the designs described in this paper. The prototype wire-wrap version of this system is operational at 40-MHz sampling rate, however, considerably higher sampling rates are expected from the printed circuit board version currently being designed.

Acknowledgment

The authors wish to thank Dr. S. Rao for valuable discussions on clock skew.

References

[1] J. V. McCanny and J. G. McWhirter, "Completely iterative, pipelined multiplier array suitable for VLSI," *Proc. Inst. Elec. Eng.*, vol. 129, pt. G, no. 2, pp. 40–46, Apr. 1982.

[2] R. A. Evans, D. Wood, K. Wood, J. V. McCanny, J. G. McWhirter, and A. P. H. McCabe," A CMOS Implementation of a Systolic Multi-Bit Convolver Chip," in *VLSI'83*, F. Anceau and E. J. Aas, Eds. Amsterdam, The Netherlands: Elsiver, 1983, pp. 227–235.

[3] J. C. White, J. V. McCanny, A. McCabe, J. McWhirter, and R. Evans, "A high speed CMOS/SOS implementation of a bit level systolic correlator," in *Proc. Int. Conf. on Acoustics, Speech, and Signal Processing* (Tokyo, Japan, Apr. 1986), pp. 1161–1164.

[4] A. Corry and K. Patel, "Architecture of a CMOS correlator," in *Proc. IEEE Int. Symp. on Circuits and Systems*, pp. 522–525, 1983.

[5] P. R. Cappello and K. Steiglitz, "Bit-level fixed architectures for signal processing," in *Proc. IEEE Int. Conf. on Circuits and Computers*, pp. 570–573, Sept. 1982.

[6] ——, "A VLSI layout for a pipelined Dadda multiplier," *ACM Trans. on Computer Systems*, vol. 1, no. 2, pp. 157–174, May 1983.

[7] ——, "Completely-pipelined architectures for digital signal processing," *IEEE Trans. Acoust., Speech, Signal Processing*, vol. ASSP-31, no. 4, pp. 1016–1023, Aug. 1983.

[8] ——, "A note on free accumulation in VLSI filter architectures," *IEEE Trans. Circuits Syst.*, vol. CAS-32, no. 3, pp. 291–296, Mar. 1985.

[9] C. E. Hauck, C. S. Bamji, and J. Allen, "The systematic exploration of pipelined array multiplier performance," in *Proc. IEEE Int. Conf. on Acoustics, Speech, and Signal Processing* (Tampa, FL, Apr. 1985), pp. 1461–1464.

[10] K. K. Parhi and D. G. Messerschmitt, "A bit parallel bit level recursive filter architecture," in *Proc. IEEE Int. Conf. on Computer Design* (Rye Brook, NY, Oct. 1986), pp. 284–289.

[11] T. G. Noll, D. Schmitt-Landsiedel, H. Klar, and G. Enders, "A pipelined 330-MHz multiplier," *IEEE J. Solid-State Circuits*, vol.

SC-21, no. 3, pp. 411–416, June 1986.
[12] M. Hatamian and G. L. Cash, "A 70-MHz 8-bit × 8-bit parallel pipelined multiplier in 2.5 micron CMOS," *IEEE J. Solid-State Circuits*, vol. SC-21, no. 4, pp. 505–513, Aug. 1986.
[13] ——, "High speed signal processing, pipelining, and VLSI," in *Proc. IEEE Int. Conf. on Acoustics, Speech, and Signal Processing* (Tokyo, Japan, Apr. 1986), pp. 1173–1176.
[14] J. W. Goodman, F. I. Leonberger, S-Y. Kung, and R. A. Athale, "Optical interconnections for VLSI systems," *Proc. IEEE*, vol. 72, no. 7, pp. 850–866, July 1984.
[15] S. Y. Kung and R. J. Gal-Ezer, "Synchronous versus asynchronous computation in very large scale integrated (VLSI) array processors," in *Proc. SPIE Symp.*, vol. 341, pp. 53–65, 1982.
[16] A. L. Fisher and H. T. Kung, "Synchronizing large VLSI processor arrays," *IEEE Trans. Comput.*, vol. C-34, no. 8, pp. 734–740, Aug. 1985.
[17] M. A. Franklin and D. F. Wann, "Asynchronous and clocked control structures for VLSI based interconnection networks," in *Proc. 9th Annu. Symp. Computer Architecture* (Austin, TX, Apr. 1982), pp. 50–59.
[18] H. B. Bakoglu, J. T. Walker, and J. D. Meindl, "A symmetric clock-distribution tree and optimized high-speed interconnections for reduced clock skew in ULSI and WSI circuits," in *Proc. IEEE Int. Conf. on Computer Design* (Rye Brook, NY, Oct. 1986), pp. 118–122.
[19] E. G. Friedman and S. Powell, "Design and analysis of a hierarchical clock distribution system for synchronous standard cell/macrocell VLSI," *IEEE J. Solid-State Circuits*, vol. SC-21, no. 2, pp. 240–246, Apr. 1986.
[20] H. B. Bakoglu and J. D. Meindl, "Optimal interconnection circuits for VLSI," *IEEE Trans. Electron Dev.*, vol. ED-32, no. 5, pp. 903–909, May 1985.
[21] P. R. Cappello, A. LaPaugh, and K. Steiglitz, "Optimal choice of intermediate latching to maximize throughput in VLSI circuits," *IEEE Trans. Acoust., Speech, Signal Processing*, vol. ASSP-32, no. 1, pp. 28–33, Feb. 1984.
[22] K. Gonoi, I. Honbori, M. Wada, K. Togashi, and Y. Kato, "A GaAs 8 × 8-bit multiplier/accumulator using JFET DCFL," *IEEE J. Solid-State Circuits*, vol. SC-21, no. 4, pp. 523–529, Aug. 1986.

CLOCK SYSTEM DESIGN

KENNETH D. WAGNER
IBM Corp.

A well-designed clock system is a fundamental requirement in high-speed computers. In this tutorial, the author provides a framework for understanding system timing and then describes how the clock system executes the timing specifications. The tutorial examines clock generation and the construction of clock-distribution networks, which are integral to any clock system. Examples from contemporary high-speed systems highlight several common methods of clock generation, distribution, and tuning. Tight control of system clock skew is essential to an effective clock system.

The careful design of clock systems is often neglected. Part of the reason is that older, slower computers had higher tolerances to variations in the clock signal and had less exacting timing requirements. Today, however, as the demand for high-speed computers grows, the design of their clock systems should become a major concern not only in achieving high performance, but also in reducing assembly and maintenance costs.

A well-planned and well-built clock system is a prerequisite to reliable long-term computer operation. Conversely, a badly designed clock system can plague a computer throughout its lifetime, affecting its operation at any speed. To make such systems function, components often have to be tuned individually at several stages of manufacturing.

Despite these costs and performance penalties, timing design is still overlooked in many systems. Although significant decisions that must be made early in computer design include such issues as clocking scheme and type of memory element, designers seldom participate. Instead, system architects may simply repeat a previously successful set of choices, despite significant changes in design specifications, technology, and environment. Of course, these systems will eventually be functional, but they will require much more maintenance and tuning—costs not always reflected back to the developers.

These attitudes prevail in part because timing design problems are rarely reported in the literature. Also, design teams tend to be secretive about their clock systems, either because they believe they are doing something new or because they are doing nothing new and are afraid to be associated with an older technique. Either way, the result is a scarcity of information on how to avoid timing problems through proper design of the clock system.

THE CLOCK SYSTEM

System timing specifications are executed using a clock system. The clock system has two main functions, clock generation and clock distribution. We use clock-generation circuitry to form highly accurate timing signals, which we then use to synchronize

Reprinted from *IEEE Design Test Comput.*, pp. 9–27, Oct. 1988.

Two types of clocked bistable elements are important in contemporary high-speed computers: the latch and the edge-triggered flip-flop.

changes in the system state. These pulsed, synchronizing signals are known as clocks. We use clock distribution to deliver the clocks to their destinations at precisely specified instants. A network, called the *clock-distribution network*, propagates clocks formed by clock generation to clocked memory elements.

Most logic design texts, such as that by McCluskey (see "Additional Reading" at the end of this article), describe bistable ele-synchronized by the clock signals. A system oscillator is the source for these periodic signals. We generate and manipulate the clock signals and precisely place clock pulses to meet the system timing requirements. We may also tune the clocks to compensate for inaccuracies in the clock pulsewidth or pulse position.

BISTABLE ELEMENTS

The focus of this article is on the timing design of systems that use static bistable elements. The techniques described can also be used in the timing design for other types of clocked memory elements, such as arrays and dynamic latches, or for precharging circuitry.

Most logic design texts, such as that by McCluskey (see "Additional Reading" at the end of this article), describe bistable elements and their characteristics in great detail. Two types of clocked bistable elements are important in contemporary high-speed computers: the latch and the edge-triggered flip-flop. The latch is transparent while its clock (control) input is active. By transparent, we mean that its outputs reflect any of its data inputs. Edge-triggered elements, such as the D flip-flop, respond to their data inputs only at either the rising or falling transition of their clock input. They do not have the transparency property of the latch.

We can describe the time-dependent behavior of a bistable element using the following parameters:

- *setup time*, the minimum time that the data input of the bistable element must be held stable before the active edge or latching level of the clock pulse occurs
- *hold time*, the minimum time that the data input of the bistable element must be held stable after the active edge or latching level of the clock pulse disappears
- *propagation delay*, the time between a change on the clock or data input of the bistable element and the corresponding change on its output

For system operation to be correct, the setup time, hold time, and minimum clock pulsewidth must be satisfied for each bistable element. Signals whose propagation delay is so long that it violates the setup time are called long-path signals. Signals whose propagation delay is so short that it violates the hold time are called short-path signals. Both conditions result in incorrect data being stored.

SYSTEM CLOCKING SCHEMES

System clocking is either single-phase, multiphase (usually two-phase), or edge-triggered. Figure 1 illustrates. The dark rectangles in the figure represent the interval during which a bistable element samples its data input. Each scheme requires a minimum clock pulsewidth.

The most widely used scheme is multiphase clocking. The multiphase clocking scheme in Figure 1b is two-phase, nonoverlapping. In this scheme, two distinct clock phases are distributed within the system, and each bistable element receives one of these two clocks. Systems that have adopted two-phase clocking include microprocessors such as the Intel 80x86 series and Motorola MC68000 family, micro-mainframes such as the HP-9000, and mainframes such as the IBM 3090 and the Univac 1100/90.

Figure 2 shows a finite-state machine, a machine that realizes sequential logic functions, with each clocking scheme. (For more on finite-state machines, see McCluskey's text.) For simplicity, primary I/O is not shown. The Amdahl 580 mainframe and Cray-1 vector processor are single-phase latch machines, such as that shown in Figure 2a. Modern high-speed microprocessors like the Bellmac-32A are two-phase latch machines with a single-latch design using nonoverlapping clock phases, such as that shown in Figure 2b. Figure 2c shows a two-phase latch machine with a double-latch design. This type of machine supports scan-path testing, since it can use LSSD latch pairs, which are hazard-free master-slave latches with a scan input port. Most contemporary IBM products, including IBM 3090 mainframes, incorporate design for testability using this structure. Systems built with catalog parts are usually flip-flop machines, such as that shown in Figure 2d, because clocked bistable elements commonly offered in bipolar and CMOS MSI chips are edge-triggered.

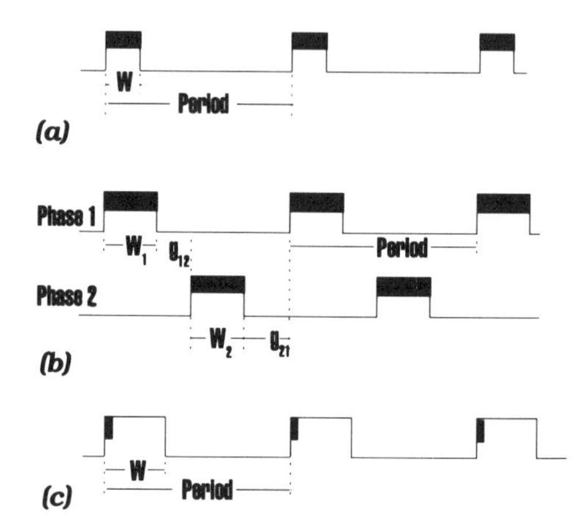

Figure 1. *System clocking waveforms; single-phase **(a)**, two-phase **(b)**, and edge-triggered **(c)**. W_i=pulsewidth of phase j and g_{ij}= interphase gap from phase i to phase j; if $g_{ij} > 0 \Rightarrow$ two-phase, nonoverlapping, if $g_{ij} < 0 \Rightarrow$ two-phase overlapping.*

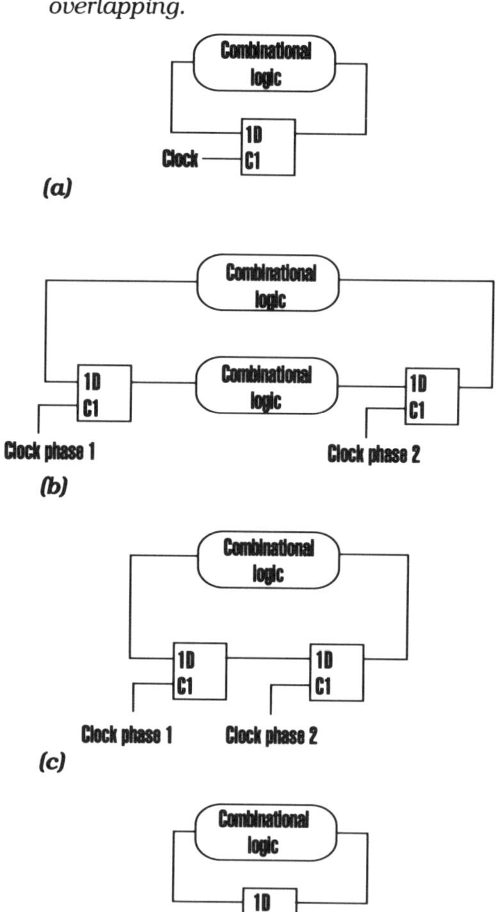

Figure 2. *General finite-state machine structures: one-phase latch machine **(a)**, two-phase latch machine with single-latch **(b)** and double latch **(c)**, and flip-flop machine **(d)**.*

THE CLOCK CYCLE

System designers characterize a computer's functionality in terms of its clock cycle, also called its machine cycle. The average number of clock cycles required per machine instruction is a measure of computer performance. Table 1 gives clock rates for some well-known systems. The designer focuses on the clock cycle because it determines the standard work interval for internal machine functions. The system state is the set of values in system memory elements at the end of a clock cycle.

A clock cycle has the following properties:

1. It consists of a sequence of one or more clock pulses.
2. The sequence of clocks generated in each cycle is identical to every other cycle.
3. No partial clock sequences can occur: clocks can only stop and start at cycle boundaries.
4. Each bistable element can be updated at most once per cycle.

These properties ensure that the transition to the next state of the system is predictable and correct. This deterministic system

In a conventional computer system, one source generates the system clock signal. Multiple processors operating synchronously may also share one signal.

behavior will hold whether clock cycles occur at the system operating rate or one at a time. We can reproduce system behavior at the operating rate by issuing single clock cycles or bursts of clock cycles, which makes system debugging much simpler.

TIMING ANALYSIS

Programs for timing analysis are used routinely to verify system timing. They can identify long or short paths, and the designer can interact with them to get estimates of signal-path delays in parts of the system. Designers can also run them after layout to get more accurate results. The delay models used for system elements are validated by circuit simulation.

Single-phase systems and multiphase overlapping systems require more extensive timing analysis than multiphase nonoverlapping and edge-triggered systems. The timing constraints of single-phase and multiphase overlapping systems are two-sided, bounded by both short paths and long paths. Figure 3 illustrates these constraints in a simplified example, where setup time and hold time are set to 0. The advantage of these systems is that they operate more quickly than their nonoverlapping counterparts.

CLOCK SIGNALS

In a conventional computer system, one source generates the system clock signal. Multiple processors operating synchronously may also share one signal. We can manipulate this clock signal in many ways before it reaches its destinations. We can divide it, delay it, shape it, buffer it, and gate it. Clocked bistable elements, either latches or flip-flops, use the signal that results from such manipulations.

Table 1. *System clock rates.*

System	Intro Date	Technology	Class	Nominal Clock Period (ns)	Nominal Clock Frequency (MHz)
Cray-X-MP	1982	MSI ECL	Vector processor	9.5	105.3
Cray-1S,-1M	1980	MSI ECL	Vector processor	12.5	80.0
CDC Cyber 180/990	1985	ECL	Mainframe	16.0	62.5
IBM 3090	1986	ECL	Mainframe	18.5	54.1
Amdahl 58	1982	LSI ECL	Mainframe	23.0	43.5
IBM 308X	1981	LSI TTL	Mainframe	24.5,26.0	40.8,38.5
Univac 1100/90	1984	LSI ECL	Mainframe	30.0	33.3
MIPS-X	1987	VLSI CMOS	Microprocessor	50.0	20.0
HP-900	1982	VLSI NMOS	Micro-mainframe	55.6	18.0
Motorola 68020	1985	VLSI CMOS	Microprocessor	60.0	16.7
Bellmac-32A	1982	VLSI CMOS	Microprocessor	125.0	8.0

For all systems, we must correctly place the leading- or trailing-edge positions of the distributed clock pulses to ensure that bistable elements switch at the correct times.

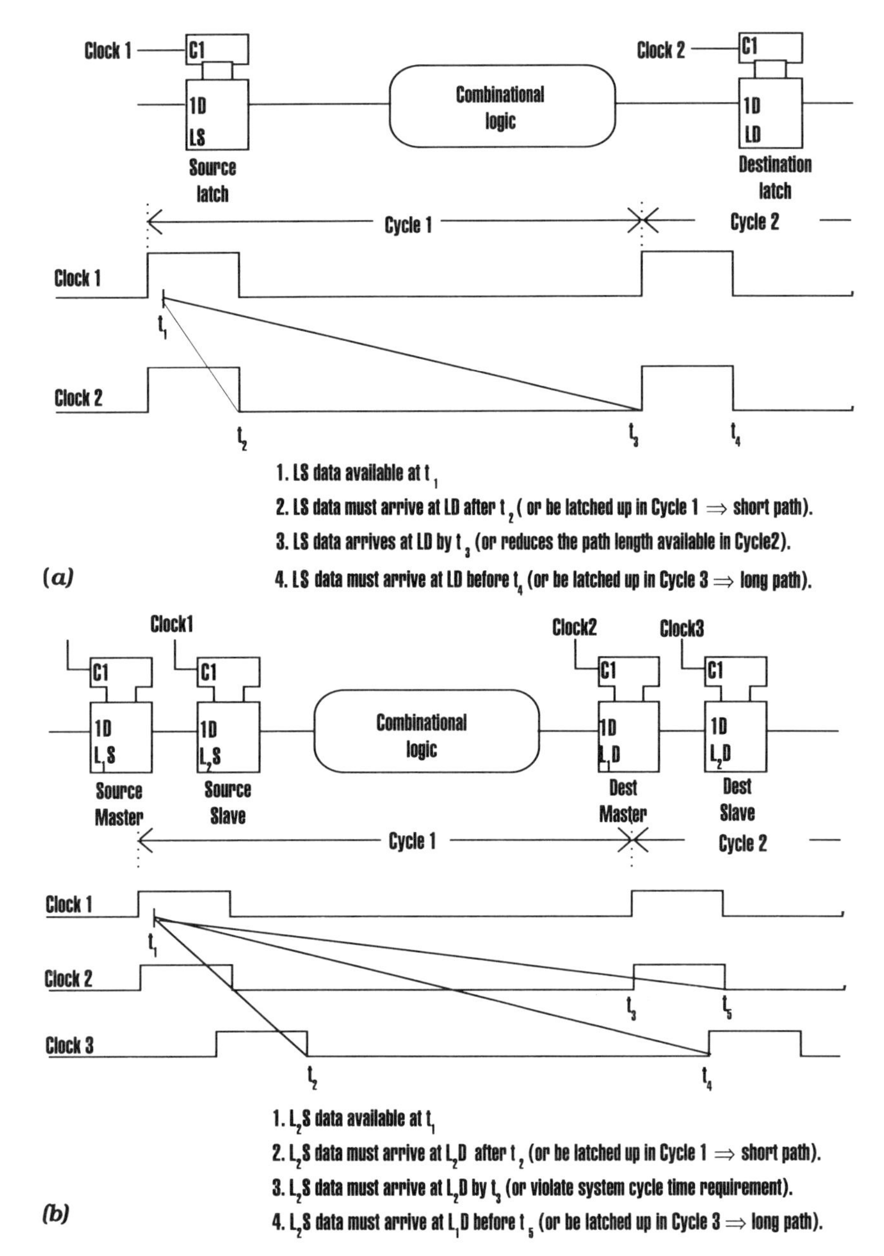

Figure 3. *Path requirements in a single-phase machine* **(a)** *and in a two-phase overlapping latch machine with a double latch* **(b)**.

SIGNAL CHARACTERISTICS

Clocked sequential logic responds to several characteristics of the clock signal: the clock period, the pulsewidth, and the leading-edge or trailing-edge position of the clock pulse. The *clock period* is the interval before the signal pattern repeats. The ideal clock signal for a bistable element is a sequence of regularly repeating pulses. Ideal pulses are rectangular with sufficient duration and amplitude to ensure the reliable operation of the bistable element. The duration of the pulse, or pulsewidth (W), can be any fraction of the clock period, but is usually less than or equal to half of it. An accurate model of a real clock pulse includes actual voltage levels and the shapes of the pulse edges.

Pulsewidth-manipulation elements have three functions: chop, shrink, and stretch.

For all systems, we must correctly place the leading- or trailing-edge positions of the distributed clock pulses to ensure that bistable elements switch at the correct times. Also, distributed clock pulses must be wide enough or they will either be filtered out in transmission or be unable to switch a bistable element because they lack the energy. *Clock-manipulation elements* reposition clock pulses and change their pulsewidths. They consist of delay elements and elements that manipulate the pulsewidth. *Delay elements* either delay a pulse, or, in a timing chain, produce a sequence of delayed pulses in response to a single pulse input. *Pulsewidth-manipulation elements* require both delay elements and logic gates.

Delay elements are available as both analog and digital circuits and are chosen according to the accuracy, flexibility, and range of signal delay required. Analog delay elements vary from simple printed or discrete wire interconnections to delay lines. Delay lines, packaged in hybrid chips, consist of lumped LC elements or distributed printed wire, which provides more accurate control. Digital delay elements include logic gates and counters. Logic gates are relatively inaccurate because of their wide delay ranges, while the time resolution of counters depends on their operating frequency.

Some delay elements are programmable, providing a range of delays. To select a particular delay, we can either connect to a particular chip output pin or tap, or control the configuration electronically by a multiplexer. A typical integrated delay line provides delays from 1 to 10 ns in 1-ns increments with a ±0.5-ns tolerance.

Pulsewidth-manipulation elements have three functions: chop, shrink, and stretch. Figure 4b shows the effect of a chopper, shrinker, and stretcher on a positive pulse. The effect of each manipulation element differs for positive and negative clock pulses. Thus, for each pulse polarity, only three of the four elements are useful. The other element has only a delay effect. Table 2 shows the values for the signal characteristics after chopping, shrinking, and stretching. AND gates have delay d_a, OR gates have delay d_o, inverters have delay d_i, delay elements have delay D, and interconnections have no delay. The signal input is a pulse of width W whose leading edge occurs at time t=0. For an element to have an effect during the pulse, the sum of d_i and D must be less than W.

Table 2. *Effect of elements that manipulate the clock pulsewidth.*

Element	**Positive Pulse** Leading Edge	Pulse-width	Function	**Negative Pulse** Leading Edge	Pulse-width	Function
A	d_a	$D+d_i$	Chopper	—	—	—
B	—	—	—	d_o	$D+d_i$	Chopper
C	$D+d_a$	W–D	Shrinker	d_a	W+D	Stretcher
D	d_a	W+D	Stretcher	$D+d_a$	W-D	Shrinker

CLOCK GENERATION

We can derive all clock signals in a synchronous machine from the system clock signal. The system clock is often a rectangular pulse train with a 50% duty cycle, called a *square wave*. The circuit that generates the system clock is at the base of the clock-distribution network. Its input is from either a voltage-controlled oscillator (VCO), a crystal oscillator (XO), or a voltage-controlled crystal oscillator (VCXO). All three sources produce a sinusoidal (single-frequency) output, which is then clamped or divided to generate the rectangular system clock. Excluding the quartz crystal, the oscillator circuit is usually packaged on a single hybrid IC.

A simple oscillator consists of an LC circuit, which we tune by carefully selecting component values that allow the circuit to resonate at the desired frequency. When we need extreme frequency stability over a wide temperature range, we use an XO. An XO consists of a tuned circuit with an embedded quartz crystal in the feedback loop. The crystal stabilizes the resonant frequency of the oscillator circuit.

When we need a larger range of selectable frequencies, we use either a VCO or a VCXO, because the XO has a very limited tunable range. A DC voltage input controls both the VCO and VCXO. The VCO could be an emitter-coupled multivibrator that produces a square wave that we can tune over a 10:1 frequency range up to 20 MHz. It could also be a capacitance-controlled oscillator that produces a sine wave tunable over a 2:1 frequency range up to microwave frequencies. If we modify the resonant frequency of an XO, we get a tuning accuracy of a few hundred parts per million in the VCXO. Thus, the XO has the most frequency stability but the least tuning flexibility, the VCXO is in the middle on both, and the VCO has the least frequency stability and the most tuning flexibility. Frequency instability in the oscillator can cause clock jitter, requiring us to assign a tolerance to the clock-edge placement in timing analysis.

From the system clock we derive the full set of clocks and clock phases that the system requires. We can generate multiphase clocks from a square-wave input in many different ways. These methods include one shots, clock choppers or shrinkers, shift-register latches, and frequency dividers, depending on the precision and flexibility required. To prevent the overlap of adjacent clock phases in a nonoverlapping clocking scheme, we use output feedback or clock choppers. If there is uneven loading on each clock phase, the relative pulse-edge positions may change, which might cause some of the clock phases to overlap. Another cause of overlap is the asymmetric rising and falling delays of contemporary devices.

Figure 5 shows two simplified circuits that create two-phase clocks. The techniques are applicable to general multiphase clock generation. The first circuit is used in the Univac 1100/90 for four-phase clock generation. It requires a fast-running square-wave clock input and a ring counter. Each stage of the ring counter enables one clock phase, and the single clock chopper

From the system clock we derive the full set of clocks and clock phases that the system requires.

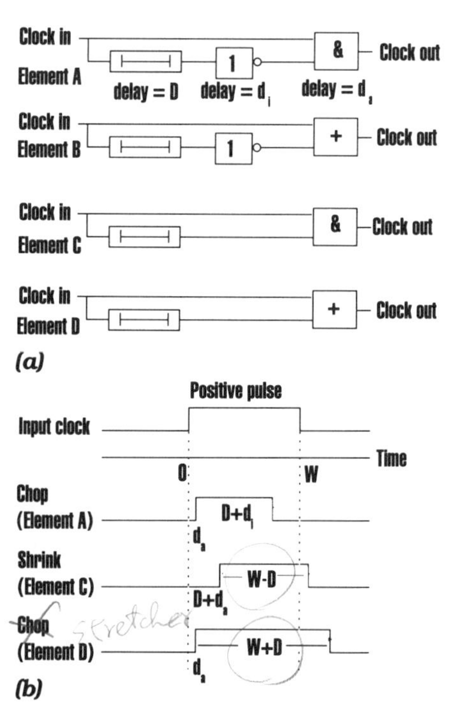

Figure 4. *Elements that manipulate the clock pulsewidth **(a)** and their effect on a positive pulse **(b)**.*

For developing, diagnosing, and producing high-speed systems, we ideally want a wide-bandwidth oscillator source that is highly accurate.

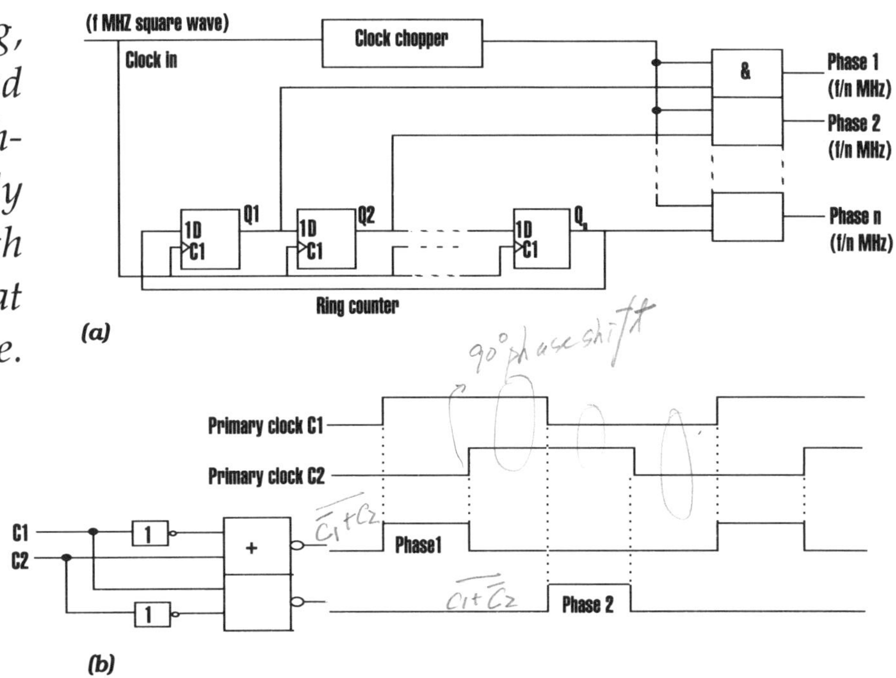

Figure 5. *Creating a two-phase clock: selecting the pulses of a fast-running clock **(a)** and decoding the primary clocks **(b)**.*

determines pulsewidth. The second circuit is used in the Bellmac-32A. It generates two-phase clocks by decoding primary clock signals. We can use a gray-code counter to produce these primary clocks, or we can use *clock shaping*. Clock shaping allows us to generate clock phases from a system clock with fixed gaps between phases (forcing pulsewidths to vary with frequency).

CLOCK SEQUENCES

The three schemes for system clocking we have looked at—single-phase, multiphase, and edge-triggered—determine the basic data flow in latch and flip-flop machines during each clock cycle. Complicating these requirements, though, are special timing considerations. For example, subsystems may require dfferent clock-arrival times so that they can communicate with each other across interfaces with large delays. Also, paths within subsystems may be too long for normal system timing. We can accommodate irregular interfaces and paths without affecting the clock cycle, although system timing becomes more complex. To handle these cases, we generate a sequence of clocks during each clock cycle and do not use normal data-path timing.

There are two timing design styles for handling the clock sequences generated during a clock cycle: multiphase design and multiclock design. Figure 6 illustrates. The dashed vertical lines represent the boundaries of the clock cycle. The solid vertical lines represent active clock edges. Time proceeds left to right across each diagram and only paths originating from the earliest (left-most) cycle are shown. In a normal multiphase (*k*-phase) design

(Figure 6a), latches clocked by phase 1 feed latches clocked by phase 2, and so on. Only the latch clocked on the last phase feeds the phase-1 latch of the succeeding cycle. All data movement proceeds phase i to phase $i+1$ modulo k.

In contrast, the multiclock design (Figure 6c) ensures that bistable elements clocked at any time T_i during one cycle feed only bistable elements clocked in the succeeding cycle. For instance, the three cycle $n-1$ clocks are early, normal and late, which correspond to the times T_0, T_1 and T_2. Each can feed any of the T_0, T_1 or T_2 bistable elements in cycle n.

In the Amdahl 580, early clocks prevent long paths between the remote channel frame and I/O processor. If we clock the source latch earlier or destination latch later than normal on a signal path, the signal has a longer interval to propagate. Of course, other signal paths between latches using normal clocks as sources and early clocks as destinations will have a shorter than normal time to propagate. Similarly, paths with latches using late clocks as sources and normal clocks as destinations will also be shorter.

Multiphase design and multiclock design can be mixed, as shown in Figure 6d. The two-phase, double-latch configuration has master latches, which feed their associated slaves in the same cycle. Each master latch is clocked at one of three timings: T_0, T_1 or T_2. The slave latch of each pair communicates with any of the master latches in the next cycle. The IBM 3033, 308X, and 3090 mainframes use similar techniques.

Figures 6b, 6e, and 6f show examples of more complex paths. Figure 6b shows the possibility of paths that skip adjacent phases in a three-phase system. The Univac 1100/90 is an example of a design with nonadjacent phase paths. Note that any phase-i-to-phase-i path in the succeeding cycle would require identical analysis to a single-phase system. Figures 6e and 6f show fractional cycle and multicycle paths. Such paths are typical of a performance-oriented design that uses two-phase latch machines.

Systems can also generate clocks that operate at several distinct cycle times, usually integer multiples of a base cycle time. We can use clocks with lower rates for parts of the system that do not need faster clocks. All clocking between subsystems must be synchronous, or else we must use techniques to reduce metastable behavior at subsystem interfaces.

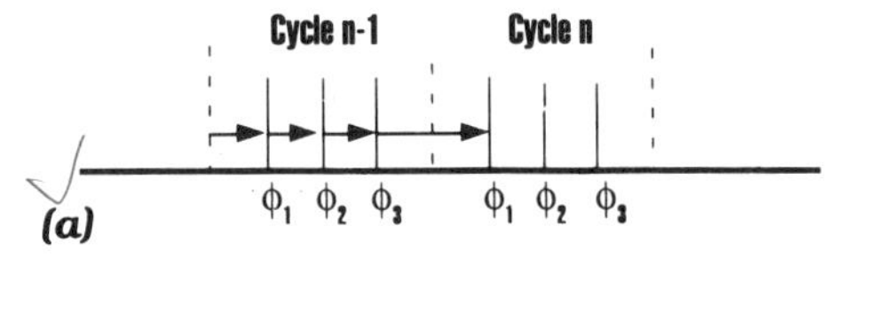

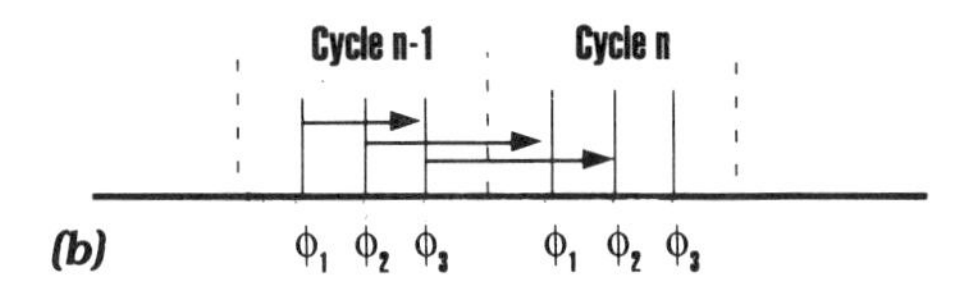

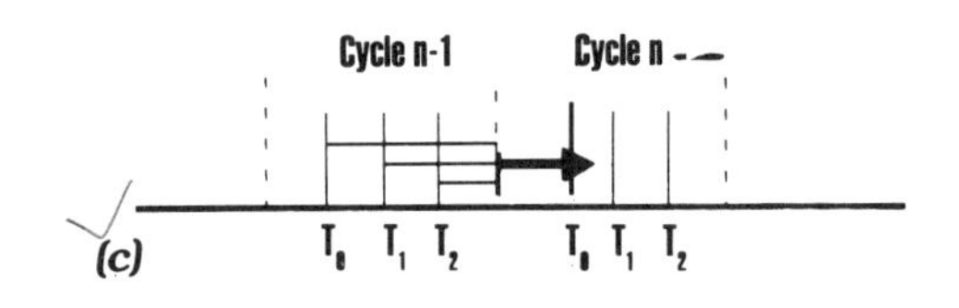

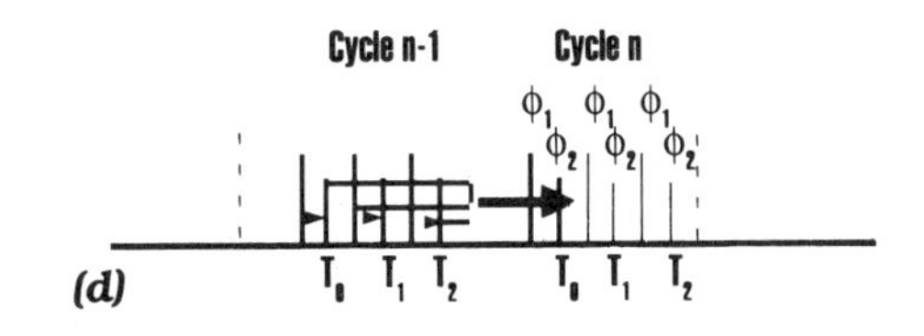

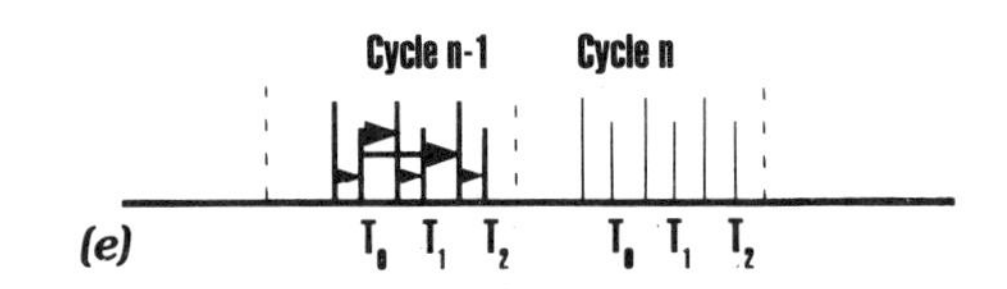

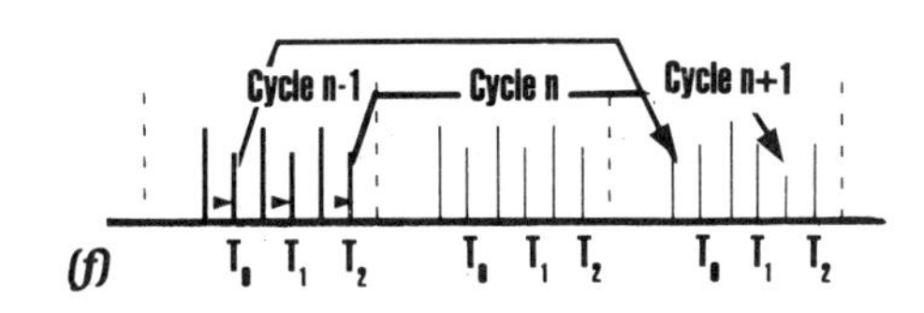

Figure 6. *Placing clock pulses; three-phase, adjacent paths* ***(a)****; three-phase, nonadjacent paths* ***(b)****; multiclock (three clocks)* ***(c)****; multiclock, two-phase* ***(d)****; multiclock, two-phase with fractional cycle paths* ***(e)****; and multiclock, two-phase with multicycle paths* ***(f)****.*

THE SYSTEM CLOCK SOURCE

For developing, diagnosing, and producing high-speed systems, we ideally want a wide-bandwidth, highly accurate oscillator source. Most systems have both a crystal-oscillator source input for production systems and a tunable source input for prototype development and AC diagnosis. During development of a multiphase system, we may need to vary the pulsewidth of any clock phase as well as to vary the relative pulse positions.

To detect marginal path-delay problems, the Amdahl 580 selects any one of three crystal oscillators as the clock source in production machines, lengthening or shortening its clock cycle.

The goal of clock distribution is to organize clocks so that the delays from the source of each clock or clock phase to its bistable elements are identical.

Operating modes are called normal, fast margin, and slow margin. These correspond to nominal clock frequency, 5% faster than nominal, and 5% slower than nominal. An external oscillator input is also available, bypassing the internal oscillators during diagnosis and development.

To detect marginal timing problems in the IBM 3090, a two-phase double-latch machine, designers made it possible to lengthen the delay between the leading edge of the slave clock and the trailing edge of the master latch clock for a selected system region (see Figure 3b). In addition, lengthening the clock cycle allows us to verify the slave-latch-to-slave-latch path delay.

We can choose between distributed or centralized clock sources to control multiprocessors synchronously. In distributed control, we let each processor or processor group in the complex use its own local oscillator, with some form of enforced synchronization between oscillators, like a phase-locked loop. Alternatively, in centralized control, we designate one oscillator as the master oscillator and have each system select this master through a local/remote switch. The second method is simpler and is common in mainframe multiprocessor models such as the Amdahl 580, IBM 3033, and IBM 370/168. Although the IBM multiprocessors use a master oscillator, other standby oscillators are phase-locked to the master oscillator and can be selected if it fails.

CLOCK DISTRIBUTION

The goal of clock distribution is to organize clocks so that the delays from the source of each clock or clock phase to its bistable elements (its destinations) are identical. In reality, however, no matter how each clock path is constructed, any two clock paths in the same machine or any two corresponding paths in different machines will always have a delay difference. Every computer operates in a different temperature, power supply, and radiation environment, and duplicate components will differ in subtle ways between computers. We must build in tolerance to these variations in any system timing design.

The most common approaches to ensure correct and reliable machine timing are worst-case analysis and statistical analysis. In worst-case analysis, we assume that all component parameters lie within some range, and the cumulative worst-case effect is still within the timing tolerance of the machine. In statistical analysis, the intent is that most machines have tolerable timing characteristics, and so we can rely on the cumulative statistical variations of component parameters to remain within the timing tolerance.

CLOCK SKEW

We specify system timing such that every system memory element has an expected arrival time for the active edge of its clock signal. Clock-edge inaccuracy is the difference between the actual and expected arrival time of this clock edge. For every pair of system memory elements that communicate, we define *path clock skew* as the sum of the clock-edge inaccuracies of the pair's source and destination. *System clock skew* is the largest path clock skew in the system. It is the value of the worst-case timing inaccuracy among all paths. We can break it into interboard skew, on-board interchip skew, and so on to the smallest timed component.

The challenge to designers of clock-distribution networks is how to control system clock skew so that it becomes an acceptably small fraction of the system clock period. As a rule, most systems cannot tolerate a clock skew of more than 10% of the system clock period. If system clock skew goes beyond the design limit, system behavior can be affected. Setup and hold times are missed, which results in long and short paths. No scheme is immune from these problems—even flip-flop machines can malfunction when clock skew is present.

Clock-powering trees, such as the one in Figure 7, are a source of clock skew. These trees are used to produce multiple copies of the clock signal for distribution. Each gate of the tree has some uncertainty associated with its delay, which is the difference between its best-case and worst-case delays. This difference is called *gate skew*. Using a worst-case timing analysis, the clock skew caused by a powering tree equals the arithmetic sum of the gate (and interconnection) skews on the path from the tree root to an output. In other words, clock skew has a cumulative effect by tree level. We can minimize this clock skew by placing all gates at a given tree level, or even the entire tree, on the same chip. In addition, we can realize elements at each tree level by using electrically matched devices and careful wiring.

Critical to efficient distribution is the clock-network layout—the physical placement of the network.

DISTRIBUTION TECHNIQUES

We must efficiently distribute the rectangular clock pulses produced through the interaction of the oscillator and clock-generation circuitry. Critical to efficient distribution is the clock-network layout—the physical placement of the network. It must conform to design rules that ensure the integrity of the clock signal by minimizing electrical coupling, switching currents, and impedance discontinuities. Other rules must prevent excessive clock skew by equalizing path delays and maintaining the quality of the signal edge. Symmetry and balanced loading at many levels of packaging, such as on the chip or on the board, are characteristic of effective clock-network layouts. To achieve these qualities, we can prearrange positions of the clock pins and make clock paths as short as possible.

It is sometimes difficult to coordinate the relative lengths of two paths that originate from a common source. To match any two paths or path segments in the clock system, we may need identical lengths of cable, wire, and interconnections; balanced loading; and equal numbers of buffer gates. A technique called *time-domain reflectometry* helps in this process by accurately measuring the line delays of long cables. In this process, line sig-

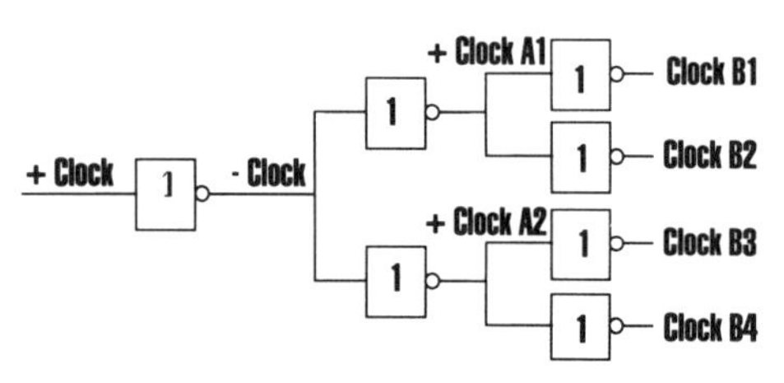

Figure 7. *Clock-powering tree.*

In practice, large systems distribute a small number of clock signals to each board or module.

nals are generated, and the signal reflections from line terminations are detected in real time. Once we measure the delays, we can equalize them by adjusting the lengths of the cables.

Duplicating the composition of two paths is not the only method of ensuring that two paths have equal delay. Another technique for matching different paths is called *padding.* In padding, we add delay elements to one or both paths. The Cray-1 uses extra interconnections and spare IC packages as padding, for example.

System designs often use a mixture of strategies. Designers might use duplication in subsections of the clock-distribution network with intermediate padding. Component screening also ensures that the performance of each system component is acceptable. Despite these techniques, some system clock skew is inevitable. However, as long as the timing analysis includes the effect of clock skew and determines that the system will function correctly, no other precautions are needed. If we detect that some system paths are failing because of excessive clock skew, then the clocks have to be tuned, or some part of the network has to be redesigned.

High-speed systems use a hierarchical structure to distribute clocks efficiently. A model for such a structure consists of *logic islands.* A logic island is a partition in the system, like a printed circuit board. Each island has a single point for clock entry, and all islands have the same line delay from their clock source, outside the island, to their clock entry point. (Tunable delay lines may precede the clock entry points.) We apply the same technique recursively to the islands themselves and to the subislands, such as chips, until we reach the individual clocked elements. Figure 8 is a diagram of the resulting structure, which resembles a star with the clock source in the center and the islands on the periphery. This model is remarkably similar to the physical organization of the Cray-1. In systems that need multiphase clocks, clock phases are generated at some convenient level in the hierarchy, and clock entry points at subsequent levels have one input for each clock phase.

In practice, large systems distribute a small number of clock signals to each board or module. Either the leading- or trailing-edge position of these clocks is very tightly controlled. When these signals reach board or module distribution, phase-generation circuits and clock choppers produce the final, well-controlled pulses. With this strategy, we can simplify clock distribution and reduce clock skew because only one clock edge has to follow a predetermined relationship to a reference clock. Using a single edge also minimizes clock skew by exploiting *common-mode action,* in which changes in power and temperature have similar effects across clock circuitry.

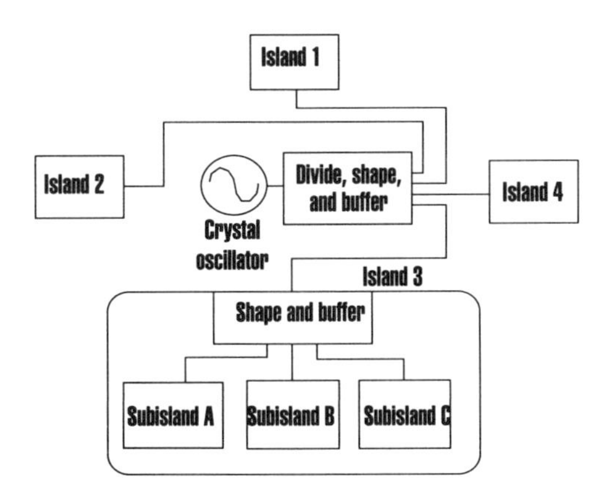

Figure 8. *Logic islands.*

Clock choppers should be close to the bistable elements that are the signal's final destination. Otherwise, asymmetric rising and falling delays in the buffer gates downstream of the clock chopper may shrink or stretch the pulse excessively. For instance, in the IBM 3090's two-phase double-latch design, the leading edge of the slave clock and trailing edge of the master clock are formed from a common input clock edge. These edges control the critical system path, so they must have minimal skew. Exploiting common-mode action in this way is also known as edge-tracking.

There are a number of simple guidelines for on-chip clock distribution to minimize asymmetries in clock-path delays and keep clock edges sharp. We balance and limit on-chip clock loading, using clock-powering trees, special buffers, symmetric layouts, and careful buffer placement. In very high speed ICs, we can select H-trees, which are symmetric, controlled-impedance clock-distribution trees composed entirely of metal. We can also reduce process-dependent clock skew in CMOS chips by adjusting clock-buffer FET parameters appropriately.

Most mainframes use gate arrays because designers need to trade off chip turnaround time with the large number of different chips required.

DESIGN DECISIONS

One of the first decisions in the design of the clock-distribution network is where to put square-wave production—the function that will produce the system clock—and where to put the function that will generate additional clocks. If a system requires many different clocks, we should distribute only a small number of clocks globally, and then generate the necessary clocks for each system section locally by manipulating (including decoding) these primary clocks.

In systems constructed from gate arrays or other semicustom chips, we cannot manipulate device characteristics individually. Most mainframes use gate arrays because the designers need to trade off chip turnaround time with the large number of different chips required. Both MOS and bipolar gate arrays have limited fanout of clock driver circuits. Whole chips or portions of chips are dedicated to controlling clock signals and buffering clock signals through powering trees (see Figure 7). The designers' intent is to produce as many copies of the clock signal as necessary to satisfy the loading requirements of the system. The more system bistable elements and associated chips and the smaller the amount of fanout allowed for each clock copy, the larger the number of required tree levels.

Figure 9 shows a simplified example of the clock-distribution network in the Amdahl 580. In this mainframe, which is based on bipolar gate arrays, the circuitry that produces the system clock is on the same card as the oscillator. Special chips on the console board then receive the system clock. These chips do clock gating and powering to provide the primary clock input for all boards. On the boards, tunable clock-distribution chips receive the primary clock and do clock chopping; derive the early, normal, and late clocks; and power these clocks sufficiently to satisfy loading requirements. The Amdahl 580 uses two ECL LSI clock-distribution chips to drive about 118 ECL LSI logic chips and static RAMs on each of its boards.

Custom systems and those constructed from catalog parts also use separate clock buffer chips when clock signals must drive large capacitive loads. Dedicated chips produce the clock copies that the system needs. Buffers on these chips must supply large

To reduce clock loading and allow for local clock gating, we can provide groups of bistable elements with their own small, local buffers.

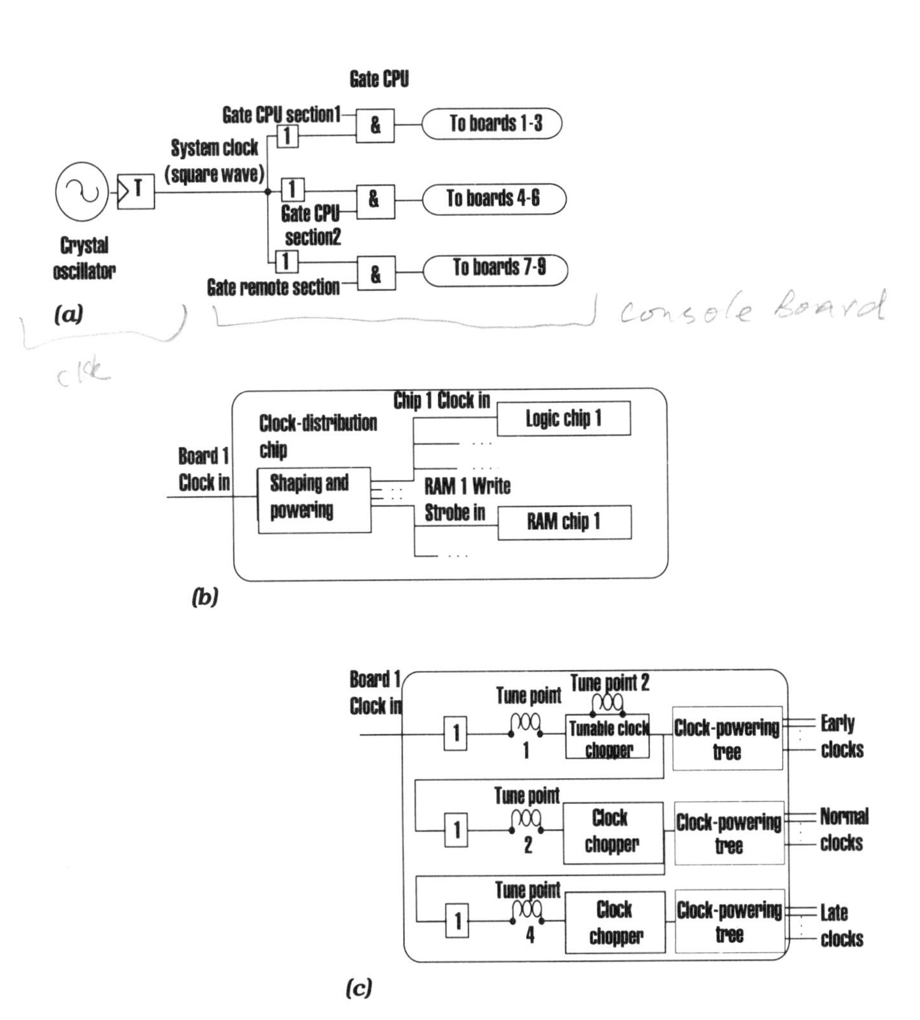

Figure 9. *Clock-distribution network with system clock gating; system-wide clock distribution* ***(a)****, board-clock distribution* ***(b)****, and clock-distribution chip* ***(c)****.*

currents, and accordingly, each buffer occupies a large area of each buffer chip. Buffer chips in the HP-9000 also generate clock phases.

In MOS custom chips or chip sets like the MC68000, we can produce clock phases on the chip and then buffer them immediately to provide adequate fanout. Most MOS single-chip systems use a single or cascaded clock buffer as a source for each distributed clock phase. Each buffer drives a very wide metal line (large load). The line composition and dimensions minimize the clock signal's power loss and voltage drop, while maximizing the speed of signal propagation. The wide metal line branches into groups of narrower metal lines which themselves may branch off, and so on. Nonmetal segments with a higher resistance, like polysilicon, should be avoided in clock-distribution lines. The distortion of the clock signal increases when it is propagated across such impedance mismatches. In these regions, clock phases may overlap and violate system timing constraints.

In single-level metal processes, power distribution has priority on the single metal layer available. Clock lines forced to cross the power lines cannot be run in metal. In the CMOS Bellmac-32A, for example, designers ran clock lines that crossed the power bus in a low-resistance silicide, and they kept the number of cross-overs to a minimum. To reach any clock load, they perform the same number of crossovers, which equalizes all the resulting path delays. In addition, they provided buffers at the crossovers to minimize these delays. In spite of the Bellmac 32-A's success, however, multilevel metal processes are a prerequisite to very high speed systems with low clock skew.

Some designers manage to avoid clock tuning by carefully designing and routing the clock network.

CLOCK GATING

Selectively deactivating the clock signal is called *clock gating*. Designers can use one of two types of clock gating, depending on the application. *Local clock gating* is a convenient way to implement many sequential circuits by locally deactivating the clock to a set of bistable elements, such as a register. To reduce clock loading and allow for local clock gating, we can provide groups of bistable elements with their own small, local buffers. The drive capabilities and numbers of on-chip clock buffers can be matched to the loading on their outputs. In *system clock gating*, the clock to an entire subsystem is deactivated. Figure 9a illustrates. In the Amdahl 580 and HP-3000, system-clock gating is done at the board level, while the Amdahl 470 does clock gating on the oscillator card. Gating signals must be valid through the entire active clock interval to prevent glitches on the gated clock line that could be sensed as valid clock pulses.

System clock gating is a tool for analyzing errors and recovering from them. It allows us to test the machine in a deterministic fashion by ensuring that a predetermined minimum number of clock cycles occurs after some machine stop condition. This type of testing makes it easier to isolate faults. For example, in the IBM 308X, the operator console deactivates the channel subsystem and logs out its contents through scan-out. It then scans in to reset the failed section. Amdahl 580 mainframes have separate console clocks, instruction and execution unit clocks, and I/O processor clocks, all of which are separately gated versions of a common, ungated system clock (also called a free-running clock).

CLOCK TUNING

High-speed computer systems with multiple boards and many chips on each board often require clock tuning after assembly. Clock tuning is calibrating the signals of the clock-distribution network. Some designers manage to avoid clock tuning by carefully designing and routing the clock network. Clock tuning during assembly and in the field is an expensive process, both in time and in the cost of the technical expertise. Because of this, designers must minimize the number of clock-tuning operations.

Tuning proceeds down the clock-distribution tree from clock source towards the clock destination.

We can determine how much clock tuning is needed by comparing clock signals probed at specified observation points. Modification of delays in the clock-distribution path then compensates for any significant inaccuracy in the clock-edge position or pulsewidth. Tuning can be manual, automatic, adaptively automatic, or a combination.

REFERENCE CLOCKS

To specify the placement of clock edges in a system, we designate one or more of the system's clocks as reference clocks. We use transitions of these timing references, or reference-clock edges, for comparison with other clock edges. We can specify the arrival time of a clock signal at any particular point in the clock-distribution network relative to these reference clocks.

TUNE POINTS

Tuning proceeds down the clock-distribution tree from clock source towards the clock destination. By tuning in this direction, we have the fewest number of tuning operations to calibrate the system because there is no backtracking. We can tune large components, such as printed circuit boards, separately to reduce the tuning requirements of the fully assembled system.

Tune points are the observation points in the clock-distribution tree where we can change the delay. Tuning methods to reposition clock edges include modifying wire lengths, selecting different taps in delay lines, or selecting one delay element from a set by controlling a multiplexer. We can tune clock choppers by adjusting their internal delay elements to control pulsewidth as well.

A tune-point hierarchy is embedded within the clock-distribution network. Figure 10 shows how we can use a tree structure to model this hierarchy. The level of the tune point in the tree is referred to as the depth in the tune-point hierarchy—the deeper the tune point, the farther away it is from the system oscillator. Chip or module primary I/O are usually the deepest accessible tune points.

Designing accessible and effective tune points may be difficult and their tuning resolution and range may be limited. One ap-

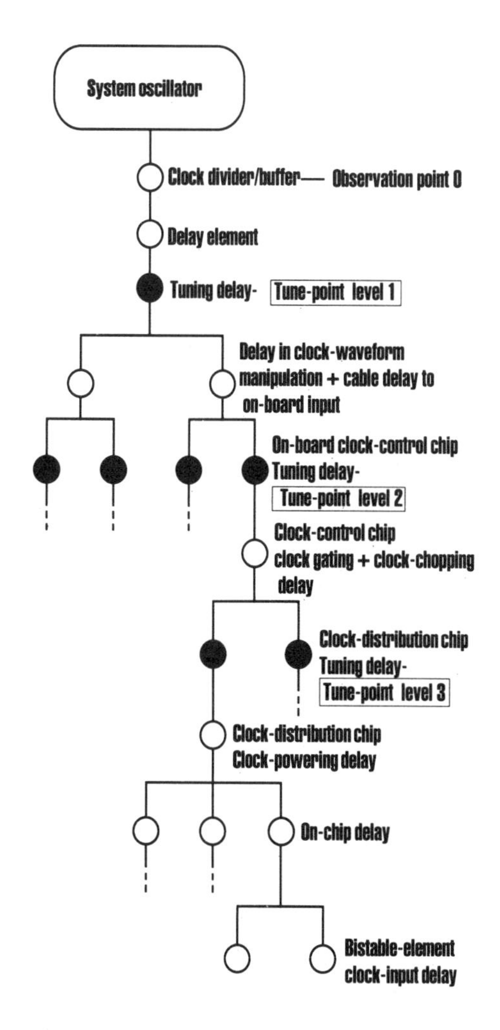

Figure 10. Tune points.

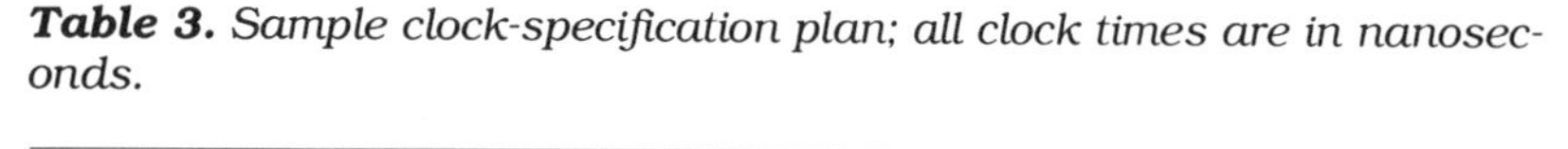

Table 3. *Sample clock-specification plan; all clock times are in nanoseconds.*

Parameters	Observation Point Level 0	Observation Point Tune-Point Level 1	Tune-Point Level 2	Tune-Point Level 3	Tune-Point Level 4
Local reference clock edge	0	+3.0	+7.0	+13.0	+20.0
Local tolerance	—	±0.5	±0.5	±0.5	±0.5
Effective clock arrival time	0	3.0±0.5	+7.0±1.0	+13.0±1.5	+20.0±2.0

proach to the problem is to determine the worst-case delay in a clock path and pad other clock paths to match this delay. For instance, in subsections of the clock-distribution network, one path may be significantly longer than all others and not have any tune points. By adding delay elements at tune points, we can pad all other comparable paths, either to have the same or shorter delay, which is fixed by design. A sophisticated physical design system can automatically design in this type of tuning by adding wire and capacitive elements as needed.

System timing is based on a clock-specification plan. The plan details the allowable ranges for clocks at each tune point, relative to a reference clock. We need to place tune points in such a way that the sum of the clock skew at the deepest tune point plus the additional skew for the signal to reach bistable elements beyond that tune point does not exceed the acceptable limit on system clock skew. Table 3 shows a sample clock-specification plan for Figure 10. The uncertainty is ±0.5 ns for each of 13 required tuning operations. In this plan, system clock skew equals 3.0 ns plus the maximum of the clock skews on the paths between the 12 tune points at Level 3 and the bistable elements that their clocks control.

We can provide extra observation points for clock tuning by distributing supplementary clock signals for use as precise reference clocks.

TUNING SCHEMES

Tuning schemes vary in sophistication. Manual tuning by a trained technician using an oscilloscope is common. The technician calibrates the tune points sequentially, starting from the one closest to the system clock source. We can provide extra observation points for clock tuning by distributing supplementary clock signals for use as precise reference clocks. No powering trees or other skew-increasing elements are allowed in the signal paths of these clocks, so they have little or no need for tuning themselves. We can thus use them safely and tune all other clock signals relative to the these references. This strategy increases tuning accuracy and decreases the number of tune points in the system.

Knowing the clock signal internal to the chip is often helpful in tuning. For MOS systems with multiple chips, on-chip clock buffering creates uncertain delays, so we must observe a representative internal clock signal. We can use this internal reference clock, which is output at a chip pin, as both a functional clock signal (for small loads), and as a reference to be compared with the corresponding references of other chips. We can then use the relative edge positions of the internal reference clocks to guide tuning.

In the clock-tuning schemes of the ETA 10 supercomputer and IBM 4341, designers provided two separate tuning resolutions, or tuning levels: rough tuning and fine tuning. Fine tuning is required for minute adjustments deep in the clock tree, while rough tuning provides the coarse adjustment earlier in the distribution and (in the ETA machine) before the system is immersed in coolant.

Untuned systems —designed with attention to component variations and to equalizing wire lengths and clock loading —eventually proved less expensive and entirely adequate.

A typical automatic tuning technique, devised for the discontinued STC CMOS mainframe, uses clock-distribution chips with many degrees of time-shifted clocks available through a large crossbar switch. In this case, a logic chip has four internal reference clocks. Each internal reference clock is a representative internal clock produced after chopping and powering one of four chip clock inputs. These representative clocks are compared with a precise reference clock. The correct skew-minimizing clock for each chip clock's primary input is then selected automatically from the crossbar switch. Registers on the clock-distribution chips are loaded to properly configure each crosspoint of the switch.

Automatic tuning often consists of closing a special feedback connection in a clock network that has an odd number of inversions in the signal path. When this connection is closed, we get oscillations with a period proportional to the total delay of the path. If necessary, we can adjust one or more tune points automatically through control signals determined by diagnostic code. Feedback tuning techniques seem attractive, and numerous patents exist for them, but their sensitivity to the clock duty cycle and the signal transition time make them complex and usually impractical.

Thus, more sophisticated tuning schemes are not necessarily better ones. Proposals for system clocking of the IBM 308X and 3090 mainframes required complex automatic tuning techniques. Eventually, untuned systems—designed with careful attention to component variations and to equalizing wire lengths and clock loading—proved less expensive and entirely adequate.

The clock system is an integral part of synchronous computers, yet it is not a widely studied aspect of their design. Early attention to system timing issues can provide benefits in system performance as well as product development time. The goal of the clock system designer is to control system clock skew at the system operating frequency as well as to minimize electrical hazards that may add undesirable components to the clock signals. Familiarity with clock-generation and clock-distribution techniques suitable for high-speed systems is essential as cycle times decrease. ■

ACKNOWLEDGMENTS

This tutorial was supported in part by the National Sciences and Engineering Research Council of the Government of Canada under its postgraduate scholarship program, in part by the National Science

Foundation under grant DCR-8200129, and in part by IBM Corp. The work was performed at Stanford University's Center for Reliable Computing and IBM.

I thank Edward McCluskey and Mark Horowitz of Stanford University, as well as Glen Langdon, Jr., and John DeFazio of IBM Corp., and Ron Kreuzenstein of Amdahl Corp. for their many helpful comments and suggestions.

ADDITIONAL READING

Bakoglu, H.B., J.T. Walker, and J.D. Meindl, "A Symmetric Clock Distribution Tree and Optimized High-Speed Interconnections for Reduced Clock Skew in ULSI and WSI Circuits," *Proc. IEEE Int'l Conf. on Computer Design*, 1986.

Domenik, S., "On-Chip Clock Buffers," *Lambda*, 1st qtr., 1981.

Friedman, E., and S. Powell, "Design and Analysis of a Hierarchical Clock Distribution System for Synchronous Standard Cell/Macrocell VLSI, *IEEE J. Solid-State Circuits*, Vol. SC-21, No. 2, 1986.

Glasser, L., and D. Dobberpuhl, *The Design and Analysis of VLSI Circuits*, chapt. 6, Addison-Wesley, Reading, Mass., 1985.

Hitchcock, R., Sr., "Timing Verification and the Timing Analysis Program," *Proc. Design Automation Conf.*, 1982.

IBM 3033 Processor Complex TO/DM, IBM Corp., Mechanicsburg, Pa., 1981.

Kogge, P., "Hardware Design and Stage Cascading," *The Architecture of Pipelined Computers*, chapt. 2, McGraw-Hill, New York, 1981.

Langdon, G., Jr., *Computer Design*, appendices C and D, Computeach Press, 1982.

Lob, C., and A. Elkins, "HP-9000: 18-Mhz Clock Distribution System," *HP J.*, Aug. 1983.

Maini, J., J. McDonald, and L. Spangler, "A Clock Distribution Circuit with a 100-ps Skew Window," *Proc. Bipolar Circuits and Technology Meeting*, 1987.

McCluskey, E.J., *Logic Design Principles: With Emphasis on Testable Semicustom Circuits*, chapts. 7-8, Prentice-Hall, Englewood Cliffs, N.J., 1986.

Seitz, C., "System Timing," *Introduction to VLSI Systems*, chapt. 7, C. Mead and L. Conway, eds., Addison-Wesley, Reading, Mass., 1980.

Shoji, M., "Electrical Design of the BELLMAC-32A Microprocessor," *Proc. Circuits and Computers Conf.*, 1982.

Shoji, M., "Elimination of Process-Dependent Clock Skew in CMOS VLSI," *IEEE J. Solid-State Circuits*, Vol. SC-21, No. 5, 1986.

Unger, S., and C.J. Tan, "Clocking Schemes for High-Speed Digital Systems," *IEEE Trans. Computers*, Vol. C-35, No. 10, 1986.

Wagner, K.D., *A Survey of Clock Distribution Techniques in High-Speed Computer Systems*, tech. rpt. 86-309, CSL Stanford Electronics Lab., CRC 86-20, Stanford Univ., Stanford, Calif., 1986.

Latch-to-Latch Timing Rules

ARTHUR F. CHAMPERNOWNE, LOUIS B. BUSHARD, JOHN T. RUSTERHOLZ,
AND JOHN R. SCHOMBURG, MEMBER, IEEE

Abstract—**The purpose of this paper is to present and analyze latch-to-latch timing rules that ensure the proper operation of synchronous systems.**

The new latch-to-latch timing rules state bounds on the amount of propagation delay for the combinational logic between consecutive latch pairs in a digital design. If the bounds are satisfied in a design, then setup and hold times are met throughout the design for each latch on each cycle of machine operation. The rules are quite general in that they apply to systems with multiple skew levels, with multiple latch parameters such as clock to data propagation delays, data to data propagation delays, setup and hold times, and with multiple clock pulse widths and clock phases.

The abstract notion of a clock skew hierarchy is introduced and characterized. The new rules apply to systems having a skew hierarchy, and most real systems have this characteristic.

The rules are established by a novel and robust method.

Index Terms—**Clock pulses, clocking, delays, digital systems, latch-to-latch, latches, multiple phase clocking, multiple skews, skew, skew hierarchy, timing.**

I. Latch-to-Latch Timing Rules

A. Introduction

THE typical process of logic design is such that it is desirable to have inequalities which, if true individually for each latch-to-latch path, will guarantee that the aggregate system does not fail due to the accumulation of effects such as skew, pulse widths, or long or short data propagation paths.

The following sections define the properties of the logic system and its components for which such rules are then proposed. This paper considers primarily the level-sensitive latch as opposed to the edge-sensitive version. Primary focus is data path, although the rule for a clock enable is also discussed. The paper treats both maximum and minimum timing concerns.

The approach that is used to prove the validity of the rules is novel in two ways. Previous work, [1], [4], [8], used a mathematical induction technique that looked across only one latch-to-latch decision, and on the first cycle of operation assumed that data were ready by setup time before the first occurrence of the trailing clock edge for each latch. This analysis is global in that it looks at delays across more than one latch-to-latch decision. Second, data are required to be stable at the input to each latch prior to the leading clock edge, but only for the first cycle of operation.

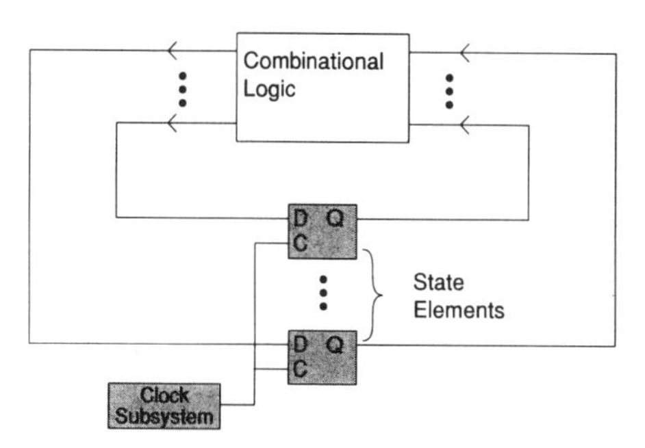

Fig. 1. Synchronous system fundamentals.

Most importantly, this work utilizes the concept of a hierarchy of clock skews. Such a notion is derived from observing that actual clock distribution systems typically exhibit this characteristic, due to the hierarchical nature of the clock distribution logic. This characteristic is exploited through the novel approach described above to obtain a set of design rules that are more general than those of previous work, and hence more generous in many applications.

B. Notation

Various temporal quantities will be defined in the paper and their values will be denoted by particular letter combinations, as in "pW" for pulse width. Values of these quantities will in some instances be estimated by "worst case" values. The worst case maximum values will be indicated by use of uppercase letters in the particular letter combination, as in "PW," and worst case minimum values will be indicated by use of lower case letters, as in "pw."

C. System Components

A synchronous digital latch system consists of a finite number of three types of components:

1) combinational logic,
2) latches (the state elements), and
3) a clock subsystem.

More complicated objects such as RAM's, ROM's, and register files generally can be modeled as collections of the first two types.

A textbook example of such a system is presented in Fig. 1. However, no system inputs or outputs are shown. Although the output of the clock subsystem is drawn as a single line, we do not rule out the possibility of multiple phases. The nature of a multiphase clock will be discussed later.

Synchronization to the system clock of primary inputs is a

Reprinted from *IEEE Trans. Comput.*, vol. 39, no. 6, pp. 798–808, June 1990.

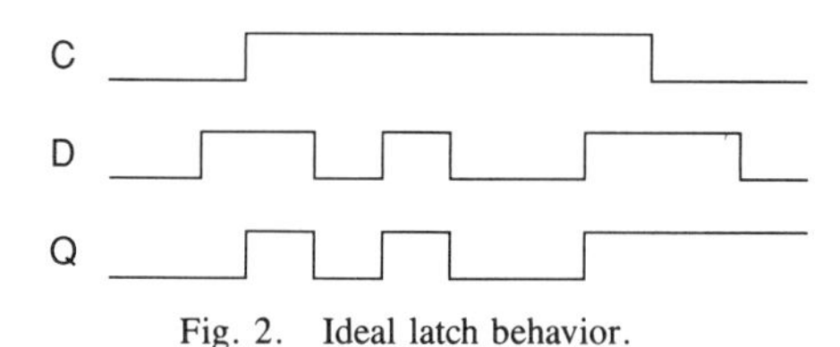

Fig. 2. Ideal latch behavior.

subject outside the scope of this paper. For our purposes it is sufficient simply to assume that they are synchronized to the system clock in some unspecified manner such that their stability characteristics are equivalent to internal signals.

1) Combinational Logic: Combinational components are devices which transform a set of input values to a set of output values according to a well-defined function. Combinational components have no memory, but instead respond directly (with some finite delay) to changes in their inputs.

For purposes of this paper, aggregates of interconnected combinational components (without feedback) are not distinguished from single components, since we are interested only in input-to-output paths through combinational logic.

2) Latches: In its simplest form the latch is a logic device having two inputs, a data input (*D*), and a clock input (*C*), and one output (*Q*). In practice there may be more terminals, but only these are needed for the current discussion.

The behavior of an ideal latch is such that if a logic one is applied to the clock input, the output value becomes the same as the data input value, changing whenever the data input value changes; if a logic zero is applied to the clock input, the output remains constant regardless of the value at the data input. This is illustrated in Fig. 2.

For real latches the polarity of the clock input may be inverted from the definition given here. In order to cover both cases, the clock input is said to have an active state, which is the value for which the output follows the data input, and an inactive state, for which the output value is held constant.

Therefore, the ideal latch behavior may be restated more generally as follows: when the clock input is active, the output *Q* of the latch follows the data input *D* of the latch. When the clock input is inactive, the output retains the value it had at the last instant the clock was active.

The value reflected at the latch output when the clock input is held inactive is called the state of the latch. Note that an active clock input does not itself cause the latch to change state, it only allows it to change.

In reality latches have nonzero delays in the response of the output to input changes, and there will be constraints on the behavior of the inputs. The following times may be defined for the real latch:

DQ, dq — The maximum and minimum times for propagating a change from the data input to the output while the clock is held active. (In case a zero–one transition propagates at a different rate from a one–zero transition, we take DQ to be the larger time, and dq to be the smaller.)

CQ, cq — The maximum and minimum delays from the leading (active-going) edge of the clock input until the input data value appears at the output, while the data input is held unchanged.

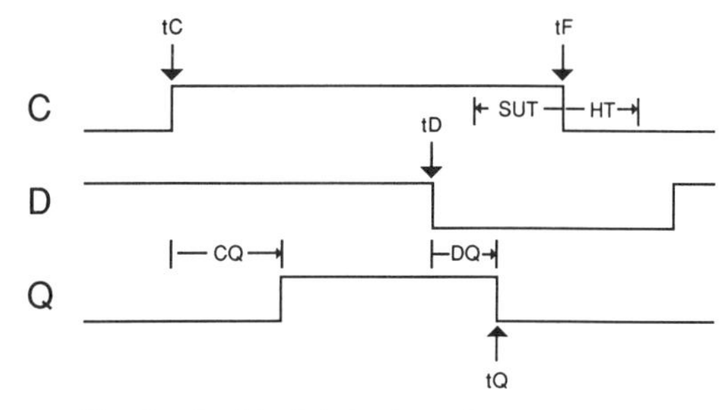

Fig. 3. Real latch behavior and parameters.

SUT — SetUp Time. The time prior to the trailing edge of the clock by which the data input must be stable in order to ensure it will be correctly latched.

HT — Hold Time. The time following the trailing edge of the clock before which the data input must be held stable in order to ensure its value will be correctly latched.

The interval between the setup time and the hold time for any clock pulse at any latch is known as the SUT_HT window. For correct operation of the latch, the data input of the latch must remain stable during this interval. If input data are allowed to change during this interval, then the behavior of the latch state and output in future time are unpredictable and unreliable, and oscillation is possible [1]–[3]. Individually SUT or HT may be negative, but the sum SUT + HT, which is the width of the SUT_HT window, must be zero or positive.

These times are illustrated in Fig. 3.

The clock input of each latch is connected by some means to the clock subsystem circuitry, which provides the latch with a stream of clock pulses. At the clock input of the latch the following times may be defined:

tC — The actual time of arrival of the leading edge of a clock pulse.

nTF — The nominal time of arrival of the trailing edge of a clock pulse.

tF — The actual time of arrival of the trailing edge of a clock pulse.

PW, pw, pW — Pulse Width. The maximum, minimum, and particular intervals of time from the leading edge of a clock pulse to its trailing edge.

It would be possible to define earliest and latest arrival times for the clock edges, (tc, TC, tf, and TF). Instead, we define a nominal time for the trailing edge, nTF, which is surrounded by a window of possible actual times, tF. The width of this window is called skew, and will be defined later. The leading edge of a clock pulse is related to its trailing edge by the pulse width ($\mathrm{pW} = tF - tC$).

It is characteristic of the latch that some minimum clock pulse width is required for correct latch functioning. Violation of the minimum pulse width may cause the latch to malfunction in ways similar to violation of the SUT_HT window, for similar reasons. In this paper, as in [8], it will be assumed that the system clock pulses are always of sufficient width to ensure correct operation, and that all latches are free of hazards.

The choice of the trailing edge, rather than the leading edge

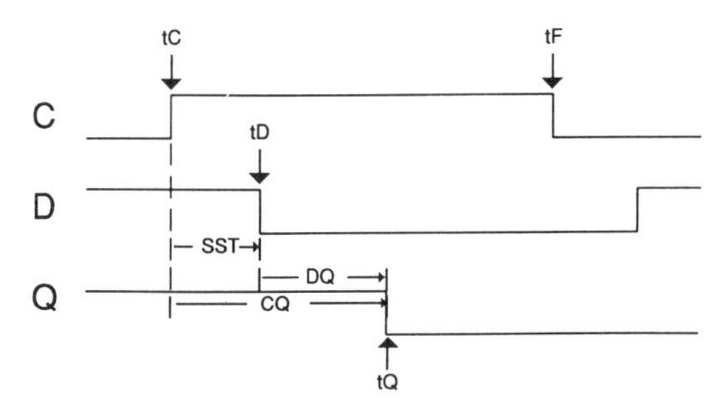

Fig. 4. SST for a real latch.

or center, as the reference point for clock pulses is arbitrary. It is selected because of its familiarity to the authors, and mathematical convenience. Choice of the other forms should lead to similar conclusions.

Fig. 4 introduces a new parameter, SST. It is called the "Sample Start Time," and is simply the difference $CQ - DQ$ as illustrated in the figure. This difference is typically positive but could be negative since the values of CQ and DQ depend upon the latch design. In effect, the data input D of the latch is not sampled until a period of time equal to SST after the leading edge of the clock pulse. The sample start time is also known as the "leading edge setup time," or the "unlatching time."

An alternate, mathematical expression of latch behavior is given in the equation

$$tQ = \max(tC + CQ, tD + DQ),$$

where tD is the time at which the input data D changes to its desired state, and tQ is the time at which the latch output Q changes to the desired state. This is the model introduced in [8]. These times are indicated in Figs. 3 and 4. In Fig. 3, $tD + DQ$ is dominant in the max, and determines tQ, and in Fig. 4 both expressions in the max are equal. In Fig. 4, if tD (the $1 \to 0$ transition of D) was moved any earlier in time and if Q was initially 1, then Q still could not change to 0 any earlier than the time tQ illustrated in Fig. 4 since $tC + CQ$ would then dominate the max to determine tQ.

It will be convenient to place latch operation modes into two categories. One is the clock controlled mode of operation. This mode occurs when $tD < tC + \text{SST}$, in which case $tQ = tC + CQ$. The other is the data controlled mode and in this mode $tD > tC + \text{SST}$, in which case $tQ = tD + DQ$. These two modes of operation are very important in the system definition given later as well as in the proof of the rule correctness.

D. Clock Definitions

The collection of circuitry that drives the clock inputs of all latches in a system is called the clock subsystem, or simply the clock. Depending upon the context, the term clock may also refer to the collective characteristics of the signals from the clock subsystem. In this paper, we are concerned only with the characteristics of the clock signals at the points where they enter the clock inputs of latches. Therefore, terms such as clock pulse, leading edge, trailing edge, active, and inactive take their meaning from the definitions given above for the clock inputs of latches.

The clock may be single phase, meaning it nominally appears at each latch of the system at the same time, or it may be multiple phase. In multiple phase systems, the fundamental clock signal is split into several versions, with each version effectively delayed by various fixed fractions (less than one) of the period. Each delayed version is called a phase, and the phases are distributed to the latches in the system with each latch getting one of the clock phases.

In Fig. 5, an illustration of the characteristics of a four-phase clock is given. The phases are usually numbered in an ascending sequence, such as 1–2–3–4 in the figure. The nominal phase separation time between an ordered pair of phases is defined as the nominal time difference between occurrences of the same event (e.g., trailing clock edges) from the first of the pair to the second, modulo the cycle time. In Fig. 5, the phase separations, denoted by S, all happen to be equal, but in general this need not be the case.

The multiple-phase clock may be completely specified by giving the phase separations between each adjacent pair of phases and the pulse widths of each phase. Note that the period of each phase is the same (T), and is equal to the sum of the phase separations. Thus, for a single-phase clock the phase separation is the same as the system clock period (cycle time).

The above defines the phase separation of the clock signals themselves. Assuming each latch is attached to exactly one of the clock phases, we can also speak of the phase separation between clocked latches. Given a combinational logic path from one latch to another, there is a nominal time from the clock pulse of the source latch to that of the destination latch. When logic is designed for this path, it is generally assumed that a signal from the source latch will be propagated in such a manner as to arrive at the destination latch in time to be captured on the next consecutive clock pulse of the destination latch's phase. In some cases, it may be desirable to have the signal captured not on the next consecutive destination phase, but on some subsequent one. In such cases the timing rules must take this into account.

In some multiphase systems such as described in [8], restrictions are placed on the selection of phases for the latches at the source and destination of logic paths. No such restrictions are required for the rules in this paper.

The above discussion of single and multiple phase clocks is provided to supply background for understanding the general application of latch-to-latch timing rules. The rule presented in Theorem 2 below is actually of a more general nature, not constrained to the types of clocks illustrated in Fig. 5. However, most applications will be of this type.

1) Clock Pulse Width and Skew: In a practical synchronous system, the clock phase(s) must be delivered to a great number of latches, requiring driver circuits to provide the clock signals. Differences in the delays through this circuitry may cause the arrival times of the clock signals to vary among latches, or possibly at the same latch over time.

The width of a clock pulse, pW, is the difference between its leading and trailing edges ($tF - tC$). At each latch s, we assume bounds of minimum pulse width, pw(s), and maximum pulse width, PW(s). The values of pw and PW are typically not observed measurements, but are engineering estimates,

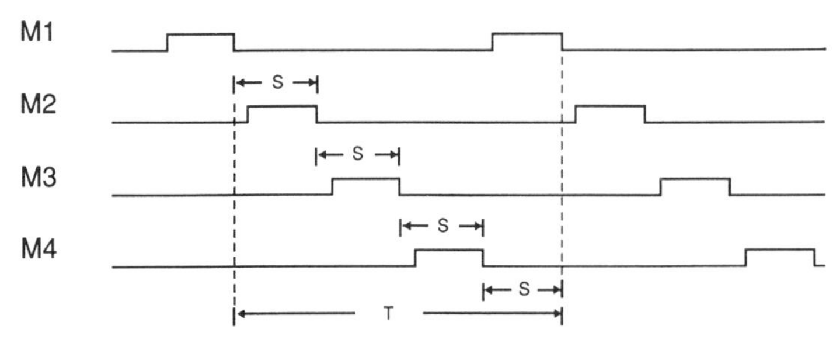

Fig. 5. Multiphase clock.

made in advance, for purposes of designing the logic of the machine.

Clock skew, or simply skew, is similarly defined as the variation in position of clock pulses. Because of its importance to latch operation, we select the trailing edge of a clock pulse as the reference. Mathematically skew is defined as follows:

Definition 1: Skew: For any pair of latches the skew value between their clock inputs is defined as

$$\mathrm{sK}(s, d, i, j) = \mathrm{ABS}((tF(s, i) - nTF(s, i)) - (tF(d, j) - nTF(d, j)))$$

where s and d are latches, i and j are positive integers, and the pairs s, i and d, j refer to the ith clock pulse at s and the jth clock pulse at d, respectively.

Essentially it is the residual separation between two trailing edges after subtracting out their nominal (intended) separation.

Again, for design purposes, it is reasonable to utilize engineering estimates for the bounds of skew. In the case of skew, only an upper bound is of practical use, and it may be any number $\mathrm{SK}(s, d)$ such that

$$\mathrm{SK}(s, d) \geq \mathrm{sK}(s, d, i, j) \qquad \text{for all } i \text{ and } j.$$

Since both the pulse width and pulse position may vary, it may be helpful to clarify how these are measured. Fig. 6 illustrates skew and pulse width for a pair of latches having the same nominal clock time.

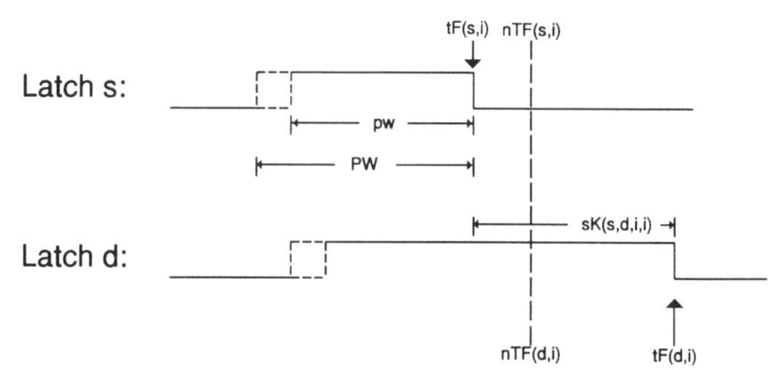

Fig. 6. Illustration of pulse width and skew.

It is important to note that for our purposes the pulse width is not constrained by the time of occurrence of the trailing edge of the clock. In particular, the largest pulse width can occur simultaneously with the earliest trailing edge, and the smallest pulse width with the latest trailing edge.

Note also that skew is the positive difference between two positions, rather than being plus or minus from some reference point. When using this information in a design, we do not know whether the clock pulse to a particular latch is skewed early or late relative to that for any other particular latch. We know only a bound on the absolute value of the largest possible difference between their trailing edges. Therefore, worst case design must consider both cases: a particular latch must be considered early when checking maximum delays to its inputs, and must be considered late when checking maximum delays from its outputs.

2) Skew Hierarchy: Practical digital systems are constructed from a variety of forms of physical packaging. For example, we may have chips residing on chip-carriers residing on daughter-boards residing on mother-boards residing on back-planes residing in racks residing in cabinets. In most cases, these levels of packaging can be seen to have the form of a hierarchy. Obviously not all systems have all of the levels mentioned above, nor are all packaging systems necessarily hierarchical. Nonetheless, most do fall into this category.

Also, practical digital systems are commonly synchronous. Implementation of a practical synchronous system is usually done by distributing clock signals from a single master oscillator. The clock distribution takes the form of a fan-out tree. Typically the clock fan-out follows the hierarchy of physical packages, with levels of buffering added to provide the needed fan-out at each level. Because each additional level of buffering increases the skew of the clock signal, the fewer levels of buffering encountered within the clock fan-out tree between any two latches, the less skew there is likely to be between these latches.

For example, a system may consist of some number of circuit boards, each with some number of chips on them. This system has three levels of packaging hierarchy: 1) system, 2) board, 3) chip. Assume the clock signal is distributed from a common point with fan-out buffers at each of these levels. It should be clear that SK values within a board can be made less than SK values between boards, since there are fewer levels of clock buffering and interconnect to cause skew.

Because skew is so detrimental to designing fast systems, designers desire to take advantage of situations that allow smaller SK values to be used. Therefore, rather than using only a single global SK value valid between any two latches in a system, engineers prefer to use the smaller SK values for paths contained within a board, or the even smaller SK values for paths contained within the same chip. A system designed with the single value of system skew, SK, would work, but would be suboptimal.

In addition to establishing that SK values become smaller as we descend the clock distribution hierarchy, it is also in-

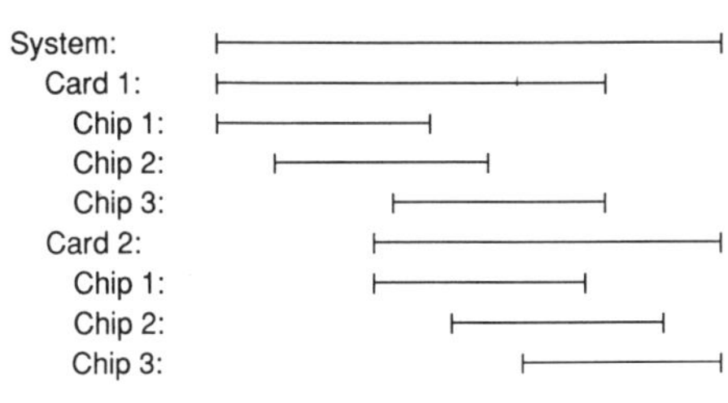

Fig. 7. Example of possible tF skew relationships.

structive to note the manner in which SK values are related among the various hierarchy levels. Because SK is defined as the maximum possible variation in clock position for the set of latches being considered, taking subsets of the latches leads to SK ranges which are no wider than the range of any containing set. This leads to viewing these relationships themselves as a form of hierarchy.

As an example, consider a single phase system consisting of two cards on a backpanel, and three chips on each of the cards, and assume that the clock distribution follows the packaging hierarchy. Fig. 7 suggests how the trailing edge positions (modulo the period) within this system might be related, and consequently, how the skew values at the various hierarchy levels would be related.

In this figure, the horizontal axis represents the time domain, and each line represents the range of possible positions for the SAME trailing clock edge. Thus, if there were no skew, all lines would collapse to points at the same t coordinate.

A formal definition of a skew hierarchy is now stated.

Definition 2: Skew Hierarchy: A skew hierarchy for a system is a collection of sets of latches in the system. These sets are called nodes of the hierarchy. With each node A is associated a number $\mathrm{NSK}(A)$ which bears the following relation to SK: if s and d are (possibly identical) latches of the system and A is the smallest node which has both s and d as elements then $\mathrm{SK}(s, d) = \mathrm{NSK}(A)$. That is, $\mathrm{NSK}(A)$ serves as the upper bound, $\mathrm{SK}(s, d)$, for all latch pairs, s, d, in A. The nodes are subject to the following constraints:

If A and B are nodes, then either B contains A, or A contains B, or A and B are disjoint. (1)

The set of all latches in the system is a node. (2)

If node A contains node B, then $\mathrm{NSK}(A) \geq \mathrm{NSK}(B)$. (3)

$\mathrm{NSK}(A)$ is intended to be the maximum skew value possible between any pair of latches in the node A.

The following theorem and corollary will be quite useful later in the timing rule analysis.

Theorem 1: Triangular Skew Theorem: If x, y, and z are any three latches in a skew hierarchy, then

$$\mathrm{SK}(x, z) \leq \max(\mathrm{SK}(x, y), \mathrm{SK}(y, z)). \tag{4}$$

The latches do not have to be distinct.

Proof: For any pair of latches v and w, let $N(v, w)$ be the smallest node containing v and w. Then, by definition,

$$\mathrm{SK}(v, w) = \mathrm{NSK}(N(v, w)).$$

$N(x, y)$ and $N(y, z)$ both contain y. So by (1) either $N(x, y)$ contains $N(y, z)$ or $N(y, z)$ contains $N(x, y)$. Without loss of generality, assume that $N(y, z)$ contains $N(x, y)$. The relation

$$\max(\mathrm{SK}(x, y), \mathrm{SK}(y, z)) = \mathrm{NSK}(N(y, z))$$

follows from (3). Now $N(y, z)$ also contains $N(x, z)$ because x and z are in $N(y, z)$. It follows from (3) that

$$\mathrm{NSK}(N(x, z)) \leq \mathrm{NSK}(N(y, z)),$$

from which the theorem follows.

Corollary: For any sequence of latches, $L(1)$, $L(2), \cdots$, $L(n)$, $n \geq 2$, in a system with a skew hierarchy, there is some pair of consecutive latches within the sequence having a skew value at least as large as the skew between the two end latches, i.e., there is some k, $1 \leq k < n$, such that

$$\mathrm{SK}(L(1), L(n)) \leq \mathrm{SK}(L(k), L(k+1)).$$

The latches, $L(1), L(2), \cdots, L(n)$, do not have to be distinct, nor do they have to be connected, other than through the clock subsystem.

Proof: For $n = 2$ choose $k = 1$. For $n = 3$ Theorem 1 applies directly. For $n > 3$, we proceed by induction. Suppose that the corollary holds for $n-1$. It follows that we can choose a j, $1 \leq j < n-1$, such that

$$\mathrm{SK}(L(1), L(n-1)) \leq \mathrm{SK}(L(j), L(j+1)).$$

By (4)

$$\begin{aligned}\mathrm{SK}(L(1), L(n)) &\leq \max(\mathrm{SK}(L(1), L(n-1)), \mathrm{SK}(L(n-1), L(n))) \\ &\leq \max(\mathrm{SK}(L(j), L(j+1)), \mathrm{SK}(L(n-1), L(n))),\end{aligned}$$

and so one of the values $k = j$ or $k = n-1$ satisfies the corollary. Hence, the corollary holds for n.

The converse of the Triangular Skew Theorem can also be proved [9], but it is not needed for the derivation of the latch-to-latch timing rules.

Note that the notion of skew hierarchy derives from the tree (hierarchical) form of the clock distribution network, rather than from the packaging hierarchy. In practice, the two generally correspond. If, in some case, they do not correspond, the skew hierarchy concept is still valid, but care must be taken to apply it according to the clock distribution tree.

It should be apparent that previous work in this field that did not utilize a skew hierarchy is not contradicted by this notion, since it represents the special case of a degenerate hierarchy having only one node.

E. The Timing Rule Problem

In order for a synchronous system to operate reliably, all setup and hold times for all system latches must be met under all conditions. This requires controlling both the maximum and minimum delays along the latch-to-latch logic paths.

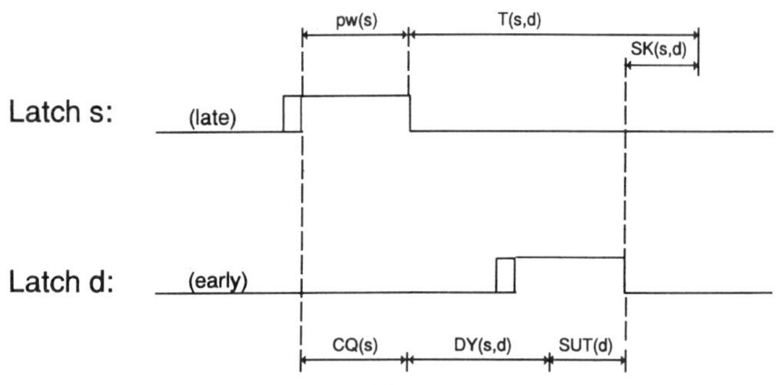

Fig. 9. Illustration of first expression of MAX.

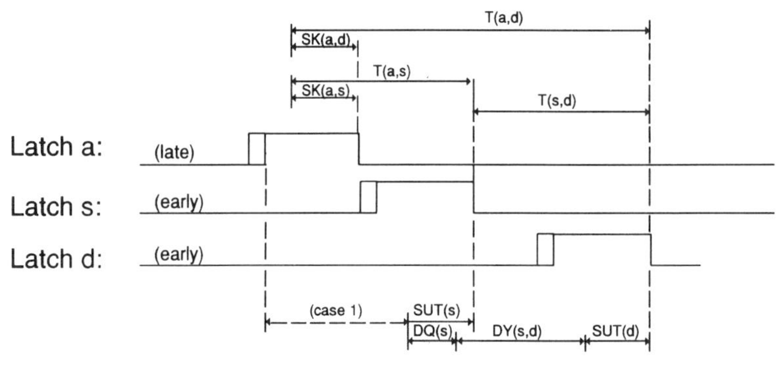

Fig. 10. Illustration of second expression of MAX.

minimum of three expressions. The three cases corresponding to these expressions are illustrated in Figs. 9–11. For sake of simplicity let us use $T(s, d)$ to represent the nominal clock separation between latch s and latch d, instead of $nTF(d, k) - nTF(s, j)$, dropping the j and k since only one pulse of each is drawn.

In the first case the clock for s is skewed as late as possible, and the clock for d is skewed as early as possible. Then the difference between the two trailing edges is $T(s, d) - \mathrm{SK}(s, d)$. Given that the data into latch s is stable by the latest leading clock edge plus SST, then it can be seen from the figure that

$$CQ(s) + DY(s, d, p) + \mathrm{SUT}(d) \leq \mathrm{pw}(s) + T(s, d) - \mathrm{SK}(s, d)$$

in order for data to be correctly captured at latch d. Rearranging terms and substituting $DQ(s) + \mathrm{SST}(s) = CQ(s)$, yields

$$DQ(s) + DY(s, d, p) - T(s, d) \leq \mathrm{pw}(s) - \mathrm{SST}(s) - \mathrm{SK}(s, d) - \mathrm{SUT}(d)$$

which is the first case of the MAX rule.

The second case considers three latches, a, s, and d, with the first skewed as late as possible, and the other two both skewed as early as possible. In other words, all of the skew appears between the first two latches leaving no skew between latches s and d. Therefore, their trailing clock edges are separated by exactly $T(s, d)$. This is clearly possible using a single skew value, and can be shown to be the worst situation for this case when the three latches are distributed in a skew hierarchy.

The path from a to s is governed by the results of the first case above, and we now consider the requirement to set up the third latch correctly. In this case, data become stable at the input to s as late as $\mathrm{SUT}(s)$ before the trailing edge, so for correct operation of latch d the figure shows that

$$DQ(s) + DY(s, d, p) + \mathrm{SUT}(d) \leq \mathrm{SUT}(s) + T(s, d)$$

which can be arranged to give the second expression in MAX:

$$DQ(s) + DY(s, d, p) - T(s, d) \leq \mathrm{SUT}(s) - \mathrm{SUT}(d).$$

The third case considers three latches, s, d, and b. This time the first two are skewed as late as possible and the third is skewed as early as possible. (This is the mirror image of case 2.) As a result there will be no skew between the first two latches s and d. In this case, the goal is to determine the condition on the path from s to d such that the path from d to b will be covered by the first case above. This means that data must be stable at d by the latest leading edge of d's clock pulse plus $\mathrm{SST}(d)$. From the figure, this is seen to be

$$CQ(s) + DY(s, d, p) + DQ(d) - CQ(d) + \mathrm{pw}(d) \leq \mathrm{pw}(s) + T(s, d).$$

Using the relation $CQ = DQ + \mathrm{SST}$ twice and rearranging terms, we get

$$DQ(s) + DY(s, d, p) - T(s, d) \leq \mathrm{pw}(s) - \mathrm{pw}(d) + \mathrm{SST}(d) - \mathrm{SST}(s)$$

which is the third term of MAX.

In [9] three examples are constructed in a two skew system which show that if any one, but only one, of the three terms of MAX is replaced by zero, then a setup failure results.

There is only one case to be considered for the MIN inequality. The task is to prevent data into d from changing before hold time has expired. The worst situation is clearly

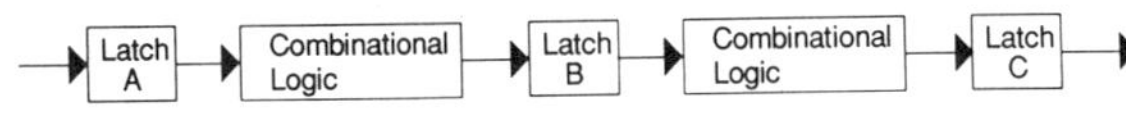

Fig. 8. Example latch path.

The behavior of a latch output in following its data input when its clock input is active is called "latch fall-through." Latch-to-latch timing rules must account for this property, which allows data arriving at the input of a latch during a clock pulse (between SST and SUT) to fall through the latch as if it were a simple combinational delay with a value of DQ.

For example, Fig. 8 shows a path from latch A through latch B to latch C. To ensure proper operation of these latches, not only must the maximum and minimum timing from A to B and from B to C be controlled, but if there is a possibility that the signals from A could arrive at B during the clock pulse, then the timing from A through B (using DQ of B) to C must also be controlled. The path from A to C in this example is called a fall-through path.

It is clearly impractical directly to check all possible fall-through paths between every pair of latches in a system for timing rule violations.

The purpose of this paper is to present timing rules which, when applied only to consecutive latch pairs, have the effect of guaranteeing that the system will not malfunction due to the effects of latch fall through. In the example of Fig. 8, this would mean that once the paths from A to B and from B to C were designed to this criterion, the A–B–C path would necessarily function correctly also.

The challenge is to obtain timing rules that have the property of being applied locally, while allowing the designer the greatest range of design freedom. We believe that the timing rules presented below meet this challenge with a greater degree of generality than any previously published, particularly those in [1], [4], [7], and [8].

The remainder of this paper consists primarily of presenting these rules, and proving their sufficiency. The data path timing rule will be presented first, and its proof given. The enable path timing rule follows it.

F. The New Timing Rule

Before stating the timing rule, it is necessary to define more carefully the conditions under which it applies.

1) Additional Definitions:

Definition 3: Synchronous Pulse Conditioned Latch System: A synchronous pulse conditioned latch system consists of a clock subsystem, combinational logic components, latches, system inputs, and system outputs. These may be interconnected subject to the following constraints.

1) Signals from the clock subsystem may not drive the data inputs of latches either directly or through combinational logic. That is, they may drive latches via only their clock inputs, whether directly or through combinational logic.

2) For some time before the first leading clock edge is applied at any latch clock input, the system is stable, i.e., no signal is changing. Specifically the required amount of time is

$$-\min(\mathrm{SST}(i)) \qquad \text{for all latches } i.$$

(The negative of the most negative SST.)

3) Each primary input to the system is controlled so that, for any clock pulse at any latch, the system input does not cause a change at the data input of the latch during the interval of time that starts at the sample start time after the leading edge of the clock pulse and ends at the hold time after the trailing edge of the clock pulse.

Essentially, a synchronous pulse conditioned latch system has the feature that all latches are clock controlled before a certain time throughout the system, and for all time at the system inputs.

Definition 4: Adjacency, $\langle s, d, p\rangle$: A pair of latches s and d are called adjacent if there is a combinational logic path from the output of latch s (the source) to the data input of latch d (the destination).

An adjacency is a pair of adjacent latches together with a particular path p between them. This will be written $\langle s, d, p\rangle$.

In the example of Fig. 8 latches A and B are adjacent, and latches B and C are adjacent, but A and C are not adjacent.

Definition 5: Nominal Time of Fall, $nTF(l, i)$: For any latch l let $nTF(l, i)$ be the nominal time of the ith trailing edge to arrive at latch l. If there is no such edge, $nTF(l, i)$ is undefined.

Definition 6: Propagation Delay, dY: For any adjacency $\langle s, d, p\rangle$ let $dY\ (s, d, p)$ be the propagation delay from the output of s to the data input of d along the path p.

2) Statement of the Rule:

Theorem 2: Latch-to-Latch Timing Rule: Consider a synchronous pulse conditioned latch system with a skew hierarchy. Let $\langle s, d, p\rangle$ be any adjacency in the system, and let j and k be positive integers such that there is a jth clock pulse at s and a kth clock pulse at d. If for all such cases one of the following inequalities holds:

$$\begin{aligned} DQ(s) + DY(s, d, p) + nTF(s, j) - nTF(d, k) \\ \le \min(\mathrm{pw}(s) - \mathrm{SST}(s) - \mathrm{SK}(s, d) - \mathrm{SUT}(d), \\ \mathrm{SUT}(s) - \mathrm{SUT}(d), \\ \mathrm{pw}(s) - \mathrm{SST}(s) - \mathrm{pw}(d) + \mathrm{SST}(d)) \end{aligned} \qquad (\mathrm{MAX}(s, d, p, j, k))$$

$$\begin{aligned} cq(s) + dy(s, d, p) + nTF(s, j) - nTF(d, k) \\ \ge \mathrm{PW}(s) + \mathrm{SK}(s, d) + \mathrm{HT}(d) \end{aligned} \qquad (\mathrm{MIN}(s, d, p, j, k))$$

then no change at any latch data input can occur during any SUT_HT window.

The first inequality, called MAX, is sometimes referred to as the "long-path" calculation, since it provides the upper bound DY. Similarly, the MIN inequality is sometimes known as the "short-path" calculation, providing permissible lower bound dy. The parameterized forms will be used below in the proof of the theorem.

Before giving the formal proof of the theorem, some informal illustrations will be used to aid in its understanding.

3) Intuitive Illustration of the Rule: Theorem 2 contains two inequalities, MAX and MIN. The MAX timing rule gives an upper bound on the allowable delay for the logic path between two adjacent latches. This bound is in the form of the

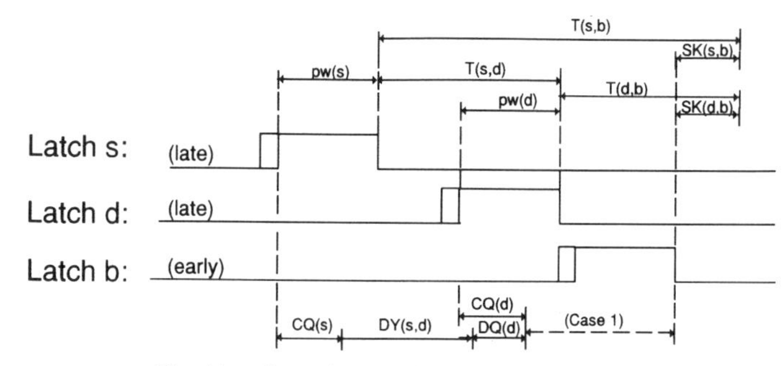

Fig. 11. Illustration of third expression of MAX.

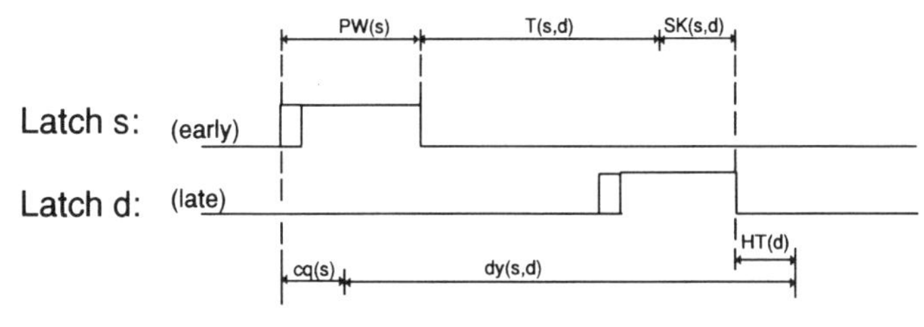

Fig. 12. Illustration of MIN.

when the clock pulses of s and d are separated by the greatest amount of time. From Fig. 12 it can be seen that

$$cq(s) + dy(s, d) - \mathrm{HT}(d) \geq \mathrm{PW}(s) + T(s, d) + \mathrm{SK}(s, d)$$

which can be rearranged into the form of MIN.

It is the use of the Triangular Skew Theorem as applied to a skew hierarchy that makes it possible to prove (as done in Theorem 2) that no other cases need to be considered.

4) Formal proof of the Theorem:

Proof of Theorem 2: Proof is by contradiction. Suppose there are some occasions on which a latch data input changes during a SUT_HT window. Choose the earliest such occurrence. If there is more than one earliest occurrence pick one of them. Call the latch d and let the clock pulse whose SUT_HT window was violated be the kth clock pulse at latch d.

The following discussion relates to events which occurred no later than the first time a latch data input changed during a SUT_HT interval. During this initial period no latch could have failed due to a SUT_HT window violation.

Notice that during this time period no combinational component or latch can generate a change of itself (i.e., any change at an output must be traceable to a change at one of the inputs with a characteristic delay).

Trace the offending change backwards through combinational components and data controlled latches. By definition of a synchronous pulse conditioned latch system, this process cannot continue indefinitely and must eventually reach the clock input of a clock controlled latch. Let the first such clock input reached be at latch s and let the clock be the jth clock pulse at s. Latch s cannot be in a failure mode. Therefore, the origin of the offending change was the leading edge of the jth clock pulse at latch s.

Reversing the sequence of the above trace yields a fall-through path $\langle l[1], l[2], p[1]\rangle$, $\langle l[2], l[3], p[2]\rangle \cdots \langle l[n], l[n+1], p[n]\rangle$ with $s = l[1]$ and $d = l[n+1]$ and a sequence of positive integers $c[2] \cdots c[n]$ identifying the clock pulsc number at each data controlled latch $l[i]$. Define $c[1] = j$ and $c[n+1] = k$.

Now we will show that for all of the adjacencies in this sequence MIN (s, d, p, j, k) must be false.

Select any one of the adjacencies $\langle l[i], l[i+1], p[i]\rangle$, $1 \leq i \leq n$. The change in question propagated through latch $l[i]$ on the $c[i]$th clock pulse. Therefore, the change must have appeared at the output of $l[i]$ at or after $tC(l[i], c[i]) + cq(l[i])$ and therefore it arrived at the data input of $l[i+1]$ at or after the time

$$tC(l[i], c[i]) + cq(l[i]) + dy(l[i], l[i+1], p[i]),$$

which, by definition of pulse width, is at or after

$$tF(l[i], c[i]) - \mathrm{PW}(l[i]) + cq(l[i]) + dy(l[i], l[i+1], p[i]). \quad (5)$$

By definition of the sequence the change propagated correctly into latch $l[i+1]$ on clock pulse $c[i+1]$ for $1 \leq i < n$. Therefore, the change must have arrived before

$$tF(l[i+1], c[i+1]) - \mathrm{SUT}(l[i+1]).$$

In the case $i = n$ (latch d) the change violated the SUT_HT window. Therefore, it arrived after the above time but before

$$tF(l[n+1], c[n+1]) + \mathrm{HT}(l[n+1]).$$

Then for all i, $1 \leq i \leq n$, it follows that the change arrived before

$$tF(l[i+1], c[i+1]) + \mathrm{HT}(l[i+1]).$$

Combining this with (5) indicates that

$$tF(l[i], c[i]) - \mathrm{PW}(l[i]) + cq(l[i]) + dy(l[i], l[i+1], p[i]) < tF(l[i+1], c[i+1]) + \mathrm{HT}(l[i+1]).$$

Applying the definition of skew (Definition 1), and rearranging terms yields

$$cq(l[i]) + dy(l[i], l[i+1], p[i]) + nTF(l[i], c[i]) - nTF(l[i+1], c[i+1]) < \mathrm{PW}(l[i]) + \mathrm{SK}(l[i], l[i+1]) + \mathrm{HT}(l[i+1]).$$

This is the inverse of MIN $(l[i], l[i+1], p[i], c[i], c[i+1])$. Thus, MIN is false for all adjacencies $1 \leq i \leq n$.

By the premise of Theorem 2, each adjacency $\langle l[i], l[i+1], p[i]\rangle$, $1 \leq i \leq n$ must have either MAX $(l[i], l[i+1], p[i], c[i], c[i+1])$ true or else MIN $(l[i], l[i+1], p[i], c[i], c[i+1])$ true. We have shown that MIN $(l[i], l[i+1], p[i], c[i], c[i+1])$ is false for $1 \leq i \leq n$. Therefore, it must be that MAX $(l[i], l[i+1], p[i], c[i], c[i+1])$ is true for $1 \leq i \leq n$. We will next show that this contradicts the supposition that a failure occurred.

Form the sum of the left-hand sides of the MAX terms for $1 \leq i \leq n$.

$$\sum_{i=1}^{n} DQ(l[i]) + DY(l[i], l[i+1], p[i]) + nTF(l[i], c[i]) - nTF(l[i+1], c[i+1]).$$

By cancelling pairs of nTF's, this simplifies to

$$nTF(s, j) - nTF(d, k) + \sum_{i=1}^{n} DQ(l[i]) + DY(l[i], l[i+1], p[i]).$$

The corresponding sum for the right-hand sides of the MAX's is

$$\sum_{i=1}^{n} \min(\mathrm{pw}(l[i]) - \mathrm{SST}(l[i]) - \mathrm{SK}(l[i], l[i+1]) - \mathrm{SUT}(l[i+1]), \mathrm{SUT}(l[i]) - \mathrm{SUT}(l[i+1]), \mathrm{pw}(l[i]) - \mathrm{SST}(l[i]) - \mathrm{pw}(l[i+1]) + \mathrm{SST}(l[i+1])).$$

By the Triangular Skew Theorem there is some integer h such that $\mathrm{SK}(l[h], l[h+1]) \geq \mathrm{SK}(s, d)$.

Now select one term from $\min(\cdots)$ expression based on whether i is less than, equal to, or greater than h. This step yields the following, possibly greater, upper bound for the left-hand side

$$\sum_{i=1}^{h-1} \mathrm{pw}(l[i]) - \mathrm{SST}(l[i]) - \mathrm{pw}(l[i+1]) + \mathrm{SST}(l[i+1]) + \mathrm{pw}(l[h]) - \mathrm{SST}(l[h]) - \mathrm{SK}(l[h], l[h+1]) - \mathrm{SUT}(l[h+1]) + \sum_{i=h+1}^{n} \mathrm{SUT}(l[i]) - \mathrm{SUT}(l[i+1])$$

which simplifies to

$$\mathrm{pw}(s) - \mathrm{SST}(s) - \mathrm{SUT}(d) - \mathrm{SK}(l[h], l[h+1]).$$

So, combining the two simplified expressions we have

$$nTF(s, j) - nTF(d, k) + \sum_{i=1}^{n} DQ(l[i]) + DY(l[i], l[i+1], p[i]) \leq \mathrm{pw}(s) - \mathrm{SST}(s) - \mathrm{SUT}(d) - \mathrm{SK}(l[h], l[h+1]).$$

By the choice of h,

$$nTF(s, j) - nTF(d, k) + \sum_{i=1}^{n} DQ(l[i]) + DY(l[i], l[i+1], p[i]) \leq \mathrm{pw}(s) - \mathrm{SST}(s) - \mathrm{SUT}(d) - \mathrm{SK}(s, d)$$

and by the definition of SK,

$$tF(s, j) - tF(d, k) + \sum_{i=1}^{n} DQ(l[i]) + DY(l[i], l[i+1], p[i]) \leq \mathrm{pw}(s) - \mathrm{SST}(s) - \mathrm{SUT}(d).$$

Hence,

$$tC(s, j) + CQ(s) + DY(s, l[2], p[1]) + \sum_{i=2}^{n} DQ(l[i]) + DY(l[i], l[i+1], p[i]) \leq tF(d, k) - \mathrm{SUT}(d).$$

The expression on the left represents the time of the leading clock edge at latch s plus the upper bound for the delays along the path from s to the data input of latch d. It is therefore the latest time that the offending change could have reached d. This is shown here to be at or before the required setup time for the kth clock pulse at d, thereby contradicting the supposition that latch d failed. Hence, the theorem is proved.

5) Interpretation of the Theorem: The theorem essentially gives the conditions for preventing any violation of any SUT_HT window. Given an adjacency $\langle s, d, p \rangle$ and a pair of clock pulses, these are the conditions that prevent a signal passing through the source latch on the first clock pulse j from violating the SUT_HT window at the destination latch on clock pulse k. To avoid that particular SUT_HT window, the data must arrive either before the setup time or after the hold time. MAX gives the conditions for ensuring the former, and MIN gives the conditions for the latter. Thus, the theorem requires either MAX or MIN for each j, k pair.

Application of this theorem to logic design is discussed below. A point to be made is that all of the variables in the theorem are local to the two latches in question and the path between them. Yet the proof shows that the possible effects of latch fall-through are accounted for, and cannot jeopardize the SUT_HT window at latch d.

G. Application of the Timing Rule

1) General Comments: In the above proof a very loose definition of a clock signal was used. In particular it was not assumed that all clocks have the same frequency and no restriction was placed on the combinational logic a clock may pass through before reaching the clock input of a latch.

In order to apply the timing rule in practice, all clocks at least must be based on subharmonics of some common frequency. Such is virtually always true.

2) Clock Enables: In practice, pulse width and timing at the clock inputs of latches are controlled by restricting the combinational logic through which a clock may pass by the following additional rules.

1) Any connection to the clock input of a latch must be a system clock or a signal derived from a system clock by ANDing it with data signals. 2) A clock may not be ORed with a data signal. 3) If a data signal is ANDed with a clock, then it must be stable while the clock is active.

This last condition may be guaranteed by observing the following rule.

Enable Timing Rule: If there is a combinational path p from the output of a latch s to the clock input of latch d then for any positive integers j and k, either there is no jth clock pulse at s or there is no kth clock pulse at d or at least one of the following inequalities holds:

$$CQ(s) + DY(s, d, p) + nTF(s, j) - nTF(d, k) \leq \min(\mathrm{pw}(s) - \mathrm{PW}(d) - \mathrm{SK}(s, d), \mathrm{SUT}(s) - \mathrm{PW}(d) + \mathrm{SST}(s))$$

$$cq(s) + dy(s, d, p) + nTF(s, j) - nTF(d, k) \geq \mathrm{PW}(s) + \mathrm{SK}(s, d)$$

where the terms $nTF(d, k)$, PW(d), and SK(s, d) are to be understood to refer to the clock pulses which would have occurred at latch d if it were always enabled.

The proof is left to the reader. It can be constructed using a form similar to that used for Theorem 2, and noting that one of the quantities in the min function of the MAX inequality is not needed. Note also that the window of stability that must be guaranteed at latch d is not the SUT_HT window about the trailing edge of clock pulse k, but the entire clock pulse itself. Hence, the SUT(d) and HT(d) terms do not appear. There are no terms replacing them because the times are measured at the latch clock input, rather than at the inputs to the ANDing circuitry.

This Enable Rule is not required for the validity of Theorem 2, but is useful in its application to logic design.

3) Application to Design: Applying Theorem 2 to the process of constructing a logic design requires understanding the meaning of the two inequalities MAX (s, d, p, j, k) and MIN (s, d, p, j, k).

In practice the latches s and d, and the starting clock pulse j, are typically treated as givens. The object is to design the logic for path p in order to reach the destination on a particular clock k (relative to j). During the process of logic design, the intended phase separation $nTF(k)-nTF(j)$ for any particular adjacency is determined, or at least known, by the designer of the logic path.

Having MAX (s, d, p, j, k) true means that data will be available to be captured in latch d on or before clock k. It can be seen that MAX (s, d, p, j, k) implies MAX $(s, d, p, j, k+1)$, and so on. Having MIN (s, d, p, j, k) true means that data will be prevented from affecting latch d on or before clock k. Similarly MIN (s, d, p, j, k) implies $(s, d, p, j, k-1)$. It should also be apparent that MAX (s, d, p, j, k) and MIN (s, d, p, j, k) cannot both be true simultaneously, since the maximum delay cannot be less than the minimum delay.

Therefore, from the designer's point of view, given a particularly k, the maximum delay of p must be controlled so that MAX (s, d, p, j, k) is true, and the minimum delay of p must ensure that MIN $(s, d, p, j, k-1)$ is true.

When designing with a single phase clock, with the destination latch always enabled, $nTF(d, k) - nTF(s, j)$ is simply the clock period T in the MAX inequality, and (because of using $k-1$) becomes zero in the MIN inequality.

4) Application to Automated Timing Analysis: The timing rule proved in Theorem 2 is quite general in its application. In particular, given any specific adjacency, it does not require information about the phase separation intended by the designer. Many present day timing analysis programs have trouble dealing with multiple-cycle situations, and generate extraneous error messages. By considering both the maximum and minimum delays together, as suggested by the theorem, this class of erroneous output can be eliminated.

5) Special Cases: In some design situations, system-wide values can be assigned to a given parameter by using the worst case possible for that parameter. In such a case, the parameter is a constant and the rules reduce to simpler forms and are easier to apply. For example, for a design with constant values of pw, SUT, CQ, and DQ, the (MAX) portion of Theorem 2 reduces to

$$DQ + DY(s, d, p) + nTF(s, j) - nTF(d, k) \leq \min(\text{pw} - \text{SST} - \text{SK}(s, d) - \text{SUT}, 0). \quad \text{(MAX1)}$$

This may be compared to the results (11) and (12) obtained in [8]. The translation of most variables from one notation to the other is straightforward. The expression in [8] for the time from the latest leading edge of the source clock pulse to the earliest trailing edge of the destination clock pulse is $P + W - TL - TT$. In our notation it is $nTF(d, k) - nTF(s, j) + \text{pw} - \text{NSK(SRS)}$. Hence, [8] becomes

$$DQ + DY(s, d, p) + nTF(s, j) - nTF(s, k) \leq \min(\text{pw} - \text{SST} - \text{NSK(SYS)} - \text{SUT}, 0) \quad \text{(MAX2)}$$

where SYS is the node that contains all latches in the system. This is identical to MAX1 except for the skew term.

It is always true that SK(s, d) $\leq$ NSK(SYS). Thus, if SK(s, d) $\geq$ pw $-$ SST $-$ SUT, then the time allowed for combinational logic (DY) in MAX1 is larger than that in MAX2 by the full difference NSK(SYS) $-$ SK(s, d). And if SK(s, d) $<$ pw $-$ SST $-$ SUT, the advantage of MAX1 over MAX2 is max(NSK(SYS) $-$ pw $+$ SST $+$ SUT, 0), which is always nonnegative. This demonstrates directly the advantage of using a skew hierarchy.

Next consider simplifying the above case still further by assuming a pulse width wide enough such that

$$\text{pw} > \text{SST} + \text{SUT} + \text{NSK(SYS)}.$$

Then the maximum allowed time reduces to

$$DQ + DY(s, d, p) + nTF(s, j) - nTF(d, k) \leq 0. \quad \text{(MAX3)}$$

(MAX3) is the best possible latch-to-latch rule in that no such rule can allow more time on every latch-to-latch decision. Essentially it is saying simply that the amount of logic delay cannot exceed the nominal clock separation.

(MAX3) is worth dwelling on. There are technologies which incorporate logic in latches [5]–[7]. If in such a technology (MAX3) can be applied and if the DQ time of each latch is utilized for logic as effectively as the combinational components use that same amount of time, then the entire phase separation time is available for useful logic, which is a very desirable situation. The validity of (MAX3) depends on pulses being of sufficient width, however, and a penalty is paid in minimum delay requirements, which are often more difficult to meet.

At the other extreme, if pw is a set to zero, and if

$$0 \leq \text{SST}(s) + \text{SUT}(s),$$

for each latch s, then the result is the edge-triggered maximum rule as given in [8]. Similarly, if PW(s) is set to zero in (MIN), the result is the edge-triggered minimum rule as given in [8]. (MAX3) and the edge-triggered rules provide an interesting spectrum as pulse width varies from relatively high values to zero.

H. Summary

A new latch-to-latch timing rule has been presented which is applicable in very general circumstances. The methods of analysis employed are powerful and robust and can be used for the development of other rules, as in [9]. Extensions of the rule to cover clock enables and edge triggered latches are discussed.

Acknowledgment

R. J. Petschauer introduced the authors to the notion of "latch fall-through." Many fruitful discussions on latch-to-latch timing rules were held with G. R. Kregness and K. L. Engelbrecht.

References

[1] A. Albicki and J. Beausang, "A method to obtain an optimal clocking scheme for a digital system," in *Proc. IEEE Int. Conf. Comput. Design: VLSI Comput.*, Oct. 1985, pp. 68–72.

[2] A. Albicki and T. Jackson, "Characterization of clocked D flip-flops," Tech. Rep. EL-84-03, Dep. Elec. Eng., Univ. of Rochester, Rochester, NY, 14627, Sept. 1984.

[3] L. B. Bushard and J. R. Schomburg, "Behavior of a simple latch model," in *Proc. Sperry Tech. Symp.*, May 1986, paper 1-19.

[4] —, "Latch to latch timing rules," in *Proc. Sperry Tech. Symp.*, May 1986, paper 1-20.

[5] J. G. Earle, "Latched carry-save adder," *IBM Tech. Disc. Bull.*, vol. 7, no. 10, pp. 909–910, Mar. 1965.

[6] D. D. Eberlein, "Custom MSI for very high speed computers," in *Dig. Papers, IEEE Spring Compcon79*, Feb. 1979, pp. 295–298.

[7] S. R. Kunkel and J. E. Smith, "Optimal pipelining in supercomputers," in *Proc. 13th Int. Symp. Comput. Architecture*, June 1986, Tokyo, pp. 404–411.

[8] C.-J. Tan and S. H. Unger, "Clocking schemes for high-speed digital systems," *IEEE Trans. Comput.*, vol. C-35, no. 10, pp. 880–895, Oct. 1986.

[9] L. B. Bushard, A. F. Champernowne, J. T. Rusterholz, and J. R. Schomburg, "Detailed latch to latch timing rules," Tech. memo. 945. Unisys, Roseville, MN, 1988.

Synchronization in Digital System Design

DAVID G. MESSERSCHMITT, FELLOW, IEEE

(Invited Paper)

Abstract—**In digital system design, synchronization ensures that operations occur in the logically correct order, and is a critical factor in ensuring the correct and reliable system operation. As the physical size of a system increases, or as the speed of operation increases, synchronization plays an increasingly dominant role in the system design. Digital communication has developed a number of techniques to deal with synchronization on a global and even cosmic scale; and as the clock speeds of chip, board, and room-sized digital systems increase, they may benefit from similar techniques. Yet, the digital system and digital communication communities have evolved synchronization techniques independently, choosing different techniques and different terminology. In this paper, we attempt to present a unified framework and terminology for synchronization design in digital systems, borrowing techniques and terminologies from both digital system and digital communication design disciplines. We then compare the throughput of synchronous and asynchronous interconnect, emphasizing how it is impacted by interconnect delay. Finally, we discuss opportunities to apply principles long employed in digital communications to the design of digital systems, with the goal of reducing this dependence on interconnect delay.**

I. INTRODUCTION

OPERATIONS in digital systems can either proceed concurrently, or they must obey a precedence relationship. If two operations obey a precedence, then the role of synchronization is to ensure that the operations follow in the correct order. Synchronization is thus a critical part of digital system design.

The most common approach to synchronization is to distribute a clock signal to all modules of the system. With the scaling of feature-sizes in VLSI design, clock speeds are increasing rapidly, but increases in complexity tend to prevent significant reductions in chip size. As a consequence of this scaling, clock speeds in digital system designs are increasing in relation to propagation delays. This is causing increasing problems with the traditional synchronous design methodologies, certainly at the system and board levels, and increasingly even within high performance chips [1]. This problem will be accentuated with the more common application of optics to system interconnection.

Problems such as large propagation delay have been faced since the earliest days of digital communication system design, and hence there are a number of opportunities to apply digital communications principles to VLSI and digital system design. However, because of the wide difference between system physical sizes in relation to clock speeds, the design styles of the communities have developed almost independently. This is in spite of the fact that digital design is a necessary element of digital communication design (for example, in modems, switches, etc.).

In this paper, we place the synchronization problem and design approaches in digital communication and digital system design in a common framework, and then examine opportunities for cross-fertilization between the two fields. In attaining the unification of design methodologies that we attempt in this paper, the first difficulty we face is the inconsistencies, and even contradictions, between terms as used in the two fields. For example, the term "self-timed" generally means "no independent clock," but as used in digital communication it means no clock at all (in the sense that if a clock is needed it can be derived from the data), and in digital system design it indicates that there is a clock signal that is slaved to the data rather than being independent. Therefore, in Section II we attempt to define a taxonomy of terminology that can apply to both fields, while retaining as much of the terminology as presently used as possible. We discuss this terminology in terms of the necessary levels of abstraction in the design process. In Section III we discuss the synchronization techniques commonly used in digital systems and digital communications, relate the two, and compare the fundamental limitations they place on throughput. In Section IV we discuss specifically how some synchronization techniques from digital communication might be beneficial in digital system design, particularly in reducing dependence of throughput on interconnect delay. Finally, in Section V we discuss briefly the interrelationship between architectural design and synchronization in digital systems. This discussion expands on an earlier conference paper [2].

II. ABSTRACTIONS IN SYNCHRONIZATION

A basic approach in system design is to define *abstractions* that enable the designer to ignore unnecessary details and focus on the essential features of the design. While every system is ultimately dependent on underlying physical laws, it is clear that if we relied on the solution of Maxwell's equations at every phase of the design, systems could never get very complex. Abstractions are often applied in a hierarchical fashion, where each layer of ab-

Reprinted from *IEEE J. Selected Areas Communi.*, vol. 8, no. 8, pp. 1404–1419, Oct. 1990.

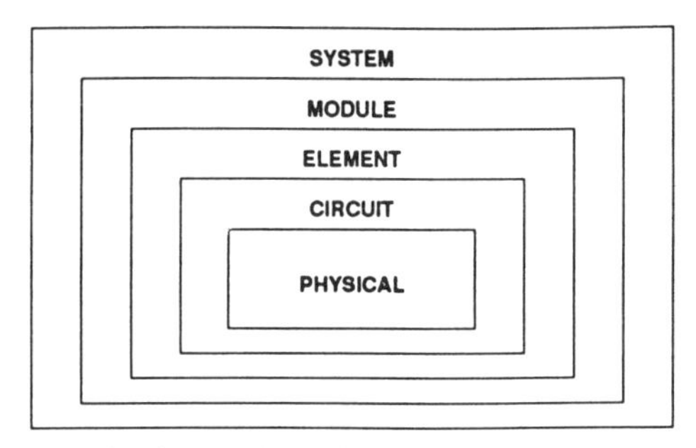

Fig. 1. An example of some abstractions applied to digital system design.

straction relies on the essential features of the abstraction level below, and hides unessential details from the higher level. This as illustrated for digital system design in Fig. 1. At the base of the design, we have the physical representation, where we have semiconductor materials, interconnect metalization, etc., and we are very concerned about the underlying physical laws that govern their properties. Above this, we have the circuit abstractions, where we deal with circuit entities such as transistors, interconnections, etc. Our descriptions of the circuit entities attempt to ignore the details of the physical laws underlying them, but rather characterize them in terms that emphasize their operational properties, such as current-voltage curves, transfer functions, etc. Above the circuit abstraction is the element, which groups circuit entities to yield slightly higher functions, such as flip-flops and gates. We describe elements in terms that deal with their operational characteristics (timing diagrams, Boolean functions, etc.) but ignore the details of how they are implemented. Above the element abstraction, we have the module (sometimes called "macrocell") abstraction, where elements are grouped together to form more complex entities (such as memories, register files, arithmetic-logic units, etc.). Finally, we group these modules together to form systems (like microcomputers).

Many other systems of abstractions are possible, depending on circumstances. For example, in digital communication system protocols, we extend the abstractions in Fig. 1, which carry us to the extreme of hardware design, to add various abstractions that hierarchically model the logical (usually software-defined) operation of the system (including many layers of protocols). Similarly, the computer industry defines other system abstractions such as instruction sets, operating system layers, etc.

In the present paper, we are concerned specifically with synchronization in the design of digital systems and digital communications. In the context of digital systems, by *synchronization* we mean the set of techniques used to ensure that operations are performed in the proper order. The following subsections define some appropriate abstractions for the synchronization design. While these abstractions are by no means new, perhaps this is the first time that a systematic treatment of them has been attempted. This systematic treatment gives us a common base of terminology for synchronization design, and is utilized in the following subsections.

While abstractions are very useful, and in fact absolutely necessary, they should be applied with care. The essence of an abstraction is that we are ignoring some details of the underlying behavior, which we hope are irrelevant to the operation, while emphasizing others that are most critical to the operation. It should always be verified that in fact the characteristics being ignored are in fact irrelevant, and with considerable margin, or else the final system may turn out to be inoperative or unreliable. This is especially true of synchronization, which is one of the most frequent causes of unreliable operation of a system. In the following, we therefore highlight behaviors that are ignored or hidden by the abstraction.

A. Some Basic Synchronization Abstractions

1) Boolean Signals: A *Boolean signal* (voltage or current) is assumed to represent, at each time, one of two possible levels. At the physical level, this signal is generated by saturating circuits and bistable memory elements. There are a couple of underlying behaviors that are deliberately ignored in this abstraction: *finite rise-time* and the *metastable* behavior of memory elements.

Finite rise-time behavior is illustrated in Fig. 2 for a simple *RC* time constant. The deleterious effects of rise-time can often be bypassed by the *sampling* of the signal. In digital systems this is often accomplished using edge-triggered memory elements. In digital communications, rise-time effects are often much more severe because of the long distances traversed, and manifest themselves in the more complex phenomenon of *intersymbol interference* [3]. In this case, one of several forms of equalization can precede the sampling operation.

Metastability is an anomalous behavior of all bistable devices, in which the device gets stuck in an unstable equilibrium midway between the two states for an indeterminate period of time [4]. Metastability is usually associated with the sampling (using an edge-triggered bistable device) of a signal whose Boolean state can change at any time, with the result that sampling at some point very near the transition will occasionally occur. Metastability is less severe a problem than rise-time in the sense that it happens only occasionally, but more severe in that the condition will persist for an indeterminate time (like a rise-time which is random in duration).

The Boolean abstraction is most valid when the rise-time is very much shorter than the interval between transitions, and metastability is avoided or carefully controlled. But the designer must be sure that these effects are negligible, or unreliable system operation will result.

2) Signal Transitions: For purposes of synchronization we are often less concerned with the signal level than with the times at which the signal changes state. In digital systems this time of change would be called an *edge* or transition of the signal. The notion of the transition time ignores rise-time effects. In fact, the transition time is subject to interpretation. For example, if we define the transition time as the instant that the waveform crosses some slicer level (using terminology from digital communication), then it will depend on the slicer level as illustrated in Fig. 3. Even this definition will fail if the

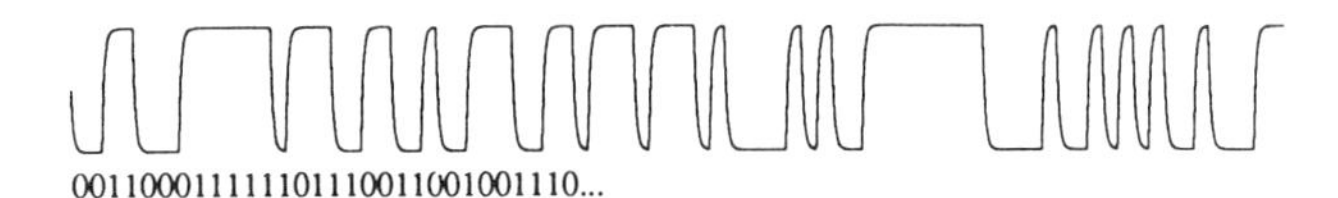

Fig. 2. Illustration of a digital signal with a finite rise-time as generated by an *RC* time-constant, where the Boolean signal abstraction is shown below.

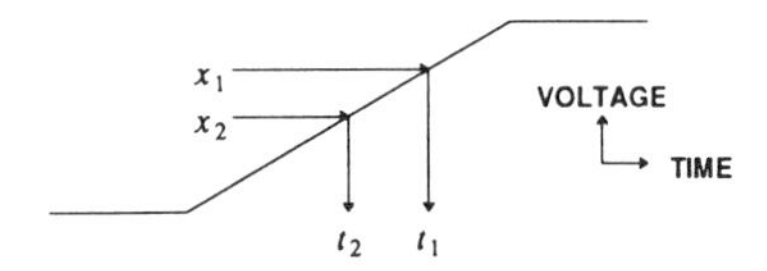

Fig. 3. The transition time t_i depends on slicer level x_i.

waveform is not monotonic in the region of the slicer level.

Transition times are a useful abstraction for the case where the rise times are very short in relation to the interval between transitions, with the result that the variation in the transition time is negligibly small over the set of all possible definitions of the transition time. Rise-time is governed by underlying physical phenomena, such as transmission line dispersion, and can be reduced by using wider bandwidth drivers or intermediate repeaters. As system clock rates increase, however, for a given interconnect style the behavior ignored by this abstraction inevitably becomes important.

The transition abstraction can be extended to the notion of *uniformly spaced transitions*. For example, a clock signal can be modeled as a square wave, in which the transitions alternate in sign, and each adjacent pair of transitions (called a *cycle*) represents a time equal to the reciprocal of the clock *frequency*. For a data signal, transitions may or may not be present depending on the Boolean data (see Fig. 2), so we have to introduce the notion of the times where transitions *might* occur, called a *transition opportunity*, whether they actually occur or not. A data signal whose transitions are slaved to a clock with uniformly spaced transitions then has uniformly spaced transition opportunities (an example is shown in Fig. 4). We can think of these transitions as being associated with a clock that has positive transitions at identical times, which we call the *associated clock*, whether or not such a clock signal exists physically.

Uniformly spaced transitions ignore possible jitter effects in the generation or transmission of the Boolean signals, which often result in small variations in the times between transitions. Hence, there is the need to define the concepts of instantaneous phase and frequency.

3) Phase and Frequency: For a Boolean signal, we can define a phase and frequency of the signal as the phase and frequency of the associated clock. It is convenient to describe mathematically a clock signal with uniformly spaced transitions as

$$x(t) = p((ft + \phi) \text{ modulo } 1), \tag{1}$$

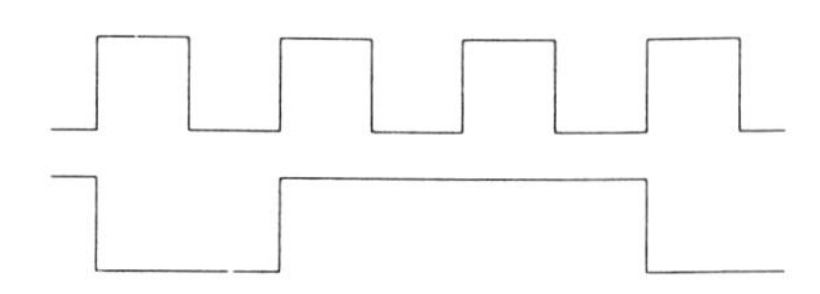

Fig. 4. A Boolean signal (below) and associated clock (above). The transitions of the Boolean signal are slaved to the positive transitions of the associated clock.

where $p(t)$ is a 50% duty cycle pulse

$$p(t) = \begin{cases} 1, & 0 \le t < 0.5 \\ 0, & 0.5 \le t \le 1, \end{cases} \tag{2}$$

f is the *nominal frequency*, and ϕ is the *phase*. As ϕ varies over the range $0 \le \phi < 1$, the transitions are shifted in time over one cycle. The phase is thus expressed as the fraction of a cycle. When we have two Boolean signals, the *relative phase* can be expressed as the phase difference $(\phi_1 - \phi_2)$ between their respective associated clocks.

A more general model that includes more possible effects replaces (1) by

$$x(t) = p(((f + \Delta f)t + \phi(t)) \text{ modulo } 1), \tag{3}$$

where f is the *nominal frequency* of the associated clock, Δf is a possible *offset* in the nominal frequency, and $\phi(t)$ is the *instantaneous phase variation* versus time. The intention here is that $\phi(t)$ does not embody a frequency offset, but rather any offset from the nominal frequency is summarized by Δf. The precise mathematical conditions for this are complicated by the modulo operation, and also depend on the model for $\phi(t)$ (deterministic signal, random process, etc.). For example, if $\phi(t)$ is assumed to be a deterministic differentiable and continuous function (with no phase jumps), then it suffices for $\phi(t)$ to be bounded,

$$\phi(t) \le \phi_{\max}, \tag{4}$$

and for such a function the derivative (instantaneous frequency) must average to zero,

$$\overline{\frac{d\phi(t)}{dt}} = 0 \tag{5}$$

(where the average is interpreted as a time average).

The model of (3) makes the assumption that the average frequency is a constant, although that average frequency may not be known *a priori* (for example, when it depends on the free-running frequency of an oscillator). Such a signal is said to be *isochronous* (from "iso," the Greek root for "equal"), whereas if the frequency is not constant (Δf is actually a function of time), the signal is said to be *anisochronous* (or "not equal"). An anisochronous signal can be modeled using (3), but the resulting phase will not be bounded. Thus, the essential difference between isochronous and anisochronous signals is the bounded phase condition of (4).

The time-varying phase in (3) is crucial where we cannot ignore small variations in the intervals between transitions, known as *phase jitter*. This jitter is usually ignored in digital system design, but becomes quite significant in digital communications, especially where the Boolean signal is passed through a chain of regenerative repeaters [3].

Directly following from the concept of phase is the *instantaneous frequency*, defined as the derivative of the instantaneous phase,

$$f(t) = f + \Delta f + \frac{d\phi(t)}{dt}. \tag{6}$$

From (5), the *instantaneous frequency deviation* $d\phi(t)/dt$ has mean value zero, so that $(f + \Delta f)$ is the *average frequency*.

Given two signals $x_i(t)$, $i = 1, 2$ with the same nominal frequency, and frequency and phase offsets Δf_i and $\phi_i(t)$, the *instantaneous phase difference* between the two signals is

$$\Delta\phi(t) = (\Delta f_1 - \Delta f_2)t + (\phi_1(t) - \phi_2(t)). \tag{7}$$

4) Synchronism: Having carefully defined some terms, we can now define some terminology related to the *synchronization* of two signals. A taxonomy of synchronization possibilities is indicated in Fig. 5 [5], [3]. As previously mentioned, a Boolean signal can be either isochronous or anisochronous. Given two data signals, if both are isochronous, the frequency offsets are the same, and the instantaneous phase difference is zero,

$$\Delta\phi(t) = 0 \tag{8}$$

then they are said to be *synchronous* (from the Greek "syn" for "together"). Common examples would be a Boolean signal that is synchronous with its associated clock (by definition), or two signals slaved to the same clock at their point of generation. Any two signals that are not synchronous are *asynchronous* (or "not together"). Some people would relax the definition of synchronous signals to allow a nonzero phase difference that is *constant* and *known*.

In practice, we have to deal with several distinct forms of asynchrony. Any signal that is anisochronous will be asynchronous to any other signals, except in the special and unusual case that two anisochronous signals have identical transitions (this can happen if they are co-generated by the same circuit). Thus, anisochrony is a form of asynchrony.

If two isochronous signals have exactly the same average frequency $f + \Delta f$, then they are called *mesochronous* (from the Greek "meso" for "middle"). For mesochronous signals, the fact that each of their phases is bounded per (4) ensures also that the phase difference is also bounded,

$$\Delta\phi(t) \le 2\phi_{\max}, \tag{9}$$

a fact of considerable practical significance. Two signals generated from the same clock (even different phases of the same clock), but suffering indeterminate interconnect delays relative to one another, are mesochronous.

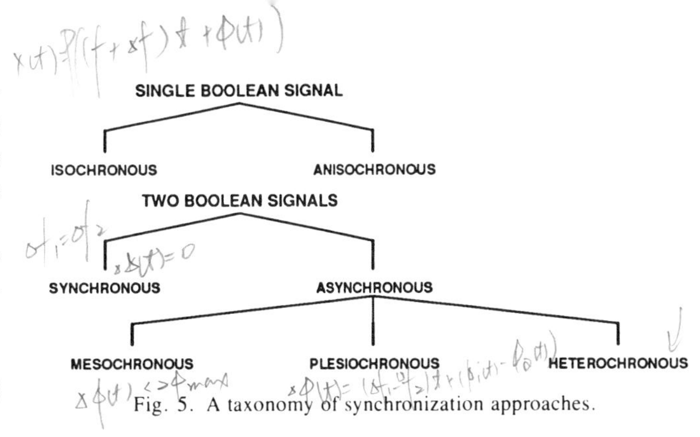

Fig. 5. A taxonomy of synchronization approaches.

Two signals that have average frequencies that are nominally the same, but not exactly the same (usually because they are derived from the independent oscillators), are *plesiochronous* (from the Greek "plesio" for "near"). Suppose the nominal frequencies are both f, but the actual frequencies are $f + \Delta f_1$ and $f + \Delta f_2$, then the instantaneous phase difference is

$$\Delta\phi(t) = (\Delta f_1 - \Delta f_2)t + (\phi_1(t) - \phi_2(t)), \tag{10}$$

where the first term increases (or decreases) linearly with time. For two plesiochronous signals, it cannot be predicted which has the higher frequency.

Finally, if two signals have *nominally different* average frequencies, they are called *heterochronous* (from the Greek "hetero" for "different"). Usually the tolerances on frequency are chosen to guarantee that one signal will have an actual frequency guaranteed higher than the other (naturally the one with the higher nominal rate). For example, if they have nominal frequencies f_1 and f_2, where $f_1 < f_2$, and the worst-case frequency offsets are $|\Delta f_1| \le \eta_1$ and $|\Delta f_2| \le \eta_2$, then we would guarantee this condition if f_1 and f_2 were chosen such that

$$f_1 + \eta_1 < f_2 - \eta_2. \tag{11}$$

B. Additional Timing Abstractions in Digital Systems

This section has covered some general considerations in the modeling of Boolean signals from a synchronization perspective. In this subsection we describe a couple of additional abstractions that are sometimes applied in digital system design, and represent simplifications due to the relatively small physical size of such systems.

1) Equipotential Region: Returning to Fig. 1, it is often assumed at the level of the element abstraction that the signal is identical at all points along a given wire. The largest region for which this is true is called (by Seitz [6]) the *equipotential region*. Like our other models, this is never strictly valid, but is useful if the actual time it takes to equalize the potential along a wire is small in relation to other aspects of the signals, such as the associated clock period or rise-time. The equipotential region is a useful concept in digital system design because of the relatively small size of such systems. However, the element dimensions for which it is valid is decreasing because of increases in clock frequency with scaling, and a single chip can generally no longer be considered an equipotential region.

2) Ordering of Signals: In the design of digital systems, it is often true that one Boolean signal is *slaved* to another, so that at the point of generation the one signal transitions can always be guaranteed to precede the other. Conversely, the correct operation of circuits is often dependent on the correct ordering of signal transitions, and quantitative measures such as the minimum time between transitions. One of the main reasons for defining the equipotential region is that if a given pair of signals obey an ordering condition at one point in a system, then that ordering will be guaranteed anywhere within the equipotential region.

III. Synchronization

The role of synchronization is to coordinate the operation of a digital system. In Section III-A and B, we review two traditional approaches to synchronization in digital system design: synchronous and anisochronous interconnection. In Section III-C, we briefly describe how synchronization is accomplished in digital communication systems. This will suggest, as discussed further in Section IV, opportunities to use digital communication techniques in digital system design.

A. Synchronous Interconnection

As shown in Fig. 6, each element (or perhaps module) is provided a clock, as well as one or more signals that were generated with transitions slaved to the clock. The common clock controls the order of operations, ensuring correct and reliable operation of the system.

We will first, in Section III-A-1), examine some fundamental limitations in the operation of synchronous interconnection, making idealistic assumptions about the ability to control the clock phases in the system and neglecting interconnect delays. In Section III-A-2), we will show how pipeline registers can be used to extend the performance of synchronous interconnect; and in Section III-A-3), we will make more realistic estimates of performance considering the effects of the inevitable variations in clock phase and interconnect delays.

1) Principle of Synchronous Interconnection: The fundamental principle of synchronous interconnection is illustrated in Fig. 7. In Fig. 7(a) a *computational block* C1 is connected to a *synchronizing register* R1 at its input. This register is *clocked* using the positive transitions of a periodic clock signal, where the assumption is that the output signal of the register changes synchronously with the positive transition of the clock. The computational block performs the same computation repeatedly on new input signals applied at each clock transition. The purpose of R1 is to control the time at which the computational block starts to perform its work, in order to synchronize it to other computational blocks in the system. This is an *edge-triggered* logic model, which we employ for its relative simplicity. There is also a more complicated *level-sensitive* model that leads to virtually identical conclusions.

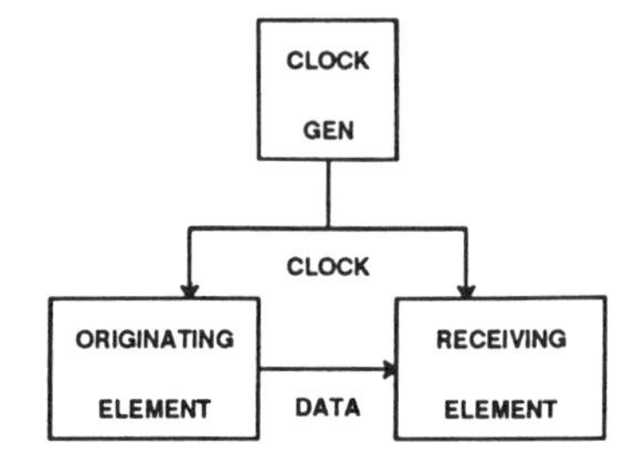

Fig. 6. Synchronous interconnection, in which a common clock is used to synchronize computational elements.

The performance measures of interest in Fig. 7 are the *throughput* (rate at which the computation is repeated) and *computational latency* (delay from the time a new input is applied until the result is available). Focusing on the latter, inevitably the computational latency is not entirely predictable. It is likely that the output signal will change more than once before it finally settles to its final correct value. For example, if the output signal actually consists of $M > 1$ Boolean signals in parallel, as is often the case, some of those Boolean signals may transition before others, or some may transition more than once before reaching a steady-state value. This behavior is an inevitable consequence of real circuit implementation of the computational block, and presents a considerable problem in the synchronization to other computational blocks. Assume the computational block has a minimum time before *any* outputs transition, called the *propagation time* t_p, and a maximum time before all the outputs reach their final and correct values, called the *settling time* t_s. Since settling time is the maximum time before the result is guaranteed to be available, it is also the computational latency. It is assumed that the propagation and settling times can be characterized and ensured over expected processing variations in the circuit fabrication.

The synchronous interconnect isolates the system behavior from these realities of circuit implementation by setting the clock period T so that there is a *certainty period* during which the output signal is guaranteed to be correct, and then samples the output signal again (using another register R2) during this certainty period as shown in Fig. 7(b). Of course, the phase of the clock for R2 must be adjusted to fall in this certainty period. This interconnect is called synchronous because the proper phase of clock to use at R2 must be known in advance—there cannot be any substantial uncertainty or time variation in this phase. With synchronous interconnect, the undesired behavior (multiple signal transitions and uncertain completion time) is hidden at the output of R2. We can then abstract the operation of the computational block, as viewed from the output of R2, as an element that completes its computation precisely at the positive transitions of the second clock, and this makes it easy to synchronize this block with others since the uncertainty period has been eliminated.

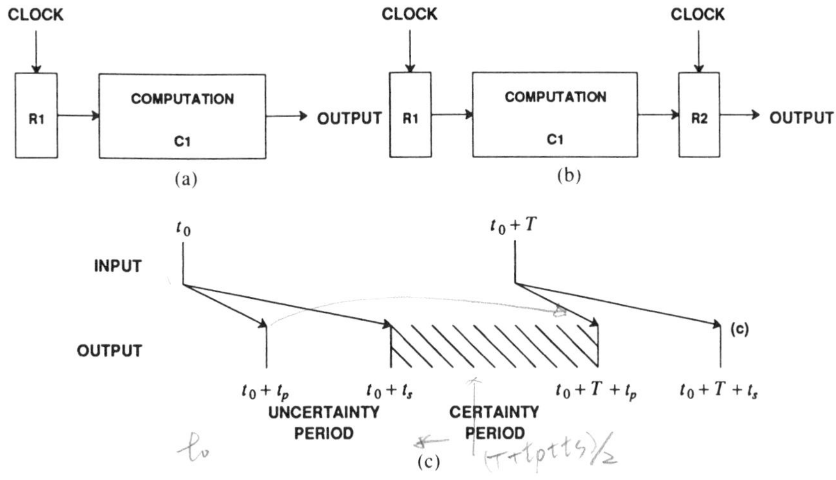

Fig. 7. Synchronizing a computation C1 by latching its input. (a) A register R1 at the input controls the starting time for the computation. (b) A second register R2 at the output samples the output signal during the certainty period. (c) Timing diagram showing the uncertainty and certainty period (crosshatched) as a function of the setting time t_s, the propagation time t_p, and the clock period T.

Given the propagation and settling times, the certainty period is shown as the crosshatched period in Fig. 7(c). The region of time not crosshatched, during which the signal may possibly be changing, is known as the *uncertainty period*. If two successive clock transitions come at times t_0 and $(t_0 + T)$, the certainty period starts at time $(t_0 + t_s)$, the time when the steady-state output due to the computation starting at time t_0 is guaranteed, and ends at time $(t_0 + T + t_p)$, which is the earliest that the output can transition due to the new computation starting at time $(t_0 + T)$. The length of the certainty period is $(T + t_p - t_s)$. An acceptable register clocking phase for R2 requires that this period must be positive in length, or

$$T > t_s - t_p. \tag{12}$$

Alternatively, we can define the *throughput* of computational block C1 as the reciprocal of the clock period, and note that this throughput is upper-bounded by

$$\frac{1}{T} < \frac{1}{t_s - t_p}. \tag{13}$$

The length of the uncertainty period $(t_s - t_p)$ is a fundamental property of the computational block, and the maximum throughput is the reciprocal of this length. In contrast, the length of the certainty period depends on the clock period T, and the goal is generally to make this length as small as practical by choosing T small.

The maximum throughput is dependent only on the length of uncertainty period, $(t_s - t_p)$, and not directly on the settling time t_s. In Fig. 8, an example is given for throughput much higher than the reciprocal of the settling time (because the uncertainty time is a small fraction of the settling time). In Fig. 8, before each computation is completed, two more computations are initiated. At any point in time, there are three concurrent computations in progress.

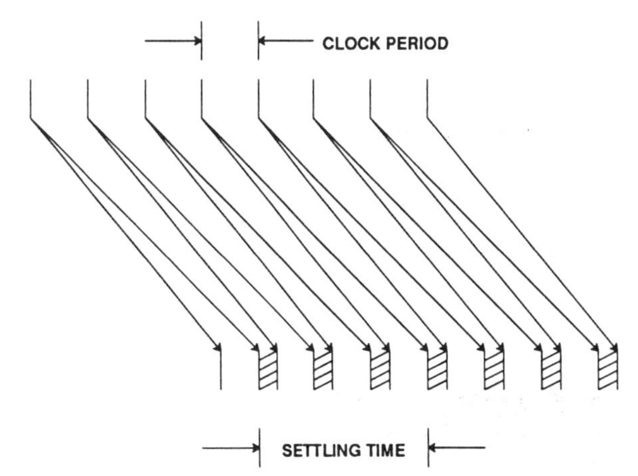

Fig. 8. An example of a synchronous interconnect with a clock period much smaller than the reciprocal of the settling time, due to a small uncertainty period.

The number of concurrent computations is limited only by the inevitable uncertainty in the computational latency. We can give three examples that illustrate a range of possibilities.

Example 1: Consider a fiber optic digital communication system, where the "computational block" is not a computation at all but rather a propagation through a guided medium. For this case, due to propagation dispersion effects, t_p and t_s are not identical, but close enough that throughputs in the range of 10^{10} bits/s are possible. Assuming a group velocity of 10^8 m/s, and a fiber length of 50 km, the settling time is 0.5 ms. At a conservative bit rate of 100 mb/s, there are 50 000 bits propagating through the fiber at any time ("concurrent computations" in the language above). At this velocity and bit rate, the maximum distance for which there is no concurrency in the communication medium is 1 m. □

Example 2: For typical practical Boolean logic circuits, designed to minimize the settling time rather than maximize the propagation time, t_p is typically very small, and concurrent computations within the computational block are not possible. The maximum throughput is the reciprocal of the settling time. □

Example 3: Consider a hypothetical (and perhaps impractical) circuit technology and logic design strategy which is designed to achieve $t_p \approx t_s$. In this case, the throughput can be much higher than the reciprocal of the settling time, and many concurrent computations within the computational block are possible. □

While Example 3 is not likely to be achieved, Examples 2 and 3 suggest the possibility of designing circuits and logic to minimize the uncertainty period $(t_s - t_p)$ (even at the expense of increasing t_s) rather than minimizing t_s as is conventional. For example, one could ensure that every path from input to output had the same number of gates, and carefully match the gate settling times. In such a design style, the throughput could be increased to exceed the reciprocal of the settling time. This has recently been considered in the literature, and is called *wave pipelining* [7].

2) Pipelining: The form of concurrency associated with Fig. 8 is known as *pipelining*. A useful definition of pipelining is the ability to *initiate* a *new* computation at the input to a computational block prior to the *completion* of the *last* computation at the output of that block. Since this results in more than a single computation in process within the block at any given time, pipelining is a form of concurrency, always available when $t_p > 0$. The number of *pipeline stages* is defined as the number of concurrent computations in process at one time. For example, if we take the liberty of calling the fiber propagation in Example 1 a "computation," then the fiber has 50 000 pipeline stages.

In conventional digital system design, t_p for computational blocks is typically small, and pipelining requires the addition of *pipeline registers*. To see this potential, make the idealistic assumption that the computational block of Fig. 7 can be split into N subblocks, the output of each connected to the input of the next, where each subblock has a propagation time of t_p/N and a settling time of t_s/N. If this is possible, then the block can be pipelined by inserting $(N - 1)$ pipeline registers between each pair of these subblocks as shown in Fig. 9. Each of these registers, according to the analysis, can use a clock frequency of $N/(t_s - t_p)$ because the uncertainty period is correspondingly smaller, assuming that each clock phase is adjusted to fall within the certainty period relative to the last clock phase.

To see that the uncertainty period is reduced for Fig. 9, in Fig. 7 the middle of the certainty period is delayed relative to the clock time t_0 by $(t_p + t_s + T)/2$. Assuming that the clock phase for each pipeline register in Fig. 9 is chosen in the middle of this certainty period, then relative to the previous clock phase the delay is N times smaller, or $(t_p + t_s + T)/2N$. The total propagation and

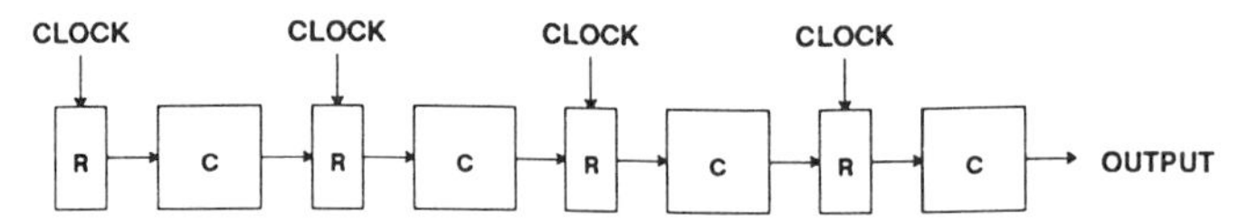

Fig. 9. Pipelining of Fig. 7 for $N = 4$ subblocks and $(N - 1 = 3)$ intermediate pipeline registers.

settling times for the pipeline are then

$$t_{p,\text{pipeline}} = (N-1) \cdot \frac{t_p + T + t_s}{2N} + \frac{t_p}{N} \tag{14}$$

$$t_{s,\text{pipeline}} = (N-1) \cdot \frac{t_p + T + t_s}{2N} + \frac{t_s}{N} \tag{15}$$

and the length of the uncertainty period is now

$$t_{s,\text{pipeline}} - t_{p,\text{pipeline}} = \frac{t_s - t_p}{N}, \tag{16}$$

a factor of N smaller. Thus, the theoretical maximum throughput is a factor of N higher. Again, the reason for this increase is that the intermediate pipeline registers have reduced the length of the uncertainty period, since the pipeline registers have rendered the uncertainty period zero for all but the last block (by controlling the computational latency with the clock).

This interpretation of the role of pipeline registers as reducing the uncertainty period is unconventional. A more common approach is to *start* with the assumption that there are pipeline registers (and, hence, pipeline stages) and a given fixed clock frequency, and then place as much of the total computation within each stage as possible, with the constraint that the settling time has to be less than the clock period. In this common viewpoint, pipelining and pipeline registers are synonymous. Within their domain of common applicability, the uncertainty period and the common viewpoints are simply different ways of expressing the same design approach. However, the uncertainty period approach we have presented here is more general, in that it includes pipelining without pipeline registers, as often occurs in communications or interconnect. More importantly, this approach serves as a unified framework under which computational blocks, communication or interconnect, and combinations of the two can be characterized—where the computational blocks often utilize pipeline registers to reduce the uncertainty period, and communications links often do not (because the uncertainty period is inherently very small).

In some situations, the *total* settling time of the computation is just as important as the throughput, for example, when the computation sits in a feedback loop. Thus, the effect of pipeline registers on this computational latency is also of interest. If we use the minimum clock period $T = t_s - t_p$ in (15), the total settling time through the pipeline is

$$t_{s,\text{pipeline}} = t_s, \tag{17}$$

and the total settling time is the *same* before and after the insertion of the pipeline registers. Thus, we have not paid a penalty in total settling time in return for the increase in throughput by a factor of N, since only the variability in settling time has been reduced.

In practice, depending on the system constraints, there are two interpretations of computational latency, as illustrated in the following examples.

Example 4: In a computer or signal processing system, the pipeline registers introduce a *logical delay*, analogous to the z^{-1} operator in Z-transforms. Expressed in terms of these logical delays, the N pipeline registers increase computational latency by N (equivalent to a z^{-N} operator). This introduces difficulties, such as unused pipeline stages immediately following a jump instruction, or additional logical delays in a feedback loop. □

Example 5: In some circumstances, the computational latency as measured in time is the critical factor. For example, in the media access controller for a local area network, the time to respond to an external stimulus is critical. For this case, as we have seen, the addition of pipeline registers need not increase the computational latency at all. With or without pipeline registers, the computational latency is bounded below by the inherent precedences in the computation as implemented by a particular technology. □

In practice, it is usually not possible to precisely divide a computational block into "equal-sized" pieces. In that case, the throughput has to be adjusted to match the *largest* uncertainty period for a block in the pipeline, resulting in a lowered throughput. There are a number of other factors, such as register setup times, which reduce the throughput and increase the overall settling time relative to the fundamental bounds that have been discussed here. One of the most important of these is the effect of interconnect delay and clock skew, which we will address next.

3) Clock Skew in Synchronous Interconnect: The effects of clock phase and *interconnect delay* (delay of signals passing between computational blocks) will now be considered. Clearly, any uncertainty in clock phase will reduce the throughput relative to (13), since earlier results required precise control of clock phase within a vanishing certainty period. Conversely, any fixed delay in the interconnect will not necessarily affect the achievable throughput, because it will increase the propagation and settling times equally and thus not affect the length of the uncertainty period. In practice, for common digital system design approaches, the effect of any uncertainty in clock phase is magnified by any interconnect delays.

In this subsection, we relax the previous assumptions, and assume that the clock phase can be controlled only within some known range (similar to the uncertainty period for the computational block). We then determine the best throughput that can be obtained following an approach similar to [8] and [9].

We can analyze clock skew with the aid of Fig. 10, in which we modify Fig.7(b) to introduce some new effects. In particular, we model the following.

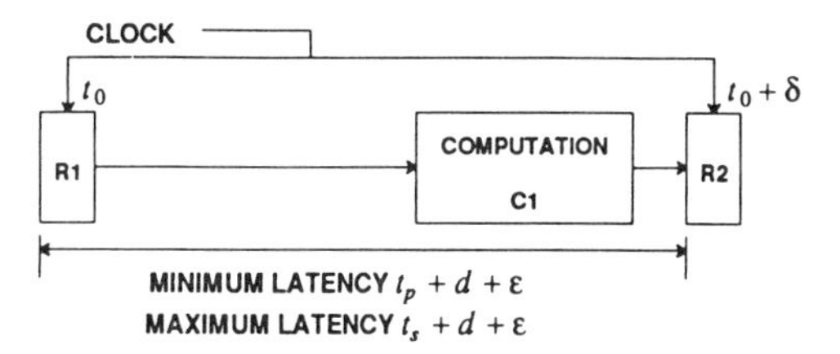

Fig. 10. Illustration of the effects of clock skew δ, where two computations C1 and C2 are synchronized using this clock.

- *Propagation and Settling Times:* As before, the propagation and settling times of the computation are t_p and t_s, except that we now include in these times any latencies relative to the clock transition imposed by the implementation of R1.
- *Interconnect Delay:* We assume interconnect delay d due to the diffusion of the signal through the interconnect wires between R1 and C1 and between C1 and R2.
- *Artificially Added Delay:* We assume that another delay ϵ is artificially added to the interconnect delay. The ϵ delay could be introduced, for example, by making R1 a double register with two clock phases, thereby introducing an artificial delay. We will find that ϵ is helpful in controlling the effects of clock skew, by effectively making the minimum interconnect delay larger than zero.
- *Clock Skew:* We assume that the clock transition at R1 occurs at time t_0 at R1, and the clock phase at R2 is $t_0 + \delta$, where δ is the clock skew. This clock skew can be either inadvertent, due, for example, to processing variations or interconnect delays in the clock distribution, or it can be deliberately controlled, for example, to adjust the R2 clock phase to fall within the certainty period. Further, it is possible for δ to be either positive or negative.

There is a certainty region of parameters $\{t_p, t_s, d, \epsilon, \delta\}$ where reliable operation is assured in Fig. 10. This is analyzed in Appendix A for the following three cases.

- *Idealistic Case:* If there is no uncertainty in d or δ and we set the delay ϵ most advantageously, then the bound of (13) is achieved. As expected, the interconnect delay need not necessarily slow the throughput.
- *Pessimistic Case:* Assume that ϵ can be precisely controlled (since it is controlled by relative clock phases) but that δ will inevitably only be controllable within a range, say, $|\delta| < \delta_{max}$. Thus, we are not attempting to set δ most advantageously. Further, we assume that all that is known about the interconnect delay is that it is bounded, $|d| < d_{max}$. The throughput is then bounded by

$$\frac{1}{T} < \frac{1}{t_s - t_p + d_{max} + 2\delta_{max}}. \tag{18}$$

For a system with a complex interconnect pattern, it would be very difficult to control the relationship of δ and d. In this case, we should expect $\delta_{max} \approx d_{max}$, and the throughput would be bounded by

$$\frac{1}{T} < \frac{1}{(t_s - t_p) + 3d_{max}}. \tag{19}$$

• *Optimistic Case:* For simple topologies like a one-dimensional pipeline, much higher throughput can be obtained by routing the signals and clocks in such a way that d and δ can be coordinated with one another [8]. Assume that the interconnect delay is known to be d_0 with variation Δd, and the skew is chosen to be δ_0 with variation $\Delta\delta$. Further assume that ϵ and δ_0 are chosen most advantageously to maximize the throughput. Then the throughput is bounded by

$$\frac{1}{T} < \frac{1}{(t_s - t_p) + 2(\Delta\delta + \Delta d)}, \qquad (20)$$

which is a considerable improvement over (19) if the delay and skew variations are small. This analysis shows that the reliable operation of the idealized synchronous interconnection of Section III-A-1) can be extended to accommodate interconnect delays and clock skew, even with variations of these parameters, albeit with some necessary reduction in throughput.

To get a feeling for the numbers, consider a couple of numerical examples.

Example 6: Consider the pessimistic case, which would be typical of a digital system with an irregular interconnection topology that prevents easy coordination of interconnect delay and clock skew. For a given clock speed or throughput, we can determine from (19) the largest interconnect delay d_{max}, that can be tolerated, namely, $(T - t_s)/3$, assuming that the interconnect delay and clock skew are not coordinated and assuming the worst-case propagation delay, $t_p = 0$. For a 100 MHz clock frequency, a clock period of 10 ns and, assuming the settling time is 80% of the clock period, the maximum interconnect delay is 667 ps. The delay of a data or clock signal on a printed circuit board is on the order of 5–10 ps/mm (as compared to a free-space speed of light of 3.3 ps/mm). The maximum interconnect distance is then 6.7–13.3 cm. Clearly, synchronous interconnect is not viable on PC boards at this clock frequency under these pessimistic assumptions. This also does not take into account the delay in passing through the pins of a package, roughly 1–3 ns (for ECL or CMOS, respectively) due to capacitive and inductive loading effects. Thus, we can see that interconnect delays become a very serious limitation in the board-level interconnection with 100 MHz clocks. □

Example 7: On a chip, the interconnect delays are much greater (about 90 ps/mm for Al-SiO2-Si interconnect), and are also somewhat variable due to dielectric and capacitive processing variations. Given the same 667 ps interconnect delay, the maximum interconnect distance is now about 8 mm. (This is optimistic since it neglects the delay due to source resistance and line capacitance—which will be dominant effects for relatively short interconnects.) Thus, we see difficulties in using synchronous interconnect on a single chip for a complex and global interconnect topology. □

Again, it should be emphasized that greater interconnect distance is possible if the clock skew and interconnect delay can be coordinated, which may be possible if the interconnect topology is simple as in one-dimensional pipeline. This statement applies at both the chip and board levels.

4) Parallel Signal Paths: An important practical importance of pipeline registers is in synchronizing the signals on parallel paths. The transition phase offset between these parallel paths tends to increase through computational blocks and interconnect, and can be reduced by a pipeline register to the order of the clock skew across the bits of this multibit register. Again, the register can be viewed as reducing the size of the uncertainty region, in this case spatially as well as temporally.

In Section III-A-1), we defined the total uncertainty region for a collection of parallel signals as the union of the uncertainty regions for the individual signals. From the preceding, the total throughput is then bounded by the reciprocal of the length of this aggregate uncertainty period. In contrast, if each signal path from among the parallel paths were treated independently (say using the mesochronous techniques to be described later), the resulting throughput could in principle be increased to the reciprocal of the maximum of the individual uncertainty periods. For many practical cases, we would expect the longest uncertainty period to include the other uncertainty periods as subsets, in which case these two bounds on throughput would be equal; that is, there is no advantage in treating the signals independently. The exception to this rule would be where the uncertainty periods were largely nonoverlapping due to a relative skew between the paths that is larger than the uncertainty period for each path, in which case there would be considerable advantage to dealing with the signals independently.

B. Anisochronous Interconnect

The synchronous interconnect approach uses isochronous signals throughout the system, since all signals are slaved to an isochronous clock. A popular alternative to synchronous interconnect has been to abandon the isochronous assumption, and further abandon the use of a global clock signal altogether. Rather, the elements of the system are chosen to fall within an equipotential region, and the interconnection between elements is designed to operate in a delay-insensitive manner; that is, operate reliably regardless of what the delays are. This is accomplished by having each element of the system generate a *completion signal*, which has a transition coincident with the settling time of that element. The completion signal is a sort of locally generated clock, and is used to synchronize the different elements. Since the completion signal depends on the settling time, which can be data-dependent, the resulting signals are anisochronous. For example, if an ALU has two instructions with different settling times, then the signal frequencies will depend on the mix of instructions, and will thus be anisochronous.

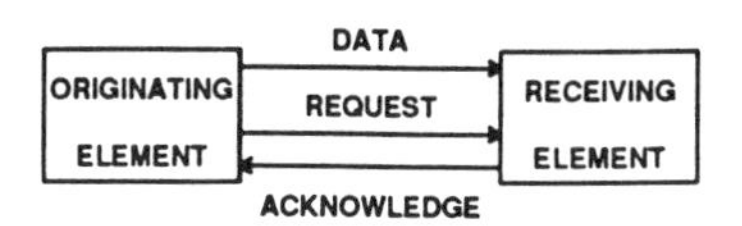

Fig. 11. Anisochronous interconnection, where coordination is performed by handshake signals.

In VLSI design, the term *self-timed interconnect* is used to describe this design approach [6]. Since this terminology conflicts with that of digital communication, we will use the term *anisochronous* instead. It is also common to refer to this method of synchronization as *asynchronous interconnect*, which is not precise, because this is only one of a number of possible asynchronous approaches (we will see some others later).

The idea behind the anisochronous interconnection is shown in Fig. 11. The originating element generates a handshake signal called a *request*, indicating that the settling time has been reached and the data are available. In addition, another handshaking signal called the *acknowledge* is generated at the destination, and indicates to the originator that the signal has been received and the originator is free to proceed with the generation of a new signal. The correct operation of the anisochronous interconnection does not depend on assumptions about the interconnect delays, other than that they be essentially identical for the data and handshake wires (which is reasonable if they are routed together). This is because the roundtrip feedback ensures that the two operations are synchronized, independent of the interconnect delay. Further details on anisochronous interconnect can be found in [6], with recent results summarized in [10]–[12].

A more detailed diagram of the anisochronous interconnect is shown in Fig. 12. The clock in Fig. 10 has been replaced by a pair of handshake blocks H1 and H2, which generate the request signal R_{out} for the next block and accept the acknowledge signal A_{in} from that same block. In addition, the calculation block now generates a completion signal C, indicating that the setting time has been completed, and accepts an input signal R_{out} which initiates its computation.

A *state transition diagram* for H1 is shown in Fig. 13 for a four-phase handshaking circuit appropriate in Fig. 12. This diagram models the order in which transitions in the H1 occur, and also the precedences that must be maintained by the circuitry in H1. For example, an arc from A_{out}^+ to R_{out}^+ indicates that the positive transition in A_{out} must precede the positive transition R_{out}.

The maximum throughput with which the system can operate is determined by the largest latency in any loop in Fig. 13. That maximum latency is the loop on the right, $R_{out}^+ \rightarrow A_{in}^+ \rightarrow R_{out}^- \rightarrow A_{in}^-$. If the settling time between the registers is the computational settling time t_s plus the interconnect delay d as shown, then the latency of the $R_{out}^+ \rightarrow A_{in}^+$ transition must be $t_s + 2d$ because this loop through H2 includes two interconnect delays plus the settling time, and the latency in the transition $R_{out}^- \rightarrow A_{in}^-$ must similarly be $2d$ because there is no computation in this loop. Additional latencies due to the handshake circuitry have been neglected. The total throughput is thus bounded by

$$\frac{1}{T} \leq \frac{1}{t_s + 4d}. \tag{22}$$

There are two-phase handshake approaches that are less reliable but reduce the $4d$ in the denominator to $2d$. Comparing (22) to (19), we see that the anisochronous interconnect throughput depends on the *actual* delay, whereas the pessimistic bound on throughput for synchronous interconnect depends on the *maximum* delay. (In the case of a pipeline, the total throughput will be dominated by the block in the pipeline with the maximum delay, so the two will be essentially the same.) On the other hand, the synchronous interconnect can take advantage of a nonzero propagation time t_p to pipeline without pipeline registers, whereas the anisochronous interconnect does not. At its best, synchronous interconnect can operate with a throughput that is independent of interconnect delay, and only limited by delay *variation*, as demonstrated in (20). Thus, depending on the circumstances, either the synchronous or anisochronous interconnect can achieve the higher throughput. However, in the presence of large interconnect delays, the synchronous interconnect clearly has a potentially higher throughput.

An important point is that the throughput of both the synchronous and anisochronous interconnect are generally adversely affected by the interconnect delay, and especially so for complicated interconnect topologies. As technologies scale, this restriction will become a more and more severe limitation on the performance of digital systems. However, as demonstrated by digital communication, which has experienced large interconnect delays from its inception, this delay-imposed throughput limitation is not fundamental, but is imposed in the synchronous case by the open-loop nature of the setting of clock phase.

An advantage of both synchronous and anisochronous interconnect is that they are free of metastability. This is avoided by ensuring through the design methodology that the clock and data are never lined up precisely, making it possible for clocked memory elements to reliably sample the signal.

C. Synchronization in Digital Communication

In digital communication, the interconnect delays are very large, so that alternative synchronization techniques are required [3]. These approaches are all isochronous, implying that the signals are all slaved to clocks, but differ as to whether a common clock distributed to each node of the network is used (mesochronous) or independent clocks are used at the nodes (plesiochronous and heterochronous). They also share a common disadvantage relative to synchronous and anisochronous interconnect—the inevitability of metastable behavior. Thus, they all have to be designed carefully with metastability in mind, keeping the probability of that condition at an acceptable level.

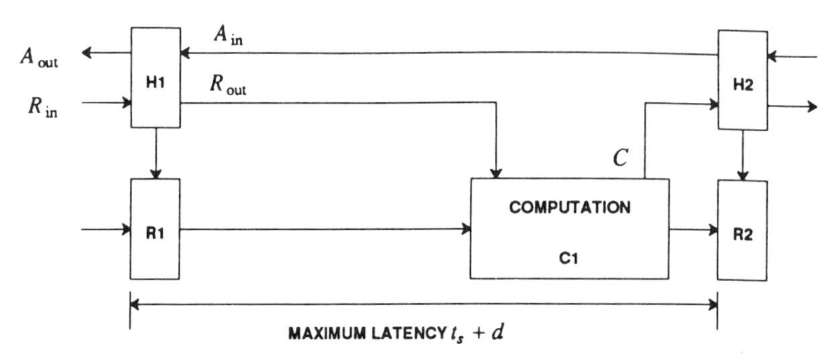

Fig. 12. An anisochronous interconnect circuit, in which relative to Fig. 10 the clock has been removed and replaced by two handshake circuits H1 and H2. H1 and H2 generate the clocks for R1 and R2, respectively. The signals R and A are, respectively, the request and acknowledge handshaking signals, essentially a locally generated clock, and C is the completion signal for the logic block.

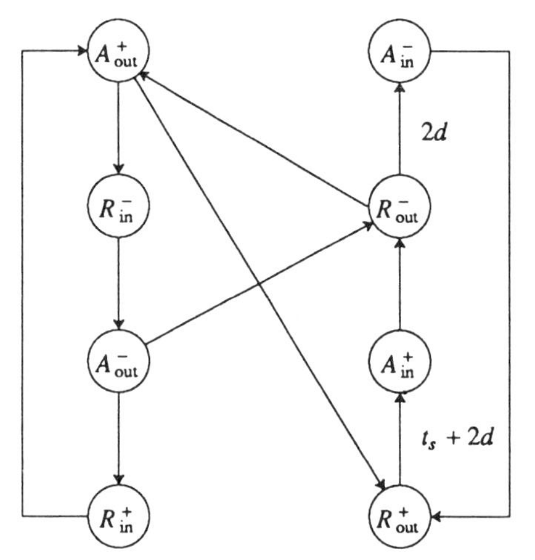

Fig. 13. A state transition diagram for a four-phase pipeline handshaking circuit H1 from [10]. The superscript "+" or "−" indicates a positive or negative transition in the corresponding signal. Arcs are labeled with the corresponding latency.

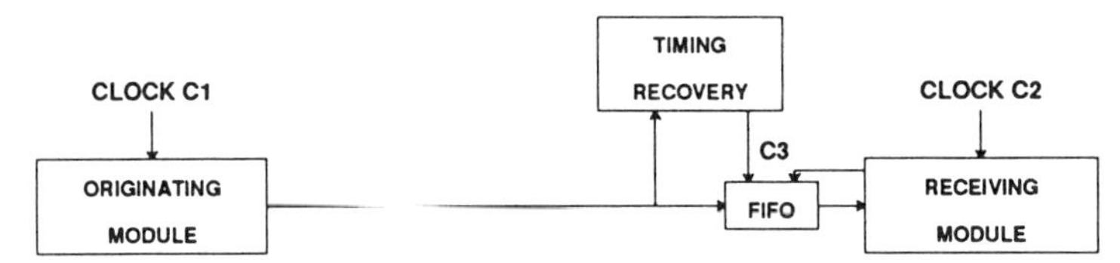

Fig. 14. An illustration of how throughput is made independent of delay in digital communication.

Limitations to throughput due to propagation delay are avoided in digital communications as shown in Fig. 14. First, two communicating modules are each provided a clock—C1 and C2. A clock is also derived from the incoming signal in the *timing recovery circuit*, and is denoted C3. These clocks have the following relationships.

• C3 is synchronous with the signal at the input to the FIFO, since it is derived from that signal.

• C3 is mesochronous to C1, since it has the same average frequency as dictated by the common signal but has an indeterminate phase due to the interconnect delay. It can also have significant phase jitter due to a number of effects in the long-distance transmission [3].

• C1 and C2 are either mesochronous, if they originated from a common source, or they are independent. In the latter case, they are either plesiochronous or heterochronous.

The purpose of the FIFO (also called an *elastic store*) is to adjust for the differences in phase, likely to be time varying, between C3 and C2. For mesochronous C1 and C2, this phase difference is guaranteed to be bounded, so that the FIFO can be chosen with sufficient storage capacity to never overflow. For heterochronous C1 and C2, where the average frequency of C1 is guaranteed to be lower than the average frequency of C2, the FIFO occupancy will decrease with time, and hence no data will ever be lost. For plesiochronous C1 and C2, the FIFO could overflow, with the loss of data, if the average frequency of C1 happens to be higher than C2. This loss of data is acceptable on an occasional basis, but may not be permissible in a digital system design.

IV. Isochronous Interconnect in Digital Systems

We found previously that the performance of both synchronous and anisochronous interconnects in digital systems are limited as a practical matter by the interconnect delays in the system. With the anisochronous approach, this limitation was fundamental to the use of roundtrip handshaking to control synchronization. In the synchronous (but not anisochronous) case, we showed that this limitation is not fundamental, but rather comes from the inability to tightly control clock phases at synchronization points in the system. The reason is the "open-loop" nature of the clock distribution, making us susceptible to processing variations in delay. If we can more precisely control clock phase using a "closed-loop" approach, the throughput of the synchronous approach can more nearly approach the fundamental limit of (13), and considerably exceed that of anisochronous interconnect in the presence of significant interconnect delays. In this section, we explore some possibilities in that direction, borrowing techniques long used in digital communication.

A. Mesochronous Interconnect

Consider the case where a signal has passed over an interconnect and experienced interconnect delay. The interconnect delay does not increase the uncertainty period, and thus does not place a fundamental limitation on throughput. If this signal has a small uncertainty period, as for example it has been resynchronized by a register, then the certainty period is likely to be a significant portion of the clock cycle, and the phase with which this signal is resampled by another register is not even very critical. The key is to avoid a sampling phase within the small uncertainty period, which in synchronous interconnect can be ensured only by reducing the throughput. But if the sampling phase can be controlled in closed-loop fashion, the interconnect delay should not be a factor, as demonstrated in digital communication systems.

Another perspective on clock skew is that it results in an indeterminate phase relationship between local clock and signal; in other words, the clock and signal are actually mesochronous. In *mesochronous interconnect*, we live with this indeterminate phase, rather than attempting to circumvent it by careful control of interconnect delays for clock and signal. This style of interconnect is illustrated in Fig. 15. Variations on this method were proposed some years ago [13] and pursued into actual chip realizations by a group at M.I.T. and BBN [14], [15] (although they did not use the term "mesochronous" to describe their technique). We have adapted our version of this approach from the mesochronous approach used worldwide in digital communication, except that in this case we can make the simplifying assumption that the phase variation of any signal or clock arriving at a node can be ignored. The primary cause of the residual phase modulation will be variations in temperature of the wires, and this should occur at very slow rates and at most requires infrequent phase adjustments. This simplification implies that clocks need not be derived from incoming signals, as in the timing recovery of Fig. 14, but rather a distributed clock can be used as a reference with which to sample the received signal. However, we must be prepared to adjust the *phase* of the clock used for sampling the incoming signal to account for indeterminate interconnect delays of both clock and signal. We thus arrive at Fig. 15.

First, we divide the digital system into functional entities called "synchronous islands." The granularity of partitioning into synchronous islands is guided by the principle that within an island the interconnect delays are small relative to logic speed, and traditional synchronous design can be used with minimal impact from interconnect delays. The maximum size of the synchronous island depends to a large extent on the interconnect topology, as we have seen previously. In near-term technology, a synchronous island might encompass a single chip within a digital system and, for the longer term, a single chip may be partitioned into two or more synchronous islands. The interconnection of a pair of synchronous islands is shown in Fig. 15(a); externally, the connection is identical to the synchronous interconnection in Fig. 6. The difference is in the relaxed assumptions on interconnect delays.

The mesochronous case requires a more complicated internal interface circuit, as illustrated in Fig. 15(b). This circuit performs the function of *mesochronous-to-synchronous conversion*, similar in function but simpler than the FIFO in Fig. 14. This conversion requires the following.

- A *clock phase generator* to create a set of discrete phases of the clock. This does not require any circuitry running at speeds greater than the clock speed, but rather can be accomplished using a circuit such as a ring oscillator phase-locked to the external clock. A single phase generator can be shared among synchronous islands (such as one per chip), although to reduce the routing overhead it can be duplicated on a per-island or even on a multiple generator per-island basis.
- A *phase detector* determines the certainty period of the signal, for example, by estimating the phase difference between a particular phase of the clock and the transitions of the signal. Generally, one phase detector is required for each signal line or group of signal lines with a common source and destination and similarly routed.
- A *sampling register* R1 with sampling time chosen well within the certainty period of the signal, as controlled by the phase detector. This register reduces the uncertainty period of the signal, for example, eliminating any finite rise-time effects due to the dispersion of the interconnect, and also controls the phase of the uncertainty period relative to the local clock. Depending on the effect of temperature variations in the system, this appropriate

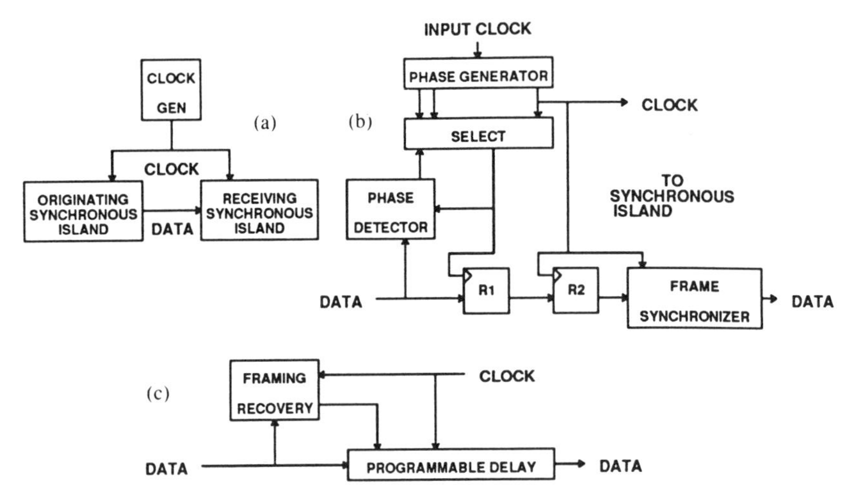

Fig. 15. A mesochronous interconnection for a digital system. (a) Clock and signal distribution are identical to the synchronous case of Fig. 6(a). Each functional unit corresponds to a synchronous island. (b) Interface circuit required for mesochronous-to-synchronous conversion. (c) Optional frame synchronizer circuit (within the synchronous island).

phase may be chosen once as power-up, or may be adjusted infrequently if the phase variations are a significant portion of a clock cycle.

• A second register R2 resamples the signal using the clock phase used internally to the synchronous island. At this point, the signal and this internal clock are synchronous. By inference, all the signals at the input to the synchronous island are also synchronous.

• If the system interconnect delays are larger than one clock cycle, it may be convenient at the architectural level to add an optional *frame synchronizer*, which is detailed in Fig. 15(c). The purpose of this block is to synchronize all incoming signal lines at the word-level (for example, at byte boundaries) by adding a programmable delay. This framing also generally requires additional framing information added to the signal.

The practicality of the mesochronous interconnection depends to a large extent on the complexity of the interconnection circuit and the speed at which it can be implemented. All the elements of this circuit are simple and pretty standard in digital systems with the possible exception of the phase detector. Thus, the practicality of mesochronous interconnection hinges largely on innovative approaches to implementing a phase detector. A phase detector can be shared over a group of signals with the same origin and destination and same path length, and it can even be time-shared over multiple such functions since we do not expect the phase to change rapidly. Nevertheless, the phase detector should be realized in a small area, at least competitive with anisochronous handshaking circuits. Further, it should be fast, not restricting the clock rate of the interconnect, since the whole point is to obtain the maximum speed. This should not be a problem on slower out-of-package interconnections, but may be difficult to achieve on-chip. Any phase detector design is also likely to display metastable properties, which must be minimized. The design of the phase detector may also place constraints on the signal, such as requiring a minimum number of transitions. In view of all these factors, the phase detector is a challenging analog circuit design problem.

B. *Heterochronous and Plesiochronous Interconnect*

Like synchronous interconnect, and unlike anisochronous interconnect, mesochronous interconnect requires distribution of a clock to all synchronous islands, although any interconnect delays in this distribution are not critical. If the power consumption of this clock is of concern, in principle it would be possible to distribute a submultiple of the clock frequency, and phase-lock to it at each synchronous island. But if the interconnect wires for clock distribution are of concern, they can be eliminated as in the digital communication system of Fig. 14, where the timing recovery is now required. There are two useful cases.

• If clocks C1 and C2 are *heterochronous*, such that C1 is guaranteed to have a lower frequency than C2, then the FIFO can be very small and overflow can never occur. However, the signal from the FIFO to the receiving module would have to contain "dummy bits," inserted whenever necessary to prevent FIFO underflow, and a protocol to signal where those bits occur.

• If clocks C1 and C2 are *plesiochronous*, then overflow of the FIFO can occur in the absence of flow control. Flow control would insert the "dummy bits" at the originating module whenever necessary to prevent FIFO overflow. Flow control could be implemented with a reverse handshaking signal that signaled the originating module, or if another interconnect link existed in the opposite direction, that could be used for this purpose.

V. Architectural Issues

While synchronous interconnect is still the most common, both anisochronous and mesochronous interconnects are more successful in abstracting the effects of interconnect delay and clock skew. Specifically, both design styles ensure reliable operation independent of interconnect delay, without the global constraints of sensitivity to clock distribution phase. Each style of interconnect has disadvantages. Mesochronous interconnect will have metastability problems and requires phase detectors, while anisochronous interconnect requires extra handshake wires, handshake circuits, and a significant silicon area for completion-signal generation.

The style of interconnection will substantially influence the associated processor architecture. The converse is also true—the architecture can be tailored to the interconnect style. As an example, if at the architectural level we can make a significant delay one stage of a pipeline, then the effects of this delay are substantially mitigated. (Consider, for example, (22) with $t_s = 0$.)

Many of these issues have been discussed with respect to anisochronous interconnection in [16]. On the surface one might presume that mesochronous interconnection architectural issues are similar to synchronous interconnection which, if true, would be a considerable advantage because of the long history and experience with synchronous design. However, the indeterminate delay in the interconnection (measured in bit intervals at the synchronous output of the mesochronous-to-synchronous conversion circuit) must be dealt with at the architectural level. For synchronous interconnection, normally the delay between modules is guaranteed to be less than one clock period (that does not have to be the case as illustrated in Fig. 8), but with mesochronous interconnection the delay can be multiples of a bit period. This has fundamental implications to the architecture. In particular, it implies a higher level of synchronization (usually called "framing" in digital communications) which line up signal word boundaries at computational and arithmetic units.

In the course of addressing this issue, the following key question must probably be answered:

> Is there a maximum propagation delay that can be assumed between synchronous islands? If so, is this delay modest?

The answer to this question is most certainly "yes" in a given chip design, but difficult to answer for a chip designed to be incorporated into a mesochronous board- or larger-level system. As an example of an architectural approach suitable for worst-case propagation delay assumptions, we can include frame synchronizers like Fig. 15(c) which "build-out" each interconnection to some worst-case delay. Every interconnection thus becomes worst-case and, more importantly, predictable in the design of the remainder of the architecture. However, this approach is probably undesirable for reasons that will soon be elaborated. Another approach is to build into the architecture adjustment for the actual propagation delays, which can easily be detected at power-up. Yet another approach is to use techniques similar to packet switching in which each interconnection carries associated synchronization information (beginning and end of message, etc.), and design the architecture to use this information. We have studied this problem in some detail in the context of interprocessor communication [17].

As previously mentioned, each data transfer in an anisochronous interconnect requires two to four propagation delays (four in the case of the most reliable four-phase handshake). In contrast, the mesochronous interconnection does not have any feedback signals and is thus able to achieve whatever throughput can be achieved by the circuitry, independent of propagation delay. This logic applies to feedforward-only communications. The more interesting case is the command-response situation or, more generally, systems with feedback, since delay will have a considerable impact on the performance of such systems. To some extent, the effect of delay is fundamental and independent of interconnect style: the command-response cycle time cannot be smaller than the roundtrip propagation delay. However, using the "delay build-out" frame synchronizer approach described earlier would have the undesirable effect of unnecessarily increasing the command-response time of many interconnections in the system. Since this issue is an interesting point of contrast between the anisochronous and mesochronous approaches, we will now discuss it in more detail.

A. Command-Response Processing

Suppose two synchronous islands are interconnected in a bilateral fashion, where one requests data and the other responds. An example of this situation would be a processor requesting data from a memory—the request is the address and the response is the data residing at that address. Assuming for the moment that there is no delay generating the response within the responding synchronous island, the command-response cycle can be modeled simply as a delay by N clock cycles, corresponding to roughly a delay of $2t_p$, as illustrated in Fig. 16.

It is interesting that this delay is precisely analogous to the delay (measured in clock cycles) introduced by pipelining—we can consider this command-response delay as being generated by N pipeline registers. As in pipelining, this delay can be deleterious in reducing the throughput of the processing, since the result of an action is not available for N clock cycles. One way of handling this delay is for the requesting synchronous island to go into a wait state for N clock cycles after each request. The analogous approach in pipelining is to launch a new data sample each N cycles, which is known as *N-slow* [18], and the throughput will be inversely proportional to N. This approach is analogous to the anisochronous interconnection (which may require considerably more delay, such as four propagation delays for the transfer in each direction).

For the mesochronous case, there are some architectural alternatives that can result in considerably improved performance under some circumstances, but only if this issue is addressed at the architectural level. Some examples of these include the following.

Fig. 16. A model for command-response in a mesochronous interconnection. The delay is with respect to cycles of the common clock.

- If we have some forward-only communications coincident with the command-response communications, we can *interleave* these feedforward communications on the same lines.
- If we have a set of N independent command-response communications, we can interleave them on the interconnection. This is analogous to *pipeline interleaving* [19] (which can make full use of the throughput capabilities of a pipelined processing element provided that we can decompose the processing into N independent streams). If the responding island cannot accommodate this high a throughput, then it can be duplicated as many times as necessary (in pipelining, an analogous example would be memory interleaving).
- If we cannot fully utilize the interconnection because of the propagation delay, then at least we should be able to allow each processing element to do useful processing (at its full throughput) while awaiting the response. This is analogous to what is sometimes done when a synchronous processor interacts asynchronously with a slower memory.

The last of these options would be available to an anisochronous system, since the communication portion could be made a separate pipeline stage. However, the first two options—two forms of interleaving of communications—are not available in anisochronous systems because the *total* throughput is bounded by the propagation delay. If the communication bottlenecks are taken into account at the architectural level, it appears that mesochronous interconnection offers considerable opportunity for improved performance.

VI. Conclusions

In this paper we have attempted to place the comparison of digital system synchronization on a firm theoretical foundation, and compare the fundamental limitations of the synchronous, anisochronous, and mesochronous approaches. A firm conclusion is that interconnect delays place a fundamental limitation on the communication throughput for anisochronous interconnect (equal to the reciprocal of two or four delays), this limitation does not exist for mesochronous interconnect. Further, mesochronous interconnect can actually achieve pipelining in the interconnect (as illustrated in Example 1) without additional pipeline registers, whereas anisochronous cannot. Further, anisochronous requires extra interconnect wires and completion signal generation. Thus, as clock speeds increase and interconnect delays become more important, mesochronous interconnect shows a great deal of promise. However, the advantages of any synchronization technique cannot be fully exploited without modifications at the architectural level.

Synchronization is the most difficult issue faced in digital system interconnect viewed as a digital communication problem, but there are some other techniques that can be considered. One of the more interesting is line code design [3]. Electrical interconnect displays dispersive properties which cause intersymbol interference at high speeds, through both bandwidth constraints and microreflections. Multilevel line codes would be an interesting possibility from the perspective of increasing the data rate within a bandwidth constraint. Also, the line code can constrain the number of transitions per unit time, easing, for example, the design of phase detectors [20].

Equalization of the intersymbol interference is a well-proven technology, but it is likely not practical at the high speeds of digital system interconnect.

Appendix A

In this Appendix, we determine the certainty region for the parameters in Fig. 10. Consider a computation initiated by the clock at R1 at time t_0. Extending our earlier results, the conditions for reliable operation are now as follows.

- The earliest the signal can change at R2 is after the clock transition $t_0 + \delta$, or

$$t_0 + \delta < t_0 + t_p + d + \epsilon. \tag{23}$$

- The latest time the signal has settled at R2 must be before the next clock transition $t_0 + \delta + T$, where T is the clock period,

$$t_0 + t_s + d + \epsilon < t_0 + \delta + T. \tag{24}$$

Simplified, these equations become

$$\delta < t_p + d + \epsilon \tag{25}$$

$$T + \delta > t_s + d + \epsilon. \tag{26}$$

Together, (25) and (26) specify a *certainty region* for $\{\delta, T\}$ where reliable operation is guaranteed.

With the aid of Fig. 17, we gain some interesting insights. First, if the skew is sufficiently positive, reliable operation cannot be guaranteed because the signal at R2 might start to change before the clock transition. If the skew is negative, reliable operation is always guaranteed if the clock period is large enough.

Idealistic Case: The most advantageous choice for the skew is $\delta = t_p + d + \epsilon$ at which point we get reliable operation for $T > t_s - t_p$. Choosing this skew, the fundamental limit of (13) would be achieved. This requires precise knowledge of the interconnect delay d as well.

Pessimistic Case: In Fig. 17 we see the beneficial effect of ϵ, because if it is large enough, reliable operation is guaranteed for any $\delta_{\max}$. In particular, the condition is $t_p + d + \epsilon > \delta_{\max}$, or

$$t_p + d > \delta_{\max} - \epsilon. \tag{27}$$

Since the interconnect delay d is always positive, (27) is guaranteed for any t_p and d so long as $\epsilon > \delta_{\max} - t_p$. Since ϵ also has the effect of increasing T, it is advantageous to ϵ as small as possible; namely, $\epsilon = (\delta_{\max} - t_p)$. Referring

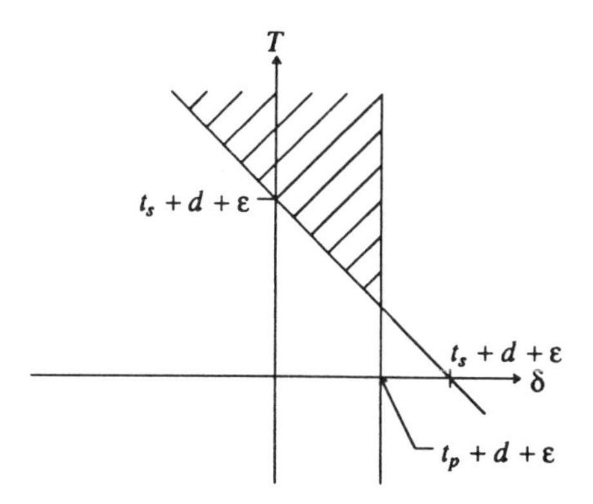

Fig. 17. The crosshatched area is the certainty region, now in terms of $\{\delta, T\}$. Note that it is not always possible to choose the clock period T sufficiently large to fall in the certainty region, but it is always possible to choose T and ϵ together large enough to accommodate any bounded clock skew.

back to Fig. 17, reliable operation is guaranteed if

$$T > (t_s - t_p) + d + 2\delta_{\max}. \tag{28}$$

Relative to the fundamental bound on clock period $(t_s - t_p)$, the clock period must be increased by $(d + 2\delta_{\max})$. Taking account of the worst case $d = d_{\max}$, we get (18).

Optimistic Case: The throughput of (20) follows easily by the same method.

Acknowledgment

The author is pleased to acknowledge helpful conversations on these issues with friends and colleagues, including T. Meng of Stanford; C.-S. Li, H.-D. Lin, S.-F. Chang, and P. Gray of U.C. Berkeley; K. Parhi of the University of Minnesota; M. Hatamian of AT&T Bell Laboratories; and H. Stone of IBM Research.

References

[1] M. Hatamian, L. A. Hornak, T. E. Little, S. K. Tewksbury, and P. Franzon, "Fundamental interconnection issues," *AT&T Tech. J.*, vol. 66, p. 134, July 1987.

[2] D. G. Messerschmitt, "Digital communication in VLSI design," in *Proc. 23rd Asilomar Conf. Signals, Syst. Comput.*, Oct. 1989.

[3] E. A. Lee and D. G. Messerschmitt, *Digital Communication.* Norwell, MA: Kluwer Academic, 1988.

[4] T. J. Chaney and C. E. Molnar, "Anomalous behavior of synchronizer and arbiter circuits," *IEEE Trans. Comput.*, vol. C-22, p. 421, Apr. 1973.

[5] G. H. Bennett, "Pulse code modulation and digital transmission," *Marconi Instruments*, Apr. 1978.

[6] C. Mead and L. Conway, *Introduction to VLSI Systems.* Reading, MA: Addison-Wesley, 1980.

[7] D. Wong, G. De Micheli, and M. Flynn, "Designing high-performance digital circuits using wave pipelining," *IEEE ICCAD-89 Dig. Tech. Papers*, Nov. 1989.

[8] M. Hatamian and G. L. Cash, "Parallel bit-level pipelined VLSI designs for high-speed signal processing," *Proc. IEEE*, vol. 75, p. 1192, Sept. 1987.

[9] M. Hatamian, "Understanding clock skew in synchronous systems," in *Concurrent Computations*, S. C. Schwartz, Ed. New York: Plenum, 1988.

[10] T. Meng, R. W. Brodersen, and D. G. Messerschmitt, "Asynchronous logic synthesis for signal processing from high-level specifications," *IEEE ICCAD 87 Dig. Tech. Papers*, Nov. 1987.

[11] T. Meng, G. Jacobs, R. Brodersen, and D. Messerschmitt, "Implementation of high sampling rate adaptive filters using asynchronous design techniques," in *Proc. IEEE Workshop VLSI Signal Processing*, Nov. 1988.

[12] T. Meng and D. G. Messerschmitt, "Automatic synthesis of asynchronous circuits from high-level specifications," *IEEE Trans. Comput. Aided Design*, accepted for publication.

[13] W. Nix, "A system for synchronous data transmission over a line of arbitrary delay," M.S. Project Report, U.C. Berkeley, 1981.

[14] P. Bassett, L. Glasser, and R. Rettberg, *Dynamic Delay Adjustment: A Technique for High-Speed Asynchronous Communication.* Cambridge, MA: M.I.T. Press, 1986.

[15] P. Bassett, "A high-speed asynchronous communication technique for MOS VLSI systems," Mass. Inst. Technol., Dec. 1985.

[16] T. Meng, G. Jacobs, R. Brodersen, and D. Messerschmitt, "Asynchronous processor design for digital signal processing," in *Proc. IEEE Int. Conf. ASSP*, Apr. 1988.

[17] M. Ilovich, "High performance programmable DSP architectures," Ph.D. dissertation, Univ. Calif., Apr. 1988.

[18] K. Parhi and D. G. Messerschmitt, "Pipeline interleaving and parallelism in recursive digital filters, Part I: Pipelining using scattered look-ahead and decomposition," *IEEE Trans. Acoust., Speech, Signal Processing*, July 1989.

[19] E. A. Lee and D. G. Messerschmitt, "A coupled hardware and software architecture for programmable digital signal processors, Part I: Hardware," *IEEE Trans. Acoust., Speech, Signal Processing*, Sept. 1987.

[20] D. G. Messerschmitt and M.-K. Liu, "A fixed transition coding for high speed timing recovery in fiber optics networks," in *Proc. IEEE Int. Conf. Commun.*, June 1987.

Performance of Synchronous and Asynchronous Schemes for VLSI Systems

Morteza Afghahi, *Member, IEEE*, and Christer Svensson

***Abstract*—Continuous advances in VLSI technology have made it possible to implement a system on a chip. One consequence of this is that the system will use a homogeneous technology for interconnections, gates, and synchronizers. Another consequence is that the system size and operation speed increase, which leads to increased problems with timing and synchronization. In this paper system and delay models necessary for the study of time performances of synchronous and asynchronous systems are developed. Clock skew is recognized as a key factor for the performance of synchronous systems. A new mode of clocking that reduces the clock skew substantially is proposed and examined. Time penalty introduced by synchronizers is recognized as a key factor for the performance of asynchronous systems. This parameter is expressed in terms of system parameters. Different techniques and recommendations concerning performance improvement of synchronous and asynchronous systems are discussed.**

***Index Terms*— Arbiter, asynchronous, clocking, delay model, digital, metastability, scaling, synchronous, synchronization, VLSI.**

I. Introduction

VLSI technology has shown spectacular capacity improvements during the last decade. Today, very powerful computers can be built from a few VLSI circuits on a chip. The introduction of VLSI systems has several important consequences. One consequence is that the system will use a homogeneous technology, that is, each part of the system is built from the same devices with the same performance. Another consequence is that the system size and operation speed increase, which leads to an increased cost of communications in terms of area and time and, thereby, increased problems with timing and synchronization.

For LSI system design, a synchronous timing network [1] is the conventional method. As the complexity of integrated circuits increase, it is believed that synchronous systems, even on a chip, will have some serious shortcomings mainly because of their limited performances due to the skew produced by the globally distributed clock signal. One of the purposes of this paper is to review the performance of synchronous VLSI systems with special attention to the effects of continuous shrinking of devices and interconnection line geometries and the effect of increasing the chip sizes.

It has long been proposed that problems associated with synchronous systems could be solved by using asynchronous [1] communication between modules or processors composing the system. In this approach the global clock distribution network in the synchronous scheme is replaced by some variety of handshaking protocols among the modules and systems. Different asynchronous schemes may be divided into two general classes, namely, totally asynchronous or self-timed systems and globally asynchronous, locally synchronous systems [1], [37].

Entirely asynchronous circuits are most suitable for simple functions with few inputs and outputs because the circuit and design complexity of these circuits increase drastically with the number of inputs and outputs [11]. Self-timed elements are also required to be contained in an equipotential or isochronic [1] region. The maximum size for an equipotential region must be sufficiently small so that the propagation delay on any wire within the area is small in comparison with the switching delay. However, the rate at which the size of this region shrinks with shrinking MOS feature sizes is higher than that at which circuit sizes decrease. Indeed, for 0.5 μm and below, the metal line propagation delay is larger than hundreds of picosecond for local distances ($\leq$ 500 μm), while high-speed devices exhibit a switching time of the order of 10 ps [6]. Therefore, less and less devices will be accommodated within each equipotential region. This will increase the interactions among the self-timed elements and call for very high bandwidth between them. But, it is characteristic of all self-timed systems that elements engaged in communication are interconnected via at least one closed loop. When dual-rail handshaking is used—which is easier and safer compared to 2-cycle encoding—signals are required to make two trips to and from the elements for each operation. Considering the relative delay of lines to gates in a scaled technology, these communications will incur a serious performance limitation for self-timed systems. These problems have led designers to consider another form of asynchronous circuit realization that resembles a conventional clocked sequential circuit, namely, globally asynchronous but locally synchronous systems.

In the globally asynchronous and locally synchronous systems, modules have their own independent local clocks but communicate asynchronously on the system level. External asynchronous signals to these modules are made internal by means of synchronizers. When the synchronizer is not properly triggered by the module clock (i.e., when its set-up or hold time is violated), the device may be set to a metastable state (MSS) and produce a logically undefined output. Consequently, the input synchronizer is a potential

Reprinted from *IEEE Trans. Comput.*, vol. 41, no. 7, pp. 858–872, July 1992.

source of random errors propagated in the system. These errors can cause module or system failures. In order to reduce the error rate (i.e., the probability of synchronizer failure), it is suggested to increase the period of the clock to allow adequate settling time after each synchronization event [1], [21]. Systems using this scheme can be called fixed period systems [13]. In another scheme a pausable or extendible clock scheme [12] is suggested. In this scheme, when the synchronizer is in MSS, the local clock pauses, but soon after the flip-flop output is logically defined, the clock generates its next pulse. Q-modules [10], [11] also use the principle of a pausable clock but in these modules provisions are provided to ensure a correct sequence of operations on the system level. Also, stoppable clocks have been suggested [12] and investigated [14] for the modules. Such a scheme, however, may only be used for some subsystems.

Each of the schemes discussed above has its merits and demerits. The system designer is often faced with the problem of selecting a scheme that will give a better performance for a given application. Synchronous and asynchronous approaches may be compared on the basis of their time performances, size, testability, extendibility, ease of design and understandability, reliability,...etc. The two approaches are, however, mainly evaluated on the basis of their time performances. There is a lack of quantitative comparisons of the two approaches except for two studies on small processor arrays [2], [3] and some basic theoretical comparisons in [4] and [5]. To investigate the performances of these approaches for high-speed applications made possible by MOS feature size shrinking will be particularly interesting. However, comparing the performance of an entirely self-timed system with a conventional system is not without fundamental incongruence. They require different techniques for synchronization and different algorithms are more suited to some than to others. In this paper, we will therefore make a quantitative comparison of time performances of the synchronous and locally synchronous, globally asynchronous schemes.

Time performance of VLSI circuits is subject to limitations from two set of factors: systematic factors, like power supply fluctuations, which can be controlled by proper techniques and factors that are random, and therefore uncontrollable by improved techniques. Thus the random factors determine the achievable performance. Our major concern in this paper is the ultimate limitations set by these factors.

The rest of the paper is organized as follows: In Section II system models for synchronous and asynchronous schemes are presented. In Section III a delay model for the synchronous system is developed. Models are developed to estimate delay variation of a single clock path and clock skew on system level. Section IV provides delay models for various asynchronous schemes. Time penalties introduced by synchronizers are expressed in terms of system parameters. In Section V we present an evaluation of the effect of scaling critical dimensions on synchronous and asynchronous time performances and provide some guidelines for VLSI system designs. In Section VI a method is proposed that makes the design of synchronous systems easier and safer. In the last section some general conclusions are drawn from previous presentations.

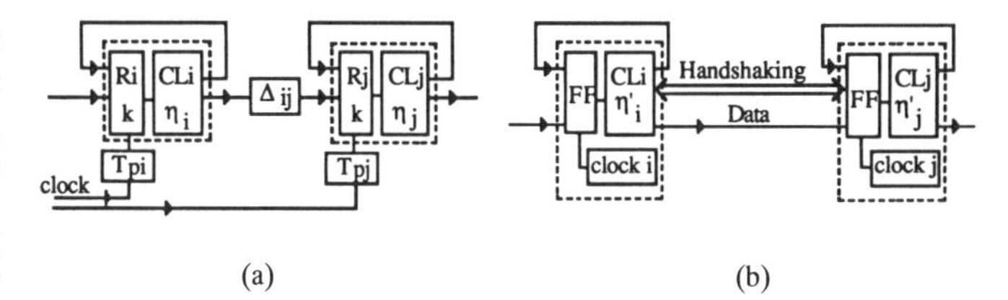

Fig. 1. A system model. (a) Synchronous system and (b) asynchronous system.

II. System Models

Time performance of synchronous and globally asynchronous, locally synchronous (hereafter we use only asynchronous) schemes, can be easily discussed in connection with a finite-state machine model of logic, Fig. 1. We consider the data rate from one module to another as the measure of time performance.

A module in a synchronous system, Fig. 1(a), may consist of registers (R) that sample and store inputs and the combinational logic (CL) that compute the task of the module. Data transfer between the modules is synchronized by a global clock signal.

An efficient and appropriate synchronous clock system should allow different modules their own local clock frequencies optimized for the corresponding module performances. Module clock frequencies as high as hundreds of megahertz or even gigzhertz for submicron technologies are not unexpected. A logic method which permits, for example, 250 MHz for an adder in a 3 μm CMOS technology has been developed [7] and [9]. The system performance is, however, determined by the data rate between modules. The clock that synchronizes the communications between modules can have a different frequency, usually lower, than the modules' clock frequencies. Thus, a suitable clock system for a synchronous scheme can be a hierarchy of clocks, one system clock which is distributed on the chip and one communication clock which synchronizes data communication between modules which is derived from the system clock. Module clocks may also be obtained from the system clock, see Fig. 2. Since both the module clock and the communication clock are derived from the same basic clock, they can be synchronized in each module. This clock system, however, when used in association with conventional clocking, results in unacceptable low module clock frequency and low data rate between modules. For example, in the method proposed by Anceau [8], the maximum module clock frequency and communication clock frequencies for a 0.3 μm technology are fixed and determined by the propagation delay of the clock metal wire rather than by the module gate delays. Assuming a skew equal to 10% of the propagation delay, the maximum module clock frequencies was estimated to be only 25 MHz and the maximum communication clock frequency to be 2.5 MHz. In Section III-C we will propose and examine another mode of clocking which makes the above clock strategy acceptable for scaled-down synchronous VLSI systems.

A module in an asynchronous system, Fig. 1(b), may be represented by an input synchronizer (FF) and the combinational logic. External asynchronous signals to these modules

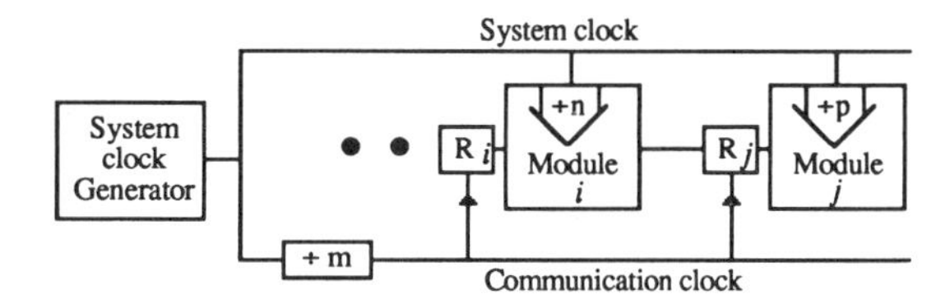

Fig. 2. A synchronous clocking strategy.

are made internal by means of synchronizers. When the synchronizer is properly triggered by the module clock, it produces a logically defined output state after its maximum normal delay time. However, if the input state changes too close in time to the triggering clock edge, the device is said to be marginally triggered and may be set to a MSS. It is then impossible to guarantee a specified time for the synchronizer to resolve from the metastable state. Consequently, the input synchronizer is a potential source of random errors propagated in the system. These errors can cause module or system failures. In order to reduce the error rate it is generally required to allocate extra time to these synchronizers to resolve to a stable state. We call this extra time, time penalty incurred by synchronizers, and will study its effects on the asynchronous system performance.

Communications between modules in an asynchronous system may be performed by some kind of handshaking between them. Depending on specific implementation of the modules, different handshaking may be used [1].

III. Synchronous Timing Model

In this section delay models necessary for evaluation of data rate between two communicating modules of the synchronous scheme are developed. In this analysis, for simplicity, edge-triggered D-flip-flops are used for the registers.

In a synchronous system, the minimum clock period T_c must be larger than the time required for a data to leave the initial flip-flop, k, the time necessary to propagate through the combinational logic and interconnection, $\eta + \Delta_{ij}$, the clock skew $\delta = (T_{pj} - T_{pi})$, and the time required to successfully latch within the final register $t_{\text{set-up}}$. That is $T_c = k + \eta + \Delta_{ij} + \delta + t_{\text{set-up}}$. Detailed analysis of this expression involves the analysis of the relations between these parameters and the clocked register parameters for the specific clock method used [15], [9], [17], [46]. The clock period is also required to be long enough so that when a clock signal is released by the clock source, it must reach the the most remote module before the next clock signal is sent. Therefore, the clock period must be greater than T_p, where T_p is the time for the clock signal to propagate from the clock source to the end of the longest clock path(s) [2]. Thus the clock period must be

$$T_S \geq \text{MAX.}\,(T_c, T_p). \tag{1}$$

In some pessimistic studies of synchronous systems [2], [3] it is assumed that the clock signal period should be constrained to be longer than the time required to charge and discharge the clock distribution tree. In this way the term T_p in (1) must be replaced by 2 T_p . In the following it is shown that

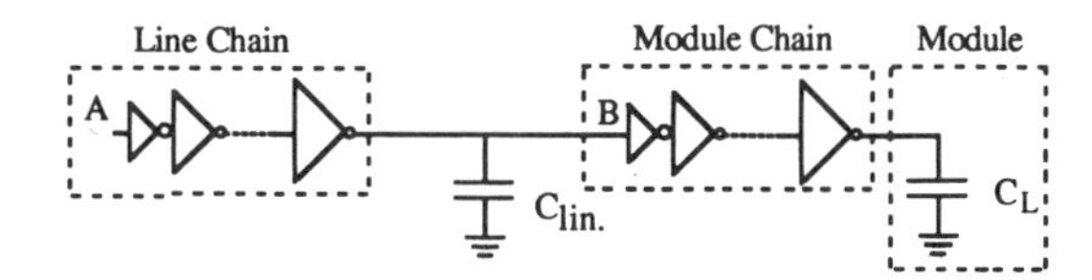

Fig. 3. A clock distribution path when the line resistance is relatively low

the clock path delay is correct in the relation (1) rather than the charging and discharging time. It is later shown that it is possible to design a system which is not even constrained by T_p. T_s is then controlled only by T_c.

Clock skew, δ, may arise from unequal clock path lengths to various modules. To equalize line length, and thus reduce the clock skew, a common practice is the use of a balanced clock network. Still process variations prevents line delays from being exactly equal, thus introducing some skew. This process induced skew is to some extent related to the clock path delay, T_p. T_p is also directly involved in relation (1). Therefore, to investigate the performance we begin with estimating the clock path delay T_p.

A. Clock Path Delay

A clock distribution path used in a synchronous system has one single input, called the clock generator or source. For practical reasons there is only one path from the source to each functional module. There are a number of different structures used for the clock distribution path, like bus, tree, hierarchical distributed buffering [16], [17], etc. However, the general organization of a clock signal path is depicted in Fig. 3. This path consists of a driver chain designed to minimize the delay introduced by the line capacitance and the driven inverter input capacitance. The path is ended with another driver chain, the module chain, to drive the load seen by the path, CL. This load is the sum of the gate capacitances of the registers included in the functional module clocked by the path. The interconnect lines are assumed to be made of metal (aluminum) and poly is used for local interconnections only.

In the above model it is assumed that the resistance of the interconnection line is considerably lower than the on-resistance of the transistor driving the line and, hence, the interconnection line can be modeled by a lumped capacitance. This is correct in a 3 μm technology. With scaling, however, the resistance of the line will soon become comparable or larger than the on-resistance of the transistor as for global lines it increases quadratically with the scaling factor and also because the size and complexity of chips increases with VLSI developments. In this case, the line driver chain minimizes the effect of the line capacitance only and will not affect the delay introduced by the line resistance. If, however, the line resistance is important, repeaters should be used along the interconnection line to reduce the effect of the line resistance and make the line delay linear with length. In Fig. 4, the interconnection line is divided into m sections.

In order to calculate the clock path delay and skew, a delay element is defined. A convenient delay element for calculation of the clock path delay is an inverter driving another inverter through an interconnection line, Fig. 5(a). This can be reduced

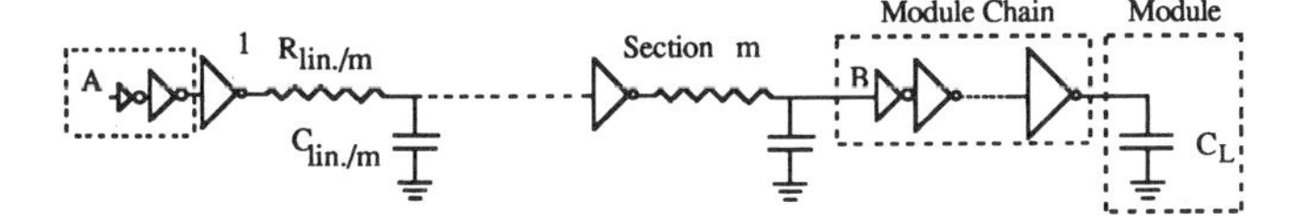

Fig. 4. Dividing the line into m sections by repeaters.

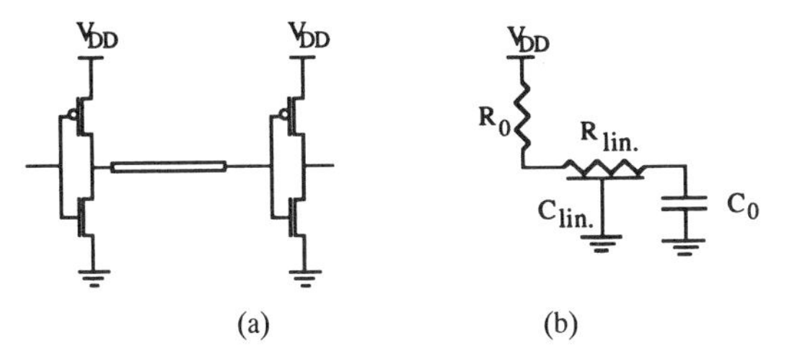

Fig. 5. The CMOS configuration (a) and the delay-model (b).

to a cascade of two inverters by putting $R_{\text{lin.}}$ and $C_{\text{lin.}}$ (i.e., resistance and capacitance of the interconnection line, respectively) equal to zero. To prevent confusions, we call an inverter driving another inverter through an interconnection line a section, and a cascade of two inverters a stage. We use the well-known empirical formula [19] for the rising (or falling) time of this element [see Fig. 5(b)].

$$T_{\text{dr}} = 1.0C_{\text{lin.}}R_{\text{lin.}} + 2.30(R_0C_{\text{lin.}} + R_0C_0 + R_{\text{lin.}}C_0) \quad (2)$$

where R_0 is the on-resistance of the driving transistor and $C_0 = C_{\text{ox}} \cdot W \cdot L$ is the input capacitance of the driven inverter. C_{ox} is the gate unit area capacitance and W and L are the width and length of the transistor, respectively. The delay of this circuit may then be obtained by averaging the rise and fall times. R_0 may be obtained from [19]:

$$R_0 \approx \frac{L/W}{\mu C_{\text{ox}}(V_{DD} - V_T)} \quad (3)$$

where is the charge carrier mobility.

B. Impact of Scaling on Clock Path Delay

In order to study the impact of scaling and technology development on the clock path delay we start with a typical 3 μm technology, see Table I. A practical minimum size transistor in this technology is assumed to have a gate width/length ratio of 7/3. Then a set of scaling rules must be chosen for the devices and the interconnections. In ideal scaling all vertical and horizontal geometries and voltages are divided by the scaling factor λ, while the substrate doping is increased by the same factor. All interconnection dimensions are also scaled down by this same scaling factor. However, for several practical reasons, this straightforward scaling rule cannot predict the performance of future technologies. A more realistic prediction of some major process parameters is presented in Table II [40]. Furthermore, the maximum chip area will also increase with the development of IC processing. Maximum chip area, A, for 3 and 1 mm feature sizes are assumed to be 0.25 and 1 cm^2, respectively [24]. The chip area corresponding to a device size of 0.3 μm can be 2 cm^2 [41]. Therefore, $\sqrt{A}$ for 3, 1, 0.5, and 0.3 μm technologies will be 0.5, 1, 1.20, and 1.40 cm, respectively. Since the maximum delay and skew that can occur on a chip is concerned, we assume that the maximum length of the clock path is $2\sqrt{A}$. This should also account for the increase in the clock path length due to the increase in chip complexity. Note that here a coupling between scaling and chip area is imposed through the expected VLSI technology development. With "scaling" we therefore mean "technological development" in the following.

TABLE I
PARAMETERS OF A TYPICAL 3 μm CMOS P-WELL PROCESS

Device Parameters		NMOS	PMOS	Unit
Threshold voltage	V_T	0.8	−0.8	V
Mobility	μ	600	220	cm^2/V s
Gate oxide thickness	t_{ox}	500	500	A_$circ$
Field oxide thickness	f_{ox}	7000	7000	A$_\circ$
Metal line				
Width		4		μm
Thickness		1		μm

TABLE II
A REALISTIC PREDICTION OF SOME CMOS PROCESS PARAMETERS

L_{eff} (μm)	V_D (V)	V_{TN} (V)	V_{TP} (V)	t_{ox} A°	μ_n (cm^2/V·s)	μ_p (cm^2/V·s)
1.0	5	0.8	−0.8	250	490	195
0.5	3	0.5	−0.5	100	433	168
0.3	3	0.5	−0.5	70	375	134

A clock path may be divided into two parts: The interconnection part from the clock source (point A in Figs. 3 and 4) to the module chain input (point B in the same figures) and the second part which is the module chain (buffer). To illustrate the effects of scaling on the interconnection line delays, the delay of a 1 cm interconnection line is first calculated. To this end, we proceed by taking a value for the number of repeaters and then iterate the transistor size of the repeater. Then the number of repeaters is incremented. This procedure is repeated until the minimum line delay is obtained. For example, the value of the minimum delay for 1 cm line in the 0.3 μm process is 4.0 ns and occurs when the number of repeaters is 13 and the size of the repeater is 16 times the minimum size inverter. The dashed line in Fig. 6 demonstrates the minimum delay of 1 cm long line for other feature sizes. The curve reveals a minimum delay at ≈ 0.7 μm. For feature sizes ≥ 0.7 μm, the feature size scaling decreases $C_{\text{lin.}}$ and increases $R_{\text{lin.}}$, but $R_{\text{lin.}}$ is still smaller or of the same order as R_0. Therefore, the delay is controlled by $C_{\text{lin.}}$ and R_0. Both of these decrease with scaling and hence the delay decreases. For feature sizes ≤ 0.7 μm $R_{\text{ lin.}}$ becomes comparable or greater than R_0. Thus, the delay is controlled by $C_{\text{lin.}}$ and $R_{\text{lin.}}$, but, the rate at which $C_{\text{lin.}}$ decreases with scaling is smaller than that at which $R_{\text{lin.}}$ increases, hence the delay increases.

To compute the minimum delay of a complete clock path, first the maximum chip size and hence the maximum line length at scaled technologies, stated above, should be considered. Then the delay of the interconnection part is added to the delay of the module chain buffer. We use a master–slave flip-flop [34, p. 219] for the edge-triggered flip-flops. We also assume that each module has sixteen flip-flops where the clock signal sees six 20/3 transistors in each device. The module chain will then be a cascade of five inverters with a tapering factor of 3. The solid line in Fig. 6 shows the minimum delay of complete paths. This curve also reveals a minimum delay at ≈ 0.7 μm. At 3 μm feature size, the interconnection part delay is 4.9 ns and the module chain delay is 6.1 ns. The total path delay is thus 11 ns. We see that the module chain contribution to the path delay is significant at this feature size. For feature sizes ≥ 0.7 μm, path delay is controlled by the module chain and line delays. Both of these decrease with scaling, hence path delay decreases. In this range, the influence of increase in line length is insignificant because the line is mainly capacitive and the rate at which the line becomes longer is smaller than that at which capacitance decreases with scaling. For feature sizes ≤ 0.7 μm, the unit line length delay increases with scaling, hence the influence of increase in the line length becomes more and more significant. In this range, the module chain contribution to the path delay becomes less and less significant. For example, at 0.3 μm, where the path length is 2.8 cm, the interconnection part delay is 10.66 ns while the module chain delay is only 0.34 ns.

Results of Fig. 6 can be used to estimate the maximum data rate of the synchronous system at different feature sizes. According to relation (1), the maximum data rate for this scheme occurs at a feature size ≈ 0.7 μm and is ≈ 200 MHz. The maximum data rate at 0.3 μm feature size will be only 90 MHz. This data rate is solely determined by the clock path delay and not by the switching time of the devices. This data rate will obviously be much lower than the high bandwidth made possible by high-speed devices at that feature size. To overcome this shortcoming, we have considered and analyzed another mode of clocking [33] that can result in a much higher data rate. This mode of clocking is described in the following.

C. Synchronous Delay Model Using Nonconventional Clocking

When inverters are inserted in a clock distribution network comparison between the total delay of the network and an RC charging time is not correct when estimating T_p, as each inverter regenerates the signal amplitude and slope with a gain much greater than the unity. The delay of each section, starting with an inverter, may then be considered separately, to the first order, from the rest of the circuit. Consider then the course of events when a signal is applied to the path shown in Figs. 3 and 4. By receiving a signal, each section will start charging or discharging its corresponding capacitance and then deliver the resulting signal to the next section. In the conventional method of clocking, implicit in expression (1), the next clock signal will be transmitted only after the first signal has reached the final point of the longest path, i.e., the module registers. In this way, sections are idle for most of the time. One may instead send a sequence of clock signals to march along the line without waiting for the first signal to reach the end of the path. In [33] it is shown, by SPICE simulation, that whatever

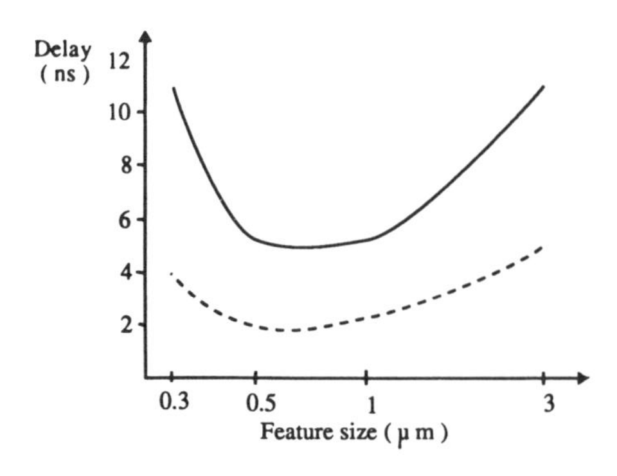

Fig. 6. Delay of 1 cm line length - - - - and a complete path ______.

the delay of a clock path may be, it is possible to send a clock on the path with a period equal to the delay of one pair of line sections. For example, simulation showed that a clock pulse with a period equal to 2 ns could be sent on a path with a total delay of 28.2 ns [33]. Using this approach for clock distribution, the data rate between communicating pairs will not be controlled by T_p in relation (1), but by T_c, thus,

$$T_S \geq T_c = k + \eta + \Delta_{ij} + \delta + t_{\text{set-up}}. \tag{4}$$

Notice that the modules tap the clock signals only from the end of the path and not from points along it. Assuming a balanced clock distribution network, the skew is minimal.

D. Clock Path Skew

There is little published material on a detailed treatment of the underlying physical causes of clock skew. Fisher and Kung [4] discuss the problem from a graph theory point of view for a processor array with an H-tree as clock distribution network. Results of some tests and simulations are presented in [17]. A probabilistic model for clock skew is also presented and examined in [43].

Clock skew may arise mainly from unequal clock path lengths to various modules. To equalize line lengths, and thus reduce the clock skew, a common practice is the use of a balanced clock network. In this way the nominal value of skew becomes zero and clock skew reduces to the variation of the clock path from the clock generator to the processing elements. These variations may originate from process variations and temperature fluctuations. Here, however, we consider only the effect of process variations. The effect of temperature fluctuations is assumed small, as local temperature variations are small in a VLSI chip. When estimating the clock path delay variation, either a worst-case or a statistical approach may be utilized. A worst-case approach can cause an unnecessarily long clock period. In a statistical approach, on the other hand, the clock parameters may be chosen so that the probability of a timing failure is very small, but not zero. This usually results in a shorter clock period. In this paper we have employed a statistical approach.

The delay of each delay element, shown in Fig. 5, can be associated with a real random variable. The delay of each clock path that is the sum of a number of such delay elements is also a real random variable. The variance of the clock path is also equal to the sum of the variances of the constituent delay elements of the path. Then, according to the central-limit theorem [22], the distribution of the path delay tends to a normal curve regardless of the actual distribution of the delay elements. Random variations of the path delay are due to both local and global variations of the parameters involved in that delay. However, since we are concerned with the differences in the paths delay on the same chip, only local variations are considered.

One approach to calculate the variance of the delay element due to the variations of process parameters is to express the relation adapted for the delay of the delay element, relation (2), in terms of independent variables. Reconsider the relation for the on-resistance of the transistor:

$$R_0 = 1/K(V_{DD} - V_T). \tag{5}$$

In a system the power supply is controlled globally and is not a random variable. Since V_T and K are both dependent on gate capacitance per unit area, they are correlated variables. However, it is theoretically and experimentally shown that their correlation factor is very close to zero, indicating that they can be considered as independent variables [44]. Consider then the gain constant, K, of the transistor. It is given by $K = \mu C_{OX} W/L$. All factors in this expression are independent variables. Now consider the variables necessary to calculate the delay of a delay element, the relation (2). These variables are geometrical dimensions, V_T (threshold voltage), μ (carrier mobility), and C_{ox} (unit area gate oxide capacitance). All these factors can be considered independent. Thus, the variance of the path delay can be determined in terms of variances of these independent random variables [22].

Analytical models were developed that related the variance of threshold voltage, gate oxide capacitance, and mobility to the device dimensions and extensive data were used to verify these models [23]. Standard deviation of geometrical parameters, W and L, for a 3 μm technology is in the range of 0.01–0.03 μm [23]. We used a value of 0.02 μm for these deviations and assume that aluminum can be deposited with $\pm 5\%$ accuracy. Using these variances and relation (2), it was possible to calculate the variance of the delay of a delay element and hence the variance of the clock path delay, see Appendix A for details.

To study the effects of scaling on the clock path delay variance, we again started from our 3 μm technology and applied the set of scaling rules to predict variances for the scaled down technologies. We assumed that the lithographic process would improve with scaling by a factor of ψ = 0.3λ [29] (this means that standard deviation of geometrical parameters, W and L, for a 0.3 μm technology, i.e., λ = 10, is 0.02/(1+0.3x10)=0.005 μm). Using the variances for scaled down technologies (see also Appendix A), the delay standard deviations, σ, for 1 cm line length and complete clock paths was calculated and depicted in Fig. 7.

These curves reveal some interesting facts. Standard deviations, σ, show the same trend with scaling as delays (see Fig. 6). At 3 μm feature size, σ is 46 ps for a 1 cm interconnection line (that includes the line driver chain) while it is 52 ps for the whole path (that consists of 1 cm interconnection line and the module chain). That is, standard deviation contribution of the module chain is only 6 ps compared to about 46 ps due to the interconnection part. This means that the variations in capacitance due to variations in line dimensions (width and thickness) is the main cause in the path delay variation and relative buffer (inverter) delay variation is minor. But, if the standard deviation is determined by variations in the line dimensions, why is the standard deviation of a 1 cm interconnection line at 0.3 μm feature size significantly less than at 3 μm feature size? To answer this question remember that in scaled technologies, repeaters were used to divide the line into sections. Then to find the whole interconnection variance, σ_2, section variances are added together. The following mathematical expression explains why σ is less for a 1 cm continuous line than for a 1 cm partitioned line.

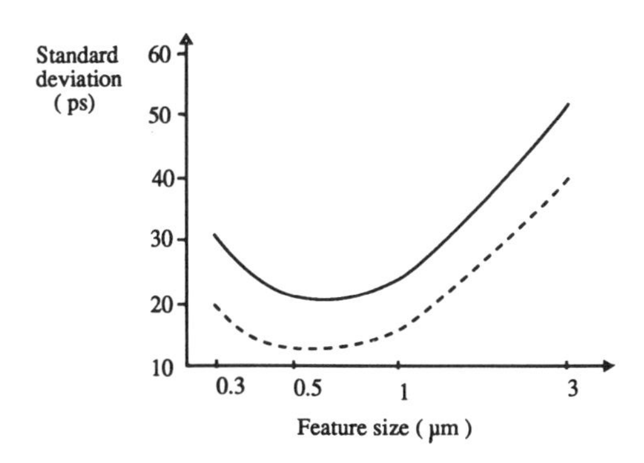

Fig. 7. Standard deviation 1 cm interconnection line - - - - - and a complete clock path ________ .

$$\sqrt{\sigma^2} > \sqrt{n \sum_{n} (\sigma/n)^2} \qquad n > 1.$$

The above discussion results in an important conclusion. Repeaters are effective in making the line delay linear with the line length in scaled technologies, but they are even more effective in reducing the standard deviation of the line delay. They make the standard deviation proportional to the square root of the line length.

In developing the expression for the variance of V_T it was assumed that V_T is independent of channel length and width. The threshold voltage of small geometry transistors is, however, a function of channel length and width. But it is demonstrated that the variations of V_T variance due to dimensional variations, in the worst case, is not more than 10% for small geometries [44]. Besides, it was noticed that the line edge variations are dominating in the path variances.

E. Maximum Skew on a Chip

Up to now we have only considered the delay variations of a single clock path. Proper operation of a single path, however, will not guarantee the proper operation of a chip containing many paths. These paths depend on each other as they may be overlapped in some parts of their length. However, it is proven that the expected range of variation of a set of random variables, that are related because they are the sums of overlapping variables, is no larger than the expected range of the corresponding set of independent random variables [43]. As we are interested in the maximum skew on a chip, we thus assume that the delay of the paths are independent random variables. Then for the set of independent paths that are normally distributed, the expected value and variance of the difference between the maximum and minimum delay (skew) is asymptotical [45]:

$$\begin{aligned} \text{Skew} &= \sigma\left(\frac{4\log R - \log\log R - \log 4\pi + 2C}{(2\log R)^{0.5}}\right) \\ \text{Variance[skew]} &\approx \frac{\sigma^2}{\log R} * \frac{\pi^2}{6} \end{aligned} \tag{6}$$

where $C \approx 0.5772$ is Euler's constant, R is the number of paths (or processing elements), and σ is the standard deviation of a single path delay. This is an asymptotic upper bound on the expected skew on a chip with R paths. To assure a very high (>99%) probability of system success, we chose a confidence level of ± 2.5 * [skew variance].

When estimating the skew in scaled technologies, notice that a unit area chip that contains R processing elements in a 3 μm technology, will have $R \times 10^2$ of the same processing elements in a 0.3 μm technology. Then for maximum skew one must also account for the increase in the area of that chip. As an example, assume that a chip in a 3 μm technology with 0.5 cm^2 area contains $R = 10^3$ of a particular processing element. According to Fig. 7 the standard deviation (σ) of the delay of a complete path on this chip is 52 ps. The maximum skew on this chip is then 0.437 ns from (6). Consider the same processing element on a 0.3 μm chip with an area of 2 cm^2. The standard deviation of the delay of a complete path on this chip is 31 ps. This chip will contain $R = 4 \times 10^5$ of the same processing element and thus the maximum skew on this chip will be 0.294 ns.

In the above discussions we assumed that the modules had suitably regular and identical structures so that a completely balanced clock distribution network could be mapped on them. Although regular structures and array type architectures are a very attractive and powerful approaches to VLSI design, cell-based designs, which are also common, rarely have regular structures and clock loading. This reduces the control over the exact placement of a clock line in automated chip assembly tools, resulting in a clock skew due to deviation of the clock network from a balanced network and different clock loads in the modules. This type of skew, that is systematic, differs from the kind of clock skew considered above which is fundamental and exists even in completely balanced clock distribution networks. In Section V it is shown that this skew can be insignificant compared to the process induced skew.

We now return to relation (1) that governs the data rate in a synchronous system. We noticed that, when the nonconventional mode of clocking is used, the data rate is solely determined by T_c. Terms involved in T_c are logic delays, interconnection delay and clock skew. Up to now we have estimated the clock skew. For a given combinational logic delay, it is only necessary to assume an interconnection delay to calculate the data rate.

In this section, we discussed the structure of a clock path. A delay element was defined. Based on this delay element the delay and variance of a clock path was estimated. It was shown that when conventional clocking is used, the data rate for systems with submicron is dictated by clock path delay rather than switching times. To alleviate this, another mode of clocking was examined. We further estimated the maximum clock skew due to the variations of process parameters. It was shown that the skew will not scale with scaling the feature size of MOS transistors. Then a probabilistic model for skew of a system on a chip was given. Based on this model and calculated variance of a single clock path, it is possible to predict the maximum skew for existing and future chips.

IV. Asynchronous Timing Model

In this section delay models for the asynchronous system are developed that serves as a basis for further analysis and discussions, althougth may seem brief.

A. The Probability of Failure

In a clocked module asynchronous system, Fig. 1(b), the external events to the modules are transferred into the modules through synchronizers by means of a sampling clock. The sampling clock frequency may in general be smaller than the module clock frequency (but synchronous with the module clock). As we will see below, the communication rate between modules can be limited by the reliability of the system, through the probability for synchronization failures. To have a low failure rate, it is generally necessary to allow extra time to these devices to resolve from metastability. The time penalty introduced in this way can be considered as an extra delay that affects the data rate. In estimating the time penalty introduced by synchronizers and therefore the data rate, one must consider the duration and the frequency with which MSS's occur.

Assuming that a positive edge triggered D-type flip-flop is used for the synchronizer, a metastability state may occur when the rising edge of the (module) sampling clock appears within a small time interval during a rising or falling edge of the input signal (i.e., when the flip-flop is marginally triggered), see Fig. 8. is equal to $t_{\text{set-up}} + t_{\text{hold}}$ of the flip-flop. The value of β is in the range of the propagation delay of the flip-flop. We assume it is the same for rising and falling edges of the input signal.

Because the input data signal is asynchronous, its rising and falling edges are distributed uniformly in time. Thus, the (conditional) probability that the rising edge of the sampling clock falls within a small region over one (average) period T2 of the data signal is

$$P = 2 * \frac{\beta}{T_2} * \frac{f_1}{f_2} = 2 * \beta * f_1 \tag{7}$$

where f_1 and f_2 are the module clock frequency and average input data frequency, respectively. It has been determined experimentally [25] that the probability that the resolution time of the flip-flop exceeds t, can be approximated by an exponential function. Thus, the average number of occurrences of MSS with duration longer than t at a clock event is

$$N(t) = P * \exp(-\frac{t-k}{\tau}) = 2 * \beta * f_1 * \exp(-\frac{t-k}{\tau}) \tag{8}$$

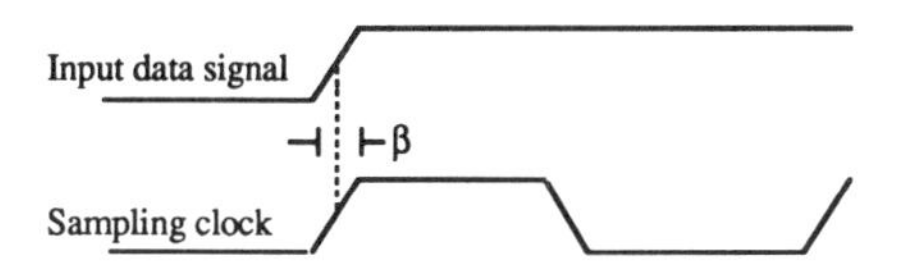

Fig. 8. Risk of metastability

where k is the intrinsic flip-flop response time and τ is the time constant of the flip flop circuit. For a set of m independent flip-flops all with the same characteristics, $N(t, m)$ is [13]

$$N(t,m) = 2 * m * \beta * f_1 * \exp(-\frac{t-k}{\tau}) \quad (9)$$

and average number of occurrences of MSS that exceeds t during n seconds is

$$N(t,m,n) = 2 * n * m * \beta * f_1 * f_2 * \exp(-\frac{t-k}{\tau}). \quad (10)$$

This failure rate is usually used as a measure of the reliability of asynchronous systems.

B. Data Rate Constraint in the Fixed-Period Clock Module

Suppose that the allowed duration of MSS is t_1, after which the information on the flip-flop output node must be valid. Those MSS exceeding t_1 can constitute a module failure. Thus, the average number of module failures during n seconds is $N(t_1, m, n)$. Notice that not all the MSS exceeding t_1 lead to a module failure, as a failure is a misinterpretation caused by a MSS. Situations exist where consistent interpretations of the input are possible even if the MSS exceeds t_1. $N(t_1, m, n)$ is the least upper bound for the expected number of failures and is thus a good measure of reliability [28]. If a chip contains R modules, the expected number of failures for the whole chip is $N(t_1, m, n, R) = R * N(t_1, m, n)$. Given the required reliability, it is now possible to calculate t_1 from $N(t_1, m, n, R)$.

There are two basic ways to provide adequate settling time after each synchronization event to achieve the required reliability. In the first technique the clock period is simply lengthened so that the excess time required by the synchronizer is covered by the clock period [1], see Fig. 9(a).

Thus, for this scheme, which can be called the simple scheme, the time interval between successive absorption of data by a module can be

$$T_{\mathrm{AS}} = t_1 + \eta + \Delta. \quad (11)$$

Clearly, the maximum data rate between these two communicative modules is $f_{\mathrm{AS}} = 1/T_{\mathrm{AS}}$. To estimate t_1, we define the chip to be specified to fail once per year. Thus, from $N(t_1, m, n, R)$, t_1 is

$$t_1 = k + \tau \ln(4 * \beta * n * m * f_1 * f_2 * R) \quad (12)$$

where $n = 31.5 \cdot 0.10^6$ s (a year) and m is the number of flip-flops in a module. In this expression we used $2N(t_1, m, n, R) = 1$ to obtain MTBF (mean time between failures). This value of t_1 gives the maximum possible value of the data rate between a pair of communicating modules.

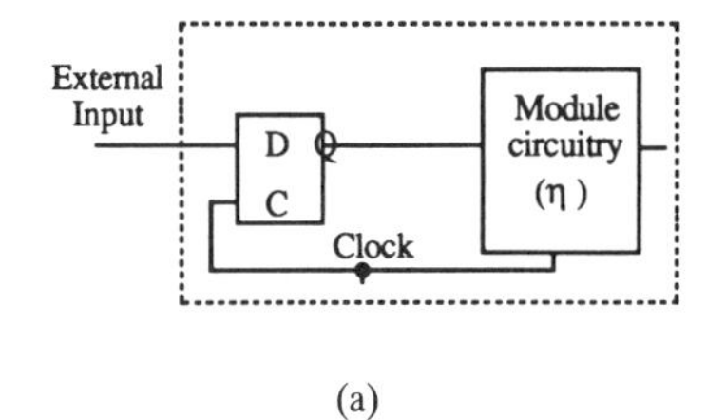

(a)

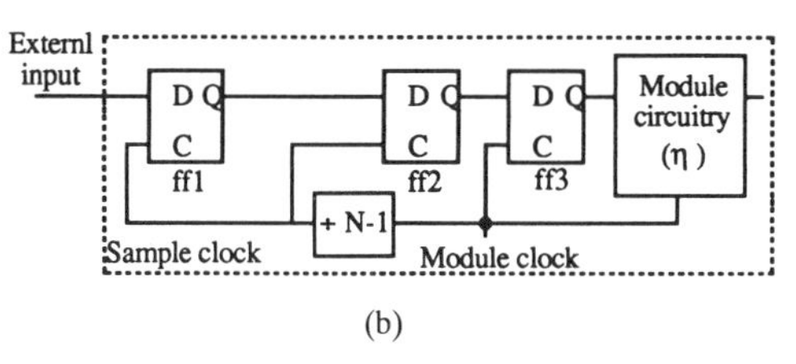

(b)

Fig. 9. Block diagrams for two basic synchronization schemes. (a) Extension of clock period. (b) Cascade of flip-flops.

Data rate is limited by $f_2 \leq f_1$. We use $f_1 = f_2 = 1/t_{\mathrm{AS}}$ to obtain t_1 and t_{AS}.

In other synchronization techniques, one may cascade $N+1$ flip-flops to obtain N clock periods settling time for the synchronizer or divide the module clock by $(N-1)$ to achieve an uninterrupted settling time of $(N-1)$ clock periods between ff1 and ff2, see Fig. 9(b), and an additional clock period between ff2 and ff3. The function of ff3 is to hide the delay of the divider and to synchronize the output with the module clock. It has been shown [39] that for $N > 2$ and $t_d \approx k$, (t_d is the delay of the divider), the divided clock synchronizer performs better than $N+1$ cascaded flip-flops both in terms of reliability and total number of hardware components. However, both these schemes introduce latency in each module which can be a serious problem in time-critical systems. For this divided clock scheme, the data rate can be obtained from

$$T_{\mathrm{AC}} = \mathrm{Max.}\ (t_1 + t_d + \Delta, \eta + \Delta). \quad (13)$$

The second term in the above expression is included for cases where the delay of the combinational logic in the module is larger than the required settling time of the input synchronizer. In this case the divider and ff3 may be omitted and the synchronizer reduces to a cascade of two flip-flops.

It is interesting to compare relation (12) to relation (6). In synchronous systems the skew grows with the square root of the logarithm of the system size ($O(\log R)^{0.5}$) while in asynchronous systems the time penalty due to synchronization grows with the logarithm of the system size ($O(\log R)$). The coefficient of these functions is proportional to the standard deviation of the path delay in the synchronous case and to the time constant of the synchronizers in the asynchronous case. These two factors are characteristic of each technology and determine the ultimate performances of systems fabricated in that technology.

C. Data Rate Constraint in the Pausable Clock Module

In a pausable clock module [12], [13], the separation between consecutive clock pulses is adapted to the state of the synchronizers. To implement this, synchronizers are monitored and, in case of occurrence of an MSS, the active phase of the clock is prolonged to any needed value, according to the longest of the detected metastable states. In this way it is guaranteed that the output of the synchronizers will not be used until they are stable and logically defined. Thus, there will be no propagated errors. The basic components of a pausable clock module are shown in Fig. 10. In this scheme edge triggered flip-flops that are provided by MSS detectors are used for synchronizers. The outputs of these detectors are ORed and can inhibit the module free-running clock generator to generate the next pulse, see Fig. 10(b).

When a metastable state occurs, it will be indicated as an inhibitive pulse I. This pulse must be ready before the clock fall edge so that no runt pulse can arise at B. Thus, the following constraint must be fulfilled:

$$t_h \geq T_F = k + t_m + t_{\mathrm{OR}}.$$

We call T_F the feedback delay. In this expression t_H is high part of the clock pulse, k is again the intrinsic delay of the flip-flop, t_M is the metastable detector delay, and t_{OR} is the delay of the multi-input OR gate. We assume that the delays are the same for both edges and, hence, the basic module clock pulse is symmetric. The multi-input OR gate may be optimized for minimum delay [34]. In this case $t_{\mathrm{OR}} = T_{\mathrm{OR3}} \ln(m)$ where T_{OR3} is the delay of a three input OR gate and m is again the number of input synchronizers (inputs and/or bits). It should mentioned that there must be mariginal conditions under which a runt pulse is produced at B causing reliability problems in the clock circuitry. However, this scheme has been practically implemented and used [11], [12].

Since the flip-flops resolution time has no absolute bound, the clock period can become unacceptably long. If the extension of the run time of the module exceeds some specified upper bound, t_c, the module is said to have failed [32]. Since the conditions that lead to a pause are the same as the conditions that lead to an error in fixed-period clocked modules, the same procedure as was used to calculate t_1 for a certain failure rate, relation (10), may be employed here to obtain t_c. However, in this case, instead of the delay k used in (11), one must use TF as this limits the the time available for the flip-flops to resolve. Although a long time may be allowed for t_c, most of the pauses tend to have a short duration. In order to estimate the average time penalty at a clock event, we calculate the average duration of pauses with t_c as an upper bound $(t_c - T_F \gg t)$ for each clock event. Using (9) this is

$$t_{\mathrm{ave.}} \approx \int_0^{t_c} tN'(t,m)dt \approx P * m\tau \approx 2\beta f_1 m\tau \qquad (15)$$

where $N'(t,m)$ is the probability density function and P is the (conditional) probability that a flip-flop causes a pause. A recommended practice in this scheme [28] is to allow several (e.g., 5) of these $t_{\mathrm{ave.}}$ in the basic module clock period to prevent most (99%) of the pauses to cause any clock period lengthening. This assumption, however, will not affect the conclusions of this study. Thus, the average data rate between two communicative modules using this scheme is

$$T_{\mathrm{AP}} = \mathrm{Max.}\ \{2(T_F + 5t_{\mathrm{ave.}}) + \Delta, T_F + 5t_{\mathrm{ave.}} + \eta + \Delta\}. \quad (16)$$

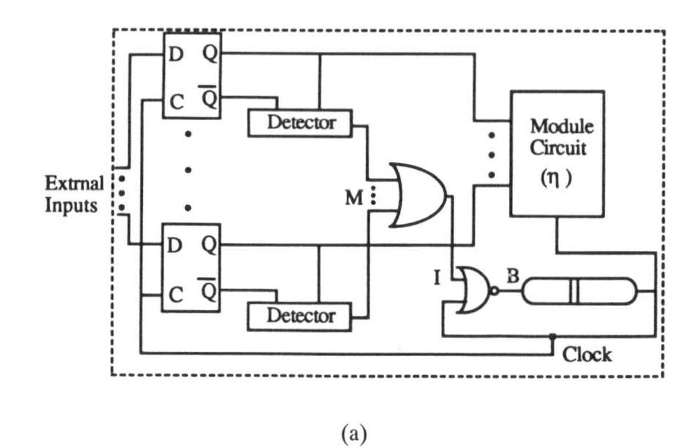

(a)

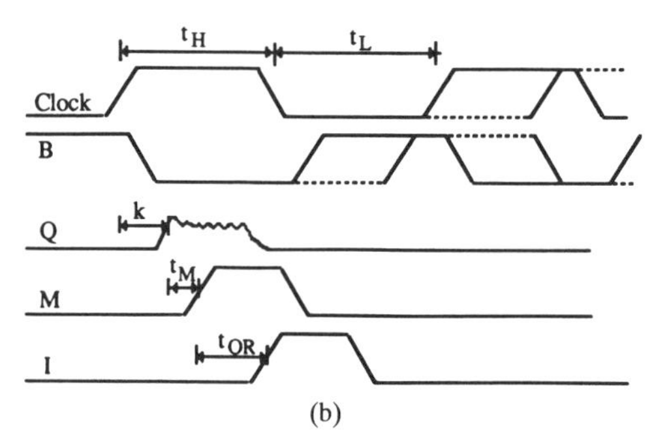

(b)

Fig. 10. A pauseable clock system scheme. (a) Block diagram. (b) Timing diagram.

The first term in the above expression is included to guarantee the minimum clock period constrained by the feedback delay. Clearly, the maximum data rate of this scheme is $f_{\mathrm{AP}} = 1/T_{\mathrm{AP}}$. Again, we assume that $f_1 = f_2 = f_{\mathrm{AP}}$. Data rate of Q-modules [11] follows the same behavior as expressed in the above relation. Notice that in the above relation the overhead time is dominated by the feedback delay and not by the average duration of pauses.

V. Comparison of Synchronous and Asynchronous Schemes

In this section our intention is to compare the performance of systems using synchronous and asynchronous schemes for their communication. In the preceding sections expressions determining the data rate in the synchronous and asynchronous schemes were developed. For the sake of easy reference these expressions are repeated below.

$$T_S = k + \eta + \Delta_{ij} + \delta + t_{\mathrm{set-up}} \quad (4)$$

$$T_{\mathrm{AS}} = t_1 + \eta + \delta \quad (11)$$

$$T_{\mathrm{AC}} = \mathrm{Max.}\ (t_1 + t_d + \Delta, \eta + \Delta) \quad (13)$$

$$T_{\mathrm{AP}} = \mathrm{Max.}\{2(T_F + 5t_{\mathrm{ave.}}) + \Delta, T_F + 5t_{\mathrm{ave.}} + \eta + \Delta\} \quad (16)$$

where T_s, T_{AS}, T_{AC} and TAP stand for data rate for synchronous, simple asynchronous, cascaded flip-flop asynchronous and pausable clock asynchronous schemes, respectively.

Handshaking protocols between modules in an asynchronous system may be realized in different ways. In the clocked module based systems, considered in this paper, handshaking may be realized by sending data only from the transmitter to the receiver (see for example [20]). This results in an upper bound for the data rate in asynchronous schemes. In this way the interconnection delay, Δ_{ij}, appears only once in the data rate expressions for asynchronous schemes. Assuming this, the terms in the above expressions may be considered equal to Δ_{ij}. Then, as we are concerned with relative performance of two schemes, the Δ_{ij} terms may be deleted from the above expressions. Comparison of the two schemes, therefore, reduces to the comparison of the effect of clock skew and delay penalty introduced by synchronizers, in synchronous and asynchronous systems, respectively. In general, these two factors are indeed characteristic for the synchronous and the asynchronous time performances. Notice that the effect of Δ_{ij} on data rate is similar to the effect of the delay of the combinational logic in the modules. Therefore, when we consider very complex modules, this is equivalent to a simpler module with a longer interconnection wire.

In order to make a general, and yet realistic, comparison between the two schemes, we considered three typical architectures for VLSI systems at an on-chip level. At one end, we considered modules with very fine granularity. This level of of granularity simulates very highly pipelined, regular (systolic type) architectures that are intended for high-speed operations required for heavy computational (real-time) demands. To consider this type of architecture is interesting not only because they exploit the power of VLSI technology, but also because at this level of granularity the effect of clock skew and synchronizers time penalty will not be disguised by the module combinational logic delay. It was assumed that the modules in this class contained a critical logic path of one flip-flop and two two-input NAND gates with a fan-out of three.

At the other extreme of modules complexity, we considered an area where VLSI systems would find large applications, that is, smart terminals, etc. Critical logic depth of these modules may be as high as 50–100 [8]. We assumed a logic depth of 100 for these functional modules, where again each step is a two-input NAND gate with a fan-out of three. The middle zone between these two extremes may be filled by modules with a complexity corresponding to a logic depth of 20. This level of complexity can be representative for logic-enhanced memories, RISC processors or computational arrays, in which modules are capable of operations such as multiplication and parallel addition [47].

If the switching function realized by a module in an asynchronous system is the same as for a module in a synchronous system, it is expected that the latter module is simpler as it does not contain extra control logic needed for the asynchronous communication. On the other hand, it is normally assumed that an asynchronous system is faster, on the average, than a synchronous system by utilizing data dependent delay (that is utilizing the fact that some operations are faster than others). We assume that these two phenomena are equal in importance so that the same function has the same delay in both schemes.

Using the scaling rules and expressions developed in the previous sections we have calculated the maximum communication rate for different schemes. Note that these rates are independent of any module clock frequency, as explained earlier. In these calculations we have used the largest chip area, defined earlier, for different process technologies. For the values of τ, k, β, t_d, t_M, t_{OR}, $t_{\text{set-up}}$, and R (the number of modules or paths on a chip) at different feature sizes see Appendix B. We also assume that the number of input bits of each module is 16. Fig. 11 shows the data rate of the synchronous and different asynchronous schemes at different feature sizes for the finest module granularity. In these results the effects of skew and synchronizers on the data rate are pronounced without being dominated by the module delay.

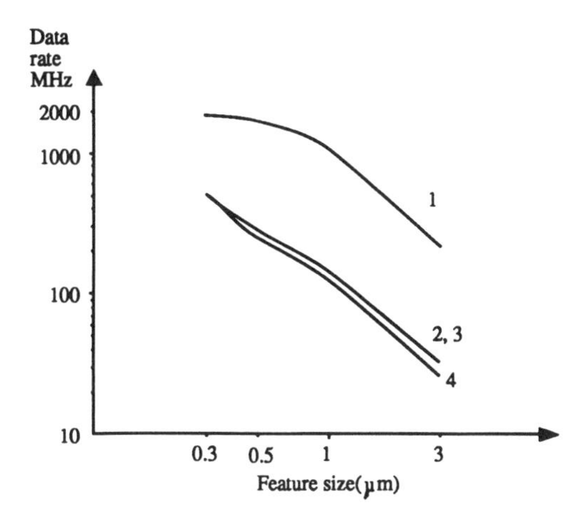

Fig. 11. Comparison of the maximum data-rate of synchronous and different asynchronous schemes for a module with a logic depth of two. 1) synchronous scheme, 2) simple asynchronous scheme, 3) cascaded flip-flop asynchronous scheme, and 4) pausable clock asynchronous scheme.

We can see from this figure that the synchronous scheme (with the improved clock distribution principle) always performs much better than all the asynchronous schemes, even for the smallest feature size. This contradicts the view that asynchronous schemes will outperform synchronous schemes in a scaled technology. The maximum data rate for the synchronous scheme in an 0.3 μm process is $\approx$ 2 GHz and for the asynchronous schemes 500 MHz. The data rate of the pausable clock asynchronous scheme is very much dependent on the number of input bits as this increases the feedback delay of this scheme (where t_{OR} dominates). For example, if the number of bits is increased from 16 (assumed above) to 32, the data rate will reduce to 380 MHz while if reduced to 8, the data rate will increase to 630 MHz. For only one bit input this scheme may provide a maximum data rate of 1.2 GHz. This scheme, however, has some other problems as it is not possible to produce pause signals without propagating runt pulses into the clock circuitry and in most applications modules rely on a constant clock period for their proper operation.

Notice that for the simple module considered here, no significant speed will be gained by going from a 0.5 μm to a 0.3 μm process. In the synchronous scheme the delay of the small module considered above plus the input flip-flop and $t_{\text{set-up}}$ in a 3 μm process is 4.2 ns. The maximum clock path skew on the system level for this level of granularity is only 0.4 ns. However, in a 0.3 μm process the delay of the module will be 0.23 ns but the maximum skew is 0.27 ns. At 2 GHz the length of the interconnection lines may seem electrically long and hence a transmission line model may be necessary to estimate line delays. However, remember that at smaller feature sizes the line is divided into more and more sections. For example in the 0.3 μm technology, the line length between two consecutive repeaters is shorter than 1 mm. This is still an electrically short line at 2 GHz.

To study the performance of the synchronous and different asynchronous schemes for more complex modules we have calculated the data rates of systems using more complex modules at different feature sizes. The results of these calculations are represented in Fig. 12.

For modules with a logic depth of up to 40, the synchronous scheme results in much higher data rates. For larger logic depths, the asynchronous scheme with cascaded flip-flops has nearly the same performance as the synchronous scheme. However, this scheme introduces a long ($\approx$ 150 τ_0 where τ_0 is the delay of a minimum size inverter driving another similar inverter) latency in each module. This can cause serious performance problems in time critical systems. The simple asynchronous scheme, in which the synchronizers are directly connected to the module's combinational logic, performs almost similar to the pausable clock scheme. Notice that in these calculations we have assumed that the number of input bits is 16. For less input bits the pausable clock scheme performs better and for more bits worse.

Another interesting conclusion can be drawn from these results. They suggest that an efficient VLSI system may use two levels of granularities. At a finer level a synchronous scheme should be employed to exploit its higher rate. At a coarser level an asynchronous scheme can be used.

We now discuss the skew introduced by unequal clock path length. It is generally preferable to exploit the transparency of modules and run the (clock) lines through the blocks rather than around them. The presence of several conducting layers makes this approach more feasible. However, in the case that a clock line must go around a module to reach another adjacent module, the maximum difference in the length of the clock line may be $\approx 2\sqrt{\text{cell area}}$. To estimate the area of modules, we have made an extensive study on the density of different logic. The results of this study may be summarized as follows. We chose an area unit = (1 poly pitch)*(1 metal pitch). For example, in our 3 μm process 1 area unit = 4*6 = 24 mm^2. Then each transistor in control logic (not very regular) occupies 44 units and for data paths (regular logic) 28 units. From the number of transistors the area of each module can be estimated. Then from the delay of 1 cm line length, given in Fig. 6, one can calculate the skew due to the displacement of modules. This skew is maximum at minimum feature size. However, at 0.3 μm technology this skew is less than 6% of the modules delay. We assumed that the area increases quadratically with logic depth. Compared to the skew induced by the process variation, this skew is negligible. Obviously, at larger feature sizes, the relative value of this skew to module delay is even smaller.

In the preceding analysis we considered the maximum clock skew only for chips with maximum area at each feature size, to show the technological trends. Most of the chips manufactured at a certain time do not necessarily use maximum area. In fact for a given application, the optimum chip sizes may be smaller than the maximum possible chip size [18]. In addition,

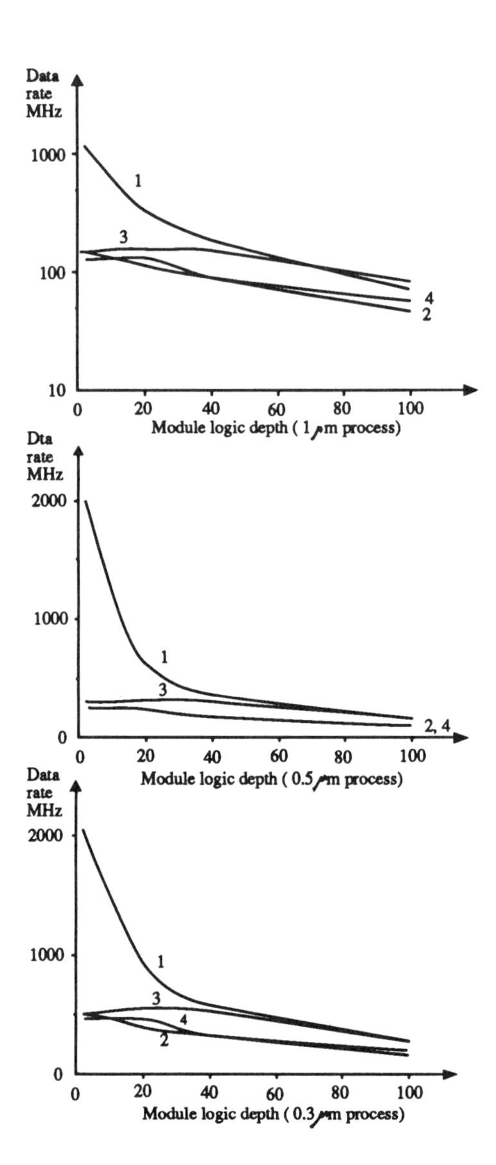

Fig. 12. Comparison of data rate at different feature sizes and logic depths. 1) Synchronous, 2) simple asynchronous, 3) cascaded flip-flop asynchronous, and 4) pausable clock asynchronous scheme.

we assumed that the aspect ratio of the interconnections (wire thickness/ wire width ratio) will remain constant with downscaling. The trends of technology, however, show an increase in this ratio. Both these factors favor data rates for a synchronous scheme which is even larger than discussed above.

VI. Management of Clock Skew and Interconnection Delay

Clock skew affects the performance of synchronous systems in two ways. It reduces the system data throughput and it makes the system failure prone. In the previous section we only considered the first effect and demonstrated that synchronous systems may result in a much higher data rate than asynchronous systems. In this section we address the second problem. For this purpose consider again a state machine representation for two communicating modules, Fig. 13(a), where a register (R, e.g., a positive edge triggered flip-flop) and a computational cell (CL) form a module. The

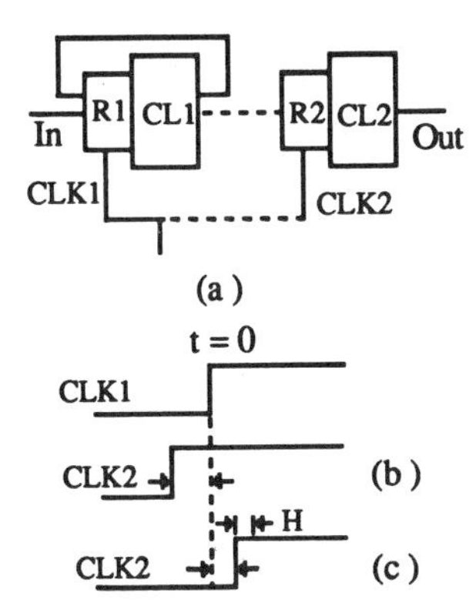

Fig. 13. Demonstration of clock skew. (a) System diagram, (b) negative skew, and (c) positive skew.

relative position of the clock signal at the site of registers R1 and R2 may be as depicted in Fig. 13(b). It is shown that [9], [17] if the relative timing of the CLK2 is as in Fig. 13(b), negative skew, the only effect of skew is to increase the clock period by the amount of skew and no failures will occur, but, consider the other situation, positive skew, Fig. 13(c). In this case the minimum delay of the register R1, the computational cell CL1, and the interconnection between the modules may be so that a data released by R1 at $t = 0$ can reach R2 before the end of the hold time (H) of this device for the present clock edge and cause race-through and failure.

It is usually expensive and difficult to deal with this problem which cannot be easily solved by increasing the clock period. To reduce the skew, great efforts must be exerted, for example, to tune the delays in the clock distribution paths. However, process induced skew cannot be tackled in this way. On the other hand, in order to increase the minimum delay of the logic circuit in the cell, it is usually necessary to add extra delays (e.g., inverters) at the register outputs as well as to increase the clock period. Designers have appealed to two phase clocking (pseudo 4 phase in CMOS) to solve this problem by increasing the gap between the phases. This not only increases the clock period, but also creates other problems, like running more clock lines over the whole design and controlling the skew between the phases. In the following a circuit technique is proposed that can handle up to half a clock period positive skew with no constraint on the minimum delays.

A single phase clocking scheme was developed that is based on N and P latches. An N latch is transparent in the high part of the clock signal and latches in the low part of the clock signal while a P type latch is transparent in the low part and latches in the high part of the clock signal [9]. Then a state machine representation of a synchronous system using the latches looks as in Fig. 14(a). This scheme seems equivalent to an ordinary 2-phased clocked system but uses a single phase clock. Positive skew causes problem because when the data are transmitted by the P1 latch at $t = 0$, CLK2 is still at high state, see Fig. 14(b). But, the data should reach the N2 latch first after the CLK2 has gone to low state to avoid set-up violation of the N2 latch. To protect this scheme against positive skew it suffice to insert an extra P latch in front of the remote N latch, Fig. 14(c). By this method a positive clock skew as long as half a clock period can be handled.

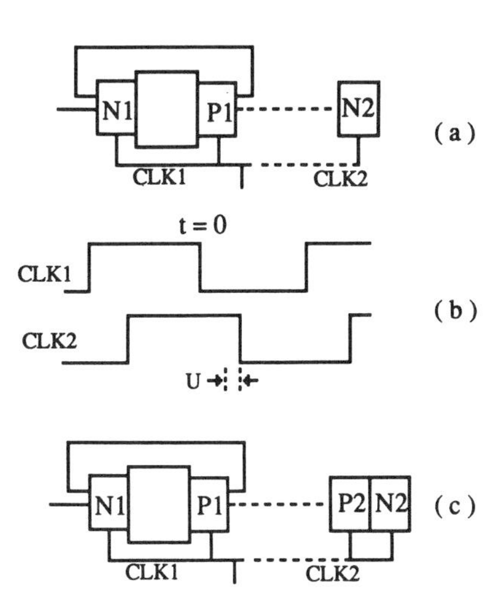

Fig. 14. A circuit technique to cancel the effect of positive skew. (a) block diagram of a synchronous system using latches, (b) relative clock signals, and (c) an extra latch is inserted on the site of the destination latch.

We now discuss the data interconnection delay. In a complex system with various functional modules on a chip or when there are some feedbacks involved, the length of inter-module wires may become important. The maximum average inter-module wire length is estimated to be one-half the size of the chip edge [48]. According to Fig. 6 this can cause a delay of about 3 ns in the 0.3 μm process that may seriously degrade the chip performance.

Using buffer amplifiers distributed along the clock lines and pipelining several clock signals on the path made the clock path delay no longer a limiting factor on the performance, instead the delay fluctuation became important. A similar approach may also be used for long range communication by "re-synchronizing" data signals along its path over the chip. Re-synchronization is done by letting the signal pass a clocked register, using the local clock, at a necessary number of positions along its path. See also [36]. By this method, data rate will not be limited by the delay of the interconnection line, but some latency will be introduced in the system [42], [49].

VII. An Example

In the 2 μm process available to us, resistance of lines is still not high enough to create any serious clocking problems. However, to examine a design in which the delay of the clock path is much larger than the delay of modules, a highly pipelined, bit-serial precharged dynamic system was implemented. This shortens the clock period in addition to increasing the number of inverters required for clock buffering. The circuit is a 512 word, 16-bit sorter on a 8×8 mm chip. For more details see [50]. Fig. 15 shows the clock distribution network and the direction of the data flow of this design. To drive the large clock capacitance of the modules, a two stage clock buffer is distributed within the individual functional modules. In this way, the clock skew due to different line length (resistance) becomes minimal, see also [16]. The initial, central clock-buffer is designed to drive the clock distribution network. The required number of buffer stages was determined to be six. The path was simulated between point A and B

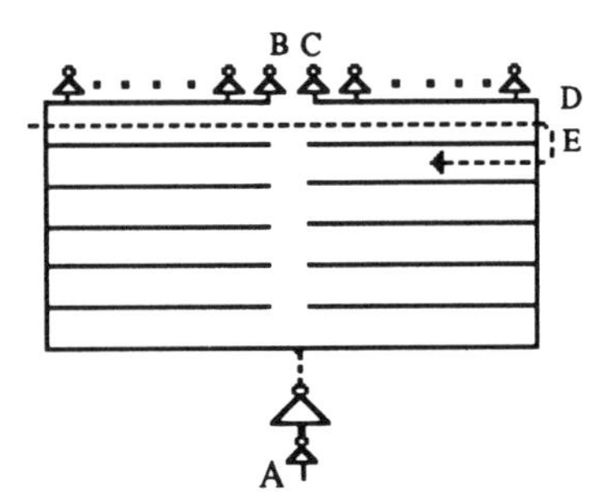

Fig. 15. A balanced clock distribution network for the sorter.

by SPICE. The delay of this path is 10 ns. However, the maximum delay of the modules is 4 ns. Thus, according to conventional clocking the maximum data rate of this system should be determined by the clock path delay and must be around 100 MHz. A 240 MHz clock pulse (determined by the module delays) was sent on the path and simulated the system function successfully.

The performance of this system is scalable, because although the delay of the path will increase with scaling, the data rate of the system will mainly be determined by the delay of the module, which decreases by scaling.

Notice that a "balanced" clock distribution network should be viewed from the points where the communicating modules see it. H-tree and binary tree are two general balanced networks. Depending on application and technology, other types of networks may appear balanced to the modules. The above network, for example, is a balanced network for our purpose, even for scaled down feature sizes. The skew due to differing path lengths, between points D and E, is a small fraction of the module delay also for a 0.3 μm process. The process induced skew will be most pronounced between points B and C, although, these points have the same distance from the clock source.

VIII. Conclusions

In this paper we compared the time performance of synchronous and globally asynchronous and locally synchronous systems. Our comparisons were based on realistic data (for existing technologies) and realistic circuit examples rather than on formal theories. We have studied feature sizes of 3 to 0.3 μm (masked transistor gate length), assuming that 0.3 μm is close to a practical lower limit [35]. The most significant findings are:

1) Clock skew introduced by process parameter variations is a key factor in synchronous systems time performance. This skew is mainly due to interconnection lines width and thickness variations.
2) Resolution time required by synchronizers to achieve a sufficiently low failure rate is a key factor in time performance of asynchronous systems.
3) It was shown that the time complexity of synchronous systems is $(O(\log R)^{0.5})$ while the time complexity of asynchronous systems is $(O(\log R))$, where R is the size of the system. The coefficients of these functions are variance of clock skew in synchronous systems and time constant of synchronizers in asynchronous systems.
4) For high-speed applications, either a pipelined mode of clocking or smaller chip sizes must be used.
5) When pipelined mode of clocking is used, the synchronous scheme always outperforms the asynchronous schemes.
6) A synchronous scheme can allow a data rate of ≈ 2 GHz in future VLSI chips.
7) No speed benefit is gained by going from 0.5 μm to 0.3 μm.
8) It was quantitatively shown that an efficient VLSI architecture may use two levels of granularities. At a finer level a synchronous scheme should be used while at a coarser level an asynchronous scheme can be preferable.
9) A circuit technique was given that makes the design of synchronous systems safer and easier.

Appendix A

The variances of threshold voltage, carrier mobility and oxide capacitance are expressed in terms of dimensions (as these are the only parameters that can be designed) in the form of

$$\sigma_{VT}^2 = \frac{1}{LW}(A + B * A_{\text{ox}})$$
$$\sigma_{\mu}^2 = \frac{A\mu}{LW}\mu^2$$
$$\sigma_{Cox}^2 = \frac{A_{\text{ox}}}{LW}Cox^2$$

where constants A, B, A_{ox}, A_μ are experimentally determined for both n and p transistors [23]. As stated in the text, we also assumed that $\sigma_W = \sigma_L = 0.02$ μm in a 3 μm technology. The variance of a random variable z that is a function of some independent random variables, $z = g(x, y,)$ may be obtained from

$$\sigma_z^2 = (\frac{\partial g}{\partial x})^2\sigma_x^2 + (\frac{\partial g}{\partial y})^2\sigma_y^2 + \cdots$$

When the number of sections in a clock path is incremented to obtain the minimum delay of the path, the minimum of the delay and σ for the path do not occur for the same number of sections. We choose the value of σ at the point minimum delay, as this is easier to capture during the simulation phase of design. For more details see [26].

Appendix B

In this Appendix the time parameters of the synchronizer are first calculated. Parameters shown in Table I are used for the 3 μm process. As stated in the text, an edge-triggered flip-flop [34, p. 219] was used for synchronizers in asynchronous schemes and registers in the synchronous scheme. The central part of this device that determines the time constant of the synchronizers is two cross-coupled NOR gates as in Fig. 16. The equivalent circuit for this circuit under metastable operation is depicted in Fig. 17. Parameters of this circuit can be expressed in terms of transistors geometry and fabrication process parameters [30]. Following the same procedure, given in the above reference in association with the data given in Table I and assuming $W_p/W_n = 2$, $W_n = 20$ mm, and

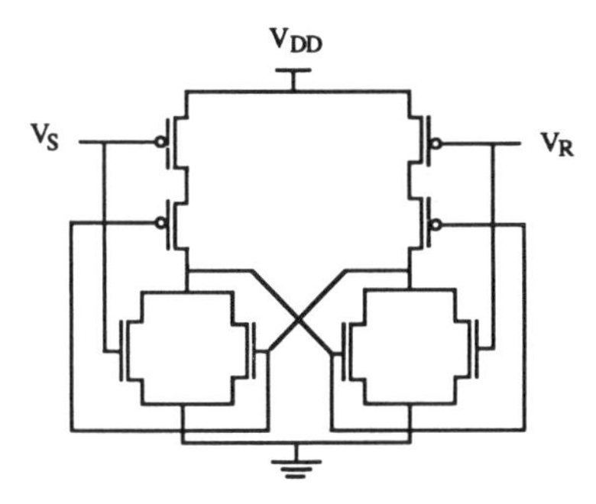

Fig. 16. Two CMOS cross-coupled NOR gates determining the flip-flop time constant.

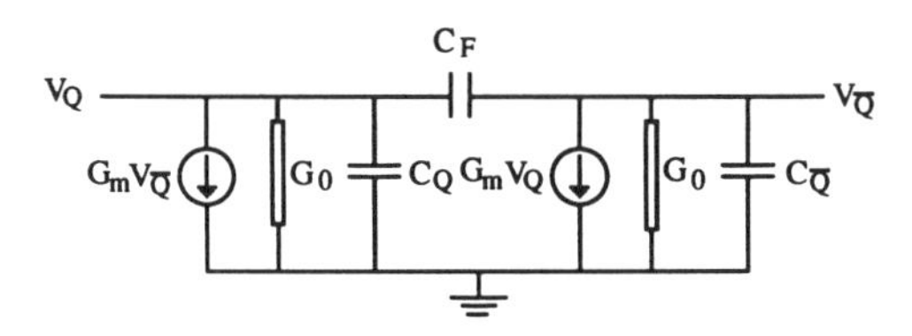

Fig. 17. The equivalent circuit for the flip-flop

$L_{dn} = L_{dp} = 8$ mm, a value of $\tau = 0.40$ ns is obtained for the time constant.

In literature (as far as we know), when a flip-flop circuit is used to calculate τ, the flip-flop is considered as it stands alone. The performance of flip-flops to be used within a module are often quite different from those of stand-alone flip-flops. For example, if the outputs of the above flip-flop are connected to a 20/3 CMOS inverter, the value of τ becomes 0.6 ns.

In our case where it is assumed that the flip-flops are connected to a two input NAND, we estimate that the maximum value of τ is 0.6 ns.

Furthermore, we have estimated k, the flip-flop intrinsic delay, by simulating the flip-flop. k is then defined as the propagation delay from the clock input to the Q output measured at 0.5 V_{DD}. A value of 2 ns was found for k. We also assume that $t_{\text{set-up}} = t_{\text{hold}} = 1$ ns. To estimate t_M, we assume that an EXOR gate with a delay 2 ns is used for this purpose. The delay of two input NAND and three input OR are estimated by SPICE and are ≈ 0.6 and ≈ 2.4 ns, respectively. To calculate the delay of these gates at scaled-down feature sizes, parameters of Table II are used. To include the effect of carrier velocity saturation, we use an equivalent mobility $\mu_e = F_v * m$ [24], where $F_v = (1 + v_{\text{ave.}}/v_0)^{-1}$, v_0 is the scatter-limited velocity, and $v_{\text{ave.}} = \mu(V_{\text{DD}})/L_{\text{eff.}}$. Using these relations and the parameters of the Table II, the delay of gates in the 1, 0.5, and 0.3 μm process are 1/5, 1/10, and 1/18 of the corresponding delay of the gates in the 3 μm process. To see how realistic the estimated delay of the scaled-down gates is, we simulated a minimum inverter, driving another similar inverter by SPICE and obtained an average delay of 0.5 ns at 3 μm. According to our scaling rules, the delay of this gate in the 0.5 μm process is 50 ps. Measured delay of this gate is reported to be 65 ps [38].

Finally, to estimate the number of processors or paths on the chips with different module granularities, we assumed that a module with one register and two logic levels, where each level is a two input NAND, will include 30 transistors. According to the figures given in the text for the size of transistors, the area of this module is 25 000 μm^2. Therefore, a 3 μm chip with an area of 0.25 cm^2 can contain 1000 of these modules and a 2 cm^2 0.3 μm chip 8 x 10^5 modules. For other module granularities, we assume that the area of the modules increases quadratically with the module logic depth.

REFERENCES

[1] C. Mead and L. Conway, *Introduction to VLSI systems.* Reading, MA: Addison-Wesley, 1980.

[2] D. F. Wann and M. A. Franklin, "Asynchronous and clocked control structures of VLSI-based interconnection networks," *IEEE Trans. Comput.*, vol. C-32, pp. 284–293, Mar. 1983.

[3] S. Y. Kung and R. J. Gal-Ezar, "Synchronous versus asynchronous computation in VLSI array processors," *SPIE*, vol. 341, pp. 53–65, 1982.

[4] A. L. Fisher and H. T. Kung, "Synchronizing large VLSI processor arrays," *IEEE Trans. Comput.*, vol. C-34, pp. 734–740, Aug. 1985.

[5] E. Arjomandi, M. J. Fisher, and N. A. Lynch, "Efficiency of synchronous versus asynchronous distributed systems," *J. ACM*, vol. 30, pp. 449–456, July 1983.

[6] D. Zhou, F. P. Preparata, and S. M. Kang, "Interconnection delay in very high-speed VLSI," *IEEE Trans. Circuit Syst.*, vol. 38, no. 7, pp. 779–790, July 1991.

[7] J. Yuan and C. Svensson, "High-speed CMOS circuit technique," *IEEE J. Solid-State Circuits*, vol. 24, no. 1, pp. 62–70, Feb. 1989.

[8] F. Anceau, "A synchronous approach for clocking VLSI systems," *IEEE J. Solid-State Circuits*, vol. SC-17, pp. 51–56, Feb. 1982.

[9] M. Afghahi and C. Svensson, "A unified single phase clocking scheme for VLSI systems," *IEEE J. Solid-State Circuits*, vol. 25, no. 1, pp. 225–233, Feb. 1990.

[10] C. F. Molnar, T. Fang, and F. U. Rosenberger, "Synthesis of delay-insensitive modules," in *Proc. Chapel-Hill Conf. VLSI*, 1985, pp. 67–86.

[11] F. U. Rosenberger, C. E. Molnar, T. J. Chaney, and T. Fanf, "Q-modules: Internally clocked delay-insensitive modules," *IEEE Trans. Comput.*, vol. 37, no. 9, pp. 1005–1018, Sept. 1988.

[12] M. Pchoucek, "Anomalous response time of input synchronizers," *IEEE Trans. Comput.* vol. C-25, pp. 133–139, 1975.

[13] W. Y. Lim and J. R. Cox, "Clocks and the performance of synchronizers," *Proc. IEE*, vol. 130, pt. E, pp. 57–64, Mar. 1983.

[14] W. Lim, "Design methodology for stopable clock systems," *Proc. IEE*, vol. 133, pt.E. pp. 65–69, Jan. 1986.

[15] S. H. Unger and C. Tan, "Clocking schemes for high-speed digital systems," *IEEE Trans. Comput.*, vol. C-35, pp. 880–895, 1986.

[16] E. B. Friedman and S. Powel, "Design and analysis of a hierarchial clock distribution system for synchronous standard cell/macrocell VLSI," *IEEE J. Solid-State Circuits*, vol. SC-21, pp. 240–246, Apr. 1986.

[17] M. Hatamian and G.L. Cash, " Parallel bit-level VLSI designs for high-speed signal processing," *Proc. IEEE*, vol. 75, no. 9, pp. 1192–1202, Sept. 1987.

[18] H. B. Bakoglu, *Circuits, Interconnections, and Packaging for VLSI.* Reading, MA: Addison-Wesley, 1990.

[19] H. B. Bakoglu and J. D. Meindl, "Optimal interconnection circuit for VLSI," *IEEE Trans. Electron Devices*, vol. ED-32, pp. 903–909, May 1985.

[20] J. O. Tuazon, J. C. Peterson, M. Pniel, and D. Liberman, "CalTech/JPL Mark II Hypercube concurrent processor," in *Proc. 1985 Int. Conf. Parallel Processing*, Aug. 1985, pp. 666–673.

[21] F. Rosenberger and T. J. Chaney, "Flip-flop resolving time test circuit," *IEEE J. Solid-State Circuits*, vol. SC-17, pp. 731–738, Aug. 1982.

[22] A. Papoulis, *Probability, Random Variables and Stochastic Process.* Tokyo: McGraw-Hill, Kogakusha, 1965.

[23] K. R. Lakshmikumar, A. Hadaway, and M. A. Copeland, "Characterization and modeling of mismatch in MOS transistors for precision analog design," *IEEE J. Solid-State Circuits*, vol. SC-21, pp. 1057–1066, Dec. 1986.

[24] Y. El-Mansy, "MOS device and technology constraints in VLSI," *IEEE J. Solid-State Circuits*, vol. SC-17, no. 2, pp. 197–203, Apr. 1982.

[25] M. Hurtado, "Structure and performance of asymptotically bistable dynamical systems," D.Sc. dissertation, Dep. EE. Washington Univ., St., Louis, MO, May 1975.

[26] M. Afghahi and C. Svensson, "Calculation of clock path delay and skew in VLSI synchronous systems," in *Proc. IEEE Euro. Conf. Circuit Theory and Design*, Brighton, England, Sept. 1989, pp. 256–269.

[27] INMOS, *IMST424 Transputer Reference Manual*, Nov. 1984.
[28] M. J. Stucki and J. K. Cox, "Synchronization strategies," in *Proc. Caltech Conf. VLSI*, Jan. 1979, pp. 375–393.
[29] H. Masuda, M. Nakai, and M. Kubo, "Characteristics and limitation of scaled-down MOSFET's due to two-dimentional field effect," *IEEE Trans. Electron Devices*, vol. ED-26, pp. 980–986, June 1979.
[30] T. Kacprzak and A. Albicki, "Analysis of metastable operation in RS CMOS flip-flops," *IEEE J. Solid-State Circuits*, vol. SC-22, pp. 57–64, Feb. 1987.
[31] T. J. Chaney and F. U. Rosenberger, "Characterizarion and scaling of MOS flip-flop performance in synchronizer applications," in *Proc. Caltech Conf. VLSI*, Jan. 1979, pp. 357–374.
[32] B. Randell, P. A. Lee, and P. C. Treleaven, "Reliability issues in computing design," *ACM Comput. Surveys*, no. 10, pp. 123–165, 1987.
[33] M. Afghahi and C. Svensson, "A scalable synchronous system," in *Proc. IEEE Symp. Circuits Syst.*, Finland, 1988, pp. 471–474.
[34] H. Weste and K. Eshraghian, *Principles of CMOS VLSI Design–A System Perspective*. Reading, MA: Addison-Wesley 1985.
[35] C. Svensson, "VLSI physics," *Integration*, vol. 1, pp. 3–19, 1983.
[36] ——, "Signal resynchronization in VLSI systems," *Integration*, vol. 4, pp. 75–80, 1986.
[37] Chapiro, Ph.D dissertation, STANCS-84-1026, Dep. Comput. Sci., Stanford Univ., Stanford CA 94305, Oct. 1984.
[38] H. Cong, J. M. Anderews, D. M. Boulin, S. Fang, S. J. Hillenius, and J. A. Michejda, "Multigigahertz CMOS dual-modules prescalar IC," *IEEE J. Solid-State Circuits*, vol. 23, no. 5, pp. 1189–1194, Oct. 1988.
[39] L. Kleeman and A. Cantoni, "Metastable behavior in digital systems," *IEEE Design Test Comput.*, pp. 4–19, Dec. 1987.
[40] D. M. Brown, M. Ghezzo, and J. M. Pimbley, "Trends in advanced process technology–Submicron CMOS device design and process requirements," *Proc. IEEE*, vol. 74, no. 12, pp. 1678–1702, Dec. 1986.
[41] W. A. Holton and R. K. Cavin, III, "A perspective on CMOS technology trends," *Proc. IEEE*, vol. 74, no. 12, pp. 1646–1668, Dec. 1986.
[42] C. E. Leiserson and J. B. Saxe,"Optimizing synchronous systems," in *Proc. IEEE Symp. Foundations Comput. Sci.*, 1981, pp. 23–36.
[43] S. D. Kugelmass and K. Steiglitz, "A probabilistic model for clock skew," in *Proc. Int. Conf. Systolic Arrays*, San Diego, CA, 1988, pp. 545–554.
[44] K. R. Lakshmi Kumar, "Characterization and modeling of mismatch in MOS devices and application to precision analog design," Ph.D. dissertation, Carlton Univ., Ottawa, Ont., Canada, 1985.
[45] H. Cramer, *Mathematical Methods of Statistics*. Princeton, NJ: Princeton Univ. Press, 1946.
[46] E. G. Friedman, "Performance limitations in synchronous digital systems," Ph.D. dissertation, Univ. California, Irvine, CA, 1989.
[47] C. Seitz, "Concurrent VLSI architectures," *IEEE Trans. Comput.*, vol. C-33, no. 12, pp. 1247–1265, Dec. 1984.
[48] S. M. Kang, "Metal-metal matrix (M3) for high-speed MOS VLSI layout," *IEEE Trans. Comput.-Aided Design*, vol. CAD-6, no. 6, pp. 886–891, Sept. 1987.
[49] J. P. Fishburn, "Clock skew optimization," *IEEE Trans. Comput.*, vol. 39, no. 7, pp. 945–951, July 1990.
[50] M. Afghahi, "A 512, 16-b bit-serial sorter chip," *IEEE J. Solid-State Circuits*, vol. 26, pp. 1452–1457, Oct. 1991.

Synchronization of Pipelines

Karem A. Sakallah, *Senior Member, IEEE,* Trevor N. Mudge, *Senior Member, IEEE,*
Timothy M. Burks, *Student Member, IEEE,* and Edward S. Davidson, *Fellow, IEEE*

***Abstract*—In this paper we apply a recently formulated general timing model of synchronous operation to the special case of latch-controlled pipelined circuits. The model accounts for multiphase synchronous clocking, correctly captures the behavior of level-sensitive latches, handles both short- and long-path delays, accommodates wave pipelining, and leads to a comprehensive set of timing constraints. Pipeline circuits are important because of their frequent use in computer systems. We define their concurrency as a function of the clock schedule and degree of wave pipelining. We then identify a special class of clock schedules, *coincident multiphase clocks*, which provide a lower bound on the value of the optimum cycle time. We show that the region of feasible solutions for single-phase clocking can be nonconvex or even disjoint, and derive a closed-form expression for the minimum cycle time of a restricted but practical form of single-phase clocking. We compare these forms of clocking on three pipeline examples and highlight some of the issues in pipeline synchronization.**

List of Symbols

i, j	Indexes used to identify pipeline stages/synchronizers.
p, r	Indexes used to identify clock phases.
p_i	Index of clock phase used to control synchronizer i.
a_i, A_i	Early and late signal arrival times at stage i.
d_i, D_i	Early and late signal departure times from stage i.
δ_i, Δ_i	Minimum and maximum propagation delays from synchronizer $i - 1$ to synchronizer i.
C	Concurrency in the pipeline.
e_p	Time, in global frame-of-reference, at which clock phase p ends (i.e., when its latching edge occurs).
E_{pr}	Phase shift from clock phase p to clock phase r.
$\mathcal{E}_{i-1,i}$	Phase shift from stage $i - 1$ to stage i.
H_i	Hold time of synchronizer i.
k	Number of clock phases.
m	Width (in bits) of the pipeline datapath.
n	Number of pipeline stages.
ν_i	Degree of wave pipelining in stage i.
R_i^E	Region of feasibility corresponding to the early arrival (hold) constraints of pipe stage i.
R^L	Region of feasibility corresponding to the late arrival (setup) constraints.
R^P	Region of feasibility corresponding to the pulsewidth constraints.
S_i	Setup time of synchronizer i.
T_c	Clock cycle time.
T_p	Width of active interval of phase p.
U	Utilization of the pipeline.
ϕ_p	Name of clock phase whose index is p.
w	Minimum allowable clock pulsewidth.

I. Introduction

In this paper we extend the work reported in [1] which applied a recently formulated general timing model of synchronous operation [2] to the special case of pipelined circuits. The model accounts for multiphase synchronous clocking, correctly captures the behavior of level-sensitive latches, handles both short and long paths, and leads to a comprehensive, yet simple, set of timing constraints. It has been successfully applied, for general circuit topologies, to the problems of clock cycle minimization using linear programming methods [3] and timing verification using an iterative relaxation algorithm [4].

By applying this model to a simple circuit structure, this paper helps to clarify various aspects of the clocking of level-sensitive latches as a function of circuit propagation delays.[1] These include the following:

- Defining pipeline concurrency as a function of the clock schedule and the degree of wave pipelining.
- Identifying a special class of clock schedules, *coincident multiphase clocks*, which yield the smallest possible cycle times for a specified degree of wave pipelining.
- Demonstrating that the space of physically realizable single-phase clock schedules derived from this model can be nonconvex or even disjoint, complicating the search for the optimal cycle time.

[1]Using edge-triggered flip-flops rather than level-sensitive latches yields a much simpler model that is considerably easier to analyze. Latches, however, are interesting for two reasons: first, they hold the promise of lower cycle times because they allow their data signals to "flow through" unhindered during the active interval of their controlling clock phase; and second, they are typically "cheaper" than flip-flops as measured by gate count or area.

Reprinted from *IEEE Trans. Computer-Aided Desi.,* vol. 12, no. 8, pp. 1132–1146, Aug. 1993.

- Providing a closed-form solution for the optimal cycle time of a restricted, but practical, form of single-phase clocking.

In addition to helping understand the above issues, the study of pipelines is further justified by their increasing use, even in the instruction execution units of single-chip microprocessors (commonly referred to as "super-pipelining" in RISC machines) [5]. In particular, the above-mentioned closed-form expression for minimum cycle time may be directly applied in the design and optimization of high performance processors which use single-phase clocking.

The remainder of this paper is organized as follows. The pipeline model is developed in Section II. In Section III we examine the dependence of pipeline concurrency on clocking and wave pipelining, and define coincident multiphase clocks. In Section IV we derive the regions of feasibility for coincident multiphase clocking and for two modes of single-phase clocking. Section V illustrates the application of the model on three example pipelines using an experimental program, *pipe*T_c, which computes the optimal clock schedules for single-phase as well as coincident multiphase clocking. Conclusions and suggestions for future work are summarized in Section VI.

II. Pipeline Model

Pipelining is frequently used to speed up the execution of a sequence of computations by dividing each into n consecutive subcomputations and overlapping their execution. Theoretically, this should yield a factor of n performance improvement over the nonpipelined case. This maximum is rarely achieved, however, because of dependencies among the operations and overhead due to clocking [6]. Performance can be defined as the sustained number of operations per unit time, and can be expressed as:

$$MOPS = \frac{U(n) \times 10^3}{T_c(n)}$$

where *MOPS* stands for millions of operations per second, $0 \le U(n) \le 1$ is the *utilization* of the n-stage pipeline, and $T_c(n)$ is the clock cycle time, in nanoseconds, at each pipe stage. Typically, $U(n)$ is a decreasing function of n which is determined empirically through simulations or benchmarking. $T_c(n)$ is also a decreasing function of n that depends on circuit delays and clocking parameters. Optimal pipeline design seeks to find the value of n which maximizes *MOPS*. This is usually done in two steps. 1) Determining $U(n)$ for a suitable range of n by analyzing the dependencies among the operations of an appropriate set of benchmark computations. This is a purely "architectural" analysis which disregards hardware implementation details, but may consider software restructuring to decrease the dependence effects. 2) Determining the minimum $T_c(n)$ for the same range of n. Generally, this is a synthesis problem which involves examining the logic design of various pipelines, and finding those that yield the minimum cycle times.

This paper addresses one aspect of the second step, namely, determining the minimum cycle time, $T_{c,min}$, for an n-stage pipeline in terms of circuit delays. The problem has been addressed previously by a number of authors including [6]–[9]. This previous work dealt mostly with simple clocking paradigms. Furthermore, the analysis was typically based on examination of a single pipe stage. In contrast, in this paper we propose a pipeline timing model that accounts for more complex clocking and for the temporal interactions among the various pipe stages.

Our pipeline model is shown in Fig. 1. The pipe stages are numbered consecutively from 0 to $(n - 1)$. The datapath through the pipeline is assumed to be m bits wide, $m \ge 1$. Each pipe stage consists of a bank of m level-sensitive latches used as synchronizing elements followed by combinational circuitry.[2] Data flow through the pipeline is regulated by a k-phase clock, where $1 \le k \le n$. Stage i is characterized by the following parameters:

- p_i: an integer denoting the clock phase used to control the synchronizing element at the output of stage i (henceforth referred to as synchronizer i).
- S_i: nonnegative setup time of synchronizer i relative to latching edge of phase p_i.
- H_i: nonnegative hold time of synchronizer i relative to latching edge of phase p_i.
- δ_i, Δ_i: minimum and maximum propagation delays ($0 \le \delta_i \le \Delta_i$) from the input of synchronizer $i - 1$ to the input of synchronizer i. Note that this definition of stage delay lumps together the two components of signal delay, namely the synchronizer delay and the combinational logic delay.

Note that, unlike earlier open-ended pipeline formulations, such as those given in [6]–[9], our model includes a virtual pipe stage, labeled 0, which forms a simple loop with stages 1 through $n - 1$. Stage 0 is used to model the times of data arriving from, and data departing to, the environment surrounding the pipeline. For example, it can be used to model the timing attributes of the memory or register file used to supply operands for the computation, and to receive results from it. The use of such a virtual stage provides a consistent mechanism to account for the boundary conditions of pipeline operation. Furthermore, as shown in Section 2.3, the open-ended pipeline is a special case of the closed pipeline.

We base the steady-state behavior of such pipelines on the general model of synchronous operation introduced in [2]. The salient features of this model, as they relate to the pipeline, are summarized below. In addition, we extend the model to allow for *wave pipelined* [10], [11] op-

[2]The value of m may vary from stage to stage and actually has no effect on the model.

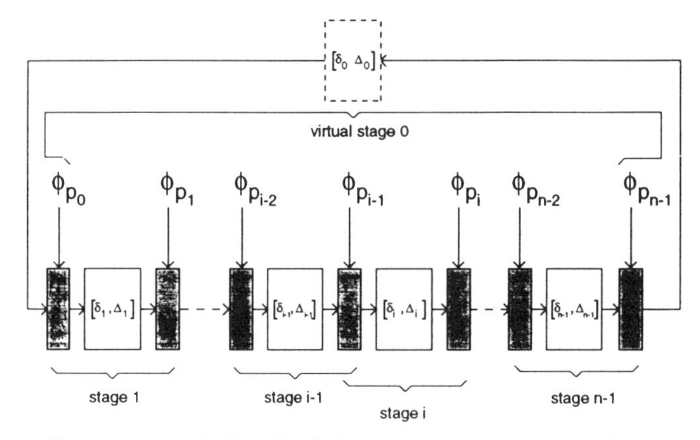

Fig. 1. n-stage pipeline. Shaded boxes represent the synchronizers.

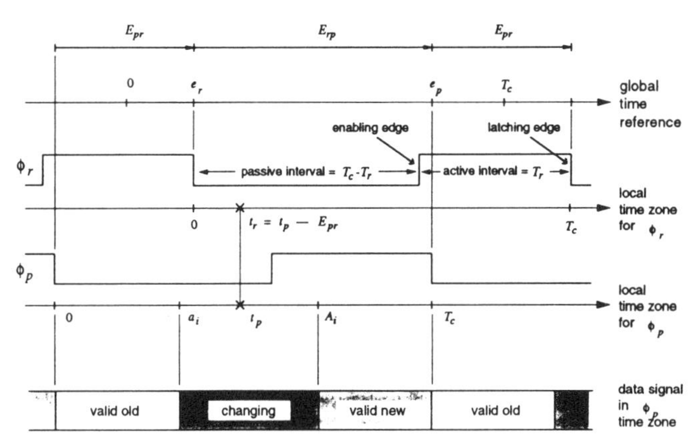

Fig. 2. Key model parameters.

eration. Fig. 2 depicts the relationships among the key parameters used in the model.

2.1. Clocking Model

The clocking model is described in terms of a *temporal* rather than a *logical* framework based on the concept of *periodic phases* which define *local time zones* related by *phase shift operators*. In this model, a k-phase clock is considered to be a collection of k periodic signals ϕ_1, ϕ_2, $\cdots$, ϕ_k—referred to as the *phases*—with a common cycle time T_c. Each phase ϕ_p divides the clock cycle into two intervals: an *active* interval of duration T_p, and a *passive* interval of duration $(T_c - T_p)$. During the active interval of a given phase, the synchronizers it controls are *enabled*; during its passive interval, they are *disabled*. The transitions into and out of the active interval are called, respectively, the *enabling* and *latching* edges of the phase. We assume, without loss of generality, that all phases are active high; thus, the enabling and latching edges correspond to the rising and falling transitions of the phase signal. Associated with the phase is a *local time zone* such that the passive interval of the phase starts at $t = 0$, its enabling edge occurs at $t = T_c - T_p$, and its latching edge occurs at $t = T_c$. The temporal relationships among the k phases (i.e., among the different time zones) are established by an arbitrary choice of a *global time reference*. We introduce e_p to denote the time, relative to this global time reference, at which phase ϕ_p *ends* (i.e., when its latching edge occurs). Finally, we define a *phase shift* operator:

$$E_{pr} \equiv \begin{cases} (e_r - e_p), & e_r > e_p \\ (T_c + e_r - e_p), & e_r \leq e_p \end{cases} \tag{1}$$

which takes on positive values in the range $(0, T_c]$. When subtracted from a time variable, t_p, in the *current* local time zone of ϕ_p, E_{pr} changes the frame of reference to the *next* local time zone of ϕ_r, taking into account a possible cycle boundary crossing.

2.2. Timing Constraints

For timing purposes, it is sufficient to characterize a data signal with respect to one clock cycle by two, possibly simultaneous, events which demark the interval during which the signal is switching between its old and new values. For the signal *arriving* at the input of synchronizer i these two events are defined to occur at $t = a_i$ and $t = A_i$ in the local time zone of phase p_i. The corresponding events of the data signal *departing* from the input of synchronizer i are defined to occur at $t = d_i$ and $t = D_i$. It will be convenient to refer to a_i and A_i as the *early* and *late* arrival times, and to d_i and D_i as the *early* and *late* departure times. The timing model of the pipeline can now be expressed by the following constraints and equations [2] for $i = 0, \cdots, n - 1$.

Clock Constraints express limitations on clock generation and distribution. This set should at least include the following minimum pulsewidth constraints:

$$T_{p_i} \geq w_{p_i} \tag{2}$$

$$T_c - T_{p_i} \geq w_{p_i} \tag{3}$$

where w_{p_i} are specified pulse width parameters. In addition, to simplify the design of the clock generator we may include "regularity" constraints such as

$$T_1 = T_2 = \cdots = T_k. \tag{4}$$

It is important to point out that the phase signals are not required to be nonoverlapping.

Latching Constraints express the conditions necessary for capturing valid data values at each of the synchronizers. They consist of two sets of requirements which, together, insure that the data signal at the input of a synchronizer is stable for a sufficient period of time before and after the occurrence of the latching edge of the corresponding clock. Mathematically,

$$a_i \geq H_i \tag{5}$$

$$A_i \leq T_c - S_i. \tag{6}$$

Synchronization Equations *macromodel* the temporal behavior of different types of synchronizing elements. Specifically, for D-type level-sensitive latches, they express the departure times of each output data signal as the later of the arrival time of the corresponding input data signal and the enabling clock edge:

$$d_i = \max (a_i, T_c - T_{p_i}) \tag{7}$$

$$D_i = \max (A_i, T_c - T_{p_i}). \tag{8}$$

Propagation Equations model the delay of the combinational stages in the pipeline, including the propagation through the input synchronizer. They express the arrival times of data at the input of synchronizer i in terms of the corresponding departure times from the input of synchronizer $(i - 1) \bmod n$, taking into account the change in the frame-of-reference from phase p_{i-1} to phase p_i:[3]

$$a_i = d_{i-1} + \delta_i - \mathcal{E}_{i-1,i} \tag{9}$$

$$A_i = D_{i-1} + \Delta_i - \mathcal{E}_{i-1,i} \tag{10}$$

where $\mathcal{E}_{i-1,i}$ is the amount of phase shift *from stage i − 1 to stage i*. In [2], this was defined to be equal to the phase shift *from clock phase* p_{i-1} to *clock phase* p_i, i.e., $\mathcal{E}_{i-1,i} \equiv E_{p_{i-1}p_i}$. This definition limited signal propagation to consecutive cycles of phases p_{i-1} and p_i, i.e., signals launched from stage $i - 1$ in any given cycle of phase p_{i-1} had to arrive and be correctly latched at stage i by the immediately following cycle of phase p_i. We extend this definition here to allow for signal propagation over multiple clock cycles by introducing the nonnegative integer parameter ν_i to indicate the number of *additional* clock cycles available for signals to propagate from stage $i - 1$ to stage i. Thus,

$$\mathcal{E}_{i-1,i} \equiv E_{p_{i-1}p_i} + \nu_i T_c. \tag{11}$$

Note that the addition of an integer number of clock cycles to the clock phase shift has the effect of changing the frame-of-reference from the current local time zone of phase p_{i-1} to the local time zone of phase p_i that begins ν_i cycles *after* its next local time zone. In particular, for $\nu_i = 0$ the phase shift reverts to its earlier definition.

2.3. Open-Ended Pipelines

Characterizing open-ended pipelines using the above model is a simple matter of replacing the departure time equations for virtual stage 0 with *specified* values that represent the pipeline boundary conditions. Specifically, the following equations for signal departure times from stage 0

$$d_0 = \max (a_0, T_c - T_{p0}) \tag{12}$$

$$D_0 = \max (A_0, T_c - T_{p0}) \tag{13}$$

are simply replaced by

$$d_0 = \hat{d}_0 \tag{14}$$

$$D_0 = \hat{D}_0 \tag{15}$$

[3]Index arithmetic in what follows will always be modulo n. To keep the equations from becoming too cluttered, the mod operator will be dropped and assumed to be implied.

where $\hat{d}_0$, and $\hat{D}_0$ denote the specified signal "departure" times from the pipeline source to its first stage. This immediately leads to the following arrival time equations at the first pipe stage:

$$a_1 = \hat{d}_0 + \delta_1 - \mathcal{E}_{0,1} \tag{16}$$

$$A_1 = \hat{D}_0 + \Delta_1 - \mathcal{E}_{0,1}. \tag{17}$$

The equations for signal arrival times at virtual stage 0

$$a_0 = d_{n-1} + \delta_0 - \mathcal{E}_{n-1,0} \tag{18}$$

$$A_0 = D_{n-1} + \Delta_0 - \mathcal{E}_{n-1,0} \tag{19}$$

which capture the signal propagation delays through the pipeline "environment" are dropped altogether. Instead, the corresponding *actual* signal departure times computed from the model equations:

$$d_{n-1} = \max (a_{n-1}, T_c - T_{p_{n-1}}) \tag{20}$$

$$D_{n-1} = \max (A_{n-1}, T_c - T_{p_{n-1}}) \tag{21}$$

are checked against the *required* signal departure times, $\hat{d}_{n-1}$ and $\hat{D}_{n-1}$ from the last pipe stage (stage $n - 1$) to the pipeline environment.

The specification of signal times entering stage 1 and leaving stage $n - 1$ represents a decoupling of the signal propagation equations around the closed pipeline and leads to an easier cycle time optimization problem. However, by explicitly including virtual stage 0 in the pipeline model, we have the added flexibility of optimizing the operation of the pipeline within its environment. Either way, it should be clear that the closed pipeline model above encompasses open-ended pipelines as a degenerate special case. The remainder of the paper focuses on studying closed pipelines.

III. Pipeline Operation Modes

Allowing multiple clock cycles for signals to propagate through a single stage has the potential of reducing the cycle time below what is possible with single-cycle propagation. However, for such operation to be feasible the minimum combinational delay of the stage must be sufficiently large to maintain adequate temporal separation between consecutive *waves* of signals (see Fig. 3). Reliance on logic delay, rather than on synchronizing elements alone, to prevent interference between consecutive data waves has been dubbed *wave pipelining* [10]. This phenomenon will occur in any pipe stage for which $\nu_i > 0$. We thus refer to ν_i as the *degree of wave pipelining* in stage i.

The number of operations concurrently in process in an n-stage pipeline need not be equal to n. Depending on the nature of the clocking scheme, the differences between the minimum and maximum delays in each stage, and the distribution of the maximum delays over all stages, it may be possible to operate the pipeline so that the number of signal waves simultaneously traveling around the closed pipeline is less than or greater than n. We capture this

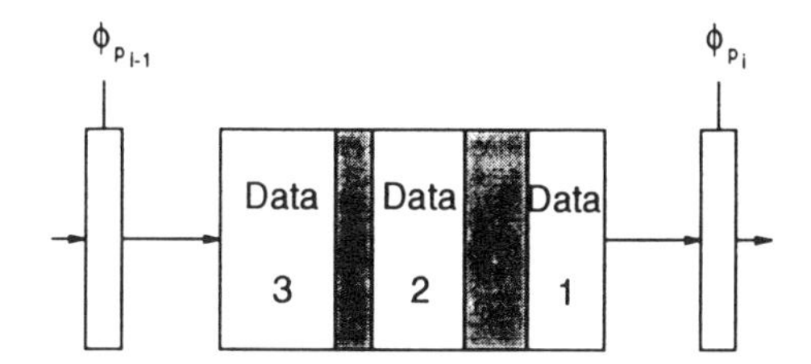

Fig. 3. Wave pipelining.

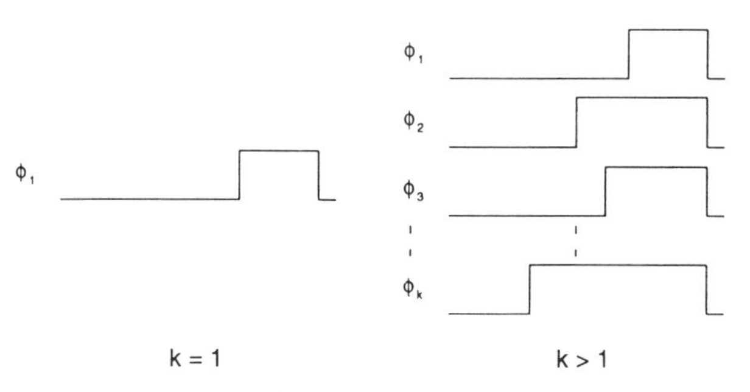

Fig. 4. Clocks with maximum possible phase shift between phases.

notion by introducing C, the *concurrency* in the pipeline, which can easily be related to the clock phase shifts and the degrees of wave pipelining by

$$C = \frac{1}{T_c} \sum_{i=0}^{n-1} E_{p_{i-1}p_i} + \sum_{i=0}^{n-1} \nu_i. \quad (22)$$

C can be thought of as the number of *virtual* pipeline stages. Note that in a closed pipeline C must be an integer; hence $\Sigma E_{p_{i-1}p_i}$ must be an integer multiple of T_c. A particular level of concurrency may be achieved by a variety of combinations of clocking schemes and wave pipelining. For example, a concurrency of 4 in a 4-stage pipeline may be obtained by a 4-phase clock where each pipe stage is allocated a fraction of the clock cycle such that $\Sigma E_{p_{i-1}p_i} = T_c$, and $\Sigma \nu_i = 3$. Alternatively, $\Sigma E_{p_{i-1}p_i} = 2T_c$, and $\Sigma \nu_i = 2$.

We limit our attention in this paper to those clocking schemes which maximize C for a given level of wave pipelining, namely those for which the sum of the clock phase shifts around the pipeline stages is equal to nT_c. Recalling that each of the individual phase shifts is at most one clock cycle, this restriction implies that $E_{p_{i-1}p_i} = T_c$ for each of the n stages. Clocking schemes for which this restriction applies include single-phase clocks and the restricted form of multiphase clocking shown in Fig. 4, which will be referred to as *coincident* multiphase clocking since the latching edges of all k phases *coincide* in time.[4] For simplicity in the equations and analysis that follows we let $\nu_i = \nu$ for all stages. The methods used, however, do handle the general case where ν_i differs from stage to stage.

With these restrictions, the concurrency C becomes

$$C = (1 + \nu)n. \quad (23)$$

[4]For general multiphase clocks, the existence of fractional phase shifts (i.e., phase shifts smaller than a full cycle) limits $\Sigma E_{p_{i-1}p_i}$ to $\leq (n - 1)$ clock cycles.

IV. Optimal Cycle Time Calculation

Subject to the simplifying assumptions made above, namely $E_{p_{i-1}p_i} = T_c$ and $\nu_i = \nu$, the phase shift from stage $i - 1$ to stage i in (11) can be expressed simply as

$$\mathcal{E}_{i-1,i} \equiv (1 + \nu) T_c. \quad (24)$$

The timing model of the pipeline can now be conveniently viewed as consisting of *three* distinct sets of constraints:

- **Pulsewidth Constraints** expressed by (2) and (3).
- **Long-Path** (*Late-Signal*) **Constraints** involving the late arrival and departure times and expressed by the setup inequalities (6), the propagation equations (10), and the synchronization equations (8).
- **Short-Path** (*Early-Signal*) **Constraints** involving the early arrival and departure times and expressed by the hold inequalities (5), the propagation equations (9), and the synchronization equations (7).

Subject to the above constraints, we outline in this section procedures for obtaining the minimum cycle time for latch-controlled pipelines for the following three clocking schemes:

1) a coincident n-phase clock,
2) a general form of single-phase clocking,
3) a restricted form of single-phase clocking.

In all three cases, the calculation of the optimal cycle time starts by finding expressions for the early and late arrival times at stage i in terms of the clock variables and circuit delays. These expressions are then combined with the hold and setup requirements to obtain the short- and long-path constraints. In one case, restricted single-phase clocking, these constraints can be solved to yield a closed-form expression for the minimum cycle time. Numerical solutions are necessary in the other two cases.

It should be noted that single-phase clocks are a special case of the more general coincident n-phase clocks. As such, the minimum cycle time possible with a coincident n-phase clock will always be less than or equal to that obtainable with a single-phase clock. Less obvious, though, is the fact that the solution space for the case of coincident n-phase clocks is convex whereas that for single-phase clocks may in fact be nonconvex or even disconnected. While we do not envision that coincident n-phase clocks are likely to be used in practice, their study is theoretically important because they provide a lower bound on the minimum cycle times possible with single-phase clocks.

4.1. Coincident n-phase Clocks

A coincident n-phase clock is obtained by setting $p_i = i$ and is characterized by $n + 1$ variables: the cycle time T_c, and the n independent phase widths $T_0, \cdots, T_{n-1}$. It is important to note that the freedom to choose a different phase width for each pipe stage is the key to the relatively simple solution procedure of the coincident n-phase case. In particular, it is always possible to choose the phase widths so that the synchronization equations (7) and (8) are simplified to:

$$d_i = D_i = T_c - T_i \tag{25}$$

This simplification can be justified as follows:

> Suppose that $D_i > T_c - T_i$ for some stage i at the optimal solution. Then $D_i = A_i > T_c - T_i$. Since changing T_i can only directly affect the departure times from stage i, it should be obvious that T_i can be decreased until $D_i = A_i = T_c - T_i$ without affecting the optimal cycle time. Note also that decreasing T_i can only increase the margin by which the hold requirement is satisfied at stage $i + 1$.

The above simplification is significant because it removes the coupling, inherent in the latch synchronization model, between the departure and arrival times. As will become apparent later, this coupling is the primary source of complexity and nonconvexity in the general single-phase case. Specifically, the optimality of the coincident n-phase solution is unchanged if we replace the synchronization equations (7) and (8) and their troublesome max function with the simple equalities (25). This in turn makes it possible to express the feasible region as a set of linear inequalities that define a convex space.

Arrival Times: Substituting (24) and (25) in (9) and (10), we can express the arrival times at stage i as:

$$a_i = T_c - T_{i-1} + \delta_i - (1 + \nu) T_c$$
$$= \delta_i - T_{i-1} - \nu T_c \tag{26}$$

and

$$A_i = T_c - T_{i-1} + \Delta_i - (1 + \nu) T_c$$
$$= \Delta_i - T_{i-1} - \nu T_c. \tag{27}$$

In addition, from (8), signals must arrive at the latest by the rising edge of the corresponding clock to satisfy (25):

$$A_i \leq T_c - T_i. \tag{28}$$

Long-Path Constraints: Combining (27) with the setup requirement (6), yields

$$(1 + \nu) T_c + T_{i-1} \geq \Delta_i + S_i. \tag{29}$$

Combining (27) with (28) leads to another constraint:

$$(1 + \nu) T_c + T_{i-1} - T_i \geq \Delta_i. \tag{30}$$

Short-Path Constraints: Substituting (26) into the hold requirement (5) yields

$$\nu T_c + T_{i-1} \leq \delta_i - H_i. \tag{31}$$

Solution Procedure: The feasible region for coincident n-phase clocking is defined in the $(n + 1)$-dimensional space of clock variables by $5n$ linear inequalities:

- $2n$ long-path inequalities (29) and (30),
- n short-path inequalities (31),
- $2n$ minimum pulse-width inequalities (2) and (3).

The minimum cycle time can be now be found by solving a linear program. Note that if $T_i \geq S_i$, as might be required for certain types of latches, (29) is subsumed by (30) and the total number of constraints in the linear program can be reduced to $4n$.

4.2. General Single-Phase Clock

When the clock phase widths at all pipe stages are forced to be equal, it may no longer be possible to satisfy the simplified latch synchronization equation (25); instead, the general model equations (7) and (8) must be invoked. It is possible under these conditions for some early signals to simply flow through the latches without having to wait for the enabling clock edge (i.e., $a_i > T_c - T_{p_i}$), effectively rendering the latches redundant. Such an operation mode has been termed "aggressive" [3] since it allows the latches to be transparent not only for the slow signals but also for the fast signals. In this case, the space of feasible solutions may become nonconvex. If we denote the feasible regions corresponding to pulse-width constraints by R^P, long-path (late-signal) constraints by R^L, and short-path (early-signal) constraints by R^E, then the overall region of feasibility is simply $R^P \cap R^L \cap R^E$. These regions are shown in Fig. 5 and are derived next, except for R^P which follows trivially from (2) and (3).

Arrival Times: The solution in the case of a single-phase clock is considerably more complicated because of the coupling between signal arrival and departure times through the latch synchronization equations (7) and (8). Unlike the coincident n-phase case, obtaining an expression for the arrival time at stage i requires the substitution of the synchronization and propagation equations of *all* pipe stages. Thus, the early arrival time at stage i is calculated by repeated application of (9) and (7) and algebraic simplification. Setting $p_i = 1$ to represent a single-phase clock, we obtain the following expression for the early arrival time at the input to synchronizer i:

$$\begin{aligned}
a_i &= d_{i-1} + \delta_i - (1 + \nu)T_c \\
&= \max (a_{i-1}, T_c - T_1) + \delta_i - (1 + \nu)T_c \\
&= \max (a_{i-1} + \delta_i - (1 + \nu)T_c, \delta_i - T_1 - \nu T_c) \\
&= \max (d_{i-2} + \delta_{i-1} + \delta_i - (2 + 2\nu)T_c, \\
&\qquad \delta_i - T_1 - \nu T_c) \\
&= \max (\max (a_{i-2}, T_c - T_1) + \delta_{i-1} + \delta_i \\
&\qquad - (2 + 2\nu)T_c, \delta_i - T_1 - \nu T_c) \\
&= \max (a_{i-2} + \delta_{i-1} + \delta_i - (2 + 2\nu)T_c, \delta_{i-1} \\
&\qquad + \delta_i - T_1 - (1 + 2\nu)T_c, \delta_i - T_1 - \nu T_c) \\
&= \cdots \\
&= \max (a_i + \delta_{i-n+1} + \cdots + \delta_i - (n + n\nu)T_c, \\
&\qquad \delta_{i-n+1} + \cdots + \delta_i - T_1 \\
&\qquad - (n - 1 + n\nu)T_c, \\
&\qquad \cdots, \delta_{i-1} + \delta_i - T_1 - (1 + 2\nu)T_c, \\
&\qquad \delta_i - T_1 - \nu T_c)
\end{aligned}$$

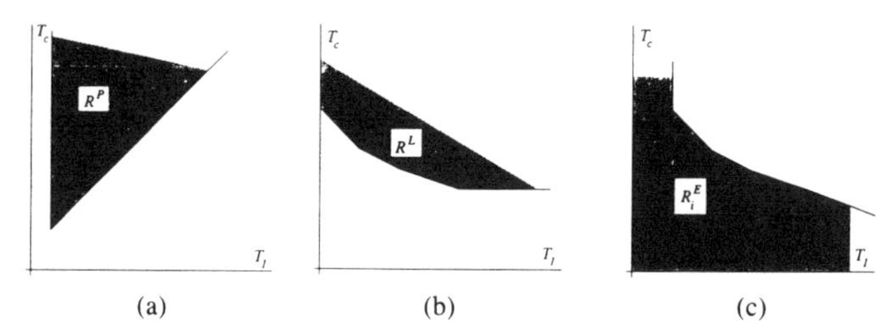

Fig. 5. Single-phase feasible regions for latches (illustrated for $\nu = 0$). (a) Pulsewidth constraints. (b) Long-path constraints. (c) Short-path constraints for stage i.

which can be expressed more conveniently as:

$$a_i = \max \left\{ a_i + \left(\sum_{j=i-n+1}^{i} \delta_j \right) - (n + n\nu)T_c, \right.$$
$$\left. \cdot \max_{0 \le l \le (n-1)} \left[\left(\sum_{j=i-l}^{i} \delta_j \right) - T_1 - (l + \nu + l\nu)T_c \right] \right\} \quad (32)$$

Similarly, the late arrival time at stage i, calculated from (10) and (8), is:

$$A_i = \max \left\{ A_i + \left(\sum_{j=i-n+1}^{i} \Delta_j \right) - (n + n\nu)T_c, \right.$$
$$\left. \cdot \max_{0 \le l \le (n-1)} \left[\left(\sum_{j=i-l}^{i} \Delta_j \right) - T_1 - (l + \nu + l\nu)T_c \right] \right\} \quad (33)$$

Note that the max functions in these expressions involve $n + 1$ arguments in which, except for the first argument, the only variables are the two clock variables T_c and T_1.

Long-Path Constraints: Expression (33) implies the following $n + 1$ inequalities:

$$A_i \ge A_i + \left(\sum_{j=i-n+1}^{i} \Delta_j \right) - (n + n\nu)T_c \quad (34)$$

$$A_i \ge \left[\left(\sum_{j=i-l}^{i} \Delta_j \right) - T_1 - (l + \nu + l\nu)T_c \right],$$
$$l = 0, \cdots, n - 1. \quad (35)$$

Eliminating A_i from the first inequality, we immediately obtain the following lower bound on T_c:

$$T_c \ge \frac{1}{n(1+\nu)} \sum_{j=i-n+1}^{i} \Delta_j$$
$$= \frac{1}{n(1+\nu)} \sum_{j=0}^{n-1} \Delta_j \equiv \frac{\overline{\Delta}}{1+\nu} \quad (36)$$

which confirms the intuition that the cycle time cannot be less than the average pipeline stage delay, $\overline{\Delta}$, when $\nu = 0$. In general, the C clock cycles during which one signal wave completes its tour of the pipeline must not comprise less total time than the sum of the maximum propagation delays around the pipeline.

Combining each of the remaining n inequalities with the setup constraint (6) we eliminate A_i to obtain:

$$(1 + l)(1 + \nu)T_c + T_1 \ge \left(\sum_{j=i-l}^{i} \Delta_j \right) + S_i,$$
$$l = 0, \cdots, n - 1. \quad (37)$$

While the physical interpretation of each of these inequalities is not as obvious as that of (36), it is still rather simple: the time available for a signal to propagate down the $(l + 1)$ pipe stages ending at stage i, and to be correctly setup for latching at stage i, is $(1 + l)(1 + \nu)$ clock cycles plus the phase width T_1 which represents the "extra" time due to the use of level-sensitive latches. Since each of these inequalities must be true for all n pipe stages, we obtain:

$$(1 + l)(1 + \nu)T_c + T_1 \ge \max_{0 \le i \le (n-1)} \left[\left(\sum_{j=i-l}^{i} \Delta_j \right) + S_i \right],$$
$$l = 0, \cdots, n - 1. \quad (38)$$

Thus the long-path constraints have been reduced to the $n + 1$ inequalities in (36) and (38) which together define a convex set in the T_c/T_1 solution space as shown in Fig. 5(b).

Short-Path Constraints: Proceeding as we did for the late arrival time at stage i, we obtain the following inequalities that must be satisfied by the early arrival time:

$$a_i \ge a_i + \left(\sum_{j=i-n+1}^{i} \delta_j \right) - (n + n\nu)T_c \quad (39)$$

$$a_i \ge \left[\left(\sum_{j=i-l}^{i} \delta_j \right) - T_1 - (l + \nu + l\nu)T_c \right],$$
$$l = 0, \cdots, n - 1. \quad (40)$$

The first of these is redundant since it is subsumed by the corresponding max-delay inequality (34). The remaining n inequalities in (40) may now be combined with the hold requirement (5) to eliminate a_i and yield the set of short-path constraints. A convenient way to obtain this set is to derive its complement, namely the set of constraints under which the hold requirements are *violated*, and then to invoke De Morgan's Law. Specifically, the hold violation region for stage i is defined by the set of n inequalities:

$$H_i > a_i \ge \left[\left(\sum_{j=i-l}^{i} \delta_j \right) - T_1 - (l + \nu + l\nu)T_c \right],$$
$$l = 0, \cdots, n - 1 \quad (41)$$

which, upon elimination of a_i, leads to the following n hold violation conditions:

$$(l + \nu + l\nu)T_c + T_1 > \left(\sum_{j=i-l}^{i} \delta_j \right) - H_i,$$
$$l = 0, \cdots, n - 1. \quad (42)$$

Introducing $\overline{R}_i^E$ to represent the hold violation region for stage i, where E stands for *early signal* (short path), (42) can equivalently be expressed as

$$\overline{R}_i^E = \overline{R}_{i,0}^E \cap \overline{R}_{i,1}^E \cap \cdots \cap \overline{R}_{i,l}^E \cap \cdots \cap \overline{R}_{i,n-1}^E \tag{43}$$

where $\overline{R}_{i,l}^E$ denotes the region of feasibility for the lth inequality in (42). By applying DeMorgan's Law to the set intersection equation (43), we obtain

$$R_i^E = R_{i,0}^E \cup R_{i,1}^E \cup \cdots \cup R_{i,l}^E \cup \cdots \cup R_{i,n-1}^E \tag{44}$$

Thus, the desired set of short-path constraints is

$$(l + \nu + l\nu)T_c + T_1 \leq \left(\sum_{j=i-l}^{i} \delta_j \right) - H_i$$
$$\text{for } at\ least \text{ one } l \in \{0, \cdots, n-1\} \tag{45}$$

Note that, unlike the corresponding long-path inequalities (38) which must all be satisfied, the above set of n short-path inequalities is satisfied if *at least one* of them is satisfied. In other words, the feasible region defined by the set of n inequalities in (38) is the *intersection* of n separate (linear bounded) convex regions, whereas that defined by the inequalities in (45) is the *union* of n separate (linear bounded) convex regions. This in turn implies that while the region defined by (38) is guaranteed to be convex, the region defined by (45), for each i, is guaranteed to be nonconvex, as shown in Fig. 5(c).

Solution Procedure: Denoting the overall region of feasibility by R, it can be conveniently expressed as:

$$R = R^P \cap R^L \cap R_0^E \cap \cdots \cap R_{n-1}^E. \tag{46}$$

Due to the nonconvexity of R_i^E, R may be nonconvex or even disconnected. Examples of these cases are illustrated in Section V. In any case, assuming that $R \neq \phi$, at the optimal solution one or more of the long-path constraints (36) and (38) must be active (satisfied as an equation). This observation forms the basis for a directed-search algorithm to find the minimum cycle time. Basically, the search begins by finding the smallest possible cycle time that satisfies the minimum pulsewidth and long-path constraints ($R^P \cap R^L$). Except for the degenerate case where the vertex of R^P lies in R^L, this point corresponds to the intersection of $T_c - T_1 = w_1$ and one of the $n + 1$ long-path constraints. This solution is now examined to see if it satisfies all of the short-path constraints. If it does, then it is optimal, otherwise we "climb" up the lower periphery of R^L until either we satisfy all the short-path constraints, or we reach the other end of the minimum pulse width region ($T_1 = w_1$) without satisfying all the short-path constraints. If the latter obtains, the problem is infeasible. A detailed description of the geometric solution approach outlined here can be found in [12].

4.3. Restricted Single-Phase Clock

A conservative application of single-phase clocking is to require that no hold times be violated even if the early signal departure from each latch occurred at the earliest possible time. This is equivalent to conservatively assigning d_i to its worst case value by using $d_i = T_c - T_1$ in place of the general early signal synchronization equation (7) even when $a_i > T_c - T_1$. This restriction restores convexity to the region of feasible solutions and leads to a closed-form expression for minimum cycle time.

Specifically, since the $d_i = T_c - T_1$ part of (25) is satisfied, the short-path constraints can be obtained from (31) by first setting $T_{i-1} = T_1$, leading to

$$\nu T_c + T_1 \leq \delta_i - H_i \tag{47}$$

which must be satisfied for all i, resulting in

$$\nu T_c + T_1 \leq \min_{0 \leq i \leq (n-1)} (\delta_i - H_i) \tag{48}$$

which corresponds to a convex region. Notice that this simplified short-path constraint can also be obtained from (45) by requiring it to be satisfied for $l = 0$, thereby shrinking R_i^E by extending the leftmost boundary edge down to the T_1 axis and removing the other edges.

When (48) is combined with the long-path constraints (36) and (38), and the pulse-width constraints (2) and (3), we obtain the following expression for the minimum cycle time:

$$T_{c,\min} = \max \left\{ \frac{\overline{\Delta}}{1 + \nu}, \max_l \frac{\max_i \left[\left(\sum_{j=i-l}^{i} \Delta_j \right) + S_i \right] - \min_i (\delta_i - H_i)}{1 + l + l\nu}, \max_l \frac{\max_i \left[\left(\sum_{j=i-l}^{i} \Delta_j \right) + S_i \right] + w_1}{2 + l + \nu + l\nu}, 2w_1 \right\}. \tag{49}$$

The feasibility of this minimum cycle time must be checked by substituting it, along with the corresponding phase width T_1, in (48). If (48) is violated, then the restricted single-phase constraints have no feasible solution. This check is necessary only if the minimum cycle time obtained in (49) is set by the third or fourth arguments of the max function; if it is determined by the first or second argument, (48) is automatically satisfied.

4.4. Observations

The solution space becomes nonconvex when the early arriving signals are allowed to flow through the latches unimpeded by the clock, i.e. when $d_i > T_c - T_i$ for one or more stages. If necessary or desired, this can be prevented by using $d_i = T_c - T_i$ instead of the actual synchronization equation $d_i = \max(a_i, T_c - T_i)$ and can be accomplished in two ways:

1) By using a restricted single-phase clock which *assumes* that the early signals always start flowing through latches on the enabling clock edge even though they may not actually start their propagation until after the clock edge. This leads to safe though not generally minimum cycle times.
2) By using a coincident multiphase clock which permits the individual phase widths to be adjusted so that $d_i = D_i = T_c - T_i$ actually occurs at every latch. This choice involves more costly clock generation and distribution, but achieves the minimum possible cycle time of any coincident clocking scheme.

V. Examples and Results

We developed a computer program *pipe*T_c which determines the optimal cycle time for n stage pipes. *pipe*T_c reads in the pipeline parameters (number of stages, stage delays, setup and hold times, and wave pipelining parameters) and produces the optimal clock schedules and signal waveforms for general single-phase, restricted single-phase, and coincident n-phase clocking using latches.

In this section we illustrate the use of *pipe*T_c on three pipeline examples to highlight some of the issues in pipeline synchronization. The results are shown in Figs. 6–12. In each figure we show:

- The pipeline parameters (minimum and maximum delays, hold and setup times, wave pipelining parameter, and minimum pulse width)
- The region of feasible solutions, in the T_c/T_1 space, for general single-phase clocking (part (a) in each figure).
- The optimal clock waveform(s), and corresponding signal waveforms at all synchronizer inputs, for:
  general single-phase clocking (part (b)),
  restricted single-phase clocking (part (c)),
  single-phase clocking with negative edge-triggered flip-flops (part (d)),
  coincident n-phase clocking (part (e)).

The clock and signal waveforms in these figures are depicted using the notation introduced in Fig. 2.

The flip-flop solutions are obtained by substituting the synchronization equations $d_i = D_i = T_c$ in the signal propagation equations and combining the results with the hold and setup constraints. This procedure is analogous to those used in Section IV for latch synchronization; however, it is much less involved due to the simplicity of the flip-flop synchronization equations, and leads to the following simple expression for minimum cycle time

$$T_{c,\min} = \max\left\{\max_i\left(\frac{\Delta_i + S_i}{1+\nu}\right),\ 2w_1\right\} \qquad (50)$$

subject to the following feasibility conditions:

$$\nu = 0:\ \delta_i \geq H_i, \qquad \text{for } i = 0, \cdots, n-1$$

$$\nu \geq 1:\ \min_i\left(\frac{\delta_i - H_i}{\nu}\right) \geq \max_i\left(\frac{\Delta_i + S_i}{1+\nu}\right).$$

The phase widths T_{p_i} can be chosen arbitrarily as long as they satisfy the minimum pulse width constraints. By comparing (49) with (50) we see that operation at the ideal latch cycle time of

$$\frac{\overline{\Delta}}{1+\nu} \qquad (51)$$

is never possible with flip-flops.

Example 1: The first example is a 4-stage pipeline with an uneven distribution of stage delays. Optimal clock schedules were computed for two cases: (a) $H_1 = 2.0$, and (b) $H_1 = 2.5$. In both cases, $\nu = 0$ and $w = 1$. The results are shown in Figs. 6 and 7 and are summarized in Table I. Examination of these results leads to the following observations:

1) The general single-phase feasible region in Fig. 6 is nonconvex. It consists of the shaded area in the T_c/T_1 plane as well as the line segment AB.
2) The optimal general single-phase cycle time is the same as the optimal coincident 4-phase cycle time reaching the ideal value ($\overline{\Delta}/(1+\nu) = 10$), and is substantially lower than the restricted single-phase optimum (16.0) and the flip-flop optimum (18.0).
3) Although there are always exactly four signal waves in the pipeline in the general single-phase and coincident 4-phase solutions, during each clock cycle there are two waves traveling in stage 0 (from $t = 2.0$ to $t = 8.0$) and in stage 2 (from $t = 2.0$ to $t = 4.0$). This limited form of wave pipelining in particular stages occurs even though $\nu = 0$ since the delays of both stages 0 and 2 are greater than the cycle time. Examination of the signal waveforms suggests another way to determine if a given stage is wave pipelining: *stage i will "contain" 2 or more waves of data in every clock cycle from* $t = D_{i-1}$ *to* $t = A_i$ *if* $D_{i-1} < A_i$*; otherwise at most one wave can be traveling in stage i.*
4) When H_1 is increased from 2.0 to 2.5, the general single-phase feasible region shrinks and becomes convex (Fig. 7). Now, the general single-phase and the restricted single-phase solutions are identical ($T_{c,\min} = 16.5$), and both are larger than the coincident 4-phase solution ($T_{c,\min} = 10.125$). Note that due to the nonconvexity of the general single-phase feasible region, a 0.5 ns change in the hold time of stage 1 causes a 6.5 ns change in the optimal cycle time. In contrast, the restricted single-phase and coincident 4-phase optima changed by 0.5 ns and 0.125 ns, respectively, in response to the same 0.5 ns change in H_1. The flip-flop solution did not change.

Example 2: The second example is a modification of the first in which the delay of stage 1 has been increased from 4.0 to 8.0. We study the effect of changing the hold time of stage 2 from 6.0 to 7.5 in 0.5 ns increments. The results are shown in Figs. 8, 9, 10, and 11, and are sum-

i	δ_i	Δ_i	H_i	S_i
0	16.0	16.0	2.0	2.0
1	4.0	4.0	2.0	2.0
2	12.0	12.0	2.0	2.0
3	8.0	8.0	2.0	2.0

$\nu = 0$, $w = 1$

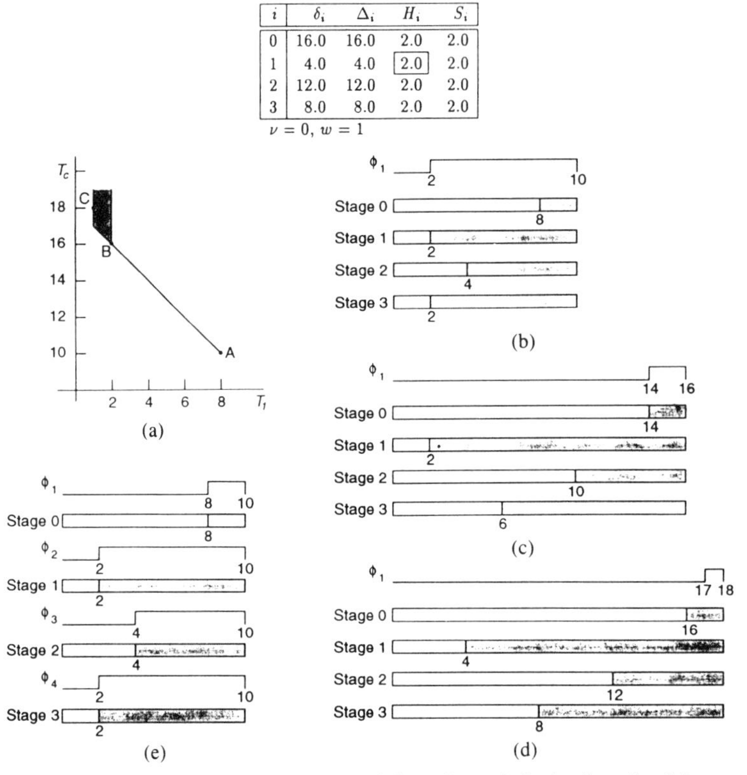

Fig. 6. Example 1, Case (a)—H_1 = 2.0 (a) General single-phase feasible region (latches). (b) Optimal general single-phase solution (point A). (c) Optimal restricted single-phase solution (point B). (d) Optimal flip-flop solution (point C). (e) Optimal 4-phase solution.

i	δ_i	Δ_i	H_i	S_i
0	16.0	16.0	2.0	2.0
1	4.0	4.0	2.5	2.0
2	12.0	12.0	2.0	2.0
3	8.0	8.0	2.0	2.0

$\nu = 0$, $w = 1$

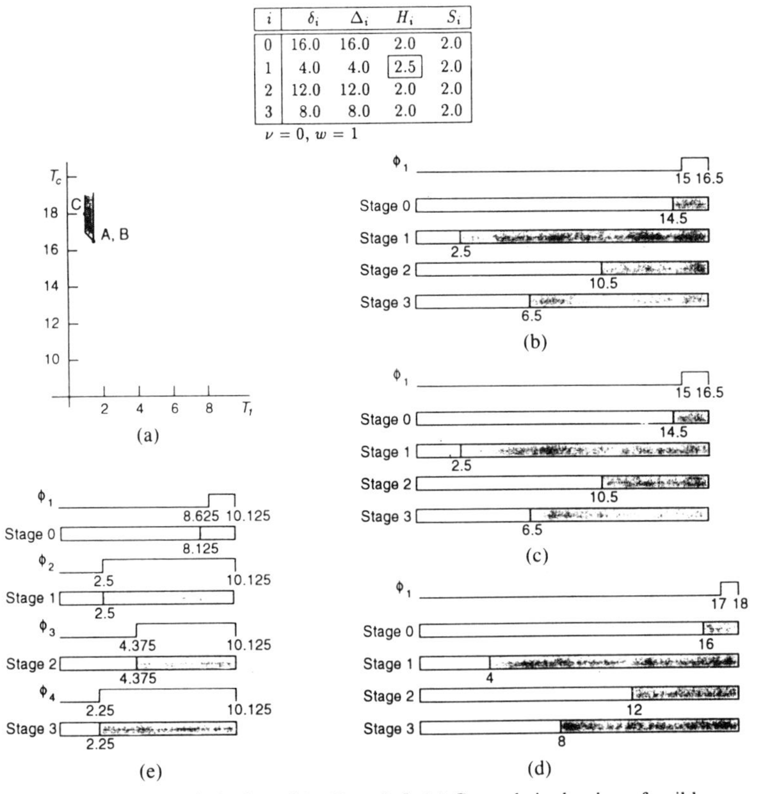

Fig. 7. Example 1, Case (b)—H_1 = 2.5. (a) General single-phase feasible region (latches). (b) Optimal general single-phase solution (point A). (c) Optimal restricted single-phase solution (point B). (d) Optimal flip-flop solution (point C). (e) Optimal 4-phase solution.

TABLE I
SUMMARY OF RESULTS FOR EXAMPLE 1 (ALL TIMES IN UNITS OF NANOSECONDS)

		G 1-Ph[1]		R 1-Ph[2]		C 4-Ph[3]	FF[4]
Case	H_1	T_c	T_1	T_c	T_1	T_c	T_c
(a)	2.0	10.000	8.000	16.000	2.000	10.000	18.000
(b)	2.5	16.500	1.500	16.500	1.500	10.125	18.000

[1]Optimal general single-phase solution
[2]Optimal restricted single-phase solution
[3]Optimal coincident 4-phase solution
[4]Optimal flip-flop solution

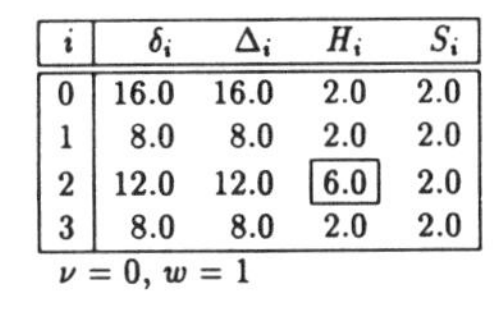

i	δ_i	Δ_i	H_i	S_i
0	16.0	16.0	2.0	2.0
1	8.0	8.0	2.0	2.0
2	12.0	12.0	6.0	2.0
3	8.0	8.0	2.0	2.0

$\nu = 0, w = 1$

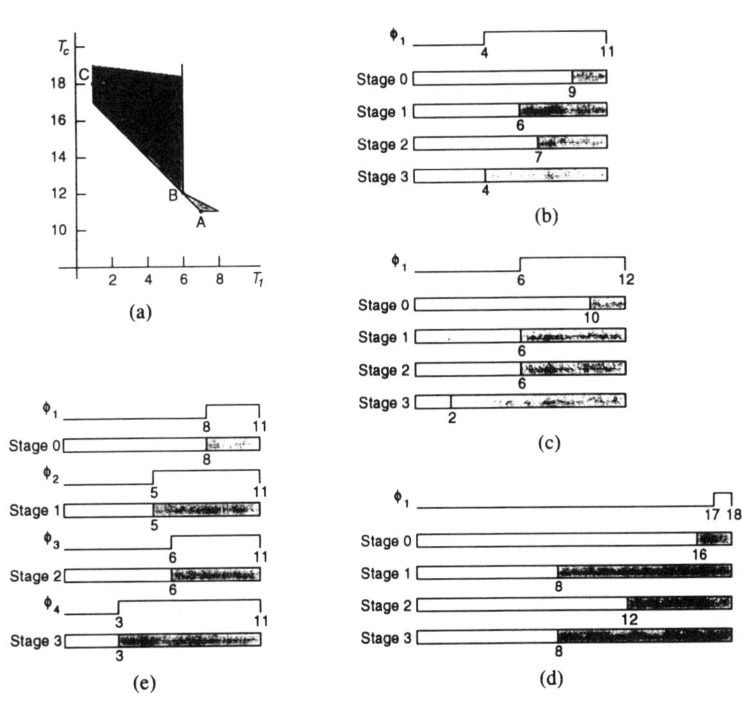

Fig. 8. Example 2, Case (a)—H_2 = 6.0. (a) General single-phase feasible region (latches). (b) Optimal general single-phase solution (point A). (c) Optimal restricted single-phase solution (point B). (d) Optimal flip-flop solution (point C). (e) Optimal 4-phase solution.

marized in Table II. The following additional observations can be made:

1) The general single-phase feasible region is nonconvex (Fig. 8), and becomes disconnected when H_2 is increased to 6.5 and 7.0 (Fig. 9 and Fig. 10). In particular, one of the disconnected subregions shrinks to a point. When H_2 is increased further to 7.5, the feasible region becomes convex (Fig. 11). Further increases in H_2 reduce the size of the feasible region, until it vanishes completely and the problem becomes infeasible.

2) The general single-phase solution is not unique. In fact, for H_1 = 6.0 and H_1 = 6.5, the same optimal cycle time (11.0) can be achieved with a range of values for T_1. This situation will arise whenever the lower bound constraint on T_c given by (36) is active (this lower bound corresponds to the horizontal line segment).

3) The restricted single-phase solution is now exhibiting wave pipelining in some stages. In fact, only the flip-flip solution is free from wave pipelining (recall that ν = 0 in these experiments).

i	δ_i	Δ_i	H_i	S_i
0	16.0	16.0	2.0	2.0
1	8.0	8.0	2.0	2.0
2	12.0	12.0	6.5	2.0
3	8.0	8.0	2.0	2.0

$\nu = 0,\ w = 1$

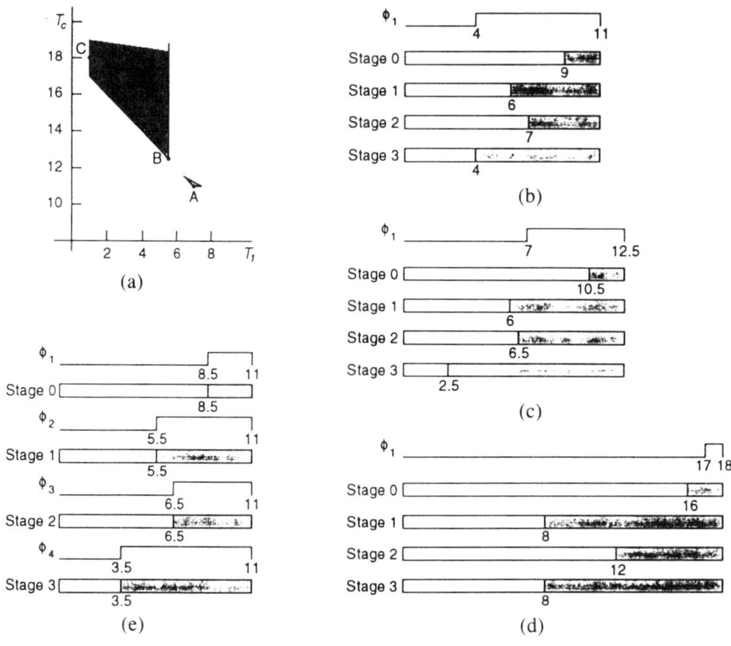

Fig. 9. Example 2, Case (b)—$H_2 = 6.5$. (a) General single-phase feasible region (latches). (b) Optimal general single-phase solution (point A). (c) Optimal restricted single-phase solution (point B). (d) Optimal flip-flop solution (point C). (e) Optimal 4-phase solution.

i	δ_i	Δ_i	H_i	S_i
0	16.0	16.0	2.0	2.0
1	8.0	8.0	2.0	2.0
2	12.0	12.0	7.0	2.0
3	8.0	8.0	2.0	2.0

$\nu = 0,\ w = 1$

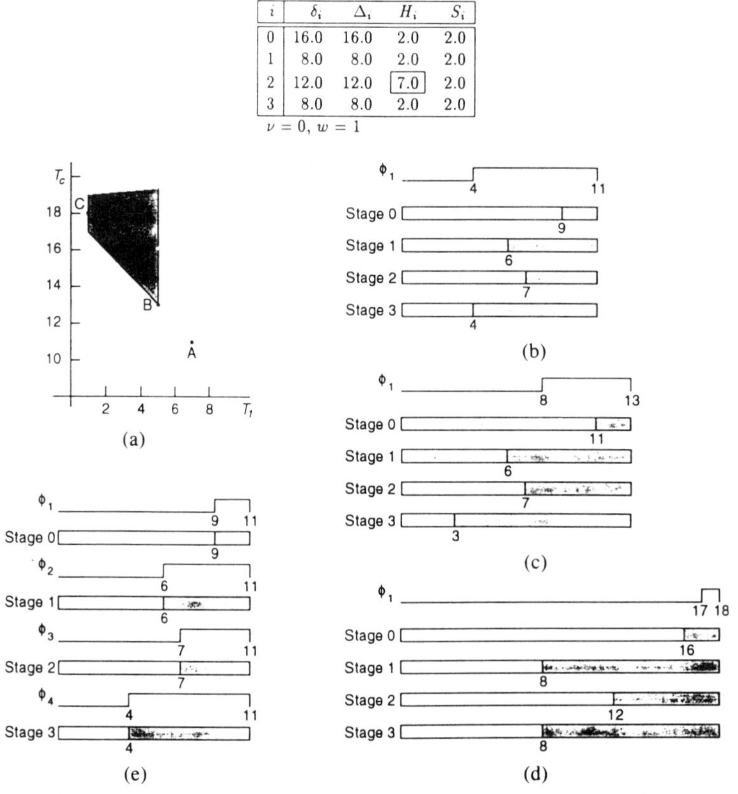

Fig. 10. Example 2, Case (c)—$H_2 = 7.0$. (a) General single-phase feasible (latches). (b) Optimal general single-phase solution (point A). (c) Optimal restricted single-phase solution (point B). (d) Optimal flip-flop solution (point C). (e) Optimal 4-phase solution.

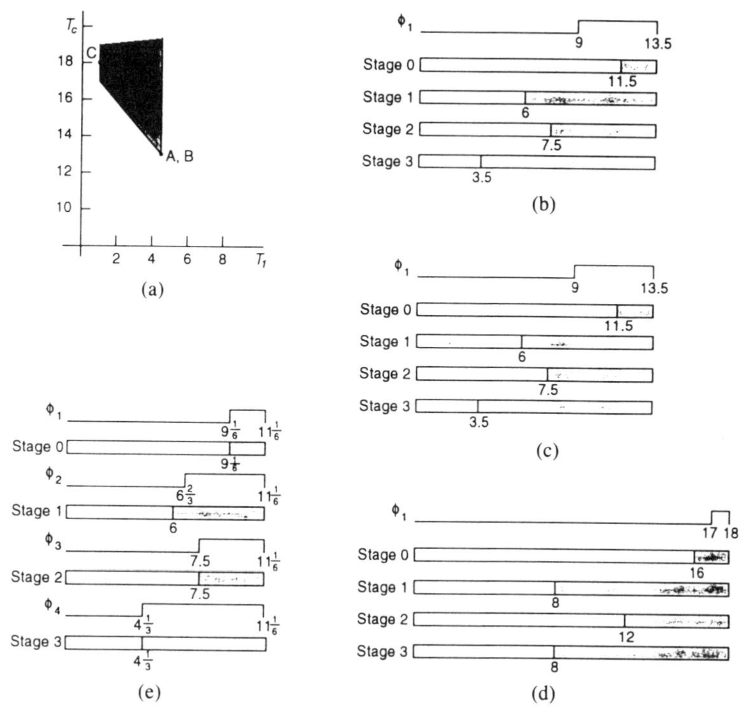

Fig. 11. Example 2, Case (d)—H_2 = 7.5. (a) General single-phase feasible region (latches). (b) Optimal general single-phase solution (point A). (c) Optimal restricted single-phase solution (point B). (d) Optimal flip-flop solution (point C). (e) Optimal 4-phase solution.

TABLE II
SUMMARY OF RESULTS FOR EXAMPLE 2 (ALL TIMES IN UNITS OF NANOSECONDS)

		G 1-Ph[1]		R 1-Ph[2]		C 4-Ph[3]	FF[4]
Case	H_2	T_c	T_1	T_c	T_1	T_c	T_c
(a)	6.0	11.000	[7.0, 8.0]	12.000	6.000	11.000	18.000
(b)	6.5	11.000	[7.0, 7.5]	12.500	5.500	11.000	18.000
(c)	7.0	11.000	7.000	13.000	5.000	11.000	18.000
(d)	7.5	13.500	4.500	13.500	4.500	11.167	18.000

[1]Optimal general single-phase solution
[2]Optimal restricted single-phase solution
[3]Optimal coincident 4-phase solution
[4]Optimal flip-flop solution

4) The 0.5 ns increase in H_2 from 7.0 to 7.5 (cases c and d) causes a 2.5 ns increase in the general single-phase optimum, a 0.5 ns increase in the restricted single-phase optimum, and a *one-sixth* ns increase in the multiphase optimum; the flip-flop optimum does not change. This is consistent with the earlier observation in example 1. The curious one-sixth ns increase in the multiphase case is readily explained in terms of the dual solution of the linear program [13] used to find the optimal cycle time.

5) Note that, for this as well as for the previous example, the optimal cycle time for general single-phase clocking is equal to either the optimal coincident multiphase cycle time or the optimal restricted single-phase cycle time. This is not true in general, as the last example demonstrates.

Example 3: The third example is for a 3-stage pipeline in which, unlike the first two examples, the minimum delays for some stages are strictly less than the corresponding maximum delays. The results are shown in Fig. 12 and suggest the following additional comments:

1) As was just noted, the optimal general single-phase cycle time (13 ns) is strictly less than the restricted single-phase optimum (14 ns) and strictly greater than the coincident 3-phase optimum (10 ns).
2) The optimal coincident 3-phase cycle time (10 ns) is greater than the average maximum stage delay ($\overline{\Delta}$ = 9.33 ns) due to the discrepancy between the minimum and maximum stage delays. Specifically, the signal arriving at stage 0 must wait 2 ns before beginning to propagate to stage 1. Averaged over the three pipe stages, this wait time accounts for the observed difference of 0.67 ns (10 − 9.33).

i	δ_i	Δ_i	H_i	S_i
0	4.0	7.0	2.0	2.0
1	5.0	7.0	2.0	2.0
2	14.0	14.0	8.0	2.0

$\nu = 0, w = 2$

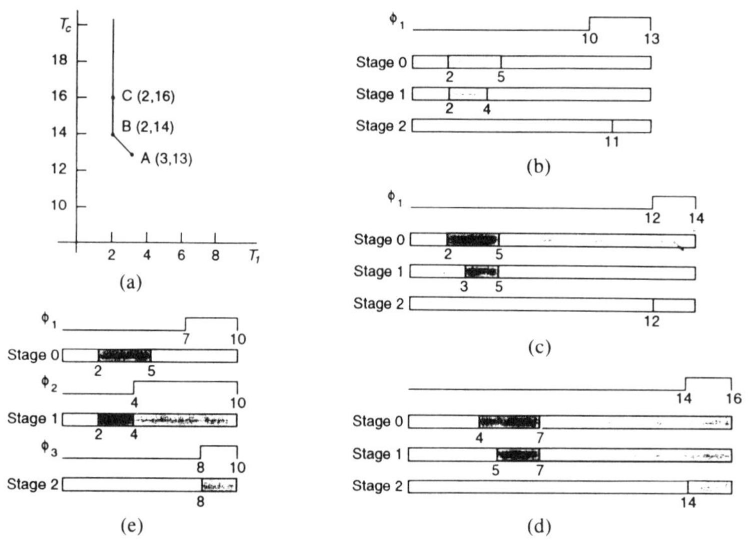

Fig. 12. Example 3. (a) General single-phase feasible region (latches). (b) Optimal general single-phase solution (point A). (c) Optimal restricted single-phase solution (point B). (d) Optimal flip-flop solution (point C). (e) Optimal 3-phase solution.

VI. Conclusions and Future Work

We have studied the problem of minimizing the cycle time for an n-stage pipeline under a variety of clocking conditions. This study clarified the relationships among concurrency, wave pipelining, and clocking. It has also led to a closed-form expression for minimum cycle time under a restricted form of single-phase clocking.

One of the important results of this study was discovering that even for single-phase clocks and simple circuit structures, the region of feasible solutions may be nonconvex or even disjoint. One must conclude, therefore, that such phenomena can also occur in the case of multiphase clocking of more complex circuits. Nonconvexity implies that slight variations in the circuit delays can cause large variations in the cycle time, leading to possible malfunction. Even more serious, a disjoint solution space poses problems of reachability (i.e., how do you start, stop and single-step the clock). In either case, trying to capitalize on the existence of such effects may lead to unreliable circuit operation and "weird" circuit behavior. Both problems can be traced to exploiting the transparency of latches for *fast* as well as slow signals to achieve lower cycle times.

With the above in mind, practical solutions should include only those clocking schemes whose feasible regions are convex, such as coincident multiphase and restricted single-phase clocks. Additional considerations, such as ease of clock generation and distribution, may further limit the range of options to just a few phases. The closed-form minimum cycle time expression for restricted single-phase clocking thus assumes greater significance as a bound on what is practically achievable.

We have incorporated wave pipelining in the model but have not addressed issues such as startability and stoppability (single-stepping). These issues remain as important open problems that must be solved before wave pipelining becomes viable. Finally, the above results do not take clock skew into account. We conjecture that clock skew, clock phase shifts, and wave pipelining can be integrated into a unified model for clock design.

References

[1] K. A. Sakallah, T. N. Mudge, T. M. Burks, and E. S. Davidson, "Synchronization of pipelines," Tech. Rep. CSE-TR-97-91, Univ. Michigan, Ann Arbor, Feb. 1991.

[2] K. A. Sakallah, T. N. Mudge, and O. A. Olukotun, "*checkT$_c$* and *minT$_c$*: Timing verification and optimal clocking of synchronous digital circuits," in *Dig. Tech. Pap. IEEE Int. Conf. on Computer-Aided Design (ICCAD)*, pp. 552–555, Nov. 1990.

[3] T. G. Szymanski, "Computing optimal clock schedules," in *Proc. IEEE/ACM Design Automation Conf. (DAC)*, pp. 399–404, 1992.

[4] T. G. Szymanski, "Verifying clock schedules," in *Dig. Tech. Pap. IEEE Int. Conf. on Computer-Aided Design (ICCAD)*, Nov. 1992, pp. 124–131.

[5] N. P. Jouppi and D. W. Wall, "Available instruction-level parallelism for superscalar and superpipelined machines," in *Proc. Conf. on Architectural Support for Programming Languages and Operating Systems*, pp. 272–282, Boston, MA, Apr. 1989.

[6] S. R. Kunkel and J. E. Smith, "Optimal pipelining in supercomputers," in *Proc. 13th Ann. Symp. on Computer Architecture*, pp. 404–411, 1986.

[7] L. W. Cotten, "Circuit implementation of high-speed pipeline systems," in *AFIPS Fall Joint Computer Conf.*, pp. 489–504, 1965.

[8] T. G. Hallin and M. J. Flynn, "Pipelining of arithmetic functions," *IEEE Trans. Computers*, vol. C-21, pp. 880–886, Aug. 1972.

[9] B. K. Fawcett, *Maximal Clocking Rates for Pipelined Digital Systems*, Master thesis, Univ. Illinois, 1975.

[10] D. Wong, G. D. Micheli, and M. Flynn, "Inserting active delay elements to achieve wave pipelining," in *ICCAD-89 Dig. Technical Pap.* pp. 270–273, 1989.

[11] D. A. Joy and M. J. Ciesielski, "Placement for clock period minimization with multiple wave propagation," in *Proc. of the 28th Design Automation Conf.*, pp. 640–643, 1991.
[12] C.-H. Chang, E. S. Davidson, and K. A. Sakallah, "Using constraint geometry to determine maximum rate pipeline clocking," in *Dig. of Tech. Pap. IEEE Int. Conf. on Computer-Aided Design (ICCAD)*, Nov. 1992, pp. 142–148.
[13] D. T. Phillips, A. Ravindran, and J. J. Solberg, *Operations Research: Principles and Practice*. New York: Wiley, 1976.

Salphasic Distribution of Clock Signals for Synchronous Systems

Vernon L. Chi

Abstract—The design of a synchronous system having a global clock must account for propagation-delay–induced phase shifts experienced by the clock signal (clock skew) in its distribution network. As clock speeds and system diameters increase, this requirement becomes increasingly constraining on system designs. This paper describes a method that exploits properties of standing waves to reduce substantially clock skews due to unequal path lengths, for distribution network diameters up to several meters. The basic principles are developed for a loaded transmission line, and then applied to an arbitrary branching tree of such lines to implement a clock distribution network. The extension of this method to two- and three-dimensional distribution media is also presented, suggesting the feasibility of implementing printed circuit board clock planes exhibiting negligible phase shift over their extents.

Index Terms—Clock plane, clock signal, distribution network, phase skew, propagation delay, salphasic clock, synchronous system, system clock.

I. Introduction

sal·pha·sic (săl-phā′zĭk) *adj.* Characterized by sudden jumps in phase between regions of relative constancy. [Lat., *salio,* to jump + NLat., *phasis,* phases of the moon.]

Synchronous system design methodology, although well developed and widely used, is limited by signal propagation delays [1]. For system diameters that are small, signal interconnections can be considered equipotential at any given instant. However, trends toward higher clock speeds and more system function are increasingly forcing designers to deal explicitly with time delays inherent in the propagation of signals on their interconnecting structures.

Some synchronous system organizations constrain data signal paths to be local in an attempt to avoid propagation delay problems. The global clock signal, however, must be propagated across the entire system in a manner that preserves the correct ordering of events throughout the system. Considerable work has been done to address this problem, ranging from investigation of various clocking disciplines [2] to tuning the distribution network conductor lengths and amplifier delays [3]–[5] to minimize clock skew across the system.

A fundamentally different approach uses self-timed and asynchronous delay-insensitive [6]–[8] disciplines. These techniques appear to promise scalability to any system size and speed, at the expense of additional hardware. For large future designs, these disciplines may become the mainstream methodology of choice, but they are substantially different from synchronous design methodology, and are neither widely understood nor practiced today.

This paper addresses the minimization of clock skew for moderately large synchronous systems, ranging up to several meters, using spatial phase properties of standing waves. This notion was first disclosed [9] for the restricted case of resonant transmission lines. The core contribution herein is the discovery that these properties do not depend on resonant structures, rather that they occur for any sinusoidal signal propagating as a wave in a finite, linear, lossless system.

The method developed is potentially quite general, applying in principle to two- and three-dimensional conducting geometries as well as to wires. This paper presents an analysis of the useful special case of loaded transmission lines, and tree structured networks composed of such lines. The extension of the method to multidimensional conductor geometries is also discussed

Section II presents the theoretical groundwork for an ideal lossless transmission line and load. The mechanism whereby salphasic behavior arises from standing waves is described, and the conditions are derived under which a finite loaded transmission line exhibits nearly salphasic behavior. In Section III, a canonical branch circuit is described that satisfies these conditions, and it is used to show that an arbitrarily branching tree composed of such circuits also satisfies these conditions, thereby demonstrating the salphasic behavior of the entire tree. Section IV describes the extension to multidimensional distribution geometries such as "clock planes." Section V reports some empirical results from application of salphasic clock distribution technology to a digital system of significant speed and size. Section VI compares salphasic technology with existing methods for distribution of clock signals, and discusses its advantages and limitations. Section VII contains a summary and conclusions, and a brief discussion of additional research needed to develop salphasic clock distribution into a broadly applicable technology.

II. Physical Principles

In an infinite lossless uniform linear transmission line, two waves V_f and V_r of equal frequencies propagating in the forward and reverse directions, respectively, are characterized as follows:

$$V_f = V_A \sin(\omega t - \beta x), \qquad (1)$$

Reprinted from *IEEE Trans. Comput.*, vol. 43, no. 5, pp. 597–602, May 1994.

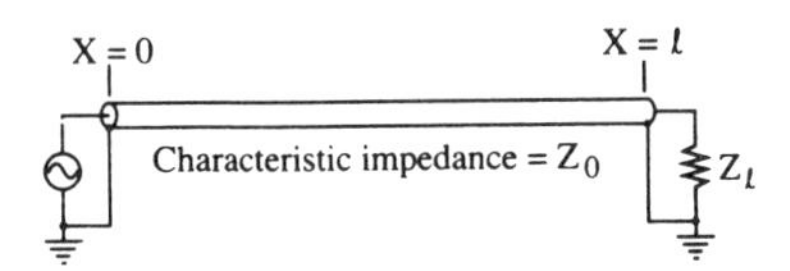

Fig. 1. Linear transmission line driven sinusoidally at $x = 0$ loaded by linear impedance Z_l at $x = l$. If the line and load are both lossless, the magnitude of the reflection coefficient is unity, resulting in a pure standing wave and ideal salphasic behavior.

$$V_r = V_B \sin(\omega t + \beta x), \tag{2}$$

where V_A and V_B represent the amplitudes, ω represents the angular frequency, t represents time, x represents the position along the transmission line, and β represents the phase constant (angular spatial frequency) of the waves. Letting $V_B = V_A$ and adding the two waves at location x and time t on the transmission line provide an instantaneous voltage

$$\begin{aligned} V &= V_A(\sin(\omega t - \beta x) + \sin(\omega t + \beta x)) \\ &= 2V_A \sin(\omega t)\cos(\beta x). \end{aligned} \tag{3}$$

According to this relationship, the temporal phase ωt of the instantaneous voltage V is independent of location x. Thus, as x varies, only the amplitude of the sinusoidal wave is affected while the temporal phase remains constant.

This behavior, which we will call *salphasic,* provides that the phase of such a wave distribution is equal for all values of x within any region in which the sign of $\cos(\beta x)$ remains constant. Salphasic behavior depends on the equality of both the amplitudes and frequencies of the forward V_f and reverse V_f traveling waves. These conditions happen to be identical with those necessary and sufficient to form standing waves.

Fig. 1 shows a finite lossless uniform linear transmission line having characteristic impedance Z_0, driven at location $x = 0$, and terminated by a load of impedance Z_l, at location $x = l$. If the transmission line and the load are both lossless, then the characteristic impedance Z_0 has no reactive component, that is , $Z_0 = R_0 + j \cdot 0$, and the load impedance Z_l has no resistive component, that is, $Z_l = 0 + jX_l$. Accordingly, the voltage reflection coefficient is [10]

$$\rho = \frac{Z_l - Z_0}{Z_l + Z_0} = -\frac{R_0 - jX_l}{R_0 + jX_l}, \tag{4}$$

which shows that $|\rho| = 1$. This satisfies the desired condition for a purely standing wave wherein the magnitude of the reflected wave V_B is given by

$$V_B = |\rho| \cdot V_A = V_A. \tag{5}$$

Therefore, a lossless, finite transmission line loaded by a pure reactance exhibits purely salphasic behavior.

Conversely, if the transmission line and its load are lossy, in general $Z_0 = R_0 + jX_0$ and $Z_l = R_l + jX_l$. In this case, the voltage varies according to the more general relationship [10]

$$V_x = V_A e^{-\gamma x} + V_B e^{\gamma x}, \tag{6}$$

where V_x is the voltage at any given location x, $\gamma = \alpha + j\beta$ is the propagation constant, and V_A and V_B are amplitudes of the forward and reverse waves, respectively. The following boundary conditions apply: At the driven end $x = 0$,

$$V_0 = V_A + V_B. \tag{7}$$

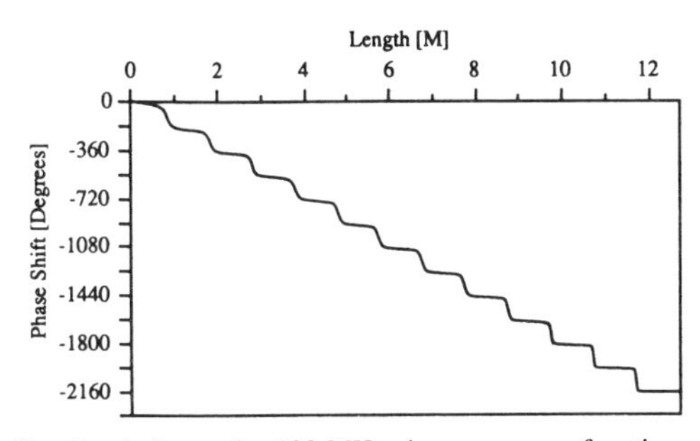

Fig. 2. Simulated phase of a 100-MHz sine wave as a function of distance from the driven end of a 12.7-m-long RG58/U coaxial cable terminated by a short circuit. Strongly salphasic behavior near the termination becomes progressively weaker toward the driven end due to the slight lossyness of the cable.

and at the loaded end $x = l$ [10],

$$\rho = \frac{V_B e^{\gamma x}}{V_A e^{-\gamma x}}. \tag{8}$$

Equations (6)–(8) can be solved for the load voltage, that is, $V_l = V_x, x = l$:

$$V_l = V_0 \frac{\rho + 1}{\rho e^{-\gamma l} + e^{\gamma l}}. \tag{9}$$

Letting $\rho = e^{\mu + j\nu}$ it is clear that the low-loss load condition $|\rho| \approx 1$ is satisfied by an equivalent condition $\mu \approx 0$. Note also that the low-loss transmission line condition is $\alpha \approx 0$. In the limit as both losses become small, (9) becomes (10), found at the bottom of the page. As α and μ approach zero, the imaginary component becomes negligible, showing that in the limit V_l and V_0 are salphasic. In particular with $|\rho| = 1$, as the real part α of the propagation constant γ becomes smaller, salphasic behavior holds for increasingly greater transmission lengths. Thus, a lower loss line can maintain salphasic behavior over greater lengths.

To demonstrate salphasic behavior for a slightly lossy transmission line, a 12.7-m length of RG58/U-type coaxial cable (Belden 9201) driven by a 100-MHz sine wave and terminated by a short circuit was simulated according to (9) and is shown in Fig. 2. Since the short circuit termination dissipates no energy, the reflected wave has the same amplitude as the incident wave at the termination. This satisfies the pure standing wave condition, which results in strongly salphasic behavior near the termination. This is illustrated in Fig. 2 by the marked steplike shape of the phase plot near the termination.

As a result of the lossiness of the cable, the incident wave is attenuated slightly as it propagates toward the termination, whereas the reflected wave is attenuated as it propagates back toward the driving point. Their amplitudes are therefore only equal at the load and become increasingly disparate toward the driving point. Thus the standing wave condition becomes progressively less well satisfied with increasing distance from the termination, resulting in progressively weaker salphasic behavior. This is evident in Fig. 2, where the steplike behavior becomes progressively softer with increasing distance from the termination.

III. Tree Topology and Its Canonical Branch Circuit

Consider a lumped constant L-section having input series impedance Z_s and output shunt admittance Y_p. Output voltage

$$\frac{V_l}{V_0} = \frac{(\cos(\nu - \beta l) + \cos(\beta l)) + j(\alpha(\sin(\nu - \beta l) - \sin(\beta l)) + \mu \sin(\beta l))}{\cos(\nu - 2\beta l) + 1}. \tag{10}$$

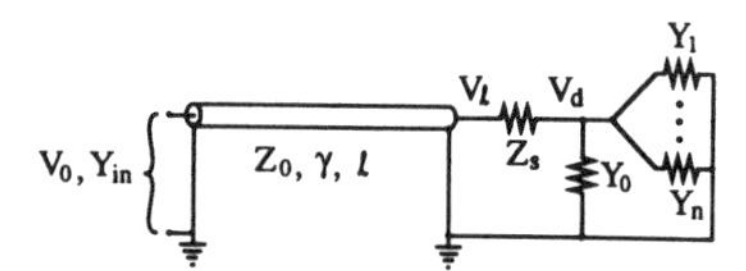

Fig. 3. Canonical branch circuit used to construct an arbitrarily branching salphasic clock distribution tree. Each admittance $Y_i, i > 0$, represents the equivalent input admittance Y_{in} of a similar canonical branch circuit. Z_s and Y_0 are lumped immitances, each of which may be optionally zero.

V_{out} is related to input voltage V_{in} by

$$V_{out} = V_{in}\frac{1}{Z_sY_p + 1}. \tag{11}$$

If Z_s and Y_p are nearly lossless, then

$$Z_sY_p \approx (jX_s)(jB_p) = -X_sB_p; \tag{12}$$

consequently the behavior of the L-section is salphasic, that is, V_{out} and V_{in} are nearly of equal or opposite phase.

Fig. 3 shows a canonical branch circuit comprising a finite linear lossy transmission line of characteristic impedance Z_0, propagation constant γ, and length l, loaded by a circuit comprising a lumped series impedance Z_s, a lumped shunt admittance Y_0, and the equivalent shunt admittances $Y_1 \cdots Y_n$ presented by n similar canonical branch circuits connected to Z_s.

The equivalent admittances $Y_i, i > 0$, are determined by using the following formula [10] for calculating the input terminal admittance presented by a loaded transmission line expressed in terms of its Z_0, γ, l, and reflection coefficient ρ due to its load,

$$Y_{in} = \frac{1}{Z_0}\frac{e^{2\gamma l} - \rho}{e^{2\gamma l} + \rho}. \tag{13}$$

The aggregate output shunt admittance connected to Z_s is the sum of the true lumped admittance Y_0 and the input admittances Y_{in} of each branch connected to Z_s. Thus, the load circuit is electrically equivalent to the L-section characterized by (11) if we let

$$Y_p = \sum_{i=0}^{n} Y_i. \tag{14}$$

Combining (9) and (11) provides a voltage transfer function for the canonical branch circuit depicted in Fig. 3,

$$V_d = V_0\frac{\rho + 1}{(Z_sY_p + 1)(\rho e^{-\gamma l} + e^{\gamma l})}. \tag{15}$$

Under sufficiently lossless conditions, it was shown in (10) that V_0 and V_l are nearly salphasic, and in (11) that V_l and V_d are nearly salphasic; hence, V_0 and V_d are also nearly salphasic. Since this holds true for each canonical branch circuit, it holds for all voltages in an arbitrarily branching tree composed of such canonical branch circuits.

A computer program based on (15) was used to simulate a model tree distribution network shown approximately to scale in Fig. 4. The model assumed an 18 × 18-in. standard multilayer glass-epoxy printed curcuit board (PCB) with 2 [oz/ft^2] copper cladding as the implementation medium for this network.

The branch circuit conductors are patterned on one layer of the PCB, separated from a ground plane by 11.8 mil of FR4 glass-epoxy dielectric. The simulated clock frequency was 40 MHz. The root branch and the vertical feeder branches were 20 mil wide, while the remaining branches were 10 mil wide. The loads, represented by a •, were each 10 pF. The numbers shown adjacent to each load represent the phase and magnitude of the voltage at the load relative to the voltage at the driving point, which was set to unit magnitude and zero phase.

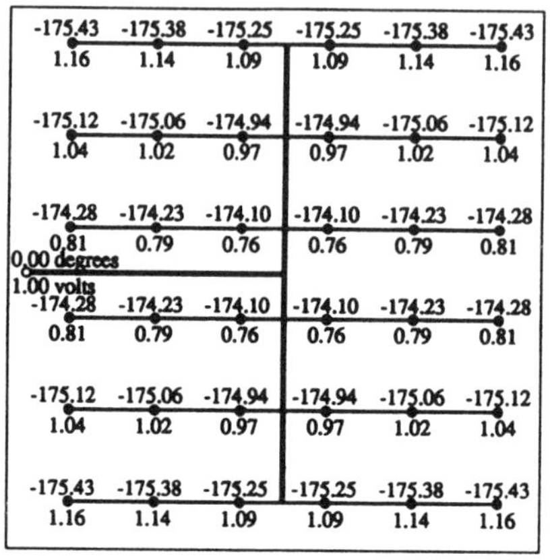

Fig. 4. Model of a salphasic clock distribution tree patterned on an 18 × 18-in. FR4 epoxy-fiberglass printed circuit board with copper conductors. Simulated phase and magnitude of the 40-MHz clock signal are shown at the driving point and at the loads.

These results show that all loads receive the clock signal in approximately the same phase, although their electrical distances from the driving point to the nearest and farthest loads is about 11 in. This distance would correspond to over 20° clock skew in conventional systems. However, due to the near salphasic behavior of the signal, clock skew between any two of the loads is only 1.33°.

IV. Generalized Conductor Geometries

The preceding development was amenable to a simple mathematical treatment due to the one-dimensional mathematical nature of conventional electrical transmission lines, and the considerable body of existing applicable knowledge.

No simple, closed-form mathematical methods are known for characterizing nonhomogeneous wave solutions for the more general two- and three-dimensional cases of standing waves with arbitrary boundary condition geometries, behaving according to the wave equation

$$\nabla^2\psi - \frac{1}{c^2}\frac{\partial^2\psi}{\partial t^2} = -4\pi f(\vec{x}, t). \tag{16}$$

Nevertheless, any linear system composed of a bounded, lossless transmission medium, with lossless linear loads, driven by a sinusoidal source exhibits pure standing wave behavior in three dimensions, and as special cases, in one and two dimensions. This can be seen through the following reasoning.

Since the system comprises purely linear components, no harmonics are produced from the sinusoidal signal. Therefore, the signal energy in the system is contained entirely in sinusoidal waves at the signal frequency.

In the steady state, no net signal energy is exchanged between the signal source and the system, because the system is bounded and lossless. Thus, in the steady state, a wave propagating from the source into the medium carries an amount of energy that must be balanced exactly by the amount of energy carried by another wave propagating back into the source from the medium.

Under any given set of boundary conditions, the wave equation admits of only two such nonhomogeneous solutions, which are identical in all respects except in their opposite directions of propagation. If the energies carried by these two waves are equal, their amplitudes must likewise be equal, thereby ensuring a pure standing wave. This remains true even after an indefinite number of reflections off the loads and boundaries of the system because the two waves are everywhere identical except for opposite directions of propagation.

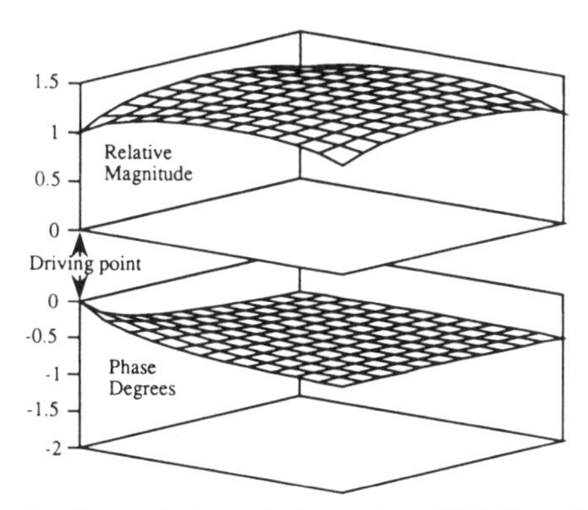

Fig. 5. Simulated magnitude and phase of a 155-MHz salphasic signal between a 8.5 × 8.5-in. pair of copper planes driven at one corner and loaded inductively at the other three.

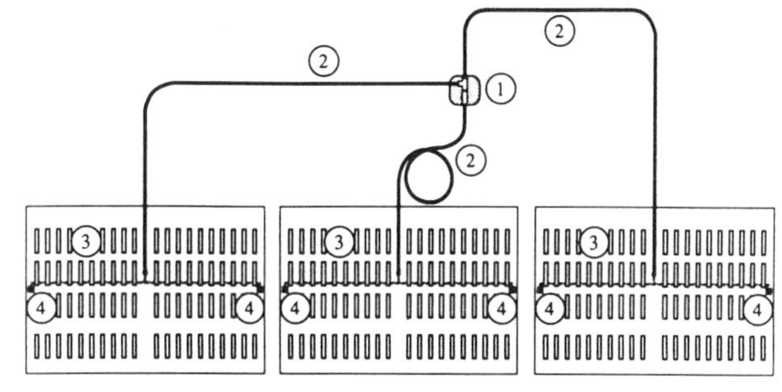

Fig. 6. Physical configuration of the Pixel-Planes 5 clock distribution tree. The clock oscillator ① is connected by the coaxial lines ② to the middles of the backplanes ③ in separate racks. On each backplane, the signal conductor branches with each branch driving half of the backplane slots. Each branch is terminated by a tunable inductor ④.

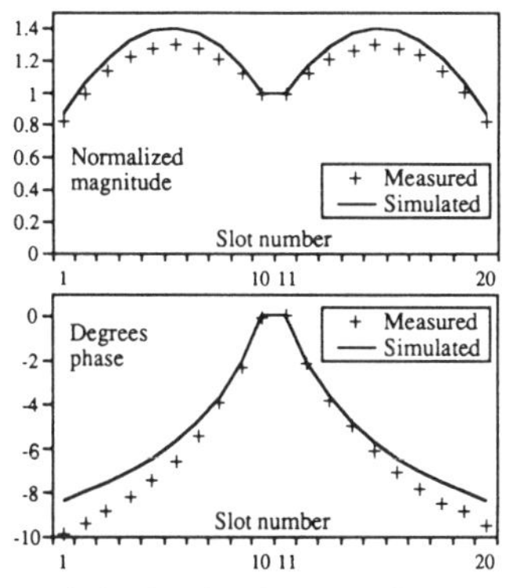

Fig. 7. Comparison of simulated and measured phase and amplitude of the *Pixel-Planes* 5 clock signal across the backplane of rack 1. This backplane was populated fully.

Since the conditions for pure standing waves are the same as for pure salphasic behavior, the notion of salphasic behavior can be generalized to propagation across surfaces or through volumes, for signals behaving according to the wave equation, although closed-form solutions corresponding to the geometric boundary conditions may not be possible.

Fig. 5 shows a simulated magnitude and phase of a 155-MHz salphasic signal between a pair of 8.5 × 8.5-in. copper planes separated by 0.028 in. of air, driven at one corner and loaded at the other three corners by 3-nH inductors having a Q of 100. The simulation was performed using finite differences, taking into account the skin effect losses of the copper planes.

V. Empirical Results

The foregoing theory and simulations fail to take into account all possible physical effects encountered in clocking a real synchronous system, such as nonlinear loads, crosstalk and noise, etc. It is therefore of interest to ascertain how salphasic technology performs in an actual system. The Pixel-Planes 5 system [11], designed and built in the Department of Computer Science at the University of North Carolina–Chapel Hill, provided a test bed for evaluating salphasic clock distribution technology.

Pixel-Planes 5 is a heterogeneous multicomputing 3-D raster graphics system having a massively parallel SIMD subsystem comprising hundreds of thousands of processors, a MIMD subsystem comprising some 40 Intel i860-based processors, and various other system functions such as video frame buffers, I/O interfaces, etc. It is currently packaged in three 19-in. equipment racks with three side-by-side backplanes each having 20 slots. This system is globally synchronous with a 160-MHz clock being distributed across all backplanes from a single crystal oscillator module. Salphasic technology is used at the backplane level, whereas conventional technology is used on the daughter boards.

The clock is distributed from a Vectron model CO-233 crystal oscillator through semirigid coaxial transmission lines to the centers of the backplanes (Fig. 6). These transmission lines were sized to transform the impedance presented by each backplane to a near open circuit, so that the oscillator voltage is substantially unaffected by connecting the backplanes. The clock circuit on each backplane branches at the driving point, with each branch driving half of the backplane in a bus-style multidrop configuration. Each branch is terminated by a tunable shunt inductor. The terminations are tuned to position the maximum voltage points of the standing wave to be located approximately at backplane slots 5 and 15.

The clock receivers on the daughter boards are AD96685 high-speed comparators, the ECL outputs of which drive conventional clock distribution trees with MC100E111 clock buffers and equal length distribution lines. On each board, the comparator is located immediately adjacent to the backplane connector, and the effective parasitic capacitance presented to the backplane is approximately nulled by a surface-mount chip inductor in shunt with the input. This minimizes the change of the standing wave distribution on the backplane with the number of populated slots.

Detailed simulations were performed in the development of the salphasic backplane clock distribution system, and detailed measurements were performed on the running system. A comparison between simulated and measured results is shown in Fig. 7 for a two-rack configuration with 24 slots populated. The data presented are for the backplane in rack 1, of which all 20 slots were populated. The measured and simulated data are mutually normalized to slot 10 of rack 1, which is the closest measurable location to the backplane driving point. Phase and amplitude measurements were performed using a Hewlett-Packard Model 4195A Network Analyzer with a pair of Tektronix P6201 1-GHz active probes with 10× attenuators to reduce probe capacitances to 1.5 pF.

The 10° skew across the backplane is equivalent to 175 pS, which is more than an order-of-magnitude improvement over the 2-nS propagation delay that would be experienced in conventional technology using the same backplane geometry.

Fig. 8 shows oscilloscope waveforms of the clock signal taken at selected slots on the fully populated backplane. These measurements were taken with a Tektronix Model DSA602 Digital Signal Analyzer with an 11A72 1-GHz preamp, again using P6201 active probes with 10× attenuators. Sampling noise above 1 GHz was suppressed by signal averaging over 16 waveform acquisitions. These waveforms demonstrate the near-sinusoidal nature of the clock signal; however, a slight second harmonic distortion is evident on close scrutiny.

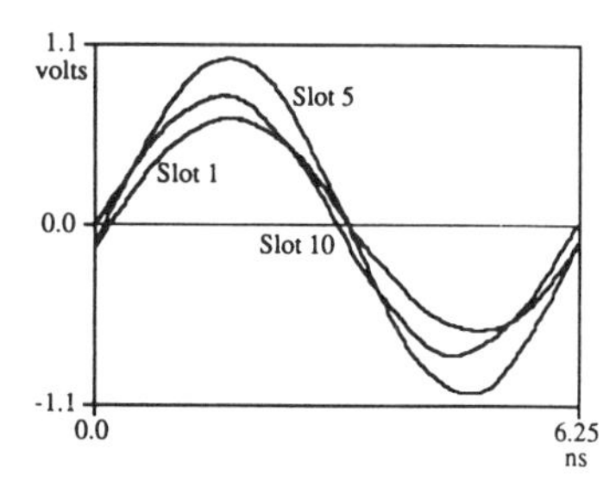

Fig. 8. Oscilloscope traces of Pixel-Planes 5 clock waveforms at selected slots of the backplane in rack 1.

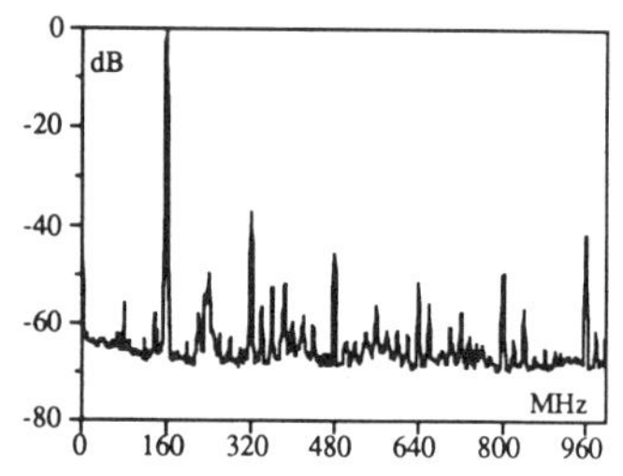

Fig. 9. Discrete Fourier transform of Pixel-Planes 5 clock waveform at slot 10 of rack 1.

A discrete Fourier transform of the slot 10 waveform using a Blackman window is shown in Fig. 9. The main spectral components are harmonics of the master clock frequency, with the second harmonic being about 2% and all others below 1%. Other spectral components appear at multiples of the 20- and 40-MHz daughter board clocks. These are all well below 0.5% of the master clock amplitude.

The salphasic clock system essentially came up on the first attempt, and, after a simple tuning procedure using the variable inductor terminations, has performed continuously without attention since August 1990. The configuration of this system, that is, the number of populated slots, has varied from 24 to 60 over this time, demonstrating the insensitivity of this design to changes in the backplane configuration.

VI. Comparison with Existing Methods

Conventional technology can, in principle, achieve as good performance as salphasic technology at the cost of extra circuitry, interconnection, and design complexity. However, salphasic technology is able to exploit a far wider range of distribution conductor geometries that not only simplifies the design, but enables more economical distribution systems. In many respects, these technologies are difficult to compare directly, and their costs in design effort are manifested in different ways. Nevertheless, the following comparisons are made.

A. Conductor Length versus Wave Extent

In conventional technology, multiple loads may be distributed along the length of any single conductor up to the point where propagation delays exceed allowable skew levels. Above 100 MHz, typically no more than two loads are permissable. Branching structures are also limited because of the need to avoid signal reflections due to impedance mismatches. Thus, high-performance designs usually employ a separate clock line for each one or possibly two clocked circuits; the lines being "tromboned" to equalize their electrical lengths, and each being individually driven by a clock buffer amplifier.

For example, consider a circuit board with 100 clocked IC's implemented in ECL technology. A typical design uses an MC100E111 1 : 9 differential clock driver IC to drive nine clock feeder lines, all tuned to the same electrical length, to distribute the clock to nine subsidiary MC100E111 IC's. In turn, these each drive nine lines, which each clock one or two IC's. These 81 lines also are tuned to the same electrical length as each other.

In salphasic technology, length considerations are manifested through their effects on the signal magnitude distribution. Clocked modules should not be connected to the distribution network near voltage nodes. Simple designs have all loads positioned well within a single half-wave voltage loop, although more ambitious designs spanning multiple loops are possible. Instead of exact length matching, usable connection regions sufficiently far from voltage nodes are observed that depend on clock frequency and propagation velocity. As indicated herein, a wide variety of conductor geometries, including clock planes, can be used to realize structures wherein these usable connection regions satisfy the clock distribution system requirements.

For example, 100 clocked IC's all could be connected to a single branching tree of conductors provided its geometry is arranged such that no voltage nodes occur, and that acceptable signal levels are presented to the clocked IC's. In practice, this translates to keeping lengths sufficiently short so that the entire structure fits well within a single half-wave loop of the standing wave.

B. Impedance Matching versus Standing Wave Control

Impedance matching is necessary in conventional technology to preserve the integrity of the clock signal. Failure to match properly can result in overshoot, ringing, and other artifacts that increase jitter and compromise noise immunity. Impedance matching requires proper termination and also limits the use of branching conductors. On a PC board, binary branching requires quartering the conductor width at each branch point to maintain forward impedance matching. For this reason, branching is not generally accepted as a viable technique in high-performance, controlled-impedance designs. Proper termination in the preceding example requires the use of 90 closely matched resistive loads.

In salphasic technology, however, a total mismatch is the desired condition. This means ensuring that the receiving loads are substantially nondissipative, typically by being high impedance. In addition, the conductor system must be sufficiently low loss. At PC board and backplane levels, this condition is satisfied easily in the subgighertz regime. In very large scale integrated circuits, however, salphasic technology will not work because relative conductor losses become too great as the geometry is scaled down to these sizes.

Optimizing salphasic networks does require selection of reactive terminations to control the position of the standing wave. Moreover, if the clock frequency is changed, the values of these terminations must be changed to reoptimize. This is in contrast to conventional technology where the terminations are insensitive to variations in clock rate. The ready availability of high-quality tunable reactances enables salphasic designs to cope reasonably with this effect.

C. Clock Power

A conventional distribution system operating at speeds requiring controlled-impedance transmission lines must dissipate at least of the order of V_{rms}^2/Z_0 per transmission line in its terminations. In the preceding example, this comes to about 2 W. The power dissipation of salphasic systems is negligible.

D. Noise and Crosstalk

Signals for very high speed systems should have quick rise times for well-defined logic intervals, yet soft rise times to minimize crosstalk and reflections of high harmonics. Thus, well-designed high-speed logic families such as MECL100E balance these conflicting requirements to produce waveforms that approximate a sine wave at their maximum operating rate. Because of this property, the crosstalk sensitivity and noise immunity of conventional and salphasic technologies at maximum logic speed are similar. At lower speeds, however, salphasic technology suffers reduced noise immunity near transition times because of the reduced slope of the clock signal at the switching threshold.

The use of phase-locked-loop clock receivers appears to offer a means to improve noise immunity. It has been known for many years that phase locked loops can be designed that are particularly robust to additive noise [12], [13]. Such designs can be used with either conventional or salphasic technology, thus making their noise performance equivalent.

Another major difference is that salphasic technology is intrinsically nondissipative except at the driving point, so that crosstalk and other noise such as harmonics generated by nonlinear loads are not damped as effectively as in conventional technology. The use of substantially linear loads and careful control of crosstalk have proven adequate in the case of the Pixel-Planes 5 system, as described herein.

E. Clock Buffer Skew versus Overdrive Dispersion

Given a carefully designed conventional distribution system, the main contribution to uncontrolled skew is component variation in the clock buffers themselves. Unfortunately, this cannot be controlled completely at design time, so the highest performance designs must often resort to component screening and matching, or even final tweaking of the buffers at system fabrication time to meet performance requirements.

In salphasic technology, no such components need be used in the distribution system, so these variations do not occur. However, the signal amplitude is not constant over the system, which introduces another source of skew due to the effect of overdrive dispersion in the clock receivers, as described in the differential comparator literature. This effect is systematic, however, and can therefore be determined at system design time.

Both technologies suffer from component variation in the propagation delays of the clock receiver circuits. Phase-locked-loop clock receivers could potentially be used with either technology to minimize or practically eliminate this variation, as well as to avoid overdrive dispersion.

VII. Summary, Conclusions, and Future Research

The notion of salphasic behavior arising naturally from pure sinusoidal standing wave signals on transmission lines was described. A canonical branch circuit comprising a low-loss transmission line with a low-loss L-section load was analyzed for overall salphasic behavior and used to implement recursively an arbitrarily branching signal distribution tree. The salphasic behavior of such a tree was shown to be well approximated by realistically low-loss circuits.

This provides sufficient evidence that salphasic behavior could be exploited to control clock skews in high-speed synchronous systems having diameters such that sufficiently lossless conditions are preserved. This depends largely on the lossyness of the transmission lines that make up the clock distribution network.

Using the salphasic approach, it is possible to build low skew clock distribution networks with a minimum of attention to adjustments and tuning, although further improvement could be achieved by doing so. Indeed, a model of an 18 × 18-in. PCB with thirty-six 10 pF loads predicts a clock skew of less than 1.33° between any two loads at a clock frequency of 40 MHz, with no tuning or adjustments whatsoever.

The arbitrarily branching tree clock distribution topology allows the use of geometries infeasible to implement in conventional technology, due to the requirement for impedance matching to control undesired signal reflections. This allows for far simpler clock distribution trees since stub length design violations need not be considered.

Clock planes similar to power and ground planes embedded in a multilayer circuit board were shown to be possible. The notion of a salphasic clock plane is very attractive both in its potential performance and in its design simplicity.

Additional research is needed to develop salphasic technology to where it is easy to use. Experiments such as those conducted on Pixel-Planes 5 should be performed for a variety of distribution geometries. Crosstalk from signal lines to such clock system geometries should be characterized more thoroughly, as should effects of nonlinear loads. The use of phase-locked-loop clock receivers to avoid overdrive dispersion and to improve noise immunity should be explored. Designs of such receivers should be developed both as stand-alone clock receivers and as standard cells for VLSI clock inputs. Where larger system diameters are to be used, techniques for dealing with phase inversions across voltage nodes will be necessary, and design tools for tuning subregions to avoid voltage node points should be developed.

References

[1] C. A. Mead and L. Conway, *Introduction to VLSI Systems.* Reading, MA: Addison-Wesley, 1980, ch. 7.

[2] S. H. Unger and C-J. Tan, "Clocking schemes for high-speed digital systems," *IEEE Trans. Comput.*, vol. C-35, pp. 880–895, Oct. 1986.

[3] K. D. Wagner and E. J. McClusky, "Tuning, clock distribution, and communication in VLSI high speed chips," Stanford University CRC Tech. Rep. 84-5, June 1984.

[4] A. L. Fischer and H. T. Kung, "Synchronizing large VLSI processor arrays," *IEEE Trans. Comput.*, vol. C-34, pp. 734–740, Aug. 1985.

[5] E. G. Friedman and S. Powell, "Design and analysis of a hierarchial clock distribution system for synchronous standard cell/macrocell VLSI," *IEEE J. Solid-State Circ.*, vol. SC-21, pp. 240–246, Apr. 1986.

[6] C. L. Seitz, "Self-timed VLSI systems," in *Proc. Caltech Conf. on VLSI*, Jan. 1979, pp. 345–355.

[7] C. E. Molnar, "Introduction to asynchronous systems," in *Proc. New Frontiers in Computer Architecture Conf.*, Mar. 1986, pp. 83–94.

[8] I. E. Sutherland, "Micropipelines," *Commun. ACM*, vol. 32, pp. 720–738, June 1989.

[9] W. Peil and H. Raillard, "Clock power distribution arrangement for high speed logic systems," U.S. Patent 3,267,394, filed 13 Feb. 1963, issued 16 Aug. 1966.

[10] R. E. Matick, *Transmission Lines for Digital and Communication Networks.* New York: McGraw-Hill, 1969, ch. 2.

[11] H. Fuchs *et al.*, "Pixel-Planes 5: A heterogeneous multiprocessor graphics system using processor-enhanced memories," *Comput. Graphics*, vol. 23, no. 3, pp. 79–88, July 1989.

[12] S. G. Margolis, "Response of a phase-locked-loop to a sinusoid plus noise," *IRE Trans. Inform. Tech.*, vol. IT-3, pp. 136–142, June 1957.

[13] P. K. Runge, "Phase-locked-loops with signal injection for increased pull-in range and reduced output phase jitter," IEEE *Trans. Commun.*, vol. COM-24, pp. 636–644, June 1976.

Timing Constraints for Wave-Pipelined Systems

C. Thomas Gray, *Member, IEEE*, Wentai Liu, *Senior Member, IEEE*, and Ralph K. Cavin, III, *Fellow, IEEE*

***Abstract*— Wave-pipelining is a timing methodology used in digital systems to achieve maximal rate operation. Using this technique, new data are applied to the inputs of a combinational block before the previous outputs are available, thus effectively pipelining the combinational logic and maximizing the utilization of the logic without inserting registers. This paper presents a timing constraint formulation for the correct clocking of wave-pipelined systems. Both single- and multiple-stage systems including feedback are considered. Based on the formulation of this paper, several important new results are presented relating to performance limits of wave-pipelined circuits. These results include the specification of distinct and disjoint regions of valid operation dependent on the clock period, intentional clock skew, and the global clock latency. Also, implications and motivations for the use of accurate delay models and exact timing analysis in the determination of combinational logic delays are given, and an analogous relationship between the multi-stage system and the single-stage system in terms of performance limits is shown. The minimum clock period is obtained by clock skew optimization formulated as a linear program. In addition, important special cases are examined and their relative performance limits are analyzed.**

I. Introduction

As INTEGRATED circuit designs continue to push the limits of current technology, the clocking methodology and the design of the clocking system is very important to overall system performance. Using conventional clocking methodologies, the minimal clock period is determined by the delay time of the latest possible transition at the circuit outputs of any combinational logic block. Thus, the designer expends major effort to reduce and optimize circuit paths that produce these delays. Still, it is obvious that individual logic gates in the combinational logic blocks are idle for the vast majority of each clock cycle. For example, in a modern CMOS process, the gate switching time (the time inputs arrive at a gate to the time at which the output of the gate is available) is typically less than 1 ns, while the overall system clock period remains greater than 10 ns. This implies less than 10% logic utilization. That is, for more than 90% of the clock cycle, individual logic gates are idle. Although it is unreasonable to assume that logic gate utilization efficiency can be improved to the point that this disparity is completely eliminated, it is possible to improve utilization by minimizing this idle time and allowing for maximal rate operation of the combinational logic. By so doing, data are presented at the inputs of the combinational logic block at a maximal rate, effectively pipelining the combinational logic block but without inserting registers. This leads to multiple "waves" of data present in the combinational logic block concurrently, hence the name *wave-pipelining*.

This mode of maximal rate operation was first formalized by Cotten [4] in 1969, though it appears that similar ideas were used in the earlier designs of the IBM System/360 Model 91[1] floating point unit [1] and a maximal rate accumulator [19]. Recently, the concept has been a subject of renewed interest as technology and design techniques have allowed for effective implementation in integrated circuits and as clock periods are being pushed to physical technology limits. To date, work in the area of wave-pipelining has included theoretical work on the timing constraints of wave-pipelined systems [4], [5], [10], [11], [15], [23], work on techniques for path equalization [5], [29], and the design of systems using the techniques of wave-pipelining [1], [6], [9], [14], [10], [20], [22], [28], [29].

In the first of recent wave-pipelining theoretical research, Ekroot [5] developed more fully a theory of the constraints for correct operation of wave-pipelined systems and presented methods for inserting combinational logic delay units into wave-pipelined systems to reduce the clock period by equalizing data delays. This optimization problem was formulated as a linear program. This work was instrumental in the resurgence of the wave-pipelining methodology. Later, Wong [29], [30] proposed more practical methods for equalizing delays through combinational logic using ECL technology. This work consisted of rough tuning circuit delays by insertion of discrete delay units and then fine tuning delays by setting individual ECL gate drive currents. Using a linear programming formulation similar to Ekroot, Fishburn [7] presented a method for insertion of delay units along the clock paths to optimize system performance by clock skew optimization. The Fishburn formulation is a special case of the formulation developed in this paper (see Section VI-D).

Concurrent with the work presented in this paper, several groups of researchers have contributed to the development of theoretical constraints of wave pipelining. Joy and Ciesielski [11], [12] have developed methods for placement of standard cells to minimize the clock period of wave-pipelined systems and have also extended the Fishburn formulation to include internal node constraints. Klass and Mulder have presented new CMOS gate structures for low delay variation [14] and have developed additional algorithms for balancing circuit delays in CMOS circuits. Srinivasan *et al.* [26], [27] extended

[1] IBM and System/360 are trademarks of the International Business Machines Corporation.

Reprinted from *IEEE Trans. Computer-Aided Desi.*, vol. 13, no. 8, pp. 987–1004, Aug. 1994.

the Fishburn formulation to include stochastic delays and also developed a methodology for cycle time optimization under certain packaging constraints. Sakallah *et al.* [2], [23] have extended their previous timing model to allow for wave-pipelined operation. Lien and Burleson [18] have developed constraints for correct clocking of wave-pipelined CMOS domino circuits based on the Sakallah timing model. Lam *et al.* [17] have derived further relationships concerning intervals of correct clocking for a class of single-stage wave-pipelined circuits. These results by Lam are analogous to the special case results developed in Section VI-C.

At North Carolina State University, several wave-pipelined circuits have been designed to demonstrate the feasibility of wave-pipelined operation in CMOS. These include a 250 MOPS 16-bit adder in 2μm CMOS [6], a 8x8 CMOS multiplier [20], and a high-speed digital sampling circuit [28]. In addition to this work, several other researchers have used the concepts of wave pipelining to increase system performance in actual designs. Qi and Peisu [22] have designed a fast experimental computer in ECL consisting of an arithmetic/logic unit and a multiplication unit. Wong [29], [31], using his CAD tools developed for path equalization, designed and implemented a 63-bit population counter in ECL with a speed-up of 2.5. Also, Ju and Jen have recently designed a high-speed CMOS wave-pipelined multiplier [13]. Using concepts similar to wave pipelining, Chappell *et al.* [3] designed a 2-ns "bubble pipelined" RAM chip. Two other recent industry examples using some degree of wave pipelining are the cache RAM access of the HP Snake workstation [16] and a crosspoint switch designed by IBM [25] for a wideband communication network.

The theoretical work to date has examined many important aspects of timing constraints for wave-pipelined circuits. However, most work concentrates on special or limiting cases. The purpose of this paper is to present a unified theory of the constraints on circuit timing parameters for both single-stage and multiple stage systems with feedback. In this work, more generalized and detailed performance limits are derived and shown, in Section VI, to be consistent with previous special case results. Based on the formulation of this paper, several important new results are presented relating to performance limits of wave-pipelined circuits. These results include the specification of distinct and disjoint regions of valid operation dependent on the clock period, intentional clock skew, and the global clock latency. Also, implications and motivations for the use of accurate delay models and exact timing analysis in the determination of combinational logic delays are given. Most importantly, an analogous relationship between the multi-stage system and the single-stage system in terms of performance limits is shown.

This paper is organized in nine sections. Sections II through V develop constraints for the single-stage wave-pipelined system. These sections present a system model, constraints for correct clocking, a method for minimizing the clock period by clock skew optimization, and a discussion of the implication of the amount of global clock latency (k). Section VI considers four important special cases of the general theory and analyzes their relative performance limits. Section VII then extends the results for the single-stage system to the multi-stage system with feedback. Lastly, Section VIII presents an example of the optimal clock period for a multi-stage system. Appendix A contains the proofs of all theorems and lemmas.

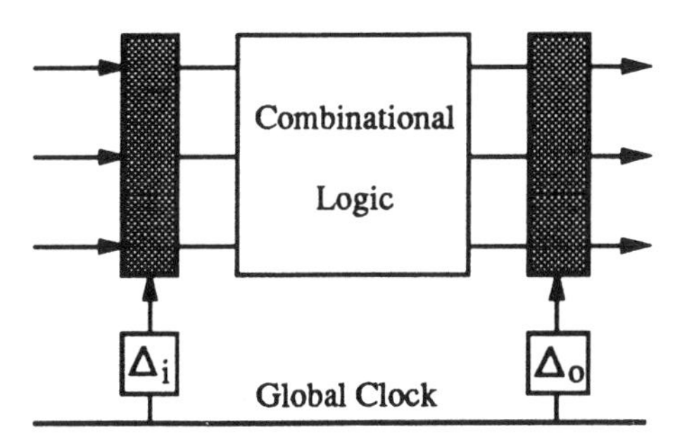

Fig. 1. Single-stage block diagram.

II. System Model

A. Circuit Model

Fig. 1 shows a block diagram of the single-stage system model to be used in this paper. As shown, the data path consists of a single block of combinational logic surrounded by input and output edge triggered registers. It is assumed that all bits of the input register are synchronized with a single input clock and all bits of the output register are synchronized with a single output clock. These clocks are generated by the clock path, which consists of delay units that produce delayed local versions of a global clock.

The combinational logic is described as a network of circuit elements. These circuit elements represent gates, collections of gates or functional circuit blocks depending on the level of detail needed. Mathematically, the logic block is described as a graph with vertex set $\mathcal{N}$ representing the set of all circuit nodes (or wires) and edge set $\mathcal{C}$ representing the set of all connections between nodes through circuit elements. For notational convenience, a set $\mathcal{G}$ is defined representing the set of all nodes that are circuit element output nodes and a set $\mathcal{O}$ is defined representing the set of all nodes that are outputs of the combinational logic block (that is, feeding into the output register). For example, in the combinational logic circuit shown in Fig. 2, $\mathcal{N} = \{1,2,3,4,5,6,7\}, \mathcal{G} = \{5,6,7\}, \mathcal{O} = \{6,7\}$, and $\mathcal{C} = \{(1,5),(2,5),(4,7),(5,6),(3,6)\}$.

In summary, the following notation will be used to represent the single-stage combinational logic block:

$\mathcal{N}$ set of all circuit nodes in logic block
$\mathcal{G}$ set of all gate output nodes in logic block
$\mathcal{O}$ set of all logic block output nodes
$\mathcal{C}$ set of all circuit connections

B. Timing Model

In addition to a physical interconnection of a circuit, timing properties of the circuit must be specified. The timing parameters can be divided into three categories: circuit timing parameters, clock timing parameters, and register timing parameters.

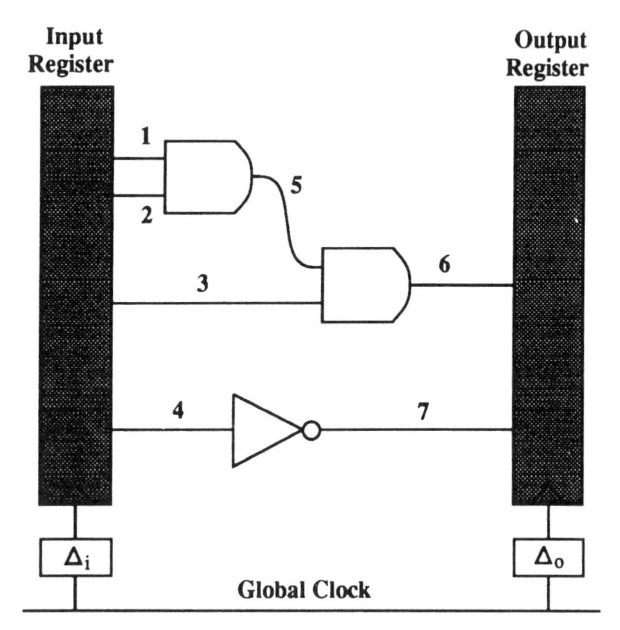

Fig. 2. Single-stage example circuit diagram.

The *circuit* timing parameters of interest are:

t_{max}	max delay from input register to output register
t_{min}	min delay from input register to output register
$t_{max}(i)$	max delay from input register to internal node i
$t_{min}(i)$	min delay from input register to internal node i
$t_{stable}(i)$	min stable time for inputs to driving gate(s) of internal node i

These parameters define the timing of data signal transitions. First, t_{max} and t_{min} refer to the maximum and minimum delay through the combinational logic between the input register and output register. These delay values are upper and lower bounds on the delay of the logic block over all expected process and environmental variation, over all circuit paths in the logic block, and for worst case input data patterns. Similarly, $t_{max}(i)$ and $t_{min}(i)$ $(\forall i \in \mathcal{G})$ represent upper and lower bounds on the delay from the input register to internal node i. The last parameter, $t_{stable}(i)$ $(\forall i \in \mathcal{G})$, represents the minimum amount of time that inputs to a circuit element or elements. Output node i must be stable together in order for node i to respond. For example, in single level CMOS gates, this parameter will depend on the worst case input rise/fall times, output loading capacitance, transistor sizes, etc.

The *clock* timing parameters of interest are:

T_{clk}	clock period
Δ_i	intentional delay added to input clock
Δ_o	intentional delay added to output clock
Δ	Δ_o-Δ_i
Δt_e^n	earliest possible clock skew at input register
Δt_l^n	latest possible clock skew at input register.
Δt_e^o	earliest possible clock skew at output register
Δt_l^o	latest possible clock skew at output register

These parameters define the timing of clock signal transitions. T_{clk} represents the period of the clock. Δ represents the effective amount of intentional skew between the local input clock and the local output clock. This is affected by the delay units shown in Fig. 1. The last four parameters, Δt_e^n, Δt_l^n, Δt_e^o, and Δt_l^o, represent various unintentional skew at the input and output registers. These skews are defined relative to a nominal arrival time and are due to variations in intentional delay units, variations in physical clock interconnect, clock jitter, etc.

The *register* timing parameters of interest are:

t_d	delay time of register
t_{setup}	setup time of register
t_{hold}	hold time of register.

These parameters define the relationship between clock and data edges as determined by the register structure being used. First, t_d defines the amount of delay between the clock edge at the register and the output data transition. t_{setup} defines the amount of time that input data must be stable *before* the clock edge arrives, and t_{hold} defines the amount of time that input data must be stable *after* the clock edge arrives.

III. Constraints for Correct Clocking

A. Register Constraints

In order to achieve proper capturing of data at the output register using the model described above, relatively stringent double-sided timing requirements must be met between clock and data edges at the inputs to the output register. The clock period must be such that the output data is latched *after* the latest data has arrived at the outputs and *before* the earliest data from the next clock period arrives.

As shown in the space-time diagram of Fig. 3, output data will be latched by the output clock at some time (t) relative to the global clock. This output clock, as described above, will be derived from the global clock delayed by a physical delay unit of length Δ_o. Given a physical delay Δ_o and a target time t, there are several options for achieving this skew. First, Δ_o could be equal to t. This would achieve the proper skew and allow for proper operation. Alternately, the cyclic nature of the clock could be relied upon to effect part of the shift. The portion of t achieved by this mechanism is kT_{clk}, where k is an integer referring to the number of clock cycles. Thus, t can be expressed as the sum of the amount of clock delay derived from the cyclic nature of the clock (kT_{clk}) and the amount of delay produced by the physical delay unit (Δ_o) [10]. That is,

$$t = kT_{clk} + \Delta_o.$$

Therefore, it is obvious that, for correct clocking, the output data must be sampled at a time (t) at which the data is stable (that is, t must fall in a nonshaded region in Fig. 3). This implies that, to ensure the latest data is latched correctly, the following constraint must be true:

$$kT_{clk} + \Delta_o - \Delta t_e^o \geq \Delta_i + \Delta t_l^n + t_d + t_{max} + t_{setup}, \quad (1)$$

where $kT_{clk} + \Delta_o = t$ as described above.

And, to ensure the earliest data is latched correctly, the following constraint must be also true:

$$\Delta_i + T_{clk} + t_d + t_{min} - t_{hold} - \Delta t_e^n \geq kT_{clk} + \Delta_o + \Delta t_l^o \quad (2)$$

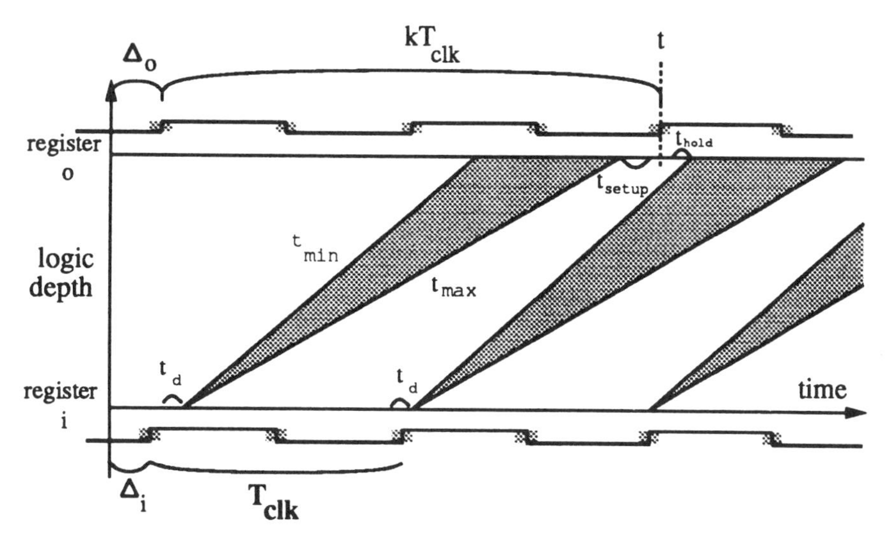

Fig. 3. Constraints at registers.

Note that when $k = 1$, constraints (1) and (2) are similar to the zero clocking and double clocking constraints developed by Fishburn [7]. This special case is described in more detail in Section VI-D.

Now, by combining constraints (1) and (2) and rearranging terms, the following condition on the number of waves is necessary for proper latching:

$$T_{clk} + t_{min} - t_{hold} - \Delta t_e^n - \Delta t_l^o \geq kT_{clk} + \Delta - t_d \geq t_{max} + t_{setup} + \Delta t_l^n + \Delta t_e^o \quad (3)$$

By transitivity, constraint (3) then implies that constraint (4) must be satisfied.

$$T_{clk} \geq (t_{max} - t_{min}) + t_{setup} + t_{hold} + \Delta t_e^n + \Delta t_e^o + \Delta t_l^n + \Delta t_l^o \quad (4)$$

Thus, the maximal clock rate is bounded by the difference of maximal and minimum delays between input and output registers as derived by Cotten [4]. It is also evident that setup and hold times along with unintentional clock skew directly increase the possible clock period.

B. Internal Node Constraints

Physical limits on switching delay of individual gates impose additional constraints on the minimal clock period. Constraint (4) indicates that the clock period is bounded only by setup/hold time and clock skew when input to output logic paths are equalized ($t_{max} = t_{min}$). However, circuit factors, such as nonzero rising/falling time, transistor sizing, capacitive loading, signal dispersion, and circuit technology, certainly can affect the clock speed. Accordingly, additional constraints due to physical properties of *internal nodes* must be considered in addition to register constraints such that fundamental circuit limitations are not exceeded and so that data waves from separate clock periods do not intermix at any of the internal nodes. In other words, the next earliest possible wave at any particular node *i* should not arrive until the latest possible wave has propagated through the node. This will be insured if all inputs to each circuit element are stable long enough for the node to respond before the next data arrives. With the amount of stable time needed at the inputs of the circuit element driving node i represented by $t_{stable}(i)$ as defined in Section II-B, these constraints at all internal nodes are expressed by constraint (5).

$$T_{clk} + \min_{(i,j)\in \mathcal{C}} (t_{min}(i)) - \Delta t_e^n \geq \max_{(i,j)\in \mathcal{C}} (t_{max}(i)) + \Delta t_l^n + t_{stable}(j), \quad \forall j \in \mathcal{G} \quad (5)$$

In order to compare the internal node constraints with the register constraints, note that constraint (5) can be rewritten as:

$$T_{clk} \geq \max_{(i,j)\in \mathcal{C}} (t_{max}(i)) - \min_{(i,j)\in \mathcal{C}} (t_{min}(i)) + t_{stable}(j) + \Delta t_e^n + \Delta t_l^n \quad \forall j \in \mathcal{G} \quad (6)$$

These constraints are similar to the internal node constraints developed by Ekroot [5].

Note that, if $t_{max}(i)$, $t_{min}(i)$, and $t_{stable}(i)$ are constant for all i in $\mathcal{G}$, the set of constraints of (6) can be written as one constraint (7).

$$T_{clk} \geq \max_{\forall j \in \mathcal{G}}(\max_{(i,j)\in \mathcal{C}} (t_{max}(i)) - \min_{(i,j)\in \mathcal{C}} (t_{min}(i)) + t_{stable}(j)) + \Delta t_e^n + \Delta t_l^n \quad (7)$$

Given the similar form of this constraint (7) and the constraint (4) derived from the register constraints, the question naturally arises as to whether the internal node constraint is needed or if is in fact implied by the register constraints. Conditions under which the constraint is not needed are expressed in the following lemma and theorem.

Lemma 3.1: The internal node constraint (7) is not active if:

$$t_{max} - t_{min} + t_{setup} + t_{hold} + \Delta t_e^o + \Delta t_l^o \geq \max_{\forall j \in \mathcal{G}}(\max_{(i,j)\in \mathcal{C}} (t_{max}(i)) - \min_{(i,j)\in \mathcal{C}} (t_{min}(i)) + t_{stable}(j)) \quad (8)$$

Note that under certain conditions the inequality of Lemma 3.1 can be guaranteed to be satisfied.

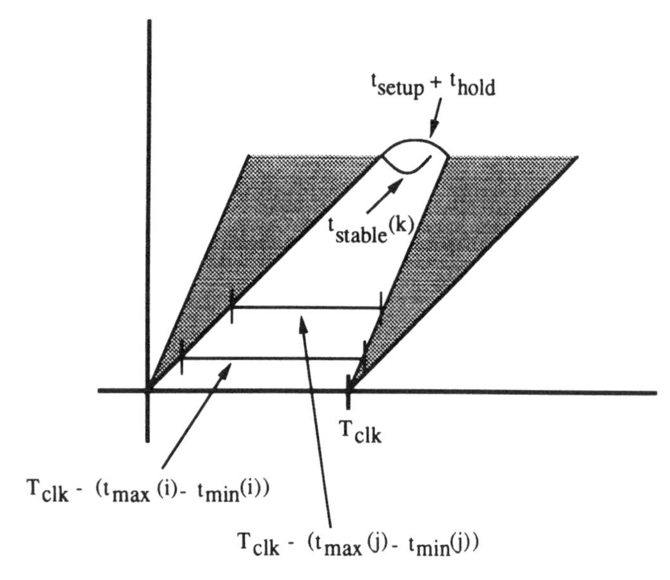

Fig. 4. Internal node timing.

Theorem 3.1: The internal node constraint (7) is not active if both of the following conditions are satisfied:

$$(i)\ t_{setup} + t_{hold} \geq t_{stable}(i) \forall i \in \mathcal{G}$$

$$(ii)\ t_{max}(j) - t_{min}(j) \geq \max_{(i,j)\in C}(t_{max}(i)) - \min_{(i,j)\in C}(t_{min}(i)) \forall j \in \mathcal{G}$$

Condition (ii) of Theorem 3.1 is satisfied when the delay difference at each node is monotonically increasing from inputs to outputs as shown in Fig. 4. That is, at each circuit element, the difference in delay is no smaller at the circuit element output than at any of its inputs. This implies that the smallest stable region between successive "waves" will be at the inputs of the output registers. Condition (i) of Theorem 3.1 is satisfied when the register performance is equal to or slower than the circuit element performance. This implies that the required stable region will be greatest at the inputs of the output register (due to the required setup and hold times). Therefore since the stable region is always larger at internal nodes than at output nodes and since the required stable time is greatest at the outputs, then, if timing requirements are met at the output register (t_{setup}+t_{hold}), the timing requirements at internal nodes ($t_{stable}(i)$) will be automatically met.

Whether or not the condition (i) is satisfied simply depends on the relative timing characteristics of the output register and the internal logic circuitry. Whether or not condition (ii) is true depends on the method used to calculate delays in circuit paths and the models used for the gate delays. In general, as shown in the following examples, this condition cannot be guaranteed to be satisfied if delay computation methods that employ accurate delay models and eliminate false delay paths are used.

First, consider the effect of different delay models. Fig. 5 shows a simple circuit consisting of a single NAND gate and a delay unit. Minimum and maximum delay values are represented by the order pairs (min,max) shown at each circuit element. In Fig. 5(a), the NAND gate is assumed to have a single delay independent of the gate input. The minimum and maximum values of this delay are shown to be 1 and 2 time units, respectively. As can be seen from the space-time diagram, condition (ii) of Theorem 3.1 is satisfied in this case. In Fig. 5(b), a more realistic model is used in which the NAND gate is assumed to have a delay dependent on the gate

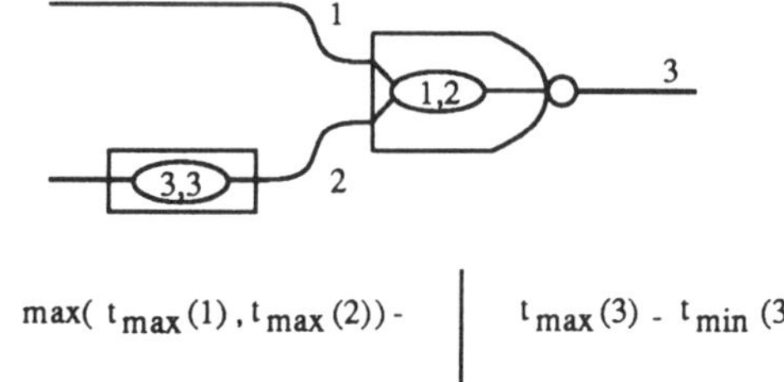

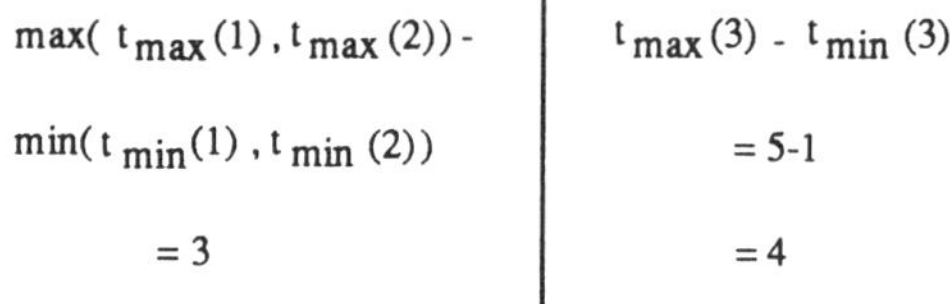

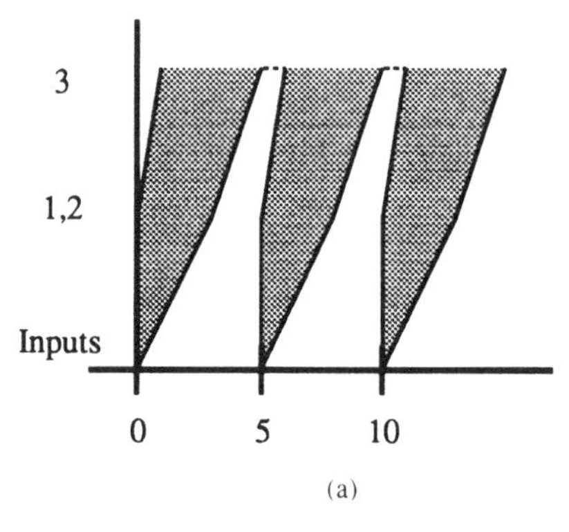

(a)

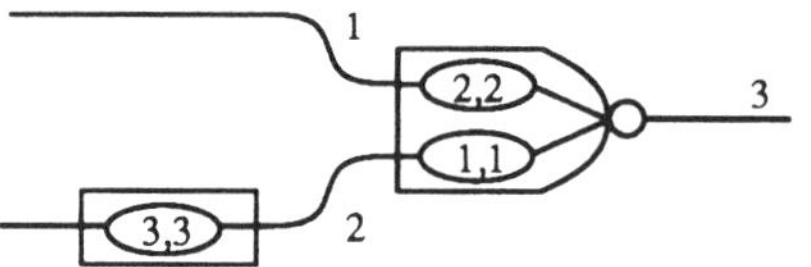

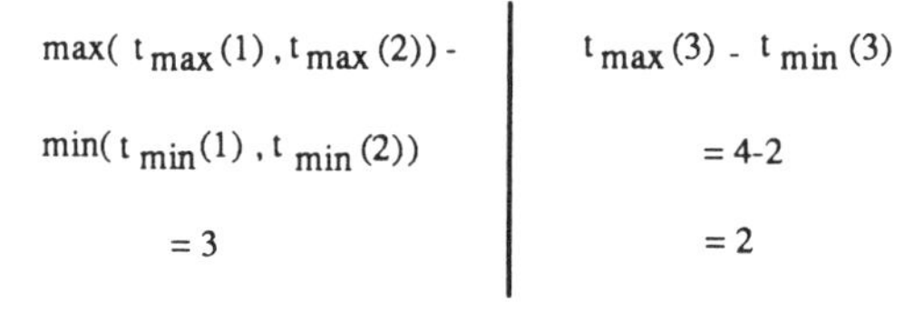

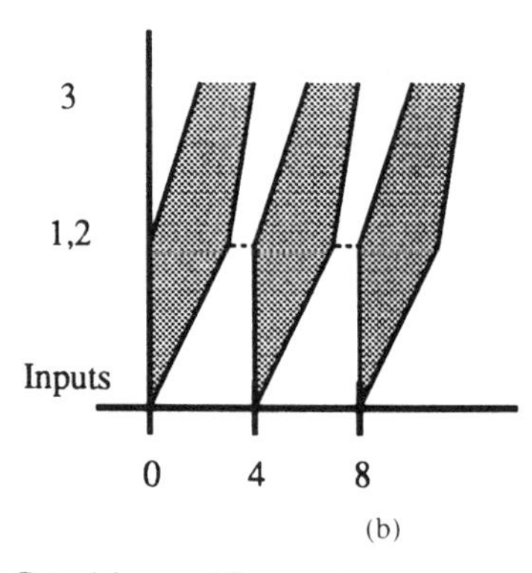

(b)

Fig. 5. Gate delay model example.

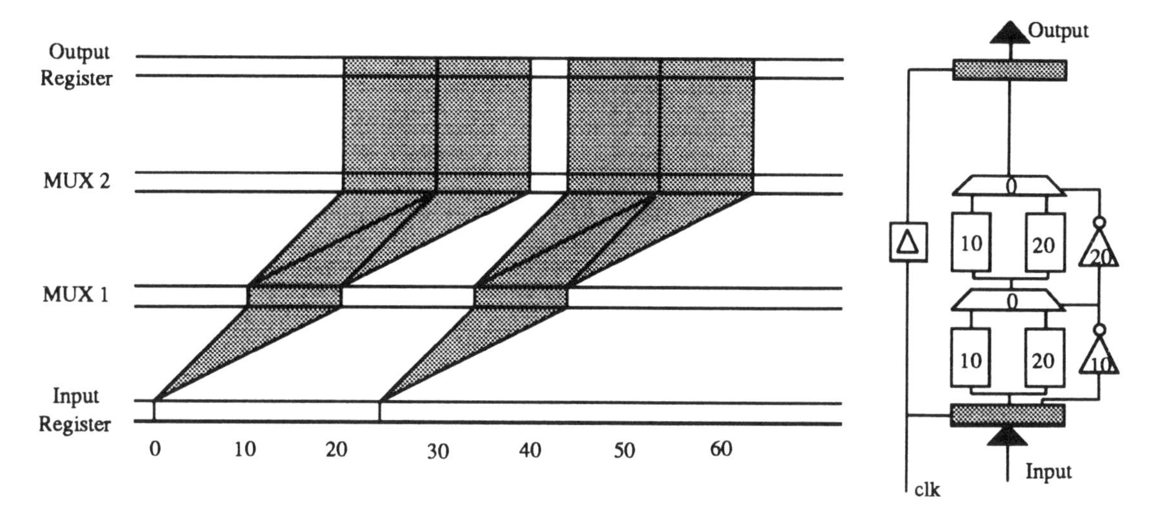

Fig. 6. Example not considering false paths ($t_{setup}=t_{hold}=t_{stable}(i)=2$).

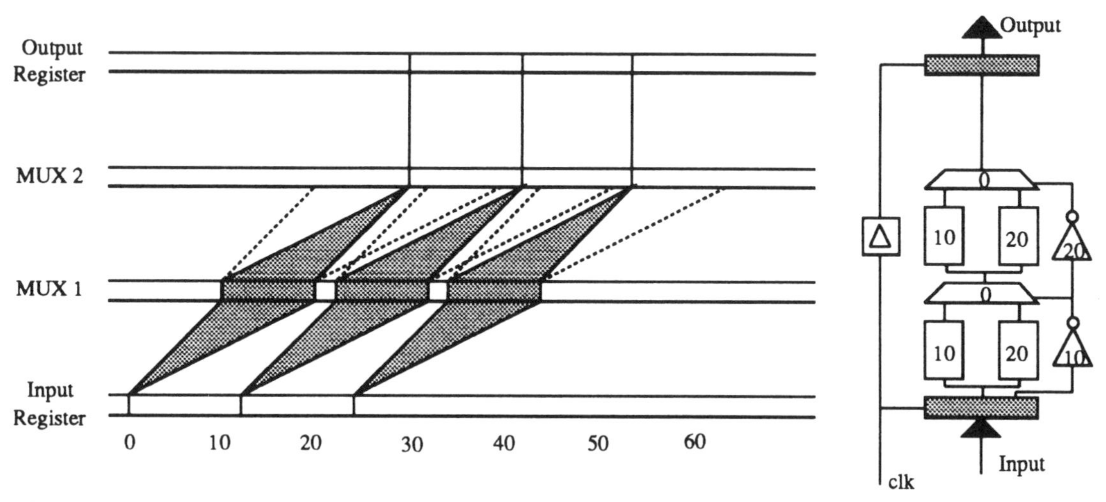

Fig. 7. Example considering false paths ($t_{setup}=t_{hold}=t_{stable}(i)=2$).

input [24]. In this case, condition (ii) of Theorem 3.1 is not satisfied.

The implications for false path elimination are demonstrated in the following example taken from [21] and shown in Fig. 6. The circuit consists of two multiplexors, four delay elements, and two inverters between the input and output registers. The numbers inside each circuit element represents the delay of that element and t_{setup} = t_{hold} = $t_{stable}(i)$ $(\forall i)$ = 2. The timing diagram in Fig. 6 graphically depicts the timing constraints at the registers and at two internal nodes (the outputs of the two multiplexors) based on a graph level timing analysis. For example, the longest delay from input register to output register (t_{max}) is 20+20=40 and the shortest delay (t_{min}) is 10+10=20. It is evident that condition (ii) of Theorem 3.1 is satisfied under this delay computation method and the figure shows that the limiting constraint is indeed at the output. The minimum clock period is 24 (t_{max}-t_{min}+t_{setup}+t_{hold}).

However, further investigation of the circuit reveals that, in fact, the path through the two elements of length 20 can never be selected due to the multiplexor control. Therefore, this path is a false path. In fact, no matter which path is selected the only true paths are of length 30 (t_{max} = t_{min} = 30). Thus, if only the constraints at the output are considered, the conclusion would be that the minimum clock period is bounded only by setup and hold times along with unintentional clock skew. This would be an erroneous conclusion, however, since this would cause data overlap at the output of the first multiplexor. The true minimum clock period is shown in Fig. 7 considering the constraint at the internal node. Note that the minimum clock period is now 12 ($t_{max}(i)-t_{min}(i)+t_{stable}(i)$). Therefore the clock period is reduced when actual circuit delays are used, but now the designer must consider the constraints at all internal nodes since condition (ii) is not satisfied.

In conclusion, unless the designer can be sure that internal node constraints are not needed, for example by using Theorem 3.1, the internal node constraints must be considered. This will depend on the relative speeds of the register and logic circuitry and the method and models used to compute circuit delays. In general, for operation at the true wave-pipelining performance limits, accurate timing analysis must be done and the internal nodes must be considered.

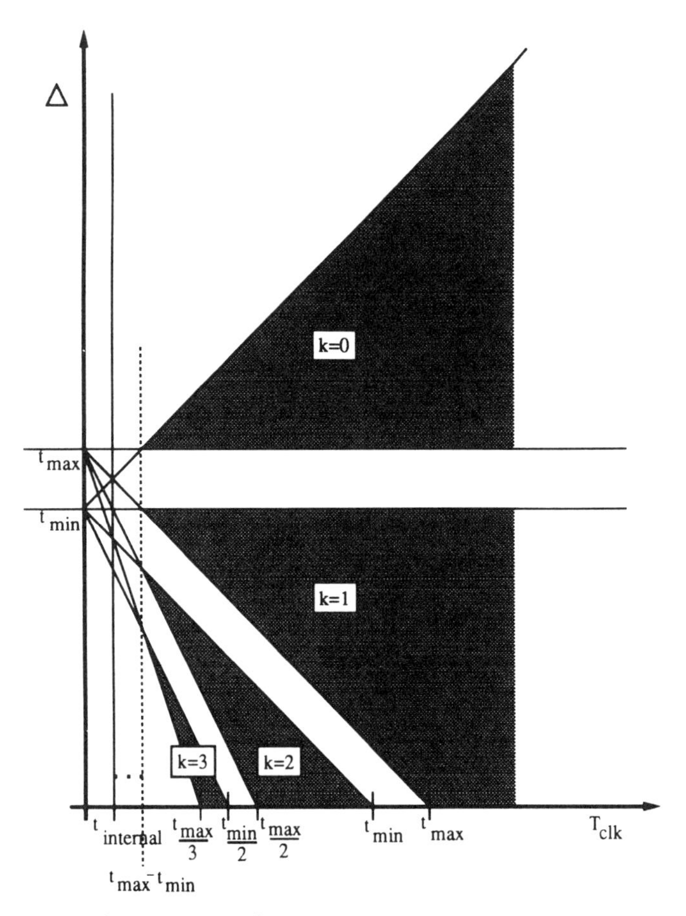

Fig. 8. Constraint space diagram.

C. Feasibility Space Diagram

Fig. 8 graphically illustrates the register and internal node timing constraints presented in the previous sections. For clearity, setup time, hold time, register delay, and unintentional clock skew have been set to zero. Also, the lower bound on clock period due to internal node constraints is defined to be $t_{internal}$:

$$t_{internal} = \max(\max_{\forall j \in \mathcal{G}} \max_{(i,j) \in \mathcal{C}}(t_{max}(i)) - \min_{(i,j) \in \mathcal{C}}(t_{min}(i)) + t_{stable}(j)) + \Delta t_e^n + \Delta t_l^n \quad (9)$$

The feasible solution space for each value of k is a convex region (shaded in Fig. 8) bounded by $T_{clk} + t_{min} = kT_{clk} + \Delta$, $kT_{clk} + \Delta = t_{max}$, and $T_{clk} = t_{internal}$ (as shown, the internal node constraint is not active). For fixed values of k, distinct feasible regions of operation are available. For example, when $k = 2$, the feasible region is bounded by $T_{clk} + t_{min} = 2T_{clk} + \Delta$ and $2T_{clk} + \Delta = t_{max}$. Note that as k is increased, the feasible region is reduced.

This figure shows an example in which the clock period is bounded below by register constraints ($t_{max} - t_{min}$). From the figure, is easy to see that this is indeed true for all values of k and that the feasible region for a particular value of k is shrunk as $t_{max} - t_{min}$ is increased. This emphasizes the necessity of delay balancing for maximal clock frequency.

IV. Minimizing the Clock Period

Given the constraints identified in the previous section, we can now proceed to minimize the clock period (T_{clk}). Given variable T_{clk} and Δ and all other parameters fixed, this can be formulated as a linear programming problem:

Min: T_{clk}**Subject to:**

$$\begin{aligned} &kT_{clk} + \Delta - \Delta t_e^o \geq \Delta t_l^n + t_d + t_{max} + t_{setup} \\ &T_{clk} + t_d + t_{min} - t_{hold} - \Delta t_e^n \geq kT_{clk} + \Delta + \Delta t_l^o \\ &T_{clk} + \min_{(i,j) \in \mathcal{C}}(t_{min}(i)) - \Delta t_e^n \geq \max_{(i,j) \in \mathcal{C}}(t_{max}(i)) \\ &\quad + \Delta t_l^n + t_{stable}(j), \\ &\quad \forall j \in \mathcal{G} \end{aligned}$$

Theorem 4.1: The minimum clock period (T_{clk}^*) of the single-stage system is

$$\max(t_{internal}, t_{max} - t_{min} + t_{setup} + t_{hold} + \Delta t_e^n + \Delta t_e^o + \Delta t_l^n + \Delta t_l^o). \quad (10)$$

Therefore, it is obvious that the minimum clock period is achievable for any value of k since T_{clk}^* does not depend on k. Note that the clock skew (Δ) needed to achieve this minimum clock period does depend on k [8].

V. The Parameter k

A. Degree of Wave Pipelining

From the constraint space diagram of Fig. 8, it is evident that the same clock period can be achieved with different values of k. It is also evident that the feasible region is different for different values of k. Therefore, it is important to investigate the physical meaning of k.

Along the data path, data that is launched by the input register at time Δ_i is captured by the output register at time $kT_{clk} + \Delta_o$ and new data is launched at time $T_{clk}+\Delta_i$. This implies the following definition for the degree of wave pipelining along the data path (W_d) relative to local clock edges.

$$W_d = \frac{kT_{clk} + \Delta}{T_{clk}}$$

This is the number of clock cycles that elapse between latching the data into the combinational logic and latching the data at the output. Note that in conventional design where $k = 1$ and $\Delta = 0, W_d = 1$ as expected.

Similarly, along the clock path, the input clock edge that arrives at the input clock at time Δ_i will arrive at the output register at time Δ_o. Thus, the following definition for the degree of wave pipelining along the clock path (W_c) is made.

$$W_c = \frac{\Delta}{T_{clk}}$$

This is the number of clock cycles elapsed while the clock edge is being delayed to generate the local clocks at the input and output of the combinational logic block.

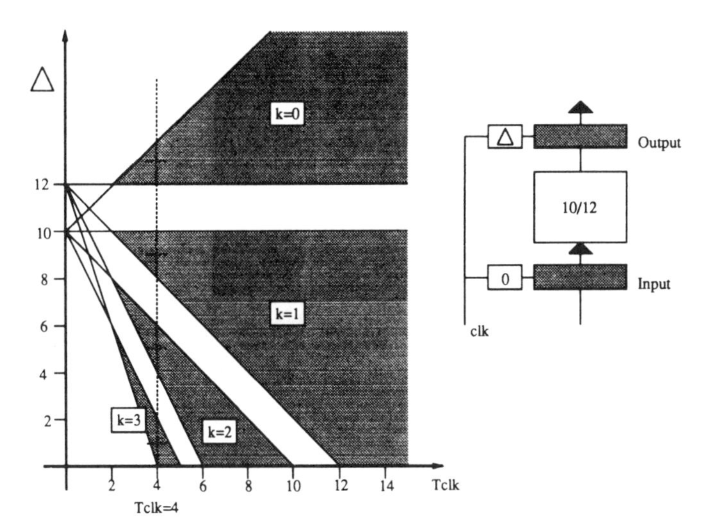

Fig. 9. Single-stage example.

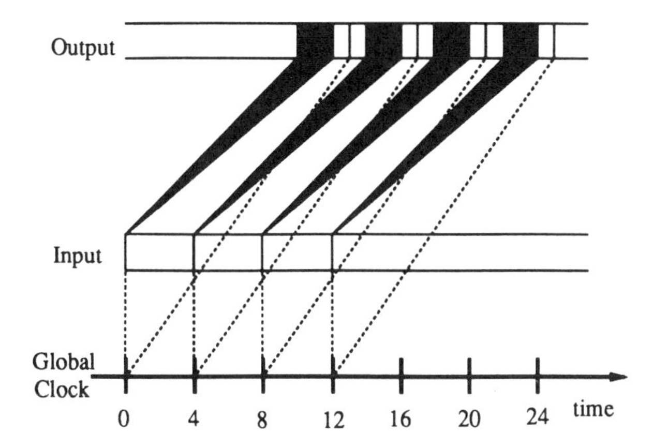

Fig. 10. Single-stage example (k=0, Δ = 13).

Using the definitions for W_d and W_c, k is the global clock latency.

$$k = W_d - W_c$$

This statement shows that k is the difference in the number of waves present in the data chain and the number of waves present in the clock chain. In other words, k is the number of global clock cycles elapsed between launching of data at the input register and capturing the data at the output register.

As an illustration, consider the single-stage system of Fig. 9 and its associated feasibility space diagram assuming setup and hold times along with unintentional clock skew are zero. The dashed vertical line indicates an intended clock period of T_{clk}=4. It is evident that, to achieve this clock period, the output register clock skew could be set to Δ=13, Δ=9, Δ=5, or Δ=1 as indicated on the feasibility space diagram. Each of these clock skews places operation in a different k region $k = 0$, $k = 1$, $k = 2$, $k = 3$ respectively. These operating points are depicted in Figs. 10–13.

Note that in these space-time logic diagrams an additional axis has been added for the global clock. Local clock edges are depicted as solid vertical lines at the input and output, and clock delay through clock delay units are depicted as dashed lines between global clock edges and local clock edges. These

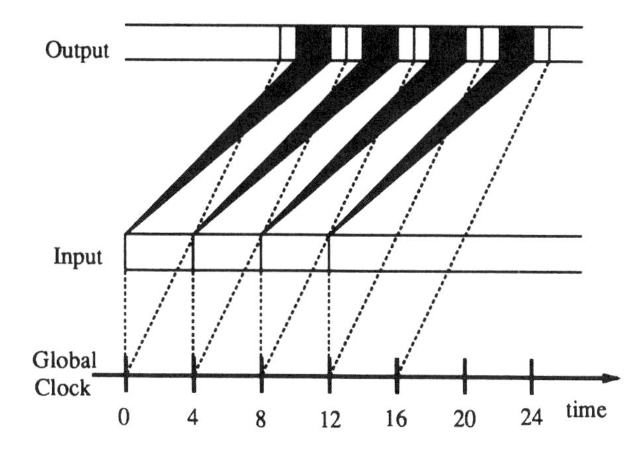

Fig. 11. Single-stage example (k=1, Δ = 9) .

figures graphically show the relative skew of both clock and data. Note also that while these figures assume ideal clock delay units indicated by single dashed lines, variation in clock delay units could be indicated by cones bounded by maximum and minimum delays as in the data path.

In Fig. 10 where $k = 0$, data is captured by a delayed version of the same global clock edge that launched the data indicating a global clock latency of zero. In Fig. 11 where $k = 1$, data is captured by a delayed version of the global clock edge of the next cycle after the launching edge, indicating a global clock latency of one. Note that if k is forced to be one and Δ is forced to be zero, as in the conventional case, the clock period T_{clk} is forced to be greater than t_{max}. In Fig. 12 where $k = 2$, it is evident that data launched by a given global clock edge will be captured by a delayed version of the global clock edge two cycles later. Fig. 13 is similar for $k = 3$.

B. Frequency Range

From the above example, the question naturally arises as to whether there is any perceivable difference in operation in different k modes. Since k is obviously not equivalent to speedup, different combinations of k and Δ for the single-stage system can produce the identical speedup, and thus the identical operation, from the system point of view.

While this is true for the single-stage system at a particular clock period (T_{clk}), there are differences in terms of the range of frequencies over which proper operation can be maintained. For example, consider the constraint space diagram in Fig. 8. It is obvious that higher k regions have a smaller range of frequencies over which proper clocking is maintained.

Mathematically, this can be expressed by considering the register constraints (1) and (2), which are rewritten below in (11) and (12) setting t_{setup}, t_{hold}, and unintentional skew to zero.

$$kT_{clk} + \Delta - \Delta t_e^o \geq \Delta t_l^n + t_d + t_{max} + t_{setup} \qquad (11)$$

$$T_{clk} + t_d + t_{min} - t_{hold} - \Delta t_e^n \geq kT_{clk} + \Delta + \Delta t_l^o \qquad (12)$$

By combining these constraints, the bounds on T_{clk} can be determined as:

$$\frac{t_{max} - \Delta}{k} \leq T_{clk} \leq \frac{t_{min} - \Delta}{k - 1} \qquad (13)$$

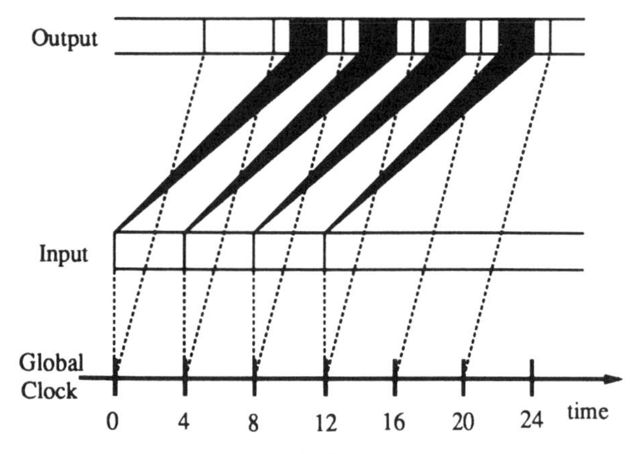

Fig. 12. Single-stage example (k=2, Δ = 5).

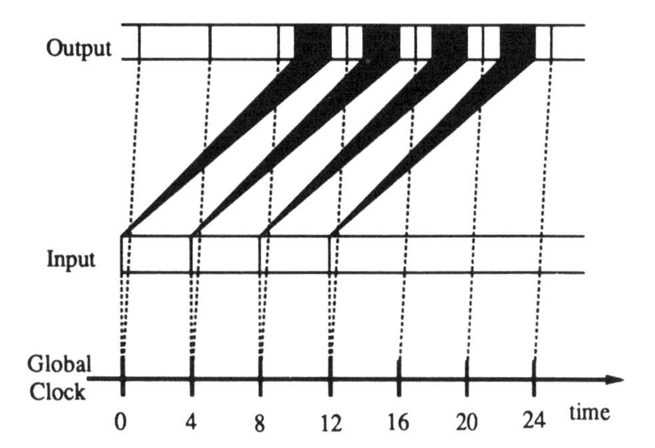

Fig. 13. Single-stage example (k=3, Δ = 1).

Thus, the range of T_{clk} is:

$$range = \frac{t_{min} - \Delta}{k - 1} - \frac{t_{max} - \Delta}{k} \quad (14)$$

Further differences in operation for different values of k will become more evident in Section VII-D when multiple stage pipelines are investigated. From a practical standpoint, k also affects the amount of process variation and environmental variation that a wave-pipelined circuit can tolerate and maintain correct operation. This is discussed in more detail in [10] and [8].

VI. Special Case

There are several special cases of the general theoretical framework that are interesting from a design point of view. In this section, the effect of each of these special case assumptions on the operation of the wave-pipelined system is analyzed. For simplicity in presentation, a single-stage system is assumed and unintentional skew, setup, and hold times are set to zero for this section. Note that these results can easily be extended to the multi-stage systems and systems with nonzero unintentional skew, setup, and hold times.

A. $\Delta \leq T_{clk}$

The addition of this constraint forces the intentional skew to, in effect, only be used for fine tuning the clock period. That is, in order to operate at the smallest clock period, the designer

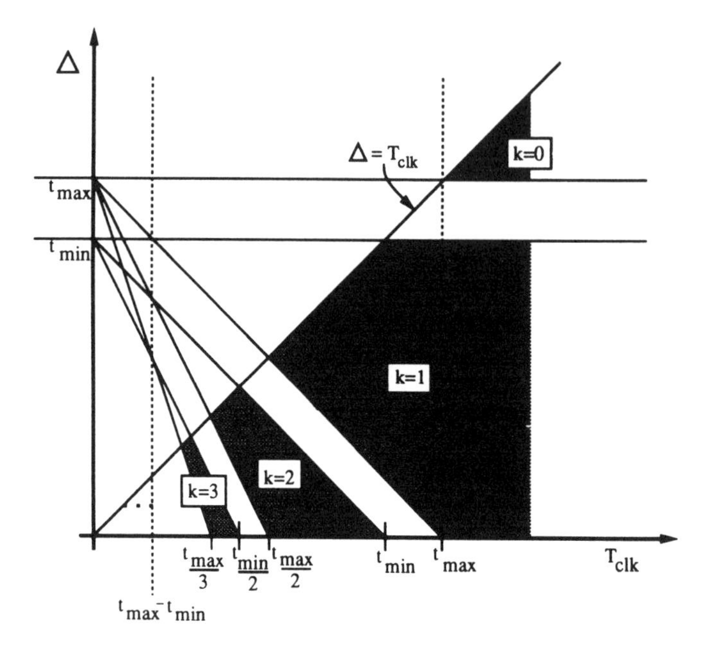

Fig. 14. Constraint Space Diagram ($\Delta \leq T_{clk}$).

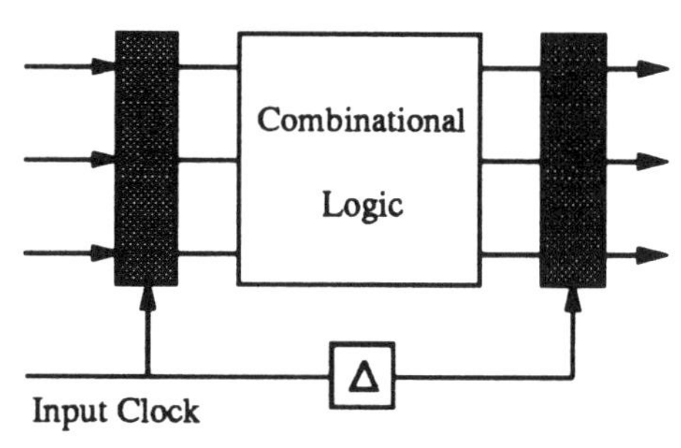

Fig. 15. Single-stage block diagram (cascaded clock).

is forced to operate in the highest k region so that, as much as possible, most of the input to output delay is derived from the cyclic nature of the clock with the remainder supplied by the physical delay. From the following theorem and the constraint space plot shown in Fig. 14, it is evident that the optimal clock period (T^*_{clk}) is now dependent on k under certain conditions.

Theorem 6.1: The minimum clock period (T^*_{clk}) of the single-stage system with $\Delta \leq T_{clk}$ is

$$\max(t_{internal}, \frac{t_{max}}{k+1}, t_{max} - t_{min}). \quad (15)$$

B. $\Delta \geq 0$

In the general set of constraints, the introduced clock delay Δ is derived from the physical delay units at the input and output registers ($\Delta = \Delta_o - \Delta_i$). An alternate system is shown in Fig. 15 in which a single clock delay is introduced in a cascade fashion between the two registers. This type of system implicitly imposes a lower bound on the introduced clock delay (Δ), and thus imposes an upper limit on the clock period for most k regions. The following theorem gives the range of valid T_{clk} for the single-stage system with this additional constraint.

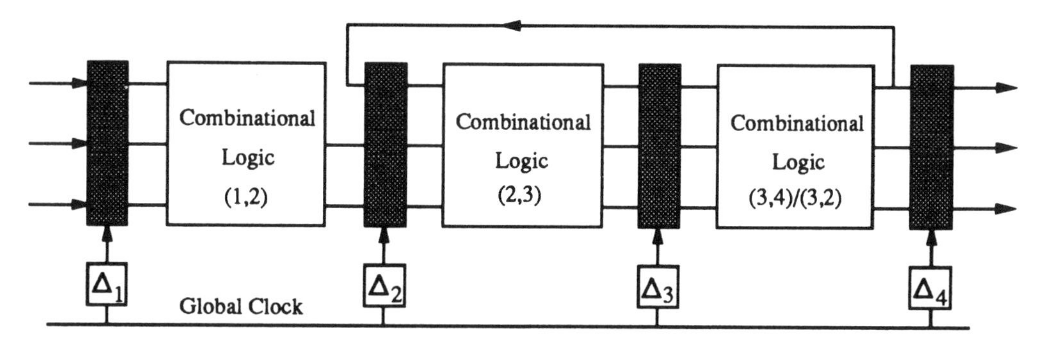

Fig. 16. Multi-stage block diagram.

Theorem 6.2: The minimum clock period (T^*_{clk}) of the single-stage system with $\Delta \geq 0$ is

$$\max(t_{internal}, t_{max} - t_{min}) \tag{16}$$

and the maximum clock period is

$$\frac{t_{min}}{k-1}. \tag{17}$$

C. $\Delta = 0$

If the clock is routed without intentional skew to both registers, an implicit constraint of $\Delta = 0$ is added. The following theorem gives the range of valid clock period for this case. Note that the valid clocking intervals identified by Lam et. al. in [17] are similar to these special case results.

Theorem 6.3: The minimum clock period (T^*_{clk}) of the single-stage system is

$$\max(t_{internal}, \frac{t_{max}}{k}), \tag{18}$$

and the maximum clock period is

$$\frac{t_{min}}{k-1}. \tag{19}$$

D. $k = 0, k = 1$

Two interesting special cases are when $k = 0$ and when $k = 1$. In these cases, the range of clock periods is not bounded above even with Δ fixed, as can be seen from Fig. 8. This implies that after Δ is fixed, the system can be run at any clock period from the minimum clock period to infinity. This kind of system is referred to as *single steppable*. The $k = 1$ case was formulated by Fishburn in [7]. Operation in this mode alleviates the debugging problems that are normally associated with wave-pipelined circuits in that the range of frequencies is continuous above the minimal clock period as in conventional circuits. However, a larger Δ is required to obtain the minimal clock period in these modes. As is evident from Fig. 8, this delay is on the order of the length of the entire combinational logic depth (t_{min} for k=1 and t_{max} for k=0).

If k=1 and Δ=0, the system is a conventional pipeline system and by Theorem 6.3 the clock period is limited by the maximum delay plus setup time and unintentional clock skew as expected.

VII. Multiple-Stage Systems

A. System Model

A representative block diagram of the multi-stage system to be considered is shown in Fig. 16. This figure shows a three-stage system surrounded by edge-triggered registers with feedback around two of the stages. In general, a system with n edge-triggered registers separating disjoint blocks of combinational logic is assumed. Each combinational logic block is labeled with an ordered pair (i, j), where i represents its input register and j represents its output register. As in the single-stage system, it is assumed that all bits of each register are synchronized by a single local clock for each register. These local clocks are derived from a global clock through delay units (represented as Δ_1 through Δ_4 in Fig. 16. Note that the common finite state machine model can be equivalently represented as a general n stage system, and therefore results derived here will be equally applicable to those systems. Also, the restriction of synchronized bits in each register can be easily relaxed to include a system with independent delay units for each register bit.

The combinational logic blocks are described as networks of circuit elements as for the single-stage system using the defined sets $\mathcal{N}_{i,j}$, $\mathcal{G}_{i,j}$, and $\mathcal{C}_{i,j}$ subscripted to indicate particular combinational logic blocks. An additional set $\mathcal{R}$ is defined to represent the set of connections between registers (equivalently, the set of combinational logic block).

In summary, the following notation will be used to represent the multi-stage system:

$\mathcal{R}$	set of all register pairs surrounding combinational logic blocks
$\mathcal{N}_{i,j}$	set of all circuit nodes in logic block between registers i and j
$\mathcal{G}_{i,j}$	set of all gate output nodes in logic block between registers i and j
$\mathcal{C}_{i,j}$	set of all circuit connections in logic block between registers i and j

Register timing parameters are superscripted to indicate logic block association. Note that for the multi-stage system, each register has a separate delayed local clock. These delays are represented by Δ_i for each register i.

The notation used for all of the timing parameters for the multi-stage system is listed below.

$t_{max}^{(i,j)}$	max delay from register i to register j
$t_{min}^{(i,j)}$	min delay from register i to register j
$t_{max}^{(i,j)}(x)$	max delay from register i to internal node x in logic block (i,j)
$t_{min}^{(i,j)}(x)$	min delay from register i to internal node x in logic block (i,j)
$t_{stable}^{(i,j)}(x)$	min stable time for inputs to driving gate(s) of internal node x in logic block (i,j)
T_{clk}	clock period
Δ_i	intentional delay added to clock at register i
Δt_e^i	earliest possible clock skew at register i
Δt_l^i	latest possible clock skew at register i
$t_d(i)$	delay time of register i
$t_{setup}(i)$	setup time of register i
$t_{hold}(i)$	hold time of register i.

B. Constraints for Correct Clocking

1) Register and Internal Node Constraints: As with the single-stage system, timing requirements must be met at all the registers of the system to insure proper operation. That is, data at the output of a combinational logic block must be captured after the latest present data arrives and before the earliest subsequent data arrives. These timing requirements must be met for all pairs of registers surrounding combinational logic blocks ($\forall(i,j) \in \mathcal{R}$). Inequalities (20) and (21) formalize these constraints.

$$k_{(i,j)}T_{clk} + \Delta_j - \Delta t_e^j \geq \Delta_i + \Delta t_l^i + t_d + t_{max}^{(i,j)} + t_{setup}(j), \quad \forall(i,j) \in \mathcal{R} \quad (20)$$

$$\Delta_i - \Delta t_e^i + t_d + T_{clk} + t_{min}^{(i,j)} - t_{hold}(j) \geq k_{(i,j)}T_{clk} + \Delta_j + \Delta t_l^j, \quad \forall(i,j) \in \mathcal{R} \quad (21)$$

By combining (20) and (21), the following statement must be true:

$$T_{clk} \geq t_{max}^{(i,j)} - t_{min}^{(i,j)} + t_{setup}(j) + t_{hold}(j) + \Delta t_e^i + \Delta t_e^j + \Delta t_l^i + \Delta t_l^j, \quad \forall(i,j) \in \mathcal{R} \quad (22)$$

$$\geq \max_{\forall(i,j)\in\mathcal{R}} (t_{max}^{(i,j)} - t_{min}^{(i,j)} + t_{setup}(j) + t_{hold}(j) + \Delta t_e^i + \Delta t_e^j + \Delta t_l^i + \Delta t_l^j) \quad (23)$$

This implies that the minimum clock period (T_{clk}) is limited by the worst case combinational block in terms of variation in delay, register parameters, and unintentional skew. This reaffirms the importance of delay balancing in *all* combinational logic blocks.

In addition to the register constraints, timing constraints must be met at all internal nodes of the system.

$$T_{clk} + \min_{(i,j)\in\mathcal{C}} (t_{min}^{(u,v)}(i)) - \Delta t_e^u \geq \max_{(i,j)\in\mathcal{C}} (t_{max}^{(u,v)}(i)) + \Delta t_l^u + t_{stable}^{(u,v)}(j), \quad \forall j \in \mathcal{G}_{u,v}, \forall(u,v) \in \mathcal{R} \quad (24)$$

Similar to the register constraints, the clock period (T_{clk}) will ultimately be limited by the worst case circuit element in any combinational block.

2) Register and Internal Node Constraints: In order to maintain proper system operation, data initiated on different clock cycles at the system inputs must never collide. In a system without feedback, this is obviously true if valid clocking constraints are satisfied in each individual stage. However, in systems with feedback, an additional constraint must be met at the feedback points. That is, data arriving from the feedback path must arrive with the proper data from the forward path. If for a given feedback loop l in the original system there are n_l registers in the loop, the number of physical pipeline stages around the loop is n_l. The goal after optimization is to increase the number of *effective* pipeline stages by some factor $\bar{k}$ so that the number of effective pipeline stages is $\bar{k}n_l$. Therefore $\bar{k}$ is the overall system speedup factor and must be constant for all loops to maintain proper operation. The following theorem formalizes these conditions under which proper operation is insured. Note that this theorem is similar to Theorem 20 of [5] arrived at by a different formulation.

Theorem 7.1: For a general system with valid clocking in each individual stage, proper system operation is maintained if

$$\frac{\sum_{\forall(i,j)\in l} k_{(i,j)}}{n_l} = \bar{k} \;\; \forall l \in \mathcal{S}$$

where $\bar{k}$ is the overall system speedup factor, $\mathcal{S}$ is the set of all loops in the system, and n_l is the number of registers in loop l.

The left side of the equation in Theorem 7.1 calculates an average value of k for each loop. This theorem implies that this average k value of every loop must be the same ($\bar{k}$) to insure the same speedup in all loops. Notice, however, that if the same k is used in all stages of the system, this condition is automatically satisfied and the overall speedup is just k. In this case satisfaction of the register constraints alone will insure proper system operation. Section VII-D will discuss the uniform k system further.

C. Minimizing the Clock Period

As in Section IV, given the constraints identified in the previous section, we can now formulate a minimization problem to minimize the clock period (T_{clk}). Given variable T_{clk} and Δ_i for all registers i and all other parameters fixed, the following linear program is formulated. As stated in Section VII-B-2), values for $k_{(i,j)}$ must be chosen to satisfy Theorem 7.1.

Min: T_{clk}

Subject to:

$$k_{(i,j)}T_{clk} + \Delta_j - \Delta t_e^j \geq \Delta_i + \Delta t_l^i + t_d + t_{max}^{(i,j)} + t_{setup}(j), \quad \forall(i,j) \in \mathcal{R}$$

$$\Delta_i - \Delta t_e^i + t_d + T_{clk} + t_{min}^{(i,j)} - t_{hold}(j) \geq k_{(i,j)}T_{clk} + \Delta_j + \Delta t_l^j, \quad \forall(i,j) \in \mathcal{R}$$

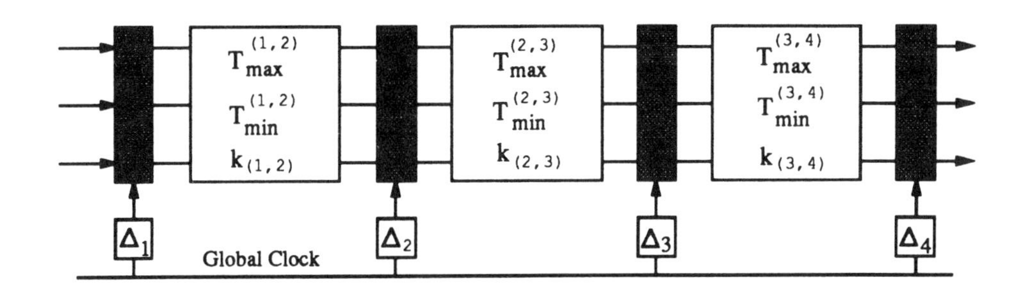

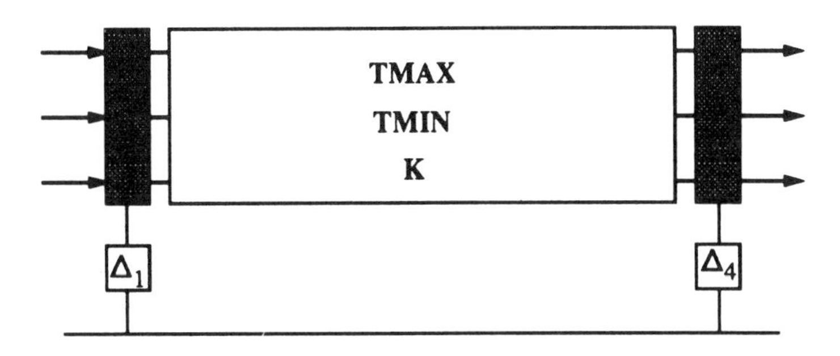

Fig. 17. Multi-stage to single-stage transformation.

$$T_{clk} + \min_{(i,j)\in C}(t^{(s,t)}_{min}(i)) - \Delta t^s_e \geq \max_{(i,j)\in C}(t^{(s,t)}_{max}(i)) + \Delta t^s_l + t^{(s,t)}_{stable}(j),$$

$$\forall j \in \mathcal{G}_{s,t}, \forall (s,t) \in \mathcal{R}$$

D. Interpretation of k

In Section V, it was shown that speedup was not directly related to k for the single-stage system. For the multi-stage system without feedback, this is also the case since the difference in clock delays to the output and input registers (Δ_j-Δ_i) is an independent variable for each combinational logic block and, thus, can be determined independently.

However, the addition of feedback imposes an additional constraint on the relationship of the difference in delays. This fact manifests itself in additional implied constraints on the clock period (T_{clk}) in systems with feedback. This additional constraint is demonstrated in the following theorem. For notational convenience, unintentional skew, setup, and hold times are set to zero.

Theorem 7.2: Let $\mathcal{S}$ be defined as the set of all loops in the system, then for multi-stage systems with feedback

$$\frac{\sum_{\forall(i,j)\in l} t^{(i,j)}_{min}}{\sum_{\forall(i,j)\in l}(k_{(i,j)} - 1)} \geq T_{clk} \geq \frac{\sum_{\forall(i,j)\in l} t^{(i,j)}_{max}}{\sum_{\forall(i,j)\in l} k_{(i,j)}} \forall l \in \mathcal{S}. \tag{25}$$

Any multi-stage system (as shown in Fig. 17) can be viewed globally as a single-stage system with timing parameters $\mathbf{TMAX} = \sum_{\forall(i,j)\in\mathcal{R}} t^{(i,j)}_{max}$, $\mathbf{TMIN} = \sum_{\forall(i,j)\in\mathcal{R}} t^{(i,j)}_{min}$, and $\mathbf{K} = \sum_{\forall(i,j)\in\mathcal{R}} k_{(i,j)}$. Constraints analogous to constraints (1) and (2) can be written for the global system where n is the number of stages in the system.

$$\mathbf{K}T_{clk} + \Delta_o \geq \Delta_i + \mathbf{TMAX} \tag{26}$$

$$\Delta_i + nT_{clk} + \mathbf{TMIN} \geq \mathbf{K}T_{clk} + \Delta_o \tag{27}$$

Combining these constraints, a constraint analogous to constraint (4) is derived.

$$T_{clk} \geq \frac{\mathbf{TMAX} - \mathbf{TMIN}}{n} \tag{28}$$

With the addition of feedback, the sum of Δ_i around the loop must be zero to insure correct clocking. Thus, the multi-stage system with feedback is analogous to the single-stage system with $\Delta = 0$ as described in Section VI-C. Rewriting the constraints of Theorem 7.2, the similarity between these constraints and the constraints of Theorem 6.3 can be seen.

$$\frac{\mathbf{TMIN}}{\mathbf{K} - n_l} \geq T_{clk} \geq \frac{\mathbf{TMAX}}{\mathbf{K}} \forall l \in \mathcal{S} \tag{29}$$

In practical terms this theorem shows that, in contrast to the single-stage system (Theorem 4.1), the minimum clock period (T_{clk}) for the multi-stage system with feedback *is* related to k. Note that in the general single-stage system, the intentional clock delays (Δ_i and Δ_o) could be set to achieve a minimum clock period of (t_{max}-t_{min}) for any k. In the multiple stage system with feedback, each individual stage (i,j) still limits the clock period to the difference in its maximum and minimum ($t^{(i,j)}_{max}$-$t^{(i,j)}_{min}$) and by its internal node constraints. But globally, since the output clock of one stage is the input clock of the next stage, a feedback loop produces a set of circular dependencies that imply a global constraint on the clock period. This limit is the total delay of the feedback loop divided by the total number of global clock cycles that it takes to get data around the loop. The total delay is just the sum of the maximum delays of each stage in the loop. The total number of global clock periods around the loop is just the sum of the $k_{(i,j)}$ of each stage of the loop, since $k_{(i,j)}$ for a particular stage is the number of globals clock cycles to get data through that stage (see Section V).

This implies that the only way to increase the speed of a wave-pipelined system with feedback in which the global constraint is the limiting constraint is to increase the sum of the $k_{(i,j)}$s around all loops. However, since the sum of $k_{(i,j)}$s also relates to the speedup of the system (Theorem 7.1), changing this sum will change the global operation of the pipeline thus producing a different computation. Whether or not this is allowable depends on overall system level considerations. Note also that increasing the sum provides no advantage in terms of clock period after the maximum rate limits of individual combination logic blocks have been reached.

Also, from Theorem 7.2, this lower bound on the clock period depends on the *sum* of the $k_{(i,j)}$, not the individual values of $k_{(i,j)}$ in each stage. This implies that different combinations of $k_{(i,j)}$ will give the same lower bound. An integer speedup (s) over the nonwave-pipelined system can be accomplished by using uniform $k_{(i,j)} = s$ in all stages. This will automatically satisfy Theorem 7.1 to insure correct operation. Note that no other combination of $k_{(i,j)}$s will give a lower bound on the clock period while achieving this speedup. Therefore, there is no advantage, in terms of clock period, to use any combination of $k_{(i,j)}$ other than uniform $k_{(i,j)}$ in all stages to yield the given speedup. However, different combinations will require different size delay units (Δ_i) and will yield different frequency ranges analogous to the single-stage system described in Sections IV and V-B. If noninteger speedup is desired or if additional limits are placed on delay unit sizes or frequency ranges, nonuniform $k_{(i,j)}$ may be necessary.

VIII. Example

As an example of the optimization of a multi-stage system with feedback, consider the system shown in Fig. 16. This is a system with three combinational logic blocks surrounded by four registers with feedback from the output of register 3 to the input of logic block B. The delay properties of each of the combinational logic blocks are shown in Table I.. Register parameters t_{setup} and t_{hold}, along with unintentional skew for all registers, is set to zero for this example and, for notational convenience, it is assumed that the internal node requirement is satisfied for all combinational logic blocks. Uniform $k_{(i,j)}$=1 operation is assumed in all stages.

The minimum clock period (T_{clk}) along with the optimal clock skew at each register can be found by the solution of a linear program as outlined in Section VII-C. The linear program is show below along with optimal T_{clk} and clock delays.

TABLE I

Register Pair	t_{min}	t_{max}
1,2	20	22
2,3	26	27
3,4	2	3

Min: T_{clk}
Subject to:

$$\Delta_2 - \Delta_1 \leq 20$$
$$\Delta_1 - \Delta_2 - T_{clk} \leq -22$$
$$\Delta_3 - \Delta_2 \leq 26$$
$$\Delta_2 - \Delta_3 - T_{clk} \leq -27$$
$$\Delta_4 - \Delta_3 \leq 2$$
$$\Delta_3 - \Delta_4 - T_{clk} \leq -3$$
$$\Delta_2 - \Delta_3 \leq 2$$
$$\Delta_3 - \Delta_2 - T_{clk} \leq -3$$

Optimal Solution:

$$\Delta_1 = 0$$
$$\Delta_2 = 7$$
$$\Delta_3 = 19$$
$$\Delta_4 = 7$$
$$T^*_{clk} = 15$$

This clock delay solution is shown graphically in Fig. 18. This timing diagram is shown with T_{clk} = 16 (for clarity, the clock period is shown longer than the optimal T^*_{clk}=15).

Since $t^{(i,j)}_{max}$ - $t^{(i,j)}_{min}$ is at most 2 (logic block (1,2)), it is evident that without feedback this system could be operated with a clock period of 2. However, with feedback, the period is limited to 15. For comparison, zero skew conventional operation would yield a minimum clock period of at least 27 due to the maximum delay of logic block (2,3), and a system in which the combinational logic between registers 2 and 4 is not pipelined at all would have a minimal clock period of 30.

Table II gives optimal clock periods and clock skews for the example system with different values of $k_{(i,j)}$. As can be seen, the theoretical limit can be reached with $k_{(i,j)}$=8 in all stages. However, this will obviously place a more stringent requirement on the precision of the delay units and the operating frequency since less variation can be tolerated.

IX. Conclusion

This paper has presented generalized timing constraints for the correct clocking of wave-pipelined circuits for both single- and multiple-stage systems with arbitrary feedback. It has been shown that distinct regions of operation are available for correct clocking of both single- and multiple-stage wave-pipelined systems. The size of these regions, in terms of the range of valid clock period (T_{clk}) and the range of valid clock skew (Δ), is dependent on the global clock latency (k) as shown in Fig. 8. It has been shown that for single-stage systems, the theoretical performance limit of $t_{max} - t_{min}$ can be achieved for any value of k.

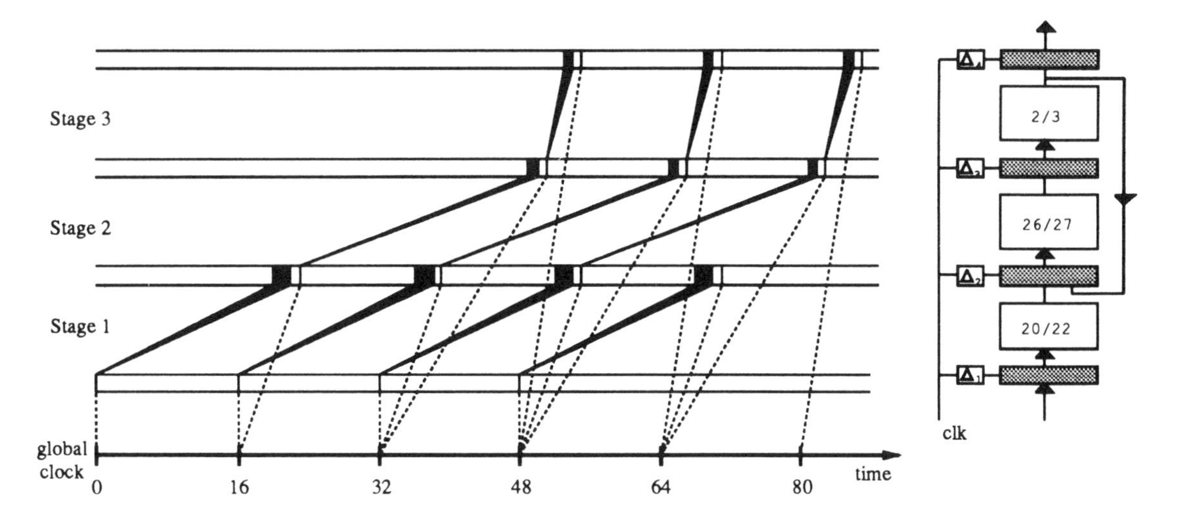

Fig. 18. Three-stage example.

TABLE II
LINEAR PROGRAM SOLUTION FOR DIFFERENT VALUES OF $k_{(i,j)}$

$k_{(i,j)}$	T^*_{clk}	Δ_1	Δ_2	Δ_3	Δ_4
1	15	0	7	19	7
2	7.5	0	7	19	7
3	5	0	7	19	7
4	3.75	0	7	19	7
5	3	0	7	19	7
6	2.5	0	7	19	7
7	2.14	0	7	19	7
8	2	0	6	18	6

For multi-stage systems with feedback, it has been shown that feedback loops impose additional implied constraints on the minimal and maximal clock period to insure proper operation at the feedback points. These constraints are dependent on the number of register stages in the feedback loop and the amount of global clock latency (k) in each stage. These constraints may limit the operation of a wave-pipelined system to significantly less than the t_{max}-t_{min} limit. Another important result of the multi-stage formulation presented in this paper is the identification of an analogous relationship between the multi-stage wave-pipelined system and the single-stage system.

The wave-pipelining constraints developed in this paper can be used by designers to make trade-offs between timing parameters such as the clock period (T_{clk}) and physical parameters such as intentional clock skew (Δ) to meet timing and reliability requirements. This paper also demonstrates the need for accurate circuit level timing analysis in detecting worst case circuit delays. Thus, system timing analysis based on the constraints for wave-pipelining coupled with accurate circuit timing analysis [8] can be used to design and verify wave-pipelined circuits making circuit operation at the limits of technology an obtainable reality.

X. APPENDIX THEOREM PROOFS

Lemma 1: The internal node constraint (7) is not active if:

$$t_{max} - t_{min} + t_{setup} + t_{hold}x + \Delta t_e^o + \Delta t_l^o \geq \max\left(\max_{\forall j \in \mathcal{G}} \max_{(i,j)\in \mathcal{C}} (t_{max}(i)) - \min_{(i,j)\in \mathcal{C}} (t_{min}(i)) + t_{stable}(j)\right) \quad (30)$$

Proof: Defining

$$t_{internal} = \max\left(\max_{\forall j \in \mathcal{G}} \max_{(i,j)\in \mathcal{C}} (t_{max}(i)) - \min_{(i,j)\in \mathcal{C}} (t_{min}(i)) + t_{stable}(j)\right) + \Delta t_e^n + \Delta t_l^n$$

and

$$t_{register} = t_{max} - t_{min} + t_{setup} + t_{hold} + \Delta t_e^o + \Delta t_l^o + \Delta t_e^n + \Delta t_l^n$$

if $t_{register} \geq t_{internal}$ as stated in the lemma and $T_{clk} \geq t_{register}$ as shown in constraint (4), by transitivity, this implies that $T_{clk} \geq t_{internal}$, which is constraint (7). Thus, constraint (7) is not an active constraint under these conditions. □

Theorem 1: The internal node constraint (7) is not active if both of the following conditions are satisfied:

$$(i)\ t_{setup} + t_{hold} \geq t_{stable}(i), \forall i \in \mathcal{G}$$

$$(ii)\ t_{max}(j) - t_{min}(j) \geq \max_{(i,j)\in \mathcal{C}} (t_{max}(i)) - \min_{(i,j)\in \mathcal{C}} (t_{min}(i)) \forall j \in \mathcal{G}$$

Proof: By the definition of t_{max} and t_{min},

$$t_{max} \equiv \max_{\forall i \in \mathcal{O}} (t_{max}(i)) \geq t_{max}(j) \forall j \in \mathcal{O} \quad (31)$$

and

$$t_{min} \equiv \min_{\forall i \in \mathcal{O}} (t_{min}(i)) \leq t_{min}(j) \forall j \in \mathcal{O}. \quad (32)$$

Now, by adding inequalities (31) and (32), the following inequality is implied:

$$t_{max} - t_{min} \geq t_{max}(j) - t_{min}(j) \forall j \in \mathcal{O} \quad (33)$$

By condition (ii),

$$t_{max}(j) - t_{min}(j) \geq \max_{(i,j)\in C}(t_{max}(i)) - \min_{(i,j)\in C}(t_{min}(i)) \, \forall j \in \mathcal{O}. \quad (34)$$

And, since

$$\max_{(i,j)\in C}(t_{max}(i)) - \min_{(i,j)\in C}(t_{min}(i)) \geq t_{max}(k) - t_{min}(k). \forall (k,j) \in C \, \forall j \in \mathcal{G} \quad (35)$$

by transitivity

$$t_{max} - t_{min} \geq t_{max}(j) - t_{min}(j) \forall j \in \mathcal{G}. \quad (36)$$

By condition (ii)

$$t_{max}(j) - t_{min}(j) \geq \max_{(i,j)\in C}(t_{max}(i)) - \min_{(i,j)\in C}(t_{min}(i)) \, \forall j \in \mathcal{G} \quad (37)$$

so, therefore,

$$t_{max} - t_{min} \geq \max_{\forall j \in \mathcal{G}}(\max_{(i,j)\in C}(t_{max}(i)) - \min_{(i,j)\in C}(t_{min}(i))). \quad (38)$$

Notice that condition (i) of Theorem 1 can be rewritten as

$$t_{setup} + t_{hold} \geq \max_{\forall i \in \mathcal{G}}(t_{stable}(i)). \quad (39)$$

Adding inequalities (38) and (39) gives condition

$$t_{max} - t_{min} + t_{setup} + t_{hold} \geq \max_{\forall j \in \mathcal{G}}(\max_{(i,j)\in C}(t_{max}(i)) - \min_{(i,j)\in C}(t_{min}(i))) + \max_{\forall j \in \mathcal{G}}(t_{stable}(j).) \quad (40)$$

which implies the following condition:

$$t_{max} - t_{min} + t_{setup} + t_{hold} + \Delta t_e^o + \Delta t_l^o \geq \max_{\forall j \in \mathcal{G}}(\max_{(i,j)\in C}(t_{max}(i)) - \min_{(i,j)\in C}(t_{min}(i)) + t_{stable}(j)). \quad (41)$$

By Lemma 1, the theorem is proved. □

Theorem 2: The minimum clock period (T_{clk}^*) of the single-stage system is shown in (42), at the bottom of this page.

Proof: Since this problem is formulated as a linear program, some vertex of the feasible region will give the optimal solution. Thus, it is only necessary to consider the intersection of feasible region boundaries. Also, as shown in in Section III-B, the last constraint may or may not be an active constraint. Thus, there are two cases for the optimal solution.

Case 1: Internal Node Constraint Active

If the last constraint is active, the feasible region is obviously bounded at

$$T_{clk} = t_{internal}$$

since this constraint does not depend on Δ or k.

Case 2: Internal Node Constraint Not Active

If the last constraint is not active, the feasible region will be defined by the first two constraints. Thus, the optimal solution will be at the intersection of the boundaries of the feasible regions defined by these two constraints. That is, the optimal solution will be at the intersection of

$$kT_{clk} + \Delta - \Delta t_e^o = \Delta t_l^n + t_d + t_{max} + t_{setup}$$

and

$$T_{clk} + t_d + t_{min} - t_{hold} - \Delta t_e^n = kT_{clk} + \Delta + \Delta t_l^o.$$

By adding these equations and rearranging terms, the intersection point, and thus the optimal solution, is

$$kT_{clk} + \Delta - \Delta t_e^o + T_{clk} + t_d + t_{min} - t_{hold} - \Delta t_e^n = \Delta t_l^n + t_d + t_{max} + t_{setup} + kT_{clk} + \Delta + \Delta t_l^o$$

$$T_{clk} = t_{max} - t_{min} + t_{setup} + t_{hold} + \Delta t_l^o + \Delta t_e^o + \Delta t_l^n + \Delta t_e^n$$

By combining Case 1 and Case 2, the theorem is proved. □

Theorem 3: The minimum clock period (T_{clk}^*) of the single-stage system with $\Delta \leq T_{clk}$ is

$$\max(t_{internal}, \frac{t_{max}}{k+1}, t_{max} - t_{min}). \quad (43)$$

Proof: For this special case, there are four constraints concerning the allowable clock period:

$$kT_{clk} + \Delta \geq t_{max} \quad (44)$$

$$t_{min} \geq (k-1)T_{clk} + \Delta \quad (45)$$

$$T_{clk} \geq t_{internal} \quad (46)$$

$$T_{clk} \geq \Delta \quad (47)$$

Constraint (46) alone implies

$$T_{clk} \geq t_{internal}$$

Constraints (44) and (45) together imply

$$T_{clk} \geq t_{max} - t_{min}$$

Constraints (44) and (47) together imply:

$$T_{clk} \geq \frac{t_{max}}{k+1}$$

Constraints (45) and (47) do not imply a bound on T_{clk}. Therefore, since there are no other lower bounds on T_{clk}, the minimum clock period (T_{clk}^*) must the maximum of the three lower bounds as stated in the theorem. □

Theorem 4: The minimum clock period (T_{clk}^*) of the single-stage system with $\Delta \geq 0$ is

$$\max(t_{internal}, t_{max} - t_{min}). \quad (48)$$

and the maximum clock period is

$$\frac{t_{min}}{k-1}. \quad (49)$$

$$\max(t_{internal}, t_{max} - t_{min} + t_{setup} + t_{hold} + \Delta t_e^n + \Delta t_e^o + \Delta t_l^n + \Delta t_l^o). \quad (42)$$

Proof:

Minimum Clock Period:

For this special case, there are four constraints concerning the allowable clock period:

$$kT_{clk} + \Delta \geq t_{max} \tag{50}$$

$$t_{min} \geq (k-1)T_{clk} + \Delta \tag{51}$$

$$T_{clk} \geq t_{internal} \tag{52}$$

$$\Delta \geq 0 \tag{53}$$

Constraint (52) alone implies

$$T_{clk} \geq t_{internal}$$

Constraints (50) and (51) together imply

$$T_{clk} \geq t_{max} - t_{min}.$$

Constraints (50) and (53) do not imply a bound on T_{clk} and constraints (51) and (53) imply an upper bound on T_{clk}. Therefore, since there are no other lower bounds on T_{clk}, the minimum clock period (T^*_{clk}) must the maximum of the two lower bounds as stated in the theorem.

Maximum Clock Period:

From the minimum clock period case, it can be seen that the only combination of constraints that imply an upper bound on T_{clk} is the combination (51) and (53). These constraint imply:

$$T_{clk} \leq \frac{t_{min}}{k-1}$$

Since there are no other upper bounds on T_{clk}, the maximum clock period is as stated in the theorem. □

Theorem 5: The minimum clock period (T^*_{clk}) of the single-stage system is

$$\max(t_{internal}, \frac{t_{max}}{k}). \tag{54}$$

and the maximum clock period is

$$\frac{t_{min}}{k-1}. \tag{55}$$

Proof:

Minimum Clock Period:

Again, if the internal node constraint is active, the minimum clock period is $t_{internal}$. Otherwise, the clock period is bounded by constraints (1) and (2). Since constraint (1) is the only lower bound constraints, substituting Δ = 0 in this constraint yields a minimum clock period of $\frac{t_{max}}{k}$.

Maximum Clock Period:

Similar to the minimum clock period case, substituting Δ = 0 in the only upper bound constraint (2) yields a maximum clock period of $\frac{t_{min}}{k-1}$. □

Theorem 6: For a general system with valid clocking in each individual stage, proper system operation is maintained if

$$\frac{\sum_{\forall(i,j)\in l} k_{(i,j)}}{n_l} = \bar{k} \;\; \forall l \in \mathcal{S}$$

where $\bar{k}$ is a constant relating to the average number of data "waves" in each stage of the feedback loop, $\mathcal{S}$ is the set of all loops in the system, and n_l is the number of registers in loop l.

Proof: Assume that the goal is to speed up the entire system by a factor $\bar{k}$. Initially, before speedup, the delay around a particular feedback loop (l) is n_l clock cycles where n_l is the number of registers in the feedback loop. In order for proper operation after speedup, from a global system point of view, the number of clock cycles around any feedback loop (l) must be $\bar{k}n_l$ so that data coherency is maintained.

The total delay around a feedback loop is the sum of the difference between the input and output clocks for all stages in the feedback loop. For stage (i, j) this delay is $k_{(i,j)}T_{clk}+\Delta_j-\Delta_i$. Therefore, around any particular feedback cycle (l) (as shown in Fig. 19) the delay (d_n) is:

$$\begin{aligned} d_n &= \sum_{i=2}^{n}(k_{(i-1,i)}T_{clk} + \Delta_i - \Delta_{i-1}) + k_{(n,1)}T_{clk} + \Delta_1 - \Delta_n \\ &= k_{(1,2)}T_{clk} + \Delta_2 - \Delta_1 + k_{(2,3)}T_{clk} + \Delta_3 - \Delta_2 + \cdots + \\ &\quad k_{(n-1,n)}T_{clk} + \Delta_n - \Delta_{n-1} + k_{(n,1)}T_{clk} + \Delta_1 - \Delta_n \\ &= \sum_{\forall(i,j)\in l} k_{(i,j)}T_{clk}. \end{aligned} \tag{56}$$

Thus, the total number of clock cycles it takes to traverse the feedback cycle is

$$\frac{d_n}{T_{clk}} = \frac{\sum_{\forall(i,j)\in l} k_{(i,j)}T_{clk}}{T_{clk}} \tag{57}$$

$$= \sum_{\forall(i,j)\in l} k_{(i,j)}. \tag{58}$$

Thus, the speedup of the new system over the original system is

$$\frac{\sum_{\forall(i,j)\in l} k_{(i,j)}}{n_l} = \bar{k}$$

or the average k value around the loop. This value must be constant for all loops to maintain proper operation. □

Theorem 7: Let $\mathcal{S}$ be defined as the set of all loops in the system. Then, for multi-stage systems with feedback,

$$\frac{\sum_{\forall(i,j)\in l} t^{(i,j)}_{min}}{\sum_{\forall(i,j)\in l}(k_{(i,j)}-1)} \geq T_{clk} \geq \frac{\sum_{\forall(i,j)\in l} t^{(i,j)}_{max}}{\sum_{\forall(i,j)\in l} k_{(i,j)}} \;\; \forall l \in \mathcal{S}. \tag{59}$$

Proof: In order to prove that the above constraints are satisfied for all loops in $\mathcal{S}$, it must be proven that the constraints are satisfied for any general loop in the system. To this end, consider the general n stage loop shown in Fig. 19. Rewriting lower bound constraint (1) for each stage in the feedback loop yields the following set of constraints:

$$k_{(1,2)}T_{clk} + \Delta_2 - \Delta_1 \geq t^{(1,2)}_{max} \tag{60}$$

$$k_{(2,3)}T_{clk} + \Delta_3 - \Delta_2 \geq t^{(2,3)}_{max} \tag{61}$$

$$k_{(3,4)}T_{clk} + \Delta_4 - \Delta_3 \geq t^{(3,4)}_{max} \tag{62}$$

$$\vdots$$

$$k_{(j-1,j)}T_{clk} + \Delta_j - \Delta_{j-1} \geq t^{(j-1,j)}_{max} \tag{63}$$

$$\vdots$$

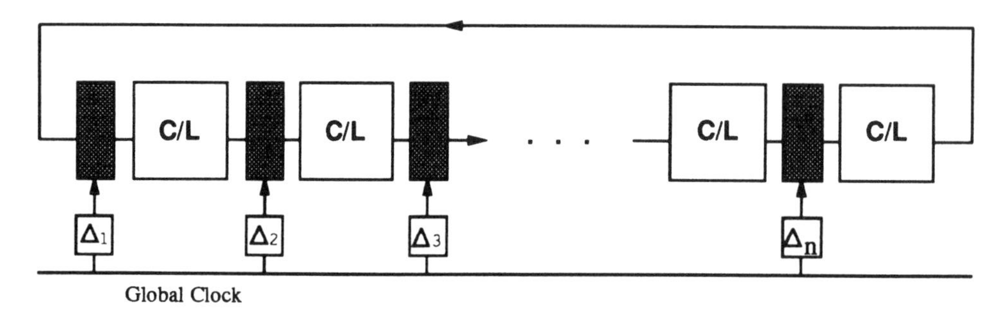

Fig. 19. n-stage feedback.

$$k_{(n-1,n)}T_{clk} + \Delta_n - \Delta_{n-1} \geq t_{max}^{(n-1,n)} \tag{64}$$

$$k_{(n,1)}T_{clk} + \Delta_1 - \Delta_n \geq t_{max}^{(n,1)} \tag{65}$$

Combining constraints (60) and (61) yields

$$\Delta_3 \geq t_{max}^{(1,2)} + t_{max}^{(2,3)} - k_{(1,2)}T_{clk} - k_{(2,3)}T_{clk} + \Delta_1 \tag{66}$$

Now, combining constraints (66) and (62) and so forth for all other constraints:

$$\Delta_4 \geq t_{max}^{(1,2)} + t_{max}^{(2,3)} + t_{max}^{(3,4)} - k_{(1,2)}T_{clk} - k_{(2,3)}T_{clk} - k_{(3,4)}T_{clk} + \Delta_1 \tag{67}$$

$$\vdots$$

$$\Delta_j \geq \sum_{i=2}^{j} t_{max}^{(i-1,i)} - T_{clk}\sum_{i=2}^{j} k_{(i-1,i)} + \Delta_1 \tag{68}$$

$$\vdots$$

$$\Delta_n \geq \sum_{i=2}^{n} t_{max}^{(i-1,i)} - T_{clk}\sum_{i=2}^{n} k_{(i-1,i)} + \Delta_1 \tag{69}$$

Finally, combining constraints (69) and (65) and rearranging terms, constraint (70) is implied.

$$T_{clk} \sum_{\forall(i,j)\in l} k_{(i,j)} \geq \sum_{\forall(i,j)\in l} t_{max}^{(i,j)} \tag{70}$$

Similarly, combining the upper bound constraints (2) for each stage in the feedback loop yields the following constraint:

$$T_{clk} \sum_{\forall(i,j)\in l} (k_{(i,j)} - 1) \leq \sum_{\forall(i,j)\in l} t_{min}^{(i,j)} \tag{71}$$

By combining constraints (70) and (71), the constraint given in Theorem 7 is implied (for $\sum(k_{(i,j)} - 1) > 0$).

$$\frac{\sum_{\forall(i,j)\in l} t_{min}^{(i,j)}}{\sum_{\forall(i,j)\in l}(k_{(i,j)} - 1)} \geq T_{clk} \geq \frac{\sum_{\forall(i,j)\in l} t_{max}^{(i,j)}}{\sum_{\forall(i,j)\in l} k_{(i,j)}}. \tag{72}$$

Since the above constraint must be satisfied for any general cycle in the system, the theorem is proved. □

REFERENCES

[1] S. Anderson, J. Earle, R. Goldschmidt, and D. Powers, "The IBM system/360 model 91 floating point execution unit," *IBM J. Research and Development*, pp. 34–53, Jan. 1967.

[2] C.-H. Chang, E. S. Davidson, and K. A. Sakallah, "Using constraint geometry to determine maximum rate pipeline clocking," in *Proc. ICCAD-92*, 1991, pp. 142–148.

[3] T. I. Chappell, B. A. Chappell, S. E. Schuster, J. W. Allan, S. P. Klepner, R. V. Joshi, and R. L. Franch, "A 2-ns cycle, 3.8 ns access 512-kb CMOS ECL SRAM with a fully pipelined architecture," *IEEE J. Solid State Circuits*, pp. 1577–1585, Nov. 1991.

[4] L. W. Cotten, "Maximum-rate pipeline systems," in *Proc. 1969 AFIPS Spring Joint Comput. Conf.* 1969, pp. 581–586.

[5] B. Ekroot, "Optimization of Pipelined Processors by Insertion of Combinational Logic Delay," Ph.D. thesis, Stanford University, 1987.

[6] D. Fan, T. Gray, W. Farlow, T. Hughes, W. Liu, and R. Cavin, "A CMOS parallel adder using wave-pipelining," *Advanced Research in VLSI and Parallel Sys.*, Providence, RI, Mar. 1992.

[7] J. Fishburn, "Clock skew optimization," *IEEE Trans. Comput.*, vol. 39, no. 7, pp. 945–951, 1990.

[8] C. T. Gray, W. Liu, and R. K. Cavin, *Wave-pipelining: Theory and CMOS Implementation.* Kluwer Academic Publishers, 1994.

[9] C. T. Gray, T. Hughes, D. Fan, G. Moyer, W. Liu, and R. Cavin, "A high speed CMOS FIFO using wave-pipelining," North Carolina State University, Tech. Rep. NCSU-VLSI-91-01, 1991.

[10] T. Gray, T. Hughes, S. Arora, W. Liu, and R. Cavin, "Theoretical and practical issues in CMOS wave-pipelining," in *IFIP Transactions A: Computer Science and Technology-VLSI '91*, pp. 397–409, Edinburgh, Scotland, 1992.

[11] D. A. Joy and M. J. Ciesielski, "Clock period minimization with wave-pipelining," *IEEE Trans. Computer-Aided Design*, pp. 461–472, Apr. 1993.

[12] D. A. Joy and M. J. Ciesielski, "Placement for clock period minimization with multiple wave propagation," in *Proc. 28th Design Automation Conf.*, 1991, pp. 640–643.

[13] S. T. Ju and C. W. Jen, "A high speed multiplier design using wave-pipelining technique," in *Proc.IEEE APCCAS*, 1992, Australia, pp. 502–506.

[14] F. Klass and J. M. Mulder, "CMOS implementation of wave-pipelining," Tech. Rep. 1-68340-44(1990)02, Delft University of Technology, Dec. 1990.

[15] P. Kogge, *The Architecture of Pipelined Computers.* McGraw-Hill, 1981.

[16] Charlie Kohlhardt, "PA-RISC processor for 'Snake' workstations," in *Proc. Hot Chips Symp.*, 1991, pp. 1.20–1.31.

[17] W. K. C. Lam, R. K. Brayton, and A. Sangiovanni-Vincentelli, "Valid clocking in wavepipelined circuits," in *Proc. ICCAD-92*, Nov. 1992, Santa Clara, CA.

[18] W. Lien and W. Burleson, "Wave-domino logic: Theory and applications," in *Proc. Int. Symp. on Circuits and Systems*, 1992.

[19] H. Loomis, "The maximum rate accumulator," *IEEE Trans. Electron. Comput.*, pp. 628–639, Aug. 1966.

[20] V. Nguyen, W. Liu, C. T. Gray, and R. K. Cavin, "A CMOS multiplier using wave-pipelining," in *Proc. Custom Integrated Circuits Conf.*, May 1993, San Diego, CA.

[21] S. Perremans, L Claesen, and H. De Man, "Static timing analysis of dynamically sensitizable paths," in *Proc. 26th Design Automation Conf.*, 1989, pp. 568–573.

[22] L. Qi and X. Peisu, "The design and implementation of a very fast experimental pipelining computer," *J. Comput. Sci. and Technol.*, vol. 3, no. 1, pp. 1–6, 1988.

[23] K. Sakallah, T. Mudge, T. Burks, and E. Davidson, "Synchronization of pipelines," Computer Science and Engineering Department, University of Michigan, Tech. Rep. CSE-TR-97-91, Feb. 1991.

[24] T. Sakurai and A. R. Newton, "Delay analysis of series-connected MOSFET circuits," *IEEE J. Solid State Circuits*, pp. 122–131, Feb. 1991.

[25] H. Shin, J. Warnock, M. Immediato, K. Chin, C.-T. Chuang, M. Cribb, D. Heidel, Y.-C. Sun, N. Mazzeo, and S. Brodskyi, "A 5Gb/s 16x16 Si-bipolar crosspoint switch," in *Proc. IEEE Int. Solid State Circuits Conf.*, pp. 228–229, Feb. 1992.

[26] A. Srinivasan and D. LaPotin, "Cycle time optimization subject to packaging constraints," in *Multichip Module Workshop*, Mar. 1991, Santa Cruz, CA.

[27] A. Srinivasan, S. Seshadri, J. G. Shanthikumar, and D. P. LaPotin, "Stochastic cycle time optimization of sequential systems," Tech. Rep., Electronics Research Laboratory, University of California, Berkeley, Dec. 1990. Tech. Memo UCB/ERL M90/91.

[28] W. v. Noije, C. T. Gray, W. Liu, T. A. Hughes, R. K. Cavin, and W. J. Farlow, "CMOS sampling circuitry with 25 ps resolution," in *Proc. Custom Integrated Circuits Conf.*, May 1993, San Diego, CA.

[29] D. Wong, "Techniques for Designing High-Performance Digital Circuits using wave-pipelining," Ph.D. Thesis, Stanford University, Sept. 1991.

[30] D. C. Wong, G. De Micheli, and M. J. Flynn, "Designing high-performance digital circuits using wave-pipelining: Algorithms and practical experiences," *IEEE Trans. Computer-Aided Design*, pp. 25–46, Jan. 1993.

[31] D. C. Wong, G. De Micheli, M. J. Flynn, and R. E. Huston, "A bipolar population counter using wave-pipelining to achieve 2.5x normal clock frequency," *IEEE J. Solid State Circuits*, pp. 745–753, May 1992.

Part 2

Clock Distribution Design of Structured Custom VLSI Circuits

Design and Analysis of a Hierarchical Clock Distribution System for Synchronous Standard Cell/Macrocell VLSI

EBY G. FRIEDMAN, MEMBER, IEEE, AND SCOTT POWELL

Abstract —**This paper describes the synchronous clock distribution problem in VLSI and techniques for its solution. In particular, a hierarchical design technique for minimizing clock skew within a VLSI circuit and its relative advantages and disadvantages is discussed. In addition, a model for clock distribution networks which considers the effects of distributed interconnect impedances on clock skew is described.**

I. Introduction

IN MOST digital systems, the transfer of data between functional elements is synchronized by a single control signal, the processing clock. This signal typically constrains the timing and performance behavior of an entire system [1]. Therefore it is imperative that a design methodology exist for distributing clocks which will permit the system to operate as fast as possible without creating any unnecessary timing uncertainties or decreasing functional chip yield. This clock distribution design methodology must also provide an environment for hierarchical design of a VLSI circuit, thus permitting the VLSI design effort to be partitionable into smaller design problems. This paper describes a design methodology for distributing clock networks in a VLSI circuit design environment. The methodology considers all hierarchical levels of design detail and permits design of a clock distribution network that provides optimal performance, such as negligible clock skew. The overall design problem remains functionally partitionable and is well-suited for a multi-person design team environment. Lastly, a clock distribution model is discussed which considers the effects of distributed interconnect impedances on clock skew.

Section II of this paper describes the requirements of a clock distribution system for a classical sequential VLSI circuit. In Section III, the particular design approach of the partitionable clock distribution scheme is discussed, while in Section IV the advantages and disadvantages of this clock distribution scheme are described. In Section V, an approach for characterizing distributed resistive and capacitive interconnect impedances for optimal performance evaluation of clocking circuitry is discussed. Section VI describes an example in which this clock distribution methodology has been implemented in a complex VLSI circuit. Finally, in Section VII, some concluding comments are made describing the relative benefits of this design approach to VLSI circuits.

II. Design Requirements of a Clock Distribution System

Clock distribution systems perform the task of synchronizing the flow of data in digital systems. As a clock signal arrives at a sequential register, it triggers the data from one bank of sequential registers to the next through a combinatorial network, which performs manipulations of the data in an appropriate functional manner (see Fig. 1) [1]. When designing the specific combinatorial logic resident between the sequential registers, careful attention must be spent on insuring that each path's timing requirements are maintained. The delay of a specific sequential signal path consists of five factors.

1) $t_{C \to Q}$—the clock-to-Q delay of the originating register in the signal path.

2) t_{logic}—the propagation delay due to the particular RC time constants of the specific path's logic and interconnect.

3) t_{setup}—the setup time of the final register in the signal path.

4) t_{skew}—the time difference between the triggering edge of the processing clock presented to two different sequential registers in the signal path. A critical race condition can result if the final register clock signal significantly leads or lags the originating register clock signal.

5) $\Delta t_{t/r}$—the additional delay caused by the effect of the variation of input transition time on each transistor within a specific sequential path. For slow rise and fall times, the propagation delay of a device can increase significantly. This delay is described as a time difference since the delay components of a synchronous signal path ($t_{c \to Q}$, t_{logic}, and t_{setup}) change as a function of input transition time.

Therefore the total propagation delay between two registers in a sequential signal path is given by

$$t_{PD} = t_{C \to Q} + t_{\text{logic}} + t_{\text{setup}} + t_{\text{skew}} + \Delta t_{t/r}. \quad (1)$$

For a design to meet its specified requirements, the largest propagation delay of any signal path being enabled

Reprinted from *IEEE J. Solid-State Circuits,* vol. SC-21, no. 2, pp. 240–246, April 1986.

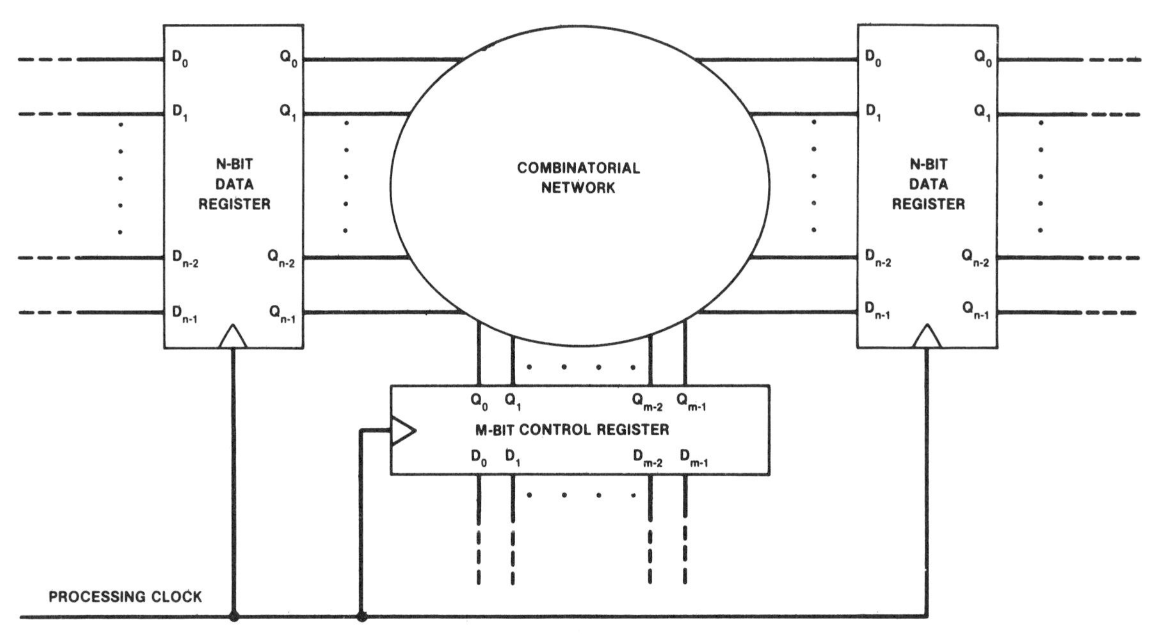

Fig. 1. Classical sequential data flow.

by the clock system must be less than the inverse of the circuit's maximum clock frequency (2) [1]–[3]

$$t_{PD_{\max}} < t_{\substack{\text{clock}\\ \text{period}}} = \frac{1}{fclk}. \tag{2}$$

Therefore, in order to permit maximum performance in a VLSI circuit, special attention must be given to decreasing nonfunctional parasitic time delays. This is commonly exhibited within the industry by an effort to decrease parasitic device and interconnect impedances [4] common in all integrated circuits. The intention of this paper is to describe a design technique for minimizing clock skew in sequential VLSI circuits while still maintaining a useful design team environment. By minimizing clock skew, the propagation delay of a path decreases, thereby permitting a higher frequency of operation and improving the data throughput of the overall system being integrated.

III. A Design Technique for Minimizing Clock Skew

When designing a clock distribution system in a VLSI circuit, it is imperative that the clock skew between each register in the circuit be limited. A common technique for minimizing the clock skew between each clock branch is the use of a general symmetric organization of the clock lines (e.g., equalizing line lengths) which, to first order, equalizes the parasitic load of each clock signal as seen by each clock buffer [2], [5], [6]. The difficulty occurs with this technique when one wants to hierarchically partition the overall chip design into separate functional elements, as is commonly done in large VLSI circuits. Ideally, each large functional element would contain its own locally optimized clock distribution system to satisfy its own particular timing constraints. However, local optimization within a functional element does not necessarily lead to global optimization of the overall chip-level clock distribution system.

If the processing clock interconnect had relatively low resistance, a chip-level centralized clock-buffer circuit would satisfy chip-level buffering requirements [7]. However, in many VLSI circuits, top-level interconnect typically is highly resistive due to the length of these connections. By centralizing the clock buffering, all sequential registers must be driven by the central clock-buffer circuit. However, due to the large capacitive fan-out of these registers, the load seen by the central clock-buffer circuit has a very large capacitive component. Also, for the central clock circuit, the combined capacitance of the registers is typically much greater than the parasitic capacitance of the clock interconnect. The delay through the final stage of the clock circuit is then proportional to $(R_{\text{buffer}} + R_{\text{line}}) \times C_{\text{load}}$. The large capacitive component of the fan-out makes the delay along a path very sensitive to line resistance. Since different branches will have different line resistances, a significant skew between processing clock signals can result [2], [8].

To decrease delay and minimize clock skew, it is advantageous to configure the centralized clock-buffering circuit so that the clock signals with higher resistances drive loads with lower capacitances. In a hierarchical VLSI circuit, due to the length of the lines, clock lines connecting the various chip-level functional elements are likely to be more resistive than clock lines within a functional element. In these situations, the final stages of the clock buffering should be distributed within the individual functional elements. In this way, the inverter stage driving the large capacitive fan-out dependent loads has lower interconnect resistance.

Therefore, in general, the initial clock-buffering stages are centralized at the chip level. Parallel connections are made from the centralized clock buffer circuit to each of

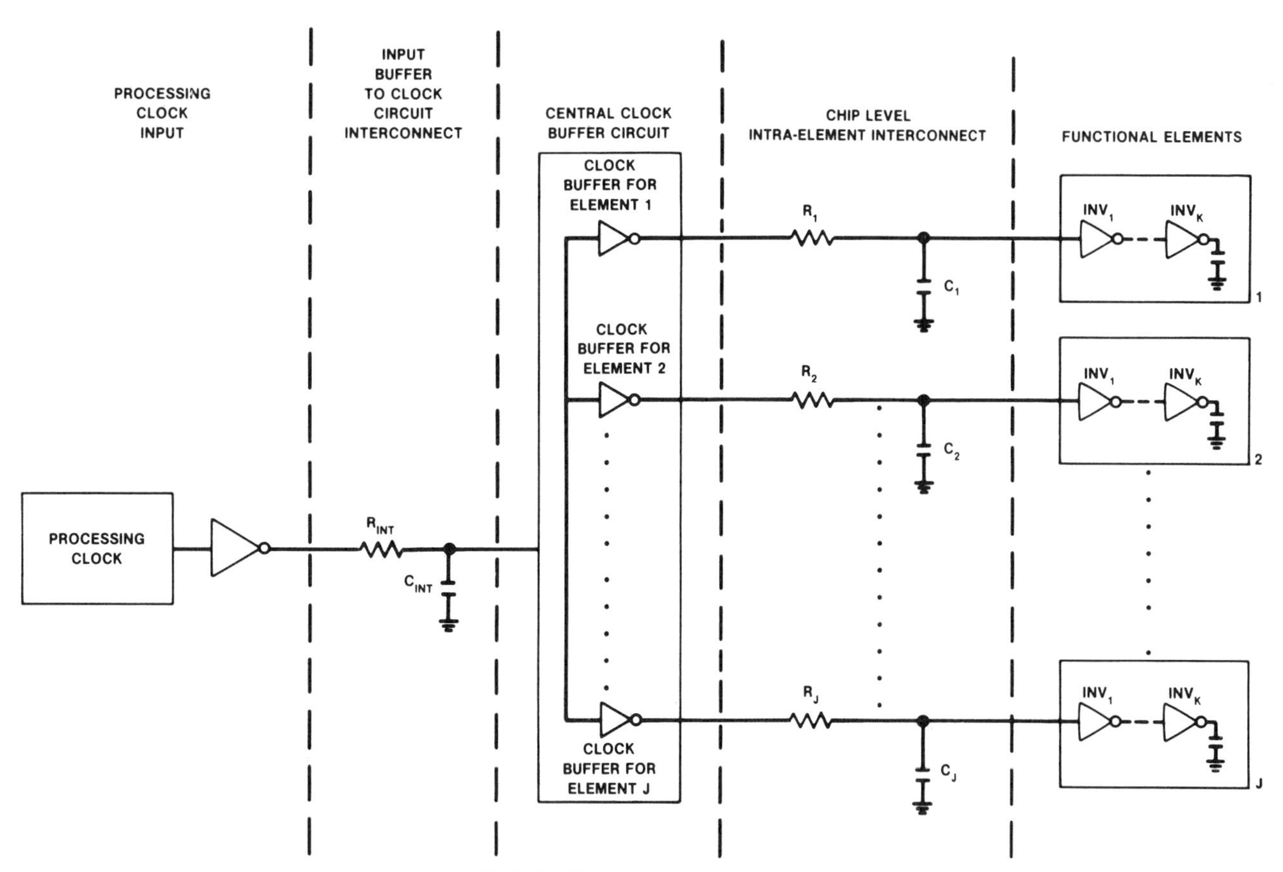

Fig. 2. Clock distribution system for minimal clock skew.

the functional elements, as shown in Fig. 2. The clock load seen at the input to each of the functional elements is relatively small, permitting these loads to be driven by the higher resistive chip-level clock interconnect. Thus clock skew due to differing line resistances is minimized by distributing the clock buffering and reducing the capacitive load driven by large chip-level resistive interconnect.

As described in [9], parameterized buffer cells can be used to geometrically size an inverter to provide an appropriate level of current drive. The effective output impedance of these devices coupled with the distributed resistive and capacitive impedances of the buffer's interconnect and fan-out define precisely the timing response of that logic stage [1], [10], [11]. Therefore one can compensate for the variation of clock propagation delay between each of the functional elements by parameterizing each of the functional elements' clock buffers resident in the central clock-buffering circuit. As shown in Fig. 2, each of the J functional elements requires its own precisely tuned clock signal to minimize the variation in clock skew between each of the J functional elements. Also shown, each of the J functional elements is constrained to a K-stage clock distribution system. This is to insure that each functional element is triggered by the same clock edge. A list of recommended constraints in the design of a minimal clock skew distribution system is as follows:

1) each of the J functional elements should utilize the same number (K stages) of clock buffering;
2) the clock signal rise and fall times seen by each sequential register within each functional element should be constrained to a maximum level; and
3) the internal functional element clock skew should be constrained to a maximum level using this same minimal clock-skew technique in a hierarchical fashion.

Accurate resistive and capacitive interconnect parasitics at the intra-element level are extracted [12], [13] and used to characterize the interconnect impedances between the functional elements and the central clock-buffer circuit. To compensate for the variation in interconnect and fan-out loading of each of the functional elements, finely tuned parameterized buffers are placed within the central clock-buffer circuit to drive each particular clock line. Each of these parameterized buffer cells is chosen with precisely the correct level of current drive (i.e., channel resistance) to compensate for the variation in interconnect and fan-out loading that each functional element requires. Thus the chip-level clock distribution system produces extremely small clock skew across the entire chip.

The number of stages necessary to implement an optimal clock distribution system is dependent upon the fan-out and interconnect loading and the application specific speed/area trade-offs. Techniques are described in the literature for determining the optimal number and geometric size of cascaded buffer stages [1], [14]–[16]. The optimal configuration of buffer stages contained within each func-

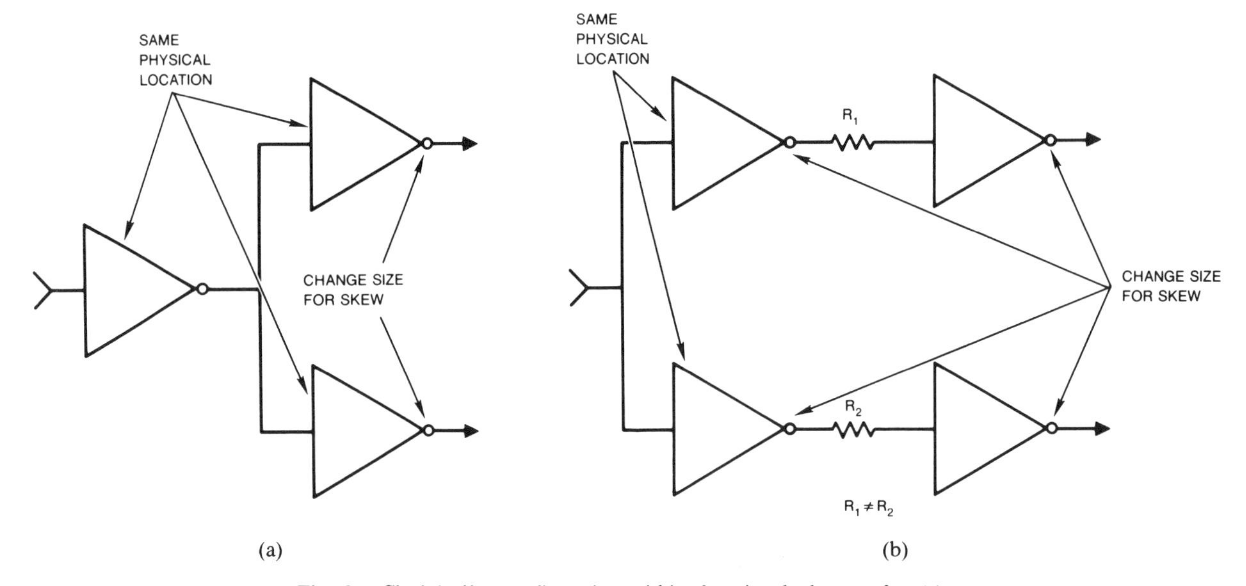

Fig. 3. Clock-buffer configuration within functional element for (a) standard cell implementation and (b) macrocell implementation.

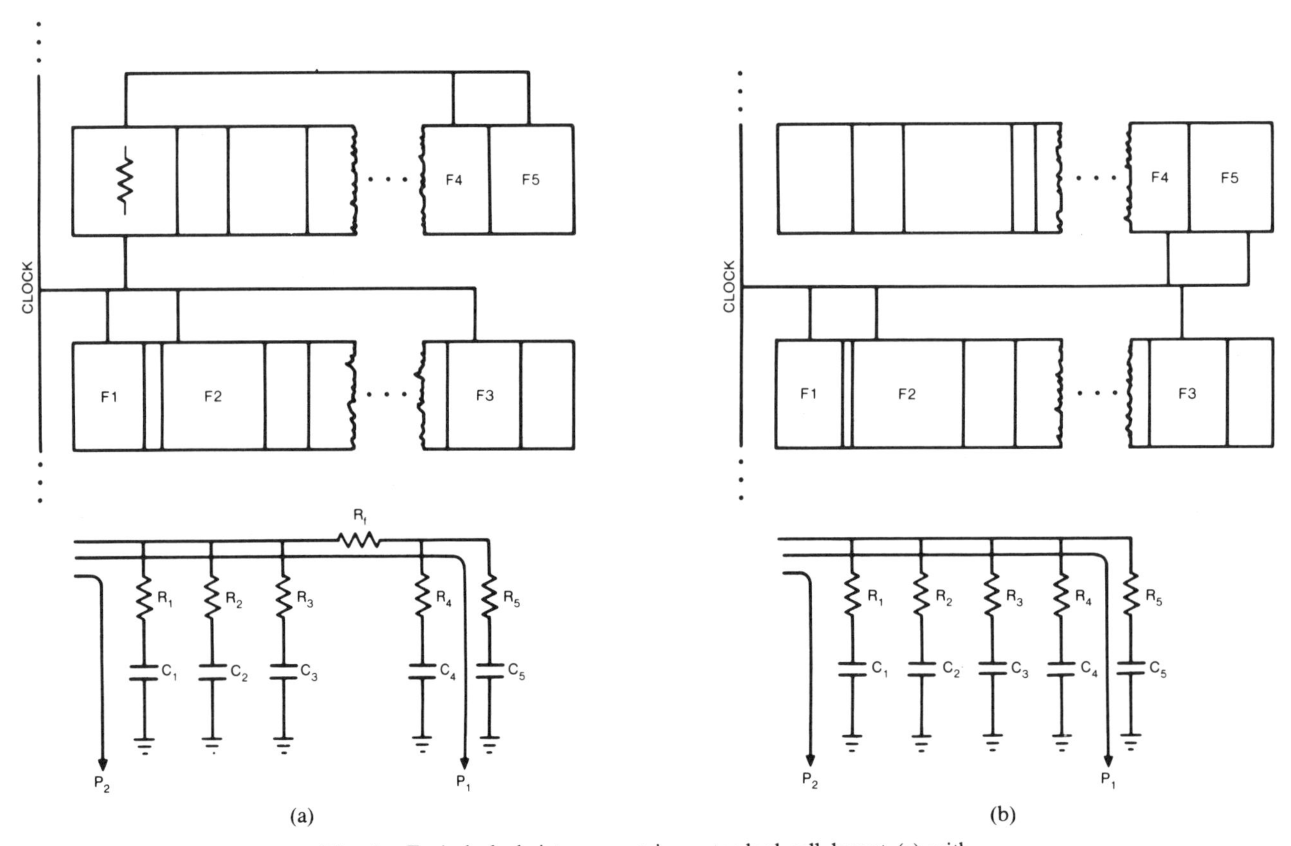

Fig. 4. Typical clock interconnect in a standard cell layout (a) with common resistance and (b) without common resistance.

tional element is dependent upon the application. The configuration illustrated in Fig. 3(a) can be used in functional elements implemented with standard cells. Since the line resistance is typically low, all inverters can be placed in close physical proximity and final adjustments for skew are made at the final stage. For functional elements implemented with macrocells, the configuration illustrated in Fig. 3(b) is more accurate since macrocell-to-macrocell connections tend to be longer than cell-to-cell connections in a standard cell environment. Therefore line resistances tend to be larger, requiring the delay through each branch to be defined independently of other branches.

For minimal clock skew, care must be taken to insure that there are no common resistive paths shared by more than one sequential gate. Fig. 4(a) illustrates a standard cell layout and the resulting clock line network. Resistance

R_f is connected to several gates and causes a difference in delay between paths $P1$ and $P2$, resulting in a clock skew between these two paths. Fig. 4(b) shows the same circuit without R_f. Notice that the resistors in the resulting network only drive one gate, as opposed to the situation in Fig. 4(a) where resistor R_f drives several gates. In Fig. 4(b), the capacitance seen by any one resistance is minimized, thereby minimizing the effect of the resistance on the overall delay (which is directly proportional to RC) and decreasing the clock skew between paths $P1$ and $P2$.

IV. Advantages and Disadvantages of the Minimal Clock-Skew Design Technique

A key advantage to this technique is in reducing chip-level clock skew. Also, overall clock delay, from the input pin to the sequential registers, is reduced. This occurs from improved partitioning of the RC loads. The inverters within each functional element drive large capacitive loads. These inverters, by placing them within their functional elements, are physically close to their loads, minimizing the interconnect impedance, particularly the resistance of these nodes [17]–[20]. The fairly long distance of the intra-element interconnect induces a large resistance between the central clock-buffer circuit and the intermediate stage clock inverters with relatively low input capacitance. Thus the RC time constants are reduced, which reduces the overall clock delay and skew.

Another important advantage in using this design technique in a synchronous VLSI circuit is the ability to partition each of the functional elements among a VLSI design team. The overall chip design can be partitioned hierarchically into a manageable domain of information with an emphasis on optimal clock distribution. Each functional element's clock distribution system is therefore optimized for its particular requirements while still maintaining overall chip-level clock integrity. The usefulness of this technique is dependent upon the ability to characterize the VLSI circuit's device and interconnect impedances. Careful attention must be spent on characterizing and modeling these resistances and capacitances in order to minimize the chip-level clock skew.

Lastly, unlike interconnect impedances, transistor transconductances tend to be very sensitive to both process and environmental variations (e.g., temperature, radiation, etc.). Therefore the performance of an optimally designed clock distribution system designed under nominal conditions would tend to fluctuate under worst-case and best-case conditions. In particular, a processing clock signal path whose performance is dominated by interconnect impedances changes differently than a clock signal path that is dominated by device impedances. For example, the delay through an inverter driving a large resistive load will tend to be fairly insensitive to variations in temperature. Conversely, the delay of an inverter driving a small resistive load will vary widely with temperature. Thus situations occur where the absolute propagation delays of clock signals significantly decrease under best-case conditions but the clock skew actually increases under these same conditions due to the different sensitivities of different signal paths to changes in device transconductance. However, if close attention is given to reducing the interconnect resistance when using the minimal clock-skew technique, the effect on clock skew is fairly minimal [17].

V. Performance Analysis

An accurate model of the clock distribution network within a VLSI circuit is necessary for proper performance evaluation. With the exception of the final stage, the clock-buffering system consists of a chain of inverters with simple interconnect between stages. The final stage consists of an inverter driving a multitude of sequential gates connected by a multibranch interconnection tree. The model for the load seen by the final clock-buffer stage is, in the general case, a large RC tree. Due to the size and complexity this tree can attain, circuit simulation of the entire network becomes prohibitive and a delay estimation technique must be used [10]. When using the clock-buffer distribution scheme described within this paper, the load seen by the final clock-buffer stage is of a simpler form, as is illustrated in Fig. 5(a). Clock skew is evaluated by comparing the delay through the branch with the largest RC time constant to the delay through the branch with the smallest RC time constant. Since the network consists entirely of parallel combinations of RC branches, several assumptions can be made to simplify the circuit used to model the load seen by the final clock-buffer stage.

The time constant R_jC_j of any particular branch j is normally one or more orders of magnitude less than the rise time of the clock signal. Approximations involving these branch time constants will have minimal effect on the accuracy of the result. The first simplifying approximation involves branch resistances. If the individual branch resistances do not vary significantly from the average branch resistance R_{av}, the impedance looking into the network of Fig. 5(a) can be approximated by the following equation (excluding the parasitic line capacitance which will be accounted for later):

$$Z_{in} \cong R_1 \| R_2 \| \cdots \| R_n \cdot \frac{(s+1/R_1C_1)(s+1/R_2C_2)\cdots(s+1/R_nC_n)}{s\left(s+1/R_{eq}C_{eq}\right)^{n-1}} \tag{3}$$

where

$$n = \text{number of sequential gates}$$

$$R_{eq} = R_1 + R_2 + \cdots + R_n = n \times R_{av}$$

$$C_{eq} = (1/C_1 + 1/C_2 + \cdots + 1/C_n)^{-1} = C_{av}/n$$

$$R_1 \| R_2 \| \cdots \| R_n = (1/R_1 + 1/R_2 + \cdots + 1/R_n) = R_{av}/n$$

$$s = j\omega = j2\pi f.$$

The capacitances $C_1, C_2, \cdots, C_n$ are all approximately equal to each other since they each represent the capacitance of a single sequential gate. Thus C can be approximated by C_{av}/n where C_{av} $(=C_1 = C_2 = \cdots C_n)$ is the input capaci-

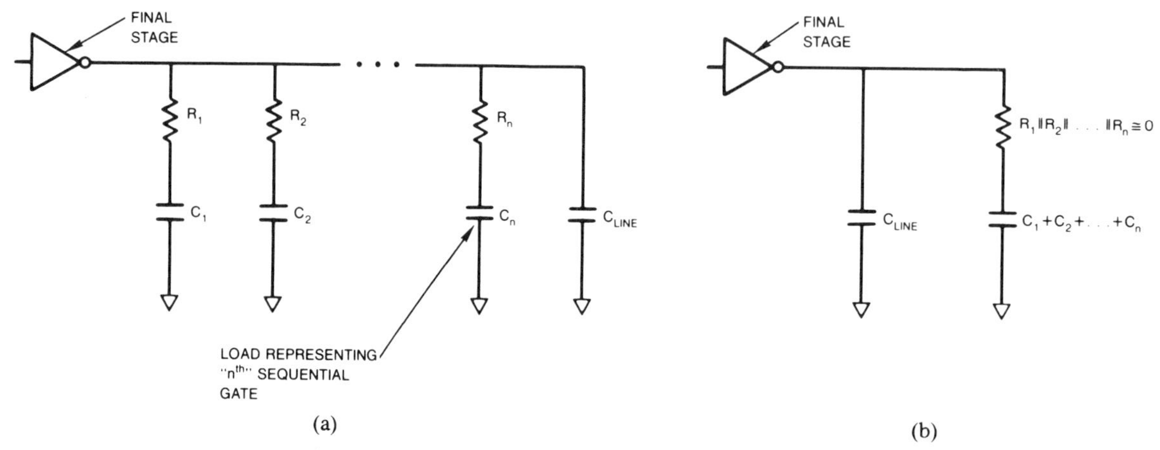

Fig. 5. (a) Complete model representing clock interconnect. (b) Simplified model representing clock interconnect.

tance of a single sequential gate. R_{eq} can be equivalently represented as $R_{av} \times n$ where R_{av} is the average of all the branch resistances. If branch resistances $R_1, R_2, \cdots, R_n$ do not vary significantly from R_{av}, the following equations hold true:

$$1/R_{eq}C_{eq} = 1/R_{av}C_{av}$$
$$\cong 1/R_1C_1 \cong 1/R_2C_2 \cong \cdots \cong 1/R_nC_n \quad (4)$$
$$R_1 \| R_2 \| \cdots \| R_n = R_{av}/n. \quad (5)$$

Using (4), the $n-1$ poles of (3) will cancel all but one of the zeros. Substituting (4) and (5) into (3), the following clock load impedance equations result:

$$Z_{\text{in}} \cong (R_{av}/n)\frac{(s + 1/R_{av}C_{av})}{s}$$
$$= R_{av}/n + \frac{1}{s(nC_{av})}. \quad (6)$$

Equation (6) suggests that the load seen by the final stage of the clock-buffering system can be modeled by a single-series RC branch where the resistance of the branch is the parallel combination of all the branch resistances and the branch capacitance is the sum of all the branch capacitances. The circuit represented by (6) is illustrated in Fig. 5(b). Additional simplification will result if a large number of sequential gates are driven by the final buffer stage and the individual branch resistances are sufficiently small such that $R_1 \| R_2 \| \cdots \| R_n$ can be taken as zero. Thus, for many cases, the clock distribution scheme presented in this paper can be adequately modeled by a simple summation of the input capacitance of all the sequential gates driven by a particular buffer with an additional parallel capacitance to account for parasitic line capacitance C_{line}.

VI. Implementation of Clock Distribution Design Technique

The clock distribution technique described in this paper has been implemented in a complex VLSI circuit. The chip was partitioned into many levels of hierarchy and required a large diverse team of VLSI designers. The VLSI design system described in [12] was used in the design of this chip.

Utilizing the aforementioned constraints listed in Section III, each functional element's clock distribution system was optimized for its particular load environment. The delay from a functional element's clock input to any internal sequential register was globally constrained to a fixed value for each functional element. Each designer insured that within their functional element, all clock skews and transition times seen by each sequential register were kept below a preset maximum limit.

Once each of the functional element's clock distribution systems was completed, interconnect parasitics were extracted [12], [13] at the top level. These were described in SPICE format and incorporated with the chip-level clock buffering in the centralized clock-buffer circuit into one large SPICE file. Considering various process and environmental conditions, each of the parameterized buffers resident within the central clock-buffer circuit was tuned for minimal chip-level clock skew. Thus, with an accurate description of device and interconnect parasitic impedances embedded within a SPICE nodal description of the complete clock distribution system, the difference of delay between each clock line, clock skew, was designed to be of almost negligible magnitude.

VII. Conclusions

An efficient design technique for a clock distribution system oriented for sequential VLSI circuits has been described. The technique can be used to minimize clock skew within a chip, thereby improving the circuit's overall performance. The design technique also provides an environment for concurrent VLSI circuit design, thereby permitting the VLSI circuit design effort to be hierarchically partitionable among a VLSI design team without impacting chip-level performance or cohesiveness. However, the technique exhibits some sensitivity to process and environmental variations, but this effect is considered to be relatively small. An accurate model for a clock distribution network in a VLSI circuit was presented. The model considers the effects of distributed interconnect impedances on clock skew. Finally, an example that utilizes this clock distribution system and the methodology for its implementation has been described.

Acknowledgment

The authors would like to acknowledge the technical contributions of G. Yacoub, W. Marking, P. Pandya, L. Tsu, D. Barach, T. Cesear, and R. Rapoza. They are also grateful to G. Persky for his sincere encouragement and advice in writing this paper.

References

[1] L. A. Glasser and D. W. Dobberphul, *The Design and Analysis of VLSI Circuits*. Reading, MA: Addison-Wesley, 1985.

[2] D. F. Wann and M. A. Franklin, "Asynchronous and clocked control structures for VLSI-based interconnection networks," *IEEE Trans. Comput.*, vol. C-32, no. 3, pp. 284–293, Mar. 1983.

[3] J. Beausang and A. Albicki, "A method to obtain an optimal clocking scheme for a digital system," in *Proc. Int. Conf. Computer Design* (ICCD), Oct. 1985, pp. 68–72.

[4] A. K. Sinha, J. A. Cooper, Jr., and H. J. Levinstein, "Speed limitations due to interconnect time constants in VLSI-integrated circuits," *IEEE Electron Device Lett.*, vol. EDL-3, no. 4, pp. 90–92, Apr. 1982.

[5] S. Dhar, M. A. Franklin, and D. Wann, "Reduction of clock delays in VLSI structures," in *Proc. Int. Conf. Computer Design* (ICCD), Oct. 1984, pp. 778–783.

[6] K. D. Wagner and E. J. McCluskey, "Tuning, clock distribution, and communication in VLSI high-speed chips," Stanford Univ., Stanford, CA, CRC Tech. Rep. 84-5, June 1984.

[7] R. Woudsma and J. M. Noteboom, "The modular design of clock-generator circuits in a CMOS building-block system," *IEEE J. Solid-State Circuits*, vol. SC-20, no. 3, pp. 770–774, June 1985.

[8] F. Anceau, "A synchronous approach for clocking VLSI systems," *IEEE J. Solid-State Circuits*, vol. SC-17, no. 1, pp. 51–56, Feb. 1982.

[9] E. Friedman, W. Marking, E. Iodice, and S. Powell, "Parameterized buffer cells integrated into an automated layout system," in *Proc. Custom Integrated Circuits Conf.* (CICC), May 1985, pp. 389–392.

[10] P. Penfield, Jr. and J. Rubinstein, "Signal delay in RC tree networks," in *Proc. 18th Design Auto. Conf.*, June 1981, pp. 613–617.

[11] C. M. Lee and H. Soukup, "An algorithm for CMOS timing and area optimization," *IEEE J. Solid-State Circuits*, vol. SC-19, no. 5, pp. 781–787, Oct. 1984.

[12] S. Powell, W. R. Smith, and G. Persky, "A parasitics extraction program for closely-spaced VLSI interconnects," in *Proc. Int. Conf. Computer-Aided Design* (ICCAD), Nov. 1985, pp. 193–195.

[13] E. Friedman, G. Yacoub, and S. Powell, "A CMOS/SOS VLSI design system," *J. Semicustom IC's*, vol. 2, no. 4, pp. 5–11, June 1985.

[14] H. C. Lin and L. Linholm, "An optimized output stage for MOS-integrated circuits," *IEEE J. Solid-State Circuits*, vol. SC-10, no. 2, pp. 106–109, Apr. 1975.

[15] R. Jaeger, "Comments on 'An optimized output stage for MOS-integrated circuits,'" *IEEE J. Solid-State Circuits*, vol. SC-10, no. 3, pp. 185–186, June 1975.

[16] H. J. M. Veendrick, "Short-circuit dissipation of static CMOS circuitry and its impact on the design of buffer circuits," *IEEE J. Solid-State Circuits*, vol. SC-19, no. 4, pp. 468–473, Aug. 1984.

[17] H-T. Yuan, Y-T. Lin, and S-Y. Chiang, "Properties of interconnect on silicon, sapphire, and semi-insulating gallium arsenide substrates," *IEEE Trans. Electron Devices*, vol. ED-29, no. 4, pp. 639–644, Apr. 1982.

[18] R. J. Antinone and G. W. Brown, "The modeling of resistive interconnects for integrated circuits," *IEEE J. Solid-State Circuits*, vol. SC-18, no. 2, pp. 200–203, Apr. 1983.

[19] G. De Mey, "A comment on 'The modeling of resistive interconnects for integrated circuits,'" *IEEE J. Solid-State Circuits*, vol. SC-19, no. 4, pp. 542–543, Aug. 1984.

[20] H. B. Bakoglu and J. D. Meindl, "Optimal interconnection circuits for VLSI," *IEEE Trans. Electron Devices*, vol. ED-32, no. 5 pp. 903–909, May 1985.

A SYMMETRIC CLOCK-DISTRIBUTION TREE AND OPTIMIZED HIGH-SPEED INTERCONNECTIONS FOR REDUCED CLOCK SKEW IN ULSI AND WSI CIRCUITS

H.B. Bakoglu[1], J.T. Walker and J.D. Meindl

Stanford University Center for Integrated Systems

Abstract

A symmetric clock-distribution tree suitable for high-speed ultra large scale integrated (ULSI) and wafer scale integrated (WSI) circuits as well as the module and board level design of high speed systems is described. In order to reduce the clock skew, the signal is delayed by an equal amount before it reaches the subblocks in the system. Design considerations regarding transmission line properties of the clock lines, reflections at the discontinuities, cross talk, termination and driving of the lines, distributed interconnection RC delay, skin effect, and the electrical design of interconnections are described. A design example in which the clock-signal is distributed over a three-inch wafer with 0.5 nsec rise time and 0.5 nsec skew (suitable for a 250 MHz signal) is also presented.

1 Introduction

Clock skew is one of the major concerns in high-speed digital-system design [1]-[10]. If two circuits clocked by the same signal are not equidistant from the clock driver, they receive the clock signal at different times. This causes clock skew, and related timing problems arise unless the clock period is much longer than the skew. This problem becomes more severe as die size increases (resulting from longer interconnection lengths) and as minimum feature size is scaled down (resulting from a higher RC constant per unit length of interconnection) [11]-[14].

Clock skew can be reduced by distributing the clock signal in such a way that the interconnections carrying the clock signal to the functional subblocks on the wafer are of equal length. If the clock signals are delayed equally before arriving at the subblocks, they will be perfectly synchronous. Using this basic idea, the distribution scheme in Fig. 1 minimizes the clock skew by repeating an H-shaped structure recursively.

2 Drive and Termination of the Clock Lines

To avoid reflections at the fan-out points, the two branches must have characteristic impedances twice the incoming line. As a result, in parallel they will act as a single line with the same impedance as the incoming line. This

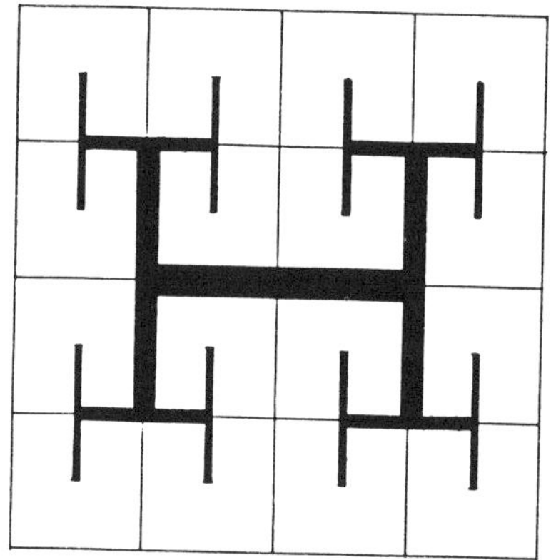

Figure 1: The H-clock tree. Clock signal is delayed by an equal amount before it reaches the subblocks; as a result, clock skew is reduced. At the branching points line width is reduced for proper impedance matching

matching condition can be accomplished by narrowing down the line width at the branching points because the characteristic impedance of a line, Z, is inversely proportional to its capacitance (Eq. 5, Fig. 1).

In addition to the reflections from the branching points, reflections from the end points must also be considered. Both the required on-resistance of the driver and the average power dissipation can be improved by terminating the line at the driving end instead of the receiving end (Fig. 2). In this scheme, the on-resistance of the driver is equal to the characteristic impedance of the line. This reduces the area required by the driver because the aim is for a $R_s = Z$ Ω source impedance rather than of 0 Ω. Since the effective on-resistance of the driver is equal to the impedance of the line, when the driver turns on, it forms a series potential divider with the transmission line and a voltage step with an amplitute of $V_{DD}/2$ starts travelling towards the receiver. The receiver acts as a capacitive termination and the voltage step doubles with a time constant of $\tau = ZC_L$, where C_L is the input capacitance of the receiver. For a 50 Ω line and 100 fF input capacitance, this risetime is 5 $psec$. Even with a 10 pF load the risetime is 500 $psec$. As a result, for all practical purposes, the line acts as if there is an open circuit at the receiving end and the doubling of the voltage step is instantenous. This doubled waveform travels back and turns the driver off when it arrives at the source end because the potential drop between the source and drain of the driving transistor becomes zero (Fig. 2). As a result,

[1]H.B. Bakoglu was supported by an IBM Graduate Fellowship. The support of DARPA is also gratefully acknowledged (contract No. MDA903-84-K-0062).

Reprinted from *IEEE Int'l Conf. Computer Design*, pp. 118–122, Oct. 1986.

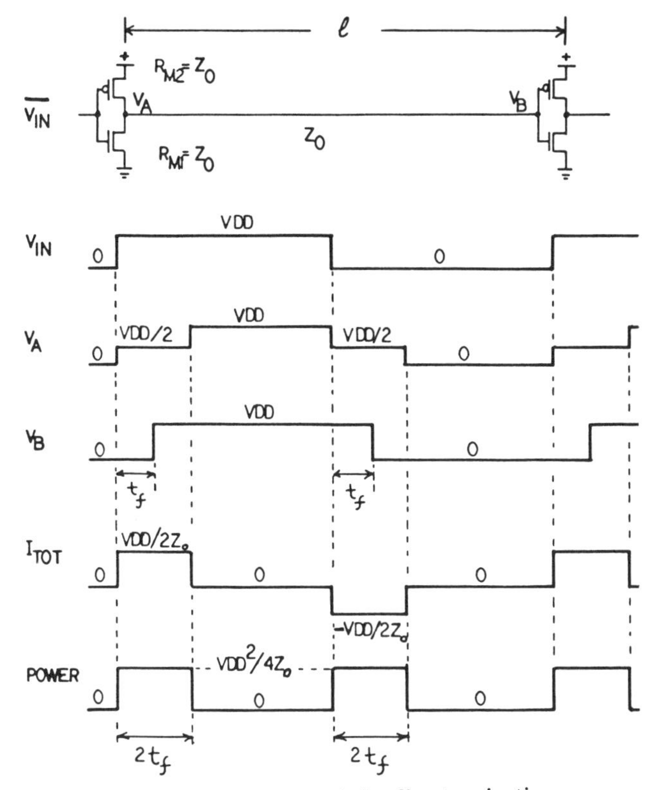

Figure 2: Low power transmission line termination

there is no DC power dissipation.

The termination condition can be calculated by setting the current supplied by the driver and into the transmission line equal. The current into the transmission line is given by

$$I = \frac{V_{DD}/2}{Z} \tag{1}$$

Since the transistor is at the linear region, it can supply a current of

$$I = \mu C_{ox} \frac{W}{L_{eff}} \left[(V_{DD} - V_T)\frac{V_{DD}}{2} - \frac{V_{DD}^2}{8} \right] \tag{2}$$

Transistor sizes are obtained by setting the currents in Eqs. 1 and 2 equal.

$$\frac{W}{L_{eff}} = \frac{1}{\mu C_{ox} Z \left(\frac{3}{4}V_{DD} - V_T\right)} \tag{3}$$

Here, μ is the mobility of the carriers, C_{ox} is the capacitance of the gate oxide per unit area, V_T is threshold and W and L_{eff} are the width and the effective length of the transistor.

3 Interconnection Resistance

When H-clock tree is used, all subblocks will receive identical clock signals; however, if the interconnections have large RC constants, the waveforms will have long rise times and a high-frequency clock signal will not be possible.

The response of a uniform lossy line to a unit step input has two components. The first one is a step function

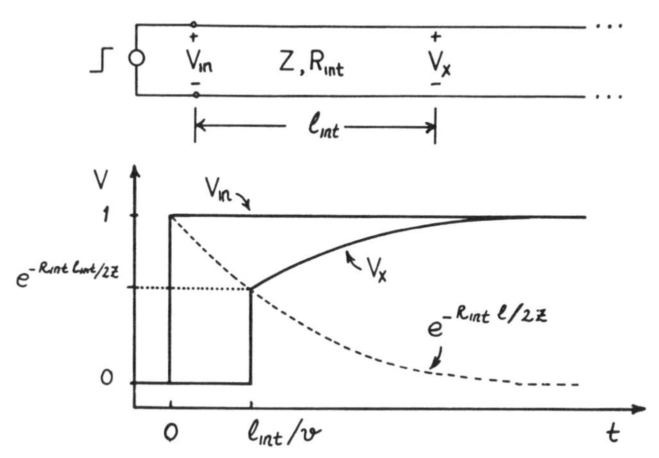

Figure 3: Waveforms in a lossy transmission line

that travels down the line with the speed of light and its magnitute is attenuated exponentially along the length of the line. The second component is a slow rising waveform that exibits an RC-like behavior (Fig. 3). If $\mathcal{R}_{int}l_{int} \gg 2Z$, the fast rising portion of the waveform is negligible and the transmission-line behaves like an RC-line. Here, $\mathcal{R}_{int}$ is the resistance of per unit length interconnection. To achieve high-speed signals, the fast rising portion of the waveform

$$(1 - loss) = e^{-\mathcal{R}_{int}l_{int}/2Z} \tag{4}$$

must be close to 1.0 (e.g. $(1 - loss) = 0.7$). Therefore, at least one of the metal levels should be scaled such that its RC constant per unit length is not too high. If the RC constant of the clock tree is not small enough, high-speed repeaters may be used [13,14]. Clock skew, however, will be determined by how well the repeaters are matched. If all the repeaters in a path are at the slow or the fast end of the process window, a relatively large clock skew may occur.

A steady current is distributed uniformly throughout the cross-section of a conductor through which it flows; however, time-varying currents concentrate near the surfaces of conductors. This is known as the skin effect [15]. As a result, at high frequencies the effective resistance of an interconnection will not decrease by making it thicker than two to four times the skin depth. The skin depth at 1 GHz for aluminum at room temperature is 2.8 μm.

4 Electrical Design of Interconnections

A major problem with interconnections fabricated on silicon wafers is their high capacitance caused by the thinness of the insulating layers, which results in a very low characteristic impedance because

$$Z = \sqrt{\frac{\mathcal{L}_{int}}{C_{int}}} = \frac{1}{vC_{int}} \tag{5}$$

where C_{int} and $\mathcal{L}_{int}$ are the capacitance and inductance per unit length of interconnection and v is the propagation speed of the electromagnetic waves in transmission line medium.

The low characteristic impedance of the transmission-line requires a greater current-drive capability from the clock driver to achieve transmission line speed.

Another major concern is the large RC time constant per unit length of interconnection. Thicker insulating layers and rectangular cross sections give the best electrical properties but they are also the most technologically challenging structures. Alternative interconnection materials can be used to reduce resistance. Au, Cu and Ag all have resistivities lower than that of Al. However, replacing Al with any of them will require a major effort because none of them are as compatible with integrated circuit processing as Al. Also, alternative dielectric materials may be used to reduce interconnection capacitance. Low diectric constant polymers (e.g. polyimide $\epsilon_r < 3.5$) are possible candidates. The capacitance and transmission-line properties of the interconnections can be further improved by fabricating air-bridges at the topmost level. Cooling the wafers to liquid-nitrogen temperature reduces interconnection resistance significantly.

In high-speed circuits, noise is also a major problem. As the number of interconnection levels increases, the capacitance and mutual inductance between neighboring lines become important. In the absence of a proper ground plane, the neighboring lines become current return paths and excessive noise and cross talk may result. Furthermore, the risetime of signals degrade because in a multiconductor environment non-TEM modes of propagation get excited. Since different modes propagate at different speeds, the waveforms get dispersed. Silicon substrate also introduces problems because it acts as a conductor for capacitive effects and as an insulator for inductive effects. This gives rise to so called "slow wave" mode in which the propagation speed is much less than $v = c_0/\sqrt{\epsilon_r \mu_r}$ [16]. Therefore, it is best not to rely on the silicon substrate as the current return path but to supply a nearby metal reference plane.

An interconnection scheme similar to the one shown in Fig. 4 would be very desirable. Here, two topmost interconnection layers are reserved for long lines in X and Y directions for ease of routing; and these planes are seperated by continuous V_{DD} and ground planes that distribute power to the system. These planes serve multiple purposes. They supply a nearby low resistivity return path for interconnections and assure good transmission line properties, eliminate cross talk between X and Y planes and reduce coupling between the lines at the same level. They also provide low resistance and low noise power distribution network by eliminating the IR drops and LdI/dt noise at power lines. Of course, successful manufacturing of these complicated structures with acceptable yields is a major challenge; however, it may be the only way to take full advantage of the performance potentials of submicron electronics.

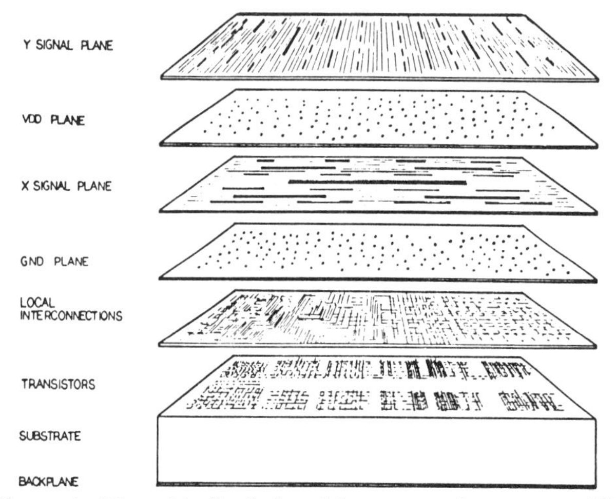

Figure 4: Hierarchically designed interconnection system. The thicker lines in the X and Y planes belong to the H-clock tree

5 Design Example

In this section a design example in which a 250 MHz clock-signal is distributed over a three-inch wafer ($D = 5.4\ cm$) is described. The specified rise time ($T_{90\%}$) of the signal distributed to the subblocks is 0.5 $nsec$ and the maximum allowable clock skew within a block is 0.5 $nsec$.

The first step is to calculate the subblock size. Assuming the local interconnections that distribute the clock signal within the subblocks have the following dimensions: $W_{int} = 2\ \mu m$, $H_{int} = 1\ \mu m$, $t_{ox} = 1\ \mu m$ and using Eq. 6 [17], the interconnection parameters are obtained as $\mathcal{R}_{int} = 150\ \Omega/cm$, $\mathcal{C}_{int} = 2\ pF/cm$, and $\mathcal{R}_{int}\mathcal{C}_{int} = 0.3\ nsec/cm^2$.

$$\mathcal{R}_{int} = \rho_{Al}\frac{1}{W_{int}H_{int}}$$

$$\frac{\mathcal{C}_{int}}{\epsilon_{ox}} = 1.15\left(\frac{W_{int}}{t_{ox}}\right) + 2.80\left(\frac{H}{t_{ox}}\right)^{0.222} + \left\{0.06\left(\frac{W_{int}}{t_{ox}}\right) + 1.66\left(\frac{H}{t_{ox}}\right) - 0.14\left(\frac{H}{t_{ox}}\right)^{0.222}\right\}\left(\frac{t_{ox}}{W_{sp}}\right)^{1.34} \quad (6)$$

Here, W_{int}, H_{int} and W_{sp} are the width, thickness and spacing of interconnections and t_{ox} is the thickness of the insulator. Because the maximum allowable clock skew within a subblock is 0.5 $nsec$, using the following delay expression for distributed RC lines [14],

$$T_{90\%} = \mathcal{R}_{int}\mathcal{C}_{int}{l_{max}}^2 = \mathcal{R}_{int}\mathcal{C}_{int}\left(\frac{d_{max}}{\sqrt{2}}\right)^2 \quad (7)$$

the largest subblock size d_{max} is calculated to be 1.83 cm. Since $d = D/2^{n/2}$ and $D = 5.4\ cm$, n is chosen to be 4 which attains $d = 1.35\ cm$ (Fig. 1).

The second step is to determine how many levels can be left unmatched at the branching points. This is important because if the final brach is a 50 Ω line and all the levels are matched, the input impedance of the clock tree be-

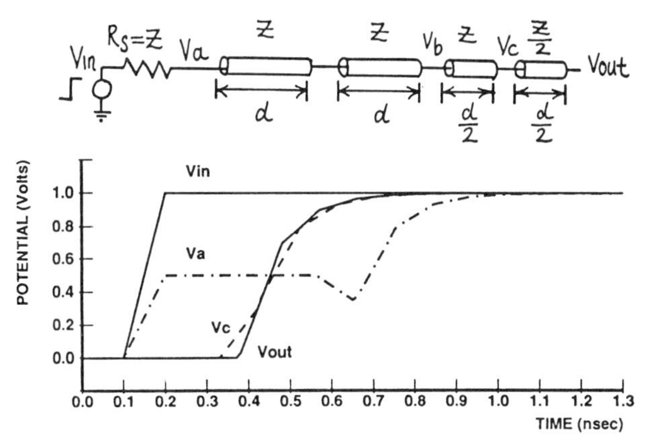

Figure 5: Simulations of the H-clock tree for three inch wafer with the last stages of branches unmatched.

Table 1: THREE INCH WAFER DESIGN PARAMETERS

$D = 5.4\ cm,\ d = 1.35\ cm,\ W_{int} = 2x\ \mu m,\ H_{int} = t_{ox} = x\ \mu m$

k	l_{int} (cm)	W_{int} (μm)	R_{int} (Ω/cm)	C_{int} (pF/cm)	$R_{int}C_{int}$ (ns/cm^2)	Z (Ω)
4	0.68	$2x$	$\frac{140}{x^2}$	1.8	$\frac{0.25}{x^2}$	37
3	0.68	$2x$	$\frac{140}{x^2}$	1.8	$\frac{0.25}{x^2}$	37
2	1.4	$7x$	$\frac{40}{x^2}$	3.6	$\frac{0.14}{x^2}$	18.5
1	1.4	$17.5x$	$\frac{16}{x^2}$	7.2	$\frac{0.12}{x^2}$	9.3
0	—	$38x$	$\frac{7.4}{x^2}$	14.4	$\frac{0.11}{x^2}$	4.6

comes 3.12 Ω. It is not easy to drive such small impedances with CMOS drivers. SPICE simulations of an equivalent transmission-line structure in Fig. 5 shows that the last set of branches can be left unmatched and still obtain a risetime less than 0.5 $nsec$. For an interconnection technology with $W_{int} = 2H_{int} = 2t_{ox}$ and the subblock size calculations of the previous paraghraphs, Table 1 is obtained.

In the third step the vertical dimensions of the clock lines must be determined to obtain a fast rising clock signal. In Table 1 the dimension x which represents the interconnection and the insulator thicknesses is left as a variable to be determined by the $loss$ requirements (Eq. 4). There are two subtleties in the following calculation: (1) The last branch is not matched. (2) Because of the fringing field effects, the clock tree will not act as a uniformly lossy line.

Since the last branch is not matched, the reflections shown in Fig. 6 must be taken into account. When all the components are summed the output potential is obtained as

$$V_{OUT} = 2x_1x_2(1+\Gamma)\left\{\frac{1-(-x_2^2\Gamma)^{N+1}}{1+x_2^2\Gamma}\right\} \tag{8}$$

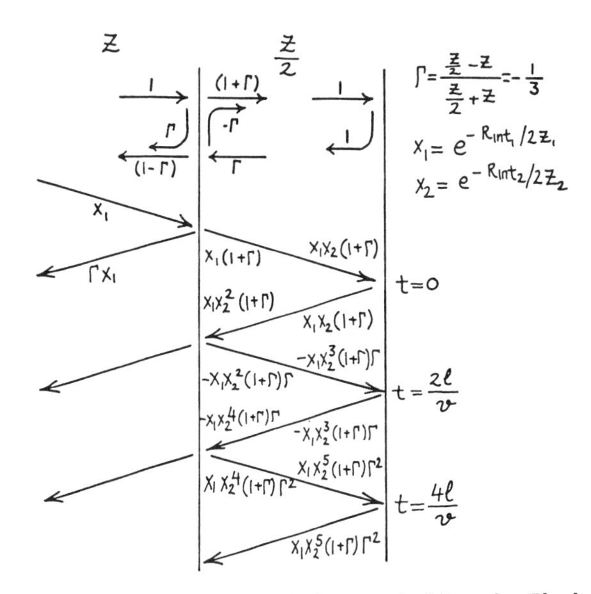

Figure 6: Reflections in the final unmatched branch. The lossy nature of the line is taken into account

Here x_1 is the portion of the step input that arrives to the final set of branches after losses and x_2 is the portion of x_1 that goes through the final branch. Both x_1 and x_2 are less that one and are given by the $(1 - loss)$ term in Eq. 4. Γ is the reflection coefficent at the discontinuity between the last set of branches and the lines that feed them. The time element is hidden in N (i.e $N(t)$). For the frequencies and dimensions that are of concern, $(-x^2\Gamma)^{N+1} \approx 0$.

Because of the fringing fields, the H-clock tree does not act as a uniformly lossy line; the lines at the end have more loss. As a result, total loss must be calculated as a product of the losses at individual levels. Using the information from Table 1 and Eq. 4, x_1, x_2 and Γ can be calculated as

$$\begin{aligned} x_1 &= exp\left\{-\sum_{k=1}^{3}\frac{l_{int}R_{int}}{2Z}\right\} = e^{-3.9/x^2} \\ x_2 &= exp\left\{-\frac{l_4R_4}{2Z_4}\right\} = e^{-1.28/x^2} \\ \Gamma &= \frac{\frac{Z}{2}-Z}{\frac{Z}{2}+Z} = -\frac{1}{3} \end{aligned} \tag{9}$$

Using Eqs. 8 and 9 and requiring that V_{OUT} is 70 percent of the full swing, $x = H_{int} = t_{ox}$ is calculated to be 4 μm. Because the slow rising portion of the waveform described in section 3. is neglected, the actual waveform will have a better risetime than the calculated 0.5 $nsec$. Finally, the skin effect limitations must be checked. Because the skin depth is 2.8 μm for aluminum at room temperature at 1 GHz, the designed interconnection thickness is within the limits of skin effect.

An on-wafer CMOS clock driver driver that is terminated at the source end is used. Since Z of the initial line is 4.6 Ω, using Eq. 3 the transistor sizes of the driver are found to be $W/L_{effNMOS} = 2456$ and $W/L_{effPMOS} = 6140$ for Stanford

2 μm CMOS process ($V_{DD} = 5\,V$, $V_T = 0.8\,V$, $k_n = \mu_n C_{ox} = 30\,\mu Amp/V^2$, $k_p = \mu_p C_{ox} = 12\,\mu Amp/V^2$). To achieve high current drive parasitic source and drain diffusion resistances and polysilicon RC delays must be carefully considered.

6 Conclusion

Clock skew can be reduced by an H-clock distribution tree that delays the signal by an equal amount before it reaches the subblocks. The following factors must be taken into account in the design of this distribution network. (1) To achieve "sharp" clock signals RC constant of the lines must be kept sufficiently small, which requires proper scaling of the vertical dimensions (interconnection and insulator thicknesses). (2) Skin effect must be taken into consideration. (3) The characteristic impedance of the lines should be properly tailored to avoid reflections at the branching points. (4) Nearby low resistivity return paths should be supplied to assure good transmission-line properties. (5) On-wafer drive and termination methods that are suitable for driving small impedances should be employed.

References

[1] C. Mead and L. Conway, *Introduction to VLSI Systems,* Reading, Mass., Addison-Wesley, 1980, Chapter 7.

[2] C.L. Seitz, "Self timed VLSI systems," *Caltech Conference on VLSI,* pp. 345–355, Jan. 1979.

[3] F. Anceau, "A synchronous approach for clocking VLSI systems," *IEEE Journal Solid-State Circuits,* vol. SC-17, pp. 51–56, Feb. 1982.

[4] B. Randell and P.C. Treleaven, eds. *VLSI Architecture,* Chap. 15 by J.A. Marques and A. Cunha, "Clocking of VLSI circuits," Prentice-Hall, New Jersey, 1983.

[5] D.F. Wann and M.A. Franklin, "Asynchronous and clocked control structures for VLSI-based interconnection networks," *IEEE Transaction on Computers,* vol. C-32, pp. 284–293, Mar. 1983.

[6] K.D. Wagner and E.J. McCluskey, "Tuning, clock distribution and communication in VLSI high-speed chips," *Center for Reliable Computing Technical Report No. 84-5,* Stanford University, Stanford, CA, June 1984.

[7] S. Dhar, M.A. Franklin and D.F. Wann, "Reduction of clock delays in VLSI structures," *IEEE Int. Conference on Computer Design: VLSI in Computers (ICCD),* pp. 778–783, 1984.

[8] J. Beausang and A. Albicki, "A method to obtain an optimal clocking scheme for a digital system," *IEEE Int. Conf. Computer Design (ICCD),* pp. 68–72, Oct. 1985.

[9] E.G. Friedman and S. Powell, "Design and analysis of a hierarchical clock distribution system for synchronous standard cell/macrocell VLSI," *IEEE Journal of Solid-State Circuits,* vol. SC-21, pp. 240–246, Apr. 1986.

[10] A.L. Fisher and H.T. Kung, "Synchronizing large VLSI processor arrays," *IEEE Trans. Computers,* vol. c-34, pp. 734–740, Aug. 1985.

[11] R.H. Dennard, F.H. Gaensslen, H.N. Yu, V.L. Rideout, E. Bassous and A.R. LeBlanc, "Design of ion implanted MOSFET's with very small physical dimensions," *IEEE J. Solid-State Circuits,* vol.SC-9, pp. 256–268, Oct. 1974.

[12] K.C. Saraswat and F. Mohammadi, "Effects of scaling of interconnections on the time delay of VLSI circuits," *IEEE J. Solid-State Circuits,* vol. SC-17, pp. 275–280, Apr. 1982.

[13] H.B. Bakoglu and J.D. Meindl, "Optimal interconnect circuits for VLSI," in *IEEE International Solid-State Circuits Conference (ISSSC) Dig. of Tech. Papers,* pp. 164–165, San Francisco, CA, Feb. 1984.

[14] H.B. Bakoglu and J.D. Meindl, "Optimal interconnection circuits for VLSI," *IEEE Trans. on Electron Devices,* vol ED–32, No.5, pp. 903–909, May 1985.

[15] B.D. Popovic, *Introductory Engineering Electromagnetics,* Reading, Mass., Adison-Wesley, 1971, Chapter 14.

[16] H. Hasegawa, M. Furukawa and H. Yanai, "Properties of microstrip line on Si-SiO_2 system," *IEEE Trans. Microwave Theory and Techniques,* vol. MTT-19, pp. 869–881, Nov. 1971.

[17] T. Sakurai and T. Tamaru, "Simple formulas for two- and three-dimensional capacitances," *IEEE Trans. on Electron Devices,* vol. ED–30, pp. 183–185, Feb. 1983.

A Unified Single-Phase Clocking Scheme for VLSI Systems

MORTEZA AFGHAHI, MEMBER, IEEE, AND CHRISTER SVENSSON

Abstract —Two of the main consequences of advances in VLSI technologies are increased cost of design and wiring. In CMOS synchronous systems, this cost is partly due to tedious synchronization of different clock phases and routing of these clock signals. In this paper a single-phase clocking scheme is described that makes the design very compact and simple. It is shown that this scheme is general, simple, and safe. It provides a structure that can contain all components of a digital VLSI system including static, dynamic, and precharged logic as well as memories and PLA's. Clock and data signals are presented in a clean way that makes VLSI circuits and systems well-suited for design compilation.

Index terms —Clocking, compilation, edge-triggered flip-flops, latches, one-phase clocking, skew, synchronous systems, timing, VLSI digital systems.

I. Introduction

VLSI technologies are continuously being developed towards an increased level of integration and higher speeds. This has several consequences. Two of the most important consequences are an increased design cost and increased cost of wiring.

The increased cost of design is due to the increased complexity of the design. One way to cope with this is to use more systematic approaches to the design, sometimes called structured design [1]. So far structured design has been mainly applied to the structural part of the design (architecture, floorplan, etc.) and not so much to the timing part. In this paper we will discuss a structural timing design. Our goal is to develop a clocking strategy that is general, simple, and safe. Generality and simplicity make design easier and therefore decrease the design cost. Safety decreases the need for careful timing analysis and will also reduce design cost. These factors also make the CMOS circuits and systems well suited for design compilation.

Part of the high wiring cost of CMOS systems is caused by the many clock phases used; often four different clock phases are to be distributed all over a design [2]. The reason for this is that multiphase clocking has been recognized as both safe and transistor-count effective. In this paper we will show that a recently proposed scheme for single-phase clocking [3], [4] is just as safe and transistor-count effective as the previous four-phase clocking techniques. This technique will be shown to be general enough to handle static, dynamic, precharged, and precharged domino logic as well as array logic like memories or PLA's. Furthermore, it allows considerably higher clock frequencies than the normal multiphase techniques [4].

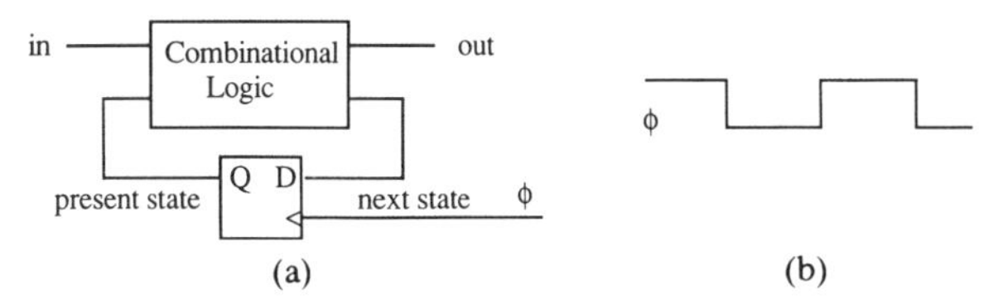

Fig. 1. (a) Single-phase state machine structure and (b) the clock signal.

II. Safe Clocking

In this paper we will limit our discussions on "safety" to timing behavior. We will thus use the term "safe" to denote safety against clock-skew and clock-slope problems, races, hazards, etc. Safe clocking is easily discussed in connection to a finite-state-machine model of logic (see Fig. 1) [1], [5]. The machine in this figure uses a positive edge-triggered D-flip-flop (ETDFF) as storage element. Input signals together with present state signals are inputs to a combinational logic block whose outputs are outputs and next-state signals of the machine. At the positive clock edge the next state becomes the present state and a new state is generated. The circuit is safe if the flip-flop is always nontransparent; that is, a state signal cannot pass the logic block more than once during one clock cycle. The discussion is easily extended to several interconnected finite state machines, a system which can be used as a model for any synchronous digital system. Designing circuits with ETDFF's is rather simple and straightforward and is used in most basic digital design texts [5].

However, if instead of edge-triggered elements data latches are used, it is possible to obtain faster and more transistor-count-effective circuits. But, it is shown that when traditional latches are used, single-phase clocking is neither safe nor fast [6] and a multiphase clock is required. This is due to the lower bound constraint on the minimum short-path delay of the combinational circuits, that could result in a data racethrough problem. For CMOS, when

Reprinted from *IEEE J. Solid-State Circuits,* vol. 25, no. 1, pp. 225–232, Feb. 1990.

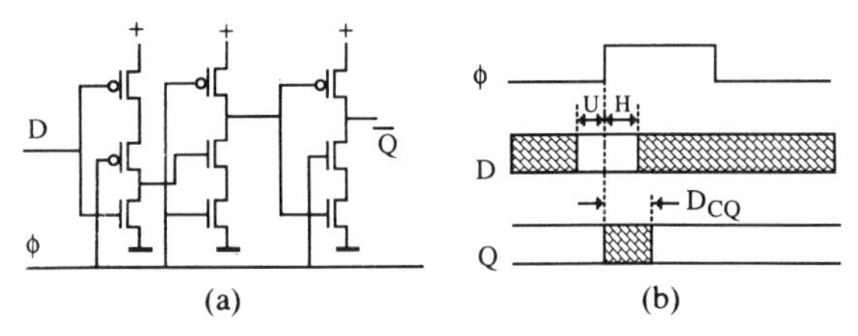

Fig. 2. Single-phase positive ETDFF: (a) circuit diagram and (b) parameters.

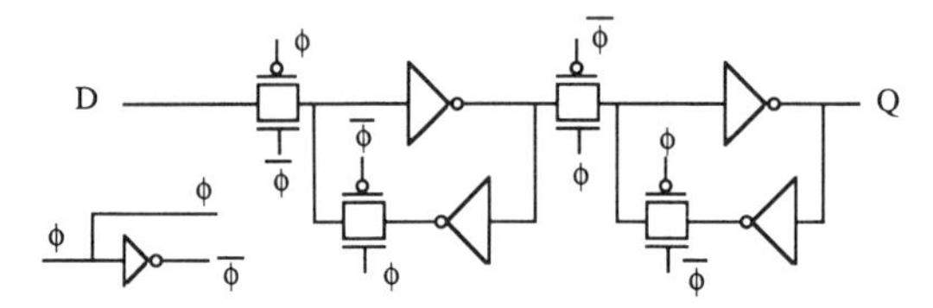
Fig. 3. Static single-phase ETDFF.

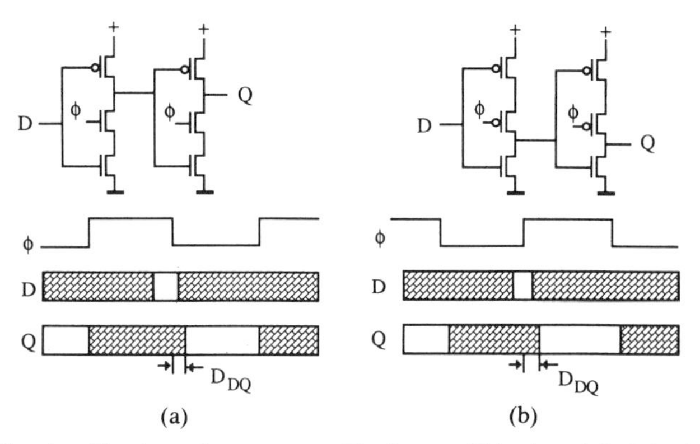
Fig. 4. Circuits and parameters of latches: (a) N-latch and (b) P-latch.

using latches, several schemes exist, in most cases requiring a four-phase clock signal (or pseudo-two-phase, which also means four clock signals) [2]. A simple example in this respect is a circuit according to Fig. 1, where the flip-flop is replaced by two dynamic latches (tristate inverters) clocked by a pseudo-two-phase clock [2]. Some of these schemes are used in connection with other techniques to save transistors, like precharged or precharged domino techniques. In the following, however, it is our intention to show that it is possible to combine the simple and safe single-phase clocking with transistor-count-effective latches. Furthermore, we propose a method that can safely handle up to half a clock period clock skew between two communicating machines without any constraint on the minimum short-path delay of the combinational logic circuit.

In this paper, we divide signals in a system into two classes, namely, clock signal and data signal. The only constraint on the clock signal is that its reference edges must occur only when specified. On the other hand, there are data signals that must obey certain specifications with respect to the clock-signal reference edges. This clear picture of the signals in a system makes the design and verification simpler. To gate the clock signal with a data signal or to delay the primary clock signal destroys this picture. Moreover, gating the clock signal usually results in considerable clock skew. Circuit and bus structures will be considered to selectively transfer data between different parts of the system without gating the clock signal with data signals.

III. Basic Components of Single-Phase Clocking

A. Storage Elements

In a synchronous system the movement of data is controlled by clocked storage elements. Thus, in this section we consider the timing behavior of storage elements of a single-phase clocking scheme.

A dynamic positive ETDFF [4] is shown in Fig. 2(a). The timing behavior of this device is the same as that of other ETDFF's, i.e., the input data must be valid during the interval given by the setup (U) and hold (H) times of the device and the output data are available and stable D_{CQ} after the active clock edge. The value of the hold time H of this kind of flip-flop is close to zero (see Section V). It is also possible to design a similar circuit which is triggered on the negative clock edge instead of the positive edge [4]. One may note that the proposed flip-flop is inverting. This is normally of little importance. The only limitation of this class of circuits is that the clock rise time must be short enough. This is, however, not a strong limitation, as a rise time of the order of 20 times the logical delay can be accepted [4], [12].

This dynamic flip-flop can thus replace the flip-flop in Fig. 1. It uses nine transistors and one clock line, which is more effective than the standard pseudo-two-phase CMOS solution using eight transistors and four clock lines (two tristate inverters as flip-flop).

If a static flip-flop is needed, a traditional CMOS master–slave flip-flop based on transmission gates [2] fits well into this timing scheme. In Fig. 3 this static ETDFF is demonstrated. This circuit is not really a true single-phase clocked circuit but is safely clocked by a single-phase clock if properly designed. By using a local inverter, as in the figure, the clock skew is under local control so that a properly designed cell will be safe in any environment. This cell uses 18 transistors.

The reason for using a two-phase nonoverlapping clock (pseudo-four-phase in CMOS) when latches are used is mainly to create two different points within each clock period to latch the data. In this way, a holding period is created before each latch that prevents the racethrough problem. This requirement may also be achieved by designing two latches that are transparent on different parts of the same clock signal, i.e., high and low part of the signal. Fig. 4(a) shows a circuit for a latch that is transparent when the clock signal is high and latches the data when the clock signal goes low. Fig. 4(b) shows the counterpart of this circuit, i.e., a circuit that is transparent when the clock signal is low and latches the data when the clock signal goes high [4]. These circuits may be called N- and P-latches, as the clock signal is applied to n and p transis-

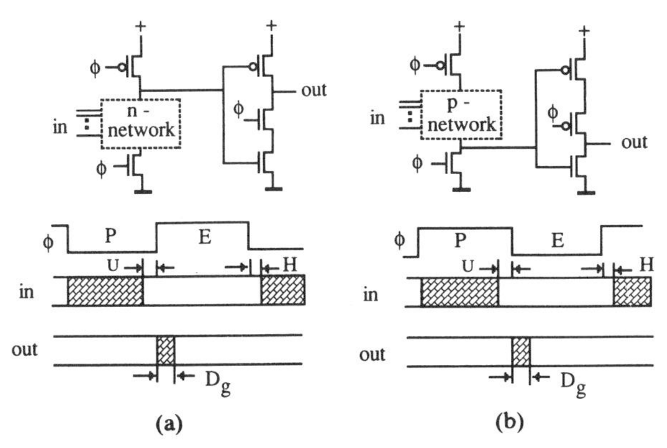

Fig. 5. Circuits and timing diagrams of (a) N-block and (b) P-block.

tors in these devices. In the same figure timing diagrams of these circuits are also depicted. Parameters of these latches are similar to other latches. Again, these devices have a forbidden window and the output signal is stable following the propagation delay of the element after the latching edge of the clock signal. As in the case of the dynamic ETDFF above, the clock edge must be fast enough.

B. Precharged Logic and Domino Logic

Precharged dynamic logic utilizes the MOS device characteristics to their best advantages. In fact, another reason for appealing to a multiphase clocking scheme was to enable several levels of nMOS dynamic logic evaluation to be performed per clock cycle [7]. This is now made possible for CMOS circuits even when a single-phase signal is used [3]. Single-phase precharged dynamic logic is based on N- and P-blocks. These blocks, with their respective timing diagrams, are demonstrated in Fig. 5(a) and (b). An N-block is precharging (P) when the clock signal is low and evaluates (E) its stable inputs when the clock signal is high. On the contrary, a P-block is in the precharge phase when the clock is high and in the evaluation phase when the clock signal is low. Therefore, a P-block using the same clock as an N-block can follow an N-block as it is in the precharging phase when the N-block is evaluating; that is, it can change its outputs. Obviously, the output of the N-block cannot be used to drive another N-block as they are in the evaluation phase at the same time. Again, these blocks can be characterized by a forbidden window during which the input data must stay stable and a delay after which the output data are stable.

As in the case of multiphase precharged logic, several levels of logic can be evaluated during one clock cycle using the domino technique [4]. This is done in the usual way, by replacing the logical stage in Fig. 5 with several cascaded logical stages separated by inverters [4], [15]. The single-phase clock thus imposes no limitations on the flexibility of the precharged/domino technique.

Comparing Fig. 2(a) and Fig. 5(a), one notices that the last stage of the positive ETDFF is identical to the last stage (the latching stage) of an N-block and, therefore, they have the same output timing. Thus a positive ETDFF

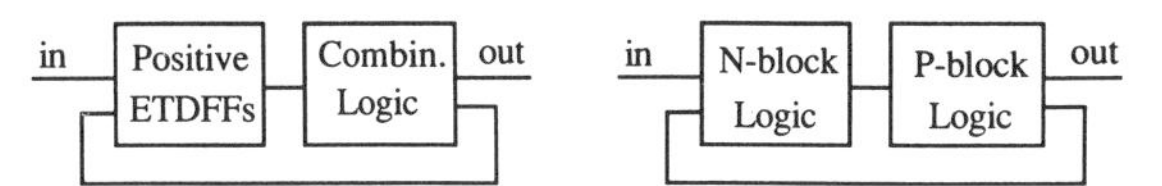

Fig. 6. Block diagram of a single-phase system with ETDFF's.

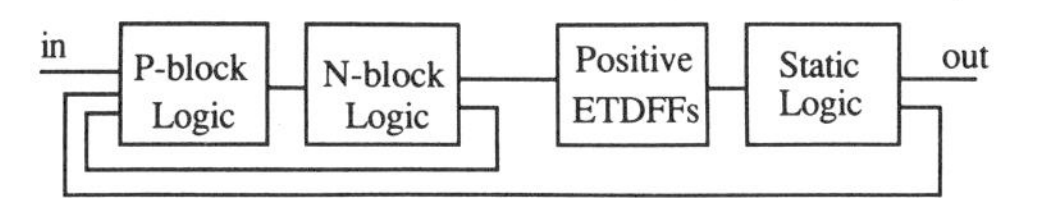

Fig. 7. A mixed technique.

TABLE I
ALLOWABLE COMBINATIONS OF FLIP-FLOPS AND PRECHARGE LOGICS

From :

To: ↓	+edge	- edge	N - Block	P - Block
+edge	ok	ok	ok	ok
- edge	ok	ok	ok	ok
N-Block		ok		ok
P-Block	ok		ok	

+edge= Positive ETDFF - edge= Negative ETDFF

can always drive a P-block and a negative ETDFF can always drive an N-block.

IV. Single-Phase Subsystem Circuits

In this section we will show that all types of circuits used in a CMOS design fit well into the single-phase clocking scheme. To this end, we employ the general concept of the finite state machine used in Section II. It is also important to include array logics, such as memory and PLA circuits, in the same regular clocking scheme.

A. Single-Phase Finite State Machine

When ETDFF's are employed as storage elements, it is relatively simple to fulfill the requirements of the state machine. The structure of the circuit becomes as shown in Fig. 6(a). In this structure, the output signal of the machine is stable after the delays of the flip-flop and the combinational circuit following the leading edge of the clock signal. This output signal can drive another state machine with a similar structure. The combinational logic is normally simple static logic in this case.

We may also form a finite state machine from precharged logic. The feedback loop must then include two precharged stages, one N-block and one P-block. Such a machine is demonstrated in Fig. 6(b).

Furthermore, we may also mix static and precharged logic. For example, static logic may be inserted between the blocks in Fig. 6(b) or a technique as depicted in Fig. 7 may be employed. The only important point to consider is that the timing of the previous latching circuit must fit the timing of the following latching or precharged circuit. In Table I a set of rules covering this problem is given.

When latches are used as storage elements, the block diagram of the state machine may take the configuration shown in Fig. 8. In the combinational logic circuit both

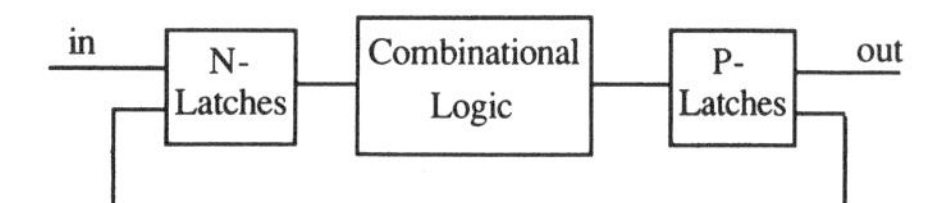

Fig. 8. Block diagram of the system with latches.

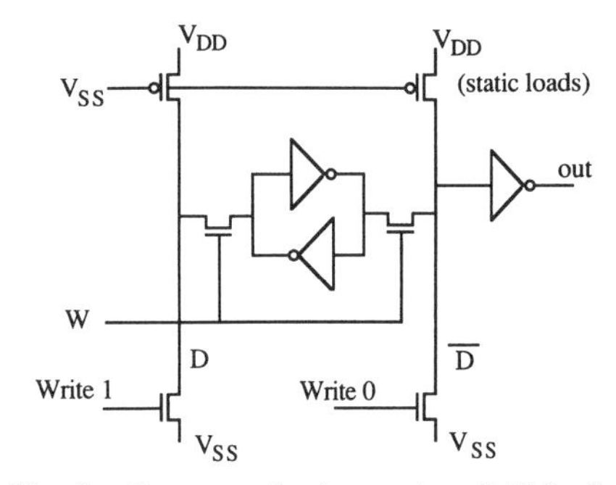

Fig. 9. Structure of a six transistor RAM cell.

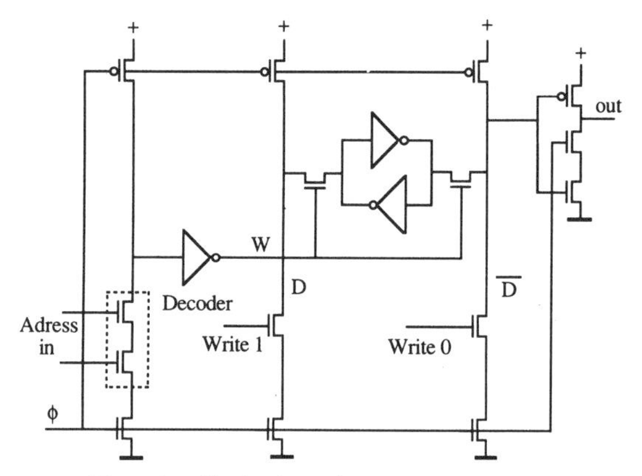

Fig. 10. Single-phase dynamic RAM circuit.

precharge p-units and static circuits can be included. This output signal can drive any system with a similar structure or systems with ETDFF's as storage element.

B. Single-Phase Memory and PLA Structures

The circuit of a normal CMOS RAM cell [2] is shown in Fig. 9. Note that this circuit is not a true CMOS circuit, since the circuit includes a static load (pseudo-nMOS). The memory cell is accessed by making the word select line W high. Often, the address decoder, generating the signals on W, is also designed in pseudo-nMOS in order to save transistors.

Now it will be shown how this memory cell and the address decoder can be designed in a two-stage domino mode with true CMOS (i.e., no static loads) and with the same standard single-phase clock as used above. This is demonstrated in Fig. 10. As shown, the address decoder is the first stage in a domino circuit and the memory cell and the bus circuitry are operated as a second stage. Both stages are then included in an N-block. As in any N-block (see Fig. 5(a)) the read data are latched when the clock signal goes low.

This proposed scheme again gives a block with the same timing behavior as an N-block (see Fig. 5(a)). Both address-in and write 1 and write 0 signals are then considered

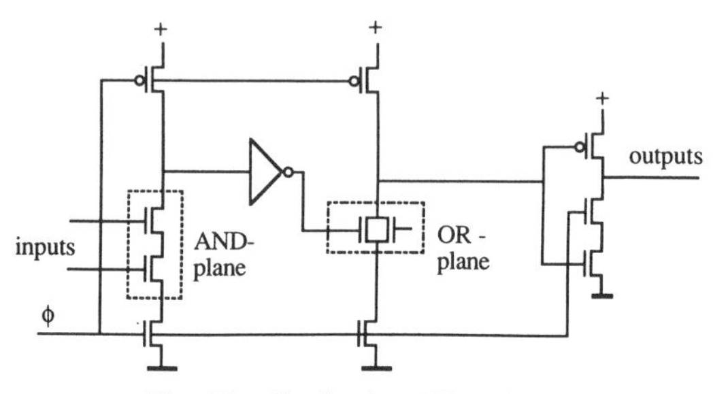

Fig. 11. Single-phase PLA circuit.

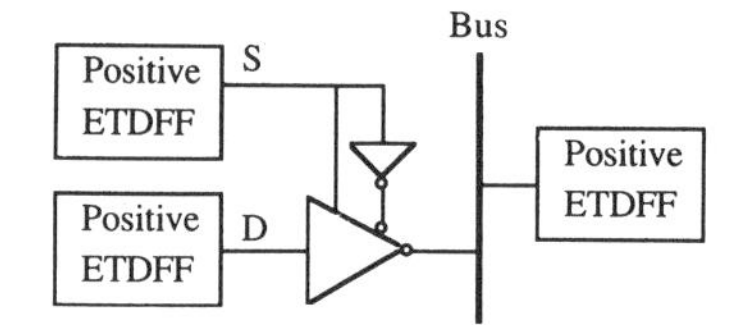

Fig. 12. Circuit for bus structures.

as input signals. The memory or the register stack thus fits exactly into the same timing scheme as any other block.

In addition, a ROM is easily designed in the same scheme. For a ROM we need only one data bus D and the memory cell includes only one transistor, connected between ground and D with its gate connected to W [2]. Data are stored by omitting some of the transistors in the array (see also Fig. 11).

The structure of a PLA also fits into the above scheme. Again, the AND plane can be considered as the first stage of a domino logic and the OR plane as the second stage (see Fig. 11). If some of the outputs are to be fed back to the inputs of the PLA to form a state machine structure, we may use the principle of Fig. 6(b). Note that the evaluation transistor (the clocked n-transistor) can be omitted in the second stage as the signal from the first stage in the domino chain always is zero during precharge. Again, the whole PLA has the same timing as an ordinary N-block.

It is also quite possible to use other variations of circuits to implement RAM's, ROM's, or PLA's. In all cases we may, for example, replace the two-stage domino principle with an N-block followed by a P-block (or vice versa). The latter principle will require more transistors but will give rise to higher speed due to less logical depth.

C. Bus Structures

The problem of selectively delivering data from flip-flops or latches to a bus, or receiving data from a bus, can give rise to timing problems [1, ch. 5]. In our structured single-phase clocking scheme we would prefer to keep the clock signal intact and not gate it or in other ways introduce an extra delay to it. The problem can be solved as shown in Fig. 12, where we access the bus with a tristate inverter, controlled by a signal with the same timing protocol as the data. The bus data will then also follow the same timing protocol and the whole system fits into the scheme already described.

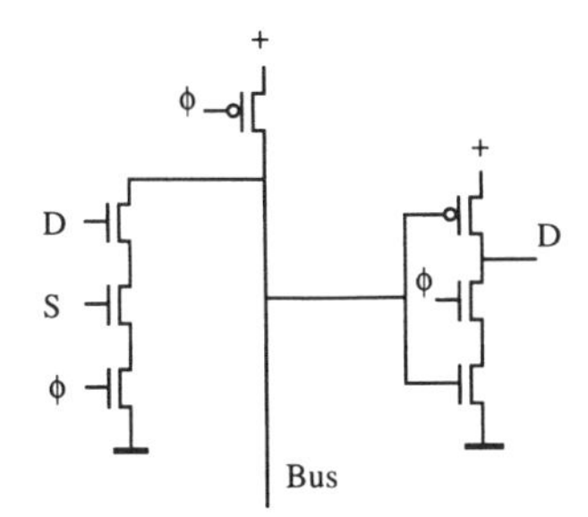

Fig. 13. Precharged bus system.

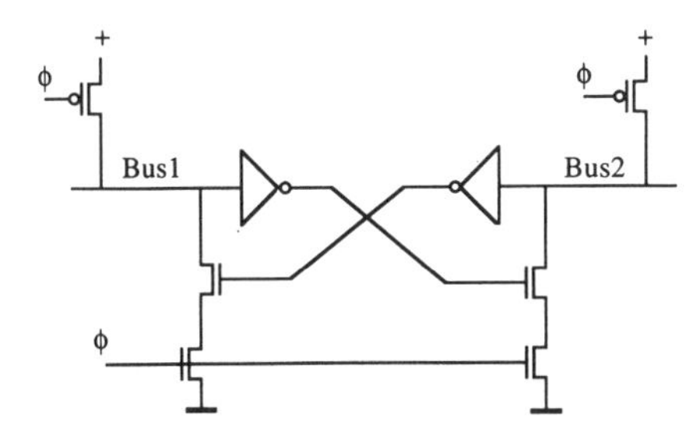

Fig. 14. Bidirectional repeater on a precharged bus.

A very useful bus structure is the precharged bus [1]. It is not only used internally on a chip but it may also be used externally, between chips. Again, we find that the single-phase clocking scheme can be applied. The precharged bus may use the same structure and timing as the internal node in a precharged circuit (Fig. 5(a)). This is demonstrated in Fig. 13, where we show a precharged bus with one input circuit and one output latch.

The input circuit is equivalent to the logic part of a precharged gate, where data and select are treated equal (and follow the same timing protocol). This circuit is thus equivalent to an open tristate when select is low. The input circuit may also include logic. If more than one input is active, the bus will perform a wired-OR function and still give valid data. The bus data can be captured either by an output latch as shown in Fig. 13 or by using it directly as input data to a second domino stage. The role of the output latch is to convert the data timing protocol from a precharged form to a normal form.

By using a precharged bus, it is also possible to introduce a bidirectional repeater circuit (which is not possible in a normal bus, see Fig. 14) [8]. This circuit is useful in long buses with large capacitive loads. It can also be used to minimize the effect of the RC delay of the line [9]. During precharge, both bus parts are precharged high. During evaluation, both bus parts will stay high unless one or both are made low, then both will become low. The bidirectional repeater is thus transparent both for data and for the wired-OR function between many input datas.

V. Timing of the Single-Phase Clocking Scheme

In this section timing considerations of the single-phase clocking scheme are covered. For this purpose we use the circuits of the finite state machine. After a brief discussion

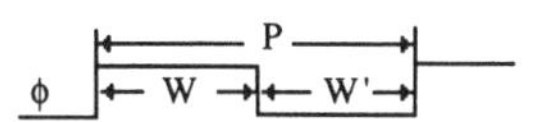

Fig. 15. Parameters of a single-phase clock signal.

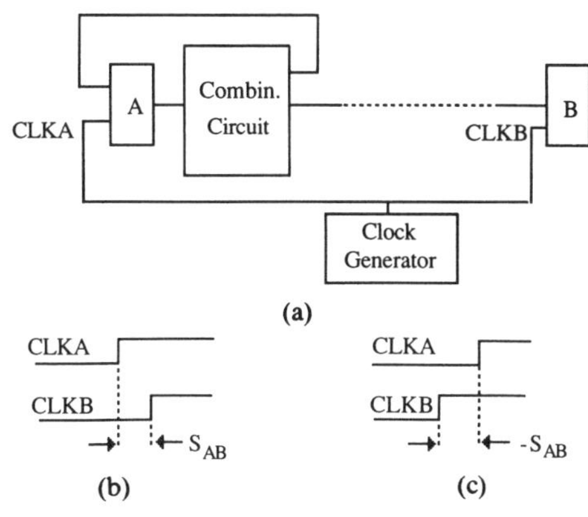

Fig. 16. Demonstration of clock skew in a single-phase system: (a) system diagram, (b) negative skew, and (c) positive skew.

about the clock signal, clock skew, and logic delays, we will consider both ETDFF's and latches as storage elements and derive the necessary constraints between the clock signal and different delays to guarantee proper operation of the machine.

The clock signal is specified by its width and period (see Fig. 15). Transition points of the clock signal, i.e., where the signal goes high or low, are also important and may be called reference edges.

In a synchronous system it is ideal if these reference edges occur at the same moment of time at all synchronizing points of the system, that is, where storage elements are placed. In reality, however, clock signals which propagate along routes with variable delays or signals may reach the same point from different paths. This variation of the delays results in clock skew. In a circuit two types of skew are encountered. They may be called self-skew and relative skew [10]. To explain the difference between these two, consider Fig. 16(a). The self-skew of $CLKA$ is the delay variation of the path from the clock generator to register A. Delay variations are caused by imprecision in processing, variation of temperature and power supply, variability among individual storage elements in the register, slopes in the clock pulse edges, etc. The effect of self-skew on timing is insignificant.

The relative skew between the clock signals $CLKA$ and $CLKB$ is the difference between the delays of the paths from the clock generator to registers A and B. $CLKA$ and $CLKB$ timing may then take the form of Fig. 16(b) where we define the skew between clock A and clock B, S_{AB}, as the time lag of $CLKB$'s clock edge after $CLKA$'s corresponding clock edge. Note that S_{AB} may take both positive and negative values. We will further use S_{ABM} for the maximum value of skew and S_{ABm} for its minimum value. Relative skew is evident in the same clock period and is present in nonlocal communications or when different flip-flops in a register receive clock signals from different paths. In estimating clock skew one may use worst-case

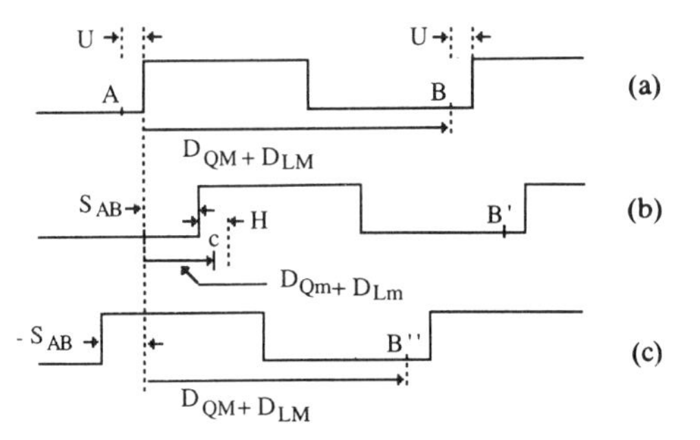

Fig. 17. Timing diagram for finite state machine with edge-triggered flip-flop.

analysis or statistical analysis [11]. In the worst-case analysis the skew is the difference between the worst delay (slowest) and the best delay (fastest) of the paths. This is assumed in Fig. 16. Statistical analysis usually results in lower skew but at the cost of a nonzero failure rate. In the following we assume that the skew associated with the leading edge and the trailing edge of the clock signal are equal. This assumption is reasonable since clock signals pass through several inverters before they reach the registers.

The combinational logic circuit included in the state machine structure can also be specified by two delay parameters. The maximum and minimum propagation delays through all circuit paths connecting any input to any output will be called D_{LM} and D_{Lm}. Flip-flops and latches are also characterized by maximum and minimum delays. We now consider the timing of single-phase clocking with edge-triggered flip-flops, latches, and precharged circuits.

A. Edge-Triggered Flip-Flop Timing

The aim of timing calculation is to specify a minimum period for the clock signal such that the state machine circuit operates correctly. For this purpose consider Fig. 16(a) and assume that registers A and B are edge-triggered flip-flops. Now consider a data stored at the input of the flip-flop A. For deterministic behavior of the state machine, this data when triggered in must have enough time to trip round the feedback loop and reach the input of the same flip-flop before the setup time of that device for the next clock cycle. This condition is graphically depicted in Fig. 17(a). An input data at point A in the current clock cycle must reach point B for the next cycle. In this case:

$$P \geqslant D_{QM} + D_{LM} + U \tag{1}$$

where D_{QM} is the maximum delay of the flip-flop. Furthermore, the data must not trip around the machine too fast, or to be more exact, it must not arrive at the flip-flop input before the end of the hold time for the present clock edge:

$$D_{Qm} + D_{Lm} \geqslant H. \tag{2}$$

As mentioned above, H is normally very close to zero, hence this condition is always fulfilled.

Now consider a situation where one of the outputs of the machine is feeding a nonlocal flip-flop, say, flip-flop B

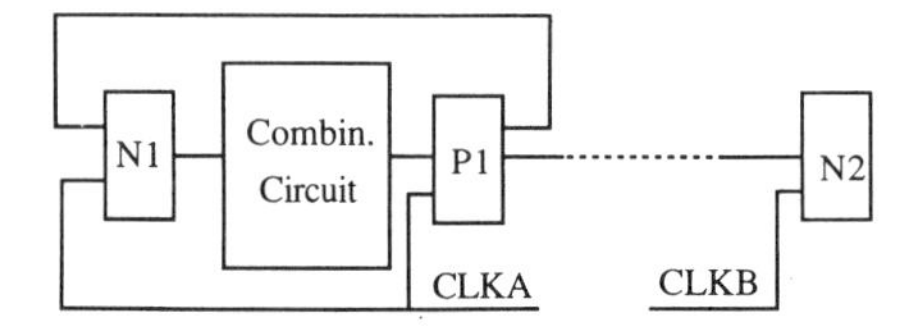

Fig. 18. Block diagram of single-phase finite state machine with latches as storage elements.

in Fig. 16(a). In this case the relative position of the clock signals at the site of flip-flop A and B may be as depicted in Fig. 16(b). Following the same principle stated above and referring to Fig. 17(b):

$$P \geqslant D_{QM} + D_{LM} + U - S_{ABm}. \tag{3}$$

The other situation must be considered. The minimum delays of the flip-flop and the combinational circuit may be such that the output signal reaches flip-flop B before the hold time of this device (see point C in Fig. 17(b)). Therefore, to avoid violation of the hold time of flip-flop B, we must have

$$D_{Lm} + D_{Qm} \geqslant H + S_{ABM} \tag{4}$$

where D_{Qm} is the minimum delay of the flip-flop. In other words, one may say that the maximum allowable clock skew is

$$S_{ABM} \leqslant D_{Lm} + D_{Qm} - H. \tag{5}$$

In the case when the skew is negative, Fig. 17(c), the clock period will increase and is obtained from the relation (3) with the sign of the skew changed. On the other hand, according to (5), the maximum positive skew is limited to the minimum delay of the flip-flop and combinational logic. This class of circuits is thus quite sensitive to positive clock skew [12], [13]. In a general system, (5) and (3) must be used for the maximum allowable skew and minimum clock period, respectively.

It should be mentioned that when a positive ETDFF drives a negative ETDFF, the minimum clock-pulse width, rather than the clock period, must fulfill the above expressions for the clock period. In this case the system will be much less sensitive to positive clock skew.

B. Data Latch Timing

The argument followed here is of the same form as used above. That is, the clock signal width and period are so determined that if data propagates along different paths and round the feedback loop, the setup and hold times of the latches must not be violated. We assume that the input data to a latch do not wait for the enabling clock edge. Rather, latches are open before input data become valid [10]. With this measure the system becomes faster and timing analysis becomes more clear. Fig. 18 shows a block diagram of the single-phase system considered here.

Now consider a data at the input of the latch $N1$. The latest arrival time of this data is the point A in Fig. 19. This data should have enough time to propagate through

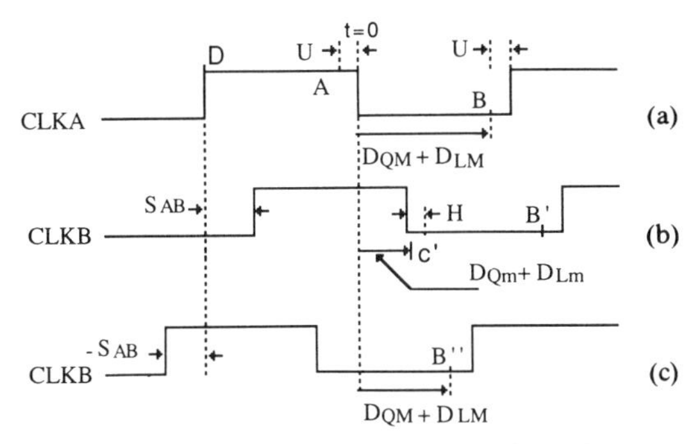

Fig. 19. Timing diagram for the state machine of Fig. 18.

the latch and the combinational circuit and reach the input of the $P1$ latch before the setup time of this latch (point B in the same figure). For the sake of clarity, we assume that the maximum and minimum delays and setup and hold times of the $N1$ latches are equal to that of the $P1$ latches. Therefore, for the above condition to hold,

$$W' \geqslant D_{QM} + D_{LM} + U \tag{6}$$

where W' is the low part of the clock signal. If another combinational circuit were included in the feedback loop between the $P1$ and the $N1$ latch, a similar relation would apply for the width of the clock signal. In the case of Fig. 18:

$$W \geqslant D_{QM} + U \tag{7}$$

where D_{QM} is the maximum delay of the $P1$ latch. Now consider a situation when one of the outputs of the $P1$ latches is sent to a remote latch $N2$ instead of to $N1$. Relative position of the $CLKA$ and $CLKB$ may then be as in Fig. 19. In this case:

$$W' \geqslant D_{QM} + D_{LM} + U - S_{ABm}. \tag{8}$$

Now we consider the effects of minimum delays of the combinational logic circuit and latches. The minimum delays of the latch and the combinational circuit must be such that the output of the $N1$ latch does not arrive to the $N2$-latch input before the end of the hold time of this latch for the present clock period (point c' in Fig. 19(b)). Therefore, to avoid violation of the hold time of the $N2$ latch, we must have

$$D_{Lm} + D_{Qm} \geqslant H + S_{ABM} \tag{9}$$

where D_{Qm} is the minimum delay of the latch. In other words, one may say that the maximum allowable clock skew is

$$S_{ABM} \leqslant D_{Lm} + D_{Qm} - H. \tag{10}$$

In the case when the skew is negative, Fig. 19(c), the W' will increase and is obtained from relation (8) with the sign of the skew changed. Again, in a general system, (10) and (8) and (7) must be used for the maximum allowable skew and minimum clock period, respectively.

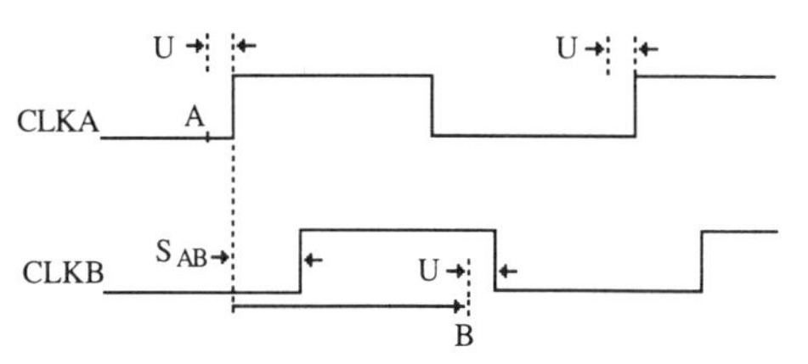

Fig. 20. Timing diagram for precharged circuits.

Traditionally, in single-phase clocking using latches, only one latch is used in the feedback loop of the state machine circuit [6]. This is equivalent to taking out latch $P1$ from Fig. 18. In this case the minimum delay of the combinational circuit and latches must be so long that a data arriving at the $N1$-latch input at point D in Fig. 19(a) cannot race through the circuit and reach the $N2$-latch input before point c'; that is:

$$D_{Lm} + D_{Qm} \geqslant W + S_{ABM}. \tag{11}$$

Fulfilling this condition makes the design very expensive and difficult. This constraint on the minimum delay of the combinational circuit and latches is a major drawback of traditional single-phase clocking with latches. This forced the designers to appeal to two-phase clocking (four phases in CMOS) in order to create two reference edges within the same clock period. Then two latches of the same kind but clocked by different phases are used in the feedback loop of the state machine.

In the single-phase scheme considered here, however, instead of using two phases we use two different kinds of latches: one that is transparent during the high part of the clock signal and the other that is transparent during the low part of the clock signal. In fact, the $P1$ latch in Fig. 18 is a barrier that stops the propagation of the data up to point $t = 0$ in Fig. 19(a). In this respect this single-phase scheme is equivalent to a two-phase scheme with no overlap or gap between the phases. Here it is only necessary to guarantee that a data released at point $t = 0$ by the $P1$ latch will not reach the $N2$-latch input before the hold time of this latch (see relation (9)). In Section VI we propose a technique that relaxes this constraint even further.

C. Precharged Circuit Timing

Here again we use the same argument as above to determine the minimum clock width and period. Assume that an N-block clocked by $CLKA$ is communicating with a P-block clocked by $CLKB$ (see Fig. 20). The latest arrival time of a data to the input of the N-block is denoted by point A in this figure. This data must have enough time to be evaluated by the N-block and reach the input of the p-block before the setup time of this block (point B in the figure). Therefore

$$W \geqslant D_{LM} + U - S_{ABm}. \tag{12}$$

W is also constrained by the precharge time T_p, so that $W \geqslant T_p$. Following the same discussion that led to (5), the

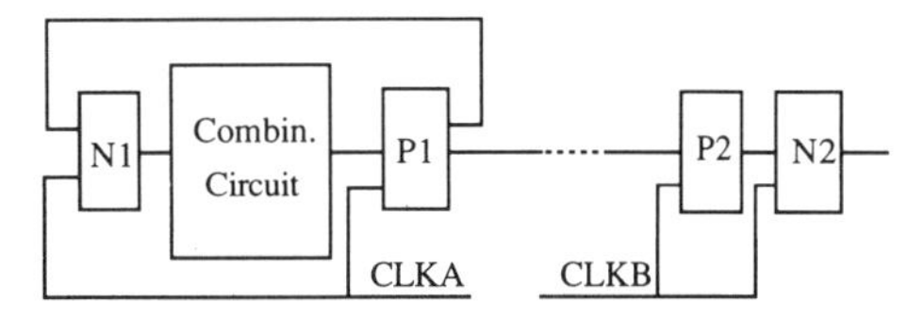

Fig. 21. Circuit technique to cancel the effects of clock skew. An extra latch is inserted on the site of the destination latch.

maximum allowable skew is

$$S_{ABM} \leqslant D_{Lm} - H. \tag{13}$$

Again, for this kind of circuit, $H \approx 0$. If there is a skew, (12) shows that a negative skew will increase the clock period. On the other hand, according to (13) the maximum positive skew is limited to the minimum delay of the block. This class of circuits is thus sensitive to positive clock skew unless some special technique is used (see [4] and [12]).

VI. Signal Resynchronization

It is usually expensive and difficult to satisfy the constraints between the clock skew and the minimum delay of the combinational logic and latches (relation (10)). To reduce the clock skew, great efforts must be exerted, for example, to hand-tune the delays in clock distribution paths, and to control other related factors. On the other hand, to increase the minimum delay of the combinational logic circuit, it is usually necessary to add extra delays (e.g., inverters) at the latch outputs as well as to increase the clock period. In the following, a method is proposed that when used can handle up to half a clock period skew with no constraint on the minimum combinational logic delay.

If we compare Fig. 19(a) and (b), it is evident that the problem is raised because when the data are transmitted by the $P1$ latch at $t = 0$, $CLKB$ is still at high state. But, the data should reach the $N2$ latch first after the $CLKB$ has gone to low state to avoid setup violation of the $N2$ latch. One, for example, can use a negative edge-triggered flip-flop instead of an $N2$ latch that receives data from the remote block. In this way, however, the setup time of the edge-triggered flip-flop may be violated. To solve this problem one can insert an extra P-latch before the $N2$ latch (see Fig. 21). By these methods a clock skew as long as half a clock period can be handled safely without putting any constraint on the minimum delay of the circuit elements (see also [14]). This same technique can also be used in association with edge-tgriggered flip-flops to relax relation (5).

When the clock skew is negative, the above racethrough problem is not present but the clock period will be longer, in accordance with relation (10).

VII. Conclusions

In this paper a unified single-phase clocking scheme was proposed. It was shown that this scheme is general, safe, and simple. The generality makes it possible to use the scheme consistently throughout a digital VLSI system. It provides a structure that can contain all components of a digital VLSI system including static, dynamic, and precharge logic as well as memories and PLA's. This also makes the CMOS circuits and systems well suited for design compilation. Timing expressions for circuits with different storage elements were developed. It was shown that a negative clock skew will just increase the clock period (this is also equivalent to signal delay) and that positive skew may give rise to failure. The problems associated with traditional one-phase clocking using latches are avoided. Techniques were proposed that relax the constraints between the clock skew and the minimum delays of the combinational logic circuits and storage elements. Using the scheme and techniques proposed in this paper should provide a simple and safe clocking for high-speed and compact VLSI digital systems. It also makes the visualization of the system simple and resembles the traditional digital systems where only edge-triggered devices were used.

References

[1] C. Mead and L. Conway, *Introduction to VLSI Systems.* Reading, MA: Addison-Wesley, 1980.
[2] N. Weste and K. Eshraghian, *Principles of CMOS VLSI Design.* Reading, MA: Addison-Wesley, 1985.
[3] Y. Ji-ren, I. Karlsson, and C. Svensson, "A true single phase clock dynamic CMOS circuit technique," *IEEE J. Solid-State Circuits*, vol. SC-22, pp. 899–901, 1987.
[4] J. Yuan and C. Svensson, "High-speed CMOS circuit technique," *IEEE J. Solid-State Circuits*, vol. 24, pp. 62–71, 1989.
[5] D. L. Dietmeyer, *Logic Design of Digital Systems.* Boston: Allyn and Bacon, 1971.
[6] S. H. Unger and C. Tan, "Clocking schemes for high-speed digital systems," *IEEE Trans. Comput.*, vol. C-35, pp. 880–895, 1986.
[7] M. Penney and L. Lau, *MOS Integrated Circuits.* Princeton, NJ: Van Nostrand, 1972.
[8] K. Chen and C. Svensson, private communication, 1989.
[9] M. Afghahi and C. Svensson, "A scalable VLSI synchronous system," in *Proc. IEEE Symp. Circuits Syst.* (Espoo, Finland), 1988, pp. 471–474.
[10] G. G. Langdon, *Computer Design.* San Jose, CA: Computeach, 1982.
[11] M. Afghahi and C. Svensson, "Calculation of clock path delay and skew in VLSI systems," in *Proc. European Conf. Circuit Theory Design* (Brighton, England), 1989, pp. 265–269.
[12] I. Karlsson, "True single phase clock dynamic CMOS circuit technique," in *Proc. IEEE Symp. Circuits Syst.* (Espoo, Finland), 1988, pp. 475–478.
[13] M. Hatamian and C. L. Cash, "Parallel bit-level VLSI design for high-speed signal processing," *Proc. IEEE*, vol. 75, pp. 1192–1202, 1987.
[14] C. Svensson, "Signal resynchronisation in VLSI systems," *Integration, The VLSI J.*, vol. 4, pp. 75–80, 1986.
[15] R. H. Krambeck, C. M. Lee, and H. S. Law, "High-speed compact circuits with CMOS," *IEEE J. Solid-State Circuits*, vol. SC-17, pp. 614–619, 1982.

Skew-free clock distribution for standard-cell VLSI designs

G.M. Blair

Indexing terms: Very large scale integration, Clock distribution, Optimal buffering

Abstract: The control of skew in globally distributed signals is essential to the safe design of VLSI. The paper describes a novel technique using Cadence EDGE software, for the creation of a skew-free distribution network in standard-cell designs which is achieved after initial placement by modification of the netlist and insertion of buffers whose drive strength is finally determined after the component is completely routed.

1 Introduction

The increase in complexity of VLSI components has led to an increase in the problems of clock-signal distribution, where the larger number of clocked subcells on a single component has resulted in a high relative fanout for the clock circuitry. If no corrective design is undertaken, this leads to long switching delays, which reduces the overall circuit speed, and poor clock edges, which cause some types of latch to fail.

The first step is to buffer the clock signal in a chain of buffers of increasing size to attain the optimum speed for driving the given capacitance [1]. However, if this final signal is distributed throughout the component on a single track, it can lead to high currents requiring wide tracks to prevent breakages due to electromigration; it can also lead to clock skew introduced in the track due to transmission line effects over the long resulting lengths (of the order of several centimetres).

A solution is to distribute the buffering at each level into several smaller buffers that drive independent nodes [2]. Thus, the current flowing from the final stage of buffering is distributed over the distinct nodes and at a level that can be supported by the technology. Similarly, the final track capacitance is distributed between the final buffers. However, the independence of these buffers may lead to different delays along the distinct distribution paths from the common, original clock signal. This may cause clock skew, with the danger of falsely clocking data transferred between latches that are adjacent but driven by different, final clock-buffers.

There are two main methods to overcome this problem. The first method is to increase the system's tolerance to possible skew by introducing multiphase clocking schemes [3]. These remove the problem by ensuring that adjacent latches are not changing at the same clock phase, and so there is no danger of falsely latching a changing value. However, a single-phased clocking scheme offers the advantages of a simpler design style and reduced routing overheads in comparison with multiphase approaches [4].

The second method is to use computer aided design (CAD) tools to minimise the clock skew by careful placement and routing (P&R) of the buffer hierarchy [5, 6], and to distribute buffers on a row-by-row basis of the standard cell placement, with buffer sizes matched according to the number of clock inputs in each row [7].

The emphasis so far has been on the use of either the design structure or on careful control of the P&R algorithms to minimise skew. The approach of this paper, however, is to focus instead upon the layout itself, as produced by existing P&R software, and to render the skew negligible by augmenting the netlist with variable sized buffers.

2 Design sequence

The automatic buffering scheme is aimed at a standard-cell design environment and provides each input for a given net with a skew-free signal by building a fully balanced distribution network. This is achieved by modification of the cell- and net-lists after placement, and modification of buffer drive strengths after routing.

From the designer's point of view, this scheme allows the completion of the design using global nets, without any considerations for the buffering or distribution of these signals, and the verification of the design by simulation (without interconnect capacitances) so that the clock distribution simply becomes part of the P&R operation. The user performs the initial placement and calls the buffering software which prompts for the name of each of the global nets. Distribution networks are then created, and the user proceeds with the normal operations to the completion of routing. Finally, the second phase of the buffering scheme is called which modifies the previously created buffers, according to the actual interconnect capacitances, without altering the existing placement or routing.

2.1 CAD prerequisites

The procedure is implemented in the Cadence EDGE software which was made available under the Eurochip initiative. The procedure relies upon several features of that suite:

(*a*) the use of the high-level database language, SKILL [8]; the Cadence compromise between LISP and C

(*b*) the ability of SKILL to interrogate and modify both the cell-list and netlist of the design during P&R

(*c*) the parameterisation of the layout of cells allowing the design of a family of buffers whose drive strength may be fine-tuned by a parameter set *after* final placement and routing.

2.2 Outline of procedure

The following describes the procedure in terms of the distribution of a single (clock) signal, although it should be repeated in a design for all widely distributed clock and control signals.

The user enters the design using schematic capture with the clock signal designated as a single, global net. Care must be taken during the design verification by simulation to ensure that the high capacitance loading on the global net does not alter the simulation results. Typically, this could be done by driving the net explicitly as an external input until the final design step, when the (unique) driver is connected to the net.

The design then enters P&R, and the placement proceeds as normal except that the priority of the global net is reduced so that it does not perturb the placement algorithm. After placement, the first part of the buffering scheme is run:

(i) The number of cells driven by the global net is counted and an appropriate buffering hierarchy is determined.

(ii) The cells driven by the net are clustered according to their position in the layout into equally sized groups.

(iii) To each group a unique buffer is assigned, and the netlist is updated so that these cells are driven by the output of that buffer.

(iv) Each new buffer is added to the placement database so that it is physically close to the cluster it is driving.

(v) Steps (ii)–(iv) are then repeated for each level of created buffers until a complete distribution tree is built.

(vi) Finally, the netlist is modified so that the last level of buffers is driven by the original global signal.

The user now continues with the standard routing program, either routing the design as normal or giving preference to the nets on the distribution network. After the design is fully routed, the second part of the buffering scheme is run:

(vii) The interconnect capacitances for the generated subnets are extracted, and a common ratio of output capacitance to input capacitance (that is, drive strength) is derived for each independence level of the distribution tree.

(viii) A parameter is then added to the instance of each buffer network which modifies the transistor sizes in accordance with the actual capacitance it drives without altering the placement of routing, so that the delays are equalised across each level of the network.

3 Implementation

The general objective is to create a buffer tree, as shown in Fig. 1, where the overall size of the buffers increases at each layer (to minimise total delay), and the sizes of individual buffers are adjusted according to the final output capacitance (to equalise the delay across each level). This section will describe the details of the procedure that rely most heavily upon the EDGE software.

3.1 Clustering

The standard cells are initially placed by the standard EDGE routines, and the automatic buffer scheme is then called.

First, a minimum spanning tree is generated for the nodes driven by the global net. This is used as the basic framework for clustering these nodes into groups to be driven by independent buffers. The generation of the minimum spanning tree using the metric of rectilinear distance corresponding to the Manhattan routing style. The algorithm of the prototype system is currently quite crude as even large bit-serial DSP chips have no more than about 3K cells driven by the same clock [9], and the procedure should not need to be repeated many times in any design.

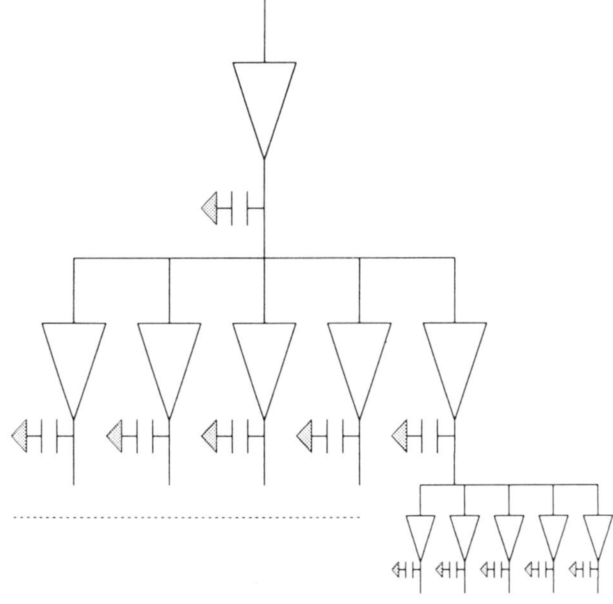

Fig. 1 *Hierarchical signal distribution tree*

The number of inputs that the designated signal must drive is used to derive the size of the groups into which the nodes are clustered and thence the expected size of the buffer at this level. The clusters are formed by coalescing the nodes in the reverse order to which they were added to the spanning tree. When a node is as large as, or larger than, the cluster size, that number of nodes are selected to form a single cluster and the rest are added back to the remaining tree.

For each cluster, a new net is created and connected to the node of each cell in the cluster. A new instance of a buffer is created and its output is connected to the new net, whereas its input is conected to the net that originally drove the cluster. Finally, the instance is added to the standard-cell array in the locality of the cluster it is driving. This whole procedure is then repeated for each level of the distribution tree until the final level alone is driven by the original signal.

The partitioning of the network by minimum spanning tree contrasts with other authors who divide the layout array into simple blocks. The difference is that the previous systems rely upon symmetric paths to each clusters to minimise the variation in interconnect capacitance; the procedure reported below compensates for any variation by modification of buffer size, and so facilitates a more general (and hierarchical) solution.

3.2 Buffer design

The P&R routines in EDGE work upon the *abstract* view of the standard cells. This is derived from the layout and is effectively the bounding box and the positions of the input and output I/O pins. After a component is completely placed and routed, the abstracts are replaced by their corresponding layout.

The buffers are actually taken from a whole family which must be designed specifically for each standard-cell library. There are two degrees of variability: the size of the abstract for P&R; and the actual size of the transistors after the layout replaces the abstract. Clearly the size of the abstract limits the size of the final transistors.

An example of a buffer abstract is shown in Fig. 2 showing the bounding box and pin position information

Fig. 2 *Example of buffer abstract*

(I/O and pwr/gnd). The height of the abstract must be the same as the common height of all the standard-cell library; the family of abstracts corresponds to the various widths which provide a range of maximum-drive options.

Once the abstract size has been selected in the procedure and it has passed through P&R, neither the size nor the positions of the I/O pins may be altered, as this would break the connections established in the component layout. However, the other features within the abstract may be freely altered. Specifically, the transistor widths of the buffer may be altered to provide any desired drive strength, within the limits imposed by the abstract, by the setting of a parameter associated with the abstract instance in the P&R environment.

Fig. 3 shows the simplified layout for three different

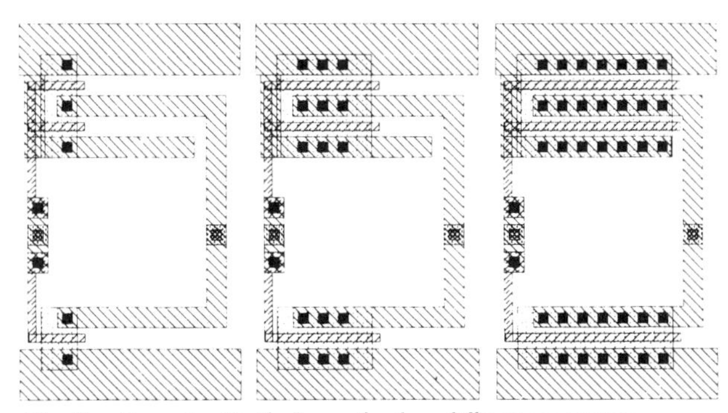

Fig. 3 *Example of buffer layout for three different parameters*

drive strengths corresponding to the same abstract and parameterised layout. The Cadence software allows both the stretching of features (as in the diffusion and polysilicon) and the conditional repetition (as with the contacts). In this example, the *p*-type transistors are simply twice the width of the *n*-type but other relations can be easily encoded. The use of horizontal polysilicon gates allows the transistor widths to be varied with as small a granularity as the layout grid.

The abstract size must be selected when it is instanced before being inserted into the array and routed. The output capacitance for a buffer is estimated from the number of nodes it must drive (the cluster size), the likely value of the interconnect capacitance, and an additional margin for error; this is then used with the chosen ratio of output-to-input capacitance to estimate the buffer transistor sizes, and the next smallest abstract to accommodate these sizes is selected. Thus, the aim is to provide the necessary drive strength while minimising area. With these buffers inserted into the array and netlist, the routing proceeds to completion.

For each buffer, the *actual* output capacitance is extracted from the routed design. For each level in the signal distribution hierarchy there is an ideal C_{out}/C_{in} ratio (R) calculated in terms of a speed/area trade-off.

At each level, the feasibility of achieving the ideal ratio is checked by the relationship:

$$\frac{C_{out(max)}}{C_{in(max)}} \leqslant R \leqslant \frac{C_{out(min)}}{C_{in(min)}}$$

The output capacitances are now fixed, and the maximum and minimum are taken over all buffers on the given level; the input capacitances are variable within the bounds of the fixed abstract. If this relationship holds then R is used, otherwise the value of the condition it violates is used as the common ratio.

The final stage is to adjust the actual transistor size in each buffer by selecting each abstract instance in turn and setting the *size* parameter to alter the transistor sizes of the actual layer, so that $C_{out(actual)}/C_{in}$ equals the common ratio for each corresponding level.

4 Concluding remarks

The elegance of this approach stems from the fact that the buffer sizes may be altered until skew is removed, without affecting the remaining layout. Thus, if simulation reveals skew on the first pass, the sizes may be altered accordingly until the skew is negligible. The only limit on this is the 'rounding error' associated with the layout's grid-size, which might typically be in the region of 1–2% of the size and hence the delay at each stage.

This buffering scheme provides a skew-free distribution network for general standard-cell VLSI designs and has been embedded within an industry standard CAD tools suite. The use of post-routing modification of the layout allows the creation of a truly balanced signal distribution network which takes full account of the variations in interconnect capacitance on the independent signal paths. With this system, simple single-phase design strategies may be adopted without the inherent problems of clock distribution.

5 References

1 HEDENSTIERNA, N., and JEPPSON, K.O.: 'CMOS circuit speed and buffer optimization', *IEEE Trans.*, March 1987, **CAD-6**, pp. 270–281

2 FRIEDMAN, E.G., and POWELL, S.: 'Design and analysis of a hierarchical clock distribution system for synchronous standard cell/macro VLSI', *IEEE J. Solid-State Circuits*, April 1986, **21**, pp. 240–246

3 WESTE, N., and ESHRAGHIAN, K.: 'Principles of CMOS VLSI design — a systems perspective' (Addison-Wesley, 1985)

4 AFGHAHI, M., and SVENSSON, C.: 'A unified single-phase clocking scheme for VLSI systems', *IEEE J. Solid-State Circuits*, Feb. 1990, **25**, (1), pp. 225–233

5 BOON, S., BUTLER, S., BYRNE, R., SETERING, B., CASALANDA, M., and SCHERF, A.: 'High performance clock distribution for CMOS asics'. IEEE Custom Integrated Circuits Conf., 1989, pp. 15.4.1-5

6 JACKSON, M.A.B., SRINIVASAN, A., and KUH, E.S.: 'Clock routing for high performance ICs'. Proc. 27th DAC, 1990, pp. 573–579

7 SAIGO, T., WATANABE, S., ICHIKAWA, Y., TAKAYAMA, S., UMETSU, T., MIMA, K., Yamamoto, T., SANTOS, J., and BUURMA, J.: 'Clock skew reduction approach for standard cell'. IEEE Custom Integrated Circuits Conf., 1990, p. 16.4.1-4

8 BARNES, T.J.: 'SKILL: a CAD system extension language'. Proc. 27th DAC, 1990, pp. 267–271

9 BLAIR, G.M.: 'Bit-serial correlator with novel clocking scheme'. Proc. ESSCIRC, 1991, pp. 157–160

Scheme for Distributing High-Speed Clock Signals in a Large Digital System

S. Padin

Indexing terms: Instrumentation, Clocks, Digital systems

The details of a clock distribution circuit for a large digital system operating at 250 MHz are presented. The arrangement provides pulse compression control and preserves the clock timing at the 100 ps level even when parts of the circuit are replaced.

Introduction: In a high-speed, synchronous digital system operating close to its maximum clock rate, pulse compression in the clock signal can be a serious problem. The effect is similar to an increase in the clock rate, and this forces the system closer to its operating limit. Pulse compression is the result of unequal rise and fall times or logic threshold errors in the clock distribution circuits, and its effects are usually avoided by using circuits which have transition times that are small compared with the clock period. To achieve this the clock circuits may have to be much faster than the rest of the system. An alternative approach is to use a feedback technique to maintain a clock mark : space ratio of unity.

Clock distribution in Caltech's digital correlator: The pulse compression problems described above were encountered during the development of a high-speed digital correlator for the Owens Valley millimetre-wave interferometer. This correlator is a 2-bit system operating at 250 MHz clock rate and involves the distribution of approximately 200 differential clock signals. The cost of distributing these clock signals using high-speed GaAs drivers was prohibitive, and so slower 100 K ECL 100113 drivers were used. These devices provide eight differential outputs, with about 1 ns transition times, and require a single-ended input. The single-ended results in a severe pulse compression effect due to threshold errors. We observed that for a cascade of clock distribution circuits using 100113 devices all from the same batch, the pulse compression effect was about 50 ps per 100113 device. However, for a distribution circuit with devices from different batches the effect was as much as 250 ps per device. The pulse width compression τ_c for a clock driver with input threshold error e_{th}, logic swing v_s and input transition time (10–90%) τ_t is

$$\tau_c \simeq \frac{1 \cdot 6 \tau_t e_{th}}{v_s} \quad (1)$$

For standard 100 K ECL circuits, $v_s = 750$ mV and $\tau_t \sim 1$ ns; hence the observed 250 ps pulse compression corresponds to a threshold error of ~ 150 mV.

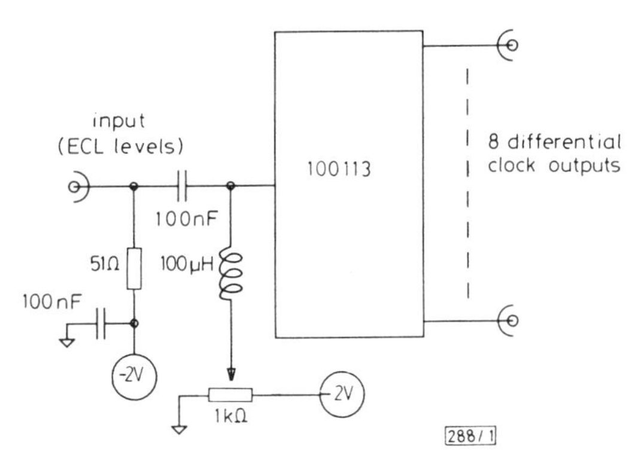

Fig. 1 *AC-coupled clock distribution circuit*

To avoid the pulse compression problem we adopted the AC-coupled circuit of Fig. 1, in which the threshold error is nulled by an external bias. Each 100113 device in the clock distribution tree[1] has its own threshold error correction circuit, so that pulse compression for the entire clock distribution system can be removed for a given operating temperature. The temperature coefficient of the threshold voltage in 100 K ECL is approximately[2] 0·1 mV/K, which translates to a pulse compression effect of about 200 fs/K for devices with 1 ns transition times.

There is a difficulty associated with the AC-coupled arrangement (and also with systems which use feedback to eliminate pulse compression), in that changes in the shape of the input waveform are converted to changes in the propagation delay of the clock distribution circuit. This affects the interchangeability of circuits, which is important for field repair of the instrument. The size of the effect was investigated by comparing the outputs of driver circuits using 100113 devices from different batches. These devices had shown inter-batch threshold differences of up to 150 mV, but the output waveform shapes were very similar, showing timing differences of no more than 100 ps. This permits the construction of a clock distribution tree in which a circuit change affects the clock timing by less than 100 ps. To achieve this, each clock distribution unit in the tree is adjusted to give zero pulse compression, and the delays of all units are made equal by inserting cable at the inputs. The effect of changing a clock distribution circuit in a cascade of two of the circuits of Fig. 1 is illustrated in Fig. 2. In this case, changing the 100113 driver in the second stage to a device from a different batch changed the timing of a clock transition at the cascade output by only

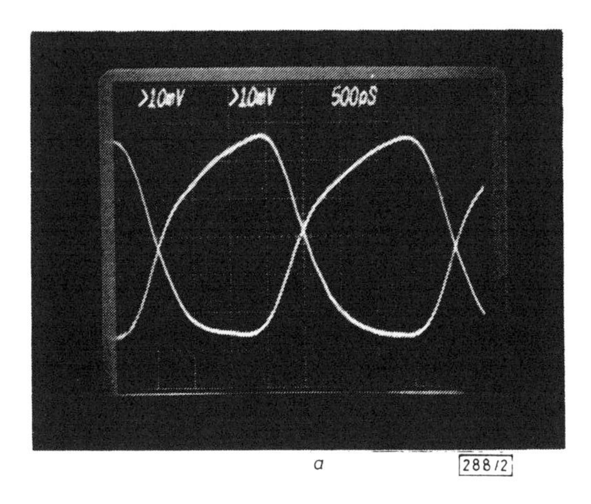

a 288/2

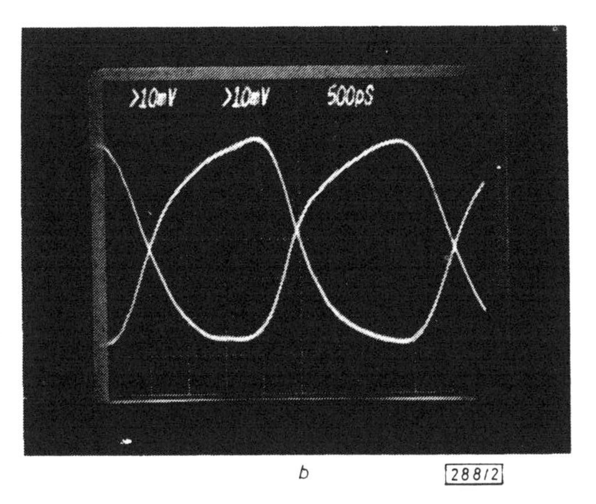

b 288/2

Fig. 2
a Complementary outputs from cascade of two of circuits shown in Fig. 1. Signals have standard 100 K ECL logic levels
b Effect of replacing second stage in cascade with circuit using 100113 device from different batch. For these measurements, oscilloscope was triggered directly from clock source so that displacement of waveform indicates change in propagation delay

about 50 ps. For situations in which the waveform shapes differ between clock distribution circuits, the propagation delay effect could be reduced by adding a high-gain differential receiver ahead of the circuit of Fig. 1. In the system described, one of the largest contributions to clock timing errors results from changes in the number of loads driven by a given 100113 driver. This effect can be as large as 100 ps.

Conclusions: We have described a simple, AC-coupled clock distribution scheme for a large, 250 MHz clock rate digital system. The scheme uses standard 100 K ECL drivers, provides pulse compression control and allows clock distribution circuits to be interchanged without affecting the system clock timing by more than 100 ps.

S. PADIN
Owens Valley Radio Observatory
California Institute of Technology
Big Pine, CA 93513, USA

17th August 1988

References

1 BLOOD, W. R.: 'MECL system design handbook' (Motorola Inc., 1983)
2 F100 K ECL user's handbook, Fairchild Camera & Instrument Corp., 1985

Clock Tree Synthesis Based on RC Delay Balancing

Fumihiro Minami and Midori Takano

ULSI Research Center, TOSHIBA Corporation

Abstract

This paper presents a novel clock tree synthesis method based on top down binary tree construction, optimal buffer insertion, and bottom up wiring with precise RC delay balancing. The proposed method has achieved near-zero clock skews with minimal clock delays while achieving similar total wire lengths in comparison with previous heuristics.

1 Introduction

The circuit speed in a synchronous VLSI design is limited by the clock skew which is defined as the maximum difference in the delays from the clock source to the synchronous components. The system speed is also limited by the inter-chip clock skew which is the clock delay difference among chips. In general, most systems require a clock skew of less than 10% of the system clock period. Thus, clock skew minimization and clock delay reduction are needed for high performance chips.

Several approaches for clock routing using a tree structure have been reported as follows: the H-tree method [1], the Method of Means and Medians (MMM) [2], and the Path Length Balancing method (PLB) [3]. However, the H-tree structure is not applicable for asymmetric distributions of synchronous components, and MMM and PLB are not aimed at path delay balancing which is the primary objective. Therefore, these techniques are insufficient in minimizing the clock skew.

This paper presents a novel approach to clock routing that minimizes the clock skew by constructing a binary tree with RC delay balancing. The proposed Path Delay Balancing method (PDB) always yields near-zero clock skews. The key point of PDB is that the branching point of a binary tree is so determined as to equilibrate the RC delay for each subtree calculated by Elmore's formula [4]. The proposed method uses a hierarchical tree structure divided by buffer cells to prevent the delay increase caused by the quadratic term of the wire length for a circuit with too many synchronous components. The synchronous components are so clustered in the lowest tree level as to balance the load capacitance within each cluster where the wire resistance is negligible, and thus the clock skew among clusters is minimized. As a result, the clock skew and clock delay are perfectly minimized with the above mentioned techniques.

This paper is organized as follows: in Section 2, the keys to minimize the skew and delay are presented. Section 3 presents the definition of the problem. The algorithm for clock tree synthesis is presented in Section 4, and practical considerations are presented in Section 5. The experimental results are presented in Section 6, and the paper is concluded in Section 7.

2 Skew and Delay Minimization

In this section, a delay calculation model is presented and then the keys to minimize the skew and delay are presented.

2.1 Delay calculation model

For a lumped RC tree network, Elmore's formula [4][6] is commonly used for delay calculation and the authors have also adopted it in this paper. According to this formula, the delay from a root to a leaf in a RC tree is defined as the sum of the products of both the resistances R_i and its downstream capacitances C_i along the path, which is defined as follows:

$$Delay = w \Sigma R_i C_i$$

where w is a constant coefficient.

In a binary tree for clock routing, Elmore's delay is calculated in the following bottom up manner.

Let the highest branch node of the tree be labeled 1 and two child nodes of node k be labeled $2k$ and $2k+1$, and the driver output pin be labeled 0, as shown in Figure 1. Let $C(k)$ be the downstream capacitance from node k and $L(k)$ be the wire length between node k and its parent node. Let r and c be the unit length resistance and capacitance, and let R_{on} and d_I be the on-resistance and the intrinsic delay of a driver cell. Then, the delay from a driver input pin to leaf node e is expressed as follows.

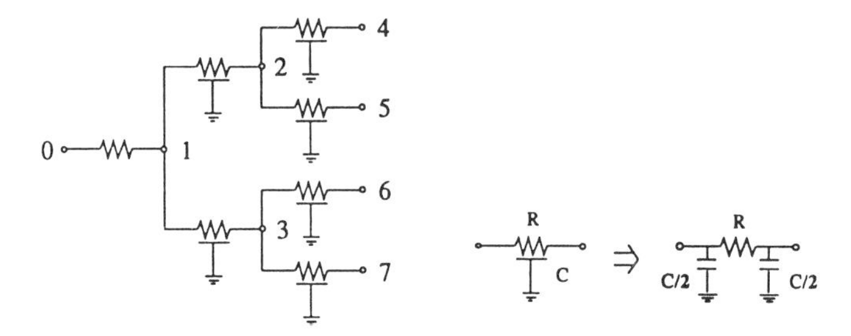

Figure 1: Binary clock tree

Figure 2: π-model subcircuit

$$Delay(e) = d_I + \Sigma d(k) \quad (1)$$
$$d(k) = w\, r\, L(k) \cdot (c\, L(k) / 2 + C(k)) \quad \text{if } k > 1$$
$$d(1) = w\, R_{on}\, C(1)$$
$$C(k) = C(2k) + C(2k+1) + c\, (L(2k) + L(2k+1))$$
$$C(e) = \text{input pin capacitance of leaf } e$$

In this expression, each wire segment which is expressed by a distributed RC line is assumed to be approximately equivalent to a π-model subcircuit as shown in Figure 2.

2.2 Skew minimization

The skew between two different paths is equal to the difference of the sum of RC products from the last common branch to each leaf because the RC products along a common path is the same. According to this, the local skew $S(k)$ at each node k can be defined as the maximum delay difference for all paths from node k to its descendant leaves. Let $T_L(k)$ and $T_S(k)$ be the maximum and minimum delays of all paths from node k to its descendant leaves, then $S(k)$ is expressed as follows:

$$\begin{aligned} S(k) &= T_L(k) - T_S(k) \\ &= \max (S(2k),\ S(2k+1), \\ &\qquad T_L(2k) + d(2k) - T_S(2k+1) - d(2k+1), \\ &\qquad T_L(2k+1) + d(2k+1) - T_S(2k) - d(2k)) \\ &= \max (S(2k),\ S(2k+1),\ \mu(k) + | \lambda(k) |) \end{aligned} \quad (2)$$

Reprinted from *Proc. IEEE Custom Integrated Circuits Conf.*, pp. 28.3.1–28.3.4, May 1992.

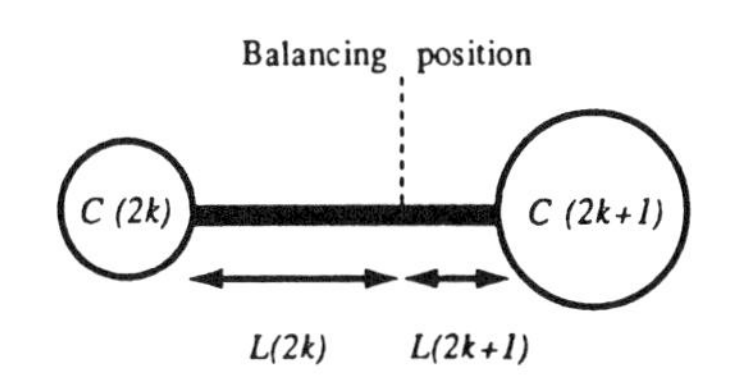

Figure 3: RC delay balancing position

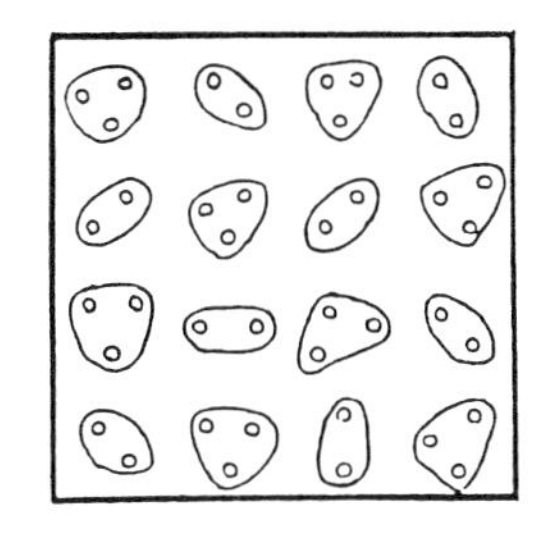

Figure 4: Clustering

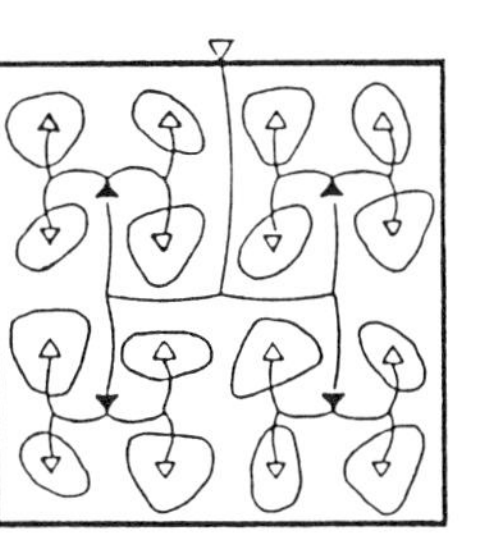

Figure 5: Tree construction

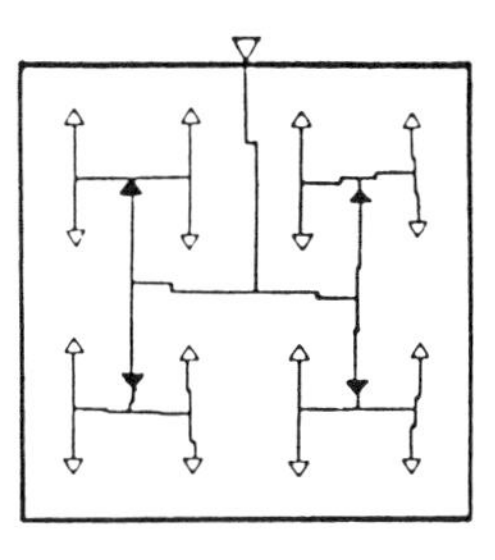

Figure 6: Balanced routing

where

$$\mu(k) = (S(2k) + S(2k+1)) / 2$$
$$\lambda(k) = d(2k) + \tau(2k) - d(2k+1) - \tau(2k+1)$$
$$\tau(j) = (T_L(j) + T_S(j)) / 2 .$$

Then, the clock skew in a tree is defined as $S(1)$.

$\lambda(k)$ means the skew between the delay median of the left side descendants and that of the right side descendants. Because $\mu(k)$ is less than either $S(2k)$ or $S(2k+1)$, if $\lambda(k)$ is forced to be equal to zero, then $S(k)$ is rewritten as follows:

$$S(k) = \max (S(2k), S(2k+1)).$$

This equation suggests that the local clock skews do not exceed the descendant clock skews. Consequently, if a position of each node k is so determined that $\lambda(k)$ is zero in a bottom up manner, the clock skew will be equal to the maximum skew at the leaves and will normally be zero. This is the key to minimize the clock skew in this paper.

The condition that $\lambda(k)$ is zero is as follows:

$$d(2k) + \tau(2k) = d(2k+1) + \tau(2k+1)$$

or

$$r L(2k) \cdot (c L(2k) / 2 + C(2k)) + \varepsilon(k) = r L(2k+1) \cdot (c L(2k+1) / 2 + C(2k+1)) \quad (3)$$

where

$$\varepsilon(k) = (\tau(2k) - \tau(2k+1)) / w .$$

As shown in this equation, delay balancing is analogous to moment balancing of a mobile where the capacitance corresponds to the weight and the resistance corresponds to the arm length. Figure 3 shows the image.

2.3 Delay minimization

For a circuit with too many synchronous components, the clock delay increases in proportion to the square of the wire length. In such a case, a hierarchical clock tree structure by multistage buffering is commonly used to reduce the clock delay.

For multistage buffering, the delay is influenced by the positions of the inserted buffer cells. Buffer insertion close to the primary driver cell causes a large delay in the buffer stage, and buffer insertion close to the leaf nodes causes a large delay in the primary driver stage, so there exists an optimal position for buffer cells to be inserted.

In the proposed method, buffer cells are so inserted as to minimize the total clock delay by a greedy method.

In addition, the clock pins in the lowest trees are routed by the Steiner-tree method which minimizes the wire length and thus reduces the delay. This is because the wire resistances in the lowest tree are very small in comparison with the on-resistances of the buffers and RC delay balanced routing is unnecessary.

3 Problem definition

Given a circuit placement, the clock pin positions, the number of buffering stages and the performance data for buffer cell types, the clock layout optimization problem can be defined as follows: construct a hierarchical clock tree by multi-stage buffering which minimizes the clock skew, delay, and wire length subject to the clock pin positions. In this paper, it is assumed that two layers are available for clock layout.

4 Algorithm

The proposed method consists of three steps, which are synchronous component clustering, hierarchical tree topology construction, and RC delay balanced routing. In tree topology construction, the optimal combination of the buffer types and levels to be inserted into the tree is obtained by a greedy method. The main process flow is shown as follows.

```
procedure clock_tree_synthesis
Begin
  B : set of sequentially inserted buffer types and
      their insertion depths in the tree
  for b ∈ B , do
    synchronous component clustering
    hierarchical tree topology construction
      /* RC delay balanced routing */
    for each level of hierarchical tree, do
      branch position determination for each subtree
      buffer positions adjustment for delay equalization
      tree level decrement
    end
    saving the routing result if the delay is the smallest
  end
End
```

4.1 Synchronous component clustering

First, synchronous components are clustered by recursive bi-sectioning until the wire resistance within each cluster becomes negligibly small, as shown in Figure 4. At each division, the dividing line is so determined as to balance the load capacitance of two regions which is defined as the sum of the estimated wire capacitance [5] and the input gate capacitance, or

$$LoadCap = c\, h\, f(fanout) + \Sigma C_g \quad (4)$$

where

- c : capacitance per unit wire length
- h : half perimeter of the minimal bounding box
- $f(fanout)$: statistical coefficient function of wire length
- C_g : input pin capacitance for each gate .

After clustering, the clock pins in each cluster are routed by the Steiner-tree method to minimize the wire length.

By this clustering, the inter-cluster skew is minimized because of the uniform load capacitance and negligible wire resistance for each cluster.

4.2 Hierarchical tree topology construction

Secondly, the binary tree topology whose root is a primary clock driver and whose leaves are clusters is constructed by top down bi-sectioning [2], and then buffer cells are inserted to construct the hierarchical tree. In each buffering stage, buffer cells of the same type are topologically inserted at the same level of the binary tree, as shown in Figure 5. The lowest buffer cells are placed at the center of each cluster region, and middle buffer cells are placed at or near the highest branch points of the subtrees, which are determined by the process explained in Section 4.3.

4.3 RC delay balanced routing

Lastly, each subtree of the hierarchical tree is recursively routed in a bottom up manner, as shown in Figure 6. In the subtree which is driven by a middle buffer cell or a primary driver cell, the branch position determination and the buffer position adjustment are so performed as to minimize skew as follows.

1) Branch position determination

In the subtree, the branch position for each tree node is so determined as to minimize its local skew $S(k)$, or to satisfy Equation (3).

Let l be $L(2k) + L(2k+1)$ and x be the ratio of $L(2k)$ to l, then x is solved as follows:

$$x = \frac{C(2k+1) + cl/2 - \varepsilon(k)/rl}{C(2k) + C(2k+1) + cl} \tag{5}$$

To minimize the wire length and clock delay, l is assumed to be the Manhattan length between node $2k$ and $2k+1$. If x is not in the range from 0 to 1, x is set at 0 or 1 so that $S(k)$ is minimized.

After each branch position determination, node k is connected to each child node with a minimum length, and $C(k)$, $T_L(k)$, and $T_S(k)$ are calculated. These are recursively processed by the following depth first procedure, as shown in Figure 7.

```
procedure delay_balance( P )
Begin
  if P is a leaf then
    F.pos = input pin position of leaf
    F.c   = input pin capacitance
    F.TL  = max delay of lower tree
    F.TS  = min delay of lower tree
    return( F )
  else
      PL : left side subtree of current node
      PR : right side subtree of current node
    FL = delay_balance( PL )
    FR = delay_balance( PR )
    decide branch position so as to minimize the λ(k)
    route from the current node to two child nodes
      l : wire length between one child node and another
    F.pos = branch position for current node
    F.c   = FL.c + FR.c + c l
    F.TL = max ( FL.TL + d(2k), FR.TL + d(2k+1) )
    F.TS = min ( FL.TS + d(2k), FR.TS + d(2k+1) )
    return( F )
  endif
End
```

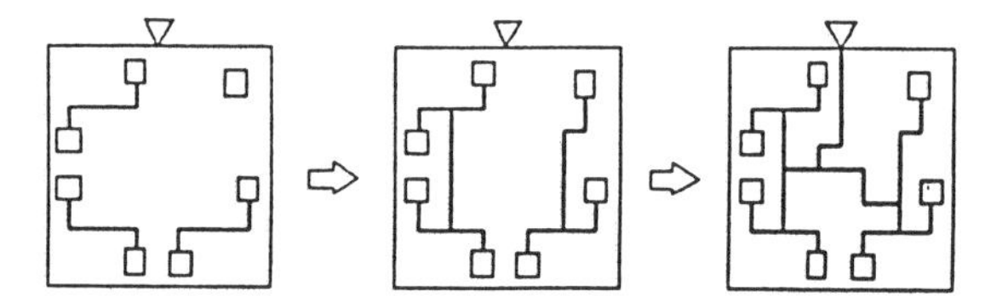

(a) Routing image

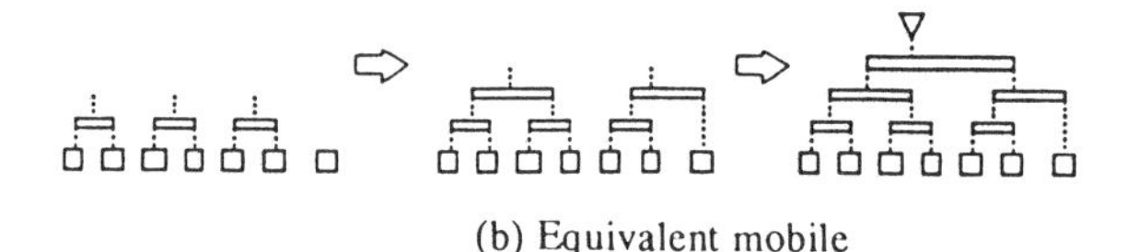

(b) Equivalent mobile

Figure 7: Bottom up routing in a single tree

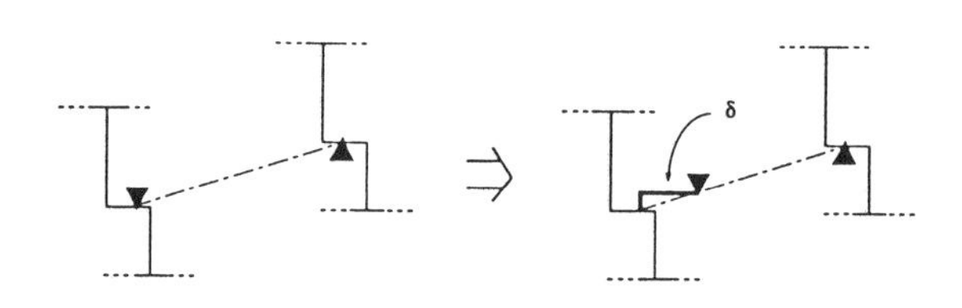

Figure 8: Buffer position adjustment

2) Buffer positions adjustment

After routing at each tree level, the buffer cell positions are so adjusted as to balance the downstream delays by the following technique.

Initially, each buffer cell is placed at the highest branch point of its subtree. Next, the buffer cell is moved to the point with a Manhattan distance δ from the initial position, where the downstream delay is equal to the largest one at the same tree level. A new position is searched along the line which connects the initial position with its brother buffer position of the upper subtree, as shown in Figure 8.

Let Δ be the difference between $\tau(1)$ of the current subtree and the largest one of the subtree at the same level, then δ is required to satisfy the following equation.

$$\Delta = w R_{on} c \delta + w r \delta (c \delta / 2 + C(1)) \tag{6}$$

Such δ is solved as follows:

$$\delta = \frac{\sqrt{b^2 + 2rc\Delta/w} - b}{rc} \tag{7}$$

where

$$b = R_{on} c + r C(1) .$$

In most cases, small δ is sufficient for equalizing the delays at the same level because $C(1)$ is relatively large. And such selection of the moving direction leads to a wire length reduction at an upper level which cancels out the wire length increase at the current level. Consequently, the adjustment for buffer cell positions is so effective that subtree delays of the same level can be balanced without elongating the total wire length.

As mentioned above, RC delay balanced routing for a hierarchical tree results in a near-zero clock skew and minimal clock delay.

5 Practical considerations

In practice, one would like to minimize the skew exactly. In this case, we can modify the RC delay balanced routing method by adding a process for the elongating the wire length in order to balance the delay exactly at each branch and at each cluster. Branch position determination just after routing between two child nodes is also effective for exact delay balancing.

Another practical concern is the many possible combinations for buffer insertion to be checked. To reduce the combinations and the computational time, the authors propose to use the upper driving limit of the tree depth for each driver type.

6 Experimental results

The proposed PDB method was implemented in C on a SUN4 system, and was tested on single and hierarchical tree examples. Table 1 shows the statistics for the test examples and the coefficient *w* was set at 0.7 for delay calculation.

In the single tree case, the routing performance for PDB was compared with that for MMM, and PLB for three industrial data and the MCNC benchmark data *Primary2*, under the condition of the same topology. Table 2 shows the computational results of the clock skew, delay, and wire length. The results showed that PDB yields a zero clock skew with a slight small delay and almost the same wire length.

In the hierarchical tree case, PDB was tested on two examples with one and two stage buffering. Table 3 shows the excellent results which had an extremely small delay compared with the single tree and a sufficiently small skew. The skew was mostly caused by the inter-cluster skew, or the maximum capacitance difference between the clusters which was always less than 10% of the average capacitance.

Figure 9 and 10 show the global and detail routing results of *Data3* by the PDB method.

7 Conclusion

This paper has presented a novel clock tree synthesis method based on top down binary tree construction, optimal buffer insertion, and bottom up wiring with precise RC delay balancing. The proposed method has achieved near-zero clock skews with minimal clock delays for a wide range of clock pin counts and distributions on industrial chips.

Future work will address wire width optimization and give consideration to internal path delay in a large macro with its blockage area during the construction of the clock tree.

References

[1] H. B. Bakoglu, J. T. Walker and J. D. Meindl, "A symmetric clock-distribution tree and optimized high-speed interconnections for reduced clock skew in ULSI and WSI circuits", *Proc. IEEE Int. Conference on Computer Design*, pp. 118-122, 1986.

[2] M. A. B. Jackson, A. Srinivasan, and E. S. Kuh, "Clock Routing for High-Performance ICs", *Proc. 27th Design Automation Conference*, pp. 573-579, 1990.

[3] A. Kahng, J. Cong, and G. Robins, "High-Performance Clock Routing Based on Recursive Geometric Matching", *Proc. 28th Design Automation Conference*, pp. 322-327, 1991.

[4] J. Rubinstein, P. Penfield and M. A. Horowitz, "Signal Delay in RC Tree Networks", *IEEE Trans. on CAD*, CAD-2(3), pp. 202-211, 1983.

[5] S. Boon, S. Butler, R. Byrne, B. Setering, M. Casalanda, and Al Scherf, "High Performance Clock Distribution for CMOS ASICs", *IEEE Custom Integrated Circuits Conference*, pp. 15.4.1-15.4.5, 1989.

[6] H. B. Bakoglu: Circuits, Interconnections, and Packaging for VLSI, Chapter 5, Addison-Wesley Publishing Co., Reading, Mass., 1990.

	dl (nsec)	*Ron* (psec/LU)	*Cg* (LU)
Driver	0.79	3.6	
Buffer1	0.70	31.4	1
Buffer2	0.68	15.7	2
Flipflops			1

(a) Performance data for used cells

	r (psec/LU mm)	*c* (LU/mm)
Horizontal wire	1.22	1.09
Vertical wire	2.14	1.50

(b) Performance data for wires

	Pin counts	Chip width (mm)	Chip height (mm)
Data1	410	6.6	6.6
Primary2	603	12.0	12.0
Data2	1026	14.7	14.7
Data3	1661	14.7	14.7

(c) Test examples

Table 1: Statistics of experiments

		MMM	PLB	PDB
Data1	Skew	0.37	0.33	0.00
	Delay	5.31	5.10	5.07
	Length	178	175	176
Primary2	Skew	0.68	0.39	0.00
	Delay	13.73	13.32	12.95
	Length	432	430	432
Data2	Skew	0.47	0.24	0.00
	Delay	33.33	32.50	32.41
	Length	726	721	723
Data3	Skew	0.71	0.49	0.00
	Delay	48.16	46.96	46.81
	Length	927	918	920

Table 2: Skew (nsec), maximum delay (nsec), and total wire length (mm) comparison in a single tree

		Without buffering	With buffering
Primary2	Skew	0.00	0.07
	Delay	12.95	3.03
	Length	432	232
Data3	Skew	0.00	0.05
	Delay	46.81	3.94
	Length	920	544

Table 3: Skew (nsec), maximum delay (nsec), and total wire length (mm) comparison between without buffering and with buffering

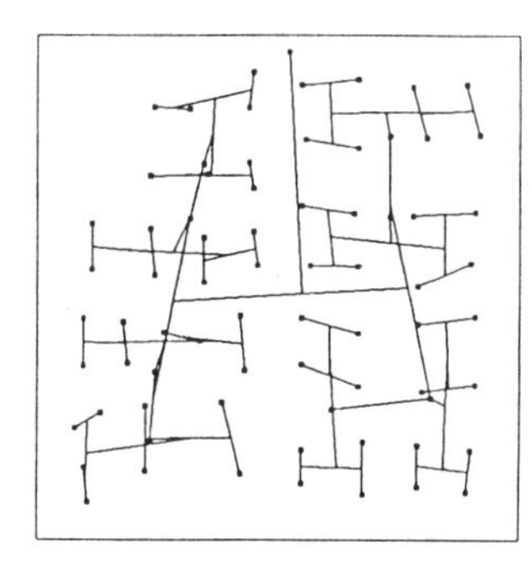

Figure 9: Global routing example

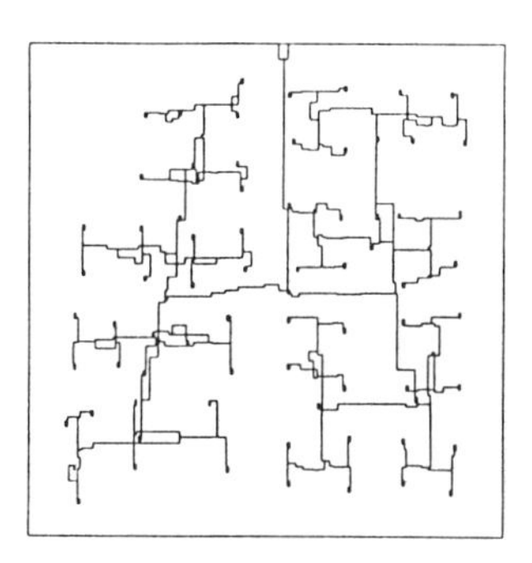

Figure 10: Detail routing example

AUTOMATIC ANALYSIS OF CIRCUITS FOR TESTER SKEW AND CLOCK DISTRIBUTION FOR VLSI CIRCUITS

Inderpal Deol

California Devices Inc., Milpitas, Ca. 95035

ABSTRACT

This paper presents TSAS (Tester Skew Analysis System) for performing automatic skew analysis and clock distribution analysis on a given circuit description. TSAS is used to detect potential tester skew or clock distribution problems before a circuit is manufactured, thereby saving production and debug costs in testing. The circuit description is read into a dynamic data structure, which is then levelized and partitioned to make the processing very efficient. The system produces appropriate messages for any skew or clock distribution errors.

INTRODUCTION

The increase in complexity and the speed of logic circuits, the rising cost of manufacturing and debugging, and the testers having a skew of several nanoseconds has brought forward a need to detect all potential problems before manufacturing and test.

Current skew analysis and clock distribution analysis programs suffer from the following deficiences :

(1) High computer resource requirements, using several hours of cpu time.

(2) Large memory requirements limiting the number of jobs run at one time.

(3) Very large learning curve because of fragmented pieces of programs with poor user interface.

(4) Large amount of data being produced which is difficult to interpret.

TSAS was developed to provide the designer with a fast, efficient tool to detect the effect of tester skew on a logic circuit and also any problems with clock distribution. The system produces easy to interpret messages with which the user can pin point the defects. The system was set up to be a general purpose tool which can handle several technologies and very large networks with ease.

TSAS was implemented on a VAX/VMS using pascal programming language. The implementation details are provided in the following sections.

SYSTEM OVERVIEW

TSAS can be viewed as having three sections, levelization and network partitioning, partition analysis, and skew analysis. Fig 1 shows the system configuration and data flow.

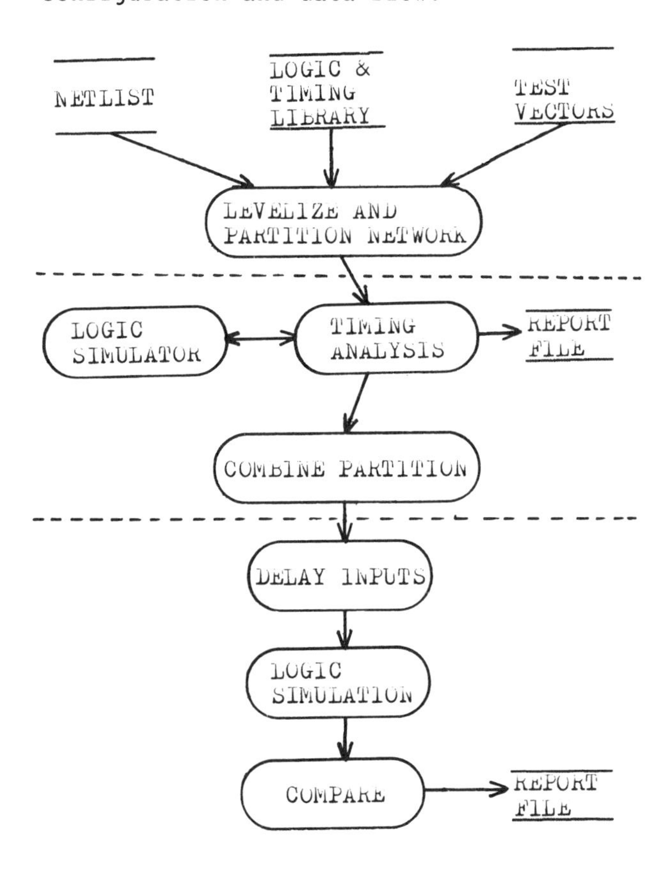

Fig 1. System Configuration

Reprinted from *Proc. IEEE Int'l Conf. Computer Design*, pp. 350–353, Oct. 1987.

LEVELIZATION AND NETWORK PARTITIONING

A network description is read in from a text file and stored in a dynamically allocated root directed tree structure for ease of forward and backward trace of nets and components. For all the primitive components used in the network a logic and timing model is read from the library. A pointer to the test vector file is established, this pointer will be used by the logic simulator to read input stimulii. The following sub sections describe the network partitioning technique :

Terminology Definition

A PPOO (primary point of observability) is defined as a net which is a primary output for a network. A SPOO (scan data point of observability) is the data net of a scannable storage device. Every SPOO has an associated CPOO (clock point of observability) which is the clock net for the storage device. A PPOC (primary point of controllability) is defined as a net which is a primary input for a network. A SPOC (scan point of controllability) is a net on which a value can be scanned in from a primary input for a network.

Partition/Levelization Implementation

The network is traversed and all the PPOO, SPOO, CPOO, PPOC and SPOC nets are marked. The network description is then partitioned using the PPOO, or the SPOO and CPOO pair as the starting point. The partition contains all the devices in the path till a PPOC or a SPOC is reached. Figures 2,3,4 show several example circuit partitions. A simple back trace recursive procedure is used to create all the partitions. Figure 5 shows the data structure used to store the partitions. At this time the original network structure is discarded and the memory freed.

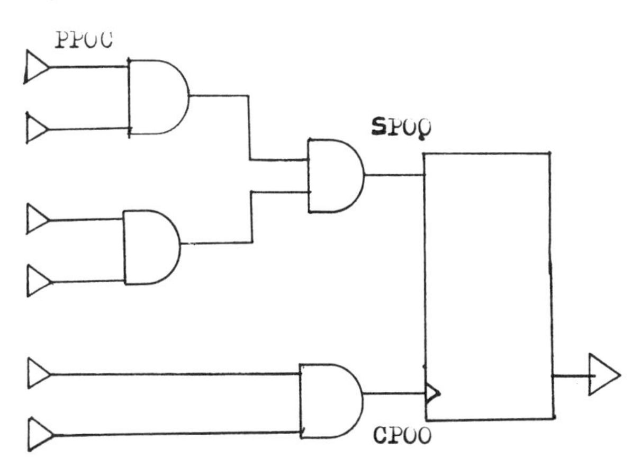

Fig. 2 Example circuit partition

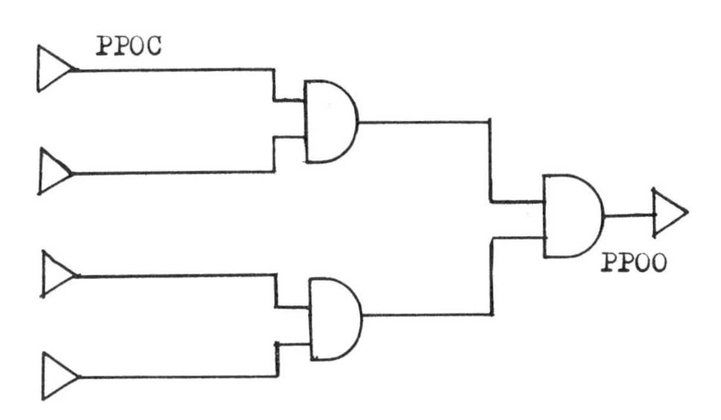

Fig. 3 Example circuit partition

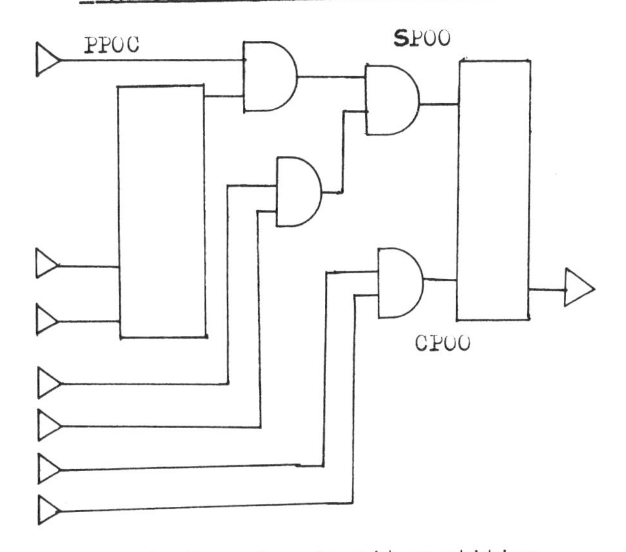

Fig. 4 Example circuit partition

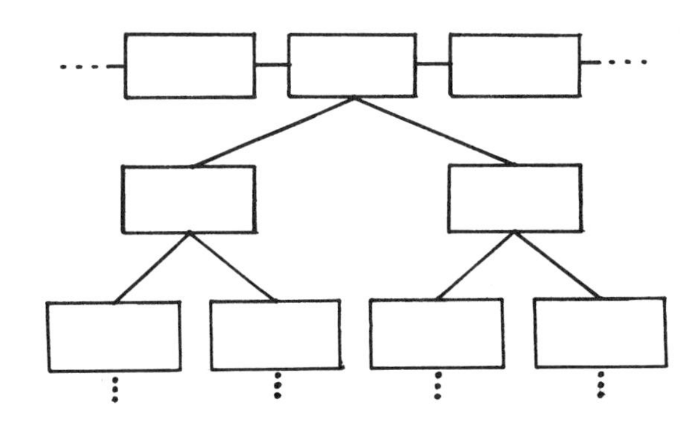

Fig. 5 Data Structure

Partition/Levelization Justification

The partitions make analysis fast and efficient, since all the partitions are made up of only combinatorial logic, evaluation of logic states and calculations are speeded up. The memory required to do logic simulation is reduced, and the number of comparisons required for analysis is reduced.

PARTITION ANALYSIS

Two types of partition analysis are performed, timing analysis and combining relevant partitions for skew analysis. The following sub sections describe the analysis technique :

Timing Analysis

For all the partitions a logic simulation is done and for each PPOO, DPOO, CPOO to PPOC, or SPOC path a minimum rise time, maximum rise time, minimum fall time and maximum fall time is calculated. These values are then compared with specified values and a list of error paths is output. The user can also specify which path delay values are to be reported. The user can then modify the circuit to fit the required specifications. Some of the other checks done are setup and hold checks, minimum clocks width checks, a error report is also generated if any check fails.

Combining relevant partitions

Skew analysis is only performed on DPOO to PPOC or SPOC partitions, and the related CPOO to PPOC or SPOC partitions. All the other partitions are dropped because the timing analysis has caught the defects. For any partition if all the PPOC or SPOC nets are a subset or equal to another partition then the two partitions are combined into one partition. This minimizes the skew analysis time, an example of two candidate partitions is shown in figure 6.

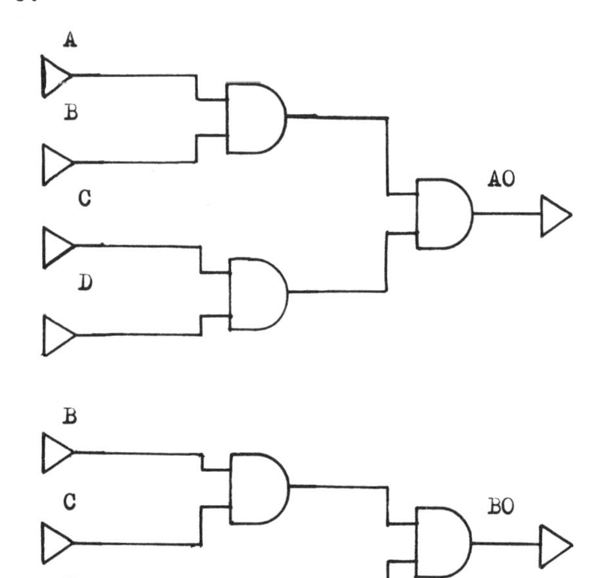

Fig. 6 Candidate Partitions For Combining

SKEW ANALYSIS

For all of the remaining partitions, each of the PPOC or SPOC net is delayed plus and minus the user defined skew time. This is done by inserting a delay buffer into the netlist. Then a logic simulation is performed and results are compared to the original simulation results at DPOO point. If any DPOO point fails the list of error path is generated and also the test vector which caused the value at DPOO to fail is written out to a error file.

Performance figures

A number of test circuits were analyzed by using TSAS running on a VAX 11/780. Each circuit had 2500 test vectors, and computer had no other active jobs running. Table 1 shows the results.

#of primitive gates	#of primary inputs	#of storage devices	CPU time minutes
900	18	56	1
2800	27	125	3
3600	27	612	13
4600	32	255	9
7200	39	227	22
10000	75	170	52
12000	79	321	68

Table 1. Performance figures for TSAS

CONCLUSION

A tester skew analysis system (TSAS) is a fast, accurate, user friendly, and powerful design analysis CAD tool using minimum computer resources. TSAS can be used to detect potential skew, timing and clock distribution problems. The reports produced can be easily interpreted to modify test vectors or the design before manufacturing and test.

ACKNOWLEDGEMENT

The author would like to acknowledge the technical contributions of Laurin Williams, Tom Ferry, and Jin Kim. The author is grateful to B. Nataraj and Sunil Mehta for there sincere encouragement in writing this paper.

REFERENCES

(1) E. Chan, "Development of a Timing Analysis Program for Multiple Clocked Network" Proceedings of the 22nd Design Automation Conference, pp. 816-819, 1985.

(2) A. Krasniewski, A. Albicki, "Simulation-Free Estimation of Speed Degredation in NMOS Self-Testing Circuits for CAD Applications" Proceedings of the 22nd Design Automation Conference, pp. 808-811, 1985.

(3) T.M. McWilliams, "Verification of Timing Constraints on Large Digital Systems" Proceedings of the 17th Design Automation Conference, pp. 139-147, 1980.

(4) R.B. Hitchcocks Sr., "Timing Verification and Timing Analysis Program" Proceedings of the 19th Design Automation Conference, pp. 594-604, 1982.

(5) E.G. Friedman, S. Powell, "Design and Analysis of a Hierarchical Clock Distribution System for Synchronous Standard Cell/Macrocell VLSI", IEEE journal of solid-state circuits vol. sc-21, no. 2, pp. 240-246, 1986.

(6) J. Beausang, A. Albicki, "A Method to Obtain an Optimal Clocking Scheme for a Digital System", Proceedings of International Conference on Computer Design", pp. 68-72, 1985.

(7) M. Muraoka, H. Iida, H. kikuchihara, M. Murakami, K. Hirakawa, "ACTAS: An Accurate Timing Analysis System for VLSI", Proceedings of the 22nd Design Automation Conference, pp. 152-158, 1985.

(8) S. Dhar, M. Franklin, D. Wann, "Reduction of clock delays in VLSI structures", Proceedings of International Conference on Computer Design", pp. 778-783, 1984.

(9) E.G. Ulrich, "The Concurrent Simulation of Nearly Identical Digital Networks", Proceedings of the 10th Design Automation Conference, pp. 145-150, 1973.

DESIGN AND VERIFICATION OF CLOCK DISTRIBUTION IN VLSI

David C. Keezer

Center for Microelectronics Research at the
University of South Florida, Tampa, Florida 33620

ABSTRACT

The availability of synchronous clock signals to internal logic elements is of paramount importance in the design of digital integrated circuits. Numerous design tools may be applied to help assure this condition. However, few verification tools exist to directly measure the effectiveness of the design approach. This paper describes some of the design considerations involved in producing a circuit with minimal clock skew. The use of the electron beam probe for verifying the effectiveness of the resulting implementation is also described. The electron beam probe allows utilization of the voltage contrast effect for measuring timing relationships between clock signals as they are distributed throughout a VLSI device. An example is shown which demonstrates subnanosecond synchronization of such signals. Another example shows how a combination of marginal design practices and slight processing deviations can lead to significant internal clock skew and device failure.

I. INTRODUCTION

The continued development of digital electronic devices has resulted in a move towards higher complexity and an emphasis on higher clock rates. Accompanying these trends is the use of large registers with distributed elements, each requiring clock signals to synchronize local activity with the global device operation.

Often many large registers must be clocked simultaneously. The physical location of these elements is distributed throughout the integrated circuit floor plan (a fact which is not readily apparent from the logical or schematic representations). Furthermore, in the case of very large registers (i.e. many tens or hundreds of elements), an individual register may itself be physically distributed.

The nodes at which clock signals are expected to be synchronized may be separated by several hundred or even a few thousand microns [1]. Each node presents a capacitive load to an already complicated distribution network made up of non-ideal transmission lines and active elements (buffers). Balancing such a network is an engineering effort in itself which usually relies on capacitance calculations and detailed circuit simulation [2-23].

Complicating this situation further is the tendency for the storage elements to exhibit shorter clock-to-output propagation delays. These delays are on the order of a nanosecond or less for some advanced technologies [24,25]. The timing requirement for internal clock distribution is in this same range.

Clock synchronization can be achieved to within subnanosecond skews across a VLSI device through the use of careful design practices. However, validating that such specifications are realized is a test problem solved most readily using an electron beam probe system operating in the voltage contrast mode. Because of the non-contact nature of the electron probe, the pitfalls of mechanical (needle) probing are avoided. These include: (1) the risk of mechanical damage; (2) the difficult prospect of locally exposing a small section of conductor which is otherwise covered with an insulating "passivation" material; and (3) the uncontrolled alteration of the measured signal by the modification of the environment surrounding the measurement point and the added capacitance of the probe itself. The mechanism whereby internal signals may be measured with the electron beam is described in Section III.

II. DESIGN CONSIDERATIONS

There are two primary sources of difficulty in designing clock distribution networks for VLSI. The first is the fact that the clock signal(s) must be made available to a large number of elements. The unavoidable high fanout requires that the clock be appropriately buffered in order to handle the drive requirements. The second difficulty is a result of the fact that the driven elements are distributed across an area of many square millimeters. The interconnect lines have intrinsic resistance and capacitance which result in significant diffusion delays.

Reprinted from *Proc. IEEE Int'l Conf. Commun. ICC '90,* vol. 3, no. 4, pp. 317.7.1–317.7.6, April 1990.

Algorithms for minimizing either total delay or silicon area for series of logic gates are presented in [2-5]. The gates are assumed to be side by side so that the effects due to wiring resistance and capacitance are neglected. However, the algorithms are useful when this condition is met. Specifically, Mohsen and Mead [5] have pointed out that the optimal fan-out for a string of successively larger drivers is *e* (=2.718...). A model is presented in [6] which can be used to determine individual transistor sizing, given the speed and capacitive load requirements. A generalized algorithm for transistor sizing is presented in [7].

Calculations of delays resulting from on-chip interconnects are given by Carter and Guise [8,9] for several interconnect materials including Aluminum, Polysilicon, and Tantalum Silicide. These are used to compare various Silicon and GaAs technologies especially as geometries are scaled down below 1um and interconnect line lengths remain at a significant fraction of the die dimension. The delay time is modelled as the sum of delays resulting from: (1) the driving gate unloaded delay; (2) line and fanout capacitance coupled with the driving gate resistance; (3) diffusion delay; and (4) RC delay due to fanout capacitance and line resistance. An important result is that the overall delay of a composite line (i.e. one made up of both high and low resistance elements) may be significantly effected by the order of placement. Specifically, high resistance elements (such as polysilicon interconnects) placed near the driving end contribute more to the delay than if placed near the receiving end. The effects of parasitic capacitances and multiple dielectric materials on interconnect RC delays are addressed in [10]. These only serve to increase the problem as geometries are reduced.

The diffusion equation has been used to estimate the size of circuit which may readily be clocked using a given technology by Anceau[11]. As device geometries scale down, so does the size of the region over which clocks can remain synchronized without careful attention to clock distribution design. Anceau estimates that this "isochronic region" is typically 7mm, 1.16mm and 0.35mm for 6um, 1um, and 0.3um NMOS technologies respectively. A proposed solution requires that the circuit be partitioned into functional elements each of which has dimensions less than these limits. A common, high speed clock is broadcast to all elements and used to drive internal processing. Inter-block communication is synchronized to a slower rate clock which is derived from the high-speed clock using a 1/N divider. A phase locked loop is used to avoid metastability problems.

Some authors have recently proposed that the system clock approach be replaced by a distributed generation of timing signals and a handshake mode of operation for transfer of data between logic blocks [12]. Nevertheless, the bulk of present day designs rely on the familiar system clock for synchronization.

Several schemes have been proposed which attempt to minimize or eliminate the circuit sensitivity to clock skew by utilizing single phase clocking. Krambeck et al [13] describe the so-called "Domino" logic approach which combines a dynamic NMOS gate with a static CMOS buffer. In the NORA technique [14], dynamic blocks are combined with a C_2MOS latch stage. Use of these design techniques requires strict adherence to a set of rules in order to assure racefree operation.

The use of a single-phase clock is described by Bazes et al [15] in conjunction with the use of single-edge logic throughout the device. This approach avoids the problems of clock-phase distortion which may result when one phase of a two-phase clock distribution encounters a longer RC delay than the other phase. The drawback is that the single-edge logic elements are more complex than those used for two-phased clocking. In this same example, the circuit design required wide distribution of signals other than clocks also with minimal timing skew. The authors describe the use of low-resistance input protection devices, routing on metal lines, and a distributed buffer tree approach to obtain the performance required. These are techniques appropriate for use in clock distribution design as well.

Recently, Lu [16] has proposed a single phased clocking scheme which utilizes the rising and falling edges of the clock to latch alternate

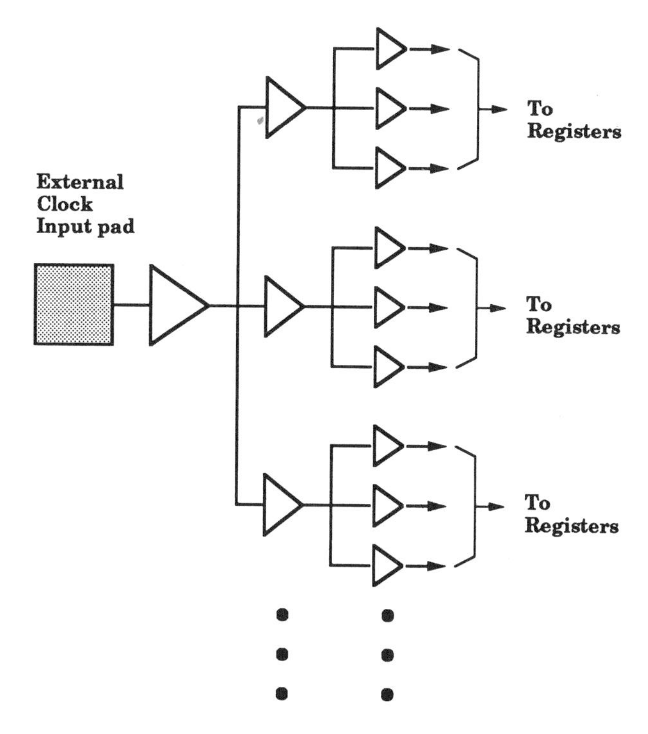

Figure 1 - Three-level buffer tree.

stages of the shift register. The approach simplifies the routing of clock signal(s) at the expense of increased cell area.

A partial example of a three level buffer tree approach is shown in Fig. 1. Here, an external source provides a single clock signal to one device pin. This signal is usually buffered immediately to minimize and/or standardize the load which the device presents to the external clock distribution network. The current drive requirement of this buffer is determined by the number of buffers in the next level of the tree and the capacitance of the interconnecting conductor lines. A similar balancing approach is used to determine the size (current capacity) of the second and third level buffers. The lowest level buffers are then directly used to clock storage elements. If the load on each node at each level is properly balanced then the clock signals reaching the storage elements will be closely synchronized.

III. ELECTRON BEAM PROBING

The electron beam probe system is made up of an electron-optical system similar to that found in standard scanning electron microscopes. The key features are depicted in Fig. 2. They are: (1) an electron "gun" designed to produce an electron beam of moderate energy (typically 1kV); (2) a system of condensing and objective magnetic lenses designed to produce a small (submicron) spot size at the sample surface; (3) an array of deflecting coils which cause the beam to scan the device surface; and (4) a secondary electron detector made up of a scintillator, a photomultiplier, and a video amplifier and display system. To this basic arrangement is added: (1) a "beam blanking" system to provide for strobing the electron beam; (2) a secondary electron (SE) "energy analyzer" with linear response; and (3) an oscilloscope-like data acquisition and display system. The theory, design, and performance of the SE energy analyzer and beam blanking system is described in great detail by Menzel, et al [26-31].

The voltage contrast measurement technique exploits the fact that the energy distribution of secondary electrons will be affected by the local surface potential. A large percentage of these SEs have energies of just a few volts and are therefore greatly affected by voltage differences typically found on the surface of integrated circuits [26]. The shift in SE energy distribution is directly proportional to these voltage differences as the beam is scanned between surfaces. What is detected by the scintillator is the integral form of the energy distribution which results in a nonlinear response. A method to linearize the measured signal through the use of a "retarding"grid and a feedback loop is described in [32-35]. The retarding grid serves as a filter which, depending on the applied voltage, permits more or less SEs from reaching the scintillator. Linearization is achieved by applying such a voltage to the retarding grid so as to

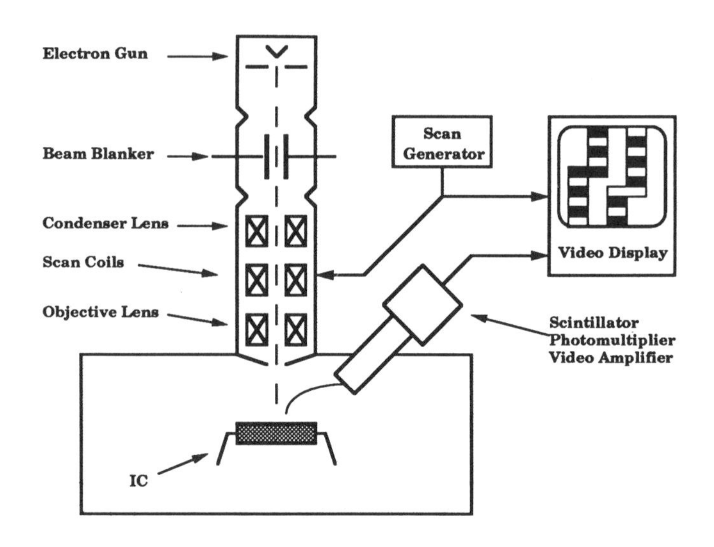

Figure 2 - Key features of the electron beam probe system.

maintain a constant signal at the scintillator. Under this condition, changes in the sample voltage are directly proportional to the changes in the required retarding grid voltage.

Using a continuous beam of electrons, this measurement system is limited by the bandwidth of the detection and feedback electronics (typically a few hundred kiloHertz). A stroboscopic sampling method is incorporated to provide for higher bandwidth measurements [28]. A beam blanking system is used to pulse the otherwise continuous beam of electrons. Pulses in the range of 100 to 1000 picoseconds are synchronized with the clock rate of the device under test. The phase relationship between the device clock edge and these pulses is then varied in order to acquire a voltage versus time waveform. The instrument scan coils may be disabled during such measurements to permit waveform acquisition from a small, submicron "spot" on the sample surface.

The voltage contrast system can be optimized to allow measurement of waveform signals from integrated circuit internal nodes with resolution and accuracy approaching that of a high performance sampling oscilloscope. Measurements taken from exposed metal conductors are typically made to 10mV to 50mV resolution. Through the use of waveform averaging techniques, 1mV resolution can be achieved as shown in Fig. 3. Here an oscilloscope-like display is shown of a 10mV (peak-to-peak) "triangle" signal obtained using the voltage contrast probe technique. The limit of resolution is approximated by the level of noise on the signal which is less than 1mV. To obtain this level of resolution, the sample must be clean and free of extraneous noise-inducing signals. The measured signal must be repetitive and system-induced noise reduced by averaging. Typically between 128 to 256 samples of the entire

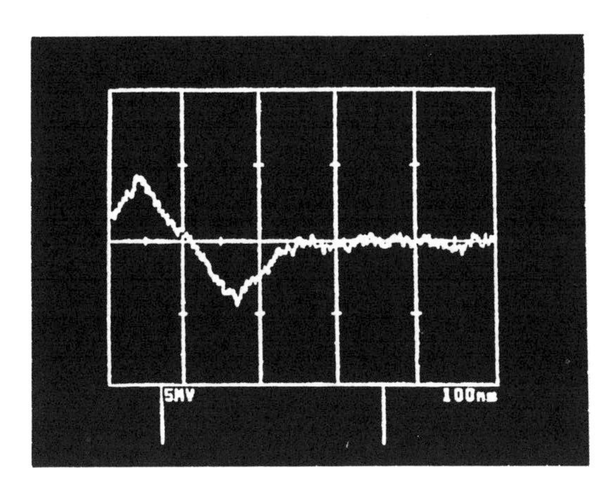

Figure 3 - Demonstration of high resolution voltage contrast measurement.

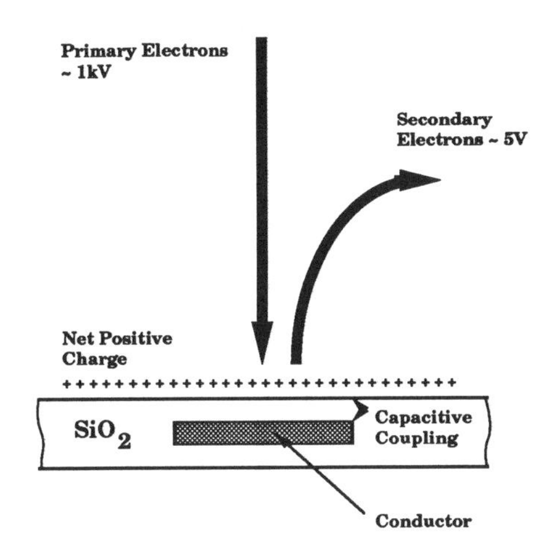

Figure 4 - Charge balance and capacitive coupling of signals through the passivation layer.

waveform are required. Linearization methods are used to obtain between one and a few percent accuracy in amplitude measurements when care is taken in calibration [33]. The use of 100ps to 200ps sample pulses provides an effective bandwidth in the gigaHertz range.

One of the key advantages of the voltage contrast measurement technique is its ability to extract signals from conductor lines which are coated with a dielectric "passivation" layer. Removal of the passivation layer (using wet chemical etch or plasma techniques) has the disadvantage of altering the capacitance and impedance of the underlying interconnect lines. It is therefore highly desirable to probe the signal without removing the passivation. This is especially important when critical measurements such as clock distribution timing are to be made.

To probe "through" the insulating passivation, the primary beam energy is adjusted (typically in the 0.8 to 1.6 kV range) so as to achieve a balance of charge injected and charge emitted (SEs) from the surface. In this range of beam energies most insulators emit more secondary electrons than are injected by the primary beam. This imbalance results in a net positive charge on the surface of the insulator which in turn creates a retarding field that reduces the net emission of SEs as depicted in Fig. 4. A value for the primary beam energy is empirically determined which results in a balancing of these effects. A changing potential on the underlying conductor is then capacitively coupled to the surface of the insulator, resulting in an observable voltage contrast signal [36]. It should be noted that this effect is only sensitive to time-varying signals on the conductor and is not useful for measuring static potential levels. This approach is ideal for measuring the typically high rate clock signals in VLSI devices.

IV. EXAMPLES

An approach similar to that depicted in Fig.1 was used to distribute the master clock throughout a particular VLSI device. In this example, the master clock was used to synchronize many hundreds of sequential elements distributed throughout the die. For the device to operate properly, the clock signals were required to be synchronous to within +/-1.0ns. The voltage contrast technique was used to measure the relative timing of these clock signals at various branches of the lowest level of the buffer tree. For these measurements, a repetitive stimulus was applied to the sample. A trigger signal, synchronous with this stimulus was used to establish the timing of the beam blanker pulses. Using 200ps electron beam pulses, phase delayed over a single clock period, voltage contrast waveforms were obtained at several different branches as shown in Fig. 5. In the figure, measurements from four nodes have each been averaged 128 times, normalized (in software), and overlaid (to facilitate comparison). It can be seen that the 50% points of all the transitions lie within a 1ns range. Using this approach at many locations within the device, the design requirement of synchronous clock distribution was validated.

In another wafer lot, a recurring failure mode was observed which pointed to a clock distribution timing error. A 42-bit shift register was observed to require only 41 cycles to "walk" a bit from input to output. The measurement problem was complicated by the fact that the 42-bit register was physically made up of seven smaller registers ranging from one to fifteen bits each as shown in Fig. 6. The voltage contrast probe was again used to measure the clock timing at the seven inputs to these registers. The first six registers were found to be receiving synchronous clock signals (within +/-1ns) as expected. However, the clock input to the

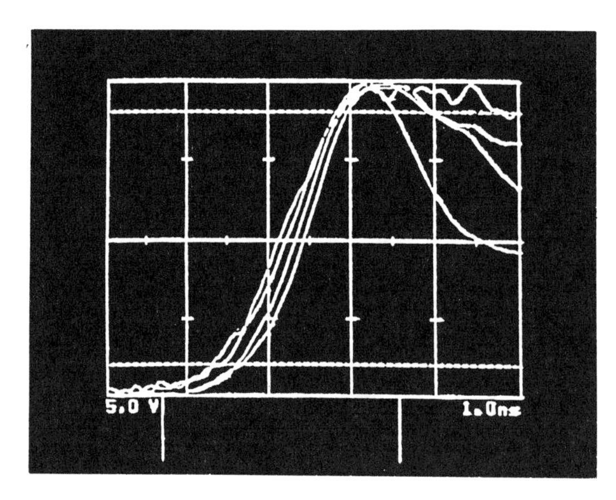

Figure 5 - Display of clock signals measured at four different nodes.

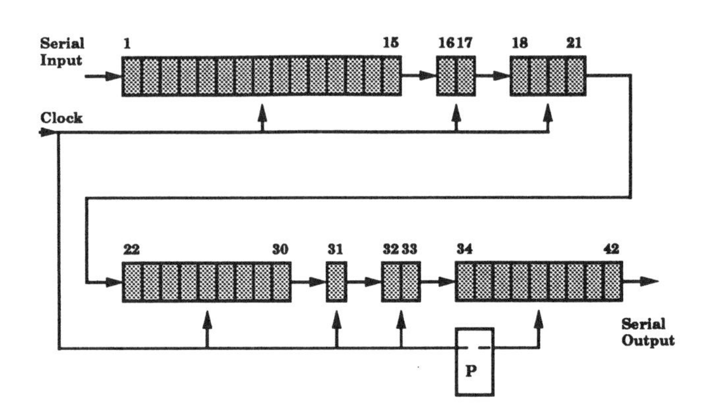

Figure 7 - Location of a passive feedthrough element causing clock skew.

last (9-bit) register was found to be delayed by 3.6ns and somewhat degraded in risetime and amplitude. This was enough to disrupt the normal shift operation of data passed from the prior registers.

The cause of the added delay in the clock signal was further traced using the electron beam probe system. It was found that the delay occurred as the signal was routed through a passive feedthrough of a logically unrelated element as depicted in Fig. 7. Such feedthroughs are incorporated in the cell layout to provide added flexibility in auto-routing. Normally the resistance of these feedthroughs is in the range of 100 Ohms or less. With a typical capacitive load of 0.5pF, the expected RC delay from input to output would be 50ps (far less than observed). In this lot of devices a processing error caused the nominal feedthrough resistance to increase to between 7,000 and 8,000 Ohms. When coupled with the 0.5pF capacitive load, the observed RC delay resulted. In this way the suspected clock skew induced error was confirmed, isolated, and traced to a specific process related error. The element feedthroughs were later redesigned to be less sensitive to such processing variations.

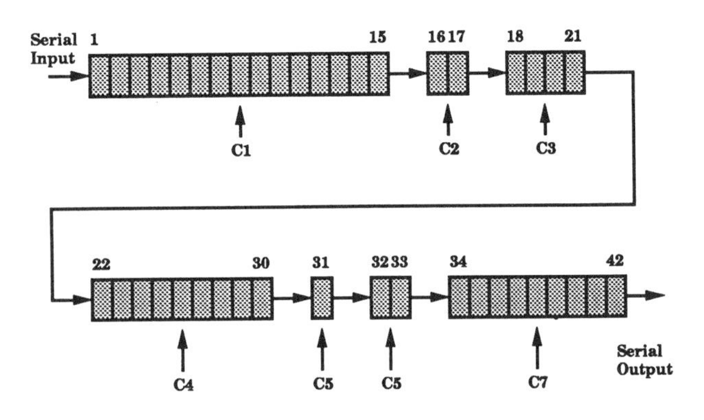

Figure 6 - Grouping of bits within a 42-bit shift register.

V. CONCLUSIONS

As illustrated by the two examples above, the voltage contrast electron beam probe is an invaluable tool for both characterizing and optimizing the clock distribution design within VLSI devices. It is also useful for pinpointing errors in the clock timing which may result not only from inaccurate circuit design but also from layout and fabrication errors or limitations. The approach provides accurate measurement of the timing relationships between internal clock signals without disrupting the physical environment surrounding the probed node or otherwise distorting the edge placement of the signals during measurement.

REFERENCES

[1] K. Yamashita, A. Kanasugi, S. Hijiya, G. Goto, N. Matsumura, T. Shirato, "A Wafer-Scale 170,000-Gate FFT Processor with Built-In Test Circuits," IEEE J. of Solid-State Circuits, VOL. 23, NO. 2, April 1988, pp.336-342.

[2] C.M. Lee, H. Soukup, "An algorithm for CMOS timing and area optimization," IEEE J. of Solid-State Circuits, VOL.19, NO.5, Oct. 1984, pp.781-787.

[3] H.C. Lin, L.W. Linholm, "An optimized output stage for MOS integrated circuits," IEEE J. of Solid-State Circuits, VOL. SC-10, 1975, pp.106-109.

[4] C. Mead, L. Conway, *Introduction to VLSI System.* Reading, MA: Addison-Wesley, 1980.

[5] A.M. Mohsen, C.A. Mead, "Delay-time optimization for driving and sensing of signals on high-capacitance paths of VLSI systems," IEEE Trans. on Electron Devices, ED-26, 1979, pp.540-548.

[6] B.A. Richman, J.E. Hansen, K. Cameron, "A deterministic algorithm for automatic CMOS transistor sizing," IEEE J. of Solid-State Circuits, VOL. 23, NO. 2, April 1988, pp.522-526.

[7] J. Shyu, A. Sangiovanni-Vincentelli, J. Fishburn, A. Dunlop, "Optimization-based transistor sizing," IEEE J. of Solid-State Circuits, VOL. 23, NO. 2, April 1988, pp.400-409.

[8] D.L. Carter, D.F. Guise "Effects of Interconnections on Submicron Chip Performance," VLSI Design, Jan. 1984, pp.63-68.

[9] D.L. Carter, D.F. Guise, "Analysis of Signal Propagation Delays and Chip-Level Performance Due to Chip Interconnections," International Conference on Computer Design: VLSI in Computers, Port Chester, NY, Nov. 1983.

[10] A.K. Sinha, J.A. Cooper, H.J. Levinstein, "Speed limitations due to interconnect time constants in VLSI Integrated Circuits,"

[11] F. Anceau, "A synchronous approach for clocking VLSI systems," IEEE J. of Solid-State Circuits, VOL. SC-17, 1982, pp.51-56.

[12] S. Komori, H Takata, T. Tamura, F. Asai, T. Ohno, O. Tomisawa, T. Yamasaki, K. Shima, K. Asada, H.Terada, "An elastic pipeline mechanism by self-timed circuits," IEEE J. of Solid-State Circuits, VOL. 23, NO. 1, Feb. 1988, pp.111-117.

[13] R.H. Krambeck, C.M. Lee, H.S. Law, "High-speed compact circuits with CMOS," IEEE J. Solid-State Circuits, VOL. SC-17, pp.614-619, June 1982.

[14] N.F. Goncalves, H.J. De Man, "NORA: A racefree dynamic CMOS technique for pipelined logic structures," IEEE J. of Solid-State Circuits, VOL. SC-18, No.3, June 1983, pp.261-266.

[15] M. Bazes, J. Nadir, D. Perlmutter, B, Mantel, O. Zak, "A programmable NMOS DRAM controller for microcomputer systems with dual-port memory and error checking and correction," IEEE J. of Solid-State Circuits, VOL. SC-18, 1983, pp.164-172.

[16] Shih-Lien Lu, "A Safe Single-Phase Clocking Scheme for CMOS Circuits," IEEE J. of Solid-State Circuits, VOL. 23, NO. 1, Feb. 1988, pp.280-283.

[17] D. Auvergne, G. Cambon, D. Deschacht, M. Robert, G. Sagnes, V. Tempier, "Delay-time evaluation in ED MOS logic LSI," IEEE J. of Solid-State Circuits, VOL. 21, April 1986, pp.337-343.

[18] L.A. Glasser, D.W. Dobberpuhl, *The Design and Analysis of VLSI Circuits*, Reading, MA: Addison-Wesley, 1985.

[19] P. Naish, P Bishop, *Designing Asics*, Chichester England: Ellis Horwood Ltd., 1988.

[20] A.E. Ruehli, P.A. Brennan, "Accurate metallization capacitances for integrated circuits and packages," IEEE J. of Solid-State Circuits, VOL. SC-8, No.4, Aug. 1973, pp.289.

[21] M. Shoji,"Electrical design of the BELLMAC-32A microprocessor," Proc. 1982 Int. Conf. Circuits and Computers, Sept. 1982, pp.112-115.

[22] J.P. Uyemura, *Fundamentals of MOS Digital Integrated Circuits*, Reading, MA: Addison-Wesley, 1988.

[23] N.H.E. Weste, K. Eshraghian, *Principles of CMOS Design, A Systems Perspective*, Reading, MA: Addison-Wesley, 1985.

[24] F.E. Anderson, J.M. Ford, "A 150k channelless gate array design in 0.5um CMOS technology," IEEE J. of Solid-State Circuits, VOL. 23, NO. 2, April 1988, pp.520-522.

[25] R.R. Cordell, "A 45-Mbit/s CMOS VLSI Digital Phase Aligner," IEEE J. of Solid-State Circuits, VOL. 23, NO. 2, April 1988, pp.323-328.

[26] E. Menzel, E. Kubalek, "Fundamentals of Electron Beam Testing of Integrated Circuits", SCANNING Vol.5, pp103-122 (1983).

[27] E. Menzel, E. Kubalek, "Electron Beam Test Techniques for Integrated Circuits", Scanning Electron Microscopy 1981/I,SEM Inc, AMF O'Hare, Chicago, pp305-322.

[28] E. Menzel, E. Kubalek, "Electron Beam Chopping Systems in the SEM", Scanning Electron Microscopy 1979/I, SEM Inc, AMF O'Hare, Chicago, pp305-318.

[29] M. Ostrow, E. Menzel, E. Postulka, S. Gorlich, E. Kubalek, "IC-Internal Electron Beam Logic State Analysis", Scanning Electron Microscopy 1982/II, SEM Inc, AMF O'Hare, Chicago, pp563-572.

[30] E. Menzel, M. Brunner, "Secondary Electron Analyzers for Voltage Measurements", Scanning Electron Microscopy 1983, SEM Inc, AMF O'Hare, Chicago, pp65-75.

[31] E. Menzel, E. Kubalek, "Secondary Electron Detection Systems for Quantitative Voltage Measurements" SCANNING Vol. 5, pp151-171 (1983).

[32] J.P. Fleming, E.W. Ward, "A technique for accurate measurement and display of applied potential distributions using the SEM", Scanning Electron Microscopy 1970, IITRI, Chicago, pp465-470.

[33] A. Gopinath, C.C. Sangen, "A Technique for Linearization of Voltage Contrast in the Scanning Electron Microscope", J Phys E 4, pp334-336 (1971).

[34] L.J. Balk, H.P. Feuerbaum, E. Kubalek, E. Menzel, "Quantitative Voltage Contrast at High Frequencies in the SEM", Scanning Electron Microscopy 1976/I, IITRI, Chicago, pp615-624.

[35] H.P. Feuerbaum, "VLSI Testing using the Electron Probe", Scanning Electron Microscopy 1979/I, SEM Inc, AMF O'Hare, pp285-296.

[36] L. Kotorman, "Non-Charging Electron Beam Pulse Prober on FET Wafers", Scanning Electron Microscopy 1980/IV, SEM Inc, AMF O'Hare, pp77-84.

Harmonic Series Analysis of Digital Clock Distribution Circuits

Clifford Kraft
AT&T Bell Laboratories
IH 2G-437
Naperville, Illinois 60566

ABSTRACT A transmission line model that can be used to analyze clock distribution nets on computer circuit boards has been derived. This model uses a finite Fourier Series expansion of a square wave clock signal with finite rise and fall times. Complex transmission and reflections coefficients are found for each harmonic at each termination and branch point in the net. Steady state solutions consisting of an infinite number of reflections of each harmonic are then found at any desired point in the circuit, and a time-domain waveform can be reconstructed at that point. This allows the designer to accurately predict such parameters as clock skew, waveshape, and ringing anywhere in the net. This information can then be used to decide the necessity of items such as termination resistors and their optimum locations. The model has been programmed in Fortran, and compares well with actual laboratory data.

I. Introduction

There are two basic nets found on computer circuit boards that profoundly affect overall system performance: the power distribution net and the clock distribution net. This paper presents a method of modeling and analyzing a clock distribution net that allows the designer to simulate the actual circuit board device clock layout. The method first expands an input clock signal into a finite Fourier Series using enough harmonics for good approximation. Each harmonic is treated separately as an independent generator input. If TEM mode propagation is assumed throughout the circuit, scattering parameters can be used to characterize the system of infinite reflections caused by complex (frequency dependent) impedances at devices. This assumption is good for harmonics up to at least 2 GHz. Load and source impedances lead to reflection coefficients at all net terminations and splits, while lands (net wiring) are treated as sections of linear transmission line with fixed, resistive characteristic impedance Z_0. Splits in lands become scattering points that lead to systems of linear equations. Finally, phasor voltages for each harmonic at a given circuit point can be combined to re-form a time Fourier Series at the point yielding the effect of an oscilloscope probe. The important questions of ringing, false clock edges and skew can then be answered at all important clock points, and decisions about possible line termination resistors, or alternate routing, can be made. At the higher harmonics, resistors must be modeled as complex impedances that include parasitic capacitance and lead inductance.

There has been some interconnect modeling of signal propagation in the past. Ghioni et al. [1] present a quasi-TEM mode analysis for fast GaAs device interconnects. Fukuoka [2] discusses higher order modes that exist on such metalization, primarily at microwave frequencies. Stripline propagation and properties are extensively analyzed by Bahl [3],[4], while the transmission line properties of parallel lands is addressed by Wheeler [5]. Fache and De Zutter [18] discuss structures consisting of a wire above a grounded plane, a geometry resembling multiwire technology, while Branin and Dvorak [6],[7] present a time-domain analysis of transmission line transients. Scattering matrices applied to n-ports was pioneered by Monaco and Tiberio [15],[17] as early as 1974. A very understandable modern description of this is given in a book by Gupta et al. [16]. Also various vendors of integrated circuits (especially ECL and fast CMOS) analyze circuit connections in the context of transmission line theory in application notes [20],[21].

II. Fourier Expansion of Clock Signal

The input signal to a clock distribution net is usually a continuous, possibly asymmetric, square wave having finite rise and fall times. These edges can be very accurately represented by exponentials of the form $1 - e^{-t/\tau}$ for the rising edge and $e^{-t/\tau}$ for the falling edge. Figure 1 shows the form of this periodic signal with different rise and fall time constants.

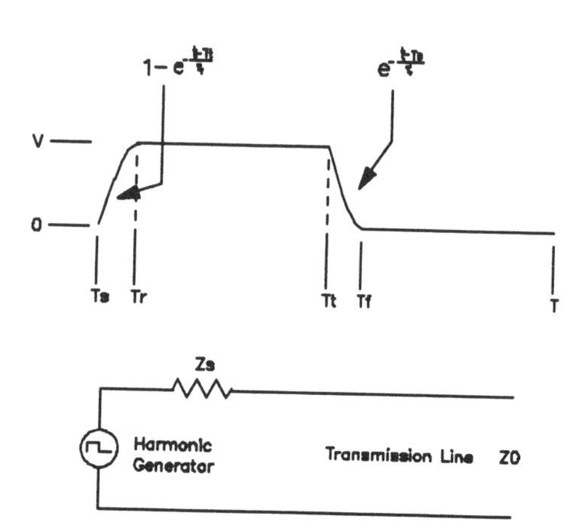

Figure 1. One Cycle of Typical Clock Generator Signal

It is well known that any periodic signal with reasonable mathematical behavior can be expanded into an infinite series of sinusoids using the following Fourier Series [8],[9]:

$$f(t) = a_0 + \sum_{n=1}^{\infty} \left[a_n \cos(n\omega_0 t) + b_n \sin(n\omega_0 t) \right] \tag{1}$$

where:

$$a_0 = \frac{1}{T} \int_{-T}^{T} f(t)\, dt \tag{2}$$

$$a_n = \frac{2}{T} \int_{-T}^{T} f(t) \cos(n\omega_0 t)\, dt \tag{3}$$

$$b_n = \frac{2}{T} \int_{-T}^{T} f(t) \sin(n\omega_0 t)\, dt \tag{4}$$

Using $f(t) = 1 - \exp[-(t - T_t)/\tau_f]$ for the rising edge, and $f(t) = \exp[-(t - T_s)/\tau_r]$ for the falling edge, with $f(t)=1$ during the rest of the high half cycle, and $f(t)=0$ during the rest of the low half cycle (the function in Figure 1), and performing the elementary integration indicated in equations (2),(3), and (4), we have first for the DC term:

$$a_0 = \frac{1}{T - T_s} \Big((T_r - T_s) - e^{T_s/\tau_r} \left[e^{-T_r/\tau_r} - e^{-T_s/\tau_r} \right] + (T_t - T_r) + e^{T_t/\tau_f} \left[e^{-T_f/\tau_f} - e^{-T_t/\tau_f} \right] \Big) \tag{5}$$

then for the cosine term:

$$a_n = \frac{2}{T - T_s} \Big\{ \frac{\sin(n\omega_0 T_r) - \sin(n\omega_0 T_s)}{n\omega_0} + \frac{\sin(n\omega_0 T_t) - \sin(n\omega_0 T_r)}{n\omega_0} - \frac{e^{T_s/\tau_r}}{1/\tau_r^{\,2} + (n\omega_0)^2} \Big[e^{-T_r/\tau_r} \Big(n\omega_0 \sin(n\omega_0 T_r) - \frac{1}{\tau_r}\cos(n\omega_0 T_r)\Big) - e^{-T_s/\tau_r} \Big(n\omega_0 \sin(n\omega_0 T_s) - \frac{1}{\tau_r}\cos(n\omega_0 T_s)\Big) \Big] +$$

Reprinted from *Proc. 32nd Midwest Symp. Circuits Syst.*, pp. 206–211, Aug. 1989.

$$\frac{e^{T_t/\tau_f}}{1/\tau_f^2 + (n\omega_0)^2}\left[e^{-T_f/\tau_f}(n\omega_0 \sin(n\omega_0 T_f) - \frac{1}{\tau_f}\cos(n\omega_0 T_f)) - e^{-T_t/\tau_f}(n\omega_0 \sin(n\omega_0 T_t) - \frac{1}{\tau_f}\cos(n\omega_0 T_t))\right]\} \quad (6)$$

and finally for the sine term:

$$b_n = \frac{-2}{T - T_s}\{\frac{\cos(n\omega_0 T_r) - \cos(n\omega_0 T_s)}{n\omega_0} + \frac{-\cos(n\omega_0 T_t) + \cos(n\omega_0 T_r)}{n\omega_0} - \frac{e^{T_s/\tau_r}}{1/\tau_r^2 + (n\omega_0)^2}\left[e^{-T_r/\tau_r}(-n\omega_0 \cos(n\omega_0 T_r) - \frac{1}{\tau_r}\sin(n\omega_0 T_r)) - e^{-T_s/\tau_r}(-n\omega_0 \cos(n\omega_0 T_s) - \frac{1}{\tau_r}\sin(n\omega_0 T_s))\right] + \frac{e^{T_t/\tau_f}}{1/\tau_f^2 + (n\omega_0)^2}\left[e^{-T_f/\tau_f}(-n\omega_0 \cos(n\omega_0 T_f) - \frac{1}{\tau_f}\sin(n\omega_0 T_s)) - e^{-T_f/\tau_f}(-n\omega_0 \cos(n\omega_0 T_t) - \frac{1}{\tau_f}\sin(n\omega_0 T_t))\right]\} \quad (7)$$

T_s, T_r, T_t, T_f, and T are the times from Figure 1, and τ_r, and τ_f are the rise and fall time constants. Usually $T_s = -T$.

Equations (5),(6) and (7) provide the spectrum of the input clock signal. Each harmonic may be thought of as a complex voltage source applied to the circuit. Reflections in the net can be computed for a harmonic alone, with resulting phasor voltage values found for that harmonic at all points in the net. After this has been done for enough harmonics for a good representation (about 40 to 50), the final phasor values at any point can be re-expanded into a time series. If V_n is the phasor value for the *n*th harmonic at a given point in the net, the time domain function at that point is

$$h(t) = a_0 + \sum_{n=1}^{N} |V_n| \cos(n\omega_0 t + Arg(V_n)) \quad , \quad (8)$$

where N is the number of harmonics being used in the approximation. It must be remembered that during phasor computation of V_n, all impedances and reflection coefficients are functions of frequency. That is, each reflection coefficient is different for each value of n.

III. Complex Scattering Parameters

A common method of modeling RF circuits uses so-called S or scattering parameters [10],[11]. This method can be used for clock distribution circuits if the assumption is made that the only important mode propagating is TEM_0. This is a very reasonable assumption for clock harmonic frequencies up to 2 GHz and possibly higher if corrections are made to the effective dielectric constant [18]. An n-port linear circuit or section of transmission line can be modeled by a system of n linear equations. The equations express the relationship between input voltage (or current) waves and outgoing ones. A typical 2-port is shown in Figure 2. The outgoing waves Q_j are functions of the incoming waves g_k.

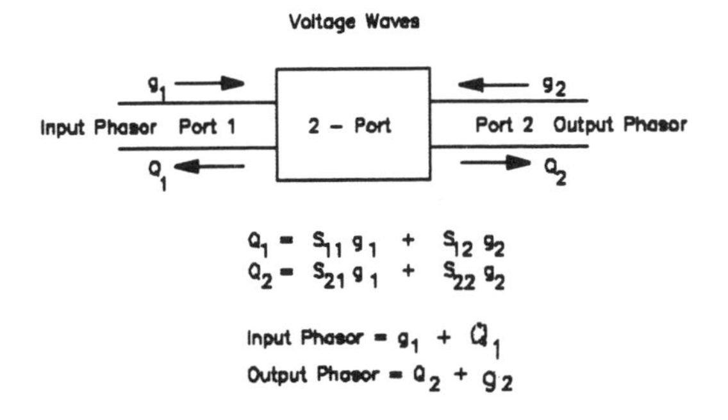

Figure 2. Typical 2-port Modeled with Scattering Parameters

The S_{ij} are complex numbers that represent transmission and reflection through the port. The diagonal elements of the S matrix (S_{11}, S_{22}, etc.) are known as reflection coefficients, while the off-diagonal elements are called transmission coefficients. A line terminator is a simple 2-port with reflection coefficient ρ and transmission coefficient 0. The general n-port S matrix for a passive transmission line split, possibly terminated in some impedance, is

$$S_z = \begin{bmatrix} \rho & \sigma & \sigma & . \\ \sigma & \rho & \sigma & . \\ \sigma & \sigma & \rho & . \\ . & . & . & . \end{bmatrix} \quad (9)$$

where $\sigma = 1 + \rho$.

The S matrix for a section of transmission line of characteristic impedance Z_0 and length z is $2 x 2$ with very simple properties.

$$S = \begin{bmatrix} 0 & e^{-\gamma z} \\ e^{-\gamma z} & 0 \end{bmatrix} , \quad (10)$$

where $\gamma = \alpha + j\beta$ is the transmission characteristic of the line. α represents line loss, and β is related to the wavelength on the line. It is well known [12],[13] that

$$\alpha = \frac{2\pi}{\lambda_0}\left(\frac{\epsilon}{2\epsilon_0}\right)^{1/2}(L-1)^{1/2} \quad (11a)$$

$$\beta = \frac{2\pi}{\lambda_0}\left(\frac{\epsilon}{2\epsilon_0}\right)^{1/2}(L+1)^{1/2} , \quad (11b)$$

where:

$L = (1 + p^2)^{1/2}$,
$p = \sigma_d / \omega\epsilon$,
ε is the permittivity of the material between the conductor and the ground plane; σ_d is its conductivity,
$\epsilon_0 = 8.854 x 10^{-12} F/m$,
$\lambda_0 = 2\pi c / \omega$ (wavelength in free space),
$c = 2.9979 x 10^8 m/s$ (speed of light in free space).

The wavelength λ in the net is $2\pi/\beta$, and the phase velocity is ω/β. The decibel attenuation per unit length is $dB/m = 10 \log_{10} e^{-\alpha}$. Equation (11b) can easily be manipulated to show that the wavelength and phase velocity of a a TEM_0 wave on a transmission line of low loss depends, to a very good approximation, only on the dielectric constant $K_e = \epsilon/\epsilon_0$ of the material between the conductor and the ground plane. Reference [14] gives a table of typical dielectric constants for materials used to make printed circuit boards; however Fache [18] points out that it is more appropriate, at least for land and multi-wire technology, to use an effective dielectric constant that is less than the material dielectric constant. This is because the system is really a layered arrangement of two different dielectrics (with air as the upper layer). The effective dielectric constant can be experimentally determined by relatively simple edge velocity measurements of single pulses or step functions. If a lossless transmission line is assumed, it will be almost non-dispersive in the frequency range of interest, and the group velocity of a pulse or edge will be very close to the phase velocity of the clock harmonics.

The simple line 2-port from equation (10) can be reduced to a transmission scalar $\Lambda = e^{-\gamma l}$ because of the form of the matrix in equation (10). If a phasor voltage V is placed on the line through source impedance Z_s, the phasor value of the voltage wave at distance z down an infinite line of impedance Z_0 is $V Z_0 e^{-\gamma z} / (Z_s + Z_0)$. If the voltage wave at the beginning of a section of line is v, its value at the end of a section of length l is $\Lambda_l v$ where $\Lambda_l = e^{-\gamma l}$. If this same line section of length l is terminated in an impedance that yields a reflection coefficient $\rho = (Z_l - Z_0)/(Z_l + Z_0)$, the reflected wave as seen back at the source is $\rho \Lambda_l^2 v$. The actual voltage phasor at the beginning of the line is the sum of the forward and reflected voltage waves, $V_s = v(1 + \rho\Lambda_l^2)$. At the end of the line it is $V_e = \Lambda_l v (1 + \rho)$. For most transmission lines found on circuit boards, γ can be replaced with $j\omega\beta$ (lossless assumption).

If a transmission line splits into n parts as in Figure 3, it can be treated as an n-port. The split point becomes a mis-match, and causes reflections. If all the outgoing branches in Figure 3 are infinitely long (or equivalently terminated in the characteristic impedance Z_0), the input impedance seen into the split is Z_0/n. The reflection coefficients ρ (diagonal) and the transmission coefficients σ (off-diagonal) of the scattering matrix (equation (9)) become

$$\rho = \frac{1-n}{1+n} \qquad (12)$$

$$\sigma = \frac{2}{1+n} \ .$$

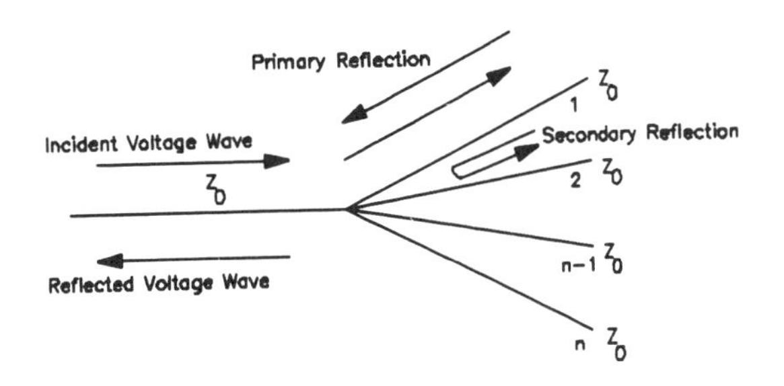

Figure 3. Transmission Line Split Point

If the split takes place at some fixed impedance point Z (such as on an integrated circuit)(Figure 4) the scattering coefficients for an n-split become

$$\rho = \frac{Z_0 Z - Z_0(nZ+Z_0)}{Z_0 Z + Z_0(nZ+Z_0)} \qquad (13)$$

$$\sigma = \frac{2ZZ_0}{Z_0 Z + Z_0(nZ+Z_0)} \ ,$$

where Z is the impedance from the split point to ground. The S matrix contains the complete scattering solution for the split. That is, even if all the n outgoing legs have reflected waves coming back, an infinite solution of all possible reflections of all orders through the split can be derived as a set of linear equations. It can be seen from equation (12) or (13) that a single split or an impedance of Z_0 placed across a transmission line causes $\rho = -1/3$, and $\sigma = 2/3$.

IV. Network Solutions

We first present the general solution to any clock distribution net containing nodes with arbitrary impedances or splits; then several interesting special cases are described. The general solution results in m linear equations in m unknowns if all net ends are terminated in some finite impedance, and there is a source sinusoidal generator feeding a single line through a source resistance Z_s. The integer m is the total number of outgoing waves in the system not counting waves from final terminations and sources. In other words, if a scattering matrix is written for each line reflection point (split or impedance point, but not line termination point or source point), m is equal to the total number of matrix rows involved with all these S matrices. The right hand side of the system of linear equations contains information about the source(s). Of course, for clock distribution, the system must be solved for every harmonic being used in the model because the impedances change with frequency. The variables in the system of equations represent outgoing voltage waves from reflection points. They can be combined to yield a voltage phasor at any circuit point. If Q_k is the outgoing wave at an impedance point, $\Lambda_p Q_k$ is the same wave down a section of line of length p. The phasors for each harmonic can be used to re-construct a time Fourier series at any circuit point to yield a waveform. A general reflection point is shown in Figure 4. The S matrix at such a point is as in equation (9) with terms like equation (13). The basic scattering matrix equation

$$\vec{Q} = S\vec{g} \qquad (14)$$

yields a set of component equations for that point. The $\vec{Q}$ are outgoing waves at split points, and the $\vec{g}$ are incoming waves at the same points. After all such sets of equations are written, they are connected by constraints

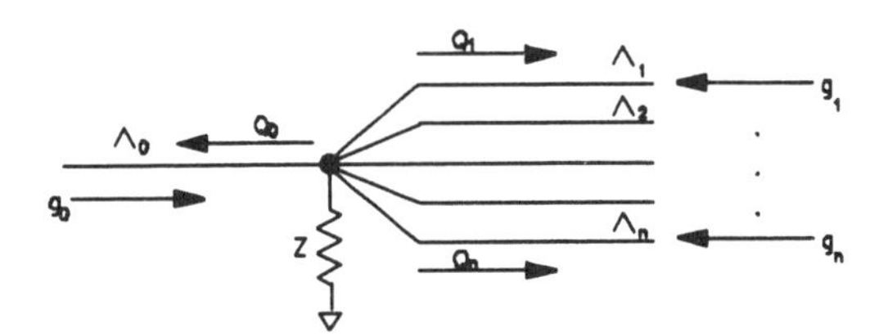

Figure 4. Generalized Reflection Point

in the following manner to yield a system that can be solved:

1) If the input to one reflection point g_i is the output of another Q_j connected only by transmission line then, $g_i = \Lambda_p Q_j$, where Λ_p is the phase constant of the connecting link. Also $g_j = \Lambda_p Q_i$. Thus two free variables g_i and g_j become bound.

2) If an output Q_j drives a line terminated in a load with reflection coefficient ρ_l, $g_j = \Lambda_s{}^2 \rho_l Q_j$. Thus a terminated line binds one free variable g_j.

3) If an input g_m is driven through a line from a generator with source impedance Z_s, $g_m = \Lambda_r[Z_0 v/(Z_0+Z_s) + \Lambda_r \rho_s Q_m]$. A driven line binds one free variable and provides a constant on the right-hand side of the final set of equations.

In all cases $\Lambda_l = e^{-\gamma l}$. Equation (14) creates m equations in $2m$ variables, the m values of $\vec{Q}$ the outgoing waves and the m values of $\vec{g}$ the incoming waves. The first situation simply connects reflection points together through sections of transmission line, where connections eliminate g variables, replacing them with Q variables. A g_k is replaced with a Q_n where n and k are not equal. The second situation replaces terminated g variables with related Q variables. In other words a g_k becomes a function of Q_k. The third supplies the source voltage and source reflection terms. The source voltage terms form the right hand side of the final system of equations. Here the g variable is replaced with a function of its related Q variable plus a constant that represents a source. A matched source or load causes no reflections; a g variable becomes zero. The result of making all such constraint connections in a complete network is a system of linear equations that can be solved for the Q variables or outgoing waves. Phasors at any point can be formed from linear combinations of these solutions. The system will be correctly determined, with non-vanishing determinant, if the network is connected, and every termination is made to a finite, non-zero impedance. Open and short-circuit terminations with $\alpha=0$ transmission lines (no loss) can lead to resonant line sections and vanishing determinants, and should be avoided. A closed solution requires at least one source. The general case allows for more than one source provided they are of the same frequency, and any phase difference is accounted for. Clock distribution nets generally have only one source.

The general matrix solution to equation (14) follows the method outlined by Monaco and Tiberio [15],[17] with sources, line sections, and terminations eliminated from the network, as just described, to greatly reduce the number of equations. First, at a given harmonic, compute the ρ and σ parts of each scattering matrix in the system. Index the incoming and outgoing waves sequentially so that the various scattering matrices can be combined into a single large S matrix for the entire net. The sub-matrices are placed on the diagonal of the larger matrix. Such a combined S matrix has the form

$$S = \begin{bmatrix} S_1 & 0 & 0 & . & . & 0 \\ 0 & S_2 & 0 & . & . & . \\ 0 & 0 & S_3 & . & . & . \\ . & . & . & . & . & . \\ . & . & . & . & . & S_m \end{bmatrix} , \qquad (15)$$

where the S_j are matrices of the form of equation (9). Form a symmetric connection matrix Γ that has Λ_p in the ith row and jth column if Q_i is connected to g_j through a length of transmission line having phase constant Λ_p. Also form a diagonal load termination matrix T having the mth diagonal element equal to $\Lambda_m^2 \rho_m$ if Q_m feeds a transmission line section of phase constant Λ_m into a load of reflection coefficient ρ_m. Also include the source reflection of any driven line with a similar term in the T matrix. The symmetric Γ matrix and the diagonal T matrix are of the form

$$\Gamma = \begin{bmatrix} 0 & \Lambda_q & 0 & . & . & . & . \\ \Lambda_q & 0 & \Lambda_p & . & . & . & . \\ 0 & \Lambda_p & 0 & . & . & . & . \\ 0 & 0 & 0 & . & . & . & . \\ . & . & . & . & . & . & . \\ . & . & . & . & . & . & . \end{bmatrix}$$

$$T = \begin{bmatrix} 0 & 0 & 0 & . & 0 \\ 0 & \Lambda_2^2\rho_2 & 0 & . & 0 \\ 0 & 0 & . & . & . \\ . & . & . & . & . \\ . & . & . & . & . \\ . & . & . & . & \Lambda_m^2\rho_m \end{bmatrix} .$$

Also include a source vector $\vec{c}$ with $\Lambda_l Z_0 v_m / (Z_0 + Z_m)$ on the mth row if g_m is driven by voltage source v_m. This source vector is of the form

$$\vec{c} = \begin{bmatrix} 0 \\ 0 \\ . \\ \dfrac{\Lambda_m Z_0 v_m}{Z_0 + Z_m} \\ 0 \\ 0 \end{bmatrix} .$$

We then have the following constraint equation:

$$\vec{g} = (\Gamma + T)\vec{Q} + \vec{c} \ . \qquad (16)$$

Substituting equation (16) into the general scattering equation (14) $\vec{Q} = S\,\vec{g}$:

$$\vec{Q} = S[(\Gamma + T)\vec{Q} + \vec{c}] \ .$$

Then finally isolating $\vec{Q}$

$$[I - S(\Gamma + T)]\vec{Q} = S\vec{c} \ , \qquad (17)$$

where I is the identity matrix. It is clear that equation (17) represents a system of linear equations in the outgoing waves Q_j. Equation (17) is the general phasor solution to the clock distribution problem. It must be solved at each clock harmonic involved in the model. Phasor solutions V_n can be formed as linear combinations of components of the Q vector at given network points, and time-domain expansions can then be found using equation (8). Equation (17) can be easily programmed to yield numerical solutions if load and source resistances, capacitances, and inductances are specified.

V. Terminations

A possible model for in-circuit loads such as integrated circuits is a resistance in parallel with a capacitance. On the other hand, termination resistors can be modeled as a resistance in parallel with a capacitance with a lead inductance in series with this. Figure 5 shows these models of terminations. Elementary circuit techniques yield the impedance for the termination resistor of the model (shown in Figure 5):

$$Z_2 = \frac{R_2}{1+q^2} + j\omega\left[\frac{L_2(1+q^2) - R_2{}^2 C_2}{1+q^2}\right] \qquad (18)$$

where $q = R_2 C_2 \omega$

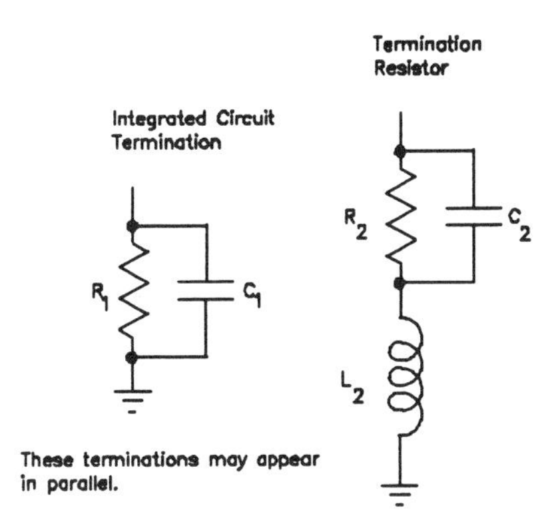

Figure 5. Typical Line Terminations

VI. Illustrative Solutions

We now present an illustrative solution to a special case. Consider the double split transmission line shown in Figure 6.

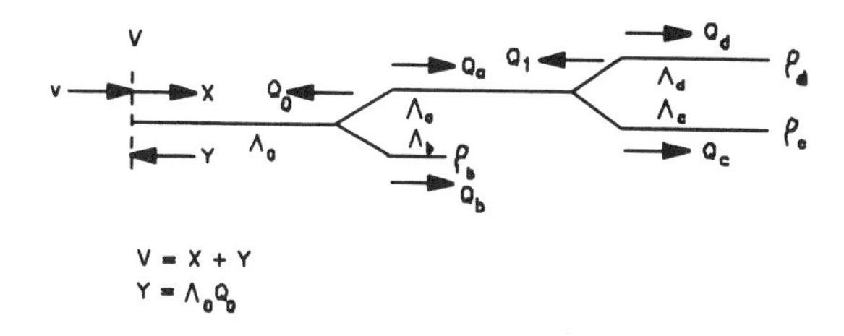

Figure 6. Special Case of Double Transmission Line Split

There are two scattering matrices involved with identical values of $\rho = -1/3$ and $\sigma = 2/3$. They are 3 x 3 matrices because there are three outgoing waves at each split. Legs b, d, and c are terminated in impedances leading to reflection coefficients ρ_b, ρ_d, and ρ_c as shown in the figure. Writing the six linear equations obtained from expanding equation (14) at each of the two nodes and substituting for the g variables in the terminated legs, we have the following system:

The equations at the first split are

$$\begin{aligned} Q_1 &= -\frac{1}{3}\Lambda_a Q_a + \frac{2}{3}\Lambda_d^2\rho_d Q_d + \frac{2}{3}\Lambda_3^2\rho_c Q_c \\ Q_d &= +\frac{2}{3}\Lambda_a Q_a - \frac{1}{3}\Lambda_d^2\rho_d Q_d + \frac{2}{3}\Lambda_3^2\rho_c Q_c \\ Q_c &= +\frac{2}{3}\Lambda_a Q_a + \frac{2}{3}\Lambda_d^2\rho_d Q_d - \frac{1}{3}\Lambda_3^2\rho_c Q_c \end{aligned} \qquad (19)$$

If x is the incoming wave from the source, the equations at the second split are

$$\begin{aligned} Q_0 &= -\frac{1}{3}\Lambda_0 x + \frac{2}{3}\Lambda_a Q_1 + \frac{2}{3}\Lambda_b^2\rho_b Q_b \\ Q_a &= +\frac{2}{3}\Lambda_0 x - \frac{1}{3}\Lambda_a Q_1 + \frac{2}{3}\Lambda_b^2\rho_b Q_b \ . \\ Q_b &= +\frac{2}{3}\Lambda_0 x + \frac{2}{3}\Lambda_a Q_1 - \frac{1}{3}\Lambda_b^2\rho_b Q_b \end{aligned}$$

If v is the generator phasor,

$$x = \frac{Z_0 v}{Z_0 + Z_s} + \rho_s \Lambda_0 Q_0 \ . \qquad (20)$$

When equation (20) is substituted into equations (19), the result is a system of six equations in six unknowns (the Q variables) that can be solved to find any phasor in the circuit. Such systems are easily solved numerically, usually with sparse matrix techniques.

The final illustrative solution will be for the degenerate case of one split (just a straight piece of line) with an infinite number of reflections. The solution to this is well known [19]. An imaginary split point is placed somewhere in the line with Q_0 emerging toward the source, and Q_1 emerging toward the load. Let Λ_1 represent the phase constant of the line toward the load, and Λ_0 the phase constant of the line toward the source. The total line would be represented by $\Lambda_0\Lambda_1$ for this special case. Equation (14) becomes:

$$
\begin{aligned} Q_0 &= g_1 \\ Q_1 &= g_0 \end{aligned} \tag{21}
$$

If x is the wave entering (from left to right) at the source, $g_0 = \Lambda_0 x$ and the constraint at the load is $g_1 = \Lambda_1^2 \rho_l Q_1$ where ρ_l is the load reflection. Combining this with equation (21) we have:

$$
\begin{aligned} Q_0 &= \Lambda_1{}^2 \rho_l Q_1 \\ Q_1 &= \Lambda_0 x \end{aligned} \tag{22}
$$

If y is the wave reflecting at the source (right to left), $y = \Lambda_0 Q_0$, and from the second equation in (22):

$$
y = \Lambda_0{}^2 \Lambda_1{}^2 \rho_l x \tag{23}
$$

The incoming wave x, at the source, must be $x = Z_0 v/(Z_0 + Z_s) + \rho_s y$ where Z_s is the source impedance, ρ_s is the source reflection, and v is the generator voltage. Combining this with (23):

$$
x = \frac{Z_0 v}{Z_0 + Z_s} \frac{1}{1 - \Lambda_0{}^2 \Lambda_1{}^2 \rho_l \rho_s} \tag{24}
$$

The total voltage phasor V at the source is the sum of x and y. Combining (23) and (24) with $\Lambda = \Lambda_0\Lambda_1$:

$$
V = x + y = \frac{Z_0 v}{Z_0 + Z_s} \frac{1 + \Lambda^2 \rho_l}{1 - \Lambda^2 \rho_l \rho_s} \tag{25}
$$

This is Chipman's solution [19] at the point $z = 0$. The total solution $V(z)$ is found by adjusting the phase factors to the point z. It should be noted that this solution was obtained without taking an infinite series or infinite limit as was done in [19]. The very powerful scattering method contains the total closed solution to an infinite number of reflections.

VI. Laboratory Data

Laboratory data were taken for a system that consisted of a 20 MHz square-wave clock entering a four-leg star section of thin, insulated wire over a large ground plane. Each leg of the star was 10 cm. in length with one leg driven by a CMOS input buffer circuit. A 100 MHz bandwidth oscilloscope took data at the center of the star through a probe. The model was adjusted to compensate for the rolloff and capacitance of the oscilloscope and probe. Two experiments were run: the first was with the three non-driven star legs open-circuit (5 megohms in the model); the second was with two of the non-driven legs terminated in 100 ohm resistors with the third still open. The oscilloscope data, as well as the model predictions, are presented in Figures 7, 8, 9 and 10.

VII. Conclusions

We have presented a model for clock distribution nets that takes into consideration infinite reflections, and leads to a system of linear phasor equations for each harmonic of the clock frequency. Reconstruction of a time series at any circuit point is simply a matter of adding the cosine terms

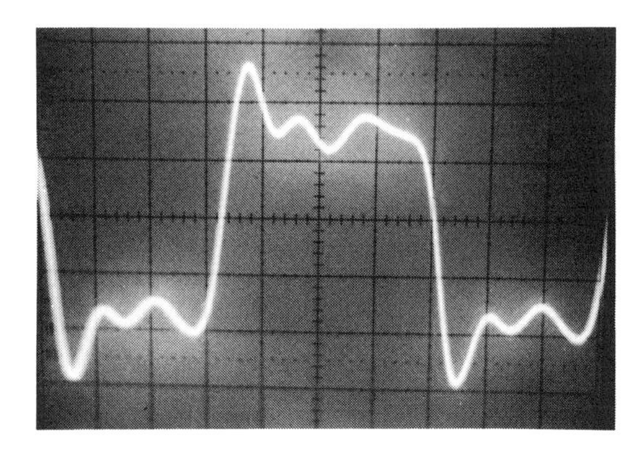

Figure 7. Laboratory data for 10 cm, four-leg star network with three open legs, and one driven leg

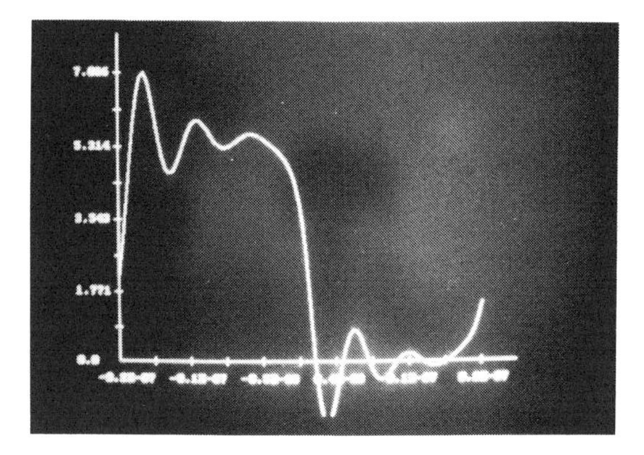

Figure 8. Model prediction for figure 7

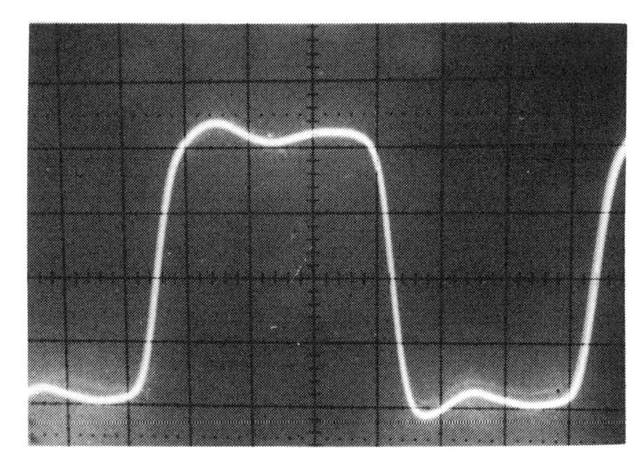

Figure 9. Laboratory data for 10 cm, four-leg star network with one open leg, two terminated legs, and one driven leg

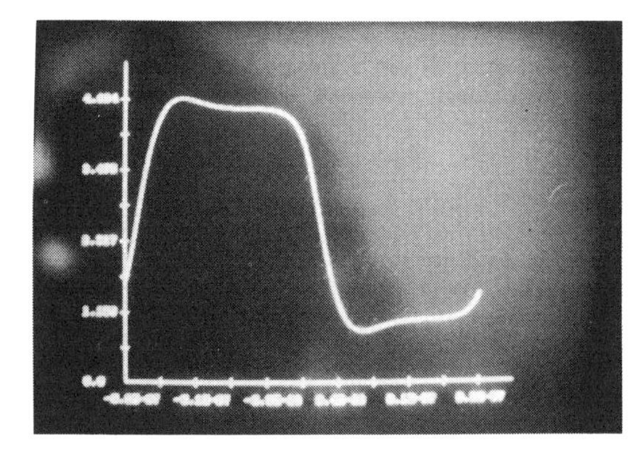

Figure 10. Model prediction for figure 9

represented by the phasor solutions. This allows construction of oscilloscope-like windows that show actual time-domain waveforms. This ability to predict the clock behavior of a real circuit allows the designer to anticipate ringing and false clock edges, and to evaluate the need and location of possible termination resistors. The model is easily programmed to facilitate numerical solution. Modified versions of it have also been used to model the effect of decoupling capacitors on power nets, and fast write strobes driving semiconductor memory devices.

References

[1] G. Ghione, I. Maio, and G. Vecchi, "Modeling of Multiconductor Buses and Analysis of Crosstalk, Propagation Delay, and Pulse Distortion in High-Speed GaAs Logic Circuits," IEEE Trans. Microwave Theory Tech., Vol. MTT-37, No. 3, Mar. 1989.

[2] Y. Fukuoka, Q. Zhang, D. Neikirk and T Itoh, "Analysis of Multilayer Interconnection Lines for a High-Speed Digital Integrated Circuit," IEEE Trans. Microwave Theory Tech., Vol. MTT-33, No. 6, June 1985.

[3] I. Bahl and R. Garg, "Designer's Guide to Stripline Circuits," Microwaves, Vol. 17, Jan. 1978.

[4] H. Howe Jr., *Stripline Circuit Design*, Artech House, Dedham, Ma., 1974.

[5] H. A. Wheeler, "Transmission Line Properties of Parallel Strips Separated by a Dielectric Sheet," IEEE Trans. Microwave Theory Tech., Vol. MTT-13, 1965.

[6] F. H. Branin, "Transient Analysis of Lossless Transmission Lines," Proc. IEEE, Vol 55, 1967.

[7] V. Dvorak, "Transient Analysis of Non-Uniform Transmission Lines," Proc. IEEE, Vol. 58, 1970.

[8] M. Protter and C. Morrey Jr., *Modern Mathematical Analysis*, Addison-Wesley, Reading Ma., 1964, Chap. 9.

[9] D. Kreider, R, Kuller, D. Ostberg, F. Perkins, *An Introduction to Linear Analysis*, Addison-Wesley, Reading Ma., 1966, Chap. 9.

[10] F. Weinert, "Scattering Parameters Speed Design of High-Frequency Transistor Circuits," Electronics, Sept. 5, 66.

[11] "S-Parameter Techniques for Faster, More Accurate Network Design," Hewlett-Packard Journal, Feb. 67.

[12] C. Kraft, "Constitutive Parameter Measurements of Fluids and Soil Between 500 kHz and 5 MHz Using a Transmission Line Technique," Journal of Geophysical Research, Vol. 92, No. B10, pp. 10,650-10,656, Sept. 1987.

[13] D. Corson and P. Lorrain, *Introduction to Electromagnetic Fields and Waves*, W. H. Freeman and Co., San Francisco, 1962.

[14] H. Ott, *Noise Reduction Techniques in Electronic Systems*, John Wiley and Sons, New York, 1976, pp 101.

[15] V. A. Monaco and P. Tiberio, "Computer-Aided Analysis of Microwave Circuits," IEEE Trans. Microwave Theory Tech., Vol. MTT-22, Mar 74, pp. 249-263.

[16] K. C. Gupta, Ramesh Garg, Rakesh Chadha, *Computer Aided Design of Microwave Circuits*, Artech House, Dedham, Ma., 1981, Chap. 11.

[17] V. A. Monaco and P. Tiberio, "Automatic Scattering Matrix Computation of Microwave Circuits," Alta Freq., Vol 39, Feb, 1970, pp. 59-64.

[18] N. Fache and D. De Zutter, "Full-Wave Analysis of a Perfectly Conducting Wire Transmission Line in a Double-Layered Conductor-Backed Medium," IEEE Trans. Microwave Theory Tech., Vol. MTT-37, No. 3, Mar. 89.

[19] R. Chipman, *Transmission Lines*, Schaum's Outline Series, McGraw-Hill, New York, 1968. pp 175.

[20] W. Blood Jr.,*MECL System Design Handbook*, Motorola Semiconductor Products Inc., 1980,83.

[21] *CMOS Data Book*, Cypress Semiconductor, San Jose, Ca., 1988, pp. 10-60.

Active Compensation of Interconnect Losses for Multi-GHz Clock Distribution Networks

Matthias Bußmann, *Student Member, IEEE*, and Ulrich Langmann, *Member, IEEE*

***Abstract*—This paper proposes an active compensation scheme for losses in heavily loaded high-speed differential clock distribution networks using Si bipolar integrated circuit technology. The compensation network consists of negative impedance converter circuitry and provides performance improvements up to multi-GHz frequencies. Even though the compensation technique is proposed in the context of Si bipolar multilevel current switch logic circuits, the analogous application with other high-speed technologies is possible where loading dominates the on-chip transmission line characteristics. The paper gives general design guidelines and reports simulation results for a 2-GHz clock distribution example problem using parameters of state-of-the-art single-poly self-aligned Si bipolar transistors.**

***Keywords*—integrated circuits, high-speed clocking, active compensation.**

I. Introduction

Clock distribution networks are one of the limiting factors for digital multi-Gbit/s circuits. Technological progress such as faster transistors can only marginally solve clock distribution problems because the physical characteristics of the interconnect remain unchanged. Technological achievements such as smaller feature sizes, larger chip area, and increased component density, are increasing the difficulties of clock distribution since they usually result in higher series resistance and higher loading per unit length for the clock distribution network. As a result, clocking of a large number of loads at multi-Gbit/s frequencies is extremely difficult.

The clock distribution problem becomes even more severe, if clocking of *large* circuits is required (c.f., [1]). The clock skew within the clock distribution network may degrade the signal delay for the critical path of the circuit. However, in the case of large circuits it may be advisable to introduce the propagation delay along the clock line as a design element in the early circuit design phase. As a simple example, consider a linear feedback shift register as shown in Fig. 1. The signal flow for the clock network and the feedback connections are running in opposite directions to decrease the influence of the propagation delay of the long feedback path on the system speed. In this case, a certain amount of clock propagation delay has a positive effect. The design goal for the clock distribution network here as in many other high-speed designs is to achieve a sufficient clock amplitude along the clock line with minimal power consumption. The clock skew, however, needs not be minimized but matched to the circuit structure.

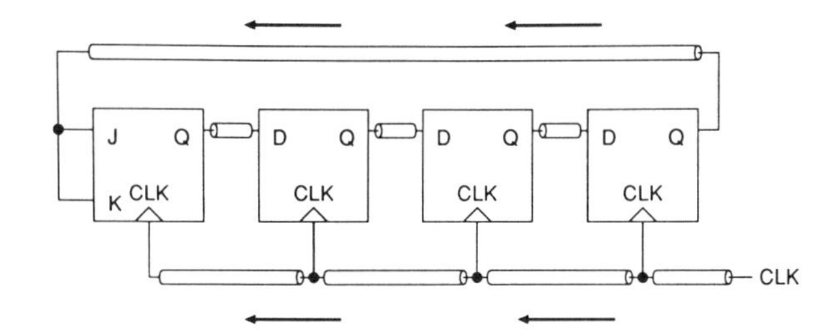

Fig. 1. Linear feedback shift register.

The amount of clock skew for optimal speed, however, is usually in a delay range which is considerably lower than a gate delay. Hence, it isn't possible to introduce additional regenerating clock buffers to reduce clock signal attenuation without drastical performance drawbacks due to clock signal propagation delay increase. A clock distribution tree would increase clock distribution overhead in regard to component count, layout area, and power consumption.

The conventional solution for the given clock distribution problem are wide clock lines and a clock driver with low output impedance and, consequently, high power dissipation. The wide clock lines are crucial 1) to lower series resistance and 2) to reduce the characteristic impedance of the clock interconnect. A smaller series resistance is lowering attenuation. A low characteristic impedance reduces the time constant existing together with a large capacitive load. Both objectives are met by using wide clock transmission lines. In turn, with increasing line width capacitances are further aggravating clock attenuation and driving problems.

This paper tries to give an alternative to the above mentioned approach by applying damping compensation techniques to the design of clock distribution networks. Damping compensation is a well-known technique which

Reprinted from *IEEE Trans. Circuits Syst.-II: Analog and Digital Signal Processing,* vol. 39, no. 11, pp. 790–798, Nov. 1992.

was used for telephone networks in the past [2]. Lately, negative resistance techniques have been applied to distributed amplifier loss compensation [3]. Influenced by this work, the present paper will show that active compensation using negative impedance converters (NIC) can improve the performance of a clock distribution network without significantly increasing power dissipation or propagation delay along the clock line. The described technique can be applied together with high-speed bipolar differential current switch logic circuits which are widely used for GHz clock frequencies. This paper reports the results obtained for a state-of-the-art single-poly self-aligned Si bipolar process [4]. The design techniques can be analogously applied with other high-speed technologies.

This paper discusses only sinusoidal clock waveforms since the described loss compensation scheme is intended for use at the speed limit for a given technology. For state-of-the-art Si bipolar technology operating at 1 GHz and above, a square wave clock input usually is converted to a sine wave by the clock driver, at least for the described heavily loaded clock distribution network. In this case, ringing and overshoot are no factor.

II. Clock Distribution Network Modeling and Analysis

For high-speed silicon bipolar technology the use of on-chip differential microstrip clock distribution networks gives considerable advantage over other clock signal wiring techniques. It minimizes on-chip coupling between adjacent signal lines and results in well-defined signal propagation characteristics since neighboring integrated circuit structures cannot influence the transmission line characteristics. Because of its differential nature, this transmission line structure is well suited for differential current switch circuits which are widely used in the Gbit/s range. In the sequel differential microstrip lines as shown in Fig. 2(a) are assumed to serve as clock wiring network.

For circuit simulations, differential microstrip wiring can be modeled using the simplified equivalent circuit model shown in Fig. 2(b) [5]. It contains the per-unit-length values of the wiring capacitances C_1', C_2', the series inductances L_1', L_2', the series resistances R_1', R_2', the coupling capacitance C_k', and the mutual inductance M' for a line segment of infinitesimal small length. For the simulation itself, a suitable segment length has to be chosen and the equivalent circuit elements have to be multiplied by the segment length l. This segment length is determined by the trade-off between computational efficiency and simulation accuracy.

Traditional transmission line theory based on the telegraph equation leads to the characteristic impedance Z_L and the propagation constant γ. They can be expressed as

$$Z_L = \sqrt{\frac{R' + j\omega L'}{G' + j\omega C'}} \tag{1}$$

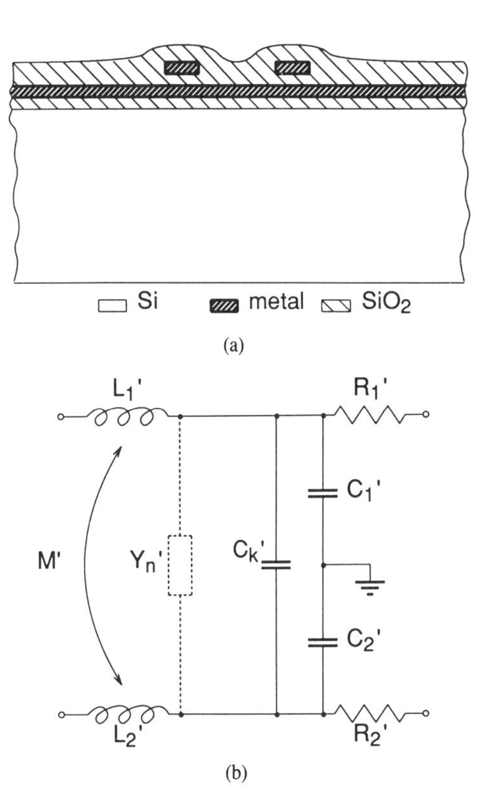

Fig. 2. Differential microstrip. (a) Cross section. (b) Simplified equivalent circuit model (active compensation dashed).

and

$$\gamma = \alpha + j\beta = \sqrt{(R' + j\omega L')(G' + j\omega C')}. \tag{2}$$

For differential microstrip transmission characteristics, mainly the odd mode values are important since for ideal differential signals only the odd mode is excited. For the application of the transmission line theory the values of R', L', C', and G' can be computed from the differential microstrip equivalent circuit model.

The characteristic impedance and the propagation constant can be obtained from the input impedances Z_S and Z_O of a transmission line for a short (index S) and an open (index O) at the receiving end:

$$Z_L = \sqrt{Z_S Z_O} \tag{3a}$$

$$\gamma = \frac{1}{2l} \ln \frac{1 + \sqrt{Z_S/Z_O}}{1 - \sqrt{Z_S/Z_O}} \tag{3b}$$

where $\beta > 0$ and $\mathrm{Im}(Z_L \gamma) > 0$ has to be considered to avoid ambiguity.

For an unloaded transmission line, a direct computation of Z_L and γ is possible. For complex transmission line structures, however, the determination of the transmission line characteristics from a small transmission line section via (3a) and (3b) is crucial for computational efficiency.

The key approach to clock distribution network analysis of periodically loaded transmission lines is to consider the load circuitry as part of the transmission line structure itself. Since loading dominates the transmission line characteristics for integrated circuits, a distributed load model results in a considerable change of Z_L and γ. The distributed load model for periodically loaded transmission lines can be applied if the distance between the loads doesn't exceed the limits set by the maximum frequency. For integrated circuit structures with maximum load distances of 50–100 μm and operating frequencies below 20 GHz, this is generally true.

III. Compensation Method

The idea behind active compensation of clock distribution network losses is to add distributed compensation circuitry to influence the propagation characteristics of the transmission line. Specifically, it is intended to reduce the attenuation constant α. Compensating the series resistance R' by a negative series resistance $-R'$ would result in an additional propagation delay through the added active element. Therefore, this paper favors an approach introducing a negative admittance $Y_n' = G_n' + sC_n'$ to the transmission line. For the differential microstrip case this can be realized as a negative admittance in parallel to the coupling capacitance C_k' (c.f., Fig. 2(b)). The advantage of a differential negative admittance consists of a simple realization using conventional circuit techniques.

The compensation method applied in this paper uses the propagation constant γ as a starting point. For the complete transmission line system including load and compensation circuitry, γ and Z_L can be expressed as

$$\gamma = \alpha + j\beta = \sqrt{(R' + sL')(x_L G_L + x_C G_C + s[C' + x_L C_L + x_C C_C])} \tag{4}$$

$$Z_L = \sqrt{\frac{R' + sL'}{x_L G_L + x_C G_C + s[C' + x_L C_L + x_C C_C]}} \tag{5}$$

where R', L', and C' denote the transmission line parameters, G_L and C_L the input conductance and capacitance of a load, G_C and C_C the input conductance and capacitance of a compensation circuit, x_L and x_C are the number of loads/compensation circuits per unit length.

Ideal loss compensation is obtained by forcing α to 0. By inspecting (4), it is obvious that there are two basic compensation techniques. The first one tries to minimize the damping constant α by adjusting the phase of γ^2 to 180°. From (4), $\alpha = 0$ can be achieved by choosing G_C and C_C under the constraints

$$\frac{C' + x_L C_L + x_C C_C}{x_L G_L + x_C G_C} = -\frac{L'}{R'} \tag{6}$$

where $x_L G_L + x_C G_C < 0$. This method potentially increases the magnitude of γ. In turn, a larger β may result in a reduced phase velocity.

The other approach tries to minimize the magnitude of γ. This implies

$$G_C = -\frac{x_L}{x_C} G_L \tag{7a}$$

and

$$C_C = -\frac{C' + x_L C_L}{x_C}. \tag{7b}$$

In this case, α as well as β can be reduced simultaneously. As a result, the phase velocity can be increased.

From the effects of the compensation method on phase velocity, it is clear that the minimization of the magnitude of γ better suits the requirements for on-chip loss compensation in integrated circuits. A large phase velocity increase would result in an increased delay for sinusoidal clock signals. With a reduction of β, however, a loss compensation together with an improvement of delay performance of the clock transmission line is possible. Furthermore, G_C and C_C can be optimized independently which results in a simple design procedure.

IV. Stability

In [3], approximate expressions for conditional stability have been established. For the compensated clock transmission line, however, this paper tries to attain unconditional stability which is obtained if passive matching at the driving and the receiving end has no influence on the circuit's stability. This is important because it is very difficult to meet the stability requirements on termination matching in a complex integrated circuit environment. For the case of a compensated clock transmission line structure unconditional stability requires (c.f., [6])

$$k = \frac{1 + |S_{22}S_{11} - S_{12}S_{21}|^2 - |S_{11}|^2 - |S_{22}|^2}{2|S_{12}S_{21}|} > 1 \tag{8a}$$

and

$$|S_{12}S_{21}| < 1 - |S_{11}|^2, \qquad |S_{12}S_{21}| < 1 - |S_{22}|^2 \tag{8b}$$

where the S_{ij} denote the commonly known S parameters. For our case of a symmetrical clock transmission line structure, this reduces to

$$k = \frac{1 + |S_{11}^2 - S_{21}^2|^2 - 2|S_{11}|^2}{2|S_{21}|^2} > 1 \tag{9a}$$

and

$$|S_{21}|^2 < 1 - |S_{11}|^2 \tag{9b}$$

where

$$S_{11} = \frac{Z_L - Z_0}{Z_L + Z_0} \quad \text{and} \quad S_{21} = e^{-\gamma 1}. \tag{9c}$$

Let $Z_0 = Z_L$, then it immediately follows from (9b) that unconditional stability requires $\alpha > 0$.

V. Negative Impedance Converter

The NIC circuit discussed here is well suited for the compensation of losses of differential transmission lines because it provides only a differential negative admittance. For the even mode, the input admittance is purely positive. This is useful for avoiding even mode oscillations. As discussed later, the circuit profits from its simple biasing and good high-frequency characteristics.

The basic NIC circuit used for active loss compensation is shown in Fig. 3 [7]–[9]. It can be classified as a voltage inverting NIC. Port 1–1′ is short-circuit stable whereas port 2–2′ is open-circuit stable.

Circuit analysis [9] yields a chain matrix

$$A \approx \begin{bmatrix} -(1+s\tau_A) & -r_p(1+s\tau_B) \\ -sC_p & K\cdot(1+s\tau_D) \end{bmatrix} \tag{10}$$

for the 2-port network of Fig. 3. The parameters can be obtained as

$$K = 2\alpha - 1 \tag{11a}$$

$$r_p = 2[r_e + (1-\alpha)r_b] \tag{11b}$$

$$C_p = 2C_\mu + C_s/2 \tag{11c}$$

$$\tau_A = 2C_\mu r_b \tag{11d}$$

$$\tau_B = r_b C_\mu - r_e C_\pi + \frac{r_e r_b}{r_e + (1-\alpha)r_b}(C_\pi + C_\mu) \tag{11e}$$

$$\tau_D = -(2C_\pi + 4C_\mu + C_s)r_e \tag{11f}$$

where α denotes the common base current gain, r_b the base resistance, r_e the emitter resistance, C_π the base-emitter capacitance, C_μ the collector-base capacitance, and C_s the collector-substrate capacitance. Please note that in contrast to [9] this analysis includes the time constant τ_B which cannot be neglected for high-frequency operation.

An approximate expression for the input admittance of an NIC terminated with a resistor R_2 in parallel to a capacitor C_2 at port 2-2′ can be obtained from the chain matrix as

$$Y_C \approx -\frac{1}{R_2 + r_p} \cdot \frac{1 + s(\tau_D + R_2[C_2 - C_p])}{1 + s\cdot\frac{R_2}{R_2 + r_p}\cdot\left(\tau_A + \frac{r_p}{R_2}\cdot[\tau_B + R_2 C_2]\right)}. \tag{12}$$

From (12), it is clear that the load at the secondary port 2-2′ significantly degrades the NIC bandwidth especially in the case of a large capacitive load and low bias currents. This bandwidth reduction imposes a constraint on the values of R_2 and C_2. Another bandwidth restriction at high frequencies is the time constant τ_D. τ_D is large for small NIC bias currents. If a large bandwidth is crucial then circuit spacing should be made larger to increase the NIC bias current while keeping the total power budget constant. Transistor saturation effects and the maximum NIC spacing imposed by the validity of the distributed analysis define an upper limit for the distance between compensation circuits.

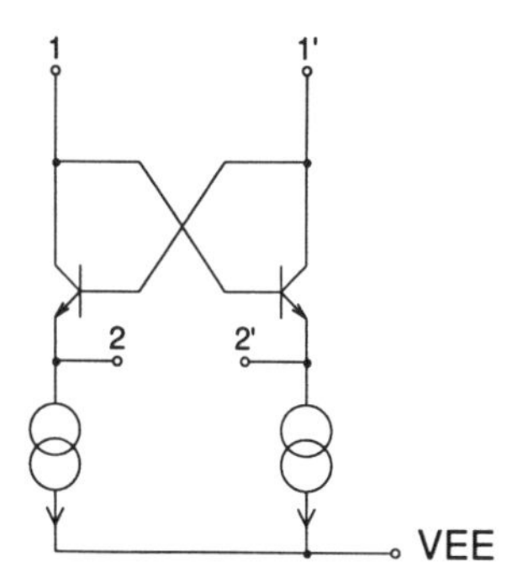

Fig. 3. Basic NIC circuit.

Fig. 4 shows a practical implementation of a negative admittance for clock transmission line loss compensation. The current sources have been replaced by resistors. The secondary port load is formed by the pull-down resistors and by the reverse biased diodes acting as a capacitance. The differential transmission line is directly connected to the primary port. Bias current can be drawn through the clock transmission line as well as through dedicated bias resistors. However, bias current drawn through the clock line acts as a substitute of the bias current of the clock driver emitter follower and thus reduces the power dissipation of the clock driver. Hence, a part of the power dissipation of the clock driver is shifted to the NIC acting as an active compensation. As a result, the total power dissipation of the clock distribution circuitry consisting of NIC's and clock driver is increased only moderately despite the active NIC circuitry. For the basic NIC circuit, one can even think of a configuration where a reduction in total power dissipation can be achieved.

The problem of NIC circuit layout can be solved quite advantageously. By placing the NIC circuit directly under the clock line, the NIC forms a homogeneous structure with the clock transmission line. The clock wiring usually is separated from other circuitry to reduce crosstalk. Therefore, an area penalty for the NIC's doesn't exist.

VI. Compensation Using NIC's

The design of the NIC compensation circuitry starts at the point where the load and the structure of the clock connections are already determined. Hence, R', L', C', G_L, C_L, and x_L are fixed. The compensation circuitry can now be optimized by selecting G_C, C_C, and x_C for minimal damping under the stability constraint $\alpha > 0$. Introducing

$$G'_{\text{tot}} = x_L G_L + x_C G_C \quad \text{and} \quad C'_{\text{tot}} = C' + x_L C_L + x_C C_C \tag{13}$$

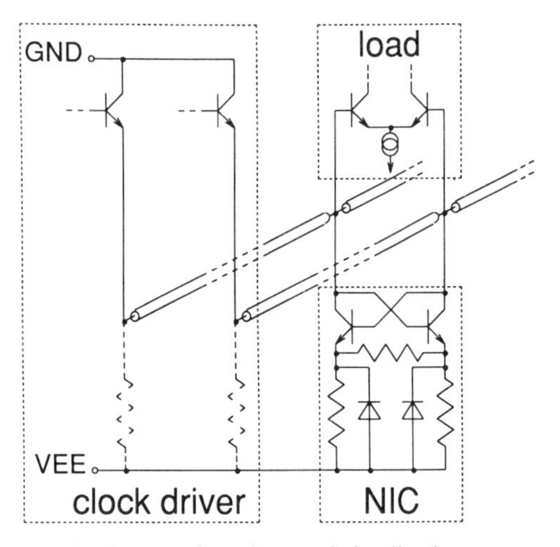

Fig. 4. NIC implementation of transmission line loss compensation.

optimal damping under the stability constraint can be achieved by minimizing G'_{tot} and C'_{tot} where $G'_{tot} > 0$ and $C'_{tot} > 0$.

As mentioned above complex transmission line structures require a considerable computational effort for the computation of Z_L and γ. Hence, a straightforward design approach is to take a short piece of the homogeneous transmission line structure. Using (3), Z_L and γ can be computed from simulated ac input impedances for the short and open circuit case. In addition to that, a direct computation of G'_{tot} and C'_{tot} is possible. By performing a reference simulation of the uncompensated transmission line structure, the improvements achieved by compensation can be related to the uncompensated values.

Fig. 5(a)–(f) show SPICE simulation results for a heavily loaded clock transmission line obtained with parameters of a state-of-the-art single-poly self-aligned Si bipolar technology [4]. A load consists of a current switch buffer with a 2-mA transistor current source. The load spacing is 100 μm. A clock transmission line of 10 μm width is used. The bias current for the compensation circuitry was chosen to be 2 mA/mm. To increase the bandwidth of the compensation circuitry, the spacing between the individual compensation circuits was doubled compared to the load distance. The simulations were obtained from an 800-μm transmission line structure section.

The most important plots for choosing R_2 and C_2 are shown in Fig. 5(a) and (b). For reference, the uncompensated values of G'_{tot} and C'_{tot} are plotted, too. Please note that a compensation circuit loading configuration similar to Fig. 4 was utilized consisting of three resistors and two reverse biased emitter-collector shorted transistors. For low frequencies, a relatively large offset to $G'_{tot} = 0$ and $C'_{tot} = 0$ of approximately 5–10% of the uncompensated values is maintained to avoid stability problems due to device parameter spread. At high frequencies G'_{tot} drops below zero. This can be tolerated as long as

$$\frac{\omega C'_{tot}}{G'_{tot}} > -\frac{\omega L'}{R'} \tag{14}$$

where $G'_{tot} < 0$. This condition is analogous to (6). That (14) holds in this case can be verified by simulations. The damping and phase constants α and β for the compensated and uncompensated case are plotted in Fig. 5(c) and (d). A damping improvement is achieved up to 12 GHz. Stability is clearly maintained with $\alpha > 0$. The phase constant β remains below the uncompensated value for low frequencies, but due to bandwidth limitations of the NIC (c.f., Fig. 5(a) and (b)) it increases beyond the uncompensated value for high frequencies. This implies that for high frequencies there exists a phase delay penalty for the compensated transmission line structure. However, in this case the phase delay penalty remains below 10 ps/mm as can be see in Fig. 5(e). For matching purposes, the characteristic impedance of the compensated transmission line structure is shown in Fig. 5(f).

The relative improvement of transmission line damping is shown in Fig. 6. More than 50% improvement is maintained up to 1.6 GHz. For verification of the ac simulations large signal transient simulations are performed at a frequency of 1 GHz. The waveform at the input and at the output of a 5-mm clock line are shown in Fig. 7. Please note that the receiving end of the transmission line is matched with the characteristic impedance at 1 GHz obtained from Fig. 5(f). For the moderate differential input voltage swing of 400 mV and very low NIC current densities, ac simulations are in good agreement with the large signal transient simulation results. The output voltage amplitude can be predicted from ac simulations with an accuracy of better than 5%. For larger voltage swings, however, amplitude differences between ac and transient simulations are caused by circuit nonlinearities and transistor saturation. Consequently, in case of saturation of the basic NIC transistors, emitter followers should be used in the compensating NIC's.

It is obvious that the good agreement between the results of ac and transient simulation can be exploited by using the ac analysis as the primary simulation tool for designing clock distribution networks. Simulation time may be cut by using only a short section of homogeneous transmission line segments for the determination of the short and open circuit impedance. Furthermore, a complex clock distribution network can be split up into homogeneous parts which can be optimized individually. As long as matching is achieved at the interfaces, a complete network synthesis can be performed without running time-consuming transient simulations of a complex clock distribution network.

VII. Design Example

The following example demonstrates the performance of the proposed loss compensation scheme for clock distribution networks at high frequencies. Consider a clock system as given in Fig. 8. A 5-mm clock line has four differently loaded sections. The clock transmission line is driven by a triple stage emitter follower clock driver. At the end of the clock line, an 8-mA clock output buffer together with a termination network matches the line.

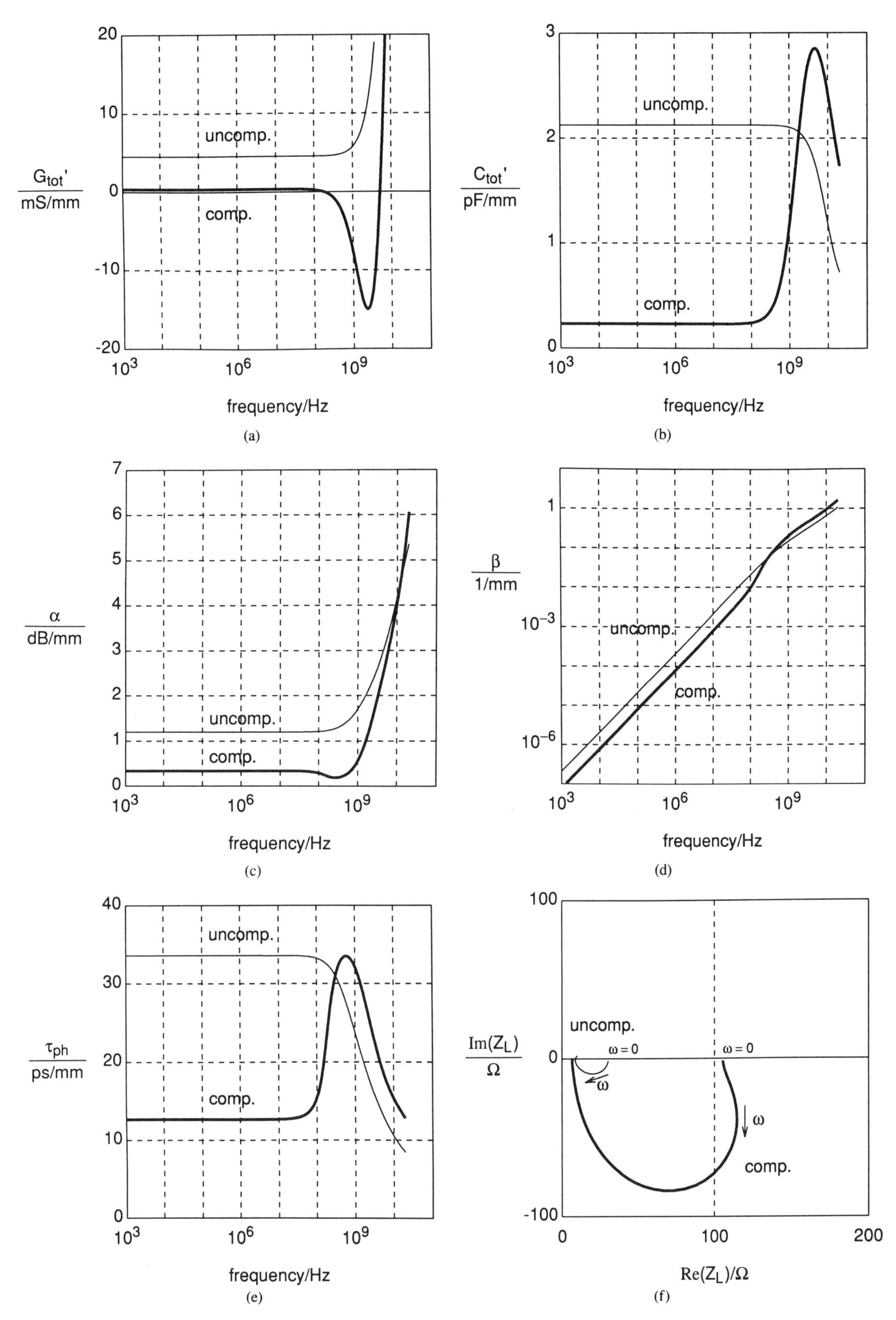

Fig. 5. Simulation results for a 10-μm clock transmission line with 2-mA buffer loads with 100-μm spacing. Compensated case (bold) and uncompensated case. (a) Total conductance G'_{tot}; (b) total capacitance C'_{tot}; (c) damping constant α; (d) phase constant β; (e) phase delay τ_{ph}; and (f) characteristic impedance Z_L.

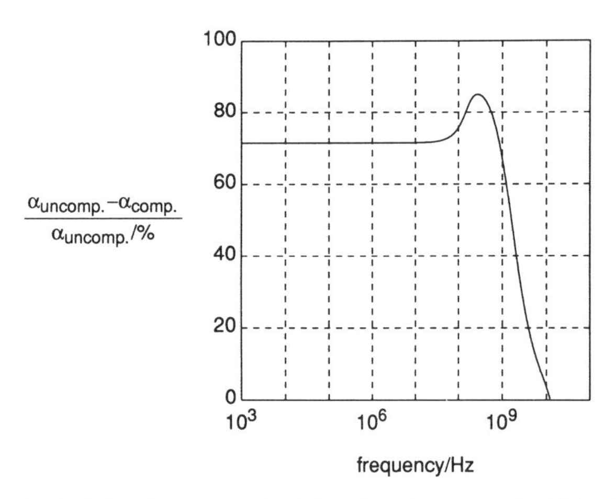

Fig. 6. Relative improvement of the damping constant obtained by active loss compensation for a 10-μm clock transmission line with 2-mA buffer loads with 100-μm spacing.

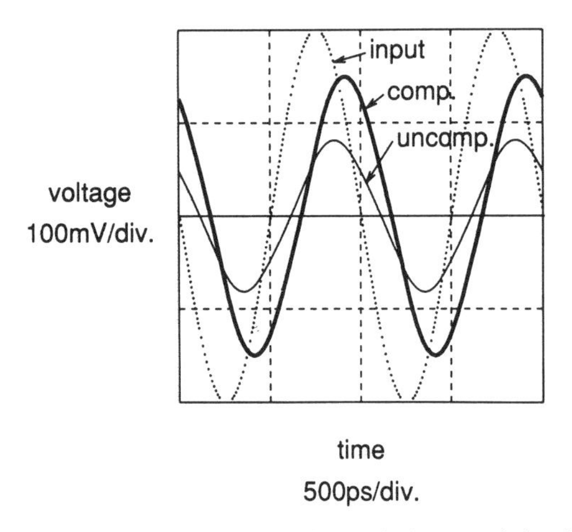

Fig. 7. Transient simulations of a 10-μm clock transmission line with 2-mA buffer loads with 100-μm spacing driven by an ideal voltage source and terminated into the characteristic impedance. The line length is 5 mm. Differential voltages are recorded at the input (dotted), the compensated output (bold), the uncompensated output.

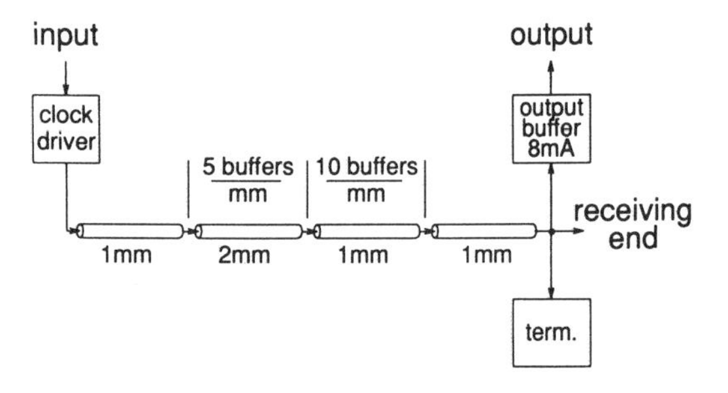

Fig. 8. Schematic of a clock distribution network showing clock driver, 5-mm clock transmission line with four differently loaded sections, an 8-mA output buffer, and a termination network (term.). A standard load consists of a buffer with a tail current of 2 mA.

The target frequency for this system is 2 GHz using the Si bipolar technology described in [4].

The problem arising in this example consists not only of the optimization of the compensation circuitry but also of avoiding reflections between the wiring segments. This requires a trade-off between minimization of C'_{tot} and G'_{tot} to obtain an optimal damping constant for the different sections and matching of the Z_L at the section interfaces. Because the increase of C'_{tot} is extremely large at the NIC's operating frequency limit (cf., Fig. 5(b)), a broadband segment match is very difficult when optimizing for maximum operating frequency.

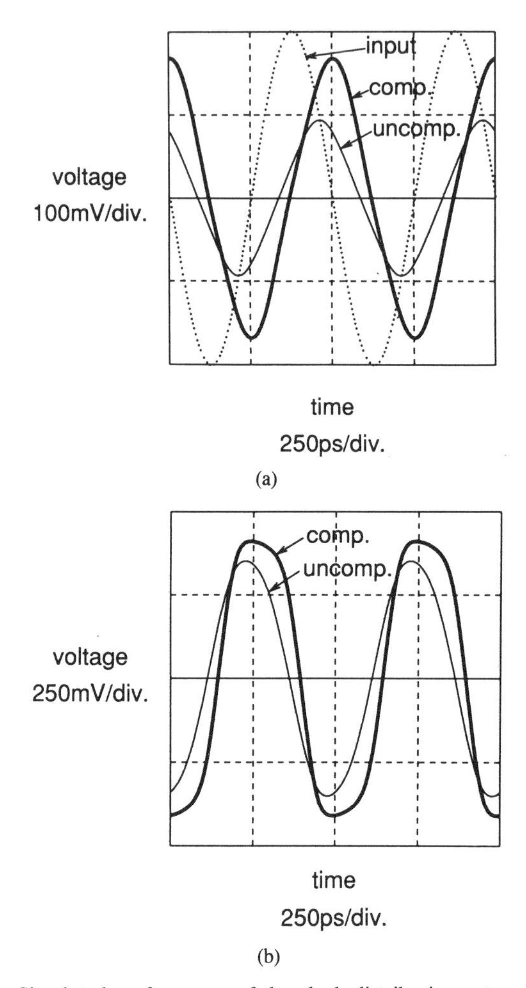

Fig. 9. Simulated performance of the clock distribution network given in Fig. 8. Differential voltage uncompenated and compensated (bold). (a) Input (dotted) and receiving end of the clock transmission line. (b) Output.

Fig. 9(a) and (b) show SPICE transient simulation results for the clock distribution system in Fig. 8. For reference, the results for the uncompensated transmission line are plotted. In Fig. 9(a), about 80% increase in amplitude at the receiving end of the clock transmission line have been achieved. As a result, the output buffer which is terminated into 50 Ω has its full output amplitude whereas for the uncompensated case, the clock voltage swing is too low to switch the buffer completely, as can be seen in Fig. 9(b). Both systems have approximately the same power consumption for the complete clock driving circuitry including NIC's. During simulation, nonlinear and saturation effects have been included by representing all building blocks by nonideal components consisting of only transistors and resistors. Generally, it is important to match the transmission lines at the receiving end to avoid voltage drops caused by standing wave effects.

The direct analogy between ac and transient simulations is difficult to observe for transmission lines with differently loaded segments since the reflection coefficients at the segment interfaces have to be taken into account. However, it is still possible to optimize damping for the individual segments and to minimize differences in characteristic impedance.

VIII. Comparison with Other Clock Distribution Techniques

Three of the other clock distribution techniques widely used in integrated circuit design are 1) the large buffer direct drive scheme, 2) the regenerating buffer design, and 3) the clock distribution tree. When comparing performance optimized clock distribution networks, amplitude loss and clock skew are the most important criteria.

The large buffer direct drive scheme with a large emitter follower driving an uncompensated transmission line, exhibits poor amplitude performance. The clock distribution scheme proposed in this paper as a direct improvement to the large buffer direct drive concept reduces this limitation. For low frequencies, clock skew or clock phase delay along the clock line can be reduced but for high frequencies a certain amount of phase delay penalty exists.

The regenerating buffer concept suffers from the additional delay introduced by the regenerators. As mentioned above, the loss compensation scheme proposed in this paper offers superior delay performance at low frequencies. At high frequencies, the delay penalty for the loss compensation scheme may theoretically result in a better performance of the regenerating buffer concept. However, in practice with buffer delays in the range of 40–50 ps, even for a relatively large delay increase, the loss compensation scheme is far from losing its advantage in delay performance. This can be verified by an expression for the maximum allowable delay $T_{RB_{max}}$ of a regenerating buffer to yield a better total clock distribution network delay than with the loss compensation scheme:

$$T_{RB_{max}} = l_{RB_{max}}(\tau_{comp.} - \tau_{uncomp.}) \tag{15a}$$

where

$$l_{RB_{max}} = \frac{\Delta_V}{\alpha_{uncomp.}}. \tag{15b}$$

$\tau_{comp.}$ and $\tau_{uncomp.}$ are the phase delays per unit length for the compensated and the uncompensated transmission lines, respectively, $l_{RB_{max}}$ denotes the maximum distance between the regenerators given by the maximum allowable voltage drop Δ_V in decibels, and $\alpha_{uncomp.}$ the damping constant of the uncompensated transmission line. For the transmission line structure in Section VI, an $l_{RB_{max}}$ of 1.66 mm can be computed using $\Delta_V = 2$ dB. For this configuration, $T_{RB_{max}}$ remains well below 10 ps, which is much lower than the actual regenerating buffer delay.

A clock distribution tree design may possibly result in an optimal clock skew and minimal amplitude loss. However, power dissipation, layout area, component count, and complexity of the resulting network must be considered as major drawbacks if compared to the loss compensation approach presented here.

IX. Conclusion

A loss compensation technique for heavily loaded clock distribution networks has been proposed. By using a differential mode NIC circuit for compensation, it is possible to adapt negative impedance loss compensation to high-speed differential current switch logic circuits. This approach only marginally increases clock phase delay at GHz frequencies. The maximum frequency where the active loss compensation leads to an amplitude improvement is basically given by the NIC bandwidth. With increasing transistor speeds, the application range for the proposed type of loss compensation is extended to higher frequencies where transmission line losses play an even larger role for system performance.

Clock distribution network design using the proposed frequency-domain approach results in very clear relationships between the basic performance parameters such as number of loads and damping. By considering the load and the compensation circuitry as part of the transmission line, simple objective functions can be derived for optimization. With low-complexity ac analyses, it is possible to optimize a complex clock distribution system in a computationally efficient way.

Acknowledgment

The authors would like to thank W. W. Brown and B. Hillery of Hewlett-Packard California Design Center, Santa Clara, CA, for making available transistor simulation parameters and T. Tzschoppe for preliminary investigations on transmission line loss compensation during his diploma thesis.

References

[1] M. Bußmann and U. Langmann, "Proposal for a versatile monolithic multi-Gbit/s m-sequence test system," *Electronics Letters*, vol. 26, pp. 1625–1626, Sept. 1990.

[2] J. L. Merrill, Jr., "Theory of the negative impedance converter," *Bell Syst. Tech. J.*, vol. 30, pp. 88–109, Jan. 1951.

[3] S. Deibele and J. B. Beyer, "Attenuation compensation in distributed amplifier design," *IEEE Trans. Microwave Theory Tech.*, vol. MTT-37, pp. 1425–1433, Sept. 1989.

[4] W. M. Huang *et al.*, "A high-speed bipolar technology featuring self-aligned single-poly base and submicrometer emitter contacts," *IEEE Electron Device Letters*, vol. EDL-11, pp. 412–414, Sept. 1990.

[5] B. Wüppermann, "Entwurf monolithisch-integrierter Registerschaltungen in Silizium-Bipolartechnologie für den Gbit/s-Bereich unter Einbeziehung von Verbindungsleitungen," dissertation, Lehrstuhl für Elektronische Bauelemente der Ruhr-Universität Bochum, 1990.

[6] J. M. Rollett, "Stability and power-gain invariants of linear two ports," *IRE Trans. Circuit Theory*, vol. CT-9, pp. 29–32, Mar. 1962.

[7] S. Takagi *et al.*, "100-MHz monolithic low-pass filters with transmission zeros using NIC integrators," *IEEE J. Solid-State Circuits*, vol. SC-26, pp. 669–671, Apr. 1991.

[8] H. Hagiwara *et al.*, "A monolithic video frequency filter using NIC-based gyrators," *IEEE J. Solid-State Circuits*, vol. SC-23, pp. 175–182, Feb. 1988.

[9] M. Kumazawa and T. Yanagisawa, "Video frequency active filters using balanced-type NIC's," *Electronics and Communications in Japan, Part 2*, vol. 69, pp. 41–50, Feb. 1986.

Power Dissipation in the Clock System of highly pipelined ULSI CMOS Circuits

Erik De Man and Matthias Schöbinger

SIEMENS A.G., Corporate Research and Development, D-81730 Munich, Germany
email : erik.de.man@zfe.siemens.de

Abstract: Pipelining is an efficient way to increase the computational throughput of a feed-forward synchronous data-path but increases the capacitive load of the clock network. Even with an optimized level of pipelining, the power dissipated in the clock system of high-throughput ULSI circuits can be a considerable part of the total power dissipation. In practical implementations complementary or complementary non-overlapping clock systems are used, requiring the distribution of two or even four different clock signals. True-single-phase registers requiring only a single clock signal enable a significant reduction of the capacitive load of the clock network and will be analyzed with respect to their potential for reduction of the total power dissipation of pipelined ULSI circuits.
Basically, the power dissipation depends quadratically on the voltage swing, so reducing the swing of the clock signals offers a potential for power dissipation savings. A novel clock driver circuit for on-chip generation of clock signals with a reduced swing will be presented, allowing a reduction of the power dissipated in the clock system of high performance ULSI circuits of more than 60%, without reducing the throughput rate of the circuit.

1. Introduction

Introducing pipelining in a data path based circuit, is an efficient and customary way to increase the throughput rate of the circuit and thus increasing the computational power of the data path. Circuits featuring a high degree of pipelining inherently have short and simple critical paths. The short paths are essential for achieving a high throughput rate, but they also allow the circuits to be realized with minimum sized devices, which is a prerequisite for low power dissipation [1]. Moreover the inherent locality of the logical paths between the pipeline registers considerably reduces the power dissipation of the logic circuits due to glitches [2].
So the principle of pipelining, applied to increase the throughput rate, may at the same time reduce the power dissipated in the logic circuitry resulting in a reduction of the total power dissipation. However, the large number of pipeline registers controlled by one or more clock signals results in a considerable capacitive load in the clock network requiring an optimization of the degree of pipelining [3].
The energy for charging the clock network each clock cycle is delivered by the clock system of the circuit. An appropriate measure for the power efficiency of a circuit is the specific power dissipation p , defined as the total power dissipation P divided by the clock frequency 1/T (which is related to the critical path delay) and divided by the number of equivalent gates G of the circuit

$$p = P \cdot T / G \qquad (1)$$

In Fig. 1 the power dissipation per gate vs. the clock frequency is compared for a number of chips realized in 1.5-µm CMOS. It shows that if the power dissipated in the clock network is not taken into account, the specific power dissipation of optimally pipelined circuits is an order of magnitude lower than the specific power dissipation of non-pipelined circuits or gate array implementations.

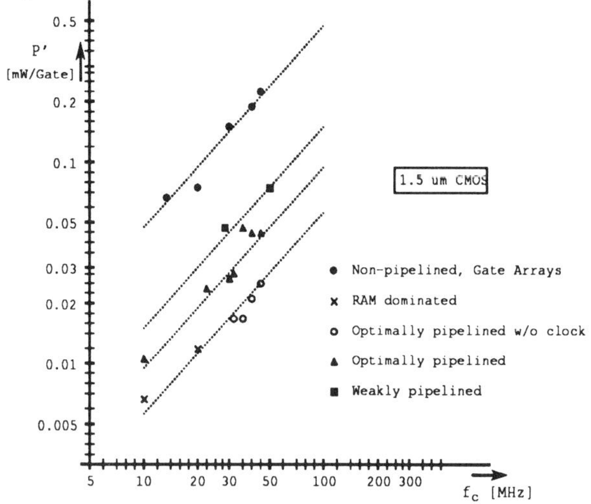

Fig. 1. Power dissipation vs clock Frequency.

However the power dissipated in the clock system of even optimally pipelined [3] high-throughput ULSI circuits may be typically as large as the power dissipated in the logic circuitry. But even if the additional power

Reprinted from *Proc. Int'l Workshop on Low-Power Design,* April 1994.

dissipation for the clock system is also taken into account (Fig. 1), the specific power dissipation of highly pipelined circuits is still about a factor 5 better than of non-pipelined circuits. Nevertheless, the clock system of such circuits still holds a great potential to reduce the total power dissipation.

2. The Clock Network

2.1. Clocking strategies

The complexity of the clock network of pipelined circuits depends on the clocking and pipelining strategy used. Because of the finite transition time of the clock signals, strategies to prevent races have to be considered in the clocking concept [4].

Complementary non-overlapping clock systems basically avoid races by the interdependent timing characteristics of the clock signals but they require the generation and routing of four different signals. But the dynamic pipeline registers of a non-overlapping clock system can be distributed over the logic blocks (Fig. 2) and can be efficiently realized as transmission gate latches exploiting the input capacitances of the pre-ceeding logic block as dynamic storage elements. Moreover, if the latches are distributed in such a way that the delay time between two successive latches is longer than the non-overlap time of the clock phases, the introduction of a non-overlap time in the clock system will not reduce the computation time of the logical circuits. The complexity in terms of capacitive clock load for non-overlapping clock systems will be analyzed in section 2.2.

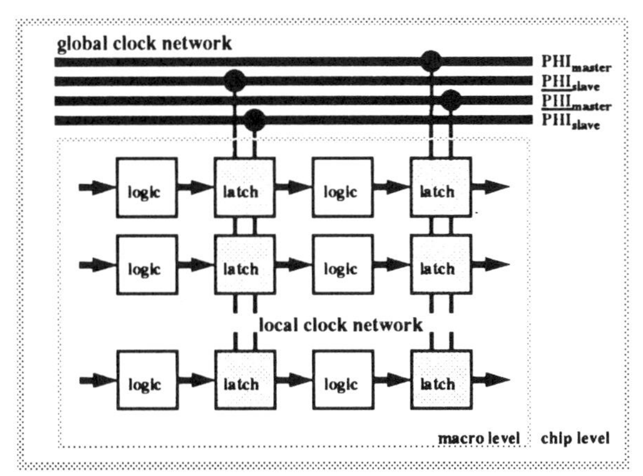

Fig.2 The clock network.

In a strict complementary clock system races must be avoided by a careful design of the clock network and pipeline registers. The registers must be optimized for delay time and clock transition requirements. The clock network must be thoroughly laid out to minimize signal delays mainly resulting from the RC-nature of the clock wiring. The advantage of strict complementary clock systems is that only two clock signals have to be routed around the chip. However this advantage can only be fully exploited if the registers are not distributed as latches over the logic blocks. The pipeline registers can be based on e.g. C^2MOS or tranmission-gate inverter latches which are more complex than the transmission based latches applicable with non-overlaping clock systems.

True single-phase clock systems [5] originally developed for high-speed CMOS circuits have the most simple clock network since only one signal has to be routed around the chip. However even more complex registers are required, which also demand a carefully designed clocking network in respect to the transition time of the clock signal. In section 3. the potential of such simple single-phase clock systems to reduce the capacitive clock load and power dissipation is evaluated. Especially the impact of the more complex register circuits on the total power dissipation will be discussed.

2.2. Capacitive load of the clock network

The expansive clock network is charged with the gate capacitances (C_g) of the clocked transistors of the pipeline registers. Another capacitive part of the clock network, is the junction capacitance of the source-drain regions of the output nodes of the clock drivers (Cj). The wiring capacitance of the clock network can be separated into the capacitance of the global clock network (C_{wg}) and the capacitance of the local clock network (C_{wl}). This is illustrated in Fig.2. The total capacitive load C_L of the clock network can be written as :

$$C_L = C_{wg} + C_{wl} + Cg + C_j \tag{2}$$

The power needed for switching the clock network 1 / T times per second is then given by :

$$P_{cn} = 1 / T \,.\, Vdd^2 \,.\, C_L \tag{3}$$

macro	gate capacitance (C_g)	local wiring capacitance (C_{wl})	global wiring capacitance (C_{wg})	junction ca-pacitamce (C_j)
FIR macro	50pF	105pF	19pF	5pF
correlator macro	16pF	38pF	12pF	2pF

Table 1 Capacitive load C_L of the clock Network

In Table 1 the relative portion of the different capacitive parts of the clock network is given for two realized examples of pipelined carry-save macros realized in a 1.0-μm technology [6]. The macros are designed for a complementary non-overlapping clock system and the registers are distributed as latches over the logic of the data path as descussed above (Fig.2).Apparently, the dominant part of the capacitive load is the capacitance of the local clock network (C_{wl}). Also for complementary clock systems for which a 50% reduction of the wiring capacitance can be expected, the local wiring capacitance still represents a major part of the clock load.

3. Single-phase registers

C^2MOS or transmission gate registers (Fig 3a) have traditionally been used as hardware efficient dynamic pipeline registers. They require complementary clock phases for the controlling of the p- and n-channel devices so that the local clock network will require the wiring of at least two different clock signals.

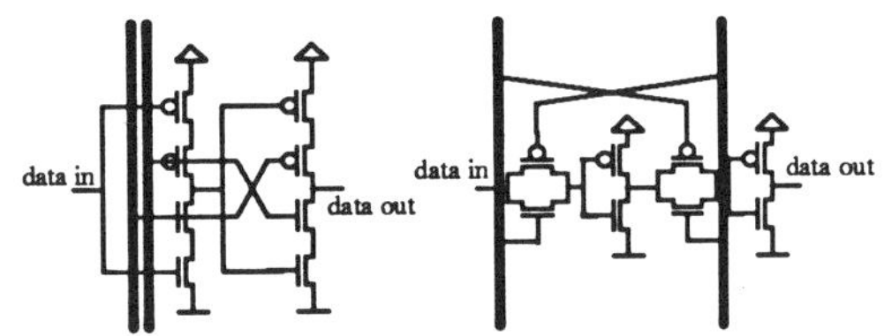

Fig. 3a. C^2MOS register and transmission-gate register.

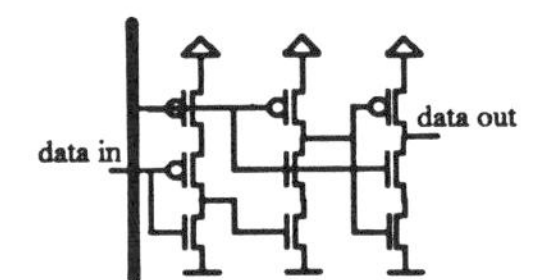

Fig. 3b. Single-phase register.

True-single-phase registers [5], originally developed for high-speed circuits, require only one clock phase to be distributed (Fig. 3b). As already mentioned, the capacitance of the global and local clock wiring can be reduced by about 50% compared to a complementary clock system . The gate capacitance will not change very much since single-phase registers also have four transistors connected to the clock signal. Since the local wiring capacitance C_{wl} represents a major part of the total capacitace of the clock network C_L , a significant saving can be expected for the power dissipation of the clock system. However, the circuits for the single-phase registers are more complex than simple C^2MOS register or pass-gate registers (Fig. 3).

In Table 2 some characteristic features of both C^2MOS and single-phase registers based on 1.0-μm CMOS designs are summarized. The capacitive load of the local clock network is reduced by 53% (from 64fF to 30fF) as expected. But for a typical input transition probability of 25% (σ=0.25), the power dissipation of the logic itself has increased by 182% (from 32μW to 102μW). The main reasons for this increase in power dissipation of the logic is that the single-phase register has more internal nodes than the C^2MOS register and it also contains a precharge inverter (central part of the register) which depending on the transition statistic of the input signal can have an unfavourable power dissipation balance.

	C^2MOS Register (Fig. 3a)	Single-phase Register (Fig. 3c)
complexity (transistors / register)	8	9
number of internal nodes	2	3
area (1.0–μm CMOS)	618 μm^2	570 μm^2
minimum clock transition time	4 ns	2 ns
clock load (local wiring and gate cap.)	64 fF	30 fF
power dissipation (@ 100 MHz)		
clock network	160 μW	75 μW
logic (σ = 0.25)	36 μW	102 μW
total	196 μW	177 μW

Table 2. Comparison C^2MOS registers and single-phase registers.

From the results of Table 2 it can be concluded, that the saving in power dissipation in the local clock network is almost completely compensated by an increase of the power dissipation of the logic of the registers. Only the saving in the global clock network remains but this is only a minor part of the complete power dissipation of the clock network (Table 1.). So applying the single-phase registers shown in Fig. 3b as pipeline registers, will bring no significant over-all saving of power dissipation of the complete circuit. Moreover, using the single-phase registers as shown in Fig. 3b results in undesirable higher demands for the clock drivers compared to conventional C^2MOS registers since faster transition times of the clock signals must be guaranteed (Table 2). However, especially for high-speed applications, the advantage of the simple clock network for circuits applying single-phase registers, of course remains unaffected.

4. The Clock Drivers

Since the power P_{cs} delivered by the clock drivers of the clock system can be a considerable part of the total power dissipation P of the circuit, the design of the drivers must also be optimized for minimum power dissipation.

CMOS clock drivers consist of a chain of cascaded drivers with increasing driving capability. An important design parameter for such driver chains is the stage ratio f defined by the ratio of the widths of the transistors of successive driver stages :

$$f = W_p^{\,i} / W_p^{\,i-1} = W_n^{\,i} / W_n^{\,i-1} \qquad (4)$$

$W_p^{\,i}$ is the width of the p-channel transistor of the driver of stage i in the chain and $W_n^{\,i}$ is the width of the associated n-channel transistor .

In a first approximation the wiring capacitances between successive driver stages are neglected. Then the parasitic capacitance of an intermediate node of the driver chain can be taken proportional to the widths of the transistors interacting with this node. Thus the capacitances of two successive intermediate nodes will also have a ratio equal to the stage ratio f and an approximate value for the power dissipation of the complete clock system can be given:

$$P_{cs} = 1 / T \cdot Vdd^2 \cdot C_L \cdot (1+1/f+1/f^2+..+1/f^N) \qquad (5)$$

C_L is the capacitive load for the clock driver, Vdd is the supply voltage, T is the clock period and N is the total number of stages in the driver chain. N is detemined by the desired input capacitance C_I of the first stage of the driver chain:

$$N = \log(C_L / C_I) / \log(f) \qquad (6)$$

The power efficiency of the clock system, can then be defined as :

$$\eta_P = P_{cn} / P_{cs} = 1 / (1+1/f+1/f^2+..+1/f^N), \qquad (7)$$

which for a given ratio of C_L / C_I is only a function of f. The factor $1-\eta_P$ characterizes the additional power dissipation overhead needed in the clock driver circuits for charging and discharging the clock network

Fig. 4 shows the power efficiency η_P calculated from (6) and (7) for C_L / C = 1000. For comparision also the results obtained from SPICE simulations of driver chains designed in a 1.5-µm CMOS technology, are shown (for C_L =100p and C_I=100fF). The small deviation between the calculated and simulated results show that the approximation of neglecting the wiring capacitance between the driver stages in (5) is justified. Moreover this also shows that the power dissipation due to short-circuit currents during clock signal transitions is negligable. In Fig. 4 also simulation results for the delay t as a function of f are shown. This well known relation [4] between t and f has a minimum for f ≅ 3. But from (7) and also from Fig. 4 it is clear that in order to miminize the power dissipation of the clock system, the stage ratio f of the drivers should be chosen as large as possible.

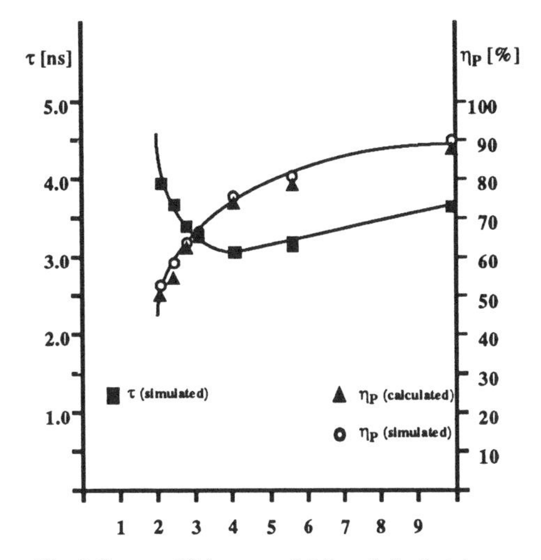

Fig. 4. Power efficiency and delay of clock drivers.

This means that the number of drivers in the chain should be minimized. However another design parameter is the transition time of the clock signals. For complementary and single-phase clock systems requiring smaller clock transition times than non-overlapping clock systems, it may not be possible to apply stage ratios much larger than f = 3. Also the delay time of the complete driver chain must be taken into account, but it is well known that the total delay increases only slightly with increasing stage ratio f (Fig. 4.). So depending on the relative importance of power dissipation and delay time of the clock drivers, the optimum for the stage ratio will be in the range f = 3..10. For example, the clock drivers for the QAM-processor of [6], driving a total clock capacitance of almost 2nF, are designed with a stage ratio of f = 8, reducing the power dissipation of the predriver stages by almost a factor of 4 with only a major delay penalty and a minor degradation of the clock signal transition time. This is also a reasonable choice considering the degradation of the transition time throughout the expansive clock network due to RC-effects

5. Reduced Clock swing

The power needed for charging and discharging the clock network 1 / T times per seconds is given by (3). If the clock swing V_{clock} can be chosen different from the supply voltage V_{dd}

$$V_{clock} = r \cdot V_{dd} \qquad \text{with } r < 1 \qquad (8)$$

then the power dissipation of a clock system supplied

from V_{dd} would be

$$P_{cs}' = 1 / T \cdot C_L \cdot V_{dd} \cdot V_{clock} \quad (9)$$

or with (8)

$$P_{cs}' = 1 / T \cdot C_L \cdot r \cdot V_{dd}^2 \quad (10)$$

A reduction of V_{clock} results in a proportional reduction of P_{cs}'. The reduced clock swing however reduces the saturation current of the transistors connected to the clock phases :

$$I_{DS} \cong W \cdot (V_{clock}-V_{th})^{\alpha} \quad \text{with } 1<\alpha<2 \quad (11)$$

(with W and V_{th} the channel width and treshold voltage of the transistor involved; $\alpha = 2$ for conventional CMOS and $\alpha = 1$ for sub-micrometer CMOS).

As a concequence of the reduced saturation current, the delay time of the pipeline registers will increase. In order to retain the original throughput rate, the channel width W of the clocked transistors will have to be resized according to :

$$W' \cong W \cdot (V_{dd}-V_{th})^{\alpha} / (V_{clock}-V_{th})^{\alpha} \quad (12)$$

This will increase the gate capacitance portion of C_L. But since this gate capacitance is only a smaller part of C_L, already for $\alpha = 2$ a reduction of the power dissipation can be obtained. For $\alpha = 1$ the required increase of the channel width and gate capacitance is smaller and the power dissipation saving correspondingly higher.

If the clock drivers could be supplied with V_{clock} instead of V_{dd} (e.g. from an external source) then the power dissipation in the clock system is :

$$P_{cs}'' = 1 / T \cdot C_L \cdot V_{clock}^2 \quad (13)$$

or with (8)

$$P_{cs}'' = 1 / T \cdot C_L \cdot r^2 \cdot V_{dd}^2 \quad (14)$$

In this case, due to the quadratic dependence of Pcs" on r , a more significant saving of power dissipation of the clock system on the chip could be obtained. However the power dissipated in the additional external source or level conversion circuit is equal to

$$P_{su} = (1 / T) \cdot C_L \cdot (V_{dd}-V_{clock}) \cdot V_{clock} \quad (15)$$

or with (8)

$$P_{su} = (1 / T) \cdot C_L \cdot (1-r) \cdot r \cdot V_{dd}^2 \quad (16)$$

is only removed off-chip but not removed from the total system power dissipation which for e.g. portable systems is an important cost factor. So even if the clock system is supplied from a reduced supply voltage, the total power dissipation on system level is still given by (10).

A novel circuit for an on-chip generation of reduced-swing clock signals, requiring no extra (external nor internal) supply voltage is shown in Fig. 5, which enables to achieve a true reduction of the power dissipation as described in (14). The circuit exploits the principle of charge charing between the capacitances of complementary clock phases and additionally reduces the swing of the clock signals to approximately half the supply voltage. In this way both principles proposed in [1] to reduce the power dissipation in the clock system were combined in an efficient circuit realization. The circuit is used in a complementary non-overlapping clock system for which the principle of distributed registers (Fig.2) can be exploited, allowing a hardware efficient implementation of the pipeline latches as simple transmission gates as described before.

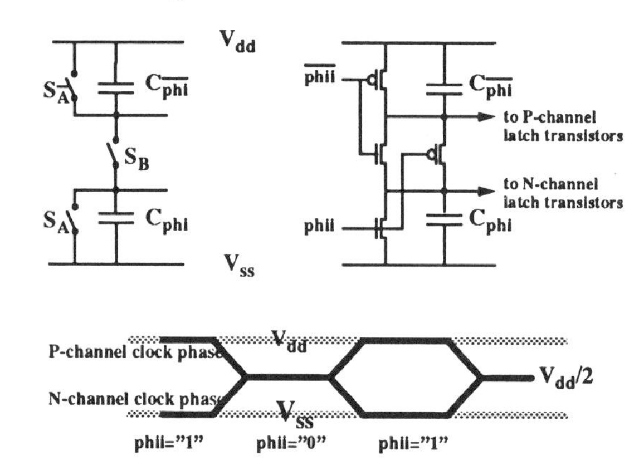

Fig. 5 Principle and CMOS implementation of complementary-phase clock driver.

The circuit of Fig. 5 requires no additional overhead compared to the standard solution. Also the widths of the clock driver transistors can be retained so that the silicon area remains unaffected. Whereas the power dissipation of the driver circuit of Fig. 5 is reduced by 75% compared to the standard clock driver circuits. The penalty in the throughput rate can be compensated by resizing the width of the clocked transisitors of the latches according to (12). This will increase the gate capacitance part of the clock load, but will only slightly increase the total clock capacitance. Simulations showed that even when maintaining the throughput rate of the circuit by resizing of the clocked transistors, a power dissipation reduction of the clock system of higly pipelined ULSI circuits of at least 60% can be achieved.

6. Conclusion

Complementary non-overlapping clock systems, complementary and true-single-phase clock systems were analyzed considering clock network complexity and register circuit complexity.
The comparison of power dissipation between C^2MOS registers and an implementation of true-single-phase registers shows that the straightforward application of single-phase registers as pipeline registers in highly pipelined ULSI circuits will yield no significant saving in power dissipation compared to complementary driven C^2MOS registers. Since the power dissipation savings due to the simpler clock networks are compensated by the increased power dissipation in the more complex register circuits, a more substantial saving of power dissipation can be obtained with a reduction of the swing of the clock signals. A special driver circuit for the on-chip generation of these reduced-swing clock signals is presented. For the conventional complementary non-overlapping clock systems, it allows a reduction of the power dissipation in the clock system of more than 60%, while maintaining the throughput rate of the circuit.

7. Acknowlegement

The ideas presented here are the result of extensive discussions with the co-workers of our staff. The authors would like to thank especially Stefan Meier, Heinz Söldner and Tobias Noll and for their valuable suggestions.

8. References

[1] T.G. Noll and E. De Man, "Pushing the Performance Limits due to Power Dissipation of future ULSI Chips", Proceedings of the ISCAS-92, pp. 1652 - 1655, San Diego, Mai 1992.

[2] C. M. Huizer, "Power Dissipation Analysis of CMOS VLSI Circuits by means of Switch-level Simulation", Proceedings of the European Solid-State Circuit Conference, pp. 61-64, Grenoble, Sept. 1990.

[3] T.G. Noll, "Carry-Save Architectures for high-speed Digital Signal Processing", Journal of VLSI Signal Processing, No. 3, pp. 121-140, 1991.

[4] C. Mead and L. Conway, "Introduction to VLSI Systems", Reading, MA: Addison-Wesley, 1980.

[5] J. Yuan and C. Svenson, "High-speed CMOS circuit techniques", IEEE Journal of solid-state circuits, Vol SC-24, No 1, pp. 62 - 70, Feb. 1989.

[6] E. De Man, M. Schöbinger, T.G. Noll and G. Sebald, "A 60-MBaud Single-Chip QAM Processor for the complete Base-Band Signal Processing of QAM Demodulators", Proceedings ISCAS-94, London, June 1994.

Part 3

Automated Layout and Synthesis of Clock Distribution Networks

EFFICIENT PLACEMENT ALGORITHMS OPTIMIZING DELAY FOR HIGH-SPEED ECL MASTERSLICE LSI'S

Yasushi Ogawa*, Tatsuki Ishii*, Yoichi Shiraishi*
Hidekazu Terai*, Tokinori Kozawa*, Kyoji Yuyama**, Kyoji Chiba***

* Central Research Laboratory, Hitachi Ltd.
Higashi-koigakubo, Kokubunji, Tokyo, 185, Japan
** Device Development Center, Hitachi Ltd.
*** Kanagawa Works, Hitachi Ltd.

Abstract

Placement algorithms optimizing signal delay as well as wirability for high-speed ECL masterslice LSI's are proposed. Equivalent constraints of wire length for clock skew, data path delay, and wired-OR are classified according to upper and lower limits. To maintain such limits, a top-down method utilizing an augmented two-dimensional clustering placement with "scope" and "zone", which are new concepts representing limits, and an iterative weighted improvement method are presented. Such algorithms are applied to hundreds of 2 K and 5 K gate ECL masterslice LSI's for a newly developed high-end mainframe computer, the Hitachi M-680H. Through such algorithms, the physical design is greatly improved by guaranteeing high wirability and improving electrical characteristics.

1. Introduction

Digital systems must be designed in accordance with logical and electrical specifications. Large computers which achieve extremely high speeds impose particularly strict timing and logic design requirements.

In designing LSI's almost all placement algorithms that have been developed and practically applied optimize wirability and chip area [1,2]. Hence, layouts using such algorithms have not necessarily satisfied electrical characteristic requirements. The LSI layouts must be repeatedly redesigned or modified manually until the electrical conditions are met. However, in the physical design of masterslice LSI's comprising a large computer, such modifications cause serious bottlenecks in design, increasing cost. Thus, a placement algorithm optimizing delays is indispensable to reducing design cost and time [3,4,5].

In general, the signal delay must be in a range between a lower and an upper limit, or below an upper limit. One of the approaches to maintaining the above conditions is to place cells closer to one another. However, this may result in lower wirability because of local wiring congestion, and it is important not to degrade wirability to ensure that electrical requirements are satisfied.

This paper presents requirements for the layout of 2 K and 5 K gate ECL LSI's used in the Hitachi M-680H. New placement algorithms optimizing the delay corresponding to each requirement are then proposed.

The electrical constraints for ECL masterslice LSI's are divided into (i) clock skew, (ii) data path delay, and (iii) resistance in wired-OR. These electrical constraints are converted to those of wire length for the placement program. The algorithm for satisfying constraints (i) and (iii) is a top-down placement algorithm into which the concepts of "scope" and "zone", representing the limits on the conditions, are introduced. Scope is the virtual region defined as the wire length threshold, and zone is a physically predetermined region on an LSI. In the algorithm for (ii), iterative weighted improvement is repeated for violated path delays. The iterative improvement is characterized by repeating weighted improvement with rapid calculation of all path delays.

Through such algorithms, performance shows marked improvement in delay and layout cost has been significantly reduced. Furthermore, an average of 99.96% wirability has been obtained for hundreds of LSI's.

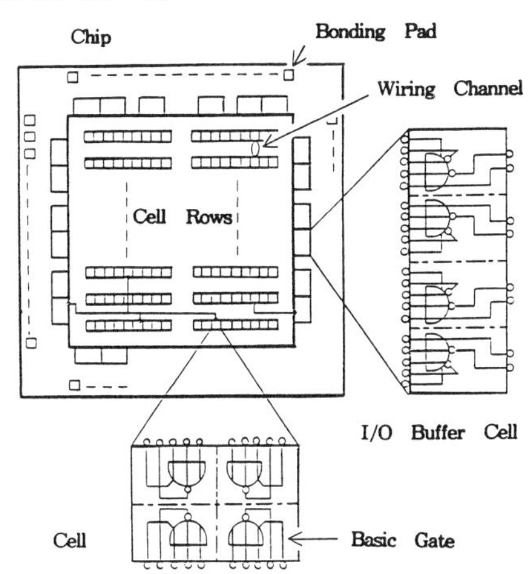

Fig. 1 Layout Model of ECL Masterslice LSI

2. Design Requirements

2.1 Overview of ECL Masterslices

The gate, cell and chip model of the high-speed

Reprinted from *Proc. ACM/IEEE 23rd Design Automation Conf.*, pp. 404–410, June 1986.

ECL masterslice LSI are shown in Fig. 1. A cell is composed of four basic gates and each gate acts as a three-input NOR or OR gate. The logic variations depend upon combinations of the basic gates. This chip has 44 rows of cells and 2 K gates. Gate delay is 0.20 ns, indicating severe constraints on the electrical characteristics.

2.2 Electrical Constraints

The performance of a computer is mainly measured by its machine cycle time. Shorter time means higher performance.

Next, the three electrical constraints on this time (clock skew, data path delay, and wired-OR) are explained.

(1) Clock Skew

The clock signal is generated by a clock generator to synchronize data transfers. The clock distribution model of the new LSI is shown in Fig. 2 (a). The clock signal is issued from a clock pin and is distributed to all latches. Although each distributed clock signal is expected to be propagated to each latch simultaneously, the propagation time actually varies according to the path length from the pin to a latch. This variation is called clock skew. Clock skew (S) is defined as:

$S = Tcmax - Tcmin,$

where Tcmax and Tcmin are the maximum and minimum propagation times, as shown in Fig. 2 (b). The clock path delay Tcp representing the propagation time from the pin to a latch must be between Tcmin and Tcmax. Naturally, the lower the clock skew, the better.

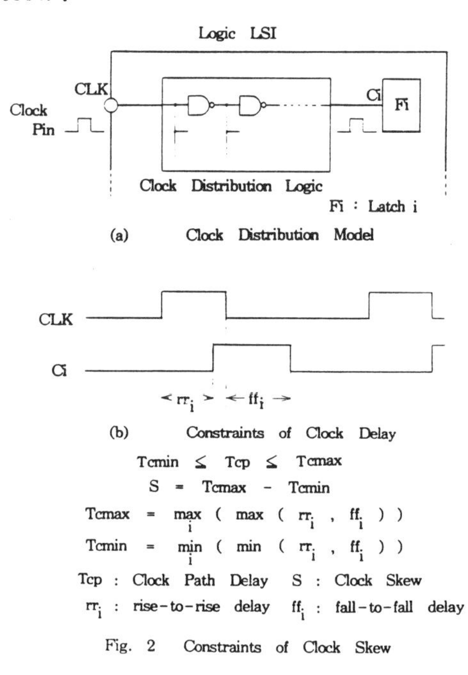

(b) Constraints of Clock Delay

$$Tcmin \le Tcp \le Tcmax$$
$$S = Tcmax - Tcmin$$
$$Tcmax = \max_i (\max (rr_i , ff_i))$$
$$Tcmin = \min_i (\min (rr_i , ff_i))$$

Tcp : Clock Path Delay S : Clock Skew

rr_i : rise-to-rise delay ff_i : fall-to-fall delay

Fig. 2 Constraints of Clock Skew

(2) Data Path Delay

The data path delay is defined as the propagation time of the data signal from a latch to the next latch. The upper and lower limits of the data path delay are illustrated by Fig. 3 (a) and (b). In order to accurately propagate a signal from a latch F1 to latch F2, the data path delay, Td, must be less than the clock cycle time, Tc, minus the clock skew, S. Also Td must exceed the maximum clock width Twmax, plus the clock skew S. Consequently, data path delay Td is defined by:

$Twmax + S \le Td \le Tc - S.$

Here, however, the lower limit is assumed to be maintained by logical design. Thus, the constraint on data path delay is $Td \le Tc - S$.

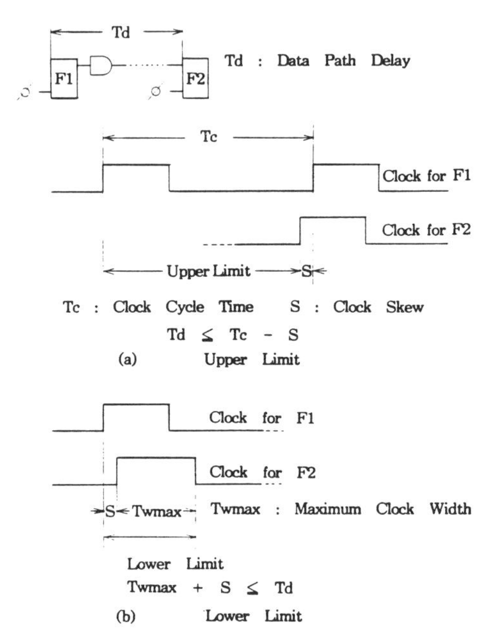

Fig. 3 Constraints of Data Path Delay

(3) Wired-OR

In ECL gates, the OR logic of the outputs can be provided by tying together the outputs without using an extra gate. This is called wired-OR. However, if the resistance value of the wire exceeds an upper limit, the wired logic doesn't work because of the large voltage drop. Thus the resistance value from the output t with terminated resistance to another output i must be restricted, as shown in Fig. 4. That is, Rmax must not exceed the upper limit Rwor.

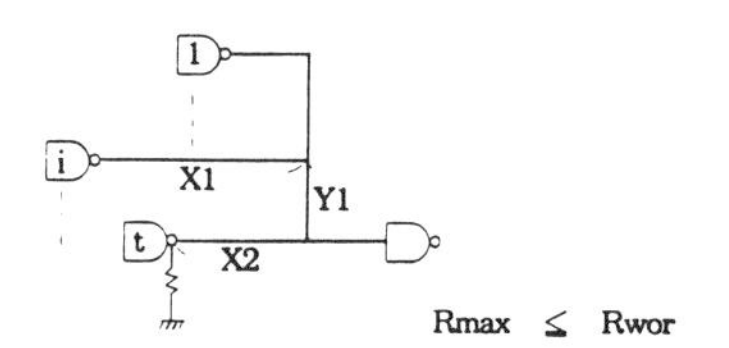

Rmax : max Rit

Rit : Resistance between Output Pins i and t

$$= rx \sum_j xj + ry \sum_k yk$$

rx : Horizontal Resistivity

ry : Vertical Resistivity

Rwor : Upper Limit of Resistance Value

Fig. 4 Constraints of Resistance in Wired-OR

3. Placement Strategy Guaranteeing Constraints

3.1 Analysis of Electrical Constraints

Since the propagation delay or voltage drop in a wired-OR is determined by the length of the net after layout, the design requirements can be converted to constraints of wire length. Thus, requirements for clock delay, data path delay, and resistance in a wired-OR correspond to wire length within a certain range or below the upper limit, as shown in Table 1. Such restrictions can be represented by combinations of two conditions: Cu, below an upper limit β, and Cℓ, above a lower limit α.

The regions of Cu and Cℓ are shown in Fig. 5. Two points, A and B, indicate two cells. When the Manhattan distance, ℓ between them is restricted to a specified range, the outer and inner contour lines show the upper and lower limits, respectively. The difference between Lx and Ly comes from wire unit capacitance or resistance, their values depending upon wire layer direction. Therefore, Cu refers to the inside of the outer contour, and Cℓ to the outside of the inner contour.

Table 1 Constraints of Length

Item	Constraint
Clock Delay	$\alpha \le \ell \le \beta$
Data Path Delay	$\ell \le \beta$
Resistance in Wired-OR	$\ell \le \beta$

ℓ : Length α : Lower Limit β : Upper Limit

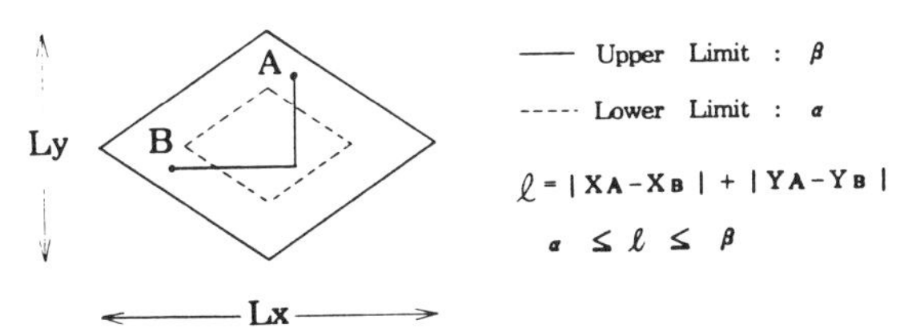

Fig. 5 Regions Corresponding to Electrical Constraints

3.2 New Concepts Representing Constraints

Three concepts are introduced to represent the above conditions: scope, zone, and weighted wire length. Scope refers to the wire length threshold defined as the virtual region to maintain condition Cu. Zone refers to a physically predetermined region on a chip to maintain condition Cℓ. The zone concept makes it simpler to implement enlongation of wires. Both zone and scope are incorporated into the top-down placement algorithm as shown in Table 2. Then, weighted wire length is used as an evaluation function in iterative improvement to maintain Cu. The above concepts are fully explained below.

Table 2 Placement Methods for Constraints

Constraint	Placement Method	
Cu : $\ell \le \beta$	Top-down Method with Scope	Iterative Weighted Improvement
Cℓ : $\alpha \le \ell$	Top-down Method with Zone	

ℓ : Length α : Lower Limit β : Upper Limit

3.3 Placement Procedure

An outline of placement procedures is shown in Fig. 6. First, in pin assignment, bonding pads and I/O buffer cells are automatically assigned according to LSI location on a card and electrical constraints such as separation of simultaneous switching signals. The second step is gate assignment to a cell. Here, a cell is a placement unit. Gates are combined to form a cell. This is performed by repeating combinations of pairs of gates on the basis of clustering. Its evaluation function is defined by geometric combination constraints and connectivity between gates. Connectivity can also represent the connectivity between gates when they are not connected directly, but through intermediate gates, in order to assign all gates to a certain number of cells. A gate is assigned to the cell type which produces the best fit. Third, initial cell placement is performed, ie, two-dimensional clustering placement [1] augmented for optimizing delay. In this step, scope and zone are applied. Fourth, iterative placement is performed on both cells and gates. Each cell location is improved iteratively through the Net Balance [1,6] and Pairwise Interchange [7] algorithms. Pairwise Interchange is also applied to gate exchange between cells. Combining these two algorithms gives higher wirability than using only one, because it helps prevent fast convergence on the local minimum value for total wire length. The effectiveness of this combination has been confirmed experimently. Finally, the weighted Net Balance and Pairwise Interchange are repeated to maintain condition Cu with the upper limit β in Table 2.

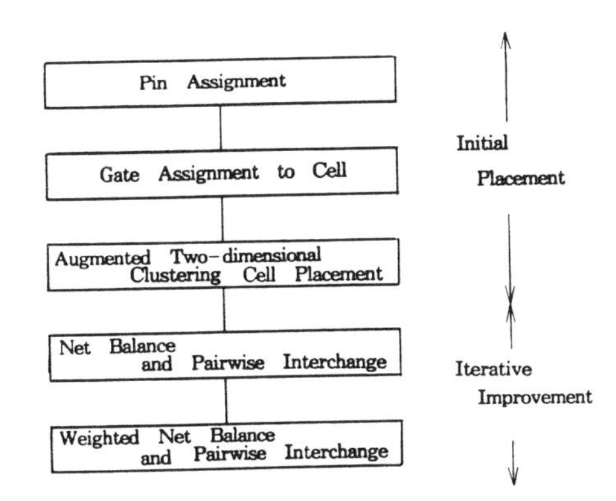

Fig. 6 Flowchart for Placement Procedures

4. Top-down Placement Algorithm

A top-down placement algorithm was proposed employing a two-dimensional clustering technique [1] to obtain good initial placement. Here, an augmented top-down placement algorithm is presented applied within the upper limit β and the lower limit α, as described in 3.1.

In this algorithm, binary partitioning is repeated until the current region to be partitioned corresponds to a cell. Thus, it permits two-dimensional cell placement.

The binary partition procedure contains the following three steps: clustering, linear placement, and partitioning. (See Fig. 7.)

(1) Clustering

A pair of cells is grouped into a cluster according to a value that represents their connectivity [8]. Similarly, a cluster is generated by grouping two clusters. Such grouping is called clustering. Here, the clustering process is repeated to construct a corresponding binary tree until all cells in the current region to be partitioned are included in a cluster.

(2) Linear Placement

Decomposing the clustering tree from top to bottom is repeated to obtain linear placement of all cells to be partitioned. Its evaluation function is to minimize the total estimated wire length, considering of the location of the external pin connections.

(3) Partitioning

A cell row is partitioned into two sets of cells. Then a partition improvement is performed to reduce the number of cuts [9]. This completes the binary current region partition process.

4.1 Top-down Placement with Scope

To maintain the upper limit β with clock delay and wired-OR constraints, the clustering value is modified. Using the clustering value weighted by nets results in closer placement of cells.

However, placement of cells too close to one another using the weighted clustering value may degrade wirability, because of local wiring congestion. Therefore, specified cells must satisfy the requirements without placing them too close to each other. Scope is then defined as a new parameter of clustering. Scope is a region threshold (whose size is determined by β in partitioning), shown in Fig. 8 (a), corresponding to the outer contour line in Fig. 5. Thus, as long as the size of a region to be partitioned is greater than that of the scope, the weighted connectivity Cel applies as shown in Fig. 8. In the other cases, conventional connectivity Clg, defined only by the logic connection, is used as a clustering value. Therefore, in the augmented clustering algorithm, to maintain high wirability, the weight in connectivity is controlled by comparing the size of a region to be partitioned with that of the scope.

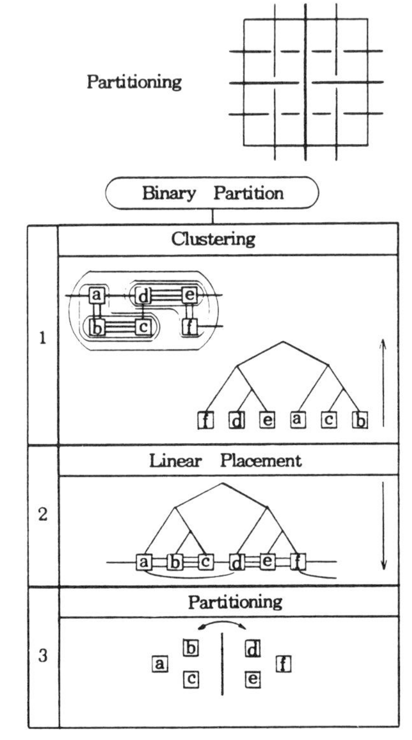

Fig. 7 Top-down Placement Algorithm

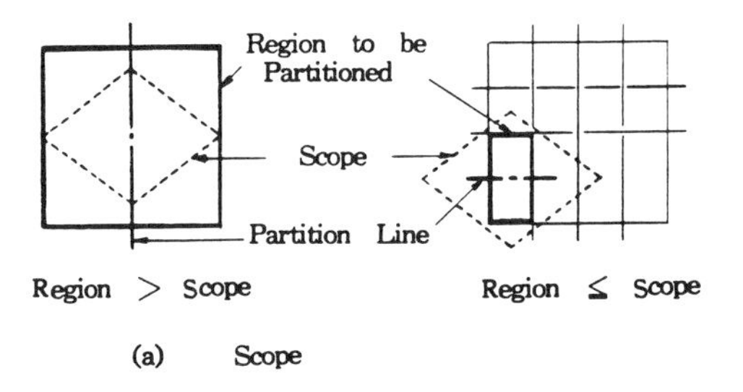

Clustering Value for Cells i and j

$$Cij = Clg + Cel$$

Clg : Connectivity from Logic Connection

Cel : Connectivity for Electrical Constraints

$$= \begin{cases} C & \text{if Region} > \text{Scope} \\ 0 & \text{else} \end{cases}$$

C : Constant for Clg, $C > Clg$

(b) Clustering Value

Fig. 8 Scope and Clustering Value

4.2 Top-down Placement with Zone

To maintain the lower limit α of clock path delay, cells are placed in a preassigned zone. For example, clock distribution gates are placed in the zone as shown in Fig. 9. Then, the clock path delay can maintain condition $C\ell$ described in Table 2 because it is larger than the path delay from the clock edge pin to the clock distribution gate, which exceeds α in all cases.

In clustering, the clustering value is added by weight for cells such as clock distribution gates to group them earlier. Also in partitioning, a set of cells is partitioned according to zone size in the current region. Examples of variations in current region zone are shown in Fig. 9. For (i), cells are partitioned into two equal size parts, and for (ii), they are not partitioned. For (iii), the binary partition in 4.1 is the same as in the top-down placement algorithm without considering zone.

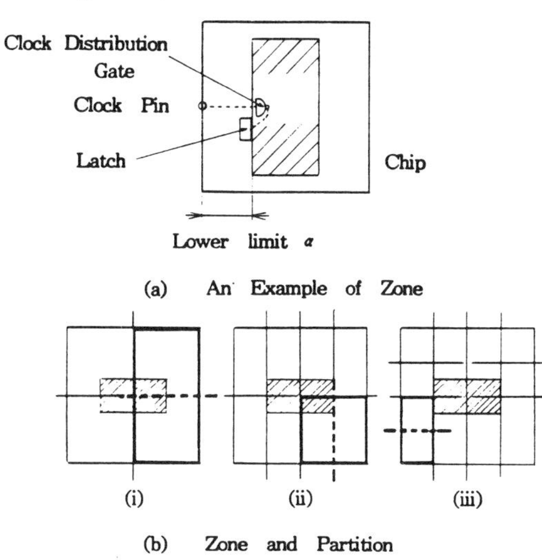

Fig. 9 Zone and Partition

5. Iterative Weighted Improvement Algorithms

An outline of iterative weighted improvement is shown in Fig. 10. The procedure consists of all path delay calculation, critical net weighting, and weighted improvement. In weighted improvement, Net Balance and Pairwise Interchange algorithms are applied as mentioned above. The aim of the procedure is to minimize path delays. It is used to calculate all path delays rapidly in the placement procedure and to repeat improvement of failed paths only in the placement improvement step. The three processes are repeated until no nets belonging to failed paths are found.

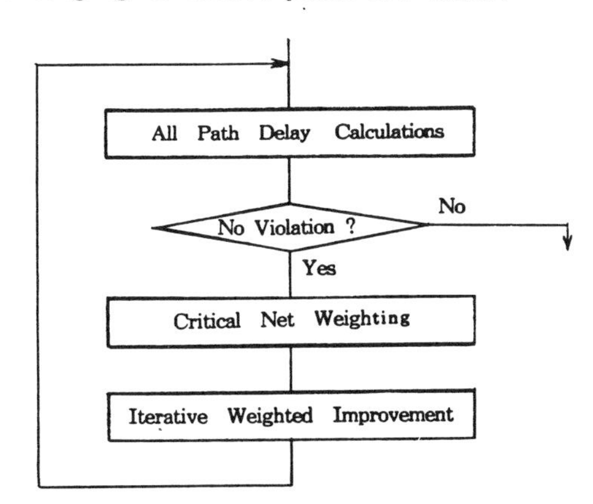

Fig. 10 Flowchart for Iterative Weighted Improvement Procedure

Normalized path delay is diagrammed in Fig. 11. Normalized path delay D is defined in the following equation.

$$D = \frac{\text{calculated path delay}}{\text{upper limit of path delay}}$$

In order to maintain the upper limit β , it is necessary to reduce both the maximum normalized path delay Dmax and the number Nf of failed paths. After calculating all normalized path delays, a weight Wpath is calculated for a path as follows.

Wpath = C*(D-1)+1, if D > 1,

unless Wpath = 1, where C is a coefficient. Wpath is assigned to nets belonging to a failed path. If the net also belongs to some other failed paths, the maximum among the weights for these paths is selected. For example, in Fig. 12 when both paths, FF0-FF1 and FF0-FF2 exceed the upper limit of the path delay, the weight of FF0-FF1 is assigned to net Z as the maximum weight. Using these weights, the evaluation function in the two improvement algorithms is defined as the total weighted wire length L as follows.

$L = \sum Wi*li,$

where li is pseudo wire length based on Steiner's tree and Wi is the weight for net i. As a result, the failed path length which appears longer than the actual length can be shortened.

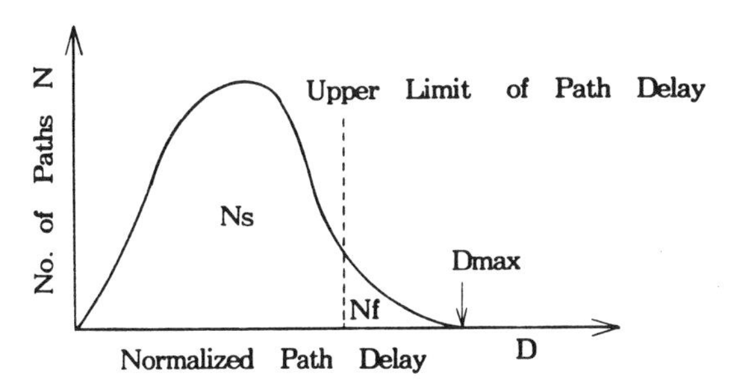

Dmax : Maximum Normalized Path Delay

Ns : No. of Paths below Upper Limit

Nf : No. of Failed Paths

Fig. 11 Definition of Normalized Path Delay

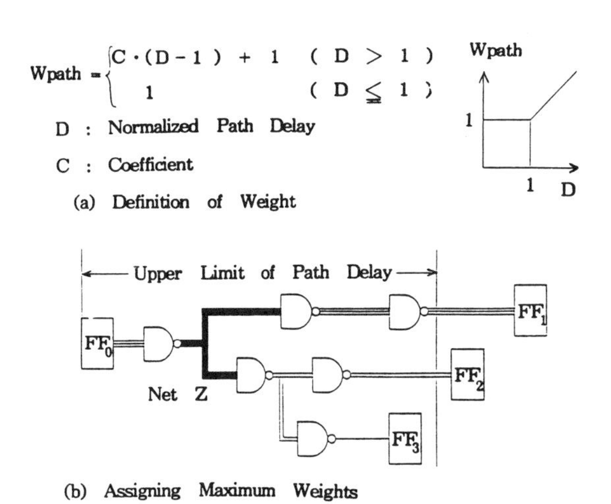

Fig. 12 Procedure for Assigning Weights

6. Routing Procedures

An outline of routing procedures is illustrated in Fig. 13. One feature is clean-up routing using a maze algorithm [10] after a channel router [11] to obtain high wirability. Another is routing nets weighted by electrical constraints by the channel router because nets unconnected by the channel router may violate the constraints due to detours by the maze router.

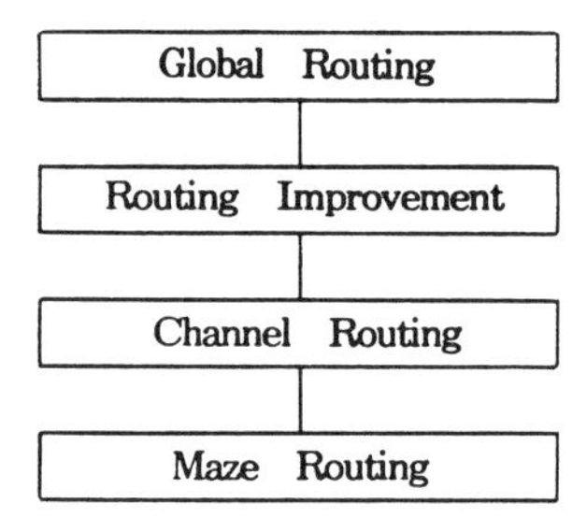

Fig. 13 Flowchart for Routing Procedures

7. Results

Experiments were carried out to evaluate the performance of the placement algorithms presented. The layout DA system incorporating such algorithms has been applied to 2 K gate ECL LSI's for the M-680H. On an 8 MIPS CPU the processing time for placement was 3.7 mins, and for routing 8.9 mins.

7.1 Electrical Characteristics Analysis

The effects of these algorithms on electrical characteristics for 9 LSI's are shown in Table 3. The number of clock path delays not within the upper and the lower limits determined by system design, decreased markedly from 134.2 to 3.3 on the average. Also clock skew improved by 0.73 ns, as shown in Fig. 14. While it is generally difficult for minimum clock path delay to be maintained above the lower limit, the ability of the top-down placement method based on the zone concept to maintain a lower limit was proven through experiments.

The maximum data path delay improved by 7.11%, and the total wire length was 0.37% greater than before. Thus, degradation in wirability was slight.

The marked effect on distribution of data path delay for an LSI is shown in Fig. 15. This effect was derived mainly from the iterative weighted improvement.

The number of resistance failures in wired-OR was reduced by 14. Thus, the effect of the proposed algorithms was confirmed.

7.2 Wirability

The aim of implementing the scope concept is to prevent degradation in wirability by not placing cells too close to one another. An experiment on wired-OR was carried out to compare cells within scope to adjacent cells. As a result, the number of unconnected pin-to-pins on the average was reduced from 2.9 to 0.8 without resistance failures, proving the effectiveness of this concept.

A histogram of unconnected pin-to-pins after optimizing all characteristics is shown in Fig. 16. On the average the number of unconnected pin-to-pins has been reduced to 0.9 for hundreds of LSI's using our layout DA. Complete wiring has been performed on approximately three-fourths of all the LSI's, proving that the algorithms proposed can guarantee high wirability.

Table 3 Effects of Delay Optimization

	Before Optinization	After Optinization
No. of Failed Path Clock Delay	134.2	3.3
Maximum Data Path Delay	107.11 %	100 %
No. of Resistance Failures in Wired-OR	14	0

On Average for 9 LSI's

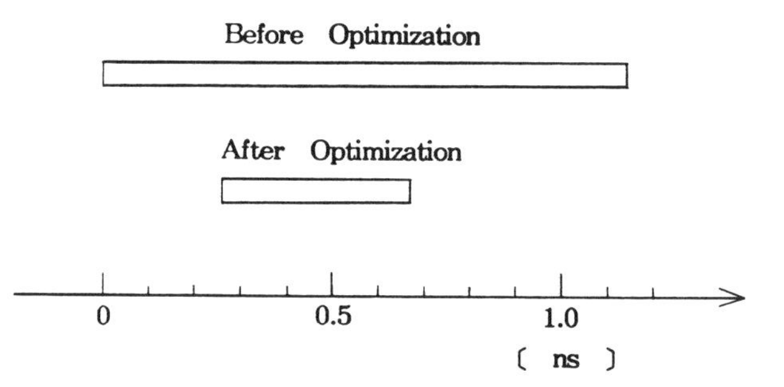

Fig. 14 Effect of Clock Skew Improvement

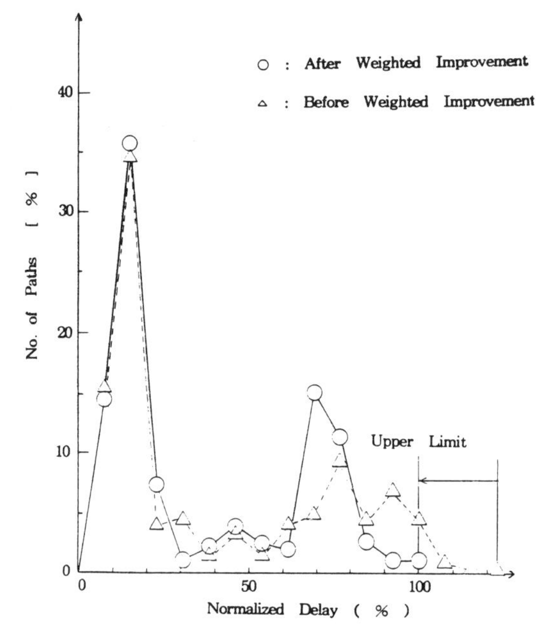

Fig. 15 An Example of Effect of Data Path Delay Optimization

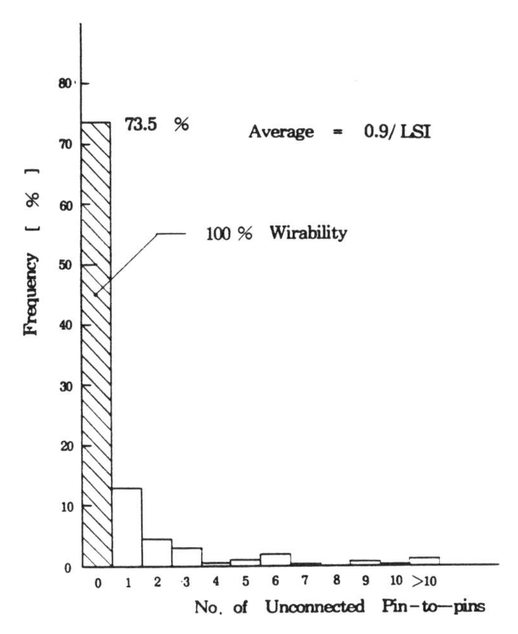

Fig. 16 Histogram of Unconnected pin-to-pins

8. Conclusion

New placement algorithms optimizing electrical characteristics for ECL masterslice LSI's have been presented. Clock skew, data path delay and resistance in wired-OR were referred to to define these characteristics. They were converted to constraints of wire length, classified into two types of conditions: Cu, below an upper limit, and Cℓ, above a lower limit. To maintain Cu, a top-down algorithm with a new concept, scope, and iterative weighted improvement were adopted. A top-down algorithm with a new concept, zone was presented for Cℓ. Scope can prevent degradation in wirablity because the cells involved are not located too close to one another. Experiments were carried out to prove the effectiveness of the algorithms. Clock skew was shown to decrease by 0.73 ns, and the maximum data path delay decreased 7.11 % on the average for 2 K and 5K gate LSI's.

The above algorithms have been applied to hundreds of 2K and 5K gate ECL masterslice LSI's for the Hitachi M-680H. As a result, electrical characteristics have been dramatically improved guaranteeing high wirability, and layout design cost has been significantly reduced. On average, there are 0.9 unconnected pin-to-pins and complete wiring has been obtained for approximately three-fourths of all the LSI's.

9. Acknowledgements

We would like to thank Drs. Yasuhiro Ohno and Katsuhiko Takizawa of Kanagawa Works for many useful discussions and comments.

References

[1] T. Kozawa, et al.; Automatic Placement Algorithms for High Packing Density VLSI, Proc. of 20th DA Conf., 1983.
[2] H. Terai, et al.; Performance Analysis of Automatic Placement and Routing for Large-scale CMOS Masterslices, ICCD, 1983.
[3] M. Burstein and M. N. Youssef,; Timing Influenced Layout Design, Proc. of 22nd DA Conf., 1985.
[4] A. E. Dunlop, et al.; Chip Layout Optimization Using Critical Path Weighting, Proc. of 21st DA Conf., 1984.
[5] K.H.Khokhani, et al.; Placement of Variable Size Circuits On LSI Masterslices, Proc. of 18th DA Conf., 1981.
[6] H.Terai, et al.; Automatic Placement and Routing Program for Logic VLSI Design based on Hierarchical layout Method, Proc. of Conf., ICCC, 1982.
[7] M. Hanan, et al.; Placement Techniques, Chap.5 in Design Automation of Digital Systems, Vol.1, Prentice-Hall, pp.213-282, 1972.
[8] D. M. Schuler, et al.; Clustering and Linear Placement, Proc. of 9th DA Workshop, 1972.
[9] B. Kernighan and S. Lin,; An Efficient Heuristic for Partitioning Graphs, Bell System Technical Journal, Vol.49, 1982.
[10] R. Kamikawai, et al.; Placement and Routing Program for Masterslice LSIs, Proc. of 13th DA Conf., 1976.
[11] T. Kozawa, et al.; Advanced LILAC - an Automated Layout Generation System for MOS/LSIs, Proc. of 11th DA Conf., 1974.

HIGH PERFORMANCE CLOCK DISTRIBUTION FOR CMOS ASICS

Scot Boon, Scott Butler, Richard Byrne, Brad Setering

VTC Incorporated
2401 East 86th Street
Bloomington, MN 55420

Marty Casalanda, Al Scherf

Control Data Corporation
8800 Queen Avenue South
Bloomington, MN 55431

ABSTRACT

An effective clock distribution system is needed for high performance CMOS standard cell designs. Commercial routers can deliver such a clocking scheme only when integrated with circuit design techniques, a customized clock network, and place and route algorithms for clock optimization. A system has been developed that can achieve clock skew of less than 500 picoseconds with phase delay under 4 nanoseconds. This system is flexible, multi-tiered, netlist specific, compatible with commercial routers, and accurately modeled. Topics of this paper include clock tree structure, interconnect constraints, buffer design methodology, netlist driven placement, localized clock assignment, simulated annealing, layout reintegration, and simulation modeling.

INTRODUCTION

ASIC designs are rapidly surpassing standard products as technology drivers. This presents a challenge to develop a clock management system capable of handling high clock rates and large gate counts. Of particular concern are clock skew and phase delay since these can lead to degraded performance or even a non-functional design.

An acceptable clock distribution scheme for standard cell designs must accomodate a wide range of design styles, performance requirements, and chip complexity. Clock concerns must be relatively transparent to the ASIC customer both from a schematic netlist and simulation perspective. VTC's clock distribution addresses these issues using a flexible, multi-tiered scheme which is netlist specific, compatible with commercial routers, and accurately modeled.

It is important to have flexibility to handle large macro blocks, uneven distributions of clock loads, and variations in the number of clock loads, core size, and performance requirements.

A multi-tiered approach allows the designer to trade off schedule and area for performance by selecting the complexity of the distribution scheme.

To satisfy the diverse demands of ASICs, it is crucial that the clocking scheme be optimized for the design netlist. A netlist specific implementation constructs the clock distribution network based on the connectivity and floorplan of the design.

Compatibility with commercial routers is needed to keep turn times to a minimum through automation of the layout task, as opposed to full custom or no clock scheme alternatives. The clock distribution method should not preclude the use of the ASIC routing features, such as floorplan flexibility or use of large macros.

It is essential for the designer to have accurate timing models for all components in his design, including clock distribution, so he can validate his design.

Finally, a high performance clock network must achieve aggressive clock skew and phase delay parameters.

Clock skew is the maximum time differential between any two registers' active clock edges. Phase delay is the propagation delay from the system clock pin to the slowest register clock pin. Figure 1 illustrates skew and phase delay on a generic schematic.

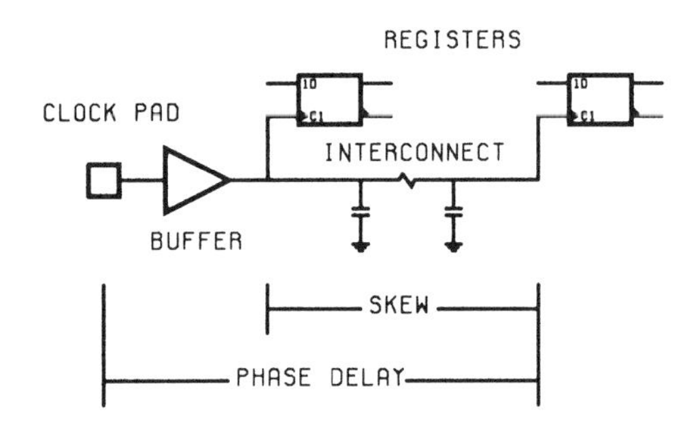

Figure 1. Clock skew and phase delay definition

Reprinted from *Proc. IEEE Custom Integrated Circuits Conf.*, pp. 15.4.1–15.4.5, May 1989.

Clock skew can be viewed as an uncertainty on the position of the active clock edge for any particular register. As skew increases, more margin is required for both setup and hold times.

The consequence of more setup time margin is either less usable cycle time or a reduction in clock frequency. With less usable cycle time fewer functions fit between registers. More pipelining may solve the problem but this increases the die size. Very high performance chips need to use all of the available cycle time, so skew is ideally targeted at 5 or 10% of the cycle time budget.

To assure a safe hold time margin more delay may be added to the fast paths. Unfortunately, adding delay to fast paths wastes chip area. Better skew performance can reduce or eliminate the need for added delay.

Clock phase delay will effect external pin setup and hold times. Chip to chip interfaces must take phase delay into account. Cycle time is wasted when two or more chips have a phase delay imbalance.

DESIGN OF THE CLOCK NETWORK

A clock distribution system has been designed for VTC using a one micron double metal CMOS process. The technology features a one micron effective channel length, 3.3 micron first metal pitch, and 3.6 micron second metal pitch.

Network Description

A multi-tiered solution to the clock distribution problem has been developed. As the skew and phase delay charts of Figures 2 and 3 indicate, a wide range of performance is available. The High Performance schemes obtain the best performance at a slightly higher cost in terms of turn time and chip area. The remainder of this paper addresses the detailed implementation of the High Performance scheme.

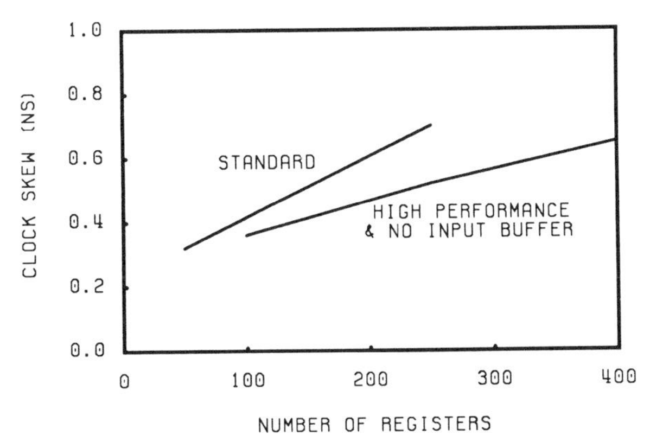

Figure 2. Skew for various clock schemes at a core size of 300 mils. The skew of the Simple scheme is about 1.5ns for 10 registers and 4ns for 50 registers, and is not shown.

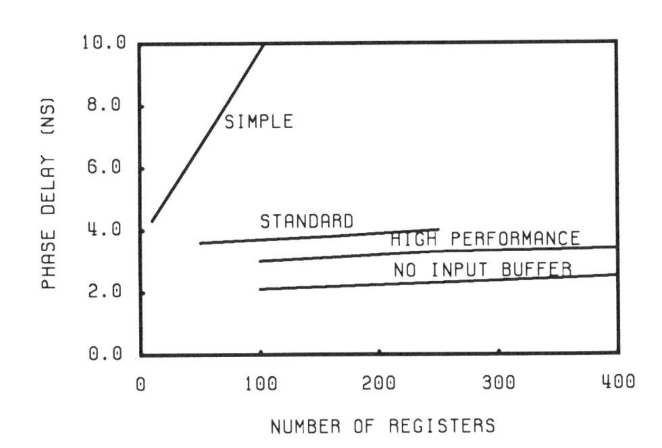

Figure 3. Phase delay for various clock schemes at a core size of 300 mils.

The High Performance scheme is constructed in the form of a distribution tree [1-3]. The external clock signal is detected at TTL or CMOS levels followed by heavy buffers within the pad cell itself. The output is routed to the center of the core in wide metal. The center component serves as a signal splitter, routing four intermediate width lines to the four quadrants of the chip. Typically five local drivers are fed from each of these lines producing local copies of the clock signal. Each local driver connects to a portion of the clocked registers. A schematic model for this network is shown in Figure 4.

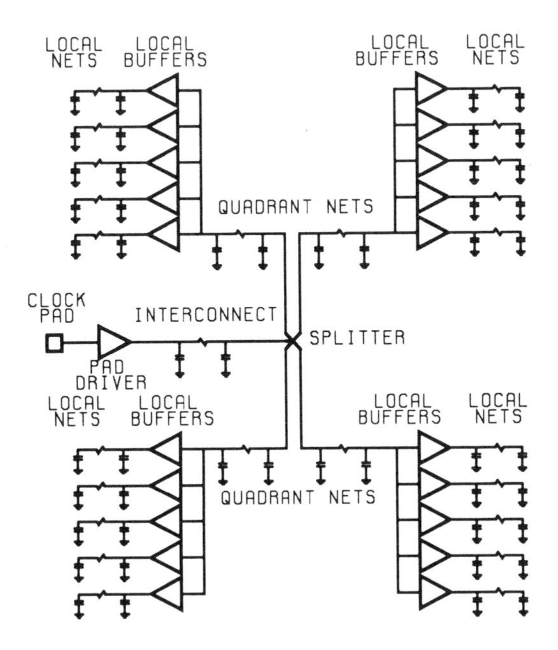

Figure 4. Schematic representation of clock network

Circuit Design Methodology

A non-CMOS level provided at the clock input pad requires level shifting and re-shaping which results in added phase delay. For this reason, CMOS interface levels are recommended. The designer may even opt to connect the external board driver directly to the input of the distribution tree. This eliminates a substantial portion of the chip phase delay.

Robust output drive on each stage helps maintain an undistorted duty cycle on the clock signal. When the transition times are slow, the duty cycle is more sensitive to logic threshold voltages. If the duty cycle is distorted, the registers' performance declines. By placing this drive capability in the pad cell the switching noise is kept out of the core power bus lines.

The High Performance scheme is targeted for up to 3200 registers. For multi-chip systems the clock network can compensate for a large variation in the number of registers so that the phase delay is more consistent. This compensation is achieved by using a fixed core size, a fixed number of local drivers, and keeping the length of the clock net segments within certain constraints.

When designing the line widths for the distribution network, keep in mind that the limiting factor may be the current density requirement rather than the RC time constant. The RMS current can be computed based on the AC power dissipation formula. The formula for AC power drawn from a supply is CV^2f, but, in the case of a signal line, current is carried during both the charge and discharge phases of a cycle, hence the RMS current is doubled. Electromigration studies by Miller [4] and Towner et. al. [5] suggest that conductor lifetime improvement is better than on-time corrections predict for pulsed current. Hence, this analysis may be conservative. If the tree capacitance is 18pf, the supply voltage 5.5v, and the clock frequency 100MHz, then

$$\begin{aligned} Irms &= 2 * C * V * f \\ &= (2)(18pf)(5.5v)(100MHz) = 20mA \end{aligned}$$

To keep current density under 500,000 amps/cm^2 requires approximately 1 micron of metal width per milliamp of current for this process. Since all 20mA is in the bus to the center splitter, at least 20 microns of bus width is required. If each quadrant bus has 3pf loading then at least 3.3 microns of width is required.

The quadrant buses are unequal in length when the center splitter is not precisely centered, so it is very important to position the splitter accurately. The local buffers are allowed to tap off of the quadrant buses anywhere, from the splitter output to the corner of the core. With a well centered splitter, the quadrant bus length can be controlled to within 10% of the averaged core dimension. The local bus length should be limited to about half of the core dimension, assuming an adequate number of local buffers, since greater lengths introduce a large RC time constant, and hence too much skew. Consider a 300 mil core with 4.2 micron wide quadrant lines. The distributed resistance and capacitance of each quadrant line is 140 ohms and 3pf, respectively. The resulting propagation delay contributes directly to clock skew. Spice simulation or the Rubinstein et. al. [6] approximation for RC networks can be used to determine the delay. Spice yields a value of 225ps.

The local buffers that drive the clocked elements have low input capacitance and high output drive capability. The low input capacitance makes the time constant from the quadrant net to the local buffer negligible. The high output drive makes the load dependent prop delay relatively insensitive to local net length and register loading variations. Four stages are used in the local buffer for the High Performance scheme, with a 0.06pf input capacitance and an 80 ohm output impedance. Using a simple pi model, the local net mismatch can be as great as the local net resistance times half the sum of the net and register capacitances. Using the 300 mil core example, a maximum length local net of 3800 microns has a resistance of 150 ohms and a net capacitance of 0.6pf. The register loading is about 3pf for 20 registers. The time constant is:

$$RC = (150)(3.6E-12)/(2) = 270ps$$

Some additional skew will occur because of local buffer prop delay fluctuations due to net and register capacitance variations. For example, with an output impedance of 80 ohms, a 1 picofarad loading discrepancy translates to about 80 picoseconds of skew.

Area Impact

Routing experiments demonstrate that the effect of going to a high performance clock distribution system, as opposed to a simplistic approach of routing a global clock signal without constraints, amounts to only about a two percent overhead.

PLACE AND ROUTE SOFTWARE

VTC proprietary software reads the ASIC database and input parameters and interfaces with the place and route tool. The place and route task is divided into three phases: netlist driven placement, automatic clock assignment, and layout reintegration.

Netlist Driven Placement

The purpose of this step is to determine the optimal placement of the clocked registers based on logical netlist constraints. To achieve this, place and route iterations are performed without attaching the clock loads to the local drivers.

At this point, power routing and various floorplan trade offs are evaluated to achieve die size and performance goals. The interconnect capacitance of the routed design is back annotated for timing simulation. The floorplan parameters are saved. Upon customer approval of the back annotated simulation these parameters are used for the final place and route following the clock distribution.

Automatic Clock Assignment

After a floorplan and placement has been optimized based on netlist constraints, VTC proprietary software, called CLOCKWISE, is used to assign the registers to local drivers based on register location. The purpose of CLOCKWISE is to partition the clock loads into a minimal but balanced capacitance configuration.

Input parameters to CLOCKWISE are:

1) Number of clock loads.
2) Number of local drivers.
3) Maximum and target capacitance imbalance on local drivers.
4) Capacitance minimization vs. balancing tradeoff.

The standard cell flexibility can create unique problems that the clock distribution must handle. The clock tree itself might be highly sensitive to regional net imbalance. Some special problems that CLOCKWISE can handle are large macro cells, loads in tight but distant clusters, local driver placement constraints, quadrant net balancing, chip to chip phase matching, and variations in input pin loading.

First CLOCKWISE performs a preliminary assignment of registers to local domains. The local drivers are assumed to be within the minimum bounding rectangle of their assigned clock loads. Initial driver locations are picked based on the clustering of the loads. Clusters that have the most distant nearest neighbors get priority. Then each unassigned load is added to the local net that will cause the least capacitance gain without exceeding the maximum imbalance goal. Hardest to fit loads have priority over easy to fit loads.

The capacitance gain is the input load capacitance plus the calculated maximum Steiner tree capacitance. The formula used for the maximum Stiener tree length is [7]:

$$\text{Length} = \tfrac{1}{2} * R * (T^{**\frac{1}{2}} + 1)$$

Where: T is the total number of pins in the net, and R is half of the perimeter of the minimal bounding rectangle.

The next step in CLOCKWISE is simulated annealing distribution improvement. After all loads have been assigned to local clock nets, zero cost dummy loads are added until each net has the same number of loads. These dummy loads allow for one way swapping of real clock loads during the simulated annealing iterative improvement phase.

The simulated annealing algorithm tries to simultaneously minimize the Steiner tree estimate of each local clock net, and the standard deviation of the capacitive loading (including interconnect) of all of the clock nets. A user supplied parameter can be adjusted to change the proportion of the net capacitance reduction versus the standard deviation in the metric formula.

Manual swapping of loads may be performed within CLOCKWISE to handle any special situations or improvements desired.

Layout Reintegration

Once the loads are assigned as desired, the layout netlist is modified accordingly and the place and route is repeated.

The pre-distribution floorplan is restored using the modified netlist. To optimize the placement, the local clock drivers need to be placed near the local clock nets. Since the clock drivers are often fairly large cells, other cells will move as well. Holding the register cells near their original location while allowing other cells to move freely during auto placement helps to maintain the pre-distribution die size and performance while allowing small changes to accomodate the clock nets.

Final routing is performed and the results are back annotated for final simulation.

SIMULATION MODELS

Clock skew and phase delay are modeled for event driven simulators and static timing analysis tools to confirm function and performance.

Prior to layout the netlist is simulated containing a dummy clock buffer that is tied to all system clock pins. The actual buffers for the selected scheme are also in the netlist but the outputs are unconnected. Consider the Mentor QuickSim modeling methodology. The target skew and phase delay values are entered by the user at the time of design expansion. The phase delay is assigned to the dummy buffer prop delay, while the skew is added to the setup and hold parameters of each register. Both values undergo derating modifications to account for process, temperature, and supply voltage effects. The resulting simulations show an averaged phase delay to all registers and any fast or slow paths attributable to skew are observed as transitions to unknown (X) states.

After layout is completed the design netlist is modified to reflect the actual local buffer assignments. The back annotation data containing the interconnect load capacitance is used during design expansion to generate a unique phase delay for each local buffer. Only the skew caused by interconnect in each local clock net is added to the specific registers' setup and hold parameters for post layout simulation. This skew is calculated assuming a worst case distribution of registers. The skew caused by quadrant interconnect mismatching is modeled in the prop delays of the central distribution component. Thus post layout simulations give the designer actual phase delays for specific circuit paths and only the skew attributed to local nets is built into the setup and hold parameters.

Another option is to use the Tangent Tansure static timing analysis tool in conjunction with the Tancell router. This tool will extract and model complex RC networks and generate a critical path report.

CONCLUSION

A multi-tiered clock distribution scheme has been designed that trades off area and schedule for high performance operation. Pad input thresholds, output drive, interconnect parasitics, input capacitance, and current density are some of the issues that must be addressed in a successful design.

Commercial place and route software, along with VTC proprietary software, uses a netlist specific implementation to optimize performance for both the clock distribution network and the customer netlist.

The clock network can be accurately modeled to simulate the effects of clock skew and phase delay using commercially available logic simulators.

The High Performance clock distribution scheme is integral to speed sensitive standard cell programs at VTC. Some of these programs are targeting 100MHz operation. Prototypes are currently in layout and fabrication.

REFERENCES

[1] K. D. Wagner, "A Survey of Clock Distribution Techniques in High-Speed Computer Systems," Stanford Computer Systems Laboratory, CRC Technical Report No. 86-20, CSL TN No. 86-309, Stanford University, Stanford, California, December 1986.

[2] H. B. Bakoglu, J. T. Walker, and J. D. Meindl, "A Symmetric Clock-Distribution Tree and Optimized High-Speed Interconnections for Reduced Clock Skew in ULSI and WSI Circuits," Proc. IEEE Int'l Conf. on Computer Design: VLSI in Computers and Processors (ICCD-86), pp. 118-122, Rye Brook, New York, October 1986.

[3] E. G. Friedman and S. Powell, "Design and Analysis of a Hierarchical Clock Distribution System for Synchronous Standard Cell/Macrocell VLSI," IEEE Journal of Solid-State Circuits, Vol. SC-21, No. 2, pp. 240-246, April 1986.

[4] R. G. Miller, "Electromigration Failure Under Pulse Test Conditions," Proc. 16th IEEE Reliability Physics Symposium, pp. 241-247, 1978.

[5] J. M. Towner and E. P. van de Ven, "Aluminum Electromigration Under Pulsed D.C. Conditions," Proc. 21st IEEE Reliability Physics Symposium, pp. 36-39, 1983.

[6] J. Rubinstein, P. Penfield Jr., and M. Horowitz, "Signal Delay in RC Tree Networks," IEEE Trans. on Computer-Aided Design, Vol. CAD-2, No. 3, pp. 202-210, July 1983.

[7] F. K. Chung and F. K. Hwang, "The Largest Minimal Rectilinear Steiner Trees for a Set of n Points Enclosed in a Rectangle with Given Perimeter", Networks Vol. 9, No. 1, pp. 19-36, Spring 1979.

A Layout Methodology for the Synthesis of High Speed Global Clock Nets

Delfin Y. Montuno and Raymond C.S. Ma

Bell-Northern Research Ltd., P.O. Box 3511, Station C,
Ottawa, Canada, K1Y 4H7

Abstract

Clock distribution in chips is important. Its importance becomes harder to overlook as the system being designed uses higher clock frequencies and larger number of flip-flops, and requires careful system partitioning into several chips and fabrication in Multi-Chip Module. In this scenario where several chips are treated as synchronous islands, we will be concerned not only with reducing intra-chip skew but also with minimizing inter-chip delay.

Related topics of this paper are system partitioning, high performance system, clocking schemes, clock distribution structures, clock synthesis and layout, buffer design, boundary scan test, intra-chip skew minimization, and inter-chip synchronization.

Introduction

Several types of *clock distribution networks* exist that try to solve the clock skew minimization problem for intra-chip clock subnets [1-4]. Most of these networks deal with the intra-chip skew and do not easily support inter-chip delay minimization. Some of them use the regularity of the routing structure to minimize clock skew but do not accommodate multiple global nets or large disparity in cell sizes [1, 2]. Some of them use clustering and balancing which could require great tree depth to minimize skew [2, 3, 4]. Some of them use hierarchy and hard macro partitioning [5]. Some use self-timing to avoid the skew problem [2].

There also exist several *clocking schemes* [2]. Clocking scheme and clock distribution structure must work together for a global clock net to work properly. The clocking scheme assumed in this paper is one with a single phase clock running at 50 MHz or 100 MHz or even higher, and each clock net can drive between 1000 and 4000 edge-triggered flip-flops.

The third component of a clock distribution system is the *design of the clock buffers and flip-flops*. However, we will not be dealing with it in this paper.

This paper will describe a clock distribution structure that is flexible enough to take into consideration both inter-chip delay and intra-chip delay problems for multiple global clock nets. Since the clock strategy considers the inter-chip delay problem, it can also be used for inter-chip boundary scan test. In the next section, we present the clock distribution structure. The software that supports the implementation of the structure is then described followed by application results.

The clock synthesis software is developed within the following context. The process technology is BiCMOS with three layers of metal for routing. The chips, composed of custom analog blocks, memory blocks, and (synthesized) logic, are laid out in gate-array style. The templates are generated using in-house developed Custom Template Generator software. The clock synthesis software is built on top of Cadence Gate Ensemble™ Release 4.0 package.

Clock Distribution Structure

Structure

The clock synthesis model consists of a *clock conditioner*, and for each global clock net, a *clock ring*, a *primary clock buffer*, and a *set of secondary clock buffers*. Figure 1 illustrates an implementation of the model.

The clock ring resides just inside the pad border region of the chip. The clock conditioner is placed just above the middle of the bottom segment of the innermost ring. It takes its clock input directly from a nearby pad along the bottom row of pads and directs its output to a primary clock buffer which is also in a nearby pad. The primary clock buffer then drives the ring, which in turn drives a set of secondary clock buffers and some I/O flip-flops. Each secondary clock buffer then drives a subset of flip-flops as shown. Note that all clock buffers are bipolar.

For symmetry, the secondary buffers are located only along the vertical sides of the ring and the primary buffers are centered at the bottom. The rings and bars are laid out in wide low resistance metal, which is essential, and the maximum length of each bar is about half the width of the chip core. The number and arrangement of secondary buffers are dependent on the load balancing criteria. The number of flip-flops used for high speed synchronous I/O and attached directly to the ring is limited.

Although only one clock net is shown, the configuration can be extended to handle multiple clock nets by allocating, for each net, a ring, a primary buffer, and a set of secondary buffers.

Reprinted from *Proc. IEEE Custom Integrated Circuits Conf.*, pp. 28.4.1–28.4.4, May 1992.

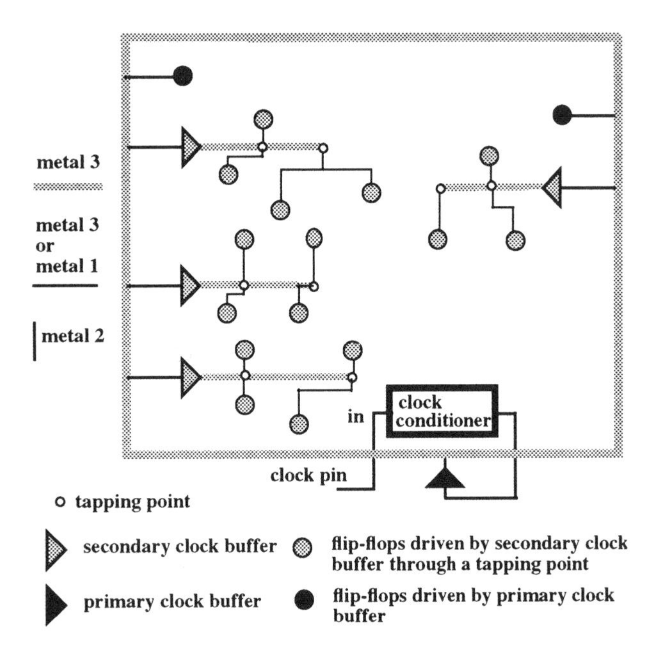

FIGURE 1. An Implementation of the Clock Synthesis Layout Model

Performance Considerations

The clock net performance is highly dependent on having low resistance metal available. We can use a simple ring structure because a metal layer with low RC is available in the target process. Let us consider our two design objectives:

Minimization of Clock Delay Although the exact inter-chip delay requirement is project specific, minimizing clock delay for high speed synchronous I/O is achieved by limiting the levels of clock buffering and by limiting interconnect length. Thus the clock input pin, the primary buffer, and the conditioning logic are all clustered near the bottom center of the chip. The flip-flop feeding a high speed output pin (or receiving from an input) is clocked from the ring and is physically clustered near the I/O pin (again to minimize interconnect length). The primary buffer size is adjusted to match the number of flip-flops attached directly to the ring. The load presented by the secondary bipolar clock buffers on the clock ring is minuscule.

Minimization of Clock Skew The minimum clock skew in the chip core is dependent on the characteristics of the single phase flip-flop. One of the primary source of skew is the RC delay around the primary clock ring; this could be reduced by driving the ring in a more symmetric fashion, although doing so results in more overall clock delay. We have not found it necessary to do this. With our particular flip-flop design, buffer characteristics, and metal RC parameters, we can obtain safe single-edge clocking with up to 50% load imbalance on the secondary buffers and still achieve the intra-chip skew requirement of below 500 ps.

Shorting all the secondary clock bars together would help to reduce skew and simplify the back annotation process, but we have not done so due to concern over the reliability of "crow-baring" several bipolar buffers.

Handling of Global Clock Net

The designer has to deal with only two logical nets per global clock net: the high speed one for inter-chip communication and the rest for intra-chip communication. Figure 2 shows the logical connectivity of a typical global clock Net_i which is connected to the input of the primary clock buffer, p_1. The output net of this buffer, *pclk*, consists of a ring and its connection to pin_1, pin_2, ..., pin_{ci} of the directly attached flip-flops for inter-chip communication and to the clock buffer macro B_i. The output net of this clock buffer macro, *cclk*, connects pins, PIN_1, PIN_2, ..., PIN_{Ci} of the flip-flops for intra-chip communication.

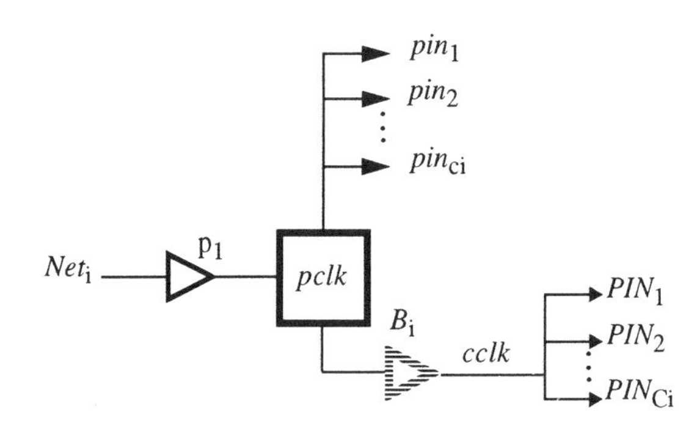

FIGURE 2. A Typical Initial Global Clock Net - The Designers' View

The partitioning and physical implementation of *cclk* is the main topic of this section. Figure 3 shows the expansion of the clock macro B_i and a logical partitioning of *cclk* as the result of a physical implementation. Here B_i is replaced by the secondary clock buffers, b_1, b_2, ..., b_{si}. The output net name of each secondary clock buffer, b_{si}, is called $cclk_{si}$. Although these nets are now not physically connected, they are considered to be logically equivalent.

The physical layout of *cclk* requires that each $cclk_{si}$ be further partitioned into groups, each of which connects to a tapping point along the wide bar coming out of a secondary clock buffer as shown in Figure 1. Each of these groups is tied to the output of a virtual clock buffer b^j_k and has a net name called $cclk^j_k$. Because of the feed through nature of the virtual clock buffers, $cclk_j$ and $cclk^j_k$ are electrically the same. These virtual buffers provide clustering capability and tapping points on the wide clock bar which is connected to a secondary buffer output.

The high speed net *pclk* has minimal delay by construction as the flip-flops are attached directly and placed very close to the ring. When partitioned into $cclk^j_k$'s, the core clock net *cclk* has minimal buffering depths and as balanced as possible partitioned capacitive loads.

After physical implementation, the replacement of the clock buffer macro is then back-annotated to the schematic. For each

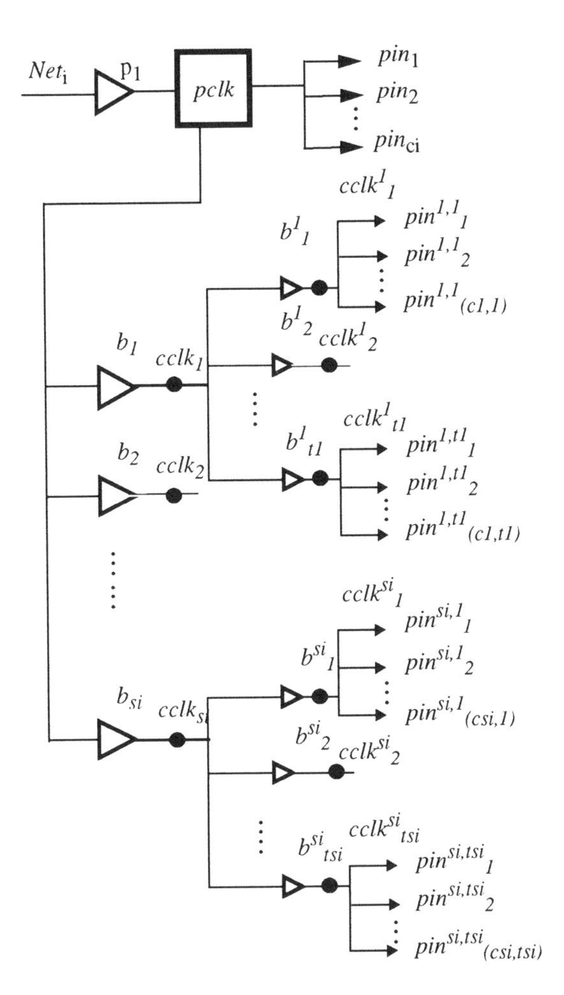

FIGURE 3. A Typical Final Global Clock Net

secondary clock buffer b_i, there is a set of t_i's tertiary virtual clock buffers which is used to partition a subset of *cclk* pins that would have been connected directly to the output of b_i. The task done here is the mapping of each pin PIN_i through partitioning into a pin $pin^{j,k}_l$ while satisfying the following relation.

$$\bigcup_{1 \le y \le Ci} PIN_y = \bigcup_{(1 \le j \le si)} \bigcup_{(1 \le k \le tj)} \bigcup_{(1 \le l \le (cj,k))} pin^{j,k}_l$$

In other words, we are solving the following problem: Given Net_i with the set of clock pins, pin_x and PIN_y, where $1 \le x \le ci$ and $1 \le y \le Ci$, and using clock net partitioning in physical layout implementation, connect pin_x directly to the clock ring and map PIN_y into $pin^{j,k}_l$ such that the above relation is satisfied and the capacitive loadings of each partition are as close as possible, from the layout point of view. The capacitive loading is in turn dependent on the following factors:

- input capacitance of clock pin of flip-flops
- number of flip-flops
- bounding area of flip-flops
- interconnect routing model

In general, there will be multiple global clock nets, and each of them can be treated, most of the time, as independent of the others. Although there may be no logical interaction among global clock nets to be transformed, there could be intense physical competition for premium buffer locations, optimal placement of flip-flops, and routing resources. In the next section, we describe the software that supports the described methodology.

Software Support

As indicated earlier, the software is built on top of the Cadence Gate Ensemble™ (GE) layout tool. The flow in the layout, illustrated in Figure 4, is generally as follows: library data and netlist data input, timing driven placement, clock synthesis, global route, final route, and if necessary, search and repair.

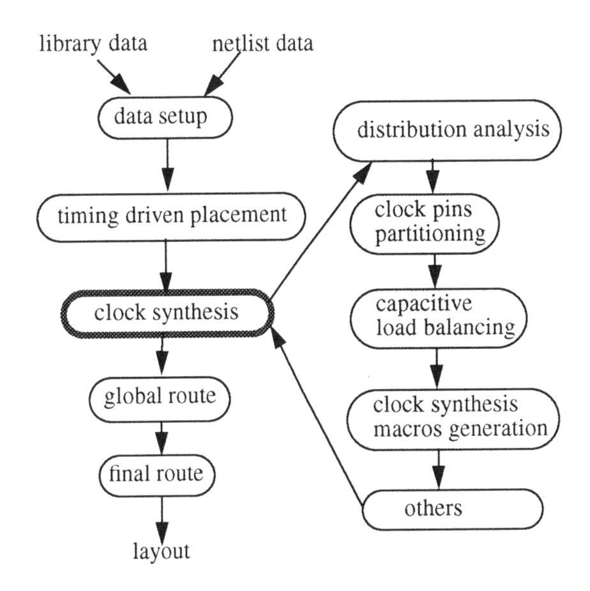

FIGURE 4. Clock Synthesis Layout Flow

The Clock Tree Synthesis (CTS) by rectangular area feature of GE can be used on a part of a chip or on an entire chip. We have developed a methodology that extends this feature to support the following steps, shaded in Figure 4: analysis of flip-flop clock pins distribution, partitioning of clock pins distribution by region, balancing of capacitive load, creation and execution of GE executable CTS macros, laying down of wide clock wires, tidying up of subnets, generation of load file, back-annotation of expanded clock macros, and others.

Analysis and Partitioning

After the timing driven placement is completed, the distribution of the flip-flop clock pins is made available to the user for interactive and graphical analysis. The analysis result provides the following information on clock pins on an areal basis: physical location, capacitive load, and count.

The above information then guides the user in interactively partitioning the clock pins into rectangular clusters while equalizing the partitioned capacitive loads as much as possible.

CTS Macro Generation and Execution

A check for completeness of partitioning and non-overlapping of partitioning is done first before GE executable CTS macros are generated. For each macro, statistics containing the following information is output: number of pins, total capacitive load, and approximate total wire load. The CTS macros are then executed in GE.

Special Clock Net Handling

Currently the wide wires are manually routed in GE. The tapping points, tertiary virtual buffers, are constrained to lie along the same cell row as that of the associated secondary clock buffer. Further splitting of a net may be necessary after CTS macro execution if the clock pin distribution is highly skewed such as in the case where large cell size disparity exists.

Load File Generation

After final routing is completed and all nets are routed successfully, a load file is generated and the post-layout simulation is performed.

Results

Table 1 shows the results of using the described methodology. The acceptable maximum load difference, as mentioned earlier, is 50%. The achieved load differences are below this tolerance for all chips laid out. In Chip A, which is about 66% memories in area, the variation is greatest. It is, however, the first chip tried and there is no second iteration made to balance the loads further. Also note that its size is about four times that of the other chips. For Chip B, a 7% load difference is achieved in the first iteration and a 0.6% in the second iteration. Similarly for Chip C, a 14.5% is achieved in the first iteration and a 2.3% in the second iteration. The main cause for the variation is due to the interconnect load which is layout tool dependent. In our case, the interconnect routing model of the layout tool is not accurately characterized yet.

Chip	Chip Size (mm^2)	Clock Frequency (MHz)	# of FFs	# of Clusters	Max. Load Difference
A	7.5x8.4	50	1228	14	32 %
		100	298	11	18 %
B	4.0x4.0	50	226	3	7 %,
B_{opt}					0.6 %
C	4.0x4.0	50	300	3	14.5 %
C_{opt}					2.3 %

Table 1: Experimental Results

Future Work

The methodology and the software tools are still in the prototypical phase. Therefore, more work remains. One conceivable concern with the methodology is that the designer has seemingly limited control in the physical implementation where more care may be needed. An example is the case of a set of tightly coupled clock pins being partitioned to multiple neighboring secondary clock buffers. This problem, however, can be handled if the designer specifies in advance the logical clustering of flip-flops, if this clustering is then preserved in the cell placement phase using the soft-grouping and region restriction features of GE, and if it is further maintained in the clock buffer assignment phase.

A number of important issues that could further improve the quality of a clock distribution are: ground bounce minimization, noise minimization, power consumption analysis, and layout tool specific interconnect model. They are also related to the design of clock buffers, flip-flops, and power net distribution.

Conclusion

This paper presents a practical clock distribution methodology. It is quite flexible because of its applicability in chips with large cell size disparity. It allows extra levels of clustering to balance load at the expense of minimal increase in total delay. It facilitates inter-chip delay minimization. Although it is implemented in gate-array style, the concept is not gate-array dependent. It can handle multiple clock nets and allows designer to cluster sensitive subset of flip-flops.

Acknowledgments

We would like to thank Phil Wilcox and Bradley Booth for their contribution on the performance requirement analysis; Gary Puukila for the discussion on the clock distribution structure; Esmond Ho for his contribution on the Custom Template Generator; Ho Nguyen for the use of and feedbacks on this methodology; John Soucie and Shawn Fitzpatrick for the laying out of the chips; and all others who are involved in one way or another.

References

[1] T. Saigo, S. Watanabe, Y. Ichikawa, et al, "Clock Skew Reduction Approach for Standard Cell," *IEEE 1990 Custom Integrated Circuit Conference*, pp. 16.4.1-16.4.4.

[2] H.B. Bakoglu, *Circuits, Interconnections and Packaging for VLSI*. Addison-Wesley Publishing Company, Inc., 1990, "Chapter 8, Clocking of High Speed Systems," pp. 338-393.

[3] *Gate Ensemble User Guide*, Release 2.0, Vol. I, March 1990, "Chapter 12 Timing," pp. 12.21-12.41.

[4] M.A.B. Jackson, A. Srinivasan, and E. S. Kuh, "Clock Routing for High Performance ICs," *27th ACM/IEEE Design Automation Conference*, 1990, pp. 573-606.

[5] E. G. Friedman and S. Powell, "Design and Analysis of a Hierarchical Clock Distribution system for Synchronous Standard Cell/Macrocell VLSI," *IEEE Journal of Solid-State Circuits*. Vol sc-21, no. 2, April 1986, pp. 240-246.

A CLOCK DISTRIBUTION SCHEME FOR NON–SYMMETRIC VLSI CIRCUITS

P. Ramanathan and Kang G. Shin

Real-Time Computing Laboratory
Department of Electrical Engineering and Computer Science
The University of Michigan
Ann Arbor, Michigan 48109–2122.

ABSTRACT

The control of clock skew to an acceptable small fraction of the clock period is essential for the correct and dependable operation of any digital system. Within a VLSI circuit, the foremost factor responsible for this clock skew is the difference in the length of clock lines to different functional elements in the circuit. In this paper we propose a clock distribution scheme that minimizes the difference in the length of clock lines.

The proposed scheme uses the hierarchy created by the clock buffers to parallelize the distribution of the clock signal. At each hierarchical level, an exhaustive search of paths with intelligent pruning is used to determine the optimal layout of clock lines at that level. Unlike other related work in this area, both delay and skew are taken into account in determining the layout.

1 Introduction

Within most VLSI circuits, data transfer between the functional elements is synchronized by a single control signal, the processing clock. This signal typically constrains the timing and performance of the circuit. It is, therefore, imperative to develop a design methodology for distributing the clock signal without creating any timing uncertainties.

There are four main factors responsible for the timing uncertainties in a VLSI circuit: (i) differences in the length of lines that deliver the clock signal to two different functional elements, (ii) differences in delays through any active elements inserted in the clock lines (e.g., buffers), (iii) differences in parameters such as resistivity and dielectric constant that determine the line time constant, and (iv) differences in threshold voltages of different elements. These factors can be eliminated by using an appropriate clock distribution scheme.

Wann and Franklin [1] proposed a distribution scheme that ensures equal lengths for all paths when the elements are identical and placed in the form of a symmetric array. This scheme, however, is not suitable for distributing the clock signal in a general VLSI circuit where the functional elements are of different sizes and where the placement of the elements is governed by the interconnection between the elements and is usually not in an array form. Likewise, the existing schemes for eliminating the other factors [2, 3, 4] are also not suitable for distributing a clock in a general VLSI circuit because they are sensitive to the fabrication process.

In this paper we propose a non-symmetric distribution scheme that will minimize the clock skew due to line lengths for functional elements with different sizes and arbitrary placement. It also accounts for the difference in delays caused by the clock buffers. Unlike the other related work in this area, both delay and skew are considered in determining the layout of the clock lines. This is important because layouts that have minimum skew might have long lines that result in large delays, and hence, degraded performance. On the other hand, layouts that have minimum delay might have large skews that could affect the correctness of the circuit. The objective used in this paper for determining the clock layout minimizes the skew subject to minimum longest delay.

The scheme proposed here assumes that a floorplan of the modules is given. This floorplan would have been determined by using a placement algorithm based on the other signals interconnecting the sub-modules that constitute that module. For this floorplan, the scheme identifies an entry point from which the clock is distributed and the route of the clock from the entry point to each of the sub-modules. The entry point is selected in such a way that the delay to the farthest sub-module from that entry point is minimum. This will minimize the maximum delay in distributing the clock signal to the sub-modules. The optimal layout of the clock lines from the selected entry point is determined by an exhaustive search of all paths with intelligent pruning. The main advantage of the scheme is that the computation of the optimal layout can be easily parallelized.

This paper is organized as follows. In Section 2, the clock layout problem is defined formally and some notations and terms are introduced. An algorithm to select the entry point is presented in Section 3. Section 4 describes an algorithm to determine the optimal clock layout from the selected entry point. An implementation of the scheme is discussed in Section 5. The paper concludes with Section 6.

2 Problem Formulation

A *module* is a group of functional elements to which the clock signal is being distributed. For clarity, the elements contained in a module will be referred to as the *sub-modules*. A *floorplan* is a relative placement of the sub-modules in the routing area. A *channel* is a rectangular zone of empty area between the sub-modules through which the signals can be routed. A *track* is a sub-unit within a channel through which a single signal can be routed. A channel is said to be *horizontal* (*vertical*) if the tracks within the channel are horizontal (vertical). A channel is said to be *peripheral* if it lies in the periphery of the floorplan. There are three possible types of intersection between a horizontal and a vertical channel: L-type, T-type, and $+$-type. A *decision point* is the point of a T-type or a $+$-type intersection.

Reprinted from *Proc. IEEE Int'l Conf. Computer-Aided Design*, pp. 398–401, Nov. 1989.

The floorplan can be represented by an undirected, weighted graph referred to as a *placement graph.* The vertices of the placement graph are the decision points in the floorplan, the input clock terminals of all the sub-modules, and the output clock terminals of all clock buffers. Clock buffers are sub-modules whose output drives the input clock signals of other sub-modules. Vertices v_l and v_k are connected by an edge if and only if there is a channel from v_l to v_k not containing any other vertex of the placement graph. In other words, there is a one-to-one correspondence between the edges of the placement graph and the channels in the floorplan. The weight of an edge $\{v_l, v_k\}$ is equal to the physical distance in the channel between the two points corresponding to v_l and v_k except when either v_l or v_k is an input clock terminal of one of the sub-modules. It is therefore proportional to the delay that will be induced on a clock signal if that channel is used as a part of the clock layout. The weight of edges $\{v_l, v_k\}$ in which one of v_l or v_k is an input clock terminal will be used to account for the delays induced by clock buffers. The assignment of weights to these edges will be discussed later in this section. One needs to consider not only the edges of the placement graph but also the line segment between two arbitrary points on the same edge. The line segment between two points u and v on the same edge will be denoted by $[u, v]$. The weight of $[u, v]$ is equal to the physical distance in the channel between the points corresponding to u and v. The weight of $[u, v]$ is sometimes referred to as the distance between u and v and is denoted by $d(u, v)$.

Let P be a floorplan of a module M that is comprised of m sub-modules M_1, M_2, ..., and M_m. There is an entry point into the floorplan from which the clock is distributed to all the sub-modules. However, all sub-modules may not receive their clock signal directly from this entry point. Instead, some of the sub-modules will receive their clock signal indirectly from the output of some other sub-module (i.e., clock buffer) in M. For simplicity of presentation, introduce a fictitious clock buffer that drives the entry point to the floorplan. For each sub-module M_i, $1 \le i \le m$, it is then possible to associate a unique level, *Level(M_i)*, that is equal to the number of clock buffers in the route from the clock entry point to the sub-module. In other words, sub-modules directly driven from the entry point are said to be in *Level 1*, while those driven by sub-modules in Level 1 are said to be in *Level 2*, and so on. Let $max\_level = \max_{1 \le i \le m} Level(M_i)$, and *Buffer($M_i$)* be the clock buffer, say, M_j that drives the clock signal of M_i. Define *Buffer(M_i)* of a sub-module M_i at level 1 as the fictitious clock buffer that drives the clock entry point to the floorplan.

Definition 1: In a placement graph, a *simple path* from s to t is a finite sequence of distinct points $u_0 u_1 \cdots u_r$, $r \ge 1$, such that: (i) $u_0 = s$ and $u_r = t$, (ii) if $r > 1$, then $u_1, u_2, \ldots, u_{r-1}$ are vertices of the placement graph, and (iii) $\forall\, j \in \{0, \ldots, r-1\}$, u_j and u_{j+1} lie on a common edge. The term "a simple path from M_i to M_j" will be used to refer to the simple path from the output clock terminal of M_i to the input clock terminal of M_j.

Definition 2: A *c-route* from an entry point e to a sub-module M is a concatenation of simple paths from e to B_1, B_1 to B_2, B_2 to B_3, ..., B_{l-1} to B_l, and B_l to M, where $B_1, B_2, \ldots, B_l$ are such that B_l = *Buffer(M)* and B_{j-1} = *Buffer(B_j)* for all $l \ge j \ge 2$.

Definition 3: The sum of the distances between adjacent points in a c-route will be referred to as its *c-length.*

Definition 4: A *c-layout* from an entry point e is a tuple of c-routes from e to all sub-modules in the floorplan.

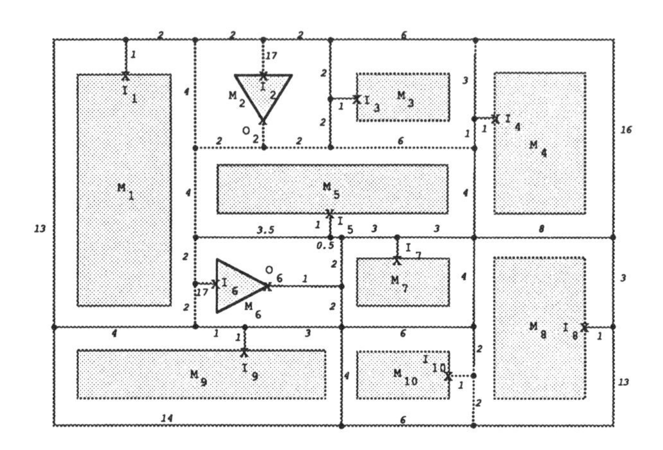

Figure 1: Floorplan of sub-modules in M.

Let SM_l denote the set of all sub-modules in level l. Let $\eta_i^{P,e}$ denote the c-length of the shortest simple path from *Buffer(M_i)* to M_i in the floorplan P when e is the clock entry point. Define $\mu_l^{P,e}$ to be $\max_{i \in SM_l} \eta_i^{P,e}$. We are now in a position to define the weights of the edges $\{v_l, v_k\}$ in the placement graph in which either v_l or v_k is an input clock terminal of a sub-module. Let I_j be the input clock terminal of sub-module M_j. If *Level(M_j)* is *max_level* or if M_j is not a clock buffer, then the weight of the edge $\{v_k, I_j\}$ is equal to the physical distance between v_k and I_j in the channel, else the weight of the edge is equal to $\mu_{l+1}^{P,e} + b + c$, where l = *Level(M_j)*, b is a constant corresponding to the delay introduced by M_j, and c is equal to the distance in the channel between the points v_k and I_j. The intuitive reason for assigning the weights to these edges in this manner is explained later in Example 1.

There are two main objectives in determining a c-layout. The first objective is to minimize the maximum difference in the c-length to the different sub-modules. The second objective is to minimize the c-length to all sub-modules. More formally, let E be the set of all points in the periphery of the floorplan. It is the set of points from which an entry point is to be selected. Let $Y^{P,e}$ denote a c-layout from e in the floorplan P. Let $L_i(Y^{P,e})$ denote the c-length of the simple path from *Buffer(M_i)* to M_i in the c-route from e to M_i in $Y^{P,e}$. Then the clock distribution problem is to determine an entry point e such that $\mu_1^{P,e} \le \mu_1^{P,s}, \forall s \in E$ and then from that entry point determine a c-layout $Y^{P,e}$ that minimizes $\sum_{i \in SM_l} \left| L_i(Y^{P,e}) - \mu_l^{P,e} \right|$, $\forall\, 1 \le l \le max\_level$. Since $\mu_1^{P,e}$ represents the delay from the entry point to the farthest (from that entry point), the first criterion minimizes the longest delay. The second criterion tries to increase the length of clock lines from the entry point to all the sub-modules to equal line length to the farthest sub-module. Consequently, the overall objective identifies a layout of clock lines that has minimum skew subject to minimum worst case delay.

Example 1: Consider the floorplan in Figure 1. The module M is comprised of ten sub-modules $M_1, M_2, \ldots, M_{10}$. Sub-modules M_2 and M_6 are clock buffers whose input is driven directly from the clock entry point of M. In addition, the input of sub-module M_8 is also driven directly from the entry point. M_2 drives the input clock signals of sub-modules M_1, M_3, M_4, and M_5 while M_6 drives the input clock signals of sub-modules M_7, M_9, and M_{10}. The sub-modules have been placed within the given area in such a way that there is ample space to route all the signals between them. The placement graph for this floorplan is also shown in Figure 1. In this graph the vertices corresponding to the decision points are indicated by • whereas those corresponding to the clock terminals of the sub-modules are indicated by ×. The weight of the edges in this graph are indicated in italics. The problem is to determine a clock entry point for the module M, the route of the clocks from this entry point to the sub-modules M_2, M_6 and M_8, and the route of the clocks from the output of M_2 and M_6 to the input clock terminals of the other sub-modules.

In this example it is easy to illustrate the reason for assigning larger weights to the edges $\{v_j, v_k\}$ in which either v_j or v_k is an input terminal of a clock buffer. Consider the sub-module M_7. The clock input to M_7 is not directly driven by the clock entry point but by the output of M_6. Therefore, total delay from the clock entry point to the clock input of M_7 is equal to the sum of the delays from the entry point to the input of M_6 ($\approx \mu_2^{P,e}$), delay introduced by M_6 ($= b$), and the delay from the output of M_6 to the input of M_7 ($= c$). By assigning $\mu_2^{P,e} + b + c$ as the weight of the edge incident on I_6 we can determine the c-route from the entry point to M_6 in parallel with the c-route from M_6 to M_7. In this example, b is assumed to be 5 and c is assumed to be 1. ■

3 Selection of the Entry Point

The goal of this section is to select a point in E that minimizes the longest delay, i.e., select $e \in E$ such that $\mu_1^{P,e} \leq \mu_1^{P,s}$ for all $s \in E$. Without loss of generality any point in E can be chosen as the origin and the other points can be represented as the distance from the origin along the periphery of the floorplan. In other words, E can be represented as a set of real numbers from 0 to $\|E\|$, where $\|E\|$ is the perimeter of the floorplan. Let $D(s_1)$ denote the distance of s_1 from the origin along the periphery.

Theorem 1: If $s_1 \in E$ and $s_2 \in E$ are two points on the same edge of a placement graph with $D(s_1) < D(s_2)$, then for all $x \in [s_1, s_2]$ and all $M_i \in SM_1$

$$\text{if } \left[D(x) - D(s_1) \leq \frac{\eta_i^{P,s_2} - \eta_i^{P,s_1} + D(s_2) - D(s_1)}{2}\right] \text{ then}$$
$$\eta_i^{P,x} = \eta_i^{P,s_1} + D(x) - D(s_1)$$
else
$$\eta_i^{P,x} = \eta_i^{P,s_2} + D(s_2) - D(x).$$

Proof: Outlined in [5]. ■

Entry, an algorithm for identifying the best entry point is based on Theorem 1. It can be informally explained as follows. Let $s_1, s_2, \ldots, s_n$ be the decision points in E such that $D(s_i) < D(s_{i+1})$ for $1 \leq i \leq n-1$. Let $s_{n+1} \equiv s_1$. First determine the shortest c-routes to the sub-modules in level 1 from each of the points $s_1, s_2, \ldots, s_n$. Then start at s_1 and move along the periphery of the floorplan from s_1 to s_2, s_2 to s_3, $\ldots s_{n-1}$ to s_n, and then back to s_1. To go from s_k to s_{k+1} it may be necessary to move through several intermediate points. Each move is comprised of determining the best entry point based on the path already traveled and then selecting the next point to move.

For example, let $u \in [s_k, s_{k+1}]$ be an intermediate point to which the algorithm has moved while going from s_k to s_{k+1}. If M_i is the farthest sub-module from u, then the next point s to which the algorithm moves is given by

$$\text{if } \left[D(s) - D(u) < \frac{\eta_i^{P,v} - \eta_i^{P,u} + D(v) - D(u)}{2}\right] \text{ then}$$
$$s = D^{-1}\left[D(u) + \frac{\eta_i^{P,v} - \eta_i^{P,u} + D(v) - D(u)}{2}\right];$$
else
$$s = \min\{z_{ij} : M_j \in SM_1, z_{ij} > s\},$$

where z_{ij} is the point of intersection between $\eta_i^{P,x}$ and $\eta_j^{P,x}$ in $[s_k, s_{k+1}]$. After moving to s the algorithm updates the current best entry point and then repeats the above steps until it reaches s_{k+1}.

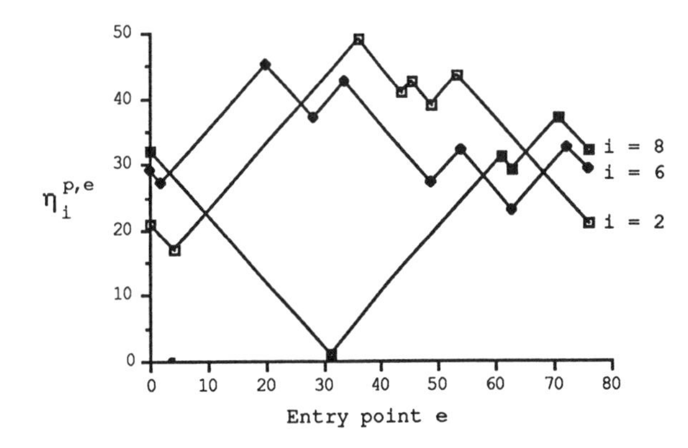

Figure 2: Variation in $\eta_i^{P,x}$ for the floorplan in Figure 1

Example 1 (cont'd.): Consider the floorplan in Figure 1. Recall that only sub-modules M_2, M_6 and M_8 are in level 1. Figure 2 shows the variation in $\eta_i^{P,x}$ for the three sub-modules in this floorplan. The plots are based on vertex 0 (see Figure 1) as the origin. Clearly, $\eta_i^{P,x}$ for $i = 2, 6$ and 8 are comprised of straight lines of slopes +1 and −1 as indicated by Theorem 1. Since in this example $\mu_1^{P,x} = \max\{\eta_2^{p,x}, \eta_6^{p,x}, \eta_8^{p,x}\}$ and $\mu_{opt} = \min_{x \in [0,76]} \mu_1^{P,x}$, it follows by either graphically or by using algorithm **Entry** that the optimal entry point is at distance 3.5 units along the periphery from the origin and $\mu_{opt} =$ 28.5 units. ■

4 Optimization Problem

The problem addressed in this section can be stated as follows: given a floorplan P and an entry point e determine a c-layout $Y^{P,e} \equiv (R_1^{P,e}, R_2^{P,e}, \ldots, R_m^{P,e})$ that minimizes $\sum_{i \in SM_l} \left|L_i(Y^{P,e}) - \mu_l^{P,e}\right|$ for all $1 \leq l \leq max\_level$. Since there is no inter-dependence between the terms of the summation,

$$\min_{Y^{P,e}} \sum_{i \in SM_l} \left|L_i(Y^{P,e}) - \mu_l^{P,e}\right| \iff \min_{R_i^{P,e}} \left|L(R_i^{P,e}) - \mu_l^{P,e}\right| \tag{4.1}$$

for all i such that $M_i \in SM_l$, and where $L(R_i^{P,e}) = L_i(Y^{P,e})$ is used just to emphasize that the right-hand side depends only on M_i.

Consider the placement graph G^P of a floorplan P. Let e be the selected entry point. Without loss of generality we can assume that e is a vertex of G^P. (If e is not a vertex of G^P, then modify G^P to include e as a vertex by augmenting the vertex set of G^P with e and replacing the edge $\{v_1, v_2\}$ containing e by two edges $\{v_1, e\}$

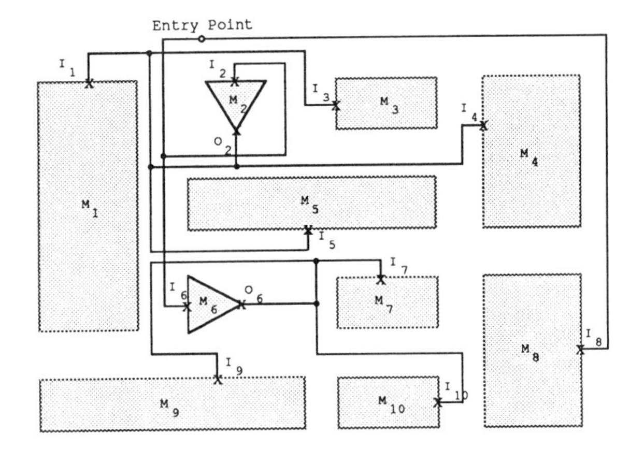

Figure 3: Optimal clock layout for floorplan in Figure 1

and $\{e, v_2\}$ of weights $d(v_1, s)$ and $d(e, v_2)$, respectively.) Since determining an $R_i^{P,e}$ that minimizes $\left|L(R_i^{P,e}) - \mu_l^{P,e}\right|$ is equivalent to finding a simple path in a graph of length closest to a given length $\mu_l^{P,e}$, the problem is NP-hard [6]. An exhaustive search (with intelligent pruning) of the paths from the entry point to the clock terminals of all the sub-modules is therefore used to find the optimal c-layout.

Route, the proposed algorithm for searching the paths, is recursive in nature. The basic idea of the algorithm is as follows. Consider a clock buffer B_l at level l. It drives the clock signals of some of the sub-modules in level $l+1$. Let *Drives(B_l)* denote the set of sub-modules driven by B_l. For clarity of presentation, it is convenient to assume that there is only one sub-module, say M_i, in *Drives(B_l)*. Extending the description to the case when *Drives(B_l)* contains more than one sub-module is relatively simple.

The algorithm starts out with a partial path that contains only the output clock terminal of B_l. At each step of the algorithm a vertex is added to the partial path until either the clock terminal of M_i is reached, or it is clear that there is no completion of the partial path that will result in a c-route with better objective value than the best known objective at that step. The algorithm terminates when every path from the output of B_l to the clock terminal of M_i has been either investigated or pruned. The pruning of the search space occurs while forming the set of possible vertices that can be appended to the current partial path. Vertices are disregarded if the shortest c-route to M_i from that vertex conclusively shows that there is no better completion of the current partial path.

5 Implementation

Entry and **Route** were implemented in C on an Apollo DN4000. The placement graph was the input to the program and the output was the optimal entry point and the route of clocks to all the sub-modules. This implementation was used to distribute clocks in several examples. In this paper we show the results for two examples.

The first example is that of the floorplan in Figure 1. As discussed in Section 3, the optimal entry point for this floorplan is at a distance 3.5 units from vertex 0. The route of the clocks from this entry point is shown in Figure 3. In this layout it is easy to see how the increased weight on the edges incident I_2 and I_6 account for the delays induced by the clock buffers M_2 and M_6, respectively. For instance, the delay to M_8 from the entry point is 28.5 units. The delay to M_1 from the entry point is the delay to M_2 (16.5 units) plus the delay introduced by M_2 (5 units) plus the delay from M_2 to M_1 (10 units), i.e., a total delay of 31.5 units. Being a small example the CPU time required to determine all these routes was less than a second.

The second example was a custom VLSI chip called the *routing controller* [7]. It has been implemented using the CONCORDE™ silicon compiler in a $2\mu m$ CMOS process and contains around 20,000 transistors. The routing of clock signals to the sub-modules were initially carried out using the placement and routing tools in the compiler. Since these placement and routing tools do not take into account the clock skew problem, the skew (in terms of difference in line lengths) was around 4000 microns between the sub-modules driven by one of the clock buffers. By using **Route**, the skew can be reduced to 20 microns for the same set of sub-modules and the algorithm requires only 4.5 seconds of CPU time on an Apollo DN4000 with 16 MB of memory. This should be contrasted to the several hours of CPU time that was required to route the other signals in the floorplan.

These two examples clearly indicate that the algorithm proposed in this paper can be used to distribute clocks even in large VLSI circuits.

6 Conclusion

A hierarchical clock distribution scheme suitable for general VLSI circuits is presented in this paper. The unique features of the proposed scheme are: (i) both delay and skew are taken into account in determining the clock layout, and (ii) the proposed scheme can be easily parallelized.

References

[1] D. F. Wann and M. A. Franklin, "Asynchronous and clocked control structures for VLSI based interconnection networks," *IEEE Trans. Comput.*, vol. C-32, no. 3, pp. 284–293, March 1983.

[2] K. D. Wagner, "A survey of clock distribution techniques in high-speed computer systems," Technical Report CRC 86-20, Center for Reliable Computing, Stanford University, December 1986.

[3] E. G. Friedman and S. Powell, "Design and analysis of a hierarchial clock distribution system for syncrhonous standard cell/macrocell VLSI," *IEEE Journal of Solid-State Circuits*, vol. SC-21, no. 2, pp. 240–246, April 1986.

[4] M. Shoji, "Elimination of process-dependent clock skew in CMOS VLSI," *IEEE Journal of Solid-State Circuits*, vol. SC-21, no. 5, pp. 869–880, October 1986.

[5] P. Ramanathan and K. G. Shin, "A clock distribution scheme for non-symmetric VLSI circuits," Technical report, Real-Time Computing Laboratory, Dept. of Electrical Engineering and Computer Science, University of Michigan, Ann Arbor, 1989.

[6] M. R. Garey and D. S. Johnson, *Computer and intractability: A guide to the theory of NP–completeness*, Freeman, San Francisco, 1979.

[7] J. W. Dolter, P. Ramanathan, and K. G. Shin, "A microprogrammable VLSI routing controller for HARTS," Technical Report CSE-TR-12-89, Dept. of Electrical Engineering and Computer Science, University of Michigan, Ann Arbor, 1989.

Matching-Based Methods for High-Performance Clock Routing

Jason Cong, *Member, IEEE,* Andrew B. Kahng, *Associate Member, IEEE,* and Gabriel Robins, *Member, IEEE*

***Abstract*—Minimizing clock skew is important in the design of high performance VLSI systems. We present a general clock routing scheme that achieves very small clock skews while still using a reasonable amount of wirelength. Our routing solution is based on the construction of a binary tree using geometric matching. For cell-based designs, the total wirelength of our clock routing tree is on average within a constant factor of the wirelength in an optimal Steiner tree, and in the worst case is bounded by $O(\sqrt{l_1 l_2} \cdot \sqrt{n})$ for n terminals arbitrarily distributed in the $l_1 \times l_2$ grid. The bottom-up construction readily extends to general cell layouts, where it also achieves essentially zero clock skew within reasonably bounded total wirelength. We have tested our algorithms on numerous random examples and also on layouts of industrial benchmark circuits. The results are promising: our clock routing yields near-zero average clock skew while using total wirelength competitive with previously known methods.**

I. Introduction

CIRCUIT speed is a major consideration in the design of high-performance VLSI systems. In a synchronous VLSI design, limitations on circuit speed are determined by two factors: the delay on the longest path through combinational logic, and the maximum clock skew among the synchronizing components. With advances in VLSI fabrication technology, the switching speed of combinational logic increases dramatically. Thus, the clock skew induced by non-symmetric clock distribution has become a more significant limitation on circuit performance.

Minimization of clock skew has been studied by a number of researchers in recent years. H-tree constructions have been used extensively for clock routing in regular systolic arrays [2], [11], [15], [34]. Although the H-tree structure can significantly reduce clock skew [11], [34], it is applicable primarily when all of the synchronizing components are identical in size and are placed in a symmetric array. Ramananathan and Shin [23] proposed a clock distribution scheme for building block design where all blocks are organized in a hierarchical structure. They assume that all clock entry points are known at each level of the hierarchy and, moreover, that the number of blocks at each level is small since an exhaustive search algorithm is used to enumerate all possible routes. Fishburn [14] gave methods to maximize the margin of error in clocking constraints, and to minimize the clock period while avoiding clock hazards, or race conditions. This is accomplished via a linear programming formulation. However, the approach assumes that the entire clock tree topology is already known.

Jackson, Srinivasan, and Kuh [18] presented a clock routing scheme for circuits with many small cells. Their algorithm recursively partitions a circuit into two equal parts, and then connects the center of mass of the whole circuit to the centers of mass of the two sub-circuits. Although it was shown that the maximum difference in path length from the root to different synchronizing components is bounded by $O(\sqrt{l_1 l_2 / n})$ in the average case, small examples exist for which the wirelengths between clock source and clock pins can vary by as much as half the chip diameter.

In this paper, we first study the problem of high-performance clock routing for cell-based designs, i.e., circuits with many small cells, such as with standard-cell or sea-of-gates design styles. We then extend our method to general cell (also known as building-block) layouts, where the wiring is restricted to specific channels. In either of these scenarios, the H-tree approach cannot be used since synchronizing components may be of different sizes and may be in arbitrary locations in the layout. The method of [23] cannot be applied either, since there is no natural hierarchical structure associated with the design and the number of synchronizing components is typically too large to allow exhaustive examination of all possible routes. The algorithm of [18] is not completely satisfactory since large skews may result even for small examples, while the approach of [14] does not construct an actual clock routing topology. With this in mind, the goal of our present work is to develop a clock routing methodology which minimizes skew while incurring little added wiring expense.

We present a basic algorithm and several variants,

Reprinted from *IEEE Trans. Computer-Aided Desi.*, vol. 12, no. 8, pp. 1157–1169, Aug. 1993.

which minimize skew by constructing a clock tree that is balanced with respect to root-leaf pathlengths in the tree (these notions will be formalized below). The approach is based on geometric matching: we start with a set of trees, each containing a single terminal of the clock signal net. At each level, we combine the trees into bigger trees using the edges of geometric matching. The end result is a binary tree whose leaves are the terminals in the clock signal net and whose root is the clock entry point. Our method is particularly suitable for designs which employ a single large buffer to drive the entire clock tree, rather than a buffer hierarchy. There are a number of reasons for such a design choice, as discussed in [2]. We note that the recently announced DEC Alpha processor uses such a single-buffer design style [12].

In the cell-based design regime, our algorithm guarantees perfect pathlength balanced trees for inputs of four or less pins. Extensive experimental results indicate that even for large clock signal nets, the maximum difference of pathlengths in the clock tree constructed by our algorithm remains essentially zero. This performance is obtained without undue sacrifice of wirelength: we prove that on average the total wirelength in our clock tree construction is within a constant factor of the wirelength in the optimal Steiner tree. Furthermore, our worst-case clock tree cost is bounded by $O(\sqrt{l_1 l_2} \cdot \sqrt{n})$ for n terminals in the $l_1 \times l_2$ grid,[1] which is the same bound as for the worst-case cost of the optimal Steiner tree.

Since the work in [18] addresses minimum-skew clock routing for cell-based designs, we implemented the algorithm of [18] for comparison purposes. For uniformly random sets of up to 1024 pins in the $l_1 \times l_2$ grid, our method produced clock routings with near-zero clock skew both in the average case and worst case, with total wirelength of the clock tree significantly lower than that produced by the method of [18]. In addition, our routing results for layouts of the MCNC Primary1 and Primary2 benchmarks are significantly better than those reported by [18]; we obtain perfectly balanced root-leaf pathlengths in the clock tree using several percent less total wire than the method of [18]. Actual clock skews for our benchmark routings, as determined by SPICE simulation, are reasonable.

We then apply our method to general cell design, by extending the notion of matching to arbitrary weighted graphs. In this scenario our algorithm produces a clock routing tree that is embedded in the channel intersection graph [10] of an arbitrary building-block layout. The clock routing trees produced by our method attain almost zero skew with only modest wirelength penalty. Experimental results show that the pathlength skew of our routing tree is less than 2% of the skew for a heuristic Steiner tree. This is achieved on average with less than 50% increase in wiring cost over the Steiner tree.

[1]The $l_1 \times l_2$ grid consists of all lattice points (x, y) with x, y integers and $0 \leq x \leq l_1$, $0 \leq y \leq l_2$.

The remainder of this paper is organized as follows. Section II defines a number of basic concepts and gives a precise formulation of our skew minimization problem. In Section III, we present the clock routing algorithm in detail for cell-based designs; Section IV extends the algorithm to general cell layouts. Experimental results of our algorithm and comparisons with previous methods are presented in Section V, and Section VI concludes with possible extensions of the method. Early versions of this paper were presented in [19] and [8].

II. Preliminaries

A synchronous VLSI circuit consists of two types of elements, synchronizing elements (such as registers) and combinational logic gates (such as NAND gates and NOR gates). The synchronizing elements are connected to one or more system-wide clock signals. Every closed path in a synchronous circuit contains at least one synchronizing element (Fig. 1). The speed of a synchronous circuit is mainly determined by the clock periods. It is well known [1], [18] that the clock period C_P of each clock signal net satisfies the inequality:

$$C_P \geq t_d + t_{skew} + t_{su} + t_{ds}$$

where t_d is the delay on the longest path through combinational logic, t_{skew} is the clock skew, t_{su} is the set up time of the synchronizing elements (assuming that the synchronizing elements are edge triggered), and t_{ds} is the propagation delay within the synchronizing elements.

The term t_d itself can be further decomposed into $t_d = t_{d\_interconnect} + t_{t\_gates}$, where $t_{d\_interconnect}$ is the delay associated with the interconnect of the longest path through combinational logic, and t_{d_gates} is the delay through the combinational logic gates on this path. As VLSI feature sizes become smaller, the terms t_{su}, t_{ds}, and t_{d_gates} all decrease significantly. Therefore, as noted above, $t_{d\_interconnect}$ and t_{skew} become more dominant factors in determining circuit performance. It was noted in [1] that t_{skew} may account for 10% or more of the system cycle time. The objective of this paper is to minimize t_{skew}, while we have subsequently addressed the problem of minimizing $t_{d\_interconnect}$ in a different work [9].

Given a routing solution for a clock signal net, the clock skew is defined to be the maximum difference among the delays from the *clock entry point* (CEP) to synchronizing elements in the net. The delay from the CEP to any synchronizing element depends on the wirelength from the CEP to the synchronizing element, the *RC* constants of wire segments in the routing, and the overall topology of the routing solution. Usually, the clock routing may be described as an *RC* tree [24], and we commonly use the first-order moment of the impulse response (also called Elmore delay) to approximate delay in an *RC* tree. The formulas derived by Rubinstein, Penfield and Horowitz [24] give both upper and lower bounds on delay in an *RC* tree.

However, although both the formula for Elmore delay

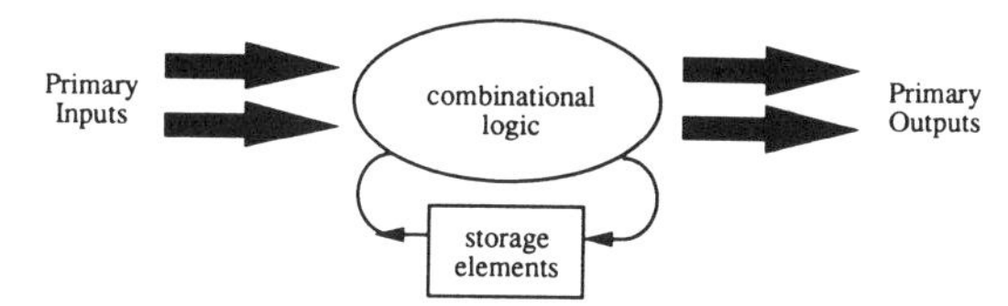

Fig. 1. A typical combinational circuit.

and those in [24] are very useful for simulation or timing verification, they involve sums of quadratic terms and are more difficult to compute and optimize during layout design. Thus, a simpler, linear *RC* model is often used (e.g., [23], [22]) so that wirelength between CEP and the synchronizing elements approximate the circuit delay. In this paper, we also use wirelength as a simple approximation of delay in a routing solution. The clock skew is hence defined to be the maximum difference in wirelength from the CEP to synchronizing elements in the clock signal net. We now give several definitions, along with a formal statement of the skew minimization problem.

A clock routing solution is represented by a rooted (Steiner) tree in the layout whose root is the CEP and whose leaves are synchronizing elements in the clock signal net. The *length*, or *cost*, of an edge in the tree is the Manhattan distance between the two endpoints of the edge, and the tree cost is the sum of all edge costs in the tree.

Definition: The *pathlength skew* of a tree is the maximum difference of the pathlengths in the tree from the root to any two leaves.

A tree is called a *perfect pathlength balanced tree* if its pathlength skew is zero. It is not difficult to construct a perfect pathlength balanced tree if we are allowed to use an arbitrary amount of wire. However, a routing tree with very high cost may distort the clock signal due to longer signal rise and fall times. Thus, we wish to construct a clock routing tree whose pathlength skew is as small as possible, without making the total tree cost too large. With this in mind, we formulate the clock routing problem as follows:

The Pathlength Balanced Tree (PBT) Problem: Given a set of n terminals, N, and real numbers B and S, find a clock routing tree connecting N such that the pathlength skew of the tree is bounded by S and the tree cost is bounded by B.

The following is immediately evident:

Theorem 1: the PBT problem is NP-hard.

Proof: Set $S = \infty$ so that the PBT problem simplifies to the minimum rectilinear Steiner tree problem, which is known to be NP-complete [17]. □

Our objective is to give a heuristic algorithm for the PBT problem. For cell-based design methodologies, we wish to construct a clock tree with pathlength skew as small as possible, using wirelength as close as possible to that in an optimal Steiner tree. Specifically we would like to obtain a clock routing solution in the $l_1 \times l_2$ grid which uses $O(\sqrt{l_1 l_2} \cdot \sqrt{n})$ total wirelength because an optimal Steiner tree will also use $O(\sqrt{l_1 l_2} \cdot \sqrt{n})$ wirelength in the average case [28].

III. A Clock Routing Algorithm for Cell-Based Design

For cell-based design, point-to-point interconnection cost is closely approximated by (Manhattan) geometric distance. Thus, in developing our clock routing algorithm for cell-based layouts, we introduce the notion of a geometric matching:

Definition: Given a set of $2k$ terminals, a *geometric matching* on this set consists of k line segments between terminals, with no two of the k segments sharing an endpoint.

Each line segment in the matching defines a *matching edge*. The cost of a geometric matching is the sum of the costs of its matching edges. A geometric matching on a set of terminals is optimal if its cost is minimum among all possible geometric matchings. An example of an optimal geometric matching over four terminals is shown in Fig. 2.

To construct a tree by iterative matching, we begin with a forest of n isolated terminals (for convenience, assume that n is a power of 2), each of which is considered to be a tree with CEP equal to the location of the terminal itself. The minimum-cost geometric matching on these n CEPs yields $n/2$ segments, each of which defines a subtree with two nodes. The optimal CEP into each subtree of two nodes is the midpoint of the corresponding segment, i.e., such that the clock signal will have zero skew between the segment endpoints.

In general, the matching operation will pair up the CEP's (roots) of all trees in the current forest. At each level, we choose the root of each new merged tree to be the *balance point* which minimizes pathlength skew to the leaves of its two subtrees (see Fig. 3). The balance point is the point p along the "straightline" connecting the roots of the two subtrees, such that the maximum difference in pathlengths from p to any two leaves in the combined tree is minimized. Computing the balance point requires constant time if we know the minimum and maximum root-leaf pathlengths in each of the two subtrees, and these values can be maintained incrementally using constant time per each node added to the clock tree.

Notice that at each level of the recursion, we only have to match half as many nodes as in the previous level. Thus, after $\lceil \log n \rceil$ matching iterations, we obtain the complete clock tree topology. In practice, we actually compute min-cost *maximum cardinality* matchings, i.e., if there are $2m + 1$ nodes, we find the optimal m-segment matching and match $m + 1$ CEPs at the next level. Fig. 4 describes of our clock routing algorithm ALG1 for cell-based design.

The following two results show that ALG1 indeed uses a reasonable amount of wirelength. We prove that our

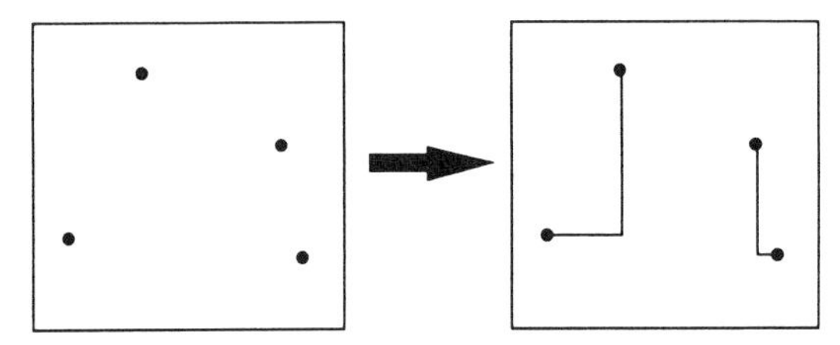

Fig. 2. An optimal geometric matching over four terminals.

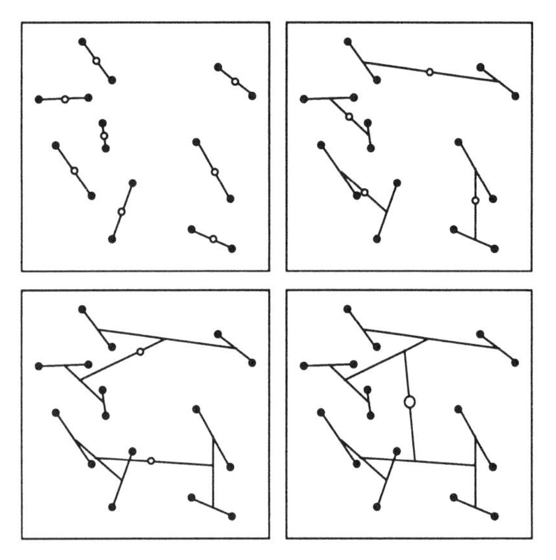

Fig. 3. ALG1 execution on a set of 16 terminals. Solid dots denote terminals, and hollow dots represent the balance points of the corresponding edges. At each level a geometric matching is computed on the balance points of the previous level. Note that although edges are depicted as straight lines, they are actually routed rectilinearly.

clock tree cost grows at the same asymptotic rate as the worst-case optimal Steiner tree cost over n terminals; we also show that our tree cost is on average within a constant factor of the optimal Steiner tree cost.

Theorem 2: For n terminals arbitrarily distributed in the $l_1 \times l_2$ grid, the maximum total wirelength of T_{ALG1} is $O(\sqrt{l_1 l_2} \cdot \sqrt{n})$.

Proof: For n terminals in the $l_1 \times l_2$ grid, the worst-case cost of an optimal matching is $O(\sqrt{l_1 l_2} \cdot \sqrt{n})$ [31]. Since the clock tree is formed by the edges of a matching on n terminals, plus the edges of a matching on $n/2$ terminals, etc., the total edgelength in the tree is

$$O(\sqrt{l_1 l_2} \cdot \sqrt{n}) + O\left(\sqrt{l_1 l_2} \cdot \sqrt{\frac{n}{2}}\right) + \cdots$$

$$= O(\sqrt{l_1 l_2} \cdot \sqrt{n}) \qquad \square$$

This is of the same order as the maximum possible total edge length for the optimal Steiner tree on n terminals [28]. Note that Theorem 2 does not directly relate the cost of our clock routing construction to the cost of the optimal Steiner tree; this is partially addressed by the following.

ALG1: A Clock Routing Algorithm for Cell-Based Designs
Input: A set of terminals N
Output: A clock tree topology T_{ALG1} with root CEP

$T = \emptyset$
$P = N$
While $|P| > 1$
  M = the edges of the optimal geometric matching over P
  $P' = \emptyset$
  For $(p_1, p_2) \in M$ **Do**
    T_1 = the subtree of T rooted at p_1
    T_2 = the subtree of T rooted at p_2
    p = a point lying *between* p_1 and p_2 on the line containing p_1 and p_2, such that p minimizes skew of the tree $T_1 \cup T_2 \cup \{(p, p_1), (p, p_2)\}$ rooted at p
    $P' = P' \cup \{p\}$
    $T = T \cup \{(p, p_1), (p, p_2)\}$
  $P = P'$ plus a possible unmatched node if $|P|$ is odd
CEP = root of T = single remaining point in P
Output clock routing tree = $T_{ALG1} = T$

Fig. 4. The matching-based clock tree routing algorithm.

Theorem 3: For random sets of terminals chosen from a uniform distribution in the $l_1 \times l_2$ grid, the total edgelength of the ALG1 clock tree will be on average within a constant factor of the total edgelength of the optimal Steiner tree.

Proof: The minimum Steiner tree cost for n terminals randomly chosen from a uniform distribution in the $l_1 \times l_2$ Manhattan grid grows as $\beta \cdot \sqrt{l_1 l_2} \cdot \sqrt{n}$ for some constant β [28]. The claim follows from the $O(\sqrt{l_1 l_2} \cdot \sqrt{n})$ worst-case bound on the minimum-cost matching at any level of the construction [31]. $\square$

The balancing operation to determine the CEP of a merged tree is necessary because the root-leaf pathlength might vary between subtrees at a given stage of the construction. In general, when we merge subtrees T_1 and T_2 into a higher level subtree T, the optimal entry point of T will not be equidistant from the entry points of T_1 and T_2 (this can be seen in the example of Fig. 3). Intuitively, balancing entails "sliding" the CEP along the "bar of the H." However, it might not always be possible to obtain perfectly balanced pathlengths in this manner (see Fig. 5).

We therefore use a further optimization, which we call *H-flipping*: for each edge e added to the layout which matches CEPs on edges e_1 and e_2, replace the "H" formed by the three edges e, e_1, and e_2 by the "H" over the same four terminals which (i) minimizes pathlength skew, and (ii) to break ties, minimizes tree cost. We now prove that for four terminals it is always possible find an "H" orientation which achieves zero clock skew, and we also bound the increase in wirelength caused by H-flipping for nets of size four. As discussed below, extensive empirical tests confirm that even for very large inputs, the H-flipping refinement almost always yields perfectly path-balanced trees with essentially no increase in wirelength.

If a net is of size two, ALG1 selects the midpoint of the segment connecting the two terminals as the balance point, and this clearly yields a perfect pathlength balanced tree. Now we show that for nets of size four, ALG1 with

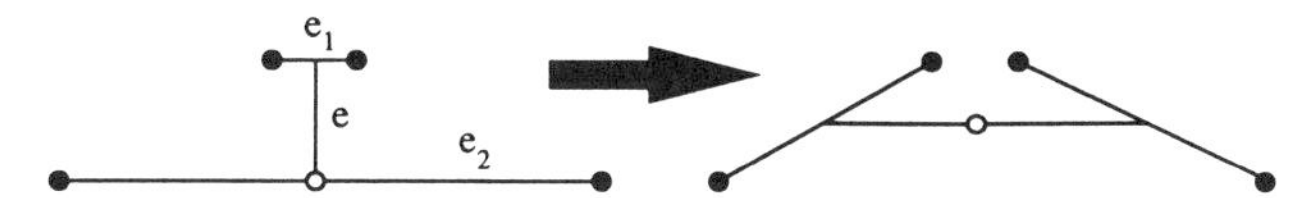

Fig. 5. Example of flipping an H to minimize clock skew: the tree on the left has *no* zero-skew balance point along the middle segment of the "H", while the tree on the right does.

the H-flipping refinement also yields perfect pathlength balanced trees (a net of size three can be treated as a net of size four in which two terminals coincide).

Let a, b, c, and d be the terminals in a net of size four. Without loss of generality, assume that ab and cd are the edges in an optimal matching and $ab \geq cd$. (for convenience, we use ab to denote both the segment ab and also its length. Let m_1 and m_2 be the midpoints of ab and cd, respectively. According to ALG1, m_1 is chosen to be the root of the subtree for a and b, and m_2 is chosen to be the root of the subtree for c and d. Then, the algorithm tries to choose the balance point p on segment $m_1 m_2$ such that

$$\frac{ab}{2} + pm_1 = \frac{cd}{2} + pm_2. \tag{1}$$

It is easy to see that if $m_1 m_2 \geq (ab - cd)/2$, we can always choose p satisfying (1). In this case, the pathlengths from p to all four terminals are the same, so that we have a perfect pathlength balanced tree. However, if $m_1 m_2 < (ab - cd)/2$, we perform H-flipping and replace ab and cd by ad and bc. Then the midpoint n_1 on bc is the root of the subtree for b and c, and the midpoint n_2 on ad is the root of the subtree for a and d. We then seek p' on $n_1 n_2$ such that

$$\frac{ad}{2} + p'n_1 = \frac{bc}{2} + p'n_2. \tag{2}$$

According to the following lemma, we are guaranteed to find p' on $n_1 n_2$ satisfying (2).

Lemma 1: If $m_1 m_2 < (ab - cd)/2$ then $n_1 n_2 \geq (bc - ad)/2$.

Proof: If we have both $m_1 m_2 < (ab - cd)/2$ and $n_1 n_2 < (bc - ad)/2$ then (see Fig. 6):

$$m_1 m_2 + n_1 n_2 < \frac{ab - cd}{2} + \frac{bc - ad}{2},$$

therefore,

$$\frac{ab + bc}{2} > m_1 m_2 + n_1 n_2 + \frac{cd + ad}{2}. \tag{3}$$

Let x be the midpoint of bd. Using similar triangles and the triangle inequality, we obtain

$$\frac{ab}{2} = xn_2 \leq n_1 n_2 + xn_1 = n_1 n_2 + \frac{cd}{2}$$

and

$$\frac{bc}{2} = xm_2 \leq m_1 m_2 + xm_1 = m_1 m_2 + \frac{ad}{2}$$

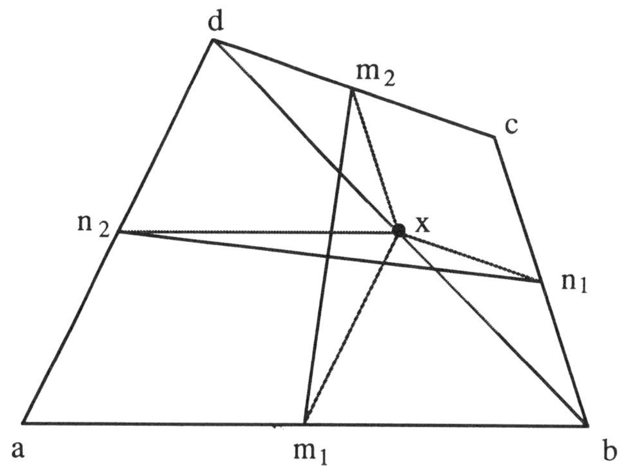

Fig. 6. Illustration for the proof of Lemma 1.

so that

$$\frac{ab + bc}{2} \leq m_1 m_2 + n_1 n_2 + \frac{cd + ad}{2},$$

contradicting (3). Therefore if $m_1 m_2 < (ab - cd)/2$ we must have $n_1 n_2 \geq (bc - ad)/2$. □

Lemma 1 implies that we can always choose the balance point p' on $n_1 n_2$ after H-flipping. Therefore, ALG1 always constructs a perfect pathlength balanced tree for a net of size four. The following lemma shows that when we replace ab and cd by ad and bc in the H-flip, the wirelength increase is bounded by a constant factor.

Lemma 2: If $m_1 m_2 < (ab - cd)/2$ then $bc + ad \leq 3(ab + cd)$.

Proof: Let x be the midpoint of bd. Again applying similar triangles and the triangle inequality, we obtain (see Fig. 7):

$$\frac{bc}{2} = xm_2 \leq xd + dm_2 = xd + \frac{cd}{2}$$

and

$$\frac{ad}{2} = xm_1 \leq xb + bm_1 = xb + \frac{ab}{2}$$

so that

$$\frac{bc + ad}{2} \leq bd + \frac{ab + cd}{2}. \tag{4}$$

Let y be the intersection of bd and $m_1 m_2$. We then have

$$by \leq m_1 y + m_1 b = m_1 y + \frac{ab}{2}$$

$$dy \leq m_2 y + m_2 d = m_2 y + \frac{cd}{2}$$

$$bd \leq m_1 m_2 + \frac{ab + cd}{2} < \frac{ab - cd}{2} + \frac{ab + cd}{2} = ab. \tag{5}$$

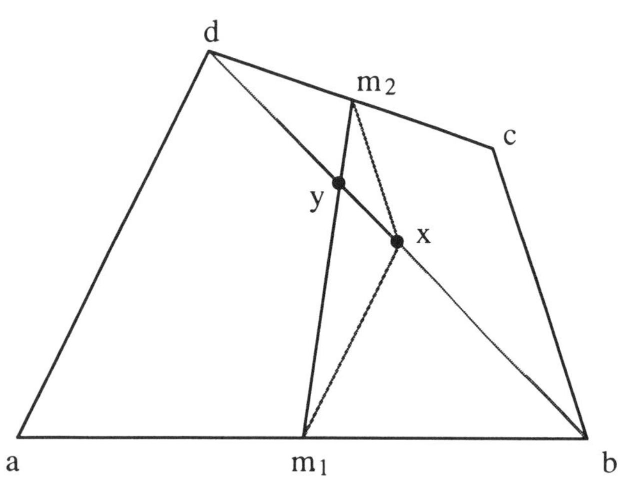

Fig. 7. Illustration for the proof of Lemma 2.

Thus, from (4) and (5) we have

$$\frac{bc + ad}{2} \leq ab + \frac{ab + cd}{2} \leq \frac{3(ab + cd)}{2}$$

or $bc + ad < 3(ab + cd)$. □

Together, these lemmas imply:

Theorem 4: It is always possible to find an "H" orientation over four terminals which achieves zero clock skew, using at most a constant factor extra wirelength.

We now briefly discuss complexity issues and the requirement of an efficient implementation. Since our method is based on geometric matching, its time complexity depends on that of the matching subroutine. A well-known algorithm for general weighted matching requires time $O(n^3)$ [16], [21]. By taking advantage of the planar geometry, the algorithmic complexity can be reduced to $O(n^{2.5} \log n)$ [33]. However, even this may be excessive for large problem instances.

In order to solve problems of practical interest, and since there is no clear relationship between the optimality of the matching and the magnitude of the skew of the resulting clock tree, we may choose to speed up the implementation by using efficient geometric matching heuristics [3], [29], [30]. Although most of these methods were designed for the Euclidean plane, they also perform well in the Manhattan metric, especially if their output is further improved by uncrossing pairs of intersecting edges in the heuristic matching (in any metric, this reduces the matching cost due to the triangle inequality; to this end, note that k intersections of n line segments may be found efficiently in time $O(n \log n + k)$ [7]).

We shall later discuss empirical results from implementation of ALG1 based on three matching methods which require time $O(n)$, $O(n \log n)$ and $O(n \log^2 n)$, respectively. Each of these three matching heuristics yields very good clock routing solutions.

The basic approach of ALG1 thus consists of $\lceil \log n \rceil$ applications of the matching algorithm. H-flipping requires constant time per node, and therefore does not add to the asymptotic time complexity. If the underlying matching algorithm runs within monotonically non-decreasing time $S(n) = \Omega(n)$, we may write $S(n) = n \cdot T(n)$ where $T(n) = S(n)/n$ is monotonically non-decreasing, and hence the total time required by ALG1 is

$$S(n) + S\left(\frac{n}{2}\right) + S\left(\frac{n}{4}\right) + \cdots$$

$$= n \cdot T(n) + \frac{n}{2} \cdot T\left(\frac{n}{2}\right) + \frac{n}{4} \cdot T\left(\frac{n}{4}\right) + \cdots$$

$$\leq n \cdot T(n) + \frac{n}{2} \cdot T(n) + \frac{n}{4} \cdot T(n) + \cdots$$

$$= T(n) \cdot \left(n + \frac{n}{2} + \frac{n}{4} + \cdots\right)$$

$$\leq 2n \cdot T(n) = 2S(n) = O(S(n))$$

i.e., the time complexity of ALG1 is asymptotically equal to the time complexity of the underlying matching algorithm.

IV. A Clock Routing Algorithm for General Cell Design

The same idea of bottom-up iterative matching which we developed in the preceding section may be easily generalized to clock routing in block layouts. In this section, we extend our method to such general cell designs, where a circuit is partitioned into a set of arbitrarily-sized rectangular cells (also referred to as blocks). Blocks may be of widely varying sizes, and are not necessarily placed in any regular arrangement. The routing is carried out in the channels between blocks, with routing over blocks prohibited. For this design style, the approximation of routing cost by geometric distance, which we used for cell-based design in the previous section, does not apply. The feasible routing regions are represented by the channel intersection graph (CIG) [10], which represents the available routing channels induced by a module layout. To capture the locations of clock pins within channels, we use the *augmented channel intersection graph* (ACIG), which is constructed as follows: for each pin incident to a routing channel, introduce a new node into the channel intersection graph which breaks the channel edge into two new edges (see the top left of Fig. 9).

Our goal is still to construct a clock signal tree with both skew and total wirelength as small as possible, except that routing of tree edges is now restricted to lie within prescribed routing channels. Given a graph G with positive edge costs, we let $minpath_G(x, y)$ denote the minimum cost path between nodes x and y, and use $dist_G(x, y)$ to denote the cost of $minpath_G(x, y)$. The notion of a matching may be extended to arbitrary weighted graphs as follows:

Definition: Given a graph $G = (V, E)$ with a positive cost function on the edges, a *generalized matching* M in G is a set of shortest paths connecting m mutually disjoint node pairs, i.e., $M = \{minpath_G(x_1, y_1), minpath_G(x_2, y_2), \cdots, minpath_G(x_m, x_m)\}$, where the x_i's and y_i's are all distinct.

A generalized matching on a set of nodes $N \subseteq V$ in G is *complete* if $m = \lfloor |N|/2 \rfloor$. The *cost* of a generalized matching M is the sum of the costs of the shortest paths in the matching, i.e., $cost(M) = \Sigma_{i=1}^{m} dist_G(x_i, y_i)$. An *optimal* complete generalized matching on $N \subseteq V$ is one with least cost. We can show the following properties of optimal complete generalized matchings:

Lemma 3: Each edge of G belongs to at most one shortest path in an optimal complete generalized matching on $N \subseteq V$ in G.

Proof: Let M be an optimal complete generalized matching on N. Suppose that edge e appears in both $minpath_G(x_i, y_i)$ and $minpath_G(x_j, y_j)$, where (x_i, y_i) and (x_j, y_j) are in M and $i \neq j$ (see Fig. 8). Because (x_i, y_i) and $(x_j, y_j) \in M$ are shortest paths in G, we have

$$dist_G(x_i, x_j) + dist_G(y_i, y_j) \leq dist_G(x_i, y_i) + dist_G(x_j, y_j) - 2 \cdot cost(e).$$

Therefore, replacing $minpath_G(x_i, y_i)$ and $minpath_G(x_j, y_j)$ by $minpath_G(x_i, x_j)$ and $minpath_G(y_i, y_j)$ would yield a complete generalized matching on N with smaller cost, a contradiction. □

Henceforth, we will assume that there are b blocks in the design. G is the underlying augmented channel intersection graph and we assume that the n clock terminals are embedded on edges of G.

Lemma 4: The routing cost between any two clock terminals in G is bounded by $l_1 + l_2$.

Proof: Let x and y be two clock terminals in G. Let P_1 be any monotone (staircase) path passing through x and connecting two opposite corners w and w' of the layout grid. Clearly, $cost(P_1) = l_1 + l_2$. Similarly, let P_2 be a monotone path passing through y and connecting w and w'. Then, $cost(P_1) + cost(P_2) = 2 \cdot (l_1 + l_2)$. Since at least one of w or w' can be reached from *both* x and y with cost at most $l_1 + l_2$, the shortest path between x and y has cost no more than $l_1 + l_2$. □

Proof: Let x and y be two clock terminals in G. Let P_1 be any monotone (staircase) path passing through x and connecting two opposite corners w and w' of the layout grid. Clearly, $cost(P_1) = l_1 + l_2$. Similarly, let P_2 be a monotone path passing through y and connecting w and w'. Then, $cost(P_1) + cost(P_2) = 2(l_1 + l_2)$. Since at least one of w or w' can be reached from *both* x and y with cost at most $l_1 + l_2$, the shortest path between x and y has cost no more than $l_1 + l_2$. □

It is clear from Lemma 4 that an optimal complete generalized matching on the clock terminals in G has cost no more than $(l_1 + l_2) \cdot \lfloor n/2 \rfloor$.

As in the previous section, our basic strategy is to construct a clock tree by computing a sequence of generalized matchings on the clock terminals. We begin with a forest of n isolated clock terminals in G (again for convenience, we assume that n is a power of 2), each of which is a degenerate tree with CEP being the terminal itself. The optimal complete generalized matching on these n terminals yields $n/2$ paths, each of which defines a subtree. The optimal CEP into each subtree is the midpoint of the corresponding path, so that the clock signal will have zero skew between the two terminals. At each level, we compute an optimal generalized matching on the set of CEPs (roots) of all subtrees in the current forest and merge each pair of subtrees into a larger subtree. As before, the root (CEP) of the resulting tree is chosen to be the *balance point* on the path connecting the two subtrees such that the pathlength skew in the resulting tree is minimized (see Fig. 9).

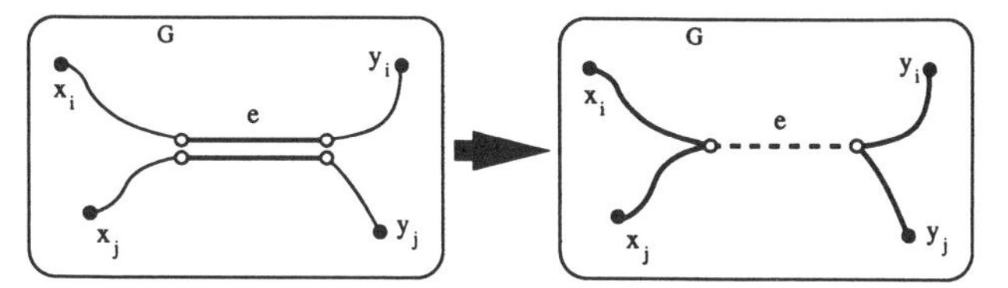

Fig. 8. Each edge belongs to at most one shortest path in an optimal complete generalized matching.

Notice that at each level of the recursion, we only have to match half as many nodes as at the previous level. Thus, in $\lceil \log n \rceil$ matching iterations, we obtain a complete clock tree topology. If n is not a power of 2, then as noted in the discussion of ALG1, there will be an odd number $2m + 1$ of nodes to match at some level. For such cases, we compute an optimal maximum-cardinality generalized matching on $2m$ nodes, and then match $m + 1$ nodes at the next level. Fig. 10 gives a formal description of our clock routing algorithm ALG2 for general cell design.

The worst-case clock tree cost produced by the algorithm can be bounded as follows:

Theorem 5: Given b blocks in the $l_1 \times l_2$ grid and n terminals of a clock signal net, the cost of the clock tree created by ALG2 is at most $(l_1 + l_2) \cdot n$.

Proof: By Lemma 4, the cost of a generalized matching on n terminals is bounded by $(l_1 + l_2) \cdot \lfloor n/2 \rfloor$. After each iteration, the number of nodes to be matched is reduced by half. Therefore, the total clock tree cost is bounded by

$$(l_1 + l_2) \cdot \frac{n}{2} + (l_1 + l_2) \cdot \frac{n}{4} + \cdots \leq (l_1 + l_2) \cdot n.$$

□

In order to compute an optimal generalized matching on a set of nodes N in G, we construct a weighted complete graph G' on N such that $weight(x, y) = dist_G(x, y)$ for each pair of nodes x and y in N. This can be accomplished by applying an $O(|E| \cdot |V| + |V|^2)$ implemen-

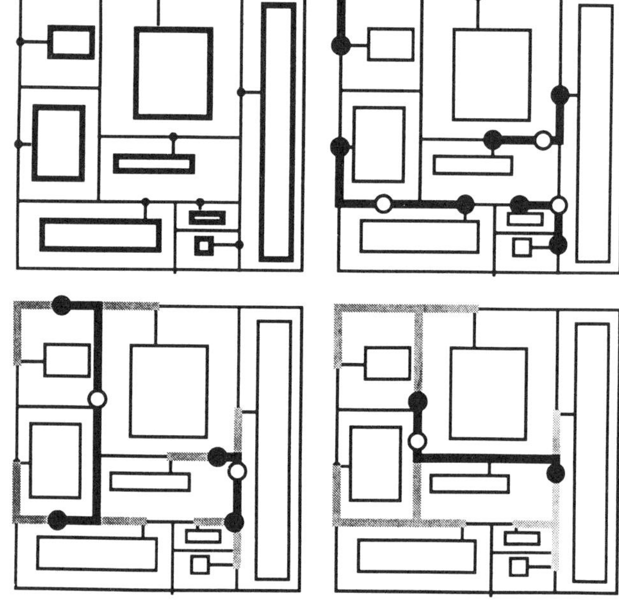

Fig. 9. ALG2 execution on a random module placement with an 8-terminal net. Solid dots are roots of subtrees in the previous level; hollow dots are roots (CEPs) of new subtrees computed at the current level. At each level an optimal generalized matching is computed on the solid points. For clarity, only the newly added wires are highlighted at each level.

ALG2: A Clock Routing Algorithm for General Cell Designs
Input: A set of terminals N embedded in a CIG G
Output: A clock tree topology T_{ALG2} with root CEP

```
T = ∅
P = N
While |P| > 1
    M = opt complete generalized matching on P
    P' = ∅
    For {p1, p2} ∈ M Do
        T1 = subtree of T rooted at p1
        T2 = subtree of T rooted at p2
        p = balance point on minpath_G(p1, p2) minimizing the
            skew of the tree T1 ∪ T2 ∪ minpath_G(p1, p2)
        P' = P' ∪ {p}
        T = T ∪ {{p, p1}, {p, p2}}
    P = P' plus an unmatched node if |P| is odd
CEP = Root of T = single remaining point in P
Output clock routing tree = T_ALG2 = T
```

Fig. 10. The matching-based clock tree algorithm for general cell design.

tation of Floyd's all-pairs shortest path algorithm [25] to the graph $G = (V, E)$. Note that G is a planar graph and therefore $|E| = O(|V|)$. Since for the augmented channel intersection graph we have $|V| = O(b + n)$, and typically $b > n$, the overall time complexity for this step is $O(b^2)$. We may then apply an $O(n^3)$ algorithm for computing an optimal complete matching in general graphs [21]. However, this complexity will result in long runtimes for large problem instances. Therefore, in order to achieve an efficient implementation, we use the greedy matching heuristic [26]. Such a heuristic matching may be improved by removing overlapping edges of shortest paths, as described in the proof of Lemma 3, so that no edge is used in more than one shortest path. The time complexity of each iteration of ALG2 is dominated by the $O(b^2)$ all-pairs shortest paths computation, which we invoke $\lceil \log n \rceil$ times, so that the overall time complexity of ALG2 is $O(b^2 \cdot \log n)$. This complexity is reasonable since the number of blocks is typically not large.

V. Experimental Results

Both ALG1 and ALG2 were implemented in ANSI C for the Sun-4, Macintosh, and IBM 3090 environments. This section summarizes the simulation results.

5.1. Empirical Data for Cell-Based Designs

We have implemented three basic heuristic variants of ALG1, corresponding to different matching subroutines. The first heuristic variant (SP) uses the linear-time space partitioning heuristic of [30] to compute an approximate matching; the second variant (GR) uses an $O(n \log^2 n)$ greedy matching heuristic [29]; and the third variant (SFC) uses an $O(n \log n)$ spacefilling curve-based method [3]. We have further tested these three variants by running each both with and without two refinements: (1) removing all edge crossings in the heuristic matching, and (2) performing "H-flipping" as necessary. Either of these optimizations can be independently added to any of the three variants, yielding a total of twelve distinct versions of the basic algorithm. The variants of the algorithm are denoted and summarized as follows:

- *SP:* Use the space-partitioning matching heuristic of [30], which induces the matching through recursive bisection of the region (rather than bisection of the set of terminal locations).
- *GR:* Use a greedy matching heuristic, which always adds the shortest edge between unmatched terminals [29].
- *SFC:* Use a space-filling curve to map the plane to a circle, then choose the better of the two embedded matchings (i.e., either all odd edges or all even edges in the induced Hamiltonian cycle through the terminal locations) [3].
- *SP+E, GR+E, SFC+E:* Same as SP, GR, and SFC, respectively, except that the heuristic matching cost is further improved by edge-uncrossing.
- *SP+H, GR+H, SFC+H:* Same as SP, GR, and SFC, respectively, except that pathlength skew is further reduced by H-flipping.
- *SP+E+H, GR+E+H, SFC+E+H:* Same as SP, GR, and SFC, respectively, except that both edge-uncrossing and H-flipping are performed.

For comparison, we also implemented

- *MMM:* The method of means and medians, similar to that of Jackson, Srinivasan and Kuh [18].

The algorithms were tested on random sets of up to 1024 terminals generated from a uniform distribution in the 1000 × 1000 grid (i.e., $l_1 = l_2 = 1000$). Results for a sample run with 50 random terminal sets at each cardinality are summarized here: Table I compares the average tree costs and Table II compares the average clock skews

TABLE I
AVERAGE TREE COSTS, IN GRID UNITS, FOR THE VARIOUS HEURISTICS

Pts	MMM	SP	GR	SFC	SP+E	GR+E	SFC+E
4	1197	1155	1136	1140	1129	1129	1130
8	2136	2075	2032	2031	1990	1990	1992
16	3506	3582	3409	3527	3343	3326	3343
32	5598	5922	5481	5788	5342	5277	5326
64	8377	9184	8526	9048	8100	8032	8068
128	12276	13793	12632	13656	11912	11725	11976
256	17874	20765	18625	20354	17573	17024	17768
512	25093	30443	27055	29618	25341	24548	25720
1024	36765	44304	38688	42750	36444	35086	37056

Pts	SP+H	GR+H	SFC+H	SP+E+H	GR+E+H	SFC+E+H	Meta
4	1125	1125	1125	1125	1125	1125	1125
8	2027	2028	1994	1971	1979	1980	1960
16	3502	3416	3428	3333	3322	3329	3268
32	5860	5628	5577	5329	5273	5304	5151
64	9226	8794	8748	8076	7982	8047	7844
128	13997	3315	13159	11871	11697	11914	11566
256	21307	19611	19713	17457	16955	17629	16919
512	31646	29175	28688	25188	24465	25483	24480
1024	46417	42110	41540	36276	34965	36814	34992

TABLE II
AVERAGE SKEW VALUES, IN GRID UNITS, FOR THE VARIOUS HEURISTICS

Pts	MMM	SP	GR	SFC	SP+E	GR+E	SFC+E
4	112.31	3.98	15.52	0.00	0.00	0.00	0.00
8	186.10	45.79	76.71	4.26	0.66	0.66	0.66
16	234.72	70.93	141.22	19.47	4.01	3.54	3.66
32	262.61	143.85	200.33	28.29	8.14	7.85	6.14
64	229.15	179.83	273.04	51.36	6.93	8.65	5.29
128	201.55	226.61	314.05	64.86	11.52	14.18	11.26
256	183.28	286.90	324.57	85.10	17.25	13.85	15.04
512	153.90	321.23	399.29	85.46	14.79	15.26	15.73
1024	125.34	339.34	402.59	89.75	17.14	16.71	15.35

Pts	SP+H	GR+H	SFC+H	SP+E+H	GR+E+H	SFC+E+H	Meta
4	0.00	0.00	0.00	0.00	0.00	0.00	0.00
8	3.38	0.12	0.00	0.00	0.00	0.00	0.00
16	1.80	3.80	0.12	0.00	0.00	0.00	0.00
32	3.53	8.64	0.00	0.00	0.00	0.00	0.00
64	13.17	27.69	1.26	0.00	0.00	0.00	0.00
128	20.79	40.34	3.18	0.00	1.02	0.24	0.00
256	41.79	51.87	7.49	0.00	0.92	0.00	0.00
512	76.35	90.66	13.51	0.39	0.62	0.39	0.00
1024	75.92	94.99	16.62	0.44	0.08	0.38	0.00

for all heuristics. The data in the tables is given in grid units.

The computational results indicate that both optimizations (edge-uncrossing and H-flipping) significantly improve both skew and total wirelength. When the refinements are combined, average pathlength skew essentially vanishes, and the wirelength of several variants is superior to the output of MMM. The best variant appears to be GR+E+H, which is based on the greedy matching heuristic together with edge-uncrossing and H-flipping. Note that the cost of the greedy matching is asymptotically as good as that of the optimal matching [26]. Tables III and IV highlight the contrast between GR+E+H and MMM, showing minimum, maximum and average values

TABLE III
MINIMUM, AVERAGE, AND MAXIMUM SKEW VALUES, IN GRID UNITS, FOR GR+E+H AND MMM

	MMM			GR+E+H		
Pts	min	ave	max	min	ave	max
4	2	112.31	379	0	0.00	0
8	46	186.10	407	0	0.00	0
16	86	234.72	416	0	0.00	0
32	118	262.61	540	0	0.00	0
64	141	229.15	337	0	0.00	0
128	120	201.55	282	0	1.02	30
256	127	183.28	250	0	0.92	46
512	103	153.90	203	0	0.62	31
1024	94	125.34	167	0	0.08	4

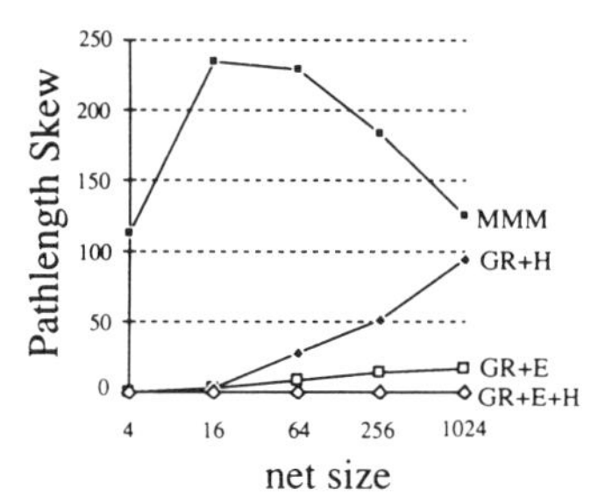

Fig. 11. Overall pathlength skew comparisons between ALG1 (GR + E + H) and MMM.

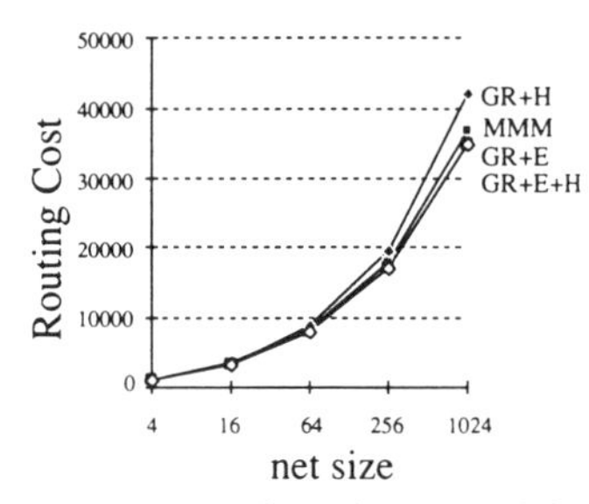

Fig. 12. Overall tree cost comparisons between ALG1 (GR + E + H) and MMM.

TABLE IV
MINIMUM, AVERAGE, AND MAXIMUM TOTAL WIRELENGTH VALUES, IN GRID UNITS, FOR GR + E + H AND MMM

	MMM			GR + E + H		
Pts	min	ave	max	min	ave	max
4	656	1197	1823	555	1125	1668
8	1089	2136	2943	1123	1979	2810
16	2841	3506	4221	2793	3322	3993
32	4813	5598	6216	4695	5273	5866
64	7624	8377	9266	7372	7982	8556
128	11439	12276	13136	11052	11697	12243
256	17220	17874	18549	16379	16955	17543
512	25093	25666	26291	23866	24465	25325
1024	36126	36765	37561	34231	34965	36179

for both total wirelength and skew. Figs. 11 and 12 depict these same comparisons graphically.

As noted in [20], any set of approximation heuristics induces a *meta-heuristic* which returns the best solution found by any heuristic in the set; we also implemented this (denoted as "Meta"), which returns the minimum-skew result from all of the other variants. Interestingly, in our experience Meta *always* returns a perfect pathlength balanced tree, i.e., for each problem instance, at least one of the other heuristic variants will yield a zero clock skew solution. This is very useful, especially when the heuristics are of similar complexity. For example, we can solve the Primary1 benchmark using all twelve methods in under two minutes on a Sun-4/60 workstation.

Fig. 13 and 14 illustrate the output of variant GR + E + H on the Primary1 and Primary2 benchmarks, using the same placement solutions as in [18]; note that although edges are depicted as straight lines in these diagrams, they are actually routed rectilinearly. Table V compares the results of GR + E + H and the results of [18] which were provided by the authors [27]: GR + E + H

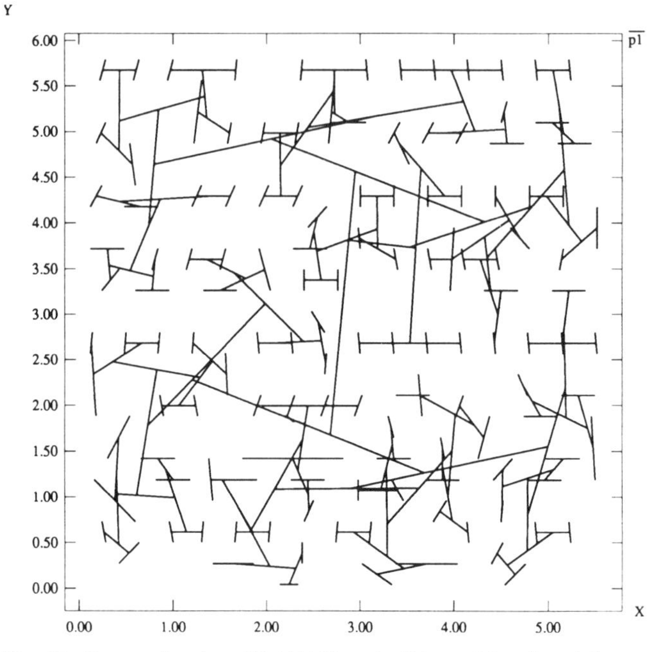

Fig. 13. Output of variant GR + E + H on the Primary1 benchmark layout.

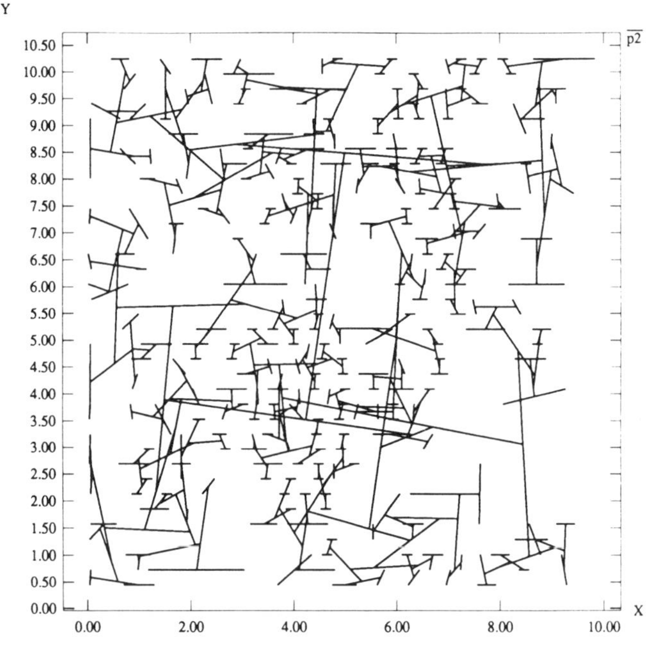

Fig. 14. Output of GR + E + H on the Primary2 benchmark layout.

completely eliminates pathlength skew while using 5%–7% less wirelength. To confirm the correlation between the linear delay model and actual delay, we ran SPICE simulations on the Primary1 and Primary2 clock trees using MOSIS 2.0-μm CMOS technology parameters and 0.3-pF sink loading capacitance); the simulated skews of our clock trees for Primary1 and Primary2 were 181 ps and 741 ps, respectively[2]. Notice that this clock skew was

[2]Vias and parasitic difference between metal layers were not considered in our simulation because detailed layer assignment has not been determined at this stage of clock routing.

TABLE V

COMPARISONS OF GR+E+H AND MMM ON PRIMARY1 AND PRIMARY2. "SKEW (STD)" DENOTES THE STANDARD DEVIATION OF THE PATH LENGTH, AND "COST" DENOTES THE TOTAL WIRELENGTH

	Skew (STD) MMM	Cost MMM	Skew (STD) GR+E+H	Cost GR+E+H	Reduction Skew (STD)	Reduction Cost (%)
Primary1	0.29	161.7	0.00	153.9	0.29	4.8
Primary2	0.74	406.3	0.00	376.7	0.74	7.3

TABLE VI

AVERAGE CLOCK TREE COSTS AND PATHLENGTH SKEWS, IN GRID UNITS, OF ALG2 AND THE STEINER TREE HEURISTIC, RESPECTIVELY.

# of Modules	Net Size	Skew		Cost	
		Steiner	ALG2	Steiner	ALG2
16	4	511.0	0.8	1537	1921
16	8	794.9	12.9	2328	3478
16	16	1101.5	22.1	3332	5873
32	4	445.0	0.4	1401	1729
32	8	804.4	4.4	2261	3407
32	16	1136.9	12.0	3357	5847

TABLE VII

AVERAGE CLOCK TREE COSTS AND PATHLENGTH SKEWS, OF ALG2 OUTPUT, NORMALIZED (PER INSTANCE) TO CORRESPONDING HEURISTIC STEINER TREE VALUES.

# of Modules	Net Size	Pathlength Skew	Tree Cost	Edge Density in Channels
16	4	0.00	1.26	1.24
16	8	0.02	1.49	1.40
16	16	0.02	1.77	1.63
32	4	0.01	1.24	1.21
32	8	0.01	1.52	1.36
32	16	0.01	1.74	1.48

obtained simply by balancing CEP-leaf pathlengths; as discussed in Section VI, more sophisticated delay models can yield a better choice of balance points in the matching-based construction.

5.2. Empirical Data for General Cell Designs

We have tested ALG2 on two sets of test cases. One set of examples contains clock nets of sizes 4, 8, and 16 on 16 blocks, and the other set contains clock nets of sizes 4, 8, and 16 on 32 blocks. Block sizes and layouts were assigned randomly in the grid by creating a fixed number of non-overlapping blocks, with length, width, and lower-left coordinates all chosen from uniform distributions on the interval [0, 1000] (i.e., $l_1 = l_2 = 1000$).

For each net size (and block number), 100 instances were generated randomly, and we compared the skew and cost of the ALG2 routing trees with those produced by the 1-Steiner heuristic [20]. Results are shown in Tables VI and VII. The skew of our clock tree is very close to zero. In no case is it more than 2% of the skew of the Steiner tree routing. The increase in total wirelength of our routing tree varies from 24% to 77% when compared with the Steiner tree. The data in the tables is given in grid units.

As with the cell-based layout benchmarks, we ran SPICE simulations on a number of examples (again using MOSIS 2.0-μm CMOS technology and 0.3-pF gate loading capacitance). The actual skew of our clock tree is consistently much smaller than that of a Steiner tree. For a typical 16-pin clock net in a 16-block design, the skews of our clock tree and the Steiner tree are 18 and 69 ps, respectively.

For the routing tree produced by ALG2, we may have overlapping edges in a channel because matching paths at different levels may use the same channel. However, by Lemma 3, no channel segment will appear in more than a single path in a matching. Therefore, there are at most $\lceil \log n \rceil$ overlapping edges in each channel. The last column in Table VII shows the average edge density in channels, computed as the average of non-zero local column densities over all columns in all channels.

VI. REMARKS AND EXTENSIONS

We recommend that the global clock routing of ALG1 or ALG2 be performed before other wiring, following standard practice. In this way, there are no wire-crossing conflicts since two layers of metal are used, one for horizontal wires, and the other for vertical wires. The exact routing of the clock tree topology may be determined in the detailed routing step.

For cell-based design, we can realize additional wirelength savings in our clock tree routing by varying the geometric embedding of individual wires in the layout. In the Manhattan metric, the "balance point" of a wire connecting two terminals is not unique but is rather a locus of many possible terminals (Fig. 15), with the extremes corresponding to the two L-shaped wire orientations. Our current implementation sets the balance point of a segment to be its "Euclidean" midpoint, but this is not necessarily an optimal choice. Using a graph-theoretic formulation, we can easily derive a polynomial-time method, based on general graph matching, for finding the optimal set of balance points within these loci.

The wire embedding at each level of our algorithm may also benefit from *lookahead* of one or more levels, i.e., when we reach a situation where pathlength skew cannot be eliminated even via the utilization of an H-flip, we can go back one or two levels in the subtrees involved and try different H-flips during previous iterations on those subtrees. In our experience, this strategy easily allows complete elimination of clock skew at the current level, and

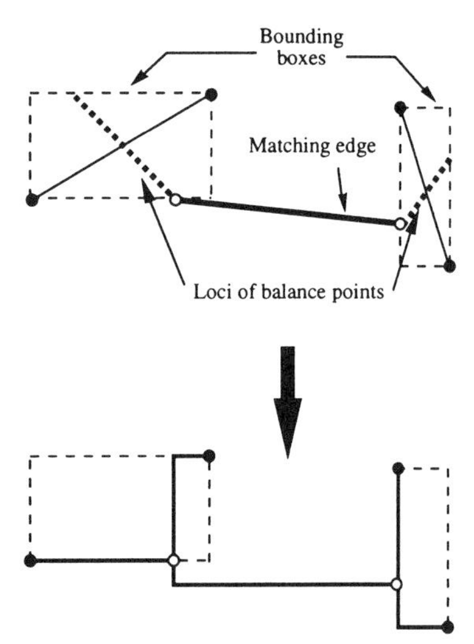

Fig. 15. Further optimizations are possible by matching over the *loci* of balance point candidates.

requires only a constant amount of computation provided the lookahead depth (i.e., number of levels) is bounded by a constant. With respect to Fig. 15, note that because the routing layers have different electrical characteristics, the choice of balance points must be optimized both with respect to locations and the actual embeddings of the wires incident to the balance point. If the layer assignment is prescribed, the balance point computation is straightforward. Alternatively, deciding between various optional embeddings may be accomplished using one level of lookahead as in [32].

Another important extension lies in the selection of the CEP at each level. Instead of using the linear delay model to select a CEP, we may use a more accurate distributed *RC* model, to select the CEP so that clock skew is reduced by as much as possible. This is a strictly local modification of our method and does not affect the execution of the rest of the algorithm (or any variant). Such an extension applies to both ALG1 and ALG2, and is particularly useful when varying capacitative loadings exist at the terminals of the clock net. Since our algorithm operates in a bottom-up fashion, and since we treat each level independently, our method is able to accommodate variable gate loading very naturally.[3]

[3]We note that Tsay [32] recently gave a clock routing algorithm which uses a bottom-up construction approach similar to the one described in this paper. Tsay's algorithm incorporates one level of look-ahead and the introduction of "extra" wire to achieve an exact zero-skew tree with respect to the Elmore delay model [13]. At each step, Tsay's method combines a pair of zero-skew trees to yield a new zero-skew tree of larger size. The linear-time "Deferred-Merge Embedding" (DME) algorithm of [4]–[6] generalizes look-ahead in maintaining all loci of CEP's that are compatible with a zero-skew tree construction. DME thus reduces the cost of an initial clock tree topology computed by any previous method, while maintaining exact zero clock skew. In regimes where the linear delay model applies, the DME method produces the *optimal* (i.e., minimum-cost) zero-skew clock tree with respect to the prescribed topology, and this tree will also enjoy optimal source-terminal delay [4], [5]. It is noteworthy that with respect to DME, our present matching-based approach yields topologies which lead to lower cost trees than such other initial topologies as those of [6], [18], [32].

Finally, we mention that the PBT problem is interesting from a theoretical standpoint: the tradeoff between pathlength balance and total edgelength appears important not only for clock skew minimization, but also for a number of applications in areas ranging from computational geometry to network design.

In summary, we have presented a bottom-up approach for constructing clock routing trees, for both cell-based and general cell designs. Skew minimization is achieved by constructing the clock tree iteratively through geometric or graph matchings, while carefully balancing the pathlengths from the root to all leaves at each level of the construction. We verified our algorithm on numerous random examples, on industry benchmark circuits, and by SPICE timing simulations; the results show near-zero average clock skew while using total wirelength that compares favorably with previous work.

References

[1] H. Bakoglu, *Circuits, Interconnections and Packaging for VLSI.* Reading, MA: Addison-Wesley, 1990.

[2] H. Bakoglu, J. T. Walker, and J. D. Meindl, "A symmetric clock-distribution tree and optimized high-speed interconnections for reduced clock skew in ulsi and wsi circuits," in *Proc. IEEE Intl. Conf. on Computer Design*, Port Chester, NY, Oct. 1986, pp. 118–122.

[3] J. J. Bartholdi and L. K. Platzman, "A fast heuristic based on spacefilling curves for minimum-weight matching in the plane," *Inf. Proc. Lett.*, vol. 17, pp. 177–180, 1983.

[4] K. D. Boese and A. B. Kahng, "Zero skew clock routing with minimum wirelength," Tech. Rep. CSD-TR-920012, Computer Sci. Dep., UCLA, Mar. 1992.

[5] —, "Zero skew clock routing with minimum wirelength," in *Proc. IEEE Intl. ASIC Conf.*, Rochester, NY, September 1992, pp. 1-1.1–1-1.5.

[6] T. H. Chao, Y. C. Hsu, and J. M. Ho, "Zero skew clock net routing," in *Proc. ACM/IEEE Design Automation Conf.*, Anaheim, CA, June 1992, pp. 518–523.

[7] B. Chazelle and H. Edelsbrunner, "An optimal algorithm for intersecting line segments," *J. Ass. Comput. Mach.*, vol. 39, pp. 177–180, 1992.

[8] J. Cong, A. B. Kahng, and G. Robins, "On clock routing for general cell layouts," in *Proc. IEEE Intl. ASIC Conf.*, Rochester, NY, Sept. 1991, pp. 14:5.1–14:5.4.

[9] J. Cong, A. B. Kahng, G. Robins, M. Sarrafzadeh, and C. K. Wong, "Provably good performance-driven global routing," *IEEE Trans. Computer-Aided Design*, vol. 11, pp. 739–752, 1992.

[10] W. M. Dai, T. Asano, and E. S. Kuh, "Routing region definition and ordering scheme for building-block layout," *IEEE Trans. Computer-Aided Design*, vol. CAD-4, pp. 189–197, 1985.

[11] S. Dhar, M. A. Franklin, and D. F. Wann, "Reduction of clock delays in vlsi structures," in *Proc. IEEE Intl. Conf. on Computer Design*, Port Chester, NY, Oct. 1984, pp. 778–783.

[12] D. Dobberpuhl, *et al.*, "A 200 mHz 64b dual-issue cmos microprocessor," in *Proc. IEEE Intl. Solid State Circuits Conf.*, San Francisco, Feb. 1992, pp. 106–107.

[13] W. C. Elmore, "The transient response of damped linear networks with particular regard to wide-band amplifiers," *J. Appl. Phys.*, vol. 19, pp. 55–63, 1948.

[14] J. Fishburn, "Clock skew optimization," *IEEE Trans. on Computers*, vol. 39, pp. 945–951, 1990.

[15] A. L. Fisher and H. T. Kung, "Synchronizing large systolic arrays," in *Proc. SPIE*, May 1982, pp. 44–52.

[16] H. Gabow, "An efficient implementation of edmonds' algorithm for maximum matching on graphs," *J. Ass. Comput. Mach.*, vol. 23, pp. 221–234, 1976.

[17] M. Garey and D. S. Johnson, "The rectilinear steiner problem is np-complete," *SIAM J. Appl. Math.*, vol. 32, pp. 826–834, 1977.

[18] M. A. B. Jackson, A. Srinivasan, and E. S. Kuh, "Clock routing for high-performance IC's," in *Proc. ACM/IEEE Design Automation Conf.*, June 1990, pp. 573–579.

[19] A. B. Kahng, J. Cong, and G. Robins, ''High-performance clock routing based on recursive geometric matching,'' in *Proc. ACM/IEEE Design Automation Conf.*, June 1991, pp. 322-327.

[20] A. B. Kahng and G. Robins, ''A new class of iterative steiner tree heuristics with good performance,'' *IEEE Trans. Computer-Aided Design*, vol. 11, pp. 893-902, 1992.

[21] E. Lawler, *Combinatorial Optimization: Networks and Matroids.* New York: Holt Rinehart and Winston, 1976.

[22] I. Lin and D. H. C. Du, ''Performance-driven constructive placement,'' in *Proc. ACM/IEEE Design Automation Conf.*, June 1990, pp. 103-106.

[23] P. Ramanathan and K. G. Shin, ''A clock distribution scheme for non-symmetric VLSI circuits,'' in *Proc. IEEE Intl. Conf. on Computer-Aided Design*, Santa Clara, CA, Nov. 1989, pp. 398-401.

[24] J. Rubinstein, P. Penfield, and M. A. Horowitz, ''Signal delay in *RC* tree networks,'' *IEEE Trans. Computer-Aided Design*, vol. CAD-2, pp. 202-211, 1983.

[25] R. Sedgewick, *Algorithms.* 2nd ed. Reading, MA: Addison-Wesley, 1988.

[26] T. L. Snyder and J. M. Steele, ''Worst-case greedy matchings in the unit d-cube,'' *Networks*, vol. 20, pp. 779-800, 1990.

[27] A. Srinivasan, private communication, Oct. 1991.

[28] J. M. Steele, ''Growth rates of euclidean minimal spanning trees with power weighted edges,'' *Ann. Probability*, vol. 16, pp. 1767-1787, 1988.

[29] K. J. Supowit, ''New techniques for some dynamic closest-point and farthest-point problems,'' in *Proc. ACM/SIAM Symp. on Discrete Algorithms*, Jan. 1990, pp. 84-90.

[30] K. J. Supowit and E. M. Reingold, ''Divide and conquer heuristics for minimum weighted euclidean matching,'' *SIAM J. Comput.*, vol. 12, pp. 118-143, 1983.

[31] K. J. Supowit, E. M. Reingold, and D. A. Plaisted, ''The travelling salesman problem and minimum matching in the unit square,'' *SIAM J. Comput.*, vol. 12, pp. 144-156, 1983.

[32] R. S. Tsay, ''Exact zero skew,'' in *Proc. IEEE Intl. Conf. on Computer-Aided Design*, Santa Clara, CA, Nov. 1991, pp. 336-339.

[33] P. Vaidya, ''Geometry helps in matching,'' in *Proc. ACM Symp. on Theory of Computing*, 1988, pp. 422-425.

[34] D. F. Wann and M. A. Franklin, ''Asynchronous and clocked control structure for VLSI based interconnection networks,'' *IEEE Trans. on Computers*, vol. 21, pp. 284-293, 1983.

Clock Routing for High-Performance ICs

Michael A. B. Jackson Arvind Srinivasan E. S. Kuh
Electronics Research Laboratory, University of California, Berkeley, CA 94720

Abstract

In this paper we focus on routing techniques for optimizing clock signals in *small-cell* (e.g., standard-cell, sea-of gate, etc...) ASICs. In previously reported work, the routing of the clock net has been performed using ordinary global routing techniques based on a minimum spanning or minimal Steiner tree that have little understanding of clock routing problems. We present a novel approach to clock routing that all but eliminates clock skew and yields excellent phase delay results for a wide range of chip sizes, net sizes (pin count), minimum feature sizes, and pin distributions on both randomly created and standard industrial benchmarks. For certain classes of pin distributions we have proven theoretically and observed experimentally a decrease in skew with an increase in net size. In practice, we have observed a two to three order magnitude reduction in skew when compared to a minimum rectilinear spanning tree.

1 Introduction

In today's highly competitive IC marketplace, company survival necessitates product differentiability. Product differentiability may be engendered in many ways, several of which include: increased performance (e.g. lower power, faster timing, etc...), lower cost, more features, or faster time to market. Thus, design techniques that enhance chip timing performance are of fundamental importance to the IC community.

The clock is the essence of a synchronous digital system. Physically, the clock is distributed from an external pad to all similarly clocked synchronizing elements through a distribution network that includes clock distribution logic and interconnects. It serves to unify the physical and temporal design representations by determining the precise instants in time that the digital machine changes state. Because the clock is important, optimization of the clock signal can have a significant impact on the chip's cycle time, especially in high-performance designs. Non-optimal clock behavior is caused by either of two phenomena: the routing to the chip's synchronizing elements, or in the non-symmetric behavior of the clock distribution logic.

Previous work in clock optimization has been contributed by several authors. H-trees have been recognized for years as a technique to help reduce the skew in synchronous systems [FK82] [KGE82] [DFW84] [BWM86] [WF83]. For regular structures such as systolic arrays the H-tree works well to reduce skew, but in the general case asymmetric distributions of clock pins are common and the H-tree is not as effective for clock routing. The large size of the clock net has led some researchers [DFW84] [Mij87] to perform buffer optimization within the clock distribution tree. More recently, [BWM86] has provided an analysis of the clock lines that considers the transmission line properties of the clock net. [BBB+89] have presented an approach for ASIC clock distribution that integrates buffer optimization into place and route algorithms, while [RS89] and [FP86] have presented approaches that consider macro-cell clock distribution. However, in all previous work the routing of the clock net is performed using ordinary routing techniques. This causes non-optimal clock behavior and as region size or the number of pins in the net increases, the undesirable behavior is exacerbated. In this paper, we focus exclusively on routing techniques for optimizing the clock signal in VLSI circuits. We demonstrate the superiority of our algorithm over standard routing techniques for examples of a wide range of ranging size.

In section two the preliminaries necessary for understanding the paper are presented. Following this, the problem is defined in section three. Section four illustrates the algorithm for clock routing and section five discusses theoretical results. Next, in section six, practical considerations are discussed. In section seven the experimental results are presented, and in section eight possible avenues for future work and conclusions regarding the approach are discussed.

2 Preliminaries

The majority of digital chips are synchronous in nature. Synchronous designs are often modeled as a Moore finite state machine for the purposes of analysis. The topological requirement imposed on such a finite state machine is that all closed signal paths must contain at least one synchronizing element. Satisfaction of this constraint has several benefits, two of which are: the assurance of deterministic behavior if the physical aspects of the design are correct, and the elmination of the requirement that the combinational logic be free of transients as long as next state sampling is performed after the longest path has settled to its final value [MC80]. For simplicity, and without loss of generality, let the synchronizing elements be edge-triggered. Furthermore, let CP denote the clock period, d_L the largest path delay through the combinational logic, t_{SKEW} the clock skew, t_{SU} the set-up time of the edge-triggered synchronizing elements, and t_{CQ} the delay from the synchronizing element's clock pins to the Q output pins. In order to guarantee that no long-path timing violations occur in the design, the following equation must be satisfied

$$CP \geq d_L + t_{SKEW} + t_{SU} + t_{CQ} \qquad (1)$$

This expression demonstrates the important relationship between the clock period, the longest path delay, and the clock skew.

The two timing related clock parameters that one must consider for high-performance design are clock skew and phase delay. Clock skew is defined to be the

Reprinted from *27th Proc. ACM/IEEE Design Automation Conf.*, pp. 573–579, June 1990.

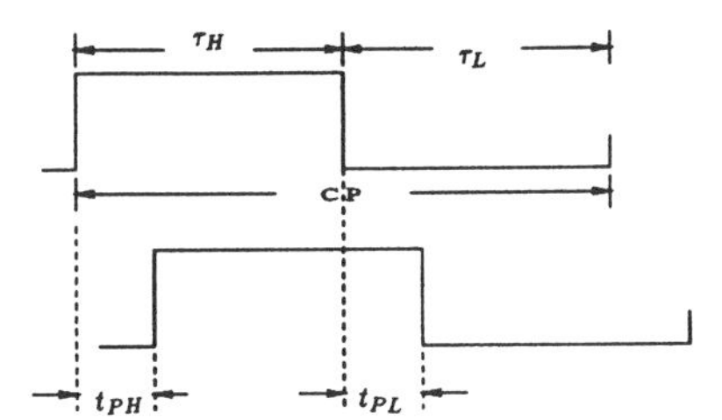

Figure 1: Relationship between τ_H, τ_L, and CP with 50 % duty cycle

maximum difference in arrival times at any two similarly clocked clock pins. Clock skew is caused by several phenomena: asymmetric routes to the clocked elements, differing interconnect line parameters, different delays through the clock distribution elements, and different device threshold voltages for the clock distribution logic. Equation 1 illustrates the important relationship between skew and the longest combinational logic path delay. As skew increases with the clock period held fixed, the efficiency of the digital system is reduced because valuable computation time is "stolen" from the total cycle time. Frequently in high-performance design environments, skew is constrained to be less than five percent of the clock period. Thus, in a 100 MHz design, skew would be constrained to be less than 500 ps. In this paper, routing techniques to help achieve this goal will be presented. Phase delay may be defined to be the maximum delay to any synchronizing clock pin. The same phenomena causing skew also contributes to phase delay. It is convenient to consider phase delay to consist of two components: an intrinsic cell delay t_{IH} or t_{IL} contributed by the externally driven clock pad and the time to charge or discharge the clock net t_{CH} or t_{CL}. Expressions for phase delay may be defined as

$$t_{PH} = t_{IH} + t_{CH} \tag{2}$$

$$t_{PL} = t_{IL} + t_{CL} \tag{3}$$

Figure 1 illustrates equations 2 and 3. Phase delay affects chip to chip interfaces by appearing as inter-chip skew and in worst-case scenarios may provide inadequate time to charge and discharge the clock net. For example, for a clock signal with a duty cycle of 50 %, the high and low portions of the clock period τ_H and τ_L, must satisfy the following constraints

$$t_{CH} < \tau_H \tag{4}$$

$$t_{CL} < \tau_L \tag{5}$$

These expressions are necessary but not sufficient conditions to guarantee proper clocking. In worst-case situations, t_{CH} and t_{CL} could conceivably constrain the clock period so it is necessary to make provisions for minimizing phase delay.

To this point, we have tacitly assumed that only one clock exists. However, in CMOS design styles it is commonplace to design with more than one clock. The ideas presented in this paper may be easily extended to the case of multiple clocks by treating the clock nets independently. When multiple clocks are present, inter-clock and intra-clock skew and phase delay must be considered. Independent treatment of the different clock nets will address intra-clock problems by minimizing the phase delay of each clock net and minimizing the skew between similarly clocked synchronizing elements. Inter-clock phase delay is minimized since intra-clock phase delay is minimized. While inter-clock skew is not explicitly minimized, should it be a problem, circuit techniques that insert delay to equalize arrival times across different clock nets will ameliorate the problem. Hereafter, for purposes of simplicity, attention will be restricted to single clock designs.

Prior to delving into the clock routing algorithm, it is necessary to define its role in the context of the overall design flow. Traditional ASIC design proceeds from logic to physical design. Physical design consists of three classical steps: placement, global routing, and detailed routing. In our proposed design flow a clock routing step is interposed between placement and global routing. Presently, during the clock routing step, the global and detailed routes of the clock net are determined and passed to the global router as blockages. The determined routes are constructed so that the clock signal behavior is optimized.

To understand the consequences of decisions made during physical design, one must model the interconnect parasitics that load the clock tree. We make the reasonable assumption that in a high-performance design environment, the interconnects are realized with aluminum due to its excellent conducting properties. Interconnect resistance R_{int} is determined using the following expression

$$R_{int} = \frac{\rho L}{WH} \tag{6}$$

where ρ is the resistivity of aluminum (3 $\mu\Omega$ cm), L is the interconnect length, W the interconnect width, and H the interconnect thickness. Interconnect capacitance C_{int} is modeled using a simple parallel-plate model given by

$$C_{int} = K_c(C_{ox} + C_I) \tag{7}$$

where

$$C_{ox} = \epsilon_{ox}\frac{WL}{t_{ox}} \tag{8}$$

and

$$C_I = \epsilon_{ox}\frac{LH}{L_s} \tag{9}$$

In these expressions K_c is a constant that is inserted to account for fringing effects, and can be calculated using the two-dimensional analysis of [DS80]. It is assumed to be 2.0 in our calculations. L_s represents the line spacing, t_{ox} represents the thickness of the field oxide, and ϵ_{ox} is the permittivity of the oxide. All lengths used to calculate resistance and capacitance are based on manhattan distances.

Based on estimates for R_{int} and C_{int}, simple and accurate interconnect delay estimates may be calculated using the first-order moment of the impulse response which has also been called Elmore's delay [RPH83] [Elm48]. The interconnects are treated as distributed RC trees and the rectilinear segments comprising the interconnect are modeled using their equivalent T-network representation.

3 Problem Definition

Given the ICs placement, the locations of blockages on the routing layers, the positions of all clock pins on the clock net, and the location of the clock pad along the periphery of the chip, the problem may be defined as follows: construct a clock tree consisting only of Manhattan segments that optimizes the clock skew, wirelength and phase delay subject to the blockages on the routing layers. In general, one seeks to minimize clock skew and wirelength, subject to constraints on phase delay and the routing. Formally, clock optimization could be formulated as follows

$$\begin{array}{ll} \min & f(t_{skew}, WL) \\ \text{subject to} & t_{CH} < \tau_H \\ & t_{CL} < \tau_L \\ & NR = 0 \end{array}$$

where WL equals the total wirelength and NR equals the number of no routes.

4 The Algorithm

4.1 The Basic Algorithm

The algorithm which we call the Method of Means and Medians (MMM) is conceptually simple and yields theoretical results which are intuitively pleasing. Let $S = \{s_1, s_2, \ldots, s_n\}$ be the set of points in the plane which represent the clock pins. Each s_i is a couple (x_i, y_i). Define

$$x_c(S) = \frac{\sum_{i=1}^{n} x_i}{n} \tag{10}$$

$$y_c(S) = \frac{\sum_{i=1}^{n} y_i}{n} \tag{11}$$

$(x_c(S), y_c(S))$ represents the center of mass of the set of points S. We shall use the notation $S_x(S)$ or simply S_x to denote the ordered set of points obtained by ordering the set S by increasing x coordinate, i.e. $x_i \leq x_j$ if $s_i, s_j \in S_x(S)$ and $i < j$. Similarly, $S_y(S)$ or simply S_y represents the ordered set of points obtained by ordering the set S by increasing y coordinate. Define

$$S_L(S) = \{s_i \in S_x \mid i \leq \lceil n/2 \rceil\} \tag{12}$$

$$S_R(S) = \{s_i \in S_x \mid \lceil n/2 \rceil < i \leq n\} \tag{13}$$

$$S_B(S) = \{s_i \in S_y \mid i \leq \lceil n/2 \rceil\} \tag{14}$$

$$S_T(S) = \{s_i \in S_y \mid \lceil n/2 \rceil < i \leq n\} \tag{15}$$

The sets S_L and S_R represent the division of S into two sets about the median x coordinate of the set of points. These sets partition the original region in the x dimension into two sub-regions with approximately equal number of elements in each sub-region. In fact, $\mid |S_L| - |S_R| \mid \leq 1$. Similarly, S_B and S_T represent the division of S into two sets about the median y coordinate of the set of points. The basic algorithm first splits S into two sets (arbitrarily in the x direction or y direction). Assume that a split of S into S_L and S_R is made. Then, the algorithm routes from the center of mass of S to each of the centers of mass of S_L and S_R respectively. The regions S_L and S_R are then recursively split in the y direction (the direction opposite to the previous one). Thus, splits alternating between x and y are introduced on the set of points recursively until there is only one point in each sub-region. The pseudo-code for the algorithm is given in Figure. 2.

4.2 Improvements

The simple algorithm described above yields good results, but there is room for further improvement. In the following discussion we define a cut in the x direction to mean a split resulting in a left region and a right region. A cut in the y direction implies a split resulting in a top and a bottom region.

```
procedure basic_MMM(S)
begin
    if |S| ≤ 1 return;
    x_0 = x_c(S); y_0 = y_c(S);
    x_left = x_c(S_L(S)); y_left = y_c(S_L(S));
    x_right = x_c(S_R(S)); x_right = y_c(S_R(S));
    route from (x_0, y_0) to (x_left, y_left) and
      (x_right, y_right);
    x_bot-left = x_c(S_B(S_L(S)));
    y_bot-left = y_c(S_B(S_L(S)));
    x_top-left = x_c(S_T(S_L(S)));
    y_top-left = y_c(S_T(S_L(S)));
    x_bot-right = x_c(S_B(S_R(S)));
    y_bot-right = y_c(S_B(S_R(S)));
    x_top-right = x_c(S_T(S_R(S)));
    y_top-right = y_c(S_T(S_R(S)));
    route from (x_left, y_left) to
      (x_bot-left, y_bot-left) and
      (x_top-left, y_top-left);
    route from (x_right, y_right) to
      (x_bot-right, y_bot-right) and
      (x_top-right, y_top-right);
    basic_MMM(S_B(S_L(S)));
    basic_MMM(S_T(S_L(S)));
    basic_MMM(S_B(S_R(S)));
    basic_MMM(S_T(S_R(S)));
end;
```

Figure 2: Pseudo-code for the basic algorithm

Delay equalization look-ahead

Consider the example shown in Figure 3, where S is the clock source. If we make a cut in the x direction and then recursively split the left and right regions in the y direction, we get the result shown in Figure 3(a). Clearly, there is skew between points P_{LT} and P_{RT}. However, if we reverse the cut directions, i.e., split in the y direction first followed by a split in the x direction, we get the result shown in Figure 3(b), which has no skew between the endpoints.

This example illustrates the need for making a good choice of cut direction at each level of the recursion tree. We make the choice by a one level look-ahead technique. Given a region to be split, the algorithm makes an x direction cut followed by a y direction cut on the resulting left and right regions. It also makes a y direction cut followed by an x direction cut. The skews for each of the

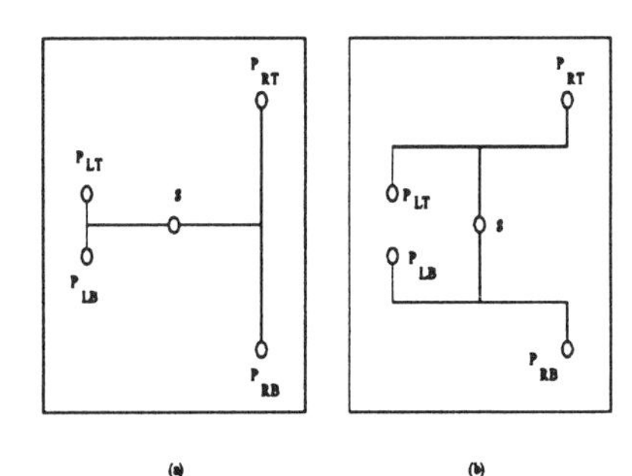

Figure 3: Clock tree (a) without look-ahead, (b) with look-ahead

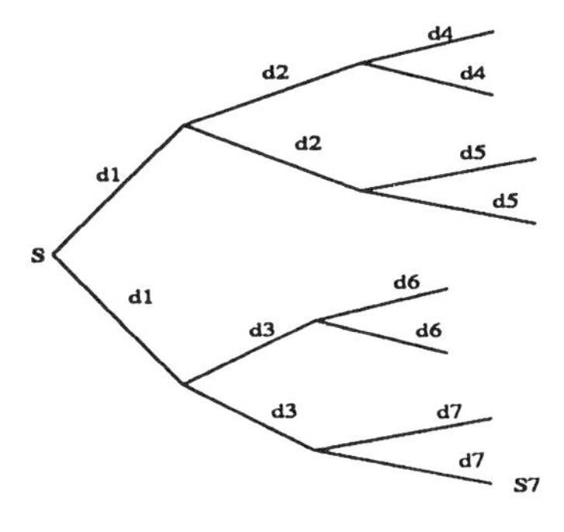

Figure 4: Clock tree representation used for delay calculations

configurations is compared and the cut direction that minimizes skew between its current endpoints is chosen. The method of estimating the skew between the endpoints is described in the following section.

Delay Calculation

We use the Penfield-Rubinstein [RPH83] algorithm for calculating delays to the endpoints in the grown clock tree. The resistance and capacitance of the tree segments are modeled by a T-network model. Consider the tree shown in Figure 4. Because of a property of the center of mass (see next section), the lengths of tree segments from the center of mass of a region to each of its two sub-regions will always be equal and symmetric. The delay from s to the endpoint $s7$ is calculated as

$$\delta_{s-s7} = R_l(0.5C_l(d_1^2 + d_3^2 + d_7^2) + C_g(4d_1 + 2d_3 + d_7) + 2C_l(d_1d_3 + d_1d_6 + d_1d_7 + d_3d_7)) \quad (16)$$

where C_g is the gate capacitance at the clock pin (assumed equal for all clock pins), C_l is the capacitance per unit length and R_l is the resistance per unit length. Note that the delay to any endpoint depends on the lengths of other segments connecting different endpoints, so it is important to use delay estimates to drive the look-ahead rather than length calculations. The complexity of the algorithm with look-ahead is $O(n \log n)$ where n is the number of clock pins. For a detailed analysis of the complexity, the interested reader is referred to [JSK90]. We shall refer to the algorithm as the Method of Means and Medians (MMM) in the following text.

5 Theoretical results

In this section we state some key results that motivate the Method of Means and Medians. For the sake of brevity we have omitted proofs of the propositions. A detailed treatment the proofs leading to the results can be found in [JSK90]. The first result is that after splitting a region into two sub-regions, the lengths of segments from the center of mass of the region to each of its sub-regions is equal and symmetric. Next, we establish a bound on the total wirelength for a gridded distribution of points and compare it with the wirelength for a minimum rectilinear Steiner tree spanning those points. We also present an interesting result which claims that increasing the number of points within a region reduces the skew. Finally, we show that the algorithm with one level of look-ahead runs in time $O(n \log n)$ where n is the number of clock pins. All our theoretical results corroborate our experimental results (Section 7).

Theorem 5.1
Given a set of points $S = \{s_1, s_2, \ldots, s_n\}$, where n is an even integer,

$$| x_c(S) - x_c(S_L(S)) | + | y_c(S) - y_c(S_L(S)) | = | x_c(S) - x_c(S_R(S)) | + | y_c(S) - y_c(S_R(S)) |$$

A similar result holds between S, $S_B(S)$ and $S_T(S)$. The significance of the above result is that at every split in the algorithm, the lengths to each of the sub-regions are always equal. Note that as we move deeper into the clock tree, the segments become shorter. Thus, at the topmost level of the clock tree when the segments are longest, we ensure exact balance and no skew.

Lemma 5.2
Given a distribution of n points on a uniform grid, where $n = 4^k, k$ is an integer ≥ 1, within a region of side 1.0 unit, the total wirelength of the tree produced by the basic algorithm grows as $\frac{3}{2}\sqrt{n}$.

Theorem 5.3
The wirelength for a minimum rectilinear Steiner tree spanning a set of n uniformly spaced points on a grid, where $n = 4^k, k$ an integer ≥ 1, within a region of side 1.0 unit, grows as $\sqrt{n}+1$. This is also the largest possible wirelength for a rectilinear Steiner tree for a distribution of n points in a unit square. Any other distribution of n points within the unit grid will yield a smaller total wirelength.

These results indicate that the wirelength of the clock tree is within a constant factor of $\frac{3}{2}$ compared to a minimum rectilinear Steiner tree for the particular distribution of points.

We conjecture that the worst-case wirelength for the Method of Means and Medians is $\frac{9}{4}\sqrt{n} - \frac{3}{2}$ [JSK90]. Thus even in this case, the total wirelength is still a constant times the wirelength for the largest minimum rectilinear Steiner tree.

We define the *sparsity p* of a distribution of points in a region to be the total number of points in the region divided by the area of the region. It is a measure of the average number of points per unit area. The next result concerns the variation of skew with sparsity.

Theorem 5.4
For a uniformly randomly distributed set of points inside a box of side n units with sparsity p, the expected maximum difference in length from the center to any endpoint for the basic algorithm is proportional to $\frac{1}{\sqrt{p}}$.

This result indicates that as the number of points within the region is increased, the skew between the endpoints is reduced. Our experimental results support this claim.

Theorem 5.5
The algorithm with one level look-ahead runs in time $O(n \log n)$, where n is the number of points in the region.

The algorithm is fast and the running time for routing a region with 4096 points on a DECStation 3100 computer (14 MIPS) was less than a second of CPU time. Therefore, speed is not an issue when running the algorithm on practical examples.

6 Practical Considerations

In practice one would like to route the clock net with minimum wirelength while satisfying a prespecified tolerance on clock skew and phase delay. We have developed a *hybrid* clock routing algorithm that performs MMM to a certain depth, i.e., until the chip has been divided into a number of regions each containing less than a certain number of clock pins. Then standard routing techniques are applied to each of the remaining sub-regions. The depth at which the transition from MMM to standard techniques occurs and the number of pins within each sub-region that are routed using standard methods are a function of the amount of skew tolerable.

Another practical concern is the degradation in the clock waveform. Usually, buffers are inserted into the clock tree to regenerate the clock waveform. We propose two strategies to deal with the placement of buffer cells. The first is to pre-place buffers at symmetric locations on the chip that coincide with expected locations of the centers of mass of the sub-regions to be driven by the buffers. Then, during the clock routing, detours are made to the pre-placed buffers so that they may drive the clock pins. These buffers then act as centers from which a clock tree is grown using the Method of Means and Medians. The second strategy is to insert buffers after placement at optimal locations determined by the Method of Means and Medians. The expected perturbation to the placement would be small considering the size and number of buffers relative to the total number of cells.

7 Experimental Results

As a test of the effectiveness of MMM it was run on twenty random examples and the MCNC industrial benchmarks Primary1 and Primary2. The twenty random examples had uniform pin distributions in a square region. For the twenty examples, four equal-sized chips with 16, 32, 64, 256 or 512 pins were generated. For comparative purposes, we routed the same pin distributions using a minimum rectilinear spanning tree (MST) algorithm. As shown in [Han66], the ratio of the length of a minimum rectilinear spanning tree and an optimal rectilinear Steiner tree is bounded by a factor of $\frac{3}{2}$. SPICE [Nag75] files were generated for all examples based on Manhattan geometries, and the interconnect was driven by a single I/O buffer pad with equivalent drive of ten times the minimum sized inverter cell in a 2 μm design style. To model gate loading, a capacitance was placed at the leaves of the clock distribution tree of value 0.3 pF.

We compared the skew, phase delay and wirelength as a function of the number of pins for MMM and MST. Additional experiments comparing skew as a function of chip size and minimum feature size may be found in [JSK90]. An extended evaluation of the results of all experiments is also given in [JSK90].

To determine the relationship between skew and the number of pins with chip size fixed at 25 mm^2, MMM and MST were compared to one another with the result appearing in Figure 5. Interestingly, the skew decreased with increasing number of pins for MMM and grew linearly for MST. Similarly, phase delay versus the number of pins for a chip size of 25 mm^2 were compared for MST and MMM. Again, MMM displayed a clear advantage with its growth in phase delay appearing to be sub-linear and MST approximating linear growth. These results can be seen in Figure 6.

The dramatic improvements in clock skew and phase delay are paid for in terms of total wirelength. To illustrate this, the average wirelength for all examples was plotted against the number of pins in Figure 7. The experimental results corroborate the theoretical $\sqrt{n}$ relationship between wirelength and number of points

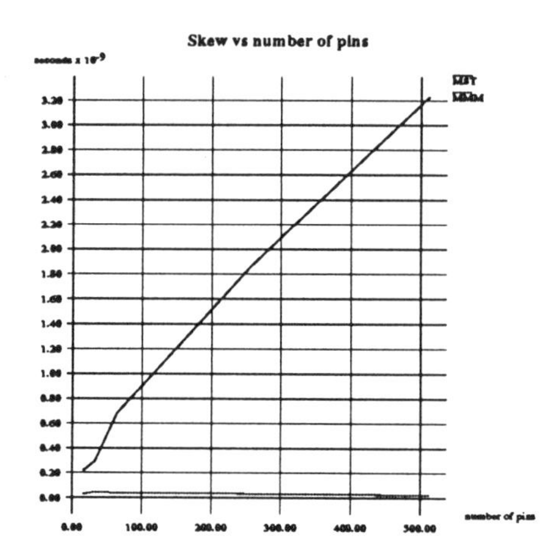

Figure 5: Skew versus no. of pins

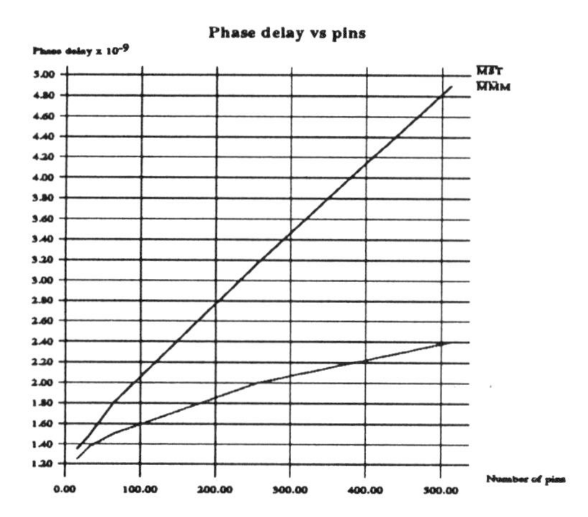

Figure 6: Phase delay versus no. of pins

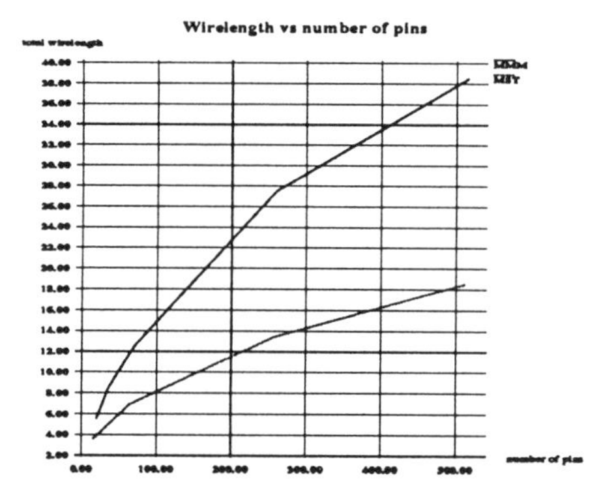

Figure 7: Wirelength versus no. of pins

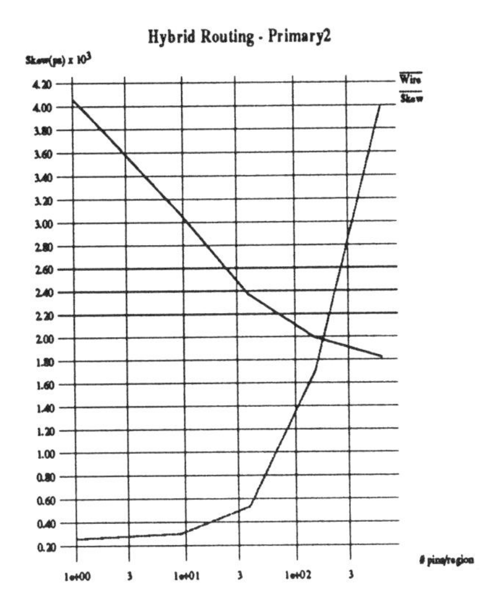

Figure 8: Hybrid routing for Primary2

Figure 9: Clock tree for example Primary1

n with the difference appearing as a constant factor. Thus, improvements in clock behavior are accompanied by an increase in the clock net's wirelength.

Figure 8 shows the results of running the hybrid algorithm to varying depths on the MCNC benchmark chip Primary2. On the x-axis we have plotted the number of pins in each region that were routed using standard techniques. The origin of the x-axis corresponds to routing using MMM for all the pins. The rightmost point on the axis corresponds to routing the clock net using a minimum spanning tree. Thus the depth to which MMM was applied decreases as we move towards the right of the figure. The solid line shows decreasing wirelength (normalized) while the dotted line shows increasing skew with decreasing depth. This provides the designer with the opportunity to arrive at a compromise between the excellent skew of MMM and the low wirelength of a minimum spanning tree.

Figure 9 and 10 show MMM's routing results for the MCNC Primary1 and Primary2 benchmarks respectively. The skew introduced for each of these examples was 31 ps and 260 ps respectively. Primary1 had 269 clock pins and Primary2 had 603 clock pins. Both placements were obtained using PROUD [TKH88]. It is interesting to note that Primary2's placement exhibited an asymmetric clock pin distribution while Primary1's remained relatively uniform. However, the asymmetry was not enough to deter MMM from yielding excellent results. Figures 11 and 12 show the voltage waveforms at the furthest and closest pins from the clock driver

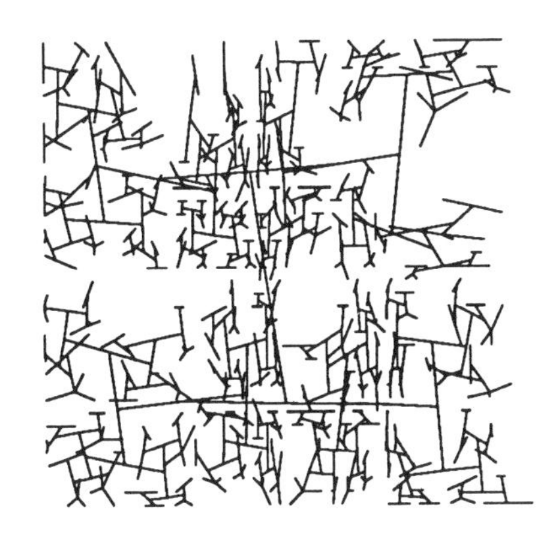

Figure 10: Clock tree for example Primary2

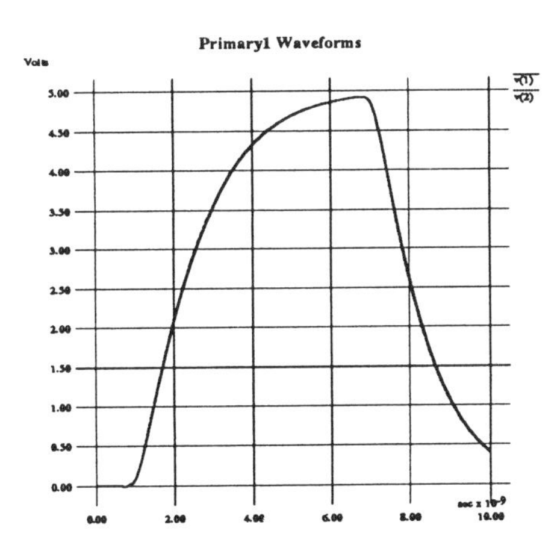

Figure 11: Clock waveforms for Primary1(MMM)

for Primary1 when routed using MST and MMM respectively. The skew introduced by MST was 4.7 ns, and the routing to the furthest point was so poor (in terms of timing behavior) that the pin was unable to charge to the supply voltage. Note that the skew generated by MMM is 31 ps and is barely distinguishable in Figure 11.

8 Conclusions and Future Work

We have presented an approach to clock routing that is clearly superior to simple minded clock routing based on a minimum spanning tree. While high-performance industrial designs are unlikely to have clock routing performed using such a simple approach as MST, the quality of the results generated by MMM are exceptional. The approach has all but eliminated clock skew and yielded excellent phase delay results for a wide range chip sizes, net sizes (pin count), technologies, and pin distributions on both randomly created and industrial benchmarks.

Future work will address clock tree buffer optimization and give consideration to blockages and routing congestion during the growth of the clock tree. Ad-

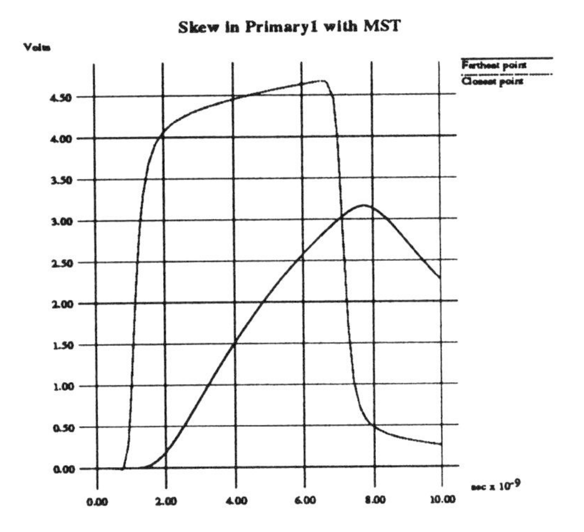

Figure 12: Clock waveforms for Primary1(MST)

ditionally, the impact of the approach on wirability and chip area will be investigated.

[Fis89] has shown that clock skew may be used to decrease the clock period of a system by introducing relative delays between the arrival times of the clock signal at the clocked pins. We are considering techniques to implement this idea.

Acknowledgements

This work was supported by SRC Grant 90-DC-008 and JSEP Grant F49620-87-C-0041.

References

[BBB+89] S. Boon, S. Butler, R. Byrne, B. Setering, M. Casalanda, and Al Scherf. High performance clock distribution for cmos asic's. *IEEE Custom Integrated Circuit Conference*, pages 15.4.1–15.4.4, 1989.

[BWM86] H. B. Bakoglu, J. T. Walker, and J. D. Meindl. A symmetric clock-distribution tree and optimized high-speed interconnections for reduced clock skew in ulsi and wsi circuits. *IEEE Int. Conference on Computer Design: VLSI in Computers and Processors (ICCD-86)*, pages 118–122, October 1986.

[DFW84] S. Dhar, M. A. Franklin, and D. F. Wann. Reduction of clock delays in vlsi structures. *IEEE Int. Conference on Computer Design: VLSI in computers (ICCD)*, pages 778–783, 1984.

[DS80] R. L. M. Dang and N. Shigyo. A two-dimensional simulation of lsi interconnect capacitance. *IEEE Electron Device Letters*, EDL-2:196–197, August 1980.

[Elm48] W. C. Elmore. The transient response of damped linear networks with particular regard to wideband amplifiers. *Journal of Applied Physics*, 19(1):55–63, January 1948.

[Fis89] J.P. Fishburn. 1989. Private Communication.

[FK82] A. L. Fisher and H. T. Kung. Synchronizing large systolic arrays. *Proceedings of SPIE*, 341:44–52, May 1982.

[FP86] E. G. Friedman and S. Powell. Design and analysis of a hierarchical clock distribution system for synchronous standard cell/macrocell vlsi. *IEEE Journal of Solid-State Circuits*, SC-21(2):240–246, 1986.

[Han66] M. Hanan. On steiner's problem with rectilinear distances. *SIAM Journal of Applied Math*, 14:255–265, 1966.

[JSK90] Michael Jackson, Arvind Srinivasan, and E.S. Kuh. Clock routing methodologies. Technical report, University of California at Berkeley, 1990. Memo ERL.

[KGE82] S. Y. Kung and R. J. Gal-Ezer. Synchronous versus asynchronous computation in vlsi array processors. *Proceedings of SPIE*, pages 53–65, May 1982.

[MC80] Carver A. Mead and Lynn A. Conway. *Introduction to VLSI Systems.* Addison-Wesley, Reading, Massachusetts, 1980.

[Mij87] D. Mijuskovic. Clock distribution in application specific integrated circuits. *Microelectronics Journal*, 18(4):15–27, 1987.

[Nag75] W. Nagel. Spice2, a computer program to simulate semiconductor circuits. *University of California, Berkeley, Memo No. ERL-M520*, May 1975.

[RPH83] Jorge Rubinstein, Paul Penfield, and Mark A. Horowitz. Signal delays in rc tree networks. *IEEE Trans. Computer-Aided Design*, CAD-2:202–211, July 1983.

[RS89] P. Ramanathan and K. G. Shin. A clock distribution scheme for non-symmetric vlsi circuits. *IEEE Int. Conference on Computer-Aided Design (ICCAD-89)*, pages 398–401, November 1989.

[TKH88] R. S. Tsay, E. S. Kuh, and C. P. Hsu. Proud: A sea-of-gates placement algorithm. *IEEE Design and Test of Computers*, pages 318–323, December 1988.

[WF83] D. F. Wann and M. A. Franklin. Asynchronous and clocked control structures for vlsi based interconnection networks. *IEEE Transactions on Computers*, C-32(3):284–293, March 1983.

Zero Skew Clock Routing with Minimum Wirelength

Ting-Hai Chao, Yu-Chin Hsu, Jan-Ming Ho, Kenneth D. Boese, *Student Member, IEEE*, and Andrew B. Kahng, *Associate Member, IEEE*

Abstract—**In the design of high performance VLSI systems, minimization of clock skew is an increasingly important objective. Additionally, wirelength of clock routing trees should be minimized in order to reduce system power requirements and deformation of the clock pulse at the synchronizing elements of the system. In this paper, we first present the deferred-merge embedding (DME) algorithm, which embeds any given connection topology to create a clock tree with zero skew while minimizing total wirelength. The algorithm always yields exact zero skew trees with respect to the appropriate delay model. Experimental results show an 8% to 15% wirelength reduction over previous constructions in [17] and [18]. The DME algorithm may be applied to either the Elmore or linear delay model, and yields *optimal* total wirelength for linear delay. DME is a very fast algorithm, running in time linear in the number of synchronizing elements. We also present a unified BB + DME algorithm, which constructs a clock tree topology using a top-down *balanced bipartition* (BB) approach, and then applies DME to that topology. Our experimental results indicate that both the topology generation and embedding components of our methodology are necessary for effective clock tree construction. The BB + DME method averages 15% wirelength savings over the previous method of [17], and also gives 10% average wirelength savings when compared to the method of [25]. The paper concludes with a number of extensions and directions for future research.**

I. Introduction

In synchronous VLSI designs, circuit speed is increasingly limited by two factors: i) delay on the longest path through combinational logic, and ii) clock skew, which is the maximum difference in arrival times of the clocking signal at the synchronizing elements of the design. This is seen from the following well-known inequality governing the clock period of a clock signal net [2], [17]:

$$\text{clock period} \geq t_d + t_{\text{skew}} + t_{su} + t_{ds}$$

where t_d is the delay on the longest path through combinational logic, t_{skew} is the clock skew, t_{su} is the set-up time of the synchronizing elements (assuming edge triggering), and t_{ds} is the propagation delay within the synchronizing elements. The term t_d can be further decomposed into $t_d = t_{d\text{-interconnect}} + t_{d\text{-gates}}$, where $t_{d\text{-interconnect}}$ is the delay associated with the interconnect of the longest path through combinational logic, and $t_{d\text{-gates}}$ is the delay through the combinational logic gates on this path. Increased switching speeds due to advances in VLSI fabrication technology will significantly decrease the terms t_{su}, t_{ds}, and $t_{d\text{-gates}}$. Therefore, $t_{d\text{-interconnect}}$ and t_{skew} become the dominant factors in determining circuit performance: Bakoglu [2] has noted that t_{skew} may account for over 10% of the system cycle time in high-performance systems. With this in mind, a number of researchers have recently studied the clock skew minimization problem.

Several results address formulations with inherently small problem size. For building block design styles, Ramananathan and Shin [21] have proposed a clock distribution scheme that applies when the blocks are hierarchically organized. The number of blocks at each level of the hierarchy is assumed to be small, since the algorithm exhaustively enumerates all possible clock routings and clock buffer optimizations. Burkis [5] and Boon *et al.* [4] have also proposed hierarchical clock tree synthesis approaches involving geometric clustering and buffer optimization at each level. More powerful clock tree resynthesis or reassignment methods were used by Fishburn [13] and Edahiro [11] to minimize the clock period while avoiding hazards or race conditions; Fishburn employed a mathematical programming formulation, while Edahiro employed a clustering-based heuristic augmented by techniques from computational geometry. All of these methods are essentially limited to small problem sizes, either by their algorithmic complexity or by their reliance on strong hierarchical clustering. In contrast, we are interested in clock tree synthesis for "flat" problem instances with many sinks (synchronizing elements), as will arise in large standard-cell, sea-of-gates, and multichip module designs.

Clock tree construction for designs with many clock sinks was first attacked by the H-tree method, which was used in regular systolic arrays by Bakoglu and other authors [1], [10], [14], [26]. The H-tree structure can significantly reduce clock skew [10], [26], but is applicable only when all of the sinks have identical loading capacitances and are placed in a symmetric array. A more

Reprinted from *IEEE Trans. Circuits Syst.-II: Analog and Digital Signal Processing,* vol. 39, no. 11, pp. 799–814, Nov. 1992.

robust clock tree construction for cell-based layouts is due to Jackson, Srinivasan, and Kuh [17]: their "method of means and medians" (MMM) algorithm generates a topology by recursively partitioning the set of sinks into two equal-sized subsets, then connecting the center of mass of the entire set to the centers of mass of the two subsets. While the MMM solution will have reasonable skew on average, Kahng *et al.* [18] gave small examples for which the source-sink pathlengths in the MMM solution may vary by as much as half of the chip diameter. In some sense, this reflects an inherent weakness in the top-down approach: it can commit to an unfortunate topology early on in the construction. Kahng *et al.* [9], [18] have proposed a bottom-up matching approach to clock tree construction: in practice their method eliminates all source-sink pathlength skew, while using 5–7% less total wirelength than the MMM algorithm. However, as the method of [9] and [18] focuses primarily on pathlength balancing, their method addresses clock skew minimization only in the sense of the *linear* delay model. Tsay [25] uses ideas similar to both [17] and [18], and achieves exact zero skew trees with respect to the Elmore delay model [12], [22]. His algorithm was the first to produce trees with exact zero skew in all cases. In the same spirit as the method of [18], Tsay's method recursively combines pairs of zero skew trees at "tapping points," analogous to the "balance points" in [18], to yield larger zero skew trees.

The primary motivation behind our work is to minimize the total wirelength of clock routing trees while maintaining exact zero skew with respect to the appropriate delay model. Total wirelength is a critical parameter of the clock routing solution since excess interconnect not only increases layout area but also results in greater tree capacitance, thus requiring more power for distribution of the clock signal. However, both the top-down method of [17] and the bottom-up methods of [9], [18], [25] concentrate on the problem of computing a clock tree *topology*, and only incompletely address the associated problem of finding a minimum-cost *embedding* of the topology. These previous methods are actually quite inflexible in that they permanently embed each internal node of the tree as soon as it becomes defined [18], or else choose the embedding with at most one level of lookahead in the tree construction [17], [25].

In this paper, we first propose a new approach that achieves exact zero skew while significantly reducing the total wirelength of the clock tree. The basic idea of our deferred-merge embedding (DME) algorithm is to *defer* the embedding of internal nodes in a given topology for as long as possible: i) a bottom-up phase computes loci of feasible locations for the roots of recursively merged subtrees, and ii) a top-down phase then resolves the exact embedding of these internal nodes of the clock tree. In practice, the DME algorithm begins with an initial clock tree computed by any previous method, then maintains exact zero clock skew while reducing the wirelength. In regimes where the linear delay model applies, our method produces the *optimal* (i.e., minimum wirelength) zero skew clock tree with respect to the prescribed topology, and this tree will also enjoy optimal source-sink delay. Experimental results in Section IV below show that the DME approach is highly effective in both the Elmore and linear delay models. We achieve average savings in total clock tree wirelength of 15% over the MMM algorithm [17] and 8% over the method of Kahng *et al.* [18]. In all cases, our clock trees have *exact* zero skew according to the appropriate delay model, and our Elmore delay computations have been confirmed by SPICE simulations which show sub-picosecond skew on all benchmark examples.

Since the DME algorithm only optimizes a prescribed topology, it cannot achieve all possible improvement of the clock tree construction. Thus, to complement this successful embedding method, we also propose a new top-down heuristic for constructing an initial clock tree topology, based on the geometric concept of a *balanced bipartition* (BB). Applying our embedding to topologies generated in this way yields a unified BB + DME algorithm which gives very promising results: we achieve 15% reduction in tree cost and as compared with the MMM algorithm [17], and we achieve 10% reduction in tree cost and a 22% reduction in Elmore delay as compared with the method of Tsay [25].[1] Again, all of our solutions have exact zero skew. Our methods are quite robust, and extend to prescribed skew formulations as well as more general optimizations of topologies for both clock routing and global routing. Furthermore, because our method implicitly maintains *all* possible minimum-cost embeddings of a topology, it may be used to reroute the clock net while preserving minimum wirelength, as may be necessary when routing density must be minimized.

The remainder of this paper is organized as follows. In Section II, we formalize the minimum-cost zero skew clock routing problem and also establish the linear and Elmore delay models that are used in the subsequent discussion. Section III presents our main results. These include: i) the DME algorithm for efficiently embedding a given topology; ii) application of the DME algorithm to both the linear and Elmore delay regimes; and iii) our unified BB + DME algorithm, which uses a top-down BB strategy to derive a good tree topology to which the DME algorithm may be applied. Section IV gives experimental results and comparisons with previous work, and Section V concludes with directions for future research.

II. Problem Formulation

The placement phase of physical layout determines positions for the synchronizing elements of a circuit, which we call the *sinks* of the clock net. A finite set of sink locations, denoted by $S = \{s_1, s_2, \cdots, s_n\} \subset \Re^2$, specifies an

[1] Note that SPICE simulations for BB + DME constructions on random sink sets (Table IV below) indicate only a 3% improvement in delay compared to the MMM algorithm. This suggests that although the Elmore model is reasonably accurate for predicting skew, it is less accurate for predicting delay.

instance of the clock routing problem. A *connection topology* is defined to be a rooted binary tree, G, which has n leaves corresponding to the set of sinks S. A *clock tree* $T(S)$ is an embedding of the connection topology in the Manhattan plane.[2] The embedding associates a *placement* in $\Re^2$ with each node $v \in G$; we will use $pl(T, v)$ or $pl(v)$ to represent this location. (When no confusion arises, we may also denote $pl(T, v)$ simply by v.) The root of the clock tree is the clock *source*, denoted by s_0. We direct all edges of the clock tree away from the source; a directed edge from v to w may be uniquely identified with w and written as e_w. We say that v is the *parent* of w, and w is a *child* of v; the set of all children of v is denoted by $children(v)$. The wirelength, or *cost*, of the edge e_w is denoted by $|e_w|$, and must be greater than or equal to the Manhattan distance between its endpoints $pl(w)$ and $pl(v)$.[3] The cost of $T(S)$, denoted $cost(T(S))$, is the total wirelength of the edges in $T(S)$.

For a given clock tree $T(S)$, let $t_d(s_0, s_i)$ denote the signal propagation time, or *delay*, on the unique path from source s_0 to sink s_i; the collection of edges in this path is denoted by $path(s_0, s_i)$. The *skew* of $T(S)$ is the maximum value of $|t_d(s_0, s_i) - t_d(s_0, s_j)|$ over all sink pairs s_i, $s_j \in S$. If $T(S)$ has zero skew then it is called a *zero skew clock tree* (ZST). Given a set S of sinks, the zero skew clock routing problem is to construct a ZST $T(S)$ of minimum cost. A variant of the zero skew clock routing problem asks for a minimum cost ZST with a prescribed connection topology:

Zero Skew Clock Routing Problem (S, G): Given a set S of sink locations, and given a connection topology G, construct a zero skew clock tree $T(S)$ with topology G and having minimum cost.

The notion of a zero skew clock tree is well defined only in the context of a method for evaluating signal delays. The delay from the source to any sink depends on the wirelength of the source-sink path, the RC constants of the wire segments in the routing, and the underlying connection topology of the clock tree.[4] Using equations such as those of Rubinstein *et al.* [22], one can achieve tight upper and lower bounds on delay in a distributed RC tree model of the clock net. However, in practice it is appropriate to apply one of two simpler RC delay approximations, either the linear model or the Elmore model, both of which are easier to compute and optimize during clock tree design.

[2]Note that the binary tree representation suffices to capture arbitrary Steiner routing topologies. Also, because the meaning is clear, we use $T(S)$ instead of $T(S, G)$ to denote a clock tree; implicitly, the embedding is always with respect to a particular topology G.

[3]To route a wire of length greater than the distance between its endpoints, the method of specified-length routing due to Hanafusa *et al.* [16] can be used.

[4]The global routing phase of layout will typically consider the clock and power/ground nets for preferential assignment to (dedicated) routing layers. We assume that the interconnect delay parameters are the same on all metal routing layers, and we ignore via resistances. Thus, wirelength becomes a valid measure of the RC parameters of interconnections.

2.1. Delay Models

2.1.1. Linear Delay: In the linear delay model, the delay along $path(s_0, s_i)$ is proportional to the length of the path and is independent of the rest of the connection topology. Normalized by an appropriate constant factor, the linear delay between any two nodes u and w in a source-sink path is

$$t_{LD}(u, w) = \sum_{e_v \in \text{path}(u, w)} |e_v|.$$

While less accurate than the distributed RC tree delay formulas of Rubinstein *et al.* [22], the linear delay model has been effectively used in clock tree synthesis [18], [21]. In general, use of the linear approximation is reasonable with older ASIC technologies, which have larger mask geometries and slower packages. Tsay [25] notes that the linear delay model is also proper for emerging optical and wave interconnect technologies. In addition, we observe that linear delay applies to hybrid packaging technologies, which have relatively large interconnect geometries [24].

2.1.2. Elmore Delay: With smaller device dimensions and higher ASIC system speeds, a distributed RC tree model for signal delay in clock nets is often required to derive accurate timing information. Typically, we use the first-order moment of the impulse response, also known as the Elmore delay [6], [8], [25]. The Elmore delay model is developed as follows. Let α and β, respectively, denote the resistance and capacitance per unit length of interconnect, so that the resistance r_{e_v} and capacitance c_{e_v} of edge e_v are given by $\alpha \cdot |e_v|$ and $\beta \cdot |e_v|$, respectively. For each sink s_i in the tree $T(S)$, there is a loading capacitance c_{L_i} which is the input capacitance of the functional unit driven by s_i.

We let T_v denote the subtree of $T(S)$ rooted at v, and let c_v denote the node capacitance of v.[5] The *tree capacitance* of T_v is denoted by C_v and equals the sum of capacitances in T_v. C_v is calculated using the following recursive formula:

$$C_v = \begin{cases} c_{L_i} & \text{if } v \text{ is a sink node } s_i \\ c_v + \sum_{w \in \text{children}(v)} (c_{e_w} + C_w) & \\ \quad \text{if } v \text{ is an internal node.} \end{cases}$$

According to [12], [22], and [23], the Elmore delay $t_{ED}(s_0, s_i)$ can be calculated by the following formula (see [25] for a discussion of underlying circuit models):

$$t_{ED}(s_0, s_i) = \sum_{e_v \in \text{path}(s_0, s_i)} r_{e_v}\left(\tfrac{1}{2} c_{e_v} + C_v\right).$$

More generally, the delay time between any two vertices u

[5]As noted earlier, we will assume that $c_v = 0$ for each internal node in all of our examples and benchmarks.

and w on a source-sink path is given by

$$t_{ED}(u, w) = \sum_{e_v \in \text{path}(u, w)} r_{e_v}\left(\tfrac{1}{2}c_{e_v} + C_v\right).$$

Elmore delay is additive: if v is a vertex on the u–w path, then $t_{ED}(u, w) = t_{ED}(u, v) + t_{ED}(v, w)$, and in particular, if v is a child of u on the u–s_i path, then $t_{ED}(u, s_i) = r_{e_v}((1/2)c_{e_v} + C_v) + t_{ED}(v, s_i)$. A sink node s_i may be treated as a trivial zero skew subtree with capacitance c_{L_i} and delay zero.

III. Main Results

This section presents our new unified approach to constructing a ZST over a given set of sinks S. At a high level, we divide the construction of the ZST into: i) generation of a connection topology, and ii) embedding of that connection topology in the Manhattan plane. Our discussion begins with the DME algorithm, which computes a wire-efficient embedding of a given topology. Next, we describe the application of the DME algorithm to both the linear and Elmore delay models. We then present a new top-down BB algorithm that creates a good connection topology, leading to the unified BB + DME algorithm.

3.1. The DME Algorithm

The DME algorithm embeds internal nodes of the topology G via a two-phase process. A bottom-up phase constructs a tree of line segments which represent loci of possible placements of the internal nodes in the ZST. A top-down phase then resolves the exact locations of all internal nodes in T. In the discussion that follows, the *distance* between two points p and q is assumed to be the Manhattan distance $d(p, q)$, and the distance between two sets of points P and Q, written $d(P, Q)$, is given by $\min\{d(p, q) | p \in P \text{ and } q \in Q\}$.

3.1.1. Bottom-Up Phase: The Tree of Merging Segments: For prescribed sink locations S and connection topology G, we construct a tree of *merging segments*. The basic idea is as follows. Each node v in G is associated with a merging segment which represents a set of possible placements of v. The merging segment of a node depends on the merging segments of its two children, so the connection topology must be processed in a bottom-up order. In building the tree of merging segments, we also assign a length to each edge in G; this length is retained in the final embedding of G as a ZST.

Let a and b be the children of node v in G. We use TS_a and TS_b to denote the subtrees of merging segments rooted at a and b, respectively. We are interested in placements of v which allow TS_a and TS_b to be merged with *minimum* added wire while preserving zero skew. Define the *merging cost* between TS_a and TS_b to be $|e_a| + |e_b|$, where $|e_a|$ and $|e_b|$ denote the lengths to be assigned to edges e_a and e_b. These lengths are chosen to minimize merging cost while balancing delays at $pl(v)$. Because delay is a monotone increasing function of wirelength, there is a unique optimal assignment of lengths to e_a and e_b.[6]

We now develop more precisely the construction of the tree of merging segments. A *Manhattan arc* is defined to be a line segment, possibly of zero length, with slope $+1$ or -1; in other words, a Manhattan arc is a line segment tilted at 45° from the wiring directions. The collection of points within a fixed distance of a Manhattan arc is called a *tilted rectangular region*, or *TRR*, whose boundary is composed of Manhattan arcs (see Fig. 1). The Manhattan arc at the center of the TRR is called its *core*. The *radius* of a TRR is the distance between its core and its boundary.

The *merging segment* of node v, $ms(v)$, is defined recursively as follows: if v is a sink s_i, then $ms(v) = \{s_i\}$. If v is an internal node, then $ms(v)$ is the set of all placements $pl(v)$ which allow minimum merging cost, that is to say, all points that are both within distance $|e_a|$ of $ms(a)$ and within distance $|e_b|$ of $ms(b)$. If $ms(a)$ and $ms(b)$ are both Manhattan arcs, then we obtain the merging segment $ms(v)$ by intersecting two TRR's, trr_a with core $ms(a)$ and radius $|e_a|$, and trr_b with core $ms(b)$ and radius $|e_b|$; i.e., $ms(v) = trr_a \cap trr_b$.

The merging cost at v has an obvious lower bound of $\kappa = d(ms(a), ms(b))$. If the merging cost is greater than κ (i.e., more wirelength is needed to balance the delays), then one edge length will equal zero and the other will equal the merging cost. Fig. 2 illustrates the algorithm for the case where the merging cost is equal to κ, and Fig. 3 illustrates the algorithm for the case where the merging cost is greater than κ. An entire tree of merging segments is illustrated by Fig. 4. The leaves of the tree of segments are all single points representing the sink locations $s_1, \cdots, s_8$, and the internal nodes are Manhattan arcs.

We prove that all merging segments are Manhattan arcs using induction and the following lemma. (Proofs of all lemmas are given in the Appendix.)

Lemma 1: The intersection of two TRR's, R_1 and R_2, is also a TRR and can be found in constant time. If $\text{radius}(R_1) + \text{radius}(R_2) = d(\text{core}(R_1), \text{core}(R_2))$, then the TRR $R_1 \cap R_2$ is also a Manhattan arc.

Lemma 1 implies that if $ms(a)$ and $ms(b)$ are both Manhattan arcs, then $ms(v)$ is a Manhattan arc, as follows: i) if the merging cost at v is equal to κ, then $d(\text{core}(trr_a), \text{core}(trr_b)) = |e_a| + |e_b| = \text{radius}(trr_a) + \text{radius}(trr_b)$, and hence, $trr_a \cap trr_b$ is a Manhattan arc; or ii) if the merging cost at v is greater than κ, then either trr_a or trr_b will be a Manhattan arc whose intersection with any convex set will also be a Manhattan arc. For each sink s_i, the merging segment $ms(s_i)$ is a single point and

[6] The uniqueness is shown as follows. Suppose the minimum merging cost is c. Define a function $f(|e_a|)$ to be the path delay from v to sinks in TS_a for edge length $|e_a|$; similarly define $g(|e_b|)$ for the path delay from v to sinks in TS_b. Define $g'(|e_a|) = g(c - |e_a|)$. A length assignment to e_a must satisfy $f(|e_a|) = g'(|e_a|)$, or alternatively, $(f - g')(|e_a|) = 0$. If both f and g are monotone increasing functions, then g' is monotone decreasing and $f - g'$ is monotone increasing. Thus $(f - g')(|e_a|) = 0$ will have at most one solution.

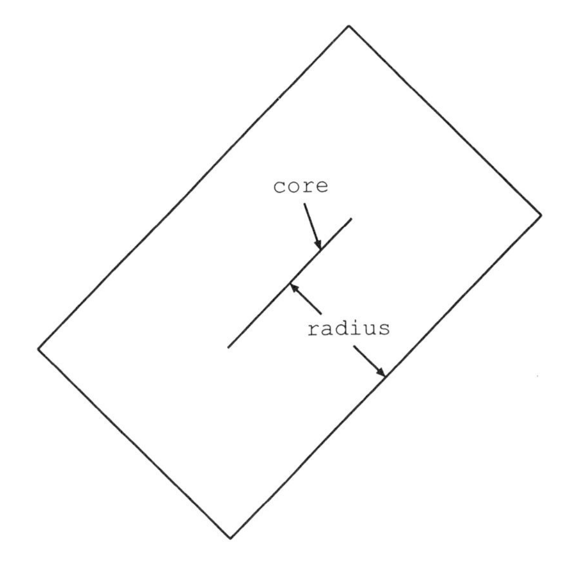

Fig. 1. An example of a TRR with core and radius as indicated.

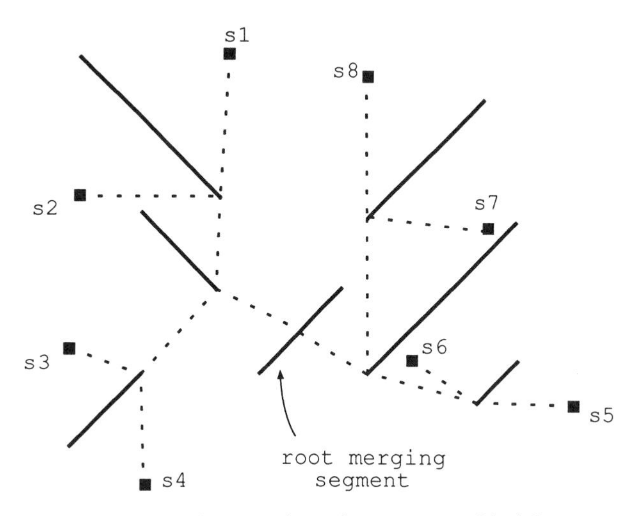

Fig. 4. An example of a tree of merging segments with sinks $s_1, \cdots, s_8$. The solid lines are merging segments and the dotted lines indicate edges between merging segments.

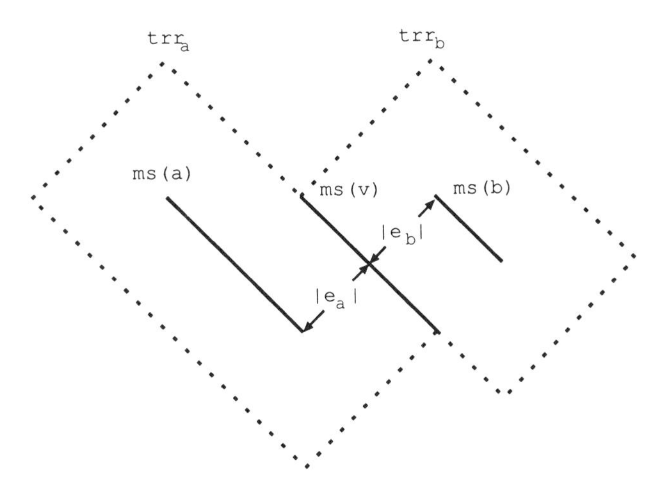

Fig. 2. Construction of merging segment $ms(v)$ when the merging cost equals κ.

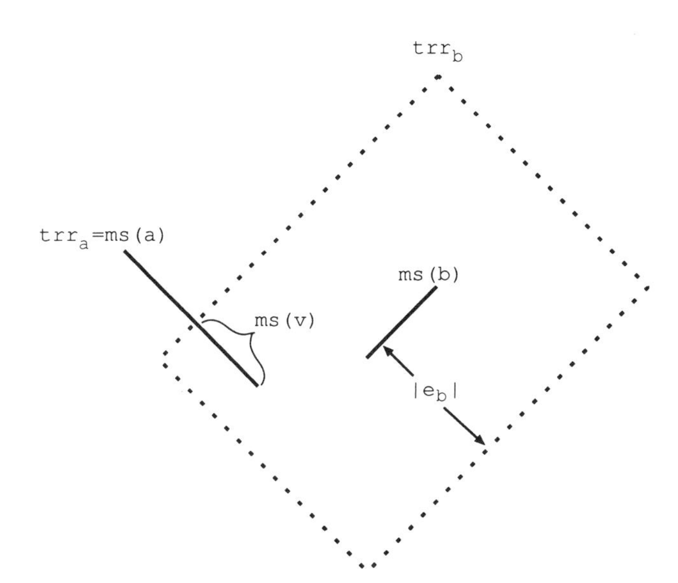

Fig. 3. Construction of merging segment $ms(v)$ when the merging cost is greater than κ. Note that in this example, radius$(trr_a) = |e_a| = 0$.

thus a Manhattan arc. By induction, therefore, all merging segments must be Manhattan arcs.

Fig. 5 gives a precise description of the procedure Build_Tree_of_Segments, which constructs the tree of merging segments. Details of the Calculate_Edge_Lengths subroutine depend on the delay model and are described in Sections 3.2.1 and 3.3.1 below.

By Lemma 1, procedure Build_Tree_of_Segments requires constant time to compute each new merging segment, and time linear in the size of S to construct the entire tree of merging segments.

3.1.2. Top-Down Phase: Embedding of Nodes: Once the tree of segments has been constructed, the exact embeddings of internal nodes in the ZST are chosen in a top-down manner. For node v in topology G, i) if v is the root node, then select any point in $ms(v)$ to be $pl(v)$;[7] or ii) if v is an internal node other than the root, choose $pl(v)$ to be any point in $ms(v)$ that is at distance $|e_v|$ or less from the placement of v's parent p (because the merging segment $ms(p)$ was constructed such that $d(ms(v), ms(p)) \leq |e_v|$, there must exist some choice of $pl(v)$ satisfying this condition). In case ii), the algorithm first creates a square TRR trr_p with radius $|e_v|$ and core equal to $\{pl(p)\}$; then, $pl(v)$ can be any point from $ms(v) \cap trr_p$ (see Fig. 6). For the tree of merging segments in Fig. 4, the resulting placements are indicated by the points at which the segments are connected by dotted lines. Fig. 7 describes the procedure Find_Exact_Placements, which uses the tree of merging segments to determine the final embedding of nodes in the ZST.

The time complexity of DME is analyzed as follows. Because each instruction in Find_Exact_Placements is executed at most once for each node in G (and the intersection of TRRs $ms(v)$ and trr_p can be found in constant time by Lemma 1), Find_Exact_Placements runs in time linear in the size of S. Because procedure

[7] If a fixed source location s_0' is specified, choose $pl(s_0) \in ms(s_0)$ with minimum distance from s_0' and connect a wire directly from s_0' to $pl(s_0)$.

```
Procedure Build_Tree_of_Segments
Input: Topology G; set of sink locations S
Output: Tree of merging segments TS containing
  ms(v) for each node v in G and edge length |e_v|
  for each v ≠ s_0
for each node v in G (bottom-up order)
  if v is a sink node,
    ms(v) ← {pl(v)}
  else
    Let a and b be the children of v
    Calculate_Edge_Lengths(|e_a|,|e_b|)
    Create TRRs trr_a and trr_b as follows:
      core(trr_a) ← ms(a)
      radius(trr_a) ← |e_a|
      core(trr_b) ← ms(b)
      radius(trr_b) ← |e_b|
    ms(v) ← trr_a ∩ trr_b
  endif
```

Fig. 5. Construction of the tree of segments.

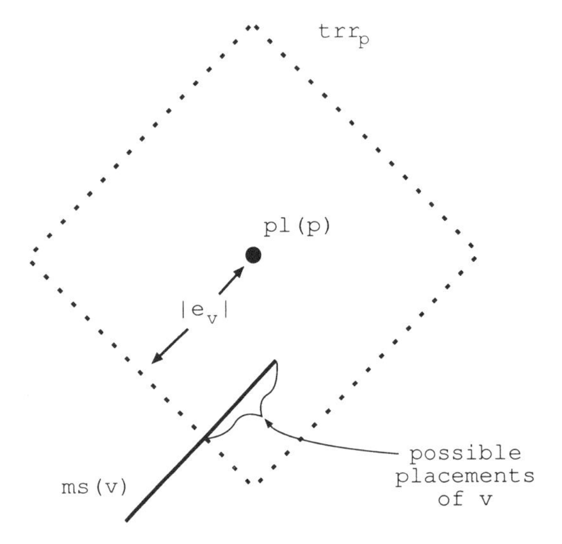

Fig. 6. Procedure Find_Exact_Placements: finding the placement of v given the placement of its parent p.

```
Procedure Find_Exact_Placements
Input: Tree of segments TS containing ms(v)
  and |e_v| for each node v in G
Output: ZST T(S)
for each internal node v in G (top-down order)
  if v is the root
    Choose any pl(v) ∈ ms(v)
  else
    Let p be the parent node of v
    Construct trr_p as follows:
      core(trr_p) ← {pl(p)}
      radius(trr_p) ← |e_v|
    Choose any pl(v) ∈ ms(v) ∩ trr_p
  endif
```

Fig. 7. Construction of the ZST by embedding internal nodes of the topology.

Build_Tree_of_Segments also runs in linear time, DME as a whole is a linear-time algorithm.

3.2. Application of DME to Linear Delay

3.2.1. Calculating Edge Lengths: Calculating the edge lengths $|e_a|$ and $|e_b|$ is straightforward in the linear delay model. Let a and b be children of v with merging segments $ms(a)$ and $ms(b)$, and let $t_{LD}(a)$ and $t_{LD}(b)$ be the delays from a and b to the sinks in their respective subtrees. Then, zero skew at v requires that

$$t_{LD}(a) + |e_a| = t_{LD}(b) + |e_b|.$$

Again, let $\kappa = d(ms(a), ms(b))$. If $|t_{LD}(a) - t_{LD}(b)| \le \kappa$, then the merging cost is minimized with $|e_a| + |e_b| = \kappa$, i.e.,

$$|e_a| = \frac{\kappa + t_{LD}(b) - t_{LD}(a)}{2}$$

and

$$|e_b| = \kappa - |e_a|.$$

On the other hand, if $|t_{LD}(a) - t_{LD}(b)| > \kappa$, then the merging cost is minimized when one of the edge lengths is equal to zero. It is easy to see that if $t_{LD}(a) > t_{LD}(b)$, then $|e_a| = 0$ and $|e_b| = t_{LD}(a) - t_{LD}(b)$; similarly, if $t_{LD}(a) < t_{LD}(b)$ then $|e_b| = 0$ and $|e_a| = t_{LD}(b) - t_{LD}(a)$.

3.2.2. Optimality of DME for Linear Delay: The following theorem states that the DME algorithm is optimal in the linear delay regime.

Theorem 1: Given a set of sink locations S and a connection topology G, the DME algorithm produces a ZST T with minimum cost over all ZST's for S having topology G.

The proof of Theorem 1 relies on Lemmas 2 and 3. Lemma 2 asserts that for any node v in an optimal ZST, $pl(v)$ is in $ms(v)$ and must therefore satisfy the constraints imposed in the bottom-up phase of the algorithm. Lemma 3 implies that the placements of two sibling nodes correspond to a closest pair of points in their respective merging segments. Together, Lemmas 2 and 3 can be used to show that placements in an optimal ZST must satisfy the top-down phase of the algorithm. Let $t_{LD}(T, x)$ denote the delay in ZST T between a point x in T and each sink which has x on its source-sink path.

Lemma 2: Given a ZST T with topology G, let v be an internal node with children a and b. Suppose the subtrees of T rooted at a and b can be generated by the DME algorithm for some placement of v on $ms(v)$, and also suppose that $q = pl(T, v) \notin ms(v)$. Then a new ZST T' with the same topology can be constructed from T by moving the placement of v so that the following hold: i) $q' = pl(T', v) \in ms(v)$; ii) $\text{cost}(T') < \text{cost}(T)$; and iii) $t_{LD}(T, q) = t_{LD}(T', q)$.

Lemma 2 is illustrated in Fig. 8. The construction of T' from T reduces the tree cost by modifying the q–a and q–b connections so that they share wire on the segment from q to q'.

Lemma 3: Suppose that a and b are two sibling nodes in ZST T with parent v, and suppose that the subtrees of T rooted at a and b can be generated using the DME algorithm. If $d(a, b) > d(ms(a), ms(b))$ and $d(a, b) > |t_{LD}(T, a) - t_{LD}(T, b)|$, then a new ZST T' can be constructed from the same topology, with $\text{cost}(T') < \text{cost}(T)$ and with $t_{LD}(T, q) = t_{LD}(T', q)$ for $q = pl(T, v)$.

Fig. 9 contains an illustration of Lemma 3. Moving the placements of nodes a and b to locations a' and b' allows the a'–q and b'–q connections to share wire on the

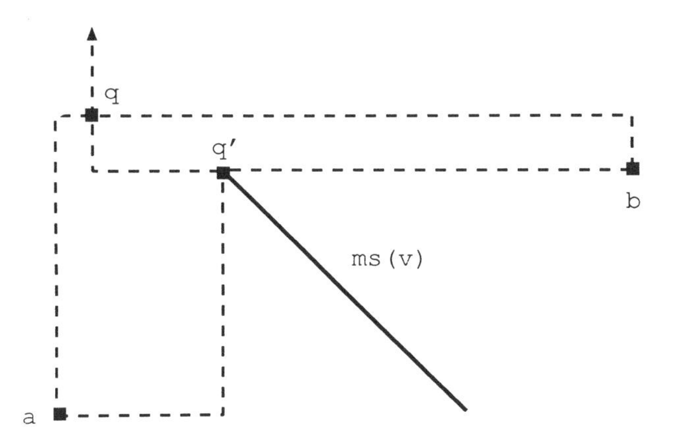

Fig. 8. Optimal placement of v must be on $ms(v)$. $pl(T, v) = q$; $pl(T', v) = q'$; and $\text{cost}(T') < \text{cost}(T)$.

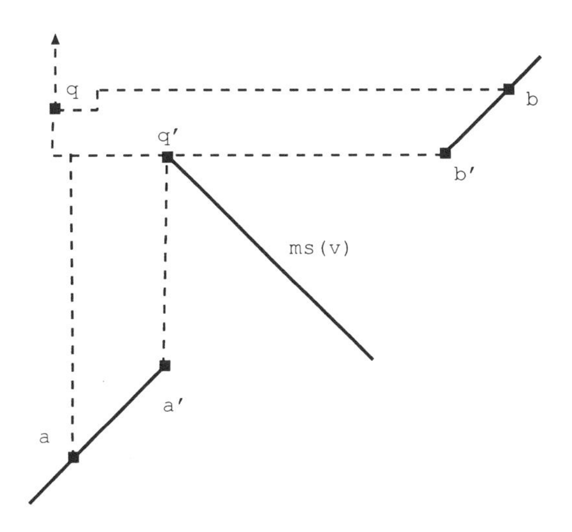

Fig. 9. Optimal placement of siblings a and b must satisfy the distance constraint in the top-down phase Find_Exact_Placements. $pl(T, a) = a$ and $pl(T', a) = a'$, etc.; and $\text{cost}(T') < \text{cost}(T)$.

segment from q' to q. The delay at point q remains unchanged.

Proof of Theorem 1: The proof is by contradiction. The DME algorithm places only two constraints on the placement of a node v in G: i) $pl(v) \in ms(v)$ and ii) $d(pl(v), pl(p)) \leq L_v$, where p is the parent of v and L_v is the edge length assigned by DME to e_v. Condition i) arises by the construction in the top-down phase of DME, and condition ii) is required by the bottom-up phase of DME. Suppose ZST T has minimum cost for point set S and topology G, but contains a node placement violating one of the two conditions. Let v be a node with greatest depth in T that violates either condition, and let w be the sibling of v. Because v has maximum depth, all of the descendants of v and w can be produced using DME. Consequently, because T has minimum cost, Lemma 2 implies that $pl(T, v)$ must be in $ms(v)$ and $pl(T, w)$ must be in $ms(w)$. Thus v does not violate condition i).

Consequently, v must violate condition ii), i.e., $d(pl(T, v), pl(T, p)) > L_v$. Let $L(T, e_v)$ denote the length of edge e_v in T. Because the length of an edge must be at least the distance between its endpoints, $L(T, e_v) > L_v$. Suppose $d(pl(T, v), pl(T, w)) \leq d(ms(v), ms(w))$. Then the subtrees of T rooted at v and w can be generated by DME for some placement of p on $ms(p)$, and by Lemma 2, $\text{cost}(T)$ can be improved by moving p to its merging segment and setting $L(T', e_v) = L_v$ and $L(T', e_w) = L_w$. If $d(pl(T, v), pl(T, w)) \leq |t_{LD}(v) - t_{LD}(w)|$, then $\text{cost}(T)$ can be reduced by moving $pl(p)$ to $pl(v)$ if $L_v = 0$, or to $pl(w)$ if $L_w = 0$. Thus, we have $d(pl(T, v), pl(T, w)) > d(ms(v), ms(w))$, and $d(pl(T, v), pl(T, w)) > |t_{LD}(v) - t_{LD}(w)|$. Then by Lemma 3 $\text{cost}(T)$ can be decreased, contradicting the assumption that t has minimum cost. □

It can be proved that in the linear model, DME also minimizes the source-sink delay in a ZST, and that this delay is equal to one-half the diameter of the sink set S. A proof of this result is contained in [3].

The DME algorithm is also optimal for any topology in the variant of the ZST problem where the source location is predefined. Suppose that $ms(s_0)$ is the merging segment for the root node s_0 of topology G and that s_0' is the prescribed source location. The DME algorithm can be modified at the beginning of the procedure Find_Exact_Placements to connect s_0' with the closest point in $ms(s_0)$. This point becomes $pl(s_0)$. Lemmas 2 and 3 can be used to prove the optimality of this method: they state that any tree rooted at a location $q \notin ms(s_0)$ will have minimum cost only if the two subtrees of G directly below the root are merged at a point $q' \in ms(s_0)$ which is then connected to s_0' by a single edge.

3.3. Application to Elmore Delay

3.3.1. Calculating Edge Lengths in the Elmore Delay Model: We use the analysis of Tsay [25] to calculate the edge lengths needed to merge two trees of merging segments TS_a and TS_b with minimum merging cost in the Elmore model. Let TS_a and TS_b, respectively, have capacitance C_1 and C_2 and delay $t_1 = t_{ED}(a)$ and $t_2 = t_{ED}(b)$; let $pl(v)$ be a merging point with minimum merging cost.

From the definition of Elmore delay, we have that $t_{ED}(v, a) = r_{e_a}((1/2)c_{e_a} + C_1)$. Thus, $pl(v)$ satisfies

$$r_{e_a}\left(\tfrac{1}{2}c_{e_a} + C_1\right) + t_1 = r_{e_b}\left(\tfrac{1}{2}c_{e_b} + C_2\right) + t_2. \quad (1)$$

Let $d(ms(a), ms(b)) = \kappa$. Suppose that TS_a and TS_b can be merged with merging cost κ; in other words, $|e_a| = x$ and $|e_b| = \kappa - x$ for $0 \leq x \leq \kappa$. Then we have resistances $r_{e_a} = \alpha x$ and $r_{e_b} = \alpha(\kappa - x)$ and capacitances $c_{e_a} = \beta x$ and $c_{e_b} = \beta(\kappa - x)$. Substituting into (1) and solving for x yields

$$x = \frac{t_2 - t_1 + \alpha\kappa(C_2 + \tfrac{1}{2}\beta\kappa)}{\alpha(C_1 + C_2 + \beta\kappa)}. \quad (2)$$

Case 1: If $0 \leq x \leq \kappa$, then there exists a feasible zero skew merging point of TS_a and TS_b with merging cost κ, $|e_a| = x$ and $|e_b| = \kappa - x$.

Case 2: If $x < 0$ or $x > \kappa$, then the assumption of merging cost κ results in a negative edge length for either e_a or e_b. In this case, an extended distance $\kappa' > \kappa$ is required to balance the delays of the two trees. If $x < 0$, which means $t_1 > t_2$, we choose $pl(a)$ as the merging point and set $|e_a| = 0$ and $|e_b| = \kappa'$. Then

$$t_1 = \alpha\kappa'(\tfrac{1}{2}\beta\kappa' + C_2) + t_2$$

and we use the quadratic formula to solve for κ':

$$\kappa' = \frac{\left((\alpha C_2)^2 + 2\alpha\beta(t_1 - t_2)\right)^{1/2} - \alpha C_2}{\alpha\beta}.$$

Similarly, if $x > \kappa$, we set $|e_b| = 0$ and

$$|e_a| = \kappa' = \frac{\left((\alpha C_1)^2 + 2\alpha\beta(t_2 - t_1)\right)^{1/2} - \alpha C_1}{\alpha\beta}.$$

The above analysis shows that a zero skew merging point between two ZST's can always be found. The merging cost depends on the distance between the roots of the ZST's, the delay of each ZST, and the tree capacitance of each ZST. Intuitively, to minimize the merging cost we should therefore choose topologies such that merged subtrees have minimum distance between their roots, along with similar capacitances and delays, so as to avoid the extra cost $\kappa' - \kappa$. This motivates our new BB algorithm, which uses the geometric notion of a *balanced bipartition* for computing a topology. Before describing this algorithm in Section 3.4 below, we observe that the DME algorithm is not optimal for all topologies in the Elmore delay model.

3.3.2. Suboptimality of DME for Elmore Delay: Recall that in the linear delay regime, the DME algorithm produces an *optimum* (minimum wirelength) ZST for *any* given topology. Our experimental results in Section IV clearly show the effectiveness of the DME algorithm in the Elmore delay model, and indeed we believe that in practice the algorithm gives solutions that are very close to optimum. However, the ZST's T in Fig. 10 and T' in Fig. 11 demonstrate that, for some sink sets and topologies, DME will not be optimal for Elmore delay. T and T' connect terminal points $s_1, \ldots, s_6$ to source s_0. Both trees are assumed to extend to the right side of s_0, with their subtrees to the right of s_0 being mirror images of the subtrees to the left of s_0 (this ensures that the source will be at s_0 in the optimal tree). In this example, we set both the unit resistance α and unit capacitance β to one, and the loading capacitance c_{L_s} of each sink node s to zero.[8]

The ZST T' in Fig. 11 is constructed so that if points s_1 and s_2 are merged at point p_1', then vertical wires from points s_3 through s_6 will merge along the horizontal wire from s_1 to s_0 with exactly zero skew. If, however, s_1 and s_2 are merged on their merging segment as shown in the tree T of Fig. 10, the delay at p_1' will increase, and jogs

[8]The example can easily be altered to have non-zero loading capacitances: shorten each edge adjacent to a terminal node by a small value $c > 0$, and then set the loading capacitance of each terminal node to c.

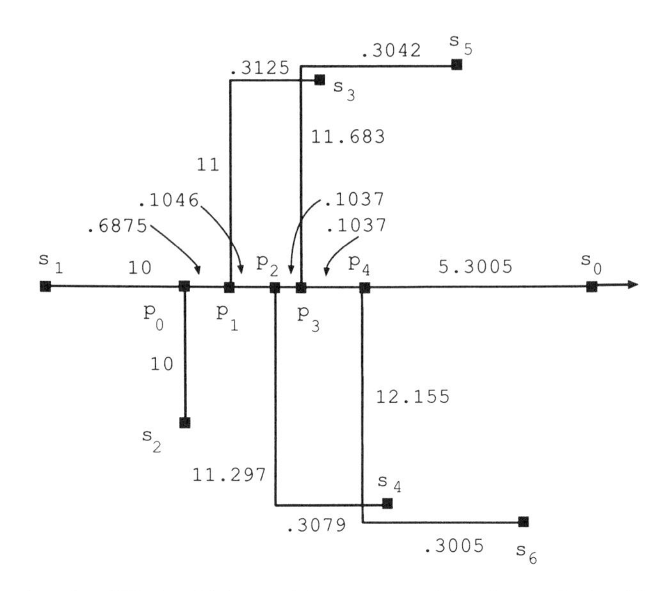

Fig. 10. ZST T, which would be constructed by the DME algorithm with suboptimal cost for its topology. (Note that the tree is not drawn to scale; lengths of horizontal and vertical segments are as indicated.)

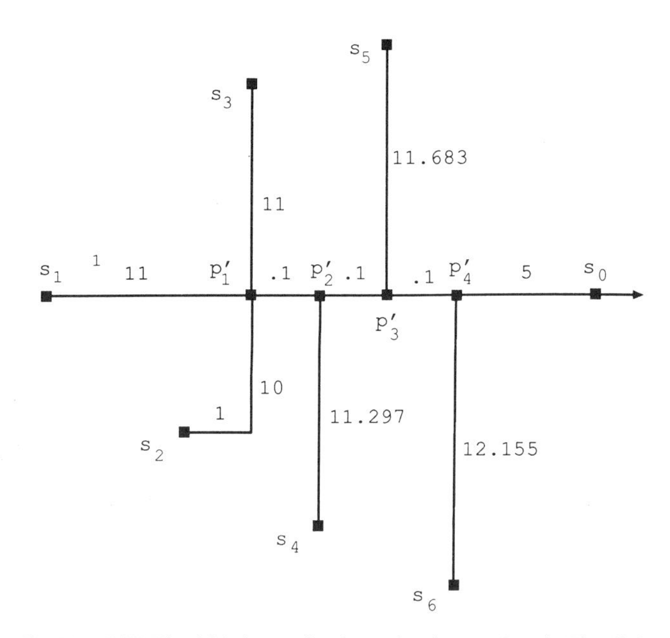

Fig. 11. ZST T', which has optimal cost for the topology in Fig. 10, but which violates the DME algorithm. In T', the internal nodes placed at p_0 and p_1 in T are placed at the same point, p_1'. (The tree is not drawn to scale; lengths of horizontal and vertical segments are as indicated.)

will be required in the edges e_{s_3} through e_{s_6}. In this example, the four required jogs are each of length greater than 0.3. Thus, their sum is greater than 1, which was the amount of wire saved initially by merging s_1 and s_2 at p_0.

Table I contains the calculated delay and capacitance at each of the internal nodes of T and T'. For example, in T' the capacitance at p_1', $C_{p_1'}$, is 33; and the delay at node p_2' is

$$t_{ED}(p_2') = t_{ED}(p_1') + 0.1 * \left(\frac{0.1}{2} + C_{p_1'}\right)$$

$$= 60.5 + 3.305 = 63.8.$$

TABLE I
DELAY AND CAPACITANCE AT EACH INTERNAL NODE IN ZST'S T AND T'

Tree T			Tree T'		
node	delay	capacitance	node	delay	capacitance
p_0	50	20			
p_1	64.0	32.0	p'_1	60.5	33.0
p_2	67.3	43.7	p'_2	63.8	44.4
p_3	71.9	55.8	p'_3	68.2	56.2
p_4	77.6	68.4	p'_4	73.9	68.4
s_0	454.0	2×73.66	s_0	428.6	2×73.44

Because unit resistance and capacitance both equal one, and because loading capacitances at the leaves are zero, the tree capacitance of a node is given by the amount of wire in its subtree. Thus, we see in Table I that cost(T') is less than cost(T) by 0.44.

3.4. Topology Generation

It is easy to see that, as hinted by the examples of Figs. 10 and 11, the choice of topology will affect the success of the DME embedding. We now present a new heuristic for generating connection topologies.[9] The heuristic works in top-down fashion, dividing the sink nodes recursively into two partitions with nearly equal total loading capacitance. We call this heuristic the BB method. The BB method offers a more powerful top-down partitioning scheme than the previous approaches of Jackson *et al.* [17] and Tsay [25], which divide the sink set recursively, using only alternating horizontal and vertical cuts.

For our description of the BB method, we introduce the following notation. Denote the diameter of S by dia(S) $= \max\{d(p, q) | p, q \in S\}$ and the number of sinks in S by $|S|$. Since the cost of any routing tree of S is greater than dia(S) and less than $|S| \cdot (\text{dia}(S))/2$, we consider dia$(S)$ to be a heuristic approximation of the cost of any ZST $T(S)$. Recall also that imbalanced loading capacitance may lead to excess edge length in the DME construction; we call a bipartition of a set of sinks S into two subsets S_1 and S_2 a *balanced bipartition* if the difference between the total loading capacitances of the two subsets is at most $\max\{c_{L_i}\}$.[10] Intuitively, we would like to find a balanced bipartition which divides set S with minimum *partition cost*, given by dia(S_1) + dia(S_2). This is the idea behind the BB heuristic. In the Euclidean metric, the problem of constructing a balanced bipartition which minimizes the sum of diameters can be solved in $O(n^2)$ time [19]. However, we are not aware of any polynomial-time algorithm that yields a minimum cost balanced bipartition in the Manhattan plane.

Let $p.x$ and $p.y$ be the x- and y-coordinates of point p. The *octagon* of set S is defined as the region formed by the intersection of eight half spaces (in clockwise order around the octagon): $y \le \max_{p \in S}\{p.y\}$, $y - x \ge \min_{p \in S}\{p.y - p.x\}$, $x \ge \min_{p \in S}\{p.x\}$, $y + x \ge \min_{p \in S}\{p.y + p.x\}$, $y \ge \min_{p \in S}\{p.y\}$, $y - x \le \max_{p \in S}\{p.y - p.x\}$, $x \le \max_{p \in S}\{p.x\}$, $y + x \le \max_{p \in S}\{p.y + p.x\}$. The *octagon set* of S, Oct(S), is the set of sink locations in S that lie on the boundary of S's octagon.

Fig. 12(a) shows the octagon for a set of 16 sink locations; the octagon set is $\{s_1, s_3, s_5, s_{10}, s_{14}, s_{16}\}$. The lines defining the octagon induce a natural circular ordering on the sinks in the octagon set. For example, $s_1 - s_5 - s_{10} - s_{16} - s_{14} - s_3 - s_1$ is the circular order of the octagon set of Fig. 12(a). Note that the octagon set construction naturally captures those parameters of the sink set which are relevant to diameter computations in the Manhattan plane. Based on extensive experimental investigations, we have found that each of the sets S_1 and S_2 in a balanced bipartition of S is likely to consist of consecutive elements in Oct(S). Based on this observation, a balanced bipartition heuristic is as follows.

1) Compute Oct(S) and sort Oct(S) in circular order.
2) Perform steps 3–5 for each set of $\lfloor 1/2 |\text{Oct}(S)| \rfloor$ consecutive sinks in Oct(S), called a *reference set* and denoted by REF$_i$, $i = 1, \cdots, |\text{Oct}(S)|$.
3) For each sink $p \in S$, compute the *weight* of p, equal to $\min_{r \in \text{REF}_i} d(p, r) + \max_{r \in \text{REF}_i} d(p, r)$.
4) Sort the sinks in ascending order of weight, then add sinks according to this order to S_1 until the difference between the sum of capacitances in S_1 and one half the total capacitance is minimized.
5) The remaining sinks are placed in S_2, and the partition cost dia(S_1) + dia(S_2) is obtained.
6) Over all reference sets REF$_i$, select the partition (S_1, S_2) with smallest partition cost.

In the example of Fig. 12(a), each set of three consecutive sinks in the octagon set will be a possible reference set: REF$_1$ = $\{s_5, s_{10}, s_{16}\}$ has partition cost 280 as shown in Fig. 12(b); REF$_2$ = $\{s_{10}, s_{16}, s_{14}\}$ has partition cost 270 as shown in Fig. 12(c); etc. After all six reference sets have been evaluated, we find that the optimal reference set is REF$_2$ with cost 270. Fig. 12(d) shows the output of the BB + DME algorithm on the instance of Fig. 12(a).[11]

The time complexity of the BB algorithm is affected by characteristics of the sink set S. The number of times that the loop over steps 3–5 must be repeated is given by

[9] No NP-completeness result has been obtained for our general minimum-cost zero skew clock tree formulation (i.e., where the topology has not been prescribed). However, [9] and [18] showed that a closely related problem (in the linear delay model), the "bounded-skew pathlength-balanced tree problem," is trivially NP-complete since it reduces the minimum rectilinear Steiner tree problem when the allowed pathlength skew is infinite. Thus, heuristics for computing promising topologies are of interest.

[10] For the linear delay model, we use uniform loading capacitances in the input to the BB algorithm, because delay depends only on the edge lengths.

[11] For the Elmore delay model, we observe that the DME algorithm is not always optimal for topologies generated by balanced bipartitioning. To see this, we modify the counter-example of Section 3.2 as follows. Let the loading capacitance of each sink be a small fixed value $\epsilon > 0$. Suppose that there are 16 sink nodes near point $s6$ within a very small radius $\delta > 0$ of each other. Similarly, suppose there are 8 sink nodes at point $s5$, 4 at $s4$, 2 at $s3$ and 1 at both $s1$ and $s2$. Then the BB algorithm will generate the topology of Fig. 10.

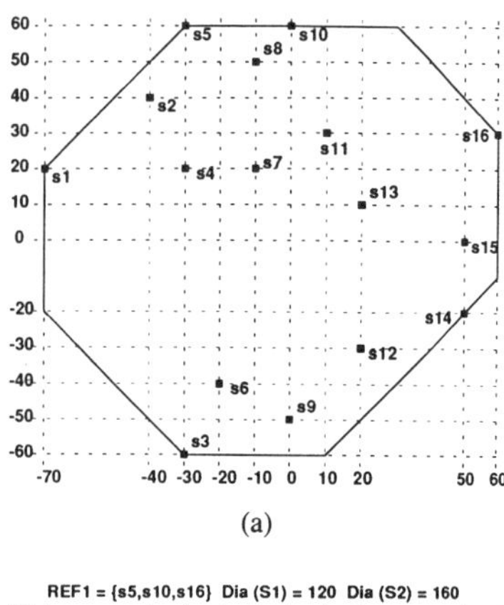

(a)

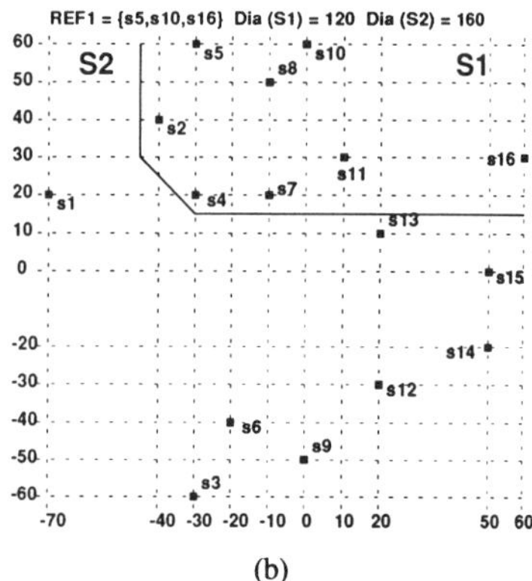

(b)

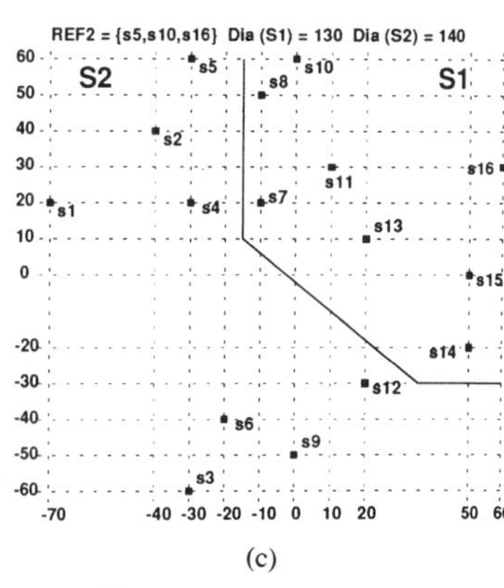

(c)

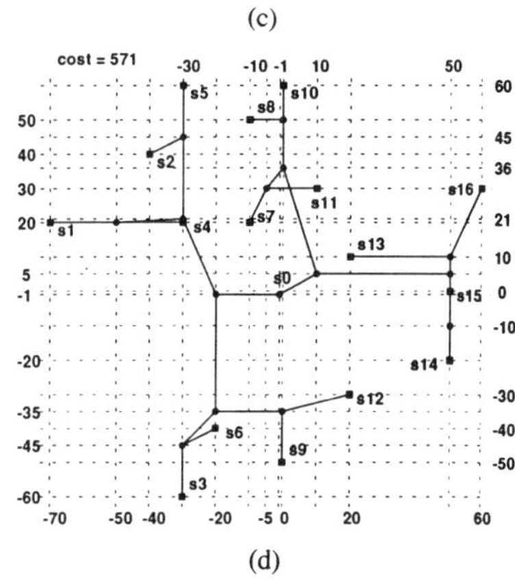

(d)

Fig. 12. (a) Octagon lines of 16 sink locations. (b) Partition result of REF_1. (c) Partition result of REF_2. (d) The ZST produced by BB + DME.

$|\mathrm{Oct}(S)|$, the number of reference sets. In the worst case this value is $\Theta(n)$, but in practice it is usually bounded by a constant. Because BB is recursive, its complexity is also affected by the relative sizes of the bipartitions. In the worst case, when loading capacitances are very unbalanced, we can have $|S_1| = 1$ and $|S_2| = |S| - 1$.

Steps 3 and 4 dominate all others in the complexity of BB and are repeated for each reference set. (The diameters in step 5 can be calculated in linear time in the Manhattan metric.) Step 4 requires $O(n \log n)$ operations each time it is run, while step 3 requires $O(n|\mathrm{Oct}(S)|)$ time. If $|\mathrm{Oct}(S)| = \Theta(n)$, then the total time used in step 3 for a single bipartition can be reduced from $O(n^3)$ to $O(n^2 \log n)$ by using a priority queue such as a Fibonacci heap.[12]

In the very worst case, we can have $|\mathrm{Oct}(S)| = \Theta(n)$ and pathologically unbalanced loading capacitances; each bipartition will require $O(n^2 \log n)$ time and the total time complexity of BB will be $O(n^3 \log n)$. If $|\mathrm{Oct}(S)| = O(1)$ but loading capacitances are still unbalanced, the time complexity will be $O(n^2 \log n)$. The time complexity is reduced when we impose very reasonable constraints on the loading capacitances, e.g., the largest and smallest capacitances can differ by at most a constant factor, or simply that the cardinalities of the partitions differ by at most a constant factor. If the loading capacitances are "balanced" and $|\mathrm{Oct}(S)| = \Theta(n)$, then the time complexity of BB is $O(n^2 \log n)$. Finally, under the most realistic circumstances, when the loading capacitances are balanced and $|\mathrm{Oct}(S)| = O(1)$, the time complexity of BB is $O(n \log^2 n)$.

IV. Experimental Results

The BB and DME algorithms were implemented on Sun SPARC workstations in the C/UNIX environment. The code can be obtained from the authors. We compared routing cost and source-sink delay of the BB + DME output with previous results of Jackson *et al.* [17], Kahng *et al.* [18], and Tsay [25], which were obtained for both the linear and Elmore delay models.

Because the DME algorithm can be applied to any prescribed topology, we also applied it to topologies obtained in previous studies. In this way, we can separate the effects of DME from the effects of complementary heuristics for generation of clock tree topologies. We used two sets of benchmarks: i) sink placements for the MCNC benchmarks Primary1 and Primary2 used in [17] and [18], and originally provided by the authors of [17] (Primary1 contains 269 sinks, and Primary2 contains 603 sinks); and ii) sink placements for the five benchmark sets r1–r5 used in [25] (the sizes of these examples range from 267 to 3,101 sinks).

4.1. Linear Delay Model

Our experimental results for linear delay are contained in Table II. We compared BB + DME with the Method of Means and Medians (MMM) of Jackson *et al.* [17] and with the bottom-up, matching based method of Kahng,

[12] The priority queue, however, will increase the worst-case space requirements from $O(n)$ to $O(n^2)$.

TABLE II
COMPARISON OF BB + DME WITH OTHER ALGORITHMS IN THE LINEAR DELAY MODEL USING MCNC BENCHMARKS PRIMARY1 AND PRIMARY2 AND BENCHMARKS R1 THROUGH R5 FROM TSAY

	number of sinks	MMM cost	KCR cost	KCR + DME cost	reduction by KCR + DME from KCR (%)	BB + DME cost	reduction by BB + DME from MMM (%)	reduction by BB + DME from KCR (%)
Primary1	269	161.7	153.9	140.3	8.8	140.5	13.1	8.7
Primary2	603	406.3	376.7	350.4	7.0	360.8	11.2	4.2
r1	267	1,815	1,627	1,497	8.0	1,500	17.4	7.8
r2	598	3,625	3,349	3,013	10.0	3,010	17.0	10.1
r3	862	4,643	4,360	3,902	10.5	3,908	15.8	10.4
r4	1,903	9,376	8,580	7,782	9.3	8,000	14.7	6.8
r5	3,101	13,805	12,928	11,665	9.8	11,757	14.8	9.1
		average			9.1		14.9	8.2

Cong, and Robins (KCR) [18]. In order to test the performance of the DME algorithm alone, we also ran DME on the topologies produced by the KCR algorithm. The combined BB + DME algorithm produced an average reduction in cost of 15% from the MMM results. We also obtained an 8% average cost reduction from the KCR algorithm. Note that in the linear model, DME also produces trees with optimal source-sink *delay* [3], and our experiments showed an average reduction of 19% from the KCR algorithm. The improvement in source-sink delay ranged from 9% for Primary1 to 23% for r3.

4.2. Elmore Delay Model

We tested the BB + DME algorithm for Elmore delay on the same benchmark sink sets. The results are contained in Table III. Again, these results indicate a significant improvement by BB + DME over previous algorithms. The average reduction in wirelength was 14% over MMM results, and 10% over the results of Tsay. It should be noted that DME alone resulted in an average improvement of only 2% over Tsay's algorithm, which can be attributed to the fact that Tsay's embedding algorithm allows deferral of the choice of placements for one level in the tree (the two endpoints of each merging segment are selected and carried to the next level, where the actual embedding is chosen to the be the point which allows the minimum connection cost).[13] Our results also indicate a very significant reduction in source-sink delay in the Elmore model: the combination of KCR + DME reduced delay over the trees of Tsay by an average of 22%.

To obtain a more complete picture of the BB + DME performance, we also tested the algorithm on sink sets with locations chosen randomly from a square grid, i.e., with coordinates $s_i.x$, $s_i.y \in [-2500, 2500]$. The size of the sink sets ranged from 8 to 64. In these experiments, we also compared our algorithm with minimum rectilinear Steiner trees (RST's) constructed by the heuristic in [7]; the BB + DME tree cost was only 64% above the heuristic RST cost. Finally, we used the circuit simulator SPICE2G.6 [20] to evaluate clock skew in the ZST's generated on the random sink sets. For both the MMM and BB + DME clock trees, SPICE decks were generated with the following specifications. The routing area was assumed to be 0.5 cm × 0.5 cm, and all the parameters were based on a 1.2-μm CMOS technology. An input clock frequency of 100 MHz and a superbuffer driven by the input clock source were assumed. The delays between the source and the sink nodes were measured at the output node of the inverter which drives the sink nodes. Table IV shows the average maximum delays, minimum delays and clock skews for the sinks sets of each size. The maximum delay of BB + DME was on average 3% less than that of MMM. The average skew of MMM was 9.2 ps while that of BB + DME was only 0.5 ps, a 93% reduction. Fig. 13 shows the output of the BB + DME algorithm on an instance containing 64 sinks. The total routing length is 50445 μm and the source-sink delay is 0.91 ns. By contrast, the MMM algorithm yielded a tree with cost 59256 μm and delay 0.94 ns for this case.

V. CONCLUSIONS AND DIRECTIONS FOR FUTURE WORK

Minimization of clock skew is critical to the design of high-performance VLSI systems. Recent research has yielded a number of heuristics which effectively eliminate skew according to either the Elmore or linear delay model. However, these previous methods concentrate on generation of the clock tree topology, and then embed the topology in the plane with little concern for the minimization of total wirelength.

Obviously, minimization of total wirelength will lead to reduction of wiring area, with the added effect of less blockage for subsequent routing phases of layout. We also note that clocking accounts for a large portion of system power requirements: wire minimization can significantly

[13]A surprising outcome of our experiments was the strong performance of topologies generated by the KCR algorithm. The combination of KCR and DME actually outperformed BB + DME by an average of 2.5% on the seven benchmarks. We expected balanced topologies to be superior in the Elmore delay model, where the amount of load on each line affects delay, but our experimental results indicate that a bottom-up approach originally designed for the linear delay model can perform as well or better. However, we note that KCR uses such techniques as H-flipping and uncrossing of matching edges; the latter has exponential worst-case time complexity. Moreover, the minimum-diameter bipartitioning approach of BB may be more useful when the distribution of sink locations is highly pathological.

TABLE III
COMPARISON OF BB + DME WITH OTHER ALGORITHMS IN THE ELMORE DELAY MODEL

	number of sinks	MMM cost	Tsay cost	Tsay + DME cost	KCR + DME cost	BB + DME cost	reduction by BB + DME from MMM (%)	reduction by BB + DME from Tsay (%)
Primary1	269	161.7	*	*	140.1	140.5	13.1	*
Primary2	603	406.3	*	*	345.2	360.8	11.1	*
r1	267	1,815	1,697	1,658	1,487	1,535	15.4	9.5
r2	598	3,625	3,432	3,368	3,020	3,065	15.4	10.7
r3	862	4,643	4,407	4,333	3,867	3,962	14.7	10.1
r4	1903	9,376	8,866	8,694	7,713	8,054	14.1	9.2
r5	3101	13,805	13,199	12,926	11,606	11,837	14.3	10.3
			average				14.0	10.0

* Results for Tsay's algorithm were obtained from Dr. Ren-Song Tsay and were not available for the Primary1 and Primary2 benchmarks.

TABLE IV
MEAN DELAY TIME AND CLOCK SKEW FOR RANDOM SINK SETS (TIME UNIT = PICOSECOND). THE RIGHT-MOST THREE COLUMNS DISPLAY RATIOS BETWEEN THE RESULTS OF BB + DME AND MMM

	MMM			BB + DME			BB + DME/MMM		
	delay		clock	delay		clock	delay		clock
#pts	max	min	skew	max	min	skew	max	min	skew
8	769.3	763.2	6.1	746.6	746.2	0.4	0.970	0.978	.07
16	801.8	797.0	4.8	783.2	782.5	0.7	0.977	0.982	.15
24	836.6	826.2	10.4	808.7	808.3	0.4	0.967	0.978	.04
32	863.5	855.6	7.9	837.3	836.5	0.8	0.970	0.978	.10
40	885.6	876.3	9.3	857.0	856.5	0.5	0.968	0.977	.05
48	908.9	896.4	12.5	876.8	876.3	0.5	0.965	0.978	.04
56	926.2	914.4	11.8	890.2	889.7	0.5	0.961	0.973	.04
64	940.6	930.1	10.5	910.7	910.2	0.5	0.968	0.979	.05
average							0.968	0.978	.07

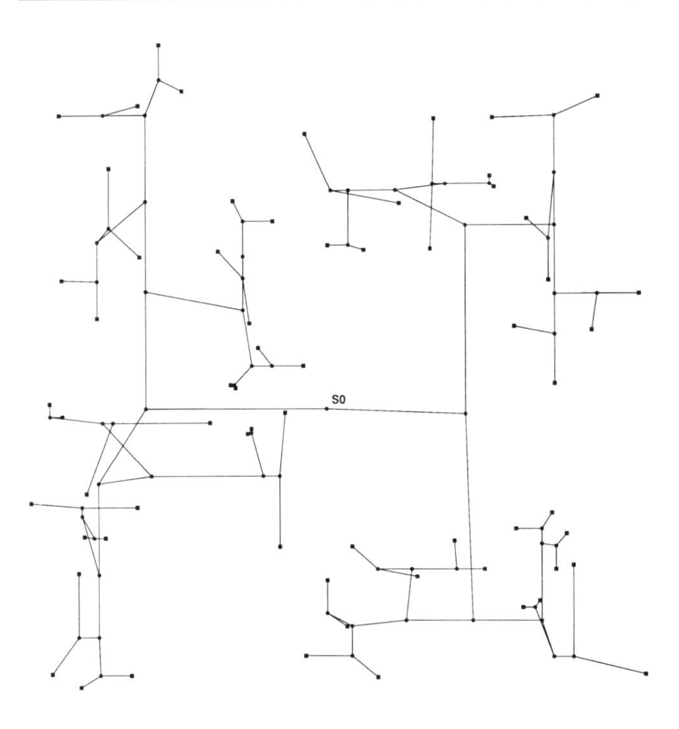

Fig. 13. An example of a ZST produced by BB + DME for 64 randomly chosen sink nodes.

reduce the power needed to drive the clock signal, thus improving system feasibility and reliability. Finally, wirelength reduction will improve performance by lessening such effects as pulse narrowing, pulse deformation, etc. Given these considerations, our work gives a unified approach to clock tree construction which combines the topology generating phase (BB) with the embedding phase (DME).

The BB heuristic generates a connection topology by recursively dividing the set of sinks into two subsets with similar total loading capacitance while at the same time minimizing the sum of diameters of the two subsets. This balance condition is a novel aspect of the method, and is useful when delay depends on both pathlength and capacitance, as in the Elmore model. The partitioning strategy based on minimizing the sum of diameters improves upon previous top-down bisection strategies of Jackson *et al.* [17] and Tsay [25], which use only horizontal or vertical cuts to partition the set of sinks.

The DME algorithm offers many improvements over previous embedding schemes. DME constructs a highly flexible tree of merging segments which allows a choice among minimum-cost zero skew clock trees. Given any connection topology over the set of sink locations, DME always produces a tree with exact zero skew, and may thus be applied to previously generated clock trees in order to improve both wirelength and delay. Experiments show that applying DME alone to the clock trees constructed by other algorithms results in wirelength reductions of 2% to 9%. The DME algorithm also extends to problem formulations where the clock source is prescribed. Finally, in the linear delay model, DME yields *optimal* total wirelength for a given connection topology, and optimal source-sink delay.

Our experimental results indicate that the BB + DME methodology yields routing solutions with exact zero skew (which we confirmed to be in the subpicosecond range using SPICE2G.6) and significantly reduced total wirelengths (8–15% less than the best previous methods). Furthermore, the superiority of BB + DME over previous methods depends on their joint application. For instance, our improvement of approximately 8% over the matching-based method of Kahng *et al.* (KCR) [18] is directly attributable to the DME embedding, since DME applied to topologies generated by KCR yields clock tree cost very similar to that obtained using BB + DME. On the other hand, DME alone can achieve only 2% out of the 15% improvement of BB + DME over Tsay [25]. Thus 13% of the cost savings can be attributed to the BB topology.

There are many promising extensions to our current approach. The DME algorithm readily applies to problems of *prescribed* skew (i.e., "useful" skew [1]), where the arrival times of the clocking signal must differ by prescribed amounts. This is handled by setting initial delays at the sinks to non-zero values. The DME algorithm can also be used for problems with *allowed skew* [1], [13], [25], where the signal must arrive at each sink within some prescribed segment of time.

Finally, the general issue of topology generation remains an important area for further investigation. A promising approach is to run DME concurrently with matching-based and other bottom-up topology generating heuristics. In general, the construction of optimal topologies appears to be very difficult (perhaps NP-hard). However, we expect further investigations in this area to have fruitful applications, for both clock tree construction and the broader area of high-performance routing.

VI. Remarks

Through independent research, the two groups of authors came up with essentially identical approaches to constructing zero skew clock routing trees with minimum wirelength for a given tree topology. The major differences between the two treatments are: i) Chao, Hsu, and Ho apply DME to the Elmore delay model, while Boese and Kahng establish the theoretical results for DME with respect to both the linear and Elmore delay models; and ii) Chao, Hsu, and Ho proposed the top-down balanced bipartition technique to generate an initial clock tree topology, while Boese and Kahng assume arbitrary existing tree topologies, e.g., those derived from the KCR method [9], [18]. The work of Chao, Hsu, and Ho [8] appeared at the 29th ACM/IEEE Design Automation Conference; the work of Boese and Kahng [3] appeared at the 5th IEEE International Conference on ASIC.

Appendix: Proofs of Lemmas 1, 2, and 3

Lemma 1: The intersection of two TRR's, R_1 and R_2, is also a TRR and can be found in constant time. If radius (R_1) + radius$(R_2) = d(\text{core}(R_1), \text{core}(R_2))$, then the TRR $R_1 \cap R_2$ is a Manhattan arc.

Proof: Rotate the plane by 45° so that the boundaries of R_1 and R_2 are vertical and horizontal line segments (see Fig. 14). Let R_1' and R_2' be the two TRR's after rotation with boundary lines given by:

- R_1': ($a_1 \le a_2$ and $b_1 \le b_2$)
$$x = a_1$$
$$x = a_2$$
$$y = b_1$$
$$y = b_2.$$
- R_2': ($a_3 \le a_4$ and $b_3 \le b_4$)
$$x = a_3$$
$$x = a_4$$
$$y = b_3$$
$$y = b_4.$$

Then $R_1' \cap R_2'$ is a rectangular region with boundary lines

$$x = \max(a_1, a_3)$$
$$x = \min(a_2, a_4)$$
$$y = \max(b_1, b_3)$$
$$y = \min(b_2, b_4).$$

Since rotating each TRR by 45° requires constant time, determining the intersection of the two TRR's $R_1 \cap R_2$ also requires only constant time.

If radius(R_1) + radius$(R_2) = d(\text{core}(R_1), \text{core}(R_2))$, then decreasing the radius of either R_1 or R_2 must cause their intersection to become empty; otherwise, we could form a path between core(R_1) and core(R_2) with length less than $d(\text{core}(R_1), \text{core}(R_2))$. Consequently, $R_1 \cap R_2$ must have zero width and be a line segment or a single point. Since $R_1 \cap R_2$ is also a TRR, it must be a Manhattan arc. □

Define a *straight-line path* between two points x and y to be any minimum-length path between them using only vertical and horizontal lines. If x and y are not on the same horizontal or vertical line, then there will be an infinite number of straight-line paths between them. Define the *projection area* $PA(x, Q)$ from a point x through a set of points Q as the set of all points p for which there exists a straight-line path from x to p that passes through Q. (Q must be between p and x.) Fig. 15 contains an example of the projection area from a point x through a Manhattan arc Q.

The next lemma about projection areas will be used to prove Lemma 2. It states that the union of two projection areas from points p and q, respectively, through a merging segment ms between them, is the entire plane.

Lemma 4: Let ms be a merging segment between the two points p and q. Then

$$PA(p, ms) \cup PA(q, ms) = \Re^2.$$

Proof: If the merging cost between p and q is greater than $d(p, q)$, then either $ms = \{p\}$ or $ms = \{q\}$. Since for

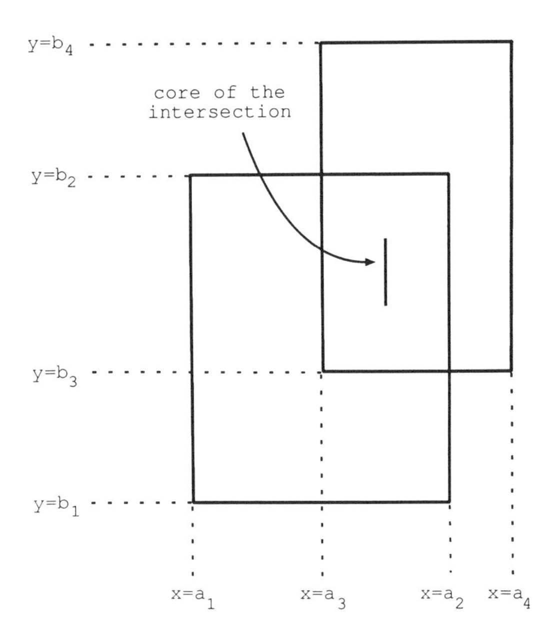

Fig. 14. Intersection of two TRR's after 45° rotation.

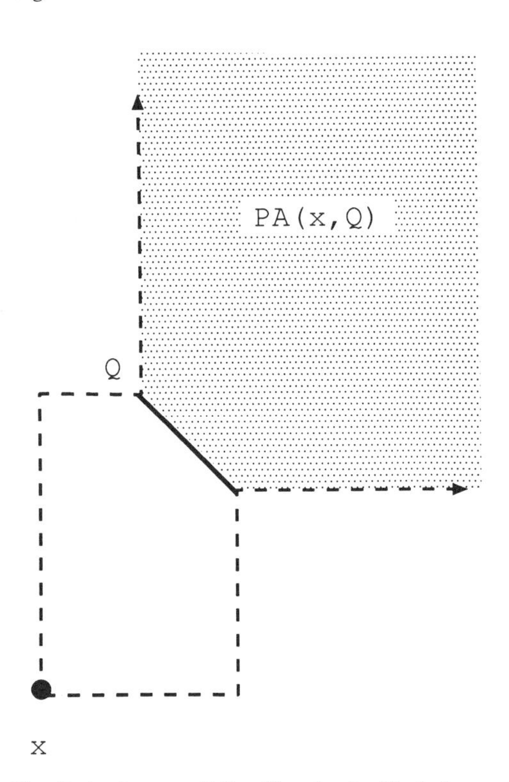

Fig. 15. Projection area $PA(x, Q)$ under the Manhattan metric.

any point x, $PA(x, \{x\}) = \Re^2$, this implies that either $PA(p, ms) = \Re^2$ or $PA(q, ms) = \Re^2$ and the proof is complete. For the case when the merging cost equals $d(p, q)$, merging segment ms is constructed as the intersection of two TRR's, trr_p and trr_q, such that core $(trr_p) = \{p\}$, core $(trr_q) = \{q\}$, and

$$\text{radius}(trr_p) = x * d(p, q)$$

$$\text{radius}(trr_q) = (1 - x) * d(p, q)$$

for some x satisfying $0 \leq x \leq 1$. If $x = 1$ or $x = 0$, the lemma is immediately true, since either $PA(p, \{p\}) = \Re^2$ or $PA(q, \{q\}) = \Re^2$ will hold. Let z_1 and z_2 be the two endpoints of merging segment ms. If $0 < x < 1$ then we need to consider the two cases depicted in Fig. 16:

a) z_1 and z_2 are both corners of the same TRR, either trr_p or trr_q. Assume without loss of generality that they are both corners of trr_p;

b) z_1 and z_2 are corners of different TRR's. Assume without loss of generality that z_1 is a corner of trr_q and z_2 is a corner of trr_p.

Define a *ray* $\overrightarrow{p_1 p_2}$ from point p_1 through point p_2 as the half-line with endpoint p_1 that extends through p_2. In case a), the straight-line path from p to z_1 is a vertical line segment and the straight-line path from p to z_2 is a horizontal segment. In Fig. 16(a) it is evident that $PA(p, \{z_1\})$ is a half plane with border line $z_1 p_1$ and $PA(p, \{z_2\})$ is a half plane bordered by line $z_2 p_2$. Furthermore, $PA(p, ms)$ is the infinite region separated from p by (and including) ray $\overrightarrow{z_1 p_1}$, segment ms, and ray $\overrightarrow{z_2 p_2}$. Similarly, $PA(q, ms)$ is the region separated from q by the same border. Consequently, $PA(p, ms) \cup PA(q, ms)$ is the entire plane.

In case b), shown in Fig. 16(b), $PA(p, ms)$ is the infinite region separated from p by (and including) $\overrightarrow{z_1 p_1}$, ms, and $\overrightarrow{z_2 p_2}$. $PA(q, ms)$ is the region separated from q by the same border. Again, $PA(p, ms) \cup PA(q, ms) = \Re^2$.

Lemma 2: Given a ZST T with topology G, let v be an internal node with children a and b. Suppose the subtrees of T rooted at a and b can be generated by the DME algorithm for some placement of v on $ms(v)$, and also suppose that $q = pl(T, v) \notin ms(v)$. Then a new ZST T' with the same topology can be constructed from T by moving the placement of v so that the following hold: i) $q' = pl(T', v) \in ms(v)$; ii) $\text{cost}(T') < \text{cost}(T)$; and iii) $t_{LD}(T, q) = t_{LD}(T', q)$.

Proof: Consider Fig. 8 of Section 3.2.2. Let a and b be the placements in T of v's children. By Lemma 4, there exists a point q' on $ms(v)$ such that there is a straight-line path either from a to q or from b to q, that passes through q'. Without loss of generality, assume that this path is from b to q. Because $bq'q$ is a straight-line path, segment bq in T can be replaced by segments bq' and $q'q$ in T' without changing the delay between b and q, and leaving the delay at point q unchanged. Moreover, the construction of $ms(v)$ ensures that zero skew is maintained by setting the edge e_a equal to the segment aq' and $pl(T', v) = q'$. Define length(T, xy) to be the edge length between points x and y in ZST T. Because the delay at q remains unchanged in T' and the a–q and b–q connections share wire between q' and q in T', we must have $\text{cost}(T') = \text{cost}(T) - \text{length}(T', q'q)$. □

Lemma 3: Suppose that a and b are two sibling nodes in ZST T with parent v, and suppose that the subtrees of T rooted at a and b can be generated using the DME algorithm. If $d(a, b) > d(ms(a), ms(b))$ and $d(a, b) >$

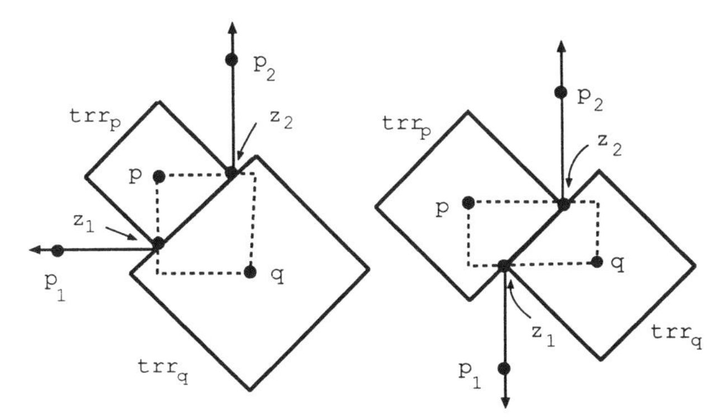

Fig. 16. Two cases to consider in the proof of Lemma 4.

$|t_{LD}(a) - t_{LD}(b)|$, then a new ZST T' can be constructed from the same topology, with $\text{cost}(T') < \text{cost}(T)$ and with $t_{LD}(T, q) = t_{LD}(T', q)$ for $q = pl(T, v)$.

Proof. (See Fig. 9 in Section 3.2.2): To prove the lemma, we will first construct a ZST T_{new} with source at $q = pl(T, v)$, and then replace the subtree of T rooted at v with part of T_{new} to create T'. Using [3, theorem 2] we show that the connections $a - q$ and $b - q$ share wire on a partial edge $c_{q'}$ in T', whereas they do not share wire in T. Because T' is also constructed so that the lengths of the $a - q$ and $b - q$ connections are the same as in T, tree T' will have lower cost than T.

Let G_v be the subtree of topology G rooted at v, and let S_v be the set of sinks in G_v. Suppose that sink s_i is the sink in S_v furthest from q. Create a new sink z that is located at a point directly opposite of q from s_i; i.e., $d(q, s_i) = d(q, z)$ and $d(s_i, z) = 2 * d(q, s_i)$. Consider a new set of sinks: $S_{new} = S_v \cup \{z\}$.

We create a topology G_{new} for S_{new} that merges G_v and z at its root, s_{new0}. We then run DME on S_{new} using topology G_{new} to create ZST T_{new}. By [3, theorem 2] T_{new} will have minimum feasible delay at each sink, equal to one-half the diameter of S_{new}, specifically $d(q, s_i)$. By the Fact used in the proof of [3, theorem 2], $ms(s_{new0})$ is the set of all points within distance $d(q, s_i)$ of every sink in S_{new}. Therefore, $q \in ms(s_{new0})$ and T_{new} can be constructed so that $q = pl(T_{new}, s_{new0})$. Let $a' = pl(T_{new}, a)$, $b' = pl(T_{new}, b)$, and $q' = pl(T_{new}, v)$. We now construct ZST T' for S by cutting off the subtree of T rooted at q and replacing it with T_{new} minus the edge between q and z. Since $t_{LD}(T', q) = d(q, s_i)$, it must be that $t_{LD}(T', q) \le t_{LD}(T, q)$. If the strict inequality holds, we add extra wire between q and q' to enforce equality, and thereby retain zero skew.

For convenience, we use $e_{a'}$ and $e_{b'}$ to represent the embeddings of edges e_a and e_b in T'. We also use e_q' to denote the partial edge between q' and q in T'. Because the subtrees of T rooted at a and b were constructed according to DME, we have $t_{LD}(T, a) = t_{LD}(T', a')$ and $t_{LD}(T, b) = t_{LD}(T, b')$. Thus, because $t_{LD}(T', q) = t_{LD}(T, q)$, it must be that

$$|e_a| = |e_{a'}| + |e_{q'}| \quad \text{and} \quad |e_b| = |e_{b'}| + |e_{q'}|. \quad (3)$$

Because $d(a, b) > d(ms(a), ms(b))$ and $d(a, b) > |t_{LD}(a) - t_{LD}(b)|$, $d(a, b)$ is strictly greater than the merging cost between $ms(a)$ and $ms(b)$. Therefore,

$$|e_a| > |e_{a'}| \quad \text{and} \quad |e_b| > |e_{b'}|. \quad (4)$$

Equations (3) and (4) imply that $|e_{q'}| > 0$, and thus

$$|e_a| + |e_b| > |e_{a'}| + |e_{b'}| + |e_{q'}|.$$

As a result, $cost(T') < cost(T)$. □

ACKNOWLEDGMENT

The authors are grateful to Dr. Ren-Song Tsay for providing benchmark data and for his communications which made this collaboration possible. We also thank Mr. Masato Edahiro for pointing out a mistake in our original of Lemma 3.

REFERENCES

[1] H. Bakoglu, J. T. Walker, and J. D. Meindl, "A symmetric clock-distribution tree and optimized high-speed interconnections for reduced clock skew in ULSI and WSI circuits," in *Proc. IEEE Int. Conf. Computer Design*, pp. 118–122, 1986.

[2] H. Bakoglu, *Circuits, Interconnections and Packaging for VLSI*. Reading, MA: Addison-Wesley, 1990.

[3] K. D. Boese and A. B. Kahng, "Zero-skew clock routing trees with minimum wirelength," in *Proc. IEEE Int. Conf. ASIC*, pp. 1.1.1–1.1.5, 1992.

[4] S. Boon, S. Butler, R. Byrne, B. Setering, M. Casalanda, and A. Scherf, "High performance clock distribution for CMOS ASICS," in *Proc. IEEE Custom Integrated Circuits Conf.*, pp. 15.4.1–15.4.4, 1989.

[5] J. Burkis, "Clock tree synthesis for high performance ASIC's," in *Proc. IEEE Int. Conf. ASIC*, pp. 9.8.1–9.8.4, 1991.

[6] P. K. Chan and K. Karplus, "Computing signal delay in general *RC* networks by tree/link partitioning," *IEEE Trans. Computer-Aided Design*, pp. 898–902, Aug. 1990.

[7] T.-H. Chao and Y.-C. Hsu, "Rectilinear Steiner tree construction by local and global refinement," in *Proc. IEEE Int. Conf. Computer-Aided Design*, pp. 432–435, 1990.

[8] T.-H. Chao, Y.-C. Hsu, and J.-M. Ho, "Zero skew clock net routing," in *Proc. ACM/IEEE Design Automation Conf.*, pp. 518–523, 1992.

[9] J. Cong, A. B. Kahng, and G. Robins, "Matching-based methods for high-performance clock routing," to appear in *IEEE Trans. Computer-Aided Design*.

[10] S. Dhar, M. A. Franklin, and D. F. Wann, "Reduction of clock delays in VLSI structures," in *Proc. IEEE Int. Conf. Computer Design*, pp. 778–783, 1984.

[11] M. Edahiro, "A clock net reassignment algorithm using Voronoi diagrams," in *Proc. IEEE Int. Conf. Computer-Aided Design*, pp. 420–423, 1990.

[12] W. C. Elmore, "The transient response of damped linear networks with particular regard to wide-band amplifiers," *J. Applied Physics*, vol. 19, pp. 55–63, Jan. 1948.

[13] J. P. Fishburn, "Clock skew optimization," *IEEE Trans. Computers*, vol. 39, pp. 945–951, July 1990.

[14] A. L. Fisher and H. T. Kung, "Synchronizing large systolic arrays," *Proc. SPIE*, pp. 44–52, May 1982.

[15] M. Garey and D. S. Johnson, "The rectilinear Steiner problem is NP-complete," *SIAM J. Applied Math.*, vol. 32, pp. 826–834, 1977.

[16] A. Hanafusa, Y. Yamashita, and M. Yasuda, "Three-dimensional routing for multilayer ceramic printed circuit boards," in *Proc. IEEE Int. Conf. Computer-Aided Design*, pp. 386–389, 1990.

[17] M. A. B. Jackson, A. Srinivasan, and E. S. Kuh, "Clock routing for high performance IC's," in *Proc. ACM / IEEE Design Automation Conf.*, pp. 573–579, 1990.

[18] A. B. Kahng, J. Cong, and G. Robins, "High-performance clock routing based on recursive geometric matching," in *Proc. ACM/IEEE Design Automation Conf.*, pp. 322–327, 1991.

[19] C. Monma and S. Suri, "Partitioning points and graphs to minimize the maximum or the sum of diameters," in *Proc. Sixth Int. Conf. Theory and Applications of Graphs*, Wiley, 1988.

[20] L. Nagel, "SPICE2: A computer program to simulate semiconductor circuits," *ERL Memo. UCB / ERL M75 / 520*, May 1975.

[21] P. Ramanathan and K. G. Shin, "A clock distribution scheme for non-symmetric VLSI circuits," in *Proc. IEEE Int. Conf. Computer-Aided Design*, pp. 398–401, 1989.

[22] J. Rubinstein, P. Penfield, and M. A. Horowitz, "Signal delay in *RC* tree networks," *IEEE Trans. Computer-Aided Design*, pp. 202–211, July 1983.

[23] T. Sakurai, "Approximation of wiring delay in MOSFET LSI," *IEEE J. Solid-State Circuits*, vol. 18, pp. 418–426, Aug. 1983.

[24] K. P. Shambrook, "An overview of multichip module technologies," in *Proc. IEEE Workshop Multichip Modules*, pp. 1–6, Mar. 1991.

[25] R. S. Tsay, "Exact zero skew," in *Proc. IEEE Int. Conf. Computer-Aided Design*, pp. 336–339, 1991.

[26] D. F. Wann and M. A. Franklin, "Asynchronous and clocked control structures for VLSI based interconnection networks," *IEEE Trans. Computers*, vol. 21, pp. 284–293, Mar. 1983.

An Exact Zero-Skew Clock Routing Algorithm

Ren-Song Tsay, *Member, IEEE*

***Abstract*—In this paper we present an exact zero-skew clock routing algorithm using the Elmore delay model. The results have been verified with accurate waveform simulation. We first review a linear time delay computation method. A recursive bottom-up algorithm is then proposed for interconnecting two zero-skewed subtrees to a new tree with zero skew. The algorithm can be applied to single-staged clock trees, multistaged clock trees, and multi-chip system clock trees. The approach is ideal for hierarchical methods of constructing large systems. All subsystems can be constructed in parallel and independently, then interconnected with exact zero skew. Extensions to the routing of optimum nonzero-skew clock trees (for cycle stealing) and multiphased clock trees are also discussed.**

I. Introduction

IN THIS PAPER we propose an exact zero-skew clock routing algorithm for optimizing the timing performance of synchronous digital systems. Clock skew is defined as the maximum difference of the delays from the clock source to the clock pins on latches. Optimization of the clock skew can dramatically reduce the system's cycle time, and, hence, the timing performance. In contrast, improper clock skew may sometimes cause clock hazard and system malfunction [6]. The following equation summarizes the relationship of the clock period P, clock skew s, worst-case data path delay $d_{\max}$, and other offset constant P_o for the condition of proper timing:

$$P = s + d_{\max} + P_o.$$

Note that P_o is a constant that includes data setup time, latch active time, and other possible offset factors such as safety margins, for example. The latch active time is the lag time for the data to be latched in after the latch is triggered by a clock signal.

It is clear from the equation that to reduce the cycle time P, it is necessary to minimize the skew s, besides the minimization of the worst-case data delay $d_{\max}$, on the combinational logics. As interconnection delay is becoming more dominating and design size is becoming larger, the clock skew is also becoming more significant in terms of performance optimization.

Many heuristics for clock routing have been proposed in the past. H-tree structures [1], [4], [10], [7] are the most widely used, especially in systolic array designs. A generalization of an H-tree that hierarchically connects the median points is proposed in [8]. A further improvement is done by bottom-up pairwise connections which construct a perfect length balanced tree [9]. However, all these heuristics focus only on wire length balancing, rather than the real objective of balancing clock delay. These approaches are not effective enough for tight skew optimization, as encountered in many high-performance designs nowadays. In contrast, what we propose is an exact algorithm that balances the clock delays directly. It is a general approach that takes into account uneven loading and buffering effects.

The outline of this paper is as follows. We first study how to compute signal delays efficiently on an *RC* tree. An *RC* tree is a connected acyclic undirected graph, with each branch associated with a resistance value and each node associated with a capacitance value.

Next, we discuss how a clock tree is modeled as an *RC* tree for delay analyses. In general, clock trees are classified into two types. The first type is *single-staged* clock trees in which clock pins are driven directly from a clock source. In order to reduce phase delays (the maximum delay from the clock source to a clock pin) and supply sufficient driving currents, usually several levels of buffers are added to create a multistaged clock tree. Thus the second type is called *multistaged* clock trees, in which the clock pins are driven from intermediate buffers, and the buffers are driven by either other buffers or the clock source. A multichip system clock tree is basically a multistaged clock tree, except that the clock pins are scattered on many chips (or cards).

The zero-skew algorithm is then presented. Based on a lumped delay model and the delay computation method, we found that any two zero-skewed subtrees can be merged into a tree with zero skew by tapping the connection to a specific location of each subtree. Basically, it is a recursive bottom-up algorithm.

Finally, we present experimental results of the zero-skew algorithm and comparisons with the wire length balancing heuristics [9]. We also discuss extensions to clock routing problems that require specific clock skew values (for cycle stealing), and to the problems of multiphase clock and optical skew.

II. Linear Time Hierarchical Delay Computation

We adopt the commonly used Elmore delay model [5], [13], [2] to calculate the signal traveling time from a clock source to each clock pin. We modify the method proposed in [13] and have a hierarchical method for computing de-

Reprinted from *IEEE Trans. Computer-Aided Desi.*, vol. 12, no. 2, pp. 242–249, Feb. 1993.

lays in a bottom-up fashion, which is the key to our zero-skew algorithm.

Rubinstein *et al.* [13] proposed a delay computation method using common path resistances of all node pairs and node capacitances. The time complexity is at least quadratic, due to the common path resistance calculation of all node pairs. Instead, we use branch resistance and total subtree capacitance for delay calculation. The new method is of linear time complexity.

To develop the algorithm, we first define a few terms. Let T represent an RC tree with every node associated with an index. We always assume the index of the root is 0. A *predecessor* of node i is a node residing on the unique path between the root and node i, but excluding node i itself. An *immediate predecessor* of node i is a predecessor of node i with no other nodes between them. Similarly, a *successor* of node i is the set of nodes which have node i as one of their predecessors. An *immediate successor* of node i is a successor of node i with no other nodes in between. The root is the node with no predecessor, and the leaf nodes are the nodes with no successors. A subtree T_i is defined as the subtree of T formed by the node i and its successors. Since T is a tree, there is only one unique edge between a node and its predecessor. So we simply define branch i as the edge between node i and its immediate predecessor.

Let c_i be the node capacitance of node i and r_i be the resistance of branch i. For convenience, if node i is the root, we set $r_i = 0$. Define $IS(i)$ as the set of all immediate successors of node i. Then the total subtree capacitance C_i of T_i is defined recursively as

$$C_i = c_i + \sum_{k \in IS(i)} C_k.$$

The above equation suggests that the subtree capacitance can be computed in a depth-first search manner. The capacitance of the subtree rooted from a node can be computed from its own node capacitance and the summation of the subtree capacitance of its immediate successors. Hence a recursive bottom-up algorithm can be used to compute the subtree capacitance of each node. The procedure of a recursive depth-first capacitance (DFC) calculation routine is outlined below. Note that a single call to *DFC*(0) will compute the subtree capacitance of each node in a bottom-up fashion.

procedure DFC(i)
1. $C_i = c_i$;
2. **for each** $j \in IS(i)$ **do**
3. DFC(j);
4. $C_i = C_i + C_j$;
5. **end for**

end procedure

The time complexity of the depth-first search procedure is linear in terms of the number of edges [3]. Since for a tree the number of edges is $|N| - 1$, hence the time complexity of computing the subtree capacitance of every node in $\theta(|N|)$, where $|N|$ is the number of nodes on T.

To calculate the delay, we first define N as the collection of all nodes on the tree T and $N(i, j)$ as the collection of nodes on the path between node i and node j, excluding i but including node j. The delay to a leaf node i can be calculated by the following formula:

$$t_{0i} = \sum_{n \in N(0,i)} r_n C_n$$

As a generalization, we can compute the "delay time" between any two nodes i and j; by the following formula, assuming i is a predecessor of j.

$$t_{ij} = \sum_{n \in N(i,j)} r_n C_n.$$

It can be shown easily that if i is an intermediate node between node k and node j, then

$$t_{kj} = t_{ki} + t_{ij}. \quad (1)$$

Suppose that node k is the root (i.e., $k = 0$); then we have

$$t_{0j} = t_{0i} + t_{ij}$$

since there is only one edge between node i and j and $t_{ij} = r_j C_j$. Hence,

$$t_{0j} = t_{0i} + r_j C_j.$$

This equation suggests that we can easily calculate the delay from the root to all leaf nodes in one depth-first search. The delay time to each node can be derived from its immediate predecessor, the branch resistance, and the subtree capacitance. Recursively, in a top-down fashion we compute the delay time to each node. The outline of the recursive depth-first delay (DFD) calculation routine is as shown below.

procedure DFD(i)
1. **for each** $j \in IS(i)$ **do**
2. $t_{0j} = t_{0i} + r_j C_j$;
3. DFD(j);
4. **end for**

end procedure

Since the algorithm again runs in a depth-first search manner, the complexity is $\theta(|N|)$. A complete hierarchical delay (HD) calculation algorithm is, simply, first an execution of DFC to obtain the capacitance information, and then an execution of DFD for delay calculation of every node. The outline of HD is as follows:

procedure HD
1. DFC(0);
2. $t_{00} = 0$;
3. DFD(0);

end procedure

Since both DFC and DFD run in linear time complexity, we easily have the following theorem.

Theorem 2.1 The delay time from the root to each node on an RC tree can be computed in $\theta(|N|)$.

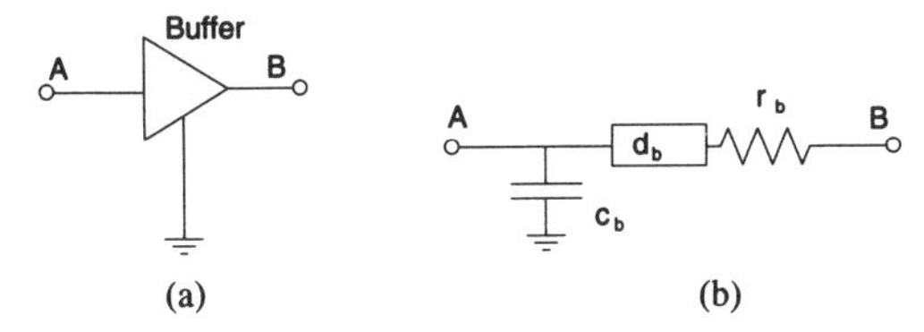

Fig. 1. (a) A clock buffer. (b) An equivalent model d_b: buffer internal delays C_b: buffer input capacitance; r_b: buffer output driving resistances; A: buffer input node; B: buffer output node.

A. Generalization to Buffered RC Trees

To handle multistaged clock trees (or buffered clock trees), we generalize the previous delay computation method for a *buffered RC tree*. Before we define what is a buffered *RC* tree, we first discuss a circuit model of the clock buffer as shown in Fig. 1(b). We specifically designate the input node of a buffer as a *buffer input node*, which is important for delay calculation. The box in Fig. 1 represents a delay element with d_b as the buffer internal delay and is connected to the buffer input node on one end and the buffer output driving resistor r_b on the other end. The buffer input capacitor c_b is on the buffer input node, and the buffer output driving resistor r_b is connected to the delay element and buffer output node. One function of buffers is to supply enough currents for driving latches. The other function of buffers is creating stages such that the subtree capacitance of the buffer output node will not be *carried over*, i.e., the equivalent total subtree capacitance as seen at the buffer input node is only c_b. Usually the buffer driving resistance and input capacitance are designed to be small values. This is why buffering usually reduces delay time.

To account for the buffering effects, we define a buffered *RC* tree to be just like a normal *RC* tree, except that each branch i is now also associated with a branch delay d_i besides the branch resistance r_i. The branch delay is always equal to zero, except in the case where it stands for a buffer. The basic delay calculation presented previously is modified as the following for buffered *RC* trees.

The calculation of the equivalent subtree capacitance at node i is now depending on whether node i is a buffer input node or not. Thus to compute the subtree capacitance of a buffered *RC* tree, we modify line 4 of DFC to

$$C_i = \begin{cases} C_i, & \text{if node } i \text{ is a buffer input node} \\ C_i + C_j, & \text{otherwise.} \end{cases}$$

We also extend the delay computation for a node i and its successor j as the following equation in order to accommodate the new branch delay situation, i.e.,

$$t_{ij} = \sum_{n \in N(i,j)} (r_n C_n + d_n).$$

Thus, line 2 of DFD is modified to be

$$t_{0j} = t_{0i} + r_j C_j + d_j$$

for delay calculation of a buffered *RC* tree, and the HD calculation routine will run exactly the same way and remain linear time complexity.

III. Delay Computation of Clock Trees

We shall discuss in this section how to model a clock tree as a buffered *RC* tree so that we can perform delay computation efficiently. Each clock tree realization consists of wiring segments, clock pins, and clock buffers. Hence, we shall first study the *RC* model of each component.

A. Equivalent π-Model for a Distributed RC Line

Distributed *RC* lines are more accurate for characterizing the circuit performance of wiring segments. A distributed *RC* line is usually represented as the symbol shown in Fig. 2(a). Either a π-model (Fig. 2(b)) or a T-model (Fig. 2(c)) is used to represent the equivalent circuit of the distributed *RC* line under the Elmore delay model.

Throughout this paper, we will use the equivalent π-model for the analysis. The equivalent π-model of a wire segment is represented by an input node, an output node, and a branch between both nodes. Let R be the total wire resistance and C the total wire capacitance. Then the equivalent input and output node capacitances are all equal to $C/2$, and the equivalent branch resistance is R.

B. Equivalent Buffered RC Tree of a Clock Tree

We use a generic example, as shown in Fig. 3(a), to illustrate how to construct an equivalent buffered RC tree from a multistaged clock tree. For this particular example, we assume a clock source is driving a buffer through wire 1, and the buffer is connected to the clock pin on a latch through wire 2. The driving resistance of the clock source is assumed to be r_s. Both wire segments 1 and 2 are represented by equivalent π-models as discussed earlier. The buffer is transformed to an equivalent circuit with buffer input capacitance c_b, buffer delay d_b, and buffer output driving resistance r_b. The end clock pin of the latch is associated with a loading capacitance c_l. The equivalent buffered *RC* tree is as shown in Fig. 3(b).

C. Lumped Delay Model

We shall introduce a *lumped delay model* based on the fact that the delay can be computed segment by segment and that the total subtree capacitance is sufficient for calculation. This model will help to ease the presentation of the zero-skew algorithm.

Recall (1):

$$t_{kj} = t_{ki} + t_{ij}.$$

Suppose i is an immediate successor of k, and j is a leaf node. Then

$$t_{kj} = d_i + r_i C_i + t_{ij}. \tag{2}$$

Consider node i as the root of the subtree T_i. To compute the delay time one level up from node k to node j, we need to know only the branch resistance r_i, the branch delay, the subtree capacitance C_i and the delay time from the root of T_i to the leaf node j, according to (1).

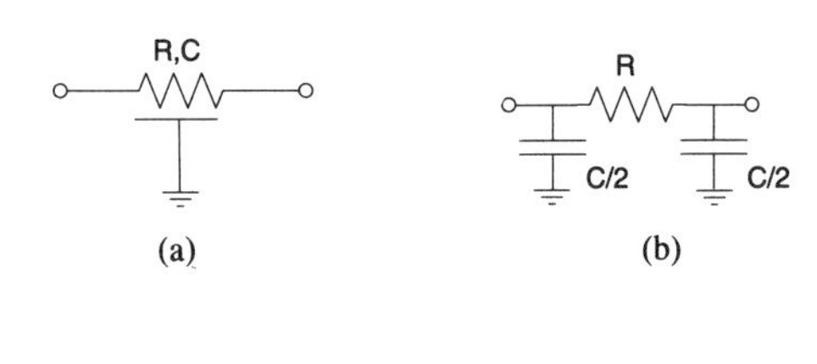

R/2 R/2 C

(c)

Fig. 2. (a) A distributed RC line. (b) The equivalent π-model. (c) The equivalent T-mode.

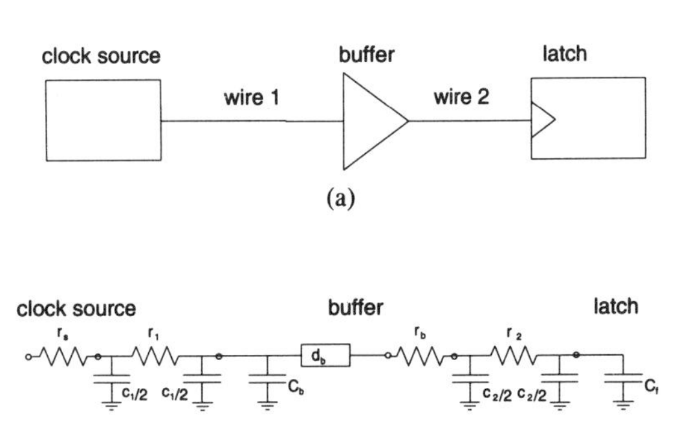

(b)

Fig. 3. (a) A generic multi-staged clock tree. (b) The equivalent buffered RC tree.

Thus we propose an equivalent lumped delay model of the subtree T_i (see Fig. 4) for simplifying the delay computation. In the equivalent circuit, the subtree T_i is replaced by an input capacitance C_i and a branch delay t_{ij} from input node i to leaf node j. We will use this lumped delay model for developing the algorithm in the next section.

IV. Zero Skew Algorithm

The zero-skew algorithm is a recursive bottom–up process. We describe only one recursive step. Repeating the process in a bottom–up fashion will construct a complete zero-skew clock tree.

We assume every subtree has achieved zero skew, which means the signal delay from the root of the subtree to its leaf nodes are equal. This is obvious if the subtree contains only one leaf node. Hence, leaf nodes are the starting subtrees of the algorithm.

To interconnect two zero-skewed subtrees with a wire and ensure zero skew of the merged tree, the problem to be solved is the decision of where on the wire the new root of the merged tree will be, such that the delay time from this new root to all leaf nodes are equal, i.e., zero skew. We will call this new *root* point on the wire a *tapping* point, and this process the *zero-skew-merge* process.

Let us discuss the example shown in Fig. 5 with two subtrees 1 and 2. First, assume the lumped delay model of each subtree is as shown in Fig. 5. The tapping point separates the interconnection wire of the two subtrees into two halves (which may not be equal). Each half wire segment is represented as a π-model as shown. To ensure the delay from the tapping point to leaf nodes of both subtrees being equal, it requires that

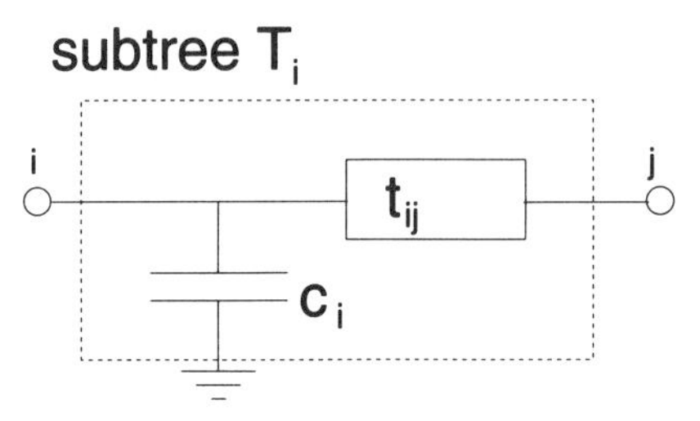

Fig. 4. An equivalent lumped delay model of a clock subtree.

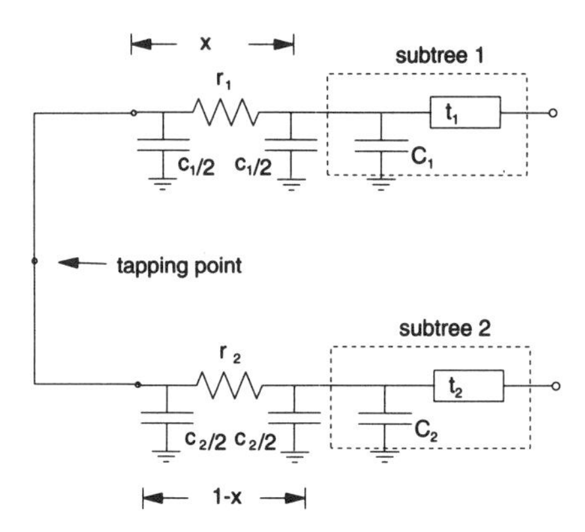

Fig. 5. *Zero-Skew Merge* of two subtrees.

$$r_1(c_1/2 + C_1) + t_1 = r_2(c_2/2 + C_2) + t_2) \quad (3)$$

according to (2). Note that r_1 and c_1 are the total wire resistance and capacitance of the wire segment 1. Similarly, r_2 and c_2 are for wire segment 2. There are no branch delays.

We assume that the total wire length of this interconnection wire segment is l. The wire length from the tapping point to the root of subtree 1 is $x \times l$. Hence, the wire length from the tapping point to the root of subtree 2 will be $(1 - x) \times l$.

Let α be the resistance per unit length of wire and β be the capacitance per unit length of wire. Then we have $r = \alpha l$, $r_1 = \alpha x l$, $r_2 = \alpha(1 - x)l$. Also, $c = \beta l$, $c_1 = \beta x l$, $c_2 = \beta(1 - x)l$.

Hence, after solving (3), we find that the zero-skew condition requires

$$x = \frac{(t_2 - t_1) + \alpha l\left(C_2 + \dfrac{\beta l}{2}\right)}{\alpha l(\beta l + C_1 + C_2)}.$$

If $0 \leq x \leq 1$, the tapping point is somewhere along the segment interconnecting the two subtrees and is legal. In case that $x < 0$ or $x > 1$, it indicates the two subtrees are too much out of balance. The interconnection wire has to be elongated. For simplicity, we discuss only the case that $x < 0$. For this case, the tapping point has to be exactly on the root of subtree 1 in order to minimize total interconnection length. Assume the elongated wire length is l'. The distributed resistance value is $\alpha l'$ and the distributed capacitance value is $\beta l'$. To determine a minimum elongated wire length l', it requires

$$t_1 = t_2 + \alpha l' \left(C_2 + \frac{\beta l'}{2} \right)$$

or

$$l' = \frac{[\sqrt{(\alpha C_2)^2 + 2\alpha\beta (t_1 - t_2)}] - \alpha C_2}{\alpha\beta}.$$

Similarly, for the case $x > 1$, the new root should be the root of subtree 2, and

$$l' = \frac{[\sqrt{(\alpha C_1)^2 + 2\alpha\beta (t_2 - t_1)}] - \alpha C_1}{\alpha\beta}.$$

It is worthwhile noting that the uneven loading effect is naturally taken care of by this approach.

A common practice for wire elongation is done by "snaking," as shown in Fig. 6. Since it is the nature of a clock wiring algorithm to *balance* the two subtrees, the *snaking* should not occur often. Real examples have shown less than 1.2% of wire length increase due to the elongation.

In case that the two subtrees are too much out of balance and the elongation severely affects the wirability, then addition of buffers, delay lines, or capacitive terminators should be considered, based on the same balancing principle. For instance, a capacitive terminator can be attached on the root of subtree 2 for case $x < 0$, instead of making a longer wire. The capacitance value, say C_t, can be determined by solving the equation $t_1 = t_2 + \alpha l (C_2 + \beta l/2 + C_t)$, or we will have $C_t = (t_1 - t_2)/(\alpha l) - (C_2 + \beta l/2)$.

Before presenting the algorithm formally, we define a few more related terms. The number of stages of a clock tree is defined as the maximum number of clock buffers on a path from the clock source to a clock pin, with the clock source counted as a buffer. A *cluster* is the collection of a clock buffer and its associated clock pins. Each cluster is tagged with a *stage number*, which is exactly the number of buffers on the path between the clock source and the clock buffer of the cluster. The number includes the clock source and the clock buffer of the cluster. In conclusion, we have the following efficient zero-skew clock routing algorithm.

Algorithm 4.1 (Zero-Skew Algorithm)

S1: Let s = number of clock tree stages.

S2: If s = 0, report results and exit: continue, otherwise.

S3: For each cluster in stage s, do

- *S3.1: Treat each clock pin in the cluster as a tapping point. Repeat steps S3.2 and S3.3 until there is only one tapping point left.*
- *S3.2: Pair up tapping points.*
- *S3.3: For each pair, perform zero-skew-merge of the two subtrees and determine the new tapping point, using the algorithm discussed in this section. If only one point in the group, then do nothing.*
- *S3.4: Connect the last tapping point directly to the clock driver output node.*

S4: Let s = s − 1. Continue from S2.

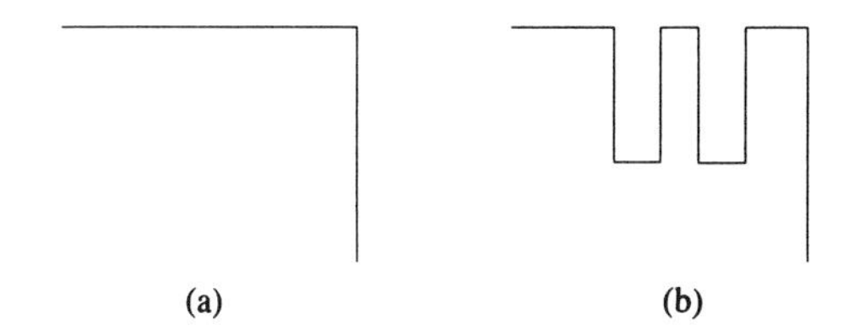

Fig. 6. (a) A regular wire. (b) Elongation of the wire by *snaking*.

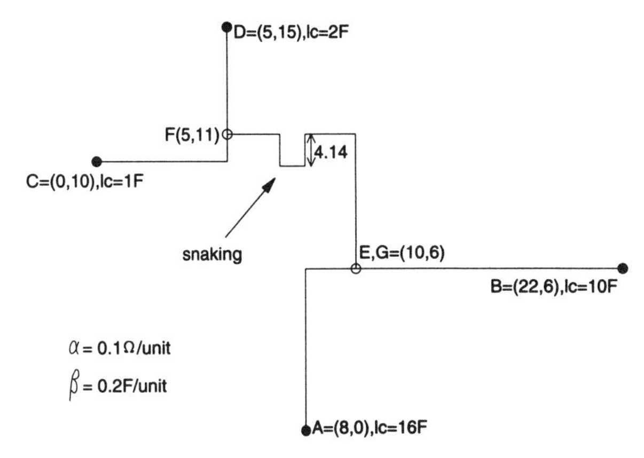

Fig. 7. A zero-skew wiring result of a simple example.

The zero-skew algorithm does not depend on the algorithm used for grouping the clock pins or tapping points into pairs. For any pairing algorithm, the zero-skew algorithm will work well. However, to optimize wirability, a minimum weighted matching algorithm may be better, or a more efficient algorithm that alternately partitions the clock pins into two equal numbered groups can be used.

When implementation occurs in real environments, we have to consider blockages and the different electric constants on different layers. The connection between any two tapping points can be done by any existing wiring algorithm that handles wiring blockages. The tapping point is then found by searching through each wiring segment of different electric constants.

To minimize the total wire length, we may construct a few possible wiring patterns (e.g., two one-bend connections) between each pair of tapping points, and pick up the one which gives shorter length at the next higher level pairing process.

Example: An example with four clock pins (Fig. 7) is used to illustrate the algorithm. Pin A is at (8, 0) with 16-F loading capacitance. Pin B is at (22, 6) with 10-F capacitance. Pin C is at (0, 10) with 1 F. Pin D is at (5, 15) with 2 F. The per unit resistance is 0.1 Ω, and the per-unit capacitance is 0.2 F. Pins A and B are in one pair and C, D in the other pair. According to the algorithm, a tapping point E is decided to be on (10, 6) so that the

TABLE I
STATISTICS OF THE TESTING EXAMPLES

Example	$R1$	$R2$	$R3$	$R4$	$R5$
No. Pins	267	598	862	1903	3101
Chip width	69 984	94 016	97 000	126 970	142 920
Chip height	70 000	93 134	98 500	126 988	145 224

TABLE II
COMPARISON BETWEEN THE ZERO-SKEW ALGORITHM AND A WIRE LENGTH BALANCING HEURISTIC

Algorithm	Zero Skew			Length Balancing	
Examples	Phase Delay (ns)	Skew (ns)	Run time(s)	Phase Delay (ns)	Skew (ps)
$R1$	1.799	0	0.1	1.798	132
$R2$	4.631	0	0.3	5.367	806
$R3$	7.055	0	0.5	7.655	702
$R4$	20.666	0	1.2	23.316	3558
$R5$	35.918	0	2.0	38.958	1931

delays to both A and B are all equal to 13.44 ns. Similarly, a tapping point F is located at (5, 11) for connection to pins C and D, with equal delay 0.96 ns. The two subtrees rooted by E and F are *very* unbalanced. We find that $x = -0.175 < 0$. Hence, the wire connecting E and F has to be elongated by 8.28 units, and the tapping point G has to coincide with E. The final wiring result is shown in Fig. 7. Note that the connections between (A, B) and (C, D) are chosen from the two one-bend connection of each pair for shorter wire length between (E, F).

V. EXPERIMENTAL RESULTS

We test our algorithm on five different sized examples. The statistics of the examples are shown in Table I. The chip width and height units are both in $1/10$ μm. We assume the per-unit resistance is 3 mΩ, and the per unit capacitance is 0.02 fF. The loading capacitances of clock pins ranged from 30 to 80 fF. For simplicity, we assume all are one-stage clock trees, i.e., no intermediate clock buffers. All experiments are conducted on an IBM 3090 machine.

We use a simple heuristic for pairing up clock pins in this experiment. We recursively partition the pins into two equal (or almost equal) halves by the median of the sorted pin list in alternate horizontal and vertical directions. This heuristic creates a binary tree for each example. Then the pins are connected, based on the zero-skew algorithm. For comparison, we also implement the wire length balancing heuristic [9] on the same binary tree. The results are shown in Table II. It is obvious that our algorithm really constructs zero-skewed clock networks according to the Elmore delay calculation. As demonstrated from the results, our algorithm shows extreme potential for cycle time improvement, especially for large chips which are becoming popular in recent products. Because of the balanced delay, the zero-skew algorithm also performs better than the wire length balancing algorithm in terms of smaller phase delay for most cases. A final clock routing result of the example $r3$ is shown in Fig. 8.

To be more convincing, we also verify the results with RICE 3.2 an accurate waveform simulation program [12]. In three test runs, we try one step input (zero rise-time ramp), and two ramp inputs with rise time 0.5 and 1.0 ns, respectively. We use 6 poles and 6 zeros for waveform simulation. The delay time at each clock pin is calculated as the difference of the time that the output is 50% of the final output value and the time that the input is 50% of the final input value. Then the skew is calculated based on the delay to each clock pin. The results are summarized in Table III. We find that the actual skew is much smaller than expected. The skew is less than 4.6 ps, even in the worst case of this experiment (see Table III).

To get better feeling on the impact of wire elongation, we also calculate the percentage of extra wire length introduced, due to the elongation, to the total wire length. As summarized in Table IV, even the worst case gives less than 1.2% penalty, which shows little effect on wirability.

VI. EXTENSIONS

In this section we outline possible extensions of the basic zero skew algorithm.

Optimum nonzero clock skew: Fishburn in [6] proposed a Linear Programming approach for an optimum clock skew assignment such that the clock period is minimized without clock hazards. This nonzero-skew assignment is mainly for the *cycle stealing* technique. The rise time of the clock signal can be adjusted to a certain time point, so that the critical paths whose delays are longer than one cycle time can be accommodated without timing violation. The result of this optimization is that the difference of the signal delays from the clock source to any two clock pins has to be a particular value. To realize the clock routing of such a nonzero-skew result requires a special routing technique to ensure an exact skew of each

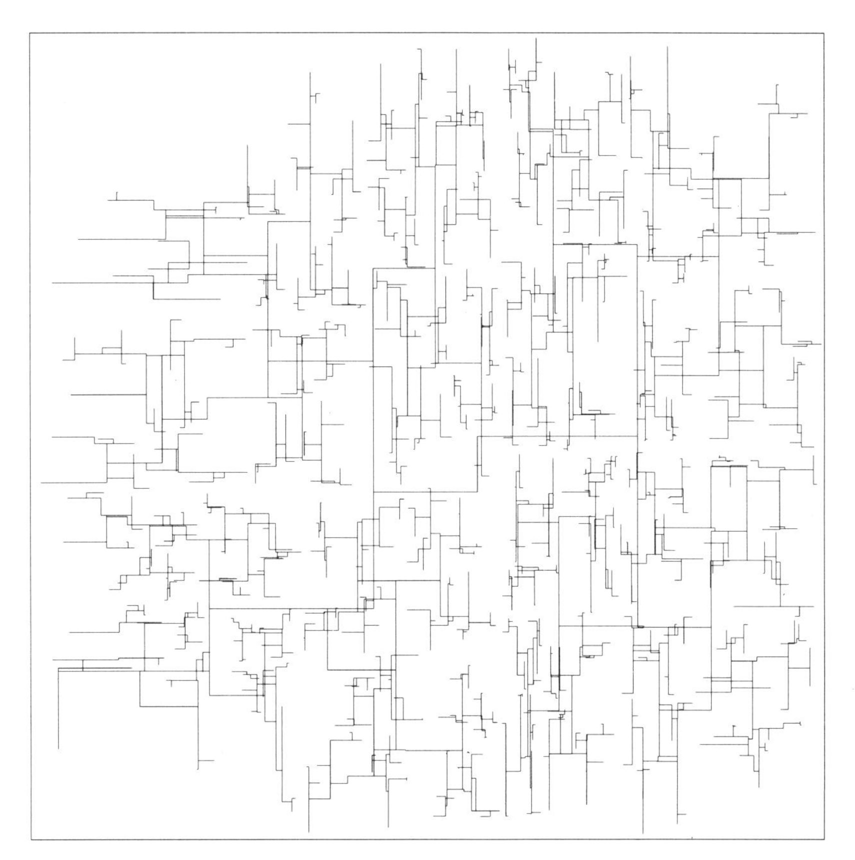

Fig. 8. A zero-skew clock routing result of the example $r3$.

TABLE III
SKEW VALUES COMPUTED FROM THE ACCURATE WAVEFORM SIMULATION PACKAGE RICE

Rise Time	0 ns	0.5 ns	1.0 ns
Examples	Skew (ps)	Skew (ps)	Skew (ps)
$R1$	0.479	0.460	0.433
$R2$	2.338	2.339	2.353
$R3$	1.304	1.304	1.302
$R4$	4.119	4.119	4.118
$R5$	4.572	4.571	4.567

TABLE IV
PERCENTAGE OF WIRE ELONGATION TO TOTAL CLOCK WIRE LENGTH

Examples	$R1$	$R2$	$R3$	$R4$	$R5$
Elongation	0.323%	0.553%	1.043%	0.415%	1.168%

pair of clock pins, as predetermined by the skew assignment algorithm.

Our zero-skew algorithm can be modified to solve this special wiring problem by adding a fictitious delay element on each clock pin. Let us assume the optimum clock delay to clock pin i is $D_0 + D_i$, where D_0 is a common offset value which is unknown until the clock routing is determined. Thus the skew between clock pin i and clock pin j is $D_i - D_j$. Let D_{max} be the maximum clock delay, i.e.,

$$D_{max} = \max_k (D_0 + D_k) = D_0 + \max_k D_k.$$

Define the fictitious delay of clock pin i as

$$d_i = D_{max} - (D_0 + D_i) = \max_k D_k - D_i.$$

In other words, each clock pin is modeled as a lumped delay model with an input loading capacitance and a branch delay, as shown in Fig. 9. Then we perform the zero-skew algorithm on this modified clock tree with the fictitious delay on each clock pin.

As pointed out by Fishburn [6], by combining this technique with the retiming technique [11], one can optimally make a tradeoff between chip area (in terms of number of logics and latches) and clock period.

Multi phase clock: To zero-skew a multiphase clock tree, we first perform the zero-skew algorithm on the clock subtree of each phase independently, but stop at the final end tapping point. Each subtree is then connected to the corresponding output of the clock divider by adding an appropriate delay element or snaking wires for balancing the delays according to the phase and the wiring distance to clock source (Fig. 10).

Optical or wave skew: For this case, the delay is proportional to the traveling distance. Then the zero-skew criterion is to equalize the traveling distance from the source to each end terminal. The wire length balancing algorithm [9], plus the wire elongation scheme, is perfect for this application.

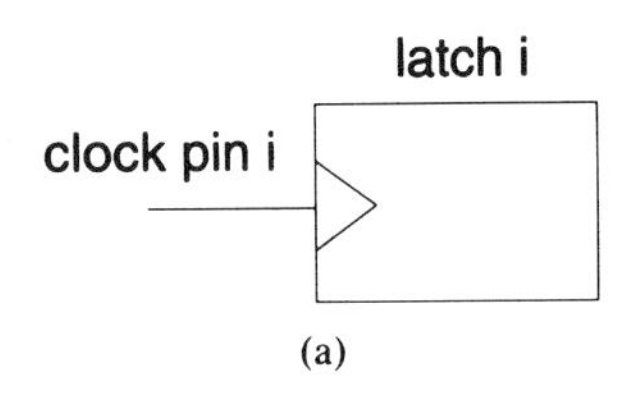

(a)

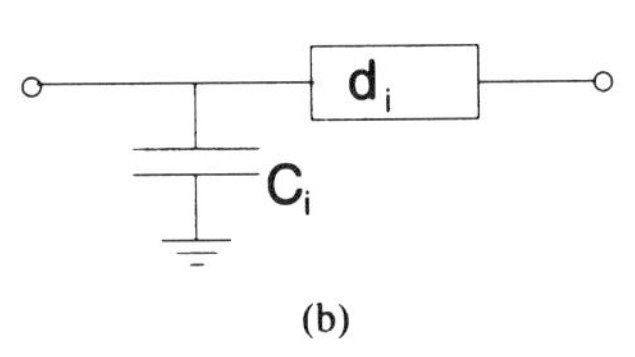

(b)

Fig. 9. (a) A clock pin on a latch. (b) The modified model of a clock pin according to the optimum skew assignment.

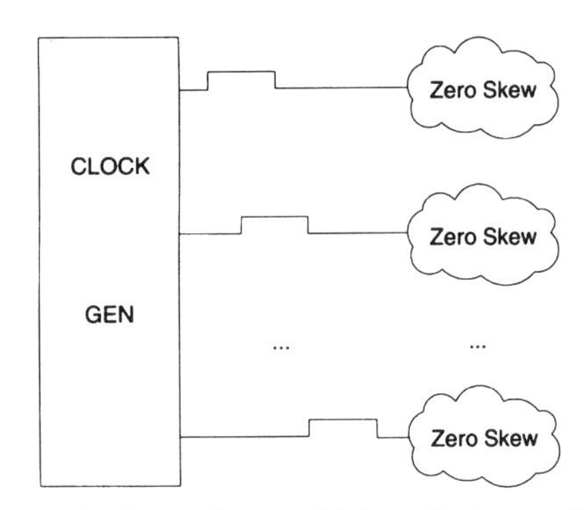

Fig. 10. Connecting a multiphase clock network.

VII. Conclusions

We have presented a novel zero-skew clock routing algorithm based on the Elmore delay calculation. The approach is ideal for hierarchical methods of constructing large systems. All subsystems can be constructed in parallel and independently, then interconnected with exact zero skew. We expect this clock routing algorithm will be widely used for performance enhancement for synchronous VLSI digital systems.

Acknowledgment

The author would like to thank Prof. Lawrence T. Pillage, Curtis L. Ratzlaff, and Nanda Gopal for their help in using the RICE package.

References

[1] H. B. Bakoglu, J. T. Walker, and J. D. Meindl, "A symmetric clock-distribution tree and optimized high speed interconnections for reduced clock skew in ULSI and WSI circuits," in *IEEE Int. Conf. Computer Design: VLSI in Computers*, 1986, pp. 118-122.

[2] P. K. Chan and K. Karplus, "Computing signal delay in general rc network by tree/link partitioning," *IEEE Trans. Computer-Aided Design*, vol. 9, pp. 898-902, Aug. 1990.

[3] T. H. Cormen, C. E. Leiserson, and R. L. Rivest. *Introduction to Algorithms*. New York: McGraw-Hill, 1990.

[4] S. Dhar, M. A. Franklin, and D. F. Wann. "Reduction of clock delays in VLSI structures," in *Proc IEEE Int. Conf. Computer Design: VLSI in Computers*, pp. 1984, pp. 778-783.

[5] W. C. Elmore, "The transient response of damped linear networks with particular regard to wide band amplifiers," *J. Appl. Phys.*, vol. 19, pp. 55-63, 1948.

[6] J. P. Fishburn, "Clock skew optimization," *IEEE Trans. Computers*, vol. 39, pp. 945-951, July 1990.

[7] A. L. Fisher and H. T. Kung, "Synchronous large systolic arrays," in *Proc. SPIE*, 1982, pp. 44-52.

[8] M. A. B. Jackson, A. Srinivasan, and E. S. Kuh, "Clock routing for high-performance IC's," in *Proc. Design Automation Conf.*, 1990, pp. 573-579.

[9] A. Kahng, J. Cong, and G. Robins, "High-performance clock routing based on recursive geometric matching," in *Proc. Design Automation Conf.*, 1991, pp. 322-327.

[10] S. Y. Kung and R. J. Gal-Ezer, "Synchronous versus asynchronous computation in very large scale integrated (VLSI) array processors," in *Proc. SPIE*. 1982, pp. 53-65.

[11] C. E. Leiserson, F. M. Rose, and J. B. Saxe, "Optimizing synchronous circuitry by retiming," in *Proc. Third Caltech Conf.*, 1983, pp. 87-116.

[12] C. L. Ratzlaff, N. Gopal, and L. T. Pillage, "RICE: Rapid interconnect circuit evaluator," in *Proc. Design Automation Conf.*, 1991, pp. 555-560.

[13] J. Rubinstein, P. Penfield, and M. A. Horowitz, "Signal delay in *rc* tree networks," *IEEE Trans. Computer-Aided Design*, vol. CAD-2 pp. 202-211, 1983.

Delay Minimization for Zero-Skew Routing

Masato Edahiro
Department of Computer Science, Princeton University
Princeton, NJ 08544-2087, USA
C&C Research Laboratories, NEC Corporation
Miyazaki, Miyamae-ku, Kawasaki 216, Japan

Abstract

Delay minimization methods are proposed for zero-skew routings. A delay-time estimation formula is derived, which can be used as an objective function to be minimized in zero-skew routing algorithms. Moreover, the optimum wire width is formulated. Experimental results show that our methods with a clustering-based algorithm achieve 50% reduction of the delay time on a benchmark data with 3000 pins.

1 Introduction

With the increase of clock rate in VLSI, clock-net routing scheme plays more critical roles. While the exact zero skew was accomplished [11], the delay minimization is still a difficult problem. In order to minimize the delay time, accurate estimation is crucial. Although the delay on wires has been thoroughly analyzed [1, 2, 8, 9], it seems very difficult to estimate the total delay time for general RC networks with a driver.

In this paper, first, we derive a simple but more accurate formula for the delay time estimation in zero-skew routings. In our formulation, we use a distributed RC model approximated by the lumped $\pi 2$-model for wires [1]. Next, we propose an approximation for the optimum wire width to minimize the delay time in zero-skew routings. We will show that our approximated delay time has the global minimum as a function of wire widths. Although there have been several researches on the wiresizing problem [1, 2], this is the first optimum wire width formulation for zero-skew routings.

Experimental results show that our delay minimization technique achieves 10%-50% reduction of the total delay time compared with a total-wire-length minimization algorithm [3] on benchmark data [6, 11].

2 Zero-Skew Routing

2.1 Definitions

Given a fan-out terminal v_r and a set of n fan-in terminals $S = \{v_1^s, v_2^s, \ldots, v_n^s\}$, a *clock tree* is defined by a tree rooted by v_r whose n leaves are S (Fig. 1). We call the fan-out terminal *root* and fan-in terminals *leaves*. A set of leaves in the subtree rooted by a node v is denoted by S_v. In this paper, clock trees are always binary, though nodes may degenerate and they do not look binary in some cases.

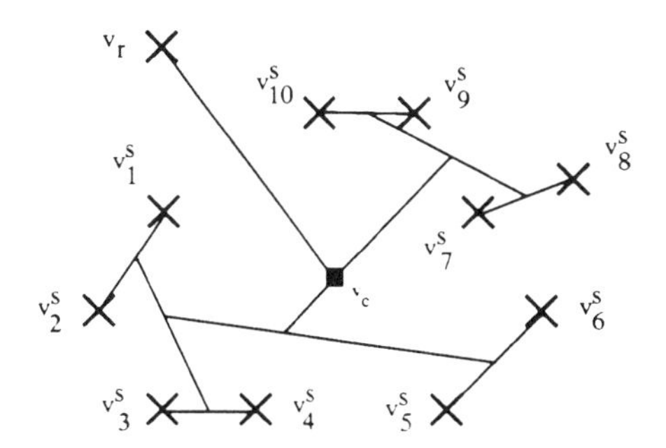

Figure 1: Clock Tree with a root v_r and leaves $S = \{v_1^s, v_2^s, \ldots, v_{10}^s\}$.

We assume that the *load capacitance* $C(v_i^s)$ is given for each leaf v_i^s. Also, the *load capacitance* $C(v)$ for an internal node v is defined by the total capacitance in S_v including wire capacitance as well as gate capacitance.

Then, a *zero-skew routing* for the given root and leaves is defined by a clock tree in which all the propagation delay time from the root to all leaves is equal. From this definition, it is clear that, for any node v in the zero-skew routing, all the propagation delay from the node v to leaves in S_v should be equal. We call this delay the *propagation delay time* $\tau(v)$ *for* v. For leaves v_i^s, $\tau(v_i^s) = 0$.

2.2 Routing Algorithm

Now, we derive equations for $\tau(v)$ and $C(v)$ to be satisfied at each internal node v on a zero-skew routing. Let v_1 and v_2 be children of v on the clock tree, and l_1 (l_2) and w_1 (w_2) be length and width of wire from v to v_1 (v_2), respectively. Also, let r and c be wire resistance and capacitance for an unit length and width wire. Then, by using $\pi 2$-model for the wiring delay [1, 9] (Fig. 2), in a zero-skew routing, $\tau(v)$ and $C(v)$ should satisfy the following equations [11]:

$$\begin{aligned}\tau(v) &= \frac{rl_1}{w_1}\left(\frac{cl_1w_1}{2} + C(v_1)\right) + \tau(v_1) \\ &= \frac{rl_2}{w_2}\left(\frac{cl_2w_2}{2} + C(v_2)\right) + \tau(v_2), \quad (1) \\ C(v) &= C(v_1) + C(v_2) + c(l_1w_1 + l_2w_2). \quad (2)\end{aligned}$$

Reprinted from *Proc. IEEE Int'l Conf. Computer-Aided Design,* pp. 563–566, Nov. 1993.

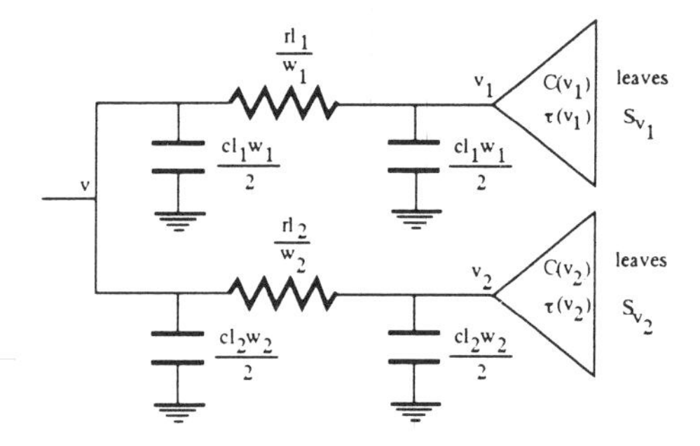

Figure 2: π2-Model for Zero-Skew Merge at v.

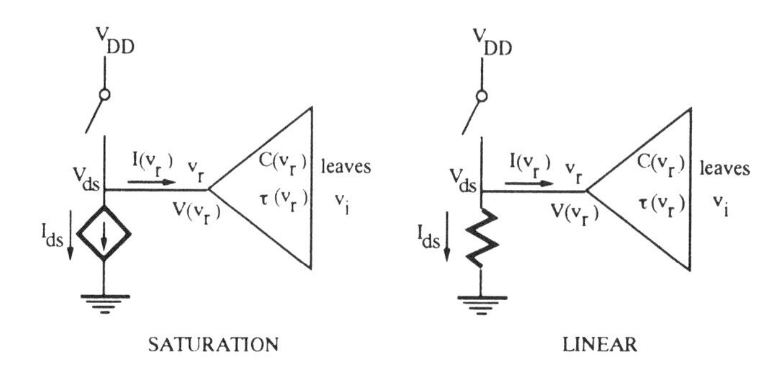

Figure 3: Equivalent Inverter Model for Fall Time Estimation.

It is important to note that, given two zero-skew routings rooted by v_1, v_2 and wire widths w_1, w_2, using above equations, the physical location of v can be calculated in constant time such that v is the parent of v_1 and v_2, the tree rooted by v is a zero-skew routing, and $l_1 + l_2$ is minimized [5]. This operation to determine the location of v is called the *zero-skew merge*.

A zero-skew routing for given a root and leaves can be constructed by $n-1$ zero-skew merges in a bottom-up fashion [3, 11].

3 Delay Time Estimation

Suppose that a pre-constructed zero-skew routing, in which all $C(\cdot)$'s and $\tau(\cdot)$'s have been calculated for the root v_r and internal nodes using the equations (1) and (2), is driven by a clock driver whose process parameters are known.

In this section, we propose a delay-time estimation formula for zero-skew routings. In order to estimate the delay time, first, we show a circuit equation at each internal node v, which is expressed by the load capacitance $C(v)$ and the propagation delay time $\tau(v)$.

Theorem 1 *For each node v and leaves $v_i^s \in S_v$ in a zero-skew routing,*

$$V(v) = V(v_i^s) + \tau(v)\frac{dV(v_i^s)}{dt}, \quad (3)$$

$$I(v) = C(v)\frac{dV(v_i^s)}{dt}, \quad (4)$$

where $V(v)$ and $I(v)$ are voltage and current values for the node v.

A proof of this theorem is shown in [4]. Then, estimated fall time t_f using above equations and the ideal (first order) equations [10] for an nMOS transistor model shown in Fig. 3 is:

$$t_f \approx 3.7\frac{C(v_r)}{\beta_n V_{DD}} + 1.4\tau(v_r), \quad (5)$$

where β_n is the nMOS transistor gain factor of the clock driver [4].

In CMOS circuits, since the output fall/rise time dominates the delay time, the delay time can be estimated by half of the fall/rise time [10]. Consequently, the delay time for the fall is estimated by:

$$t_d \approx 1.85\frac{C(v_r)}{\beta_n V_{DD}} + 0.7\tau(v_r). \quad (6)$$

The delay time for the rise is estimated similarly.

4 Wire Width Optimization

In this section, we derive an approximation formula for the optimum wire width to minimize the delay time. For a wire segment on a clock net, large wire width may reduce the propagation delay when the wire segment shared by many leaves, while smaller wire width may make the gate switching faster because of smaller wire capacitance. Since the delay time is estimated by a linear function of the propagation delay and the load capacitance, there should exist an optimum wire width. We show that our approximated delay time has the global minimum as a function of wire widths.

4.1 Variable Wire Width Model

Consider an internal node v with children v_1 and v_2 in a zero-skew routing. Let l_1 (l_2) and w_1 (w_2) be the length and width of the wire from v to v_1 (v_2). Suppose that we are constructing the zero-skew routing in a bottom-up fashion. That is, the positions of v_1 and v_2 have already been determined, and $C(v_1)$, $C(v_2)$, $\tau(v_1)$ and $\tau(v_2)$ have been calculated. Thus, these values are constants in the optimization.

Now, at the next step, the position of v is to be determined. *The wire-width optimization* is to determine the values w_1 and w_2 so as to minimize the delay time. Note that l_1, l_2 and the position of v are expressed as functions of w_1 and w_2 in this optimization.

The length l is defined by the distance between v_1 and v_2, and w_{min} denotes the minimum width of a wire. These values are also constants. In addition, we define the sum R_v of ratio (length/width) over all wire segments p from the root v_r to v, i.e.,

$$R_v = \sum_{p \in \{\text{path from } v_r \text{ to } v\}} \frac{\text{length of } p}{\text{width of } p}. \quad (7)$$

Since we are constructing the routing in a bottom-up fashion, the value R_v can not be calculated. However, in this paper, we assume that R_v can be approximated from the structure of the clock tree and positions of v_1 and v_2, which are independent from w_1 and w_2. We discuss the approximation method in [4].

Now, suppose that the delay time t_d can be expressed by the formula

$$t_d = \alpha_1 \frac{C(v_r)}{\beta V_{DD}} + \alpha_2 \tau(v_r) + \alpha_3, \quad (8)$$

where α_1, α_2, and α_3 are constants. In our estimation in equation (6), $\alpha_1 = 1.85$, $\alpha_2 = 0.7$, and $\alpha_3 = 0$. We assume $\beta = \beta_n = \beta_p$ in this section. Using the equations (1) and (2), the delay time t_d is rewritten in a function of w_1 and w_2 as follows:

$$t_d(w_1, w_2) = \left(\frac{\alpha_1 c}{\beta V_{DD}} + \alpha_2 r c R_v\right)(l_1 w_1 + l_2 w_2) + \alpha_2 r \left(\frac{c l_1^2}{2} + \frac{l_1 C(v_1)}{w_1}\right) + const. \quad (9)$$

4.2 Optimum Wire Width

In this section, we show the optimum width formulae derived from equation (9). Before deriving the formulae, we show a global optimality theorem for the equation. We use two conditions in the theorem. First, wire widths w_1 and w_2 are not less than the minimum wire width w_{min}. Next, $l_1 + l_2 = l$. Since the sum $l_1 + l_2$ of wire lengths from v to v_1 and v_2 can not be less than the distance l between v_1 and v_2, the optimum solution under this condition is the global optimum if the solution is feasible, i.e., both l_1 and l_2 are non-negative. We will classify into several cases based on this feasibility.

Theorem 2 (Wire Width Optimization) *On the domain $w_1, w_2 \geq w_{min}$ and under $l = l_1 + l_2$, the function $t_d(w_1, w_2)$ has the global minimum when:*

$$w_1 = w_2 = w^* = \max\left(w_{min}, \sqrt{\frac{\alpha_2 r}{\frac{\alpha_1 c}{\beta V_{DD}} + \alpha_2 r c R_v}\left(\frac{C(v_1)C(v_2)}{C(v_1)+C(v_2)}\right)}\right). \quad (10)$$

The wire lengths for this optimum width are:

$$l_1^* = \frac{(\tau(v_2) - \tau(v_1)) + rl(\frac{C(v_2)}{w^*} + \frac{cl}{2})}{r(cl + \frac{C(v_1)}{w^*} + \frac{C(v_2)}{w^*})}, \quad (11)$$

$$l_2^* = l - l_1^*. \quad (12)$$

Again, the proof is shown in [4]. It is important to note that this global optimum solution also minimizes *the delay sensitivity* introduced in [7]. The delay sensitivity is defined by the change in delay with respect to the change in wire widths. Since our solution satisfies $\frac{\partial t_d}{\partial w_1} = \frac{\partial t_d}{\partial w_2} = 0$ [4], the sensitivity is theoretically *zero*. This means that our wire width optimization achieves the most reliable zero-skew routing as well as the minimum delay time.

Now, we show the optimum width formulae for zero-skew routings. Since, in the above theorem, we did not care about the condition that l_1 and l_2 should be non-negative values, we need to classify into the following three cases.

Case 1: $[l_1^* > l \ (l_2^* < 0)]$
In this case, we need to fix $l_2 = 0$, i.e., $v \equiv v_2$, for the delay minimization. From equations (11) and (12), for $\exists w_1^+ \leq w^*$,

$$\tau(v_2) - \tau(v_1) = rl(\frac{cl}{2} + \frac{C(v_1)}{w_1^+}).$$

There are two subcases.

Case 1.1: $[w_1^+ < w_{min}]$
In this case, whichever wire width ($\geq w_{min}$) is used, $\tau(v_2) - \tau(v_1)$ is too large for a zero-skew routing under $l = l_1 + l_2$. Thus, we need to use a *detour* wire from v_1 to v ($\equiv v_2$), whose length is larger than l. In order to minimize the delay, w_1 should be w_{min} because the length of the detour wire needs to be minimized. From equation (1), the length of the detour wire should be $l_1 = l_1^+$ such that

$$\tau(v_2) = \frac{rc}{2}(l_1^+)^2 + \frac{rC(v_1)}{w_{min}} l_1^+ + \tau(v_1).$$

Case 1.2: $[w_1^+ \geq w_{min}]$
In this case, we do not need any detour edge if $w_1 = w_1^+$ is used. In addition, it is easy to see that this is the case to minimize the delay. Therefore, $l_1 = l$ and $w_1 = w_1^+$.

Case 2: $[0 \leq l_1^* \leq l$ and $0 \leq l_2^* \leq l]$
Since our global optimum specified in Theorem 2 is feasible, the optimum wire width and corresponding wire lengths are $w_1 = w_2 = w^*$, $l_1 = l_1^*$, and $l_2 = l_2^*$.

Case 3: $[l_2^* > l \ (l_1^* < 0)]$ Similar to Case 1.

5 Experimental Results

The proposed methods have been implemented with a clustering-based zero-skew routing algorithm [3]. The delay time estimation (6) is utilized for an objective function in the algorithm, and the wire width optimization is applied to each wire segment where $\alpha_1 = 1.85$, $\alpha_2 = 0.7$, and $\alpha_3 = 0$.

In our test circuit model, an inverter is added to the root v_r of generated zero-skew routings. The total delay time was measured by time difference between the input transition (50% level) and the 50% level at leaves, which is estimated from the equation (6).[1] Also, the propagation delay time was measured by the time difference between the root v_r and leaves, which is estimated by $\tau(v_r)$. For the drive inverter, we used $\beta_p = \beta_n = 5.6 \ [mA/V^2]$ for all experiments. Also, $V_{DD} = 5[V]$.

[1] Although the estimated delay time from (6) is smaller than that from existing estimation as discussed in [4], the delay time by SPICE simulation is still smaller. We used 60% of the value from (6) to fit with SPICE which uses more accurate but complicated model.

Table 1: Comparison between Zero-Skew Routings generated by a total-wire-length minimization algorithm [3] and delay minimization algorithm.

	#pins		[3] estimated	[3] simulated	Delay Minimization estimated	Delay Minimization sim. (ratio to [3])
prim1	269	TWL	131427		131877	
		PD	2.34ns	2.63ns	0.36ns	0.37ns (0.14)
		TD	6.35ns	6.31ns	5.60ns	5.63ns (0.89)
		SKW	0.00ns	0.00ns	0.00ns	0.00ns
prim2	603	TWL	306053		317296	
		PD	8.97ns	10.89ns	0.99ns	1.01ns (0.09)
		TD	15.90ns	15.93ns	12.77ns	12.60ns (0.79)
		SKW	0.00ns	0.03ns	0.00ns	0.00ns
r1	267	TWL	1289004		1288597	
		PD	1.13ns	1.35ns	0.71ns	0.80ns (0.59)
		TD	2.05ns	2.19ns	1.91ns	2.04ns (0.93)
		SKW	0.00ns	0.00ns	0.00ns	0.00ns
r2	598	TWL	2537488		2559898	
		PD	3.58ns	4.58ns	1.51ns	1.69ns (0.37)
		TD	4.79ns	5.08ns	4.06ns	4.15ns (0.82)
		SKW	0.00ns	0.00ns	0.00ns	0.00ns
r3	862	TWL	3227150		3266236	
		PD	4.70ns	6.07ns	1.79ns	1.98ns (0.37)
		TD	6.39ns	6.71ns	5.40ns	5.44ns (0.81)
		SKW	0.00ns	0.01ns	0.00ns	0.00ns
r4	1903	TWL	6588826		6657174	
		PD	14.92ns	16.73ns	4.00ns	4.44ns (0.27)
		TD	15.70ns	17.10ns	11.58ns	11.47ns (0.67)
		SKW	0.00ns	0.02ns	0.00ns	0.00ns
r5	3101	TWL	9867854		9952239	
		PD	33.42ns	32.92ns	5.32ns	5.83ns (0.18)
		TD	28.89ns	33.23ns	17.60ns	17.30ns (0.52)
		SKW	0.00ns	0.05ns	0.00ns	0.01ns

In our experiments, we compared our delay minimization methods with a total-wire-length minimization algorithm [3] by the SPICE circuit simulator using benchmark data prim1-prim2 [6] and r1-r5 [11]. In the algorithm [3], minimum width wires are used.

Table 1 shows the estimated and simulated values for the total wire length [TWL], the propagation delay time [PD], the total delay time [TD], and the clock skew [SKW]. Table 1 shows that our estimation fits with the simulation within 10% for most data. In addition, our methods accomplished 10%-50% shorter total delay time than the algorithm [3].

6 Conclusions

Delay-time estimation formulae and wire-width optimization methods have been proposed for zero-skew routings. Computational experiments showed that our methods with a clustering-based zero-skew routing algorithm achieved 10%-50% reduction of the total delay time compared with a total-wire-length minimization algorithm.

Acknowledgment

The author would like to thank A. B. Kahng and K. Boese of UCLA and R. S. Tsay of ArcSys Inc. for providing benchmark data. Also the author is grateful to R. J. Lipton and T. Takahashi of Princeton University for valuable discussion.

References

[1] H. B. Bakoglu: *Circuits, Interconnections, and Packaging for VLSI.* Addison-Wesley, Reading, Massachusetts, 1990.

[2] J. Cong, K.-S. Leung, and D. Zhou: Performance-Driven Interconnect Design Based on Distributed RC Delay Model. *CSD-920043*, CS Dept., UCLA, 1992.

[3] M. Edahiro: A Clustering-Based Optimization Algorithm in Zero-Skew Routings. *Proc. of 30th DAC*, 1993, pp.612-616.

[4] M. Edahiro: Delay Minimization for Zero-Skew Routing. *CS-TR-415-93*, Dept. of CS, Princeton University, 1993.

[5] M. Edahiro: Equi-spreading tree in Manhattan distance, unpublished.

[6] M. A. B. Jackson, A. Srinivasan, and E. S. Kuh: Clock Routing for High-Performance ICs. *Proc. of 27th DAC*, 1990, pp.573-579.

[7] S. Pullela, N. Menezes, and L. T. Pillage: Reliable Non-Zero Skew Clock Trees Using Wire Width Optimization. *Proc. of 30th DAC*, 1993, pp.165-170.

[8] J. Rubinstein, P. Penfield, and M. A. Horowitz: Signal Delay in RC Tree Networks. *IEEE Trans. on CAD*, Vol. CAD-2 (1983), No. 3, pp.202-211.

[9] T. Sakurai: Approximation of Wiring Delay in MOSFET LSI. *IEEE J. of Solid-State Circuits*, Vol. SC-18 (1983), No. 4, pp.418-426.

[10] N. Weste and K. Eshraghian: *Principles of CMOS VLSI Design: A Systems Perspective.* Addison-Wesley, Reading, Massachusetts, 1985.

[11] R. S. Tsay: Exact Zero Skew. *Proc. of ICCAD91*, 1991, pp.336-339.

Reliable Non-Zero Skew Clock Trees Using Wire Width Optimization

Satyamurthy Pullela, Noel Menezes, and Lawrence T. Pillage
Department of Electrical and Computer Engineering
The University of Texas at Austin
Austin, TX 78712

Abstract

Recognizing that routing constraints and process variations make non-zero skew inevitable, this paper describes a novel methodology for constructing reliable low-skew clock trees. The algorithm efficiently calculates clock-tree delay sensitivities to achieve a target delay and a target skew. Moreover, the sensitivities also show that wires should be widened as opposed to lengthened to reduce skew since the former improves reliability while the latter reduces it. This paper introduces the concept of designing *reliable* clock nets with process-insensitive skew.

1 Introduction

Clock skew is defined as the maximum difference in the delays from the output of the clock buffer to the inputs of the clocked elements on a chip. With the increasing density of VLSI and the use of several pipeline stages in hardware design, clock skew becomes a dominating factor in determining the clock period of synchronous digital systems.

While reducing clock skew should be the main focus of clock net algorithms, clock-signal delay should be considered as well since it affects system-level skew[5]. Moreover, for reliable results, it is also important to consider that the actual width of a line on a fabricated chip may differ from the expected width due to intra-chip process variations such as over- and under-etching, mask misalignment, spot defects, etc. This paper shows that these wire width process variations can significantly impact the delay and skew.

Recent algorithms [4, 6] proposed for clock-skew reduction construct binary tree-like structures with the clock pins at the leaf nodes. "Zero-skew" trees are created in a recursive bottom-up manner from the clock pins upwards by tapping the connection between two zero-skew subtrees at such a location so as to create a parent zero-skew tree rooted at the tapping point. Earlier algorithms [5, 8] assume that the wire length is a valid measure of the delay and try to equalize the lengths from the root of the clock tree to the leaf nodes. These algorithms, in effect, minimize skew while routing the clock net. Our algorithm, on the other hand, routes the tree first and then minimizes skew by varying the wire widths of the tree branches. This gives the router additional flexibility in routing around possible blockages. Since the algorithms described in [5, 8] can easily be modified to route around blockages, they serve as excellent starting points for the algorithm described here.

Our approach uses *sensitivities* to address the most important issues in clock tree synthesis. Firstly, skew and delay are optimized using sensitivity information. In addition, sensitivities are used to increase the reliability of synthesized clock nets. The sensitivity approach allows us to specify which wire lengths and/or widths to vary for clock tree synthesis. This paper, however, uses the latter approach since widening is shown to increase reliability (decrease delay sensitivity to process variation) while lengthening is shown decrease it.

Our paper is organized as follows: In Section 2, we detail a simple RC delay model [2] and the inaccuracies associated with using it. The futility of "zero-skew" clock nets is also introduced along with the concept of *reliable* clock nets which are insensitive to process variation. In section 3, we propose an efficient algorithm to calculate the sensitivities of the leaf-node signal delays to variations in the width of the tree branches. Techniques for delay reduction, reliability, and skew reduction based on these sensitivities are described in section 4. We then present results, conclusions, and possible extensions of our approach.

2 Practical Considerations

Each branch of a clock tree can be represented by a distributed resistance-capacitance segment. Each distributed RC line is modeled as an equivalent lumped L- or π-circuit. The lumped resistance of the L-circuit model of an RC line of length l_i is usually approximated by rl_i and the capacitance by cl_i where r and c are the resistance and capacitance per unit length. Figure 1 shows a route for eight clock pins. The equivalent RC tree with every segment replaced by an L-model is shown in Figure 2.

We present our notation first: The branch connecting node i to its parent node in the RC tree is labeled i, and the width of this branch is denoted by w_i. The resistance of branch i is denoted by R_i. The sum of the resistances of the branches on the path from node i to the root is called the upstream resistance, R_{u_i}. $R_{c_{ij}}$ is used to refer to the sum of the resistances along the

*This work was supported in part by IBM Corp., the National Science Foundation, and the Semiconductor Research Corporation under contract # 92-DP-142.

Reprinted with permission from *30th Proc. ACM/IEEE Design Automation Conf.*, S. Pullela, N. Menezes, and L. T. Pillage, "Reliable Non-Zero Clock Trees Using Wire Width Optimization," pp. 165–170, June 1993.

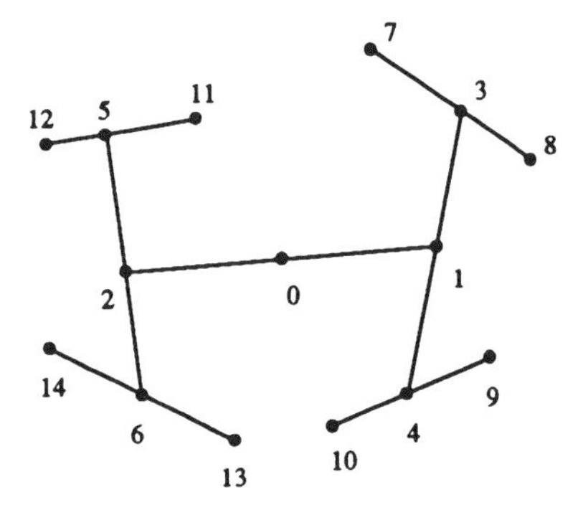

Figure 1: Example clock route on 8 pins.

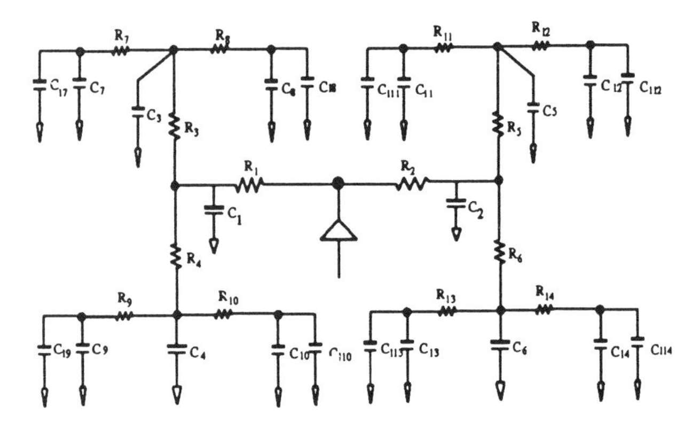

Figure 2: Equivalent RC tree for route in Figure 1.

common path from node i to the root and from node j to the root. The capacitance at node i is denoted by C_i. There are additional capacitances C_{li} at the leaf nodes. The downstream capacitance, C_{d_i}, at a node i is defined as the sum of the capacitances at node i and all its descendant nodes. $U(j)$ is the set of all branches that lie on the path from node j to the root of the tree. R_{u_j} (the upstream resistance) denotes the sum of resistances of the branches in set $U(j)$. If node i lies on the path from node j to the root, branch j (i) as well as node j (i) are said to lie *downstream* (*upstream*) of branch i (j) and node i (j).

2.1 Elmore Delay Approximations

The Elmore delay model [2, 7] is commonly used to approximate the signal delay in RC tree networks. This model approximates the delay from the root to any node n in a lumped RC tree by the sum of the products of the branch resistance, R_i, and the downstream capacitance, C_{d_i}, for every branch on the path from the root to the node, i.e.

$$T_{D_n} = \sum_{i \in P(n)} R_i C_{d_i} \tag{1}$$

where $P(n)$ is the set of nodes that lie on the path from the root to node n excluding the root.

However, it should be noted that the Elmore delay is a first-order approximation which is generally applied as a dominant time constant, or step-response delay approximation for the circuit. In reality, for slow input transition times (inputs to the RC net, which are outputs of the clock driver), the signal is a ramp follower with an actual delay T_D[11]. For faster transition times, the Elmore delay is an approximation.

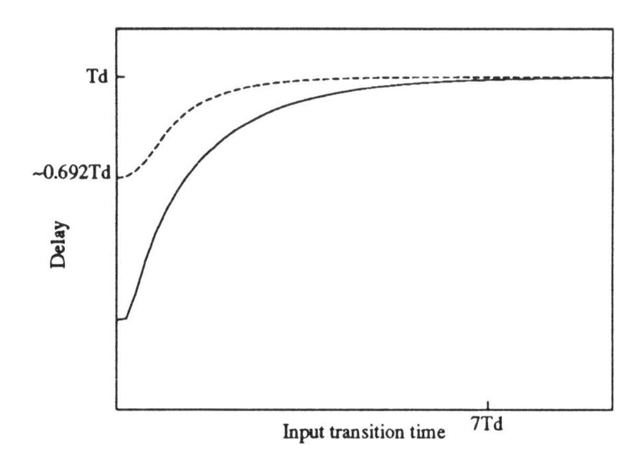

Figure 3: Plots of delay versus input transition time for two nodes in an RC tree with equal T_Ds.

Figure 3 shows the variation in the actual delay for two nodes in an RC tree as a function of the input rise time. A highly-accurate third-order model was used to generate these curves[1]. For a step input, the 50% delay is approximated by $0.692T_D$. However, from Figure 3, we observe that this approximation may, at times, be inaccurate. Roughly speaking, when the transition time is less than $7T_D$, the response is no longer a ramp follower and the first order model may have significant error[11]. Therefore, a clock net designed for zero skew using the Elmore delay model may not yield zero skew for small input rise times. We expect, however, that the non-zero skew due to the inaccuracy of the Elmore delay model for small rise times is only a second-order effect when compared to the skew due to process-related wire width variations. Our approach does not preclude the use of more accurate delay models[3]; however, the first-order model does provide excellent efficiency.

In order for an algorithm to attain zero skew for any input rise time it would have to match the delay curves at all the leaf nodes in the circuit – a seemingly impossible task. However, reducing the delay will reduce the transition-time induced skew.

2.2 Reliable Clock Routing

During fabrication, the width of a line on a chip may differ from the expected width because of process variations. Ideally, the clock net should guarantee a certain skew and delay taking these process variations into account. It is essential for a robust clock net to exhibit a certain degree of insensitivity to these process variations.

Consider a change in the width of branch 1 of Figure 1 from a specified value of w_1 to $w_1 + \Delta w$ due to process variation. Δw is a random variable which does not depend on the branch width. Its statistical characteristics are determined by the process. If the new resistance and capacitance of branch 1 due to this variation are $R_1 + \Delta R_1$ and $C_1 + \Delta C_1$ respectively, the change in the Elmore delay to any node n downstream of branch 1 is given by

$$\Delta T_{D_n} = \Delta R_1(C_{d_1} - C_1) \tag{2}$$

From (2), we see that the skew of binary tree-like clock nets is extremely sensitive to the changes in the widths of the branches closest to the root of the tree. Small changes in the widths of such branches can, therefore, have a large effect on skew. As a result, a carefully designed low-skew clock net may yield high skew values during actual fabrication due to process-related wire width variations, if these variations are not taken into account during design.

3 Sensitivity of Elmore Delays to Circuit Parameters

For an RC tree, the first moment of the impulse response at any node is the Elmore delay. It has been shown that this first moment for all the nodes in the circuit can be obtained from the solution of a dc-equivalent circuit in which all the capacitors are replaced by current sources of value equal to the capacitance[3]. Since dc-circuit sensitivities can be readily determined by using the *adjoint* sensitivity technique[9], the sensitivity of the Elmore delays to circuit parameters can be obtained with similar ease.

The adjoint of an RC circuit [9] is topologically equivalent to the original circuit except that the independent sources are set to zero and the output node of interest is driven by a unit current source. An adjoint analysis of node i in an RC tree which has a nominal delay T_{D_i} can be shown to yield the following sensitivity equations for circuit parameters R_j and C_j:

$$\begin{aligned} \frac{\partial T_{d_i}}{\partial R_j} &= C_{d_j} \qquad R_j \in U(i) \\ &= 0 \qquad \textit{otherwise} \end{aligned} \tag{3}$$

$$\frac{\partial T_{d_i}}{\partial C_j} = R_{c_{ij}} \tag{4}$$

To compute the sensitivities for a set of nodes, a different adjoint network would have to be set up and analyzed for each node of interest. Instead, the delay sensitivities for all nodes with respect to all the resistances and capacitances in the circuit can be calculated using (3) and (4) in $O(n^2 log(n))$ time by path-tracing the tree. These sensitivities can be used to obtain the sensitivities with respect to the wire widths.

3.1 Delay Sensitivity to Wire Width

The change in delay to a pin with respect to a wire can be obtained from the sensitivity of delay with respect to the wire capacitance and resistance by applying the chain rule. The sensitivity coefficient for pin i with respect to wire j is expressed as follows:

$$\frac{\partial T_{D_i}}{\partial w_j} = \frac{\partial T_{D_i}}{\partial R_j}\frac{\partial R_j}{\partial w_j} + \frac{\partial T_{D_i}}{\partial C_j}\frac{\partial C_j}{\partial w_j} \tag{5}$$

The terms $\partial T_{Di}/\partial R_j$ and $\partial T_{Di}/\partial C_j$ can be evaluated as described previously. $\partial R_j/\partial w_j$ and $\partial R_j/\partial w_j$ can be computed from the following equations for the resistance and capacitance per unit length:

$$R = \frac{K_R}{w} \tag{6}$$

$$C = K_C w + C_{fr} \tag{7}$$

where K_R and K_C are independent of the width, $K_C w$ is the area capacitance and C_{fr} refers to the fringe capacitance. Differentiating (6) and (7) with respect to w yields

$$\frac{\partial R}{\partial w} = -\frac{R}{w} \tag{8}$$

$$\frac{\partial C}{\partial w} = \frac{C - C_{fr}}{w} \tag{9}$$

whence,

$$\frac{\partial T_{D_i}}{\partial w_j} = \frac{C_j - C_{fr_j}}{w_j}\frac{\partial T_{D_i}}{\partial C_j} - \frac{R_j}{w_j}\frac{\partial T_{D_i}}{\partial R_j} \tag{10}$$

A positive value of sensitivity obtained from (10) indicates a case where widening wire j increases the delay to leaf node i while a negative value indicates that the delay decreases.

It is clear from (10) that the sensitivities are a function of the wire widths (similar expressions can be derived in terms of wire lengths). This implies that the sensitivities need to be recomputed whenever the width of a wire is modified. The next section demonstrates the simplicity of this task.

3.2 Updating Sensitivities

The computation of the sensitivities is a one-time process. Whenever the width of a wire changes, the sensitivities can be updated in $O(n^2)$ time as opposed to the $O(n^2 log(n))$ time needed for the initial calculation of sensitivities.

Changing the width of branch j by ΔW_j causes the resistance and capacitance to change by ΔR_j and ΔC_j respectively. The effect of this change on the sensitivities of the delay of the leaf nodes with respect to the widths of all branches in the tree can be summarized as follows:

$$\Delta\frac{\partial T_{D_i}}{\partial w_k} = \begin{cases} \frac{C_k}{w_k}\Delta R_j & k \in D(j) \text{ and } j \in U(i) \\ -\frac{R_k}{w_k}\Delta C_j & k \in U(i) \cap U(j) \\ \frac{2C_{d_j}}{w_j}\Delta R_j & j = k \\ 0 & \text{otherwise} \end{cases} \tag{11}$$

4 Clock Tree Synthesis

With efficient algorithms to calculate the delay sensitivities of nodes in an RC tree, we can propose the following steps for clock-tree synthesis:

4.1 Delay Reduction

Figure 4(a) shows the distribution of the delays to the leaf nodes of a typical clock tree. The extremal points of this distribution determine the skew while the average determines the delay. During delay reduction, we bring the average of this distribution to or slightly below a specified target delay, T_{target}.

Sensitivities provide a means of selecting the right wires to widen in order to bring the average delay of an RC tree as close as possible to a specified target delay. The delay to a single node can be reduced trivially by widening the wire that has the highest *negative* sensitivity to this node. However, widening this wire may have a detrimental effect on delays to nodes which have *positive* sensitivities with respect to this wire. The ideal candidate for wire widening should, therefore, be selected using a more global metric.

A cost D_j is assigned to every branch j in the tree. This cost takes into account the effect of widening wire j on all the leaf nodes. If S_{ij} denotes the sensitivity of the delay to pin i with respect to wire j then the cost D_j is given by:

$$D_j = \sum_{i=1}^{N} S_{ij}(T_{D_i} - T_{target}) \tag{12}$$

where N denotes the number of leaf nodes (clock pins) in the tree.

We observe the following from the above equation:

- If $T_{D_i} > T_{target}$, and $S_{ij} < 0$ (widening decreases delay), D_j decreases.
- If $T_{D_i} < T_{target}$, and $S_{ij} > 0$, D_j decreases.
- If $T_{D_i} > T_{target}$, and $S_{ij} > 0$, D_j increases.
- If $T_{D_i} < T_{target}$, and $S_{ij} < 0$, D_j increases.

The first two cases clearly aid delay reduction by bringing the distribution of the delays closer to the target delay. The last two cases tend to disturb this distribution by increasing the delay of the pins with already high delays and lowering those with already low delays. Hence, at each iteration the wire with the *least* cost is widened by a constant amount which is based upon the minimum grid size. The sensitivities are recomputed and this procedure continues until the specified target delay is achieved.

4.2 Skew Reduction

Assuming that the distribution of delays in Figure 4(a) is normal, we might define the clock-tree skew as $6\sigma_{nom}$, where σ_{nom} is the standard deviation of this normal distribution. Referring to Figure 4(b), the *nominal skew* may be approximated by the difference between the nominal delay at pin i, T_{D_i}, and the nominal delay at pin j, T_{D_j}, since these delays lie near the boundaries of the nominal distribution.

As shown in Figure 4(b), however, delays T_{D_j} and T_{D_i} can vary significantly due to process disturbances. In fact, when all the wires in the clock tree are thin, statistical variations in the widths of wires closest to

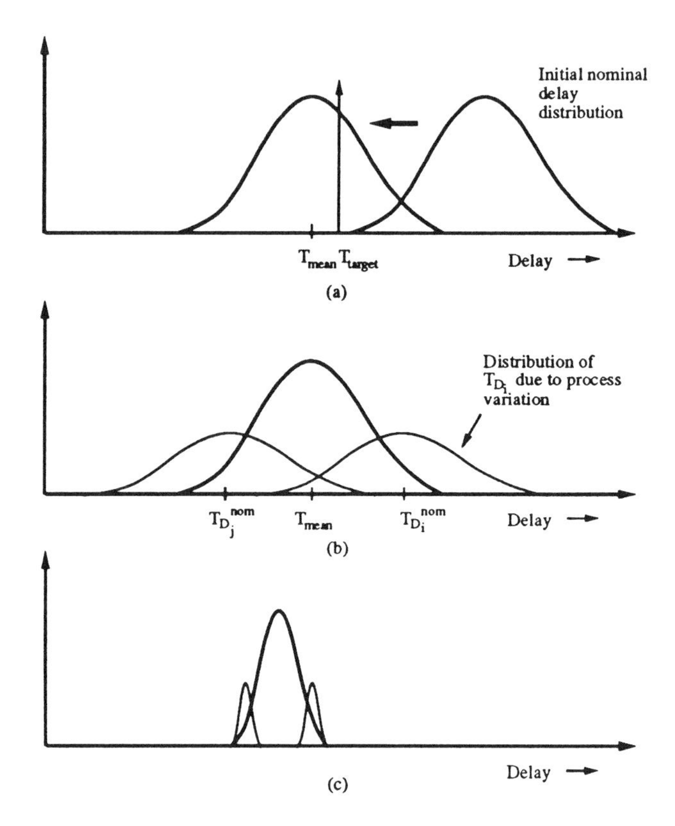

Figure 4: Distribution of delays before and after desensitization and skew minimization

the driver may have an enormous impact on the *actual* value of skew. In other words, if the delays T_{D_i} and T_{D_j} are also modeled as normal distributions with standard deviations of $\sigma_{T_{D_i}}$ and $\sigma_{T_{D_j}}$ respectively, the $\sigma_{T_{D_i}}$'s may be significantly larger than the σ_{nom}. A worst-case *bound* on the skew may, therefore, be expressed as

$$skew_{bound} = 6\sigma_{nom} + 3\sigma_{T_{D_i}} + 3\sigma_{T_{D_j}} \tag{13}$$

To reduce the skew in a reliable manner, we must reduce the variance of the nominal delay distribution, as well as the variances of all the nominal delays due to process variations, as shown in Figure 4(c). Toward this end, we begin by first attempting to reduce all $\sigma_{T_{D_i}}$'s to much less than σ_{nom}. Then, once all wires are de-sensitized in this manner, we attempt to reduce σ_{nom} while maintaining the de-sensitization conditions on all wires.

4.2.1 De-sensitization

Low values of absolute sensitivities imply negligible change in delays for small changes in wire widths. This information can be used to guide the reduction of the $\sigma_{T_{D_i}}$'s due to process variations. It is well understood that increasing the widths of the wires should reduce the delay sensitivities at the leaf nodes since small process variations of the widths would result in smaller changes in the overall delays, and consequently, smaller skew. The objective then is to make the delay sensitivities at the leaf nodes *small* enough so that the upper bound on the skew is acceptable without widening wires excessively.

Consider, as an example, that the width of a single wire j in the clock tree varies from its expected value due to a single process disturbance (e.g. a spot defect). Under such circumstances, we would want to make the delay to all leaf nodes insensitive to changes in the width of wire j. The maximum process variation in the width of wire j corresponds to a maximum change in delay, at one of the leaf nodes, $\Delta T^j_{D_{max}}$. Therefore, to de-sensitize wire j, we might increase w_j, so that $\Delta T^j_{D_{max}}$ is less than the maximum allowable change in skew, ΔT. Or, increase w_j until

$$\Delta w_{max} S_{ij} < \Delta T \tag{14}$$

for all leaf nodes i, where Δw_{max} represents the maximum change in w_j due to process variations.

Under the assumption of a single-defect model, we would de-sensitize all of the wires in the tree as follows: Starting at the leaf nodes, widen all of the wires in a bottom-up manner so that the maximum change in delay at any pin due to process variations at wire j is less than some acceptable change in skew, ΔT. In this paper, we have chosen ΔT to be $\sigma_{nom}/2$.

It is important to point out that the wires are widened in a bottom-up manner from the leaf-nodes in order to properly consider the possible changes in the upstream sensitivities which tend to increase the $\sigma_{T_{D_i}}$s. Note that it can be shown that increasing the width of a wire j, is guaranteed to make all positive sensitivities smaller, and most negative sensitivities smaller in magnitude. However, those wires which lie along the unique path from j to the root node may have the magnitude of their negative sensitivities increased by the increase in their downstream capacitance. Thus, the bottom-up traversal from the leaf-nodes will use the correct value of downstream capacitance.

Of course, all of this assumes a single defect model which is far from accurate. Under- and over-etching, for example, influence a large number of wires simultaneously in a manner which is difficult to predict. Therefore, we establish a more stringent condition for de-sensitization by widening wires until the following holds:

$$\Delta w_{max} S_{ij} < \Delta T / L \tag{15}$$

where L is the depth of the tree. In other words, we do not allow any wire in the tree to cause a change in delay of more than $\Delta T/L$ at a leaf node thereby ensuring that the maximum possible change in delay from the root to a leaf node is less than ΔT.

4.2.2 Reducing Nominal Skew

After the wires are de-sensitized, we attempt to narrow the distribution of delays about its mean value, thereby reducing the nominal skew. This is accomplished by attempting to make the delays of the leaf nodes as close to the mean of the distribution as possible. In this paper, the sensitivities are used to select wires which narrow the distribution while disturbing the mean of the distribution as little as possible. Let ΔT_{ij} denote the change in the delay of node i when the width of wire j is changed by an amount Δw_j. The skew would be reduced to zero if

$$\Delta T_{ij} = T_{mean} - T_{D_i} \tag{16}$$

for all leaf nodes i in the tree. However, it seems highly improbable that skew could be eliminated by varying the width of a single wire. Therefore, we select a wire that forces the new delays of as many pins as possible, closer to the current mean delay. To choose the best wire to widen for skew reduction we assign a cost

$$D_j = \sum_{i=1}^{N} (|T_{D_i} + \Delta T_{ij} - T_{mean}|) \tag{17}$$

to each wire j. Skew could be eliminated if we could find a wire with zero cost. Alternatively, we select to widen the wire with the least cost. The change in delay ΔT_{ij} is estimated using sensitivities by

$$\Delta T_{D_i} = S_{ij} \Delta w_j \tag{18}$$

It should be noted that (18) is valid only for small changes in width. The wire width increment, Δw_j, is constant for all iterations and is based upon the grid size.

To summarize, the complete clock tree synthesis approach is to reduce delay, de-sensitize, then reduce skew. However, we recall from Section 4.2.1 that widening a wire j may increase the magnitude of the sensitivity of wires upstream of j. In order to avoid undoing the de-sensitization process, we add an additional step (4) to the clock tree synthesis procedure as follows:

1. Meet RC delay requirement. (Delay reduction phase).
2. Starting at the leaf nodes, widen wires until their $\Delta T_{D_{max}}$ is less than $\sigma_{nom}/2L$. (Desensitization phase).
3. Use (17) to select the best wire to reduce skew. Widen this wire by a predetermined increment. Update σ_{nom}.
4. Starting at the modified wire, trace back to root widening wires as necessary to ensure that $\Delta T_{D_{max}}$ is still less than $\sigma_{nom}/2L$.
5. Go to step 3 until skew objectives are met.

5 Results

We applied the clock tree synthesis procedure to the 5 examples in [4]. The per micron resistance and capacitance are assumed to be $3m\Omega$ and $0.02fF$ respectively. These examples were routed using the methods

Example	# pins	Initial		Final		Worst case skew (ns)	w_{max}
		Skew (ns)	Delay (ns)	Skew (ns)	Delay (ns)		
r1	267	0.325	2.23	0.080	1.270	0.203	4.5
r2	598	1.888	5.56	0.179	4.016	0.572	7.0
r3	862	1.529	7.55	0.149	4.295	0.498	11.8
r4	1903	4.893	20.25	0.579	11.506	1.252	7.9
r5	3101	4.370	40.85	0.898	16.978	2.141	6.4

Table 1: Delay and skew results for examples in [4].

of means and medians[5]. The initial skew and average delay are shown in columns 3 and 4 of Table 1.

Selecting a target delay of one-half the initial delay and a target skew of 5% of the initial delay, we applied the clock tree synthesis procedure described above. Using wire width increments of 0.3 microns during the skew reduction phase (steps 3 – 5) these targets were achieved in less than 100 iterations. As expected, the quality of the results increases with the number of iterations. Columns 5 and 6 show the nominal skew and delay after 100 iterations. To verify the effectiveness of the desensitizaton phase, a Monte Carlo simulation of 500 trials was performed on the final clock tree by varying the widths of all wires using a normal distribution with $\sigma = 0.05$ microns (which is a reasonable assumption for today's submicron processes). The worst case skew results of the Monte Carlo simulation are shown in Column 7. Column 8 shows the maximum wire widths in the final clock tree. Note that, as expected, the widest wires appear near the root – they were widened during the desensitization phase.

6 Extensions to Clock Meshes

The first moment computation for an RC tree configuration has been shown to be a simple task [10]. We have also demonstrated the ease of calculation of sensitivities for RC tree structures. Most clock routes have tree topologies; however, recently there has been a great deal of interest in clock meshes. The calculation of the moments for RC meshes can be accomplished in a very efficient manner [1]. As for the sensitivities, we know that they can be calculated for RC meshes since the adjoint method applies to generalized circuits. We are, however, currently working on ways to calculate the sensitivities of RC meshes with acceptable efficiency. Such a capability would be extremely useful, since along with enabling wire-width optimization for RC clock meshes, an efficient adjoint sensitivity approach for meshes could be used to calculate the sensitivities with respect to adding loops of metal. In words, the sensitivities could aid in predicting the effect of adding a loop where there was none to begin with, to improve skew and/or reliability.

7 Conclusion

We have presented a technique to perform clock tree routing using wire width adjustment. Our algorithm yields low values for delay and skew for any type of tree configuration. Most importantly, we introduce the concept of reliable clock net design. Our approach increases the reliability of the clock net by considering the effect of process variations on the skew.

8 Acknowledgment

We wish to thank Ashok Balivada for his participation in the initial stages of this project. We also wish to thank Dr. Ren-Song Tsay for providing us with the example circuits.

References

[1] C. L. Ratzlaff, N. Gopal, and L. T. Pillage. "RICE: Rapid Interconnect Circuit Evaluator," *Proc. 28th ACM/IEEE Des. Auto. Conf.*, Jun 1991.

[2] W. C. Elmore. "The transient response of damped linear networks with particular regard to wideband amplifiers," *J. Applied Physics*, 19(1), 1948.

[3] L. T. Pillage and R. A. Rohrer. Asymptotic waveform evaluation for timing analysis. *IEEE Trans. Comp. Aided Design*, 9, 1990.

[4] Ren-Song Tsay. "Exact zero skew," *Proc. IEEE Int'l. Conf. Computer-Aided Des.*, Nov. 1991.

[5] M. A. B. Jackson, A. Srinivasan, and E. S. Kuh. "Clock routing for high performance ICs," *Proc. 27th ACM/IEEE Des. Auto. Conf.*, Jun 1990.

[6] Ting-Hai Chao, Yu-Chin Hsu, and Jan-Ming Ho. "Zero skew clock net routing," *Proc. 29th ACM/IEEE Des. Auto. Conf.*, Jun 1992.

[7] P. Penfield and J. Rubinstein. "Signal delay in RC tree networks," *Proc. 19th ACM/IEEE Des. Auto. Conf.*, 1981.

[8] A. Kahng, J. Cong and G. Robins. "High-performance clock routing based on recursive geometric matching," *Proc. 28th Des. Auto. Conf.*, 1991.

[9] S. W. Director and R. A. Rohrer. "The generalized adjoint network sensitivities," *IEEE Trans. Circuit Theory*, Vol. CT-16, no. 3, Aug. 1969.

[10] C. J. Terman. "Simulation tools for digital LSI design," PhD thesis, Massachusetts Institute of Technology, Sept. 1983.

[11] C. L. Ratzlaff, S. Pullela, and L. T. Pillage. "Modeling the RC-interconnect effects in a hierarchical timing analyzer," *Proc. IEEE Custom Integrated Circuits Conf.*, 1992

Topological Design of Clock Distribution Networks Based on Non-Zero Clock Skew Specifications

Jose Luis Neves and Eby G. Friedman

Department of Electrical Engineering
University of Rochester
Rochester, New York 14627

Abstract - A methodology is presented in this paper for designing clock distribution networks based on application dependent clock skew information. Contrary to previous approaches, the clock distribution network is designed such that the clock skew between two sequentially adjacent registers can be non-zero in order to maximize synchronous performance. The clock skew information is used to define the network topology and determine the delay values of the network branches. The design strategy for determining the topology of the clock distribution network considers the hierarchical description of the circuit to define the branching points of the clock tree. Algorithms to determine the optimal clock delay to each register and the network topology, given the non-zero clock skew specifications, are described and clock distribution networks are developed for example circuits.

1. INTRODUCTION

The distribution of clock signals is one of the primary limitations to maximizing the performance of synchronous integrated circuits. With the continuing reduction of feature size concurrent with increasing chip dimensions, interconnect delay has become increasingly significant, perhaps of greater importance than active device delay. Therefore, globally distributed signals, such as clock signals, can dominate and limit system performance. In this paper, focus is placed on the topological design of clock distribution networks, using a specified clock skew schedule to define the optimal clock delay to each register.

Several techniques have been developed to improve the performance of clock distribution networks, such as repeater insertion [1] to convert highly resistive-capacitive networks into effectively capacitive networks, symmetric distribution networks such as H-tree structures [2] to ensure minimal clock skew, and zero skew clock routing algorithms [3,4] to automatically layout these high speed networks in cell-based designs. These and similar approaches implement the clock distribution network so as to minimize the clock skew between each leaf of the clock distribution tree. However, these approaches neglect two fundamental properties of clock distribution networks. First, clock skew is only meaningful between sequentially adjacent registers, therefore there is no need to eliminate the skew between two clock signals that do not belong to the same data path. Second, it has been shown that clock skew [5] can be used to improve circuit performance by evening out the delay between slow and fast paths. This strategy permits the system clock frequency to be determined by the average path delay, instead of by the critical worst case path.

Clock skew is manifested by a lead/lag relationship between the clock signals that control a local data path, where a local data path is composed of two sequentially adjacent registers with, typically, combinational logic between them. If the clock delay to the initial register is less than the clock delay to the final register, the clock skew is described as negative. Likewise, if the clock delay to the initial register is greater than the clock delay to the final register, the clock skew is described as positive [6].

In this paper a methodology is proposed for designing synchronous clock distribution networks based on the optimal scheduling of the clock signals [7,8]. Given technology related information and the timing requirements of the circuit, such as the maximum permissible clock path delay and the aforementioned clock skew information, a topology of the clock distribution network is produced. The overall structure of the clock distribution network and the minimum delay values of each branch of the network are defined. The topological design of the clock distribution network exploits the hierarchical description of the circuit, although an approach is also presented for those circuits with minimal or no hierarchical information.

A theoretical background and an algorithm for determining the minimum clock delay of each clock signal, based on the optimal scheduling of the clock skew, is presented in section 2. In section 3, the delay values and the hierarchical information contained in the description of the circuit are used to design the topology of the clock distribution network. Experimental data describing several examples and discussion of these results are presented in section 4. Finally, section 5 summarizes the primary achievements of this paper and outlines further research.

2. DETERMINATION OF CLOCK PATH DELAY

The design of a clock distribution network requires a description of the circuit at the register transfer level and the desired clock skew between each pair of sequentially adjacent registers. The temporal characteristics of the functional data between each application-specific local data path are implicitly known from the clock skew information. The minimum clock delay of each register based on the clock skew data is obtained by a process called *negative clock skew* or *cycle stealing*, investigated independently by [9] and [10]. This approach shifts delay from the faster neighboring local data paths into the slower critical paths, thereby reducing the system-wide clock period and improving overall circuit performance.

2.1 Theoretical Background

To determine the minimum clock delay from the clock source to two sequentially adjacent registers, it is important to investigate if a relationship exists between the clock skews of the sequentially adjacent registers occurring within a global data path. Furthermore, it is necessary to describe the clock skew of global data paths which contain feedback paths. Before these issues are discussed, the concept of clock skew is first defined.

Definition 1: Given two sequentially adjacent registers, R_i and R_j, the clock skew between these two registers is defined as

$$T_{SKEW_{ij}} = T_{CD_i} - T_{CD_j}, \quad (1)$$

where T_{CD_i} and T_{CD_j} are the clock delays from the clock source to the registers R_i and R_j, respectively.

The path between two sequentially adjacent registers is described in this paper as a *local data path*; this is compared to a *global data path*, where a global data path can consist of one or more local data paths. The relationship between the clock skew of sequentially adjacent registers in a global data path is called conservation of clock skew and is presented below.

Theorem 1: For any given global data path, clock skew is conserved. Alternatively, the clock skew between any two registers which are not necessarily sequentially adjacent is the sum of the clock skews between each pair of registers along the global data path between those two same registers.

Although clock skew is defined between two sequentially adjacent registers, Theorem 1 shows that clock skew can exist between any two registers in a global data path. Therefore, it extends the definition of clock skew introduced by Definition 1 to any two non-sequentially adjacent registers belonging to the same data path. It also illustrates that the clock skew between any two non-sequentially adjacent registers which do not belong to the same global data path has no physical meaning.

A typical sequential circuit may contain sequential feedback paths, as illustrated in Figure 1. It is possible to establish a relationship between the

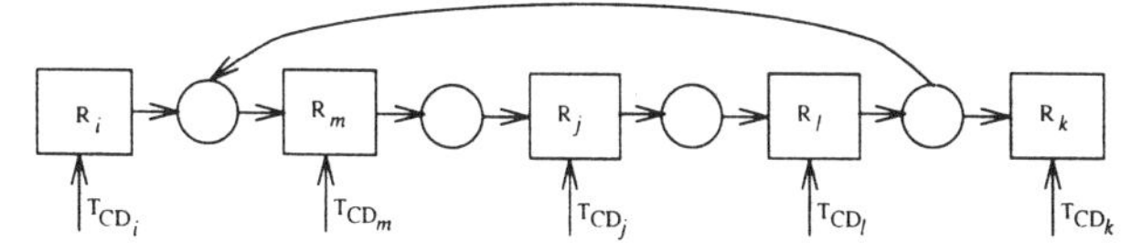

Figure 1: Data path with feedback paths

clock skew in the forward path and the clock skew in the feedback path, because the initial and final registers in the feedback path are also registers in the forward path. As shown in Figure 1, the initial and final registers in

Reprinted from *Proc. 36th Midwest Symp. Circuits Syst.*, pp. 468–471, Aug. 1993.

the feedback path R_l-R_j are the final and initial registers of the forward path R_j-R_k-R_l. This relationship is formalized below.

Theorem 2: For any given global data path containing feedback paths, the clock skew in a feedback path between any two registers, say R_l and R_j, is related to the clock skew of the forward path by the following relationship,

$$T_{SKEWfeedback,lj} = - T_{SKEWforward,jl} \quad (2)$$

2.2 Clock Path Delay Algorithm

A synchronous circuit is formed by one or more global data paths. For each global data path, the clock delay of the registers is calculated by first choosing the local data path with the largest clock skew. If the clock skew is positive (negative), the clock delay of the final (initial) register of the local data path is assigned a constant K, and the clock delay of the initial (final) register is assigned the constant K plus (minus) the clock skew of the final (initial) register, where the constant K is the minimum clock delay of the circuit. The clock delay of the following register connected to the final register of the local data path is the clock delay of the final register minus the clock skew between them. Similarly, the clock delay of the previous register connected to the initial register is the clock delay of the initial register plus the clock skew between them. Proceeding in this fashion for the other registers, the clock delay of each of the registers in the global data path is calculated in terms of K.

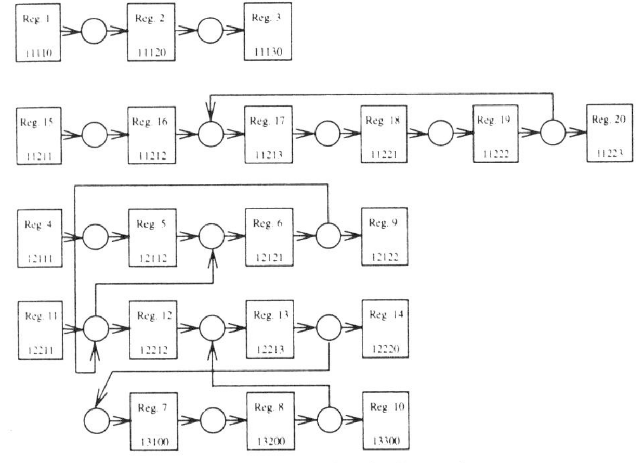

Figure 2: Example of a data path

To illustrate how clock delay is calculated, consider the synchronous circuit illustrated in Figure 2, where the rectangles represent registers and the circles represent combinational logic. Figure 3 illustrates the graph representation of this circuit, where a vertex represents a register instance, a directed edge represents a physical path between a pair of registers, and the edge weight represents the clock skew of each local data path. Consider, for example, the global data path, R_{15} to R_{20}. The maximum absolute clock skew is three for both local data paths (R_{15}, R_{16}) and (R_{16}, R_{17}), respectively, in Figure 2. Therefore, the clock delay to R_{16} from the clock source is the constant K, and the remaining clock delays are given as a function of K plus the local clock skew.

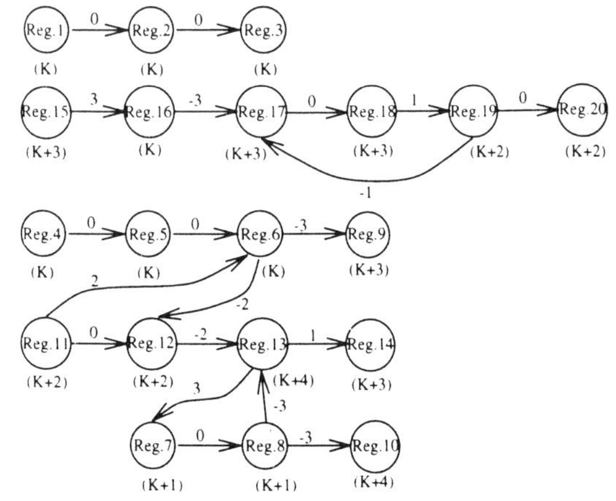

Figure 3: Optimal delay assignment for the data path example

The procedure to calculate the minimum clock delay is formalized in an algorithm and summarized in Figure 4 and called *Path_Delay*. After finding the local data path with the greatest clock skew, each of the clock delays to the other registers is calculated by traversing the global data path and attributing to each node the clock delay values that enforce the desired clock skew specification. If the initial clock skew specification satisfies Definition 1 and Theorems 1 and 2, each node is visited only once and the time complexity of the algorithm is $O(n)$, where n is the total number of nodes in the graph. The clock delay of each register obtained with this algorithm is represented by an expression in terms of K attached to each vertex, as illustrated in Figure 3.

Algorithm *Path_Delay*()
begin
  Circuit = $G(v,e)$;
  for each Datapath D **in** Circuit G **do**
  find MAX(T_{SKEWij});
  if $T_{SKEWij} < 0$ $v_i = K$; $v_j = K - T_{SKEWij}$;
  else $v_j = K$; $v_i = K + T_{SKEWij}$;
  while (v_i,v_j not visited) **and** (v_i,v_j in Datapath D) **do**
    Delay v_{j+1} = delay $v_j + T_{SKEWj,j+1}$;
    Delay v_{i-1} = delay $v_i + T_{SKEWi-1,i}$;
    Mark v_{i-1}, v_{j+1} as visited;
end *Path_Delay*

Figure 4: Algorithm to find the optimal clock delay to each register

Assuming a single clock source and a clock distribution network that produces the delay values found from the previous procedure, the clock skew requirements of the circuit can be satisfied. Such a network is illustrated in Figure 5 for the example shown in Figure 2, where the numerals inside the rectangles represent clock delays.

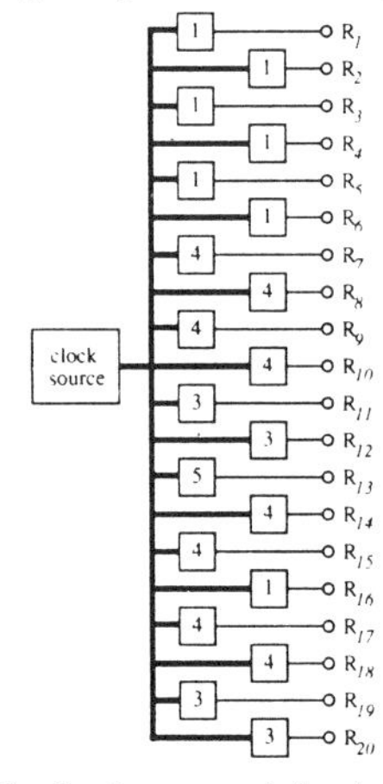

Figure 5: Clock distribution network for the data path example

3. TOPOLOGY OF CLOCK DISTRIBUTION NETWORK

In this section a methodology is presented for designing the topology of a clock distribution network that implements the delay values obtained from the algorithm *Path_Delay* introduced in section 2. A reasonable initial strategy is to provide an independent path delay for each register. This approach has the advantage of isolating each clock signal and requires a simple circuit composed of cascaded inverters and interconnect segments for which the behavior and delay models have been previously studied in great detail [1,11]. However, having independent clock paths drive each register is impractical because a typical VLSI circuit may contain many thousands of registers and, if each clock signal requires an independent clock path, significant chip area would be expended.

In this paper the clock distribution network for a specific circuit is designed in a tree structure using the hierarchical information of the original circuit description and the clock delay values obtained from algorithm *Path_Delay*. Subsection 3.1 illustrates the construction of the clock tree based on the hierarchical description of the circuit. The techniques to calculate the branch delays are presented in subsection 3.2. Subsection 3.3 describes a methodology for constructing clock distribution networks when minimal or no hierarchical information is available.

3.1 Construction of the Clock Tree Structure

Almost any large VLSI circuit is composed of several levels of hierarchy, where the number of levels is defined by the size and organization of the system and related factors such as the complexity of the circuit, the number of transistors, and the design methodology. During circuit placement, the elements of each hierarchical module are typically placed physically close to each other. Based on this reasoning, the hierarchical description of the circuit can be used to constrain the structure of the clock distribution network.

To build the clock distribution network, the hierarchy of the circuit is extracted from the circuit netlist and represented as a tree structure. The root vertex is the clock source, the internal vertices are the branching points derived from the hierarchy, and the leaf vertices are the registers. How the tree is built depends upon the unique code assigned to each register, which locates a register within the hierarchy of the circuit netlist. Figure 6 illustrates the hierarchical tree structure for the circuit example shown in

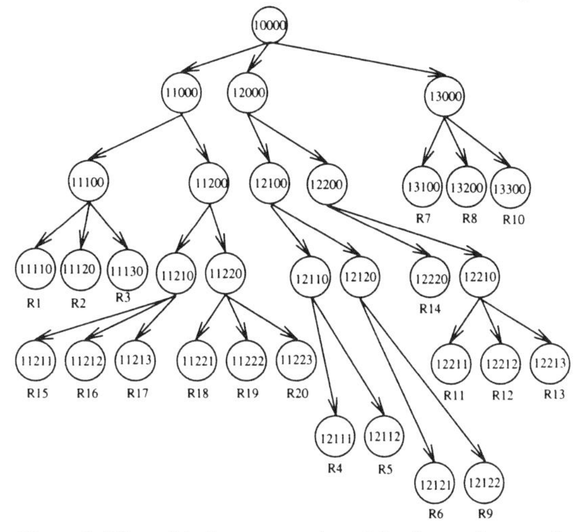

Figure 6: Hierarchical representation of the data path example

Figure 2. For the purpose of calculating the individual clock delays, the branches of the tree are classified as either external or internal. The external branches are the branches connected directly to the registers. All other branches are classified as internal.

3.2 Calculation of Branch Delay

The minimum clock delays obtained from the algorithm *Path_Delay* are calculated without regard to hierarchical information. However, when the hierarchy is considered, these delay values are insufficient to completely determine the delay of each branch of the clock distribution network because of two characteristics of the hierarchical representation:

1) The clock skew between sequentially adjacent registers may not depend on the delay of the internal branches. Consider, for example, the global data path R_1 to R_3 in Figure 2. From Figure 6, these registers are driven by the same branching point, meaning that the clock skew specifications are satisfied solely by the delay of the individual external branches, driving R_1, R_2, and R_3.

2) The clock skew specifications may be satisfied if the delay of some of the internal branches is known. As an example, the global data path formed by R_{15} to R_{20} contains three registers, R_{15} to R_{17}, driven by one branching point while the remaining three registers, R_{18} to R_{20}, are driven by a second branching point. The clock skew between registers R_{17} and R_{18} can only be satisfied if the delay of the internal and external branches driving these two registers is known.

The process of determining the individual branch delays is divided into three steps:

Delay of External Branches - When both registers within a local data path are driven by the same branching point, the clock skew specifications are completely satisfied by the delay values assigned to the external branches. These delay values are provided by *Path_Delay* and are expressed in absolute terms, provided that a value is assigned to the constant *K*. Since the delay of a branch cannot be zero, unit delay is assigned to *K* without loss of generality.

Delay of Internal Branches - It is possible however, to have a global data path driven by more than one branching point. For example, the global data path R_{15} to R_{20} is driven by two separate branches of the clock distribution network. Starting with the common vertex that drives the data path (vertex V_{11200}), variables are assigned to each of the branches, as illustrated in Figure 7(a). The clock delay equations are chosen to satisfy the clock skew specifications. The set of equations are

$$c - d = 3$$
$$d - e = -3$$
$$a + e = b + f$$
$$f - g = 1$$
$$g = h$$

This system of equations has multiple solutions, since more unknowns exist than constraints. To obtain a solution, the delay of the external branches is calculated initially, providing a solution for variables *c* to *h*, respectively. Rewriting these equations, it is found that $a = b$. Attributing unit delay to these variables, a solution, shown in Figure 7(b), satisfying these clock skew specifications is found.

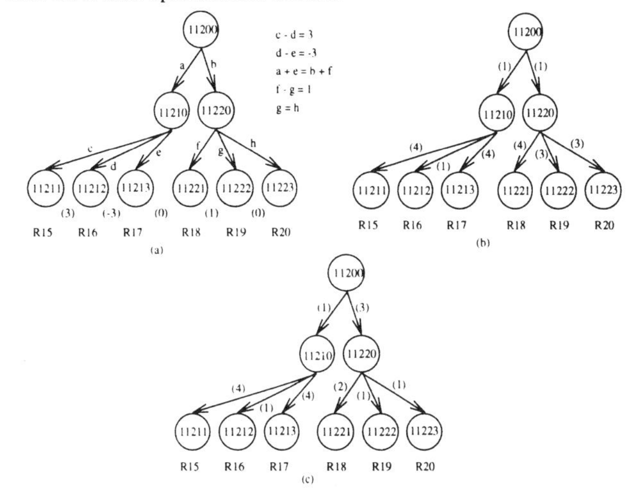

Figure 7: Calculation of the internal branch delays

Delay Shifting - It is possible to reorganize the delay of the external branches to reduce the total number of delay units needed to build the clock distribution network, thereby reducing the die area. Consider, for example, Figure 7(b). The external branches connected to registers R_{18} to R_{20} may each have their branch delay reduced by two, if the delay of the internal branch driving each of these registers is increased by two. Another advantage of shifting the delay is the increased flexibility of the circuit implementation, since the delay can be shifted between branches to accommodate for variations in the layout placement. An example of delay shifting is illustrated in Figure 7(c).

Extending this procedure to the other global data paths in the circuit of Figure 2, it is possible to define the minimum delay values for each branch of the clock distribution network. The complete delay specification for the clock distribution network required to synchronize the system depicted in Figure 2 is illustrated in Figure 8.

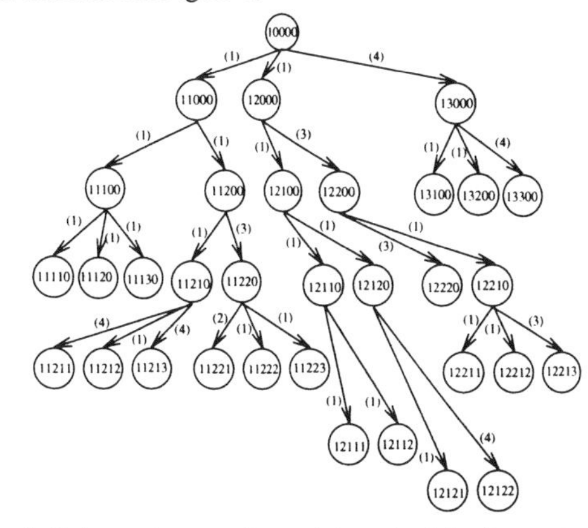

Figure 8: Delay assignment for each branch of the data path example

3.3 Reorganization of the Clock Tree

The methodology for constructing the clock distribution network presented in section 3 assumes that the circuit netlist is described hierarchically. It is possible, however, to have circuit descriptions which are completely or partially flat. For those circuits described non-hierarchically, the output of the previous section is a clock distribution network with independent clock paths for each register, similar to that shown in Figure 5. In this subsection, an approach for transforming these types of inefficient clock distribution networks into tree structures is presented. As an example, the clock distribution network for the flat representation of the circuit in Figure 2 is transformed below in Figure 9 into a clock tree with several levels of hierarchy.

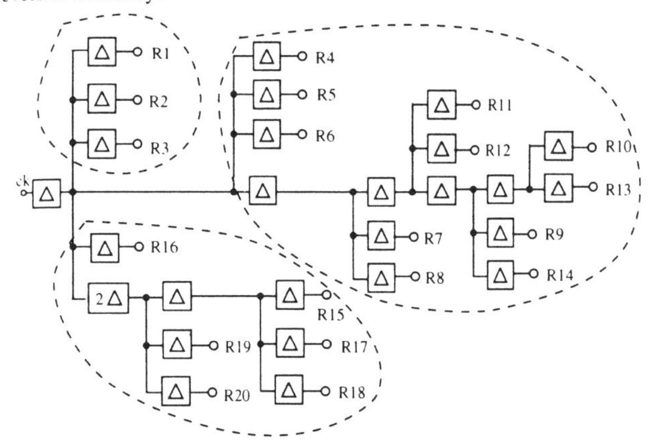

Figure 9: Delay assignment of non-hierarchically defined clock distribution network

Without regard to placement information, the clock distribution tree can be built by placing the fast clock paths in branches close to the clock source and the slow clock paths in branches farther from the clock source. For this purpose, the branch with the longest delay is partitioned into a series of segments, where each segment emulates a precise quantized delay value, Δ. Between any two segments, there is a branching point to other registers or sub-trees of the clock tree, where several segments with delay greater or equal to Δ are cascaded to provide the appropriate final delay at each leaf node. The result of this approach is illustrated in Figure 9, where a significant saving of segment delays, as compared to Figure 5, is obtained (from 56 to 28).

4. RESULTS

The algorithms for determining the minimum clock delay of each register and for calculating the clock delay and circuit topology of the clock distribution network have been implemented in C. Table 1 illustrates some of the features of the algorithms for several circuit examples. The circuit illustrated in Figure 2 is example *cir1*, cited in Table 1. The second example, *cir2*, illustrates a clock distribution network in which each of the registers of a global data path are all interconnected. The final circuits, *cir3* and *cir4*, exemplify the effects of having large number of registers driven by the same branching point.

Circuit	Registers	Minimum clock delay		Total delay units	
		No hierarchy	Hierarchy	No hierarchy	Hierarchy
cir1	20	5	8	56	57
cir2	15	7.2	9.2	57	53
cir3	56	11.1	13.2	184	115
cir4	37	8	10.6	163	124

Table 1: Minimum clock delay and total number of delay units for several example circuits

The second column in Table 1 shows the number of registers within each circuit. The third and fourth columns depict the maximum clock delay of the clock distribution network based on a non-zero clock skew schedule. The fourth column describes the maximum delay once the hierarchy of the circuit description is considered. The final two columns describe the total number of delay units required to implement the circuit without and with using the hierarchical information of the circuit. The total number of delay units is an indirect measure of the layout area required to implement the clock distribution network.

5. CONCLUSIONS

In this paper a strategy for designing the topology of a clock distribution network based on non-zero clock skew is presented. The primary result of this work is a description of the topology of the network with minimum delay values assigned to each branch of the clock tree, such that the clock skew specifications between any two sequentially adjacent registers are satisfied. The design of the topology exploits the hierarchical description of the circuit, although an approach for developing the topology of clock distribution networks for circuits with minimal or no hierarchical information is also described.

It is shown that the clock skew between any two registers belonging to the same global data path is the sum of the individual clock skews of the sequentially adjacent registers between those two registers. It is also shown that the clock skew of any feedback path between two registers is the negative of the clock skew of the forward path between the same two registers. These results are used in an algorithm that provides the optimal clock delay values for each individual clock path. Examples are described which depict the significant area advantages of tree structured clock distribution networks with a minimal increase in clock delay.

Future research includes both determining the optimal non-zero clocking characteristics and the design of circuit structures to emulate the delay values of the network branches. The circuit design process requires accurate delay models for both the buffers and the interconnect lines. The buffer delay model must consider the effects of ramp shaped waveforms and short-channel effects, such as velocity saturation and channel length modulation. The buffer delay model must also be integrated with an interconnect delay model which includes the effects of both parasitic capacitance and resistance. These results will permit the efficient synthesis of high speed clock distribution networks for application to high performance VLSI/ULSI-based synchronous systems.

REFERENCES

[1] H. B. Bakoglu, *Circuits, Interconnections and Packaging for VLSI*, Addison-Wesley, 1990.

[2] H. B. Bakoglu, J. T. Walker, and J. D. Meindl, "A Symmetric Clock-Distribution Tree and Optimized High-Speed Interconnections for Reduced Clock Skew in ULSI and WSI Circuits," *Proceedings of the IEEE International Conference on Computer Design*, pp. 118-122, October 1986.

[3] T.-H. Chao, Y.-C. Hsu, J.-M. Ho, K. D. Boese, and A. B. Kahng, "Zero Skew Clock Routing with Minimum Wirelength," *IEEE Transactions on Circuits and Systems-II: Analog and Digital Signal Processing*, Vol. CAS-39, No. 11, pp. 799-814, November 1992.

[4] R.-S. Tsay, "An Exact Zero-Skew Clock Routing Algorithm," *IEEE Transactions on Computer-Aided Design of Integrated Circuits and Systems*, Vol. CAD-12, No. 2, pp. 242-249, February 1993.

[5] E. G. Friedman and J. H. Mulligan Jr., "Clock Frequency and Latency in Synchronous Systems," *IEEE Transactions on Signal Processing*, Vol. SP-39, pp. 930-934, April 1991.

[6] E. G. Friedman, "Clock Distribution Design in VLSI Circuits - an Overview," *Proceedings of the IEEE International Symposium on Circuits and Systems*, pp. 1475-1478, May 1993.

[7] K. A. Sakallah, T. N. Mudge, O. A. Olukotun, "*checkTc* and *minTc*: Timing Verification and Optimal Clocking of Synchronous Digital Circuits," *Proceedings of the IEEE/ACM Design Automation Conference*, pp. 111-117, June 1990.

[8] T. G. Szymanski, "Computing Optimal Clock Schedules," *Proceedings of the IEEE/ACM Design Automation Conference*, pp. 399-404, June 1992.

[9] E. G. Friedman, *Performance Limitations in Synchronous Digital Systems*, Ph. D. Dissertation, University of California, Irvine, California, June 1989.

[10] J. P. Fishburn, "Clock Skew Optimization," *IEEE Transactions on Computers*, Vol. C-39, No. 7, pp. 945-951, July 1990.

[11] T. Sakurai, "Closed-Form Expressions for Interconnection Delay, Coupling, and Crosstalk in VLSI's," *IEEE Transactions on Electron Devices*, Vol. ED-40, No. 1, pp. 118-124, January 1993.

Circuit Synthesis of Clock Distribution Networks based on Non-Zero Clock Skew

José Luis Neves and Eby G. Friedman

Department of Electrical Engineering
University of Rochester
Rochester, NY 14627

ABSTRACT

A methodology is presented in this paper for synthesizing clock distribution networks by inserting circuit structures to emulate the delay values assigned to specified branches of the clock tree. These clock distribution networks are designed with localized non-zero clock skew [1] so as to improve circuit performance and reliability. The design methodology is targeted for CMOS technology.

The clock lines are transformed from distributed resistive-capacitive interconnect lines into purely capacitive interconnect lines by partitioning the RC interconnect lines with inverting repeaters. The inverters are specified by the geometric size of the transistors, the slope of the ramp shaped input/output waveform, and the output load capacitance. The branch delay model integrates both an inverter delay model and an interconnect delay model. Maximum errors of less than 3% for the delay of the clock paths and 6% for the clock skew between any two registers belonging to the same global data path are obtained as compared to SPICE Level-3.

1. INTRODUCTION

Most existing digital systems utilize fully synchronous timing, requiring a reference signal to control the temporal sequence of operations. Globally distributed signals, such as clock signals, are used to provide this synchronous time reference. These signals can dominate and limit the performance of VLSI-based systems. This is, in part, due to the continuing reduction of feature size concurrent with increasing chip dimensions. Thus interconnect delay has become increasingly significant, perhaps of greater importance than active device delay. Furthermore, the design of the clock distribution network, particularly in high speed applications, requires significant amounts of time, inconsistent with the high design turnaround of the more common data flow portion of a VLSI circuit.

Several techniques have been developed to improve the performance and design efficiency of clock distribution networks, such as repeater insertion [2] to convert highly resistive-capacitive networks into effectively capacitive networks, symmetric distribution networks [3], such as H-tree structures, to ensure minimal clock skew, and zero-skew clock routing algorithms [e.g., 4,5], to automatically layout clock nets. A common weakness of these approaches is that the clock distribution network is designed so as to minimize the clock skew between each register, not recognizing that localized clock skew [6,7] can be used to improve synchronous circuit performance and minimize the likelihood of any race conditions. Furthermore, no known techniques exist today that can automatically synthesize high speed and robust clock distribution techniques while including distributed buffers along the clock path. A novel methodology is therefore presented in this paper for efficiently synthesizing distributed buffer tree-structured clock distribution networks.

The methodology is divided into four major phases. The first phase is the determination of an optimal clock schedule [8,9], defining the localized clock skew schedule which maximizes circuit performance and reliability. In the second phase, a topological design of the clock distribution network is obtained [1], producing a clock tree with minimum delay values assigned to each branch. In the third phase, circuit structures are designed to implement the branch delay values. The final phase is the geometric layout of the clock distribution network. The third phase is the focus of this paper, although the second phase is briefly reviewed to motivate and provide background to the circuit synthesis of clock distribution networks.

In Section 2, the topological design of clock distribution networks is briefly reviewed. The process of implementing branch delay values with CMOS inverters is described in Section 3. Clock path delays derived from this approach are compared to SPICE and presented in Section 4. Finally, the primary results described in this paper are summarized in Section 5.

2. TOPOLOGICAL DESIGN

The topological synthesis of clock distribution networks is divided into three steps. In the first step, the minimum clock path delay of each register in the circuit is determined from the specified clock skew schedule. This task is made possible by recognizing that the clock skew between any two registers in the same global data path is the sum of the clock skews between each pair of registers along the data path formed by the two registers. Furthermore, the clock skew in a feedback path between any two registers is the negative of the clock skew in the forward path. In the next step, the topology of the clock distribution network is determined from the hierarchical description of the circuit. In the final step, delay values are attached to each branch of the clock distribution network, satisfying the initial clock skew assignment. A detailed explanation of each step can be found in [1].

Figure 1 illustrates the topology of a clock distribution network of an example circuit composed of twenty registers.

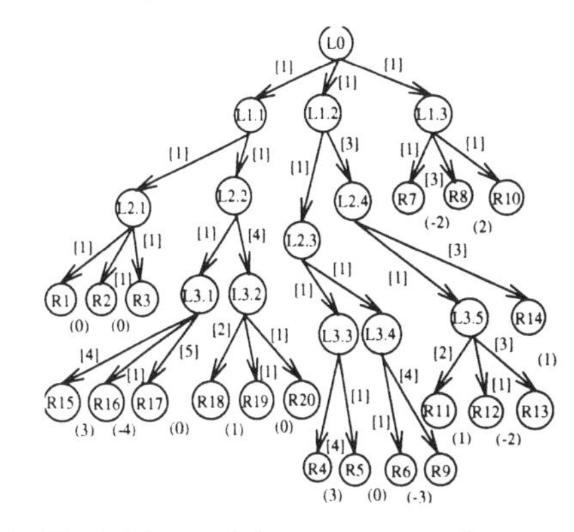

Figure 1: Clock delay and skew assignment for an example clock distribution network

Reprinted from *Proc. IEEE Int'l Symp. Circuits Syst.*, pp. 4.175–4.178, June 1994.

The numbers in parenthesis are the original clock skew specifications, derived from the optimal clock scheduling phase, while the numbers in brackets are the minimum delay values assigned to each branch.

The clock skew between two sequentially adjacent registers is defined in this paper as positive if the clock signal arrives at the final register before arriving at the initial register. The clock skew is negative if the clock signal arrives at the initial register before arriving at the final register. The clock skew is zero if the clock signal arrives at both registers at the same time [6].

3. DESIGN OF CIRCUIT DELAY ELEMENTS

The delay of the circuit structures that emulates the delay values associated with each branch of the network requires high precision, because variations in the delays of the internal branches are propagated throughout the network, causing unacceptable variations in the desired clock skew. It is important to note that it is much more difficult and significant to satisfy the clock skew *between* any two clock paths rather than to satisfy each individual clock path delay.

Delay elements can either be composed of passive or active circuit elements. Implementing the delay elements within the clock distribution network as a passive RC network is unacceptable for several reasons; 1) the delay of each branch is highly dependent on the delay of every other branch, 2) the clock signal waveform would degrade, limiting system performance and reliability, 3) an accurate delay model of a passive clock distribution network for a circuit with thousands of registers is difficult to obtain, and 4) the layout of the passive RC network is highly sensitive to small variations in position or length of the clock lines, producing unacceptable variations in the localized clock skew. Therefore, two criteria must be met to successfully synthesize a clock distribution network. First, the delay of each branch must be implemented such that each branch is independent of the delay of the other branches. Second, the clock branches must be designed such that no physical layout constraints are created which are difficult to implement.

To satisfy both of these criteria, the strategy adopted is to implement the delay segments with active elements, specifically CMOS inverters. Due to the high input impedance of a CMOS inverter, the inverter effectively isolates each clock branch from each other. Additionally, the interconnect lines can be modeled as purely capacitive lines by properly inserting these distributed CMOS inverters as repeaters along the clock signal path [2]. The insertion points are chosen such that the output impedance of each inverter is much greater than the resistance of that portion of the driven interconnect line. This strategy permits the length of a single interconnect line to be accurately modeled as a lumped capacitance with negligible resistance. However, the strategy also places a maximum constraint on the length of an individual portion of an interconnect line between inverting repeaters, thereby limiting the placement of a clock branch within the circuit layout.

3.1 Preserving Clock Signal Polarity

Using a single inverter to produce a specific branch delay may invert the polarity of the clock signal for those clock paths consisting of an odd number of branches. To maintain the proper signal polarity, the tree structured graph representing the topology of the clock distribution network is searched to identify those branches requiring two inverters, ensuring that the number of inverters from the clock source to every register remains even (or odd), while utilizing a minimum number of inverters. As shown in Figure 1, the clock paths that require an extra inverter are those branches which drive R_1, R_2, R_3, and R_{14}, since these paths have an odd number of branches.

3.2 Implementation of Clock Path Delay

Since a signal path in a clock distribution network is typically composed of more than one branch, the total delay of the clock path t_{cpd} is given by the summation of the delays of each individual branch along the clock path τ_{bi}, as

$$t_{cpd} = \sum_{i=1}^{n} \tau_{bi} \tag{1}$$

In order to accurately sum the individual delay components along a clock signal path, four circuit characteristics must be considered: the output impedance of the inverter as compared to the resistance of the interconnect lines driven by the inverter, the capacitance of the interconnect line as compared to the input capacitance of the branches and/or registers driven by the inverter, the output waveform shape of the driving inverter, and the input waveform shape of the driven inverters. These circuit characteristics are considered in order to 1) isolate each branch delay, 2) determine the geometric size of the transistors and load capacitance of each branch, and 3) integrate the inverter and interconnect delay equations used to calculate the delay of each clock path.

The capacitive load at the output of the inverting buffer is composed of the capacitance of an interconnect line plus the input capacitance of each inverter driven by that interconnect line, shown as

$$C_L = C_{line} + \#\ of\ branches * C_g, \tag{2}$$

where Cg is the input gate capacitance of an inverter and the inverters are assumed to be of equal size. If the inverters are not of equal size, the second term in (2) becomes a summation among all the branches connected to the driving point.

The output waveform of an inverter belonging to one branch is the input waveform of the inverter(s) driving the following branch. This circuit configuration is illustrated in Figure 2, where s_i is the slope of the input signal, s_o is the slope of the output signal, and C_{Li} is the capacitive load of each branch i.

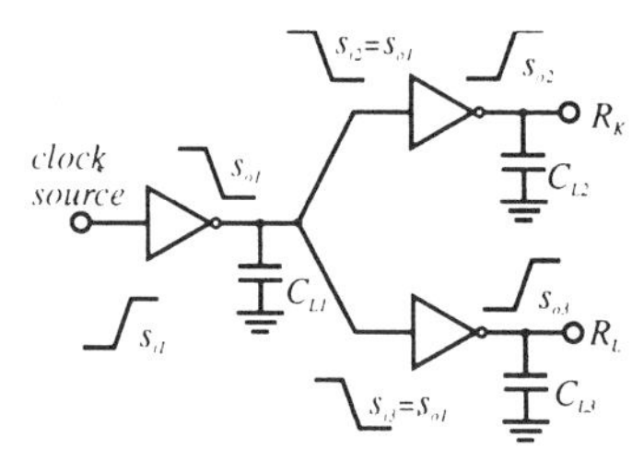

Figure 2: Integration of input/output waveform in a clock path

Under the assumption that the output impedance of the inverter is much larger than the interconnect resistance, the interconnect lines are modeled as lumped capacitors and the slope of the output waveform of the driving inverter is equal to the slope of the input waveform of the driven inverter(s). The slope of the input/output waveform can be characterized by two parameters, the geometric size and the capacitive output load of the driving inverter. Since the capacitive load includes the interconnect line capacitance and the input gate capacitance of each driven inverter, the transistor size and output load of each branch are each determined such that the branch delay and the total delay of a clock path are both satisfied.

3.3 Branch Delay Modeling

The delay equations describing the inverting repeater, shown in (3)-(5), are used to determine the geometric dimensions of the transistors and are based on the MOSFET α-power law I-V model developed by Sakurai and Newton [10].

$$I_{DS} = \begin{cases} 0 & (V_{GS} \le V_{th} \quad : cutoff) \\ \left(\frac{I'_{DO}}{V'_{DO}}\right)V_{DS} & (V_{DS} < V'_{DO} \quad : linear) \\ I'_{DO} & (V_{DS} \ge V'_{DO} \quad : saturation) \end{cases} \quad , (3)$$

where

$$I'_{DO} = I_{DO}\left(\frac{V_{GS} - V_{th}}{V_{DD} - V_{th}}\right)^{\alpha} = \frac{W}{L_{eff}} P_C(V_{GS} - V_{th})^{\alpha} \quad , (4)$$

$$V'_{DO} = V_{DO}\left(\frac{V_{GS} - V_{th}}{V_{DD} - V_{th}}\right)^{\alpha/2} = \frac{W}{L_{eff}} P_V(V_{GS} - V_{th})^{\alpha/2} \quad , (5)$$

and where I_{DO} is the drain current at $V_{GS} = V_{DS} = V_{DD}$, V_{DO} is the drain saturation voltage at $V_{GS} = V_{DD}$, V_{th} is the threshold voltage, α is the velocity saturation index, and V_{DD} is the supply voltage. The parameters α, V_{DO}, I_{DO}, and V_{th} are calculated as explained in [10].

The input of each clock branch is assumed to be driven by a ramp signal with rising and falling slopes, s_r and s_f, respectively, selected such that during discharge (charge), the effects of the PMOS (NMOS) transistor can be neglected [11]. The circuit structure of the inverter is illustrated in Figure 3a, with the shape of the input and output waveforms illustrated in Figure 3b.

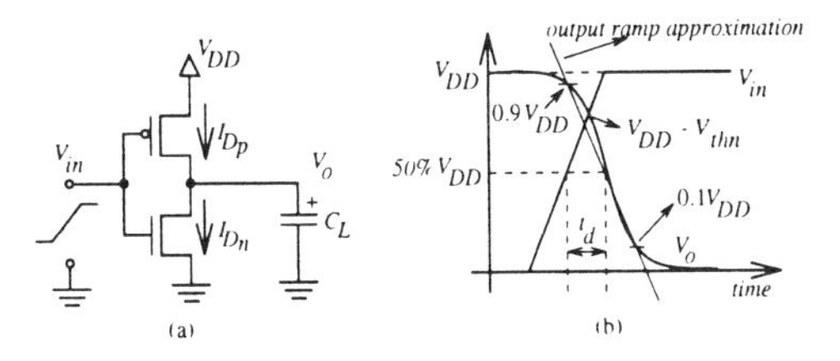

Figure 3: (a) Delay element; (b) Input/output waveforms of the delay element

The time t_d from the 50% V_{DD} point of the input waveform to the 50% V_{DD} point of the output waveform is defined as the delay of the circuit element. Equation (6), derived from (3)-(5) [10], describes the delay of a CMOS inverter in terms of its load capacitance C_L.

$$C_L = \frac{2I_{DO}}{V_{DO}}\left[t_d - \left(\frac{1}{2} - \frac{1 - v_T}{1 + \alpha}\right)s_r\right], \quad where\ v_T = \frac{V_{th}}{V_{DD}} \quad (6)$$

The output waveform of the driving inverter is the input signal to all the branches connected to this inverter and is approximated by a ramp shaped waveform. This approximation is achieved by linearly connecting the points $0.1V_{DD}$ and $0.9V_{DD}$ of the output waveform and is accurate as long as the interconnect resistance is negligible as compared to the inverter output impedance. The slope is expressed as

$$s_r = \frac{t_{0.9} - t_{0.1}}{0.8} = \frac{C_L V_{DD}}{I_{DO}}\left(\frac{0.9}{0.8} + \frac{V_{DO}}{0.8V_{DD}} ln\frac{10V_{DO}}{eV_{DD}}\right) \quad . (7)$$

Equations (6) and (7) provide the necessary relationships to design the circuit elements of a clock signal path, as explained below:

Design of clock signal path - The initial branch of each clock signal path is the branch connected directly to the clock source. The input slope is assumed zero for this branch, since the source of the clock signal is assumed to behave as a step function. Given the branch delay, the capacitive load of the inverter is determined from (6). The effective interconnect capacitance is the capacitance derived from (6) minus the gate capacitance of the branches connected to the driving buffer. The slope of the output signal is obtained from (7) and is the input slope of the following stage. To design the circuit structures of the following branches, the previous steps are used recursively by first calculating the required load capacitance, and then using this value of capacitance to calculate the slope of the signal waveform of the following stage. This procedure is continued until each register driven by the clock distribution network is reached.

For a certain range of branch delay, the load capacitance obtained with (6) may be small enough to be of the same order of magnitude as the drain capacitance of the buffer, compromising the accuracy of the buffer delay. To overcome this uncertainty, the load capacitance is increased, requiring the driving buffer to be resized by the same order of magnitude. Assuming that the geometric channel length remains fixed, the width of a resized transistor is [10]

$$I_{DO(new)} = \frac{W_{new}}{W_{measured}} I_{DO(measured)} \quad (8)$$

Equations (6)-(8) are sufficient to determine the geometry and load capacitance of every inverter along the branches of the clock distribution network.

4. EXPERIMENTAL RESULTS

Table 1 compares the difference between the calculated and measured clock path delays for the circuit shown in Figure 1. The second column depicts the desired delay obtained from the topological and circuit design of the clock distribution network. The third column shows the delay values of each clock path derived from SPICE circuit simulation using Level-3 device models, while the fourth column depicts the per cent error between the calculated and the numerically derived delay, where the maximum error is less than 3%.

Table 1: Comparison between calculated and measured clock path delay

Clock Path	Delay (ns)	SPICE (ns)	Error (%)
R_1, R_2, R_3	3.0	3.07	2.3
R_{15}, R_4, R_9	7.0	7.11	1.6
R_{18}	8.0	8.10	1.3
R_{19}, R_{20}	7.0	7.04	<1
R_5, R_6, R_{16}	4.0	4.06	1.5
R_{17}	8.0	8.06	<1
R_{14}	7.0	7.17	2.4
R_{11}	7.0	7.16	2.3
R_{12}	6.0	6.14	2.3
R_{13}	8.0	8.20	2.5
R_7, R_{10}	2.0	1.98	1.0
R_8	4.0	4.07	1.8

A more significant measure of the effectiveness of this clock distribution network design methodology is to guarantee that the clock skew between any pair of registers in the same global data path is accurately satisfied, rather than the delay of each individual clock path. Table 2 illustrates the clock skew between registers for the circuit illustrated in Figure 1. Column two shows the scheduled clock skew implemented with the design methodology described in this paper for the pair of registers presented in column one. Column three depicts the values obtained from SPICE circuit simulation, while column four shows the per cent error between both measurements. Note that the maximum error is 6%, a number well within practical and useful limits.

Table 2: Comparison between specified and measured clock skew values

Registers	Specified Skew (ns)	Measured (ns)	Error (%)
R_1 - R_3	0.0	0.0	0.0
R_{15} - R_{16}	3.0	3.0	0.0
R_{12} - R_{13}	-2.0	-2.06	3.0
R_{17} - R_{19}	1.0	1.03	3.0
R_4 - R_{14}	0.0	0.06	6.0
R_6 - R_{12}	-2.0	-2.08	4.0
R_5 - R_{13}	-4.0	-4.14	3.5

The individual data paths have been selected to illustrate several types of clock skew situations, such as non-zero clock skew between registers in the same data path or in separate data paths. More specifically, zero clock skew between registers in the same data path is illustrated by the path between registers R_1 and R_3. The path between registers R_{15} and R_{16} illustrates positive clock skew for registers in the same data path, while the path between registers R_{12} and R_{13} illustrates negative clock skew for registers in the same data path. In these three examples, the clock skew is only dependent upon the delay of the external branches, and therefore these clock paths are independent of the delay variations of the internal branches of the clock distribution network. Other examples are more illustrative of the possible effects of internal branch delays within the clock path. The path between registers R_{17} and R_{19} illustrates non-zero clock skew in a data path with feedback which is dependent on the delay of its internal branches. The last three examples illustrate the clock skew between two registers belonging to interconnected data paths. The first example, the path between registers R_4 and R_{14}, illustrates zero clock skew between two registers. The final examples, the path between registers R_5 and R_{13} and registers R_6 and R_{12}, illustrate the effects of negative clock skew on two registers. Observe that in the last examples, the error tolerance of the clock skew is still within acceptable margins, exhibiting good accuracy, even for those paths in which the clock skew between two registers have a significant portion of the clock distribution network in common.

5. CONCLUSIONS

VLSI/ULSI-based synchronous systems require the efficient synthesis of high speed clock distribution networks in order to obtain higher levels of circuit performance. In this paper, circuit performance is improved by using non-zero localized clock skew to reduce the minimum clock period. An integrated methodology is presented for synthesizing clock distribution networks and is divided into four phases, 1) optimal clock scheduling, 2) topological design of the clock distribution network, 3) design and modeling of the circuit delay elements, and 4) layout implementation. The focus of this paper is the third phase, the design and modeling of the circuit delay elements.

A strategy for implementing the branch delay values of the clock distribution network is presented, using CMOS inverters for the delay elements. The minimum number of inverters to satisfy the branch delay is obtained, while preserving the polarity of the signal driving the clock input of each register. Delay equations of an inverter, derived from the α-power law I-V model, are described. The inverter delay model accurately determines the delay of each clock path by considering the fanout, interconnect capacitance, and the slope of the input and output waveforms of each branch along the clock path. Comparisons between expected and simulated delays of each individual clock signal path produce circuits with a maximum error of less than 3%. Furthermore, the maximum error between scheduled and simulated clock skew for any two registers belonging to the same global data path is 6%.

Thus, an integrated methodology for synthesizing clock distribution networks for high performance VLSI circuits is presented. This methodology, based on inserted delay elements, accurately synthesizes localized clock skews. Furthermore, this methodology utilizes non-zero clock skew to improve overall system performance.

REFERENCES

[1] J. L. Neves and E. G. Friedman, "Topological Design of Clock Distribution Networks Based on Non-Zero Clock Skew Specifications," *Proceedings of the IEEE 36th Midwest Conference on Circuits and Systems*, August 1993 (in press).

[2] H. B. Bakoglu, *Circuits, Interconnections and Packaging for VLSI*, Addison-Wesley, 1990.

[3] H. B. Bakoglu, J. T. Walker, and J. D. Meindl, "A Symmetric Clock-Distribution Tree and Optimized High-Speed Interconnections for Reduced Clock Skew in ULSI and WSI Circuits," *Proceedings of the IEEE International Conference on Computer Design*, pp. 118-122, October 1986.

[4] T.-H. Chao, Y.-C. Hsu, J.-M. Ho, K. D. Boese, and A. B. Kahng, "Zero Skew Clock Routing with Minimum Wirelength," *IEEE Transactions on Circuits and Systems-II: Analog and Digital Signal Processing*, Vol. CAS-39, No. 11, pp. 799-814, November 1992.

[5] R.-S. Tsay, "An Exact Zero-Skew Clock Routing Algorithm," *IEEE Transactions on Computer-Aided Design of Integrated Circuits and Systems*, Vol. CAD-12, No. 2, pp. 242-249, February 1993.

[6] E. G. Friedman, "Clock Distribution Design in VLSI Circuits - an Overview," *Proceedings of the IEEE International Symposium on Circuits and Systems*, pp. 1475-1478, May 1993.

[7] J. P. Fishburn, "Clock Skew Optimization," *IEEE Transactions on Computers*, Vol. C-39, No. 7, pp. 945-951, July 1990.

[8] K. A. Sakallah, T. N. Mudge, O. A. Olukotun, "*checkTc* and *minTc*: Timing Verification and Optimal Clocking of Synchronous Digital Circuits," *Proceedings of the IEEE/ACM Design Automation Conference*, pp. 111-117, June 1990.

[9] T. G. Szymanski, "Computing Optimal Clock Schedules," *Proceedings of the IEEE/ACM Design Automation Conference*, pp. 399-404, June 1992.

[10] T. Sakurai and A. R. Newton, "Alpha-Power Law MOSFET Model and its Applications to CMOS Inverter Delay and Other Formulas," *IEEE Journal of Solid State Circuits*, Vol. SC-25, No. 2, pp. 584-594, April 1990.

[11] N. Hedenstierna and K. O. Jeppson, "CMOS Circuit Speed and Buffer Optimization," *IEEE Transactions on Computer-Aided Design*, Vol. CAD-6, No. 2, pp. 270-281, March 1987.

The Application of Localized Clock Distribution Design to Improving the Performance of Retimed Sequential Circuits *

E G Friedman
Department of Electrical Engineering
University of Rochester, New York, USA

Clock distribution networks synchronize the flow of data signals between data paths and the design of these networks can significantly affect system performance. By exploiting specific lead/lag clock waveform relationships, the delay of the worst case paths can be minimized by shifting time from sequentially adjacent less critical paths. Thus, as retiming is performed on pipeline registers by shifting their temporal locations, local clock distribution networks are used to tune the effective path delays of each local data path. Thus, retiming can be considered a two-phase process, logical partitioning for gross temporal allocation followed by local clock distribution design. By applying localized clock distribution design to the critical path delays of a large system, both maximum and minimum delay paths can be adjusted such that these paths will be of approximately equal delay, improving both system performance and reliability and permitting the realization of higher performance retimed circuits.

1. INTRODUCTION

Retiming is a behavioral synthesis methodology for improving synchronous performance in VLSI-based digital systems by automating the process of pipelining for parallel concurrency such that logic elements are shifted across sequentially adjacent registers so as to minimize the delay of the critical worst case paths [1-3]. In this synthesis technique, the total latency is maintained constant; the objective is to choose an optimal temporal location for each of the pipeline registers so as to minimize the maximum clock period of any local data path. Thus, the system operating clock frequency is maximized by adjusting the relative temporal placement of the pipeline registers. The approach presented in this paper integrates localized clock distribution design with retiming in order to improve overall synchronous performance as well as provide greater temporal resolution of the retiming process.

In presenting these results, the paper is composed of five principal sections. A brief overview of a synchronous system and its component parts is presented in section 2. The nature of the clock distribution network directly affects the performance and reliable operation of a synchronous digital system. Minimum and maximum constraint relationships as well as design techniques to improve performance are described in Section 3. In Section 4, latency and clock frequency relationships are developed in terms of the circuit and timing characteristics of a data path. Practical design equations are provided for analyzing the performance of retimed circuits. An example of applying localized clock distribution design to a pipelined system is provided in section 5 and some conclusions are presented in section 6.

2. HIGH PERFORMANCE SYNCHRONOUS DATA PATHS

A synchronous digital system is composed of data paths in which data are moved from a register through some logic functions and into a second register. The synchronization of the data flow between the initial and final registers is typically coordinated by a single control signal, commonly called the clock signal. Thus, a synchronous digital system, as depicted in Figure 1, is composed of three interrelated systems [4,5]:

1. the combinatorial network which performs the logical function of the digital system,
2. the registers which store the data signals awaiting the synchronizing clock pulse, and

*This research was supported in part by NSF Grant MIP-9208165

Reprinted from *Proc. IEEE Asia-Pacific Conf. Circuits Syst.*, pp. 12–17, Dec. 1992.

3. the clock distribution network which generates the synchronizing clock pulse and defines when data can flow from one register to the next.

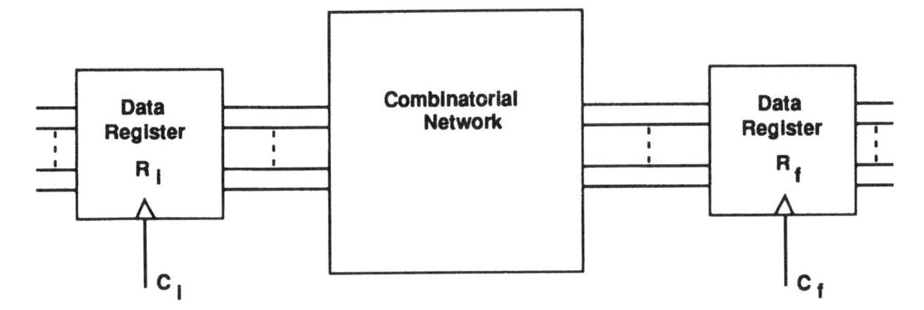

Figure1: Synchronous Data Path

The total delay of a data path is determined by the time required to leave the initial register once the clock signal C_i arrives, T_{c-Q}, the time necessary to propagate through the logic and interconnect, $T_{logic} + T_{int}$, and the time required to successfully propagate to and latch within the final register of the data path, T_{set-up} [6].

$$T_{PD} = T_{c-Q} + T_{logic} + T_{int} + T_{set-up} \ . \qquad (1)$$

In (1), the individual delay components making up the total delay T_{PD} of a data path are summed. Those data paths whose total delay plus any clock skew are greatest represent the critical worst case timing requirements of a digital system, and the delay and clock skew of these paths must be minimized in order to maximize the performance of the entire digital system. Thus, in a high performance synchronous digital system, the critical paths constrain and define the maximum performance of the entire system. Therefore, the goal in designing a high performance system is to minimize each delay component in (1) as well as to utilize any possible advantages (and minimize any possible disadvantages) of the clock distribution circuitry which will increase the speed of operation of the critical data paths.

A general form of a data path is shown in Figure 2, where an initial register R_i begins the data path and is followed by N stages of logic and N+1 sections of interconnect, ending in a final register R_f. Each interconnect section is represented as a single pole time constant and designated as T_i, where i represents each logic and interconnect stage and N is the total number of logic stages. Thus, in a data path composed of N logic stages, there are N+1 interconnect sections.

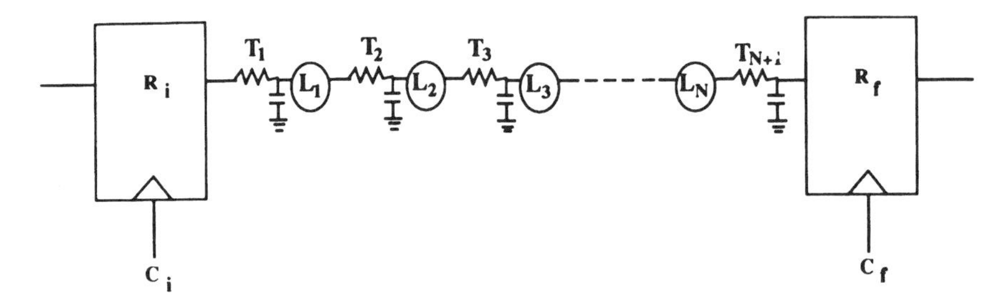

Figure 2: Synchronous Data Path with N Stages of Logic

Since logic paths are composed of only logic stages and interconnect sections, the total delay through a logic path can be modeled as the sum of the delay through the individual logic stages and interconnect sections. For convenience in representing the delay through the system, it is desirable to combine each individual RC interconnect section and logic stage as a single delay component of the logic path. This permits one to define the time required for the data signal to propagate through the ith distributed RC interconnect section T_i and logic stage L_i as T_{fi}.

The minimum possible latency L_{min} of a data path, for a given function, occurs when it is completely unpipelined. For N logic stages traversed between the input and output of the system,

L_{min} can be expressed as

$$L_{min} = \sum_{i=1}^{N} T_{fi} \quad . \tag{2}$$

When registers are inserted into the data path from system input to output (defined as the global data path), the minimum clock period can be decreased (providing a higher maximum clock frequency), albeit with an increase in latency. Each global data path is composed of individual cascaded register-to-register data paths and these are defined as local data paths. Each local data path is composed of an initial and final register and typically, n logic stages between them. Note that each register within a local data path performs double duty, serving as the initial (final) and final (initial) register of the current and previous (next) local data path, respectively.

In order to improve computational concurrency, the process of pipelining and retiming is applied to these systems. In the pipelining process, registers are inserted into the data path in order to provide temporal concurency of the data flow and increased system clock frequency. Retiming then follows the initial pipelining process by shifting the location of the pipeline registers in order to minimize the maximum clock period.

3. CLOCK DISRIBUTION NETWORKS

In a synchronous digital system, the global clock signal is used to define a relative time reference for all movement of data within that system. Because this function is vital to the operation of a synchronous system, much attention has been given to the characteristics of these clock signals and the networks used in their distribution [4,5,7-17]. Most synchronous digital systems consist of cascaded banks of sequential registers with combinatorial logic between each set of registers. The functional requirements of the digital system are satisfied by the logic stages, while the global performance and local timing requirements are satisfied by the careful insertion of pipeline registers into equally spaced time windows to satisfy critical worst case timing constraints [4,5,7-10,13] and the proper design of the clock distribution network to satisfy critical timing requirements as well as to ensure that no race conditions exist [7,9,10].

Each data signal typically is stored in a latched state within a bistable register awaiting the incoming clock signal to define when the data should leave the register. Once the enabling clock signal reaches the register, the data signal leaves the bistable register and propagates through the combinatorial network, and for a properly working system, enters the next register and is fully latched into that register before the next clock signal appears [6,8]. C_i and C_f represent the clock signals driving the initial register and the final register, respectively, and both originate from the same clock signal source. The clock delay of the initial clock signal T_{Ci} and the final clock signal T_{Cf} define the time reference when the data signals begin to leave their respective registers. These clock signals originate from a clock distribution network which is designed to generate a specific clock signal waveform which synchronizes each register [4-10]. The difference in delay between two sequentially adjacent clock paths, as shown in (3), is the clock skew T_{SKEW}. If the clock signals C_i and C_f are in complete synchronism (i.e., the clock signals arrive at their respective registers at exactly the same time), the clock skew is zero.

$$T_{SKEW} = T_{Ci} - T_{Cf} \quad . \tag{3}$$

The minimum allowable clock period T_{cpmin} between two registers in a sequentially adjacent data path is given by [4,5]

$$\frac{1}{f_{clkMAX}} = T_{cp_{min}} \geq T_{PD} + T_{SKEW} \quad , \tag{4}$$

where T_{PD} is defined in (1) and T_{SKEW} can be positive or negative depending on whether C_f leads or lags C_i, respectively. A timing diagram depicting each delay component in (1) in terms of the clock period is shown in Figure 3. These waveforms show the timing requirement of (4) being barely satisfied.

3.1 Maximum Data Path/Clock Skew Constraint Relationship

For a design to meet its specified timing requirements, the greatest collective propagation delay of any data path between a pair of data registers, R_i and R_f, being synchronized by a clock distribution network must be less than the inverse of the maximum clock frequency of the circuit as shown in (4). If the time of arrival of the clock signal at the final register of a data path T_{Cf} leads that of the time of arrival of the clock signal at the initial register of the same sequential data path T_{Ci} (see Figure 4A), the clock skew is referred to as positive clock skew and, under this condition, the maximum attainable operating frequency is decreased [4,5,7-10]. Positive clock skew is the additional amount of time which must be added to the minimum clock period to reliably apply a new clock signal at the final register, where reliable operation implies that the system will function correctly at low as well as at high frequencies.

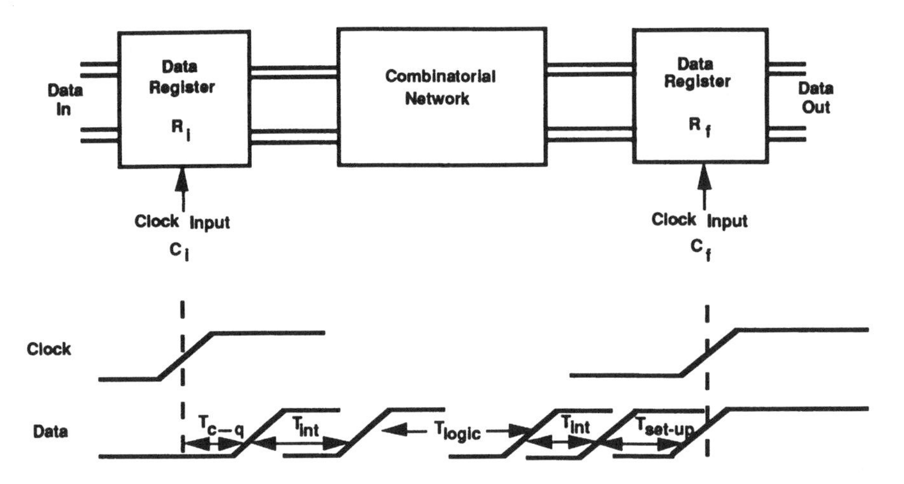

Figure 3: Timing Diagram of Clocked Data Path

In the positive clock skew case, the clock signal arrives at R_f before it reaches R_i. From (1) and (4), the maximum permissible positive clock skew can be expressed as

$$T_{SKEW} \leq T_{cp} - (T_{c\text{-}Q} + T_{logic} + T_{int} + T_{set\text{-}up}) \quad \text{for } T_{Ci} > T_{Cf} \quad . \qquad (5)$$

This situation is the typical critical path timing analysis requirement commonly seen in most high performance synchronous digital systems. In circuits where positive clock skew is significant and (5) is not satisfied, the clock and data signals should be run in the same direction, thereby forcing C_f to lag C_i and making the clock skew negative.

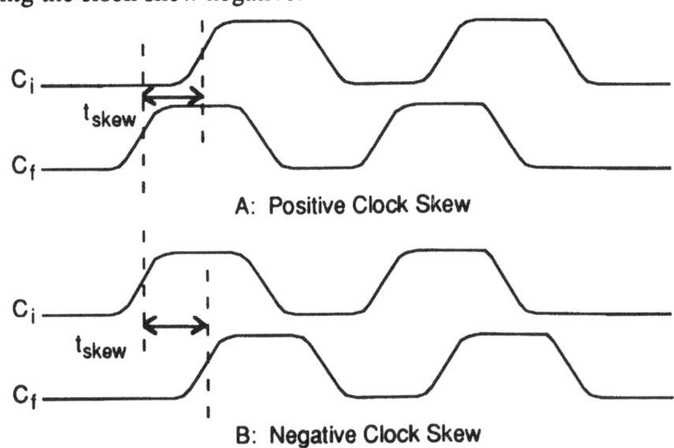

Figure 4: Clock Timing Diagrams

3.2 Minimum Data Path/Clock Skew Constraint Relationship

If the clock signal arrives at R_i before it reaches R_f (see Figure 4B), the clock skew is defined as being negative. Negative clock skew can be used to improve the maximum performance of a synchronous system by decreasing the delay of a critical path; however, a potential minimum constraint can occur [4,5,7,9,10], creating a race condition. In this case, the clock skew, when C_f lags C_i, must be less than the time required for the data to leave the initial register, propagate through the interconnect and combinatorial logic, and set-up in the final register (see Figure 3). If this condition is not met before the data stored in register R_f can be shifted out of R_f, it is overwritten by the data that had been stored in register R_i and had propagated through the combinatorial logic. Correct operation requires that R_f latches data which corresponds to the data R_i latched during the previous clock period. This constraint on clock skew is

$$|T_{SKEW}| \leq T_{PD} = T_{c\text{-}Q} + T_{logic} + T_{int} + T_{set\text{-}up} \quad \text{for } T_{Cf} > T_{Ci} \quad . \tag{6}$$

An important example in which this minimum constraint can easily occur is in those designs which use cascaded registers, such as a serial shift register or a k-bit counter. As shown in Figure 5, T_{logic} is equal to zero and T_{int} approaches zero (since cascaded registers are typically designed, at the geometric level, to abut). If $T_{Cf} > T_{Ci}$ (i.e., negative clock skew), then the minimum constraint becomes

$$|T_{SKEW}| \leq T_{c\text{-}Q} + T_{set\text{-}up} \quad \text{for } T_{Cf} > T_{Ci} \quad , \tag{7}$$

and all that is necessary for the system to malfunction is a poor relative placement of the flip flops or a highly resistive connection between C_i and C_f. In a circuit configuration such as a shift register or counter, where negative clock skew is a more serious problem than positive clock skew, provisions should be made to force C_f to lead C_i, as in the example circuit of Figure 5.

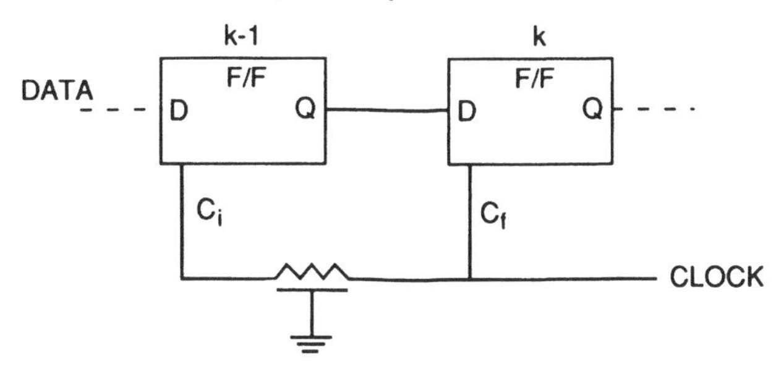

Figure 5: k-Bit Shift Register with Positive Clock Skew

As higher levels of integration are achieved, on-chip testability becomes necessary. Data registers, configured in the form of serial set/scan chains when operating in the test mode, are a common example of a built-in test design technique [18]. The placement of these circuits is typically optimized around the functional flow of the data. When the system is reconfigured to use the registers in the role of the set/scan function, different path delays are possible. In particular, the clock skew of the local data path can be negative and greater in magnitude than the local register delays. Therefore, with increased negative clock skew, (7) may not be satisfied and the incorrect data will latch into the final register of the reconfigured local data path.

Finally, in ideal scaling of MOS devices, all linear dimensions and voltages are multiplied by the factor 1/S, where S > 1. Device dependent delays, such as $T_{c\text{-}Q}$, $T_{set\text{-}up}$, and T_{logic}, scale as 1/S while interconnect dominated delays such as T_{SKEW} remain constant to first order, and if fringing capacitance is considered, actually increase with decreasing dimensions [16,19]. Therefore, when examining dimensional scaling, (6) and (7) should be considered carefully.

3.3 Enhancing Synchronous Performance by Applying Negative Clock Skew

Negative clock skew, as previously mentioned and as shown in (4), can be used to improve synchronous performance by minimizing the delay of the critical worst case data paths [4,5,10,20].

By applying negative clock skew to the critical paths, excess time is shifted from the neighboring less critical local data paths to the critical local data paths by forcing C_i to lead C_f at each critical local data path. This negative clock skew represents the additional amount of time that the data signal at R_i has to propagate through the n stages of logic and n+1 sections of interconnect and into the final register. Negative clock skew subtracts from the logic path delay, thereby decreasing the minimum clock period. This, in effect, increases the total time that a given critical data path has to accomplish its functional requirements by giving the data signal released from R_i more time to propagate through the logic and interconnect stages and latch into R_f. Thus, the differences in delay between each local data path is minimized, thereby compensating for any inefficient partitioning of the global data path into local data paths common in most practical systems.

The maximum permissible negative clock skew of any data path, however, is dependent upon the clock period itself as well as the time delay of the previous data paths. This results from the structure of the serially cascaded local data paths making up the global data path. Since a particular clock signal synchronizes a register which functions in a dual role, as the initial register of the next local data path and as the final register of the previous data path, the earlier C_i is for a given data path, the earlier that same clock signal, now C_f, is for the previous data path. Thus, the use of negative clock skew in the ith path results in a positive clock skew for the preceding path, which may then establish the new upper limit for the system clock frequency. It should be emphasized that in [7,9,10], the authors describe many of these characteristics of clock skew and its effects on the maximum clock frequency and, in [7,9], designate the lead/lag clock skew polarity (positive/ negative clock skew) notation as the opposite of that used here.

4. PIPELINING AND RETIMING OF SYNCHRONOUS SEQUENTIAL SYSTEMS

In digital systems, the minimum latency, as described in (2), occurs when the data path consists entirely of logic stages; it is the time required for propagation of a data signal through these logic stages. The clock period for this system, which is also the latency, is equal to the time required to process one data sample. If new data appear at the input of a system at time intervals smaller than the latency for this simple configuration, registers can be inserted into the data path to increase the frequency at which new data signals are processed through the system and appear at the system output. This degrades the latency, however.

The total delay from the output of the initial register R_i to the output of the nth logic stage is the sum of the individual T_{fi} terms along that local data path as shown below:

$$T_{logic} + T_{int} = \sum_{i=1}^{n} T_{fi} \quad . \tag{8}$$

Thus, for a local data path consisting of n logic stages, the time delay through the path T_{PD} can be expressed as

$$T_{PD} = T_{c\text{-}Q} + \sum_{i=1}^{n} T_{fi} + T_{set\text{-}up} \quad . \tag{9}$$

Equation (9) is composed of the delay required to get out of and into the initial and final registers, respectively, and the time required to propagate through n stages of logic and n + 1 sections of interconnect. If T_{REG} represents the total register related delay of both R_i and R_f, (9) can be written as

$$T_{PD} = T_{REG} + \sum_{i=1}^{n} T_{fi} \quad . \tag{10}$$

Thus, the total time to move a data signal through a data path is composed of the overhead requirements to get in and out of the final and initial registers as well as the time required to perform the logical operations.

The latency L is defined as the time required to move a data signal from the input of the system to its output. For the special case of no pipelining, the latency equals the minimum clock period. Each added register increases the latency by a register delay. Thus, the latency of a pipelined data path is the summation of the total delay through the global data path.

$$L = \sum_{i=1}^{N} T_{fi} + \sum_{k=1}^{M} T_{ek} \quad , \tag{11}$$

where N is the number of logic stages per global data path, M is the number of local data paths (and clock distribution networks) per global data path, and the total number of clock periods (and registers) required to move a particular data signal from the input of the system to its output is M+1. The maximum permissible negative clock skew T_{ek} in (11) is given by (12). In (12), T_{ek} is the aggregate delay of the kth local data path due to the initial and final registers (T_{REGk}) and the

$$T_{ek} = T_{REGk} + T_{SKEWk} \quad , \tag{12}$$

clock distribution network (T_{SKEWk}). T_{ek} can be used to represent the acceptable tolerance of negative clock skew for the kth local data path and can be described as the total effective delay of the registers and clock distribution network per local data path. Note that T_{ek} is typically positive for most circuit configurations.

In a retimed system, the latency L is constant. The objective is to shift the temporal location of each of the pipeline registers so as to minimize the maximum clock period. Thus, for a given system,

$$\Delta L = 0 \quad . \tag{13}$$

Since, from (13), M is held constant and, in (11), N and T_{fi} are assumed constant, the only two remaining opportunities to improve system wide performance are to temporally shift the pipeline registers keeping M constant (i. e., classical retiming), or to vary T_{ek}. In (12), T_{REGk} is assumed constant for a specific circuit implementation, thereby leaving T_{SKEWk} as the sole parameter to vary in (4) and (11) in order to improve system performance.

The clock period T_{cp} can be expressed as

$$T_{cp} \geq Max \; \{ T_{REG_k} + \sum_{i=1}^{n} T_{fi} + T_{SKEW_k} \} \tag{14}$$

$$T_{cp} = \begin{cases} \sum_{i=1}^{N} T_{fi} & \text{for } M = 0 \quad (15) \\ T_{ek} + \sum_{i=1}^{n} T_{fi} & \text{for } M \geq 1 \quad (16) \end{cases}$$

where Max is used to designate the critical worst case path. For a pipelined global data path with registers placed at both its input and output, the total latency L can also be described by the relation

$$L = \frac{M+1}{f_{clk}} \quad . \tag{17}$$

The primary objective of applying localized clock distribution design to retimed circuits is to partition a global data path into local data paths of approximately equal delay. By applying the concept of local negative and positive clock skew, one can, in effect, even out the delays of each pipelined data path, making each local data path of approximately equal delay.

Thus, this approach can be used to increase both the clock frequency and the temporal resolution of retimed circuits. Typically, a pipeline register is repositioned within a Boolean network in order to minimize overall clock period, thereby providing a temporal resolution defined by the minimum number of logic elements shifted within the Boolean network [1-3,21,22]. By using the approach of positive and negative clock skew, in addition to increasing the maximum system clock frequency, one can shift time between local data paths at a resolution much smaller than is possible by standard retiming algorithms [10]. Therefore, the maximum resolution of a retimed circuit is defined by the precision of the clock distribution network. An order of magnitude improvement in resolution is possible by applying the aforementioned local clock distribution design methods instead of relying solely on adjusting the relative temporal placement of the logic elements. For a 1 μm CMOS technology, precision of the clock distribution network on the order of 200 to 300 picoseconds [8] are possible instead of 2 to 5 nanoseconds typical of the logical transformations of many retiming algorithms [21,22].

5. AN EXAMPLE OF APPLYING LOCAL CLOCK SKEW TO RETIMED CIRCUIT

Consider the non-recursive retimed system shown in Figure 6 where the horizontal oval represents a logic delay and the vertical oval a clock delay. Since the local data path from R_2 to R_3 represents the worst case path (assuming the register delays are equal), by delaying C_3 with respect to C_2, negative clock skew is added to the R_2 - R_3 local data path. If C_1 is synchronized with C_3, then the R_1 - R_2 local data path receives some positive clock skew. Thus, if the register delays are 2 ns., C_2 would be designed to lead C_3 by 1.5 ns., forcing both paths to be the same total $T_{PD} + T_{SKEW} = 7.5$ ns. Thus, the critical path of the retimed circuit is further temporally refined to the precision of the clock distribution network and the entire system (for this simple example) could operate at a clock frequency of 133.3 MHz. instead of 111.1 MHz. Note that $|T_{SKEW}| < T_{PD}$ (|-1.5 ns.| < 9 ns.) for the R_2-R_3 local data path; therefore, the correct data signal is successfully latched into R_3 and no minimum data path/clock skew constraint relationship exists. This design technique of applying localized clock skew is particularly affective in sequentially adjacent temporally irregular local data paths; however, it is applicable to any type of synchronous sequential system and for certain architectures, significant improvement in performance and reliability is both possible and likely.

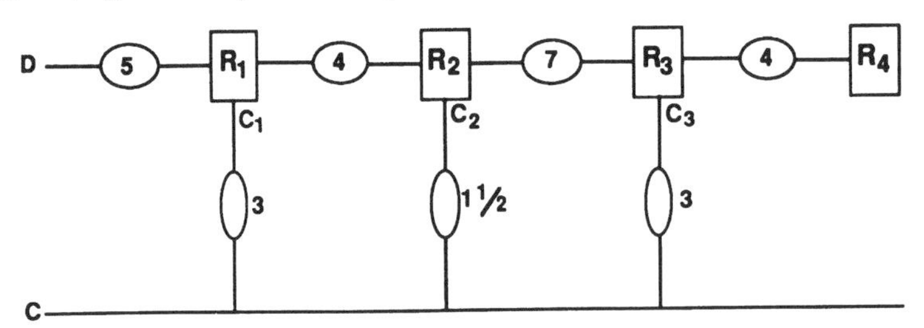

Figure 6: Example of Localized Clock Distribution Design Technique
Applied to Retimed Circuit

6. CONCLUSIONS

Clock distribution networks have been described in terms of their data path timing requirements. Local clock skew can be used to improve performance by locally varying its magnitude and lead/lag relationship with its sequentially adjacent local data path. Data path/clock skew constraint relationships were developed for both the positive clock skew case and the negative clock skew case. From these specific constraint relationships, recommended design procedures were offered to eliminate the deleterious effects of clock skew on both the maximum performance and the reliable operation of a synchronous digital system. Techniques were described for improving synchronous performance by applying negative clock skew for equalizing path delays between local data paths, thereby minimizing the delay of the critical worst case paths. Thus, local clock distribution networks can be used to tune the effective path delays of each local data path. This provides logically retimed circuits with retiming precision consistent with the resolution of the clock distribution network. Thus, retiming becomes a two-phase process; 1) logical retiming for gross temporal partitioning and 2) local clock distribution design for fine readjustment of the clock period. An order of magnitude improvement in resolution is attainable in retimed circuits by applying local clock distribution design methods instead of relying solely on moving logic elements across sequentially adjacent data paths.

7. REFERENCES

[1] Leiserson, C. E. and Saxe, J. B., "Optimizing Synchronous Systems," *Proceedings of 22nd Annual Symposium on Foundations of Computer Science*, October 1981, pp. 23-26.

[2] Leiserson, C. E. and Saxe, J. B., "Optimizing Synchronous Circuitry by Retiming," *Proceedings of the 3rd Caltech Conference on Very Large Scale Integration*, 1983, pp. 87-116.

[3] Leiserson, C. E. and Saxe, J. B., "Retiming Synchronous Circuitry," *Algorithmica*, Vol. 6, January 1991, pp. 5-35.

[4] Friedman, E. G. and Mulligan, Jr., J. H., "Clock Frequency and Latency in Synchronous Digital Systems," *IEEE Transactions on Signal Processing*, Vol. SP-39, No. 4, April 1991, pp. 930-934.

[5] Friedman, E. G. and Mulligan, Jr., J. H., "Pipelining of High Performance Synchronous Digital Systems," *International Journal of Electronics* , Volume 70, Number 5, May 1991, pp. 917-935.

[6] Friedman, E. G., "The Limiting Performance of a CMOS Bistable Register based on Waveform Considerations," *International Journal of Electronics* (in press).

[7] Hatamian, M. and Cash, G. L., "Parallel Bit-Level Pipelined VLSI Designs for High Speed Signal Processing," *Proceedings of the IEEE*, Vol. 75, No. 9, September 1987, pp. 1192-1202.

[8] Friedman, E. G. and Powell, S., "Design and Analysis of a Hierarchical Clock Distribution System for Synchronous Standard Cell/Macrocell VLSI," *IEEE Journal of Solid-State Circuits*, Vol. SC-21, No. 2, April 1986, pp. 240-246.

[9] Hatamian, M., "Understanding Clock Skew in Synchronous Systems," Chapter 6 of *Concurrent Computations (Algorithms, Architecture, and Technology)*, S. K. Tewksbury, B. W. Dickinson, and S. C. Schwartz (Eds.), New York, New York: Plenum Publishing, 1988, ch. 6.

[10] Fishburn, J. P., "Clock Skew Optimization," *IEEE Transactions on Computers*, Vol. C-39, No. 7, July 1990, pp. 945-951.

[11] Anceau, F., "A Synchronous Approach for Clocking VLSI Systems," *IEEE Journal of Solid-State Circuits*, Vol. SC-17, No. 1, February 1982, pp. 51-56.

[12] Wann, D. and Franklin, M., "Asynchronous and Clocked Control Structures for VLSI Based Interconnection Networks," *IEEE Transactions on Computers*, Vol. C-32, No. 3, March 1983, pp. 284-293.

[13] Alves Marques, J. and Cunha, A., "Clocking of VLSI Circuits," *VLSI Architecture*, Randell and Treleaven (Eds.), Englewood Cliffs, New Jersey:Prentice-Hall, 1983, pp. 165-178.

[14] Unger, S. H. and Tan, C-J., "Clocking Schemes for High-Speed Digital Systems," *IEEE Transactions on Computers*, Vol. C-35, No. 10, October 1986, pp. 880-895.

[15] Bakoglu, H. B., Walker, J. T., and Meindl, J. D., "A Symmetric Clock-Distribution Tree and Optimized High-Speed Interconnections for Reduced Clock Skew in ULSI and WSI Circuits," *Proceedings of IEEE International Conference on Computer Design*, October 1986, pp. 118-122.

[16] Gura, C. V., "Analysis of Clock Skew in Distributed Resistive-Capacitive Interconnects," University of Illinois, Urbana, Illinois, SRC Technical Report No. T87053, June 1987.

[17] Wagner, K. D., "Clock System Design," *IEEE Design & Test of Computers*, October 1987, pp. 9-27.

[18] Williams, T. and Parker, K., "Design for Testability-A Survey," *Proceedings of the IEEE*, Vol. 71, No. 1, January 1983, pp. 98-112.

[19] Bakoglu, H. B. and Meindl, J. D., "Optimal Interconnection Circuits for VLSI," *IEEE Transactions on Electron Devices*, Vol. ED-32, No. 5, May 1985, pp. 903-909.

[20] E. G. Friedman, *Performance Limitations in Synchronous Digital Systems*, Ph.D. Dissertation, University of California, Irvine, June 1989.

[21] De Micheli, G., "Synchronous Logic Synthesis: Algorithms for Cycle-Time Minimization," *IEEE Transactions on Computer-Aided Design*, Vol. CAD-10, No. 1, January 1991, pp. 63-73.

[22] Malik, S., Sentovich, E. M., Brayton, R. K., and Sangiovanni-Vincentelli, A., "Retiming and Resynthesis: Optimizing Sequential Networks with Combinational Techniques," *IEEE Transactions on Computer-Aided Design*, Vol. CAD-10, No. 1, January 1991, pp. 74-84.

Retiming With Non-Zero Clock Skew, Variable Register, and Interconnect Delay

Tolga Soyata and Eby G. Friedman

Department of Electrical Engineering
University of Rochester, Rochester, NY 14627

Abstract

A retiming algorithm is presented which includes the effects of variable register, clock distribution, and interconnect delay. These delay components are incorporated into retiming by assigning Register Electrical Characteristics (RECs) to each edge in the graph representation of the synchronous circuit. A matrix (called the Sequential Adjacency Matrix or SAM) is presented that contains all path delays. Timing constraints for each data path are derived from this matrix. Vertex lags are assigned ranges rather than single values as in standard retiming algorithms. The approach used in the proposed algorithm is to initialize these ranges with unbounded values and continuously tighten these ranges using localized timing constraints until an optimal solution is obtained. The algorithm is demonstrated on modified MCNC benchmark circuits and both increased clock frequencies and elimination of all race conditions are observed.

1 Introduction

Retiming is a sequential optimization technique used to increase the clock frequency of synchronous circuits by relocating the registers in the circuit while maintaining the original function and latency of the system. In retiming algorithms proposed to date, variable register, clock distribution, and interconnect delays are essentially ignored. Without including these delay components, standard retiming algorithms are not sufficiently accurate for building practical high speed circuits. For this reason, clock distribution, variable register, and interconnect delay must be integrated into the retiming process in order to ensure that retiming becomes a practical and useful design methodology.

Both register and interconnect delay are similar in magnitude to the delay of the logic elements. Also, variations in clock delay between widely separated registers may create clock skews which can drastically affect circuit operation. Undesirable clock skew can produce a net negative delay within a local data path. This implies the existence of a race condition, which must be avoided as a condition imposed on the retiming process.

In most retiming algorithms proposed to date, registers are assumed to have zero delay (e.g., [1, 2]) or equal delay (e.g., [3]). In [3], the set-up (t_s) and hold (t_p) times are non-zero constant values, creating an effective clock period of $T_{PD} + t_s + t_p$, where T_{PD} is the worst case path delay of the synchronous circuit. Since constant register delays are assumed throughout the circuit, $t_s + t_p$ is added to each individual local data path, biasing the clock period by this amount. However, this simple summation is not sufficiently accurate since each local data path may have a different register delay.

Integrating clock skew into the retiming process was first proposed in [4, 5]. The authors of this paper originally introduced the strategy of integrating clock skew and variable register delays into retiming by attaching electrical information to the edges of the graph representing the synchronous circuit [6]. These delay parameters are defined as Register Electrical Characteristics (RECs) in this original work and are adhered to herein. Following this work, the integration of clock skew into retiming was discussed in [7]. In this paper, constraints are placed on the clock skew to permit the use of standard linear programming methods. Variable register and interconnect delays were not considered. In [8], a branch and bound algorithm is briefly introduced to solve the general retiming problem while considering non-zero clock skew, variable register, and interconnect delay. In general, there has been a growing interest in making retiming into a more practical and useful design methodology, evidenced by [1–10].

In this paper, a retiming algorithm is presented which incorporates variable register and interconnect delay and non-zero localized clock skew. Either rising edge or falling edge triggered D flip flops and a single phase clock is assumed throughout the synchronous digital circuit. To accomplish the integration of variable clock distribution, interconnect, and register delays into the retiming process, a path between logic elements is defined in this paper as the traversal from weighted edge to weighted edge, an edge being interpreted as a connection between logic elements containing zero, one, or more registers. With this definition, clock, register, and interconnect delays are assigned to each edge. Thus, as registers are shifted from edge to edge, different clock skews and register delays are considered in each of the local path delays. This permits

Reprinted with permission from *Proc. IEEE Int'l Conf. Computer-Aided Design,* T. Soyata and E. G. Friedman, "Retiming with Non-Zero Clock Skew," pp. 234–241, Nov. 1994.

both maximum clock periods and race conditions to be detected on a path-by-path basis. This approach, therefore, initially requires approximate (or estimated) values of register, clock distribution, and interconnect delays which can be replaced with more accurate values as the exploratory retiming process becomes better specified [6, 8].

The paper is organized as follows. Background and definitions of important terms used throughout the paper are provided in Section 2. In Section 3, models of non-zero clock skew, variable register, and interconnect delay are presented. In Section 4, the Sequential Adjacency Matrix (SAM) is introduced. Timing constraints, derived from the SAM, are described in Section 5. The proposed retiming algorithm RETSAM is presented in Section 6. Results of applying the proposed algorithm to MCNC benchmark circuits are presented in Section 7 and finally some conclusions are drawn in Section 8.

2 Background and Definitions

The absolute delay of the clock signal from the global clock source to a specific register (or memory element) is the **clock delay** and is denoted as T_{CD}. The difference between the clock delay of any two registers is the **clock skew** between these registers, denoted as T_{Skew}. The notion of **localized clock skew** and its application to increasing the clock frequency within pipelined systems was introduced by Friedman and Mulligan in [11]. They show that only clock skew between **sequentially adjacent registers** (registers that receive information at successive clock intervals and are either directly connected or connected by logic elements) is significant in pipelined systems. A **local data path** is formed between two sequentially adjacent registers. The local data path with the greatest delay is the **critical data path**, whose delay defines the minimum clock period of the circuit.

The definition of sequential adjacency is extended in this paper to edges on a graph. **Sequentially adjacent edges** are those edges that are connected via a fully combinatorial path. The last register of the initial edge and the first register of the final edge are sequentially adjacent, thereby making the path sequentially adjacent.

The clock skew T_{Skew} between two sequentially adjacent edges i and j is defined as

$$T_{Skew}(i,j) = T_{CD}(i) - T_{CD}(j). \tag{1}$$

If $T_{CD}(j) > T_{CD}(i)$, the clock skew between registers i and j is defined as being negative. Negative clock skew occurs if the initial clock signal leads the final clock signal of a local data path. If $T_{CD}(j) < T_{CD}(i)$, the clock skew between registers i and j is positive. Positive clock skew occurs if the initial clock signal lags the final clock signal of a local data path. In the case that $T_{CD}(j)$ equals $T_{CD}(i)$, i.e., the clock signal reaches the clock input of the two registers at precisely the same time, the clock skew is zero.

Positive clock skew increases the path delay of a local data path, potentially making its local data path a critical path, whereas **negative clock skew** may improve circuit speed in critical paths [5, 12], however it may also create negative path delays, resulting in **race conditions**. Race conditions are caused by *early-clocking*, i.e., clocking of registers before the relevant data is successfully latched. A race condition occurs if the skew is negative and greater in magnitude than the total local data path delay [5, 11, 12]. Those paths with negative delay are called **short paths** [13]. Similarly, a **long path** designates those paths with a delay greater than the desired clock period of the circuit.

A synchronous circuit can be modeled by a graph composed of a vertex set V and an edge set E. $|V|$ and $|E|$ refer to the cardinalities of these sets, i.e., the number of vertices and edges in the graph, respectively. Vertices denote logic elements and edges denote the connection between vertices. v_n and e_k represent vertex n and edge k, respectively. Every edge e_k connects two vertices. These two vertices are called the start vertex and the end vertex of e_k and are denoted $e_k.start$ and $e_k.end$.

The logic element delay represented by v_n is $d(v_n)$ and is measured in **time units (tu)**. The number of registers on an edge between two vertices is represented by the weight of the corresponding edge e_k and is denoted by $w(e_k)$. Edge-to-edge and vertex-to-vertex paths are represented by $e_i \rightsquigarrow e_j$ and $v_i \rightsquigarrow v_j$, respectively. The lag of a vertex v, $r(v)$, is defined in [1] and adhered to in this paper. Using this definition, retiming can be defined as assigning a lag to each vertex using the following formula,

$$w_r(e) = w(e) + r(u) - r(v), \tag{2}$$

where e is an edge connecting vertices u and v and $w(e)$ and $w_r(e)$ are the weight of edge e before and after retiming, respectively. A W matrix, defined in [1], contains all vertex-to-vertex path weights. The elements of this matrix, $W(i,j)$, can be calculated as

$$W(i,j) = \min\{w(p) : \ p : v_i \rightsquigarrow v_j\}. \tag{3}$$

3 Register Electrical Characteristics (RECs)

In order to consider the effects of clock distribution, variable register, and interconnect delay, a number set, the **Register Electrical Characteristic** (REC), is assigned to each edge of the graph in the following form: $T_{CD} : T_{Set-up}/T_{C\to Q} — T_{Int1}/T_{Int2}$. T_{CD} is the clock delay from the global clock source to each register, T_{Set-up} is the time required for the data at the input of a register to latch, $T_{C \to Q}$ is the time required for the data to appear at the output of the register upon arrival of the clock signal, and T_{Int} is the total interconnect delay along that edge and can be considered as being composed of two parts, T_{Int1} and T_{Int2}, if there is more than one register along that edge.

By attaching delay components to registers located on edges (connections between logic elements), the local path must be defined from edge-to-edge [6, 8] rather than vertex-to-vertex, as in standard retiming algorithms [1]. A modified version of the graph introduced in [1] is shown in Figure 1, in which an REC is assigned to each edge. By assigning a clock delay to each edge, the circuit is assumed to be partitioned into regions of similar clock delay, i.e., registers that are located on the same edge are physically located within the same clock delay region. Therefore, registers that end up on the same edge after retiming are assumed to have similar clock delay. Registers that move to different edges are assumed to have the clock and register delays of the new edge. Since registers on different edges may be considered to have different clock and register related delays, moving a register from one edge to another edge during retiming will not only create different local data paths with different logic, register, and interconnect delays, but may also change the localized clock skew of the new local data paths.

The local data path delay $T_{PD}(i,j)$ from edges e_i to e_j is

$$T_{PD}(i,j) = T_{C \to Q}(i) + T_{Int2}(i) + T_{Logic}(i,j) + T_{Int1}(j) + T_{Set-up}(j) + T_{Skew}(i,j), \tag{4}$$

where $T_{Logic}(i,j)$ is the delay of the logic elements between e_i and e_j including the interconnect delay of the zero weight edges along the path between these edges. If parallel paths exist, minimum and maximum local data path delays, $T_{PD_{min}}$ and $T_{PD_{max}}$, are defined. If $T_{PD_{min}}(i,j) < 0$, a race condition between e_i and e_j exists since in this local data path the final register is clocked before the data signal arrives and is successfully latched.

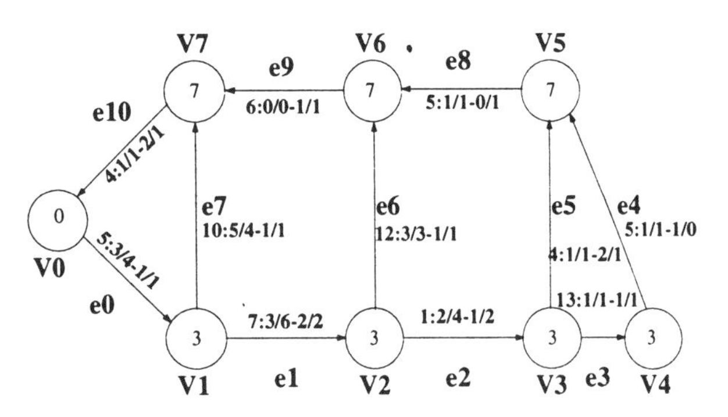

Figure 1: Graph of the digital correlator [1] with added REC values

If registers R_i and R_j are located on the same edge e_k and are sequentially adjacent, then, according to the definition of the RECs, the clock skew between R_i and R_j is zero from (1) since the clock delays of both registers are the same. This assumption is made since registers on the same edge would typically be physically close, and therefore the difference in clock delay to each register and the interconnect delay between these registers would be negligible. Furthermore, since no vertices (logic elements) exist between registers R_i and R_j when both are on the same edge, the logic delay between the two registers is zero. Since all registers located on the same edge are defined to have the same timing characteristics (REC values), all sequentially adjacent registers located on the same edge have a similar internal path delay. A path composed of multiple registers on an edge could possibly be the critical worst case path of the overall circuit and its delay is defined as $T_{PD_{Internal}}(e_k)$, given by

$$T_{PD_{Internal}}(e_k) = T_{C \to Q}(e_k) + T_{Set-up}(e_k). \tag{5}$$

4 Sequential Adjacency Matrix (SAM)

The Sequential Adjacency Matrix (SAM or the S matrix) is an $|E| \times |E|$ matrix whose element $S(i,j)$ is the path delay from e_i to e_j. The S matrix element, $S(i,j)$, is calculated from

$$S(i,j) = \max\{T_{PD}(i,j) : p : e_i \rightsquigarrow e_j \wedge w(p) = W(i,j)\}, \tag{6}$$

If parallel paths exist between any two edges, the S matrix is composed of two matrices, S_{min} and S_{max}. Equations (7) and (8) are used to calculate the values of these two matrices. In order to reduce the number of matrices, a combined matrix, S', is used. $S'(i,j)$ contains $S_{min}(i,j)$ if $S_{min}(i,j)$ contains a zero or negative entry, and contains $S_{max}(i,j)$ if no zero or negative entry exists. The importance of $S_{min}(i,j)$ is determined by whether a zero or negative entry exists, thereby denoting a race condition. If $S_{min}(i,j)$ is completely positive, the maximum valued entries in $S_{max}(i,j)$ limit the maximum speed of the circuit. Equation (9) is used to calculate the combined matrix, S'.

$$S_{min}(i,j) = \min\{T_{PD}(i,j) : p : e_i \rightsquigarrow e_j \wedge w(p) = W(i,j)\}, \tag{7}$$

$$S_{max}(i,j) = \max\{T_{PD}(i,j) : p : e_i \rightsquigarrow e_j \wedge w(p) = W(i,j)\}, \tag{8}$$

$$S'(i,j) = \begin{cases} S_{min}(i,j), & if\ S_{min}(i,j) \le 0 \\ S_{max}(i,j), & if\ S_{min}(i,j) > 0. \end{cases} \tag{9}$$

Note that the S' matrix contains information for only those paths that can potentially cause the circuit to function improperly. Therefore, the negative entries in the S_{min} matrix override the positive entries in the corresponding S_{max} matrix during the calculation of the S' matrix. This occurs since these negative entries flag catastrophic race conditions. For the remainder of the paper, the notation for the combined matrix S' is denoted as S for simplicity. The S matrix of the graph of Figure 1 is shown in Table 1. The light shaded elements of the table indicate those paths with race conditions (negative values) and the dark shaded elements indicate those paths with a path delay greater than the desired clock period. In this example, a target clock period of 23 tu is assumed. Paths with zero delay are marginal race conditions that are not permitted and would appear as light shaded. The unshaded elements of the table indicate those paths that neither limit the maximum performance of the circuit nor create a catastrophic race condition.

5 Timing Constraints

In this paper, a branch and bound algorithm is used in which unbounded values are initially assumed for the lag ranges. These lag ranges are tightened using timing constraints derived from the SAM. There are four different types of timing constraints: negative edge weight, long path, short path, and internal path constraints. These different types of constraints are explained in greater detail in the following subsections.

5.1 Negative edge weight constraints

As introduced in [1], a properly retimed graph contains no negative edge weights. Negative edge weights are permitted for peripheral edges in [14] in order to shift the registers to the periphery of a synchronous circuit. This approach permits combinatorial optimization to be performed on the circuitry placed between the peripheral edges. However, since the retiming algorithm described in this paper does not exploit this feature of **resynthesis**, negative edge weights are disallowed. The negative edge weight constraint can be written as

$$w(e) \geq 0, \; \forall \, e \in E \,. \tag{10}$$

Table 1: SAM for the graph of Figure 1. Light shaded entries represent short paths, whereas dark shaded entries represent long paths for $c = 23$ tu. Unshaded entries denote permissible paths.

SAM		to										
		e0	e1	e2	e3	e4	e5	e6	e7	e8	e9	e10
from	e0	24	11	22	15	28	25	12	9	32	24	21
	e1	38	42	20	13	26	23	10	40	30	22	35
	e2	39	43	54	-1	12	9	44	41	16	23	36
	e3	45	49	60	53	15	63	50	47	22	26	42
	e4	32	36	47	40	53	50	37	34	9	16	29
	e5	32	36	47	40	53	50	37	34	9	16	29
	e6	34	38	49	42	55	52	39	36	59	18	31
	e7	24	28	39	32	45	42	29	26	49	41	21
	e8	25	29	40	33	46	43	30	27	50	9	22
	e9	16	20	31	24	37	34	21	18	41	33	13
	e10	5	9	20	13	26	23	10	7	30	22	19

5.2 Long path constraints

If a clock period c is desired, then all paths with a delay greater than c must be eliminated. Long paths are represented by entries in the S matrix that exceed a desired clock period c. In Table 1, long paths for $c = 23$ tu are depicted using dark shaded elements. In order to eliminate these long paths, the two edges that create the long path are made nonsequentially adjacent.

Two registers are sequentially adjacent if there exists a zero weight path between them. According to this definition, in order to make two edges, e_i and e_j, nonsequentially adjacent, three approaches are possible: 1) the source or 2) the destination edges can be made zero weight, i.e., all registers can be removed from these edges, or 3) one or more registers can be placed within each zero weight path between the source and destination registers. The first two conditions exist since by eliminating the initial and/or final register of a local data path, a longer path is created which may have a smaller delay (due to negative clock skew). These three conditions can be written in terms of path and edge weights as follows:

$$w(e_i) = 0, \tag{11}$$

$$w(e_j) = 0, \tag{12}$$

$$W(e_i.end, e_j.start) > 0. \tag{13}$$

If (11) or (12) is satisfied, then no registers exist on edge i or j, respectively, and therefore all local data paths between edges i and j are eliminated. If (13) is satisfied, all possible paths between edges i and j have a weight of at least one. This violates the definition of sequential adjacency, i.e., no paths exist with a zero weight between these two edges. Intuitively, it is stated in (11), (12), and (13) that either the initial or final edge does not have any register located on it or there is at least one register along every path between these two edges.

5.3 Short path constraints

Short paths appear as zero or negative entries in the S matrix. $S(i,j) \leq 0$ indicates a short path originating at e_i and terminating at e_j. If e_i and e_j form a short path, then the initial and the final registers of this path must be made nonsequentially adjacent. Equations (11), (12), and (13) are used to eliminate any catastrophic short paths (or race conditions).

5.4 Internal path constraints

Internal long paths are created between two sequentially adjacent registers on the same edge when the edge weight is greater than one. Internal long path constraints can be formulated using (5) as

$$w(e_i) \leq 1, \ \forall\, i : T_{PD_{internal}}(e_i) > c. \tag{14}$$

Note that **internal short paths** are not possible since the clock skew between any two registers on the same edge cannot be negative (it must be zero). Therefore, internal short paths are not considered in this paper.

5.5 Constraints due to vertex lags

Constraints (10), (11), (12), (13), and (14) are written in terms of edge weights. These constraints can be rewritten as (15), (16), (17), (18), and (19), respectively, to reduce the number of necessary operations.

$$r(e.start) - r(e.end) \leq w(e), \ \forall e \in E, \tag{15}$$

$$r(e_i.start) - r(e_i.end) = w(e_i), \tag{16}$$

$$r(e_j.start) - r(e_j.end) = w(e_j), \tag{17}$$

$$r(e_i.end) - r(e_j.start) \leq W(i,j) - 1, \tag{18}$$

$$r(e_i.start) - r(e_i.end) \geq w(e_i) - 1, \quad \forall i : T_{PD_{internal}}(e_i) > c. \tag{19}$$

In order to provide some intuition to (15), (16), (17), (18), and (19), note that, given two vertices u and v, the value $r(u) - r(v)$ can be thought of as "the number of registers taken out of the path $p : u \rightsquigarrow v$." Given this interpretation, it is implied in (15) that "the number of registers taken from an edge e cannot be greater than the original weight of the edge," i.e., none of the edge weights can be negative. In a similar manner, it is stated in (16) and (17) that "the number of registers taken from edge e_i and e_j, respectively, must be equal to the original weight of this edge," implicitly stating that this edge should be made zero weight. In (18) it is stated that "the registers taken from the path $p : e_i \rightsquigarrow e_j$ must be less than the original weight of this path minus one," implicitly stating that at least one register should be left along any path between registers e_i and e_j, thereby making this path nonsequentially adjacent. Finally, in (19) it is implied that either zero or one register should be left on an edge e that contains an internal long (worst case) path.

6 Retiming Algorithm

In this section two algorithms are introduced: 1) Algorithm *RETSAM* to perform retiming of synchronous circuits and 2) Algorithm *CHECKCP* to check the feasibility of a specific clock period. These two algorithms are explained in the following subsections.

6.1 *RETSAM*: Algorithm for Retiming

Retiming a synchronous circuit is achieved by performing a binary search of all possible clock periods of the circuit. The pseudo-code of the retiming algorithm is shown in Figure 2. The lower and upper bounds of the binary search are CP_{min} and CP_{max}, respectively. Initially the lower bound is zero (Step 1). If the original graph does not contain any race conditions, the critical path delay of the original graph defines the upper bound of the binary search (Step 2). The SAM is calculated in Step 3 and used throughout the algorithm. If the original graph contains one or more race conditions, the maximum value in the SAM is used as the upper bound (Step 4). During the binary search, a specific clock period, CP_{target}, is checked for feasibility using algorithm *CHECKCP* (Steps 5 and 6). Depending on whether a solution exists (Step 7) or not (Step 8), the lower and upper search bounds are adjusted and the binary search continues until the minimum clock period is determined (Step 9).

6.2 *CHECKCP*: Clock Period Feasibility Check

A feasibility check for a specific clock period CP_{target} is achieved by solving the set of nonlinear inequalities for the vertex lag ranges. If all the constraints are satisfied for every path in the graph, the clock period is considered feasible. Pseudo-code for the algorithm that determines the feasibility of a clock period is shown in Figure 3. Lag ranges are stored in an array called $r[]$. The timing constraints are derived from the SAM. An approximate solution can be obtained for the minimum clock period if the clock period feasibility test *CHECKCP* is terminated once the binary search bounds become sufficiently tight.

The most important step in *CHECKCP* is solving for the vertex lags, $r()$. The objective of the retiming algorithm is to yield a set of vertex lags that satisfy (15) through (19). To achieve this, the vertex lag ranges are initialized with unbounded values ($[-\infty \ldots \infty]$). Timing constraints are continuously applied to these vertex lags in order to tighten

1. $CP_{min} = 0$
2. CP_{max} = clock period of the original graph
3. Calculate SAM
4. If the original graph has race conditions $CP_{max} = \max\{S(i,j), \forall i,j\}$,
5. Choose $CP_{target} = \lfloor \frac{CP_{max}+CP_{min}}{2} \rfloor$,
6. Check for feasibility of $c = CP_{target}$
7. If set of inequalities can be successfully solved, then $CP_{max} = CP_{target}$
8. If not, then $CP_{min} = CP_{target}$
9. Continue this process until $CP_{min} = CP_{max}$

Figure 2: Pseudo-code for *RETSAM*

the ranges until eventually all the constraints are satisfied. Once the vertex lags are each defined, these lag values are used to determine the edge weights of the retimed graph according to (2). Table 2 exemplifies the solution method for the vertex lags of the graph shown in Figure 1.

The following types of equality and inequalities are created from the aforementioned timing constraints:

$$r(v_a) - r(v_b) = k, \tag{20}$$

$$r(v_a) - r(v_b) \leq k, \tag{21}$$

$$r(v_a) - r(v_b) = k_1 \;\; or \;\; r(v_c) - r(v_d) = k_2, \tag{22}$$

$$r(v_a) - r(v_b) = k_1 \; or \; r(v_c) - r(v_d) \leq k_2, \tag{23}$$

$$r(v_a) - r(v_b) = k_1 \;\; or \;\; r(v_c) - r(v_d) = k_2 \;\; or \;\; r(v_e) - r(v_f) \leq k_3, \tag{24}$$

where $r()$ are vertex lags and k_n are constant values. The *or* statements that appear in (22), (23), and (24) prohibit the use of standard linear programming methods [15] and

1. $r[0] = [0 \ldots 0]$
2. $r[k] = [-\infty \ldots +\infty], k = 1, \ldots, E-1$
3. Create a constraint list for clock period c
4. Adjust lags to satisfy all constraints
5. If all lags fixed and all constraints unsatisfied

   → Clock period is not feasible
6. If all constraints are satisfied

   → Clock period is feasible
7. If c is feasible and all lags are not fixed

   → Use lower bounds of the unfixed lags

Figure 3: Pseudo-code for clock period feasibility test, CHECKCP

necessitate the use of branch and bound techniques. Note the existence of *multiple choices* in each inequality.

To gain insight into how these multiple choice inequalities are created, consider retiming the graph of Figure 1, for which the SAM is shown in Table 1. To achieve a clock period of $c = 23$ tu, the dark shaded and light shaded paths must be avoided, since they represent long paths for $c = 23$ tu and short paths, respectively. To avoid, for example, the path $p : e_3 \rightsquigarrow e_0$, there exists three possible choices, derived from (11), (12), and (13), resulting in the multiple choice inequality,

$$r(3) - r(4) = 1 \; or \; r(0) - r(1) = 1 \; or \; r(4) - r(0) \leq -1, \tag{25}$$

which states that to eliminate the path starting at e_3 and terminating at e_0, either e_3 or e_0 must be zero weight, thereby making the path $p : e_3 \rightsquigarrow e_0$ non-existent, or at least one register must be placed between the initial and terminating vertices of the path p.

To solve for a set of vertex lags that provide a proper retiming, an inequality similar to (25) is written for each short or long path shown in Table 1. In this algorithm, the unbounded value $[-\infty \ldots \infty]$ is initially assigned to each vertex lag range. Only one vertex lag ($r(0)$ for simplicity) is initialized to 0 and the other lags are calculated relative to $r(0)$. The vertex lag non-negativity constraint from (15) is applied to each vertex, shown in the first three rows of Table 2, thereby further tightening the vertex lag ranges. It is shown in row 4 that the constraint from (15) cannot be used to further tighten the vertex lag ranges. The long path constraints from (16), (17), and (18) are therefore used in row 4 to further tighten the bounds. Short paths are also eliminated using (16), (17), and (18). Each time the bounds are tightened by applying long (or short) path constraints, (15) is applied to the new set and the neighboring vertex lag ranges to ensure non-negative edge weights on each edge. The algorithm may reach a point where the application of the constraints can no longer tighten the bounds any further. This situation is indicated in Table 2 by a dark shaded row, where the application of each of the long path, short path, and non-negativity rules cannot tighten any further the vertex lag ranges. Once this occurs, all possible values for each vertex lag are tested. On the first dark shaded row in Table 2, there are two unfixed lags with cardinalities two and three, respectively. Therefore, $2 * 3 = 6$ possible solutions exist and must be evaluated. If a solution is reached, the algorithm is terminated and the resulting vertex lag ranges are used to determine the edge weights of the retimed graph. If all possible solutions are considered and a set of vertex lag ranges cannot be determined that satisfy all constraints, a solution for that specific clock period does not exist.

Table 2: Example solution for $c = 23$. A single value is shown for equal lower and upper bounds.

Constraint	Type	r(0)	r(1)	r(2)	r(3)	r(4)	r(5)	r(6)	r(7)
		0	−∞..∞	−∞..∞	−∞..∞	−∞..∞	−∞..∞	−∞..∞	−∞..∞
r(0)-r(1)≤1	Negativity on e_0	0	−1..∞	−∞..∞	−∞..∞	−∞..∞	−∞..∞	−∞..∞	−∞..∞
r(1)-r(7)≤0	Negativity on e_7	0	−1..∞	−∞..∞	−∞..∞	−∞..∞	−∞..∞	−∞..∞	−1..∞
Negativity on e_1, e_2, e_3, e_4, e_5, e_6, e_7, e_8, e_9		0	−1..0	−2..0	−3..1	−4..0	−3..0	−2..0	−1..0
r(1)-r(7)=0 or r(7)-r(1)≤0	Long path: $e_7 \rightarrow e_7$	0	-1	−2..0	−3..1	−4..0	−3..0	−2..0	-1
r(6)-r(7)≤0	Negativity on e_9	0	-1	−2..0	−3..1	−4..0	−3..0	−2.. − 1	-1
r(5)-r(6)≤0	Negativity on e_8	0	-1	−2..0	−3..1	−4..0	−3.. − 1	−2.. − 1	-1
r(4)-r(5)≤0	Negativity on e_4	0	-1	−2..0	−3..1	−4.. − 1	−3.. − 1	−2.. − 1	-1
r(3)-r(5)≤0	Negativity on e_5	0	-1	−2..0	−3.. − 1	−4.. − 1	−3.. − 1	−2.. − 1	-1
r(2)-r(6)≤0	Negativity on e_6	0	-1	−2..−1	−3.. − 1	−4.. − 1	−3.. − 1	−2.. − 1	-1
r(2)-r(6)=0 or r(7)-r(0)=0 or r(6)-r(7)≤-1	Long path: $e_6 \rightarrow e_{10}$	0	-1	−2..−1	−3.. − 1	−4.. − 1	−3.. − 1	-2	-1
r(5)-r(6)≤0	Negativity on e_8	0	-1	−2..−1	−3.. − 1	−4.. − 1	−3.. − 2	-2	-1
r(4)-r(5)≤0	Negativity on e_4	0	-1	−2..−1	−3.. − 1	−4.. − 2	−3.. − 2	-2	-1
r(3)-r(5)≤0	Negativity on e_5	0	-1	−2..−1	−3.. − 2	−4.. − 2	−3.. − 2	-2	-1
r(2)-r(6)≤0	Negativity on e_6	0	-1	-2	−3.. − 2	−4.. − 2	−3.. − 2	-2	-1
Iteration #1: choose r(5)=-2		0	-1	-2	−3.. − 2	−4.. − 2	-2	-2	-1
Iteration #1: choose r(3)=-3		0	-1	-2	-3	−4.. − 2	-2	-2	-1
r(2)-r(3)=1 or r(3)-r(4)=1	Short path: $e_2 \rightarrow e_3$	Condition 1 satisfied							
r(3)-r(4)≤1	Negativity on e_3	0	-1	-2	-3	-2	-2	-2	-1
r(3)-r(4)=1 or r(6)-r(7)=0 or r(4)-r(6)≤-1	Long path: $e_3 \rightarrow e_9$	Constraints are not satisfiable							
Iteration #2: choose r(5)=-3		0	-1	-2	−3.. − 2	−4.. − 2	-3	-2	-1
⋮									

7 Experimental Results

The retiming algorithm *RETSAM* is implemented in C on a SUN 4 workstation. To permit evaluating the proposed retiming algorithm, modified MCNC benchmark circuits [16, 17] have been analyzed using this algorithm and compared with an implementation of the Leiserson-Saxe retiming algorithm [1]. The resulting clock periods of the retimed benchmark circuits are reported.

The application of RETSAM to the example MCNC benchmark circuits are described in Table 3. The initial five columns describe the properties of the modified benchmark circuits. These properties are 1) the name of the benchmark example as it appears in the MCNC archive, 2) the number of edges and 3) vertices in the graph of each circuit, 4) the latency of the circuit, and 5) the original clock period. The sixth column contains the minimum clock period of the retimed circuit using RETSAM. The final column lists the clock period of the benchmark circuits that were retimed using standard retiming algorithms without considering RECs.

In these circuits, the average register delay ($T_{C \rightarrow Q} + T_{Set-up}$) of each circuit is added to each local data path to compensate for the effects of the variable register delays, i.e., the register delay of each local data path is assumed to be constant and equal to the average register delay of the retimed circuit with variable REC values. As shown in Table 3, the minimum clock period of the majority circuit from LGSynth89 derived from RETSAM is less than from existing retiming algorithms. This occurs since localized negative clock skew [5, 12] subtracts delay from the critical path such that the worst case path delay is smaller, thereby causing the minimum clock period to be less. Also, note that no race conditions are exant in those circuits retimed by RETSAM, a conclusion that cannot be drawn with other retiming algorithms.

8 Conclusions

A retiming algorithm is presented which considers variable clock distribution, register, and interconnect delay. To permit the consideration of these delay components, register electrical characteristics (RECs) are attached to each edge and the original path delays are redefined to be from edge-to-edge rather than vertex-to-vertex. A set of inequalities are created based on these edge-to-edge path delays, permitting the determination of the minimum clock period of the retimed synchronous circuit. An iterative method using ranges of vertex lags rather than constant vertex lags is presented to solve for the edge weights.

The limitations and advantages of the retiming algorithm are compared using a set of modified MCNC benchmark circuits. The results of applying RETSAM to the benchmark circuits show that a more accurate and generalized retiming can be performed than with existing retiming algorithms which do not consider variable clock distribution, register, and interconnect delay. Additionally, the clock period can be further minimized due to localized negative clock skew. Finally, catastrophic clock skew induced race conditions are detected and eliminated.

Summarizing, modified MCNC benchmark circuits have been retimed using a new algorithm which considers the effects of variable clock distribution, register, and interconnect delay on local data paths. This algorithm represents a significant generalization of existing retiming algorithms, permitting the use of retiming for the automated synthesis of higher speed, more reliable pipelined digital systems.

Table 3: Results of the application of the retiming algorithm to MCNC benchmark circuits

Example	Graph properties				T_{CP} After Retiming (tu)	T_{CP} $T_{SKEW}=0$ T_{REG}=const (tu)
	Edges	Vertices	Latency	Initial T_{CP} (tu)		
LGSynth89 - multi-level (netblif)						
C17	26	19	6	92	29	25
b1	34	19	5	80	33	28
cm138a	62	29	4	110	43	38
cm42a	65	34	3	101	46	38
cm82a	59	37	4	139	47	40
majority	26	17	5	103	33	35
LGSynth91 - multi level (blif)						
C17	19	12	5	63	32	26
b1	16	10	2	50	33	24
cm42a	49	18	4	78	35	32
cm82a	22	12	3	49	28	26
cm85a	70	36	3	102	51	40
cm150a	69	38	2	90	56	48
cm151a	38	22	3	93	41	36
decod	89	24	4	69	43	37
parity	47	32	2	77	49	36
tcon	65	34	2	40	34	24
Figure 1 with the added REC values						
Is91	11	8	4	45*	23	19

* denotes a graph that has race conditions before retiming

References

[1] C. E. Leiserson and J. B. Saxe, "Retiming Synchronous Circuitry," *Algorithmica*, Vol. 6, pp. 5–35, January 1991.

[2] A. T. Ishii, C. E. Leiserson, and M. C. Papaefthymiou, "Optimizing Two-Phase, Level-Clocked Circuitry," *Proceedings of the 1992 Brown/MIT Conference on Advanced Research in VLSI and Parallel Systems*, pp. 245–264, March 1992.

[3] G. De Micheli, "Synchronous Logic Synthesis: Algorithms for Cycle-Time Minimization," *IEEE Transactions on Computer-Aided Design*, Vol. CAD-10, No. 1, pp. 63–73, January 1991.

[4] J. P. Fishburn, "Clock Skew Optimization," *IEEE Transactions on Computers*, Vol. 39, No. 7, pp. 945–951, July 1990.

[5] E. G. Friedman, "The Application of Localized Clock Distribution Design to Improving the Performance of Retimed Sequential Circuits," *Proceedings of the IEEE Asia-Pacific Conference on Circuits and Systems*, pp. 12–17, December 1992.

[6] T. Soyata, E. G. Friedman, and J. H. Mulligan, Jr., "Integration of Clock Skew and Register Delays into a Retiming Algorithm," *Proceedings of the IEEE International Symposium on Circuits and Systems*, pp. 1483–1486, May 1993.

[7] B. Lockyear and C. Ebeling, "The Practical Application of Retiming to the Design of High-Performance Systems," *Proceedings of the IEEE International Conference on Computer-Aided Design*, pp. 288–295, November 1993.

[8] T. Soyata and E. G. Friedman, "Synchronous Performance and Reliability Improvement in Pipelined ASICs," *Proceedings of the ASIC Conference*, September 1994.

[9] S. Simon, E. Bernard, M. Sauer, and J. A. Nossek, "A New Retiming Algorithm for Circuit Design," *Proceedings of the IEEE International Symposium on Circuits and Systems*, pp. 4.35–4.38, May/June 1994.

[10] L. Chao and E. H. Sha, "Retiming and Clock Skew for Synchronous Systems," *Proceedings of the IEEE International Symposium on Circuits and Systems*, pp. 1.283–1.286, May/June 1994.

[11] E. G. Friedman and J. H. Mulligan, "Clock Frequency and Latency in Synchronous Digital Systems," *IEEE Transactions on Signal Processing*, Vol. 39, No. 4, pp. 930–934, April 1991.

[12] E. G. Friedman, "Clock Distribution Design in VLSI Circuits — an Overview," *Proceedings of the IEEE International Symposium on Circuits and Systems*, pp. 1475–1478, May 1993.

[13] K. A. Sakallah, T. N. Mudge, T. M. Burks, and E. S. Davidson, "Synchronization of Pipelines," *IEEE Transactions on Computer-Aided Design of Integrated Circuits and Systems*, Vol. CAD–12, No. 8, pp. 1132–1146, August 1993.

[14] S. Malik, E. M. Sentovich, R. K. Brayton, and A. Sangiovanni-Vincentelli, "Retiming and Resynthesis: Optimizing Sequential Networks with Combinatorial Techniques," *IEEE Transactions on Computer-Aided Design*, Vol. CAD-10, No. 1, pp. 74–84, January 1991.

[15] E. L. Lawler, *Combinatorial Optimization: Networks and Matroids*. Holt, Rinehart and Winston, NewYork, 1976.

[16] R. Lisanke, "Logic Synthesis and Optimization Benchmarks User Guide: Version 2.0," Tech. Rep., Microelectronics Center of North Carolina, December 1988.

[17] S. Yang, "Logic Synthesis and Optimization Benchmarks User Guide: Version 3.0," Tech. Rep., Microelectronics Center of North Carolina, January 1991.

Resynthesis of Multi-Phase Pipelines

Narendra V. Shenoy* Robert K. Brayton Alberto L. Sangiovanni-Vincentelli

University of California - Berkeley CA 94720

Abstract

This paper describes an algorithm for deriving *necessary and sufficient* constraints for a multi-phase sequential pipeline to operate at a target clock cycle. Constraints on delays of the pipeline stages are used to drive a combinational logic delay optimizer to resynthesize the pipeline stages for improved performance. A main advantage of such an approach is that a global picture of the distribution of delays in the circuit is obtained. It also permits safe cycle stealing through level-sensitive latches across pipeline stages.

1 Introduction

The past decade has seen automated logic synthesis play a dominant role in IC design. Synthesis tools have concentrated on the optimization of combinational logic for area, for delay or for a weighted combination of the two. But for practical use, these algorithms need to be incorporated in the design of sequential circuits. We focus on synchronous sequential circuits. A consequence of designing in the sequential domain is that cycle time constraint has to be met.

Previous work in performance optimization for sequential circuits falls into two classes. The first relies on using resynthesis techniques for combinational logic, such as critical-path restructuring [14] [1], gate decomposition [9], and buffer optimization [2] [13]. The specified timing constraints translate to a set of performance constraints on one or more "pieces" of combinational logic. In the single phase edge-triggered case, the pipeline performance optimization problem is equivalent to a combinational "speedup" problem [8]. The approach in [1] is based on approximating level-sensitive latches by edge triggered flip-flops and can handle arbitrary multi-phase circuits. Slack is used to direct resynthesis and logic movement across memory elements repeatedly to find the best clock cycle at which the circuit can operate. The slack based approach is myopic in its optimization. To overcome this, simulated annealing is used to guide the optimization. However this may result in much larger circuits than necessary, especially when a target clock cycle is given.

Our approach differs from [1] in several ways;

1. we focus on the problem of satisfying a target clock cycle,
2. only combinational regions are optimized, with no movement of logic across memory elements,
3. flip-flops and latches are handled without any approximations,
4. we restrict the algorithm to acyclic pipeline circuits.

The second approach for performance optimization involves techniques that reposition memory elements (*retiming*) to meet the clock cycle constraint. Details may be found in [6] [5] [7] [3]. A notable effort to merge the two approaches was by DeMicheli [4]. Using the notion of synchronous logic operations, DeMicheli, suggested using combinational synthesis techniques coupled with register movement to optimize the performance of single phase edge-triggered designs.

The paper is organized as follows. Definitions of terms are provided in Section 2. Section 3 describes the approach in detail. An example is presented in Section 4. The implementation and results are given in Section 5.

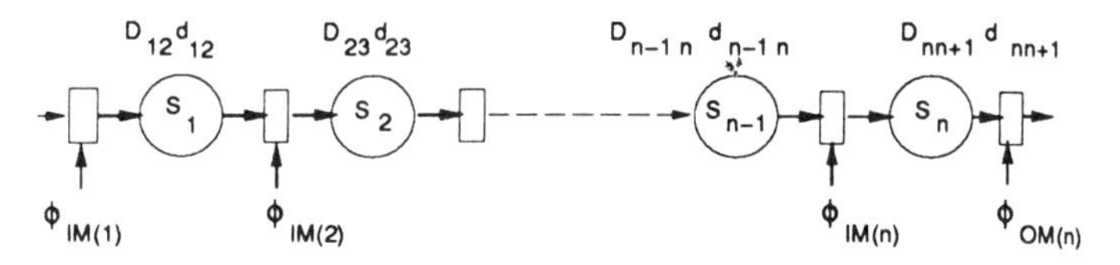

Figure 1: Multi-Phase Pipeline Circuit

2 Definitions

Memory elements are assumed to be level-sensitive [1] (also called a "latch"). A latch has a data input, a clock input and a single output. The data at the input is transmitted to the output as soon as the active period of the latch begins. The output is held at the data value from the time the active period ends until the start of the next active period and fresh data from the input arrives. The input and output are not isolated during the active period and time can be borrowed by the logic preceding the latch, in order to complete the computation. For correct operation, we need the data signal to be stable at the data input before the latching edge occurs by an amount called the *set-up* time (S). It is also required that the signal be stable after the latching edge by an amount called the *hold* time (H). We assume, without loss of generality, that level-sensitive latches are active when the phase is high. Thus, the falling edge of each phase is the critical edge with respect to which set-up and hold constraints must be satisfied.

A clocking scheme, Φ is a collection of l periodic signals with a common period c, and is represented by $\Phi = (\phi_1, \phi_2, \cdots, \phi_l)$. Associated with each phase ϕ_i are two real numbers r_i and e_i, the time of occurrence of the rising and falling edges of ϕ_i $(0 \leq (r_i, e_i) \leq c)$. The clocking scheme specifies a an ordering of the rise and fall of the phases. Let $0 \leq e_1 \leq e_2 \cdots \leq e_l = c$.

Each gate in the circuit has a delay associated with every input-output pin pair. A multi-phase pipeline ($\mathcal{P}$) consists of n stages of combinational logic separated by latches. Each stage of combinational block has inputs from the previous stage (*i.e.* from the latches) and perhaps some inputs from the external world (called *primary inputs*). It has outputs feeding the next stage and outputs to the external world (called *primary outputs*). To simplify the discussion, we make the following restrictions throughout the paper:

1. stage S_1 has primary inputs through latches,
2. stage S_n has primary outputs through latches, and
3. all other stages to have no primary inputs/outputs [2](see Figure 1).

The uth stage denoted by S_u has its inputs latched by a bank of latches denoted by u and its outputs latched by a bank of latches denoted by $u+1$. The delay of each stage S_u is characterized by two parameters, the longest delay (D_{uu+1}) and the shortest delay (d_{uu+1}) from an input to an output. An implicit map $IM : \{1, 2, \cdots, n\} \rightarrow \{1, 2, \cdots l\}$ maps an input of stage S_k to the phase that the input is latched. Similarly $OM : \{1, 2, \cdots, n\} \rightarrow \{1, 2, \cdots l\}$ maps the output of stage S_k to the phase that the output is latched. All inputs (outputs) of a stage are latched on the same phase. Also note $OM(k) = IM(k+1)$.

[1]The extension to include edge-triggered elements is simple.

[2]The extension to handle primary outputs at intermediate stages is straightforward.

Reprinted with permission from *Proc. ACM/IEEE Design Automation Conf.*, N. V. Shenoy, R. K. Brayton, and A. L. Sangiovanni-Vincentelli, "Resynthesis of Multi Phase Pipelines," pp. 490–496, June 1993. © Association for Computing Machinery.

For any path of purely combinational elements from a latch u clocked on phase $\phi(u)$ to a latch v clocked on phase $\phi(v)$ define

$$K_{uv} = 0 \quad \text{if } e_{\phi(u)} < e_{\phi(v)}$$
$$= 1 \quad \text{otherwise.}$$

For a path $p : u_1 \rightarrow u_2 \rightarrow \cdots \rightarrow u_k$, where u_i is a latch on the path and each sub-path $u_i \rightarrow u_{i+1}$ contains no latches on it, the number of clock cycles available for computation along it is $\sum_{i=1}^{k-1}(K_{u_i u_{i+1}})$.

The problem statement is:

Given a pipeline $\mathcal{P}$ with n stages, using a clocking scheme with l phases, find an implementation that meets a given cycle time constraint c.

The problem implicitly expects the rise and fall times for the phases to be determined, consistent with the ordering provided by the clocking scheme.

3 Approach

3.1 Clocking Constraints

We are interested in necessary and sufficient conditions for correct clocking for an arbitrary multi-phase pipeline. The topological structure of a circuit and the distribution of delays within it give rise to a set of constraints called the *internal clocking constraints*. A more detailed discussion on these constraints can be found in [15],[11]. The internal clocking constraints may be classified into two categories.

1. Long path constraints: Let $p : u_1 \rightarrow u_2 \rightarrow \cdots \rightarrow u_k$ be a path with latches $u_1, \cdots u_k$ on it, with each sub-path $u_i \rightarrow u_{i+1}$ containing no latches and the path p containing no repeated latches. We require $e_{\phi(u_k)} - r_{\phi(u_1)} \geq \sum_{i=1}^{k-1}(D_{u_i u_{i+1}} - K_{u_i u_{i+1}} c) + S$. These are also known as the set-up constraints.

2. Short path constraints: For every path $p : u \rightarrow v$, with only combinational elements on p, we require $e_{\phi(v)} - r_{\phi(u)} \leq d_{uv} + (1 - K_{uv})c - H$. These are also called the hold constraints.

It was shown in [11] that these constraints can be represented on a graph, which leads to an efficient algorithm to determine the optimum clock cycle c. In this paper, we are given a target value of c. The problem is to determine bounds on the D_{kk+1}'s and the d_{kk+1}'s. There is an implicit constraint hidden in this formulation, namely $D_{kk+1} \geq d_{kk+1}$.

Let us make the following assumptions for ease of presentation :

1. *the longest path from an input of stage S_i to an output of stage S_j, $j \geq i$, has a delay $\sum_{k=i}^{j}(D_{kk+1})$,*

2. *all gates have identical rise and fall times,*

3. *the clock skew is negligible and all memory elements have the same set-up and the same hold times,*

4. *the shortest path from an input of stage S_k to an output is larger than the hold time H, i.e. $d_{kk+1} \geq H$,* [3]

5. *the target clock cycle is greater than $S + H$.*

The first assumption will be relaxed in section 3.3.1. A method to overcome the second assumption can be found in [12]. The extension to include clock skew and different set-up/hold times is easy and will not be detailed.

[3] This is true because the hold time for most of the latches in present day technology is 0. Arguably, even if it has a finite hold, the combinational regions in most designs have short paths that satisfy this constraint.

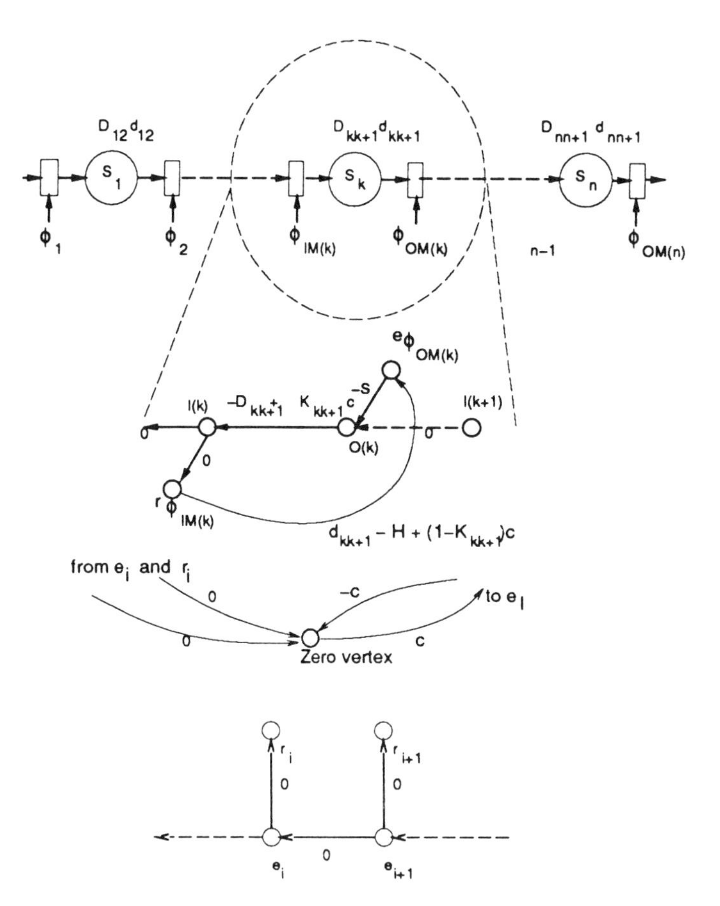

Figure 2: Graph construction: the kth stage

We now describe the construction of a finite, edge-weighted, directed graph $\mathcal{G} = (V, E)$, with vertex set V and edge set E. It is used to derive bounds for the resynthesis procedure. Whenever we say "add an edge of weight w from u to v ", we mean the following: if no edge exists from u to v create an edge with weight w, if such an edge exists, append w to the list of weights on the edge. An edge weight can be either a constant or a linear function of the clock cycle c. For every phase i, there are two vertices r_i and e_i. Stage S_k is split into two vertices $O(k)$ and $I(k)$ and an edge from the former to the latter (called a long path edge), with weight $-D_{kk+1} + K_{kk+1}c$, is added. There is an edge from $e_{\phi_{OM(k)}}$ to $O(k)$ with weight $-S$ and an edge from $I(k)$ to $r_{\phi_{IM(k)}}$ with weight 0 . Place an edge (called a short-path-edge) from $r_{\phi_{IM(k)}}$ to $e_{\phi_{OM(k)}}$ with weight $d_{kk+1} - H + (1 - K_{kk+1})c$. In addition place edges of 0 weight from $I(k+1)$ to $O(k)$ ($k = 1, \cdots, n-1$). To force the rise and fall times to be consistent with the clocking scheme, we construct a zero vertex (z). An edge from z to e_l (with weight c) and an edge from e_l to z with weight $-c$ are added. Also edges from $r_i (i = 1, \cdots l-1)$ and $e_i (i = 1, \cdots, l-1)$ to z (with weight 0) are added. Edges from e_i to r_i ($i = 1, \cdots, l$) and from e_{i+1} to e_i ($i = 1, \cdots, l-1$) with weight 0 are added. A portion of the graph for the kth stage is shown in Figure 2. We now present a theorem on the necessary and sufficient condition for c to be a valid clock cycle for the pipeline [12]:

Theorem 3.1 *The specified clock cycle c is feasible* if and only if *there is no negative cycle in $\mathcal{G}$.*

Theorem 1 leads to a set of linear inequalities between the D_{kk+1}'s, d_{kk+1}'s and c. This is equivalent to eliminating the e_i's and the r_i's from the system of inequalities that need to be satisfied for correct clocking. A simple bound on the size of the set of cycles is [12]:

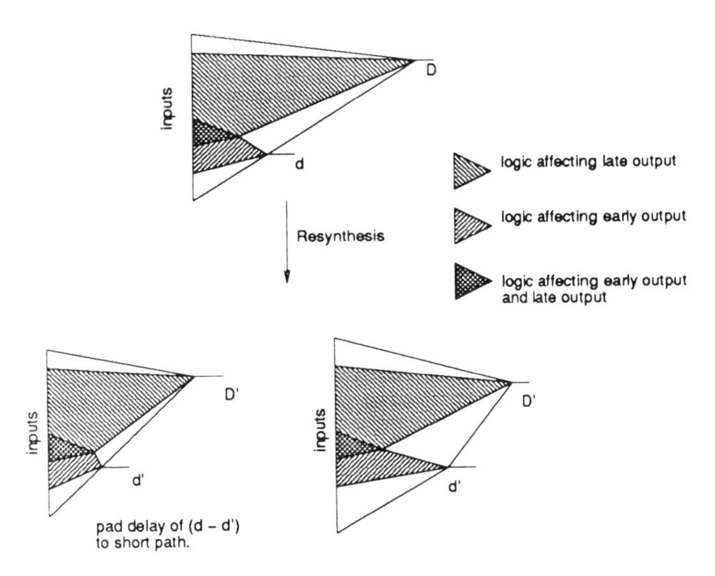

Figure 3: Combinational Optimization: effect on long/short paths

Corollary 3.1 *For an n stage pipeline, there are $O(n^2)$ relevant cycles.*

3.2 Resynthesis

To understand the implications of resynthesis on regions of the pipeline, we model the algorithm for combinational optimization $\mathcal{C}$, as follows.

- input: a piece of combinational logic G, with arrival times specified at the inputs and required times specified at the outputs. Let the slack at each output be the difference between the required and arrival times. If the slack is positive for all outputs then G satisfies the performance constraint. If the slack is negative for an output then it is a critical output.
- output $\mathcal{C}(G)$: a combinational circuit, logically equivalent to G. $\mathcal{C}$ will never cause an output with positive slack in G to have a negative slack in $\mathcal{C}(G)$, *i.e.* it does a careful restructuring to guarantee that non-critical paths do not become critical.

Assuming equal rise and fall times for gates in the library, we seek to understand the effect of $\mathcal{C}$ on long and short paths of a combinational region. Let D and d be the long and short paths in G. We denote the long and short paths after resynthesis by primes (refer to Figure 3). The algorithm $\mathcal{C}$ ensures that $D' < D$. If $d' \geq d$, there is no cause for concern because all the short path lengths appear as positive weights in $\mathcal{G}$: so increasing d can only help eliminate negative cycles. On the other hand if $d' < d$, we could possibly be introducing negative cycles. We make the assumption that delay padding (by discrete amounts) is permitted so that the shortest path in $\mathcal{C}(G)$ is just greater than d. The complete issue of repadding is discussed in [12]. However this requires $d \leq D'$ to be consistent with our model. Note that this requirement only simplifies the discussion, *i.e.* we assume that the d for each stage is fixed to the value of the current implementation. We need to append the constraint above ($D'_{kk+1} \geq d_{kk+1}$ for each stage S_k) [4] to the set of constraints that force all cycles in $\mathcal{G}$ to have non-negative weight. This implies that the target clock is never so small as to require speeding up short paths (if so we must let the short path delays be variables).

We have the following set of constraints

1. those arising from cycles in $\mathcal{G}$ and

[4] We can choose to let d'_{kk+1} be a variable rather than fix it to d_{kk+1}, and add the constraint $d'_{kk+1} \geq H$. After combinational optimization we need to pad delays so that all short paths are greater than d'_{kk+1} and less than D'_{kk+1}.

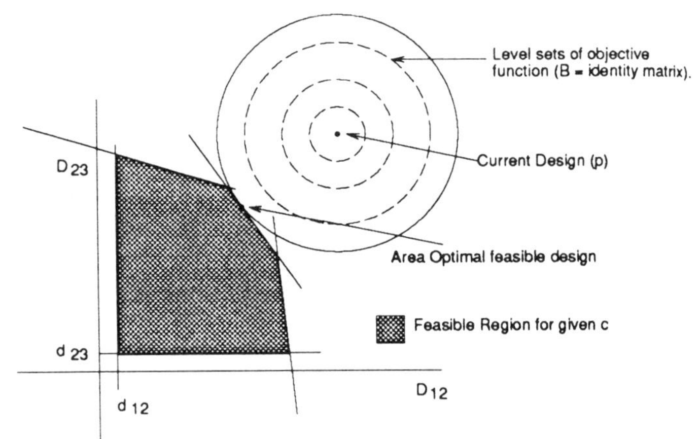

Figure 4: Design Space: feasible region and initial design ($n = 2$)

2. $D'_{kk+1} \geq d_{kk+1}$.

This gives a polytope P in R^n, which lies in the positive orthant, and represents all feasible delay assignments that meet the target cycle. The initial design is represented by the point $p = (D_{12}, D_{23}, \cdots, D_{nn+1})$ (Figure 4 shows the feasible region for a 2 stage pipeline). If the design is feasible, then the point $p \in P$; and there is no resynthesis to be done. Since $c > S + H$, we find that setting $D'_{ii+1} = d_{ii+1}$ is a feasible solution. This is a consequence of the fact that the feasible region is a cone with its vertex at $(d_{12}, d_{23}, \cdots, d_{nn+1})$.

3.3 Deriving Constraints for Resynthesis from Clocking Constraints

Let us assume that there is an area penalty associated with each speedup, which is directly proportional to the deviation from the initial point p. This suggests the formulation of the following optimization problem. Let $D = (D_{12}, \cdots, D_{nn+1})$ be a vector of known delays of the n stages. Let $D' = (D'_{12}, \cdots, D'_{nn+1})$ be the vector of unknown target delays. We find a D' which is a solution to

$$\min(D' - D)^{\mathrm{T}}(D' - D)$$
$$D' \in P.$$

This is an example of constrained optimization with a positive definite (convex) objective function over a convex region P. There are well known algorithms to solve this problem. An algorithm tailored for our formulation is described in [12]. Note that we could have chosen some other objective function like-

1. $\min_{D' \in P} \max_{k=1,\cdots n} |D'_{kk+1} - D_{kk+1})|$
2. $\min_{D' \in P} \sum_{k=1}^{n} |D'_{kk+1} - D_{kk+1}|$.

The quadratic objective function is selected since it is the simplest and guarantees a unique optimum (if feasible). If the d'_{kk+1}'s are chosen to be variables then they too must be entered in the objective function in a manner similar to the D'_{kk+1}'s.

The optimization problem can be generalized as

$$OPT: \quad \min(D' - D)^{\mathrm{T}} B(D' - D)$$
$$D' \in P.$$

The simplest choice for B is the identity matrix. However, we may choose to weigh different stages of the pipeline depending on their

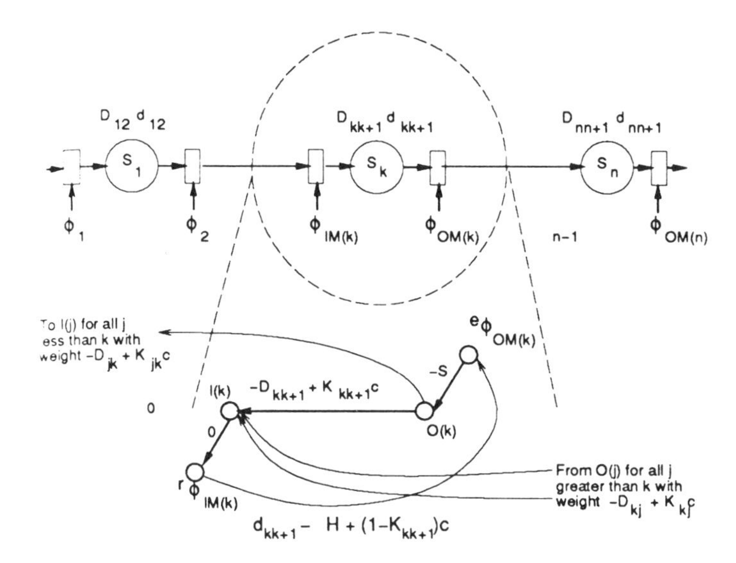

Figure 5: Graph without Assumption 1: the kth stage

resynthesis "potential". The potential of a stage is a positive number which represents the ease of resynthesizing the stage to meet an arbitrary target delay. In such a case B is a diagonal matrix with $[B]_{kk} = b_k$. . The "potential" is computed as a heuristic function of

1. The difference between the longest and shortest paths of the stage. For a stage with small difference, resynthesis for a target delay of the stage may cause all outputs to be critical.

2. The size of the stage. If the stage has a large size then the resynthesis algorithm has a better chance of identifying nodes on critical cuts [14].

The actual function used to compute this figure will depend to a large extent on the algorithm used for combinational resynthesis, and the factors that aid the particular algorithm. The resynthesis algorithm in our implementation is the path restructuring approach presented in [14].

Theorem 3.2 *Any optimum solution to OPT will have* $D'_{kk+1} \leq D_{kk+1}$ *for* $k = 1, \cdots, n$.

We need to resynthesize the the stages S_i for which $D'_{ii+1} < D_{ii+1}$. We set the arrival times at the inputs to stage S_i to be 0, and the required times at all the outputs to be D'_{ii+1}. After combinational resynthesis, we pad delays if there are any short path violations(see next subsection). We can have n such resynthesis steps (one for each stage). The resynthesized regions are mutually disjoint.

3.3.1 Relaxing Assumption 1

So far it was assumed that the longest path in the circuit from (an input of) stage S_i to (an output of) stage S_j ($j > i$), was the sum of the longest paths in each of the stages. This need not be true, and relaxing this assumption will permit the resynthesis to take advantage of cycle-stealing across several latches. However, we must be willing to do a complicated resynthesis procedure. Let $D_{ij}(d_{ij})$ denote the longest (shortest) path from an input of stage S_i to an output of stage S_{j-1}. Note

$$\sum_{k=i}^{j-1} d_{kk+1} \leq d_{ij} \quad , \quad \sum_{k=i}^{j-1} D_{kk+1} \geq D_{ij} \ .$$

The construction of the graph $\mathcal{G}$ is modified as follows (Figure 5). The edges from $I(k+1)$ to $O(k)$ are no longer present. Instead of n variables previously, we now have $\frac{(n+1)n}{2}$ variables. The objective function also changes appropriately, *i.e.*, the unknown vector in OPT is $(D'_{12}, D'_{13}, \cdots, D'_{jk}, \cdots, D'_{nn+1})$, where $j < k$. We have additional constraints of the form

$$D'_{ij} \leq D'_{ik} + D'_{kj} \left\{ \begin{array}{l} i = 1, \cdots, n-1 \\ i+2 \leq j \leq n+1 \\ i+1 \leq k \leq j-1 . \end{array} \right. \quad (1)$$

The number of edges in the graph increases and but the number of cycles remains unchanged. From (1), we obtain the following result.

Lemma 1: $D'_{ij} \leq \sum_{k=i}^{j-1} D'_{kk+1}, j > i+1.$

In order to guarantee that short path delays remain consistent after padding delays if any, we need to add the constraints

$$D'_{ij} \geq D'_{ik} + d_{kj} \left\{ \begin{array}{l} i = 1, \cdots, n-1 \\ i+2 \leq j \leq n+1 \\ i+1 \leq k \leq j-1 \end{array} \right. \quad (2)$$

$$D'_{ii+1} \geq d_{ii+1} i = 1, \cdots, n \ . \quad (3)$$

The feasible region is a polytope in the positive orthant of the D'_{ij}'s, but it is no longer a cone, *e.g.* consider the set of constraints

$$\begin{aligned} d_{12} &= d_{23} = 0.2 \\ D'_{13} &\leq D'_{12} + D'_{23} \\ D'_{13} &\geq D'_{12} + 0.2, D'_{23} + 0.2 \\ D'_{12}, D'_{23} &\geq 0.2 \\ d_{13} &= 0.6 \\ D'_{13} &\geq 0.6. \end{aligned}$$

If the last 2 constraints are ignored, we have a cone with the vertex at $(D'_{12} = 0.2, D'_{23} = 0.2, D'_{13} = 0.4)$. In order to make the feasibility check easier we expand the feasible region as follows. Let

$$\hat{d}_{ij} = \sum_{k=i}^{j-1} d_{kk+1} \qquad j-1 \geq i \quad (4)$$

and replace the constraints in (2) by

$$D'_{ij} \geq D'_{ik} + \hat{d}_{kj}.$$

Note that $\hat{d}_{ij} \leq d_{ij}$, so we have enlarged the feasible region to be a cone, with its vertex defined by the point $(\hat{d}_{12}, \hat{d}_{23}, \ldots, \hat{d}_{nn+1})$. As before, we could choose to let d_{ij}'s be variables and add the constraints

$$d'_{ij} \geq (j-i)H$$

The resynthesis is now done for each stage with arrival and required times placed on its inputs and outputs. The resynthesis is done in a "forward" manner, *i.e.* stage S_k is resynthesized before stage S_{k+1}. Let D''_{ij} denote the longest path from an input of stage S_i to an output of stage S_{j-1}, after stage S_{j-1} (and all stages S_i, $i < j$) has been resynthesized. We set the arrival times at the inputs to stage S_k to

$$a_x = 0 \qquad \forall x \in I(k) \quad (5)$$

Assume that stage S_k has a single output y. The extension to multiple outputs is easy and will be shown later. We shall specify a pair of required times for each output; a *hard* required time (r_y^h) and a *soft* (r_y^s) required time. The latter is less than the former. If the circuit is to operate correctly it *must* meet the hard required time. However we would like the current output of stage S_k to attain the soft required time, in order to help the resynthesis of the stages that follow it. The required times at the output of stage k are set to

$$r_y^h = \min(D'_{kk+1}, \min_{i<k}(D'_{ik+1} - D''_{ik})) \quad (6)$$

$$r_y^s = \min_{j>k+1}(D'_{kj} - \sum_{i=k+1}^{j-1} \min(\tilde{D}_{ii+1}, D'_{ii+1}))). \quad (7)$$

The hard required time is a function of

1. The maximum delay permitted through stage k (first term) and
2. The effect of the stages that have been resynthesized (second term)

The soft required time uses an *as fast as possible* heuristic to ease the resynthesis of future stages. The required time to the delay optimizer is r_y^s. The output of the optimizer is accepted if the hard constraint is met. For the last stage S_n, r_y^s is undefined, since there is no $j > n+1$ and consequently, there are only hard constraints. This guarantees the following after resynthesis:

1. From the first term for r_y^h we get $D''_{kk+1} \leq D'_{kk+1}$.
2. The second term in the required time constraint for r_y^h yields

$$D''_{kk+1} \leq (D'_{ik+1} - D''_{ik}) \quad \forall i < k$$
$$\Rightarrow D''_{kk+1} + D''_{ik} \leq D'_{ik+1}$$
$$\Rightarrow D''_{ik+1} \leq D'_{ik+1}.$$

3. The intuitive reason for defining r_y^s is now explained. This term distributes slacks over the regions that have not yet been resynthesized and hence is a heuristic.

$$D''_{kk+1} \leq (D'_{kj} - \sum_{i=k+1}^{j-1} \min(\tilde{D}_{ii+1}, D'_{ii+1})) \quad \forall j > k+1$$

$$\Rightarrow D''_{kk+1} + \sum_{i=k+1}^{j-1} \min(\tilde{D}_{ii+1}, D'_{ii+1}) \leq D'_{kj}.$$

The term $\tilde{D}_{ii+1}$ in the min constraint is the longest sub-path in stage S_i contained in any path from an input of stage S_k to an output of stage S_{j-1}. If $\tilde{D}_{ii+1} < D'_{ii+1}$ then this stage is not critical for the constraint during the resynthesis of stage S_k. We assume that resynthesis of stage S_i will not make it critical. Should it become critical later, it will be reflected in the computation of r_y^h for S_i. If $D'_{ii+1} < \tilde{D}_{ii+1}$ then the first term during resynthesis of stage S_i will guarantee that its final delay is less than D'_{ii+1}.

$$D''_{kk+1} + D''_{k+1j} \leq D'_{kj}$$
$$\Rightarrow D''_{kj} \leq D'_{kj}.$$

We get the following result:

Proposition 3.1 *If the required time constraints (as described in (6) and (7)) are met for all stages, then $D''_{kj} \leq D'_{kj}$ and $D''_{ik+1} \leq D'_{ik+1}$, for all k, $i < k$ and $j > k+1$.*

The resulting circuit (after resynthesis) will have a feasible clock cycle c. As a side note, should the combinational re-synthesis fail at any stage, say S_j, we can repeat the optimization problem, giving the cost of resynthesizing stage S_j a large weight. This ability to restart may be used repeatedly to find a good final solution.

To extend the discussion above to stages with multiple outputs, we must take care to ensure that D''_{ik} (in (6)) only includes the delay along paths from an input of previous stages S_i that reach the output of S_k in question. Similarly $\tilde{D}_{ii+1}$ (in 7) must be interpreted as the longest path in stage S_i from the output of S_k to an output of S_{j-1}.

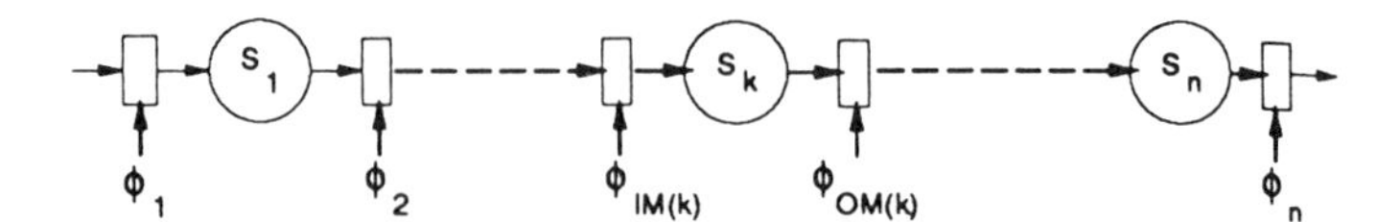

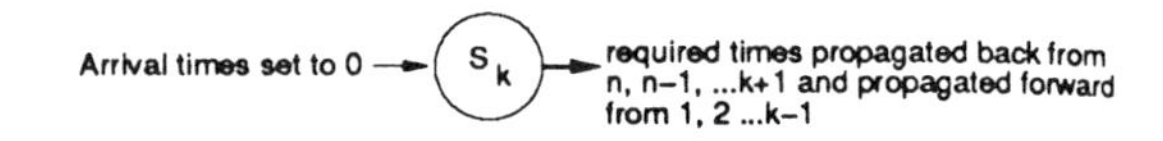

Figure 6: k^{th} Resynthesis Region

4 An Example

Consider the pipeline shown in Figure 7a. Set-up and hold times are assumed to be zero. Let x and y be bounds for stage 1 and stage

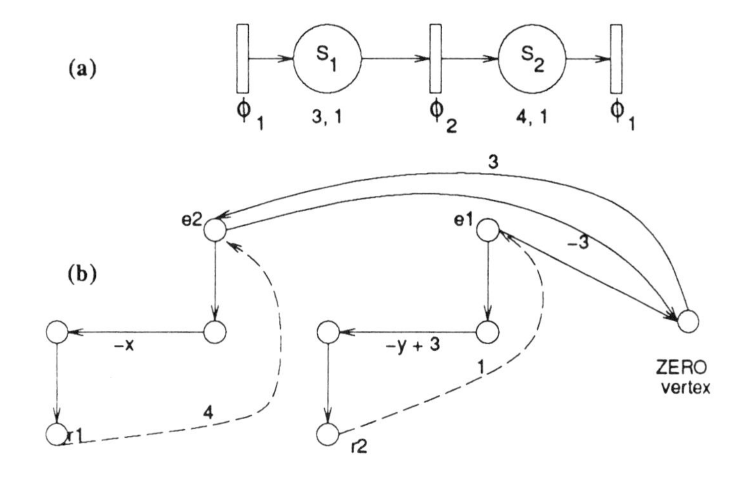

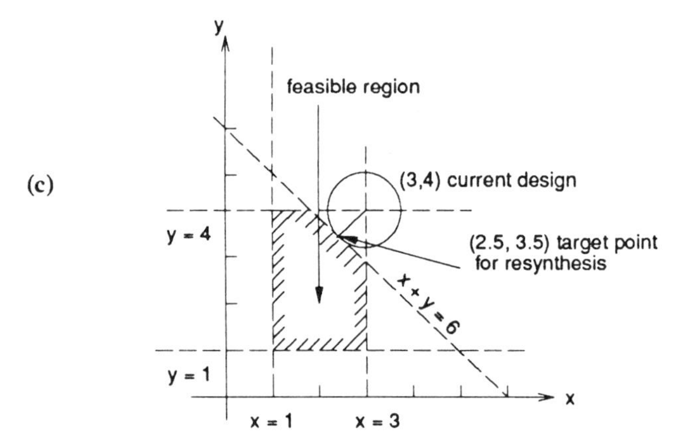

Figure 7: Example 1: pipeline , its graph and solution

2 respectively. The graph with non-zero edge weights is shown in Figure 7b. Let the target clock cycle be 3. Enumerating all cycles that could be negative we obtain
$x \leq 3, x+y \leq 6, y \leq 6, x \leq 4, y \leq 4$.
The constraint $D'_i \geq d_i$ leads to
$x \geq 1, y \geq 1$.
The objective function is $\min((x-3)^2 + (y-4)^2)$. This problem can be graphically solved (see Figure 7c), to obtain a solution of $x = 2.5, y = 3.5$.

To see the need for the extended model, consider the pipeline with multiple outputs (Lines 1 and 2) for each stage as shown in Figure 8a. We introduce a new variable z (longest delay from stage 1 to stage 2) the graph gets modified as shown in Figure 8b. Note that D_{23} has a value 4 along Line 2 and value 2 along Line 1. Though we need to use the value of 4 for the quadratic optimization, we must take care to use the correct value (of 2) when we compute the required time at the output of Stage 1 in Line 1. If the current longest path from stage 1 to

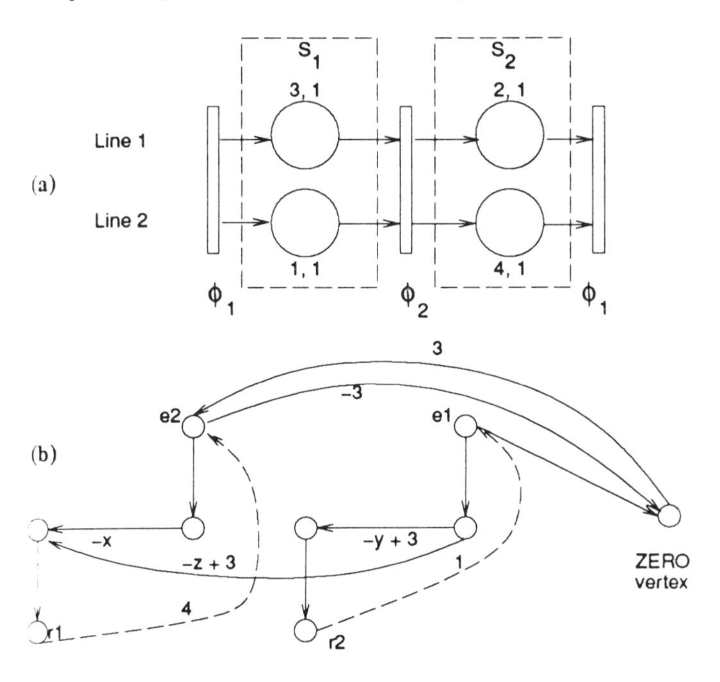

Figure 8: Example 2: pipeline and its graph

stage 2 is 5 (instead of 7, as in the simplified model), the constraints for a clock cycle of 3 are
$x \leq 3, z \leq 6, y \leq 6, x \leq 4, y \leq 4, z - x - y \leq 0, x \geq 1, y \geq 1, z \geq x + 1$.
The objective function is $\min((x-3)^2 + (y-4)^2 + (z-5)^2)$. The optimum value is 0 and the solution point is (3, 4, 5), *i.e.* the initial design is feasible. Thus, in the simplified model, the target delays would have been unnecessarily computed for a circuit with the longest delay from stage 1 to stage 2 less than the sum of the longest individual stage delays, . Let us use a target clock cycle of 2.5; the constraints are
$x \leq 2.5, z \leq 5, y \leq 5, x \leq 3.5, y \leq 3.5, z - x - y \leq 0, x \geq 1, y \geq 1, z \geq x + 1$.
The optimum value is attained at $x = 2.5, y = 3.5, z = 5$. The arrival and required times for the stages are shown in Table 1. So

stage #	*Line*	*input arrival*	*output required*
1	1	0	$\min(2.5, 5-2) = 2.5$
	2	0	$\min(2.5, 5-3.5) = 1.5$
2	1	0	$\min(3.5, 5-2.5) = 2.5$
	2	0	$\min(3.5, 5-1) = 3.5$

Table 1: Arrival and Required times for resynthesis

only the regions in Line 1 - Stage 1 and Line 2 - Stage 2 need to be resynthesized.

5 Implementation and Results

We have implemented a version of the algorithm described in section 5 in the sequential synthesis system SIS, developed at UC Berkeley [10]. Since there are no pipeline benchmark circuits, we constructed a set of circuits which are described below. All circuits use level-sensitive memory elements clocked using 2 phases. The outputs of successive stages are clocked on alternate phases. The five pipeline circuits are:

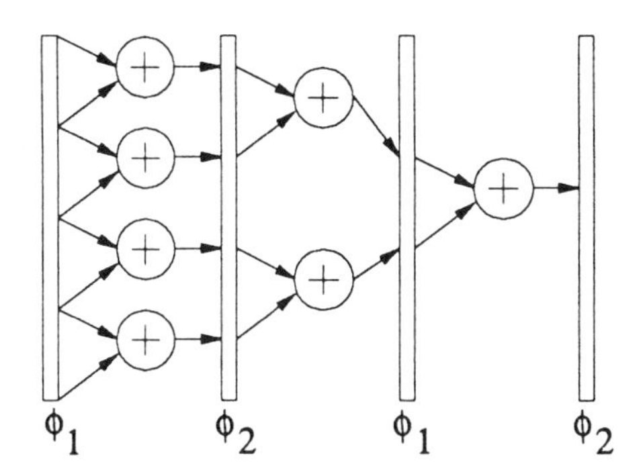

Figure 9: addT3- a 3 stage pipeline

1. adder: A 10 bit adder with 1 stage.
2. mcnc: This is a 2 stage cascade of 2 benchmarks from the MCNC suite of examples. The first stage is **alu2** and the second stage is **cm138**.
3. parity: Computes the parity of a 8 bit input, the parity of the even inputs and the parity of the odd inputs. It has 2 stages.
4. addTi: Described later in this section.
5. pop: Counts the number of ones in a 32 bit input using 4 stages.

Table 2 summarizes results of the algorithm applied to the pipeline circuits described above. We use a simple delay model. We first decompose the circuit into 2 input and/or gates and assign a delay of 1 unit for each gate and 0.2 units per fanout of a gate. The optimal clock for these circuits is computed using algorithms described in [11] (column 2). We use the resynthesis algorithm repeatedly with different target clock cycles. The target clock cycle given in column 3 is the best result after a few iterations. Column 4 gives the results of using the optimal clock algorithm on the resynthesized pipeline. The area of these circuits is measured in terms of the number of 2-input and/or gates. Columns 5 and 6 give the initial and final area. Column 7 gives the total time taken by the procedure. The last column gives the number of memory elements in the circuit. In the first 6 circuits, the algorithm for computing the target clock cycles, is able to take advantage of cycle stealing and a substantial reduction in the clock cycle is observed for a small area penalty.

name	*clock*			*area*		*time*	*latches*
	initial	*target*	*final*	*initial*	*final*	(sec.)	
adder	21.4	15.3	15.2	103	135	6.0	45
addT2	10.0	7.2	7.2	26	29	1.1	10
addT3	12.1	8.2	8.0	89	106	20.5	25
addT4	21.3	19.2	19.2	226	240	10.9	56
addT5	26.2	22.0	22.0	511	535	43.0	119
mcnc	54.6	47.7	47.6	384	412	12.8	24
parity	16.0	13	12.2	39	54	6.7	11
pop	27.3	25.4	25.4	249	320	44.1	59

Table 2: Table of Results for simple delay model

The complexity of the quadratic program grows as the number of stages. Recall that the number of variables and the number of constraints for a pipeline grow as the square of the number of stages. In order to study the efficiency of the quadratic program we constructed a pipeline (addTn) with n stages as follows: the pipeline computes the sum of 2^n inputs (each of width 3 bits). Stage S_i has 2^{n-i+1} inputs and 2^{n-i} outputs. A 3 stage pipeline is shown in Figure 9. The time taken by the algorithm to analyze the circuit and come up with the stage delays (D'_{kk+1}) is summarized in Table 3 below. The second and third columns give the initial and final clock cycles. The next two

n	*clock*		*time*(sec.)	
	initial	*target*	*QP*	*RSY*
2	10.00	7.2	0.44 (3, 8)	0.6
3	12.10	8.72	1.15 (6, 20)	7.8
4	21.30	19.2	2.87 (10, 43)	8.0
5	26.20	22.0	11.44 (15, 76)	31

Table 3: Table of Results for addTn

columns give the time required to solve the quadratic program (QP) and the time taken for resynthesis (RSY). The size of the quadratic program, *i.e.* number of variables and number of constraints is shown alongside the entry in column QP in brackets. The quadratic program seems to display a behavior which is nearly cubic in the number of stages.

We now study the effectiveness of the algorithm in distributing slack. In Table 4, we attempt to do resynthesis for upto 4 target clock cycles, until the resynthesis algorithm is unable to produce a faster clock. In the first set of experiments (Table 3), we gave parameters to the combinational speed-up algorithm that would best meet the requirements. In this case we used the default settings to the combinational speed-up algorithm in all the examples and for all the clock cycles. A "-" means that the speed-up algorithm failed to produce any improvement for lower target clock cycles with the default settings. We see that the area overhead can be controlled by specifying a target clock cycle. The only anomaly in the table is *parity* (row 7), where setting a faster target clock resulted in an area saving. We attribute this to the fact that decreasing the target clock cycle increases the ϵ critical network (see [14] for details) which allows the optimizer to explore a large space for delay optimization.

name	*initial*	*target*			
	(clock/area)	*(clock/area)*			
adder	21.4/103	20/104	19/104	18/123	17/127
addT2	10.0/26	9/28	8/29	-	-
addT3	12.1/89	11/95	10/101	9/101	-
addT4	21.3/226	20/238	-	-	-
addT5	26.2/511	25/551	24/563	22/569	-
mcnc	54.6/384	52/391	50/413	48/412	-
parity	16.0/39	15/54	14/45	-	-
pop	27.3/249	26/252	-	-	-

Table 4: Area-Delay trade-off

6 Conclusions

We have described a systematic procedure for resynthesis of pipelines to meet a target clock cycle. The constraints extracted using Theorem 1 are necessary and sufficient for correct clocking. Since there can be many feasible solutions, we chose to pick the solution which requires minimum delay deviation from the current design. This result is then translated into a set of arrival and required time constraints and path-based restructuring techniques are used to resynthesize the combinational stages to satisfy these constraints. An implementation of the algorithm shows delay improvement ranging from 7.3% to 42.2% at an area penalty ranging from 4.7% to 38.4%, on a set of pipeline circuits (Table 2).

Acknowledgements

The authors wish to thank Dr. K. J. Singh for providing and supporting the `speed-up` algorithm, and for several interesting discussions. Thanks to the anonymous reviewers who gave constructive comments.

References

[1] K. A. Bartlett, G. Boriello, and S. Raju. Timing optimization of multi-phase sequential logic. *IEEE Transactions on Computer-Aided Design*, pages 51–62, 1991.

[2] C. L. Berman, J. L. Carter, and K. F. Day. The Fanout Problem: From Theory to Practice. In *Advanced Research in VLSI: Proceedings of the 1989 Decennial Caltech Conference*, pages 69–99, 1989.

[3] T. M. Burks, K. A. Sakallah, and T. N. Mudge. Multi-phase retiming using $minT_c$. In *ACM/SIGDA Workshop on Timing Issues*, 1992.

[4] G. De Micheli. Synchronous Logic Synthesis: Algorithms for Cycle-Time Minimization. In *IEEE Transactions on Computer-Aided Design*, pages 63–73, 1991.

[5] A. Ishii, C. E. Leiserson, and M. C. Papaefthymiou. Optimizing Two-Phase Level-Clocked Circuitry. In *Advanced Research in VLSI*, 1992.

[6] C. E. Leiserson and J. B. Saxe. Optimizing Synchronous Systems. In *Journal of VLSI and Computer Systems*, pages 41–67, 1983.

[7] B. Lockyear and C. Ebeling. Optimal Retiming of Multi-Phase Level-Clocked Circuits. In *Advanced Research in VLSI*, 1992.

[8] S. Malik, K. J. Singh, R. K. Brayton, and A. Sangiovanni-Vincentelli. Performance optimization of pipelined circuits. In *Proceedings of the International Conference on Computer-Aided Design*, pages 410–413. IEEE, 1990.

[9] P. G. Paulin and F. Poirot. Logic Decompostion Algorithms for the Timing Optimization of Multi-Level Logic. In *Proceedings of the International Conference on Computer Design*, pages 329–33, 1989.

[10] E. Sentovich *et al.* Sequential Circuit Design Using Synthesis and Optimization. In *Proceedings of the International Conference on Computer Design*, pages 328–333, 1992.

[11] N. Shenoy, R. K. Brayton, and A. Sangiovanni-Vincentelli. Graph Algorithms for Efficient Clock Schedule Optimization. In *Proceedings of the International Conference on Computer-Aided Design*, pages 132–136, 1992.

[12] N. Shenoy, R. K. Brayton, and A. Sangiovanni-Vincentelli. Resynthesis of Multi-Phase Pipelines. Technical report, U. C. Berkeley, UCB/ERL memo, 1993.

[13] K. J. Singh and A. Sangiovanni-Vincentelli. A Heuristic Algorithm for the Fanout Problem. In *Proceedings of the Design Automation Conference*, pages 357–360, 1990.

[14] K. J. Singh, A. R. Wang, R. K. Brayton, and A. Sangiovanni-Vincentelli. Timing Optimization of Combinational Logic. In *Proceedings of the International Conference on Computer-Aided Design*, pages 282–285, 1988.

[15] T. G. Szymanski. Computing Optimal Clock Schedules. In *Proceedings of the Design Automation Conference*, pages 399–404, 1992.

Part 4

Analysis and Modeling of the Timing Characteristics of Clock Distribution Networks

Elimination of Process-Dependent Clock Skew in CMOS VLSI

MASAKAZU SHOJI, SENIOR MEMBER, IEEE

Abstract—Delays of two clock signals propagating along their respective CMOS logic circuit paths can be matched against all processing variations if the sum of the pull-up delays of PFET's along the first signal path is matched to that of the second path, and if the sum of the pull-down delays of NFET's along the first path is matched to that of the second path. The new design technique allows generation of a skewless pair of upgoing and downgoing CMOS clocks, and the technique allows one to design CMOS VLSI free from process-induced race conditions. The technique is flexible for light or heavy clock load and for the choice of decoder logic. The technique has a wide application in MOS circuits other than clock decoders.

I. Introduction

ONE OF THE fundamental design considerations of CMOS VLSI is how to distribute, decode, and buffer clock signals. Clock skew is the difference between the time when a clock edge actually arrives and the time when the clock edge should have arrived (as determined from the working principles of the circuit in question). VLSI designers spend a lot of time to contain clock skew under tight control. The distributed clock system of the WE32000 CPU chip set [1] was the most fundamental, and simultaneously practical solution to minimizing the clock skew originating from differing delays associated with clock distribution over a large chip. As for the clock skew originating from clock distribution delay there is no alternative except to compensate for the delay at each destination, as was done in the distributed clock system. Therefore a distributed clock system, with sophistication to compensate for distribution delays, is expected to play an important role in future VLSI design.

Clock skew originating from the different delays of clock distribution, however, does not account for all of the clock skew problem. How to use the "primary" clocks after they arrive at each destination of the chip without introducing additional skew is equally important. How to buffer or decode the primary clocks before they are used to clock latches has not been studied in depth. Use of four equally spaced phases within a machine cycle to effect master-to-slave latch data transfer guaranteed that the WE32000 chips are fundamentally free from race conditions [1], [2]. In such a four-phase timing scheme, a small amount of skew originating from decoding and buffering the primary clocks does not matter. In future CMOS VLSI chips having higher FET counts and higher operating frequencies, however, edge-triggered master–slave latches will be used more and more. Absence of a "dead phase" in the edge-triggered timing offers an advantage of more data processing time and less hardware, but the timing scheme demands higher precision in clock generation: in a scaled-down CMOS technology where a typical gate delay is well under 1 ns, that much clock skew may create a data flowthrough across the master–slave latch pair. However, a 1-ns clock decoding skew is not an easy objective to achieve: 1.75-μm WE32100 CPU was designed to a clock decoding skew of plus or minus 2.5 ns. The required high-precision clock design must be achieved under a difficult condition of increasing scale of integration. The objective of this paper is to provide a flexible solution to this problem.

II. Definition of the Problem

The problem is to equalize delays of all the clock decoders and clock buffers, and to maintain the clock edges matched under all conditions. Typical clock decoder and clock buffer circuits are shown in Fig. 1(a), and the clock waveforms with and without skew are schematically shown in Fig. 1(b).

Delay of a CMOS gate is defined as the difference between the time the input voltage crosses $V_{DD}/2$ and the time the output voltage crosses $V_{DD}/2$. This definition may give an unreasonable delay time (for example, a negative delay time) if the static switching threshold of the gate is not at $V_{DD}/2$, and if the rate of change of the input voltage is very small. This definition gives valid and reasonable delay time, however, when the transition time of the input signals (on a 10–90-percent basis) is less than several times (typically five times) the delay of the gate that is defined for the input signal that switches instantly. In a clock circuit the definition has clear meanings.

Delay associated with decoding or buffering the primary clock signal of fabricated devices depends on the particular fabrication process the chip underwent, and a designer cannot intervene in the process to control the delay. If a designer is able to keep the relationship of the buffered or decoded clock edges such as *NI*, *I*, and *D* of Fig. 1(b),

Reprinted from *IEEE J. Solid-State Circuits,* vol. SC-21, no. 5, pp. 875–880, Oct. 1986.

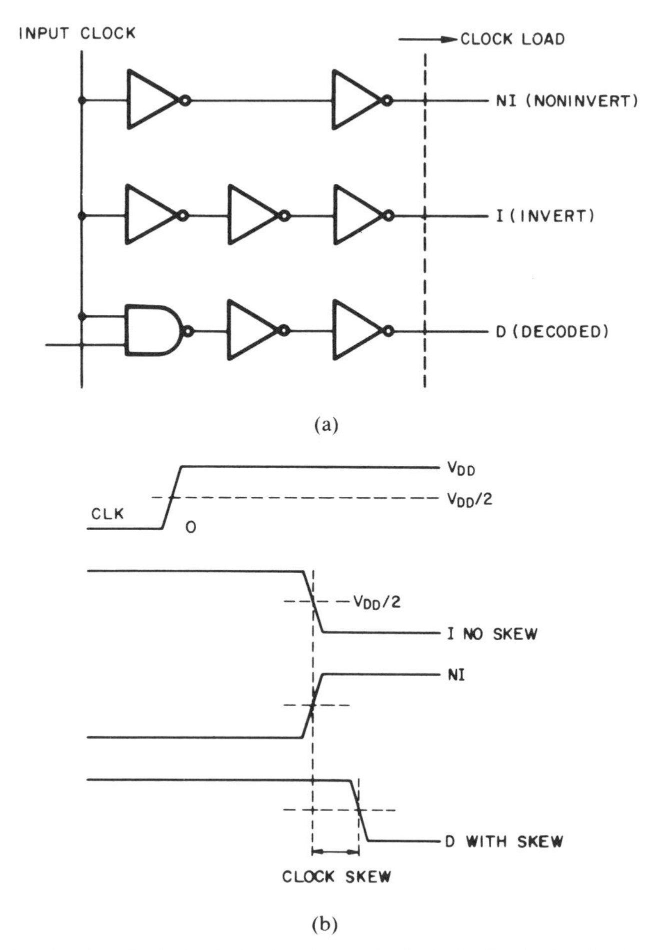

Fig. 1. (a) Typical clock decoder and clock buffer logic circuits. (b) Clock waveforms and skew.

tracking each other accurately, he will be able to design a high-performance VLSI chip.

Variation in propagation delay of the clock edges caused by processing spread is usually characterized by extracting three sets of process parameters of CMOS FET's for a given CMOS process. They are: high-speed (or high current) process parameters; typical (or medium current) process parameters; and low-speed (or low current) process parameters. The high-current process gives typically a half to a third of the gate delay of the low-current process. Even if all the clock edges are accurately aligned for typical processing, they may not be aligned any more in the deviation from the typical processing.

A misleading practice often followed by the "worst-case" design methodology is to simulate designed circuits on a combined low-PFET low-NFET current (abbreviated to PL–NL) process: when the information about the typical and the best performance of the chip is required, the circuits are simulated on a combined medium-PFET medium-NFET current (PM–NM) process and on a combined high-PFET high-NFET current (PH–NH) process, respectively. In these restricted combinations the current drive capabilities of a PFET and an NFET vary approximately proportionally, and therefore the delays vary proportionally also. Clock skew generated by this type of variation is relatively small. However, this is not what happens in an actual chip-production environment. In fact, PFET and NFET drive can vary somewhat independently. Such a process unbalance may occur through failure to control the threshold voltages of the PFET's and NFET's. If the PFET current drive capability decreases while that of the NFET increases, a larger clock skew emerges among different clock decoder-buffer circuits. It is important to note that by unbalancing a process a larger clock skew is created, more than by speeding up or by slowing down the process in a balanced way. The worst "unbalanced" process is either a combination of PFET high current and NFET low current (PH–NL process) or a combination of PFET low current and NFET high current (PL-NH process). Although considering this combination could be very pessimistic, a designer must be prepared to face the possibility of such a variation, and besides, variations intermediate between the worst unbalanced processing and the best balanced processing are not unlikely. Therefore the problem is to produce a design that withstands this type of variation, as well as more balanced process variations.

III. Variation in FET Characteristics by Process

High-speed and low-speed process parameters are defined such that only 2 percent of the processed devices have FET drive capabilities beyond the respective specifications: 96 percent of the entire population should fall within the high and the low FET speed limits. A convenient parameter to define the drive capability is the time constant of an FET, T_D, defined by

$$T_D = (C_G + C_D)(V_{DD}/I_{\mathrm{sat}}) \tag{1}$$

where C_G is the gate capacitance, C_D is the drain diffused island capacitance, and I_{sat} is the drain current when both gate and the drain voltages are set for the maximum channel current of a digital operation. T_D is a parameter proportional to gate delay: typically five times the averaged T_D for PFET and for NFET is the delay of an inverter with three times fan-out. Table I lists T_D, for the three process conditions, for the twin-tub 1.75-μm CMOS process [3]. The time constants were determined by using ADVICE circuit simulator [4] for the minimum size FET (4 μm wide) at 65 °C.

In Table I, "ratio" means the ratio to the typical processing. The drive capability is the inverse of the delay time. This result shows that ratio of drive capabilities of NFET to PFET varies from 6.02 for NH–PL process to 0.73 for NL–PH process. The variation is the practical limit, since parameters like the NFET channel length and the thin oxide thickness approximately track the respective parameters of the PFET. Thus a designer must be prepared that the parameters could vary in the production environment, within a remarkably wide range. The realistic process parameter variations never occur such that all the parameters specifying the drive capability of an FET fluctuate simultaneously in their respective directions to

TABLE I
TIME CONSTANT OF FET'S

Process	NFET Picosecond	ratio	PFET Picosecond	ratio
High current (H)	88.0	0.556	201.0	0.620
Typical (M)	158.0	1.000	324.0	1.000
Low current (L)	273.0	1.730	530.0	1.630

increase (or decrease) the drive capability. There are positive and negative statistical correlations among the parameters. If the set of the fastest possible parameter values was used by neglecting the correlations, the device would be unrealistically fast. The high-current and the low-current processes used in the present analysis include the correlations to represent our realistic fast and slow devices [3].

IV. MECHANISM OF CLOCK SKEW

Fig. 2(a) shows CMOS inverter chains used to generate a pair of inverting and noninverting clocks. When the input node I makes a low-to-high transition, the waveforms at the intermediate nodes A, B, C, 1, and 2 are as shown schematically in Fig. 2(b) and (c). In order to make the clocks skew-free, a conventional designer designs the circuit by scaling the FET's to satisfy

$$T_I = T_{NI} = T_0 \tag{2}$$

where

$$T_I = T_A + T_B + T_C$$
$$T_{NI} = T_1 + T_2.$$

This design is carried out assuming the typical process conditions. When the same circuit is processed by the high-current process, delays $T_A, T_B, \cdots, T_2$ will be different. Delay T_A depends on the drive capability of an NFET (MNA of Fig. 2(a)) which changed by a factor f_N, where $f_N > 1$. Then

$$T_A(H) = T_A(1/f_N)(< T_A).$$

In this equation, $(1/f_N)$ can be found from the NFET ratio column of Table I as 0.556. Delay T_B depends, however, on the driving capability of a PFET (MPB in Fig. 2(a)), which changed by a factor f_P. Then

$$T_B(H) = T_B(1/f_P)$$

and

$$(1/f_P) = 0.620$$

from Table I. In this way we obtain

$$T_I(H) = (T_A + T_C)(1/f_N) + T_B(1/f_P)$$
$$T_{NI}(H) = T_1(1/f_N) + T_2(1/f_P). \tag{3}$$

Since the clocks were assumed skew-free for the medium current process, (2) holds. Then

$$T_A + T_B + T_C = T_1 + T_2$$

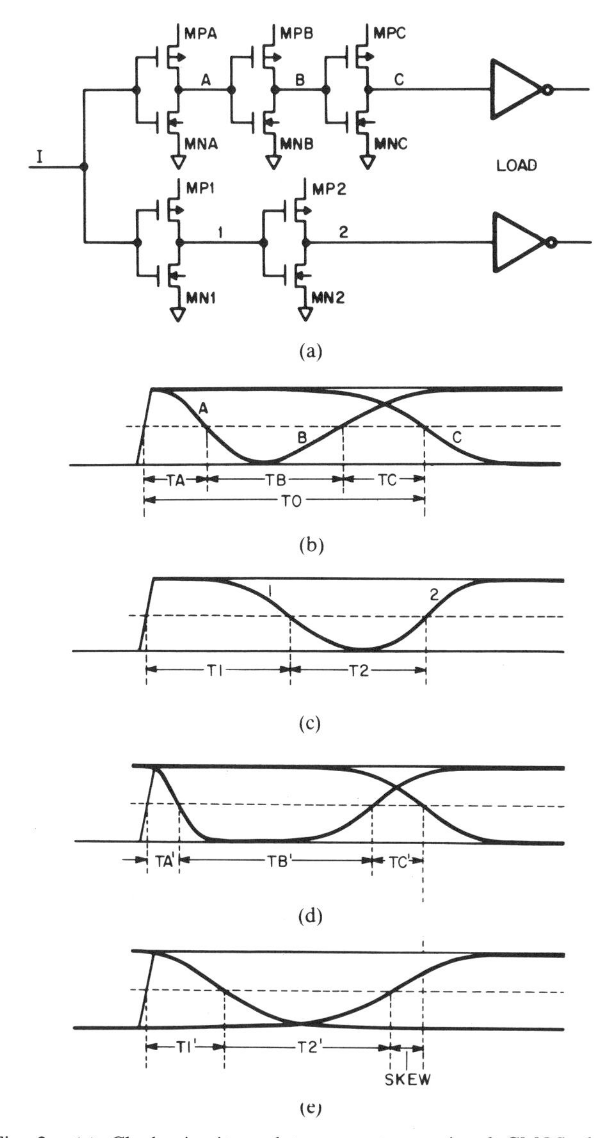

Fig. 2. (a) Clock circuit used to generate a pair of CMOS clocks. (b)–(e) Clock waveforms.

or

$$T_A + T_C - T_1 = -(T_B - T_2)$$

and therefore

$$T_1(H) - T_{NI}(H) = (T_B - T_2)((1/f_P) - (1/f_N)). \tag{4}$$

From (2) and (3), an ideal skewless condition

$$T_I(H) = T_{NI}(H)$$

is met when $f_N = f_P$ or $T_2 = T_B$, or when both are satisfied. If the circuit is processed by the PH–NH or PL–NL process, $f_P = f_N$ is approximately satisfied (within 10 percent, from Table I). Then the clocks remain skew-free against this type of variation. If the process varies from the typical PM–NM process to the PL–NH process, however, $f_N > 1$ and $f_P < 1$. Then the circuit will deliver clocks with large skew. The mechanism is schematically shown in Fig. 2(d) and (e).

TABLE II
CLOCK CHAIN SKEW DATA

Design FET size	(TA+TC)/TB nanosecond	T1/T2 nanosecond	Skew (nanosecond) PH–NH	PL–NL	PM–NM	PH–NL	PL–NH
I A(10/04),1(07/03) B(10/04),2(50/20) C(40/16) LOAD(100/40)	1.34/0.88	1.22/1.01	0.02	0.02	0.00	0.02	0.00
II A(10/04),1(05/02) B(19/08),2(20/08) C(32/13) LOAD(100/40)	1.50/0.62	0.82/1.31	0.00	0.08	0.00	0.38	0.07
III A(10/04),1(05/02) B(05/02),2(09/07) C(40/16) LOAD(100/40)	1.46/1.38	0.61/2.23	0.04	0.00	0.00	0.46	1.30

Equation (4) shows, however, that if the circuit is originally designed to satisfy

$$T_B = T_2 \tag{5}$$

in addition to the condition of zero skew at the typical process (2), then the circuit remains skew-free, whatever the process may be. From (2) and (5) we obtain

$$T_A + T_C = T_1. \tag{6}$$

Equations (5) and (6) represent the following: when input node I of Fig. 2(a) pulls up, gates A, C, and 1 pull down, by NFET MNA, MNC, and MN1, respectively. Equation (6) means that the sum of the pull-down delays by NFET MNA (gate A) and MNC (gate C) of the first logic chain should be equal to the pull-down delay of NFET MN1 (gate 1) of the second logic chain. Further, gate B of the first chain and gate 2 of the second chain pull up. Equation (5) means that the pull-up delays of PFET MPB (gate B) and that of MP2 (gate 2) should be equal.

By (5) and (6) the delays contributed by NFET and PFET are separately matched. If the precision matching is exercised the delays of the two chains are robustly matched whatever the process variations. If only total delays of the two chains are matched (by (2)) the delay will not be matched if the process varies.

This design procedure is always possible. Delays are determined by the size of FET's in Fig. 2(a) (MPA, MNA,$\cdots$,MN2). Since there are ten FET's, there are ten adjustable FET sizes, and there are only two equations. There are many solutions to the problem. The way to choose the most desirable solution is discussed in the later sections.

V. SIMULATION RESULTS

Several models of the circuit shown in Fig. 2(a) were designed and their skew characteristics were examined by ADVICE [4] simulations. Table II summarizes the results. The inverter chains had fixed, 40 times the minimum size clock load, and had delays at the typical process ranging from 2 to 3 ns. Wiring parasitics were ignored. In the distributed clock system of WE32000 CPU a pair of well-synchronized primary clocks is distributed over the entire chip of about 1 cm^2. Each local clock decoder taps off the primary clocks, decodes, buffers, and drives the clock loads within the area typically 1 mm^2. Therefore the wiring parasitic capacitance is not primarily important. Since the effects of wiring capacitance can be made small by a proper local clock decoder design, and since inclusion of wiring parasitics never brings about anything essentially new, the parasitics were not included.

In Table II, FET size was measured by the unit of 1/2 grid of Gate-Matrix style layout [5]. A Gate-Matrix layout is a symbolic layout that allows only quantized FET sizes. The unit FET in 1.75-μm CMOS is $W_m = 3.25$ μm wide, that equals half of the pitch of the metal grid (6.5 μm). The FET size must be an integral multiple of W_m. Although the layout style allows only discrete FET sizes, a fine adjustment of the circuit delay is feasible if relatively large FET's are used, as in Table II. Design I approximately follows the principle of the last section, and therefore the skew is quite small: (5) and (6) of the last section are satisfied with an accuracy of 13 percent. This design provided the best skew compensation. The best skew compensation occurs when (5) and (6) are closely satisfied, but not necessarily when they are exactly satisfied. The theory of the last section assumes that the delay of an inverter chain is a sum of the delays of the component inverters. This is only an approximation, and therefore (5) and (6) are closely satisfied, but not exactly. That there is a design of inverter chains that offers the minimum skew against process variations remains true independent of the accuracy of the approximation.

The delays and the skews listed in Table II are both small. This is because the simplest design example was chosen to demonstrate the principle of the skew elimination. What is important is the maximum skew relative to the delay. Design I is much better than designs II and III.

Design II, which is an average design before the new design principle was found, and design III, which is a very unbalanced design, show increasingly more skew as extreme process variation occurs. An unbalanced process like

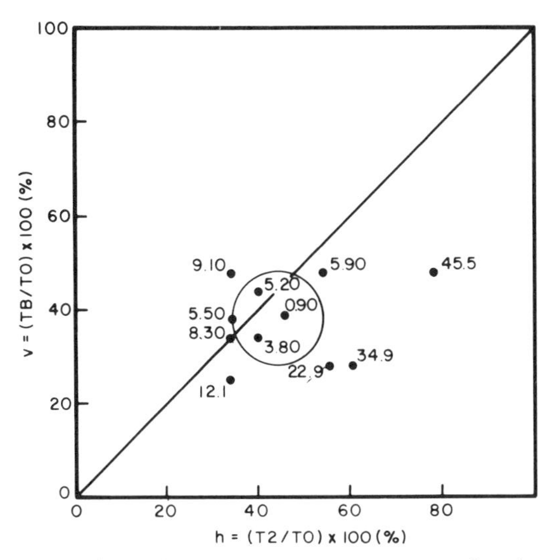

Fig. 3. Clock skew in percent versus parameters h and v.

PH–NL or PL–NH generates consistently more skew than a balanced process like PH–NH or PL–NL. The worst-case skew of design I (0.02 ns) is one order of magnitude smaller than the worst-case skew of design II (0.38 ns), showing how effective the design procedure is. Input voltage to the chain was a pulse voltage source that had a constant rise time of 0.5 ns. This was convenient to determine very small skew accurately by simulation. Experimental verification of the skew compensation is very desirable, but it is very difficult to attain the required measurement accuracy (0.01 ns or less). The circuit simulations are less accurate than experiments to determine the absolute delay, but the simulations are more accurate to determine the design parameter dependence of the clock skew. By this reason the results of Table II prove the effect of skew compensation.

Fig. 3 shows a map of skew versus percentage of the pull-up delay of the two chains. With reference to Fig. 2(b) and (c), the horizontal axis is the quantity h defined by

$$h = (T_2/T_0) * 100 \text{ (percent)}$$

and the vertical axis is the quantity v defined by

$$v = (T_B/T_0) * 100 \text{ (percent)}$$

where T_0 is the delay of the chains, matched at the typical process. The numbers associated with each point in Fig. 3 are the percentage skew defined by

$$\left(\frac{|\text{Skew}|}{T_0}\right) * 100 \text{ (percent)}.$$

Fig. 3 illustrates that minimum skew occurs when $h = v$ is approximately satisfied, but not exactly. For very accurate compensation, the range of h is limited practically to 35–50 percent. The theory of the last section assumes that the delay of the cascaded inverter chain is equal to the sum of the individual inverter delays. In reality the delay depends on the waveform, and the theory does not hold when an excessive waveform distortion occurs by heavy capacitive loading or by the Miller effect. In spite of this complication the skew compensation does work: the range of h for the best compensation falls within the range where the design of the chains is the easiest. Practical application never suffers very much from the limited range of h.

CMOS gates do not always have the switching threshold voltage exactly at $V_{DD}/2$, used to define delay time. Practically, it is impossible to scale FET's of each gate to attain the uniform threshold. The switching threshold of a gate changes by process variation. It may be argued that an ideal CMOS clock is the one that generates controlled skew to cancel the threshold variation by the process variation. This idea is impractical, as long as the switching thresholds of the gates at the typical process are not adjusted to $V_{DD}/2$. The proposed technique in this paper is not intended to design such ultraprecision clock circuits. Fig. 3 shows, however, that a clock circuit that is not designed carefully generates 20–30-percent skew by process variation. Since the total delay of the chain is 2–3 ns, the skew amounts to 1 ns, which is enough to create the race conditions. Our objective is a modest one: to prevent such a timing disaster by clearly recognizing the mechanism of clock skew and the technique to prevent the race conditions.

The design data of this section assume that the input clock of Fig. 2(a) makes a low-to-high transition. Design of a pair of perfectly symmetrical CMOS clock generators whose outputs are matched for both input low-to-high and high-to-low transition is more time consuming, since we have no convenient computer aids: the design process by interpolating the FET sizes from the simulation data is very tedious. Our best design is summarized in Table III. Although the skew is not as good as design I (0.06-ns worst case), this inverter chain generates a practically perfect pair of CMOS clocks against all the process variations. Although the design of a clock generator based on this technique is tedious, the basic idea of the technique should be quite obvious to any designer.

VI. Additional Considerations

Application of the basic concept of this paper is wide. For example, a pair of clock generator logic chains, one of which includes a NAND2 decoder and the other a straight chain of inverters, can be matched by replacing the first stage of the inverter chain by a NAND2 gate, whose one input is permanently connected to V_{DD}, as shown in Fig. 4(a). This is necessary because process dependence of delay of series-connected NFET's is different from that of a single NFET. In general, if one of the chains includes a gate other than an inverter, the other chain must include the same kind of gate at any one of the equivalent locations. A NAND2 gate of Fig. 4(a) can be either at location 1 or 3, but not at 2. Pull-down delays T_x and T_y (by the series-connected NFET's of NAND gates) must be separately matched. In this precision application, a difference in pull-down delay in response to a transition from input A and delay in response to a transition from input B of a

TABLE III
Clock Chain Skew Data

Design FET size	(TA+TC)/TB nanosecond	T1/T2 nanosecond	Skew (nanosecond) PH–NH	PL–NL	PM–NM	PH–NL	PL–NH
IV A(15/06), 1(07/03) B(13/06), 2(44, 18) C(66/27) LOAD(132/54)	1.34/1.03	1.27/1.10	0.00	0.06	0.00	0.00	0.07

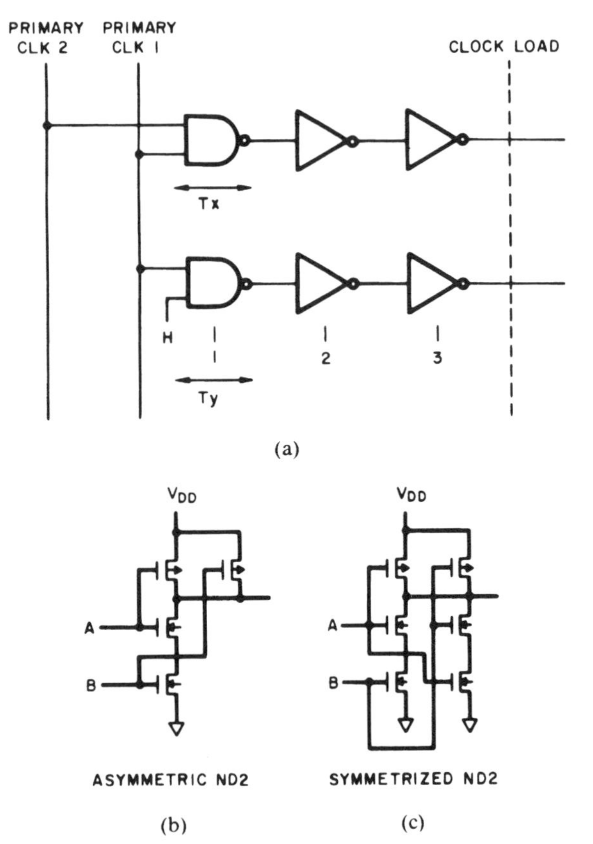

Fig. 4. (a) Delay matching of phase clock decoders. (b) Conventional NAND2 gate. (c) Symmetrized NAND2 gate.

NAND2 gate shown in Fig. 4(b) is important. Delay from input *A* is shorter than that from input *B*. The delays can be made equal by using a symmetrized NAND2 gate shown in Fig. 4(c). Delays of two logic chains that have different numbers of cascaded stages of gates, or that have different types of gates, can be matched by this technique. The degree of matching, however, may not be so good as shown in Table II. In a four-times countdown circuit with strobe, the skew between the counted down clock and the strobe was reduced at least by a factor of two using this technique.

VII. Conclusion

Delays of any pair of CMOS logic paths can be matched despite process variations. The technique is a flexible alternative and a supplement to various circuit techniques to reduce skew, including the use of feedback. To arrive at a design by this technique, however, many simulations must be run. This is particularly true when both the input pull-up and the input pull-down delays of more than two circuits must be all matched. A computer-aided design tool that would execute the design procedure can be written, but only after very accurate models of the FET's and of the parasitics become available. The present models provide good ideas, but are not accurate enough without experimental calibration. The problem will be solved eventually, and this technique of clock skew compensation should become practical, since submicrometer CMOS circuits will require more advanced design techniques.

Acknowledgment

The author wishes to thank T. G. Szymanski and J. H. Condon for stimulating discussions and P. J. Weinberger, H. K. Gummel, D. Caplan, P. W. Diodato, W. W. Troutman, and the four reviewers for valuable suggestions on the manuscript.

References

[1] M. Shoji, "Electrical design of BellMac-32A microprocessor," in *ICCC 82 Dig.*, Oct. 1982, pp. 112–115.
[2] N. Weste and K. Eshraghian, *Principles of CMOS VLSI Design*. Reading, MA: Addison-Wesley, 1985.
[3] L. C. Parrillo, L. K. Wang, R. D. Swenumson, R. L. Field, R. C. Melin, and R. A. Levy, "Twin-tub CMOS II-advanced VLSI technology," in *IEEE Tech. Dig. Int. Electron Device Meeting*, 1982, p. 706.
[4] L. W. Nagel, "Spice 2: A computer program to simulate semiconductor circuits," Univ. of Calif., Berkeley, Memo ERI-M520, May 1975.
[5] A. D. Lopez and H. F. Law, "A dense gate-matrix layout style for MOS LSI," in *ISSCC 80 Dig.*, Feb. 1980, pp. 212–213.

An Upper Bound of Expected Clock Skew in Synchronous Systems

Steven D. Kugelmass and Kenneth Steiglitz

Abstract—**We consider a statistical model for clock skew in which the propagation delays on every source-to-processor path are sums of independent contributions, and are identically distributed. Upper bounds are derived for expected skew, and its variance, in tree distribution systems with N synchronously clocked processing elements.**

We apply these results to two special cases of clock distribution. In the first, the *metric-free* model, the total delay in each buffer stage is Gaussian with a variance independent of stage number. In this case, the upper bound on skew grows as $\Theta(\log N)$. The second, *metric*, model, is meant to reflect VLSI constraints. Here, the clock delay in a stage is Gaussian with a variance proportional to wire length, and the distribution tree is an H-tree embedded in the plane. In this case, the upper bound on expected skew is $\Theta(N^{1/4}(\log N)^{1/2})$.

Index Terms—**Clock distribution, clock skew, processor arrays, synchronization, VLSI clocking.**

I. Introduction

Previous studies of the effect of system size on synchronization have used worst case analysis (see, for example, [1]–[6]), which may be overly pessimistic as systems get very large. Fisher and Kung [3], for example, used two basic models: one in which clock skew is proportional to the difference in path length from two nodes to their nearest common ancestor in the distribution tree; and one in which it is proportional to the sum of those lengths. The purpose of this note is to introduce a statistical model for clock skew, and to study its asymptotic growth rate for two important special cases. This model is meant to describe those systems where the uncertainty in clock arrival time is due to the accumulated effect of a large number of independent random deviations from nominal values. One main result will be an upper bound on expected clock skew that grows as $\Theta(\log N)$ in the case that variation in delay of a buffer stage is independent of wire length. Another result will be a somewhat more optimistic estimate than the one for the planar H-tree using the summation model of Fisher and Kung.

II. The Model

The basic limitation on clocking speed in a synchronous system is the difference in arrival times of a particular waveform edge at communicating processors. In our analysis, we make three basic assumptions about the clock distribution method.

1) The paths from the clock source to the processing elements are identical, in the sense that the paths are of equal length, contain the same number of wires and buffer stages, and are equally loaded.

2) The clock arrival times are random variables, and are the sums of uncertain independent delays through many wires and buffer stages. It is reasonable to expect, therefore, by the central limit theorem, that the arrival time of a clocking signal at a processor is well modeled asymptotically as a Gaussian random variable.

3) For the purpose of arriving at an upper bound on clock skew that is independent of topology, we make the additional assumption that any processor can communicate with any other.

Thus, our basic model applies to the situation where wire lengths are equalized in a multistage clock distribution system. It does not apply to situations where source-destination clock delays are tuned, where loading is nonuniform, or where wire lengths are disparate.

III. Cramér's Result and a Lemma

We use the following classical result of Cramér [7]. The notation $n(\mu, \sigma^2)$ is used for a normal distribution with mean μ and variance σ^2.

Theorem 1 [Cramér]: Let x_i, $i = 1, \cdots, N$ be independent samples from an $n(\mu, \sigma^2)$ distribution, and let the *range* $R = x_{\max} - x_{\min}$ be the difference between the largest and smallest x_i. Then the expected value of R is asymptotically:

$$E[R] = \sigma\left[\frac{4\ln N - \ln\ln N - \ln 4\pi + 2C}{(2\ln N)^{1/2}} + O\left(\frac{1}{\log N}\right)\right] \tag{1}$$

where $C \approx 0.5772\cdots$ is Euler's constant. The variance of R is given by

$$\operatorname{Var}[R] = \frac{\sigma^2}{\ln N}\frac{\pi^2}{6} + O\left(\frac{1}{\log^2 N}\right). \tag{2}$$

In general, the arrival times of clock signals in a distributed system will not be independent, but we can use Cramér's result in the case of general dependencies by the following lemma (for a proof, see [8]). Let y_i, $i = 1, 2, \cdots, N$ be independent identically distributed real random variables, with $N = kn$, and let the sets σ_j, $j = 1, 2, \cdots, n$ be n disjoint subsets of distinct y_i, each of cardinality k. Let the τ_i be similarly defined, with k distinct elements each, except that they are not necessarily pairwise disjoint. Define the corresponding sums of the y_i by

$$s_j = \sum_{y\in\sigma_j} y \tag{3}$$

and

$$t_j = \sum_{y\in\tau_j} y. \tag{4}$$

Assume further that the probability distribution functions of the component processes are differentiable with finite means and variances.

Lemma 1 [8]: The expected range of the t_j is no greater than the expected range of the s_j.

This allows us to use Theorem 1 to bound the range even when the samples are dependent. We now use these results to analyze two examples of clock distribution systems.

Reprinted from *IEEE Trans. Comput.*, vol. 39, no. 12, pp. 1475–1477, Dec. 1990.

IV. The Metric-Free Tree

In the *metric-free tree* model, we assume that the clock is distributed to processors at the leaves of a binary tree, and that the delays in every stage have the same probability distribution, and are independent. A *stage* here includes the buffers and transmission wires. This models the situation where all wire lengths in the distribution tree can be made equal in length, such as might be the case in clock distribution to many chips on a board, or to many boards in a system.

The arrival time of a clock signal to any processor is the sum of the delays along the path from the root of the tree to the leaf containing the processor. The arrival time is therefore the sum of $\log_2 N$ Gaussian random variables, each with some variance independent of stage number, say σ_0^2. It therefore is Gaussian with variance $\sigma_0^2 \log_2 N$. Applying Theorem 1 with this distribution, we find that the expected skew is

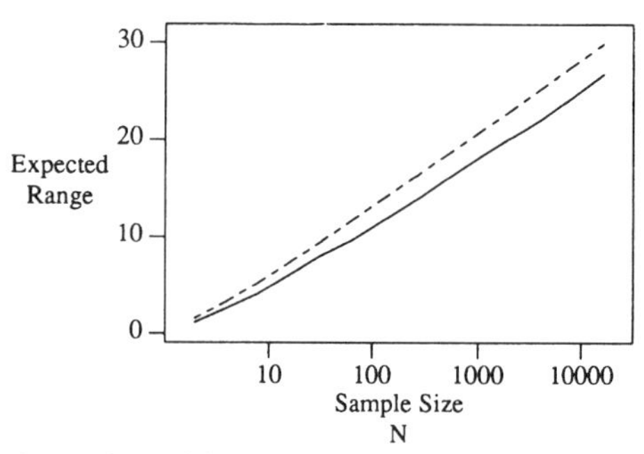

Fig. 1. Comparison of the metric-free tree asymptotic bound (dashed) and Monte Carlo simulations (solid), 100 trials.

$$
\begin{aligned}
E[\text{Skew}] &= \sigma_0 \sqrt{\log_2 N}\left[\frac{4\ln N - \ln\ln N - \ln 4\pi + 2C}{(2\ln N)^{1/2}} + O\left(\frac{1}{\log N}\right)\right] \\
&= \sigma_0 \frac{4}{\sqrt{2\ln 2}} \ln N + \text{lower order terms} \\
&= \Theta(\log N). \qquad (5)
\end{aligned}
$$

Fig. 1 shows the results of a Monte Carlo simulation and shows that the bound is relatively tight, despite the fact that dependence is ignored; the gap between the asymptote and simulation results decreases steadily from about 20% for $N = 2^2$ to about 10% for $N = 2^{14}$.

V. The Metric *H*-Tree

The second example, the *H*-tree, is a *metric tree* in the sense that the wire length, and hence delay through a clock distribution stage, depends on its location in the tree. This dependence of wire length on stage number is a consequence of the requirement that the topology must be both area-efficient and planar [9].

There are two distinct views of the effects of increasing system size (number of leaves, or processors) under the metric assumption. The first is to assume that a tree with an arbitrary number of leaves can be embedded in the fixed area of the integrated circuit. The alternative to this view sets a lower limit on the size of the smallest feature, in this case the size of a wire at the tree's leaves. Each preceding level is progressively larger and the area of the entire clock tree grows with increasing system size. This view ignores the effect of shrinking feature size but is compatible with increases in chip die size. We will adopt the second model, as did Fisher and Kung [3].

The total delay due to the buffer delays is, as before, Gaussian with variance $\sigma_b^2 \log_2 N$, where σ_b^2 is the variance associated with each buffer.

Next, we consider the delays due to wires. For this analysis, we assume that a wire delay is Gaussian with variance σ_w^2 proportional to its length. The linear relationship for the variance can be justified by considering a long wire to be equivalent to two shorter wires placed end to end. The propagation delay of the long wire is equal to the sum of the propagation delays of the two shorter wires, and the variances add because the delays of the short wires are independent.

In an *H*-tree, the wire length doubles every two levels, so the root-to-leave distance, starting with a unit length at the leaves, is

$$
1 + 1 + 2 + 2 + 4 + 4 + \cdots + \sqrt{N}/2 + \sqrt{N}/2 = 2(\sqrt{N} - 1). \qquad (6)
$$

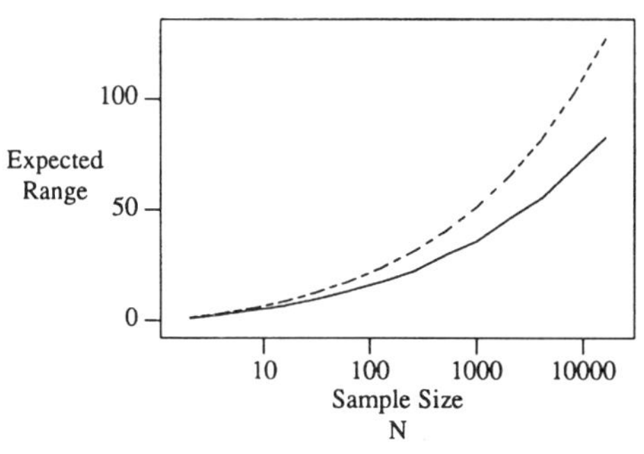

Fig. 2. Comparison of the *H*-tree asymptotic bound (dashed) and Monte Carlo simulations (solid), 100 trials.

Thus, the total delay due to the buffers and the wires has variance

$$
\sigma_b^2 \log_2 N + \sigma_w^2 2(\sqrt{N} - 1). \qquad (7)
$$

For large N the contribution due to the wires dominates. Applying Theorem 1, the expected skew is therefore

$$
\begin{aligned}
E[\text{Skew}] &= \sigma_w \sqrt{2}\sqrt{\sqrt{N} - 1}\left[\frac{4\ln N - \ln\ln N - \ln 4\pi + 2C}{(2\ln N)^{1/2}} + O\left(\frac{1}{\log N}\right)\right] \\
&= \sigma_w 4\left((\sqrt{N} - 1)\ln N\right)^{1/2} + \text{lower order terms} \\
&= \Theta\left(N^{1/4}(\log N)^{1/2}\right). \qquad (8)
\end{aligned}
$$

Fig. 2 shows the results of a Monte Carlo comparison between the asymptotic bound and simulation. In this case, the bound is not as tight as in the metric-free case, because the shared variates, representing deviation from the independence assumption, are near the root of the tree, where the wire lengths are longer.

VI. Variance

It is now possible to give an estimate on the probability that the sample value of the skew exceeds a certain value. Assume that X is a random variable with mean μ and variance σ^2. Then the one-sided Chebyshev inequality [10] yields an upper bound on the probability of exceeding the mean skew by an amount a:

$$
P(X > (\mu + a)) \le \frac{\sigma^2}{\sigma^2 + a^2} = \frac{1}{1 + \alpha^2\mu^2/\sigma^2} \qquad (9)
$$

letting $\alpha = a/\mu$, the fractional deviation from the expected value of skew.

The proof of Lemma 1 in [8] shows that if R_d and R_I are the ranges of dependent and independent variates, respectively, then for every z

$$P(R_d > z) \le P(R_I > z). \tag{10}$$

Thus, the Chebyshev inequality shows that

$$P(R_d > \mu + a) \le \frac{1}{1 + \alpha^2\mu^2/\sigma^2} \tag{11}$$

and Theorem 1 then yields

$$P(R_d > \mu + a) \le \frac{1}{1 + \alpha^2 48 (\ln N)^2/\pi^2} \tag{12}$$

in both the metric and metric-free cases.

VII. Conclusions

We have analyzed the situation where the clock arrival times at N processors are the sum of many uncertain components, but are identically distributed. The resulting expected range of arrival times reflects statistical dispersion, and is in this sense more realistic than worst case analysis when the model is valid.

Our Θ (log N) bound on skew is much more optimistic than the $\Theta(N^3)$ estimate given by Kung and Gal-Ezer [1]. As Hatamian [5] points out, however, their analysis estimates the delay from the clock source to the processors, and not the difference in arrival times.

The summation model of Fisher and Kung [3] predicts both an upper and lower bound of $\Theta(\sqrt{N})$ for the skew of an H-tree, which is, not surprisingly, a bit more pessimistic than our asymptotic upper bound of $\Theta(N^{1/4} (\log N)^{1/2})$ for expected skew.

The constants of proportionality and the variance estimate allow us to apply this bound in practical situations where the clock delay is accurately modeled as the sum of many independent contributions.

Acknowledgment

The authors wish to thank J. P. Singh for helpful discussions and literature survey, and to the referees for thoughtful comments and for catching an error.

References

[1] S. Y. Kung and R. J. Gal-Ezer, "Synchronous versus asynchronous computation in very large scale integrated (VLSI) array processors," *Proc. SPIE,* vol. 341, Real Time Signal Processing V, 1982.

[2] D. F. Wann and M. A. Franklin, "Asynchronous and clocked control structures for VLSI based interconnection networks," *IEEE Trans. Comput.,* vol. C-32, no. 3, pp. 284-293, Mar. 1983.

[3] A. L. Fisher and H. T. Kung, "Synchronizing large VLSI processcr arrays," *IEEE Trans. Comput.,* vol. C-34, no. 8, pp. 734-740, Aug. 1985.

[4] M. Hatamian and G. L. Cash, "Parallel bit-level pipelined VLSI designs for high-speed signal processing," *Proc. IEEE,* vol. 75, no. 9, pp. 1192-1201, Sept. 1987.

[5] M. Hatamian, "Understanding clock skew in synchronous systems," Princeton Workshop on Algorithm, Architecture and Technology Issues in Models of Concurrent Computations, Princeton University, 1987.

[6] H. Kopetz and W. Ochsenreiter, "Clock synchronization in distributed real-time systems," *IEEE Trans. Comput.,* vol. C-36, no. 8, pp. 933-940, Aug. 1987.

[7] H. Cramér, *Mathematical Methods of Statistics.* Princeton, NJ: Princeton University Press, 1946.

[8] S. D. Kugelmass and K. Steiglitz, "A probabilistic model for clock skew," in *Proc. Int. Conf. Systolic Arrays,* San Diego, CA, May 1988, pp. 545-554.

[9] M. S. Paterson, W. L. Ruzzo, and L. Snyder, "Bounds on minimax edge length for complete binary trees," in *Proc. Thirteenth Annu. ACM Symp. Theory Comput.,* May 1981, pp. 293-299.

[10] S. Ross, *A First Course in Probability.* New York: Macmillan, 1984.

On the Calculation of Optimal Clocking Parameters in Synchronous Circuits with Level-Sensitive Latches

MICHEL R. DAGENAIS AND NICHOLAS C. RUMIN

Abstract—**A new algorithm has been developed for the automatic determination of the optimal clock waveforms for synchronous circuits containing level-sensitive latches. From a specification of only the number of clock phases, the rise and fall times of the clock phase transitions, and the order in which they occur, the algorithm computes the minimum time interval between the transitions, while accounting for the clock skew. Timing errors, such as incorrect hold times, are also detected. Existing procedures, in contrast, either verify if a circuit meets a given specification of these clock intervals, or they work with a very restricted set of clocking schemes. The procedure is iterative, and can be formulated as a linear programming problem. It yields an upper bound on the shortest valid clock period at each iteration. Results are presented for a simplified form of this algorithm, implemented in the transistor-level, timing analysis program TAMIA.**

I. Introduction

ONE important problem faced by designers of MOS VLSI circuits is that of specifying optimal clocking parameters (i.e., clock period, duty cycle and phase relationship) for synchronous systems. This is a problem because they rely on level-sensitive latches for their clocked elements which, in turn, necessitates the use of multi-phase clocking methodologies. These can be quite complex in high performance circuits. In contrast, it is much easier to analyze the timing of circuits designed with MOS MSI components or those using the bipolar technology. The reason lies in the fact that such designs are based on edge-triggered latches.

What simplifies the timing analysis of designs based on edge-triggered latches is the fact that the timing equations are decoupled at the latches, since the input and output signals do not depend directly on each other. This has permitted the development of powerful timing analysis tools, of which probably the best known is SCALD [1]. An excellent review of this field is found in [2].

In a level-sensitive latch, on the other hand, a signal can propagate from its input to its output at any time during the entire period when the latch is *on*. The additional unknown that is added to the timing equations is what leads to the difficulty of computing the optimal clocking parameters [3]. One of the best known attempts to analyze MOS VLSI circuits is that of Ousterhout [4]. Although his timing analyzer CRYSTAL represents important pioneering work, its weakness lies in the fact that it cannot handle properly complex clocking schemes. Most important, it cannot handle signals which cross phases, which is a very common situation in MOS circuits. Jouppi's TV [5] is in many respects similar to CRYSTAL and again it is hampered by the fact that it is tied to a very specific clocking scheme.

An important contribution to the timing analysis of synchronous circuits containing level-sensitive latches was made by Agrawal [6]. In his procedure the user specifies the clock waveform, the frequency and the duty cycle. The analysis exploits the borrowing of time between clock phases, but not between consecutive clock cycles. The result is a list of timing violations, if any. Agrawal also described an iterative procedure in which the user enters a minimum and a maximum clock frequency, and the analysis is repeated until the highest frequency without timing violations is found through a binary search between these two frequencies.

More recently, Szymanski [7], [8] presented the timing analysis program LEADOUT. This procedure also verifies if a circuit containing level-sensitive latches will function correctly with a specified set of clocking parameters. The user can determine the maximum frequency by repeated analysis. This procedure uses a switch level model and allows for time borrowing between phases as well as between cycles. Because the circuit is compiled before the analysis, many sets of different clocking parameters can be verified at an impressive speed.

The main difference between the approach of Agrawal and Szymanski and the procedure presented here is that the former analyze a specified set of clocking parameters, whereas this new procedure synthesizes the clock waveforms. All the clocking parameters, including the duty cycle and the phase relationship, are determined automatically. Moreover, the proposed iterative algorithm produces a near optimal solution at each step, and typi-

Reprinted from *IEEE Trans. Computer-Aided Desi.*, vol. 8, no. 3, pp. 268–278, March 1989.

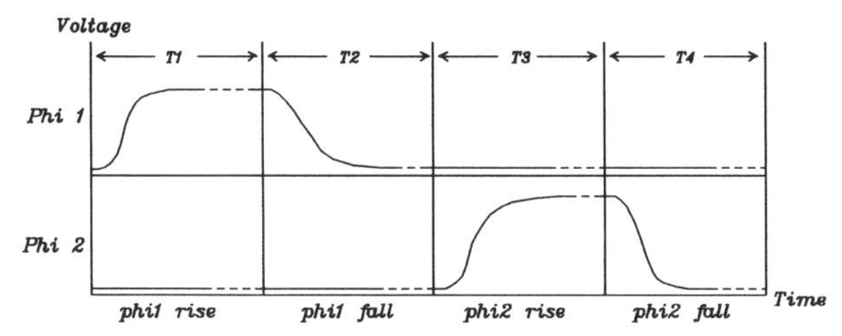

Fig. 1. Two phase non-overlapping clocking scheme decomposed into four timing intervals.

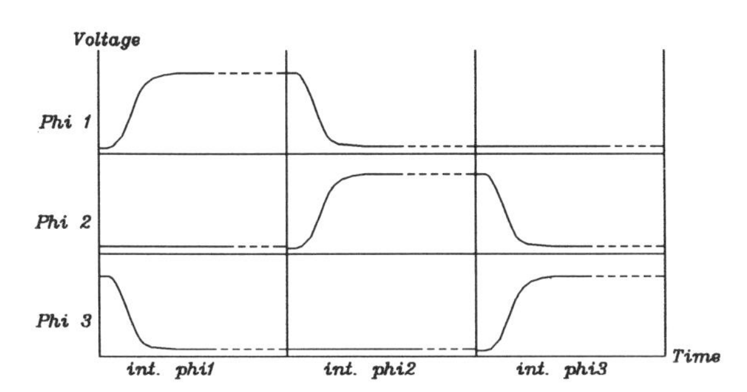

Fig. 2. Three phase overlapping clocking scheme decomposed into three timing intervals.

cally requires few iterations. This is in contrast to the exhaustive binary search method used in MOTIS [6]. A simplified version of this procedure has been implemented and tested in the transistor-level timing analysis program TAMIA [9]–[11].

The notion of *timing intervals* is first introduced. Then, the general methodology is illustrated with the conceptually simpler edge-triggered latches. Next, the timing requirements of level-sensitive latches are described and formalized. This is followed by an examination of several simplified ways to find the clocking parameters for circuits using level-sensitive latches. An iterative algorithm to find the optimal clocking parameters for level-sensitive latches is then proposed. Finally, results obtained with a simplified implementation are discussed.

Timing Intervals and Synchronizing Events

The approach to the computation of optimal clocking parameters which is presented below is based on the idea of dividing the clock cycle into *timing intervals*. The clock transitions are the *synchronizing events*, and these can be either *dependent or independent*. Every timing interval may contain only one independent event. A clocking methodology is specified by enumerating the timing intervals and the synchronizing events within each interval. The clocking parameters are then determined by computing the durations of the timing intervals.

In the two-phase, non-overlapping clocking scheme shown in Fig. 1 all four clock transitions are independent events and, therefore, there are four timing intervals. The duration of the *phi1 rise* and *phi2 rise* intervals are determined by the set-up times of the *phi1* and *phi2* latches. On the other hand, clock skew on the two clock phases determines the durations of the other two intervals, *phi1 fall* and *phi2 fall*. These non-overlap intervals are needed to ensure that the *phi1* and *phi2* latches are never on at the same time.

The three-phase, overlapping scheme shown in Fig. 2 illustrates the idea of dependent events. Here the third clock phase ensures that no more than two out of three latches in a feedback loop can conduct simultaneously. Thus, only three timing intervals are needed. For example, the *phi2 rise* and *phi1 fall* events can be dependent and therefore, put in the same interval.

Timing Requirements

The timing requirements of both edge-triggered and level-sensitive latches are examined in this section. The input to a timing analysis procedure consists of a circuit description, a definition of the signals at the primary inputs, and a specification of the clocking methodology. Each input signal is defined in time coordinates relative to the beginning of the interval in which that signal originates, i.e., the interval containing the synchronizing event that caused a transition of that signal. Indeed, the absolute time in the clock period at which the signal changes is not known since the duration of the timing intervals is yet to be determined.

The circuit elements are defined at a high level, namely combinational blocks and latches. Nevertheless, the discussion can be applied equally well to timing analysis at the gate, switch or transistor level. Fig. 3 shows the circuit elements, that are assumed to be unidirectional with a minimum and maximum delay associated with each input–output pair.

Definitions

A number of quantities used to express the timing requirements that the latches impose are defined below. Unless otherwise stated, the times are relative to the beginning of the clock cycle. For the sake of brevity, it is assumed that the latches are either positive-edge-triggered or are sensitive to the clock's high level. Also, for simplicity, the clocks are assumed to have sharp edges (zero rise and fall times).

Input and output of the latch:

in_stable: latest time at which the latch input stabilizes

in_change: earliest time at which the latch input is no longer stable

out_stable: latest time at which the latch output stabilizes

retard: the delay in the arrival time of a signal at the input of a level-sensitive latch, relative to the clock edge which puts the latch into the *on* (transparent) state

Latch delay parameters

$delay_{et}$: Maximum propagation delay from rising clock edge to output;

$delay_{ls}$: Maximum propagation delay from latch input to output;

setup: Maximum set-up time for an edge-triggered latch.

hold_time: Minimum hold time for an edge-triggered latch;

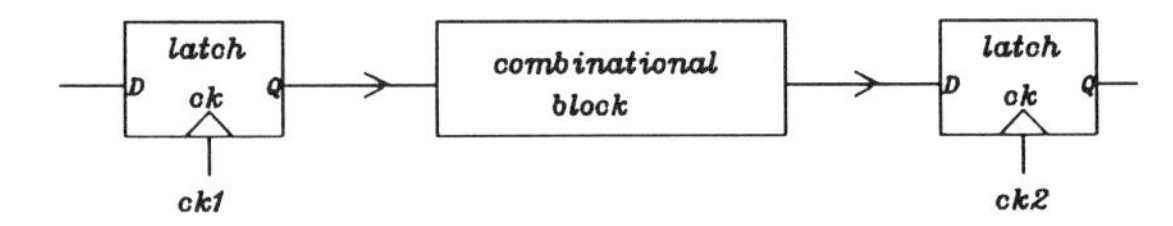

edge-triggered (et) latch: $ck \uparrow$, $Q = D$ else $Q = Q_0$
level-sensitive (ls) latch: $ck = 1$, $Q = D$ else $Q = Q_0$

Fig. 3. Circuit decomposed into combinational blocks and latches.

min_pulse_width: Minimum pulse width for an edge-triggered latch.

Interval boundaries

$start_a$, end_a: Start and end time of interval a;

$start_b$, end_b: Start and end time of interval b ($end_a = start_b$);

end: End time of the complete clock cycle.

Position of clock transitions relative to timing intervals

rise_min: Minimum delay between $start_a$ and clock rise;

rise_max: Maximum delay between $start_a$ and clock rise;

fall_min: Minimum delay between $start_b$ and clock fall;

fall_max: Maximum delay between $start_b$ and clock fall.

Edge-Triggered Latches

In this section, the timing requirements generated by edge-triggered latches are examined, and the calculation of clock timing parameters is discussed. The detailed procedure is given in Appendix A.

The analysis begins with interval a during which the clock signal controlling a latch rises and latching occurs. The latest time, relative to the beginning of interval a, when the latch output will be stable, is given by

$$out\_stable = rise\_max + delay_{et}. \quad (1)$$

Several timing requirements must be satisfied to ensure proper latching of data. First of all, a time equal to at least the *setup* time of the latch must elapse between the latest moment when the input becomes stable, *in_stable*, and the the earliest time when latching can begin:

$$in\_stable \leq start_a + rise\_min - setup. \quad (2)$$

The latch's *minimum pulsewidth* requirement is satisfied by requiring that the clock's latest rising edge and its earliest falling edge, which occurs during interval b, are separated by an interval given by

$$min\_pulse\_width \leq (start_b + fall\_min) - (start_a + rise\_max). \quad (3)$$

Finally the *hold time* of the latch must be satisfied. This involves ensuring that latching is complete before the latch input becomes undefined, and can be expressed by the following inequality:

$$hold\_time \leq in\_change - (start_a + rise\_max). \quad (4)$$

To insure correct operation of the circuit, these inequalities must be verified at every latch that is active during the interval considered. The analysis proceeds one interval at a time, starting with the first interval in the clock cycle. Thus when an interval a is reached, all the signals originating in preceding intervals of the clock cycle have already been processed. Therefore, when such a signal reaches the input of an active latch, the *in_stable* value is known. The timing requirement of the latch is therefore met by adjusting the value of $start_a$ through (2). The value of $start_a$ that satisfies the timing requirements at all the n latches which are active in interval a is computed by expressing (2) as an equality, and taking the maximum value over the n latches:

$$start_a = max_{i=1,n}(in\_stable_i + setup - rise\_min_i). \quad (5)$$

A problem arises when the signal on the input to a latch that is active during interval a depends on the outcome of a subsequent interval. Such a signal is said to *cross the cycle boundary*. This occurs, for example, at the start of the analysis, when interval a is the first one in the clock cycle, and the signal inputs to the active latches are generated at the end of the clock cycle. The *in_stable* value for these latches is not know. Therefore, their timing requirements are not examined until the end where they are satisfied by adjusting the cycle duration.

After the last interval has been examined, the time at which every signal that crosses the cycle boundary stabilizes is known. However, (2) can not be used to test the set-up requirements for a latch driven by such a signal. This is because $start_a$ represents the start time of interval a in the current clock cycle, whereas the *in_stable* that has been computed for the signals crossing the cycle boundary is the time at which these signals are stable in the next clock cycle. Therefore, the setup requirement for such a latch is met by specifying the time at which the clock cycle ends. Hence, (2) becomes

$$(in\_stable - end) \leq start_a + rise\_min - setup. \quad (2a)$$

Again, an equation similar to (5) is needed to compute the clock period that will satisfy the set-up requirements of all the n latches driven by signals crossing the cycle boundary. From (5) and (2a) this is

$$end = max_{i=1,n}(in\_stable_i - (start_a + rise\_min_i - setup)). \quad (5a)$$

The discussion is completed by considering the *min pulsewidth* and *hold time* requirements. The former must also be considered when computing the minimum duration of an interval. Once the analysis has been performed in interval a, a minimum value of $start_b$ is computed to satisfy the *min pulsewidth* requirement at all the latches active in a. This is done by modifying (3) to the form

$$start_b = max_{i=1,n}(rise\_max_i + start_a + min\_pulse\_width_i - fall\_min_i). \quad (6)$$

The *hold time* requirement is different. Notice that (4) compares $start_a + rise\_max$ with *in_change* which are dependent. The value of $start_a$ has already been adjusted to insure that any signal originating in a previous interval becomes completely stable before latching occurs. Therefore, the input of the latch can only change due to a signal originating in interval a. Consequently, the difference between $start_a$ and in_change is a fixed value which is circuit dependent, and cannot be adjusted by varying the duration of the timing intervals. Thus the *hold time* can only be checked. If any violation is found, the circuit has to be modified.

The analysis of synchronous circuits using edge-triggered latches is relatively simple because the output of the latch, out_stable, does not depend on its input, in_stable, as can be seen from (1). For this reason, the timing requirements at the input of a latch are decoupled from the requirements at other latches which are active in subsequent intervals, and which depend on the value of out_stable. Thus, even when the in_stable value at the input of a latch cannot be computed until later in the cycle because the signal crosses the cycle boundary, the out_stable value can still be computed and the analysis can proceed. This is not the case, however, for level-sensitive latches, where nonlinear simultaneous equations exist along each feedback loop in the circuit. Synchronous circuits based on such latches are discussed next.

The Timing Requirements Generated by Level-Sensitive Latches

The level-sensitive latch turns *on* with the rising edge of the clock in interval a, and turns *off* with the clock's falling edge during interval b. The time at which the output of the latch stabilizes depends on both the clock and the latch input signal. If the input of the latch is stable before the latch turns *on*, the out_stable value depends on the time of the clock's latest rising transition, $start_a + rise\_max$. On the other hand, if the input of the latch stabilizes after the latch turns *on*, the out_stable value depends on the time when the input signal stabilizes, in_stable. This double dependency can be treated by introducing the notion of *retardation*.

The retardation is null when the input signal is stable before the latch turns *on* with the rising clock, in which case the output does not depend on the value in_stable. When the input signal settles after the latch turns on, the retardation indicates by how much the input signal arrives after the clock's rising transition. Thus, the retardation is expressed as follows:

$$retard = max\big(0, in\_stable - (rise\_max + start_a)\big). \tag{7}$$

The time at which the output stabilizes, out_stable, is then easily computed using the retardation value, $retard$, the clock's latest rising transition, $start_a + rise\_max$, and the delay through the latch, $delay_{ls}$. The resulting equation has the form

$$out\_stable = start_a + rise\_max + retard + delay_{ls}. \tag{8}$$

Just as in the case of edge-triggered latches, proper latching is ensured by imposing set-up and hold time requirements. The former is specified by requiring that the output of the latch be stable before the latch starts to turn off. Under worst-case conditions, this amounts to the following inequality:

$$out\_stable \leq fall\_min + start_b. \tag{9}$$

The hold time requirement is that the input of the latch remain stable until after the latch is completely turned off. The inequality expressing the worst-case condition is

$$in\_change \geq fall\_max + start_b. \tag{10}$$

Finally, it is necessary to consider a requirement analogous to the minimum pulsewidth requirement of edge-triggered latches. Notice that (10) sets the beginning of the b interval, thereby ensuring that the contents of the latch cannot be corrupted by a signal originating in the *current* interval. However, in the presence of severe clock skew, a latch could take a long time to turn off, allowing its contents to be corrupted by a transition which originates in the *following* interval. Hence it is necessary to impose an additional requirement to insure that the latch is completely off before the b interval ends. This requirement is expressed by the following inequality:

$$end_b \geq start_b + fall\_max. \tag{11}$$

As in the case of the edge-triggered latches, these inequalities insure correct operation of the level-sensitive latch. They can be used to determine the minimum acceptable duration of each individual timing interval for this latch. The problem, however, of computing an acceptable set of clocking parameters in a circuit containing many latches is much more complicated.

Computing Timing Intervals for Level-Sensitive Latches

Consider interval a during which the clock makes a positive transition. If the signal at the input of a latch originated during a preceding interval, the time at which it stabilized, in_stable, is known. Hence, the time at which the latch's output stabilizes, out_stable, can be determined from (7) and (8). To compute the minimum time at which interval b can start, the setup time, given by (9), is expressed as an equality, and the maximum value over all latches turning *off* in interval b is retained. This yields

$$start_b = max_{i=1,n}(out\_stable_i - fall\_min_i). \tag{12}$$

However, when the signal at the input of the latch originates in a subsequent interval, this equation can not be used. As in the case of edge-triggered latches, the *setup* requirement can then be satisfied by adjusting the duration of the clock cycle, in a manner analogous to (5a). Thus the minimum duration for the clock cycle which satisfies the *setup* time at such latches is computed by modifying (12) as follows:

$$end = max_{i=1,n}(out\_stable_i - fall\_min_i - start_b). \tag{12a}$$

Unfortunately, this does not solve the problem of computing the circuit's clocking parameters. This is because, for a signal crossing the cycle boundary, the value of *in_stable* is not known and, therefore, *out_stable* can not be determined from (7) and (8). Unlike the case of edge-triggered latches, these two quantities are not decoupled and, as a result, the outputs of latches driven by signals generated in a subsequent interval can not computed. In fact, every clocked feedback loop in the circuit, yields a set of simultaneous equations for the time at which each signal in the loop becomes stable.

A solution can be found by assuming a bound on the retardation. Thus, if latch i is assumed to have associated with it a maximum retardation of max_retard_i, the maximum time at which the output of the latch becomes stable can be computed from (8). Even though the resulting *out_stable* value will be pessimistic, all the interval durations can now be computed. Evidently, in general, these durations will not represent the shortest possible clock interval durations.

At the end of the analysis, the duration of the clock cycle must be adjusted to satisfy the bounds that were imposed on the retardation at these latches. At those latches where the input signal crosses the cycle boundary, the value *in_stable* represents the time at which the signal settles one clock cycle later. Thus the value for the current clock cycle is obtained by subtracting the cycle duration *end* from *in_stable*. In that case, (7) is replaced by

$$retard = max\big(0, (in\_stable - end) - (start_a + rise\_max)\big). \qquad (7a)$$

This retardation value must be smaller than the maximum retardation allowed for that latch. Thus the following inequality is derived from (7a):

$$max\_retard \geq max\big(0, (in\_stable - end) - (start_a + rise\_max)\big) \qquad (7b)$$

which can be simplified to

$$max\_retard \geq (in\_stable - end) - (start_a + rise\_max). \qquad (7c)$$

The minimum duration of the clock cycle is obtained by taking the maximum value of *end* which barely satisfies (7c) at all the relevant latches:

$$end = max_{i=1,n}\big(in\_stable_i - (start_a + rise\_max_i) - max\_retard_i\big). \qquad (12b)$$

Before concluding, it should be noted that (11), which is needed to compute the duration of the b interval, must be modified to account for the beginning of this interval $start_b$ and the slowest clock fall time. Thus, in its place the following expression is used:

$$end_b = max_{i=1,n}(start_b + fall\_max_i). \qquad (13)$$

As in the case of edge-triggered latches, the hold time requirement given by (10) is not used in the calculation of clocking parameters and can only be used to check whether the clock skew is small enough.

Some simple techniques for assigning retardation values, and thereby making the problem of computing clocking parameters tractable are discussed in the next section.

Approximate Solutions for Level Sensitive Latches

It is useful at this point to examine some simple methods for solving for the clocking parameters of synchronous circuits designed with level-sensitive latches. These amount to different approaches to assigning retardation values, and are interesting from the point of view of the extent to which they approach the optimal clocking parameters.

Fig. 4 shows a simple circuit using the two-phase non-overlapping clocking scheme of Fig. 1. It contains two complete stages of clocked logic, connected in a single closed loop. To keep the example simple, the effect of clock skew is not considered here and the non-overlap intervals *phi1 fall* and *phi2 fall* are assumed to have zero duration. Also, it is assumed that the transitions on the clocks are sharp and occur at the very beginning of each interval, i.e., *rise_max*, etc., are all zero.

The simplest solution method requires a null retardation at every latch, such that the input of the latch is stable before the latch turns on with the clock's rising transition. It amounts to treating the latches as edge-triggered ones. This scheme will be referred to as *null retardation in every phase*, and is in fact used in CRYSTAL [4].

Using this simplified solution method, the interval durations can be determined easily. Suppose that the analysis begins at the inputs to latches *1a* and *1b*. Thus the interval being considered is *phi1 rise* as defined in Fig. 1. Since *retard* is zero, (8) yields a value of *out_stable* of 10 ns for both these latches. From (9), the *phi1 fall* interval for these latches could begin immediately after, in view of the simplifying assumptions made above. These assumptions also lead to the result that the duration of this interval could be zero, as can be seen from (11). However, for a zero retardation at the inputs to latches *2a* and *2b*, the next interval, *phi2 rise*, must begin after the outputs of both combinational blocks *L1a* and *L1b* stabilize. This will be at 30 and 70 ns, respectively. Hence, the clock interval *phi2 rise* can begin not earlier than at 70 ns.

The signals at the outputs of latches *2a* and *2b* stabilize at 80 ns since they have equal delays. By a reasoning similar to that used in the case of the other two latches, the *phi2 rise* interval could end at this time and the duration of the *phi2 fall* interval could be zero. The signal at the output of latch *2a* propagates through a combinational block before reaching the input of latch *1b* at 80 + *L2a* = 100 ns. In the case of latch *2b* the output signal follows a slower path and settles at the input of latch *1a* at 80 +

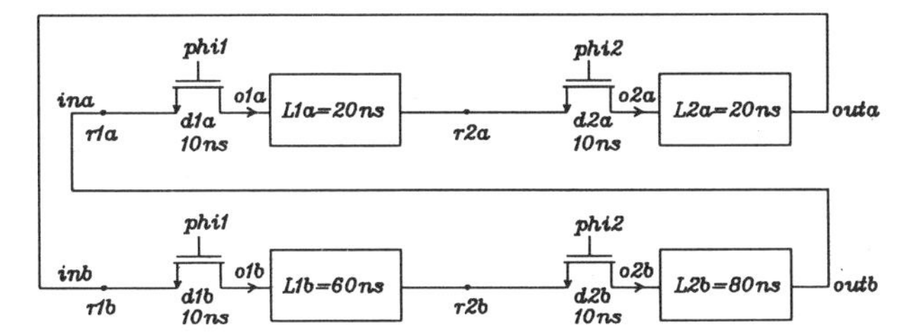

Fig. 4. Simple synchronous circuit with associated delays through the level-sensitive latches and combinational blocks.

L2b = 160 ns. Since the retardation of these latches has been assumed to be zero, the cycle duration must be 160 ns, so that the signals at the *1a* and *1b* latch inputs settle before the next cycle begins.

These clocking parameters insure correct operation of the circuit. They lead to the shortest cycle duration, ensuring a null retardation at every latch in the circuit. However, the circuit may, in fact, function properly at a higher frequency with some retardation at the latch inputs. Therefore, the simplification in this solution comes at the cost of a pessimistic clock cycle duration.

Instead of requiring a null retardation in every phase, one can use a method called *null retardation in initial phase* to avoid the problem of solving simultaneous equations. In that case, a null retardation is assumed only when the input of the latch is a signal that crosses the cycle boundary. The cycle boundary will change depending on whether the analysis starts with the *phi1* or with the *phi2* latches. Both cases are examined below and, interestingly, give different results.

Suppose that the analysis starts with the *phi1* latches. Since initially the signals at the input of the *1a* and *1b* latches are unknown, a null retardation is assumed at these latches. Thus, the first interval *phi1 rise* can last long enough to allow for propagation through these latches, namely 10 ns. From (12), interval *phi1 fall* can start right after, at 10 ns in the clock cycle. The interval *phi2 rise* can start at the same time since clock skew is neglected here and, therefore, the duration of *phi1 fall* can be essentially zero.

The inputs of latches *2a* and *2b* settle at 10 + *L1a* = 30 ns and 10 + *L2b* = 70 ns, respectively. Since the beginning of interval *a* for these latches was just set at 10 ns, from Eq. (7), the retardation at these latches is, respectively, 30 − 10 = 20 ns and 70 − 10 = 60 ns. Thus, according to (8), their outputs settle at 10 + $retard_{2a}$ + *d2a* = 40 ns and 10 + $retard_{2b}$ + *d2b* = 80 ns. Consequently, according to (12), the interval *phi2 rise* cannot end earlier than 80 ns. Finally, the inputs of latches *1a* and *1b* settle at 80 + *L2b* = 160 ns and 40 + *L2a* = 60 ns, respectively. Since the maximum retardation allowed at these latches was 0, from (12b), the cycle duration must be 160 ns.

A different solution may be obtained when the clock cycle boundary is set between *phi1* and *phi2*. In that case, the analysis starts with the *phi2* latches and requires a null retardation for them. Thus the first interval, *phi2 rise*, lasts 10 ns to allow for the propagation through the latches. Interval *phi2 fall* can have a null duration and *phi1 rise* thus starts at 10 ns. The inputs of latches *1a* and *1b* then settle at 10 + *L2b* = 90 ns and 10 + *L2a* = 30 ns. Consequently, the retardation at latches *1a* and *1b*, given by (7), is 90 − $start_3$ = 80 ns and 30 − $start_3$ = 20 ns, respectively, where $start_3$ is the beginning of the *phi1 rise* interval. Hence, from (8), the outputs of these latches settle at $start_3$ + $retard_{1a}$ + *d1a* = 100 ns and $start_3$ + $retard_{1b}$ + *d1b* = 40 ns, respectively. The inputs of latches *2a* and *2b* then settle at 100 + *L1a* = 120 ns and 40 + *L1b* = 100 ns, respectively. Therefore, according to (12b), the clock cycle must end at 120 ns to be consistent with the null retardation assumed at the *phi2* latches.

It is interesting to note that the cycle duration varies with the phase in which the null retardation is assumed. Both sets of clocking parameters insure correct operation, so that the set which yields the shorter cycle time can safely be retained. However, the results obtained are still pessimistic because of the restriction imposed on the retardation, and the circuit may in fact work with an even shorter cycle duration.

In the analysis based on an assumed maximum retardation of 0 ns at latches *1a* and *1b*, the resulting worst-case delays show that the signals feeding back to the inputs of latches *1a* and *1b* settle at 160 and 60 ns, respectively. The resulting clock period is determined by the late arrival of the signal at the input of latch *1a*. As a result, the signal at the input of latch *1b* has spare time. This suggests that by increasing the allowed retardation at the input of latch *1a*, it may be possible to borrow some of the spare time from latch *1b*, thereby shortening the clock period. Since the input of latch *1a* settles 100 ns after the input of latch *1b*, its maximum retardation may be increased by, say, 100/2 = 50 ns.

Since the maximum retardation allowed at latches *1a* and *1b* is now 50 ns and 0 ns, respectively, the output of these latches settle at 50 + *d1a* = 60 ns and 0 + *d1b* = 10 ns. Thus the interval *phi1 rise* can end and interval *phi2 rise* can begin at $start_3$ = 60 ns. The input of latches *2a* and *2b* settle at 60 + *L1a* = 80 ns and 10 + *L1b* = 70 ns, respectively, for a retardation of 80 − $start_3$ = 20 ns and 70 − $start_3$ = 10 ns, respectively. The outputs of latches *2a* and *2b* then settle at 20 + $start_3$ + *d2a* = 90 ns and 10 + $start_3$ + *d2b* = 80 ns, respectively. Therefore, the inputs of latches *1a* and *1b* are stable at 80 + *L2b* = 160 ns and 90 + *L2a* = 110 ns. Since at latch *1a* the allowed retardation is 50 ns the required cycle duration is 160 − 50 = 110 ns according to (12b). On the other hand, the maximum retardation allowed at latch *1b* is 0 and its input settles at 110 ns for a cycle duration of 110 ns. Thus there is no spare time at either latch and the optimal cycle duration has been reached.

In general, the circuits are much more complex than the one in Fig. 4 and, as a result, the determination of the optimal clocking parameters is much more difficult. This is the subject of the following section.

Optimal Clocking Parameters for Level-Sensitive Latches

The problem of computing the optimal clocking parameters for level-sensitive latches can be seen as an optimization problem. Interestingly, this problem is very similar to the problem of toroidal compaction [12] and can be formulated as a linear programming problem.

The variables are the interval durations and the stabilization time at the output of each latch. Thus, there are $(n + m)$ variables if n is the number of latches and m the number of intervals. Inequalities are set to insure that any signal leaving a latch and reaching another latch does so before the destination latch turns *off*. There can be up to (n^2) such inequalities. Other inequalities are required to insure that the signal has time to propagate through the destination latch while the latch is *on* (n inequalities). The cost function to minimize is the sum of the interval durations.

Standard techniques for solving the linear programming optimization problem could be used. However, the large number of variables and inequalities make this approach very costly. Moreover, such algorithms would not suit well the different levels of circuit modeling which take into account not only a fixed delay but also the slope of the waveforms. Finally, this problem has certain features which can be exploited to obtain a more efficient procedure.

We propose a new iterative procedure to determine the optimal clocking parameters. A pseudocode description is available in Appendix B. Each iteration step in the procedure can be performed in a time linearly proportional to the circuit size. At each step, a near-optimal solution is available. Indeed, the solution produced always satisfies the timing requirements and leads to the shortest cycle time, given the current retardation values set at the latches. At the end of the iterative process, the retardation values producing the optimal clocking parameters are found, and the shortest possible cycle time is obtained.

The analysis proceeds one interval at a time in a breath-first-search manner. The latches turning *off* are processed first and determine the start time of the interval examined using (12). Then, the latches turning *on* are examined, once the start time of the interval is computed. A problem arises when the first latch in a clocked loop is reached. Indeed, the input signal to such a latch crosses the cycle boundary and cannot be determined until later in the cycle, since it originates in a subsequent interval.

Initially, the maximum retardation at the first latch in each loop is set to 0 and the analysis proceeds to find a set of clocking parameters for this initial iteration. Then, at the end of each iteration, the retardation value is slightly increased at each such latch which is on the critical path. The fact that the retardation is increased delays the *out_stable* value computed at the corresponding latch from (8). This change in the value of *out_stable* can only delay the time at which a certain number of nodes become stable. Therefore, it cannot shorten the minimum cycle time, *min_cycle_end*, which is required by the latches turning off from (12), (12a), (13). However, the increase in the retardation value at the latches on the critical path can, according to (12b), shorten the cycle time required at these latches. If the increased retardation at a latch on the critical path causes an additional delay around the loop, but increases *in_stable* by a smaller amount, a shorter cycle time is possible according to (12b). Thus the following theorems hold:

Theorem 1: The only portion of the cycle which can be shortened is that caused by the *retardation* requirements of (12b). Thus in any given iteration, a lower bound on the shortest possible cycle time is the minimum cycle time allowed by the *setup time* and *minimum pulsewidth* requirements of (12), (12a), (13).

Corollary 1: The difference between the minimum cycle time imposed by the *retardation* requirements and that imposed by the *setup* and *min pulsewidth* requirements is the maximum amount by which the cycle time can be shortened. Thus, when this difference goes to 0, the cycle time cannot be shortened and the optimal solution is reached.

Proof: The iterations start with a null retardation at the first latches in each loop. The minimum cycle duration required by the *setup* and *min pulsewidth* requirements, (eqs. (12) (12a), (13)), cannot be shortened since the retardation cannot be negative. Thus the only way to obtain a better upper bound on the minimum valid cycle duration is by increasing the retardation allowed at these latches. Indeed, changing the retardation values can affect the minimum cycle time imposed by the *retardation* (see (12b)).

Increasing the retardation at a latch which is not on the critical path cannot shorten the cycle duration. The retardation at all the latches on the critical path must be increased in order to get any reduction in the cycle time. Increasing this retardation by δ can delay the time at which nodes stabilize by up to δ and correspondingly increase the cycle time required by the *setup* and *min pulsewidth* requirements by up to δ. The signals on the feedback paths can then be delayed by a value between 0 and δ, which is less than or equal to the additional retardation allowed on the critical paths. Consequently, the added retardation cannot increase the cycle time imposed by the *retardation* requirements.

Theorem 2: The cycle duration imposed by the *setup* and *min pulsewidth* requirements cannot be shortened. It increases monotonically with the retardation at the latches.

Proof: Equations (12), (12a), (13) represent nonlinear functions which increase monotonically with the retardation. The retardation starts at null in the first iteration and can only increase at each iteration. Thus, the lower bound on the cycle duration provided by the *setup* and *min pulsewidth* can only increase as the iterative solution progresses.

Theorem 3: The only way to shorten the cycle duration is by increasing the retardation allowed at the latches in the critical feedback paths. The increase in retardation at these latches cannot increase the cycle duration imposed by the *retardation* requirements. This insures that the iterative solution converges monotonically towards the optimal solution.

Proof: The time at which the output of a latch stabilizes is at its minimum when the retardation is null at all the latches. The cycle duration imposed by the *setup time* and *min pulsewidth* requirements cannot be shortened according to *Theorem 2*. Thus only the *retardation* requirements could possibly be relaxed to allow a shorter cycle duration. In fact, the only term in this requirement which can be changed in the retardation value. When the retardation is increased by δ at some latches, the time at which the latch input becomes stable can increase by at most δ. Consequently, the minimum cycle time required by (12b) at a latch cannot increase as a result of increasing the retardation at this and other latches by a given amount δ. If the *in_stable* time at the latch increases by less than δ, the minimum cycle duration required is shorter. It should be noted that the retardation at all the latches on critical feedback loops must be increased in order to get any reduction on the cycle time.

Theorem 4: The optimal solution is reached when the minimum cycle duration imposed by the *retardation* requirements reaches the duration imposed by the other requirements.

Proof: Indeed, increasing the retardation is the only way to shorten the cycle duration. It cannot increase the duration imposed by the *retardation* requirements, but does increase the duration imposed by other requirements. Thus until the two durations become equal, the solution can only improve. When the difference between the two durations becomes zero, no more improvement is possible by *Corollary 1*. Thus, the solution obtained at that point must be the optimal solution.

It should be noted that the "first" and "last" intervals in a cycle are an arbitrary notion because of the cyclic nature of the phenomenon. Indeed, one can in fact decide to break the cycle and start the iterative analysis in any interval. This would change the latches at which null retardation values will be initially assumed, and correspondingly the related signals which cross the new cycle boundary. Consequently, the near-optimal solution obtained in the first iterations may change depending on how the cycle is broken. However, the shortest cycle duration reached at the end of the iterative process does not vary.

Discussion

This algorithm has been implemented in TAMIA, a transistor-level timing analysis procedure developed by the authors [9]–[11]. However, in its present implementation, it performs a single iteration which corresponds to the *null retardation in the initial phase* method and thus produces near-optimal clocking parameters. This choice was dictated by the fact that a transistor-level procedure requires a long processing time, not suitable for multiple iterations.

TABLE I
CPU Time and Accuracy of TAMIA. The Circuit Bridge Contains Bridges, and Number in Parentheses is Number of Enumerated Transistors

Circuit		CPU Time on Micro VAX II			Critical Path Delay	
Name	# of Trans.	i/o	Circuit Decomp	Timing Analysis	TAMIA	SPICE
largsp	20	16.1 s	0.23 s	8.2 s	42.1 ns	42.5 ns
bridge	18(60)	16.8 s	0.29 s	17.1 s	87.3 ns	81.5 ns
treeph	94	31.1 s	0.76 s	35.9 s	61.8 ns	57.0 ns
twolps	32	14.9 s	0.26 s	11.6 s	23.3 ns	25.6 ns
treadd	1696	95.5 s	11.4 s	848 s	469 ns	442 ns

Circuits using various clocking strategies, (two-phase overlapping and nonoverlapping, three-phase, . . .), have been successfully analyzed using TAMIA. Some results are presented in Table I, where it can be seen that the accuracy obtained with the transistor-level timing model is within 5 to 10 percent of SPICE, and the processing time for the single iteration grows linearly with the circuit size, and amounts to approximately 0.4 s per transistor on a micro VAX II computer. The delays shown are for the critical path.

The switch-level timing analysis procedure LEADOUT, developed by Szymanski [7], also uses an iterative algorithm. Although LEADOUT does not automatically determine the optimal clocking parameters, it does allow time borrowing between cycles, and has to cope with the circular dependencies in order to verify if the circuit can operate correctly under specified clocking parameters. Interestingly, Szymanski reports that the number of iterations required is usually small (e.g., 2 or 3) in most cases but can be very large for some pathological cases [8].

Conclusion

A new procedure has been presented for automatically determining the optimal clocking parameters for synchronous circuits using level-sensitive latches and an arbitrary number of clock phases. This added functionality, not available previously, is obtained through an iterative process. To overcome the problems presented by level-sensitive latches, two major techniques are used. First, within each interval, the nodes are ordered so that the latches turning *on* are evaluated after those which are turning *off*, (and on which the former depend through the start time of the interval). Secondly, to break the circular dependencies that exist along clocked loops, initial retardation values are assumed at some latches, which impose corresponding requirements on the cycle duration.

This procedure is not bound to a particular delay model and can be used for gate, switch and transistor-level static timing analysis. An algorithm that iteratively refines the values of retardation has been proposed and its convergence demonstrated. However, further experiments will be required to evaluate various means to obtain a rapid convergence. When a costly timing model is used and iterating many times is not practical, diverse heuristics may rapidly lead to a near-optimal solution. The interval in the cycle at which the analysis starts can change the initial solutions obtained. Also, the lower bound on cycle duration provided by *Corollary 1* can be used to decide whether further iteration is desirable, and to determine the augmentation in the retardation at latches on the critical path.

Appendix A
Solving for Edge-Triggered Latches

The procedure to determine the optimal clocking parameters for circuits using edge-triggered latches is examined in this section. The analysis performs a breathfirst search in the circuit, starting at the primary inputs, for each timing interval. The minimum duration of each timing interval is determined in the process. The timing requirements generated at latches which receive feedback signals are used to determine the clock cycle duration.

The analysis in an interval starts at the primary inputs. The worst-case delays at the primary inputs have been supplied with the circuit description. Messages are sent to all the elements connected to these primary inputs to notify them that the delay value is available. At each element, the number of such messages received is monitored. When one message has been received for each input terminal of the element, the worst-case delays at the output of the element can be determined using the selected timing model. Then, all the elements connected to this output node are notified that the delay on this node has been computed. The computation of worst-case delays thus advances as a wavefront that propagates from the primary inputs to the primary outputs of the circuit.

When no transition can occur on a node during the interval considered, the node has the status *continue*. This occurs at primary inputs which have no transition in the interval considered, at the output of latches when they are in latching state or at the output of elements driven by nodes which have the *continue* state. In such a case, no delay needs to be computed and the elements connected to such nodes can readily be examined.

When the element to be analyzed is an edge-triggered latch, the worst-case delays at the output are obtained using (1). In addition, timing requirements are generated at latches when they encounter a latching transition in the interval considered. If the signal on the latch input node originates in a previous interval in the clock cycle, the associated worst-case delays are available and (5) is used.

However, when the signal originates in a subsequent interval, the latch input signal is in the *continue* state but with no worst-case delays yet associated with it. In that case, the *setup* time requirement is satisfied by adjusting the cycle duration, since the signals considered here do cross the cycle boundary. Moreover, adjusting the cycle duration does not affect the worst-case delay of other signals and keeps the current interval as short as possible.

The cycle duration is adjusted at the end of the analysis, once all the worst-case delays have been computed. The cycle *end* is determined by applying (5a) to all the latches encountered in the analysis which had input signals originating in intervals subsequent to the latching interval.

The procedure is described below in pseudo-code. It computes the optimal clocking parameters for circuits containing edge-triggered latches. That is the parameters which lead to the shortest cycle time while insuring correct operation under worst-case conditions. Indeed, this procedure verifies that the timing requirements are met under worst-case conditions at all the latches in the circuit. Furthermore, the start time of each timing interval is set as early as possible in the cycle while not violating the timing requirements, thus ensuring the optimality of the clocking parameters.

The procedure can be summarized as follows; the computation of the *hold time* and *min pulsewidth* requirements was left out of this pseudo-code procedure for brevity.

```
{

/* The circuit is traversed in a breath-first search
   in each interval                                  */

  for each timing interval i, i=1 to last
   { interval_end[i] = 0
     for each node primary input and clock signal
      { propagate(node)
      }

     until the queue of nodes is empty
      { node = pop_first(queue)
        propagate(node)
      }
   }

/* At the end, the requirements generated by signals that cross
   the cycle boundary are used to determine the minimum cycle
   duration using equation (5a). */

  until the stable_node_queue is empty
   { node = pop_first(stable_node_queue)
     min_rise = pop_first(stable_time_queue)
     int = pop_first(stable_int_queue)
     min_end = in_stable - min_rise - interval_end[int]
     interval_end[last] = max(interval_end[last],min_end)
   }
}


/* The elements attached to a node are searched to find which are
   ready to analyze */

propagate(node)
{
  for each element attached to node
   { note that node is analyzed
     if(element ready to analyze) analyze(element)
   }
}


/* An element is processed, the status of its output is found */
analyze(element)
{

/* For a combinational element, the output node is analyzed and
   put on the queue to affect the elements which depend on it */

  if(element is combinational)
   { compute the status and delay at the output_node
     push_last(queue,output_node)
   }
  else
   { if(clock input is rising)
      { if(input_node status is INIT)

/* The in_stable value is not available, it must be a signal
   that wraps around the cycle boundary. */
```

```
        { push_last(stable_queue_node,input_node)
          push_last(stable_queue_time,rise_min)
          push_last(stable_int_queue,i - 1)
        }
        else

/* It is a normal setup time requirement (5) which determines the
   start time of the interval */

        { min_start = in_stable - rise_min
          interval_end[i - 1] = max(min_start,interval_end[i - 1])
        }
      }

/* The delay at the output of the latch is computed (1) */

    compute the status and delay at the output_node
    push_last(queue,output_node)
  }
}
```

Appendix B
Solving for Level-Sensitive Latches

The following procedure summarizes the computation of optimal clocking parameters for synchronous circuits which use level-sensitive latches. The *hold time* computation was left out for brevity.

Each iteration step in the above procedure can be performed in time linearly proportional to the circuit size. At each step, a near-optimal solution is available. Indeed, the solution produced always satisfies the timing requirements and leads to the shortest cycle time, given the current retardation values set at the latches. At the end of the iterative process, the retardation values producing the optimal clocking parameters are found and the shortest possible cycle time is obtained.

```
{
  the retardation at each latch is set to 0
  min_cycle_end = -1
  interval_end[last] = 0
/* Iterate the solution until no further improvement is possible */

  for(min_cycle_end < interval_end[last])
  { for each timing interval i, i=1 to last
    { for each node primary input and clock signal
      { propagate(node)
      }

/* The output of latches turning off are processed in the first
   queue */

      min_pulse_width = 0
      until the first_queue of nodes ready is empty
       { node = pop_first(first_queue)
         propagate(node)
       }

/* The interval duration is affected by the min pulse width
   requirement */

      interval_end[i] = interval_end[i - 1] + min_pulse_width

/* At this point, the end time of the previous interval is known.
   The output of latches turning on can now be processed */

      until the first_queue and second_queue are empty
       { if(first_queue is not empty)
          { node = pop_first(first_queue)
          }
         else
          { node = pop_first(second_queue)
          }
         propagate(node)
       }
    }

/* The setup time requirement for signals crossing the cycle
   boundary are processed and affect the cycle duration (12a) */

    until the stable_node_queue is empty
     { node = pop_first(stable_node_queue)
       min_rise = pop_first(stable_time_queue)
       int = pop_first(stable_int_queue)
       min_end = in_stable - min_rise - interval_end[int]
       interval_end[last] = max(interval_end[last],min_end)
     }

/* The cycle duration required by the setup time and the minimum
   pulse width is obtained */

     min_cycle_end = interval_end[last]
/* The cycle duration imposed by the retardation requirements is
   computed (12b) */

     until the retard_node_queue is empty
      { node = pop_first(retard_node_queue)
        min_rise = pop_first(retard_time_queue)
        min_end = in_stable - min_rise
        if(min_end < interval_end[last])
         { interval_end[last] = min_end
           min_end_queue set empty
           push_first(min_end_queue,node)
         }
        else if(min_end = interval_end[last])
         { push_first(min_end_queue,node)
         }
      }
/* The retardation at the first latches on each critical feedback
   path is slightly increased to converge towards a shorter cycle
   time in the next iteration. */

     until the min_end_queue is empty
      { node = pop_first(min_end_queue)
        node->retardation = node->retardation + retard_step
      }
  }
}

/* The status and delay at a node is computed */

propagate(node)
{
  if(node is output of latch turning off)
   { compute the status and delay at the node

/* The in_stable value is not available, the latch is driven
   by a signal which crosses the cycle boundary. The requirement
   is stored and checked at the end of the cycle */

     if(node status is INIT)
      { push_last(stable_node_queue,node)
        push_last(stable_time_queue,fall_min)
        push_last(stable_int_queue,i - 1)
      }
     else

/* At normal latches, out_stable can be computed (7)(8) and
   the setup time requirement affects the start time of the
   current interval (12) */

      { min_start = out_stable - fall_min
        interval_end[i - 1] = max(interval_end[i - 1],min_start)
      }
     min_pulse_width = max(min_pulse_width,fall_max)
   }
  else if(node is output of latch turning on)
   { if(input_node status is INIT)

/* The retardation value (7) cannot be computed since
   in_stable is not available. A value is assumed and the
   node is remembered in the retard queue to adjust the cycle time
   to satisfy this retardation. */

      { retard = node->retard
        push_last(retard_node_queue,node)
        push_last(retard_time_queue,retard + interval_end[i - 1])
      }

/* The in_stable value is available and the retardation is
   computed */

     else retard = max(0,in_stable - rise_max - interval_end[i - 1])

     compute the status and delay at the node
   }
/* It is not a latch. The status and delay is simply computed */

  else compute the status and delay at the node

/* The elements attached to the node are examined to find those
   which are ready to analyze */

  for each element attached to node
   { mark that node is analyzed
     if(element ready to analyze) analyze(element)
```

```
    }
}

/* When an element is ready to analyze, its output node is put in
   the first queue for latches turning off and for combinational
   blocks. The latches turning on are put in the second queue. */

analyze(element)
{
  if(element is combinational)
  { push_last(first_queue,output_node)
  }
  else
  { if(element is latch turning on)
    { push_last(second_queue,output_node)
    }
    else push_last(first_queue,output_node)
  }
{
```

ACKNOWLEDGMENT

The authors wish to thank Tom Szymanski for stimulating discussions. Suggestions from Mark Horowitz are gratefully acknowledged, and contributed to the Linear Programming formulation of this problem.

REFERENCES

[1] T. M. McWilliams, ''Verification of timing constraints on large digital systems,'' in *Proc. 17th Design Automation Conf.*, Minneapolis, MN, pp. 139-147, June 1980.

[2] R. B. Hitchcock, ''Timing verification and the timing analysis program,'' in *Proc. 19th IEEE Design Automation Conf.*, Las Vegas, NV, pp. 594-604, June 1982.

[3] L. A. Glasser and D. W. Dobberpuhl, *The Design and Analysis of VLSI Circuits*. Reading, MA: Addison-Wesley, 1985.

[4] J. K. Ousterhout, ''Switch-level timing verifier for digital MOS VLSI,'' *IEEE Trans. Computer-Aided Design*, vol. CAD-4, pp. 336-349, July 1985.

[5] N. P. Jouppi, ''Timing analysis for NMOS VLSI,'' *Proc. 20th Design Automation Conf.*, Miami Beach, FL, pp. 411-418, June 1983.

[6] V. D. Agrawal, ''Synchronous path analysis in MOS circuit simulator,'' *Proc. 19th IEEE Design Automation Conf.*, Las Vegas, NV, pp. 629-635, June 1982.

[7] T. G. Szymanski, ''LEADOUT: A static timing analyzer for MOS circuits,'' *Proc. 1986 IEEE Int. Conf. on CAD*, Santa Clara, CA, pp. 130-133, Nov. 1986.

[8] —, ''LEADOUT: A static timing analyzer for MOS circuits,'' private communication.

[9] M. R. Dagenais and N. C. Rumin, ''Circuit level timing analysis and design verification of High Performance MOS computer circuits,'' *Proc. 1986 Int. Conf. on Computer Design*, Port Chester, NY, pp. 356-359, Oct. 1986.

[10] M. R. Dagenais and N. C. Rumin, ''Timing Analysis and Verification of Digital MOS Circuits,'' *Proc. CompEuro 1987*, Hamburg, Germany, pp. 242-245, May 1987.

[11] M. R. Dagenais, ''Timing analysis for MOSFETs, an integrated approach,'' Dep. Elect. Eng., McGill Univ., Montreal, July 1987.

[12] P. Eichenberger, and M. Horowitz, ''Toroidal compaction of symbolic layouts for regular structures,'' *Proc. 1987 IEEE Int. Conf. on CAD*, Santa Clara, CA, pp. 142-145, Nov. 1987.

On *RC* Line Delays and Scaling in VLSI Systems

C. Svensson and M. Afghahi

Indexing terms: Integrated circuits, Large-scale integrations, Clocks, Metal-oxide-semiconductor structures and devices

We propose that a special interconnection metal layer is introduced into the VLSI process and used for global clock distribution and communication. We show that by this method the interconnection delay will not be a limiting factor for the speed of synchronous circuits in CMOS down to 0·3 μm feature sizes and up to 20 mm chip sizes.

Introduction: It is well known that the *RC* delay of long interconnection lines in VLSI circuits may limit the performance of the circuit, and that this effect becomes worse with scaling (for smaller feature sizes).[1] Several methods have been proposed to deal with this problem: asynchronous communication,[2] semi-asynchronous communication,[3] special clock distribution schemes[4] etc. We propose a simple technical solution to this problem in this letter.

Scaling of interconnection lines: The basic problem is that interconnection lines do not scale in time as do transistors (or as does the logic). Whereas the delay of logic decreases linearly with the scaling factor, the delay of a line tends to increase quadratically (for a fixed line length and assuming constant field scaling). This is easily modelled. Consider an interconnection line with width w, length l and thickness t on an oxide layer of thickness t_{ox}. It will have a resistance of $R = \rho l/wt$ and a capacitance of $C = \varepsilon wl/t_{ox}$, where ρ is the metal resistivity and ε is the oxide dielectric constant. Its delay is, to first order:

$$t_d = RC = \rho\varepsilon l^2/tt_{ox}$$

made up of two parts, one material $\rho\varepsilon$, and one geometrical l^2/tt_{ox}. In scaling, t and t_{ox} are expected to decrease linearly with the scaling factor, whereas l stays constant (as the chip sizes are not expected to be reduced). In fact, the tendency is instead that l increases with time (and thus along with scaling) as chip sizes increase. The result is that t_d increases at least quadratically with scaling. Using the example of aluminium and silicon dioxide gives $\rho\varepsilon = 0{\cdot}95 \times 10^{-18}$ s. We may then estimate t_d for a cross-chip interconnection line in some assumed scaled VLSI chips, see Table 1. We have also given a typical logical delay for a 3 μm process (delay of a fast full adder[5]) and corresponding linearly scaled values. As can be seen from the table, the interconnection delay already starts to dominate in a 1 μm process and becomes very large (100 times the logic delay) in a 0·3 μm process.

Table 1

Feature size	t, t_{ox}	l	t_d	logic
μm	μm	mm	ns	ns
3	1	10	0·1	4
1	0·3	15	2·4	1·2
0·3	0·1	20	38	0·4

Introduction of an extra communication layer: The solution to the problem we propose is that a special long range interconnection layer is introduced. This layer may be a third metal layer, and it is not scaled, instead it also maintains $t = t_{ox} = 1\,\mu$m in the scaled processes. This means that it will maintain a width of 3 μm, and that it therefore must be routed in a coarser grid than the normal metal layers. Assuming that this is possible, the delay of a 20 mm long line of this type will be 0·4 ns, comparable to the logic delay. We may even be somewhat more optimistic; we may for example use a polymer as insulator instead of oxide, with about 50% lower dielectric constant. With typical logical and interconnection delays of the order of 0·5 ns we may expect that clock frequencies of 1 GHz are feasible in synchronous VLSI systems using a 0·3 μm process.

The low density of such an interconnection layer is not considered to be a problem, as we need it only for global communication and clock distribution. Locally, we can always use normal lines (with a maximum local line length of say 0·5 mm, their delay is limited to 0·02 ns) so the need for global interconnections is limited.

This extra communication layer resembles proposed extra metal layers for supply distribution.[6] Thus both global communication and supply distribution may be solved in a similar way in future multimetal layer processes.

Driving the global lines: Another possible problem may be the driving of these lines. A line driver can be considered as an impedance transformer, from the internal logic capacitance level to the line capacitance level. An internal capacitance level can be considered to be six minimum transistor input capacitances (fanout of 3), equal to about 0·1 pF in a 3 μm process. This capacitance scales linearly with the scaling factor. The line capacitance of a global line of the proposed type is given by $\varepsilon wl/t_{ox}$, and is thus not scaled, as w, l and t_{ox} are not assumed to scale (instead we consider l to grow as in Table 1). A typical 3 μm value of line capacitance is 1 pF, increasing to 2 pF when it reaches 20 mm long. An optimum driver consists of several inverters in a chain, each with a 'capacitance gain' of about e.[7] In our case the number of stages becomes ln (line capacitance/internal capacitance), that is 2, 6 and 8 stages for the three examples, respectively. The delay of one such stage is about $e\tau_0$, where τ_0 is the delay of one standard inverter. Our worst case occurs in the 0·3 μm process, with a driver delay of $8e\tau_0$. We estimate τ_0 to be 0·025 ns for this process, thus giving a line driver delay of 0·5 ns. Again, we obtain the same value as the logical delay, indicating that the global interconnection line can be driven with a delay of the same order of magnitude as the logic delays.

Conclusion: We propose that a special interconnection metal layer, which is coarser than the ordinary layers, is introduced into the VLSI process and used for global clock distribution and communication. We have shown that by this method the interconnection delay will not be a limiting factor for the speed of synchronous circuits in CMOS down to 0·3 μm feature sizes and up to 20 mm chip sizes. This result indicates that clock frequencies of up to 1 GHz are feasible in fully synchronous VLSI CMOS chips.

C. SVENSSON
M. AFGHAHI

10th March 1988

Dept. of Physics and Measurement Technology
Linköping University, S-581 83 Linköping, Sweden

References

1 MEAD, C., and CONWAY, L.: 'Introduction to VLSI systems' (Addison-Wesley, Reading, 1980)

2 MOLNAR, C. F., FANG, T., and ROSENBERGER, F. U.: 'Synthesis of delay-insensitive modules'. Chapel-Hill conference on VLSI, 1985, pp. 67–86

3 ANCEAU, F.: 'A synchronous approach for clocking VLSI systems', *IEEE J. Solid-State Circ.*, 1982, **SC-17**, pp. 51–56

4 AFGHAHI, M., and SVENSSON, C.: 'A scalable synchronous system'. 1988 IEEE Symposium on Circuits and Systems, Finland, June 8–10

5 JI-REN, Y., KARLSSON, I., and SVENSSON, C.: 'True single clock dynamic CMOS circuit technique', *IEEE J. Solid-State Circ.*, 1982, **SC-22**, pp. 899–901

6 SONG, W. S., and GLASSER, L. A.: 'Power distribution techniques for VLSI circuits'. Proc. of 1984 Conference on Advanced Research in VLSI, MIT, 1984, pp. 45–52

7 MOHSEN, A. M., and MEAD, C. A.: 'Delay-time optimization for driving and sensing of signals on high-capacitance path of VLSI systems', *IEEE J. Solid-State Circ.*, 1979, **SC-14**, pp. 462–470

Analysis and Design of Latch-Controlled Synchronous Digital Circuits

Karem A. Sakallah, *Member, IEEE,* Trevor N. Mudge, *Senior Member, IEEE,* and Oyekunle A. Olukotun

Abstract—**We present a succinct formulation of the timing constraints for latch-controlled synchronous digital circuits. We show that the constraints are mildly nonlinear and prove the equivalence of the nonlinear optimal cycle time calculation problem to an associated and simpler linear programming (LP) problem. We present an LP-based algorithm which is guaranteed to obtain the optimal cycle time for arbitrary circuits controlled by a general class of multiphase overlapped clocks. We illustrate the formulation and an initial implementation of the algorithm on some example circuits.**

Nomenclature

k	Number of clock phases.
l	Number of latches.
ϕ_i	Clock phase i.
T_i	Duration of active interval of ϕ_i.
s_i	Start time, within the common clock cycle, of the active interval of ϕ_i.
C_{ij}	ϕ_i to ϕ_j phase-ordering flag.
K_{ij}	ϕ_i to ϕ_j input/output relationship flag.
S_{ij}	ϕ_i to ϕ_j phase-shift operator.
p_i	Clock phase used to control latch i.
A_i	Arrival time of valid data signal at input of latch i.
D_i	Departure time of valid data signal from input of latch i.
Q_i	Departure time of valid data signal from output of latch i.
Δ_{DCi}	Setup time for latch i.
Δ_{DQi}	Propagation delay of latch i.
Δ_{ij}	Propagation delay from an input latch i through a combinational logic block to an output latch j.

I. Introduction

THE analysis and design of synchronous digital circuits which are controlled by level-sensitive latches is generally acknowledged to be a difficult problem [1]–[3]. The difficulty is due mainly to the coupling between the input and output terminals of a latch while the latch is enabled. This in turn leads to a set of cyclic timing constraints which must be satisfied by a properly designed circuit. The analysis problem seeks to determine if these constraints are indeed satisfied for a given circuit and a given clocking scheme. The design problem, on the other hand, attempts to find, for a given circuit and clocking scheme, the minimum clock cycle time which would not violate these constraints. In both cases the cyclic nature of the constraints frustrates intuitive solution approaches based on simple graph traversal methods, such as CPM [4], which require the constraint set to be acyclic.

This paper has two main goals. The first is to present a succinct, yet complete, formulation of the timing constraints for latch-controlled synchronous digital circuits. The constraints in this formulation are easily constructed, almost by inspection, for any circuit topology and clocking scheme, and are clearly seen to be nonlinear, though mildly so. The second goal of the paper is to formally prove the equivalence of the nonlinear design problem (i.e., the minimum clock cycle optimization problem with nonlinear timing constraints) to an associated and simpler linear programming (LP) problem. This proof forms the basis of an LP-based algorithm for finding the optimal cycle time.

Most current methods for the analysis and design of level-sensitive synchronous digital circuits assume edge triggering to simplify the analysis and then apply some heuristics to approximate the level-sensitive constraints. As a consequence, in the analysis case, they may declare a design to be in violation of timing constraints when in fact it is not, and in the design case, they may not produce the minimum cycle time. Our modeling of the level-sensitive constraints is not an approximation and so avoids both of these problems.

This paper is organized as follows. A review of previous work in the area is given in Section II. Section III presents a formulation of the timing constraints for level-sensitive synchronous digital circuits. They are seen to be nonlinear. Section IV shows that the solution of the apparently nonlinear optimal cycle time design problem can in fact be found by solving an associated LP problem, and therefore that the entire body of LP theory can be brought to bear on design and analysis problems. An initial implementation of this LP-based solution procedure is illustrated in Section V for several example circuits. Finally, Section VI contains some concluding remarks and a discussion of future directions.

Reprinted from *IEEE Trans. Computer-Aided Desi.*, vol. 11, no. 3, pp. 322–333, March 1992.

II. Previous Work

The timing analysis of digital logic circuits goes back at least to the work of Kirkpatrick in the 1960's [5]. However, this and much subsequent work was concerned with timing analysis for edge-triggered logic. It has only been during this decade, with MOS VLSI emerging as the leading technology for implementing digital systems, that the timing analysis and design of level-sensitive logic have become important. In this period several authors have addressed the question of level-sensitive latches, including Agrawal [6], Jouppi [1], Ousterhout [2], Unger [7], Szymanski [8], Cherry [9], Wallace [10], and Dagenais [11], [3].

One of the earliest, Agrawal, attempted to find the maximum frequency of operation of a logic circuit through a bounded binary search algorithm. Jouppi proposed an iterative scheme based on the concept of "borrowing." In the first iteration, the critical path(s) in the circuit are determined by pretending that the latches are edge triggered. This approximation is removed in subsequent iterations. In each of these iterations an attempt is made to reduce the clock cycle time to a value determined by the second most critical path. This is accomplished by trading (borrowing) the slack time in the subcritical path. In practice, only one borrowing iteration is performed to limit the computation cost. The TV program incorporates this borrowing algorithm and has been effectively applied for the verification of several large commercial chips. Ousterhout developed Crystal, a MOS timing verification program similar in many respects to TV. However, Crystal makes no attempt at dealing with clocking issues and confines its attention to the proper modeling of signal delay through trees of MOS transistors. In fact, Ousterhout acknowledged the inherent difficulty of dealing with level-sensitive latches.

Unger developed a set of timing constraints for a limited form of two-phase clocking with level-sensitive latches. He considered both the short-path (early arrival) as well as the long-path (late arrival) problems and presented a heuristic procedure for computing the minimum cycle time subject to these constraints. This, to our knowledge, was the first explicit formulation of the timing constraints of latch-controlled circuits as a system of linear inequalities. The LEADOUT program, developed by Szymanski, is an equation-based MOS timing analysis tool which handles multiphase clocking and level-sensitive latches. The temporal behavior of latches is specified by "max" constraints similar to those encountered in CPM graphs. To eliminate the inevitable cyclic dependencies among these constraints, the circuit is first partitioned into its strongest-connected components, and constraints are generated for each cycle-free path within the components. The reference does not, however, provide enough detail about LEADOUT's clocking model to explain how the cyclic dependencies induced by clock periodicity are removed. A unique feature of LEADOUT is the compilation of the timing constraints into a fast-executing program which allows repeated analysis of a circuit with different clocking or device parameters.

More recently, several authors proposed linear programming formulations for the minimum cycle time problem. The first such formulation appears to be due to Cherry in the Pearl CMOS timing analyzer. Like LEADOUT, Pearl starts out by constructing a causality graph which captures the dependencies among data and clock signals. The data signals at the inputs and outputs of latches are then *forced* to satisfy the appropriate setup and hold times, and a set of linear inequalities which determine the "spacing" between pairs of clock edges is derived. The minimum cycle time satisfying these spacing requirements is then found by solving a linear program. The simplification of the system timing constraints to a small set of clock spacing requirements seems, however, to be based on certain implicit assumptions about when data signals at the latch inputs are available.

In ATV (Abstract Timing Verifier), Wallace introduces the notion of local and common (absolute) frames of reference for time and uses it to characterize the temporal relations among clock phases through a set of *translations*. Then, in a process similar to that used in Pearl, ATV calculates bounds on the difference between pairs of translations in the form of linear inequalities, suggesting a linear programming solution. The program, however, requires the ordering of the clock edges to be specified by the user prior to the actual optimization step. To determine the optimal clock schedule, therefore, it would be necessary to run ATV several times, with a different ordering of the clock edges each time (e.g., nonoverlapped or overlapped). Another limitation in the ATV approach is its handling of loops of level-sensitive latches. The algorithm used by the program to deal with feedback and level-sensitive latches is to unroll (replicate) the circuit graph a user-specified number of clock cycles, n_c. Beside the attendant inefficiency of such an algorithm when n_c is large, if n_c is *smaller* than the number of cycles covered by any loop of latches in the circuit, the solution generated by the program will only be an approximation to the true solution.

The most recent effort at addressing this problem is due to Dagenais. He developed a MOS timing analysis and design tool, TAMIA, which represents the timing behavior of general multi-phase-clocked latch-controlled circuits by a set of nonlinear coupled relations. The design problem, which aims at finding the optimal clock parameters for a given circuit, is then solved approximately by an iterative graph-based algorithm. Because it models a circuit at the transistor level, however, in its current implementation TAMIA carries out just one iteration of this algorithm.

III. Problem Formulation

We consider synchronous digital circuits controlled by arbitrary k-phase clocks (to be defined shortly). We assume (see Fig. 1) that the circuits can be decomposed into

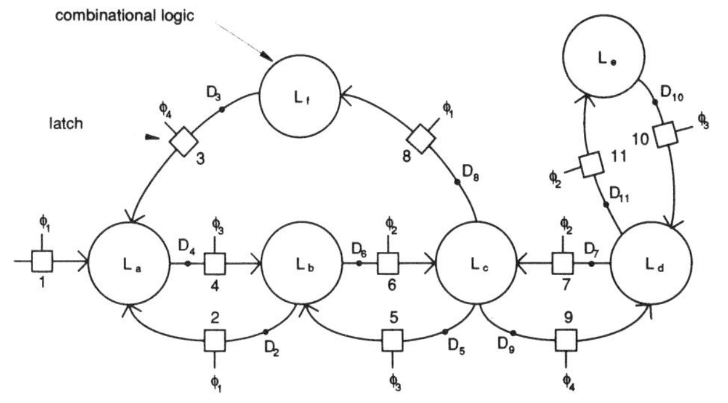

Fig. 1. Generalized logic model.

stages of feedback-free combinational logic blocks with clocked inputs and outputs.[1] The clocked elements at the inputs and outputs of the combinational stages are level-sensitive latches which provide temporary storage of data and act as *synchronizers*. These latches can be either static (for example, cross-coupled NAND gates) or dynamic (for example, MOS pass transistors). Regardless of the implementation, the functional and timing behavior of these latches is similar, and the formulation presented here applies to both types.

We place no restrictions on the combinations of clock phases used to control the input and output latches of a combinational stage other than a requirement that the set of clock phases controlling each feedback loop in the circuit be nonoverlapping; specifically, we require the logical AND of this set of phases to be identically equal to 0 at all times.

A. Clocking Methodology

An arbitrary k-phase clock is defined to be a collection of k periodic signals, $\phi_1, \phi_2, \cdots, \phi_k$, referred to as phases, with the same period. Each phase consists of two intervals: an *active* interval, during which the latches controlled by the phase are enabled; and a *passive* interval, when the latches are disabled. Without loss of generality we assume that all clock phases are active high; i.e., their active intervals occur when the phase signal assumes the logic 1 state.

We define the following clock variables (see Fig. 2):

- T_c: the clock cycle time, or period.
- s_i: the start time, relative to the beginning of the common clock cycle, of the active interval of ϕ_i.
- T_i: the duration of the active interval of ϕ_i.

For brevity, we will identify ϕ_i with its active interval, and simply refer to s_i as the start of ϕ_i and to T_i as the duration or width of ϕ_i. In addition, if ϕ_i and ϕ_j control an input latch and an output latch, respectively, of a combinational logic block L, we will simply refer to ϕ_i as an input phase, ϕ_j as an output phase, and ϕ_i/ϕ_j as an input/output-phase pair of L. We also introduce two $k \times k$ matrices C and K with elements C_{ij} and K_{ij}, defined as follows:

[1]Fig. 1 is adapted from Glasser and Dobberpuhl [13, fig. 6.7, p. 335].

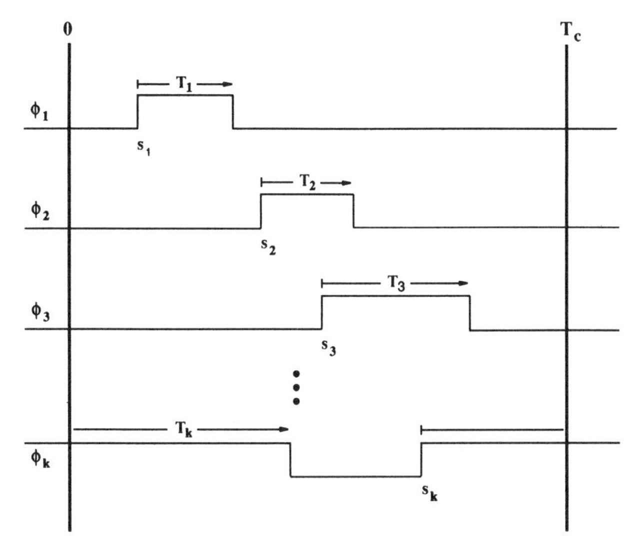

Fig. 2. Clock signal variables.

$$C_{ij} \equiv \begin{cases} 0, & i < j \\ 1, & i \geq j \end{cases} \tag{1}$$

$$K_{ij} \equiv \begin{cases} 1 & \text{if } \phi_i/\phi_j \text{ is an I/O phase pair of any logic block} \\ 0 & \text{otherwise.} \end{cases} \tag{2}$$

The K matrix identifies all I/O phase pairs for a particular circuit. The C matrix is used to determine if a clock cycle boundary must be crossed when going between an I/O phase pair ϕ_i/ϕ_j (see eqs. (6) and (12) below).

We can now state the relations among the various clock variables as a set of inequalities which are collectively referred to as the *clock constraints*:

C1. Periodicity Constraints:

$$T_i \leq T_c, \qquad i = 1, \cdots, k \tag{3}$$

$$s_i \leq T_c, \qquad i = 1, \cdots, k. \tag{4}$$

C2. Phase Ordering Constraints:

$$s_i \leq s_{i+1}, \qquad i = 1, \cdots, k-1. \tag{5}$$

C3. Phase Nonoverlap Constraints:

$$s_i \geq s_j + T_j - C_{ji}T_c \qquad \forall (i, j) \ni K_{ij} = 1. \tag{6}$$

C4. Clock Nonnegativity Constraints:

$$T_c \geq 0 \tag{7}$$

$$T_i \geq 0, \qquad i = 1, \cdots, k \tag{8}$$

$$s_i \geq 0, \qquad i = 1, \cdots, k. \tag{9}$$

These inequalities, except for the nonoverlap constraints C3, are intuitively obvious. Constraints C3 ensure that the output phase ϕ_j of every I/O phase pair must end before the input phase ϕ_i starts. This in turn guarantees that the clock phases controlling any feedback loop in the circuit are never simultaneously overlapping.

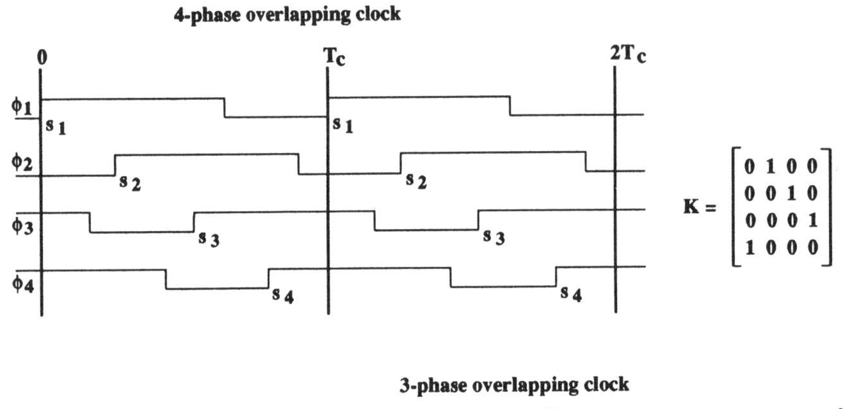

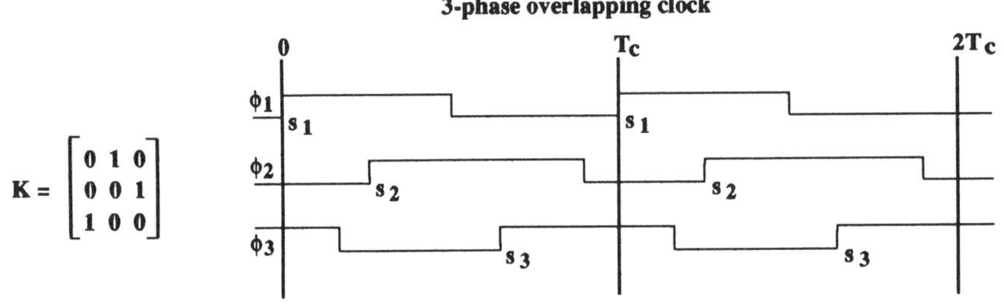

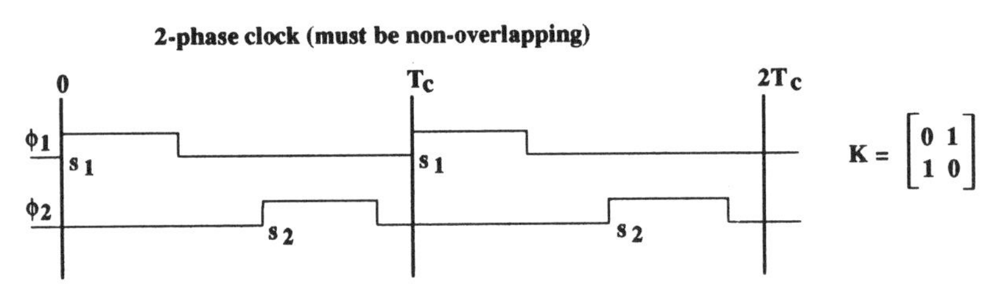

Fig. 3. Clocks with two, three, and four phases.

Constraints C1–C4 should be viewed as the minimum set of requirements that must be satisfied by a k-phase clock. Further requirements, such as minimum phase width, minimum phase separation, and clock skew, can be easily added to this minimum set but will not be treated here.

It is important to point out that the clock model introduced here is a temporal and not a logical one. The clock phases in this model are *not assumed to* have any logical relationships to one another; they are *not prevented from* being logically related either. This separation of the logical (functional) and timing aspects makes it possible to *map* any clocking discipline to our temporal framework by a suitable preprocessing step. This would include, for example, identifying derived and qualified clocks, followed by relabeling and ordering the clock phases according to (5). The generality of our clock model is demonstrated in Fig. 3 by showing how it applies to commonly used two-, three-, and four-phase clocking schemes. Note in particular that, for $k = 2$, the clock constraints ensure that the two phases are nonoverlapping, as they should be.

B. Latch Constraints

As will become evident, the simplicity of the formulation we present stems from a careful choice of time variables and naturally leads to a solution by linear programming.

We describe here the timing constraints necessary for the correct operation of D-type latches. Such latches have three terminals, representing data input, data output, and clock input (see Fig. 1). The circuit is assumed to contain l latches, numbered from 1 to l. For each of these latches we define the following variables and parameters:

- p_i: denotes the clock phase used to control latch i (e.g., latch 3 in Fig. 1 has $p_3 = 4$).
- A_i: denotes the arrival time, relative to the beginning of phase p_i, of a valid data signal at the input to latch i.
- D_i: denotes the departure time, which is the earliest time, relative to the beginning of phase p_i, when the signal available at the data input of latch i starts to propagate through the latch.
- Q_i: denotes the earliest time, relative to the beginning of phase p_i, when the signal at the data output of latch i starts to propagate through the succeeding stages of combinational logic.
- Δ_{DCi}: denotes the setup time for latch i required between the data input and the trailing edge of the clock input.
- Δ_{DQi}: denotes the propagation delay of latch i from the data input to the data output of the latch while

the clock input is high. It is assumed that $\Delta_{DQi} \geq \Delta_{DCi}$.

- Δ_{ij}: denotes the propagation delay from an input latch i through a combinational logic block to an output latch j. If latches i and j are not directly connected by a combinational block, then $\Delta_{ij} \equiv -\infty$.

Notice that both D_i and Q_i will always be nonnegative quantities, whereas A_i is unrestricted in sign.

The constraints governing latch operation fall into two categories: *setup* constraints and *propagation* constraints. The setup constraints guarantee that a latch has sufficient time to lock (store) the signal at the data input before that signal is allowed to change again. Thus,

$$A_i + \Delta_{DCi} \leq T_{p_i} \qquad i = 1, \cdots, l. \tag{10}$$

Since A_i can be negative, signifying that valid data has arrived *before* the onset of phase p_i, (10) may sometimes be satisfiable by a clock phase whose width T_{p_i} is 0! A more realistic setup constraint is obtained if A_i is replaced by D_i, yielding

$$D_i + \Delta_{DCi} \leq T_{p_i}, \qquad i = 1, \cdots, l. \tag{11}$$

In this case, since D_i is always nonnegative, the constraint places a lower bound on the width of phase p_i equal to the required setup time. We will adopt this more realistic constraint in our analysis.

Unlike the setup constraints which are local, the propagation constraints are global. They relate the departure times of signals at different latches in the circuit using the combinational propagation delay parameters. Since latch variables are referenced to the beginning of their corresponding clock phase, it is convenient to define the following phase-shift operator:

$$S_{ij} \equiv s_i - (s_j + C_{ij} T_c). \tag{12}$$

Adding S_{ij} to a time variable moves its referenced point (origin) ahead from s_i to s_j.

The propagation of signals from the inputs to the outputs of latches is simply described by

$$Q_i = D_i + \Delta_{DQi}, \qquad i = 1, \cdots, l. \tag{13}$$

Now consider a combinational path which starts at latch j and ends at latch i. The data signal at latch j starts to propagate at time Q_j and thus reaches latch i at $Q_j + \Delta_{ji}$, all times being referenced to the beginning of phase p_j. Therefore, relative to the beginning of phase p_i, the signal *arrives* at latch i at time $Q_j + \Delta_{ji} + S_{p_j p_i}$. The data signal at latch i becomes valid when *all* relevant input signals have had sufficient time to propagate through the combinational circuitry leading to latch i. Thus, the *arrival time* of a valid signal at latch i becomes

$$A_i = \max_j \, (Q_j + \Delta_{ji} + S_{p_j p_i}), \qquad j = 1, \cdots, l. \tag{14}$$

The propagation constraints through the combinational logic can now be expressed as follows:

$$D_i = \max \, (0, A_i), \qquad i = 1, \cdots, l \tag{15}$$

which express the fact that if a valid signal arrives at latch i before the start of phase p_i, then the departure time of that signal must be delayed to the beginning of phase p_i.

By eliminating the Q and A variables using (13) and (14), the latch constraints can be written exclusively in terms of signal departure times D_i along with the various clock variables, as follows:

L1. Setup Constraints:

$$D_i + \Delta_{DCi} \leq T_{p_i}, \qquad i = 1, \cdots, l. \tag{16}$$

L2. Propagation Constraints:

$$D_i = \max \, (0, \max_j \, (D_j + \Delta_{DQj} + \Delta_{ji} + S_{p_j p_i})), \quad i, j = 1, \cdots, l. \tag{17}$$

L3. Latch Nonnegativity Constraints:

$$D_i \geq 0, \qquad i = 1, \cdots, l. \tag{18}$$

Using the notation scheme defined in this section, it is now possible to write down the set of timing constraints for arbitrary circuits by inspection. It is assumed that the circuit has been decomposed into clocked combinational stages, and that the various delay parameters have been calculated. We illustrate this process in the Appendix for the circuit shown in Fig. 1.

IV. Optimal Clock Cycle Calculation

The minimum clock cycle time can be calculated by solving the following optimization problem, denoted by P1:

P1. Optimal Cycle Time:

Minimize T_c

Subject to Clock Constraints C1, C2, C3, and C4

Latch Constraints L1, L2, and L3.

P1 is a nonlinear optimization problem. Let R_1 denote the feasible region of P1, i.e., the set of solutions to P1 which satisfy the specified constraints. The nonlinearity of P1 is due to the max functions in the latch propagation constraints L2. A linear optimization problem is obtained if these propagation constraints are *relaxed* as follows:

L2R. Relaxed Propagation Constraints:

$$D_i \geq D_j + \Delta_{DQj} + \Delta_{ji} + S_{p_j p_i}, \qquad i, j = 1, \cdots, l. \tag{19}$$

Thus, we define the following linear program:

P2. Modified Optimal Cycle Time:

Minimize T_c

Subject to Clock Constraints C1, C2, C3, and C4

Latch Constraints L1, L2R, and L3

If R_2 is defined as the feasible region of P2, it should be obvious that $R_1 \subseteq R_2$.

Our objective in this section is to show that the minimum cycle time obtained by solving the linear program P2 is the same as the minimum cycle time of the original nonlinear problem P1. If we denote the optimal value for P1 and P2 by $T_{c,\min}^{(P1)}$ and $T_{c,\min}^{(P2)}$, respectively, then this result can be expressed by the following theorem:

Theorem 1:

$$T_{c,\min}^{(P1)} = T_{c,\min}^{(P2)}.$$

Proof: The critical element of this proof is the observation that the optimal value of a linear program cannot be improved with the addition of extra constraints [4, p. 170]. It folllows that, since P2 is a relaxed version of P1, $T_{c,\min}^{(P1)} \geq T_{c,\min}^{(P2)}$. Therefore the theorem is proved if we can show that P2 can be augmented with constraints such that the following two stipulations are true:

1) The optimal value of the augmented problem is not *greater* (worse) than $T_{c,\min}^{(P2)}$.
2) The constraints of the augmented problem are equivalent to those of P1.

Thus the augmented problem, which we will call P3, has the same optimal solution (i.e., the same values for all variables) as the original problem P1. Since the only difference between P1 and P2 are the latch propagation constraints, we need to examine the following two cases:

1) All D variables in the optimal solution to P2 are at their minimum values. Thus, the optimal solution to P2 satisfies all the constraints of P1, including the latch propagation constraints L2. In this case, P2 is equivalent to P1, and their optimal values are the same.
2) One or more of the D variables is not at its minimum value. Thus the optimal solution to P2 violates some of the latch propagation constraints L2. To force the solution to satisfy the L2 constraints, we augment the constraints of P2 with equality constraints as follows:
   (a) If $A_i \leq 0$ and $D_i > 0$ for some latch i, add the equality constraint $D_i = 0$.
   (b) If $A_i > 0$ and $D_i > A_i$ for some latch i, add the equality constraint $D_i = A_i$.
   The addition of these equality constraints may in some cases cause the departure time at some other latch j which previously satisfied constraints L2 to now violate them. In such cases we add further equality constraints (either (a) or (b), as appropriate) for all such affected latches, and repeat the procedure as often as necessary, until the constraints of P3 become equivalent to those of P1. It should be obvious that the addition of such equality constraints does not increase the cycle time. Thus the two stipulations stated above are true, and the theorem is proved. □

A geometric interpretation of this theorem is shown in Fig. 4. The figure clearly shows the following relationships between the original problem P1 and the modified (relaxed) problem P2:

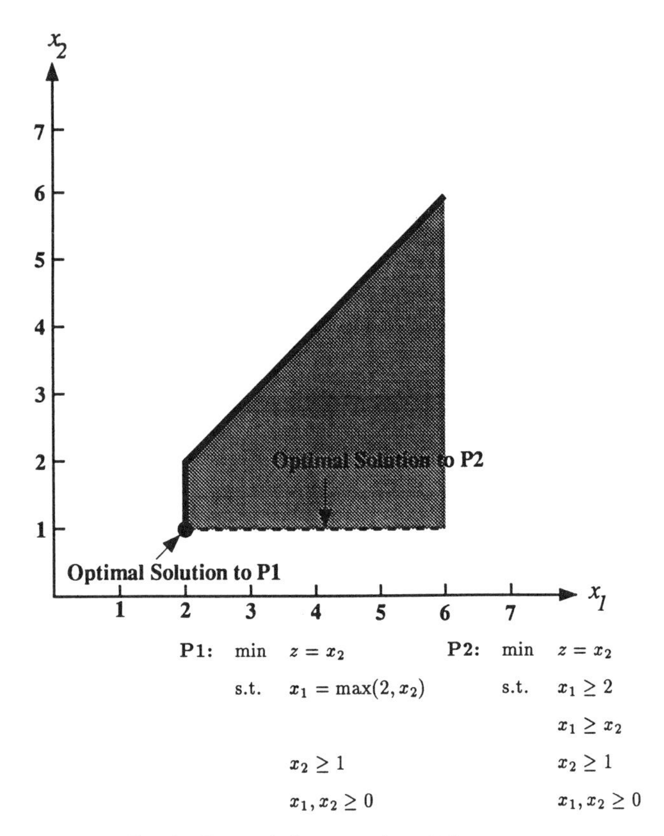

Fig. 4. Geometric interpretation of Theorem 1.

- The feasible region of P1 (the two heavy line segments) is a subset of the feasible region of P2 (the shaded area, including the bordering line segments.)
- The optimal values of P1 and P2 are equal: $z_{\min} = 1$.
- P2 does not have a unique optimal solution. In fact any point on the horizontal dashed line segment is an optimal solution to P2. On the other hand, P1 has a unique optimal solution: the single point (2, 1). If X_1 and X_2 are used to denote, respectively, the optimal solution sets to P1 and P2, then the figure makes it clear that $X_1 \subseteq X_2$. This relationship, in fact, holds regardless of the uniqueness or nonuniqueness of the optimal solutions, as can be easily verified by trying other objective functions. For example, if $z = x_1$, then $X_1 = X_2 = \{(2, x_2) | 1 \leq x_2 \leq 2\}$; and if $z = x_1 + x_2$, then $X_1 = X_2 = \{(2, 1)\}$.
- An optimal solution to P2 may not be a feasible solution to P1. For example, the point (4, 1) is an optimal solution to P2 but is clearly infeasible for P1. The figure suggests how such a solution may be made feasible for P1: *minimize* x_1 *until it satisfies* $x_1 = \max(2, x_2)$. Applying this to the point (4, 1) yields the correct optimal solution (2, 1) to P1. The minimization step can, in fact, be carried out by "sliding" x_1 to the left until the max constraint in P1 is satisfied.

Theorem 1 forms the basis for the following algorithm to find the optimal solution of P1 by linear programming:

Algorithm MLP: Optimal Cycle Time Calculation by Modified LP

(*Comment:* m is an iteration counter; v is a convergence flag.)

1) Solve P2. Denote the optimal solution found by D_i^0 for $i = 1, \cdots, l$.
2) Set $m = 0$, $v =$ TRUE.
3) For $i, j = 1, \cdots, l$, evaluate $D_i^{m+1} = \max(0, \max_j (D_j^m + \Delta_{DQj} + \Delta_{ji} + S_{p_j p_i}))$.
4) If $D_i^{m+1} \neq D_i^m$ for any i, set $v =$ FALSE, and increment m.
5) If $v ==$ TRUE stop; otherwise set $v =$ TRUE and go to (3).

Several observations should be made about this algorithm:

- The overall performance of the algorithm depends on how efficiently we can solve the LP P2 in step 1. The most commonly used method for solving linear programs is the simplex algorithm, which, on average, takes between n and $3n$ steps to reach the optimum, where n is the number of problem constraints [4, p. 54]. For P2, it is easy to show that the number of constraints is bounded from above by $4k + (F + 1)l$, where F is the *maximum* fan-in to any latch in the circuit and is usually a small number, such as 3 or 5. Thus, the number of constraints is *linear* in the number of latches, l. Additionally, by lumping latches corresponding to vector signals with similar timing (e.g., 32-bit data buses), the number l can be reasonably small even for large circuits. The complexity of step 1, therefore, grows only linearly with l. The cost of this step can be reduced further by taking advantage of the special properties of this particular LP. For instance, by treating all latches as though they were positive-edge-triggered flip-flops, a very good initial guess can be quickly generated and used as the starting point for optimization.
- The algorithm involves a Jacobi-style iteration (steps 3 to 5) which updates the values of the departure times until all the max propagation constraints are satisfied. A more efficient Gauss–Seidel-style iteration is obviously possible. In fact, an event-driven update mechanism which only calculates the departure times which have changed from the previous iteration can be easily implemented. With such an enhancement, the cost of the iterative steps is greatly reduced for large circuits.
- When the departure times are udpated in step (3), the clock variables are held fixed at the optimal values that were found by solving the LP in step (1). Thus, the update process fixes the clock "schedule" and "slides" the departure variables toward the time origin.
- The update iteration is guaranteed to terminate because the departure times are bounded from below. In the examples we have attempted, the update process usually terminated in two to three iterations (in some cases no iterations were even necessary.)

V. Examples

In this section we illustrate our proposed formulation with three examples. The examples were solved using an initial implementation of the MLP algorithm which incorporates a simple parser, a dense-matrix LP solver which implements the standard simplex algorithm, and graphical output routines.

The first example, adapted from [3], is shown in Fig. 5. It is a simple two-stage system connected in a loop and controlled by a two-phase clock. To facilitate comparison with [3] we assume that all latches have equal setup and propagation delays of 10 ns. We also assume the same values for the combinational logic delays, except for block L_d, whose delay, Δ_{41}, will be varied to study its effect on the optimal cycle time. The resulting set of timing constraints are:

- Periodicity constraints: $T_i, s_i \le T_c$, $i = 1, 2$.
- Phase ordering constraints: $s_1 \le s_2$.
- Phase nonoverlap constraints: $s_1 \ge s_2 + T_2 - T_c$ and $s_2 \ge s_1 + T_1$.
- Latch setup constraints:
  $D_1 + 10 \le T_1 \quad D_2 + 10 \le T_2$
  $D_3 + 10 \le T_1 \quad D_4 + 10 \le T_2$.
- Latch propagation constraints:
  $D_1 = \max(0, D_4 + 10 + \Delta_{41} + s_2 - s_1 - T_c)$
  $D_2 = \max(0, D_1 + 10 + 20 + s_1 - s_2)$
  $D_3 = \max(0, D_2 + 10 + 20 + s_2 - s_1 - T_c)$
  $D_4 = \max(0, D_3 + 10 + 60 + s_1 - s_2)$.
- Nonnegativity constraints: $T_c \ge 0$; $T_{p_i}, s_{p_i} \ge 0$, $p_i = 1, 2$; and $D_i \ge 0$, $i = 1, \cdots, 4$.

Figs. 6 and 7 compare the results obtained by the MLP algorithm with the *null retardation in the initial phase* (NRIP) algorithm, described in [3]. The diagrams in Fig. 6 show, for each experiment, two complete cycles of the resulting clock schedule along with a "strip" which identifies the names and delay values for each of the combinational blocks and the times at which the data signals depart from each of the four latches. The shaded portions in these strips represent propagation through the latches themselves (Δ_{DQi}), whereas gaps in the strips indicate signals that arrive *earlier* than (and must thus wait for) the enabling edge of the corresponding clock phase. For example, the MLP solution for the $\Delta_{41} = 120$ ns case, shown in part (c) of the figure, has a cycle time of 140 ns with signals departing from latches 1 through 4, respectively, at 60 ns, 90 ns, 140 ns, and 210 ns. Furthermore, the input to latch 3 becomes valid at 120 ns, 20 ns earlier than the rising edge of ϕ_1; thus departure from latch 3 must wait until ϕ_1 rises at 140 ns.

These results lead to several observations:

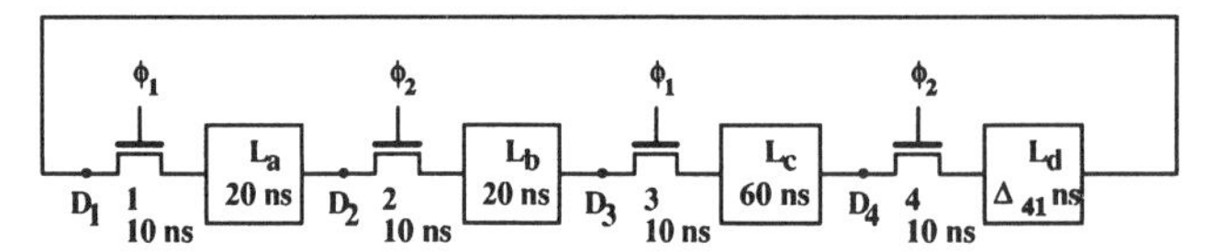

Fig. 5. Block diagram for example 1.

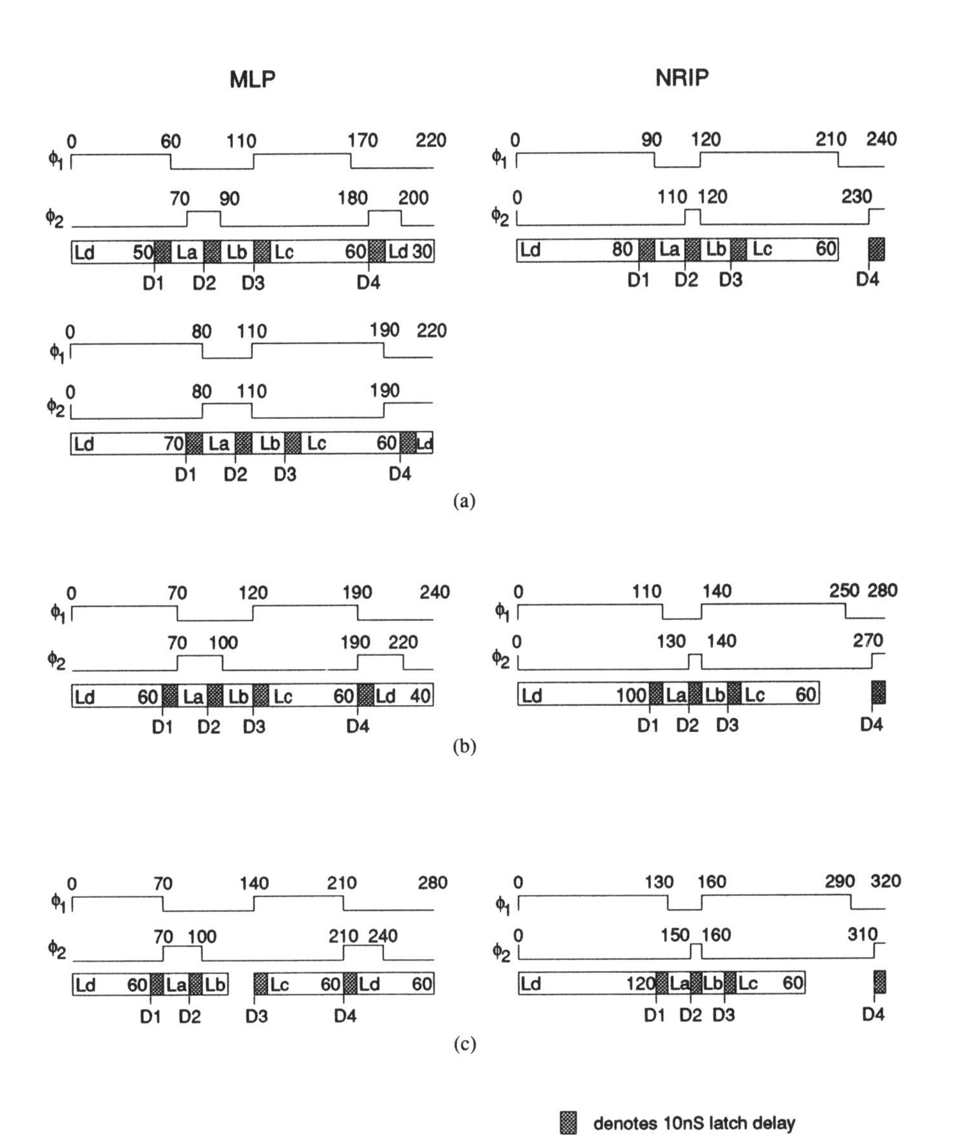

Fig. 6. Timing diagrams for example 1. (a) $\Delta_{41} = 80$ ns. (b) $\Delta_{41} = 100$ ns. (c) $\Delta_{41} = 120$ ns.

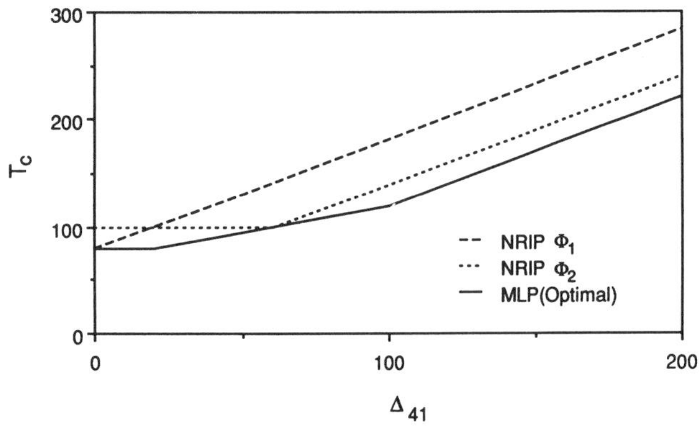

Fig. 7. T_c versus Δ_{41} for example 1.

- Unless additional constraints are placed on the minimum widths and separations of clock phases, the optimal solution will not be unique. This is illustrated with two such solutions for the $\Delta_{41} = 80$ ns case (see the top of Fig. 6). Each of these solutions has a cycle time of 110 ns, even though their phase signals are quite different. Physically, this means that the timing constraints of a given circuit may be satisfied by a number of different clock schedules which share a common cycle time. Additional requirements, such as minimum duty cycle, may be applied to select one of these different solutions. The apparent uniqueness of the solution found by the NRIP

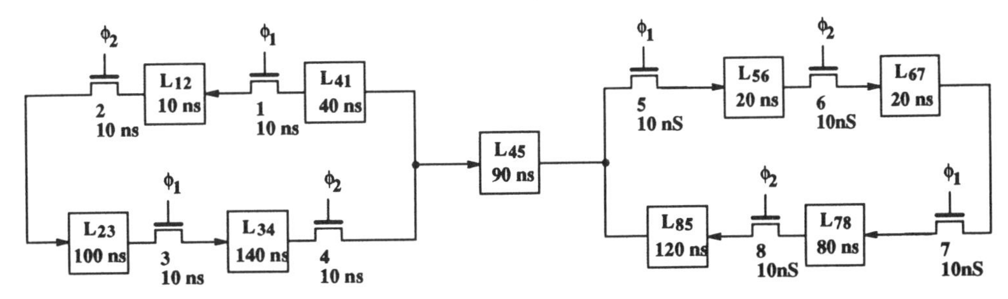

Fig. 8. Block diagram for example 2.

algorithm is due to its implicit *minimum* constraints on phase widths and separations.

- The NRIP algorithm produces an optimal solution for $\Delta_{41} = 60$ ns. For all other values of Δ_{41}, the cycle time found by NRIP is suboptimal (see Fig. 7).
- The piecewise-linear dependence of T_c on Δ_{41} in the optimal solution has three distinct segments. For $0 \leq \Delta_{41} \leq 20$ ns, T_c is independent of Δ_{41}, and is set by some other delay in the circuit. When $\Delta_{41} \geq 20$ ns, block L_d becomes critical, and any increase in Δ_{41} causes T_c to also increase. For $20 \leq \Delta_{41} \leq 100$ ns, T_c increases by 1 ns for every 2-ns increase in Δ_{41} because the added delay is shared between the two clock cycles ("borrowed" from ϕ_1). For $\Delta_{41} \geq 100$, T_c increases in direct proportion to Δ_{41}, since the additional delay can no longer be shared between the two clock cycles, and slack is inevitably introduced in the cycle with shorter delay. The rather simple dependence of the optimal cycle time on Δ_{41} in this case is due to the simplicity of the topology of this particular circuit. In fact, one can show that since the feedback loop consists of two complete clock cycles, the optimal cycle time is the maximum of the average delay around the loop and the difference between the delays for each of the cycles making up the loop.

The cycle time calculations using the MLP and NRIP algorithms for a more complicated example are shown in Figs. 8 and 9.

The following additional observations can be made:

- Unlike the previous example, the cycle time found by the NRIP algorithm is significantly higher (35%) than the optimal cycle time. While this result cannot be generalized for other circuits, it does point out that the approximate solution found by NRIP may deviate appreciably from the exact solution, and additional iterations might be necessary. Because it can be found fairly quickly, however, the NRIP solution may be used in the LP step of the MLP algorithm as a starting point.
- Because of the coupling of the timing constraints through the feedback loops in the circuit as well as through the periodic clock signals, the notion of a *critical path* is clearly inadequate as a basis for discussing the optimality of the solution, and the dependence of that solution on the circuit's delays. Instead of a single critical path, the circuit has several critical combinational delay *segments* which may be disjoint. The criticality of these segments, and the subcriticality of others, are directly related to associated slack variables in the inequality constraints. The techniques of parametric analysis in linear programming can be usefully applied here to study the effects of varying the circuit delays on the optimal cycle time.

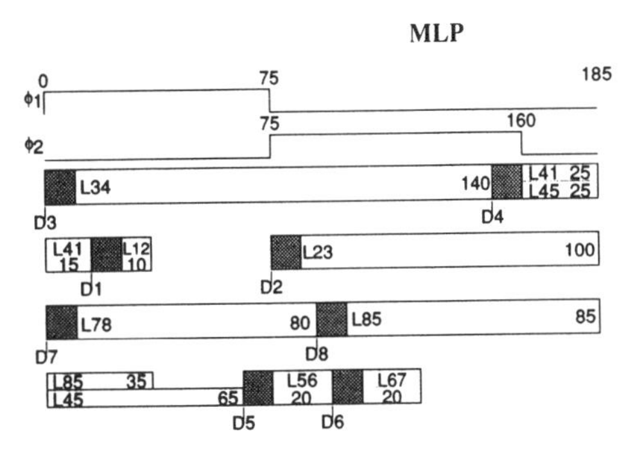

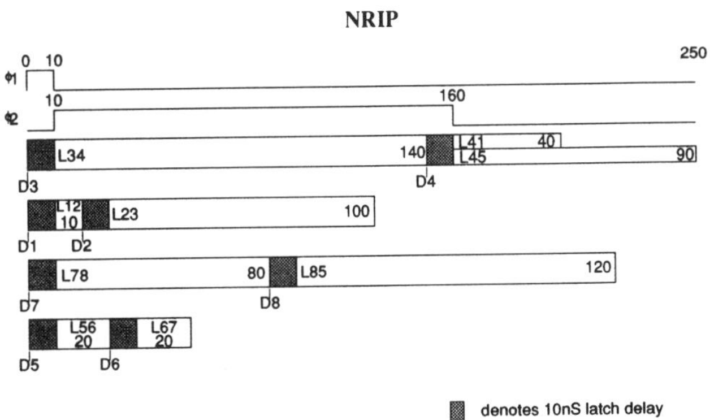

Fig. 9. Timing diagrams for example 2.

The final example illustrates the application of the MLP algorithm to study and optimize the timing of a 250 MHz gallium arsenide microcomputer currently under development at the University of Michigan [12]. Fig. 10 shows a simplified block diagram of the microcomputer's CPU and its primary cache subsystem. The CPU implements an existing instruction-set architecture, the MIPS R6000, and has as its main components a register file of 32 32-bit registers, an ALU, a shifter, and an integer multiply and divide unit. The data path is 32 bits wide and is timed

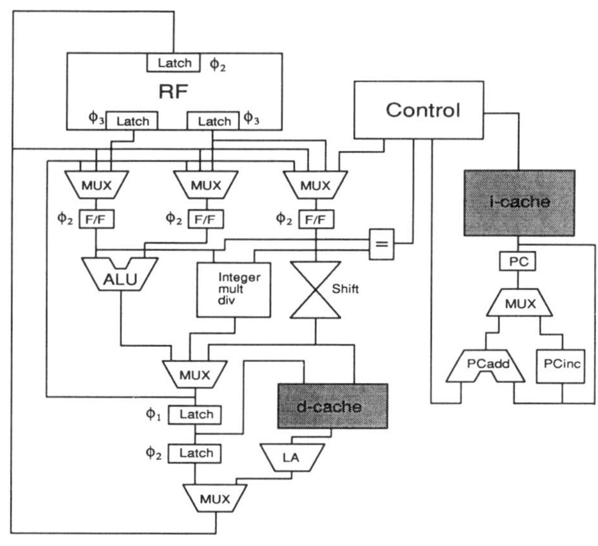

Fig. 10. GaAs MIPS system diagram.

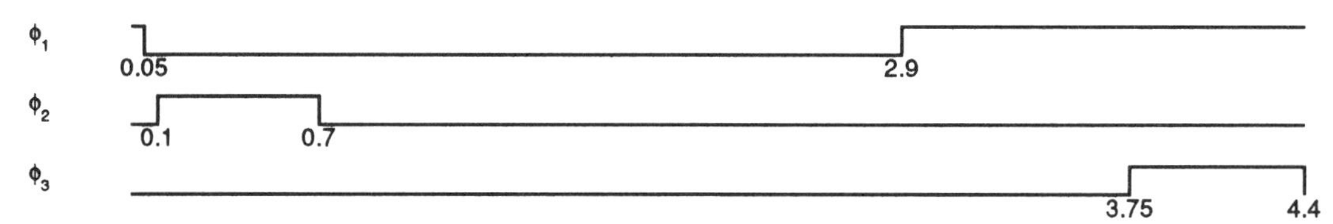

Fig. 11. MLP-generated clock schedule for the GaAs MIPS data path.

with a three-phase clock. The diagram shows the major logic blocks in the CPU data path, which is implemented as a single chip. The shaded blocks show the instruction cache and the data cache, which are implemented as a set of 15 high-speed GaAs 1 K × 32 SRAM chips. The synchronizing elements are a combination of latches and flip-flops (F/F's, see Fig. 10). To reduce the effects of chip crossings the CPU and the primary caches are integrated into a single multichip module (MCM).

We applied the MLP algorithm to a timing model of the data path to obtain its optimal clock schedule, assuming that the cache subsystem could be designed to match the speed of the CPU. The data path contains roughly 30 000 transistors, the majority of which are in the register file (see Table I for a breakdown of the transistor count amoung the major data path blocks). The timing model was abstracted from the transistor level, and consists of 18 synchronizing elements, 15 of which are level-sensitive latches. Each of the synchronizers in the model represents a 32-bit-wide data bus. The timing parameters (propagation delays and setup times) were extracted from circuit simulations using SPICE. The resulting clock schedule is shown in Fig. 11. It is interesting to note the following:

- The number of constraints for this example was 91. Even though the current implementation of the MLP algorithm uses a dense-matrix solver, its execution time (on a DECStation 3100) was hardly noticeable (on the order of a few seconds).

TABLE I
TRANSISTOR COUNT FOR MAJOR BLOCKS OF THE GaAs MIPS DATAPATH

Block Name	No. of Transistors
Register File (RF)	16,085
Arithmetic/Logic Unit (ALU)	3419
Shifter	1848
Integer Multiply/Divide (IMD)	6874
Load Aligner	1922
Total	30,148

- The optimal cycle time found by MLP (4.4 ns) is 10% higher than the target cycle time of 4 ns. We are continuing to refine the delay parameters of the model from additional circuit simulations as well as actual measurements on prototype chips, and to apply the MLP algorithm throughout the design process in order to monitor any changes in the optimal cycle time.
- Phase ϕ_3 in the optimal clock schedule is completely overlapped by ϕ_1. While this relationship might seem odd at first, it is easy to explain once the function of ϕ_3 in the circuit is recognized. This third phase is used as a precharge clock for the register file storage cells to speed up the readout of data into the ALU. The total overlap of ϕ_3 by ϕ_1 is not a problem since there are no direct paths in the circuit between these two phases (i.e., $K_{13} = K_{31} = 0$). This result also

points out some of the generality of the timing model proposed in this paper, namely that it is able to overlap clock phases if necessary to produce a shorter cycle time.

VI. Conclusions

We have shown in this paper that the optimal cycle time for circuits controlled by level-sensitive latches can be determined by solving a linear program. In contrast to earlier models, the timing constraints presented here are quite general, being based on very few assumptions, allowing them to accommodate a much wider class of clocking schemes. In addition, these constraints are very easy to generate for arbitrary circuit structures. The LP formulation provides a convenient theoretical foundation for analyzing such constraints and for developing algorithms that are potentially more efficient than the simplex algorithm. We are currently investigating just such algorithms, noting that the entries of the constraint matrix for this problem are exclusively topological (i.e., 0, ± 1). We also intend to use parametric programming techniques to quantify the notion of critical path segments and to study the effects on the optimal cycle time of varying the circuit delays.

Appendix

To illustrate the notation developed in Section III, we present here the complete set of timing constraints for the circuit shown in Fig. 1. The circuit has 11 latches and is controlled by a four-phase clock with the following K matrix:

$$K = \begin{bmatrix} 0 & 0 & 1 & 1 \\ 1 & 0 & 1 & 1 \\ 1 & 1 & 0 & 0 \\ 0 & 1 & 1 & 0 \end{bmatrix}.$$

Thus there are nine I/O phase pairs; the corresponding phase-shift operators (used in the latch propagation constraints) are

$$S_{13} = s_1 - s_3$$
$$S_{14} = s_1 - s_4$$
$$S_{21} = s_2 - s_1 - T_c$$
$$S_{23} = s_2 - s_3$$
$$S_{24} = s_2 - s_4$$
$$S_{31} = s_3 - s_1 - T_c$$
$$S_{32} = s_3 - s_2 - T_c$$
$$S_{42} = s_4 - s_2 - T_c$$
$$S_{43} = s_4 - s_3 - T_c.$$

The timing constraints can now be stated as follows:

- Periodicity constraints:

$$T_i \le T_c, \; s_i \le T_c, \qquad i = 1, 2, 3, 4.$$

- Phase-ordering constraints:

$$s_1 \le s_2 \le s_3 \le s_4.$$

- Phase nonoverlap constraints:

$$s_1 \ge s_3 + T_3 - T_c \qquad s_2 \ge s_1 + T_1 \qquad s_3 \ge s_1 + T_1$$
$$s_4 \ge s_2 + T_2 \qquad s_1 \ge s_4 + T_4 - T_c$$
$$s_2 \ge s_3 + T_3 - T_c$$
$$s_3 \ge s_2 + T_2 \qquad s_4 \ge s_3 + T_3 \qquad s_2 \ge s_4 + T_4 - T_c.$$

- Latch setup constraints:

$$D_i + \Delta_{DCi} \le T_1, \qquad i = 1, 2, 8$$
$$D_i + \Delta_{DCi} \le T_2, \qquad i = 6, 7, 11$$
$$D_i + \Delta_{DCi} \le T_3, \qquad i = 4, 5, 10$$
$$D_i + \Delta_{DCi} \le T_4, \qquad i = 3, 9.$$

- Latch propagation constraints:

$$D_2 = \max(0, D_4 + \Delta_{DQ4} + \Delta_{42} + S_{31}, D_5 + \Delta_{DQ5} + \Delta_{52} + S_{31})$$
$$D_3 = \max(0, D_8 + \Delta_{DQ8} + \Delta_{83} + S_{14})$$
$$D_4 = \max(0, D_1 + \Delta_{DQ1} + \Delta_{14} + S_{13}, D_2 + \Delta_{DQ2} + \Delta_{24} + S_{13}, D_3 + \Delta_{DQ3} + \Delta_{34} + S_{43})$$
$$D_5 = \max(0, D_6 + \Delta_{DQ6} + \Delta_{65} + S_{23}, D_7 + \Delta_{DQ7} + \Delta_{75} + S_{23})$$
$$D_6 = \max(0, D_4 + \Delta_{DQ4} + \Delta_{46} + S_{32}, D_5 + \Delta_{DQ5} + \Delta_{56} + S_{32})$$
$$D_7 = \max(0, D_9 + \Delta_{DQ9} + \Delta_{97} + S_{42}, D_{10} + \Delta_{DQ10} + \Delta_{10,7} + S_{32})$$
$$D_8 = \max(0, D_6 + \Delta_{DQ6} + \Delta_{68} + S_{21}, D_7 + \Delta_{DQ7} + \Delta_{78} + S_{21})$$
$$D_9 = \max(0, D_6 + \Delta_{DQ6} + \Delta_{69} + S_{24}, D_7 + \Delta_{DQ7} + \Delta_{79} + S_{24})$$
$$D_{10} = \max(0, D_{11} + \Delta_{DQ11} + \Delta_{11,10} + S_{23})$$
$$D_{11} = \max(0, D_9 + \Delta_{DQ9} + \Delta_{9,11} + S_{42}, D_{10} + \Delta_{DQ10} + \Delta_{10,11} + S_{32}).$$

- Nonnegativity constraints:

$$T_c \ge 0$$
$$T_{p_i} \ge 0, \; s_{p_i} \ge 0, \qquad p_i = 1, 2, 3, 4$$
$$D_i \ge 0, \qquad i = 1, \cdots, 11.$$

REFERENCES

[1] N. P. Jouppi, "Timing verification and performance improvement of MOS VLSI designs," Ph.D. thesis, Stanford University, Stanford, CA 94305-2192, Oct. 1984.

[2] J. K. Ousterhout, "A switch-level timing verifier for digital MOS VLSI," *IEEE Trans. Computer-Aided Design*, vol. CAD-4, no. 3, pp. 336–349, 1985.

[3] M. R. Dagenais and N. C. Rumin, "On the calculation of optimal clocking parameters in synchronous circuits with level-sensitive latches," *IEEE Trans. Computer-Aided Design*, vol. 8, pp. 268–278, Mar. 1989.

[4] D. T. Phillips, A. Ravindran, and J. J. Solberg, *Operations Research: Principles and Practice.* New York: Wiley, 1976.

[5] T. I. Kirkpatrick and N. R. Clark, "PERT as an aid to logic design," *IBM J. Res. Develop.*, vol. 10, no. 2, pp. 135-141, Mar. 1966.

[6] V. D. Agrawal, "Synchronous path analysis in MOS circuit simulator," in *Proc. 19th Design Automat. Conf.*, 1982, pp. 629-635.

[7] S. H. Unger and C.-J. Tan, "Clocking schemes for high-speed digital systems," *IEEE Trans. Comput.*, vol. C-35, pp. 880–895, Oct. 1986.

[8] T. G. Szymanski, "LEADOUT: A static timing analyzer for MOS circuits," in *ICCAD-86 Dig. Tech. Papers*, 1986, pp. 130-133.

[9] J. J. Cherry, "Pearl: A CMOS timing analyzer," in *Proc. 25th Design Automat. Conf.*, 1988, pp. 148–153.

[10] D. E. Wallace and C. H. Sequin, "ATV: An abstract timing verifier," in *Proc. 25th Design Automat. Conf.*, 1988, pp. 154–159.

[11] M. R. Dagenais, "Timing analysis for MOSFETs; An integrated approach," Ph.D. thesis, McGill University, Montreal, Quebec, Canada H3A 2A7, June 1987.

[12] R. B. Brown, J. A. Dykstra, T. N. Mudge, and R. Milano, "A GaAs microsupercomputer: Rationale and design," Tech. Rep. CSE-TR-42-90, University of Michigan, Dept of EECS, Ann Arbor, MI 48109-2122, 1980.

[13] L. A. Glasser and D. W. Dobberpuhl, *The Design and Analysis of VLSI Circuits.* Reading, MA: Addison Wesley, 1985.

Robin Hood: A System Timing Verifier for Multi-Phase Level-Sensitive Clock Designs

Ren-Song Tsay   Ichiang Lin
IBM, T. J. Watson Research Center

Abstract

We report an efficient system timing verification approach that employs latch graph representation. It is a general approach that can handle most clocking schemes. We name the tool Robin Hood *for its ability of analyzing cycle stealing. Both early and late mode timing constraints are considered. The tool generates a detailed slack report to help designers identify where and how much the correction should be, in case of violations. The algorithm has been implemented and tested on several real designs. An example with 1926 latches and dense interconnections is easily verified in 0.62 seconds on an IBM 3090 [1] machine.*

1 Introduction

As the demand for high-speed digital systems grows, the design of clocking schemes becomes a major concern. Mainly, the clock cycle time determines the speed of computation and data processing. Besides, correct clocking is most important for proper system operation. Therefore, it is extremely important to have both a well-planned clocking strategy, and an adequate system timing verifier. A system timing verifier can identify timing problems at an early design stage and ensure reliable system operation. This is essential for the speedup of product design time and the reduction of the costs of possible after-production problem fixing and tuning.

In this paper, we develop verification methods for multi-phase level-sensitive latched clocking schemes. Edge-triggered latch is treated as a special case of level-sensitive latch, with its active interval width set to *zero*. The algorithm proposed in this paper is very general that it can handle most clocking schemes, which include multi-phase edge-triggered, level-sensitive, master-slave, or mixed schemes.

Design and verification of level-sensitive latched systems have long been recognized as difficult problems [1]. The difficulty is mainly due to the cyclic timing constraints. Because signals can be launched at any point in the active interval, to check whether signals arrive illegally after the setup time of a latch, it depends on when the signals are launched at the preceding latches. A loop may well lead the trace back to the same latch to provide the launch time, which unfortunately depends on the arrival time to be determined. This cyclic constraint phenomenon has failed many researchers to provide a successful timing verifier.

Despite the complication, level-sensitive latched designs do provide an attractive advantage that the cycle time can be made as small as possible by widening the active interval (of course, without violating timing constraints). This is due to the *cycle stealing* technique, which allows certain combinational logic path delay to stretch longer than the given cycle time while it still satisfies early and late mode timing constraints. It is because, in the extreme case, signals can travel from the opening edge of a latch to the closing edge of the succeeding latch, and that duration usually is longer than one cycle time. However, the stealing is valid only if the preceding or succeeding stage takes less time than expected.

[1] IBM 3090 is a trademark of International Business Machines Corporation.

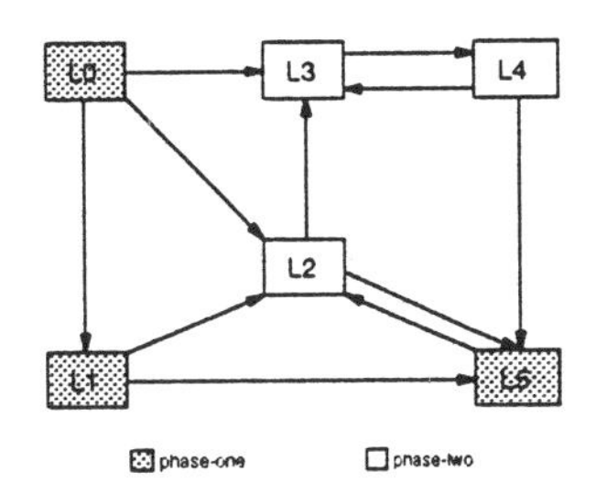

Figure 1: The latch graph of a two-phase example.

This calls for the necessity of a verification tool to examine all such necessary and sufficient conditions.

In our approach, we deviate from other timing analysis work by exploiting latch graph for system timing analysis. The delay analysis of combinational blocks is performed independently and summarized in the latch graph. It will be demonstrated that the latch graph greatly simplifies the analysis algorithm and facilitate the possibility of early system timing verification.

Since loops are mainly due to the existence of latches (each combinational logic block is an acyclic directed graph), the latch graph also helps to expose loop conditions more explicitly. With a special arrival time updating technique, loops are managed successfully in our algorithm. Quick convergence of a proposed iterative algorithm is guaranteed.

2 System Timing Abstraction

We abstract a synchronous digital system by a Latch Graph $G(L, E)$, which is a directed graph with each node representing a latch controlled by a clock signal. Each edge e_{ij} is directed from a latch L_i to latch L_j and is associated with two parameters d_{ij} and D_{ij}, the minimum and maximum path delays. For better timing accuracy. false path analysis can be used to calculate d_{ij} and D_{ij} These parameters summarize the timing property of the combinational logic block connecting the two latches. A latch graph example is as shown in Fig. (1). Later in this paper, we will use the notation $L_i \rightarrow L_j$ to indicate that there is a directed edge e_{ij} from L_i to L_j.

The separation of system information and combinational logic block information has several advantages. First, we do not have to simulate the whole system at one time. Whole system simulation may be infeasible due to the complexity. In our case, the timing simulation is done on each combinational logic block, which is much smaller as compared to the whole system. Secondly, this separation concept fits very well with hierarchical design methodology.

Thirdly, depending on the design progress, the minimum and maximum delays of the combinational logic blocks can be either estimated by designers or by some specific tools at early design stages, when detailed information is not available. Then at the final design phase when details are available, an accurate analysis is then performed for latch graph

Reprinted from *Proc. IEEE Int'l Conf. ASICs*, pp. 516–519, Sept. 1992.

extraction.

3 Timing Constraints

To simplify the discussion but without loss of generality, we ignore skew, latch internal propagation delays, setup, hold time, and assume the clocks to all latches are with same active interval. A more detailed formulation can be found in [2].

We shall first study the clock waveform on each latch, since all timing constraints are relative to clock. Let us take a latch, say L_i, as an example. Assume that the clock opens at time ϕ_i (the clocking phase) and after an *active interval* (or pulse width), the clock closes at time Φ_i. Any data arrives within active interval can freely propagate to next stages. The data arrives before the opening edge is latched until the clock opens for propagation. Hence, the data launch time equals to $\max[\phi_i + t_i]$, with t_i being the data arrival time.

To have a correct logic operation occur at certain cycle time, it requires that the input data arrives after the closing edge of previous cycle (otherwise, the data will go to different cycle), *but* definitely before the closing edge of current cycle. These two constraints are usually refered to as *Early Mode Timing Constraint* (EMTC) and *Late Mode Timing Constraint* (LMTC), respectively.

Assume that the latest data arrival time is A_i and the earliest is a_i. Then the legal timing constraints can be formulated as the LMTC

$$A_i \leq \Phi_i \tag{1}$$

and the EMTC

$$a_i \geq \Phi_i - P \tag{2}$$

where P is the given cycle time.

Therefore, if we can compute A_i and a_i, essentially we are done with the timing verification, because the clock waveform is pre-specified and Φ_i is known. So let us first see how A_i can be calculated. Suppose that $L_j \rightarrow L_i$. The latest data arrival time at L_i depends on the latest data launch time (which in turn depends on the latest data arrival time at L_j) plus the worst case path delay, D_{ji}. Note that this calculated value is refered to a time point certain number of cycles later. Let us say the number is δ_{ji}. Then we have to offset the calculated value by $\delta_{ji}P$ such that we are comparing the arrival time at the correct cycle. Henceforth, δ_{ji} is sometimes refered to as a *cycle offset* number. Obviously, A_i is the latest among the worst case data arrival time from all preceding latches, i.e.

$$A_i = \max_{L_j \rightarrow L_i} \{\max[\phi_j, A_j] + D_{ji} - \delta_{ji}P\} \tag{3}$$

Similarly, a_i is the minimum among the earliest from all preceding latches, i.e.

$$a_i = \min_{L_j \rightarrow L_i} \{\max[\phi_j, a_j] + d_{ji} - \delta_{ji}P\} \tag{4}$$

where d_{ji} is the best case path delay between L_j to L_i.

These two recursive equations reveals the complete timing relationships among the latches. The recursive form also suggests that iterative algorithms may be used for solving the problem. We will elaborate the algorithm in Sec. 5.

Note that wave pipelining (or maximum rate pipelining) designs can be verified by the same algorithm with appropriate δ_{ji} setting. The formulation can also be generalized to cover edge-triggered designs and two-phase master-slave latched systems [2].

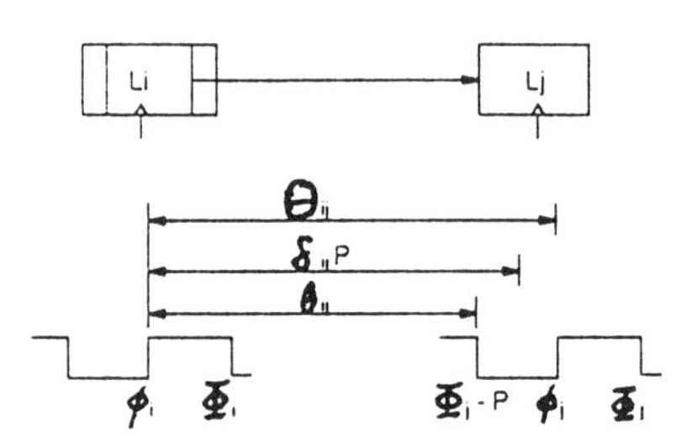

Figure 2: Use the max and min data travel times to identify cycle stealing edges

4 Cycle Stealing

The key difference of level-sensitive to edge-triggered designs is the capability of doing cycle stealing. Although cycle stealing has been mentioned in many papers, it is never been formally defined. In this section, we will try to define what is cycle stealing and use it for design characterization in our timing verifier.

Cycle stealing can be understood better by considering an edge-triggered design with certain paths that has timing violations. The violations occur at the situation that data launched at the opening edge of an edge-triggered arrives after the opening edge of the succeeding one. Since the timing constraints concern more to the closing edge, we can think of the above case as data launch at the closing edge and run over the closing edge of the succeeding one. If now level-sensitive latches are used and the preceding stage has an appropriate short path delay, then the data will arrive earlier than the closing edge and launch earlier; henceforth, it may avoid the supposed to be violation. This is like *stealing* cycle time from preceding stages, and that is why the name. Yet, *cycle stealing* is only true for late mode timing. In early mode, we are doing the contrary to stealing but actually *giving* cycle time to preceding stages. But since *cycle stealing* is more a prevalent term, we will use the same term for both late and early modes.

In essence, the *cycle stealing* edges are the edges that give timing violations if all level-sensitive latches are replaced by edge-triggered latches. To detect the violation in edge-triggered case, we shall first quantify the max and min data travel time, Θ_{ij} and θ_{ij}. If the path delay falls out of the window defined by $[\theta_{ij}, \Theta_{ij}]$, then it violates timing and hence is a *cycle stealing edge*. Depending on whether $D_{ij} > \Theta_{ij}$ or $d_{ij} < \theta_{ij}$, we will name it a *late mode cycle stealing edge* (LCSE) or an *early mode cycle stealing edge* (ECSE).

To help understand the calculation of max and min data travel times, please refer the following formulations to Fig. 2. The max data travel time Θ_{ij} is calculated from the opening edge of a latch to the opening edge of the succeeding latch after certain prespecified clock cycles. Mathematically, the calculation is as follows.

$$\Theta_{ij} \equiv \delta_{ij}P + \phi_j - \phi_i$$

Similarly, the min data travel time θ_{ij} is calculated to the closing edge of the succeeding latch but one cycle earlier, i.e.

$$\theta_{ij} \equiv (\delta_{ij} - 1)P + \Phi_j - \phi_i$$

For an LCSE to have a successful cycle stealing when level-sensitive latches are used, it requires that the data have to arrive, in worst case, before the closing edge with an amount that is greater than the violation, $D_{ij} - \Theta_{ij}$, i.e.

$$\Phi_i - A_i > D_{ij} - \Theta_{ij}$$

Note that for simplification, we have assumed that the data always arrives in the clock active interval. The analysis can be extended to more general cases.

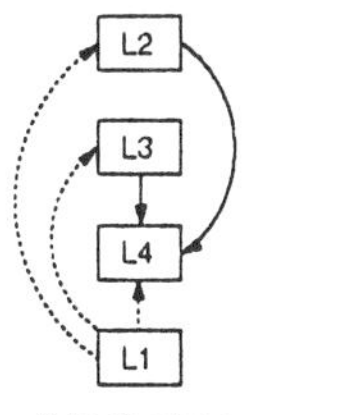

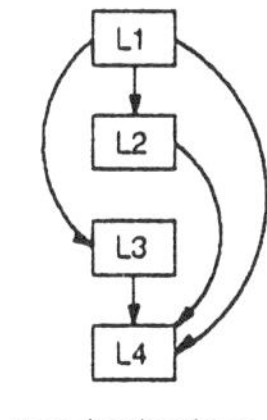

Figure 3: Appropriate latch ordering can reduce the number of back edges and shorten runtime.

i	ϕ_i	a_i	S_{e_i}	A_i	S_{l_i}
0	0	-4.0	0.0	0.0	10.0
1	0	-3.6	0.4	-0.2	6.2
2	5	3.4	2.4	12.1	-1.1
3	5	5.3	4.3	11.4	-0.4
4	5	1.1	0.1	9.1	1.9
5	0	-4.4	-0.4	10.8	-4.8

Table 1: The clocking phase, arrival interval, and slack of each latch.

Similarly, for an ECSE, a successful early mode cycle stealing requires

$$a_i - \phi_i > \theta_{ij} - d_{ij}$$

which means the earliest data should arrive after the opening edge by an amount more than the violation.

In general, the clock design quality can be improved or deteriorated by varying pulse width, phase assignment, etc. .... We can use the number of successful cycle stealing edges to measure design quality. The more unsuccessful cycle stealing edges imply more design effort to make timing correction.

5 Algorithm

As hinted in last section, we shall discuss how to devise an efficient iterative algorithm. From Eq. 3 and Eq. 4, if the earliest and latest data arrival times of all preceding latches are known then the arrival time of a latch can be calculated exactly. However, due to possible loops we may not be able to have all predecessors with known arrival time.

But we can execute the equations in a different way. Instead of waiting for all predecessors to have data arrival time ready, we start from a latch whose arrival time has been calculated or set and use it to update the arrival time of all succeeding latches. Once finishing the update of all succeeding latches, then we mark the status of this latch as DONE. If any successor whose status was DONE but its arrival time is being updated, then we change its status to PERTURBED.

i	j	d_{ij}	D_{ij}	δ_{ij}	θ_{ij}	Θ_{ij}	early CS	late CS
0	1	16.4	19.8	2	16	20	-	-
0	2	3.4	4.8	0	1	5	-	-
0	3	15.4	15.9	1	11	15	-	success
1	2	3.5	5.5	0	1	5	-	success
1	5	8.6	9.7	1	6	10	-	-
2	3	10.3	10.4	1	6	10	-	failure
2	5	4.2	9.8	1	1	5	-	failure
3	4	5.8	8.1	1	6	10	success	-
4	3	10.4	10.8	1	6	10	-	success
4	5	10.6	15.5	2	11	15	failure	success
5	2	5.6	6.1	0	1	5	-	failure

Table 2: Minimum and maximum path delays, data travel times, and cycle offsets

Then we continue the same procedure on the next latch that is not DONE or PERTURBED. At the end, if no latch is perturbed then we should have the answer. Otherwise, another iteration is executed with NEW latch statuses.

Apparently, if we can reduce, or eliminate, the number of PERTURBED latches, we can reduce the number of iterations needed to have the final solution. Let us take the ordering of the latches as the sequence of the latch scanned in each iteration. We found the PERTURBED status is caused by the *back* edges, which point back to some latches which have been DONE before. Thus, we should reduce the number of back edges to shorten runtime. To do so, intuitively, we must follow the data flow direction.

To achieve this, we perform a depth-first search on the latch graph $G(L, E)$ similar to *topological sort* algorithm. During the search, whenever a latch is *finished* (all outgoing edges are traversed), then we push it onto the front of a linked list. After the depth-first search, we then label the latches from 1 to $|L|$, the total number of latches, according to the sequence in the linked list. The complexity of this ordering algorithm is $O(|L| + |E|)$.

It can be easily shown that if the latch graph is *acyclic*, then the order we obtained is exactly the topological order which gives no *back edges*. An example with different ordering is shown in Fig. 3 for illustration. For cyclic latch graphs, the number of back edges of such an ordering is minimized.

To make the result useful, a slack report is important to designers to learn where is timing violation and how much correction is enough. Slack basically is the difference of the actual arrival time to the required time. In our case, *late mode slack* $S_{l_i} = \Phi_i - A_i$, and *early mode slack* $S_{e_i} = a_i - (\Phi_i - P)$. A negative slack indicates timing violation.

In practice, the designers will fix the timing violation whenever it presents. Thus, if the worst case data arrives after the closing edge (hence a violation), we should assume the data launch at (rather than after) the clock closing edge. To early mode violation, we don't have to do anything special since data will launch correctly at the clock opening edge according to calculation.

Now we outline the algorithm below to conclude the discussion.

Algorithm 5.1 (System Timing Verification)

Input: *latch graph* $G(L, E)$.

Output: *early and late mode slacks of each latch.*

S1: index latches from 1 to $|L|$ *as discussed.*

S2: For each L_i, *except primary input, set* $a_i = \Phi_i$, *and* $A_i = \Phi_i - P$. *For primary inputs, set* a_i *and* A_i *according to specification.*

S3: repeat S4 until no PERTURBED latches.

S4:

S4.1: for each latch L_i, *set status*$[L_i]$ = *NEW;*

S4.2: for i=1 to $|L|$ *do*

S4.2.1: for each L_j *that* $L_i \rightarrow L_j$,

S4.2.1.1: if $A_i \leq \Phi_i$ *(no late mode timing violation)*
$X = \max[\phi_i, A_i] + D_{ij} - \delta_{ij} P$ *(see Eq. (3))*
*else (*cut-off *if late mode timing violation)*
$X = \Phi_i + D_{ij} - \delta_{ij} P$

S4.2.1.2: $x = \max[\phi_i, a_i] + d_{ij} - \delta_{ij} P$ *(see Eq. (4))*

S4.2.1.3: if $X > A_j$ *then* $A_j = X$; *if* $x < a_j$ *then* $a_j = x$.

S4.2.1.4: if a_j *or* A_j *is modified and status*$[L_j]$ == *DONE then set status*$[L_j]$ = *PERTURBED;*

S4.2.2: status$[L_i]$ = *DONE;*

S5: compute late and early mode slacks of each latch.

example	#latches	#edges	cycle time (ns)	early mode		late mode		#iter	runtime (sec)
				#CS success	#CS failure	#CS success	#CS failure		
S1	62	518	22	0	0	4	16	1	0.0042
S2	101	1862	22	0	2	61	188	1	0.0139
S3	457	2630	40	0	0	28	156	1	0.0210
S4	467	5814	80	0	0	0	145	1	0.0450
S5	472	12811	80	0	3	94	132	1	0.0984
S6	782	12993	33	0	1	9	47	2	0.1573
S7	1926	53114	67	0	0	318	164	2	0.6177

Table 3: timing analysis results of real examples. (CS: cycle stealing)

Convergence

For any iterative algorithm to be useful, we need to know whether it will converge in finite steps. Our proposed algorithm is guaranteed to converge according to the following arguments. Due to the max and min functions in Eq. 3 and Eq. 4, the value of A_i (and a_i) is monotonically increasing (decreasing) after each iteration. Since we also limit the latest launch time to the clock closing edge Φ_i, there is a upper limit to A_i. Similarly, there is a lower limit to a_i. Combining these two properties, we can easily show that the algorithm will converge in finite number of steps.

We are going to show show that the *loop gains* determine the number of iterations. For simplicity, we discuss only the late mode case. The early mode case is discussed in [2].

For a loop $LOOP_l$, the maximum allowable loop delay Γ is the summation of the max data travel time allocated on each edge, i.e.

$$\Gamma \equiv \sum_{e_{ij} \in LOOP_l} \Theta_{ij}$$

Then the loop gain Δ_l is the difference of the actual total path delays to the maximum allowable loop delay, i.e.

$$\Delta_l = \left(\sum_{e_{ij} \in LOOP_l} D_{ij} \right) - \Gamma$$

Any positive loop gain is an indication of timing violation. This is because the latest arrival time will keep increasing by the same amount after each iteration until run out of bound, and then the algorithm terminates. The number of iterations can be easily calculated by dividing the width of the active interval by the loop gain. Obviously, the maximum number of iterations is determined by the minimum positive loop gain.

6 An Example

We now use an example to illustrate the features of our algorithm. The example consists of 6 latches with 11 edges as shown in Fig. (1). It is a two-phase clocking design with 10 ns cycle time. Every latch has 6 ns of active interval. The clocking phase ϕ_i of L_0, L_1, and L_5 are all equal to 0 (phase-one clock). That of L_2, L_3, L_4 are all equal to 5 ns (phase-two clock). The parameters of latches are summarized in Table (1).

The edge information is displayed on Table (2). Note that e_{02}, e_{12}, and e_{52} are from phase-one clocking to phase-two clocking on the same cycle. Therefore, the corresponding cycle offsets are equal to zero. The edges e_{15}, e_{23}, e_{34}, and e_{43} are from the same phase clocking to the same phase clocking on the next cycle, hence cycle offset is one. The edge e_{03} is from phase-one to phase-two clocking on the next cycle; hence, the cycle offset is one. The edge e_{25} is from phase-two to phase-one clocking on the next cycle; hence, the cycle offset is one. The edge e_{01} is from phase-one clocking to phase-one clocking on two cycles latter to accommodate longer path delays. Therefore, the cycle offset is two. Similarly, the edge e_{45} is from phase-two clocking to phase-one clocking on two cycles latter. Hence, the cycle offset is two. The complication of the clocking scheme is made to demonstrate the generality of the algorithm.

Among the 11 edges, there are two early mode cycle stealing edges and seven late mode cycle stealing edges. Of the two early mode cycle stealing edges, one succeeds and one fails. Of the seven late mode cycle stealing edges, four are successful and three are failed (see Table (2)).

The latch order determined by a depth-first search is as indicated by the index. Although in general, it is difficult to list all feedback loops, this example has only three loops: $(LOOP_1)$ $L_2 \rightarrow L_5 \rightarrow L_2$, $(LOOP_2)$ $L_3 \rightarrow L_4 \rightarrow L_3$, $(LOOP_3)$ $L_2 \rightarrow L_3 \rightarrow L_4 \rightarrow L_5 \rightarrow L_2$. Within these three loops, only $LOOP_1$ has positive loop gain.

The loops mainly affect the success or failure of late mode cycle stealing. The failure of e_{25} and e_{52} is due to the positive loop gain of $LOOP_1$. The failure of e_{23} is because of the saturation of latch L_2. The successes of late mode stealing of e_{43} and e_{45} are contributed by the shorter path delay of e_{34}. Similarly, the success of the late mode stealing of e_{12} is because of shorter path delay of e_{01}. Edge e_{03} is the only late mode successful edge due to the early data launch time from the primary input L_0.

For the early mode, the success of edge e_{34} is because of the excessive minimum path delays of both e_{03} and e_{23}. The failure of e_{45} is because the preceding stage e_{34} does not provide the necessary excessive delay.

For this example, it takes two iterations of the algorithm to converge a to a final solution.

7 Experimental Results

We have implemented the algorithm and tested on many real examples. All examples are with certain number of timing loops. All early mode computations are finished in one iteration, and the late mode analysis takes only slightly more than one iteration in average. The runtime is very short. It takes less than 0.62 seconds for the most complicated examples. This facilitates the possibility of interactive system timing designs. The designers can iteratively modify clocking scheme or do slack assignment for least effort timing correction. For a summary of the experimental results, please see Table (3). In fact, latch graph generation consumes more cpu time, in range of one to ten minutes, which we did not include here. More efficient methods are under investigation.

References

[1] J. K. Ousterhout. A switch-level timing verifier for digital MOS VLSI. *IEEE Transactions on Computer-Aided Design of Integrated Circuits and Systems*, CAD-4, 1985.

[2] Ren-Song Tsay and Ichiang Lin. A system timing verifier for multi-phase level-sensitive clock designs. Technical Report RC 17272, IBM Yorktown Research Center, 1991.

Analyzing Cycle Stealing on Synchronous Circuits with Level-Sensitive Latches

Ichiang Lin John A. Ludwig Kwok Eng
IBM Corporation
Enterprise Systems, Kingston, New York 12401

Abstract

A new analytic method for timing analysis on level-sensitive synchronous circuits is presented. Because a level-sensitive latch acts as combinational logic when the clock signal is active, a data signal arriving at the latch during the clock active interval may propagate through the latch to the next latch stage. In a single phase clocking scheme, a signal propagation that starts at the leading edge of the clock may end before the trailing clock edge but after the leading clock edge of the next cycle. Therefore, a signal propagation delay through combinational logic can be longer than a clock cycle (called cycle stealing) if the next cycle's data launch time can be delayed. Cycle stealing may propagate through several latches without any timing problem if a delay of the data launch time in each stage is allowed.

The proposed method can fully discover the cycle stealing effect on level-sensitive synchronous designs. The algorithm first constructs a latch graph from a timing analysis on the combinational logic, then it analyzes cycle stealing based on overlay timing relationships among latch nodes. A breadth-first search examines all possible cycle stealing among latches. The algorithm also considers the fact that cycle stealing is topology dependent.

The timing analysis program also takes into account the variation of clock width and leading and trailing clock edges in each latch, so the data can be used to assist physical design. The accurate timing information benefits both clock tree and combinational circuit design using level-sensitive clock schemes. The program has been implemented on an IBM RISC System/6000 [1] coupled with an IBM integrated circuit (IC) design system. The results show the benefit of using cycle steal opportunities in the design.

1 Introduction

An accurate timing analysis tool allows the IC design process to detect timing errors and to make proper design decisions. It is vital for a timing-driven design environment to have such tools to capture timing problems in the early design stage. One main concern of current timing analysis tools on level-sensitive synchronous circuits is the inaccurate timing analysis on transparent latches. As a signal arrives at a latch during its active interval, the signal may pass through that latch as combinational logic. Because its arrival time is after the next cycle leading clock edge, the signal propagation delay can be longer than the clock cycle time. A delay of the next cycle signal launch time is allowed if the signal propagation delay time in the next cycle is not critical. Various terms can be found in literature to describe the extension of arrival times, such as retardation, slack borrowing, slack steal, or transparent latch. We use the term "cycle stealing" to emphasize the signal propagation which steals a portion of the next clock cycle time. With the cycle stealing technique employed, a signal propagation which needs more than one clock cycle time is not a late path if its latch time can be extended to the next cycle. Note that other design issues, such as fast paths or early mode delays, need be considered when level-sensitive latches are used in the design.

A cycle stealing may propagate through several latch stages, as shown in a two-stage cycle stealing in Figure 1. Because there is a delay on the latch time in the first clock cycle, the trigger time in the second clock cycle is postponed and so is the latch time in the second clock cycle. Hence, the second clock cycle may steal some clock cycle time from the following clock cycle to allow the signal propagation.

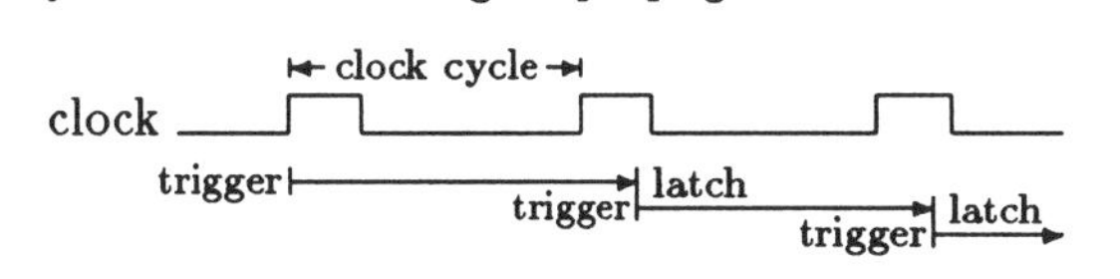

Figure 1: A two-stage cycle stealing.

In the existing timing analysis tools, which tend to look at only a single cycle, all signal propagation delay times between any two latches are bounded in some fixed time. The analysis results tend to be conservative because the cycle steal opportunities are not explored at all. For example, a timing verification, called "trigger-to-trigger test", computes path slacks as the differences between path delays and the cycle time, which is the time between the clock leading edge and its next cycle clock leading edge. The result of this analysis is too pessimistic because some failed paths under this test do not actually violate timing constraints because of the cycle steal effect. Another test, called "trigger-to-latch test", computes path slacks as the difference between the path delays

[1]RISC System/6000 is a trademark of International Business Machines Corporation.

Reprinted with permission from *29th Proc. ACM/IEEE Design Automation Conf.*, I. Lin, J. A. Ludwig, and K. Eng, "Analyzing Cycle Stealing on Synchronous Circuits with Level-Sensitive Latches," pp. 393–398, June 1992.

and the longest possible capture time, which is the time between the clock leading edge and the capturing clock trailing edge. The results tend to be too optimistic because some safe paths under this test actually violate timing constraints. Using both tests, two slack reports may expose these safe paths which pass the trigger-to-trigger test and those failed paths which fail the trigger-to-latch test. However, the results cannot discriminate timing violations from non-violations on these paths which fail the trigger-to-trigger test but pass the trigger-to-latch test.

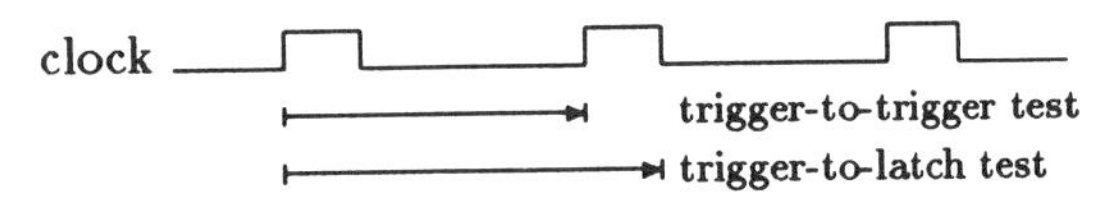

Figure 2: Trigger-to-trigger and trigger-to-latch tests

The difficulty of timing analysis on level-sensitive synchronous circuits is a challenge to both the industry and academia. Earlier study on cycle stealing can be found in [1][2]. The timing rules and properties are investigated in [3][4]. An event propagation method is one which treats a signal propagation in the circuit as an event. Each event propagates through logic gates and then compares its arrival time against the clock constraints. The variations of using this approach can be found in [5]-[8].

An arrival time assertion method is used in several timing analysis tools [9]-[11]. In this method, the arrival time on each latch is initially asserted. The asserted arrival time can be set, for example, at the middle of the active clock interval. Then the asserted arrival times are corrected by an iterative procedure. The change in each latch arrival time assertion is determined by the slack associated with the latch. The slack information is then decided by the asserted arrival times of the latch and its neighboring latches, and the arrival time assertions are adjusted by the slack information on the paths connected to the latch. The iteration may terminate if there is no further change in the assertions or other threshold conditions that are met. So far, no theoretical proof is provided to show that this approach does not sometimes treat a safe path as a timing violation.

The features of this paper are the following:

1. The block-oriented timing analysis method derives a slack report based on the given timing assertions on sources (arrival times) and sinks (required times) of the combinational logic network [12]. Because the accuracy of assertions needs to be ensured, the clock skews among all clock pins have to be minimized. Due to the cycle steal effect, the assertions at the beginning of timing analysis may no longer be valid. Arrival times through combinational logic need to be evaluated for every assertion adjustment. To avoid this problem, we use a latch graph to depict delays among latches. The latch graph contains combinational logic path delays in addition to the slack information. Any change in the arrival time at one latch is directly reflected on its neighboring latch nodes by the latch graph without re-computing combinational logic delays.

2. Most approaches mix the combinational logic path delay calculation with the cycle steal analysis at the same time [5]-[8]. We distinguish the data paths from the clock distribution tree. The path delays from latch to latch through combinational logic are first calculated to construct the latch graph. The cycle steal analysis which follows is based on the given latch graph. This provides a convenient method of synchronous circuit design with cycle steal analysis because the clock waveforms on latches may be adjusted by various design techniques on the clock distribution tree, and the latch graph can serve this need on a higher level basis.

3. We view the cycle steal problem as active interval propagation in contrast to the event-driven or arrival time assertion methods mentioned above. Overlay timing relationships which are used to depict the cycle steal behaviors among latches are derived for latch nodes of the latch graph. The cycle steal behavior on each latch can be clearly observed by its overlay timing diagram. A multiple stage cycle steal is also easily analyzed by this approach. The breadth-first tracing technique is applied for this purpose.

4. This work considers the nature of cyclic graphs which exist in synchronous circuits. In fact, the cycle steal timing analysis is topology dependent [6][13]. The feedback loop structure in synchronous circuits has a timing property which can be reflected in the overlay timing relationship. The tracing procedure analyzes feedback loops in circuits according to this property.

In this paper, the concept of an overlay timing diagram and latch graph is introduced in Sections 2 and 3. The property of feedback loops and their relation to overlay timing diagrams are discussed in Section 4. A breadth-first cycle steal algorithm and its experimental results are also presented in Sections 5 and 6.

2 Overlay Timing Diagram

Consider the latch $L(x)$ which receives the signal from the incoming latch $L(i)$, and sends the signal to the outgoing latch $L(j)$ as shown in Figure 3. Each clock waveform may be different. We assume that the setup time and the hold time hereinafter have been considered in the clock waveforms. A signal which starts at time A in latch $L(i)$ and passes through the combinational logic with delay $d(i,x)$ arrives at time B in latch $L(x)$. The signal which is triggered at time B in latch $L(x)$ passes through the combinational logic with delay $d(x,j)$ and arrives at time C in latch $L(j)$.

In order to illustrate the timing relationships between $L(x)$ and $L(i)$, the incoming latch of $L(x)$, the $L(i)$ clock waveform is shifted as

$t'(i) = t(i) + d(i,x)$,

where $t(i)$ is the time the signal is launched at latch

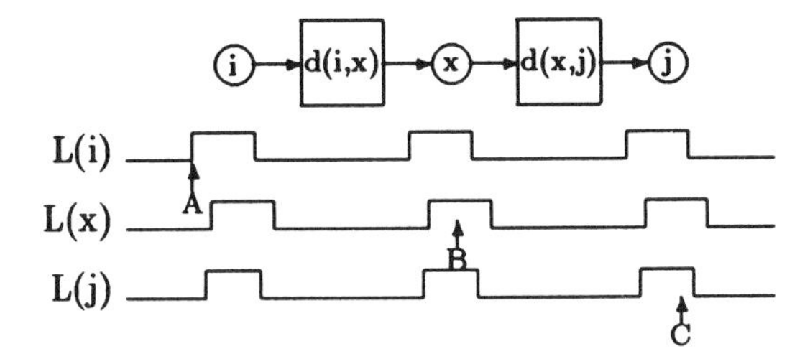

Figure 3: Three latches and their clock waveforms.

$L(i)$, $t'(i)$ is the time the signal arrives at latch $L(x)$, and $d(i,x)$ is the worst case signal propagation delay time from $L(i)$ to $L(x)$ through the combinational logic. In Figure 3, $t'(i)$ is the time B if $t(i)$ is the time A. Figure 4 shows the clock waveform of $L(i)$ overlaid with the clock waveform of $L(x)$ with this shift. Because of the shift of the clock waveform of $L(i)$, the time A in $L(i)$ is aligned with the time B in $L(x)$. We may observe that the amount of time $CS1$ is stolen by the signal launched at time A. In the overlay timing diagram, we define a window which is the clock active-interval of $L(x)$. The launch time D which is located outside of the window fails to arrive before $L(x)$ closes.

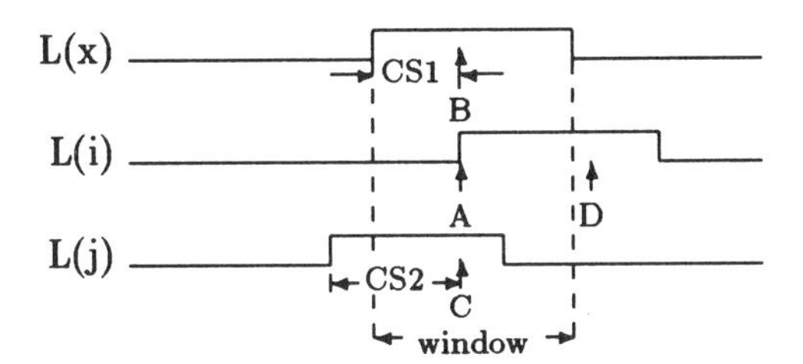

Figure 4: Overlay clock waveforms.

We may also construct the overlay timing relationships between the clock waveforms of $L(x)$ and its outgoing latch $L(j)$ with
$t'(j) = t(j) - d(x,j)$,
where the definitions of $t'(j)$, $t(j)$ and $d(x,j)$ are similar to the above. In Figure 3, $t'(j)$ is the time B if $t(j)$ is the time C. The time C, which is the arrival time of the signal propagation to $L(j)$, is aligned with the times A and B. The signal which launches at the time A in $L(i)$ must steal $CS2$ time from the clock cycle in $L(j)$. Note that the information of delay times and clock waveforms has been embedded in the overlay timing diagram.

For any incoming latch $L(i)$, the overlay clock waveform of $L(i)$ moves to the right with respect to the clock window of $L(x)$ if the delay $d(i,x)$ increases. On the other hand, the overlay clock waveform of the outgoing latch $L(j)$ moves to the left if the delay $d(x,j)$ increases. We can identify the case in which the clock active-interval of $L(i)$ appears to the left of the window of $L(x)$. Figure 5(a) shows no timing violation, labeled *Safe*, between $L(i)$ and $L(x)$, and no cycle steal is necessary. We may also identify the case of a timing violation if the clock active-interval of $L(i)$ appears to the right of the window of $L(x)$. Figure 5(b) shows a late-mode timing violation, labeled *Failed*, between $L(i)$ and $L(x)$, and the cycle stealing doesn't help. A similar way of identifying the timing status between $L(x)$ and $L(j)$ is also shown in Figure 5.

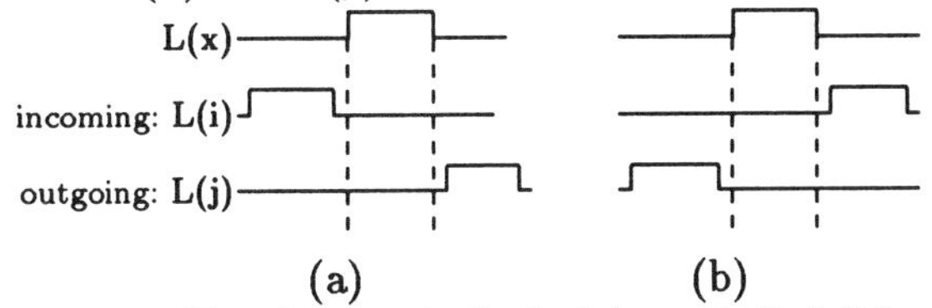

Figure 5: Signal delay is *Safe* (a) or *Failed* (b).

If the $L(i)$ clock active-interval overlaps with the window of $L(x)$, it may or may not violate the timing constraints depending on the relationship of $L(i)$ with the $L(j)$ clock waveform. Figure 6(a) shows that $L(i)$ can do cycle stealing from $L(x)$ and does not need to steal any cycle time from $L(j)$. Figure 6(b) shows that $L(i)$ can do cycle stealing from both $L(x)$ and $L(j)$. Figure 6(c) shows that even though $L(i)$ can steal cycle time from $L(x)$, the signal fails to steal cycle time from $L(j)$ because the signal arrival time on $L(j)$ is after the trailing clock edge of $L(j)$.

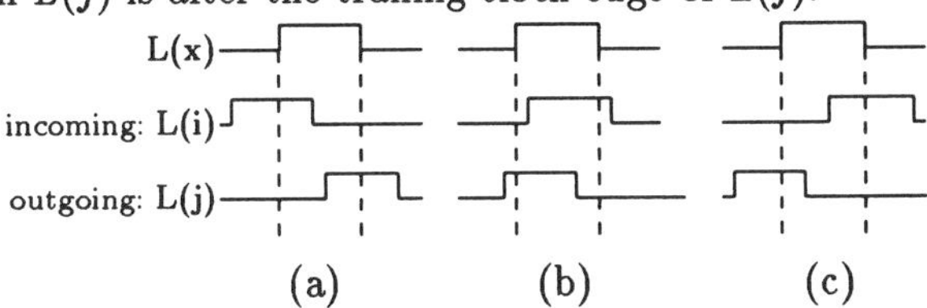

Figure 6: Clock active-intervals overlap with window.

3 Latch Graph Construction

A latch graph is first constructed before applying the cycle stealing timing analysis. The latch graph is a directed graph in which the vertices are the latches, primary inputs and primary outputs, and the edges are the delay times among them. Each edge represents a path from a latch (or primary input), to a latch (or primary output) through the combinational logic network. For each edge, we compute the late-mode worst case delay time from the output pin of the latch (or primary input) to the input pin of the latch (or primary output) through the combinational logic network. For each vertex, we compute the times of the leading and trailing clock edges if the information is available.

From a latch graph, we can construct the overlay timing relationships with incoming latches $L(i)$'s and outgoing latches $L(j)$'s in each latch $L(x)$. Consider the latch graph in Figure 7(a), in which latch $L(b)$ has three incoming latches, $L(a)$, $L(d)$, $L(e)$, and two outgoing latches, $L(c)$ and $L(e)$. Let the clock cycle time be 10 ns. The overlay timing relationships in $L(b)$ are constructed as shown in Figure 7(b). We may label the edge of $L(a)$ to $L(b)$ to be *Safe*, as we mentioned previously. Since the trailing clock edge of $L(c)$ is the earliest time among all trailing clock edges of outgoing latch waveforms, we define a new restricted window as shown in Figure 7(b). We can not determine the status of the edges, $L(d)$ to $L(b)$

and $L(e)$ to $L(b)$, from the overlay timing relationships because their clock active-intervals overlap with the restricted window. It is possible that the trigger time (or latch time) may be set at any point of the overlap.

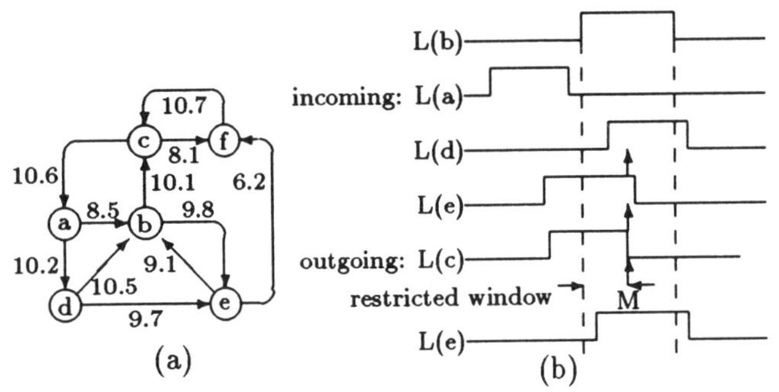

Figure 7: Latch graph and overlay timing diagram.

A backtracking method using the breadth-first search technique is applied to verify the overlay relationships through multiple latch stages. In the backtracking process, the restricted window information, i.e. the time M in Figure 7(b), propagates backward in the graph in order to verify the feasibility of cycle stealing. The backtracking process will verify one latch at a time. The latch under investigation, latch $L(b)$ in Figure 7, is called the root latch.

Consider the example in Figure 7 where a signal launched from $L(d)$ through $L(b)$ to $L(c)$ may be captured by $L(c)$, depending on the trigger time on $L(d)$. Figure 8 shows that the signal launched from the leading clock edge on $L(a)$ arrives at $L(d)$ during the clock active interval, and hence it arrives at $L(c)$ after the trailing clock edge on $L(c)$. The multiple stage cycle steal will be verified by the backtracking process.

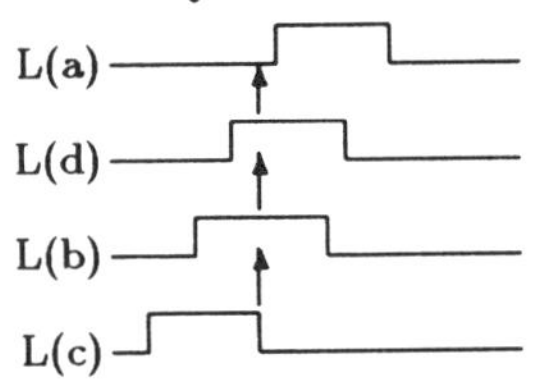

Figure 8: Multiple-stage cycle steal and backtracking process.

4 Feedback Loop Detection in Latch Graph

In the backtracking process, a feedback loop starts at a root latch $L(x)$, traverses the latch graph, and ends at $L(x)$. Every latch in the feedback loop is visited once but the root latch $L(x)$ is visited twice. The delays in some feedback loops cannot satisfy the timing requirements, and the backward propagation of the overlay timing information can detect these timing violations. Let us first examine the case of cycle steal in a single-stage feedback loop. Consider the latch graph in Figure 9(a). Assume the clock cycle time, clk, is known. If the delay time $d(x,x)$ is less than clk, the signal propagation through the feedback loop can be captured by $L(x)$ itself, as shown in Figure 9(b). Figure 9(c) shows the overlay timing relationship if $d(x,x)$ equals clk. Figure 9(d) illustrates the timing violation in which $d(x,x)$ is greater than clk. In Figure 9(d), the restricted window is initially set at

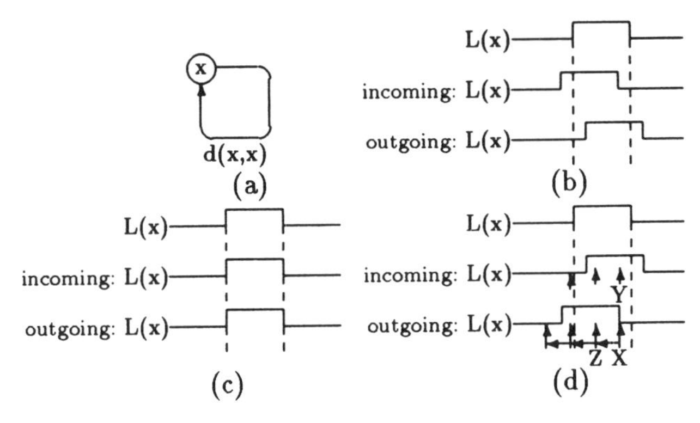

Figure 9: Overlay timing diagrams with a feedback loop.

the time X. The trigger time for the incoming latch $L(x)$ has to be set before the time Y. Because the incoming and outgoing latches actually are the same latch $L(x)$, the latch time for the outgoing latch $L(x)$ follows the time Y to be set on the time Z. We call the new setting on the restricted window 'feedback change'. Hence the restricted window becomes narrow, and it forces the trigger time of the incoming latch to occur earlier. After several iterations, the signal from the incoming latch will not arrive before the restricted window closes.

We may classify the timing violations into two types. The first type is the violation labeled *Failed*, as shown in Figure 5(b). The second type of timing violation is caused by signal propagation in a feedback loop. If the total signal propagation delay time in an N-stage feedback loop exceeds $N \times clk$, where clk stands for clock cycle time, it cannot satisfy the timing constraints.

Lemma 1 When the backtracking process detects a feedback loop, the total delay time in the feedback loop exceeds $N \times clk$ if a feedback change occurs in the overlay timing diagram of the root latch.

Proof: In the backtracking process, the restricted window limit, the time M in Figure 7(b) or the time X in Figure 9(d), is propagated backward through the feedback loop with one clock cycle time ahead through each latch stage. A feedback change implies that the $L(x)$ trigger time at N clock cycle time ahead must launch the signal earlier than the time of the window limit. For instance, the latch $L(x)$ in Figure 9(d) in the previous cycle must launch the signal at the time Z in order to let $L(x)$ capture the signal in the next cycle time. Hence the total delay time through a feedback loop is greater than $N \times clk$.

Lemma 2 For a feedback loop found in the backtracking process, a feedback change occurs in the root latch if the total delay time in the feedback loop exceeds $N \times clk$.

Proof: If the total delay time in a feedback loop is greater than $N \times clk$, the root latch $L(x)$ at N clock cycle ahead must set the trigger time earlier than the time of the window limit, such as the time Z in Figure 9(d). The backtracking calculates the trigger time after the N stage backward search and reports the feedback change.

Theorem 1 For a feedback loop detected by the backtracking process, the total delay time in the feedback loop exceeds $N \times clk$ if and only if a feedback change occurs.

It follows from the Lemmas 1 and 2.

The significance of the above theorem is as follows. The decrement of the feedback change, X to Z in Figure 9, may be very small so that the backtracking may theoretically explore the same feedback loop infinite times before the timing violation occurs. It is unnecessary to actually find the timing violation in which the window limit is set before the leading edge of the incoming clock waveform. In our algorithm, the backtracking process stops when a feedback change in a feedback loop is detected.

5 Cycle Stealing Based Timing Analysis Algorithm

The breadth first backtracking technique is applied to the algorithm. We defined *Safe* and *Failed* for the timing status of edges as mentioned. Here we define three other labels for them. If the signal propagation delay on an edge satisfies the timing requirement by stealing cycle time from the succeeding latch stages, we label it *CS*. If the signal delay can steal cycle time from one latch stage but not multiple latch stages, we label it *CSfailed*. If the timing violation is because of its feedback loop delay exceeding $N \times clk$, we label it *CSloop*. *CSloop*, *CSfailed* and *Failed* violate timing constraints.

Theorem 2 If neither the first type of the timing violations, *Failed* or *CSfailed*, nor the second type of the timing violations, *CSloop*, is found by the algorithm, no timing violation occurs.

Proof: Since both latches and primary outputs will be investigated by the algorithm once, all the timing violations will be explored and labeled. If there is a timing violation at latch $L(x)$, the timing violation will be detected when $L(x)$ is under investigation by the algorithm.

The backtracking process should explore all the feedback loops which are associated with the root latch, and discard the feedback loops which do not belong to the root latch. But in the backtracking process, it may be trapped in the feedback loops which do not go through the root latch, and result in tracing the same feedback loop infinite times. To avoid such a situation, each latch vertex x has a descendant list which records all latches on the paths from latch x to the root latch. If a new explored latch vertex i

```
Read in data file, initialize data array.
Construct overlay timing information.
Check each edge, label Safe and Failed if found.
Label other edges CS.
Initialize descendant list in each vertex latch.
For each latch ID, called root, do the following:
  Put root into FIFO queue.
  While the queue is not null, do the following:
     Get latch ID, called x, from the queue.
     For each incoming i of latch x, do the following:
       If latch i is not in the descendant list of x,
       do the following:
         Check if (i, x) is CSfailed,
                 update negative slack if necessary.
         If the window limit of i is changed,
           update the window limit and tracing info.
           If latch i is root, found feedback loop.
           Else put latch i in the queue.
              Maintain the descendant list of i.
     If found feedback loop
       report feedback loop information,
               update CSloop and slack information.
  Reset descendant lists.
Generate slack report.
```

is in the descendant list of x, a feedback loop which does not belong to the root latch is found, and it is excluded from the search process.

The number of feedback loops is theoretically exponential to the size of the latch graph. In our current implementation, the program does not report all feedback loops in the design. Instead, it averages the negative slack at each latch stage in the feedback loop. For each edge, only the maximum negative slack is recorded.

6 Implementation and Results

The algorithm has been implemented on an IBM RISC System/6000 in C code. The timing analysis program consists of two steps. In the first step, timing information is gathered from the circuit design, and a latch graph is constructed. All circuit design data are retrieved from a design database, and timing information is computed by an IBM timing tool. The second step takes the data, performs the cycle stealing timing analysis and generates a slack report.

For the available test circuits, the experimental results are based on various clock cycles and widths of the clock active-interval. Table 1 shows that the number of failed paths decreases when the width of the clock active-interval increases. As the width of the clock active-interval increases, the number of paths which successfully steal cycle time increases. Although the experiments do not include the early mode timing violations, we can expect that the number of early mode violations increases as the width of the clock active-interval increases. The decision on the width of the clock active-interval needs to take into account

ckt name	# of latches	# of PI's/PO's	clock cycle (ns)	active interval (ns)	# of Safe	# of CS	# of CS loop	# of CS failed	# of Failed	1st step runtime (sec)	2nd step runtime (sec)
decred	10	46/6	20.0	1.0	262	101	0	42	113	6.0	0.38
				2.0	200	243	1	50	24	6.0	0.38
				4.0	178	275	1	50	14	6.0	0.37
bbbtest	53	11/37	25.0	1.0	1778	64	0	5	15	10.0	1.33
				2.0	1717	134	0	1	10	10.0	1.31
				4.0	1514	345	1	2	0	10.0	1.31
aa2gor	237	123/107	80.0	4.0	5669	81	0	0	64	59.0	4.75
				8.0	5641	145	0	0	28	59.0	4.21
				16.0	5580	234	0	0	0	60.0	4.22
sma	251	100/121	50.0	4.0	11643	488	0	3	677	114.0	9.23
				8.0	11022	1250	0	2	537	114.0	9.25
				16.0	9071	3309	0	2	429	113.0	9.42

Table 1: Experimental Results.

both early- and late-mode timing violations.

7 Conclusions

We have presented a new method to fully explore cycle steal opportunities in the timing analysis for level-sensitive synchronous circuit designs. In this paper, we use a single-phase clock discipline to explain the proposed analytic method to the cycle steal problem. It is true that two-phase overlapping clock designs can be analyzed in a similar way. For the application of multiple-phase clocking designs, it is understandable that the information of the combinational logic path delay between any two latches and the interval between the two phases can be reflected on the overlay timing relationships as we mentioned. The parameters, such as clock skew, hold time and setup time, can be represented in the clock waveform, and other parameters, such as clock delay and latch delay can be lumped with combinational delay in the latch graph. The timing analysis also considers the variation of clock width and leading/trailing edges in each latch, so the program is useful in the physical design environment. The scope of this paper does not cover the false path problem in which some critical paths in design are never sensitized by input patterns. The paper does not distinguish the types of signal transitions, such as rising, falling, stable and change. The accurate timing analysis comes from a latch graph construction which may take a higher order computation time than the one of slack computation in the block-oriented approach. This approach, however, is efficient for the dynamic change in circuit designs.

Acknowledgements

The authors would like to thank the referees for their valuable suggestions. The support from Donald R. Thompson, Robert F. Hatch, Cyril A. Price, and our colleagues is gratefully acknowledged.

References

[1] V. D. Agrawal, "Synchronous Path Analysis in MOS Circuit Simulator", Proc. 19th Design Automation Conf., pp. 629-635, 1982.

[2] J. K. Ousterhout, "A Switch-Level Timing Verifier for Digital MOS VLSI", IEEE Trans. Computer-Aided Design, vol. CAD-4, no. 3, pp. 336-349, July 1985.

[3] S. H. Unger and C. Tan, "Clocking Schemes for High-Speed Digital Systems", IEEE Trans. on Computers, vol. C-35, no. 10, pp. 880-895, October 1986.

[4] M. R. Dagenais and N. C. Rumin, " On the Calculation of Optimal Clocking Parameters in Synchronous Circuits with Level-Sensitive Latches", IEEE Trans. Computer-Aided Design, vol. 8, no. 3, pp. 268-278, March 1989.

[5] T. G. Szymanski, "LEADOUT: A Static Timing Analyzer for MOS Circuits", Proc. 1986 IEEE Int. Conf. on CAD, pp. 130-133, Nov. 1986.

[6] M. Glesner, J. Schuck, R. B. Steck, "SCAT - A New Statistical Timing Verifier In a Silicon Compiler System", Proc. 23rd Design Automation Conf., pp. 220-226, 1986.

[7] D. E. Wallace and C. H. Sequin, "ATV: An Abstract Timing Verifier", Proc 25th Design Automation Conf., pp. 154-159, 1988.

[8] J. J. Cherry, "Pearl: A CMOS Timing Analyzer", Proc. 25th Design Automation Conf., pp. 148-153, 1988.

[9] N. Weiner and A. Sangiovanni-Vincentelli, "Timing Analysis In A Logic Synthesis Environment", Proc. 26th Design Automation Conf., pp. 655-661, 1989.

[10] K. A. Sakallah, T. N. Mudge and O. A. Olukotun, "Analysis and Design of Latch-Controlled Synchronous Digital Circuits", Proc. 27th Design Automation Conf., pp. 111-117, 1990.

[11] D. Hathaway and J. Goetz, "Timing Analysis Using Slack Stealing", IBM Invention Disclosure, 1990.

[12] R. B. Hitchcock, "Timing Verification and the Timing Analysis Program", Proc. 19th Design Automation Conf., pp. 605-615, 1982.

[13] J. Beausang and A. Albicki, "A Method to Obtain An Optimal Clocking Scheme For A Digital System", Proc. Int. Conf. Computer Design: VLSI Comput., pp. 68-72, Oct. 1985.

Part 5

Specification of the Optimal Timing Characteristics of Clock Distribution Networks

Clock Skew Optimization

John P. Fishburn

Abstract—**This paper investigates the problem of improving the performance of a synchronous digital system by adjusting the path delays of the clock signal from the central clock source to individual flip-flops. Through the use of a model to detect clocking hazards, two linear programs are investigated: 1) Minimize the clock period, while avoiding clock hazards. 2) For a given period, maximize the minimum safety margin against clock hazard. These programs are solved for a simple example, and circuit simulation is used to contrast the operation of a resulting circuit with the conventionally clocked version. The method is extended to account for clock skew caused by relative variations in the drive capabilities of N-channel versus P-channel transistors in CMOS.**

Index Terms—**Clocking, clock skew, finite-state machines, linear programming, optimization, synchronous circuits.**

I. Introduction

Synchronous circuit designers ordinarily try to eliminate clock skew, which may be defined as variations in the delays from the central clock source to the flip-flops (FF's) of the system. This effort can involve equalization of wire lengths, careful screening of off-the-shelf parts, symmetric design of the distribution network, and design guidelines to eliminate skew due to process variations [1], [2]. Clock skew can limit the clocking rate of a synchronous system or cause malfunction at any clock rate. Some static timing analyzers [3] detect incorrect operation in the presence of skew, and allow user tuning of clock and data paths until correct operation is verified.

In this paper, we examine the following question: How can a synchronous system be improved by adjusting the delays between the central clock and individual FF's? We have two goals in mind: 1) To speed up the clock rate at which the circuit will function correctly. 2) For a given clocking rate, to maximize the margin of safety against circuit malfunction due to clocking hazards. The analysis will use, for simplicity, positive-edge-triggered D-flip-flops.

II. Clock Hazards

A. Double-Clocking and Zero-Clocking

In 1965, Cotten [4] described a "data race" mechanism in which clock skew can cause a synchronous system to fail. In Fig. 1, if $x_j > x_i + d_{ij}$, then when the positive clock edge arrives at FF_i, the data "race ahead" through the fast path, destroying the data at the input to FF_j before the clock gets there. When the clock edge finally arrives at FF_j, the wrong data are clocked through. Since the data are clocked through two FF's with one clock edge, this has also been called *double-clocking*.

Analogously, *zero-clocking* can be used to designate the case when the data reach the FF too late relative to the *next* clock edge. This occurs in Fig. 1 when $x_i + d_{ij} > x_j + P$, where P is the clock period.

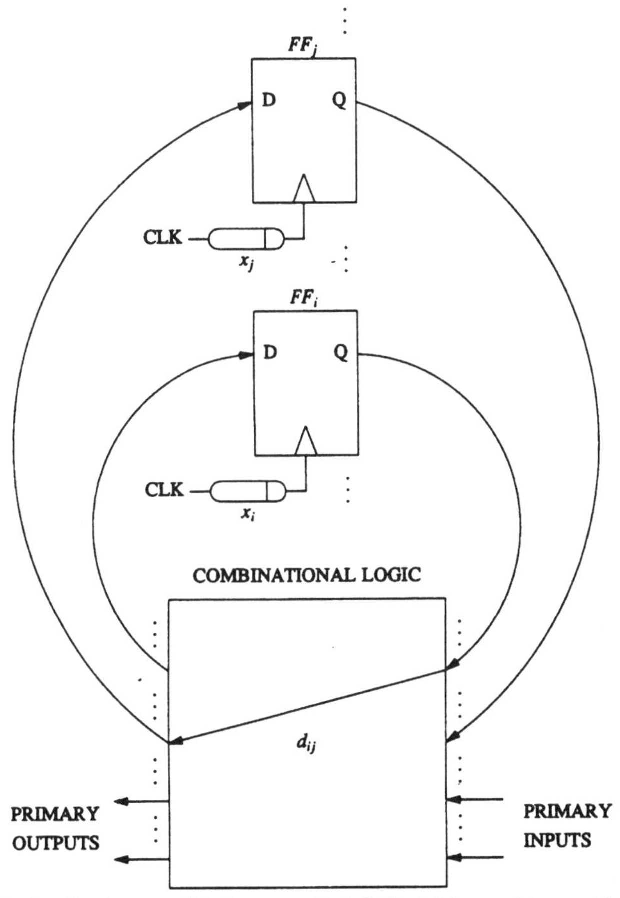

Fig. 1. Synchronous digital system with individual delays x_i interposed between the central clock source and the FF's of the synchronous system. For each FF pair (FF_i, FF_j), bounds MIN(i, j) and MAX(i, j) are computed for the time-of-flight d_{ij} from FF_i to FF_j.

B. General Model for Detecting Clock Hazards

This section develops a set of inequalities that can tell us, in general, whether either of the above hazards is present. This development is similar to that found in [1] and [5], except that a minimum and maximum delay is calculated for every source/destination FF

Reprinted from *IEEE Trans. Comput.*, vol. 39, no. 7, pp. 945–951, July 1990.

pair, as opposed to assuming overall delay bounds between banks of FF's. This is necessary for the linear programs that follow. Fig. 1 illustrates the basic components of the model:

1) FF_i receives the central clock delayed x_i by its own delay element. A later section will outline a method for delay line construction. Depending on the technology, there is some minimum delay MIN_DEL that can be generated: $x_i \geq \text{MIN\_DEL}$. We will assume some uncertainty in the clock delays. There will be two constants, $0 < \alpha \leq 1 \leq \beta$, with the property that if the nominal clock delay is x_i, then the actual clock delay x can vary from clock edge to clock edge, but must always fall in the interval $\alpha x_i \leq x \leq \beta x_i$.

2) In order for a FF to operate correctly when the clock edge arrives at time x, it is assumed that there are constants SETUP and HOLD such that correct input data must be present and stable during the time interval $(x - \text{SETUP}, x + \text{HOLD})$.

3) Timing conditions attach to each primary input and output. It is assumed for simplicity that these inputs and outputs are connected to FF's outside the synchronous system, each of which is controlled by the central clock source through its own delay line. Thus, all timing paths begin and end at a FF, making unnecessary a separate terminology for the I/O constraints. $FF_1, \cdots, FF_K$ denote the internal FF's, and $FF_{K+1}, \cdots, FF_L$ denote the external FF's. We do not have the ability to vary the clock delays to FF's external to the synchronous system, so while $x_1, \cdots, x_K$ are variables, $x_{K+1}, \cdots, x_L$ are constants determined by the circuit and its environment.

4) For $1 \leq i,\ j \leq L$, we compute lower and upper bounds $\text{MIN}(i, j)$ and $\text{MAX}(i, j)$ for the time that is required for a signal edge to propagate from FF_i to FF_j. Since it is possible that multiple paths exist from FF_i to FF_j, $\text{MIN}(i, j)$ and $\text{MAX}(i, j)$ must be computed as the minimum and maximum of these path delays. If no such path exists, we define $\text{MIN}(i, j) = \infty$ and $\text{MAX}(i, j) = -\infty$ for notational convenience. Although the data delay internal to the FF itself could be included in SETUP and HOLD, we choose to include it in MIN and MAX. Besides simplifying the notation, this is desirable because the FF internal delay can be variable due to data-dependent delay, or due to the construction and output loading of the FF's.

To avoid double-clocking between FF_i and FF_j, the data edge generated at FF_i by a clock edge must arrive at FF_j no sooner than a period of time HOLD after the latest possible arrival of the same clock edge. The earliest that the clock edge can arrive at FF_i is αx_i, the fastest propagation from FF_i to FF_j is $\text{MIN}(i, j)$. The latest arrival time of the clock at FF_j is βx_j. Thus, we have

$$\alpha x_i + \text{MIN}(i, j) \geq \beta x_j + \text{HOLD}. \tag{1}$$

To avoid zero-clocking, the data edge generated at FF_i by a clock edge must arrive at FF_j no later than SETUP amount of time before the earliest arrival of the next clock edge. The latest that the clock edge can arrive at FF_i is βx_i, the slowest propagation from FF_i to FF_j is $\text{MAX}(i, j)$, the clock period is P, and the earliest arrival time of the next clock edge at FF_j is $\alpha x_j + P$. Hence,

$$\beta x_i + \text{SETUP} + \text{MAX}(i, j) \leq \alpha x_j + P. \tag{2}$$

III. Two Linear Programs

A. Minimize P Subject to Clocking Constraints

If we desire to make the period P as short as possible while satisfying the system of inequalities (1) and (2), and if the values SETUP, HOLD, α, β, $\text{MAX}(i, j)$, $\text{MIN}(i, j)$ and the x_i for $i = K+1, \cdots, L$ are assumed to be constant, while P and the x_i for $i = 1, \cdots, K$ are variable, then what we have is a *linear program*. In a standard form [6], this system is as follows:

LP_SPEED: minimize P subject to

$$\alpha x_i - \beta x_j \geq \text{HOLD} - \text{MIN}(i, j), \qquad \text{for } i, j = 1, \cdots, L;$$

$$\alpha x_j - \beta x_i + P \geq \text{SETUP} + \text{MAX}(i, j), \qquad \text{for } i, j = 1, \cdots, L;$$

$$x_i \geq \text{MIN\_DEL}, \qquad \text{for } i = 1, \cdots, K.$$

B. Maximize Minimum Margin for Error in Clocking Constraints

Clocking hazards can be particularly vexatious because they are potentially intermittent. A system on the verge of double-clocking, for instance, might pass system diagnostics but malfunction at unpredictable times due to fluctuations in ambient temperature or power supply voltage. One way to armor a system against this problem

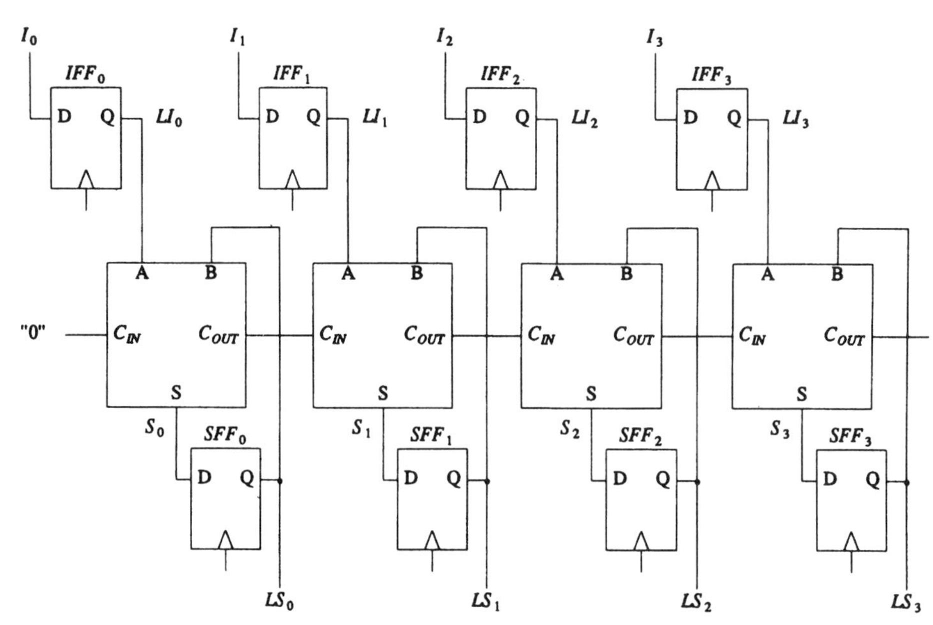

Fig. 2. A ripple-carry adder with input and accumulation register. Various delays are interposed between the central clock source and the clock inputs to the individual FF's.

would be to increase the values of the constants SETUP and HOLD in LP_SPEED, at the cost of an increase in the clock period P. If, however, P is a fixed quantity it would be desirable to maximize the minimum over all the constraints of the *slack*, or amount by which the inequality is satisfied. This converts the problem into a *maximin* problem [6], which can be stated as a linear program in the following way. Introduce a new variable M, which is added to each of the main constraint inequalities so that when it is maximized by the program, it will be the minimum slack over all the inequalities. P is now a constant. The variables M and x_i, $i = 1, \cdots, K$ are to be determined.

LP_SAFETY: maximize M subject to

$$\alpha x_i - \beta x_j - M \geq \text{HOLD} - \text{MIN}(i, j), \quad \text{for } i, j = 1, \cdots, L;$$

$$\alpha x_j - \beta x_i - M \geq \text{SETUP} + \text{MAX}(i, j) - P, \quad \text{for } i, j = 1, \cdots, L;$$

$$x_i \geq \text{MIN\_DEL}, \quad \text{for } i = 1, \cdots, K.$$

An additional benefit of LP_SAFETY is that it maximizes the tolerance of a system to variations in the speed of its parts, thereby lowering manufacturing costs. For a given reliability that is desired in a machine built from off-the-shelf parts, less stringent screening is required. For VLSI systems, LP_SAFETY would serve to improve yield in the face of process variations by centering the design away from clocking hazards.

IV. A Simple Example

At the current time, there exists no CAD tool that can automatically perform clock skew optimization. However, with the help of the circuit timing simulator ADVICE [8] and the PORT Linear Programming package [10], a simple circuit has been optimized by hand. This is a 4-bit ripple-carry adder with accumulation and input register in 1.25-μm CMOS (Fig. 2). For simplicity, the carry-in is held at zero, and the carry-out is ignored. The four primary inputs I_{0-3} feed the input FF's IFF_{0-3}, whose outputs LI_{0-3} are the A inputs of the adder. The four adder outputs S_{0-3} are fed to the sum FF's SFF_{0-3}, whose outputs LS_{0-3} are primary outputs. LS_{0-3} are also fed back into the B inputs of the adder.

A. Circuit Characterization

Numerous ADVICE runs yielded estimates for MAX and MIN. It was found that under various conditions, the delay from IFF_i or SFF_i to SFF_j was always bounded by $\text{MIN}(i, j) = 2.1 + 1.5^*(j - i)$ ns and $\text{MAX}(i, j) = 3.2 + 2.7^*(j - i)$ ns. It was assumed that FF's external to the circuit had no clock delay. Delay from an external FF to an input FF was assumed to be exactly MAX = MIN = 6 ns, while from a sum FF to an external FF was exactly MAX = MIN = 3 ns. SETUP and HOLD were both set equal to 1 ns. The delay lines were constructed by the method described in Section VII-C, with fine-tuning by iterative simulations with the delay-lines driving their actual loads in the final circuit. The *unskewed adder* was the accumulator described above, with all eight FF's connected directly to the central clock source. The *skewed adder* was the same accumulator with delay lines interposed between the central clock and the individual FF's.

B. Linear Program Solutions

With the measured circuit parameters as given above, LP_SPEED and LP_SAFETY were solved by the PORT linear programming software. The program's solution, in nanoseconds, for LP_SPEED was as follows. The delays to IFF_{0-3} were 0.00, 0.05, 1.55, and 3.05, and to SFF_{0-3} were 0.00, 0.05, 1.55, and 4.15. P was 8.15. For $P = 13.0$, the solution for LP_SAFETY was: The delays to IFF_{0-3} were 0.82, 1.50, 2.18, and 2.86, and to SFF_{0-3} were 0.00, 0.68, 1.36, and 2.04. M, the safety margin, was 1.92. For both LP_SAFETY and LP_SPEED, the limiting constraints at the optimum points came from paths beginning or ending outside the circuit.

C. Performance of the Resulting Circuit

Since LP_SAFETY also speeds up the circuit, its solution was selected to construct the delay lines in the skewed adder. An ADVICE simulation at nominal conditions exhibited correct behavior by both adders. Temperature was 25°C, and the clock period was 10 ns. The clock was initially stopped then cycled for three ticks separated by the given clock period, and then stopped again. The sum register was initialized to 0001, the input register to 1111. The values 1111, 0001, and 0001 were made available at the primary inputs for loading into the input register on the three ticks. This caused a signal to travel the length of the carry chain on both the first and second clock ticks, with values 0000, 1111, and 0000 appearing after the three ticks of the clock on the LS_{0-3} outputs.

A series of simulations was then performed with progressively shorter clock periods, but holding constant the other conditions, to determine the minimum feasible clock period for each circuit. The unskewed adder worked correctly down to a clock period of 9.5 ns before zero-clocking. The skewed adder was able to work correctly at a clock period of 7.5 ns before zero-clocking, or 2.0 ns less than the critical path delay.

How is it possible that the circuit can be run at a clock period less than the critical path delay? The answer, of course, is that a logic path can act as a delay line, containing more than one signal wavefront at a single instant. The simulation of the skewed adder showed that this was in fact happening in the carry chain at the 7.5 ns clock period. This phenomenon has been exploited in a pipelining technique known as *maximum-rate pipelining* [1], [5], [9], [13]. In maximum-rate pipelining, the clock period is determined not by the maximum path delay through the logic, but by the difference between the maximum and minimum delays. When the clock runs at this maximum rate, the pipeline contains more bits of information than FF's. For this reason, the clock in a maximum-rate pipeline cannot be single-stepped or even slowed down significantly. In the present scheme, by contrast, single-stepping is always possible. Any sequence $P_1, P_2, \cdots$ of intervals between clock edges will drive the circuit correctly as long as each P_i is large enough to satisfy (2).

A second series of simulations stressed the adders in the direction of double-clocking by adding variable amounts of additional clock delay to SFF_3 (in the case of the skewed adder, in addition to the 2.04 ns already present). Other conditions were kept unchanged from the nominal case. With clock period held constant at 10 ns, the unskewed and skewed adders were able to tolerate 3.0 and 4.1 ns clock delay to SFF_3, respectively, before double-clocking took place between IFF_3 and SFF_3. The extra resilience of the skewed adder was due to the fact that, in each bit position i, LP_SAFETY had assigned more clock delay to IFF_i than to SFF_i, thus centering the circuit away from double-clocking.

V. Some Practical Considerations

A CAD tool that optimizes clock delays would include a static timing analyzer that could compute $\text{MIN}(i, j)$ and $\text{MAX}(i, j)$ between an input i and an output j of combinational logic. In this regard, most static timing analyzers that exist today are deficient in two respects. First, lower bounds on delay are usually not computed, although formulas for lower bounds on RC network delays are available [16]. Second, delays are not computed per input–output pair. Rather, the user specifies particular data-ready times at inputs, and the analyzer computes resulting output times. A static timing analyzer that can compute maximum delays in this manner could be used as a subroutine in an algorithm to compute MAX. For any input i, $\text{MAX}(i, j)$ could be computed for all j by setting the data-ready time of input i to zero, and all other inputs to $-\infty$. $\text{MAX}(i, j)$ can then be assigned the resulting data-ready time at output j. A similar procedure could compute MIN using a static timing analyzer that could compute minimum delays.

If a system is to be built from off-the-shelf parts, only one clock delay variable can be attached to all the FF's controlled by a package pin. This may force a suboptimal solution, but the linear program can still correctly model the situation by coalescing into one the variables associated with these FF's.

Calculation of logic delays involves many uncertainties. The delays of both combinational logic and the clock distribution network are affected by process, temperature, noise, and voltage variation. Even if these physical parameters could be held constant, inaccuracies remain in the models used by the static timing analyzer. These uncertainties limit the performance that can be obtained from optimizing clock delays. It should be stressed, however, that these uncertainties have been explicitly taken into account in the linear programs presented here, through the use of upper and lower bounds on delay. The only delay uncertainty that does not equally affect the performance of conventionally clocked systems is that associated with the added clock delay lines. One scheme to generate these delays might be as follows. Between the central clock source and the FF's, the distribution network would be similar to that in conventional clock distribution networks. Every effort would be made to equalize the clock delay in this network. This network would have a certain distribution delay DD, and the linear program would have the constraints $x_i \geq$ MIN_DEL replaced by $x_i \geq$ DD + MIN_DELl. The variable delay elements (Section VII-C) would be inserted between the leaves of this network and the actual FF's. This scheme allows the variable part of each delay to be in one spot (at the FF), avoiding the additional uncertainties that would be introduced if it were distributed.

In general, partitioning a system and optimizing the pieces separately can give suboptimal results. We saw this with the adder, where the speedup was limited not by the maximum-rate pipelining limit, but by boundary constraints. The best results can be obtained by encompassing as much as possible in the synchronous system that is to be optimized. In general computer systems, the boundary might be pushed out to include all of the most tightly coupled, delay-critical parts of the system. At the boundary, additional FF's could be used to decouple the optimized system from the external world.

Adding FF's to a system before optimizing might result in increased throughput as a result of a kind of "poor man's pipelining." For example, one might double the number of FF's in a system by replacing every FF with two connected in series. The optimization procedure would tend to assign more delay to the first one of the pair. If conditions are right (i.e., if MIN is close to MAX in the right places), the end result would be to almost double the clock rate, and hence the throughput. The effect is similar to conventional pipelining because although the transit time of a single datum through the system is not decreased, the throughput is increased. Conventional pipelining requires the designer to partition combinational logic into stages of comparable delay. In poor man's pipelining, on the other hand, the clock skew serves to compensate for inequalities among stage delays. This relaxed constraint on stage delays might reduce the number of FF's required by allowing partitions with fewer interpartition signals.

VI. Clock Skew Versus Retiming

As Fig. 3 illustrates, adding clock delay to an FF has an effect similar to movement of the FF backwards across combinational-logic module boundaries [11]. This movement, called *retiming*, can be controlled by a mixed-integer linear program [12] to minimize clock period. In this sense, clock skew and retiming are continuous and discrete optimizations with the same effect. Although Fig. 3 illustrates a situation in which the designer can choose between the two transformations, the two methods can in general complement each other:

1) Since retiming moves FF's across discrete (and perhaps large) amounts of logic delay, the resulting system can still benefit from (smaller amounts of) clock skew.

2) Retiming to minimize clock period may cause an unacceptable increase in the number of FF's. Retiming to minimize the number of FF's [12] may be preferred if speed can be bought back more cheaply with clock skew.

3) Retiming does not address the problem of double-clocking or process-dependent timing variation (although presumably it could be extended to do so).

4) Efficient linear programming software packages are widely available, but mixed-integer linear programming packages are not.

5) FF's whose inputs are the primary inputs of a system can be skewed but not retimed.

6) FF's integrated with combinational logic in off-the-shelf parts can be skewed (in tandem, if controlled by a single pin) but not retimed.

VII. Process-Dependent Clock Skew in VLSI

It is assumed in this section that the synchronous system to be optimized is completely contained on a single chip. As has been mentioned, the model for detecting clock hazards has a built-in allowance for uncertainty in both logic and clock delays. This allowance and the design centering given by LP_SAFETY can make a system more robust in the face of delay variations. Unfortunately, this allowance is more conservative than necessary for some of the sources of delay variance. For example, the speed of logic gates can vary considerably due to unavoidable variations in the manufacturing process. But a single silicon chip will tend to have very little process variation across its surface. For example in CMOS, the drive capability of a fixed-size *N*-channel or *P*-channel MOS transistor (NFET or PFET)

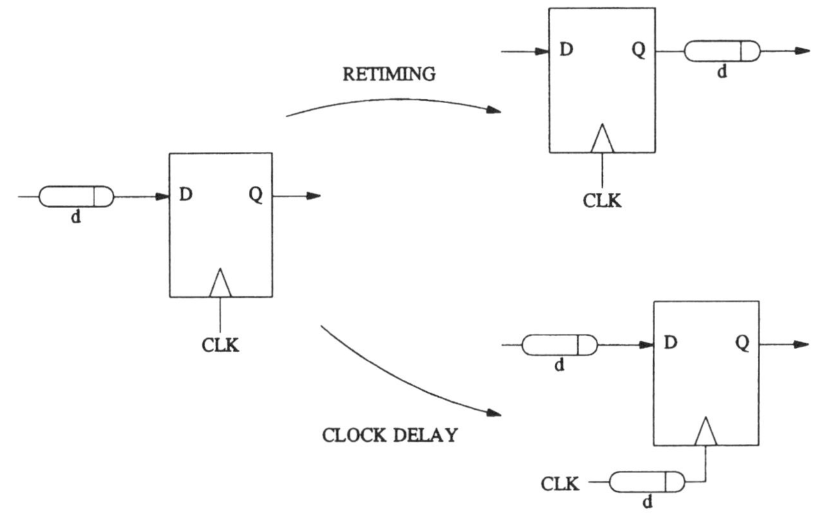

Fig. 3. Retiming and clock delay transformations applied to the same circuit.

varies insignificantly as a function of its position on the chip. The clock hazard model, on the other hand, allows the delay of a fixed-size logic gate with fixed output load to vary from place to place, and even from clock edge to clock edge. A more reasonable model would account explicitly for process variation, and restrict the uncertainties modeled by gate delays to the more nearly random causes, such as temperature variation or noise. We will now briefly sketch how the linear programs might be modified to take into account process variation.

A. Uniform Variation of Gate Delays

If all the gate delays on a chip vary uniformly, then there is no problem. It is easily seen that if a chip is free of clock hazards, if the delays of all the gates on a chip increase or decrease by some factor, and if the clock rate changes by the same factor, then the resulting chip will also be free of clock hazards.

B. Independent Variance of PFET and NFET Drive Capability in CMOS

A more complicated situation occurs in CMOS, where the current drive of NFET's varies somewhat independently of the current drive of PFET's. Shoji [2] solves the problem of the resulting clock skew in CMOS VLSI by means of a more detailed accounting and control of the sources of clock delay. We will call this the "NP-matched-clock solution." Instead of assigning a single delay x_i to the passage of an input-rising edge through a given chain of inverters i in the clock distribution network, the delay is separated into two parts: the pulldown delay n_i and the pullup delay p_i. n_i is the sum of the delays of the odd inverters in the chain, each of whose NFET's is pulling down its output, and p_i is the sum of the delays of the even inverters, each of whose PFET's is pulling up its output. The delay of a gate is defined to be the difference in time between when the input and output cross 50% of voltage swing. Rather than size the transistors in all the chains to equalize $n_i + p_i$, as is done conventionally, a more stringent matching is performed. The transistors in all the chains are sized to make all the n_i equal, and to simultaneously make all the p_i equal. The result is that all the chain delays track each other in the face of independent variance in the NFET and PFET current drive. If the NFET and PFET current drives change by factors of ND and PD, respectively, all the chain delays remain equal:

$$\frac{n_i}{\mathrm{ND}} + \frac{p_i}{\mathrm{PD}}. \quad (4)$$

Many ADVICE runs for a 1.75-μm CMOS process showed [2] that (4) was reasonably accurate as long as the pullup and pulldown delays were kept balanced by satisfying the inequalities

$$0.35 \leq \frac{p_i}{n_i + p_i} \leq 0.5 \quad (5)$$

which are equivalent to the linear inequalities $n_i - p_i \geq 0$ and $13p_i - 7n_i \geq 0$.

By replacing the clock delay variables x_i with the pulldown and pullup variables n_i and p_i, we can modify LP_SAFETY to take into account the NFET/PFET process variation. The resulting program, called CMOS_LP_SAFETY, will allow a finer control over the generation of clock delay, since n_i and p_i are controlled separately, rather than their sum. A similar transformation can be used to convert LP_SPEED into a program CMOS_LP_SPEED that minimizes clock period while avoiding clock hazards across all CMOS processes. It is necessary to sample the NFET/PFET process parameter space at a finite number of points: ND takes on A values $\mathrm{ND}_1, \mathrm{ND}_2, \cdots, \mathrm{ND}_A$, and PD independently takes on B values $\mathrm{PD}_1, \mathrm{PD}_2, \cdots, \mathrm{PD}_B$. Each sample (a, b) can be considered a separate process, and is characterized by its NFET and PFET relative drive parameters, ND_a and PD_b. In [2], for example, the process parameter space is broken up into nine processes, with ND taking on the values 0.556, 1.000, and 1.730, and PD independently taking on the values 0.620, 1.000, and 1.630.

The constraint inequalities for CMOS_LP_SAFETY can now be written down by repeating the constraint inequalities of LP_SAFETY for each process. Since SETUP, HOLD, MAX, and MIN will have different values in different processes, they are now functions of a and b: SETUP(a, b), HOLD(a, b), MAX(i, j, a, b) and MIN(i, j, a, b). MIN_DEL will be replaced by minimums for the delay line pulldown and pullup delays $\mathrm{MIN\_DEL}_D$ and $\mathrm{MIN\_DEL}_U$. For a given P, we wish to maximize M by adjusting the pulldown and pullup variables n_i and p_i, for $i = 1, \cdots, K$:

CMOS_LP_SAFETY: maximize M subject to

$$\alpha\left(\frac{n_i}{\mathrm{ND}_a} + \frac{p_i}{\mathrm{PD}_b}\right) - \beta\left(\frac{n_j}{\mathrm{ND}_a} + \frac{p_j}{\mathrm{PD}_b}\right) - M \geq \mathrm{HOLD}(a, b) - \mathrm{MIN}(i, j, a, b) \quad (6)$$

$$\alpha\left(\frac{n_j}{\mathrm{ND}_a} + \frac{p_j}{\mathrm{PD}_b}\right) - \beta\left(\frac{n_i}{\mathrm{ND}_a} + \frac{p_i}{\mathrm{PD}_b}\right) - M \geq \mathrm{SETUP}(a, b) + \mathrm{MAX}(i, j, a, b) - P \quad (7)$$

for all processes (a, b) and for $i, j = 1, \cdots, L$;

$$n_i \geq \mathrm{MIN\_DEL}_D \text{ and } p_i \geq \mathrm{MIN\_DEL}_U, \quad \text{for } i = 1, \cdots, K;$$

$$n_i - p_i \geq 0 \text{ and } 13p_i - 7n_i \geq 0, \quad \text{for } i = 1, \cdots, K.$$

Both α and β in the above program can be much closer to 1 than their counterparts in the original LP_SAFETY program, since they no longer have to account for process variation in the clock delay lines. If in fact $\alpha = \beta = 1$, and as long as $\mathrm{MIN}(i, j, a, b) \geq \mathrm{HOLD}(a, b)$ for all processes (a, b) and FF's i and j, and if P is big enough, the feasible region of CMOS_LP_SAFETY is nonempty. Simply set all the pulldown delays n_i equal to a single nonnegative constant, and all pullup delays p_i to some other nonnegative constant, such that inequality (5) is satisfied. This equalization of pulldown and pullup delays across all clock delay lines is in fact the NP-matched-clock solution. Unlike the NP-matched clock solution, the solution to CMOS_LP_SAFETY does not necessarily have the property that all clock delays track each other across all process variations. However, both the solution to CMOS_LP_SAFETY and the NP-matched-clock solution are in the feasible region of CMOS_LP_SAFETY, and so both represent solutions that avoid clocking hazards in the face of all NFET/PFET process variations. In general, however, the solution to CMOS_LP_SAFETY enjoys a greater margin-of-safety M, and hence is relatively more immune to other kinds of delay variation. Likewise, the solution to CMOS_LP_SPEED allows a higher clock rate than the NP-matched-clock solution.

C. Construction of Clock Delay Lines with Given Pullup and Pulldown Delays

This section demonstrates the construction of delay lines in 1.25-μm CMOS with various values of n and p, the pulldown and pullup delays. It is not claimed that this method is in any sense optimal. A more refined procedure involving transistor sizing has been investigated [7]. The purpose here is simply to demonstrate that the range of achievable values is continuous above acceptably small lower bounds for n and p, and within bounds on n/p. The delay line consists of a chain of an even number of inverters, with a capacitor attached to the output of each of the first two inverters. The inverters are identical in size. This size is made large enough to reduce to an acceptable level the delay variation due to data-dependent capacitance variation in the

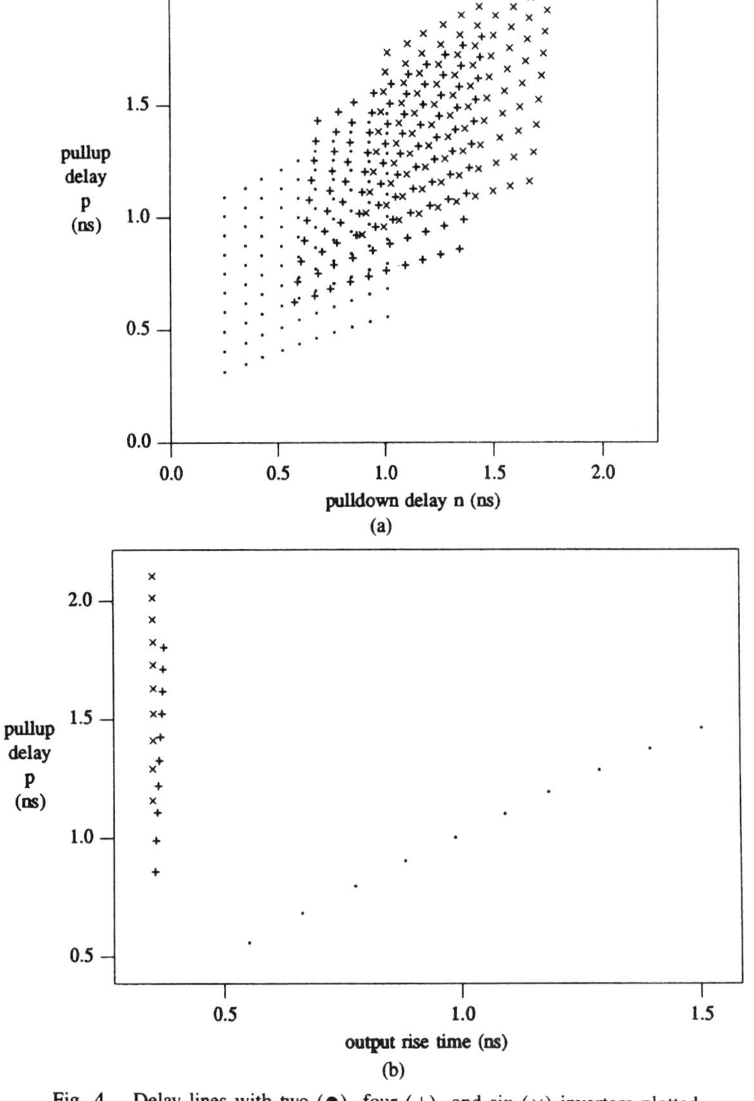

Fig. 4. Delay lines with two (●), four (+), and six (×) inverters plotted with respect to (a) pullup versus pulldown delays, and (b) pullup delay versus output rise time.

FF. The inverter PFET/NFET size ratio was set to equalize pullup and pulldown delays. The capacitor is constructed from the gate of a FET, with source and drain tied to ground. If only the total delay is to be controlled, then only the capacitor on the first inverter is necessary. Inverter chains with two, four, and six inverters were simulated by ADVICE, with each capacitor constructed with FET gate areas taking on the values 0, 100, $\cdots$, 900 μm^2. Each simulation measured not only n and p, but also the rise time (time from crossing 10% to 90% of voltage swing) of the delay line output. Rise time is of interest because too large a value can increase FF internal delay. Fig. 4(a) plots these 300 delay lines with respect to the n and p values that were achieved. Fig. 4(b) plots the delay lines with respect to p and the rise time of the delay line output, for the 30 delay lines with maximum capacitance attached to the first inverter. With two inverters, the delay line exhibits quite a large variation in rise time, as the second capacitance is varied. Depending on the technology and application, this might not be acceptable, and it may be necessary to use at least four inverters. Four inverters with no extra capacitance yield $n = 0.58$ and $p = 0.62$ ns.

VIII. Conclusions and Directions for Future Research

Inequalities (1) and (2) govern the correct operation of synchronous systems. The conventional approach of eliminating clock skew is a *feasible point* of the linear programs generated by these inequalities. In general, however, this point is neither the fastest nor the safest. These can be discovered by solving the linear programs. The optimized system can be constructed with little extra cost, and will provide faster and more reliable operation than a conventionally-clocked system.

It is not known how fast the number of constraints grows with circuit size. This number cannot be greater than twice the square of the number of FF's, but for practical circuits may be much smaller. If the linear program becomes too large, it may become necessary to investigate solution procedures that take advantage of the special form of the constraints.

The current approach should be extended to higher performance clocking schemes, such as one-phase level-sensitive latches [1], [5]. This higher performance brings with it an increased susceptibility to double-clocking, but the current approach explicitly guards against this danger.

Ideally, all variables should be considered jointly when optimizing a design. Although this is usually impossible, progress is made when two or more formerly separate optimization steps are joined into one. Clock skew and retiming are both of a linear character. It is likely that efficient procedures could be given for optimizing systems by jointly considering both sets of variables.

A third linear optimization, insertion of delay lines in *combinational logic,* has also been studied [13]. Although this work was in the context of maximum-rate pipelining, logic delay lines can provide additional safety margin against double-clocking, as well as enhance the ability of clock skew to reduce the clock period of a single-steppable machine, by reducing $\mathrm{MAX}(i, j) - \mathrm{MIN}(i, j)$.

Transistor sizing in CMOS has been shown [14] to be a *posynomial* [15] programming problem. Posynomial programs, though generally nonlinear, can always be transformed into *convex* programs, and thus enjoy the property that a local minimum is guaranteed to be a global minimum. Unfortunately, when both transistor sizes and clock delays are varied, the result is a *signomial program* [15]. A signomial program is not necessarily equivalent to a convex program. A proof of equivalence for this problem would open up the possibility of efficiently optimizing all four classes of variables jointly: clock delays, FF positions, logic delay lines, and logic-gate transistor sizes.

Acknowledgment

Thanks to E. Rosenberg for enlightening me about maxi-min programs, to N. Schryer and L. Kaufman for providing the PORT linear programming software, and to J. Chandross, R. Finkel, L. Fishburn, J. Javitt, T. Szymanski, V. Visvanathan, S. Vogel and the four reviewers for helpful comments.

References

[1] P. M. Kogge, *The Architecture of Pipelined Computers.* New York: McGraw-Hill, 1981, pp. 21-39.

[2] M. Shoji, "Elimination of process-dependent clock skew in CMOS VLSI," *IEEE J. Solid-State Circuits,* vol. SC-21, no. 5, pp. 875-880, Oct. 1986.

[3] T. M. McWilliams, "Verification of timing constraints on large digital systems," *J. Digital Syst.*, vol. 5, no. 4, pp. 401-427, 1981.

[4] L. W. Cotten, "Circuit implementation of high-speed pipeline systems," in *AFIPS Proc. 1965 Fall Joint Comput. Conf.*, vol. 27, pp. 489-504.

[5] M. J. Flynn and S. Waser, *Introduction to Arithmetic for Digital Systems Designers.* CBS College Publishing, 1982, pp. 215-222.

[6] P. E. Gill, W. Murray, and M. H. Wright, *Practical Optimization.* New York: Academic, 1981.

[7] P. V. Argade, "Sizing an inverter with a precise delay: Generation of complementary signals with minimal skew and pulse width distortion in CMOS," *IEEE Trans. Comput.-Aided Design*, vol. CAD-8, no. 1, pp. 33-40, Jan. 1989.

[8] L. W. Nagel, "SPICE2: A computer program to simulate semiconductor circuits," Memo ERL-M520, Univ. of California, Berkeley, May 9, 1975.

[9] L. W. Cotten, "Maximum-rate pipeline systems," in *AFIPS Proc. 1969 Spring Joint Comput. Conf.*, vol. 34, pp. 581-586.

[10] P. A. Fox and N. L. Schryer, "The PORT mathematical subroutine library," *ACM Trans. Math. Software,* vol. 4, no. 2, pp. 104-126, June 1978.

[11] L. A. Glasser and D. W. Dobberpuhl, *The Design and Analysis of VLSI Circuits.* Reading, MA: Addison-Wesley, 1985, pp. 345-347.

[12] C. E. Leiserson, F. M. Rose, and J. B. Saxe, "Optimizing synchronous circuitry by retiming," in *Proc. Third Caltech Conf. Very Large Scale Integration,* R. Bryant, Ed., 1983, pp. 87-116.

[13] B. C. Ekroot, "Optimization of pipelined processors by insertion of combinational logic delay," Ph.D. dissertation, Dep. Elec. Eng., Stanford Univ., Sept. 1987.

[14] J. P. Fishburn and A. E. Dunlop, "TILOS: A posynomial programming approach to transistor sizing," in *Proc. IEEE Int. Conf. Comput.-Aided Design (ICCAD-85),* Santa Clara, CA, Nov. 1985, pp. 326-328.

[15] J. G. Ecker, "Geometric programming: Methods, computations and applications," *SIAM Rev.*, vol. 22, no. 3, pp. 338-362, July 1980.

[16] J. Rubinstein, P. Penfield, and M. Horowitz, "Signal delay in RC tree networks," *IEEE Trans. Comput.-Aided Design,* vol. CAD-2, no. 3, pp. 202-211, July 1983.

Clock Frequency and Latency in Synchronous Digital Systems

Eby G. Friedman, *Senior Member, IEEE*, and J. H. Mulligan, Jr., *Fellow, IEEE*

Abstract—**This paper describes the tradeoff in the design of synchronous digital systems between clock frequency and latency in terms of the circuit characteristics of a pipelined data path. A design paradigm relating latency and clock frequency as a function of the level of pipelining is developed for studying the performance of a synchronous system. This perspective permits the development of design equations for constrained and unconstrained design problems which describe these performance parameters in terms of the delays of the logic, interconnect, and registers, clock skew, and the number of logic stages.**

These results provide a new approach to the design of those synchronous digital systems in which latency and clock frequency are of primary importance. From the behavioral specifications for the proposed system, the designer can use these results to select the best logic architecture and the best available device technology to determine if the performance specifications can be satisfied, and if so, what design options are available for optimization of other system attributes, such as area. Furthermore, the results provide a systematic procedure for the design of the synchronous system once the logic architecture and technology have been selected by the designer.

I. Introduction

In a synchronous digital system, the latency is defined as the total time required to process a signal by moving a particular data signal from the input of a system to its output. The minimum latency occurs when the data path is composed entirely of logic stages; it is the time required to propagate a data signal through these logic stages. The clock period for this system is equal to the time required to process one data sample; namely, the latency. If the system requirement for the time interval at which data is sampled at the input (i.e., the clock period) is less than the latency for this simple configuration, registers can be inserted into the data path to increase the frequency at which new data signals are processed and appear at the output of the system, thereby degrading the latency. This process is spoken of as pipelining.

Different applications of synchronous digital systems suggest different criteria for use in the optimization of their performance. For example, for a broad class of systems, optimization is done on the basis of a speed/area product. On the other hand, there are applications which are particularly sensitive to the latency of the system implementation. The results discussed in this paper are primarily intended for feedforward nonrecursive systems and describe a design approach for choosing the appropriate level of pipelining, thereby defining the system clock frequency and latency based on application specific performance requirements and architectural and technological limitations.

Systems can be designed which minimize the latency, maximize the clock frequency, or achieve tradeoffs between minimum latency and maximum clock frequency. In Section II of this paper, relations between latency and clock frequency are developed in terms of the delay components and circuit characteristics of a data path. In Section III, a graphical interpretation of the performance tradeoffs of a pipelined system is presented, illustrating the constraints and limitations of the design space.

Most synchronous digital systems are designed to satisfy specific performance requirements such as minimum clock frequency or maximum latency. Thus, in these systems the design problem becomes either one of maximizing the clock frequency while not exceeding a maximum latency or minimizing the latency while meeting a specified clock frequency. In certain systems, neither the latency nor the clock frequency ultimately constrains the design problem. In these unconstrained design problems, the level of pipelining can be chosen to tradeoff the latency with the clock frequency. These constrained and unconstrained systems are investigated in Section IV. Finally, some conclusions are presented in Section V.

II. Relation Between Latency and Clock Frequency

Since logic paths are composed of only logic stages and interconnect sections, the total delay through a logic path can be modeled as the sum of the delay through the individual logic stages and interconnect sections. For convenience in representing the delay through the system, it is desirable to define the time required for the data signal to propagate through the ith distributed RC interconnect section T_i and logic stage L_i (see Fig. 1) as T_{fi} and the average delay of all the logic and interconnect stages per data path as T_{fN}. Thus, an unpipelined data path provides the minimum latency $L_{\min}$ of a data path and, for N logic stages traversed between the input and output of the system, $L_{\min}$ can be expressed as

$$L_{\min} = \sum_{i=1}^{N} T_{fi} = NT_{fN}. \tag{1}$$

Once registers are inserted into the data path representing the complete system, the minimum clock period is decreased (higher maximum clock frequency), albeit with an increase in latency. If the original system data path is defined to be the global data path, then each individual data path between inserted registers within a global data path can be described as a local data path. Each local data path is composed of an initial and final register and typically, n logic stages between them. Note that each register within each local data path performs double duty, serving as the initial (final) and final (initial) register of the current and previous (next) local data path, respectively.

For each pipeline register, additional register related delay

Reprinted from *IEEE Trans. Signal Processing*, vol. 39, no. 4, pp. 930–934, April 1991.

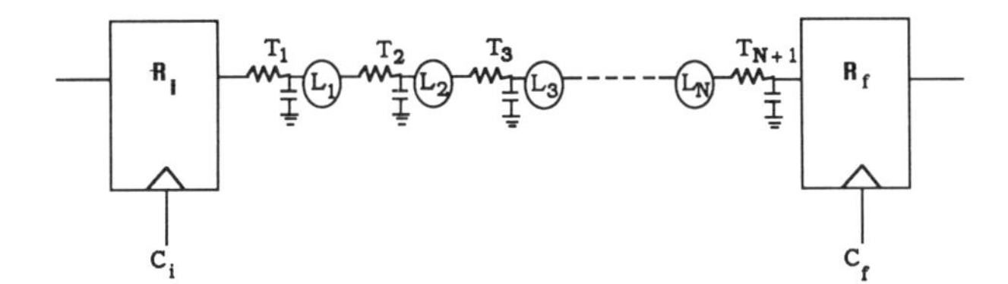

Fig. 1. Synchronous data path with N stages of logic.

components are added to the logic and interconnect delays. These register related delay components, as observed in Fig. 1, originate in R_i and R_f. T_{c-Q} is the time interval between the arrival of the clock signal at R_i and the appearance of the data signal at the register output. The time required for the signal at the output of the final logic stage to propagate through the $N + 1$st interconnect section and latch into the final register R_f is the set-up time $T_{\text{set-up}}$.

Thus, for a local data path consisting of n logic stages, the time delay through the path T_{PD} can be expressed as

$$T_{\text{PD}} = T_{c-Q} + \sum_{i=1}^{n} T_{fi} + T_{\text{set-up}}. \tag{2}$$

If T_{REG} represents the total register related delay, then

$$T_{\text{REG}} = T_{c-Q} + T_{\text{set-up}} \tag{3}$$

and (2) can be written as

$$T_{\text{PD}} = T_{\text{REG}} + \sum_{i=1}^{n} T_{fi}. \tag{4}$$

As shown in Fig. 1, the times of arrival of the initial clock signal C_i and the final clock signal C_f define the time reference when the data signals begin to leave their respective registers. These clock signals originate from a clock distribution network which is typically designed to generate a specific clock signal waveform synchronizing each register [1]–[3]. The difference in delay between two sequentially adjacent clock paths is described as the clock skew T_{SKEW}. If the clock signals C_i and C_f are in complete synchronism (i.e., the clock signals arrive at their respective registers at exactly the same time), the clock skew is zero. If the time of arrival of the clock signal at the final register of a data path (C_f) leads that of the clock signal at the initial register of the same sequential data path (C_i), then the clock skew is defined as positive; this condition degrades the maximum attainable operating frequency. This positive clock skew represents the additional amount of time which must be added to the minumum clock period to reliably apply a new clock signal at the final register. If C_f lags C_i, the clock skew is defined to be negative; this can be used to improve the maximum performance of a synchronous system. This negative clock skew represents additional amount of time for the data signal at R_i to propagate through the n stages of logic and $n + 1$ sections of interconnect into the final register. This clock skew is subtracted from the logic path delay, thereby decreasing the minimum clock period. The maximum permissible negative clock skew of any data path, however, is dependent upon the clock period itself as well as the time delay of the previous data paths. This occurs because the use of negative clock skew in the ith path results in a positive clock skew for the preceding path, which may establish the upper limit for the system clock frequency, as discussed below. It should be noted that in [1], [3], Hatamian and Cash describe these characteristics of clock skew and its effects on the maximum clock frequency and designate the lead/lag clock skew polarity (positive/negative clock skew) opposite to that described herein.

The maximum clock frequency at which a synchronous digital system can move data is defined in (5) below:

$$f_{\text{clk}} = \frac{1}{T_{\text{cp}}} \leq \frac{1}{T_{\text{PD}} + T_{\text{SKEW}}} \tag{5}$$

where T_{cp} is the clock period, T_{PD} is defined in (4), and the local data path with the greatest $T_{\text{PD}} + T_{\text{SKEW}}$ represents the critical path of the system, i.e., establishes the maximum clock frequency. Note that for positive clock skew, the maximum clock frequency decreases while for negative clock skew, the maximum clock frequency increases.

If a single register is inserted into the data path, registers external to the system would be required to synchronize the external data flow. Two registers, one at the input and the other at the output of the global data path, represents a self-contained synchronous system (as shown in Fig. 1). With each additional register, the latency increases. For a pipelined data path, the latency is the summation of the total delay through the global data path as shown below in (6) and (7):

$$L = \sum_{i=1}^{N} T_{fi} + \sum_{k=1}^{M} T_{ek} \tag{6}$$

$$L = NT_{fN} + MT_{eM} \tag{7}$$

where N is the number of logic stages per global data path, M is the number of local data paths (and clock distribution networks) per global data path, and $M + 1$ is the number of clock periods (and registers) required to move a particular data signal from the input of the system to its output. T_{ek} can be represented by (8) where T_{ek} is the aggregate delay of the kth local data path due to the initial and final registers (T_{REG}) and the clock distribution network (T_{SKEW}):

$$T_{ek} = T_{\text{REG}} + T_{\text{SKEW}}. \tag{8}$$

The average of the individual T_{ek} values over all the M serially connected cascaded data paths is defined as T_{eM}. Since T_{fN} is the average delay of all of the logic and interconnect stages (between the input and the output of the system), for convenience and improved interpretation, (7) represents the latency in terms of average delays instead of individual summations.

The average number of logic stages per local data path n is given by (9) below:

$$n = \frac{N}{M}. \tag{9}$$

The clock period T_{cp} can be expressed as

$$T_{cp} \geq T_{\text{REG}} + nT_{fN} + T_{\text{SKEW}} \tag{10}$$

$$T_{cp} = \begin{cases} NT_{fN} & \text{for } M = 0 \quad (11) \\ T_{eM} + \dfrac{NT_{fN}}{M} & \text{for } M \geq 1. \quad (12) \end{cases}$$

III. Design Paradigm for Pipelined Synchronous Systems

Registers are inserted into global data paths in order to increase the clock frequency of a digital system with, albeit, an increase in the latency. This tradeoff between clock frequency and latency is graphically described in Fig. 2. In this figure both the latency and the clock period are shown as a function of the number of pipeline registers M inserted into a global data path.

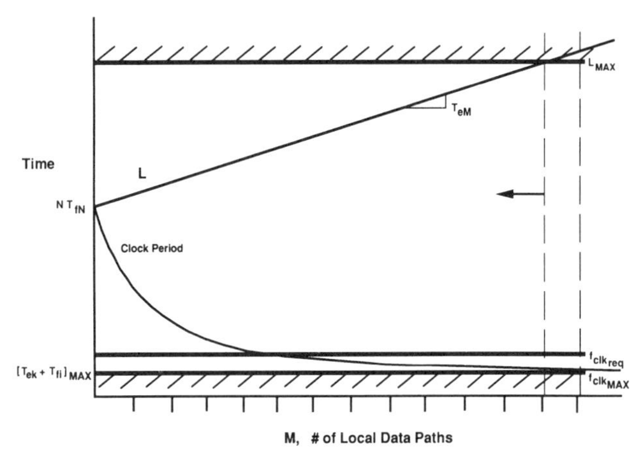

Fig. 2. Design paradigm for pipelined synchronous systems.

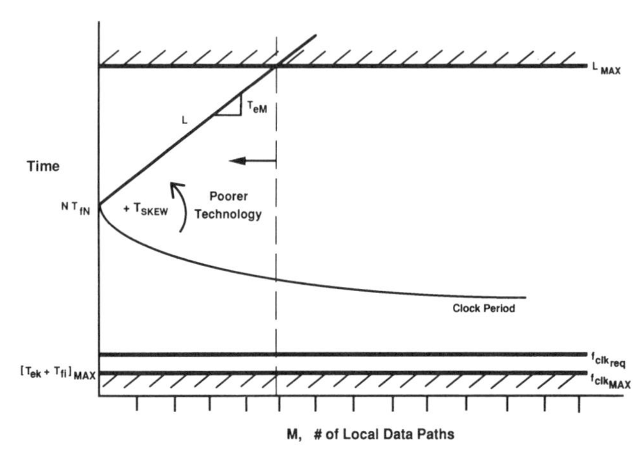

Fig. 3. Effect of positive clock skew and technology on design paradigm.

If no registers are inserted into the data path, the minimum latency $L_{\min}$ is the summation of the individual logic delays, NT_{fN}, as shown by (1) or by (7) with $M = 0$. As each register is inserted into the global data path, L increases by T_{eM}. Thus L increases linearly with M; this is depicted in Fig. 2.

As seen from (12), the expression for the clock period contains a term which varies inversely with M; this behavior is shown in Fig. 2. From (9) and (12), it is seen that the minimum clock period occurs when n equals one. The local data path having the largest value of $T_{ek} + T_{fi}$, defined as the critical data path, establishes the maximum clock frequency f_{clkMAX} for the system. This assumes that logical operations are being performed (i.e., the function is not a simple shift register). The MAX subscript in Fig. 2 is used to emphasize that the critical local data path constrains the minimum clock period (maximum clock frequency) of the total global data path.

Most design requirements must satisfy some specified maximum time for latency while satisfying or surpassing a required clock frequency. The design constraints due to $L_{\max}$ and f_{clkMAX} are shown in Fig. 2 by the vertical dashed lines. Thus, for a given $L_{\max}$, the recommended maximum clock frequency and level of pipelining is defined by the intersection of the L curve and the $L_{\max}$ line. If $L_{\max}$ is not specified and the desire is to make the clock frequency as high as possible, then the recommended f_{clk} is defined by the intersection of the clock period curve and the f_{clkMAX} line. Thus, for a specified L and f_{clk}, the extent of the possible design space is indicated by the horizontal arrow. If L and f_{clk} are both of importance and no $L_{\max}$ or f_{clkMAX} is specified, then some optimal level of pipelining is required to provide a "reasonably high" frequency while maintaining a "reasonable" latency. This design choice is represented by a particular value of M, defining an application specific f_{clk} and L.

The effects of clock skew, technology, and logic architecture on latency and clock period are graphically demonstrated in Figs. 3 and 4. If the clock skew is positive or if a poorer technology (i.e., slower) is used, as shown in Fig. 3, then T_{eM} increases and L reaches $L_{\max}$ at a smaller value of M than previously. In addition, the minimum clock period increases (decreasing the maximum clock frequency) which, for large positive clock skew or a very poor technology, eliminates any possibility of satisfying a specified clock frequency f_{clkreq} and decreases the entire design space as defined by the intersection of L and $L_{\max}$. Also, for a poorer technology or logic architecture, the intersection between either the latency or the clock period curve and the ordinate shifts upwards since T_{fN} increases due to the slower technology and N increases for the less optimal architecture.

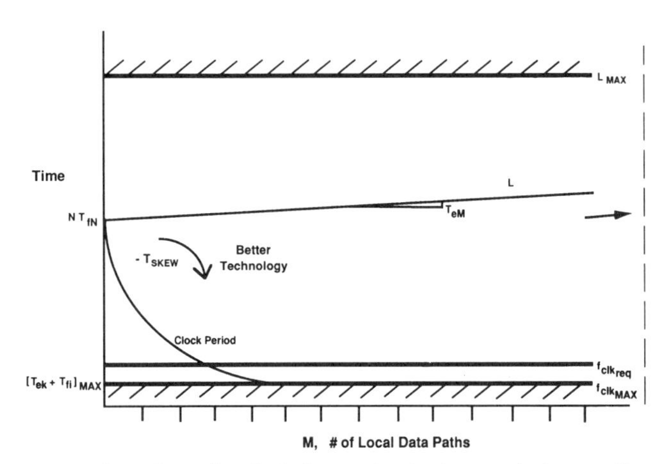

Fig. 4. Effect of negative clock skew and technology on design paradigm

If the clock skew is negative or a better technology (i.e., faster) is used, as shown in Fig. 4, T_{eM} decreases and the latency is less dependent on M. In addition, the minimum clock period decreases, satisfying f_{clkreq} and f_{clkMAX} with less pipelining. The optional design space, represented by the intersection of L and $L_{\max}$, is much larger, permitting higher levels of pipelining if very high clock rates are desired. Also, for a better technology or logic architecture, the intersection between either the latency or the clock period curve and the ordinate shifts downward since T_{fN} decreases due to the faster technology and N decreases for a more optimal architecture. Thus, Figs. 3 and 4 graphically describe how clock skew, technology, and logic architecture affect both the latency and the maximum clock frequency of a pipelined synchronous digital system.

IV. Formulation of Design Equations

A. Design Objectives

Three types of design problems are considered using this approach: 1) the maximum latency constrains the design problem, 2) the required clock frequency constrains the design problem, or 3) the problem is unconstrained and a tradeoff between L and f_{clk} must be made.

1) Maximum Latency: In applications where the maximum latency of a system is specified and $L_{\max}$ constrains the design space, the degree of pipelining can be determined from (13), where T_{eM} is taken as the estimate of an average $T_{\text{REG}} + T_{\text{SKEW}}$:

$$M \leq \frac{L_{\max} - NT_{fN}}{T_{eM}}. \qquad (13)$$

A range of possible values of clock frequency is defined in (14), assuming a value of M from (13), where the lower bound on clock frequency is due to the constraint on maximum latency and the upper bound ensures correct system operation:

$$\frac{M}{L_{\max}} \leq f_{\text{clk}} \leq \frac{M}{MT_{eM} + NT_{fN}}. \qquad (14)$$

Thus, as shown in Fig. 2, for a given maximum latency and knowledge of the average logic, register, and clock delay characteristics of a global data path, the degree of pipelining and range of clock frequency can be directly determined.

2) Required Clock Frequency: In applications where the maximum clock frequency is specified and f_{clkMAX} constrains the design space, the number of registers and the latency can be determined from (15) and (7), respectively,

$$M = \frac{NT_{fN}}{T_{\text{CP}} - T_{eM}}. \qquad (15)$$

Thus, as shown in Fig. 2, for a given maximum or required clock frequency and knowledge of the average logic, register, and clock delay characteristics of global data path, the minimum latency and the required level of pipelining can be directly determined.

3) Unconstrained Design Requirement: Each additional register increases L by T_{eM} and decreases the maximum clock period by the decreased logic delay [4]–[11]. There exists a level of pipelining where the increase in latency costs the system more than the increase in clock frequency benefits the system. In order to quantify this, an arbitrary performance criterion (the pipelining efficiency, P_e) is defined to describe the performance cost of latency. P_e is a measure of the relative performance penalty incurred by the insertion of a single additional pipeline register to an existing global data path. It is a normalized function which is the ratio of the total local logic delay to the total local data path delay, after the register has been inserted. It defines what percentage of the local data path delay is logic related and what percentage is register related. As n increases, the ratio of the total local logic delay to the total local data path delay increases toward unity, reaching it when n is infinite (or practically, when the total local logic delay is much greater than the register delay).

The benefit of inserting a register into a data path is decreased clock period as described by (12). A measure of the cost/benefit of inserting registers into an N stage global data path is the function $P_e f_{\text{clk}}$, where P_e increases for increasing n and f_{clk} decreases for increasing n. A different function could be applied if the effects of increased area, for example, were also of significant importance [7]–[9], [12]. Cappello *et al.* [8], [9] describe an AP product, where A is the chip area and P is the clock period, and use this figure of merit to optimize the speed/area performance of a pipeline system. However, the results described in this paper emphasize optimal latency and clock frequency over area/speed optimization. These results provide an approach to the design of those systems in which both latency and clock frequency are of primary importance.

If one assumes that P_e and f_{clk} are assigned equal importance and there are no constraints placed on $L_{\max}$ or f_{clkMAX} in achieving the design of the system, then an optimal value of the number of logic stages between pipeline registers N_{opt} can be obtained by determining where the product $P_e f_{\text{clk}}$ is maximized or where

$$\frac{d(P_e f_{\text{clk}})}{dn} = 0. \qquad (16)$$

By the use of (16), N_{opt}, the optimal number of logic stages per local data path, is obtained as

$$N_{\text{opt}} = \frac{1}{T_{fN}} \sqrt{T_{\text{REG}}(T_{\text{REG}} + T_{\text{SKEW}})}. \qquad (17)$$

Under the condition of an ideal clock distribution network with zero clock skew, (17) simplifies to (18):

$$N_{\text{opt}} = \frac{T_{\text{REG}}}{T_{fN}}. \qquad (18)$$

N_{opt}, in (18), is the ratio of the register delay overhead to the average stage delay of the data path. If $T_{\text{REG}} \ll T_{fN}$, which occurs when the stage performs a large high level function, then the cost of inserting registers is small and N_{opt} should be as small as feasible (since N_{opt} must be an integer, its smallest realizable value is one) or one should pipeline as often as the system permits. If $T_{\text{REG}} \gg T_{fN}$, which often exists when operating at the level of individual logic stages, then the cost of inserting registers is high and N_{opt} is some large number specified by (18). Another interpretation of (18) is that the optimal number of logic stages between registers occurs when the total logic path delay NT_{fn} equals the total register delay T_{REG}, thereby maximizing $P_e f_{\text{clk}}$.

T_{SKEW} in (17) can be zero, negative, or positive with the constraint that if T_{SKEW} is negative, its magnitude must be less than T_{REG}. It is interesting to note that the effect of clock skew on N_{opt} is relative to T_{REG} and T_{fN}. Thus, if T_{REG} is large with respect to T_{SKEW}, the relation essentially reduces to (18). Also, positive clock skew adds directly to T_{REG} and increases the cost of pipelining, thereby increasing the recommended number of logic stages between registers and quantifying how the clock distribution network affects the optimal design of a high speed data path.

V. Conclusions

Latency and clock frequency are convenient parameters on which to base the design of high speed synchronous digital systems. The results of this paper deal directly with the systematic design of those systems based upon these two performance attributes.

In system design, global data paths are often partitioned into local pipelined data paths, thereby decreasing the delay of the critical paths and increasing the clock frequency, albeit with an increase in latency. The results presented deal specifically with three types of design options; namely, that in which 1) $L_{\max}$ constrains the design space, 2) f_{clkMAX} constrains the design space, and 3) the design space is unconstrained and a tradeoff must be made between L and f_{clk}. The solution suggested for the unconstrained design problem is the use of an algorithm which considers the effects of increased latency and increased clock frequency for increasing levels of pipelining.

There is an important class of practical systems which require both high clock frequency and minimal latency (e.g., radar, sonar, high speed computing) for which there is a need for developing a design approach which satisfies their application specific performance objectives. The results presented in this paper permit the implementation of a systematic strategy for designing high performance synchronous digital systems in which both clock frequency and latency are of primary interest.

Acknowledgment

The authors thank the reviewers for comments which were helpful in the revision of the manuscript.

References

[1] M. Hatamian and G. L. Cash, "Parallel bit-level pipelined VLSI designs for high speed signal processing," *Proc. IEEE*, vol. 75, no. 9, pp. 1192–1202, Sept. 1987.

[2] E. G. Friedman and S. Powell, "Design and analysis of a hierarchical clock distribution system for synchronous standard cell/macrocell VLSI," *IEEE J. Solid-State Circuits*, vol. SC-21, no. 2, pp. 240–246, Apr. 1986.

[3] M. Hatamian, "Understanding clock skew in synchronous systems," in *Concurrent Computations (Algorithms, Architecture, and Technology)*, S. K. Tewksbury, B. W. Dickinson, and S. C. Schwartz, Eds. New York: Plenum, 1988, ch. 6.

[4] L. W. Cotton, "Circuit implementation of high-speed pipeline systems," in *Proc. Fall Joint Comput. Conf.*, 1965, pp. 489–504.

[5] P. R. Cappello and K. Steiglitz, "Bit-level fixed-flow architectures for signal processing," in *Proc. IEEE Int. Conf. Circuits Comput.*, Sept. 1982, pp. 570–573.

[6] P. R. Cappello and K. Steiglitz, "Completely-pipelined architectures for digital signal processing," *IEEE Trans. Acoust., Speech, Signal Processing*, vol. ASSP-31, no. 4, pp. 1016–1023, Aug. 1983.

[7] C. E. Leiserson and J. B. Saxe, "Optimizing synchronous systems," in *Proc. 22nd Annu. Symp. Foundations Comput. Sci.*, Oct. 1981, pp. 23–26.

[8] P. R. Cappello, A. LePaugh, and K. Steiglitz, "Optimal choice of intermediate latching to maximize throughput in VLSI circuits," in *Proc. IEEE Int. Conf. Acoust., Speech, Signal Processing*, Apr. 1983, pp. 935–938.

[9] P. R. Cappello A. LaPaugh, and K. Steiglitz, "Optimal choice of intermediate latching to maximize throughput in VLSI circuits," *IEEE Trans. Acoust., Speech, Signal Processing*, vol. ASSP-32, no. 1, pp. 28–33, Feb. 1984.

[10] J. R. Jump and S. R. Ahuja, "Effective pipelining of digital systems," *IEEE Trans. Comput.*, vol. C-27, no. 9, pp. 855–865, Sept. 1978.

[11] M. Hatamian, L. A. Hornak, T. E. Little, S. K. Tewksbury, and P. Franzon, "Fundamental interconnection issues," *AT&T Tech. J.*, vol. 66, no. 4, pp. 13–30, July/Aug. 1987.

[12] K. O. Siomalas and B. A. Bowen, "Synthesis of efficient pipelined architectures for implementing DSP operations," *IEEE Trans. Acoust., Speech, Signal Processing*, vol. ASSP-33, no. 6, pp. 1499–1508, Dec. 1985.

checkT_c and *minT_c* : Timing Verification and Optimal Clocking of Synchronous Digital Circuits

Karem A. Sakallah, Trevor N. Mudge and Oyekunle A. Olukotun
Department of Electrical Engineering and Computer Science
University of Michigan
Ann Arbor, MI 48109-2122

Abstract

We introduce two CAD tools, *checkT_c* and *minT_c* , for timing verification and optimal clocking. Both tools are based on a new timing model of synchronous digital circuits which is: 1) general enough to handle arbitrary multi-phase clocking; 2) complete, in the sense that it captures signal propagation along short as well as long paths in the logic; 3) extensible to make it relatively easy to incorporate "complex" latching structures; and 4) notationally simple to make it amenable to analytic treatment in some important special cases. We are currently using these tools to help in the design of a 4ns gallium arsenide micro-supercomputer.

1 Introduction

We present algorithms for timing verification and optimal clocking based on a new *timing model* of synchronous digital circuits. The new model extends our earlier work in [1] to handle short- as well as long-path propagation. It assumes that the circuits are logically (functionally) correct, and focuses only on capturing their latching, synchronization and propagation properties. The model can be viewed as a synthesis of ideas from earlier work in the field of timing analysis of synchronous systems. On the other hand, it also includes several key concepts which make it significantly more versatile than previous models. In particular, on the issue of level-sensitive latches which has received a great deal of attention recently (e.g. [2]–[6]), the model offers a general and accurate, yet simple, treatment. Briefly, the salient features of the new model are:

- A general treatment of *multi-phase* clocks which underscores the centrality of the common clock cycle T_c and emphasizes the *temporal* rather than the *logical* relations among clock *phases*; in particular, the notions of phase-relative *time zones* and of a *phase-shift operator* eliminate much of the notational clutter which has plagued previous efforts in dealing with "complex" clocking schemes.
- An extensible framework which allows the incorporation of arbitrary *synchronizing* structures, i.e. circuit structures where clock and data signals converge and interact, as long as a timing *macromodel* of these interactions can be provided; in this paper we only consider timing macromodels for D-type level-sensitive latches.
- Completeness, in the sense that the model accounts for signal propagation along the *shortest* as well as the *longest* paths in a circuit.

and, finally,

- Simplicity, through careful attention to notation and the use of variable and parameter symbols with mnemonic value.

The model applies to a synchronous sequential circuit containing l synchronizers (latches or flip-flops) controlled by a k-phase clock. Space limitations preclude a detailed development of the model here. Instead, a summary of its parameters, variables, and constraints is given in Table 1, and interested readers are referred to [7, 8] for a more comprehensive treatment. We devote the remainder of this paper to a discussion of the model's application to timing verification (Sec. 2) and optimal clocking (Sec. 3) and its use in the design of a micro-supercomputer (Sec. 4).

2 Timing Verification

The goal of timing verification is to determine if a given circuit can be operated with a specified clock schedule without any setup or hold time violations. This can be conveniently achieved by first finding a solution to the synchronization and propagation constraints, and then checking if this solution satisfies the latching constraints. With a fixed clock schedule, such a solution can be found by *several* passes of a CPM-like traversal of the circuit graph. Iteration is necessary if the circuit contains latches connected in feedback loops; if the only latches in the circuit are flip-flops, a single traversal would be sufficient.

Defining V_i^S and V_i^H to be the setup and hold time violations at latch i, Alg. 2.1 incorporates the above approach and forms the basis of the timing verification procedure

Reprinted from *Proc. IEEE Int'l Conf. Computer-Aided Design*, pp. 552–555, Nov. 1990.

Parameters	
p_i :	clock phase controlling latch i
S_i :	setup time of latch i
H_i :	hold time of latch i
δ_i :	min delay of latch i
Δ_i :	max delay of latch i
δ_{ij} :	min delay from latch i to latch j
Δ_{ij} :	max delay from latch i to latch j

Variables	
T_c :	cycle time
T_{p_i} :	width of active interval of phase p_i
$E_{p_j p_i}$:	forward phase shift from phase p_i to phase p_j
a_i :	earliest signal arrival time at latch i
A_i :	latest signal arrival time at latch i
d_i :	earliest signal departure time from latch i
D_i :	latest signal departure time from latch i

General System Timing Constraints (GSTC)	
Latching:	$a_i \geq H_i$
	$A_i \leq T_c - S_i$
Synchronization:	$d_i = \max(a_i, T_c - T_{p_i})$
	$D_i = \max(A_i, T_c - T_{p_i})$
Propagation:	$a_i = \min_j(d_j + \delta_j + \delta_{ji} - E_{p_j p_i})$
	$A_i = \max_j(D_j + \Delta_j + \Delta_{ji} - E_{p_j p_i})$

Table 1: Timing Model

Algorithm 2.1: Timing Verification in *checkT*

```
1.  for (i = 1,...,l) {
2.      V_i^S = V_i^H = 0 ;
3.      D_i^0 = T_c - T_{p_i} ;
4.      d_i^0 = T_c - S_i ;
5.  } /* for */
6.  m = 0;
7.  repeat
8.      done = true ;
9.      m = m + 1 ;
10.     for (i = 1,...,l) {
11.         for (j = 1,...,l) {
12.             A_i^m = max_j(D_j^{m-1} + Δ_j + Δ_ji - E_{p_j p_i}) ;
13.             a_i^m = min_j(d_j^{m-1} + δ_j + δ_ji - E_{p_j p_i})
14.         } /* for */
15.         if (A_i^m ≥ T_c - S_i) {
16.             V_i^S = A_i^m - T_c + S_i ;
17.             A_i^m = T_c - S_i
18.         }
19.         if (a_i^m ≤ H_i) {
20.             V_i^H = H_i - a_i^m ;
21.             a_i^m = H_i
22.         } /* if */
23.         D_i^m = max(A_i^m, T_c - T_{p_i}) ;
24.         d_i^m = max(a_i^m, T_c - T_{p_i}) ;
25.         if ((D_i^m ≠ D_i^{m-1}) or (d_i^m ≠ d_i^{m-1}))
26.             done = false ;
27.     } /* for */
28. until done;
```

in the *checkT*$_c$ tool. Several observations should be made about this algorithm:

- The late and early departure times are initialized, respectively, to their minimum (line 3) and maximum (line 4) possible values. With these starting values, successive updates of each D_i and A_i (d_i and a_i) during the iterative portion of the algorithm (lines 7–28) are guaranteed to be non-decreasing (non-increasing). The iteration can thus be conveniently viewed as a process of "sliding" each D_i and A_i to the right and each d_i and a_i to the left.
- The late and early arrival times are *clipped*, respectively, at their maximum (line 17) and minimum (line 21) possible values. This clipping has the desireable effect of localizing the corresponding setup or hold violation to the latch in question for better diagnostics.
- The iteration in the algorithm is guaranteed to terminate because: 1) successive updates of the late departure times are non-decreasing, and are bound from above by $T_c - S_i$; and, 2) successive updates of the early departure times are non-increasing, and are bound from below by $T_c - T_{p_i}$.
- The update formulas in the algorithm (lines 12,13,23,24) have the flavor of a mixed Jacobi/Gauss-Seidel iteration. Other variants are obviously possible. In fact an event-driven update mechanism which only calculates those arrival and departure times which have changed from the previous iteration can be easily implemented. With such an enhancement the cost of the iterative steps can be greatly reduced for large circuits.
- In addition to computing setup and hold time violations, it is possible to extend Alg. 2.1 to identify the *critical* path segments in the circuit. The long path from latch j to latch i is considered critical if $D_j + \Delta_j + \Delta_{ji} - E_{p_j p_i} = T_c - S_i$. Similarly, the short path from latch j to latch i is critical if $d_j + \delta_j + \delta_{ji} - E_{p_j p_i} = H_i$. Furthermore, the *slacks* in the other (non-critical) path segments can be easily found.

3 Calculation of Optimal Cycle Time

The minimum clock cycle time can be found by solving the following optimization problem:

Program PG: Optimal Cycle Time—GSTC

Minimize T_c

Subject to GSTC (Table 1)

Following the approach suggested in [1], we solve this problem by first transforming the nonlinear program PG to an associated linear program PXG by *relaxing* the min and max constraints. However, unlike the more restrictive case in [1], the procedure of finding the minimum cycle time for the GSTC case is complicated by the presence of both min and max functions in the constraints, and an approach similar to Algorithm MLP in [1] cannot be guaranteed to yield an optimal solution to program PG under all circumstances. In fact applying such an approach may produce a solution that violates one or more hold time constraints. Fortunately, there is also enough additional information to indicate how this infeasible solution should be modified to make it both feasible and optimal.

Assume, therefore, that we carry out the following two steps:

1. Solve program PXG.
2. Disregarding the hold time constraints, "slide" the solution found in step (1) so that the synchronization and propagation constraints are satisfied.

The solution obtained by this procedure can, then, be interpreted as that of a linear program—call it PXG′—which is identical to program PXG except that the hold time at each latch i is *reduced* by the corresponding violation amount V_i^H. Specifically, the hold time constraints, $a_i \geq H_i$, are replaced by:

$$a_i \geq H_i - V_i^H \qquad (1)$$

Finally, With these constructions, it is now possible to obtain the optimal solution of program PG by applying the sensitivity analysis techniques of linear programming [9, Sec. 4.4, p. 160] to the solution of program PXG′. In particular, program PG can be "obtained" from program PXG′ by *increasing* the hold time at each latch i by V_i^H. Denoting the optimal values of programs PG and PXG by $T_{c,min}^{(G)}$ and $T_{c,min}^{(XG)}$, we may therefore conclude that:

Conjecture 3.1 $T_{c,min}^{(G)} = T_{c,min}^{(XG)} + \sum_{i=1}^{l} \pi_i V_i^H$.
where π_i denotes the optimal value of the *dual* variable corresponding to the ith modified hold time constraint in (1). This result is stated as a conjecture because we have not yet worked out the detailed proof steps. It assumes that the solution of program PXG′ remains basic feasible after the right-hand-side vector is modified.

The detailed algorithmic steps for computing the minimum cycle time in the GSTC case are given in Alg. 3.1.

Algorithm 3.1: Optimal Clocking in ***minT*** $_c$ ***1***

1. *Solve PXG; Denote solution by* D_i^0 *and* d_i^0
2. $T_{c,min}^G = T_{c,min}^{XG}$;
3. $m = 0$;
4. **repeat**
5. $\quad done =$ **true** ;
6. $\quad m = m + 1$;
7. $\quad$ **for** $(i = 1, \cdots, l)$ {
8. $\quad\quad$ **for** $(j = 1, \cdots, l)$ {
9. $\quad\quad\quad A_i^m = \max_j(D_j^{m-1} + \Delta_j + \Delta_{ji} - E_{p_j p_i})$;
10. $\quad\quad\quad a_i^m = \min_j(d_j^{m-1} + \delta_j + \delta_{ji} - E_{p_j p_i})$
11. $\quad\quad$ } /* **for** */
12. $\quad\quad$ **if** $(a_i^m \leq H_i)$ {
13. $\quad\quad\quad V_i^H = H_i - a_i^m$;
14. $\quad\quad\quad a_i^m = H_i$
15. $\quad\quad$ } /* **if** */
16. $\quad\quad D_i^m = \max(A_i^m, T_c - T_{p_i})$;
17. $\quad\quad d_i^m = \max(a_i^m, T_c - T_{p_i})$;
18. $\quad\quad$ **if** $((D_i^m \neq D_i^{m-1})$ **or** $(d_i^m \neq d_i^{m-1}))$
19. $\quad\quad\quad done =$ **false** ;
20. $\quad$ } /* **for** */
21. **until** *done;*
22. **for** $(i = 1, \cdots, l)$
23. $\quad T_{c,min}^G = T_{c,min}^G + \pi_i V_i^H$;

4 Example

The timing model and algorithms introduced in this paper have been implemented in two experimental CAD tools: *checkT*$_c$ which examines a circuit for adherence to a specified clock schedule, and reports on setup and hold time violations; and *minT*$_c$ which determines the optimal clock schedule. We are using these tools to help in the design of a 250 MHz galluim arsenide micro-supercomputer currently under development at the University of Michigan[10]. Figure 1 illustrates the use of *minT*$_c$ in the design procedure. At the top of the figure is a simplified block diagram of the micro-supercomputer's CPU and its primary cache subsystem. The CPU implements an existing instruction-set architecture, the MIPS R6000, and has as its main components, a register file of 32 32-bit registers, an ALU, a shifter, and an integer multiply and divide unit. The datapath is 32-bits wide and it is timed with a 3- phase clock. The diagram shows the major logic blocks in the CPU datapath which is implemented as a single chip. The shaded blocks show the instruction cache and the data cache which are implemented as a set of 15 highspeed GaAs 1 K x 32 SRAM chips. The synchronizing elements are a combination of latches and flip-flops. To reduce the effects of chip crossings the CPU and the primary caches are integrated into a single multi-chip module (MCM). To avoid slow inter-chip bus protocols, the MCM is timed as one logic entity using $minT_c$ to obtain the optimal clock schedule. To apply $minT_c$ the MCM was first macromodeled to obtain the short- and long-path delays for the chip crossings using SPICE simulations.

The optimal clock schedule obtained by $minT_c$ can be seen below the block diagram. The time units are nanoseconds. To simplify the analysis performed by $minT_c$ each of 32-bit wide latches and flip-flops were modeled as a single latch. The output of $minT_c$ also includes waveforms of the signals arriving at the inputs of each latch in the MCM. Five of these waveforms, corresponding to the labeled latches in the block diagram, are shown below the clock waveforms.

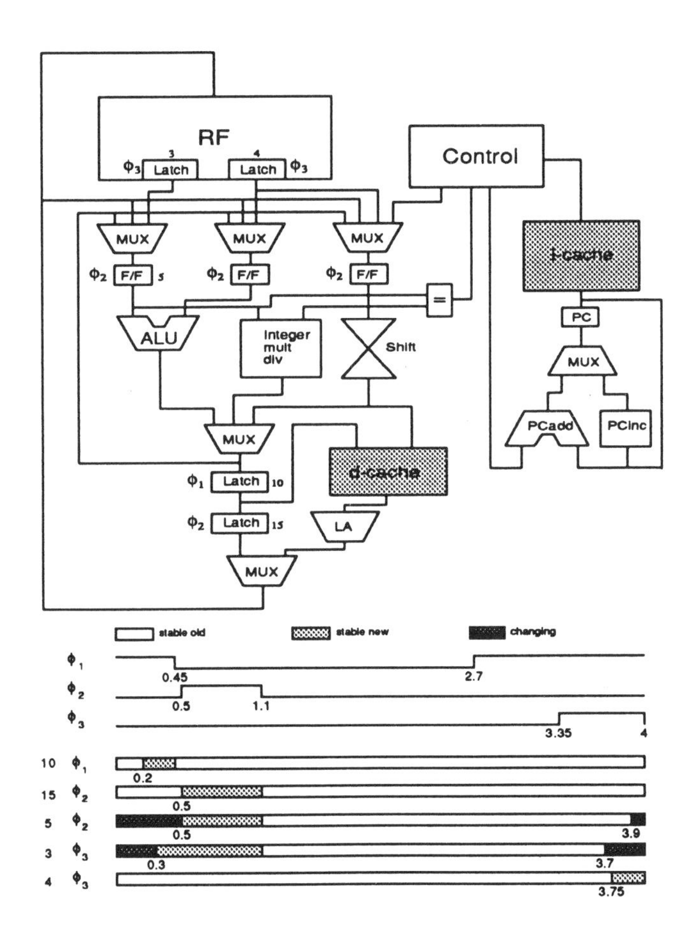

Figure 1: GaAs MIPS Datapath and Timing Waveforms

5 Conclusions and Future Work

The timing model presented in this paper establishes a framework for the study of the dynamic behavior of synchronous digital systems. The model requires that a circuit be decomposed into combinational and sequential parts, a generally difficult task. Furthermore, the dynamic behavior of the sequential structures must be captured in timing "macromodels" similar to those of the D-type latches presented in this paper. Both of these steps, decomposition and macromodeling, require further investigation. Other extensions to the model, for example to handle clock skew, are relatively straightforward and are currently being incorporated.

In parallel with the model development effort, we are investigating efficient implementations of algorithms 2.1 and 3.1. We are also looking into the derivation of closed-form design criteria for special circuit structures such as pipelines and CPU data paths.

References

[1] K. A. Sakallah, T. N. Mudge, and O. A. Olukotun, "Analysis and Design of Latch-Controlled Synchronous Digital Circuits ," in *Proceedings of the 27th Design Automation Conference*, 1990.

[2] N. P. Jouppi, *Timing Verification and Performance Improvement of MOS VLSI Designs*, PhD thesis, Stanford University, Stanford, CA 94305-2192, October 1984.

[3] T. G. Szymanski, "LEADOUT : A Static Timing Analyzer for MOS Cicuits," in *ICCAD-86 Digest of Technical Papers*, pp. 130–133, 1986.

[4] J. J. Cherry, "Pearl : A CMOS Timing Analyzer," in *Proceedings of the 25th Design Automation Conference*, pp. 148–153, 1988.

[5] D. E. Wallace and C. H. Sequin, "ATV: An Abstract Timing Verifier," in *Proceedings of the 25th Design Automation Conference*, pp. 154–159, 1988.

[6] M. R. Dagenais and N. C. Rumin, "On the Calculation of Optimal Clocking Parameters in Synchronous Circuits with Level-Sensitive Latches," *IEEE Transactions on Computer-Aided Design*, vol. 8, no. 3, pp. 268–278, March 1989.

[7] K. A. Sakallah, T. N. Mudge, and O. A. Olukotun, "A Timing Model of Synchronous Digital Circuits," Technical Report CSE-TR-47-90, University of Michigan, Dept of EECS, Ann Arbor, MI 48109-2122, 1990.

[8] K. A. Sakallah, T. N. Mudge, and O. A. Olukotun, "Optimal Clocking of Synchronous Systems," Technical Report CSE-TR-65-90, University of Michigan, Dept of EECS, Ann Arbor, MI 48109-2122, 1990.

[9] D. T. Phillips, A. Ravindran, and J. J. Solberg, *Operations Research: Principles and Practice*, John Wiley and Sons, Inc., 1976.

[10] R. B. Brown, J. A. Dykstra, T. N. Mudge, and R. Milano, "A GaAs Micro-Supercomputer: Rationale and Design," Technical Report CSE-TR-42-90, University of Michigan, Dept of EECS, Ann Arbor, MI 48109-2122, 1990.

Computing Optimal Clock Schedules

Thomas G. Szymanski
AT&T Bell Laboratories, Murray Hill, NJ 07974
Email: tgs@research.att.com

Abstract

Consider the problem of optimizing the parameters of a multiphase clock for a circuit containing both edge-triggered flipflops and level-sensitive latches. We demonstrate that recently proposed linear programming approaches to this problem require excessive computation time. We give an alternative method in which LP constraints are generated selectively, thus allowing fast solution. Various formulations of short path constraints are discussed, as are experimental results for large circuits.

1 Introduction

Consider a circuit containing edge-triggered flip-flops and level-sensitive latches, collectively referred to as *memory elements*, interconnected with combinational logic. The circuit is driven by a multiphase clock. We want to determine those values of the various clock parameters that minimize the overall clock period.

Many previous papers, including [1, 2, 3, 4, 5, 6], have demonstrated fast methods for verifying the correctness of clock schedules. Armed with such a verification algorithm, a user might then hunt for an optimum schedule by trial-and-error or multidimensional binary search. Trial-and-error has obvious disadvantages, and binary search is prohibitively expensive. If there are k clock parameters and we need to perform σ probes along each dimension, then σ^k verification runs would be needed.

A more elegant and systematic approach treats the problem as a mathematical optimization problem. Several papers, including [7, 1], have mentioned this possibility without pursuing it. More recently, Sakallah and his colleagues have written a series of papers [8, 9, 6] exploring and refining the approach.

In this paper, we describe our experiences using linear programming methods to optimize clock schedules. Initially, we were disappointed with the resulting computation times. Too many constraints were needed to model a circuit and the bulk of these constraints were superfluous. Eventually we discovered ways to determine just which constraints are actually necessary. A minimal such set of constraints can be generated quite rapidly; our implementation spends more time reading the circuit description than generating constraints. The generated constraints may then be solved directly, or combined with additional constraints arising at the system level and then used for timing analyses at higher levels in the design hierarchy. Our methods are built on those of [1, 9, 3].

2 The Circuit Model

Our model borrows heavily from [8]. We assume that a circuit is clocked by a linearly-ordered sequence of k clock events, $E_1, \ldots, E_k$. A *clock schedule* with *period* π consists of a vector of k real numbers $0 \le t_1 \le t_2 \ldots \le t_k = \pi$. When the circuit is run according to the schedule, clock event E_i occurs at time $t_i + j\pi$ for every integer j. A clock schedule is said to be *feasible* if the circuit operates correctly under these conditions. Given a schedule $\langle t_1, \ldots, t_k \rangle$, denote the *separation* between clock events within the same cycle by $\phi_{i,j} = t_j - t_i$. Note that $\phi_{i,j}$ is nonpositive when $i \ge j$ and that $\phi_{i,j} = -\phi_{j,i}$.

Each memory element M has setup and hold delays which are denoted by *setup*(M) and *hold*(M) respectively. Each flip-flop F is triggered *skew*(F) time units after its associated clock edge, denoted $\updownarrow F$. Each latch L is associated with two clock edges, an opening edge and a closing edge, denoted $\uparrow L$ and $\downarrow L$ respectively, which specify that portion of the cycle when the latch is transparent. The corresponding skews are denoted by *oskew*(L) and *cskew*(L). The combinational logic interconnecting the memory elements is assumed to be acyclic with $\Delta(M, M')$ and $\delta(M, M')$ denoting the maximum and minimum combinational delay between memory elements M and M'.

Next we must define the intended behavior of the circuit. Accordingly, we stipulate that each memory element M *assumes* exactly one value per clock cycle. This value is assumed at $\downarrow M$ for latches and $\updownarrow M$ for flipflops and is a combinational function of the values last assumed by those memory elements M' for which $\Delta(M', M)$ is defined. This provides us with a delay-independent simulation model; it is now the job of timing analysis to find those conditions under which the real circuit will have equivalent behavior.

Our clocking model differs somewhat from that of [8] in that we bind opening edges of latches to specific events in the linear clock order, whereas they allow opening edges to float free and solve for their position. In our environment, multi-phase clocks are usually implemented by distributing a few master clocks in quadrature from which specific local clocks are decoded. As such, one latch's closing edge quite naturally becomes another latch's opening edge. This practice is prevalent in precisely those leading edge CMOS designs which make most heavy use of level-sensitive latches. In such an environment, *all* clock events must be linearly ordered. In any case, it is easy to extend either model to accommodate the other's perspective.

Many circuits will require preprocessing to fit either

Reprinted from *29th Proc. ACM/IEEE Design Automation Conf.*, pp. 399–404, June 1992.

model, especially circuits in which the clock signals themselves are generated by combinational logic. Latches that can open more than once per cycle, or that can produce multiple values during a single open period, must be detected and modeled as distinct events using symbolic simulation techniques, e.g., [4]. Qualified clocks must be detected, and appropriate setup and hold requirements fabricated to check stability of qualifying logic signals while the clock is active.

For the sake of clarity, we have omitted such complicating details as distinct rising and falling delays or different "flavors" of internal latch delay (e.g., different delays from clock to output, or data to output). Treatment of these details is straightforward. Circuits employing wave pipelining fall beyond the scope of our work.

3 Short Path Constraints

Two different formulations of short path or hold constraints were presented in [8]. Suppose that a_L and d_L respectively represent the earliest time when data arrives at, and departs from, latch L. Omitting details of propagation times through latches and time zone translation, the more aggressive approach is given by

$$\begin{aligned} d_L &= \max(a_L, t_{\uparrow L}) \\ a_L &\geq hold(L) \\ a_L &= \min_{M \to L}(d_M + \delta(M, L)) \end{aligned}$$

and the more conservative approach by

$$\begin{aligned} d_L &= t_{\uparrow L} \\ a_L &\geq hold(L) \\ a_L &= \min_{M \to L}(d_M + \delta(M, L)) \end{aligned}$$

where $\min_{M \to L}$ ranges over all M for which $\delta(M, L)$ is defined. The aggressive formulation yields solutions in which the a_i's and d_i's can be larger than in the conservative formulation, and therefore less likely to violate hold constraints. In principle, the aggressive formulation might yield a solution with a shorter period, although this did not happen in any of the circuits we studied.

Unfortunately, there are common situations, such as a latch driven by a qualified clock signal, in which the aggressive formulation is incorrect as stated. A similar problem arises in circuits which permit the clock to be stopped between adjacent E_i's to save power. A "fix" is possible, but it involves modifying all constraints for signals that might be "in flight" during the pause. Another difficulty with the aggressive approach is that the set of feasible solutions for the equations might not be convex or even connected. Indeed, we can show that clock schedule optimization under aggressive hold constraints is NP-complete.

We therefore chose the conservative approach. Within this model, each pair of memory elements gives rise to a short path constraint of either the form $\phi_{i,j} \geq c$ with $i < j$ or $\phi_{j,i} \leq c$ otherwise. The constant c depends only on the minimum delay between the elements and the hold time requirement of the destination. Thus all the short path constraints may be reduced to roughly k^2 inequalities.

4 The Lower Bounds Graph

We can express lower bounds on the required separation between circuit events using a graph we call the *lower bound graph*, or *LBG*. The primary use of the graph is to encode long path or setup constraints, but we will also use it to express system level requirements imposed by the user, some of the short path constraints of §3, minimum pulse width requirements on clock lines, etc. The LBG contains one vertex e_i for each clock event E_i, and one vertex l_i for each level sensitive latch L_i. Vertices thus correspond to circuit events, specifically, the occurrence of a clock edge or the assumption of a latch value.

Each arc in the graph has a weight and a color. An arc weight has units of time and corresponds to a requirement that two events must be separated by at least that much time for the circuit to operate correctly. The color of an arc is either black or red, with black meaning that the connected events occur in the same (abstract) clock cycle, and red meaning they occur in adjacent cycles.

Arcs are placed in the LBG as described in Figure 1. The graph may contain several arcs between the same pair of vertices, but for computational purposes we can discard all but the heaviest such arc. Note that an LBG is strongly connected, and the subgraph consisting of just the black arcs is acyclic. Note further that although flipflops are not represented explicitly as vertices in the LPG, they are represented in the arc set.

Figure 2 shows a simple circuit and its corresponding LBG. Delays and skews are shown on the circuit's wires and we assume that all latches have a setup time of 1.5 time units. There are two events per clock cycle, E_1 when signal CK rises, and E_2 when it falls. Event E_1 causes latches A and C to open, and B to close, whereas E_2 does the opposite. Latch A assumes a new value at E_2 of each clock cycle, and latch B then uses this value to compute its next value at E_1 of the subsequent cycle. Accordingly, there is a red arc of weight 10 from a to b in the LBG. Because of clock skew and setup requirements, the data assumed by B at E_1 must arrive at B no more than 0.5 time units after E_1 occurs. We therefore place a black arc of weight -0.5 between b and e_1 in the LBG.

Now we need an additional definition and some notation. A *path* P is a sequence of alternating vertices and arcs, $v_0, a_0, v_1, a_1, \ldots, v_n$, $n \geq 0$, such that each a_i leads from vertex v_i to v_{i+1}. This is denoted by $P\colon v_0 \rightsquigarrow v_n$. Paths are allowed to pass through the same vertex more than once. The sum of the arc weights in P is denoted $w(P)$, and $r(P)$ is the number of red arcs in P.

5 Relevant Constraints

In this section we shall show that every path in a LBG gives rise to a constraint, more specifically a linear inequality, involving the variables of a clock schedule for the circuit. In order for the circuit to operate correctly, every constraint implied by every path must be satisfied by the clock schedule. Fortunately, only those paths that we call *relevant* matter for purposes of testing whether a schedule is feasible. The set of relevant paths turns out to be

Arc	Condition	Black?	Weight
$l_i \rightarrow l_j$	true	$\downarrow L_i < \downarrow L_j$	$\Delta(L_i, L_j)$
$e_i \rightarrow l_j$	$E_i = \updownarrow F$	$i < \downarrow L_j$	$\Delta(F, L_j) + skew(F)$
$e_i \rightarrow l_j$	$E_i = \uparrow L_j$	$i < \downarrow L_j$	$oskew(L_j)$
$l_i \rightarrow e_j$	$E_j = \updownarrow F$	$\downarrow L_i < j$	$\Delta(L_i, F) + setup(F) - skew(F)$
$l_i \rightarrow e_j$	$E_j = \downarrow L_i$	true	$setup(L_i) - cskew(L_i)$
$e_i \rightarrow e_j$	$E_i = \updownarrow F$ & $E_j = \updownarrow F'$	$i < j$	$\Delta(F, F') + skew(F) - skew(F') + setup(F)$
$e_i \rightarrow e_j$	true	$i < j$	any lower bound on time between E_i and next E_j

Figure 1: Lower Bounds Graph Construction Rules. The indicated arc is added to the graph whenever flipflops F and F' can be found that satisfy the specified condition. The arc's color and weight are then given by the two remaining columns.

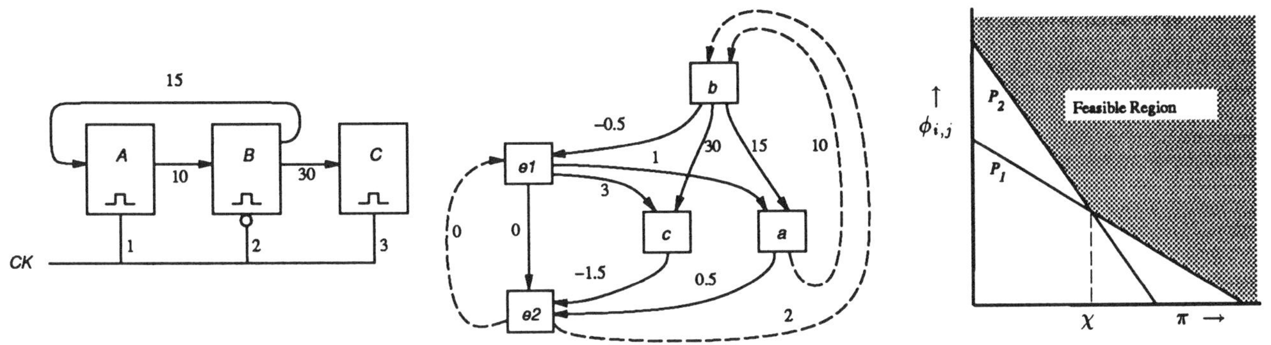

Figure 2: A circuit and its LBG. All setup times are 1.5. Red arcs shown dashed.

Figure 3: Two Constraints.

a function of the clock period, with more and more paths becoming relevant as the period decreases. However, the period can never decrease below the critical value ψ defined below. The rest of the section develops those properties of relevant paths that make them computationally useful.

Theorem 5.1 *Suppose that P: $e_i \rightsquigarrow e_j$ is a path in an LBG. Let S be a feasible clock schedule with period π. Then $w(P) \leq r(P)\pi + \phi_{i,j}$.*

We henceforth refer to the inequality $w(P) \leq r(P)\pi + \phi_{i,j}$ as the *constraint implied by path P*. For example, in Figure 2, the path from e_1 to a to b to c to e_2 has weight 39.5 and 1 red arc, so the corresponding constraint is $39.5 \leq \pi + \phi_{1,2}$. In linear programming terminology, such a constraint constitutes a bounding hyperplane on the set of feasible clock schedules. Figure 3 relates $\phi_{i,j}$ to π for the constraints implied by two paths, P_1 and P_2, with the same endpoints, e_i and e_j, but differing numbers of red arcs. For $\pi < \chi$, P_1 is less constraining than P_2, and the reverse is true for $\pi > \chi$. This notion that certain paths are most constraining for certain ranges of the clock period motivates the next definitions.

A path P is said to be *relevant for period λ* if for any other path P' with the same endpoints as P, $w(P') - r(P')\lambda < w(P) - r(P)\lambda$ if $r(P') < r(P)$ and $w(P') \leq w(P)$ if $r(P') = r(P)$. The set of all paths relevant for period λ will be denoted by $\mathcal{R}(\lambda)$ and the set of all paths in an LBG will be denoted by $\mathcal{P}$. Finally, define ψ to be the maximum of $w(C)/r(C)$ taken over all cycles C in the LBG.

Lemma 5.2 *If path P is relevant for period λ, so is every prefix of P.*

Theorem 5.3 *Let S be a clock schedule with period π. Then S satisfies all constraints implied by $\mathcal{P}$ if and only if S satisfies all constraints implied by $\mathcal{R}(\pi)$.*

Lemma 5.4 *$\mathcal{R}(\rho) \supseteq \mathcal{R}(\lambda)$ if $\rho \leq \lambda$.*

Theorem 5.5 *A clock schedule with period π is not feasible unless $\pi \geq \psi$.*

Corollary 5.6 *No path in $\mathcal{R}(\psi)$ contains a cycle.*

Theorem 5.7 *Given a schedule S with period π and any λ with $\lambda \leq \pi$. Then the following are equivalent:*
(1) *S satisfies all constraints implied by $\mathcal{P}$,*
(2) *S satisfies all constraints implied by $\mathcal{R}(\pi)$,*
(3) *S satisfies all constraints implied by $\mathcal{R}(\lambda)$,*
(4) *S satisfies all constraints implied by $\mathcal{R}(\psi)$.*

6 Algorithms

To test whether a proposed clock schedule S with period π is feasible, it suffices, according to Theorem 5.7, to check whether S satisfies all constraints implied by $\mathcal{R}(\lambda)$ for any $\lambda \leq \pi$. By Lemma 5.4, picking λ as large as possible will yield the smallest constraint set, so we should generate the constraints implied by $\mathcal{R}(\pi)$.

On the other hand, suppose that we are trying to find an optimum clock schedule for the circuit, possibly constrained by additional requirements arising from system level considerations. The set of constraints U implied by $\mathcal{R}(\psi)$ is ideal for this purpose because, again referring to Theorem 5.7, a clock schedule is feasible if and only if it satisfies U. Thus U represents *all* the timing constraints

```
subroutine update (m, w, r)
  if w - rλ > w_m - r_m λ then
    w_m ← w;   r_m ← r;
end subroutine;

for each i with 1 ≤ i ≤ k
  for each vertex n
    w_n ← -∞;   r_n ← 0;
  w_{e_i} ← 0;
  r ← 0;
  while r_n = r for at least one vertex n
    for each vertex n with r_n = r in topological
          order according to the black arcs
      prevw_n ← w_n;
      for each black arc a from vertex n to any m
        update (m, w_n + w(a), r);
    for each j with 1 ≤ j ≤ k and r_{e_j} = r
      emit constraint w_{e_j} ≤ rπ + φ_{i,j} using the
            current values of w_{e_j} and r;
    for each vertex n with r_n = r
      for each red arc a from vertex n to any m
        update (m, prevw_n + w(a), r+1);
    r ← r + 1;
```

Figure 4: Generate all constraints implied by $\mathcal{R}(\lambda)$.

implicit in the LBG. Typically U is a small set, as we shall see in §7.

6.1 Generating Relevant Constraints

As we have seen, constraints are implied by paths between the e_i's in the LBG. For each e_i, we use a single source, longest path algorithm to successively consider all relevant paths from e_i that contain exactly 0 red arcs, then 1 red arc, then 2 red arcs, etc. The correctness of this algorithm is based on Lemma 5.2 and the following assertion, which holds at the start of the for loop on the variable j, and which states the relationship amongst the program variables r, i, w_n and r_n:

> w_n and r_n give the weight and number of red arcs in that path for which $w(P) - r(P)\lambda$ is a maximum, taken over all paths P: $e_i \rightsquigarrow n$ with P relevant for λ, and $r(P) \leq r$.

Suppose that v and a are the number of vertices and arcs in the LBG. By Theorem 5.5, we can assume $\lambda \geq \psi$. Corollary 5.6 tells us the algorithm will terminate with $r \leq v$, and so the worst case running time is $O(kva)$. At most vk^2 constraints are generated.

6.2 Computing ψ

The problem of computing ψ is a variant of the minimum mean-weight cycle problem studied by Lawler [10] and Karp [11]. Karp's algorithm produces an exact answer in $\Theta(va)$ time, whereas Lawler's algorithm produces an answer with b bits of precision in $O(bva)$ time. Lawler's algorithm was better suited to our purposes because it could be augmented with heuristics to accelerate convergence.

```
subroutine checkforcycles (λ)
  for each vertex n
    d_n ← -∞;
    b_n ← a dummy arc from e_1 to n with weight -∞;
  d_{e_1} ← 0;
  for pass ← 0 to v
    if d_{e_1} > 0 then
      return "positive cycle through e_1";
    if no d_n changed during the last pass then
      return "no positive cycles";
    for each vertex n for which d_n did change
      for each arc a from vertex n to any m
        if a is black then
          t ← w(a) else t ← w(a) - λ;
        if d_n + t > d_m then
          d_m ← d_n + t;   b_m ← a;
    every so often
      examine the graph B whose arcs consist
            of {b_n | n is a vertex};
      if B contains a positive weight cycle then
        return "positive cycle";
  return "positive cycle exists";
end subroutine;

u ← large number;
l ← 0;
while u - l > ε
  ψ ← (u - l)/2;
  checkforcycles (ψ);
  if a positive cycle was found then
    l ← ψ (or something larger: see text);
  else
    u ← ψ (or something smaller: see text);
```

Figure 5: Algorithm to determine ψ.

Lemma 6.1 (Lawler) *Let G be a LBG and let $G(\lambda)$ be a directed graph constructed from G by subtracting λ from the weight of every red arc and leaving the black arcs unmodified. Then G contains a cycle C with $w(C)/r(C) > \lambda$ if and only if $G(\lambda)$ contains a positive weight cycle.*

It is easy to test whether a directed graph contains a cycle of positive weight using the Bellman-Ford algorithm. Calculating ψ then becomes a simple binary search for the maximum value of λ for which $G(\lambda)$ contains a positive weight cycle, as shown in Figure 5.

The subroutine checkforcycles computes the length of the longest path to each vertex in $G(\lambda)$ starting from a single source vertex. The code uses "selective tracing," that is, it relaxes just those arcs that leave a vertex whose distance increased during the previous pass. This helps because most of the d_n assume their ultimate values during the first few passes. More importantly, keeping track of the active vertices allows early detection of convergence.

We use e_1 as the source. Once per relaxation pass we check whether a positive weight path to e_1 has been discovered. This test is cheap, and often effective because the

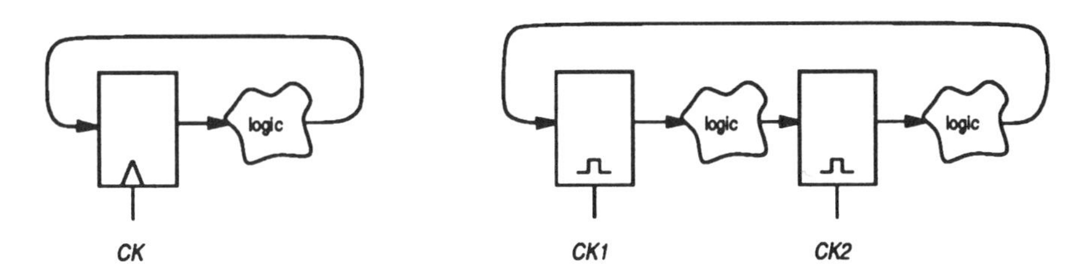

Figure 6: Bipartite transformation to convert edge-triggered circuit into two phase, level-sensitive circuit.

e_i's are so tightly coupled to the rest of the graph.

Every time a new longest path to a vertex n is discovered, we store in b_n the last arc on that path. Thus, the b_n together define a subgraph in which each vertex has indegree exactly one. It is easy to find and examine all cycles in this graph in $O(v)$ time using the following method. Associate the null tag with each vertex. Pick an untagged vertex m and trace back through the b_n's tagging vertices with m as you go. If you encounter a vertex already tagged with m, you have found a cycle and can compute its weight. Otherwise, pick some other untagged vertex and continue. If this process uncovers a positive weight cycle, then the subroutine can terminate immediately.

Whenever a positive weight cycle is detected by `checkforcycles`, we can use that cycle to accelerate the convergence of the surrounding binary search for ψ. In a normal binary search, detecting a cycle in $G(\lambda)$ would cause λ to become a new lower bound on ψ. However, if the cycle detected in $G(\lambda)$ had weight w with r red arcs, then a better lower bound on ψ is provided by $(w + r\lambda)/r$.

The movement of the upper bound on ψ can also be accelerated by noting the minimum decrease in λ that would be needed for the longest path computation to have a different outcome. More specifically, in the case where `checkforcycles` terminates without finding any positive cycles, the B graph is a spanning tree rooted at e_1 with one non-tree arc added, namely the dummy arc from e_1 to e_1. In this case we calculate the maximum value of λ for which the routine would have produced a different tree.

7 Experimental Results

We wrote three programs. The program `lps` optimizes clock schedules using the linear programming framework of [8]. The program `solve` uses the reduced constraints set approached introduced in this paper. The program `check` is a clock schedule verifier using relaxation, as described in [6]. This program was used to double check the other two. An option on `check` determines whether aggressive or conservative short path constraints are used.

7.1 The Data

Since flipflops essentially disappear during the construction of the LBG, we needed circuits with lots of level-sensitive latches to stress our techniques. The natural place to look was in full-custom designs. Although we had access to several such chips, they had all been captured in ways which were unsuitable for our needs.

transformed circuit	D	B	s38417	s38584
# of memory elements	512	1890	3272	2904
# of connections	20496	30122	67704	35956
cycletime using flipflops	39	26	94	112
cycletime using latches	23	17	63	70
time to read circuit	4.9	7.2	15.5	8.6
aggressive `lps` time	235	1822	7718	4880
conservative `lps` time	48	284	1432	761
`solve` time	1.4	2.1	6.0	4.1
# generated constraints	19	13	21	21

Figure 7: Experimental results.

We then turned to standard cell designs. Since we did not want to create software to perform the extensive preprocessing mentioned in §2 until we were sure that our algorithms were practical, we obtained some netlists for circuits built entirely using edge-triggered flipflops with a single clock. Two such circuits had been designed in house, and two were taken from the ISCAS'89 benchmark set. We used a unit delay model for the ISCAS circuits. We converted these designs to level-sensitive circuits using the transformation shown in Figure 6. The transformation consists of duplicating the circuit and cross-connecting the data inputs of just the memory elements. Thus if the output of circuit element x is input to element y, then x_i is an input to y_i unless y is a memory element, in which case x_i is an input to y_{1-i}. Each copy of the circuit is given its own clock and all flipflops are changed to latches. The characteristics of the test circuits are shown in Figure 7. Note that these circuits all run substantially faster after transformation, which indicates that the optimal clocking schedule is making good use of the transparent phase of the latches.

7.2 The Programs

The program `lps` takes a circuit description and a linear order for clock events and generates a linear program for the optimal clock schedule. An option specifies whether aggressive or conservative constraints should be generated. The max and min operators in the constraints are, of course, *relaxed* to linear inequalities in the process. We tried three different LP packages (two using the simplex algorithm and one using an interior point method) on the resulting linear programs. We soon learned that solving the dual rather than primal problem was orders of magnitude faster regardless of which solver was used. Although one solver was generally the fastest of the three, it was not uniformly so; on occasional problem instances it was as much as five times slower than one of the other solvers.

The program `solve` reads in a circuit description, a linear order for the clock events, and a set of user provided constraints. User constraints can be any linear inequalities involving the variables in the clock schedule, for example, "no interval between events is more than 25% of the period," or "certain intervals must be equal." `Solve` then constructs a set of short path constraints using the conservative criteria in §3. The LBG is then built incorporating any short path or user constraints that give lower bounds on the separation between clock events. Then the program calculates ψ and generates long path constraints as described in §6. The resulting constraints are combined with any user or short path constraints that give upper bounds on event separation, as well as a few constraints that are implicit in our formulation, such as $\pi \geq \psi$. This final constraint set is then passed on to an LP solver. The whole program is about 1300 lines of C.

7.3 Results

The performance of the algorithms is shown in Figure 7 using one processor on a Silicon Graphics 4D/440 processor. Both programs `lps` and `solve` use a common routine to read a circuit description into memory and build a minimal data structure, so we list the time for that task separately. We then show the time used by `lps` to solve the full set of constraints prescribed in [8] under both the aggressive and conservative formulations of hold constraints. The time shown is actually the best time over the three LP solvers we tried. Finally, we list the time used by `solve` to compute ψ, generate $\mathcal{R}(\psi)$, and solve the resulting LP, along with the size of $\mathcal{R}(\psi)$. Quite clearly, `solve` is orders of magnitude faster than `lps` and spends more of its time absorbing input than producing answers.

Another experiment was to determine how much the optimal clock period of a circuit was affected by the use of conservative rather than aggressive constraints. It has been shown[9] that relaxing the conservative constraints will always yield the optimum answer, but the relaxed aggressive constraints might produce an overly optimistic (i.e., wrong) answer. We used the `lps` program to generate and solve both sets of constraints for each of our test circuits, as well as bipartite-transformed versions of every circuit in the ISCAS'89 suite. In all but one case the answers were the same. The exception was s15850, for which the relaxed aggressive constraints yielded a period that was 0.3% faster than the optimal conservative schedule. The aggressive schedule, when verified with `check`, turned out to have several hold violations that exceeded this "improvement." So for our test data, the use of aggressive constraints did not yield any faster schedules than the conservative ones. They did, however, slow the LP solver down dramatically. For this and the other reasons listed in §3, we suggest that the conservative formulation is more useful in practice than the aggressive formulation.

Acknowledgments

David Gay was an invaluable source of help regarding both linear and non-linear programming. I used his AMPL mathematical programming language as a front end to the various solvers used in the experimental phase of this research. Rae McLellan taught me a lot about sophisticated clocking schemes.

References

[1] D. E. Wallace and C. H. Sequin, "ATV: An abstract timing verifier," in *Proceedings of the IEEE/ACM Design Automation Conference*, vol. 25, pp. 154–159, June 1988.

[2] N. Weiner and A. Sangiovanni-Vincentelli, "Timing analysis in a logic synthesis environment," in *Proceedings of the IEEE/ACM Design Automation Conference*, vol. 26, pp. 655–661, June 1989.

[3] A. T. Ishii and C. E. Leiserson, "A timing analysis of level-clocked circuitry," in *Proceedings of the Sixth MIT Conference* (W. J. Dally, ed.), Advanced Research in VLSI, pp. 57–69, MIT Press, 1990.

[4] T. G. Szymanski, "LEADOUT: A static timing analyzer for MOS circuits," in *Digest of Technical Papers of the IEEE International Conference on Computer-Aided Design*, pp. 130–133, Nov. 1986.

[5] J. J. Cherry, "Pearl: A CMOS timing analyzer," in *Proceedings of the IEEE/ACM Design Automation Conference*, vol. 25, pp. 148–153, June 1988.

[6] K. A. Sakallah, T. N. Mudge, and O. A. Olukotun, "*check*T_c and *mint*$_c$: Timing verification and optimal clocking of synchronous digital circuits," in *Digest of Technical Papers of the IEEE International Conference on Computer-Aided Design*, pp. 552–555, Nov. 1990.

[7] M. R. Dagenais and N. C. Rumin, "On the calculation of optimal clocking parameters in synchronous circuits with level-sensitive latches," *IEEE Trans. Computer-Aided Design*, vol. 8, pp. 268–278, Mar. 1989.

[8] K. A. Sakallah, T. N. Mudge, and O. A. Olukotun, "Analysis and design of latch-controlled synchronous digital circuits," in *Proceedings of the IEEE/ACM Design Automation Conference*, vol. 27, pp. 111–117, June 1990.

[9] K. A. Sakallah, T. N. Mudge, and O. A. Olukotun, "Algorithms for timing verification and optimal clocking of synchronous digital circuits," Tech. Rep. CSE-TR-71-90, University of Michigan, Ann Arbor, Sept. 1990.

[10] E. L. Lawler, *Combinatorial Optimization: Networks and Matroids*. Holt, Rinehart and Winston, 1976.

[11] R. M. Karp, "A characterization of the minimum cycle mean in a digraph," *Discrete Mathematics*, vol. 23, pp. 309–311, 1978.

Verifying Clock Schedules

Thomas G. Szymanski
AT&T Bell Laboratories
Murray Hill, NJ 07974
email: tgs@research.att.com

Narendra Shenoy
University of California
Berkeley, CA 94720
email: shenoy@penny.berkeley.edu

Abstract

Many recent papers have formulated both timing verification and optimization as mathematical programming problems. Such formulations correctly handle level-sensitive latches, long and short path considerations, and sophisticated multi-phase clocking schemes.

This paper deals with the computational aspects of using such a formulation for verifying clock schedules. We show that the formulation can have multiple solutions, and that these extraneous solutions can cause previously published algorithms to produce incorrect or misleading results. We characterize the conditions under which multiple solutions exist, and show that even when the solution is unique, the running times of these previous algorithms can be unbounded. By contrast, we exhibit a simple polynomial time algorithm for clock schedule verification. The algorithm was implemented and used to check the timing of all the circuits in the ISCAS-89 suite. Observed running times are linear in circuit size and quite practical.

1 Introduction

An elegant mathematical programming formulation for timing verification and clock schedule optimization was presented in [SMO90]. The formulation, henceforth referred to as the *SMO formulation*, handles both long and short path constraints, deals correctly with both edge-triggered and level-sensitive latches, handles circuits with complex, multi-phase clocking schemes, and is sufficiently flexible to handle various system constraints that might be externally imposed upon the circuit being analyzed. It also provides a rigorous framework upon which one can discuss convexity of solution spaces, robustness of solutions, etc. For these reasons, several recent papers have built upon this model to treat more advanced timing problems including retiming [BSM92] and wave-pipelining [LB92].

In the SMO formulation, the timing properties of a circuit are modeled using the parameters and variables appearing in Figure 1. In this paper, we shall deal only with level-sensitive latches, the extension to edge-triggered flip-flops being a straightforward exercise. We identify the latches of the circuit with the integers from 1 to n. The symbol $\rightarrow$ denotes the "fans out to" relation on latches, that is, $i \rightarrow j$ if and only if there is a path of strictly combinational logic elements extending from the output of latch i to the input of latch j. Without loss of generality, we assume that each latch has at least one fanin, that is, for every j, there exists an i such that $i \rightarrow j$. The term *path* henceforth means a sequence of $k \geq 0$ latches, $j_0, \ldots, j_k$, such that $j_i \rightarrow j_{i+1}$ for $0 \leq i < k$.

The *clock schedule optimization problem* asks us to find the minimum value of π for which there is an assignment to all variables consistent with the constraints in Figure 1. The *clock schedule verification problem*, on the other hand, presents us with values for π, w and e and asks us to find values for the rest of the variables so as to satisfy the constraints. Since this paper deals exclusively with the schedule verification problem, we can simplify the SMO constraints to better suit our purpose. We introduce the auxiliary variables B, λ, and Λ as shown in Figure 2. It is easy to see that B, λ, and Λ are uniquely determined by the clock schedule, as is E. We may therefore recast the clock schedule verification problem as asking whether there exist a, d, A, and D that obey the *equations* and satisfy the *constraints* appearing in Figure 2.

The following lemma will be invoked many times throughout this paper.

Lemma 1.1 *Let (a, d, A, D) be a solution to the equations. Let $j_0, \ldots, j_k$ be a path with $k \geq 0$. Then $D_{j_k} \geq A_{j_k} \geq D_{j_0} + \sum_{i=0}^{k-1} \Lambda_{j_i,j_{i+1}}$. Moreover, if $j_0 = j_k$, then $0 \geq \sum_{i=0}^{k-1} \Lambda_{j_i,j_{i+1}}$.*

Proof. ¿From the equations, $D_i \geq A_i \geq D_j + \Lambda_{j,i}$ for any i and j for which $j \rightarrow i$. In particular, $D_{j_k} \geq A_{j_k} \geq D_{j_{k-1}} + \Lambda_{j_{k-1},j_k}$. Continuing the substitution, $D_{j_k} \geq A_{j_k} \geq D_{j_0} + \sum_{i=0}^{k-1} \Lambda_{j_i,j_{i+1}}$. Setting $D_{j_k} = D_{j_0}$ and simple manipulation complete the result. □

The following observation, although elementary, is used so frequently throughout this paper that it is worth stating explicitly.

Fact 1.2 *The functions* min *and* max *are both monotonic in their arguments, that is, increasing the value of any argument cannot decrease the value of the function, nor can decreasing the value of any argument cause an increase in the value of the function.*

Finally, a solution (a^*, d^*, A^*, D^*) is called a *minimum* solution, if for any other solution $(\tilde{a}, \tilde{d}, \tilde{A}, \tilde{D})$, we have $a_i^* \leq \tilde{a}_i$, $d_i^* \leq \tilde{d}_i$, $A_i^* \leq \tilde{A}_i$, and $D_i^* \leq \tilde{D}_i$, for all i.

Reprinted from *Proc. IEEE/ACM Int'l Conf. Computer-Aided Design*, pp. 124–131, Nov. 1992.

	Parameters
n	number of latches in circuit
p_i	clock phase controlling latch i
S_i	setup time of latch i
H_i	hold time of latch i
$\delta_{i,j}$	minimum combinational delay from latch i to latch j
$\Delta_{i,j}$	maximum combinational delay from latch i to latch j
	Variables defining the clock schedule
π	clock period
w_i	length of time that clock phase i is active
e_i	absolute time within period when phase i begins
	Other variables
$E_{i,j}$	time between start of phase i and *next* phase j
a_i	earliest signal arrival time at latch i
A_i	latest signal arrival time at latch i
d_i	earliest signal departure time from latch i
D_i	latest signal departure time from latch i
	Equations and constraints
	$E_{i,j} = \begin{cases} e_j - e_i & \text{if } i < j \\ \pi + e_j - e_i & \text{otherwise} \end{cases}$
	$d_i = \max(a_i, \pi - w_{p_i})$
	$D_i = \max(A_i, \pi - w_{p_i})$
	$a_i = \min_{j \to i}(d_j + \delta_{j,i} - E_{p_j,p_i})$
	$A_i = \max_{j \to i}(D_j + \Delta_{j,i} - E_{p_j,p_i})$
	$a_i \geq H_i$
	$A_i \leq \pi - S_i$

Figure 1: The SMO timing model for level-sensitive latches. It may easily be extended to accomodate edge-triggered flip-flops.

Variables determined directly by clock chedule
$E_{i,j} = \begin{cases} e_j - e_i & \text{if } i < j \\ \pi + e_j - e_i & \text{otherwise} \end{cases}$
$B_i = \pi - w_i$
$\lambda_{j,i} = \delta_{j,i} - E_{p_j,p_i}$
$\Lambda_{j,i} = \Delta_{j,i} - E_{p_j,p_i}$
Equations defining arrival and departure times
$d_i = \max(a_i, B_i)$
$D_i = \max(A_i, B_i)$
$a_i = \min_{j \to i}(d_j + \lambda_{j,i})$
$A_i = \max_{j \to i}(D_j + \Lambda_{j,i})$
Constraints for correct operation
$a_i \geq H_i$
$A_i \leq \pi - S_i$

Figure 2: The SMO timing model as simplified for clock schedule verification.

2 Solving the Equations

Iteration is a popular method for solving a set of equations $V = \mathcal{E}(V)$ over a set of variables $V = \{V_j \mid 1 \leq j \leq n\}$. One begins with a trial "solution" V^0, and computes a succession of iterates, $V^{i+1} = \mathcal{E}(V^i)$, until convergence occurs, $V^{i+1} = V^i$, at which point V^i is a solution to $\mathcal{E}$. When the equations are formed from monotonic operators, as in our case, the entire sequence of iterates will be monotonic nondecreasing or nonincreasing, provided we choose the initial iterate properly. Thus, when $V^0 \leq V^1$ componentwise (that is, $V_j^0 \leq V_j^1$ for all j) we are guaranteed that $V^0 \leq V^1 \leq V^2 \cdots$. Similarly, when $V^0 \geq V^1$, we will have $V^0 \geq V^1 \geq V^2 \cdots$. The monotonicity of the sequence of iterates can be used to prove convergence provided a suitable upper (or lower) bound is known for the solution.

For clock schedule verification, we can partition the equations into two independent subsets, one subset (the *earliest time equations*) involving the variables a_i and d_i, and the other subset (the *latest time equations*) involving the variables A_i and D_i. By choosing appropriate starting points for either subset, we can choose whether we converge upwards or downwards to a solution. The choice of starting point is crucial; it affects both the correctness of the result and the time needed for the process to converge. We shall show that the best approach is to converge upwards in both sets of equations to what will turn out to be the minimum fixed point of the verification equations. The rest of this section defines this technique more precisely, proves that it yields a minimum solution if any solution exists at all, and establishes that the method runs in polynomial time.

Construction 2.1 *For a given clock schedule, compute E, B, λ, Λ as described above. Then, for all $m \geq 0$, define*

$$\begin{array}{rcll} a_i^0 & = & -\infty & \\ A_i^0 & = & -\infty & \\ a_i^m & = & \min_{j \to i}(d_j^{m-1} + \lambda_{j,i}) & \text{if } m > 0 \\ A_i^m & = & \max_{j \to i}(D_j^{m-1} + \Lambda_{j,i}) & \text{if } m > 0 \\ d_i^m & = & \max(a_i^m, B_i) & \\ D_i^m & = & \max(A_i^m, B_i). & \end{array}$$

It is easy to see that this process is well defined. Simply compute A^0 and a^0, then D^0 and d^0, then A^1 and a^1, etc. The values of the variables turn out to be monotonic nondecreasing in m, as shown in the following lemma.

Lemma 2.2 (Monotonicity) *For all i and $m \geq 1$, $a_i^m \geq a_i^{m-1}$, $d_i^m \geq d_i^{m-1}$, $A_i^m \geq A_i^{m-1}$, and $D_i^m \geq D_i^{m-1}$.*

Proof. By induction on m. We will only provide the argument for a_i^m and d_i^m, the argument for A_i^m and D_i^m being similar. For the basis with $m = 1$, we have $a_i^1 \geq a_i^0$ because $a_i^0 = -\infty$. Moreover, $d_i^0 = B_i$, and $d_i^1 \geq B_i$, so $d_i^1 \geq d_i^0$.

For the inductive step, assume the lemma is true for $m - 1$ with $m > 1$. By the monotonicity of min and the inductive assertion, we have $\min_{j\to i}(d_j^{m-1} + \lambda_{j,i}) \geq \min_{j\to i}(d_j^{m-2} + \lambda_{j,i})$. By definition, $a_i^m = \min_{j\to i}(d_j^{m-1} + \lambda_{j,i})$ and $a_i^{m-1} = \min_{j\to i}(d_j^{m-2} + \lambda_{j,i})$, so $a_i^m \geq a_i^{m-1}$. The monotonicity of max then implies that $\max(a_i^m, B_i) \geq \max(a_i^{m-1}, B_i)$, and so $d_i^m \geq d_i^{m-1}$. □

The next lemma shows that the iterated variables approach a solution from below. As a consequence, if the process converges, it will converge to a solution which is a minimum solution.

Lemma 2.3 *Let $(\tilde{a}, \tilde{d}, \tilde{A}, \tilde{D})$ be a solution to the equations. Then for all i and $m \geq 0$, $a_i^m \leq \tilde{a}_i$, $d_i^m \leq \tilde{d}_i$, $A_i^m \leq \tilde{A}_i$, and $D_i^m \leq \tilde{D}_i$.*

Proof. By induction on m using an argument nearly identical to Lemma 2.2. We will only provide the argument for a_i^m and d_i^m, the argument for A_i^m and D_i^m being similar. For the basis with $m = 0$, we have $a_i^0 \leq \tilde{a}_i$ because $a_i^0 = -\infty$. Moreover, $d_i^0 = B_i$, and $\tilde{d}_i \geq B_i$ directly from the equations, and so $d_i^0 \leq \tilde{d}_i$.

For the inductive step, assume the lemma is true for $m - 1$. By the monotonicity of min and the inductive assertion, we have $\min_{j\to i}(d_j^{m-1} + \lambda_{j,i}) \leq \min_{j\to i}(\tilde{d}_j + \lambda_{j,i})$. By definition, $a_i^m = \min_{j\to i}(d_j^{m-1} + \lambda_{j,i})$ and $\tilde{a}_i = \min_{j\to i}(\tilde{d}_j + \lambda_{j,i})$, so $a_i^m \leq \tilde{a}_i$. The monotonicity of max then implies that $\max(a_i^m, B_i) \leq \max(\tilde{a}_i, B_i)$, and so $d_i^m \leq \tilde{d}_i$. □

As the iteration proceeds, the values of the variables increase. The next lemma will be used to relate each such change to some latch which is "to blame."

Lemma 2.4 *For any i, if $m \geq 1$ and $d_i^m > d_i^{m-1}$, then there exists a j such that $j \to i$, and $d_i^m = a_i^m \leq d_j^{m-1} + \lambda_{j,i}$. Moreover, if $m > 1$, then $d_j^{m-1} > d_j^{m-2}$.*

Proof. Since $d_i^m > d_i^{m-1}$ by hypothesis, $d_i^m = \max(a_i^m, B_i) > \max(a_i^{m-1}, B_i) = d_i^{m-1}$. Lemma 2.2 then implies that $d_i^m = a_i^m$ and $a_i^m > a_i^{m-1}$.

Case 1: $m = 1$. By definition, $a_i^1 = \min_{p\to i}(d_p^0 + \lambda_{p,i})$. Pick j such that $j \to i$ and $d_j^0 + \lambda_{j,i} = \min_{p\to i}(d_p^0 + \lambda_{p,i})$. Then $a_i^1 \leq d_j^0 + \lambda_{j,i}$, as required.

Case 2: $m > 1$. Since $a_i^m > a_i^{m-1}$, we have $\min_{p\to i}(d_p^{m-1}+\lambda_{p,i}) > \min_{p\to i}(d_p^{m-2}+\lambda_{p,i})$. Clearly, there must exist at least one p for which $p \to i$ and $d_p^{m-1} > d_p^{m-2}$. Take $j = p$. Then $a_i^m = \min_{p\to i}(d_p^{m-1} + \lambda_{p,i}) \leq d_j^{m-1} + \lambda_{j,i}$, satisfying the lemma. □

The next lemma is the key technical result of this section. It shows that the iterations need not be continued indefinitely. More specifically, it shows that if the earliest time equations do not converge within n iterations, then the latest time equations do not converge at all.

Lemma 2.5 *If $d_i^n \neq d_i^{n-1}$ for some i, then the equations have no solution.*

Proof. By Lemma 2.2, $d_i^n > d_i^{n-1}$. Construct a path $j_0, \ldots, j_n$ by setting $j_n = i$ and then applying Lemma 2.4 n times to obtain $j_{n-1}, \ldots, j_0$. Then, for each k, $0 \leq k < n$, we have $j_k \to j_{k+1}$ and $d_{j_{k+1}}^{k+1} \leq d_{j_k}^k + \lambda_{j_k,j_{k+1}}$. Moreover, for $1 \leq k < n$, we have $d_{j_k}^k > d_{j_k}^{k-1}$. Accordingly, for any p and q with $p < q$, we have $d_{j_q}^q \leq d_{j_p}^p + \sum_{k=p}^{q-1} \lambda_{j_k,j_{k+1}}$. We also have $d_{j_q}^q > d_{j_q}^{q-1}$, which implies by Lemma 2.2 that $d_{j_q}^q > d_{j_q}^p$. Since each j_i is an integer between 1 and n inclusive, we can pick p and q so that $j_p = j_q$. Thus $d_{j_q}^q \leq d_{j_q}^p + \sum_{k=p}^{q-1} \lambda_{j_k,j_{k+1}}$. Hence, $0 < \sum_{k=p}^{q-1} \lambda_{j_k,j_{k+1}}$, implying that $0 < \sum_{k=p}^{q-1} \Lambda_{j_k,j_{k+1}}$. However, Lemma 1.1 applied to the subsequence $j_p, \ldots, j_q$, says that if a solution exists, we must have $0 \geq \sum_{k=p}^{q-1} \Lambda_{j_k,j_{k+1}}$. Therefore the equations are unsolvable. □

The next two lemmas do for the variables A and D what the previous two lemmas did for a and d. Note that the statements of the lemmas are very close, but not identical, to the earlier lemmas.

Lemma 2.6 *For any i, if $m \geq 1$ and $D_i^m > D_i^{m-1}$, then there exists a j such that $j \to i$, and $D_i^m = A_i^m = D_j^{m-1} + \Lambda_{j,i}$. Moreover, if $m > 1$, then $D_j^{m-1} > D_j^{m-2}$.*

Proof. Similar to Lemma 2.4. □

Lemma 2.7 *If $D_i^n \neq D_i^{n-1}$ for some i, then the equations have no solution.*

Proof. Similar to Lemma 2.5. □

Next we show that Construction 2.1 converges to a solution if it converges at all.

Lemma 2.8 *Let p and q be integers such that $d_i^{p-1} = d_i^p$ for every i, and $D_i^{q-1} = D_i^q$ for every i. Then (a^p, d^p, A^q, D^q) is a minimum solution to the equations.*

Proof. It is trivial to verify that (a^p, d^p, A^q, D^q) satisfies the equations. Lemma 2.3 shows that the solution is a minimum solution. □

Finally we arrive at the key result of this section, namely, a polynomial time algorithm for solving the equations.

Theorem 2.9 *If the equations have a solution, then they have a minimum solution. Moreover, this minimum solution can be found in $O(n^3)$ time.*

Proof. Clearly we can compute the vectors d^n and D^n in $O(n^3)$ time. In $O(n)$ additional time, we can apply Lemmas 2.5 and 2.7 to test whether any solution exists. If one does, then (a^n, d^n, A^n, D^n) is a solution. Moreover, by Lemma 2.8, it is a minimum solution. □

The reader will note that our suggested solution to the latest time equations is equivalent to the Bellman-Ford algorithm for finding shortest paths in a directed graph with positive and negative edge weights. Ishii[IL90] has previously advocated this approach as well.

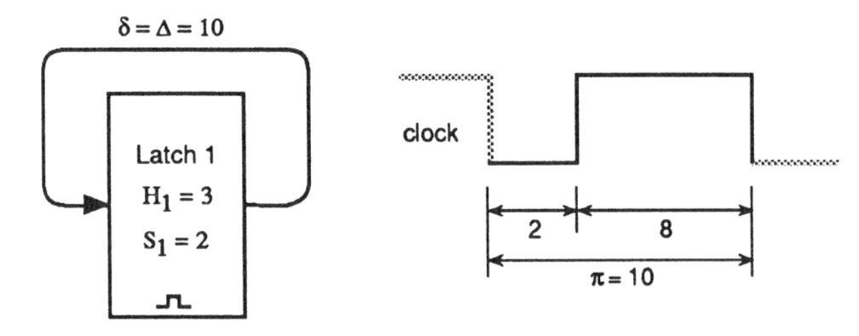

Figure 3: A circuit and clock schedule for which the SMO equations have multiple solutions.

3 Uniqueness

The algorithm of the previous section can be used to find a solution to the equations. Clock schedule verification can then be performed by checking whether the solution obeys the constraints on setup and hold times listed in Figure 2. This of course raises an important question, namely, what if the equations have multiple solutions? Some of the solutions might satisfy the constraints, while other solutions violate them. For example, consider the circuit and clock schedule shown in Figure 3. The corresponding SMO equations become

$$\lambda_{1,1} = 0 \qquad \Lambda_{1,1} = 0$$
$$a_1 = d_1 \qquad A_1 = D_1$$
$$d_1 = \max(a_1, 2) \qquad D_1 = \max(A_1, 2)$$

and the SMO constraints are simply

$$a_1 \geq 3 \qquad A_1 \leq 8.$$

For any $x \geq 2$, $A_1 = D_1 = a_1 = d_1 = x$ is a valid solution to these equations. .Solutions having $x < 3$ exhibit a hold violation, solutions having $x > 8$ exhibit a setup violation, and solutions with $3 \leq x \leq 8$ are free of violations.

Thus we see that solutions to the SMO equations are not, in general, unique, and different solutions can give differing results vis-a-vis the timing constraints. In this section, we will characterize the circumstances under which multiple solutions exist. More specifically, we will show that multiple solutions can only exist at the optimal clock period, and are due to cycles of zero aggregate delay in the circuit.

Throughout this section, we will use $(\tilde{a}, \tilde{d}, \tilde{A}, \tilde{D})$ to denote an arbitrary solution to the equations, and (a^*, d^*, A^*, D^*) to denote the minimum solution to the equations. We begin with a pair of lemmas needed in the proof of the theorem which is to follow.

Lemma 3.1 *If $\tilde{d}_i > d_i^*$, then there exists a j such that $\tilde{d}_j > d_j^*$, $j \rightarrow i$ and $\tilde{d}_i \leq \tilde{d}_j + \lambda_{j,i}$.*

Proof. By hypothesis, $\tilde{d}_i = \max(\tilde{a}_i, B_i) > \max(a_i^*, B_i) = d_i^*$, implying $\tilde{a}_i > a_i^*$. Thus, $\tilde{a}_i = \min_{p \rightarrow i}(\tilde{d}_p + \lambda_{p,i}) > \min_{p \rightarrow i}(d_p^* + \lambda_{p,i}) = a_i^*$. Hence we can find a j such that $j \rightarrow i$ and $\tilde{d}_j > d_j^*$. Moreover, $\tilde{a}_i \leq \tilde{d}_j + \lambda_{j,i}$. □

Lemma 3.2 *If $\tilde{D}_i > D_i^*$, then there exists a j such that $\tilde{D}_j > D_j^*$, $j \rightarrow i$ and $\tilde{D}_i = \tilde{D}_j + \Lambda_{j,i}$.*

Proof. Similar to Lemma 3.1. By hypothesis, $\tilde{D}_i = \max(\tilde{A}_i, B_i) > \max(A_i^*, B_i) = D_i^*$, implying $\tilde{A}_i > A_i^*$. Thus, $\tilde{A}_i = \max_{p \rightarrow i}(\tilde{D}_p + \Lambda_{p,i}) > \max_{p \rightarrow i}(D_p^* + \Lambda_{p,i}) = A_i^*$. Take j such that $j \rightarrow i$ and $\tilde{D}_j + \Lambda_{j,i} = \max_{p \rightarrow i}(\tilde{D}_p + \Lambda_{p,i})$. Certainly $\tilde{D}_j > D_j^*$. Moreover, $\tilde{A}_i = \tilde{D}_j + \Lambda_{j,i}$. □

We are now ready to show that multiple solutions to the equations are due to the presence of zero weight cycles in the circuit. Such a cycle represents a path through the circuit in which a signal returns to some latch at precisely the same time (relative to the clock period) as it left that latch.

Theorem 3.3 *If the equations have more than one solution, then there exists a path $j_0, \ldots, j_k$, $k \geq 1$, such that $j_0 = j_k$ and $0 = \sum_{i=0}^{k-1} \Lambda_{j_i,j_{i+1}}$.*

Proof. Let $(\tilde{a}, \tilde{d}, \tilde{A}, \tilde{D})$ be a solution other than the minimum solution (a^*, d^*, A^*, D^*).

Case 1: $\tilde{d}_m > d_m^*$ for some m. Construct a path $j_0, \ldots, j_n$ by setting $j_n = m$ and then applying Lemma 3.1 n times to obtain $j_{n-1}, \ldots, j_0$. Then, for each i, $0 \leq i < n$, we have $j_i \rightarrow j_{i+1}$ and $\tilde{d}_{j_{i+1}} \leq \tilde{d}_{j_i} + \lambda_{j_i,j_{i+1}}$. Clearly, this path must contain repeated indices, so pick p and q such that $p < q$ and $j_p = j_q$. The path that will satisfy the lemma is $j_p, \ldots, j_q$. Substituting, $\tilde{d}_{j_q} \leq \tilde{d}_{j_p} + \sum_{i=p}^{q-1} \lambda_{j_i,j_{i+1}}$. Thus $0 \leq \sum_{i=p}^{q-1} \lambda_{j_i,j_{i+1}} \leq \sum_{i=p}^{q-1} \Lambda_{j_i,j_{i+1}}$. By Lemma 1.1, we conclude that $0 = \sum_{i=p}^{q-1} \Lambda_{j_i,j_{i+1}}$.

Case 2: $\tilde{D}_m > D_m^*$ for some m. Use Lemma 3.2 to construct a path $j_0, \ldots, j_n$ with $j_n = m$, such that $j_i \rightarrow j_{i+1}$ and $\tilde{D}_{j_{i+1}} = \tilde{D}_{j_i} + \Lambda_{j_i,j_{i+1}}$ for each i, $0 \leq i < n$. Pick p and q such that $p < q$ and $j_p = j_q$. Now consider the subpath $j_p, \ldots, j_q$. Clearly, $\tilde{D}_{j_q} = \tilde{D}_{j_p} + \sum_{i=p}^{q-1} \Lambda_{j_i,j_{i+1}}$, and we have $0 = \sum_{i=p}^{q-1} \Lambda_{j_i,j_{i+1}}$.

Case 3: $\tilde{d}_m = d_m^*$ and $\tilde{D}_m = D_m^*$ for every m. This implies that $\tilde{a}_m = a_m^*$ and $\tilde{A}_m = A_m^*$ for every m, in contradiction to the assumption that the two solutions are distinct. □

An immediate consequence is that non-unique solutions are a phenomenon that can only occur at the optimal clock period. Said another way, the equations have a unique solution whenever the clock schedule has a suboptimal period.

Corollary 3.4 *If the equations have more than one solution for a clock schedule with period π, then π is optimal, that is, no valid schedule has a period less than π.*

Proof. Let $j_0, \ldots, j_k$ fulfill the conditions of Theorem 3.3. Then

$$0 = \sum_{i=0}^{k-1} \Lambda_{j_i,j_{i+1}} = \sum_{i=0}^{k-1} \Delta_{j_i,j_{i+1}} - \sum_{i=0}^{k-1} E_{p_{j_i},p_{j_{i+1}}},$$

which may be further rewritten as

$$0 = \sum_{i=0}^{k-1} \Delta_{j_i,j_{i+1}} - \sum_{i=0}^{k-1} (e_{p_{j_{i+1}}} - e_{p_{j_i}}) - c\pi,$$

where c is the number of i, $0 \le i < k$, for which $p_{j_{i+1}} \le p_{j_i}$. Observing that

$$\sum_{i=0}^{k-1} (e_{p_{j_{i+1}}} - e_{p_{j_i}}) = \sum_{i=1}^{k} e_{p_{j_i}} - \sum_{i=0}^{k-1} e_{p_{j_i}} = e_{p_{j_k}} - e_{p_{j_0}} = 0,$$

we can conclude that

$$0 = \sum_{i=0}^{k-1} \Lambda_{j_i,j_{i+1}} = \sum_{i=0}^{k-1} \Delta_{j_i,j_{i+1}} - c\pi.$$

Reducing π would certainly violate Lemma 1.1, and so no solutions exist for smaller values of π. □

Having shown that zero-weight cycles are a necessary condition for multiple solutions to the SMO equations, we shall next show that they are also sufficient. We'll do this by showing how such a cycle in a solution can be used to construct other solutions as well. The basic idea is to add some ϵ to all the A_i's and D_i's along the cycle. In general, this alone will not yield a solution. However, the "perturbed" values can be used to reinitialize the algorithm of §2, which can then be used to converge upwards to a new solution. Before proceeding, we note a consequence of the previous theorem.

Corollary 3.5 *Let $(\tilde{a}, \tilde{d}, \tilde{A}, \tilde{D})$ be a solution to the equations and let $j_0, \ldots, j_k$ be as in Theorem 3.3. Then $\tilde{D}_{j_{i+1}} = \tilde{A}_{j_{i+1}} = \tilde{D}_{j_i} + \Lambda_{j_i,j_{i+1}}$, for $0 \le i < k$.*

Proof. First, suppose that some $D_{j_i} \ne A_{j_i}$. Renumbering the j's if necessary, assume that it is $D_{j_k} \ne A_{j_k}$. By Lemma 1.1, $D_{j_k} \ge A_{j_k} \ge D_{j_0} + \sum_{i=0}^{k-1} \Lambda_{j_i,j_{i+1}} = D_{j_0} = D_{j_k}$. This clearly implies that $D_{j_k} = A_{j_k}$.

Second, suppose that some $D_{j_i} = A_{j_i} \ne D_{j_{i-1}} + \Lambda_{j_{i-1},j_i}$. Once again, renumbering if necessary, assume that $D_{j_1} = A_{j_1} \ne D_{j_0} + \Lambda_{j_0,j_1}$. By Lemma 1.1, $D_{j_k} \ge D_{j_1} + \sum_{i=1}^{k-1} \Lambda_{j_i,j_{i+1}}$. Thus $D_{j_k} > D_{j_0} + \sum_{i=0}^{k-1} \Lambda_{j_i,j_{i+1}} = D_{j_0}$. This is impossible since $j_k = j_0$. □

Construction 3.6 *Suppose that $(\tilde{a}, \tilde{d}, \tilde{A}, \tilde{D})$ is a solution to the equations and that $j_0, \ldots, j_k$, $k \ge 1$, is a path for which $j_0 = j_k$ and $0 = \sum_{i=0}^{k-1} \Lambda_{j_i,j_{i+1}}$. Let C be the set $\{j_i | 0 \le i < k\}$ and let ϵ be any positive real. For all $m \ge 0$, define*

$$\begin{aligned} \hat{A}_i^0 &= \begin{cases} \tilde{A}_i + \epsilon & \text{if } i \in C \\ \tilde{A}_i & \text{otherwise} \end{cases} \\ \hat{D}_i^m &= \max(\hat{A}_i^m, B_i) \qquad \text{if } m \ge 0 \\ \hat{A}_i^m &= \max\nolimits_{j \to i}(\hat{D}_j^{m-1} + \Lambda_{j,i}) \qquad \text{if } m \ge 1. \end{aligned}$$

Lemma 3.7

$$\hat{D}_i^0 = \begin{cases} \hat{A}_i^0 = \tilde{D}_i + \epsilon = \tilde{A}_i + \epsilon & \text{if } i \in C \\ \tilde{D}_i & \text{if } i \notin C \end{cases}$$

Proof. Suppose that $i \in C$. Corollary 3.5 tells us that $\tilde{D}_i = \max(\tilde{A}_i, B_i) = \tilde{A}_i \ge B_i$. Hence, $\hat{D}_i^0 = \max(\hat{A}_i^0, B_i) = \max(\tilde{A}_i + \epsilon, B_i) = \tilde{A}_i + \epsilon$. Since $\tilde{D}_i = \tilde{A}_i$, we have $\hat{D}_i^0 = \tilde{D}_i + \epsilon$.

Next suppose that $i \notin C$. Then $\hat{D}_i^0 = \max(\hat{A}_i^0, B_i) = \max(\tilde{A}, B_i) = \tilde{D}_i$. □

The next several lemmas show that the iterations performed on the perturbed solution yield monotonically non-decreasing values, must converge to another solution within n steps, and does not affect the A_i and D_i values for those latches i that are on the cycle C.

Lemma 3.8 *For all i and $m \ge 1$, $\hat{A}_i^m \ge \hat{A}_i^{m-1}$, and $\hat{D}_i^m \ge \hat{D}_i^{m-1}$.*

Proof. By induction on m. For the basis, $m = 1$. First consider $\hat{A}_i^1$. There are two cases, depending on whether i lies on the cycle.

Case 1: $i \notin C$. Lemma 3.7 tells us that $\hat{D}_j^0 \ge \tilde{D}_j$ for all j. Then $\hat{A}_i^1 = \max_{j \to i}(\hat{D}_j^0 + \Lambda_{j,i}) \ge \max_{j \to i}(\tilde{D}_j + \Lambda_{j,i}) = \tilde{A}_i = \hat{A}_i^0$.

Case 2: $i \in C$. By Corollary 3.5, there exists some $p \in C$ such that

$$\tilde{A}_i = \max_{j \to i}(\tilde{D}_j + \Lambda_{j,i}) = \tilde{D}_p + \Lambda_{p,i}.$$

By Lemma 3.7,

$$\hat{A}_i^1 = \max_{j \to i}(\hat{D}_j^0 + \Lambda_{j,i}) \ge \hat{D}_p^0 + \Lambda p, i = \tilde{D}_p + \epsilon + \Lambda_{p,i}.$$

Together, these imply that

$$\hat{A}_i^1 \ge \tilde{A}_i + \epsilon = \hat{A}_i^0,$$

as was to be shown.

Having established that $\hat{A}_i^1 \ge \hat{A}_i^0$, it is easy to see that $\hat{D}_i^1 = \max(\hat{A}_i^1, B_i) \ge \max(\hat{A}_i^0, B_i) = \hat{D}_i^0$ by the monotonicity of max. Thus $\hat{D}_i^1 \ge \hat{D}_i^0$.

For the inductive step, assume the lemma is true for $m - 1$ with $m > 1$. By the monotonicity of max and the inductive assertion, we have $\max_{j \to i}(\hat{D}_j^{m-1} + \Lambda_{j,i}) \ge \max_{j \to i}(\hat{D}_j^{m-2} + \Lambda_{j,i})$. Recall that $\hat{A}_i^m = \max_{j \to i}(\hat{D}_j^{m-1} + \Lambda_{j,i})$ and $\hat{A}_i^{m-1} = \max_{j \to i}(\hat{D}_j^{m-2} + \Lambda_{j,i})$, so substitution gives $\hat{A}_i^m \ge \hat{A}_i^{m-1}$. The monotonicity of max then implies that $\max(\hat{A}_i^m, B_i) \ge \max(\hat{A}_i^{m-1}, B_i)$, and so $\hat{D}_i^m \ge \hat{D}_i^{m-1}$ as was to be shown. □

Lemma 3.9 *For any i, if $m \ge 1$ and $\hat{D}_i^m > \hat{D}_i^{m-1}$, then there exists a j such that $j \to i$, and $\hat{D}_i^m = \hat{A}_i^m = \hat{D}_j^{m-1} + \Lambda_{j,i}$. Moreover, if $m > 1$, then $\hat{D}_j^{m-1} > \hat{D}_j^{m-2}$.*

Proof. Analogous to Lemma 2.6. Since $\hat{D}_i^m > \hat{D}_i^{m-1}$ by hypothesis, $\hat{D}_i^m = \max(\hat{A}_i^m, B_i) > \max(\hat{A}_i^{m-1}, B_i) = \hat{D}_i^{m-1}$. Lemma 3.8 then implies that $\hat{D}_i^m = \hat{A}_i^m$ and $\hat{A}_i^m > \hat{A}_i^{m-1}$.

Case 1: $m = 1$. By definition, $\hat{A}_i^1 = \max_{p \to i}(\hat{D}_p^0 + \Lambda_{p,i})$. Pick any j such that $j \to i$ and $\hat{D}_j^0 + \Lambda_{j,i} = \max_{p \to i}(\hat{D}_p^0 + \Lambda_{p,i})$. Hence $\hat{A}_i^1 = \hat{D}_j^0 + \Lambda_{j,i}$.

Case 2: $m > 1$. Since $\hat{A}_i^m > \hat{A}_i^{m-1}$, we have $\max_{p\to i}(\hat{D}_p^{m-1} + \Lambda_{p,i}) > \max_{p\to i}(\hat{D}_p^{m-2} + \Lambda_{p,i})$. Pick any j be such that $j \to i$ and $\hat{D}_j^{m-1} + \Lambda_{j,i} = \max_{p\to i}(\hat{D}_p^{m-1} + \Lambda_{p,i})$. By our choice of j, $\hat{D}_i^m = \hat{A}_i^m = \hat{D}_j^{m-1} + \Lambda_{j,i}$. Moreover, $\hat{D}_j^{m-1} > \hat{D}_j^{m-2}$. □

Lemma 3.10 *For any $i \in C$, if $m \geq 1$ then $\hat{D}_i^m = \tilde{D}_i + \epsilon$.*

Proof. Since $\tilde{D}_i + \epsilon = \hat{D}_i^0$ by Lemma 3.7, it suffices to show that $\hat{D}_i^m = \hat{D}_i^0$ for all m. Suppose the contrary, that is, suppose that $\hat{D}_i^m > \hat{D}_i^0$ for some $m \geq 1$. Accordingly, there is some $k \geq 1$ such that $\hat{D}_i^k > \hat{D}_i^{k-1}$. Apply Lemma 3.9 to construct a path $j_0, \ldots, j_k$ such that

$$\hat{D}_i^k = \hat{D}_{j_k}^k = \hat{D}_{j_0}^0 + \sum_{p=0}^{k-1} \Lambda_{j_p,j_{p+1}}.$$

Moreover, Lemmas 3.8 and 3.7 tell us that $\hat{D}_i^k > \hat{D}_i^0 = \tilde{D}_i + \epsilon$, and so we have,

$$\tilde{D}_i + \epsilon < \hat{D}_{j_0}^0 + \sum_{p=0}^{k-1} \Lambda_{j_p,j_{p+1}}.$$

Next, apply Lemma 1.1 to path $j_0, \ldots, j_k$ to see that

$$\tilde{D}_i \geq \tilde{D}_{j_0} + \sum_{p=0}^{k-1} \Lambda_{j_p,j_{p+1}}.$$

Thus, $\hat{D}_{j_0}^0 > \tilde{D}_{j_0} + \epsilon$, contradicting Lemma 3.7. □

Theorem 3.11 (converse of Theorem 3.3) *Let $j_0, \ldots, j_k$, $k \geq 1$, be a path such that $j_0 = j_k$ and $0 = \sum_{i=0}^{k-1} \Lambda_{j_i,j_{i+1}}$. Then the equations either have no solutions, or else they have infinitely many solutions, and the solutions are unboundedly large.*

Proof. If the equations have any solution at all, then we can apply Construction 3.6 to that solution. We claim that $\hat{D}_i^n = \hat{D}_i^{n-1}$ for all i. It is easy to show this using an argument along the same lines as the proof of Lemma 2.7. That is, assume the contrary, apply Lemma 3.9 to produce a path passing through the same node twice, and from there show the existence of a cycle of positive weight contradicting Lemma 1.1.

It is now straightforward to verify that $(\tilde{a}, \tilde{d}, \hat{A}^n, \hat{D}^n)$ is a solution to the equations. Moreover, each value of ϵ used in Construction 3.6 gives rise to a different solution, as readily seen from Lemma 3.10. We conclude that there are infinitely many solutions to the equations. □

Corollary 3.12 *If there are more than one solution to the equations, then there are solutions that violate the setup constraints.*

Proof. Simply pick ϵ large enough. □

Theorem 3.13 *The uniqueness of a solution can be determined in $O(n^2)$ time.*

Proof. Let $(\tilde{a}, \tilde{d}, \tilde{A}, \tilde{D})$ be a solution to the equations. Define the relation *determines* (symbolized $\triangleright$) by $j \triangleright i$ if and only if $j \to i$ and $\tilde{D}_i = \tilde{D}_j + \Lambda_{j,i}$. We claim that the equations have multiple solutions if and only $\triangleright$ is cyclic, that is, if and only if there exists some i for which $i \overset{+}{\triangleright} i$.

For the "if" part, suppose that $\triangleright$ is cyclic. Then we can find a sequence $j_0, \ldots, j_k$ with $k \geq 1$, such that $j_0 = j_k$ and $j_i \triangleright j_{i+1}$ for each i with $0 \leq i < k$. By definition of $\triangleright$, this means that $\tilde{D}_{j_{i+1}} = \tilde{D}_{j_i} + \Lambda_{j_i,j_{i+1}}$ for $0 \leq i < k$. Substituting, we have $\tilde{D}_{j_k} = \tilde{D}_{j_0} + \sum_{i=0}^{k-1} \Lambda_{j_i,j_{i+1}}$. Since $j_0 = j_k$, it must be that $0 = \sum_{i=0}^{k-1} \Lambda_{j_i,j_{i+1}}$. Since we hypothesized that the equations have at least one solution, Theorem 3.11 tells us that multiple solutions exist.

For the "only if" part, suppose that the equations have multiple solutions. By Theorem 3.3, there exists a path, $j_0, \ldots, j_k$, $k \geq 1$, such that $j_0 = j_k$ and $0 = \sum_{i=0}^{k-1} \Lambda_{j_i,j_{i+1}}$. We shall show that $j_i \triangleright j_{i+1}$ for each i with $0 \leq i < k$, by assuming the contrary and deriving a contradiction. Accordingly, after renumbering the j's if necessary, assume that $j_{k-1} \not\triangleright j_k$. Then $\tilde{D}_{j_k} \neq \tilde{D}_{j_{k-1}} + \Lambda_{j_{k-1},j_k}$. Since the equations require $\tilde{D}_{j_k} \geq \tilde{D}_{j_{k-1}} + \Lambda_{j_{k-1},j_k}$, we must have $\tilde{D}_{j_k} > \tilde{D}_{j_{k-1}} + \Lambda_{j_{k-1},j_k}$. If we apply Lemma 1.1 to the path $j_0, \ldots, j_{k-1}$, we see that $\tilde{D}_{j_{k-1}} \geq \tilde{D}_{j_0} + \sum_{i=0}^{k-2} \Lambda_{j_i,j_{i+1}}$. Combining these, we get $\tilde{D}_{j_k} > \tilde{D}_{j_0} + \sum_{i=0}^{k-1} \Lambda_{j_i,j_{i+1}}$. This implies that $0 > \sum_{i=0}^{k-1} \Lambda_{j_i,j_{i+1}}$, in contradiction to our earlier statement that this sum was exactly 0. We conclude that $j_0 \overset{+}{\triangleright} j_0$ and hence that $\triangleright$ is cyclic.

Having established the claim, we can present the desired algorithm. Given a solution, construct the $\triangleright$ relation in $O(n^2)$ time. Then use a topological sort algorithm or a strong components algorithm to determine whether $\triangleright$ is cyclic. □

Having determined that the SMO equations might have multiple solutions, we are faced with a question of interpretation, namely, which is the "right" solution? At least two viewpoints are possible.

One viewpoint is that the minimum fixed point of the equations is the preferred solution because this is what the actual circuit "does" when started from an arbitrary stable initial state. When the circuit first begins to be clocked, signals leave latches at the opening edge of the latches. As operation continues, departure times get pushed out later and later into the transparent interval of the latches. Eventually the circuit reaches a steady state and the arrival and departure times stay constant. Like Construction 2.1, the actual circuit begins with arrival and departure times as small as possible and converges upwards to an equilibrium position. Indeed, we can interpret variable A_i^m in that construction as the latest arrival time at latch i of any signal that began at the opening edge of some latch and subsequently passed through m or fewer latches during their transparent phases.

Another viewpoint is that it is not permissible to operate a circuit under a clock schedule that has multiple solutions. The rationale here is that each solution represents a possible operating point of the circuit, and all

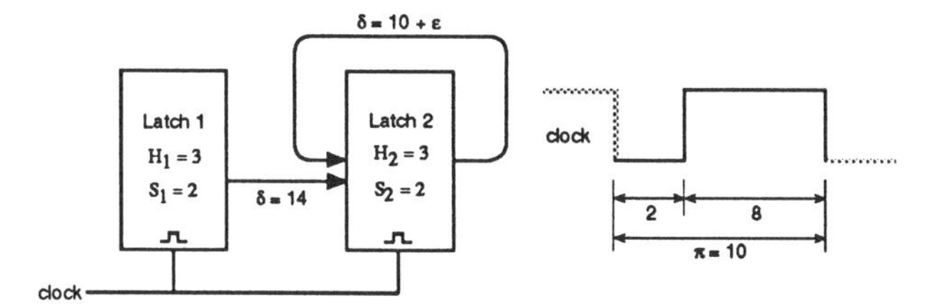

Figure 4: A circuit and schedule in which upwards convergence of early arrival times takes arbitrarily long. For $m < 4/\epsilon$, $d_2^m = 2 + m\epsilon$, and for $m \geq 4/\epsilon$, $d_2^m = 6$.

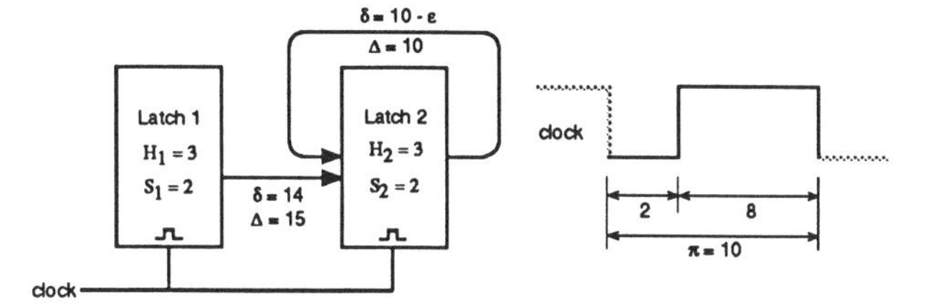

Figure 5: A circuit and schedule in which downwards convergence of early arrival times takes arbitrarily long. The least fixed point of the latest equations has $A_2 = D_2 = 7$. Taking this as a starting point for iterating the earliest time equations, we have $a_2^m = 7 - m\epsilon$ for $m < 1/\epsilon$, and $a_2^m = 6$ for all larger m.

the operating points are in equilibrium with each other. The slightest perturbation of a delay or of the arrival of a clock signal (possibly caused by a power supply fluctuation or other environmental effect) can cause the circuit to switch from operating point to another. Since we know that some of the solutions violate setup requirements, we conclude that it is too dangerous to operate the circuit this close to the "edge." Said another way, there is no margin for error under such a schedule.

4 Traps for the Unwary

Even in the absence of multiple solutions, iterative methods for solving the a and d equations can take arbitrarily many iterations regardless of the choice of starting point and the direction of convergence. In Figure 4 we see a circuit and schedule which takes $4/\epsilon$ iterations to reach a solution under upwards convergence. Note however that the A and D equations for this same circuit and schedule have no solution. A close rereading of the proof of Lemma 2.5 reveals that the a and d equations are only guaranteed to converge in n iterations when the A and D equations are known to be solvable.

Earlier approaches [SMO90, TL91] initialize the A and D equations using lower bounds, and the a and d equations using upper bounds. The convergence for the A and D equations is upwards and for the a and d equations downwards. It is easy to show that the latter process must converge downwards because the first iterate is componentwise less than or equal to the zeroth iterate and the solutions are bounded below. Another approach [SBSV92] attempts to first solve the A and D equations by upward convergence and then uses this solution to provide initial values for the a and d equations which then converge downwards to a solution. Unfortunately, both these methods can take arbitrarily long as illustrated in Figure 5. Moreover, neither process necessarily finds a minimum fixed point for the equations, so we might not get the "right" answer vis-a-vis the hold constraints (it is easy to see that if any solution has a hold violation, then the minimum solution will too).

When the SMO formulation is used for finding an optimal clock schedule, a mathematical optimization program is used to find a clock schedule and a set of arrival and departure times, (a, d, A, D). Unfortunately, as we saw in Corollary 3.4, it is likely that multiple solutions exist for an optimal schedule, and so one must carefully interpret whether such a schedule is in fact correct. Moreover, if the min and max operators in the equations were relaxed to inequalities before being passed to the solver as advocated in [SMO92], then the schedule returned by the optimizer is correct but the arrival and departure times returned by the solver might not even satisfy the original SMO equations. It has been suggested that iterating the equations using the times from the solver as starting values is will yield a solution. This is true, but probably unwise. First, it might take a long time to converge if the a and d values decrease. Second, you might converge to a solution other than the minimum solution. It probably makes more sense to simply discard the arrival and departure times found by the solver and use the verification algorithm of §2 to calculate a minimum solution directly from the optimal schedule. Depending on one's view on the multiple solution phenomenon, one might then test whether the solution satisfies the setup and hold constraints, or else test the solution for uniqueness using Theorem 3.13.

An amusing insight is provided by considering the operation of the algorithm of §2 as a simulation of the operation of the circuit during the first n clock cycles after power is turned on. As the early arrival and departure times increase monotonically to their steady-state values, they might very well violate hold requirements at various latches. This can happen even if the equations have a unique solution and no hold constraints are violated at that solution. This implies that any reset operations intended to initialize the circuit to a consistent state should persist for at least as many cycles as it takes the algorithm to converge.

5 Implementation Experience

We implemented the algorithm shown in Figure 6. It is easy to show that this algorithm gives the same answer as Construction 2.1, converges at least as fast, and only uses $O(n)$ storage. Not shown are a number of features that make the algorithm more useful in practice. First, critical short and long paths can be recovered after execution by storing with each A_i or a_i the value of j which last caused it to be increased. Second, this same backtracking can be used to print out a critical long path if the algorithm

```
for each i with 1 ≤ i ≤ n do
      A_i ← a_i ← −∞;
for m ← 1 step 1 until m > n
      for each i with 1 ≤ i ≤ n do
            D_i ← max(A_i, B_i);
            d_i ← max(a_i, B_i);
      for each i with 1 ≤ i ≤ n do
            A_i ← max(A_i, max_{j→i}(D_j + Λ_{j,i}));
            a_i ← max(a_i, min_{j→i}(d_j + λ_{j,i}));
      if no A_i or a_i changed during this pass then
            return "algorithm converged";
return "algorithm diverged";
```

Figure 6: Pseudo-code for the algorithm as implemented.

diverges. Finally, the A_i and a_i should be "clipped" to $\pi - S_i$ in order to prevent the propagation of false errors which might otherwise occur when an overly long path continues on to other latches. Users usually only want to see a diagnostic at the first late latch in each such path. Various heuristics may be employed to make the algorithm run faster, for example, only evaluating those equations for which an argument has changed during the previous iteration.

The program itself is only 558 lines of C, most of which are concerned with reading the circuit description or formatting the output. We ran the program on transformed versions of all the ISCAS '89 benchmarks as described in [Szy92] using clock schedules that had been found to be optimal by other means. The largest such circuit had 3272 latches and 67704 edges in the $\rightarrow$ relation. In all cases the running time of the algorithm was less than 20 seconds, almost all of which was consumed reading the circuit description and building data structures. Moreover, only a few iterations were ever necessary for the algorithm to converge, implying (for these circuits anyway) that signals do not usually flow continuously through very long chains of transparent latches without having to stop and wait.

6 Summary

Although the solution to the SMO equations is not necessarily unique, multiple solutions can only occur at the optimal clock period. The presence of a critical cycle in the circuit is both a necessary and sufficient condition for these multiple solutions to exist. We offered some viewpoints on the physical significance of non-unique solutions, and pointed out some of the complications that they cause in both timing verification and optimization.

Simple iterative techniques can be used to solve the equations, but one must use care in picking the starting point. The wrong starting point can lead to arbitrarily long running times and incorrect results, but the correct starting point is guaranteed to converge in n iterations, where n is the number of latches in the circuit. The total running time is at most $O(n^3)$ time, or $O(ne)$ time where e now is the number of edges in the circuit graph. Given a solution, we can test in $O(n^2)$ time (alternatively, $O(e)$ time) whether it is a unique solution.

The algorithm is simple to implement, and we ran it on enough test circuits to conclude that it takes more time in practice to read the circuit than to run the algorithm.

References

[BSM92] Timothy M. Burks, Karem A. Sakallah, and Trevor N. Mudge. Multi-phase retiming using *min*T_c. In *ACM/SIGDA Workshop on Timing Issues in the Specification and Synthesis of Digital Systems*, March 1992.

[IL90] Alexander T. Ishii and Charles E. Leiserson. A timing analysis of level-clocked circuitry. In William J. Dally, editor, *Proceedings of the Sixth MIT Conference*, Advanced Research in VLSI, pages 57–69. MIT Press, 1990.

[LB92] Wei-Han Lien and Wayne Burleson. Wave-domino logic: Timing analysis and applications. In *ACM/SIGDA Workshop on Timing Issues in the Specification and Synthesis of Digital Systems*, March 1992.

[SBSV92] Narendra Shenoy, Robert K. Brayton, and Alberto L. Sangiovanni-Vincentelli. A pseudo-polynomial algorithm for verification of clocking schemes. In *ACM/SIGDA Workshop on Timing Issues in the Specification and Synthesis of Digital Systems*, March 1992.

[SMO90] Karem A. Sakallah, Trevor N. Mudge, and Oyekunle A. Olukotun. *check*T_c and *min*T_c: timing verification and optimal clocking of synchronous digital circuits. In *Digest of Technical Papers of the IEEE International Conference on Computer-Aided Design*, pages 552–555, November 1990.

[SMO92] Karem A. Sakallah, Trevor N. Mudge, and Oyekunle A. Olukotun. Analysis and design of latch-controlled synchronous digital circuits. *IEEE Transactions on Computer-Aided Design of Integrated Circuits*, 11(3):322–333, March 1992.

[Szy92] Thomas G. Szymanski. Computing optimal clock schedules. In *Proceedings of the IEEE/ACM Design Automation Conference*, volume 29, pages 399–404, June 1992.

[TL91] Ren-Song Tsay and Ichiang Lin. A system timing verifier for multi-phase level-sensitive clock designs. Technical Report RC 17272, IBM, October 1991.

Part 6

Clock Distribution Networks for Targeted VLSI/WSI Architectures

Synchronous versus asynchronous computation in very large scale integrated (VLSI) array processors

S. Y. Kung, R. J. Gal-Ezer
Department of Electrical Engineering—Systems
University of Southern California, Los Angeles, California 90007

Abstract

This paper compares timing and other aspects of a synchronous and asynchronous square array of processing elements, fabricated by means of VLSI technology. Timing models are developed for interprocessor communications and data transfer for both cases. The synchronous timing model emphasizes the clock skew phenomenon, and enables derivation of the dependence of the global clock period on the size of the array. This O(N**3) dependence, along with the limited flexiblity with regards to programmability and extendability, call for a serious consideration of the asynchronous configuration. A self timed (asynchronous) model, based on the concept of wavefront oriented propagation of computation, is presented as an attractive alternative to the synchronous scheme. Some potential hazards, unique to the asynchronous model presented, and their solutions are also noted.

1. Introduction

The availability of low cost, fast VLSI (Very Large Scale Integration) devices promises the practice of cost-effective, high speed, parallel processing of large volumes of data. The traditional design of parallel computers is becoming unsuitable for the design of highly concurrent VLSI computing processors. It usually suffers from heavy supervisory overhead incurred by synchronization, communication and scheduling tasks, which severely hamper the throughput rate which is critical to real-time signal processing. In fact, these are the key barriers inherent in very large scale computing structure design. Moreover, though VLSI provides the capability of implementing a large array of processors on one chip it imposes its own constraints on the system. Large design and layout costs [1] suggest the utilization of a repetitive modular structure. In addition, communication, which costs the most in VLSI chips, in terms of area, time and energy, has to be restricted (to localized communication). In general, highly concurrent systems require this locality property in order to reduce interdependence and ensuing waiting delays that result from excessive communication [2]. Moreover, this locality constraint may further render the utilization of centralized control and global synchronization less appealing. As a result, the use of asynchronous, distributed control and localized data flow may become a more effective approach to the design of very large scale, highly concurrent computing structures. This paper will attempt to present some key factors affecting the tradeoff between the two timing schemes.

The timing framework is a very critical issue in designing the system, especially when one considers large scale computational tasks. Two opposite timing schemes come to mind, namely the Synchronous and the Asynchronous timing approaches. In the synchronous scheme, there is a global clock network which distributes the clocking signals over the entire chip. The global clock beats out the rythm to which all the processing elements in the array execute their sequential tasks. All the PEs operate in unison, all performing the same, identical, operation. In contrast, the asynchronous scheme involves no global clock, and information transfer is by mutual convenience and agreement between each processing element and its immediate neighbors. Whenever the data is available, the transmitting PE informs the receiver of that fact, and the receiver accepts the data whenever it is convenient for it do so. This scheme can be implemented by means of a simple handshaking protocol.

A proper timing model, which includes all lines of communication between adjacent processing elements, is essential for timing analysis and comparison between the synchronous and the asynchronous approaches. Numerous studies on the tradeoffs between the general two schemes have been carried out. Seitz [5], [2, chap.7] observes that, for large scale systems, synchronous timing approaches pose increasing difficulties. An interesting analysis is carried out by Franklin [6] as far as comparision of the two timing schemes is

* Research supported in part by the Office of Naval Research under contract N00014-81-K-0191 and by National Science Foundation under Grant ECS-80-16581.

Reprinted with permission from *Proc. of SPIE,* S. Y. Kung and R. J. Gal-Ezer, "Synchronous Versus Asynchronous Computation in Very Large Scale Integrated (VLSI) Array Processors," vol. 341, pp. 53–65, May 1982. © SPIE.

concerned. Franklin's models represent multiple VLSI chip interconnection schemes. In-depth studies of the physical processes involved in transfer of information between adjacent transistors in neighbouring processing elements on the chip, which are applicable to VLSI oriented timing discussions, have been implemented [7].

A general representation of all VLSI systems is to view them as a computation graph. The graph's nodes represent devices, which compute boolean functions, and its arcs are wires, which are responsible for information transfer and distribution of power and timing waveforms.

Several computing structures, which are designed to meet the constraints and goals of VLSI technology, have been suggested. Since the VLSI technological constraints render the general purpose array processor rather inefficient, a special class of applications, i.e. recursive and local data dependent algorithms, has become a focus of discussion [3,4]. Accordingly, we shall limit ourselves to matrix related configurations, similar to that of Fig. 1, and we narrow down the structures to two major groups. The systolic array is an example of a totally synchronous system, and is typical of the first group. The second group, consisting of asynchronous timing scheme configurations, is represented by data flow machines.

The *Systolic Array* , introduced by H.T.Kung and C.E.Leiserson [2,chap.8.3], was one of the first systematic attempts to harness VLSI power in the service of computationaly intensive tasks. Their design consists of a regular array of identical processors in a linearly connected or a hexagonally mesh-connected geometry. The linear array is suited (and dedicated) to matrix-vector multiplication and solution of triangular linear systems, while the hex-connected array is used for matrix multiplication and LU decomposition type problems. From the timing point of view, the systolic array is wholly synchronous, and requires global clock distribution, therefore suffering the inherent penalty of all synchronous systems, that of being able to clock at only the rate determined by the slowest element in the array and by the system clock skew.

Another feature of the systolic array is its' total dedication to implementing a given algorithm. The array is not programmable, and each algorithm requires a seperate and distinct array configuration. Also, because of the strict synchronized timing, all of the processors, except perhaps some special peripheral elements, must be performing the same task in unison. There is no room for multi-tasking, even if the tasks are serial and not interwoven one into the other. The systolic array concept, therefore, involves an inflexibility which might reduce its scope of applications.

The *Data Flow Machine* is asynchronous and consists of numerous processing elements (PEs) which can operate independently. A main feature of the data flow machine is that an instruction is ready for execution when its operands are available. There is no concept of control flow, nor is there a program location counter. A consequence of this philosophy is that many instructions may be available for execution at once, and the rate of throughput will depend on the availability of processing resources. A weakness of the data flow concept lies in the fact that much managing and bookkeeping are required to allocate the resources efficiently. The decomposition process, for example, which decomposes a program into its concurrent components, is time consuming, requires an extensive operating system and, above all, demands that each PE be aware of the global systems' resources, how many PEs are available, the sizes of their respective local memories and their workload. This is the penalty of pursuing general purpose computing features while attempting efficient parallelism and concurrency. It also limits the ability of the general purpose data flow machine to be implemented in VLSI.

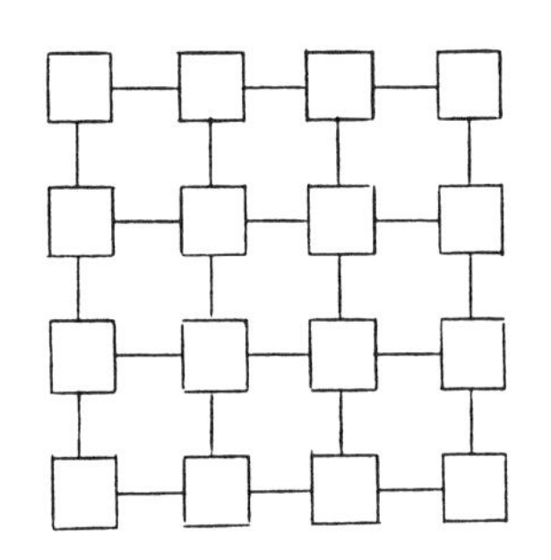

Fig. 1: Square Array of Processors

```
FETCH    A(k);                 (* From the left *)
FETCH    B(k);                 (* From above    *)
MULTIPLY  D := A(k)*B(k);
ADD      C(k) := C(k-1) + D;
FLOW     A(k);                 (* To the right *)
FLOW     B(k);                 (* Downward     *)
```

Fig. 2: Operation Sequence of Matrix Multiplication

As matrix operations form the backbone for the parallel array structures dealt with in this paper, we will, throughout this paper, employ the basic matrix multiplication,

C := A*B, as a bench test for evaluation and comparison of the timing characteristics of our synchronous and asynchronous models. The nature of matrix multiplication, as implemented via the wavefront computation propagation concept, is described below. It suffices, at this point, to provide the basic operations executed by each PE within the processor array. In the kth recursion, a PE executes: C(k) := C(k-1) + A(k)*B(k), where C(k-1) is the result of the k-1 recursion and already resides within the PE, while A(k) and B(k) are supplied by the PE's immediate neighbors, as a preliminary to the actual accumulating product operation. The sequence of operations is, therefore, as shown in Fig. 2. Some of the operations in the matrix multiplication sequence can be merged, through pipelining and/or concurrency, but all must be carried out as part of the recursive task.

2. Timing Analysis of the Synchronous Configuration

In the synchronous scheme, we assume a global clock distribution network, distributing the clocking signals over the entire VLSI chip. In the asynchronous approach, information transfer is implemented by means of a handshaking protocol. The nature of the handshaking protocol varies slightly from one system to the next. We will base our presentation on a particular "wavefront" oriented computation scheme, described in detail below.

The essential difference between a synchronous system and an asynchronous one is their timing; therefore, the timing issues constitute the major factor in defining the performance evaluation criteria of the processing array. In addition to the delay due to timing considerations, the timing constraints also commonly impose constraints on other system parameters, such as area allocation, layout rules, power consumption and, of course, overall performance of the array.

Timing aspects in the array environment can, in general, be divided into three groups: logic transition factors, including propagation delays within the basic logic elements (gates, flip flops, registers, etc.), time involved in charging the capacitances of the data and control distribution networks and time involved in charging the clock distribution network. The first two groups are dominant in the asynchronous systems, whereas all three groups play a role in the synchronous configurations, especially the first and the third. For the purpose of this paper, we will assume that the time factors due to the basic logic elements, from the processing element down to individual gates, is of the same order for both synchronous and asynchronous configurations. Our main target will, therefore, be the timing aspects that differentiate the two schemes: control and clock timing involved in interprocessor communications and data transfer.

2.1 Clock Skew Phenomena

In the synchronous global clock distribution network, one encounters the "Clock Skew" phenomenon, which arises due to three factors. The first source of clock skew is the RC of the global distribution line. The second is due to unequal clock paths to various PEs in the array. Both factors are a function of the layout scheme of the chip. The third factor contributing to clock skew is the variance of values of the gate threshold voltage, Vt, of the PE gate which receives and generates the global clock signal to the interior of the PE, thus serving as a buffer between the global clock distribution network and the local clock distribution paths. All three factors are dependent on the fabrication process [6]. The following analysis describes the essence of the clock skew problem.

The timing analysis, associated with the propagation of a signal along a conducting path between two MOS switches within a chip is similar to that of the transmission lines one encounters in power systems. The only difference is that the line inductance is negligible, as the rate of change of currents is negligible. A detailed analysis of solving the partial differential equations can be found in [7]. An important conclusion reached there is that the line capacitance plays a dominant role in the line delay. In fact, when the signal of interest is the clocking signal, it will, in many cases, create disastrous syncronization problems, particularly if the clock traverses a large distance on the chip. This could happen quite commonly in a large VLSI chip.

As crucial as the capacitive effect of the clock transmission line is the resistance through which that capacitance must charge. It, too, is dependent on the length of the line. Due to constraints involved in the layout of the VLSI chip, a major factor in the line resistance is the distribution of the material of which the line is made. As the resistance of diffusion is of the order of 100 times that of metal, the length of diffusion paths in the clock line is a predominant factor in that lines' time constant.

The equation which governs the exponential waveform of the clocking signal on the line is:

$$V_t = V_{DD} * [1 - \exp(-t/RC)] \quad (1)$$

The time t, when the gate switches its logic state, is dependent on both RC and the gate threshhold voltage, Vt. This is best shown by the following graph depicting the dependence of line voltage and time. This time is dependent on the line length, and is closely related to the ratio of diffusion and metal of the line.

The points marked on figure 3 clearly indicate the total clock skew, dt, which consists of the skew contributions due to value variations of both Vt and RC. The uncertainty in Vt is a fabrication phenomenon, e.g. Vt may be 20% higher or lower than the typical threshold voltage, Vt=2.5 volts, as given in manufacturers' data sheets. The above figure also yields the equation:

$$\begin{aligned} dt &= \text{clock skew} \\ &= t2 - t1 \\ &= RC(max)*\ln[Vt(max)] - RC(min)*\ln[Vt(min)]. \end{aligned} \quad (2)$$

This establishes the clock skew as a function of R,C and Vt.

2.2 A Timing Model for a Synchronous Array

H Tree Clocking Distribution — The clock skew due to unequal clock path lengths to the PEs may become potentially hazardous. It is, therefore, very desirable to elliminate this skew contributor. To this end, we assume that the-array clocking network is of an H-tree nature [6]. This may be implemented by placing the global clock generator at the root of a binary distribution tree. All the processing elements are at the various levels of the tree as children of preceding nodes (see fig. 4). Every node represents a processor. This ensures the equal lengths of the clock paths to all the processors. Thus, for the purpose of reducing clock skew, this clock distribution scheme appears to be optimal. The clock skew between adjacent PEs will, therefore, be determined solely by eq. (2).

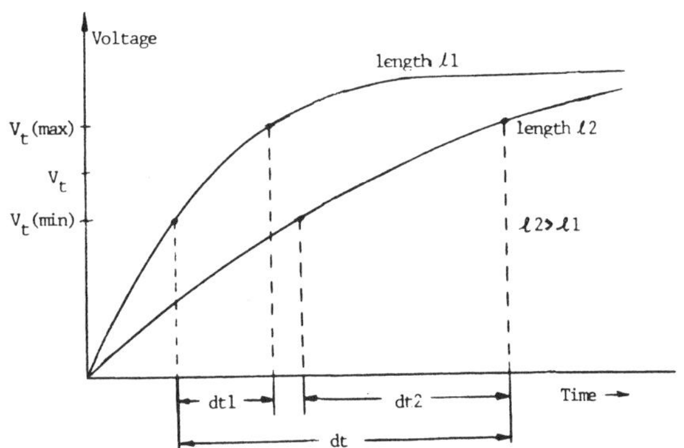

Fig. 3: Clock Skew Timing Diagram

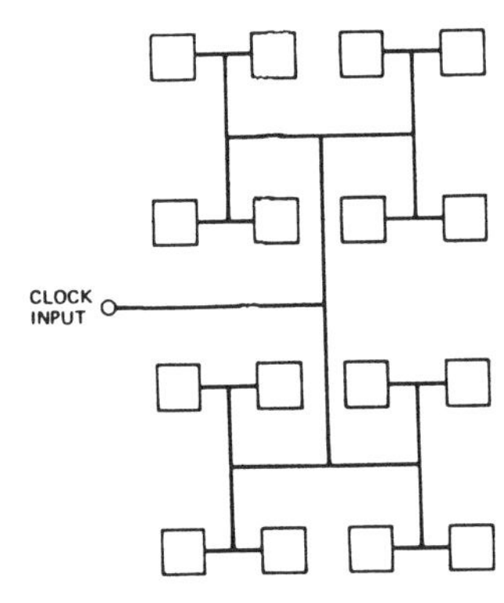

Fig. 4: H-Tree Clock Distribution Network

Timing Analysis — The complete circuit analysis of the clock distribution network is rather involved, so we shall only summarize the final results here, with the details elaborated in Appendix A. Let L be the dimension of a PE. Then the array dimension will be N*L. By appropriately modelling the H-tree distribution network as a binary tree configuration of RC branches (see Appendix A), it can be shown that the time constants, the rise time and the clock skew associated with the clocking signal distributed to the PEs, are of O(N**3)! This is corroberated by the simulation results provided below. These results place a severe restriction on the ability to generate a global, synchronous clock signal in a system with large N. It should also be noted that these conclusions are based on a full and homogenous binary distribution network. Implementing a synchronous array with an incomplete H-tree distribution, or non-homogenous clock signal paths, will add to this clock skew.

For the purpose of simulating the distribution network, a value of L=0.1 mm. was chosen. Typical values for the various physical parameters were applied. Simulation was carried out via the SPICE program, and relevant data extracted from the timing graphs. The results are depicted in Fig. 5, which graphs the clock skew as a function of N and of the conducting material distribution factor, r. To further verify the results, a two-pole approximation of the distribution network, as described in Appendix A, was executed. Here, too, the O(N**3) dependency is evident.

System Considerations Several system level assumptions are made with regards to the synchronous array in order to enable a realistic evaluation of its performance. First, the algorithm implemented by the array consists of identical recursions, where each recursion involves several states of the PE executing it. For example, we assume that addition requires K{add} clock cycles, multiplication K{mult} cycles, data flow K{flow} cycles, etc. Thus, the matrix multiplication algorithm, described above, which calls for a multiplication, an addition and two flow/fetch tasks would require K{mult}+K{add}+2*K{flow} clock cycles. Note that, whereas in the asynchronous scheme flow and fetch are seperate tasks, in a synchronous array with a two phase clock they may be combined into one task that involves a simultaneous flow-to-left and fetch-from-right or flow-down and fetch-from-up. This combination is clearly an attribute and is, therefore, taken into consideration in the synchronous model.

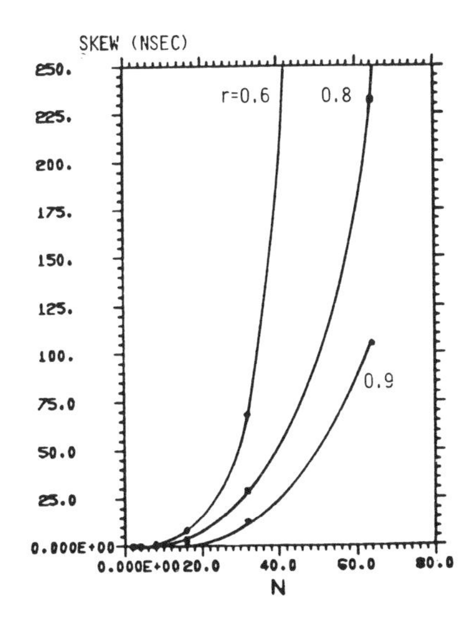

Fig. 5: Clock Skew vs. N

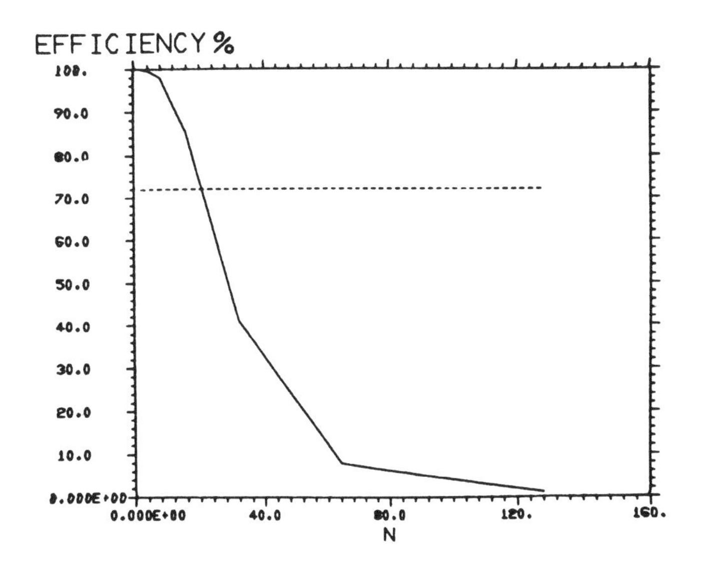

Fig. 6: Efficiency vs. N

A single, independent PE allows operation with a nominal minimal clock pulse period of T(pe). To perform the tasks of one recursion, the processor would require an execution time of [K{mult}+K{add}+2*K{flow}]*T(pe). In contrast, the same PE within the array will be clocked at intervals of T(ck), where: T(ck) - T(pe) > T(skew). We introduce an efficiency factor, EFF, which describes the array's performance efficiency based on its timing capabilities:

$$\mathrm{EFF} = \frac{\text{independent PE execution time of one recursion}}{\text{PE execution time of one recursion within the array}} \qquad (3)$$

Clearly, in the case of the synchronous configuration, EFF = T(pe)/T(ck). Also very evident is the strong dependency of the efficiency factor on the clock skew. The relationship between EFF and N is given in Fig. 6, for r=0.8 and T(pe)=20 Nsec. It should be noted that the model presented is general enough to include either a programmable or an array dedicated to the algorithm.

3. Asynchronous Configuration

The problem of clock skew in the synchronous parallel processor array configuration evidently calls for serious consideration of the asynchronous alternative. Systems which are classified as asynchronous, self-timing or data flow do not suffer from the clock skew problem. We, therefore, turn out attention to the asynchronous timing scheme.

It has been shown that Data Flow Machines suffer from several key disadvantages, due to their general-purpose nature. The solution hinges upon taking advantage of special data structures imposed by the class of algorithms considered. More precisely, for the matrix operation class, we should address the question of wether it is possible to generalize the synchronous (systolic) array to cope with with the asynchronous (self-timed) computation.

The answer to this query is yes. However, the conceptual framework needs to be changed. For this purpose, the so-called wavefront notion becomes useful and essential. The *Wavefront Array Processor* (WAP) combines the merits of both schemes. In other words, the WAP can function as a Systolic Array in one extreme and as a Data Flow Machine in the other extreme. We use the model of the WAP to provide a comparision of the timing schemes.

3.1 Computational Wavefront

For the purpose of illustrating the WAP concept, let us again use multiplication as an example. Let A = {a(ij)} and B = {b(ij)} and C = AxB = {C(ij)} all be NxN matrices. The matrix A can be decomposed into columns Ai and matrix B into rows Bj, and, therefore,

$$C = A_1B_1 + A_2B_2 + \cdots + A_NB_N \tag{4}$$

The matrix multiplication can then be carried out in N recursions, executing

$$C(k) = C(k-1) + A_kB_k \tag{5}$$

recursively for k = 1, 2, . . . , N.

The topology of the matrix multiplication algorithm can be mapped naturally onto the square, orthogonal NxN matrix array of the WAP (cf. Fig. 7). To create a smooth data movement in a localized communication network, we make use of the computational wavefront concept. For the purpose of this example, a wavefront in the processing array will correspond to a mathematical recursion in the algorithm. Succesive pipelining of the wavefronts will accomplish the computation of all recursions.

As an example, the computational wavefront for the first recursion in matrix multiplication will now be examined. Suppose that the registers of all the processing elements (PEs) are initially set to zero:

$$C(0)_{i,j} = 0 \quad \text{for all } (i,j). \tag{6}$$

The entries of A are stored in the memory modules to the left (in columns), and those of B in the memory modules on the top (in rows). The process starts with PE (1,1):

$$C(1)_{1,1} = C(0)_{1,1} + a_{1,1}b_{1,1} \tag{7}$$

is computed. The computational activity then propagates to the neighboring PE's (1,2) and (2,1), which will execute:

$$C(1)_{1,2} = C(0)_{1,2} + a_{1,1}b_{1,2} \tag{8}$$

and

$$C(1)_{2,1} = C(0)_{2,1} + a_{2,1}b_{1,1} \tag{9}$$

The next front of activity will be at PE's (3,1), (2,2) and (1,3), thus creating a computation wavefront traveling down the processor array. It may be noted that wave-propagation implies localized data flow. Once the wavefront sweeps through all the cells, the first recursion is over. As the first wave propagates, we can execute an identical second recursion in parallel by pipelining a second wavefront immediately after the first one. For example, the (1,1) processor will execute

$$C(2)_{1,1} = C(1)_{1,1} + a_{1,2}b_{2,1} \tag{10}$$

$$C(k)_{ij} = a_{i1}b_{1j} + a_{i2}b_{2j} + \ldots + a_{ik}b_{kj} \tag{11}$$

and so on.

The pipelining is feasible because the wavefronts of two successive recursions will never intersect (Huygen's wavefront principle), as the processors executing the recursions at any given instant will be different, thus avoiding any contention problems.

3.2 Asynchronous (Self-Timing) Wavefront propagation

As is evident, the pipelining of the wavefronts can be easily implemented in a highly synchronous fashion, using global clock distribution as discussed above. The result of such an implementation is referred to as the synchronous wavefront array processor, which is, in fact, a variant of the systolic array. However, a much more intriguing aspect of the wavefront notion lies in the fact that it is amenable to asynchronous computation. This is explained below.

To accomplish a self-timed, asynchronous computation, the processors in the array, must *wait* for a primary wavefront (of data), then perform the computation it calls for and, finally, act as a *secondary source* of new wavefronts. For example, operations (8) and (9) will not be executed until PEs (1,2) and (2,1) confirm inputting of (a_{11},b_{12}), and (a_{21},b_{11}) respectively. By the same token, in the next front of the wave, cells (1,3), (2,2) and (1,3) will be involved. PE (2,2), for example, has to *wait* until PEs (1,2) and (2,1) flow their data, b_{12} and a_{21}, respectively. Only after the arrival of that data will the (2,2) cell execute its own operation: $C(1)_{2,2} = C(0)_{2,2} + a_{21}b_{12}$ and activate its own successors, PEs (2,3), and (3,2).

The self-timed processor array essentially simulates a phenomenon of wavefront propagation. In particular, it exhibits the self-generating mechanism of the wavefront property. From a hardware point of view, in order to implement this wait, processors are provided with data transfer buffers. Hence, a FETCHing of data involves an inherent WAITing for the buffer to be filled (DATA SENT) by the adjacent, data sourcing, processor. Thus, the processing will not be initiated until the arrival of the data wavefront (this is similar to the concept of data flow machines [10-18]). Each processor can FLOW data to the input buffers of the neighboring PE's, thus acting as a secondary source of data-wavefronts (Huygen's principle). To avoid the overrunning of data wavefronts (in conformation with Huygen's principle), the processor hardware ensures that a processor cannot send new data to the buffer unless the old data has been used by the neighbor. Thus, the wavefront concept suggests that interprocessor communication employ buffers and "DATA SENT/DATA USED" flags between adjacent processors.

In short, the handshaking (wait) for wavefronts of data allow for globally asynchronous operation of processors i.e. there is no need for global synchronization.

3.3 Asynchronous Wavefront Array Processor Configuration

To implement the asynchronous computation, special attention is given to the PE's interfacing with its neighbors. Every interior element has a uni-directional buffer and independent status and control flags to each of its four adjacent elements. These buffers can be supported by an appropriate multiplexing subsystem, under the control units' supervision.

The concept of the asynchronous approach, as implemented in the WAP architecture, are described above and in [8,9]. This scheme involves a handshaking protocol between the adjacent processing elements. The protocol ensures the regularity and the continuity of the flow of information through the processor array. This calls for an additional number of input and output signal lines for each processor element. The global clock of the synchronous scheme is now replaced by the Data Sent and the Data Used lines which establish an exchange between the adjacent processors with regards to data transmission timing. These signals form a part of the Self timed systems [5], each processor signalling to its neighbor whenever it is ready to take an action.

3.4 Timing Analysis

As the PE itself may be internally synchronous, with an internal clock period, T(pe), which is not affected by factors outside the processors' bounds, we shall assume such a model. Note also that, in contrast with the synchronous configuration, the asynchronous PE will have seperate flow and fetch tasks. Therefore, the number of tasks per recursion, for the same matrix multiplication algorithm applied to the synchronous model, would now be [K{mult}+K{add}+2*K{flow}+2*K{fetch}]*T(pe).

The transfer of data, in the basic model, from PE(i) to PE(j) calls for PE(i) to apply the appropriate data to the interprocessor data bus, and a pulse generated on the Data-Sent control line (see figure 8). The width of that pulse, t(1), must be greater than the data setup time of the Data Input Buffer of PE(j). After the pulse has been generated,

PE(i) can turn to the next task of the recursion. It is of interest to note that, although the Data Input Buffer and the Data Transfer Control flip flop, described below, are physically within the domain of PE(j), from a timing point of view they must concede to the timing constraints of PE(i).

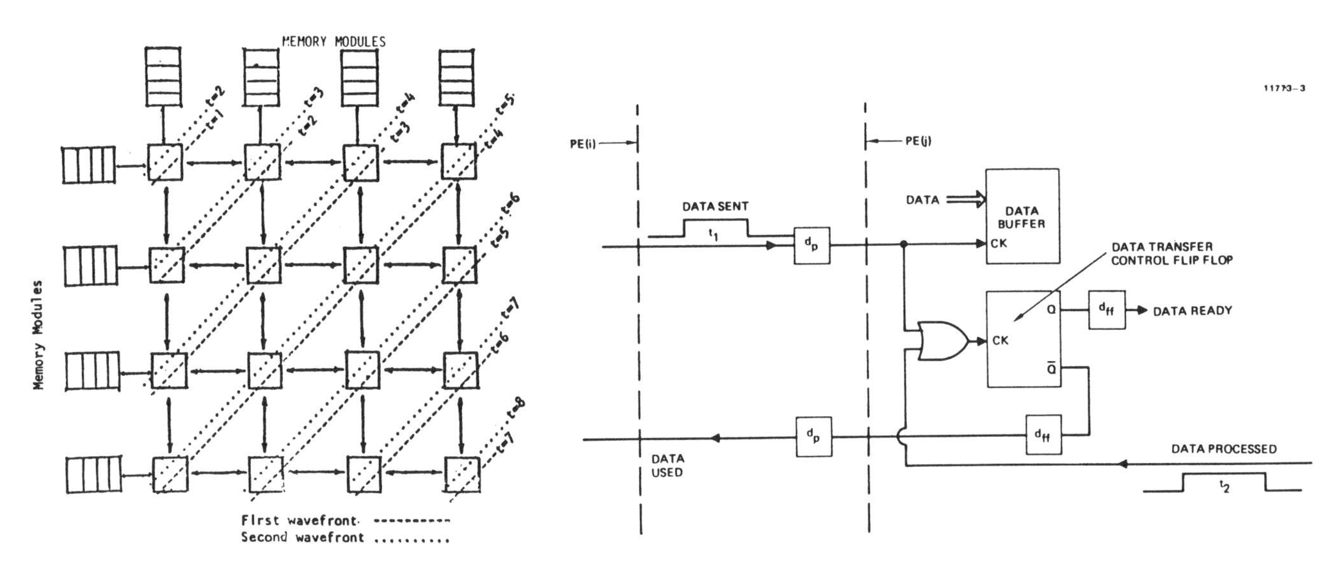

Fig. 7: Configuration for NxN WAP

Fig. 8: Interprocessor Handshaking Scheme

The negative edge of the Data-Sent pulse enters the data into the Data Input Buffer of PE(j), and also toggles the Data Transfer Control flip flop, thereby notifying PE(j) of the availability of data. In general, PE(j) will be waiting for that data, and will immediately execute a FETCH instruction. The time lost in this transaction is, in the worst case, T(pe(j))+d(ff), where d(ff) is the input to output delay involved in toggling the flip flop. Upon completion of the data FETCH, PE(j) issues a Data-Processed pulse of duration t(2), where t(2) must be larger than the clock pulse width required by the flip flop. The flip flop transition, in the form of the Data-Used signal, is then propagated back to PE(i). This involves a time delay of d(ff)+d(p), where d(p) is the propagation delay of the interprocessor flagging signal.

In none of the applicational algorithms implemented todate has there been a necessity to have two consecutive FLOW tasks in the same direction. Thus, after a FLOW has been executed, PE(i) is implementing its next task concurrently with the FETCH executed by PE(j). By the time PE(i) has to carry out the next FLOW to PE(j), the Data-Used signal will have already been set, and there is, therefore, no waiting involved. The timing penalty for this situation is, at worst, t(2)+d(ff). This penalty is paid only once, and does not multiply by the number of recursions, nor by the number of FLOW (FETCH) tasks, provided they are not consecutive. To this delay, one must add the PE time added for seperate FLOW/FETCH instructions, which is f*T(pe). Evaluation of the efficiency factor for the asynchronous model is highly dependent on both the architecture of the PE and on the algorithm that is being implemented. Based on some reasonable assumptions relating to the execution time ratio between multiplication and the other operations involved in the bench test matrix multiplication algorithm, EFF is calculated to be of the order of 72%. As has been mentioned, a higher computation intensity within the PE will improve the efficiency factor in this model.

4. Comparison of Synchronous and Asynchronous Configurations

There are numerous criteria in evaluating the timing schemes of the processing array. The performance tradeoffs between the two general approaches are highly dependent on the internal architectural features of the individual PE, such as pipelining and concurrency, the number of PEs required in the array, the processing technology employed, as well as the requirements and nature of the algorithms implemented by the array. Timing analysis can be applied to any given system configuration, resulting in curves of the type provided in Fig. 6. The crossover points between attractivity of an asynchronous scheme to that of a synchronous design can then be found from these curves. Since the O(N**3) relationship between clock skew and number of PEs is not technology dependent, it seems evident that, for large array sizes, the asynchronous scheme, provided by the WAP, will prevail.

There are also other considerations which may be crucial in the favoring of one configuration over the other. The globally synchronous array will severely limit the programmability and somewhat hamper the system extendability. This lack of flexibilty will diminish its cost-effectiveness. In this respect, the powerful notion of computational wavefront propagation, together with the asynchronous timing capability, enable us to deal with the issues of programmability and extendability quite readily. On the other hand, synchronous arrays may provide more facile chip testability , due to their regular timing features.

The above discussion indicates the need for establishing a careful tradeoff between the two timing schemes. It also opens the door to the concept of GALS (Globally Asynchronous and Locally Synchronous) configurations, which is a logical compromising outcome of the previous analysis. The GALS approach, which will incorporate blocks of synchronous PEs in a globally asynchronous system, will allow a merging of the merits of both timing schemes, establishing a justifiable tradeoff between the two. However, the outcome of the analysis of GALS is has not yet become clear at this point.

5. Timing Aspects of the WAP

As the WAP has a globally asynchronous configuration, timing problems arise only as a result of the processor-to-processor asynchronous communication scheme. The major considerations, from a system-level point of view, include inhibiting deadlock and racing situations within the array. The elimination of these hazardous conditions is carried out by means of both hardware features and software syntax rules and constraints. We will deal, here, only with racing problems, as they are are caused solely by the self timing feature.

In order to retain the wavefront oriented propagation of computation, each front of the wave must be tied to its preceding and succeeding fronts in such a manner so as to eliminate the possibility of a processor's "running away" without transferring the required data and activity parameters to its neighbors. To clarify this issue, assume that PE1 is the left-hand predecessor of PE2, and that the current algorithm requires r recursions, each of which includes the instruction sequence of Fig. 9.

If PE1 is not tied to PE2 by some restrictive rules, it can, conceivably, complete its recursion in much less time than PE2. Thus, PE1 may be flowing data resulting from its third recursion when PE2 has just completed its first recursion and is expecting second recursion data from PE1.

The ensuing conclusion is that PE1 and PE2 must be chained together. Originally, this chaining was carried out by the concept of activity movement. Once PE1 finished the tasks associated with one recursion, it propagated activity via an Activate signal to PE2. The concept of transfer of activity does not, however, resolve all of the communication problems within the array. One outstanding example of this occurs when PE1 transmits data to PE2 twice within a recursion. If PE2 is slow, and has not managed to input the first data word transmitted, that data will be written over by PE1 before it was accessed by PE2, and is therefore lost. A possible solution to this dilemma would be to create a FIFO input buffer to accumulate incoming data until it has been digested by the receiving processor. This imposes further system constraints, such as defining the depth of the buffer, which, in turn, will call for additional program bookkeeping and resource allocation management.

```
REPEAT
   WHILE WAVEFRONT IN ARRAY DO
      BEGIN
      FETCH <From Left>;
               .
               .
      FLOW  <To Right>;
               .
               .
      END;
 UNTIL TERMINATED;
```

Fig. 9: Instruction Sequence Relating to One Recursion.

```
FLOW (PE1):    WAIT UNTIL Data Used
               IF Data Used THEN:
                  TRANSFER DATA TO THE DATA BUS
                  SET Data Sent TO "1"
                  RESET Data Used TO "0"

FETCH (PE2):   WAIT UNTIL Data Ready
               IF Data Ready THEN:
                  INPUT DATA FROM THE INPUT BUFFER
                  SET Data Used TO "1"
```

Fig. 10: Data Transfer Sequence

As the source of the "run-away" problem appears to be the need for a "wait" state which will hold PE1 from proceeding in its tasks without waiting for PE2, it seems natural to tie the "wait" state concept to the data transfer itself, rather than to the wavefront propagation phenomenon, the integrity of which it is helping to retain. It is therefore suggested to replace the Activate/Active signal pair with data transfer handshaking signals, as described above. Each processor will have four input buffer registers, one from each direction, and two pairs of handshaking signals (one pair for each direction of data transfer) to each of its four neighbors to provide the communicating relationship necessary. These two signals are the Data Sent (D.S.) and Data Used (D.U.) signals described above. The sequence of events invoked by a data transfer is given in Fig. 10. Data Ready is a derivative of Data Sent.

This sequence ensures that a transmitting PE will never overwrite previously sent data before it has been used, and that a receiving PE will always be reading new, unused information. Thus, the data transfer mechanism "chains" the two adjacent PEs together in a timewise loose, yet sequentially determined manner.

Another timing problem created by the asynchronous nature of WAP communications involves the propagation of the Terminated flag. Each PE has its independent Terminated flag, which it propagates to its down- and right-hand neighbors. Each processor also receives two incoming Terminated signals from its up- and left-hand neighbors. When a processor has finished the tasks called for within one recursion, it checks for the conditions of terminating the current phase of instructions. In every PE except for the (1,1) cell, termination is dependent upon the termination of the PE's predecessors. Here, too, a "runaway" condition is possible. The solution to this situation is quite simple. Details of the solution are postponed to a later paper.

Conclusion

We have presented a comparison between the synchronous and the asynchronous timing analysis of a square, NxN array of processors. Whereas the asynchronous model incurs a fixed time delay overhead due to the handshaking processes, the synchronous time-delay is due primarily to the clock skew which changes dramatically with the size of the array, N. More precisely, our analysis indicates that the clock skew grows with array size at a significant rate of O(N**3). An immediate conclusion from this analysis is that, while for small N a globally synchronized processor array may be easier to implement, for larger N a self-timed (asynchronous) system may become much more favorable. However, there are - and we have briefly looked into - some other important factors, such as programmability, extendability, testing, racing, etc. which must be taken into consideration. In short, the only definitive conclusion is that the ultimate decision has to hinge upon the final hardware performance evaluation as well as the system applicational requirements.

Acknowledgement

The authors wish to acknowledge the contribution of Mr. S. Shenoy of USC, Los Angeles to the discussions relating to the timing analysis and especially to the SPICE simulation efforts.

References

1. I.E.Sutherland and C.Mead, "Microelectronics and Computer Science", Scientific American, Vol. 237, no. 9, Sept. 1977, pp.210-228.

2. C.Mead and L.Conway, "Introduction to VLSI systems", Addison Wesley, 1980.

3. S.Y.Kung, "VLSI Array Processor for Signal Processing", Conference on Advanced Research in Integrated Circuits, MIT, Cambridge, MA., Jan. 28-30, 1980.

4. J.M.Speiser and H.J.Whitehouse, "Architectures for Real Time Matrix Operations", Proc., GOMAC, Nov., 1980.

5. C.L.Seitz, "Self Timed VLSI Systems", Prc. of the CALTECH Conference on VLSI, Jan. 1979.

6. M.Franklin, D.Wann, "Asynchronous and Clocked Control Structures for VLSI Based Interconnection Networks", 9th Ann. Symposium on Computer Architecture, Austin, Texas. Apr. 1982.

7. G.D.Bilardi, M.D.Pracci, F.P.Preparata, "A critique and an Appraisal of VLSI Models of Computation", CMU Conf. on VLSI Systems and Computations, Computer Science Press, Oct. 1981.

8. S.Y.Kung, R.J.Gal-Ezer and K.S.Arun, "Wavefront Array Processor: Architecture, Language and Applications", MIT Conf. on Advanced Research in VLSI, MIT, Cambridge, Ma., Jan. 1982.

9. S.Y.Kung, K.S.Arun, R.J.Gal-Ezer, and D.V.Bhaskar Rao, "Wavefront Array Processor: Language, Architecture and Applications", submitted to IEEE Trans. Computers

10. J.B.Dennis and D.P.Misunas, "A Preliminary Architecture for a Basic Data Flow Processor", Proc. 2nd Ann. IEEE Symposium on Computer Architecture, pp. 126, Jan. 1974.

11. J.E.Rumbaugh, "A Data Flow Multiprocessor", IEEE Trans. Computers, Vol. c-26, No. 2, pp. 138-146, Feb. 1977.

12. J.B.Dennis, "The Varieties of Data Flow Computers", Proc. 1st Intl. Conf. on Distributed Computing Systems, pp. 430-439, Oct. 1979.

13. J.B.Dennis, "Data Flow Supercomputers", IEEE Computer, pp. 48-56, Nov. 1980.

14. K.P.Gostelow and R.E.Thomas, "Performance of a Simulated Data Flow Computer", IEEE Trans. Computers, Vol. C-29, No. 10, pp. 905-919, Oct. 1980.

15. I.Watson and J.Gurd, "A Prototype Data Flow Computer with Token Labelling", Proc. AFIPS Conf., Vol. 48, NCC, New York, pp. 623-628, Jun. 1979.

16. A.Davis, "A Data Flow Evaluation System based on the concept of recursive locality", Proc. AFIPS Conf., Vol. 48, NCC, New York, pp 1079-1086, Jun. 1979.

17. Arvind, K.P.Gostelow and W.Plouffe, "An Asynchronous Programming Language and Computing Machine", Dept. of Information and Computer Science, Univ. of Calif., Irvine, Tech. Report 114a, Dec. 1978.

18. W.B.Ackerman, "Data Flow Languages", Proc. AFIPS Conf., Vol. 48, NCC, New York, pp. 1087-1095, Jun. 1979.

APPENDIX A

Let L be the dimension of a PE. Then the array dimension will be N*L. Taking each segment of the H-tree as a lumped R-C branch provides the equivalent circuit of Fig. A1. Note that the innermost branch of the H tree, which provides the final clock signal to the PE, corresponds to the rightmost segment of the circuit, consisting of R_0 and C_0. Each "H" of the H-tree is replaced by two levels of RC segments, as shown. As the arms of any H structure are of equal length, the values of R and C representing both levels of the H structure are equal. Also, as the lengths of the H arms double from one H-level to the next, the values of both R and C at each H-level will be twice those of their successor H-levels. The levels of the H-tree structure are denoted by q, where $1 \leq q \leq \log_2 N$. The values of R and C in the root of the distribution tree will, therefore, be $N*R_0$ and $N*C_0$, respectively.

The clock distribution network is divided into paths of metal conductor and those of diffusion (or polysilicon). The metal will have a capacitance/area of C(m) and a resistance/square of R(m) associated with it, and the diffusion will have C(d) and R(d), respectively. The clock path to each PE consists of r parts metal and (1-r) parts diffusion. We assume that this distribution of the conducting material holds true for each and every segment of the H-tree.

By the simple circuit equivalence shown in Fig. A2(a), the system of Fig. A1 can be replaced by N**2 parallel branches, all of which are identical to that provided in Fig.

A2(b). Each serial branch represents the equivalent circuit, as seen by any individual PE, between the clock input at the root of the H-tree and that PE.

As a first order approximation of the equivalent circuit, we lump all the resistances and all the capacitances of Fig. A2(b) together. The sum of the resistances is:

$$R'_{eq} = \{3/8 * \mathrm{SUM}_{q=1 \text{ to } \log_2 N} (8^q) + N^3\} * R_0 = \{10*N^3 - 3\}*R_0/7 \quad (a.1)$$

and of the capacitances is:

$$C_{eq} = \{1 + 2 * \mathrm{SUM}_{q=1 \text{ to } \log_2 N} (2^{-q})\} * C_0 = \{3 - 2/N\} * C_0 \quad (a.2)$$

The equivalent time constant of the distribution network is, therefore,

$$\tau_{eq} = \{10*N^3 - 3\}*\{3 - 2/N\} *R_0*C_0/7 \quad (a.3)$$

It should be noted that this time constant, and the clock pulse rise time and the clock skew associated with it, are of O(N**3). This is corroberated by the simulation results provided in the main text.

To further verify the results, a two-pole approximation of the distribution network, as shown in Fig. A3, was executed. Here, $C_{eq(1)}$ represents the parallel combination of all the intermediate capacitances, implying that all the intermediate resistors have been short circuited. Thus, $C_{eq(1)} = 2*C_0*(1-1/N)$. The resulting transfer function is then:

$$T_1 = \frac{1}{1 + (3N^3 - 2N^2 + 1)*R_0*C_0*s + 2(N^3 - N^2)*(R_0*C_0*s)^2} \quad (a.4)$$

The negative reciprocal of the poles for this model, for N>2, are:

$$\tau_1 = (3N^3 - 2N^2 + 1)*R_0*C_0 \quad \text{and} \quad \tau_2 = 2*R_0*C_0/3 \quad (a.5)$$

which are, of course the time constants of the step input response. Again, the O(N**3) dependency is evident.

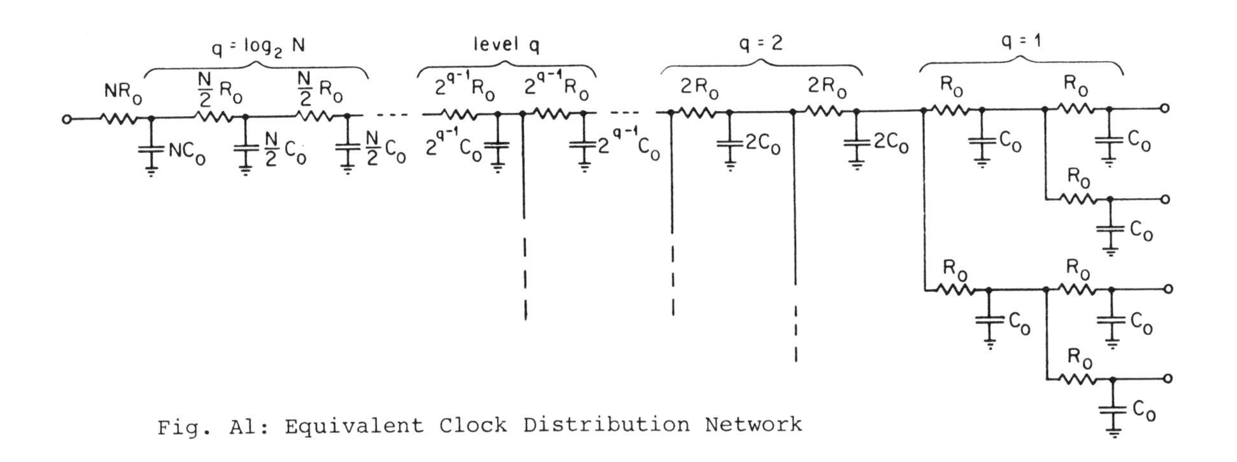

Fig. A1: Equivalent Clock Distribution Network

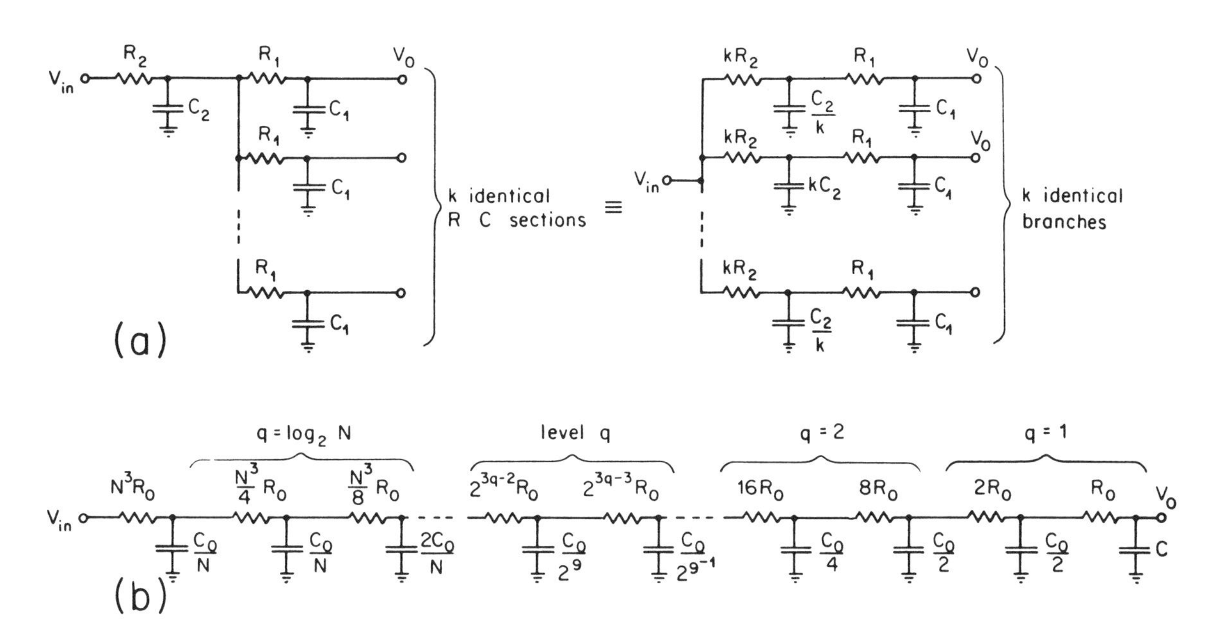

Fig. A2: (a) Circuit Equivalence
(b) Equivalent Clock Distribution Circuit for Individual PE.

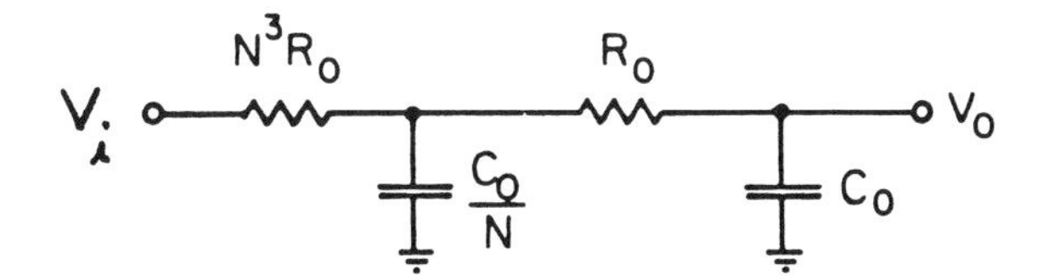

Fig. A3: Two Pole Approximation of Individual PE Clock Path.

Synchronizing Large VLSI Processor Arrays

ALLAN L. FISHER, MEMBER, IEEE, AND H. T. KUNG

Abstract —**Highly parallel VLSI computing structures consist of many processing elements operating simultaneously. In order for such processing elements to communicate among themselves, some provision must be made for synchronization of data transfer. The simplest means of synchronization is the use of a global clock. Unfortunately, large clocked systems can be difficult to implement because of the inevitable problem of clock skews and delays, which can be especially acute in VLSI systems as feature sizes shrink. For the near term, good engineering and technology improvements can be expected to maintain the feasibility of clocking in such systems; however, clock distribution problems crop up in any technology as systems grow. An alternative means of enforcing necessary synchronization is the use of self-timed asynchronous schemes, at the cost of increased design complexity and hardware cost. Realizing that different circumstances call for different synchronization methods, this paper provides a spectrum of synchronization models; based on the assumptions made for each model, theoretical lower bounds on clock skew are derived, and appropriate or best possible synchronization schemes for large processor arrays are proposed.**

One set of models is based on assumptions that allow the use of a pipelined clocking scheme where more than one clock event is propagated at a time. In this case, it is shown that even assuming that physical variations along clock lines can produce skews between wires of the *same* length, any one-dimensional processor array can be correctly synchronized by a global pipelined clock while enjoying desirable properties such as modularity, expandability, and robustness. This result cannot be extended to two-dimensional arrays, however; the paper shows that under this assumption, it is impossible to run a clock such that the maximum clock skew between two communicating cells will be bounded by a constant as systems grow. For such cases, or where pipelined clocking is unworkable, a synchronization scheme incorporating both clocked and "asynchronous" elements is proposed.

Index Terms —**Clock skew, processor arrays, synchronization, systolic arrays, VLSI complexity.**

I. INTRODUCTION

IN describing a processor array algorithm, it is often convenient to picture the processors as running in lock step. This *synchronized* view, for example, often makes the definition of the structure and its correctness relatively easy to follow: computations proceed in discrete steps which may be cleanly characterized. Perhaps the simplest means of synchronizing an ensemble of cells is the use of broadcast clocks. A clocked system in general consists of a collection of functional units whose communication is synchronized by external clock signals. A variety of clocking implementations are possible; the essential point is that by referring to the global time standard represented by the clock, communicating cells can agree on when a cell's outputs should be held constant and when a cell should be sensitive to its input wires. When different cells receive clock signals by different paths, they may not receive clocking events at the same time, potentially causing synchronization failure. These synchronization errors due to clock skews can be avoided by lowering clock rates and/or adding delay to circuits, thereby slowing the computation. The usual clocking schemes are also limited in performance by the time needed to drive clock lines, which will grow as circuit feature size shrinks relative to total circuit size. Therefore, unless operating at possibly unacceptable speeds, very large systems controlled by global clocks can be difficult to implement because of the inevitable problem of clock skews and delays.

As a practical aside, we should note that at current LSI circuit densities, clock distribution is still a solvable problem. Two somewhat pessimistic studies of which we are aware [3], [5] do not take into account either the tricks that a circuit designer can use to reduce the RC constant of his clock tree or the promise of multiple-layer metallization and low-resistance silicides. Given these factors, it seems that the usual clocking schemes should remain feasible for on-chip synchronization in the near term. Moreover, for some specific structures, such as one-dimensional arrays, clocking can be effectively used even in the presence of large signal propagation delays (see Section V-A).

An alternative approach to clocking is self-timing [10], in which cells synchronize their communication locally with some variety of "handshaking" protocol. It is easy to convince oneself that any synchronized parallel system where cells operate in lock step can be converted into a corresponding asynchronous system of this type that computes the same output — the asynchronous system is obtained by simply letting each cell start computing as soon as its inputs become available from other cells. The self-timed asynchronous scheme can be costly in terms of extra hardware and delay in each cell, but it has the advantage that the time required for a communication event between two cells is independent of the size of the entire processor array. A serious disadvantage of fully self-timed systems is that, given current digital de-

Reprinted from *IEEE Trans. Comput.*, vol. C-34, no. 8, pp. 734–740, Aug. 1985.

sign methodology, they can be difficult and expensive to design and validate.

An advantage that self-timed systems often enjoy, in addition to the absence of clock skew problems, is a performance advantage that results from each cell being able to start computing as soon as its inputs are ready and to make its outputs available as soon as it is finished computing. This allows a machine to take advantage of variations in component speed or data-dependent conditions allowing faster computation. This advantage will seldom exist in regular arrays such as systolic systems, however, for two reasons.

1) Usually, each cell in a regular array performs the same kind of computation as every other cell; thus, there is little opportunity for speed variation.

2) In cases where variations do exist, the throughput of computation along a path in an array is limited by the slowest computation on that path. The probability that a worst case computation will appear on a path with k cells is $1 - p^k$ where p is the probability that any given cell will *not* be performing a worst case computation. This quantity approaches unity as k grows, so large arrays will usually be forced to operate at worst case speeds.

The result of these considerations is that clocking is generally preferable to self-timing in the synchronization of highly regular arrays. This paper derives techniques for synchronizing large arrays, using clocking where possible and preserving some of the advantages of clocked schemes where clocking breaks down.

II. Basic Assumptions

The basic model that we will use for considering synchronization of VLSI processor arrays is as follows.

A1) Intercell data communications in an *ideally synchronized* processor array, in which all processors operate in lock step, are defined by a directed graph COMM, which is laid out in the plane. Each node of COMM, also called a *cell,* represents a cell of the array, and each directed edge of COMM, called a *communication edge,* represents a wire capable of sending a data item from the source cell to the target cell in every cycle of the system. Any two cells connected by a communication edge are called *communicating cells.*

A2) A cell occupies unit area.

A3) A communication edge has unit width.

We now add assumptions which provide the basis for clocked implementations of ideally synchronized arrays.

A4) A clock for a clocked processor array is distributed by a rooted binary tree CLK, which is also laid out in the plane. A cell of COMM can be clocked if the cell is also a node of CLK.

A5) A clocked system may be driven with clock period $\sigma + \delta + \tau$ where σ is the maximum clock skew between any two communicating cells, δ is the maximum time for a cell's outputs to be computed and propagated to a communicating cell, and τ is the time to distribute a clocking event on CLK.

This assumption is an abstraction of properties common to all clocking schemes. The detailed relationships between these parameters and other more specific parameters such as flip-flop setup and hold times depend on the exact clocking method used. An exact representation of minimum clock period might be something like $\max(\tau, 2\sigma + \delta)$ in a particular case, but such formulas will exhibit the same type of growth with respect to system size as the simple sum used here.

Note that if we adopt the usual convention that the clock tree is brought to an equipotential state before a new clock event is transmitted, eliminating clock skew can lead only to a constant factor increase in performance since it must always be true that $\sigma \leq \tau$. In particular, speed of light considerations impose the following condition.

A6) The time τ required to distribute a clocking event on a clock tree CLK in a particular layout is bounded below by $\alpha \cdot P$ where $\alpha > 0$ is a constant and P is the (physical) length of a longest root-to-leaf path in CLK.

Thus, since the clock tree must reach each cell in the array, large arrays which are synchronized by equipotential clocking must have clock periods at least proportional to their layouts' diameters. Note that in the remainder of this paper we will relate transmission delays to wire length; delays are caused by other factors, of course, but we choose to treat them together as a "distance" metric.

In the case where an array grows too big for its clock tree to be driven at the desired speeds due to the time needed to bring long wires to an equipotential state, it is possible to take advantage of the propagation delay down a long wire by having several clock cycles in progress along its length. This mode of clocking is often used in large mainframe computers, but not, to our knowledge, on chips. The electrical problems of passing a clean signal on a chip in this fashion are severe, due to analog phenomena such as damping and reflections. We can instead simulate this behavior by replacing long wires with strings of buffers, which will restore signal levels and prevent backward noise propagation. These buffers are spaced a constant distance apart; a good candidate is that distance which will cause wire delays between buffers to be of the same size as a buffer's propagation delay. This allows us to replace assumption A6) with the following.

A7) If CLK is a buffered clock tree, the time τ required to distribute a clocking event on a particular unbuffered segment of CLK is the maximum delay through a buffer and its output wire. Thus, τ is a constant independent of the size of the array.

To ensure that successive clock events remain correctly spaced along the clock path, we make the following assumption.

A8) The time for a signal to travel on a particular path through a buffered clock tree is invariant over time.

The following section describes two clock skew models based on the above assumptions, and Sections IV and V explore the problem of clocking under these models. Section VI considers the case where assumption A8) does not hold, and hence condition A6) holds rather than condition A7). Section VII briefly discusses the practicality of the models and the results obtained.

III. Two Models of Clock Skew

Given a basic model consisting of conditions A1) through A5), plus A7) and A8), the following sections consider the implications of two models of clock skew. First, in Section IV we consider the case where clock skew between two cells depends on the difference in their physical distances from the root of the clock tree. This *difference model* corresponds reasonably well to the practical situation in high-speed systems made of discrete components, where clock trees are often wired so that delay from the root is the same for all cells. Formally, we assume the following.

A9) The clock skew between two nodes of CLK, with respect to a given layout, is bounded above by $f(d)$ where f is some monotonically increasing function and d is the positive difference between the (physical) lengths of the paths on CLK that connect the two nodes to the root.

This assumption is illustrated in Fig. 1. The two circles connected by the dashed line have clock skew between them which is no more than $f(d)$ where d is the length of the crosshatched segment. This segment represents the difference between the cells' distances to their nearest common ancestor in the clock tree.

As systems grow, small variations in electrical characteristics along clock lines can build up unpredictably to produce skews even between wires of the same length. In the worst case, two wires can have propagation delays which differ in proportion to the sum of their lengths. Especially since it is not possible to tune the clock network of a system on a single chip, Section V considers a model in which the skew between two nodes depends on the distance between them along the clock tree. Formally, the *summation model* (so called because the distance between two nodes is the sum of their distances from their nearest common ancestor, while the difference measure used above is the difference between those distances) uses the following upper and lower bound assumptions.

A10) The clock skew between two nodes of CLK, with respect to a given layout, is bounded above by $g(s)$ where g is some monotonically increasing function and s is the (physical) length of the path on CLK that connects the two nodes.

A11) The clock skew between two nodes of CLK, with respect to a given layout, is bounded below by $\beta \cdot s$ where $\beta > 0$ is some constant and s is the (physical) length of the path on CLK that connects the two nodes.

Fig. 2 illustrates these assumptions; here both the upper and lower bounds on the skew between the two communicating cells depend on the entire length of the path between them, which is the sum of their distances to their nearest common ancestor in the tree.

The two models of clock skew introduced above can be formally derived as follows, for the case when both functions f and g are linear. Let h_1 and h_2, with $h_1 \geq h_2$, be the distances of any two cells to their nearest common ancestor in the clock tree. Let $m + \varepsilon$ and $m - \varepsilon$ be the maximum and minimum time, respectively, to transmit a clock signal across a wire of unit length where ε corresponds to the variations in electrical characteristics along clock lines. Then the clock skew σ between the two cells can be as large as

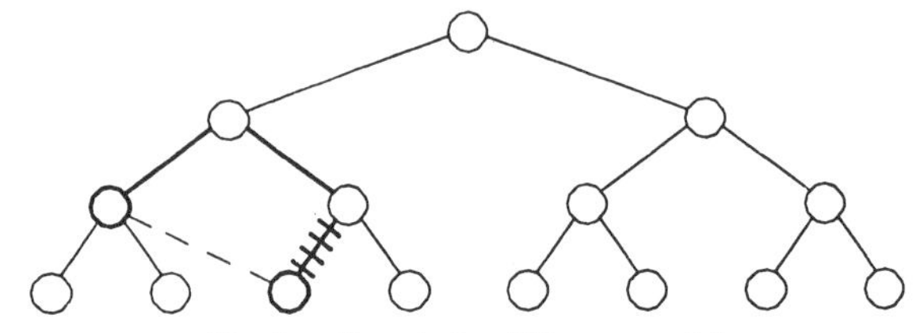

Fig. 1. Skew in the difference model.

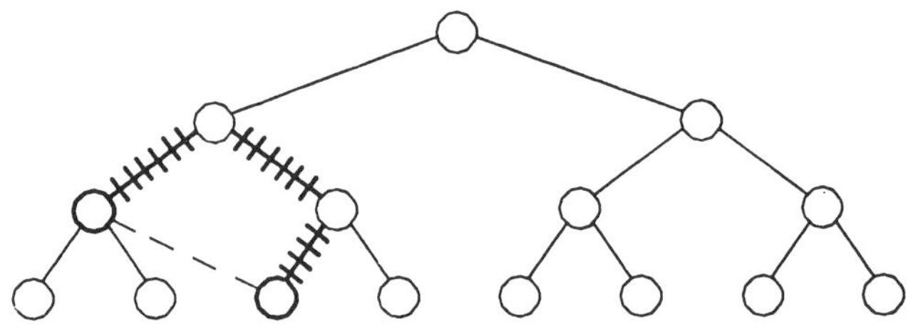

Fig. 2. Skew in the summation model.

$$\sigma = h_1(m + \varepsilon) - h_2(m - \varepsilon) = (h_1 - h_2)m + (h_1 + h_2)\varepsilon.$$

Noticing that $d = h_1 - h_2$, $s = h_1 + h_2$, and $s \geq d > 0$, we have

$$(m + \varepsilon) \cdot s \geq \sigma = m \cdot d + \varepsilon \cdot s \geq \varepsilon \cdot s.$$

We see that the upper and lower bounds correspond directly to assumptions A10) and A11) used in the summation model, while the difference model covers the case when terms involving ε can be ignored.

IV. Clocking Under the Difference Model

Assuming the basic model defined in Section II, along with condition A9), which states that the skew between two cells is bounded above by a function of the difference between their distances from the root, we note that only a bounded amount of clock skew will occur if we ensure that all nodes in COMM are equidistant (with respect to the clock layout) from the root of CLK. This can be achieved for any layout for COMM of bounded aspect ratio, without increasing the areas of the layout by more than a small constant factor, by distributing the clock through an H-tree [8]. This scheme is illustrated for linear, square, and hexagonal arrays in Fig. 3, in which heavy lines represent clock edges and thin lines represent communication edges.

More precisely, we have the following result.

Lemma 1: For any given layout of bounded aspect ratio, it is possible to run a clock tree such that all nodes in the original layout are equidistant (with respect to the clock tree) from the root of the tree, and the clock tree takes an area no more than a constant times the area of the original layout.

By a theoretical result [1] that any rectangular grid (for example, an $n^{2/3} \times n^{1/3}$ grid) can be embedded in a square grid by stretching the edges and the area of the source grid by at most a constant factor, we have the following theorem.

Theorem 2: Under the difference model of clock skew, any ideally synchronized processor array with computation and communication delay δ bounded by a constant can be simulated by a corresponding clocked system operating with a clock period independent of the size of the array, with no more than a constant factor increase in layout area.

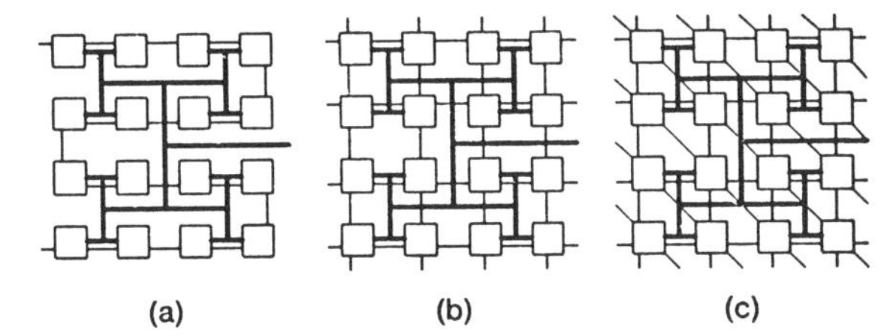

Fig. 3. H-tree layouts for clocking. (a) Linear arrays. (b) Square arrays. (c) Hexagonal arrays.

V. Clocking Under the Summation Model

This section relaxes the assumption of the previous section by using the summation model rather than the difference model for clock skews. The clock skew between two nodes of CLK, with respect to a given layout, is related to the (physical) length of the path on CLK that connects the two nodes. Note that because the summation model is weaker than the difference model, any clocking scheme working under the summation model must also work under the difference model. The reverse of the statement is not true, however. For example, the clocking scheme illustrated in Fig. 3(a) for linear arrays may not work under the summation model since two communicating cells (such as the two middle cells on the left-hand side of the layout) could be connected by a path on CLK whose length can be arbitrarily large as the size of the array grows. In the following we give another clocking scheme for linear arrays that works even under the summation model for clock skew; in addition, we show that it is impossible, under this model, to clock a two-dimensional array in time independent of its size. In this sense, linear arrays are especially suitable for clocked implementation.

A. *Clocking One-Dimensional Processor Arrays*

Given any ideally synchronized one-dimensional array [Fig. 4(a)], we propose a corresponding clocked array [Fig. 4(b)] obtained by running a clock wire along the length of the one-dimensional array. By A10) the maximum clock skew between any two neighbors is bounded above by a constant $g(s)$ where s is the center-to-center distance between neighboring cells. Thus, we have the following result.

Theorem 3: Under the summation model of clock skew, any ideally synchronized one-dimensional processor array with computation and communication delay δ bounded by a constant can be simulated by a corresponding clocked system, as illustrated in Fig. 4, operating at a clock period independent of the size of the array.

Skew between the host and the ends of the array can be handled similarly by folding the array in the middle (Fig. 5), and the array can be laid out with any desired aspect ratio by using a comb-shaped layout (Fig. 6).

With the clocking schemes illustrated, we see that the clock period for any one-dimensional processor array can be made independent of the size of the array. As a result, the clocked array may be extended to contain any number of cells using the same clocked cell design. These clocked schemes are probably the most suitable for synchronizing one-dimensional arrays due to their simplicity, modularity, and expandability. Note that one-dimensional arrays are especially important in practice because of their wide applicabilities and their bounded I/O requirements [4].

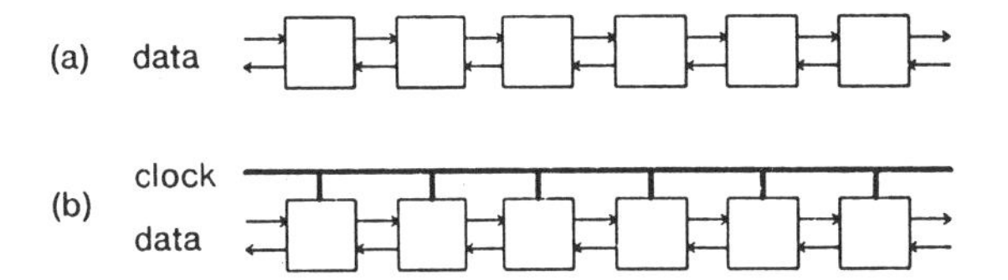

Fig. 4. (a) Ideally synchronized one-dimensional array. (b) Corresponding clocked array.

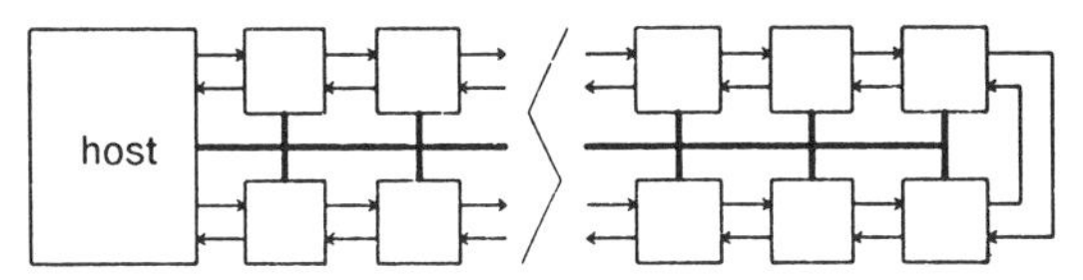

Fig. 5. Array folded to bound skew with host.

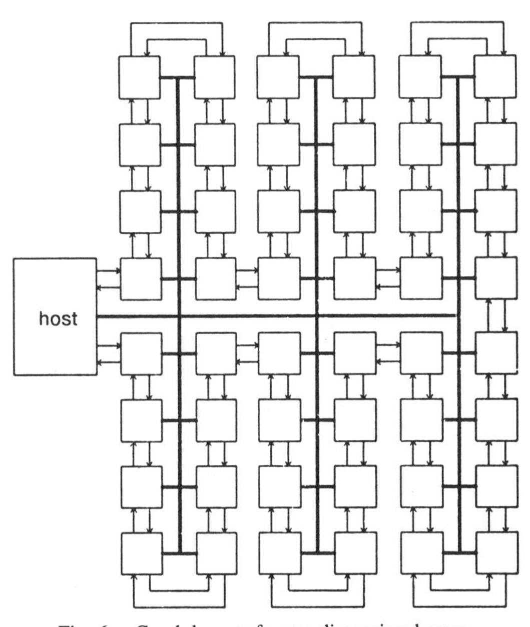

Fig. 6. Comb layout of a one-dimensional array.

B. *A Lower Bound Result on Clock Skew*

We show here that the result of Theorem 3 for the one-dimensional array cannot be extended to two-dimensional structures. Consider any layout of an $n \times n$ array clocked by a global clock tree CLK; the nodes of CLK include all cells of the array. Let σ be the *maximum clock skew* between two communicating cells of the array. We want to prove that σ cannot be bounded above by any constant independent of n. We use the following well-known result [6].

Lemma 4: To bisect an $n \times n$ mesh-connected graph at least $c \cdot n$ edges have to be removed, where $c > 0$ is a constant independent of n.

Bisecting a graph means partitioning the graph into two subgraphs, each containing about half of the nodes of the original graph. Here, for the $n \times n$ mesh-connected graph we assume that none of the subgraphs contain more than $(23/30) \cdot n^2$ nodes. We also use the following simple lemma without giving a proof.

Lemma 5: For any subset M of at least two nodes of a binary tree, there exists an edge of the tree such that its

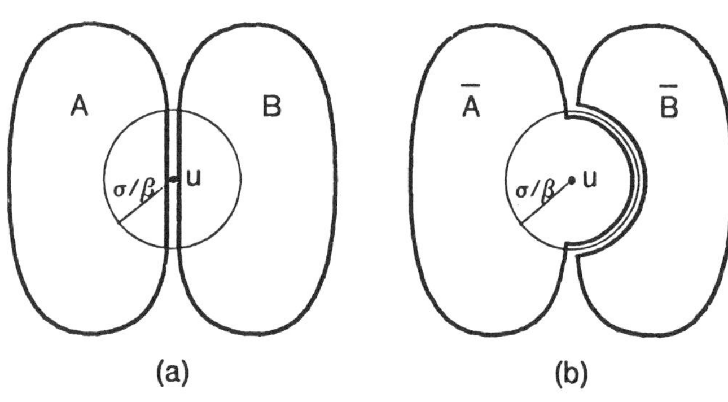

Fig. 7. (a) Original partition of the communication graph. (b) New partition of the communication graph.

removal from the tree will result in two disjoint subtrees, each having no more than two-thirds of the nodes in M.

The n^2 cells of the $n \times n$ array form a subset of nodes of CLK. By Lemma 5 we know that by removing a single edge, CLK can be partitioned into two disjoint subtrees such that each subtree has no more than $(2/3) \cdot n^2$ cells. Denote by A and B the sets of cells in the two subtrees. Let u be the root of the subtree that contains cells in A. Consider the circle centered at u and with radius of σ/β where β is defined in A11) [Fig. 7(a)]. If there are $\geq(1/10) \cdot n^2$ cells inside the circle, then by A2)

$$\pi(\sigma/\beta)^2 \geq n^2/10, \qquad \text{or } \sigma = \Omega(n),$$

and thus σ cannot be bounded above by any constant independent of n. Suppose now that there are fewer than $(1/10) \cdot n^2$ cells inside the circle. Note that any of those cells in A which are outside the circle cannot reach any cell in B by a path on CLK with (physical) length $\leq \sigma/\beta$. Thus, these cells cannot have any communicating cells in B (with respect to the $n \times n$ array) since by A11) the clock skew between these cells and any cell in B would be greater than $\beta \cdot \sigma/\beta = \sigma$, and the clock skew between any two neighboring cells is assumed to be no more than σ. Now let $\bar{A}$ be the union of A and the set of cells in the circle, and $\bar{B}$ be B minus the set of cells in the circle, as in Fig. 7(b). Then $\bar{A}$ and $\bar{B}$ form a partition of the $n \times n$ array, and each of them has no more than $(1/10) \cdot n^2 + (2/3) \cdot n^2 = (23/30) \cdot n^2$ cells. From Fig. 7(b), we see that any edge in the $n \times n$ array connecting a cell in $\bar{A}$ and a cell in $\bar{B}$ must cross the boundary of the circle. Since the length of the boundary is $2\pi\sigma/\beta$, by A3) $\bar{A}$ and $\bar{B}$ are connected by no more than $2\pi\sigma/\beta$ edges. By Lemma 4 we have $2\pi\sigma/\beta \geq c \cdot n$, or

$$\sigma = \Omega(n).$$

Therefore, as n increases, σ grows at least at the rate of n; we see that it is impossible to run a global clock for the $n \times n$ array such that the maximum clock skew σ between communicating cells will be bounded above by a constant, independent of n.

The above proof for two-dimensional mesh graphs can be generalized to deal with other classes of graphs. For the generalization, we need to define the *minimum bisection width* of a graph [11], which is the number of edge cuts needed to bisect the graph. For example, by Lemma 4 the minimum bisection width of an $n \times n$ mesh-connected graph is $\mu(n)$. We have the following general result.

Theorem 6: Suppose that the minimum bisection width of an N-node graph is $\Omega(W(N))$ and $W(N) = O(\sqrt{N})$. Then

$$\sigma = \Omega(W(N)).$$

Since under the summation model of clock skew, two-dimensional $n \times n$ processor arrays cannot be efficiently implemented by clocked controls, their implementation should be assisted by some self-timed scheme, as discussed in the next section.

VI. Hybrid Synchronization

In the absence of the invariance condition A8), in which case pipelined clocking fails, or for communication graphs with asymptotically growing clock skews under the summation model, global clocking is unable to provide constant clock rates as a system grows. In this case, a scheme similar to one described by Seitz [10], where local clocks are controlled by a self-timed handshaking synchronization network, can be used.

In this approach, we break up the layout into bounded-size segments called *elements* and provide each element with a local clock distribution node. The clock distribution nodes employ a handshaking protocol to synchronize among themselves and then distribute clock signals to the cells in their elements. Given assumptions about the maximum delay of a computation node and its communication wires, we can clock the cells in each element in constant time. This structure is illustrated in Fig. 8, in which the heavy lines and black boxes represent the self-timed synchronization network and the narrow lines represent local clock distribution to the cells in each element. Note that the subordination of the local clocks to the self-timed network avoids the possibility of synchronization failure due to a flip-flop entering a metastable state since an element stops its clock synchronously and has its clock started asynchronously.

This provides the asymptotic performance of a self-timed system by making all synchronization paths local while isolating the self-timed logic to a small subsystem and allowing the cells to be designed as if the entire system were globally clocked. The hybrid approach has the additional advantage that a single synchronization design can be used for many different structures. This simplification of the usual fully self-timed scheme is made possible by the fact that we are willing to assume a maximum delay for the cells; this is the same assumption made in ordinary clocked schemes. Note that we are willing to let the entire array operate at worst case cell speed since even a fully self-timed array would usually wind up operating at that speed regardless.

VII. Practical Implications

This section addresses some practical aspects of the material of this paper. Perhaps the most important practical issue raised in this study is the question of the use of pipelined clocking on chips; this section discusses some of the potential limits to this technique and presents some simple

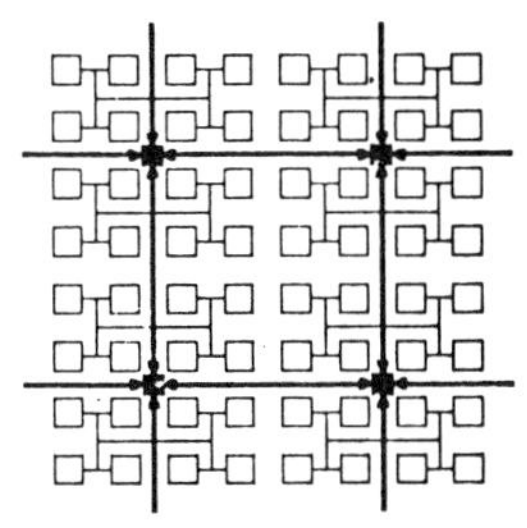

Fig. 8. Hybrid synchronization scheme.

experimental evidence suggesting its practicability. It also discusses the practical relevance of the difference and summation models of clock skew.

The practicality of pipelined clocking hinges on two issues: the limitations of the uniformity assumption A7), and the delay associated with distributing clocks in a conventional fashion. Pipelined clocking only makes sense if clock event transmission is uniform enough to gain an advantage over equipotential clocking.

This relationship, in turn, depends on the relative speeds of logic and interconnect. Pipelined clock trees, with short interconnection paths, will run at logic switching speeds (to the extent that uniformity obtains). The speed at which equipotential clock trees can run is determined by the impedance of the interconnect and by its physical dimensions. We would thus expect pipelined clocking to be most applicable where switching speeds are high and interconnect is long and has high impedance; for example, wafer-scale gallium arsenide may be a likely candidate.

The uniformity of transmission of clock events is subject to a number of factors. One obvious limitation is the uniformity of a buffer in passing rising and falling edges. For an nMOS superbuffer, for example, making transit times for rising and falling edges the same requires careful circuit tuning, and the resulting circuit will be very sensitive to manufacturing process parameters. One solution to this problem is to make each buffer respond only to rising edges on its input and to generate its own falling edges with a one-shot pulse generator. This solution has the disadvantage that the pulse width must be either wired into the circuit or programmable by some means. This may actually be convenient in some cases, if the pulse generating circuitry can be designed to model the delay of the logic circuitry. Beyond static considerations, however, the transmission of clock events can also be affected by noise, for example, internet capacitive coupling in MOS circuits. This problem can only be avoided by careful design, and further research is needed in estimating its magnitude.

Another, simpler approach to the rising/falling edge problem is to build a distribution line as a string of inverters. If the impedance of the outputs of the odd inverters is the same as that of the even inverters, rising and falling edges should traverse the string at essentially the same speed. Although this approach eliminates any inherent bias in favor of one type of edge, it does not result in speed independent of the length of the inverter string. Assume that the discrepancy between rising and falling transit times for a pair of inverters is normally distributed with a mean of zero and variance V. The sum of the discrepancies of n inverter pairs will be similarly distributed, with variance nV. If a fixed yield, independent of n, is desired, chips with a discrepancy sum proportional to the standard deviation, hence proportional to $\sqrt{n}$, must be accepted. Since a minimum condition for a given chip to run with cycle time T is that the sum of discrepancies be no greater than T, some chips will run with cycle times at least proportional to $\sqrt{n}$. The constant factors involved may still be small enough, however, to make this scheme feasible in practice.

As a simple trial of practical issues, an nMOS chip consisting of a string of 2048 minimum inverters was designed, without any special attention paid to making interconnect impedance uniform. An equipotential single phase clock signal could be run through the entire string with a cycle time of approximately 34 μs; even with the disadvantage of a slight bias in the circuit design toward falling edges, a pipelined clock could be run with a cycle time of 500 ns, 68 times faster. The same speedup was observed on five separate chips, indicating that the effect of the bias in the circuit design dominated the type of probabilistic effects described above. Assuming that transit times and any discrepancy between rising and falling edges scale linearly, a similar inverter string of any length could be clocked 68 times faster in pipeline mode than in equipotential mode. This figure does not indicate that pipelined clocking is actually applicable in this case; a chip of this size in this technology could easily be clocked with a 50 ns cycle time with a well-designed low-resistance equipotential clock. However, it does suggest that pipelined clocking may well be feasible where switches are fast and wires are slow.

The second practical issue related to this work is the question of the applicability of the difference and summation models of clock skew. First, both models apply only where pipelined clock distribution is possible; otherwise, clock period inevitably grows with the system, and the only means of improving performance are technology improvement, clever design, and self-timing. For the difference model to apply and for H-tree or other equidistant clocking schemes to be useful, it must be possible to closely control the "length" (that is, the delay characteristics) of the clock tree. This is possible in systems where wires are discrete entities that can be tuned, and indeed this is common practice in such systems. Whether this is true for integrated circuits is another question, hinging on the variability of the fabrication process and on noise characteristics.

The summation model is much more robust. Given the possibility of pipelined clocking, almost any imaginable means of transmitting clock events will have the property that cells close together on the clock tree will have bounded skew between them. We can thus be confident that linear arrays and similar structures will work as well as pipelined clocking can work.

VIII. Concluding Remarks

In this paper, we have analyzed the effect of clocked synchronization on the performance of large processor arrays. We have identified the key issues on which this depends (clock delay and clock skew) and have proposed means of implementing pipelined clocking for integrated circuits. We have considered two models of the dependence of clock skew on layout properties and have derived upper and lower bounds for the performance of processor arrays of varying topologies. The key results here are that one-dimensional arrays can be clocked at a rate independent of their size under fairly robust assumptions, while two-dimensional arrays and other graphs with similar properties cannot. We have also discussed some of the practical implications of these theoretical results.

This study has concentrated on the interaction of clock skew models with the communication structure of arrays with bounded communication delay; future work should also examine cases where asymptotically growing delays occur. One interesting such case is that where the communication graph COMM, neglecting edge directions, is a binary tree. It has been shown that a planar layout of a tree with N nodes of unit area must have an edge of length $\Omega(\sqrt{N}/\log N)$ [9]. Under the summation model of Section V then, if we make the additional assumption that communication delays grow with path length in the same way as clocking delays, a tree may be clocked at no loss in asymptotic performance simply by distributing clock events along the data paths.

Furthermore, if COMM is acyclic, as in the tree machine algorithms described in a paper by Bentley and Kung [2], and the ratio between lengths (in the layout) of any two edges at the same level in the graph is bounded, pipeline registers can be added on the long edges, with the same number of registers on all of the edges in a given level. This makes all wires have bounded length, thus causing the time needed for a cell to operate and pass on its results to be independent of the size of the tree. Adding the registers increases the layout area by at most a constant factor since they, in effect, just make wires thicker. For example, an H-tree layout has this property and allows a tree machine of N nodes to be laid out in area $O(N)$ with delay through the tree of $O(\sqrt{N})$ and constant pipeline interval.

References

[1] R. Aleliunas and A. L. Rosenberg, "On embedding rectangular grids in square grids," *IEEE Trans. Comput.*, vol. C-31, pp. 907–913, Sept. 1982.

[2] J. L. Bentley and H. T. Kung, "A tree machine for searching problems," in *Proc. 1979 IEEE Int. Conf. Parallel Processing*, Aug. 1979, pp. 257–266.

[3] M. A. Franklin and D. F. Wann, "Asynchronous and clocked control structures for VLSI based interconnection networks," in *Proc. Ninth Annu. Symp. Comput. Architecture*, Apr. 1982, pp. 50–59.

[4] H. T. Kung, "Why systolic architectures?" *IEEE Comput. Mag.*, vol. 15, pp. 37–46, Jan. 1982.

[5] S. Y. Kung and R. J. Gal-Ezer, "Synchronous vs. asynchronous computation in VLSI array processors," in *Proc. SPIE Symp., Vol. 341, Real-Time Signal Processing V*, Soc. Photo-Opt. Instrument. Eng., May 1982.

[6] R. J. Lipton, S. C. Eisenstat, and R. A. DeMillo, "Space and time hierarchies for classes of control structures and data structures," *J. ACM*, vol. 23, no. 4, pp. 720–732, Oct. 1976.

[7] C. A. Mead and L. A. Conway, *Introduction to VLSI Systems*. Reading, MA: Addison-Wesley, 1980.

[8] C. A. Mead and M. Rem, "Cost and performance of VLSI computing structures," *IEEE J. Solid-State Circuits*, vol. SC-14, pp. 455–462, Apr. 1979.

[9] M. S. Paterson, W. L. Ruzzo, and L. Snyder, "Bounds on minimax edge length for complete binary trees," in *Proc. Thirteenth Annu. ACM Symp. Theory Comput.*, May 1981, ACM SIGACT, pp. 293–299.

[10] C. L. Seitz, "System timing," in *Introduction to VLSI Systems*, C. A. Mead and L. A. Conway. Reading, MA: Addison Wesley, 1980, ch. 7.

[11] C. D. Thompson, "A complexity theory for VLSI," Ph.D. dissertation, Dep. Comput. Sci., Carnegie-Mellon Univ., Pittsburgh, PA, Aug. 1980.

Comparison of Tree and Straight-Line Clocking for Long Systolic Arrays

Marios D. Dikaiakos and Kenneth Steiglitz*

Department of Computer Science
Princeton University

Abstract

Achieving efficient and reliable synchronization is a critical problem in building long systolic arrays. We address this problem in the context of synchronous systems by introducing probabilistic models for two alternative clock distribution schemes: tree and straight-line clocking. We present analytic bounds for the Probability of Failure and examine the trade-offs between reliability and throughput in both schemes. Our basic conclusion is that as the one-dimensional systolic array gets very long, tree clocking becomes preferable to straight-line clocking.

1 Introduction

Several problems in scientific computation and signal processing can be solved efficiently by special-purpose one-dimensional systolic architectures [MMJ89]. Such solutions may have significant practical importance if they perform and scale well for large problem sizes. This implies that ultimately they should comprise many processing elements to achieve a high degree of parallelism. It appears therefore that some future special-purpose machines will be built as very long systolic arrays of fine-grained components.

One of the limiting factors in building long pipelines is the difficulty in achieving proper and reliable synchronization. In this paper we investigate clock synchronization failures in such systems, in terms of their length and parameters that characterize clocking circuitry, such as delays in buffers and wires, and variance in buffer response time. Since the one-dimensional pipeline is the simplest topology for inter-processor communication, our results also provide some insight into the problem of synchronizing large parallel systems in general. First, we concentrate on the case where the clock distribution network is implemented as a regular f-ary tree (*tree clocking*) [FK84], and analyze the effect of clock skew on system performance and reliability using a probabilistic model for clock skew. In addition to tree clocking, *straight-line clocking* is addressed. In this scheme the clock is propagated alongside the pipeline, in parallel with the data-flow. In [FK84] it is suggested that this scheme is effective because skew between adjacent processing elements is bounded, and building or extending such a distribution network is fairly easy. In that case, we focus on clock synchronization failures due to the lack of uniformity of clocking buffers in passing rising and falling edges. Again, we use a probabilistic approach and derive asymptotic results for the probability of clock synchronization failure.

*This work was supported in part by NSF Grant MIP-8912100, and U.S. Army Research Office-Durham Contract DAAL03-89-K-0074.

2 Tree Clocking

We examine first the clocking of long systolic pipelines where the clock is distributed to pipeline stages (PEs) via a symmetric regular f-ary CLOCK tree. Nodes and edges in the CLOCK tree correspond to buffers and wires in the clock distribution network respectively, and the root of CLOCK corresponds to the clock source. The clock source has the responsibility to drive the entire CLOCK tree and wait for a clock pulse to arrive at all destinations before sending the next pulse (*equipotential* clocking). A one-phase clocking scheme is adopted.

The pipeline stages are attached to the leaves of the CLOCK tree. Their interconnection is serial, as shown in figure 1. Each PE is formed by two sub-cells: **CL** which is a combinational circuit, and register **R** which is an edge-triggered flip-flop.

We are interested only in clock skew between adjacent stages of the pipeline because the data transfer occurs only between contiguous PEs. Let the node j in CLOCK be the closest common ancestor of leaves i and $i+1$, and let t_j be the departure time of a clock

Reprinted with permission from *J. VLSI Signal Processing,* M. D. Dikaiakos & K. Steiglitz, "Comparison of Tree & Straight-Line Clocking for Long Systolic Arrays," pp. 1177–1180, 1991. © Kluwer Academic Press.

edge from j. If the arrival times of that clock edge at PEs i and $i+1$ are t_i and t_{i+1} respectively, the clock skew between PEs i and $i+1$ equals $t_{i+1} - t_i$, and is attributed to two causes: The temporal fluctuations in clock-buffer delays, called *run-time skew* and denoted by δ_i^r; and the variations in delay characteristics of different components (because of different chip characteristics), called *build-time skew* and denoted by δ_i^b. In other words, the cumulative clock skew can be expressed as follows :

$$\delta_i^r + \delta_i^b = t_{i+1} - t_i \quad (1)$$

where δ_i^r, δ_i^b may be either nonnegative or negative. Build-time skew remains constant after selecting the clock buffers off the shelf, and building the clock distribution network. In real designs, the clock distribution network is tuned so as to minimize the effects of build-time skew. The tuning procedure usually involves the adjustment of delay elements, buffers, and wires and can guarantee a "negligible" build-time clock skew [Wag88].

For the calculation of *run-time clock skew* we assume that CLOCK follows the *metric-free* tree model [KS90]. In this model, all buffers (nodes) are identical and add to the clock signal a delay modeled by the same probability distribution. Wires (edges) which propagate the clock have equal lengths. Therefore, every wire has the same probability distribution for delay, which can be lumped with the delay of the buffer that follows it. The metric-free tree presents a reasonable abstraction for distribution networks which provide a clock signal to chips on a number of boards in a system.

Assumption 2.1 *The delay inserted in the clock signal by buffer k, and the wire leading to it, is considered to be a random deviate τ_k, distributed normally with zero mean and finite variance, i.e., $\tau_k \sim N(0, \sigma^2)$, and independent from clock edge to clock edge.*

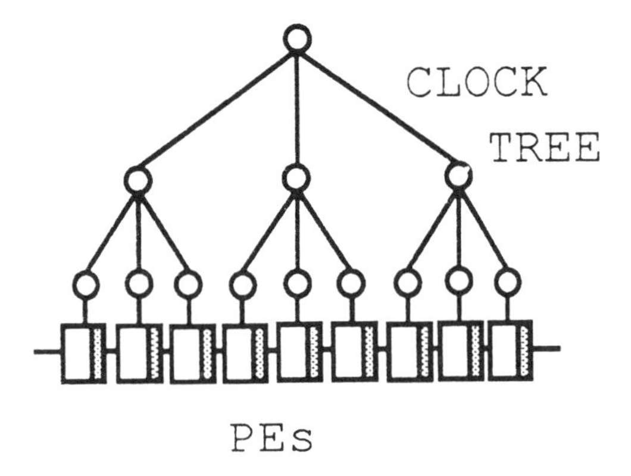

Figure 1: Tree clocking scheme.

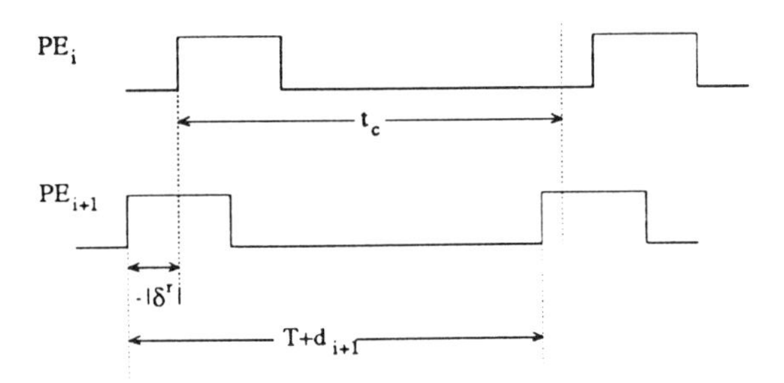

Figure 2: Type-1 Failure: the output of PE i appears too late to be latched by PE i+1.

Actually, each buffer adds a positive delay to the clock signal, and therefore there is a non-zero mean for the delay distribution. If we consider the distinct paths that route the clock to two adjacent PEs, the difference of their cumulative means is equal to the *build-time skew* between them. Nevertheless, because of CLOCK tree symmetricity, our *run-time skew* analysis is independent of the build-time skew value, and may proceed as if the cumulative means along the two distinct paths cancel out (i.e. as if the build-time skew were zero). Therefore the mean values of τ_k's may be considered zero.

In the case where the clock signal arrives earlier at PE $i+1$ than at PE i (negative clock skew), and the clock period is not large enough, the data computed in PE i may arrive at PE $i+1$ after the arrival of the next pulse's leading edge (see figure 2). In that case, proper synchronization is not guaranteed and data may be lost [HC87]. Therefore, the following condition must be satisfied:

$$d_{i+1} + \delta_i^r > t_c - \delta_i^b - T \quad (2)$$

where t_c is the total computation delay. The random variable d_l is the difference in rising-clock-edge arrival times at PE l that compensates for the fact that two successive rising clock edges arrive with different delays at the same PE l. Although the two successive pulses have been emitted by the central clock source within time T, they arrive at PE l with a time difference of $T + \sum_{k \in K_{s,l}} (\tau_k' - \tau_k)$ between each other. The random variable τ_k corresponds to the delay inserted in one clock pulse by some buffer in the path from the clock source to PE l, τ_k' corresponds to the delay inserted in the next clock pulse by the same buffer k,

K_{ji} denotes the set of CLOCK nodes (buffers) along the path from node j to node i, not including j, and S corresponds to the clock source (root of CLOCK).

In the following sections, we lump δ_i^b with t_c, denote the sum as t_c^b, and pursue our analysis as if the run-time skew δ_i^r were the only cause of clock failures. Failures due to violations of (2), will be referred as *type-1 failures* in contrast to *type-2 failures* which do not influence the clock period (see [HC87, DS90]).

We restrict our attention to type-1 failures, in order to examine the relation between reliability and clock speed in very long pipelines. We denote as $G(1, N-1)$ the probability that clock skew does not cause type-1 synchronization failures. We proved that ([DS90]):

$$G(1, N-1) \geq \left(1 - \frac{\sqrt{\log_f N}}{\sqrt{\pi} \cdot \tau} \cdot e^{-\tau^2/4\log_f N}\right)^{N-1} \quad (3)$$

where τ is the *normalized margin* $\tau = (T - t_c^b)/\sigma$. This relation provides some insight into the tradeoff between clock period and reliability. In [DS90] it is shown that there is a tight range of clock period values where the guaranteed reliability (i.e., the lower bound on the probability of success) changes sharply from low to very high. The design challenge is to satisfy two conflicting goals: first, the achievement of an acceptable reliability level, which requires the increase of T. Second, the maximization of pipeline throughput and PE utilization, achieved by making T as close as possible to t_c^b.

We notice that if we demand that the lower bound in (3) be equal to $1-\epsilon$, where ϵ is a very small positive real number, the probability of success would be close to one. To achieve this as $N \to \infty$, it is sufficient that:

$$T = t_c^b + \frac{2}{\sqrt{\log_f e}} \cdot \log_f N \quad (4)$$

From this, we conclude that as the length of the one-dimensional systolic array increases, an increase in the clock period proportional to $\log_f N$, is sufficient to guarantee negligible failure probability.

Equation (4) can be easily extended to cover the existence of build-time skew ([DS90]). In fact, it can be shown that build-time skew does not affect the pipeline throughput asymptotically. In practice however, it might require substantially higher values of T to guarantee highly reliable functioning of the systolic array.

3 Straight-Line Clocking

Straight-line (pipelined) clocking represents an alternative to equipotential clocking for synchronous systems. Under the straight-line scheme, the clock distribution network is composed of a series of buffers (repeaters) which carry successive pulses from the global clock source, so that several clock pulses are simultaneously active in the system (figure 3). As Fisher and Kung point out [FK84], straight-line clocking is most applicable in cases where PE speeds are very high, and interconnect is long and has high impedance. A potential cause of clock synchronization failure is the lack of uniformity in repeaters when passing falling and rising clock edges. Worst case analysis has shown that *differences in the delay between leading and falling edges may cause the disappearance of clock pulses and thus synchronization failure* [GS90]. Thus, the failure mode of straight-line clocking is different from that of tree clocking.

Our principal assumption is that whenever a pulse of width w goes through a buffer, its width is changed and becomes $w + \epsilon$, where ϵ is a normal random variable with zero mean and finite variance. We model the pulse width as a random process, where the state of the process corresponds to pulse width, and the discrete time variable corresponds to the buffer stage where the pulse is currently in. If w ever reaches zero, the pulse disappears and the process is *absorbed*. The assumption about the random changes (increments) ϵ of the pulse width and the process absorption at zero leads us to a thoroughly analyzed form of random process, *Brownian Motion with Absorbing Barrier at zero* [KT75].

We chose the mean value of the pulse's random increments to be zero, which means that we implicitly assume that repeaters are designed to respond uniformly to rising and falling edges. Failures occur because of the variance in the response time. Assuming that the central clock source generates a waveform with a *duty cycle* $\alpha = w/T$, the probability that a clock pulse reaches the end of the pipe when N is very large (long pipelines) has the following behavior

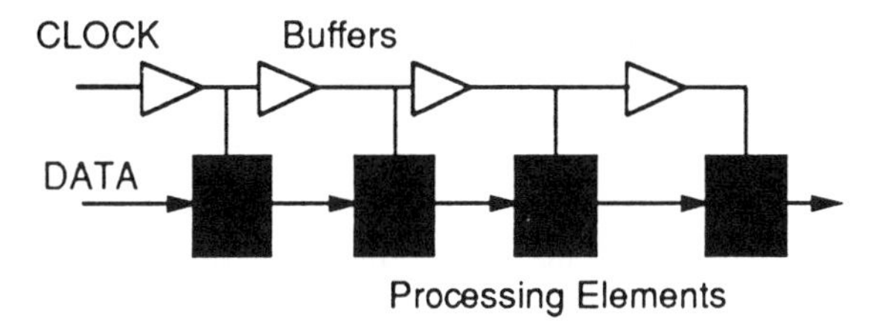

Figure 3: Straight-Line clocking scheme.

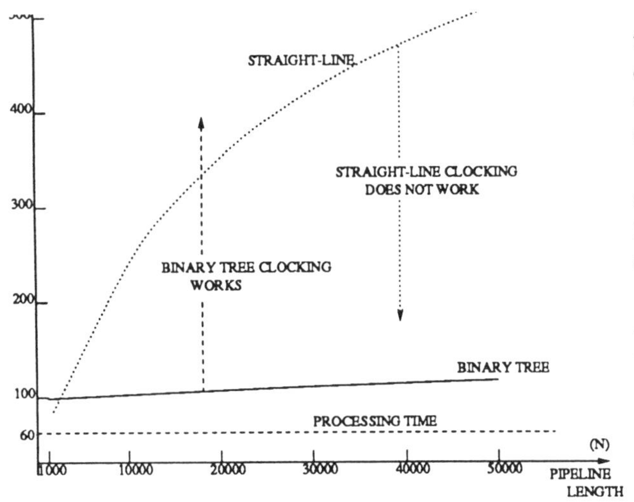

Figure 4: Straight-Line vs Tree Clocking Schemes: $N = 1000$ to 50000.

([DS90]):

$$\alpha_1 \cdot T/\sigma\sqrt{N} \ \geq \ Pr[D > N] \qquad (5)$$

where $a_1 = 2 \cdot \alpha/\sqrt{2\pi}$. If $Pr[D > N] \approx 1$, we can easily obtain: $T > \sigma \cdot \sqrt{N}/\alpha_1$. In addition to the previous inequality, the clock period T should always be greater than the computation time t_c. Therefore :

$$T \ > \ max\{t_c, \frac{\sigma \cdot \sqrt{N}}{\alpha_1}\} \qquad (6)$$

Whenever straight-line clocking is adopted, this presents a *necessary* but not *sufficient* condition for the systolic array to function with a very high reliability. We therefore conclude that as N gets larger, the clock period should grow faster than the square root of the pipeline length. Otherwise, a frequent occurrence of clock failures is expected. It is interesting to note that this conclusion agrees with the heuristic argument presented in [FK84].

4 Conclusions

Our first conclusion refers to the tree clocking scheme: we proved that for very long one-dimensional systolic arrays of length N, a growth in the clock period proportional to $\log_f N$ is sufficient to guarantee very high reliability with regard to synchronization failure.

The second conclusion relates to straight-line clocking: in that case we showed that a *necessary* condition for a systolic pipeline to function with very high reliability, is that its clock period grow proportionally to $\sqrt{N}$. In both cases, the acceptable reliability levels are determined by corresponding estimates of the Mean Time to Failure ([DS90]). Given these conclusions, we can see that as the systolic array gets very long, tree clocking is preferable to straight-line clocking.

As a concrete example, figure 4 shows a plot of the clock period T for values of pipeline length $N = 1000$ to $N = 50000$ respectively, when $t_c = 60$, $\sigma = 1$, and $\alpha = 0.5$. The probability of failure is at most 10^{-30}.

References

[DS90] M.D. Dikaiakos and K. Steiglitz. Comparison of Tree and Straight-Line Clocking for Long Systolic Arrays. Technical report, CS-TR-271-90, Dept. of Computer Science, Princeton University, June 1990.

[FK84] A.L. Fisher and H.T. Kung. Synchronizing Large VLSI Processor Arrays. *IEEE Transactions on Computers*, c-34:734–740, August 1984.

[GS90] M. Greenstreet and K. Steiglitz. Bubbles Can Make Self-Timed Pipelines Fast. *Journal of VLSI Signal Processing*, vol. 2:139–148, 1990.

[HC87] M. Hatamian and G. L. Cash. Parallel Bit-Level Pipelined VLSI Design for High-Speed Signal Processing. In *Proceedings of the IEEE*, volume 75, No. 9, pages 1192–1202, September 1987.

[KS90] S. Kugelmass and K. Steiglitz. An Upper Bound on Expected Clock Skew in Synchronous Systems. *IEEE Transactions on Computers*, vol. 39, no. 12:1475–1477, December 1990.

[KT75] S. Karlin and H. Taylor. *A First Course in Stochastic Processes*. Academic Press, 1975.

[MMJ89] J. McCanny, J. McWhirter, and E. Swartzlander Jr, editors. *Systolic Array Processors*. Prentice Hall, 1989.

[Wag88] K. Wagner. Clock System Design. *IEEE Design and Test of Computers*, pages 9–21, October 1988.

Clocking Arbitrarily Large Computing Structures Under Constant Skew Bound

Ahmed El-Amawy, *Senior Member, IEEE*

***Abstract*—We describe a new scheme for global synchronization of arbitrarily large computing structures such that clock skew between any two communicating cells is bounded above by a constant. The new clocking scheme does not rely on distribution trees, phase-locked loops or handshake protocols. Instead it utilizes clock nodes which perform simple processing on clock signals to maintain a constant skew bound irrespective of the size of the computing structure. Among the salient features of the new scheme is the interdependence between network topology, skew upper bound, and maximum clocking rate achievable.**

We use a 2-D mesh framework to present the concepts, to introduce three network designs and to prove some basic results. For each network we establish the (constant) upper bound on clock skew between any two communicating processors, and show its independence of network size. Let F be the fan-in/fan-out of a clock node and Δ be the maximum (node + link) delay. It will be shown that any of the three networks complies with the following:

a) **Maximum skew between communicating cells** $= (5 - F)\Delta;\ 2 \leq F \leq 4$
b) **Maximum skew between any two node inputs** $= (6 - F)\Delta;\ 2 \leq F \leq 4$.

The second result is important in setting up timing constraints on clock signals for each respective network. The constraints are simple and easy to implement.

Besides theoretical proofs, simulations have been carried out to verify correctness and to check the workability of the scheme. Also a 4×4 network has been built and successfully tested for stability. Other issues such as node design, clocking of nonplanar structures such as hypercubes, and the new concept of fuse programmed clock networks are addressed. A discussion on practical implementation issues is also given. The discussion shows that hardware overhead incurred by the proposed scheme is comparable to that associated with current asynchronous control schemes.

***Index Terms*— Clock distribution, clock skew, constant skew bound, fuse programmed network, hypercube, large size system, multiprocessor, synchronous system, VLSI.**

I. Introduction

THE simplest way of controlling pipelined and parallel computing structures is the use of a global clock to connect the sequence of computation with time [1]. This design discipline is usually referred to as the synchronous discipline. Synchronous systems are by far the best known and most widely used in VLSI implementations and could also be attractive for multichip/multiboard implementations. [2]

The clock skew problem has long been recognized as a major obstacle in implementing very large synchronous systems possibly employing thousands of processors [2]–[6]. The term clock skew refers to the variation in effective arrival times of a clocking event at different clocked cells. Ideally, a clocking event should arrive at (and affect) all the cells at the same time. Due to a combination of several effects: variation in threshold voltages, variation in signal propagation delays on wires, and different buffer delays (in buffered clock networks), the effective clock arrival times might well vary from cell to cell. Clock skew can be a problem even within a single chip [1].

The problem clock skew introduces is that it directly relates to the maximum clocking rate achievable [6], [2], [3], [4]. Larger skews would necessarily impose lower clocking rates if the synchronous system is to operate correctly. To demonstrate, consider a simplified model in which a source processor P_s sends data to a receiving processor P_r as shown in Fig. 1. Assume that both processors perform similar functions and that in every clock cycle P_s accepts new data, processes it, and sends the result to P_r. In an *ideally synchronized* system, the clock will arrive at both processors simultaneously. The minimum clock period, in such a case, is equal to sum of the times needed to: 1) write[1] into R (of P_s), t_w, 2) process the data (combinational logic delay), t_p; and 3) transmit the result (propagation delay from P_s to P_r), t_t.

In a real-life system, however, clock skew must be brought into the picture. Consider, for example, what happens when P_r receives its clocking event τ units of time before P_s does. The amount of time available for the execution of the three basic operations mentioned above will effectively be reduced by τ, the amount of clock skew. This could cause errors in the data transfer and could therefore lead to system malfunction. Hence, it is necessary to add the amount of worst case skew between any two communicating processors to the minimum clock period. Consequently a larger skew will automatically lower the maximum allowable clocking rate. S-Y. Kung states [3]: "For very large systems, the clock skew incurred in global clock distribution is a nontrivial factor causing unnecessary slow down in clock rate." Some designers try to minimize the effects of clock skew by equalization of wire lengths, careful screening of off-the-shelf parts, symmetric design of the distribution network, and design guidelines to reduce skew due to process variations [7]–[9].

The way clocking signals are distributed is another factor in determining the maximum clocking rates. Basically, there are two schemes for clock distribution: *equipotential clocking* [2], [4], [6] and *buffered clocking* [4], [10]. In equipotential clocking, the entire clock network is viewed as a surface which

[1] We assume that setup and hold times are included in t_w.

Reprinted from *IEEE Trans. Parallel and Distributed Syst.*, vol. 4, no. 3, pp. 241–255, March 1993.

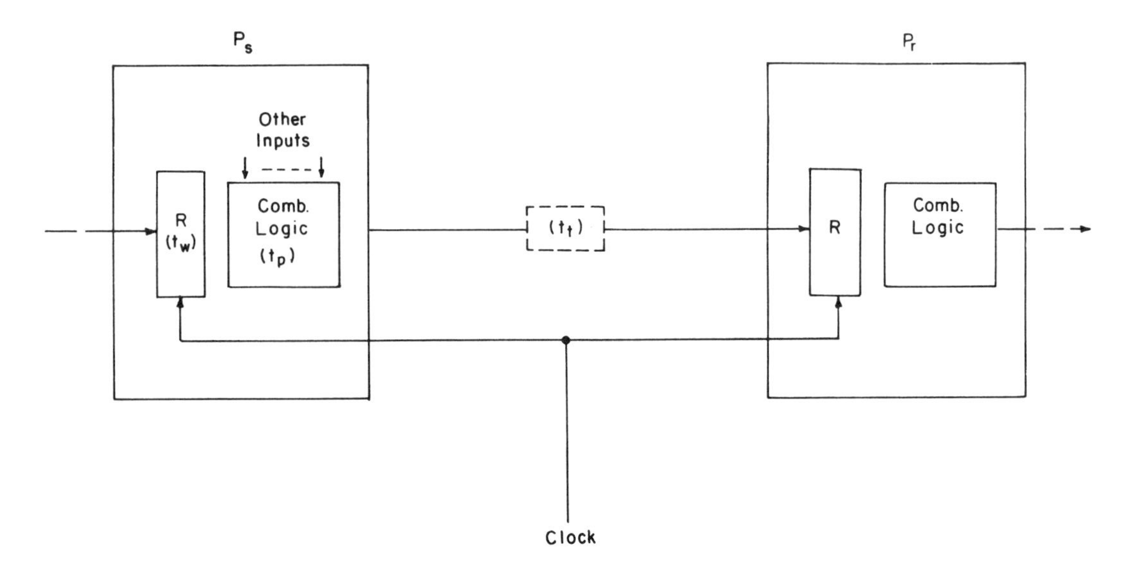

Fig. 1. A simplified model for synchronous interprocessor communication.

must be brought to a single final voltage (charged/discharged) in each half of the clock cycle. For these networks, it is clear that the time to distribute the clock is directly proportional to the network's diameter (longest path from source to leaf) [2], [4]. This time is a limiting factor in determining the clocking rate and is clearly more severe in larger-sized networks.

The other scheme, buffered clocking, utilizes strings of buffers to replace long wires [4], [10]. The buffering effect allows clocking events to be pipelined through the network such that several events could co-exist in the network at the same time. In such a case, consecutive events need only be separated by the maximum delay through a buffer and its output wire [4], which is also considered the time to distribute the clock. Notice that this is much less limiting, particularly for larger-sized networks, insofar as clocking rates are concerned. Generally, the minimum clock period $T_{\min}$ is determined by [2],

$$T_{\min} = \text{Max}\{t_w + t_p + t_t + \tau, 2t_d\}$$

where t_d is the time to distribute the clock. In most cases of interest, particularly with buffered clock networks, $t_w + t_p + t_t + \tau > 2t_d$ and therefore, $T_{\min} = t_w + t_p + t_t + \tau$.

Fisher and Kung [4] investigated clock skew problems in large 2-D arrays clocked by an H-Tree. The H-Tree guarantees that every computing cell is at the same physical distance from the root. Using two models in which all sources of skew are treated together as a distance metric, they have shown that if small delay variations are not ignored (summation model) clock skew tends to grow with system size. We directly quote Fisher and Kung [4]: "The paper shows that under this assumption (summation model, 2-D case) it is impossible to run a clock such that the maximum skew between two communicating cells will be bounded by a constant as systems grow" . . . "Therefore, unless operating at possibly unacceptable speeds, very large systems controlled by global clocks can be difficult to implement because of the inevitable problem of clock skews and delays."

Some researchers investigated alternative methods for controlling very large computing structures. The methods are either based on the asynchronous mode or on a hybrid synchronous/asynchronous approach [1]–[6]. In asynchronous control, the requirement of correct "timing" is substituted by correct "sequencing" [3], [6]. There is no global timing reference in an asynchronous system. Processors work asynchronously but information transfer between processors is by mutual convenience. Whenever data are available, the transmitting processor informs the receiver which accepts data whenever possible. The receiver then acknowledges that the data have been consumed. The scheme can be implemented using a handshake protocol that ensures that computational steps proceed in an orderly manner.

Seitz [1] described an alternative discipline known as "self-timing" for controlling large systems. Self-timed systems are interconnections of parts called elements. Elements can be thought of as performing computational steps whose initiation is caused by signal events at their inputs and whose completion is indicated by signal events at their outputs. Self-timed elements can be designed as synchronous systems with internal clocks that can be stopped synchronously and restarted asynchronously. This scheme allows synchronous elements to communicate reliably because their clocks are partly dependent. Elements may also be designed as asynchronous circuits. A similar scheme was described by Fisher and Kung [4] for systolic array control.

This paper describes a novel clock distribution scheme which guarantees tight synchronization between any two communicating processors regardless of the size of the computing network. More precisely we describe different but related clock distribution network designs such that in each case clock skew is bounded above by a constant (property of the network) independent of network size. Unlike all previous alternative methods a single clock source is needed for the whole network. This is very attractive for VLSI/WSI implementations since the single crystal can be located external to the chip/wafer.

The new approach makes it possible to extend the best known and most popular timing discipline (the synchronous discipline) to cases where it was previously considered too impractical and/or too inefficient. In the design of modular systems, where expandability and reconfiguration are essential, it is difficult if not impossible to determine the clock period using any current clocking scheme [2]. This is due to the lack of *a priori* knowledge about the final size or configuration of the system. The proposed scheme may be attractive for such systems since skew bound is known *a priori*. Further, the new scheme eliminates the possibility of metastable failure associated with most alternatives of the synchronous discipline. We will show that the hardware overhead associated with our scheme is, in the worst case, comparable to that associated with asynchronous schemes. Thus, the designer of a large computing structure is being offered a new alternative that is simple and efficient.

In Section II we describe the proposed scheme and introduce a distribution network capable of utilizing the principles of the scheme to provide constant upper-bound on skew. Timing constraints on clock signals are given and theoretical bounds are derived. Section III discusses design issues such as network and node design, clocking of nonplanar data networks and practical implementation aspects. Two other "valid" networks are introduced and simulation results are summarized. We also discuss the new concept of fuse programmed clock networks, a spin-off of the new scheme. Section IV is a conclusion.

II. The Clocking Scheme

The clocking scheme we now describe is based on the premise of clock processing. Clock signals are distributed, combined, and processed much the same way data signals are processed [11], [12]. Simple timing constraints, that are easily realizable, are imposed on the timing of clock signals. To distinguish between a clock processing entity and a data processing entity we shall hereafter reserve the terms "clock node" or "node" for the former and "data cell," "cell," or "processor" for the latter. Generally an important function of a clock node is to enforce the adherence of clock signals to the timing constraints. A node also creates a new local reference for neighboring nodes in the clock network.

Here, a 2-D mesh framework is utilized to introduce the underlying concepts and to prove some basic results. We consider a 2-D computing structure in which cells are connected in a mesh fashion such that each cell is connected to its four neighbors with the exception of boundary cells which have fewer neighbors. As a preliminary, we associate a clock node to each data cell such that each node clocks its associated cell, as illustrated in Fig. 2. We will assume that the delay from a node to its associated cell is negligible. These assumptions can be relaxed as we demonstrate in Section III.

The principle underlying the new scheme can be termed "branch-and-combine." Starting from a single clock source, the clock signal is distributed to neighboring nodes which in turn act as sources to neighboring nodes down the network. Each node (except that connected to the source) receives at least two inputs from neighboring nodes, which are then combined into one signal in accordance with the function and timing constraints enforced by the node. Copies of this signal are then sent to neighboring nodes and used to clock the associated cell.

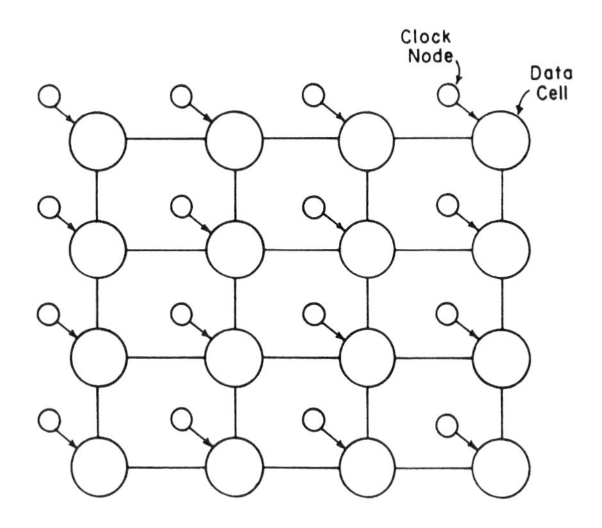

Fig. 2. Clock node per data cell.

In the process, the scheme adaptively selects the path through nodes such that skew is kept within fixed and known bounds.

A. Preliminaries

For clarity we omit data cells from the figures to follow keeping in mind the node per cell assumption, unless otherwise noted. We presume thus that clock nodes are located at mesh points and are identified by their coordinates. It is also presumed, for each network discussed, that nodes are of the same type and that the degree = fan-in = fan-out = F. Positive logic and positive edge triggering are assumed throughout the paper.

1) Notation: The following notation will be used in the remainder of this paper:

t = time

t^+ = time of $0 \rightarrow 1$ transition

$\check{t}, \hat{t}$ = lower and upper bounds on time t, respectively

$N(i,j)$ = node (i,j)

$ND(i,j) = g \pm \delta$ = delay through $N(i,j)$

$\hat{ND}(i,j) = g + \delta$ = maximum delay through $N(i,j)$

$\check{ND}(i,j) = g - \delta$ = minimum delay through $N(i,j)$

$LD(i,j,k,l) = m \pm \epsilon$ = link delay from $N(i,j)$ to $N(k,l)$

$\check{LD} = m - \epsilon$ = minimum link delay

$\hat{LD} = m + \epsilon$ = maximum link delay

$P[(i,j),(k,l),\cdots,(u,v)]$ = A path from $N(i,j)$ to $N(u,v)$ which goes through the nodes listed in between, in the given order.

$PD[(i,j),(k,l),\cdots,(u,v)]$ = is the delay through the given path (nodes and links) excluding the delay through the end node.

$t^+_{i,j}(I_x)$ = time at which input I_x of $N(i,j)$ makes a $0 \rightarrow 1$ transition.

$\check{t}^+_{i,j} = \text{Min}\{t^+_{i,j}(I_x), t^+_{i,j}(I_y), \cdots\}$, where $I_x, I_y, \cdots$ are inputs to $N(i,j)$

$t^+_{i,j}(o)$ = time at which $N(i,j)$ outputs a $0 \rightarrow 1$ transition.

Generally, $\check{\theta}$ will be used to denote minimum value (lower-bound) on variable θ, and $\hat{\theta}$ will denote the maximum value (upper-bound) on θ. A * superscript (e.g., θ^*) will denote a realilzation (actual value for some case). Also the following

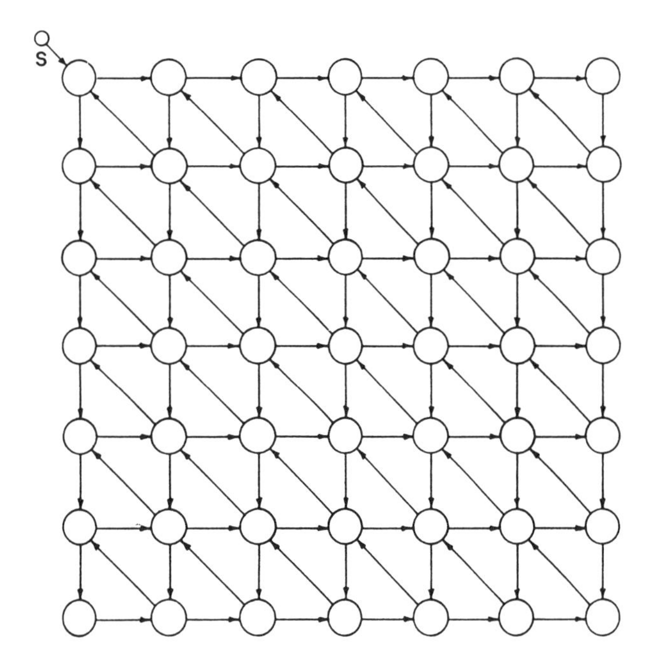

Fig. 3. Clock network with $F = 3$.

assumptions will be used to simplify expressions:

(node + link) delay = $g \pm \delta + m \pm \epsilon = \alpha \pm \beta$

$\alpha = g + m, \beta = \delta + \epsilon, \Delta = \alpha + \beta.$

B. Network Analysis

As mentioned earlier three clock network designs will be introduced. But since the basic principles are identical in all three cases, we arbitrarily elected to detail the analysis of the network shown in Fig. 3 ($F = 3$). Deriving results for the other two networks ($F = 2$ and $F = 4$) becomes straightforward after the analysis of the Fig. 3 network is completed. The results for the other two networks will be summarized in Section III. For convenience we shall, henceforth, use the notation FX to denote a network with $F = X$. Thus $F2, F3$, and $F4$ will be used to refer to the networks with $F = 2, 3$, and 4, respectively.

Consider the clock network in Fig. 3. Inputs to $N(i, j)$ will be labeled I_x, I_y, and I_z, for the x, y, and diagonal directions, respectively. Because of the cyclic nature of the network, the reader may suspect that the network could oscillate. Indeed this could happen if appropriate timing constraints are not imposed on network signals. In fact, the very cyclic nature of the network is the reason skew can be bounded by a constant! [13].

An attempt to analyze the behavior of the final network design could easily lead to circular arguments. Therefore we follow an alternative approach in which we start with a very simple (and unrealistic) model to establish some facts and derive some results. We then modify the model (to make it more realistic) and show that results based on the first model still hold true under the new set of conditions. We then present the final (practical) model and prove that results still hold.

1) Model 1: The model consists of the following conditions. Those conditions marked with * will be progressively relaxed as the analysis progresses.

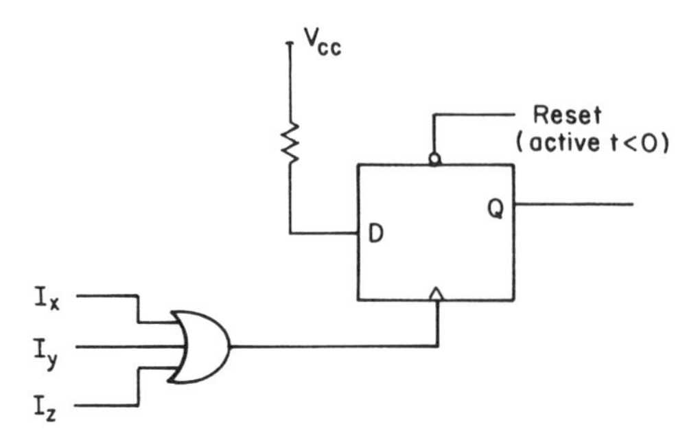

Fig. 4. Node structure assumed for Model 1.

1) Network is reset for all $t < 0$. All node outputs are at logic zero for $t < 0$.

2)* Clock source S produces a step signal at $t = 0$.

3) a) The clock distribution network topology is that shown in Fig. 3.

   b) All outputs of a given node are copies of the same signal.

4) $\alpha \geq 2\beta$. This means that (node + link) delays are allowed to vary by 50%, but not more. For analysis purposes we notice here that since $\beta = \delta + \epsilon$, this condition implies $\alpha > 2\epsilon$ and also $\Delta = \alpha + \beta > 2\beta$. The condition is merely intended to simplify following the proofs but is not necessary otherwise. Formally, we could relax the condition to require delays to only be nonnegative.

5) Network delays are time-invariant. This means delays are not functions of time but could vary from node to node ($g \pm \delta$) and link to link ($m \pm \epsilon$). The condition will be useful with subsequent models but not this one, however.

6)* Each node functions as follows: In response to the first arriving $0 \rightarrow 1$ transition on any of its inputs, it produces a step output after one node delay, i.e., $t_{i,j}^{+}(o) = \check{t}_{i,j}^{+} + g \pm \delta \forall i, j$.

Conditions 6 suggests a node structure such as that in Fig. 4, which is adopted for this model. Notice that the node cannot make a $0 \rightarrow 1$ output transition unless triggered. Notice also that we implicitly presume (for this model) that each data cell requires a step clock signal to perform its function. Hence clock skew between two communicating data cells is nothing but the difference between $t^{+}(o)$ of their corresponding nodes.

Lemma 1: Under Model 1, the clock network will not oscillate and every node will produce a single $0 \rightarrow 1$ output transition.

Proof: The proof is straightforward. Since each node can only make a single $0 \rightarrow 1$ transition, and since each node has at least one input connected to S via a network path, each node is guaranteed a trigger condition. Thus once all nodes have been triggered, their outputs will remain high indefinitely. Thus the network will not oscillate and every node will output exactly one $0 \rightarrow 1$ transition. □

Theorem 1: The clock network described by Model 1 guarantees clock skew between any two communicating cells to be bounded above by 2Δ (constant), regardless of network size.

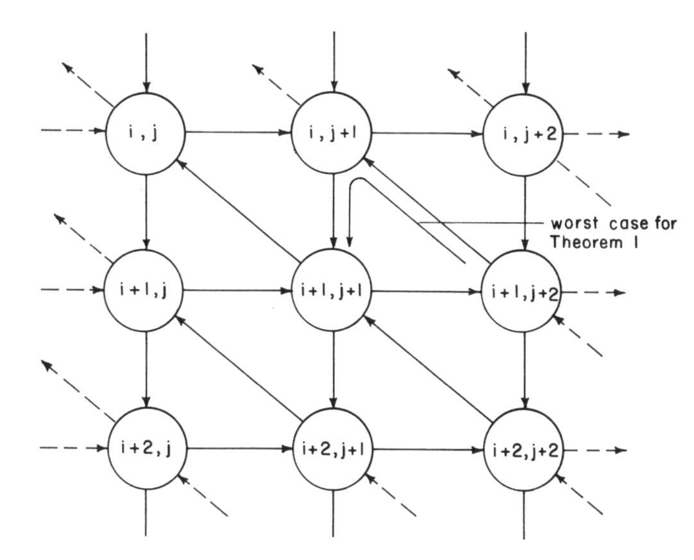

Fig. 5. An arbitrary section of a large $F3$ network.

Proof: Consider any arbitrary[2] section of an arbitrarily large network adhering to Model 1, such as that in Fig. 5. Because of symmetry we only need to consider any two adjacent nodes, differing in only one coordinate. Thus, without loss of generality, consider nodes $N(i+1, j+1)$ and $N(i+1, j+2)$ of Fig. 5. Two cases of interest can be distinguished. Either node could receive its trigger before the other. The case when both nodes are triggered simultaneously is trivial since skew $\leq 2\delta < 2\Delta$.

Case 1: $\check{t}^+_{i+1,j+1} < \check{t}^+_{i+1,j+2}$.

In this case,

$$\begin{aligned}\check{t}^+_{i+1,j+2} &= \check{t}_{i+1,j+1} + ND(i+1, j+1) \\ &\quad + LD(i+1, j+1, i+1, j+2) \\ &= t^+_{i+1,j+1}(o) + LD(i+1, j+1, i+1, j+2).\end{aligned}$$

Thus,

$$\begin{aligned}t^+_{i+1,j+2}(o) \leq t^+_{i+1,j+1}(o) &+ \hat{LD}(i+1, j+1, i+1, j+2) \\ &+ \hat{ND}(i+1, j+2)\end{aligned}$$

or

$$t^+_{i+1,j+2}(o) \dot{-} t^+_{i+1,j+1}(o) \leq m + \epsilon + g + \delta.$$

Hence for this case maximum skew is $\leq \Delta$.

Case 2: $\check{t}^+_{i+1,j+1} > \check{t}_{i+1,j+2}$.

In the worst case $N(i+1, j+2)$ produces an output $0 \rightarrow 1$ transition that reaches $N(i, j+1)$ before the other two inputs, and $N(i, j+1)$ produces an output that reaches $N(i+1, j+1)$ also before the two other inputs (see illustration in Fig. 5). To determine maximum skew we only need to consider this worst case scenario.

$$\check{t}^+_{i,j+1} = t^+_{i,j+1}(I_z) \leq t^+_{i+1,j+2}(o) + \hat{LD}(i+1, j+2, i, j+1).$$

[2] Sections containing top or left boundary nodes can be included by assuming any unused input to be permanently grounded. For bottom and right boundary nodes, simply ignore the unused outputs.

By simple extension we have

$$\begin{aligned}t^+_{i+1,j+1}(o) \leq t^+_{i+1,j+2}(o) &+ \hat{LD}(i+1, j+2, i, j+1) \\ &+ \hat{ND}(i, j+1) + \hat{LD}(i, j+1, i+1, j+1) \\ &+ \hat{ND}(i+1, j+1).\end{aligned}$$

Therefore maximum skew is

$$t^+_{i+1,j+1}(o) - t^+_{i+1,j+2}(o) \leq 2(m + \epsilon + g + \delta) = 2\Delta.$$

Hence in all cases, skew between any two communicating cells is $\leq 2\Delta$. □

The following result is very useful for future developments.

Theorem 2: Under Model 1, maximum skew between any two inputs to the same node is bounded above by 3Δ.

Proof: Define a hop as signal propagation through a node and one of its output links. Clearly a hop delay = $\alpha \pm \beta$. The proof is very simple since an input to a certain node can be reached from any of the two other inputs, to the same node, in at most three hops. The only exception is top and left boundary cells, where unused inputs are grounded. But those inputs do not trigger their respective nodes and can be ignored. Arbitrarily let

$$\check{t}^+_{i,j} = t^+_{i,j}(I_x) \text{and } \hat{t}^+_{i,j} = t^+_{i,j}(l_z).$$

It is clear that in the worst case

$$\begin{aligned}t^+_{i,j}(I_z) &\leq t^+_{i,j}(I_x) + \hat{PD}[(i,j), (i, j+1), (i+1, j+1), (i, j)] \\ &\leq t^+_{i,j}(I_x) + 3\Delta.\end{aligned}$$

Hence maximum skew between $I_x(i,j)$ and $I_z(i,j)$ is 3Δ. Recall that in our notation path delay does not include the delay through the end node.

For the other cases, we observe the similarity by tracing signal path from the first arriving input to the last arriving input to the same node, which takes at most 3 hops in each case. Fig. 6 illustrates all six possible cases. Observe the worst case assumption, that the signal sourced at $N(i,j)$ arrives before other inputs to any node in the path, in every case. □

2) Model 2: We change two of Model 1's conditions to arrive at a slightly more realistic model. Simple further changes to Model 2 will lead to our final model. Model 2 differs from Model 1 in Conditions 2 and 6. The new conditions are:

2′) The clock source S produces a single clock pulse at $t = 0$, whose duration T_h (high portion time) is bounded as follows

$$3\Delta + 2\beta + \mu \leq T_h \leq 3\Delta + 2\beta + \mu'$$

where μ and μ' are finite numbers such that $\mu' > \mu > 0$. We shall use $\check{T}_h$ and $\hat{T}_h$ to denote the lower and upper bounds on T_h, respectively.

6′) Each node functions as follows:

a) A node will not produce an output transition unless triggered.

b) When triggered by a $0 \rightarrow 1$ input transition, the node produces a pulse whose duration T_h is a defined in Condition 2′.

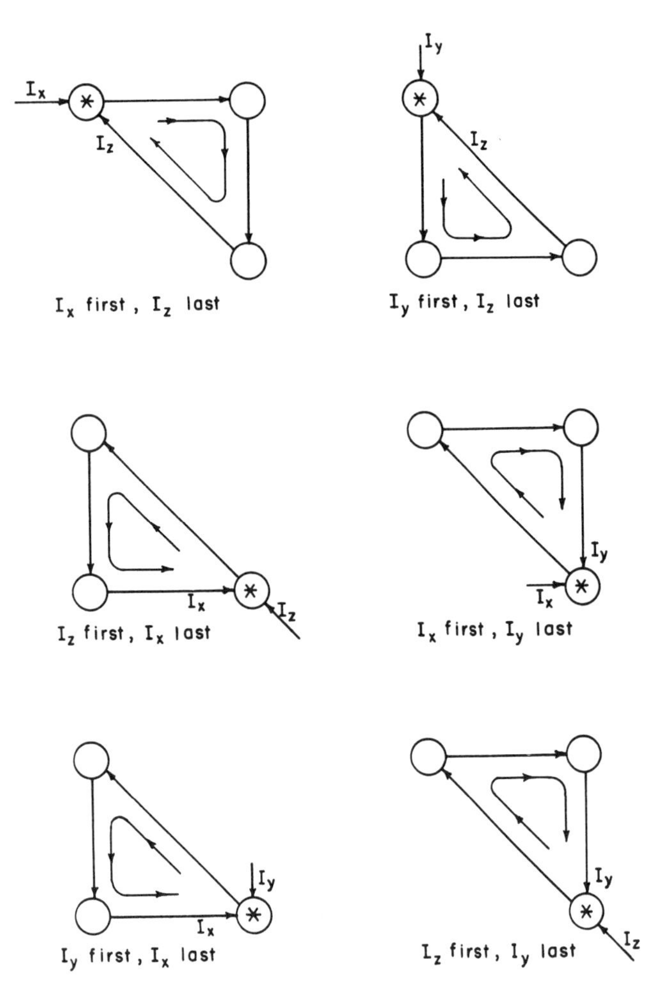

Fig. 6. Illustration of proof to Theorem 2.

c) Any $0 \rightarrow 1$ input transition arriving within $\check{T}_h$ units of time after the node is triggered will be ignored. However any two input transitions separated by $\check{T}_h$ or more may trigger the node twice, thus producing two pulses. It will definitely produce two pulses if any two consecutive inputs are separated by $\geq T_h^*$ (the actual duration of the first of the two). This latter situation must be avoided.

Condition 6′ suggests a node whose structure and function are exemplified in Figs. 7(a) and (b), respectively. The condition implies that a processing cell only requires one pulse of duration $\geq \check{T}_h$ to complete its function. For this model, the upper bound on T_h is only needed to emphasize bounded pulse duration but is otherwise arbitrarily $> \check{T}_h$. During the course of the analysis we shall see that T_h bounds can be reduced by 2β each without changing the results. We initially need the bounds as defined to help develop simpler proofs.

Thus Model 2 consists of Conditions 1,2′, 3, 4, 5, and 6′. We observe here, that (unlike Model 1) one could suspect the network to oscillate under this model. We shall now prove that it will not.

Theorem 3: Under Model 2, the clock network will not oscillate and each node will produce exactly one pulse.

Proof: We use induction on network size to prove the theorem.

Induction hypothesis: If a network of size $i \times i$ does not oscillate and each of its nodes produces exactly one pulse, then a network of size $i+1 \times i+1$ will not oscillate and each of its nodes will produce exactly one pulse.

Basis: 2 × 2 network.

Consider the 2 × 2 network shown in Fig. 8(a) and assume compliance with Model 2. Temporarily snip open the diagonal link at $N(1,1)$ as shown in Fig. 8(b). We assume that the delay from point a to $N(1,1)$ is negligible. Because of Condition 6′, and since the network in Fig. 8(b) does not have feedback it could not oscillate.

Since $N(1,1)$ has a single input connected to S, it will be triggered only once, and will thus produce a single pulse. This pulse will reach $N(1,2)$ and $N(2,1)$ such that $\alpha - \beta \leq t_{1,2}^+(I_x), t_{2,1}^+(I_y) \leq \alpha + \beta$. Since each of $N(1,2)$ and $N(2,1)$ has a single input, they will be triggered only once producing one pulse, each, such that

$$2(\alpha - \beta) \leq t_{2,2}^+(I_x) \leq 2(\alpha + \beta); \text{and,}$$
$$2(\alpha - \beta) \leq t_{2,2}^+(I_y) \leq 2(\alpha + \beta).$$

Hence,

$$|t_{2,2}^+(I_x) - t_{2,2}^+(I_y)| \leq 4\beta < 3\Delta < \check{T}_h.$$

It follows then that $N(2,2)$ will also produce a single pulse which reaches point b [Fig. 8(b)] at, say t_b^+, such that $3(\alpha - \beta) \leq t_b < 3(\alpha + \beta)$.

To study the effect of connecting points a and b, we divide time into four intervals such that

$I_1 : t < 0$
$I_2 : 0 \leq t < t_b^+$
$I_3 : t_b^+ \leq t < t_b^+ + T_h(b)$
$I_4 : t_b^+ + T_h(b) < t$ where $T_h(b)$ denotes the pulse duration at point b.

From Condition 1, we know that during I_1 all nodes are reset and therefore connecting a and b will not alter the situation. Assume a, b connected during I_1, open otherwise. From Condition 6′ a node will not make an output transition unless triggered by a $0 \rightarrow 1$ input transition. Therefore $N(2,2)$ will not trigger before $\check{t}_{2,2}^+$ which requires $N(1,2)$ [or $N(2,1)$] to have triggered earlier. This is the case regardless of a, b connection status. Hence point b will not receive a $0 \rightarrow 1$ transition before t_b^+, regardless. Therefore connecting points a and b during I_2 will not alter the behavior of the network. At this point we assume a and b connected during I_1 and I_2, disconnected otherwise.

Remark: In the remainder of the proof, the term *network behavior* will be used strictly to describe oscillation potential, and the one output pulse per node condition.

Once a node is triggered, it produces a pulse that remains high for T_h units of time (Condition 6′). Therefore point b will be at logic 1 during I_3 irrespective of its connection status to point a. Since $t_b^+ < \check{T}_h$, the pulse on b and that from S will overlap in time and by Condition 6′$N(1,1)$ will only trigger once if point b is connected to point a during I_3. Consequently

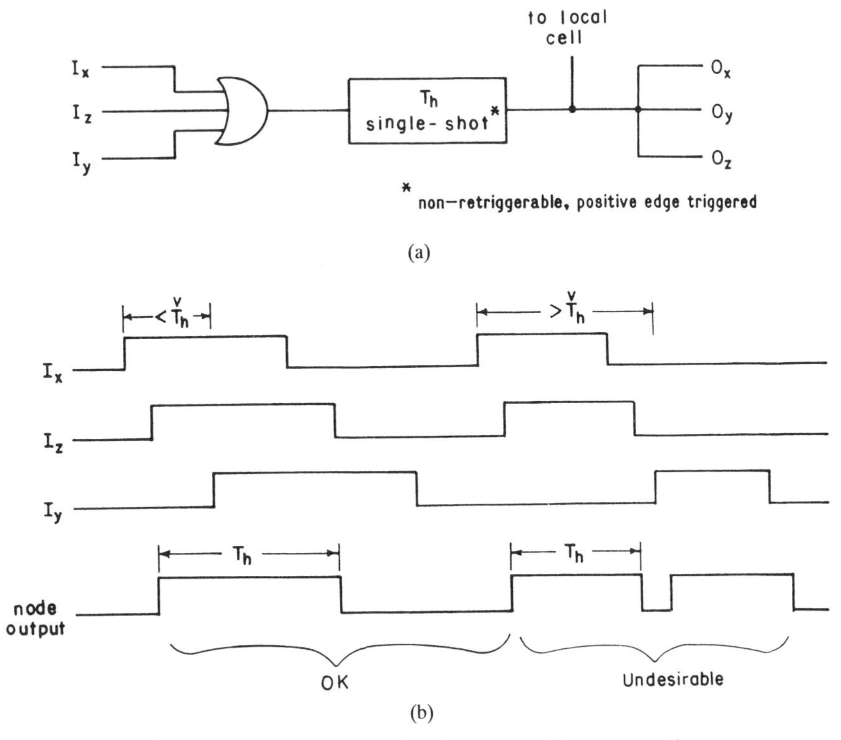

Fig. 7. Model 2 (a) assumed node structure (b) node function.

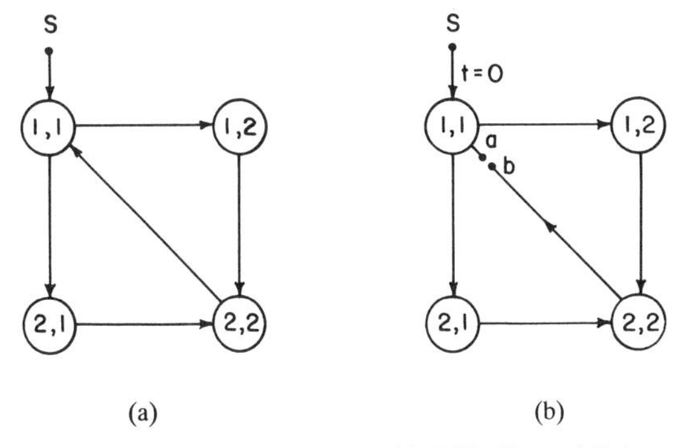

Fig. 8. 2×2 network under Model 2. (a) With diagonal link connected. (b) With diagonal link open.

$N(2,1)$, $N(1,2)$, and $N(2,2)$ will each trigger once. Hence connecting b to a during I_3 will not change network behavior, and we now assume a and b connected prior to I_4.

With point b disconnected from point a during I_4, point b is at logic 0. Moreover, after timing out its T_h, each node will drop its output to logic 0 and remain their indefinitely. Connecting b to a during I_4 can alter network behavior only if it could retrigger $N(1,1)$. But this cannot happen since b is at logic 0 during that interval. Therefore by connecting a to b at all times the 2×2 network in Fig. 8(a) will not oscillate and every node will produce exactly one pulse.

Induction Step: Assume true for $i \times i$ and show true for $i+1 \times i+1$.

Our approach here is similar to our approach in the basis part. We shall initially disconnect feedback (diagonal) connections from row $i+1$ and column $i+1$ to the $i \times i$ subnetwork. We then reconnect these links one by one, cumulatively, showing that, in each case, the connection will not alter network behavior until all feedback links are connected.

Consider the network shown in Fig. 9 and assume its adherence to Model 2, except that links carrying $I_z(m,i), 1 \leq m \leq i$, and $I_z(i,r), 1 \leq r < i$, are snipped open at the receiving nodes, such that the delay between the open point and the receiving node is negligible. Clearly this network cannot oscillate since:

a) the complete subnetwork of size $i \times i$ (containing $N(k,l)$; $1 \leq k,l \leq i$) does not oscillate, by hypothesis.
b) There is no feedback from the extra row and/or column to the $i \times i$ subnetwork.
c) There is no feedback (closed loops) in row $i+1$ and/or column $i+1$.
d) No node could produce a pulse without being triggered (Condition $6'$).

For the same reasons mentioned above, we observe that each node in the $i \times i$ subnetwork will produce exactly one pulse. However, the same cannot be said automatically (without proof) about the nodes in row $i+1$ or column $i+1$.

Consider $N(1,i+1)$. This node has one input coming from the $i \times i$ subnetwork. Thus this node will produce exactly one pulse. Because of symmetry, one can conclude that $N(i+1,1)$ also will produce one pulse. Now consider $N(2,i+1)$. This node has two inputs emanating at $N(2,i)$ and $N(1,i+1)$ which produce one pulse each. If the timing is such that $|t^+_{2,i+1}(I_x) - t^+_{2,i+1}(I_y)| < \check{T}_h$, it will produce a single output pulse. Since $I_x(2,i+1)$ emanates at $N(2,i)$ and $I_y(2,i+1)$ at $N(1,i+1)$, we distinguish two cases.

Case 1: $\check{t}^+_{2,i} \leq \check{t}^+_{1,i+1}$.

In this case,

$$\begin{aligned} t^+_{2,i+1}(I_x) &= \check{t}^+_{2,i} + ND(2,i) + LD(2,i,2,i+1) \\ &= \check{t}^+_{2,i} + ND(2,i) + m \pm \epsilon. \end{aligned}$$

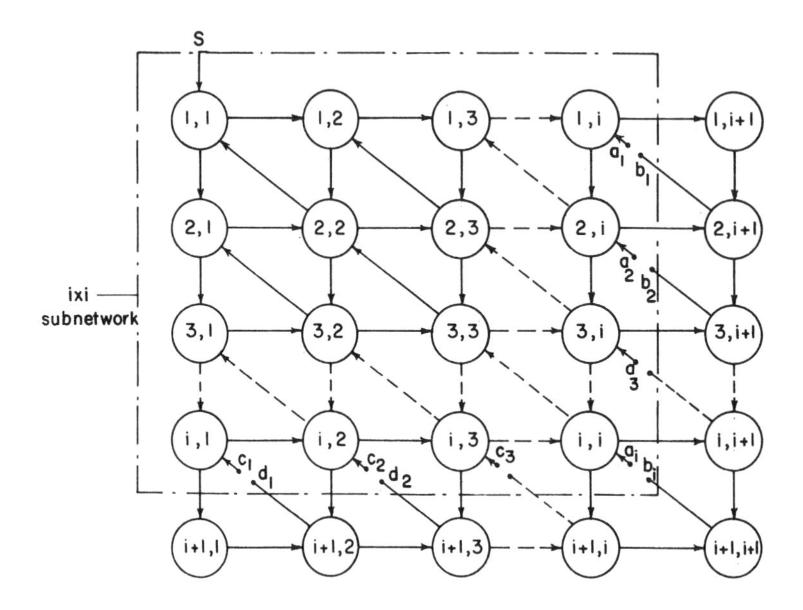

Fig. 9. The $(i+1) \times (i+1)$ network with feedback links to row i and column i opened.

Any signal reaching $I_y(2, i+1)$ must go through $N(1, i-1)$ because that node is on any path leading to $I_y(2, i+1)$. Node $(1, i-1)$ could be triggered from $N(2, i)$ or from $N(1, i-2)$. Assume that $I_z(1, i-1) < I_x(1, i-1)$. For this case

$$t^+_{2,i+1}(I_y) \leq \check{t}^+_{2,i} + ND(2,i) + \hat{LD}(2,i,1,i-1) + \hat{PD}[(1,i-1),(1,i),(1,i+1),(2,i+1)] \leq \check{t}^+_{2,i} + ND(2,i) + (m+\epsilon) + 3\Delta.$$

If $I_x(1, i-1) \leq I_z(1, i-1)$, the above relationship will still hold true because $I_y(2, i+1)$ will surely arrive earlier or at the same time as $I_x(1, i-1)$. Specifically, this is true because $N(1, i-1)$ will only produce one pulse (by hypothesis). It is clear from Condition 4 that $t^+_{2,i+1}(I_y) > t^+_{2,i+1}(I_x)$. Hence,

$t^+_{2,i+1}(I_y) - t^+_{2,i+1}(I_x) \leq 3\Delta + 2\epsilon < \check{T}_h$.

Thus, for this case $N(2, i+1)$ produces a single pulse (see Condition 6′).

Case 2: $\check{t}^+_{2,i} > \check{t}^+_{1,i+1}$.

Here, the worst case is when $I_x(2, i+1)$ is sourced at $N(1, i)$:

$$t^+_{2,i+1}(I_x) = \check{t}^+_{1,i} + PD[(1,i),(2,i),(2,i+1)], \text{and}$$
$$t^+_{2,i+1}(I_y) = \check{t}^+_{1,i} + PD[(1,i),(1,i+1),(2,i+1)].$$

Thus for this case

$$|t^+_{2,i+1}(I_x) - t^+_{2,i+1}(I_y)| = (m \pm \epsilon) + ND(2,i) + (m \pm \epsilon) - [(m \pm \epsilon) + ND(1,i+1) + (m \pm \epsilon)].$$

In the extreme we have

$$|t^+_{2,i+1}(I_x) - t^+_{2,i+1}(I_y)| = [(\alpha+\beta) + (m+\epsilon)] - [(\alpha-\beta) + (m-\epsilon)] = 2\beta + 2\epsilon < \check{T}_h.$$

It follows that for this case also $N(2, i+1)$ will produce a single pulse which will reach point b_1 (still disconnected from a_1 as in Fig. 9) at say $t = t^+_{b_1}$. To simplify the rest of the proof we introduce the following lemma.

Lemma 2: For the network in Fig. 9, assume and fix arbitrarily the connection status for the feedback pairs (a_k, b_k) and $(c_r, d_r), \forall 1 \leq k \leq i$ and $1 \leq r \leq i-1$ except for one pair (a_p, b_p), $1 \leq p \leq i$. The time at which the first $0 \rightarrow 1$ transition reaches point b_p is independent of (a_p, b_p) connection status, provided that, the connection status of all other feedback links [(a_k, b_k) and (c_r, d_r) pairs] remains unchanged.

Proof: In order for the a_p to b_p connection to alter the timing at b_p, some transition has to reach point a_p, otherwise $I_z(p, i)$ will have no effect. Also since all other feedback links are assumed fixed to a certain connection status, only a transition on a_p could alter the timing at b_p. Because we are only concerned with the first $0 \rightarrow 1$ transition at b_p, and since a node will not produce an output unless triggered, it follows that the timing of the first $0 \rightarrow 1$ transition at b_p is independent of (a_p, b_p) connection status. □

We shall now show that connecting points a_1 and b_1 will not alter network behavior. Two cases for the timing of $t^+_{b_1}$ with respect to $t^+_{1,i}(I_x)$ can be distinguished (the case $t^+_{b_1} = t^+_{1,i}(I_x)$ is trivial).

Case 1): $t^+_{b_1} > t^+_{1,i}(I_x)$.

For this case, the worst case arises when $t^+_{b_1} = t^+_{1,i}(I_x) + \hat{PD}[(1,i),(1,i+1),(2,i+1),(1,i)]$. Therefore, $t^+_{b_1} - t^+_{1,i}(I_x) \leq 3\Delta < \check{T}_h$.

To study the effect of connecting points a_1 and b_1 (Fig. 9) we divide time into three intervals as follows

$J_1 : t < t^+_{b_1}$

$J_2 : t^+_{b_1} \leq t < t^+_{b_1} + T_h(b_1)$

$J_3 : t^+_{b_1} + T_h(b_1)$.

During J_1, point b_1 is at logic 0 whether or not it is connected to a_1 (from Condition 1 and Lemma 2). Therefore connecting a_1 to b_1 during J_1 will not alter network behavior. If during J_2, a_1 is connected to b_1 a pulse will apply to $I_z(1,i)$ at $t^+_{b_1}$. But this pulse will (in the worst case) have the timing given above and $N(1,i)$ will not retrigger because of a_1, b_1 connection. Thus connecting a_1, b_1 during J_1 and J_2 will not change network behavior. Moreover, in J_3 point b_1 is at logic 0, thus connecting it to a_1 could not cause $N(1,i)$ to retrigger, and thus could not alter network behavior. Hence for this case connecting a_1 to b_1 all the time is assumed since it will not cause any change in network behavior.

Case 2): $t^+_{b_1} < t^+_{1,i}(I_x)$.

In the worst case both $I_x(1,i)$ and the signal at b_1 are sourced at $N(2,i)$. In that case,

$$t^+_{1,i}(I_x) \le \check{t}^+_{2,i} + \hat{PD}[(2,i),(1,i-1),(1,i)], \text{and}$$
$$t^+_{b_1} \ge \check{t}^+_{2,i} + \check{PD}[(2,i),(2,i+1),(1,i)].$$

Hence in the worst case,

$$\begin{aligned} t^+_{1,i}(I_x) - t^+_{b_1} &\le ((\alpha+\beta)+(m+\epsilon)) - ((m-\epsilon)+(\alpha-\beta)) \\ &\le 2\beta + 2\epsilon < \check{T}_h. \end{aligned}$$

Dividing time into three intervals as in Case 1), one can, following similar argument, connect a_1 to b_1 during J_1 without altering network behavior. However, since in this case $t^+_{b_1} < t^+_{1,i}(I_x)$, with a_1 connected to b_1 during J_2, $N(1,i)$ will be triggered by $I_z(1,i)$ not $I_x(1,i)$. But since the time difference between the $0 \rightarrow 1$ transitions on these inputs is $< \check{T}_h$, it follows that $N(1,i)$ will still produce a single pulse despite the a_1, b_1 connection. Hence connecting a_1 to b_1 during J_1 and J_2 will not alter network behavior. Point b_1 would remain at logic 0 during J_3 if disconnected from a_1. Hence like in J_1, connecting a_1 to b_1 during J_3 (and thus all the time) will not alter network behavior.

Thus, for both cases connecting a_1 and b_1 will not alter network behavior. Because of symmetry, will also connect points c_1 and d_1 safely. Notice that once a feedback link has been connected, it will remain so for the rest of the proof.

Remark: With a_1 and b_1 connected, the maximum skew between $I_x(2,i+1)$ and $I_y(2,i+1)$ reduces to 3Δ as can be verified by tracing the closed path containing $N(2,i+1)$, $N(1,i)$ [or $N(2,i)$], and then $N(2,i+1)$ again.

Now consider $N(3,i+1)$, and points a_2 and b_2. Following an argument similar to that followed with $N(2,i+1)$, two cases can be distinguished.

Case 1): $\check{t}^+_{3,i} < \check{t}^+_{2,i+1}$.

For the worst case scenario,

$$t^+_{3,i+1}(I_x) = \check{t}^+_{3,i} + ND(3,i) + LD(3,i,3,i+1), \text{ and}$$
$$\begin{aligned} t^+_{3,i+1}(I_y) \le \check{t}^+_{3,i} + PD[(3,i),(2,i-1),(2,i), \\ (2,i+1),(3,i+1)]. \end{aligned}$$

From Condition 4, we have

$$t^+_{3,i+1}(I_y) > t^+_{3,i+1}(I_x).$$

In the extreme

$$t^+_{3,i+1}(I_y) - t^+_{3,i+1}(I_x) \le 3\Delta + 2\epsilon < \check{T}_h.$$

Case 2): $\check{t}^+_{3,i} \ge \check{t}^+_{2,i+1}$.

In the worst case $I_x(3,i+1)$ originates at $N(2,i+1)$. In such a case, the link containing a_1 and b_1 (now connected) will be used.

$$\begin{aligned} t^+_{3,i+1}(I_x) = \check{t}^+_{2,i+1} + PD[(2,i+1),(1,i), \\ (2,i),(3,i),(3,i+1)], \text{ and} \end{aligned}$$
$$t^+_{3,i+1}(I_y) = \check{t}^+_{2,i+1} + ND(2,i+1) + LD(2,i+1,3,i+1).$$

Thus $t^+_{3,i+1}(I_x) > t^+_{3,i+1}(I_y)$ in this case, and

$$t^+_{3,i+1}(I_x) - t^+_{3,i+1}(I_y) \le 3\Delta + 2\epsilon < \check{T}_h.$$

Therefore in both cases

$$|t_{3,i+1}(I_x) - t^+_{3,i+1}(I_y)| < \check{T}_h$$

which implies that with points a_2 and b_2 disconnected $N(3,i+1)$ produces exactly one pulse.

As for the relative timing of $\check{t}^+_{2,i}$ and $t^+_{b_2}$, the reader can easily verify that if $t^+_{b_2} > \check{t}^+_{2,i}$ then by inspecting the path connecting $N(2,i), N(2,i+1), N(3,i+1), b_2$ we have

$$t^+_{b2} - \check{t}^+_{2,i} \le 3\Delta < \check{T}_h.$$

If $\check{t}^+_{b2} < t^+_{2,i}$, then by comparing the path connecting $N(3,i)$, $N(2,i-1)$, $N(2,i)$ to the path $N(3,i)$, $N(3,i+1)$, b_2 (or the path $N(2,i+1)$, $N(1,i)$, $N(2,i)$ to the path $N(2,i+1)$, $N(3,i+1)$, b_2) one concludes

$$\check{t}^+_{2,i} - t^+_{b_2} \le 2\beta + 2\epsilon < \check{T}_h.$$

It follows that in both cases $|\check{t}^+_{2,i} - t^+_{b_2}| < \check{T}_h$ and it is safe to also connect point a_2 to b_2 without altering network behavior. This can be verified following an argument similar to that used with points a_1 and b_1, above. Similarly and because of symmetry one can connect points c_2 and d_2 safely.

Proceeding similarly in steps (cumulatively) we can connect all feedback links from row $i+1$ and column $i+1$ to the $i \times i$ subnetwork and still have no oscillations with every node producing exactly one pulse. In the final step only points a_i and b_i will be reconnected, and a stable $i+1 \times i+1$ network will be obtained. □

With all the feedback links connected, the maximum skew between $I_x(r,i+1)$ and $I_y(r,i+1)$ [and $I_x(i+1,r)$ and $I_y(i+1,r)$], $1 \le r \le i+1$ reduces to 3Δ. Therefore, the constraint on signal timing can be relaxed to

$3\Delta + \mu \le T_h \le 3\Delta + \mu'$ where μ and μ' are finite numbers such that $\mu' > \mu > 0$. Therefore, we have,

Lemma 3: By changing bounds on T_h in Model 2 to $3\Delta + \mu \le T_h \le 3\Delta + \mu'$, the network under consideration will not oscillate.

Lemma 4: The clock network described by Model 2 guarantees the maximum skew between any two communicating cells to be bounded above by $2\,\Delta$, where Δ is maximum (node + link) delay, regardless of network size.

Since every clock node produces exactly one pulse, we can consider the skew to be the time difference between arrival times of $0 \rightarrow 1$ transitions at the two data cells in concern. The rest of the proof is hence identical to the proof of Theorem 1. □

Lemma 5: Under Model 2, maximum skew between any two inputs to the same node is bounded above by 3Δ, where Δ is maximum (node + link) delay.

Proof: The proof is identical to the proof of Theorem 2. □

While proving Theorem 3, an extra row and an extra column were added to a stable $i \times i$ network. The resulting $i+1 \times i+1$ network was shown to be stable. We could have extended the $i \times i$ network by an extra column (or row) only to obtain a stable $i \times i+1$ (or $i+1 \times i$) network. The proof to Theorem 3 does not depend on simultaneous addition of a row and a column. Moreover, we could have only added a subset of nodes in column $i+1$ [such as nodes $N(1, i+1)$ and $N(2, i+1)$] and still have a stable network capable of providing a constant skew bound. In either case the proof to Theorem 3 could be trivially modified to support the above. This leads to the following result.

Theorem 4: A network need not be square or even regular to be viable for implementing the proposed scheme. □

3) Model 3: This final model must permit a free running clock, to be practical. We observe that under Model 2 when S produces a single pulse so will every node in the network. After the output pulse expires, a node outputs logic 0 permanently (becomes idle). The exact sequence in which different nodes become idle depends on the specific delays of different links and nodes. For example $N(1,2)$ would turn idle before $N(2,1)$ iff $LD(1,1,1,2)+ND\ (1,2) < LD(1,1,2,1)+ND\ (2,1)$. If we wait until all nodes in the network become idle (all inputs and outputs at logic 0) and then apply a second pulse at S, the nodes will produce pulses (a second pulse each) and then turn idle in a manner similar to the response to the first pulse, due to the time-invariance assumption (Condition 5).

However, to apply a second pulse, we do not have to wait until all nodes turn idle. All we have to consider is worst case timing at any node in the network, and determine a safe minimum separation time between pulses such that clocking events resulting from consecutive S pulses would not *intersect*. This means that we must not allow any input to $N(i,j)$ belonging to clocking event k to reach the node while the node is still processing signals belonging to event $k-1$, for all i and j. This *intersection* of events could cause malfunction, just as using a clock period whose duration is less than the minimum required by data path requirements in a synchronous data transfer could cause failure (see Section I). In our case the malfunction would occur if the next clocking event arrives too early such that an input belonging to event k is treated like it belonged to event $k-1$.

Fortunately, we know that across the network and regardless of the size, the skew between node inputs is $\leq 3\Delta < \hat{T}_h$. Hence, if we assume the same node design associated with Model 2 the OR's output pulse duration will always be $< 2\hat{T}_h$. Since network delays are assumed time-invariant (Condition 5), applying a clocking event no earlier than $2\hat{T}_h$ units since the application of the previous one guarantees no intersection in the above described sense. From a logic design viewpoint, however, we must allow the OR output to drop for a period at least equal to the singleshot's setup time λ, in every cycle. Worst case timing for consecutive clocking events at a node is illustrated in Fig. 10.

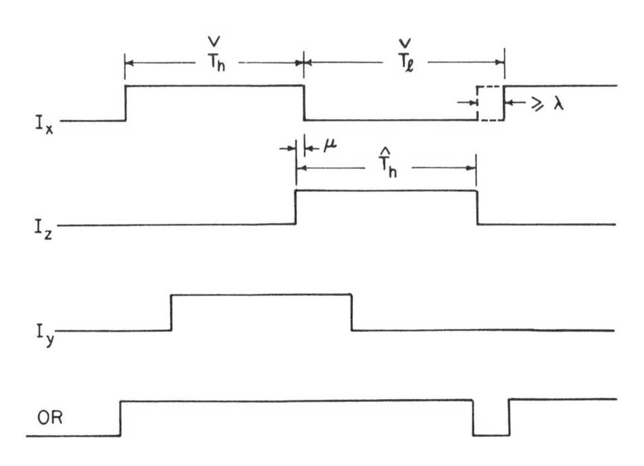

Fig. 10. Worst case timing for consecutive clocking of a node.

It is clear that we must also establish a lower bound on the duration of the low part of a pulse T_l. This is given by

$$T_l \geq \hat{T}_h + 3\Delta + \lambda - \check{T}_h = 3\Delta + \mu' - \mu + \lambda$$

where λ is the minimum setup time for the singleshot, and where $\mu' > \mu > 0$. Thus for a free running clock the minimum clock period is

$$\check{T}_h + \check{T}_l = 6\Delta + \mu + \mu' + \lambda - \mu = 6\Delta + \mu' + \lambda.$$

In the next section we discuss ways to allow for faster clocking rates.

To arrive at Model 3 we only have to modify Conditions $2'$ and $6'$ to become:

$2''$) The clock source S produces a continuous stream of pulses starting at $t = 0$. The timing of these pulses adheres to the following bounds. For the high portion

$$3\Delta + \mu \leq T_h \leq 3\Delta + \mu'$$

and for the low portion

$$T_l \geq 3\Delta + \mu' + \lambda - \mu$$

where $\mu' > \mu > 0$, and λ is the minimum setup time for the singleshot.

$6''$) Each node functions as follows. Whenever triggered, the node will produce an output pulse that adheres to the timing constraints defined in $2''$. All input $0 \rightarrow 1$ transitions separated in time by $< \check{T}_h$ will be ignored except the first. Any two consecutive input transitions separated in time by $> T_h^* + \lambda$ will trigger the node twice, where T_h^* is the actual duration of the first arriving of these transitions. This must be avoided. We can hence assume a structure identical to that in Fig. 7(a).

Thus Model 3 consists of Conditions 1, 2$''$, 3, 4, 5, and 6$''$.

Because delays are assumed time-invariant and because of Conditions 2$''$ and 6$''$, clocking events will not intersect. Thus we have.

Theorem 5: Under Model 3, clocking events will not intersect and thus the network will not oscillate. □

Due to Theorem 5, skew results for the previous models apply to this model as well. We thus have

Theorem 6: Under Model 3, the maximum clock skew between any two communicating cells is bounded above by 2Δ, where Δ is the maximum (node + link) delay. □

Theorem 7: Under Model 3, maximum skew between any two inputs to the same node is bounded above by 3Δ, where Δ is the maximum (node + link) delay. □

III. Design Issues

Designing clock networks based on the proposed scheme involves network topological issues and node design issues. We will discuss node design issues in this section. However, network topological issues are interesting and worthy enough of a separate and detailed study. Based on graph-theoretical models, a preliminary study [13] reveals that for a certain topology to be able to guarantee constant skew upper bound, regardless of the size, the corresponding graph must satisfy some necessary conditions. The key requirement is that the graph must contain cycles such that any pair of communicating processors is clocked from nodes belonging to the same cycle directly, or indirectly via finite length paths. These paths may consist of nodes or buffers or both nodes and buffers. It is also possible to relate network properties to the skew bound and timing constraints [13]. In this section we demonstrate the existence of alternative *valid* network topologies for clocking a 2-D mesh of processors. We will also present an *invalid* topology. This section will also introduce the new concept of *fuse programmed* clock networks, a spin-off of the proposed clocking scheme. It will also be demonstrated that the scheme is capable of clocking nonplanar data networks such as hypercubes. A discussion on the practicality of the proposed scheme is also given.

A. Alternative Network Topologies

Here we demonstrate the existence of alternative clock network designs for the 2-D mesh case. We also show through an example that not every network topology can support the new scheme. The results reported in this subsection are supported by the theoretical work in [13] and by simulation results, to be presented shortly.

1) An $F4$ Network: Consider the ($F4$) clock network shown in Fig. 11. It can be observed that any pair of nodes associated with two communicating cells is included in a cycle of length 2. Hence, skew between the two inputs to a node is $\leq 2\Delta$. This can be verified by noticing that any node input can be reached from the other input in two hops. Also since any node can be a local source to a neighboring node with one hop delay, maximum skew between communicating cells is Δ. Model 3 can be easily modified to accommodate the $F4$ network (assuming similar node structure) by specifying [13]:

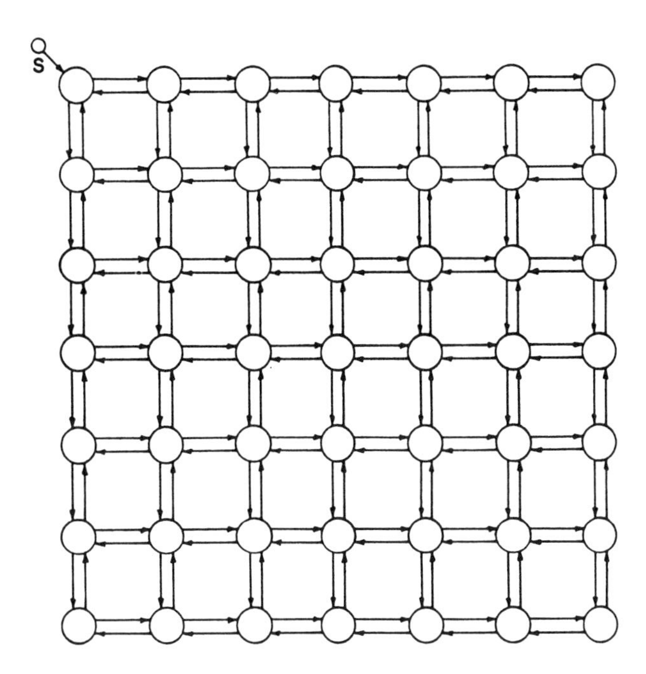

Fig. 11. A valid $F4$ clock network.

— $\mu + 2\Delta \leq T_h \leq 2\Delta + \mu'$
  $T_l > 2\Delta + \mu' + \lambda - \mu$

— Minimum clock period = $4\Delta + \mu' + \lambda$ where μ, μ', Δ, and λ are as defined earlier.

Clearly this network allows for a faster clock and provides smaller skew bound compared to the $F3$ network, but requires 33% additional connections.

2) An $F2$ Network: The ($F2$) network shown in Fig. 12 is also valid and incurs less link cost than the two other networks considered thus far. Every pair of nodes associated with two communicating cells is in a cycle of length 4. Following arguments similar to those above, it can be easily verified that maximum skew between any pair of communicating cells is 3Δ and maximum skew between any two inputs to a node is 4Δ [13]. It can similarly be verified that the minimum clock period in this case (assuming an adapted Model 3) is $8\Delta + \mu' + \lambda$ where Δ, μ, μ', and λ are as defined earlier.

Table I summarizes the main parameters for the three networks discussed. From the table it is clear that, for any of the three networks presented, one can state:

a) maximum skew between communicating cells = $(5 - F)\Delta, 2 \leq F \leq 4$

b) maximum skew between any two node inputs = $(6 - F)\Delta, 2 \leq F \leq 4$.

3) Example of an Invalid Network: Consider the network shown in Fig. 13. In this network skew can grow unchecked. This can be observed by considering the two rightmost nodes in the bottom row. These nodes have their common reference at $N(1,1)$ (top left node). As network size grows, delay differences can add up with no known limit. Indeed, we simulated this network and found that skew grew without apparent limit as the size increased.

4) Simulation and Test Results: We simulated the three valid networks ($F = 2,3,4$) on Encore's Multimax

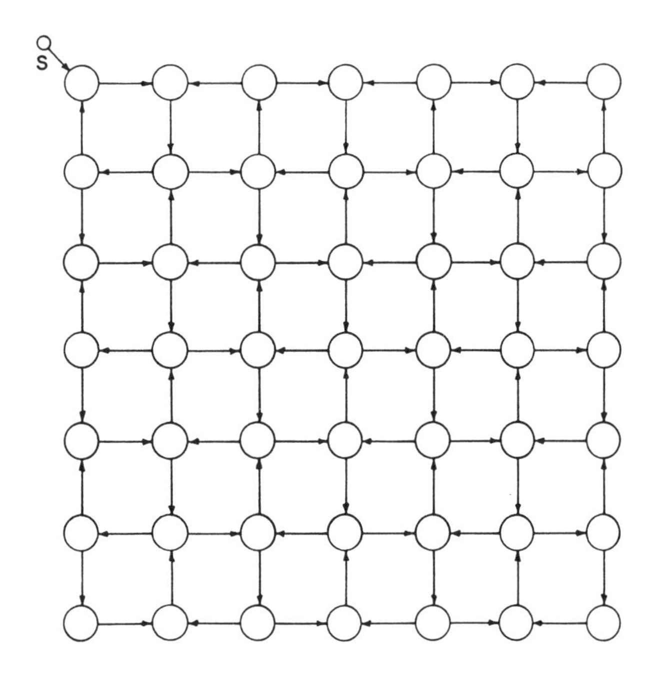

Fig. 12. A valid $F2$ clock network.

TABLE I
THE MAJOR PARAMETERS FOR THE F_2, F_2, AND F_4 NETWORKS

Network	Maximum Skew Between Communicating Cells	Maximum Skew Between Any Two Inputs to a Node	Minimum Clock Period
$F2$	3Δ	4Δ	$8\Delta + \mu' + \lambda$
$F3$	2Δ	3Δ	$6\Delta + \mu' + \lambda$
$F4$	Δ	2Δ	$4\Delta + \mu' + \lambda$

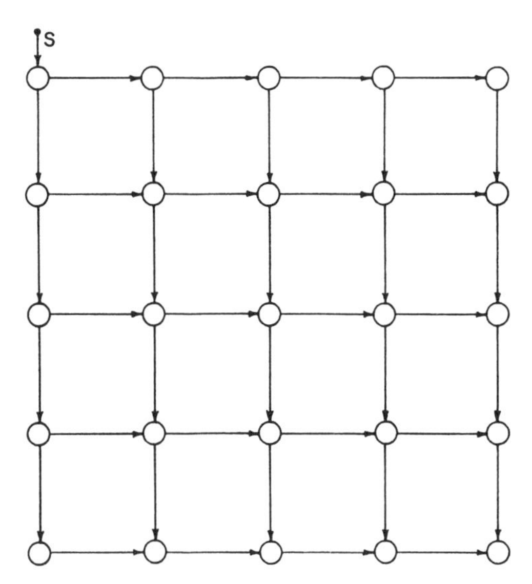

Fig. 13. An invalid clock distribution network.

multiprocessor using the C programming language. We assumed a 500 × 500 network in each case. Node and link delays were lumped together and assigned as (output) link delays. A random link delay ranging from 7 to 15 (corresponding to $\alpha-\beta$ and $\alpha+\beta$) units of time was assigned to different links using a random number generator. (The values 7–15 and their units are actually irrelevant here.) For each network we checked the following.

1) Lack of oscillation. This was checked by simulating the network's response to 10 consecutive clock pulses. Model 3 was assumed for the $F3$ network, whereas adaptations of the model were used for the $F4$, and $F2$ networks as discussed in Sections III-A1 and III-A2, respectively. For each event, arrival times at all inputs were recorded, for each node. The times at which the output of each node started and ended were also recorded. Then we successfully checked for lack of event intersections (in the sense described in Section II-B3). We also successfully checked for lack of multiple pulses per node per event.
2) Maximum skew between node inputs. The simulation runs mentioned above were also utilized to check maximum skew between node inputs. Input arrival times at each node were compared and the maximum difference recorded for each clocking event. In no case did that skew exceed the expected bound listed in Table I for any of the three networks, with $\Delta = 15$.
3) Maximum skew between any two communicating processors. This was done by recording the arrival time of the first input to each node for every clocking event. These times were compared for each pair of neighboring processors and the difference was found. In no case did that skew exceed the value listed in Table I with $\Delta = 15$, for each network.

A 4 × 4, $F3$ network was built of SSI (Schottky TTL) chips using a sequential node design (to be discussed shortly). The network was tested successfully at 9.5 MHz for lack of oscillation.

B. Node Design Issues

The node design suggested earlier [Fig. 7(a)] for Model 3 is obviously not unique. Other simple designs without singleshots can be developed. To exemplify, two alternative designs will be described briefly with their relative merits (or demerits) pointed out. All node designs described in this paper exhibit approximately two gate delays.

It is tempting to consider a purely combinational logic design such as that in Fig. 14. The OR performs the combining function. The limiter logic is utilized to limit the length of output pulse width to T_h. In the figure we use an inverter and a cascade of buffers to invert signal P and provide the necessary delay ($= T_h$). Notice that for this design node delay is only that through the OR and AND gates. However the design in Fig. 14 requires a minimum clock cycle > 3 T_h because the signal $P(R)$ could remain high (low) for approximately $2T_h$ which rules out another input pulse for at least that long (2 T_h) after the earliest input drops. Therefore although this design is very simple, it is not as efficient as the one in Fig. 7(a) in terms of clocking rate.

To avoid the problem pointed out above consider the slightly more complex node design in Fig. 15(a). For simplicity, assume flip-flop switching time $= d$. A $0 \rightarrow 1$ transition on P at $t = 0$ sets FF1 at $t = d$. At $t = T_h - d$, FF2 is triggered.

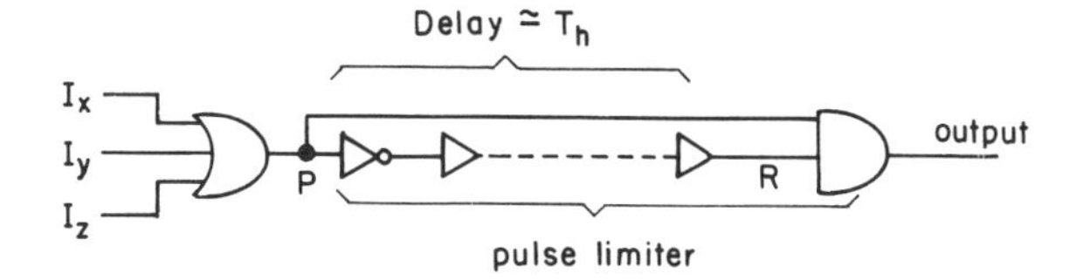

Fig. 14. Purely combinational logic node design.

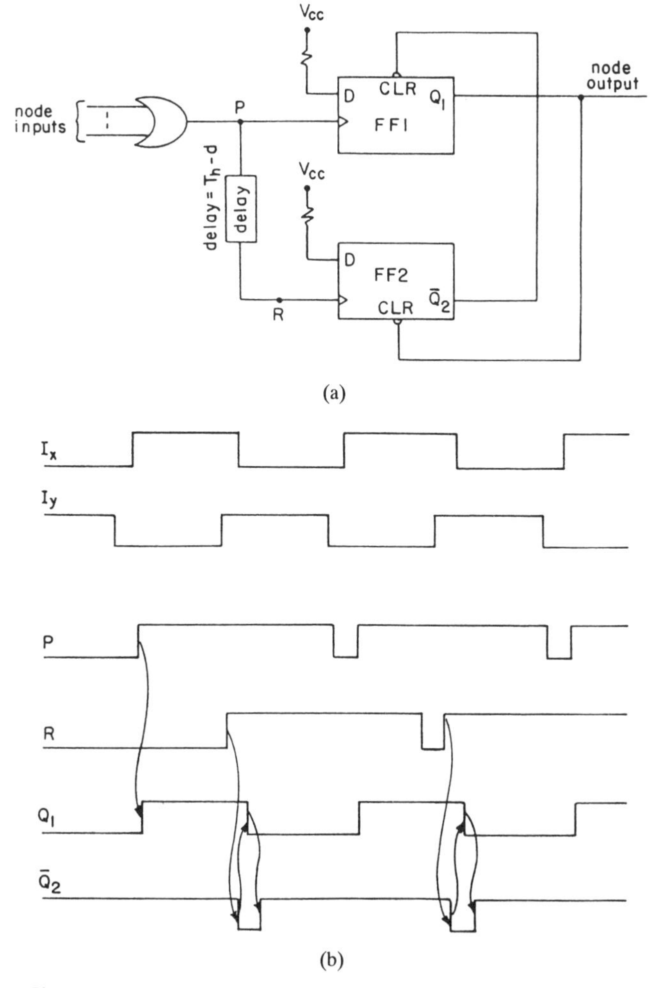

Fig. 15. A more efficient node design. (a) Logic diagram. (b) Timing.

A $t = T_h, Q_2$ drops causing FF1 to clear at $T_h + d$. Thus the output Q_1 is high for T_h as desired. The design in Fig. 15(a) allows the low portion of the clock signal to be as short as that achievable with the circuit in Fig. 7(a). Fig. 15(b) illustrates the timing for the design in Fig. 15(a). We utilized this design in the test circuit mentioned in the previous subsection.

C. Fuse Programmed Networks

It is evident that combining inputs at a clock node necessitates imposing a lower bound on the low portion of clock signals. An ORing function at the input of a node has been assumed, thus far, to implement combining which in turn necessitates a pulse limiter after the OR. This latter function can be eliminated if the order in which input signals arrive at the node is known. This requires 1) delays to be time invariant; 2) logic at each node to detect the arrival order of inputs and, 3) all node inputs, except the first arriving, to be disabled permanently; by fusing them out for instance.

The first requirement in not possible in a strict sense. However delay variations with time are usually small and take place relatively slowly. Moreover, in VLSI/WSI delays tend to increase or decrease uniformly over the network [14]. The second requirement (fusible connections) is feasible given the ready existence of the technology such as in programmable logic arrays PLA's. As for the third requirement, an RS latch and a pair of multiplexers at each node can be utilized to decide on the relative arrival times of different pairs of inputs by observing the status of the latch for each combination of the multiplexers' select values.

After programming the connections, the potential for lengthening clock pulse duration would no longer exist. Thus, in theory, the timing constraints can be dropped altogether, and the network can be treated liked a buffered clock network. Observe that with this method, the OR naturally serves the buffering function often employed in other clocking schemes. The delay through this OR becomes the only node delay with this scheme. A network programmed as such will combine the merits of buffered networks and the new scheme.

D. Clocking Nonplanar Data Networks

The new scheme can be utilized to control data processing networks of nonplanar topologies such as hypercubes and m-D meshes [13]. One straightforward way is to assign a clock node to each processor and connect the nodes in the same way data processors are connected. Equivalently, the node can be placed within the processor and two extra wires are added to the data links to carry clock signals back and forth. The resulting network will then be similar in concept to the $F4$ network in the sense that every processor exchanges clock signals with all other processors with which it directly communicates. Clearly the same timing constraints applied to the $F4$ network would have to be applied in this case [13]. Since in the worst case any processor will be clocked no later that Δ units of time after any of its neighbors is clocked, skew will be bounded above by Δ, just as in the $F4$ network case.

We now provide an example that demonstrates the applicability of the new scheme to hypercube processor networks. The example relaxes the assumption on node to cell delay and also shows how several processors can be safely clocked from a single node.

1) A Hypercube Example: An n-dimensional hypercube consists of 2^n cells labeled 0 through 2^{n-1}. Two cells are directly connected iff their binary labels differ in exactly one bit. An n-dimensional hypercube can be partitioned into 2^{n-k} subcubes of dimension k each, $k \leq n$. In such a case each subcube would contain 2^k cells. One method of partitioning is to consider the binary address of a cell, $A = a_{n-1}a_{n-2} \cdots a_k a_{k-1} \cdots a_1 a_0$ as having two parts. Bits 0 through $k-1$ represent the in-subcube label, whereas bits k through $n-1$ represent the subcube label.

For this example we assign a single clock node to each subcube, and give it the same label as the associated subcube. Thus we employ 2^{n-k} clock nodes labeled from 0 to $2^{n-k}-1$.

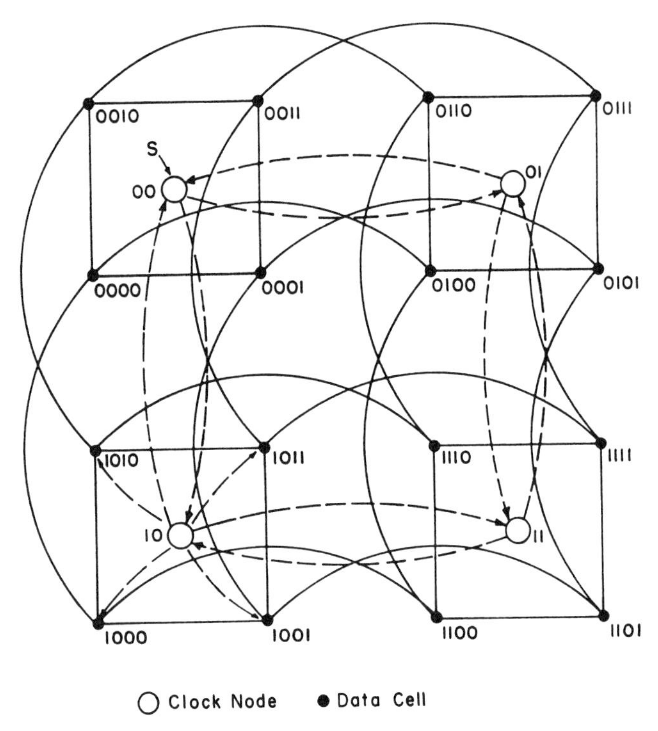

Fig. 16. Clocking a hypercube using the proposed scheme.

Two clock nodes are connected together in a 2-cycle (i.e., there is a link form each node to the other) if and only if their labels differ in exactly one bit. The clock node associated with a subcube is responsible for clocking all the cells in that subcube. The network containing the clock nodes is itself an $(n-k)$-cube. Fig. 16 illustrates a 4-cube partitioned into 4 subcubes of dimension 2, and the connections among the clock nodes. The figure also illustrates clock distribution within subcube 10, which contains nodes 1000, 1001, 1010, 1011. Notice that the clock source S can be connected to any node. This is due to the symmetry of the clock network. In fact this is also the case with any of the valid networks discussed earlier for the 2-D mesh case!

The six conditions of Model 3 (as applied to the $F4$ network) are imposed on this network. In addition, we assume that the delay from a clock node to a cell in its associated subcube is $v \pm \eta$. Thus, the upper skew bound is $\Delta \pm 2\eta$ in this case because clocks at two processors associated with a certain node can be skewed by $\leq 2\eta$, and outputs of two adjacent nodes can be skewed by $\leq \Delta$. Note that η depends on k implicitly.

E. Discussion

We now attempt to address some questions that may have occurred to the reader regarding the practicality of the new scheme.

1) Timing Constraints: The timing constraints are actually easy to implement. There are basically two requirements. First the two (high and low) parts of the cycle must not be of identical length, which is likely to be the case anyway. In systems triggering on the high part, that part must be shorter than the low part. One can always condition the source by using either the output directly or an inverted version of it.

The second requirement is that each of the low and high parts must be longer than the maximum skew between any two inputs to a node. That skew depends on the chosen network topology and the maximum delay values of nodes and links. Obviously this requirement is trivial to implement (by stretching the clock), but raises the more interesting question: what is the effect of the timing constraints on clock speed? We will attempt to answer this question next.

2) Clock speed: The skew upper bound for the networks presented earlier ranges form Δ to 3Δ (see Table I). Clearly a tradeoff between hardware cost and skew bound exists and must be dealt with. We believe that for the scheme to be attractive one has to favor networks offering lower skew bounds. As we discuss later hardware overhead may not be excessive. The most hardware extensive clocking topology would incur hardware overhead comparable to that incurred by current asynchronous schemes.

The next question is: how large Δ is? All node designs presented in this paper offer approximately two gate delays each. Link delay will depend on system layout/architecture. We assume here that inter-node delay can be kept below or about one gate delay as suggested by Fisher and Kung in [4] in the context of buffered clock networks. Thus one can estimate Δ to be about three gate delays, which is one gate delay more than its counterpart in buffered networks. For networks with characteristics like those of the $F4$ network, this translates into a minimum clock period of about 12 gate delays.

The clock cycle for a typical processor is about 20 gate delays [15]. Hence a skew of Δ corresponds to about 15% of a typical clock cycle. Hoshino [10] suggested that skew of about 10% of the clock cycle is acceptable. He actually used 10% as an example without providing rationale. We do not know whether 15% is an acceptable skew level in current systems. However, it has been demonstrated by Hoshino [10] that for many scientific applications the *linear scaling law* holds; that is, speed (or throughput) is proportional to the number of processors. He also predicted that the law will hold into the future. For systems catering to such applications, there will always be a size threshold beyond which the new clocking scheme offers better performance than any other scheme. This is due to the independence of skew bound from system size. It is therefore possible that the benefits of easy control of large systems with virtually unlimited parallelism, and the ease of scaling (expandability) could outweigh some slow down in the basic clock rate, if that is necessary.

3) Hardware Overhead: The overhead incurred by the new scheme can be divided into node costs and link costs. We estimate that a typical node design would perhaps require 10–20 gates. Given that a typical processor consists of thousands to hundreds of thousands of gates [15], it is obvious that node hardware overhead is insignificant, even if we associate a node with each processor.

For link costs consider the most extensive approach where each processor is assigned a node which exchanges clock signals (a pair of wires) with all the nodes associated with processors which directly communicate with its assigned processor (as in the $F4$ network). In this approach we have increased the number of wires between a pair of communicating processors by two. Clearly, relative link cost will depend on the level of parallelism of inter-processor communications and could be excessive for bit serial systems. However, we have demonstrated (see Section III-D1) a scheme where each node clocks several processors. That scheme trades link and node costs for some fixed increase in skew bound. The point is, with this new technique the designer will always have some flexibility in trading off hardware cost and performance.

We finally point out that the hardware overhead in any asynchronously controlled system is about the same as the overhead incurred by the most hardware extensive design based on the proposed scheme. In an asynchronous case, at least two handshake signals are needed between every pair of communicating processors and logic is needed in every processor to process these signals. Thus the amount of hardware per processor is comparable to that of the $F4$ network and is perhaps more than that needed to clock the hypercube in Fig. 16.

IV. Conclusion

We have presented a novel clocking scheme which enables global synchronization of computing networks with unlimited size such that skew is bounded by a constant. We have shown theoretically that, if clock signals follow a set of simple constraints, the proposed clock networks will not oscilliate despite the cyclic nature of the networks. Three "valid" clock networks for 2-D mesh-based topologies have been proposed and their constant skew bounds given. We have described a straightforward method for synchronizing other topologies such as hypercubes under constant clock skew bound, based on the proposed scheme. The new concept of fuse programmed clock networks has also been introduced and a discussion on practical implementation issues of the new technique has been presented.

Acknowledgment

The author would like to thank the anonymous reviewers for their helpful suggestions. He would also like to thank V. Angrish for writing the simulation software and J. Altemus for assembling the test circuit.

References

[1] C.L. Seitz, "System timing," in *Introduction to VLSI Systems,* C. Mead and L. Conway, Eds. Reading, MA: Addison Wesley, 1980, ch. 7.

[2] D. Wann and M. Franklin, "Asynchronous and clocked control structures for VLSI based interconnection networks," *IEEE Trans. Comput.,* vol. C-32, no. 3, pp. 284–293, Mar. 1983.

[3] S-Y. Kung, "On supercomputing with systolic/wavefront arrays," *Proc. IEEE,* vol. 72, no. 7, pp. 867–884, July 1984.

[4] A. Fisher and H. T. Kung, "Synchronizing large VLSI processor arrays," *IEEE Trans. Comput.,* vol. C-34, no. 8, pp. 734–740, Aug. 1985.

[5] M. Franklin and S. Dhar, "On designing interconnection networks for multiprocessors," in *Proc Int. Conf. Parallel Processing,* 1986, pp. 208–215.

[6] S. Y. Kung, S. C. Lo, S. N. Jean, and J. N. Hwang, "Wavefront array processors—Concept to implementation," *IEEE Comput. Mag.,* vol. 20, no. 7, pp. 18–33, July 1987.

[7] J. P. Fishburn, "Clock skew optimization," *IEEE Trans. Comput.,* vol. C-39, no. 7, pp. 945–951, July 1990.

[8] P. M. Kogge, *The Architecture of Pipelined Computers.* New York: McGraw-Hill, 1981, pp. 21–39.

[9] M. Shoji, "Elimination of process-dependent clock skew in CMOS VLSI," *IEEE J. Solid-State Circuits,* vol. SC-21, no. 5, pp. 875–880, Oct. 1988.

[10] T. Hoshino, *PAX Computer: High Speed Parallel Processing and Scientific Computing.* Reading, MA: Addison-Wesley, 1989, ch. 8.

[11] A. El-Amawy, "Arbitrarily large clock networks with constant skew bound," U.S. Patent 5 163 068, 1992.

[12] ——, "Branch-and-combine clocking of arbitrarily large computing networks," in *Proc. Int. Conf. Parallel Processing,* Aug. 1991, pp. I-409–417.

[13] ——, "Properties of branch-and-combine clock networks," Tech. Report #01-91C-AE, ECE Dep., Louisiana State Univ..

[14] F. M. Klaasen, "Review of physical models for MOS transistors," in *Proc. NATO Advanced Study Institute on Process and Device Modeling for Integrated Circuit Design,* Louvain-la-Neuve, Belgium, July 1977, pp. 541–571.

[15] G.S. Almasi and A. Gottlieb, *Highly Parallel Computing.* Redwood City, CA: Benjamin/Cummings, 1989, ch. 3.

Design and Evaluation of Wafer Scale Clock Distribution

D.C. Keezer V.K. Jain

Center for Microelectronics Research and the
Department of Electrical Engineering at the
University of South Florida, Tampa, Florida 33620

Abstract

Realization of high performance wafer scale systems depends heavily on the ability to synchronize local and global activity. The primary mechanism for accomplishing this is through the generation and distribution of clock signals to the physically separated elements of the wafer scale system. In this paper we describe a computer program which is used for simulating the performance of large (WSI) clock distribution networks, themselves made up of both passive transmission line elements and active buffers. The theoretical basis for the model is first briefly reviewed. Next, we present several example networks for 4 inch wafers, comparing the AC performance as a function of power dissipation, layout area, and harvesting yield. The sensitivity of performance (as measured by propagation delays and clock skew) to random or systematic process-related fluctuations and defects is calculated using the model. Finally, a new measurement system is introduced for electrically characterizing the performance of clock distribution networks on fabricated wafers and within multichip modules (MCMs).

I. Theoretical background

Techniques for WSI synchronous clock distribution often utilize a tree-like structure with several levels of hierarchy [1-5]. This is required to cope with the inherently large fanout associated with the clock signal, . The clock input is connected to the "trunk" and the signal is subsequently split and distributed to each major "branch". Each major branch feeds the signal to two or more smaller branches. This process continues until enough minor branches are formed so that the signals can be used to directly clock storage or logic elements (the "leaves"). Differences in clock distribution design methods involve: (1) the lengths, widths, and physical arrangement of the interconnections (branches), and (2) the location and type of buffer elements interspersed throughout the tree. The most prevalent approach remains the basic "tree" structure or modifications of it. In fact, Dikaiakos and Steiglitz [5] have recently shown that the "tree" approach offers better control for signal distribution within long systolic arrays as compared with pipelined clocking schemes.

A buffered clock tree network utilizes an external source which is buffered to minimize and standardize the load presented to the external system. The current requirement for the first, "trunk" buffer is determined by the number and size of buffers in the next level of the tree and the capacitance of the interconnecting conductor lines. Optimization of buffer sizes has been the focus of much research [6-10]. A similar balancing approach is used to determine the size of the second and third level buffers. The lowest level buffers (leaves) are used to clock storage and logic elements. If the load on each of the nodes at each level is closely matched then the clock signals reaching the storage elements will be closely synchronized. If the interconnect lengths are not matched then the propagation delays for signals following different paths will be dissimilar. These differences manifest themselves as clock skew. Significant clock skew can severely limit the maximum operating frequency of the system.

Delays due to signal pathlengths and interconnect RC values are significant in WSI and MCM systems. Careful attention must therefore be paid to the physical layout of the conductor lines which link clock buffers and fanout the clock signal. The H-clock tree distribution method which distributes the clock signal via a hierarchy of "H" structures has been reviewed by Bakaglu [5]. An example, passive H-clock distribution network is shown in Fig.1 for the case of a four inch wafer. A clock driver is connected to the center of the main "H" structure from which the signal is transmitted to the four corners. The four signals arrive synchronously at the corners because the signal pathlengths are equal and loads are balances. They form the inputs to the next level of hierarchy, which each are again redistributed to the four corners of smaller "H" structures. The process may continue through several levels of hierarchy. The ends of the lowest-level H's are used to clock local logic or to supply signals to local buffers. In a passive, controlled impedance arrangement, the widths of the conductor lines forming the H's become progressively smaller as the signal propagates to lower levels in the hierarchy. These narrowing widths have been exaggerated in the figure. Bakoglu [5] has demonstrated the use of a passive H-clock tree on a 3 inch diameter wafer where clock skew was limited to less than 30ps.

Reprinted from *Proc. IEEE Int'l Conf. Wafer Scale Integration,* pp. 168–175, Jan. 1992.

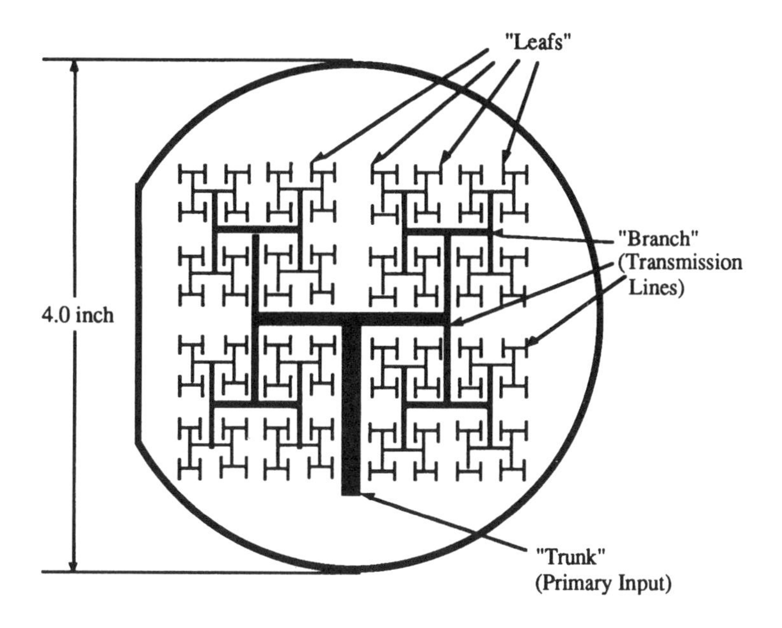

Figure 1 - A passive, four level, WSI H-Clock distribution network. (Linewidths are exaggerated to show detail.)

A buffered version of the H-clock tree is shown in Fig.2. Here, active buffers are used between hierarchy levels. This approach minimizes the problem associated with the potentially large RC time constants of the interconnect lines. However, clock skew is now influenced by process variations across the wafer which effect the speed of the individual buffers.

Signal delays due to interconnections have been calculated for VLSI by Carter and Guise [11,12]. The delay time is modeled as the sum of contributions from: (1) the driving gate unloaded delay, (2) line and fanout capacitance coupled with the driving gate resistance, (3) diffusion delay, and (4) RC delay due to fanout capacitance and line resistance. The total delay is given by [11]:

$$T_d = R_g (C_g + lc + C_0) + 1/2rcl^2 + rlC_0 \quad (1)$$

where R_g and C_g are the resistance and parasitic capacitance of the driving gate, l is the length of the interconnect line, r is the line resistance per unit length, c is the line capacitance per unit length, and C_0 is the input capacitance of the receiving gate. Multi-stage buffers may be used to provide high current drive capability while maintaining low and/or standardizing input capacitance.

For multi-stage buffers, additional time must be added to account for the internal delay. Equation (1) then becomes:

$$T_d' = R_g (C_g + lc + C_0) + 1/2rcl^2 + rlC_0 + T_i \quad (2)$$

Here, T_i can be determined [6,8,9,10] for a series of closely spaced buffers by neglecting the effects due to the short interconnects between them. For a network of buffers linked by patterned conductors, this expression may be applied to each net and the contributions added to determine the delay for a particular signal path. By tailoring the buffers and interconnect geometries, the signal path time delays can be matched to provide a synchronous clock distribution network.

One of the major thrust directions in high performance computation today is in the use of massively parallel processor arrays. For example, at the recent High Speed Digital System Interconnections Workshop [13] , a major challenge (aimed primarily at the interconnection aspects) was posed by one of the IBM representatives for use on a future 4096 processor machine with a multi-stage switch to connect these processors to 4096 output nodes. Such an arrangement is depicted in Fig.3. Notice, in particular, that in addition to the complexity of the switch, there is the added requirement of extremely high speed (i.e. 2 Gbps). The issues of clocking in such massively parallel systems are of major concern.

II. Examples of Clock Distribution Networks

An H-clock distribution network for a 4 inch WSI network is shown in Fig.4. This arrangement splits a single clock signal into 256 minor branches (leaves). Because of the symmetry of the structure, synchronous clock signals will be distributed to all 256 leaf points. To match impedances, the widths are approximately halved at each branch point. Alternatively, impedance matching might be sacrificed at the lower levels where the path lengths are small, allowing constant width traces to be used at these levels. For instance, if the main truck is split four time (rather than eight) with the lower level conductor widths fixed at 10 um, then the trunk width would be 160 um.

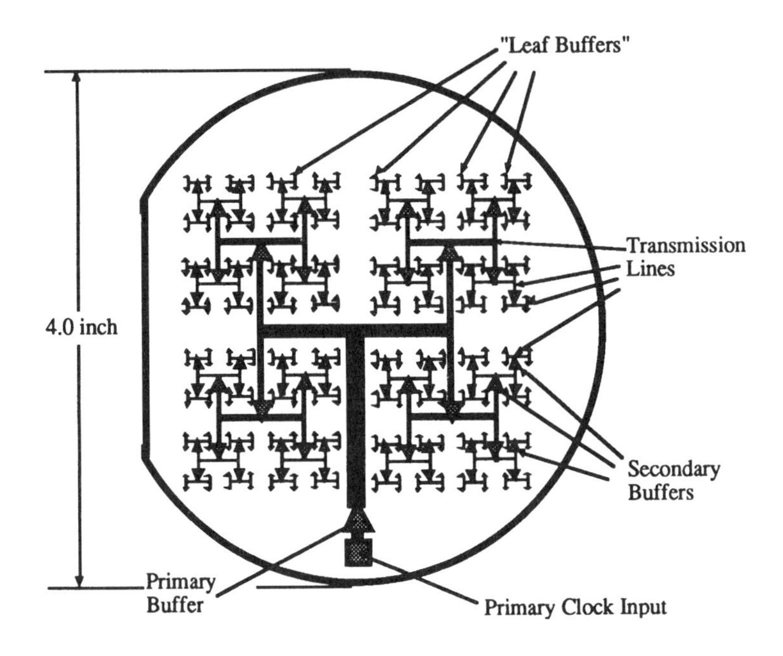

Figure 2 - A buffered, four level, WSI H-Clock distribution network.

Active buffers may be inserted at the input to each H structure (as was described in Fig.2). This approach has the advantage of reducing the load on any one buffer at the expense of adding multiple buffers. The overall diffusion delay is also reduced because it is proportional to the *square* of the interconnect length. However, additional delays (typically on the order 1 or 2 ns) are added as the signal propagates through each buffer. Of even more concern is the fact that the internal delay, T_i, of the buffers will be a sensitive function of the fabrication parameters. This will result in significant clock skew across a wafer or from one wafer to another.

A computer model has been developed to calculate the overall propagation delay of a clock signal (from wafer input to the leaf buffers). This model is based on equation (4) and is readily adapted to a wide variety of clock distribution networks. The terms in equation (4) form the input parameters for the program with the lengths and widths of the interconnections being specified as a series of values depending on the network layout. By adjusting these values we can determine the sensitivity of the design to process variations which may occur across a wafer, or from wafer to wafer.

Using nominal values of C_0=0.1pf, C_g=0.1pf, R_g=20 Ohms, T_i=2ns, conductor sheet resistance = 0.04 Ohms / Sq., conductor capacitance = 9.47 pf / mm^2, an initial trunk width of 150 um, and limiting the minimum linewidth to 4.5um, we have calculated a nominal delay of 10.024 ns. Table 1 gives delay

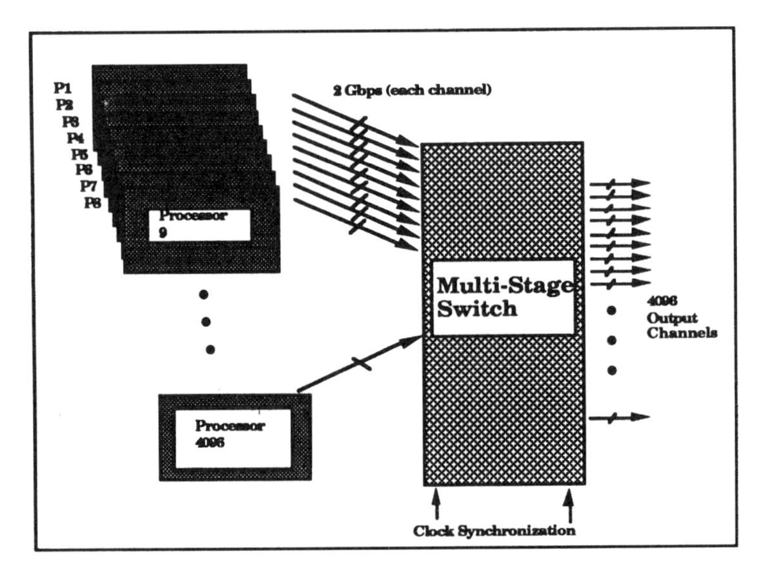

Figure 3 - A massively parallel, 4096 processor system utilizing a high speed multi-stage switch.

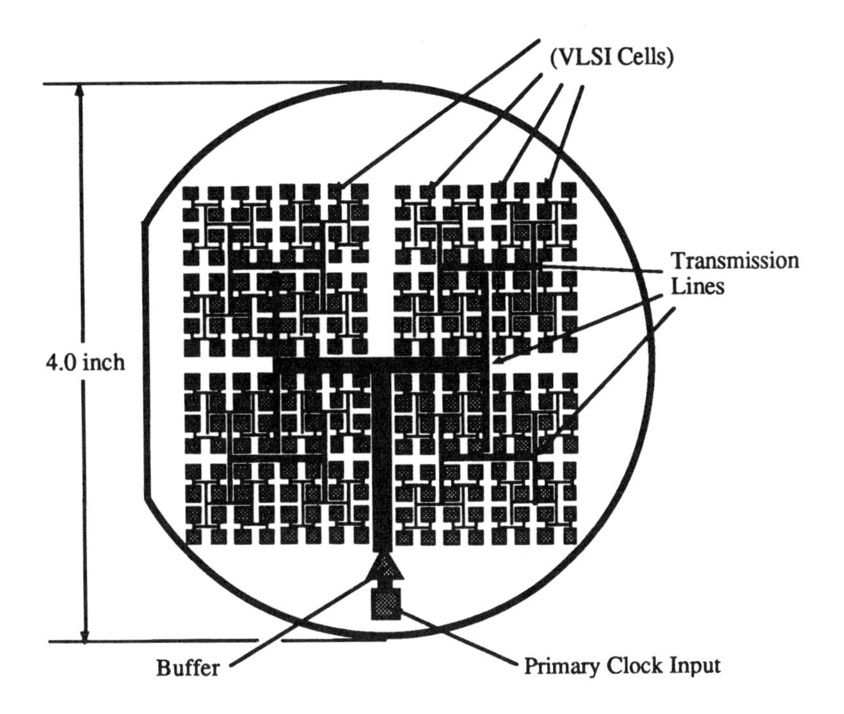

Figure 4 - 256 cell, WSI network with clock connectivity enhanced. (Clock buffers are not illustrated.)

values for this arrangement when each of the process parameters is varied by ±20% of nominal. As can be seen in the Table, the internal buffer delay (T_i) is by far the most sensitive parameter, with T_d' varying by 3.2ns or 32% over this range. In the worst case, if all parameters are varied together (rather than separately) by ±20% then T_d' is found to range from 7.539 ns to 12.878 ns. This represents a potential clock skew of over 5 ns across a poorly processed wafer.

	T_d (ns)			
	+20%	-20%	delta(ns)	delta(%)
Buffer input capacitance (C_0)	10.028	10.020	0.0008	0.080 %
Internal buffer delay (T_i)	11.624	8.424	3.200	32.000 %
Buffer output resistance (R_g)	10.276	9.772	0.504	5.000 %
Buffer output capacitance (C_g)	10.025	10.022	0.003	0.030 %
Conductor linewidth (w)	10.276	9.779	0.497	5.000 %
Conductor sheet resistance	10.177	9.871	0.306	3.000 %
Conductor capacitance / mm^2	10.423	9.625	0.798	8.000 %

Table 1 - Results of buffered H-Clock sensitivity analysis for a network with narrowing linewidths. $T_{d(Nom)}$ = 10.024 ns

Table 2 shows the results of similar calculations for a buffered H-clock distribution network where the conductor linewidths are constant. In this example we have set w=20um, with all other nominal values the same as used for Table 1. In this case the nominal delay is 9.054 ns. Again, the most sensitive parameter is T_d'. This arrangement is less sensitive to variations in R_g and w because the lower overall line capacitance results in a reduced delay contribution to T_d'. It should also be noted that if, say, 1 ns buffers were used, then each of the values given in the tables would be reduced by 4 ns (as would the nominal delays).

	T_d (ns)			
	+20%	-20%	delta(ns)	delta(%)
Buffer input capacitance (C_0)	10.028	10.020	0.0008	0.08 %
Internal buffer delay (T_i)	11.624	8.424	3.200	32.0 %
Buffer output resistance (R_g)	10.276	9.772	0.504	5.0 %
Buffer output capacitance (C_g)	10.025	10.022	0.003	0.03 %
Conductor linewidth (w)	10.276	9.779	0.497	5.0 %
Conductor sheet resistance	10.177	9.871	0.306	3.0 %
Conductor capacitance / mm^2	10.423	9.625	0.798	8.0 %

Table 2 - Buffered H-Clock sensitivity analysis for a network with constant, 20um linewidths. $T_{d(Nom)}$ = 9.054 ns

By increasing the size of the buffer transistors, shorter delays can be realized. In Fig.5 the total propagation delay (input to leaf node) is plotted as a function of the total silicon area required by the clock distribution circuitry (including the buffers and the interconnect lines). This example assumes a typical 2μm CMOS process technology. It is clear that when the buffer transistors are enlarged, the propagation delay is reduced. However, this graph illustrates clearly the principle of diminishing returns. Beyond a certain point (about 100 to 200 mm^2), reduction in propagation delay carries an enormous area penalty. For example, to reduce the propagation delay by 6% from 13.25ns to 12.4ns requires a 400% increase in silicon area for the clock distribution network. Furthermore, in this analysis, all of the additional area is in active silicon (transistor sizing) so that yield and reliability are reduced while power dissipation is increased. Such trade-off analyses are crucial to the success of ULSI systems.

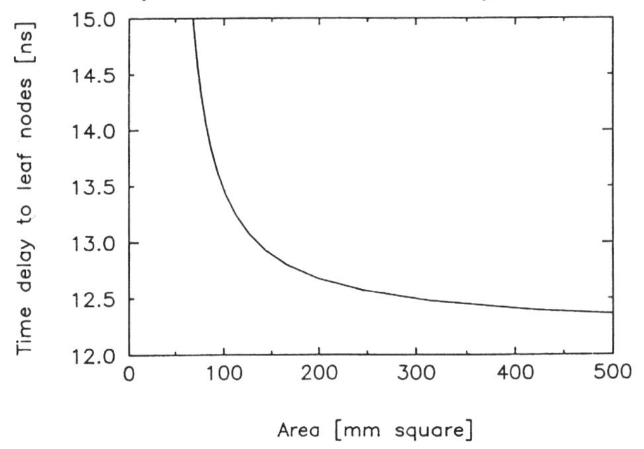

Figure 5 - Calculation of the trade-off between clock signal delay time and the total silicon area devoted to the distribution network. Values are for a hypothetical 2μm WSI system.

III. WSI / MCM Characterization Probe System

Instrumentation for characterizing how well the clock signal is actually synchronized in the fabricated system includes: (1) a stable high frequency clock source supplied to the ULSI/MCM system; (2) a non-invasive method for extracting internal signals (probing); and (3) a suitable waveform analysis instrument. We utilize a Hewlett Packard model HP82000, 200 MHz digital test system to provide both a stable source of high frequency data as well as to capture and characterize signals. When a clock signal faster than 200 MHz is required, GaAs multiplexers are used to generate signals up to about 1 GHz. A high speed (34 GHz) sampling oscilloscope is used for measurements above 1 GHz. A photograph of this arrangement is shown in Fig.6. Non-invasive probing of high speed clock signals on ULSI/MCM circuits requires the use of high impedance, low capacitance, active FET probes or non-contact methods such as voltage-contrast electron beam probing. With a stroboscopic electron beam probe system, clock synchronization has been verified to subnanosecond precision [1]. An active FET probe is currently being incorporated into the wafer probe test system at USF. This will allow detailed and precise measurements to be made of clock signals at various points within the WSI or MCM system.

IV. Conclusions

A computer model for the design of clock distribution networks for WSI and MCM systems have been described. For these methods to be effective, the extreme conditions found in WSI systems must be carefully considered. The computer model may be used to determine the clock delay at various stages within the distribution network. By adjusting process parameters in the model, sensitivity analyses have been performed to determine the likely clock skew across the wafer or from wafer to wafer.

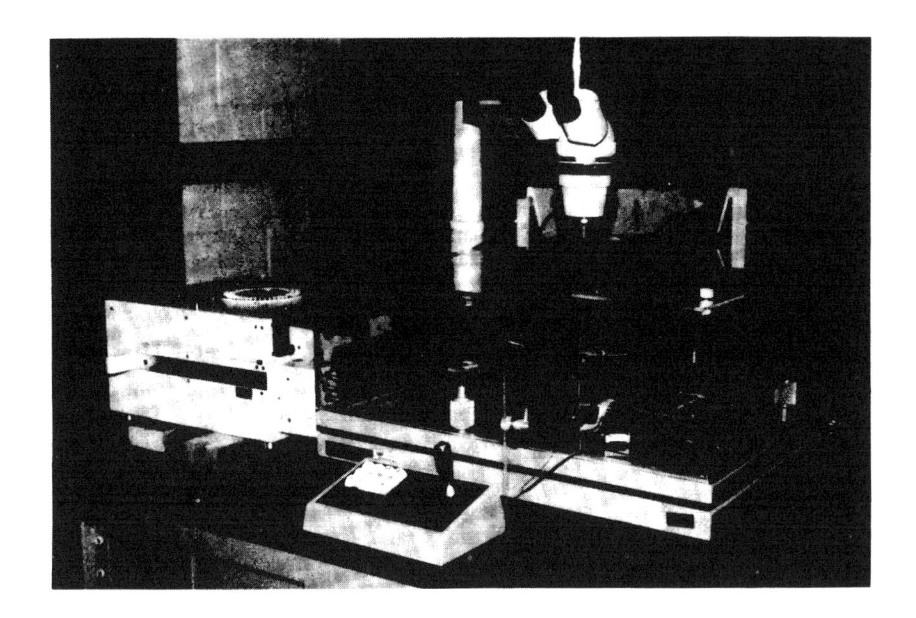

Figure 6 - Test instrumentation for measuring clock distribution for WSI and MCM systems.

V. Acknowledgements

The authors are grateful to the Defense Advanced Research Projects Agency (DARPA) which provided partial funding for this work under contract No. MDA 972-88-J-1006.

VI. References

[1] D.C. Keezer, "Design and Verification of Clock Distribution in VLSI", IEEE Intl. Conf. on Communications, ICC'90, vol.3, pp.811-816, April 1990.
[2] D.C. Keezer, V.K. Jain, "Clock Distribution Strategies for WSI: A Critical Survey," Proc. of the Intl. Conf. on Wafer Scale Integration Conf., pp.277-283, 1991.
[3] D.C. Keezer, V.K. Jain, "Neural Network Clock Distribution," Proc. of the IFIP Workshop on Silicon Architectures for Neural Nets, Nov. 1990.
[4] M. Dikaiakos and K. Steiglitz, "Tree vs. Pipelined Clocking for Long Systolic Arrays," Proc. Intl. Conf. on Acoustics Speech and Signal Processing, May 1991.
[5] H.B. Bakaglu, *Circuits, Interconnections, and Packaging for VLSI*, pp.338-393, Reading, MA: Addison-Wesley, 1990.
[6] C.M. Lee, H. Soukup, "An algorithm for CMOS timing and area optimization," IEEE J. of Solid-State Circuits, VOL.19, NO.5, Oct. 1984, pp.781-787.
[7] C. Mead, L. Conway, *Introduction to VLSI System*. Reading, MA: Addison-Wesley, 1980.
[8] A.M. Mohsen, C.A. Mead, "Delay-time optimization for driving and sensing of signals on high-capacitance paths of VLSI systems," IEEE Trans. on Electron Devices, ED-26, 1979, pp.540-548.
[9] B.A. Richman, J.E. Hansen, K. Cameron, "A deterministic algorithm for automatic CMOS transistor sizing," IEEE J. of Solid-State Circuits, VOL. 23, NO. 2, April 1988, pp.522-526.
[10] J. Shyu, A. Sangiovanni-Vincentelli, J. Fishburn, A. Dunlop, "Optimization-based transistor sizing," IEEE J. of Solid-State Circuits, VOL. 23, NO. 2, April 1988, pp.400-409.
[11] D.L.Carter, D.F. Guise "Effects of Interconnections on Submicron Chip Performance," VLSI Design, Jan. 1984, pp.63-68.
[12] D.L. Carter, D.F. Guise, "Analysis of Signal Propagation Delays and Chip-Level Performance Due to Chip Interconnections," International Conference on Computer Design: VLSI in Computers, Port Chester, NY, Nov. 1983.
[13] IEEE Workshop on Interconnections Within High-Speed Digital Systems, Santa Fe New Mexico, May 22-24 1991.

OPTICAL DISTRIBUTION OF CLOCK SIGNALS IN WAFER SCALE DIGITAL CIRCUITS*

H. U. Chou and M. A. Franklin

Computer and Communications Research Center
Washington University
St. Louis, Missouri 63130

ABSTRACT

In present large VLSI circuits a single global clock source is typically used and the clock signal is distributed from a central point to many different points on the chip via metal lines. Due to variations in signal line delays, clock skew may occur. This clock skew will increase as technology moves towards wafer scale integration. One possible solution to this problem is to distribute the clock optically. A laser can be used to provide light pulses which are distributed to detectors by means of an optical fiber network. The detectors can be spread uniformly across the wafer area. This paper considers the overall design of such an optical clock distribution system, and in particular the alternatives available for on-chip detector fabrication. Photoresistor, photodiode and phototransistor based detectors are compared in terms of the light power requirements and transient response.

1. INTRODUCTION

In recent years, a good deal of interest and research has been focussed on the area of wafer scale integration. This is a continuation of the work in VLSI (Very Large Scale Integration) that has dominated MOS semiconductor research over the last dozen or so years. VLSI technology has decreased the channel length of MOS transistors to less than a micrometer, allowing large numbers of transistors to reside on a single chip. Current VLSI chips may contain several hundred thousands of transistors on a square centimeter, with several megabit memory chips soon becoming available. Wafer scale integration (WSI) advances this technology one step further, permitting integration of an entire digital system on a single silicon wafer[1]. Although the use of wafer scale integration will open the way for design of high performance systems, new problems will also be introduced. With the miniaturization of devices and the increase in the number of devices per system, communications delay, as opposed to device switching delay, can become a performance bottleneck[2]. To take advantage of faster device speeds it is therefore important to find ways to keep communications delays reasonably small (i.e., of the same order as device delays).

One important delay problem relates to the distribution of the clocking signal. In most synchronous systems, a clock signal is centrally generated and then distributed by electrical lines to various areas of the circuit. One delay parameter (clock skew) related to clock distribution is the variance in the times clock pulses arrive at different physical locations in a given system. While ideally clock pulses should appear simultaneously throughout the system, their arrival times instead vary across the system. Though clock skew can be reduced through proper design techniques such as the use of equal length clock lines and optimized buffer design[3], it can still be a problem as systems grow in space.

Consider now an approach to reducing clock skew. Begin by subdividing the system into small area subsections, where the physical distance across each subsection is limited so that there is negligible clock delay (and hence skew) within each block[4]. If the various subsections can be synchronized by broadcasting an ideal single global clock (i.e., zero clock skew clock signals) to each, then the clock skew problem will be eliminated. One possible method to distribute the clock signal in a nearly ideal manner utilizes optical communication. In traveling a short distance in a single medium, light moves at a constant speed. Unlike electrical signals traveling in metal, there is no dependence on any resistive or capacitive effects.[5]

One optical clocking approach is to broadcast a single pulsed light source to the entire wafer. Within the wafer optical detectors would be present with a single detector placed near the center of each of the subsections discussed above. Present laser and LED technologies provide high powered semiconductor sources that can switch on and off in less than a few picoseconds with power outputs of over 10 milliwatts[6]. If light pulses were broadcast from such a source to the optical detectors, the clock skew between these detectors would be significantly smaller than if metal clock lines were used. The only delay dependence would be the distance between the light source and the light detectors and the detector response. If distance differences are under a few millimeters then free space delays of under 10 picoseconds are possible. In systems with a cycle time of a few nanoseconds, the clock skew associated with this delay would be negligible. Thus, each subsection would receive a nearly ideal clock signal.

2. LIGHTING IMPLEMENTATIONS

The design of an optical clocking system such as discussed above requires the presence of a dependable and power efficient lighting scheme. Goodman presented

* This research was sponsored in part by funding from NSF Grant DCR-8417709 and ONR Contract N00014-8D-C-0761.

Reprinted from *IEEE Int'l Conf. Computer Design*, pp. 117–122, Oct. 1987.

several lighting implementation approaches appropriate to this clocking application[2]. These included holograms, integrated waveguides, fiber optics and free air light broadcasting. Hologram and integrated waveguide technologies are still in the experimental stages of development and will not be discussed in this paper. In the fiber approach, a single laser directs a stream of light pulses to the detectors with separate single fibers of equal length going from the source to each detector. The free air, light broadcasting or "flooding" uses appropriate optics to spread the beam across the wafer using air as the waveguide to send the light pulses to each detector. Such an approach eliminates the need for a tricky fiber interconnection scheme. The "flooding" approach, however, requires a total light output far in excess of normal light source outputs since the entire surface of the wafer must be illuminated with a high intensity light even though only the small area covered by the detectors need the light. Due to this, the power requirements of such a system are impractical. Therefore, in this paper the approach considered uses fiber optics or a similar directed waveguide approach. The general structure of this system is illustrated in Figure 1.

A single laser will be current modulated to distribute the light through many equal length glass fibers, each terminating at a light detector at the surface of the MOS circuit. Glass fibers are chosen because the high quality of the present fiber technologies, and the efficient, straightforward implementation scheme that fibers promise. Current glass fibers have attenuations under 1 dB/km and core diameters of less than 10 micrometers, capable of transmitting data rates of over a gigabyte per second over spans of several kilometers without a repeater[5].

3. COMPARISON OF POSSIBLE PHOTODETECTORS

Given a lighting scheme, the next step in designing an optical clock distribution scheme concerns selecting and designing an on-chip semiconductor photodetector which can switch between high and low logic signals at high speeds in response to relatively low incident light power. The most common high speed detector devices today are photodiodes, photoresistors, and phototransistors. In the remainder of this paper each of these devices is analyzed and then compared. Initially the devices are analyzed under steady state low logic output conditions in a CMOS p-well technology. This allows one to calculate the minimum magnitude of light power that is necessary to drive the devices and yields a base requirement for the incident power needed from the light sources. The devices used in this analysis of this section are based on a 1 micrometer λ fabrication dimension with a five volt power supply. The value of λ represents ½ the minimum line width allowed in a typical technology. Later in this paper a detector transient analysis is performed and some conclusions about the most appropriate detector device for this application presented.

For the steady state analysis, it is assumed that the light comes from a single point source with a constant wavelength of 850 nm. This wavelength has proven to be sensitive for silicon devices and is also readily available from various light emitting diodes and lasers[7]. The light source is assumed to illuminate the various detector areas with the same intensity light. Silicon dioxide windows are layered over the detectors to allow a better interface to the silicon.

The three devices, the photodiode, the photoresistor and the phototransistor are all analyzed following the same approach with the devices embedded in the same overall circuit. The load conditions for the photodetectors are determined by the circuit in which the device is implemented. For the CMOS case described this circuit could be designed using an enhancement mode PMOS load device. Each circuit drives a CMOS inverter which acts as a capacitive load to the circuit (see Figure 2). In all cases the gate of the load device is connected to ground, and the device is driven to saturation. The load currents are therefore fairly constant. The operation of the circuits under steady state conditions is fairly straightforward. The value of load current under the steady state conditions is easily calculated using the standard MOSFET square law equation[8]. If a minimum sized load transistor with a W/L ratio of 1 is used, a load current of 76.8 µA is found for a power voltage of 5 volts and a threshold voltage of 1 volt. This value of load current is used as the basis for the steady state analysis. That is, 76.8 µA is the amount of current that the conducting (illuminated) photodetector is required to sink. Note that more than this current will be necessary in the non-steady state situation when charging the capacitance of the following CMOS gate.

Details of the comparative steady state analysis may be found in reference 9. From the known characteristics of each component type (i.e., photodiode, photoresistor and phototransistor) and their associated photocurrent equations the analysis yields the results of Table 1.

Table 1: Steady State Power Requirements

Device	Power Density (mW/cm^2) for 100 x 100 µm device area	Area (μm^2) for 500 mW/cm^2 power density
Photodiode	13400	268000
Photoresistor	1780	*
Phototransistor		
(β** = 14)	270	5390
(β = 140)	27	539

* requires 1780 mW/cm^2, independent of area.

** β is the transistor current gain.

From the table above, one sees that the power requirements between the various devices differ significantly. The phototransistor requires light densities much less than the photodiode due to its gain characteristics. The photoresistor lies midway between the other two for light density. To further compare the devices, transient characteristics of the devices are obtained.

For the photoresistor, the transient response depends on the recombination times of the material. This time specifies the time required for the light energized carriers to return to their expected lower energy positions (i.e. to recombine within the semiconductor lattice). In typical silicon, the recombination times, τ_n and τ_p, are approximately a few microseconds. For fast response times, the carrier lifetimes have to be sharply reduced by forcing in recombination traps, greatly increasing the necessary incident light requirements. For a gate to switch in the one nanosecond range, one would want a lifetime on the order of 100 picoseconds. For a τ_n and τ_p of this size, $(PW)/L = 17.8$ kW/cm^2 is required. This does not take into account the effect of the traps on the carrier mobilities. For carrier lifetimes of 10 to 100 picoseconds, the mobilities can fall to 5 or 10 $cm^2/(V\ sec)$. These numbers represent mobilities of approximately 1/100 of normal values.[10] Therefore, the power density would have to be increased by a factor of 100 to $P = 1.78\ MW/cm^2$. For a 100 by 100 μm detector, this corresponds to a total power of 178 watts instead of 178 μW previously calculated. Such a high light density as well as total power would be highly impractical. Thus, the photoresistor is eliminated from our choices of possible photodetectors.

The two remaining devices are both CMOS junction photodevices, which benefit because recombination-generation (R-G) delays have very little effect on transient responses. The reason the R-G delays have so little effect on the junction devices is that generation occurs in the depletion region, and recombination occurs in the bulk material. When the light generates the carriers in the depletion region, the electric field due to the reverse biasing sweeps the carriers out of the depletion to the bulk areas where they slowly recombine. When the light turns off, the recombination continues slowly in the bulk, but the depletion area is free of carriers, so the current quickly dies. The only current flowing across the junction is the reverse saturation current which is increased due to the injected carriers in the bulk materials, but not so much as to change the voltage level across the diode. Due to the moderate electric fields across the depletion areas, the transit times are also quite small. Therefore, the transient response of the devices is due primarily to delays needed to charge and discharge the various system capacitances.

For the phototransistor, the high to low delay comes not only from the collector current discharging the load capacitance, but also from the photocurrent discharging the base-collector (B-C) capacitance. If one looks at the discharging of the B-C capacitance, one sees that the photocurrent must discharge this capacitance itself. This B-C capacitance charging time tends to be the limiting performance factor in most phototransistors[10], with the B-C capacitance being a value of a few femtofarads. Given a collector (load) current of 76.8 μA, the incident power requirements in Table 1 were decreased by using a higher β to reduce the necessary photocurrents. Unfortunately with the small photocurrents into the base calculated previously, calculations indicate[8] that it would take the phototransistor 10 to 100 nanoseconds to discharge the initial B-C capacitance. Therefore, the lower power requirements shown in Table 1 for phototransistors correspond to slower devices as β increases. The transient response ultimately limits the amount of light power reduction possible through current gain to the circuit. Therefore it appears that a high gain phototransistor, while decreasing power requirements, would not be fast enough to respond. A lower gain phototransistor or a photodiode would therefore be desired in an optical clocking system. For simplicity of analysis, the photodiode is considered for further analysis. A low gain phototransistor could be analyzed in a similar manner with a factor of β added to the photocurrent equation.

4. ITERATIVE PHOTODIODE ANALYSIS

A more detailed transient analysis of the photodiode circuit is given in this section. To obtain actual values of photodiode area and power density required, the circuit of Figure 2 was simulated on a digital computer. Using an iterative $i(t)=c(dv(t)/dt)$ approach, the functions of the currents, voltages, and capacitances versus time were evaluated for various input parameters[9]. This is described in more detail later in this section. A three phase algorithm was employed to simulate the circuit. The first phase of the algorithm calculates the maximum photodiode area, A_{max}, (and hence the photodiode capacitance) that I_{DS} can charge for a specified rise time. Given this maximum area as an upper bound, the second phase of the algorithm uses a binary search to find the optimum photodiode area, A_{opt}, that provides a fall time equal to the rise time. On the first pass of this algorithm, an estimated photocurrent is initially used. Using A_{opt}, the third phase of the algorithm calculates the optimum light power density (and hence photocurrent) required to discharge the capacitances during the fall time. The last two phases are then repeated, optimizing both detector area and power density to achieve a minimum area-density product for a given rise and fall time.

To use the algorithm, the currents of the circuit must be known at every time point. These currents include the FET current, I_{DS}, the generated photocurrent, I_P, and the reverse bias leakage current, I_O. The currents are determined using standard circuit and device equations.[8] In the non-illuminated case, the total current, I_T, is the FET current since the photocurrent is normally zero and the leakage current is minimal. In the illuminated case, however, the total current is the the FET current minus the leakage current and photocurrent.

The FET current is specified by the distributed charge FET current equation[8] using both the linear and saturation modes of the transistor. With the FET current clearly defined, the photocurrent is next specified. When the light pulses on, the light intensity is assumed to be constant, and the pulse is assumed to be perfectly square. The photocurrent flowing through the n-p junction is defined as described in the standard photojunction equation[7].

Although the FET current (I_{DS}) and the photoelectric current (I_P) are the primary currents present in the analysis other currents also need to be analyzed. The FET has a small leakage current through the diffusion region, but since the device is always on, the drain to source current dominates there and this current is not considered. The reverse biased photodiode has a reverse saturation current (I_O), which is normally insignificant, but since carriers remain in high doped materials outside of the depletion region (where slow recombination times exist), the current needs to be considered. The carriers are both generated in these regions and are swept out from the depletion region due to the electric field. The recombination lifetime in typical silicon ranges from 0.1 to 1.0 microseconds[11], so carriers generated in the p-well and the n diffusion region remain active long after a light pulse is turned off. These carriers increase the number of minority carriers present in the device, greatly increasing the diffusion current that flows from the n to p regions. Since the distribution of the free carriers in the diffusion and p-well cannot be precisely defined, a uniform distribution was assumed to calculate the approximate minority densities in both regions as the number of generated carriers in the region divided by the the total volume of the region.

Given that the currents have been obtained, we next consider evaluating the capacitances needed for the algorithm. The main components of the systems capacitance come from the load capacitance, the source diffusion capacitance of the FET, and the junction capacitance of the reverse-biased photodiode. The load capacitance can be approximated by the oxide capacitance of a 2 x 4 λ NMOS gate plus that of a 2 x 4 λ PMOS gate. This assumes that the FET and photodiode is driving a typical CMOS inverter. Thus the load capacitance is defined as simply $C_L = 16 C_{ox} \lambda^2$, if diffusion overlap and sidewall capacitances are ignored. The diffusion and device capacitances consist of a junction width capacitance which can be calculated using standard bipolar techniques.

Results for λ ranging from 0.25 to 1.5 μm are given in Figures 3 and 4. Figure 3 demonstrates the relationship between maximum photodiode area and lambda for three rise times of 1, 2, and 5 nanoseconds. Figure 4 shows how the minimum incident device light power depends on lambda. The power voltage, V_{DD}, was assumed to be 5 V for λ greater or equal to a micrometer, 2 V for 0.5 μm and 1 V for 0.25 μm. Likewise the doping densities are assumed to decrease as λ decreases to decrease the electrostatic and built-in voltages, ϕ_p and V_{bi}, to match the power voltages. Despite the scaling, however, the FET current values remain fairly constant.

From Figure 3 one sees that the maximum photodiode area requirement increases for decreasing λ for the rise time case. This can be understood by viewing the total capacitance driven by the load FET as being the sum of the detector capacitance (C_1) plus the following CMOS gate capacitance (C_2). As λ and gate thickness decrease, C_2 decreases. As V_{DD} decreases, I_{DS} decreases thus effectively maintaining a constant driving resistance, R. Thus for a constant delay, $R\,(C_1 + C_2)$, as λ and hence C_2 decrease, C_1 increases so the detector area increases.

From Figure 4 one sees that as lambda decreases the total light power requirement decreases dramatically. This rapid decrease occurs independently of switching delay. This is due to the decreasing gate capacitances, power voltages, and doping densities. The device light power varied only slightly for varying delay values of 1 to 5 nanoseconds, so only the average line is included in Figure 4.

5. DESIGN OF AN OPTICALLY CLOCKED SYSTEM

To further verify the model predictions found above, a small photodiode circuit (Figure 5) was designed using the MAGIC layout system.[12] The circuit included a RESET pulldown from the output of the diode to ground to reset the node low after a short time delay from the output inverter. Two CMOS inverters were used to buffer the output of the photodiode and provide higher capacitive drive capabilities. The final dimensions of the circuit are 52.5 micrometers wide by 135 micrometers long. The photodiode cell by itself is 36 by 48 micrometers, a small fraction of the total area. The photodiode area was roughly determined from the analysis described in section 4. For further details on the design of the circuit see reference 9.

To determine the circuit current, one must first know the power output of the laser which drives the detectors. The necessary power per device was projected for various delay times. This value times the number of detectors in the system decides the total output necessary from the laser. Current GaAs lasers have been reported with outputs as high as 400 milliwatts for coupled multiple stripe phased array lasers, but outputs of 10 to 40 milliwatts are more common[6]. For the example given in this paper, a maximum output power is desired, so the 400 milliwatts for the array laser is assumed.

Given a 400 milliwatt laser directing light towards 16 detectors, and assuming a 3 dB loss in fiber connections, incident light power per device is 12 milliwatts. Such a system could clock 16 different subsections of the wafer. If each of these regions represents a one square centimeter area, the maximum distance between detectors centered in the regions would be one centimeter. The RC delay of a four micrometer by one centimeter metal line would be slightly less than one nanosecond. This is for a single line. Driving a clock tree over the 1 cm^2 area could have a delay which is a factor of 10 higher. In this context the optical delays are insignificant. Note also that tree delay grows as $O(L^3)$ where L is the length of a side of the square area.[3] If the 4 cm by 4 cm area was clocked by a single clock tree, the tree delay would be about 64 times greater than for the 1 cm^2 area.

The circuit was extracted from its MAGIC file and simulated on SPICE to find transient curves. The results from SPICE indicated rises and falls at (a), (b), and (c) in Figure 5 to all be under 2 nanoseconds. These transient times are all acceptable considering that the current speed of MOS gates are of the same magnitude.

6. CONCLUSION

From the analysis of the previous section one sees that a laser pulsed fiber system can be used to clock digital MOS circuit at frequencies near 100 MHz. Such a system requires a relatively small area of silicon for each optical to digital cell.

One large obstacle in present technologies stems from the low efficiencies of the detectors. When integrated in silicon with doping densities as required for CMOS gates, silicon photodetectors tend to be narrow, and therefore do not absorb the majority of the incident light for current generation. The incident photons instead penetrate deep into the substrate generating current to be removed by substrate contacts. Another major problem is the low power outputs of current lighting sources. The GaAs light sources have limited output power (typically < 1 watt) due to high input current requirements, low reliability at high power, and temperature instability.

More research will have to be performed to verify the models and concepts presented in this paper. Two photodetector circuits have been designed and fabricated by MOSIS for such a purpose, but they have yet to be tested using high power light sources.

7. BIBLIOGRAPHY

[1] McDonald, J. F., E. H. Rogers, K. Rose, A. Steckl, "The Trials of Wafer-Scale Integration," *IEEE Spectrum,* Vol 21, No. 10, pp. 32-39, October 1984.

[2] Goodman, J. W., F. I. Leonberger, S. Y. Kung, R. A. Athale, "Optical Interconnections for VLSI Systems," *Proceedings of the IEEE,* Vol. 72, No. 7, pp. 850-866, July 1984.

[3] Dhar, S., M. A. Franklin, D. F. Wann, "Reduction of Clock Delays in VLSI Structures," *Proceedings of the International Conference on Computer Design,* Port Chester, New York, 1984.

[4] Dhar, S., *Performance Analysis of Regular VLSI Arrays,* M. S. Thesis, Department of Electrical Engineering, Washington University, St. Louis, December 1984.

[5] Guterl, F., G. Zorpette, "Fiber Optics: Poised to Displace Satellites," *IEEE Spectrum,* Vol. 22, No. 8, pp. 30-37, August 1985.

[6] Newman, D. H., "GaAs Optoelectronic Devices," in *Gallium Arsenide: Materials, Devices and Circuits,* ed. M. J. Howes, D. V. Morgan, pp. 429-451, John Wiley & Sons, New York, 1985.

[7] Deboo. G. J., C. N. Burrous, *Integrated Circuits and Semiconductor Devices: Theory and Application,* McGraw-Hill, New York, 1971.

[8] Muller. R. S., T. I. Kamins, *Device Electronics for Integrated Circuits,* John Wiley & Sons, New York, 1977.

[9] Chou, H. U., *Optical Distribution of Clock Signals in Wafer Scale Digital Circuits,* M. S. Thesis, Department of Electrical Engineering, Washington University, St. Louis, Missouri, December 1986.

[10] Mead, C., "A Sensitive Electronic Photoreceptor," *Chapel Hill Conference on Very Large Scale Integration,* ed. Henry Fuchs, pp. 463-471, Computer Science Press, Inc., Rockville, Maryland, 1985.

[11] Pierret, R. F., *Modular Series on Solid State Devices: Semiconductor Fundamentals, Volume I* Addison-Wesley Publishing, Reading, Massachusetts, 1983.

[12] Scott, W. S., R. N. Mayo, G. Hamachi, J. K. Ousterhout, *1986 VLSI Tools: Still More Works by the Original Artists,* Report No. UCB/CSD 86/272, Computer Science Division (EECS), University of California, Berkeley, California, December 1985.

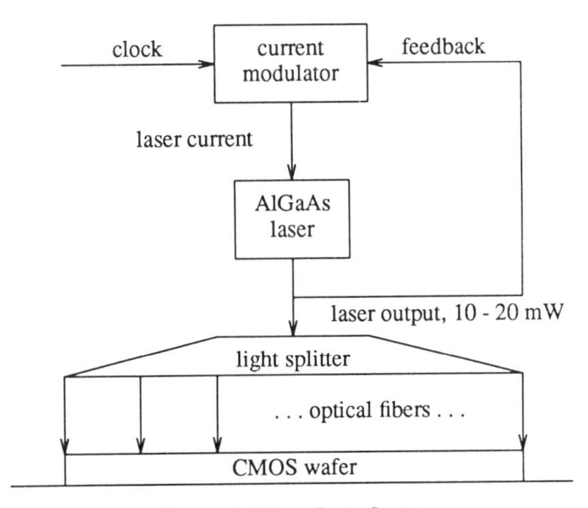

Figure 1: Fiber Optic System

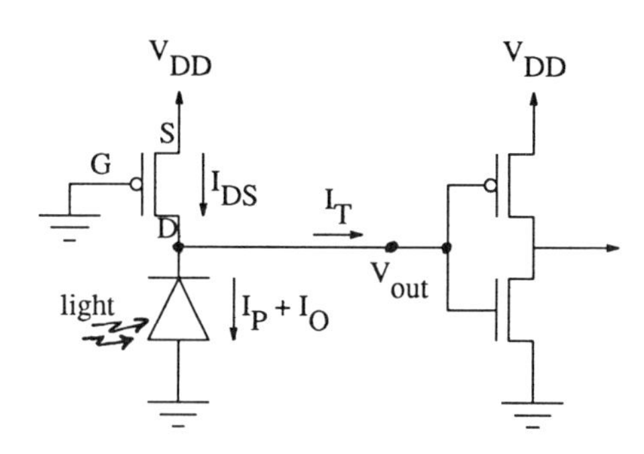

Figure 2: Currents and Voltages of Photodiode Circuit

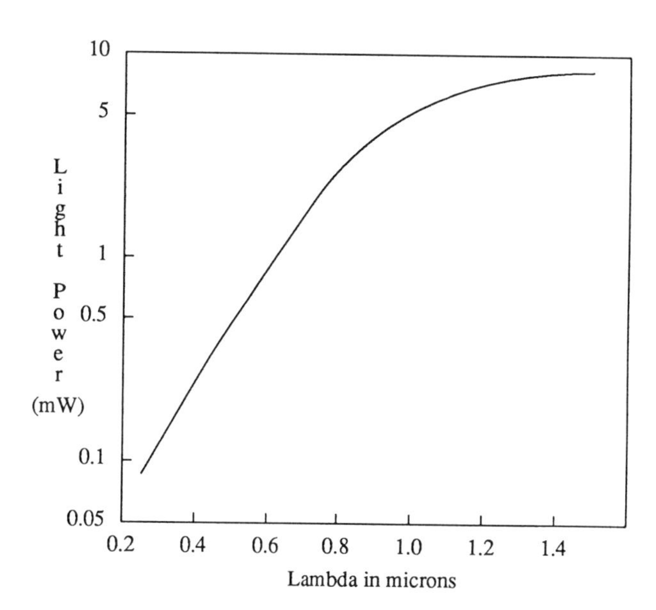

Figure 4: Optimum Light Power for One Photdiode versus Lambda
(W/L = 1/8, average over all fall times = 1.0, 2.0 and 5.0 ns)

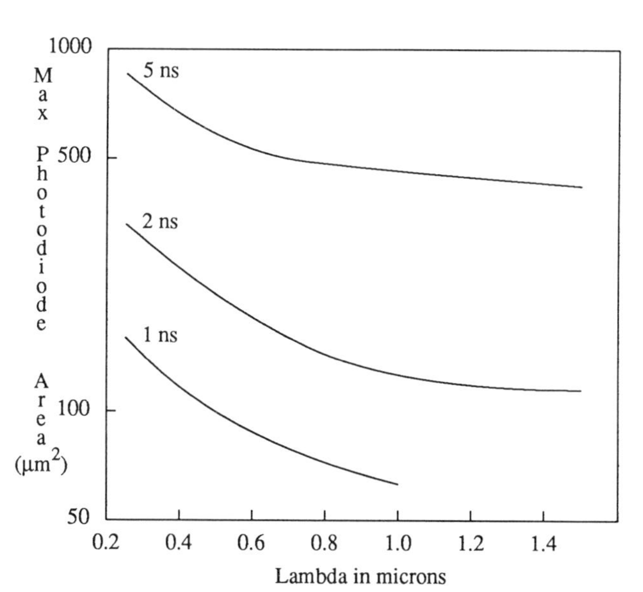

Figure 3: Maximum Allowable Photodiode Area versus Lambda
(W/L = 1/8, rise times = 1.0, 2.0, and 5.0 ns)

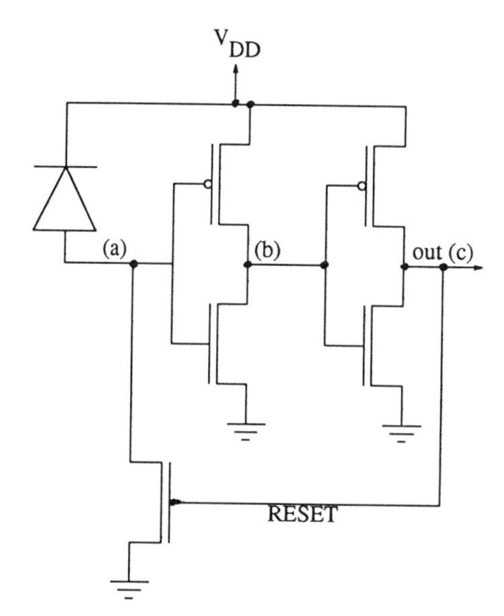

Figure 5: Circuit Diagram for Fiber Optic System

Part 7

Example Implementations of Clock Distribution Networks

Electrical Design of BELLMAC-32A Microprocessor

M. Shoji

Bell Laboratories
Murray Hill, New Jersey 07974

ABSTRACT

BELLMAC-32A chip was designed by using a layout scheme that minimizes the power bus noise and by using a distributed clock circuit that guaranteed synchronization of the entire chip within ± 3.5 NS. The design also obtains high speed by logic optimization and FET size scaling. The CPU chip has operated at the predicted frequency (8 MHz) reliably, without any noise problems.

INTRODUCTION

BELLMAC-32A is the Bell System's 32 bit CMOS VLSI microprocessor which has many advanced features.[(1)] The die area of 99 mm^2 contains 146K FET sites. Despite the complexity the chip must meet stringent noise specifications and must run at an 8 MHz clock frequency. The chip was developed using a Top-Down design method.[(2)] The circuit design and verification techniques used were so effective that all noise and frequency objectives were met in the first mask set. This paper describes the techniques used and the initial results.

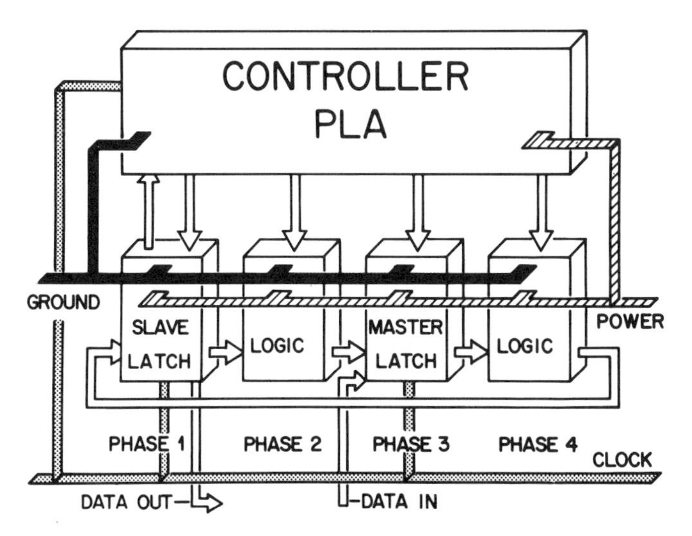

Fig. 1 Functional block diagram of BELLMAC-32A

Figure 1 shows the functional block diagram. A clock period consists of phases 1, 2, 3 and 4. The slave latches, the slave latch logic, the master latches and the master latch logic work during these phases respectively. The PLA controller generates all the control signals. The design objectives are to distribute skew-free clock signals and noise free power voltages over the entire chip, and to obtain the highest possible clock rate by speeding up logic circuits.

CLOCK DISTRIBUTION

In our 2.5 micron CMOS technology one level of metal interconnect is available, so it must be used for power distribution. When a clock line crosses the power bus, it must do so on silicide. This creates series resistance and delay. The clock distribution strategy must insure that any delay due to series resistance is equal for all clocks. To achieve this it is necessary to keep the number of clocks which are widely distributed down to a low number. In this chip two primary clocks were used.

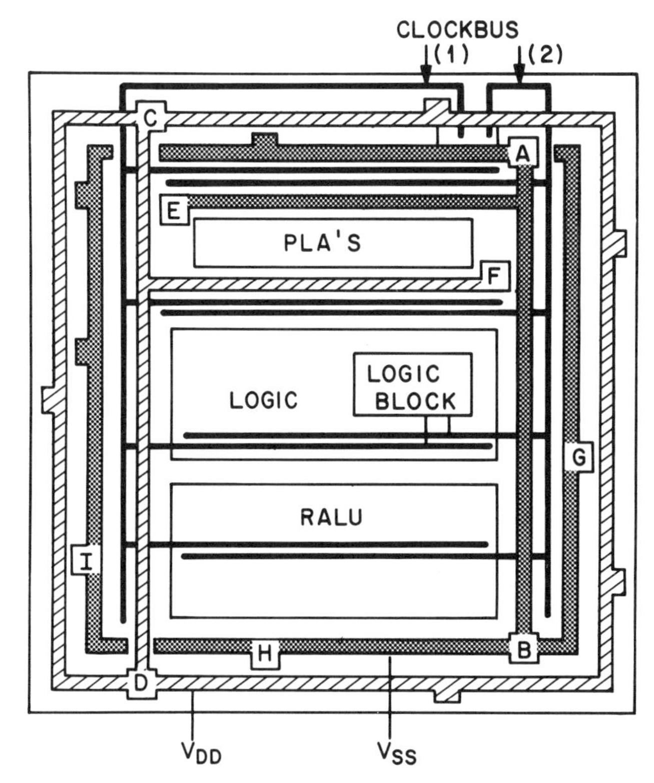

Fig. 2 Clockbus and powerbus structure of BELLMAC-32A

In Fig. 2 the clock buses originate from the upper right-hand side and cross over the V_{DD} bus twice. Then they cross over the V_{SS} or the V_{DD} bus and come into the area where the circuits to be clocked exist. To reach any clock load there are always three power bus crossovers. Therefore the internal clock edges are delayed everywhere by the same amount. Buffers were provided at crossovers to minimize this equalized delay. All the clock signals which actually clock circuits (local clocks) were created from the two primary clock signals.

All local clocks are designed to give a delay of 15 NS ± 3.5 NS

Reprinted from *Proc. IEEE Int'l Conf. Circuits and Computers,* pp. 112–115, Sept. 1982.

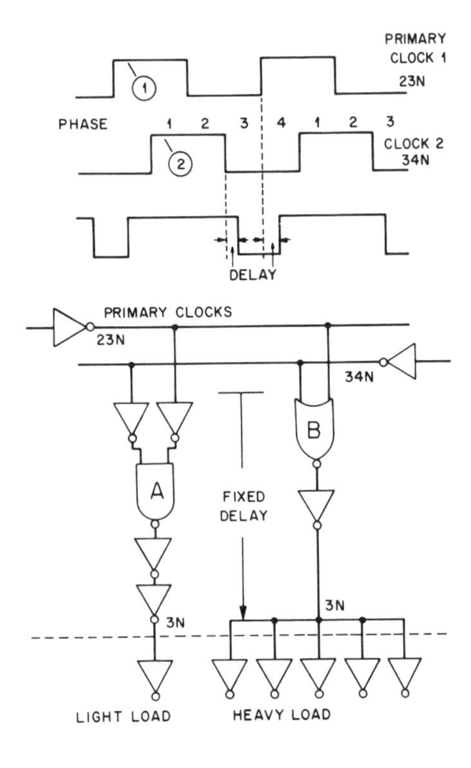

Fig. 3 Local clock decoding scheme

with respect to the primary clock. Figure 3 gives an example of how such a local clock is created. The primary clocks are combined by a NOR gate to create a local phase 3 clock. The circuits and FET sizes were adjusted to attain the 15 NS delay. For a small clockload circuit I is used. For a large clockload circuit II (with large FET's) is used.

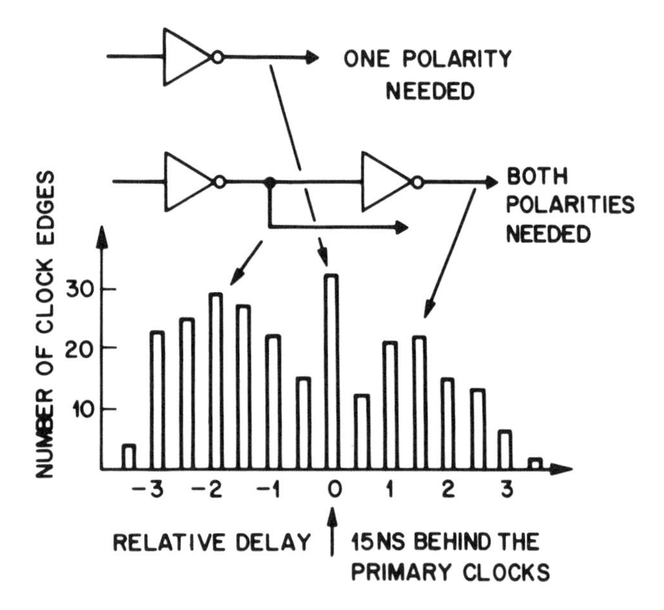

Fig. 4 Histogram of clock skew

The end result of the local clock deskewing is shown in Fig. 4 which is a histogram of delay for local clock edges based on simulations.[3] This histogram has three population peaks. The central peak is for the clocks that are needed only in one polarity. The two broad peaks are for clocks that are required for both polarities. The difference of the peak positions equals the delay of the extra buffer to create a second polarity clock from the first polarity. All clock edges are within 15 NS ± 3.5 NS from the primary clock edges as planned.

POWER BUS NOISE

As CMOS circuits switch, a spike of current flows in the power bus. Because of the resistance and the inductance of the power bus, the current spike develops noise voltage. Figure 2 shows the final structure of the power bus. Both ends of the vertical power bus (100 microns wide) have bonding pads. PLA's are served by a pair of 100 micron wide horizontal power buses. The buses end at internal bonding pads. The technique of placing pads at both ends cuts the power bus length in half. This reduces the power bus noise by a factor of 4, because both current flow and impedance are halved. The detaiis of the power bus noise were determined by simulations[3] which included the inductance of the power bus.[4]

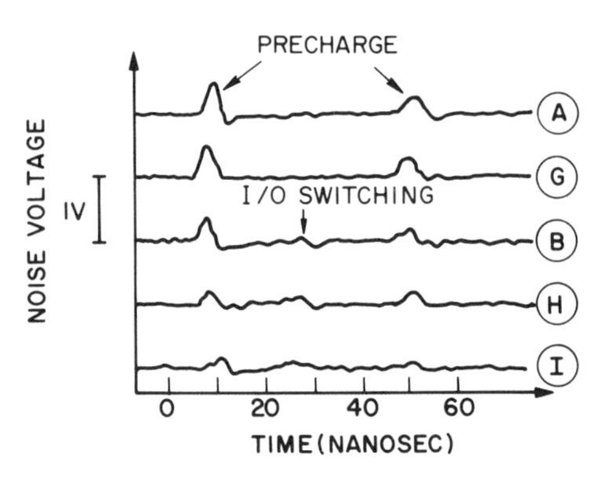

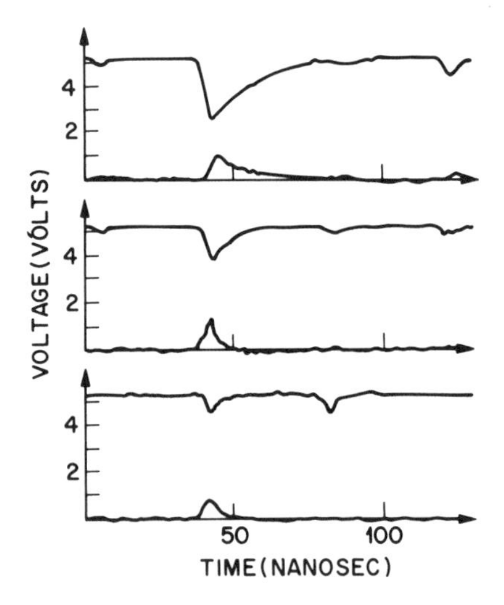

Fig. 5 (a) Ground bus noise at several I/O frame locations

(b) Noise on the power bus supplying the PLA's

Figure 5(a) shows the simulation results of the ground bus noise at many points on I/O frame (refer Fig. 2). Noise due to PLA precharge is the highest at location A, and lowest at location I. The noise due to I/O switching is highest at location I and unobservable at location A.

Interim results obtained during the design process leading to the final placement of power bus are traced in Fig. 5(b). This shows the noise on the horizontal power bus supplying the PLA's for two interim designs and for the final design. At first there were no pads terminating the buses (top figure). The noise voltage was over 2 volts. When the internal pads were added noise improved, but was still unacceptable (middle figure). Finally the precharge timing of the ROM part of the PLA was shifted by 1 phase so that it was no longer coincident with the decoder precharge. As the bottom figure shows, this split the noise pulse into two parts, both of which were acceptably small.

LOGIC SPEEDUP

The Top-Down design method requires partitioning of logic into blocks. After each logic block is laid out[5] they are wired together using LTX routing software.[6] The timing design was integrated into the block partitioning and design from the beginning. Timing also entered into design of logic and the determination of FET sizes. To carry out the partitioning process we need to maximize block size. Timing considerations enter into the choice. We assume a square logic block of size L (mm). An inverter of minimum size drives a silicide of length L, metal of length L and 3 fanouts. If we determine L such that the signal delay is doubled by the parasitics we have $L = 800$ microns. If the block size is less than 800 microns the parasitic effect is less than a factor of two, and therefore most of the blocks were made less than that limit. For a few larger blocks special design techniques were needed such as the use of multiple clock generators for the same clock signal.

Another way in which timing design was used to help determine partitioning relates to path delay. When logic blocks communicate with each other via the wiring channel, the resistance of silicide crossovers adds to the signal delay. When the wiring channel is long, and loaded by fanouts, signal transmission between the logic blocks can have long delays. When a signal is transferred from block A to B, combined with a second signal, transferred to block C, and further combined with a third signal, the two sequential block to block transfers add further to delay. The delay can be reduced by moving all logic combining the first two signals to block C.

An example of logic changes to improve timing occurs when a signal from block A drives blocks B, C, D and E and not all the blocks will require the signal at the same time. Suppose block B requires the signal immediately but blocks C, D, and E can tolerate a later signal from A. Then the signals to blocks C, D and E can be buffered, thereby reducing fanout and increasing speed.

In addition to the above techniques used to speedup block to block signal transfer, many techniques were used to speedup signal paths within blocks. Some of these have been described elsewhere, including optimization of the ratio of NFT to PFET size[7] and optimization of the FET sizes in cascaded buffer strings.[8]

An important new technique to improve timing was used for the first time on this chip, and will be described in some detail here. The circuit of Fig. 6 is a Domino CMOS 4 input AND gate.[9] This circuit has usually been designed as shown by layout I. Layout II, however, gives shorter delay. Switching of the FET chain of Fig. 6 can be equivalently described by a discharge of an RC chain circuit (low-pass filter) consisting of the resistance of the FET channel and the parasitic capacitances of the FET. If the size

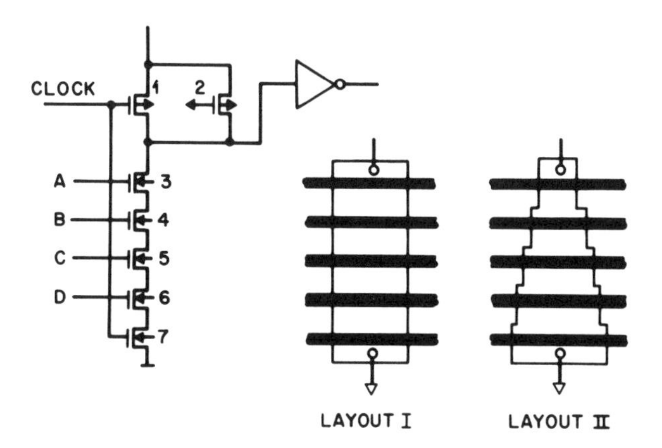

Fig. 6 Domino CMOS circuit and speedup method

of the FET nearest the output node (FET 3) is reduced the last link of the RC chain gains series resistance but loses parallel capacitance. In a long RC chain circuit the decrease of capacitance has more effect than increase in resistance, and thereby the switching delay decreases. This explains why layout II gives shorter delay than layout I. A rigorous quantitative analysis of this effect will be presented elsewhere.[10] Use of this technique to adjust transistor size reduces delay 15-30% depending on the type of gate and loading conditions.

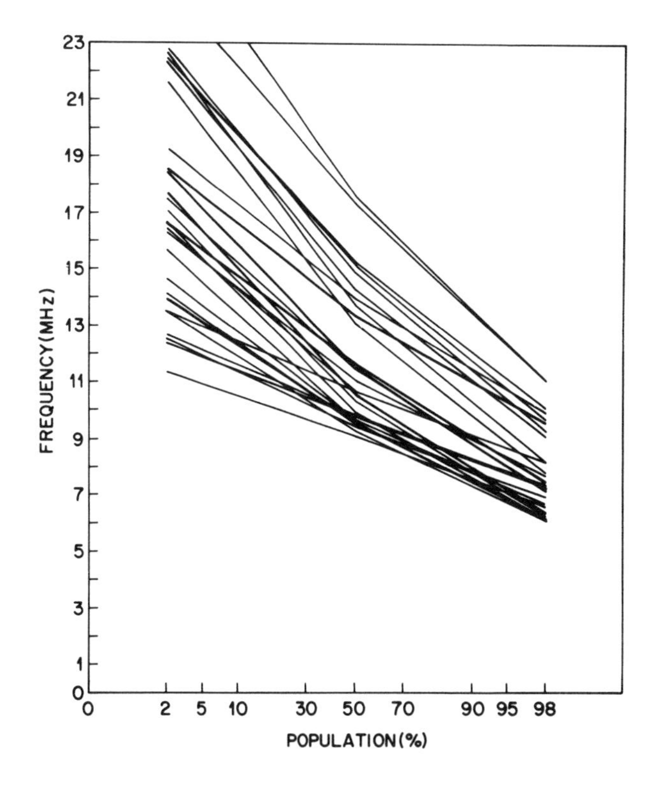

Fig. 7 Parametric yield versus frequency, determined from the critical path studies

The expected yield versus maximum operating frequency relationship was determined by simulations of many critical paths for low, medium and high current processes. The results of these simulations are given in Fig. 7, which predicts that about 50% of the functional chips will work above 8 MHz. Before the timing optimization was done there were hundreds of critical paths that were as much as a factor of two slower than the slowest path in the final design.

RESULTS

When this design was implemented in silicon most BELLMAC-32A chips worked at a frequency above 8 MHz at 70°C as planned. No clock skew in excess of 3.5 NS was measured and no adjustments to logic or FET size was needed for timing improvements. Also there were no noise effects. This was a new achievement in VLSI technology. Expensive speed improvement and noise suppression work after functional verification was unnecessary. With careful circuit design the speed of a chip was improved by a factor of 2, noise was eliminated and clock skew was reduced to tolerable level, before the chip was committed to silicon.

ACKNOWLEDGEMENT

The author wish to thank R. H. Krambeck and B. T. Murphy for their stimulating discussions on the subjects and helpful comments on the manuscript.

REFERENCES

[1] D. E. Blahut, A. K. Goksel, R. H. Krambeck, H. F. S. Law, P M. Lu, W. F. Miller, H. C. So, "The architecture and implementation of a 32 bit microprocessor with minicomputer performance," Compcon, Feb., 1982, San Francisco, Calif.

[2] R. H. Krambeck, D. E. Blahut, S. F. Law, B. W. Colbry, H. C. So and M. L. Harrison, "Top-down design of one-chip 32 bit CPU," VLSI 81, Edinbrough, Scotland, September, 1982.

[3] L. W. Nagel, "Spice 2: A computer program to simulate semiconductor circuits," Memo ERI-M520, Univ. of California, Berkeley, Calif., May, 1975.

[4] H. Hasegawa, M. Furukawa and H. Yanai, "Properties of microstrip line on Si-SiO_2 system," IEEE Trans. on Microwave Theory and Techniques, vol. MTT-19, pp. 869-881, Nov., 1971.

[5] A. D. Lopez and H. F. law, "A dense gate-matrix layout style for MOS LSI," SSCC 80 digest, pp. 212-213, Feb., 1980.

[6] B. W. Colbry and J. Soukup, "Layout aspects of the VLSI microprocessor design," International Symposium on Circuits and Systems, Rome, Italy, May, 1982.

[7] S. M. Kang, "A design of CMOS polycells for LSI circuits," IEEE Trans. on Circuits and Systems, vol. CAS-28, pp. 838-843, Aug., 1981.

[8] C. M. lee and H. Soukup, to be published.

[9] R. H. Krambeck, C. M. lee and H. F. Law, "High speed compact circuits with CMOS," to be published in IEEE J. of Solid State Circuits.

[10] M. Shoji, to be published.

Reliable Chip Design Method in High Performance CMOS VLSI

Masakazu Shoji

AT & T Bell Laboratories
Murray Hill, New Jersey, 07974

Abstract

Reliable, high performance CMOS VLSI's can be designed reproducibly, if care is exercised in the chip and circuit design. Experiences gained in the WE32100 CPU chipset design and reported elsewhere are reviewed and systematized in this paper.

Introduction

In this paper the problems of CMOS VLSI circuit design are summarized and various solutions are reviewed, thereby bringing the process of CMOS VLSI development into perspective.

A high performance CMOS VLSI chip requires (1) a noise-free power bus, (2) a skew-free, hard-driven clock source, and (3) fast logic circuits[1]. Further (4) some idea of chip performance should be known to the designer during the design process.

[1] Power bus

When a CMOS gate switches, charge-discharge current spikes flow through the power bus. By the inductance and the resistance, noise voltages are generated that affect the other circuits. A CMOS gate shown in Fig.1 is equivalent to two switches. When the one closes the other opens. The first(NFET) is in parallel to, and the second(PFET) in series with, the load capacitance(C_L). In CMOS PLA, C_L is between the output node and the ground, as shown in Fig.1. A charge current spike flows through the entire span of the power(V_{DD}, V_{SS}) buses when the series switch closes(α). When the parallel switch closes, however, the capacitor discharges via a loop including the parallel switch and therefore the discharge current closes within a localized loop(β). The noise effects of the local discharge is minimized by the circuit layout that closes the current loop within the gate(connecting nodes Γ to node G, rather than connecting to a convenient location of the ground bus). This technique minimizes noise generation at the evaluation time of the PLA[2].

The internal resistances of FET's that generate the spikes are higher than the impedance of the power bus. The noise voltage can be analyzed using an equivalent circuit consisting of current generators of the FET and the linear power bus model including resistance(R), inductance(L) and capacitance(C). Since the number of current generators and the series impedance(L,R) are both proportional to the power bus length, the worst-case power bus noise is proportional to the square of the length. Shortening the length is very effective for noise control[1].

The power bus consists of (1) the on-chip buses, (2) the bonding wires, and (3) the package wiring. We consider the most essential ones, (1) and (2).

The silicon substrate has conductivity (σ) of the order of $10^2 mho/cm$. The skin depth is $(2/\omega\mu\sigma)^{1/2}$, where $\mu = 1.25\times10^{-8} henry/cm$ and ω is the dominant frequency of the Fourier component of the current spike. The skin depth is $4 - 0.4mm$, if ω is $10^7 - 10^9 radian/sec$. The skin depth is more than the thickness of silicon. The inductance of the on-chip bus must be calculated assuming that the magnetic field penetrates into the silicon substrate, stopped only by the back metal. The inductance formula in this case is given by Hasegawa et al[3].

Let the current spike waveform be $I(t)$. The inductive and the resistive components of the power bus noise are, respectively, $\Delta V_L = LdI(t)/dt$ and $\Delta V_R = RI(t)$. The ratio of the two, $\Delta V_L/\Delta V_R$ is $2t_{LR}/t_S$, where $t_{LR}=(L/R)$ is the inductance-resistance time constant of the power bus and t_S is the width of the current spike. An estimate of $(2t_{LR})$ is made by $(2t_{LR})=10^{-11}W\log(4000/W)$ where W is the power bus width in microns and the dimensional parameters typical for 1.75 micron CMOS were used. We have $2t_{LR}=3.6 nanosec$ for $W=100\mu$ and $2t_{LR}=0.6 nanosec$ for $W=10\mu$. If a typical current spike is 1 nanosec wide, the inductive noise predominates if a power bus is wider than 20μ . Bonding wires have $2t_{LR}$ in the range 10-30 nanosec, and the effects of the inductance predominates.

The current spike $I(t)$ has a bell-shaped waveform of Fig.1: $I(t)=0$ when $t<0$, $I(t)\to 0$ when $t\to\infty$ and $I(t)$ has a maximum when $t=t_{\max}$, in between. The width of the spike is the switching time t_S . The area underneath the curve is the charge drained by the NFET upon switching, $Q=C_L V_{DD}$ where C_L is the load capacitance and V_{DD} is the power supply voltage. The inductive power bus noise is proportional to $|dI(t)/dt|$, or the slope of the curve. A complementary relationship exists between $|dI(t)/dt|_{\max}$ and the switching time t_S [4],

$$\left|\frac{dI(t)}{dt}\right|_{\max} t_S^2 \geq 4Q = 4C_L V_{DD} \qquad (1)$$

To minimize $|dI(t)/dt|_{\max}$ the slopes of $I(t)$ at the leading and at the trailing edges must be designed to the same[4]. The optimum current spike is an equilateral triangle of Fig.1 for which the equal sign holds in Eq.(1). Equation (1) is useful to evaluate output driver noise. An output driver is properly designed if it is within 20-30% of the limit of Eq.(1).

Let the number of simultaneously switching output driver be N_0 , the number of ground connections be N_g , the inductance of each bonding wire be L_b , and the inductance of the chip power bus be L_i . Then V_N , the noise voltage caused by the simultaneous pulldown of N_0 drivers, is estimated using dI/dt of Eq.(1) as

$$\frac{V_N}{V_{DD}} \geq 4\left(\frac{t_{LC}}{t_S}\right)^2 \qquad (2)$$

where $t_{LC}^2=((N_0/N_g)L_b+L_i)C_L$ is the square of the LC time constant and where t_S and C_L are defined in Fig.1. The normalized power bus noise is proportional to the square of the ratio (t_{LC}/t_S) . Time t_{LC} is determined by the packaging. Using a pin grid array package on a PC board t_{LC} is estimated as

Reprinted from *Proc. IEEE Int'l Conf. Computer Design: VLSI in Computers*, pp. 389–392, Oct. 1986.

$t_{LC} = (6 \times 2(nanohenry) \times 100(picofarad))^{1/2} = 1\ nanosec$
$V_N = 240millivolt$ if $t_S = 10nanosec$, which is less than the FET threshold voltage. If this is the slowest case the fastest case would be $t_S = 4\,nanosec$. Then $V_N = 1.51volts$, which is more than the FET threshold voltage. It is therefore difficult to design the output driver, simultaneously meeting the delay specification at the slowest conditions and the noise specification at the fastest conditions. If a conventional package like a pin grid array or a chipcarrier is used, t_{LC} in Eq.(2) stays constant, as the CMOS technology scales down. Then the power bus noise would be a very serious problem in a VHSIC VLSI circuit. An advanced flipchip package will reduce t_{LC} by an order of magnitude, thereby providing a long-term solution[5]. Techniques to ease the output driver noise, like multiple V_{SS}/V_{DD} bondings, multiple bonding wires, separating power bus of the outdrivers from the internal power bus and avoiding coincidence in I/O timing are already used[1]. The I/O protocol can also provide an effective and immediate solution: when a chip is outputting data, no input of the chip should be strobing data[1].

If the noise voltage is so large that the drain/source junctions conduct, the noise voltage is clamped. But the injection of the minority carriers is undesirable, because it increases the latchup susceptibility and can discharge dynamically held nodes.

[2] Clock bus

A perfectly synchronized clock system does not always provide the best performance. It is impossible, however, to assign a balanced time budget by adjusting the individual clock edges, since the number of clock loads in a VLSI circuit is huge, amounting to 10-15% of the random gates. The best strategy is to design a well-synchronized clock system, and then for a small number of critical paths to improve timing by shifting clock edges[1].

The objective of clock design is twofold: (1) to distribute skew-free primary clock edges all over the VLSI chip, and (2) to design the circuits that interface the primary clock to the destination skew-free.

(1) Clock signals are distributed by metal wires. The signal delay through an unloaded metal line is certainly the lower bound on clock skew. The metal line has capacitive(C), resistive(R), and inductive(L) parasitics. The RC and LC delays of a typical 2 cm long, unloaded metal wire is shown by curves RC and LC of Fig.2, respectively. The vertical axis is the delay in nanosec, and the horizontal axis is the width of the clock bus in microns. Curve LC' shows the clock delay due to the velocity of light($c_0/\sqrt{\epsilon_{OX}}$) . Because of the skin effect[3] of silicon, the propagation velocity is several times less than the light velocity(curve LC). Figure 2 shows that the signal delay of an unloaded 2cm long clockbus is not much less than 1 nanosec. In a VHSIC circuit whose clock frequency is 50-100MHz the clock distribution delay should be compensated. Curves 1-5 of Fig.2 show the delay when the bus is loaded by CMOS gates. The additional capacitive load is measured assuming a minimum size buffer of 1.75 micron CMOS, whose gate capacitance is 0.04 picofarad. If there are 10-30K gates in the random logic, 1-3K gates of them must receive clock signals directly. From Fig.2, a 20 micron wide clockbus(considered very wide) has an unloaded delay of 1 nanosec(sum of RC and LC delays). Delay is more, since the gates of FET's load the clockbus.

One technique to resolve the problem is the distributed clock system shown in Fig.3[1]. In the distributed clock system, the input clock is buffered by the clock driver, and is then distributed over the entire chip. Whenever the clockbus continues via less conductive crossunders (X1,X2), buffers are provided on the far end of the crossunder. Clock edges at nodes C1-C6 are brought into synchronization by adjusting the FET sizes of the crossunder buffers and the clock buffers at the individual logic blocks, CB1,CB2,.. This clock system allows adjustment of the distribution delays between nodes C_0 - C_6 . Clock buffers CB1-6 of Fig.3 can be clock decoders. The scheme therefore allows a nonoverlapping two-phase clock systemes used in WE32100 CPU[1]. From the flexibility in delay compensation the distributed clock system will be the most promising choice in VHSIC.

If more than one chip is clocked by the same system clock(CKIN of Fig.3) and if the chips exchange signals, the chips' internal clock edges must be well synchronized. Since the chips are processed in different lots, the internal clock delays are not the same, even if the clock circuits are identical. If the delay is T_D for the slowest chip, as much as $T_D/2$ skew may emerge when the fastest and the slowest chips communicate. The best strategy is to reduce the internal clock delay T_D to the minimum, and to introduce a selection procedure to match chips.

(2) Process variations influence the delays of the clock buffers. A new technique that eliminates the process induced clock skew by FET scaling will be presented using a simple example, but the concept is quite general. The circuit of Fig.4(a) generates a pair of up- and down-going CMOS clocks at nodes B and 3, respectively. We want to keep the two clock edges matched even if the processing conditions should vary. Figure 4(b) shows the voltage waveforms of nodes I,1,2,3,A and B. Delay times T_1,T_3 and T_A depend on the conductances of NFET's MN1, MN3 and MNA, respectively, while delay times T_2 and T_B depend on those of PFET's MP2 and MPB , respectively. The conductance is proportional to the FET size(width). In conventional circuit design method, FET sizes were adjusted to attain

$$T_A + T_B = T_1 + T_2 + T_3 \qquad (3)$$

In a design satisfying Eq.(1), the right-hand side of Eq.(3) may not equal the left-hand side if the ratio of the conductances of PFET and NFET varies. If the same circuit is designed instead to satisfy[6],

$$T_1 + T_3 = T_A \qquad T_2 = T_B \qquad (4)$$

Eq.(3) is also satisfied. Furthermore, the circuit is matched even if the ratio of the conductances of PFET and NFET might vary. Computer simulations show that the circuit designed following the principle has the worst skew less than 1/10 of the conventionally designed circuit. The design method improves the circuit even when gates other than inverters are involved, and even when the wiring parasitics are included.

[3] Gate delay control

A crucial circuit design decision is whether or not a logic function block should be implemented in faster CMOS dynamic logic[7]. It is a common practice to implement the statevector generation logic and the datapath logic with dynamic circuits. Concomitant to the choice the precharge time slot must be allocated, and the way of generating the precharge clock must be resolved. The latches that drive the dynamic logic are transparent when the precharge clock is on. A CMOS static latch is preferred, since the logic levels snap quickly upon closure of the latches.

A node in a critical path often looks like node X of Fig.5. The node is connected to a large number(N) of NFET's whose equivalent internal resistance is R_C and that has output capacitance C_D . Therefore the delay time of the node cannot be reduced below NR_CC_D by FET scaling alone. There are limited varieties of techniques to cope with the problem: (a) Combine the signals in several levels of hierarchy, (b) Use a sense amplifier that effectively adds a series negative resistance, and (c) Use a negative capacitance generator to reduce the effective load[8].

Domino CMOS dynamic gates are noninverting gates. Implementation of arbitrary logic requires slow AND dynamic gates as well as fast OR dynamic gates. The delay of a long NFET chain of an AND dynamic gate can be reduced by scaling the FET's in a pyramid shape, thereby inducing the drift motion of the potential profile within the FET chain[9].

The technique of speeding up the CMOS static logic chain is well established. Scaleup of a gate chain, choice of NFET/PFET size ratio, use of symmetric and asymmetric gates and careful layout techniques are now routinely practiced in CMOS VLSI design[1][10].

[4] Chip performance evaluation

Delays of PLA's and registers can be evaluated using a circuit simulator. The repetitive structure allows simple coding of the circuit into a circuit simulator code. Delays of datapath components like adders and shifters are often calibrated using a test pattern.

Delays of the random static logic is the most tedious to evaluate. If the technology, process and the environmental conditions are defined, the total delay of a logic chain whose FET sizes are optimized is approximately proportional to the sum of the number of logic signals combined along the logic path(an inverter combines one signal, a NAND2 gate two signals etc.). Figure 6 shows the simulation results for 1.75 micron CMOS gates in a typical compact circuit environment, at the typical process, power and temperature conditions. The delay is indeed proportional to the number of signals combined by the gate. Screening the logic design before layout, using the simple criteria, is very useful. A logic design that fails the simple criteria practically never produces a circuit meeting the delay specification. If the logic design is pre-qualified, the rest is to scale the FET's properly.

[5] Conclusion

Our experience in the development of WE32100 CPU chipset showed that a high performance CMOS VLSI circuits can be developed by a combination of the top-down design method[11] and the CMOS circuit design method summarized in this paper.

[6] Acknowledgement

The author wish to thank T.Szymanski and J.R.Brews for valuable discussions on the subjects and for many improvements of the manuscript.

References

[1] M.Shoji,"Electrical design of BellMac32A microprocessor", ICCC82 digest,pp.112-115,October,1982

[2] H.F.S.Law and M.Shoji,"PLA design for the BellMac32A microprocessor" ICCC82 digest,pp.161-164,October,1982

[3] H.Hasegawa,M.Furukawa and H Yanai,"Properties of microstrip line on $Si-SiO_2$ system",IEEE Trans. on Microwave Theory and Techniques, vol.MTT-19,pp.869-881,Nov.,1971

[4] M.Shoji,C.M.Lee and R.L.Gordon,"A complementary relation between the maximum $dI_0(t)/dt$ and the switching time of a CMOS output driver" to be published

[5] Papers in IBM J. Research and Development,vol.26,No.3,May,1982

[6] M.Shoji,"Elimination of process-dependent clock skew in CMOS VLSI", to be published in IEEE Journal of Solid State Circuits, October,1986

[7] R.H.Krambeck,C.M.Lee and H.F.Law,"High speed compact circuits with CMOS", IEEE J. Solid-State Circuits,vol.SC-17,pp.614-619,June,1982
N.P.Goncalves and H.J.de Man,"NORA: a racefree dynamic CMOS technique for pipelined logic structures",IEEE Journal of Solid State Circuits, vol.SC-18,pp.261-268,June,1983

[8] M.Shoji and R.M Rolfe,"Negative capacitance bus terminator for improving the switching speed of a microcomputer databus",IEEE J. Solid-State Circuits, vol.SC-20,pp.828-832,August,1985

[9] M.Shoji,"FET scaling in Domino CMOS gates" IEEE Journal of Solid State Circuits,SC-20,pp.1067-1071,1985

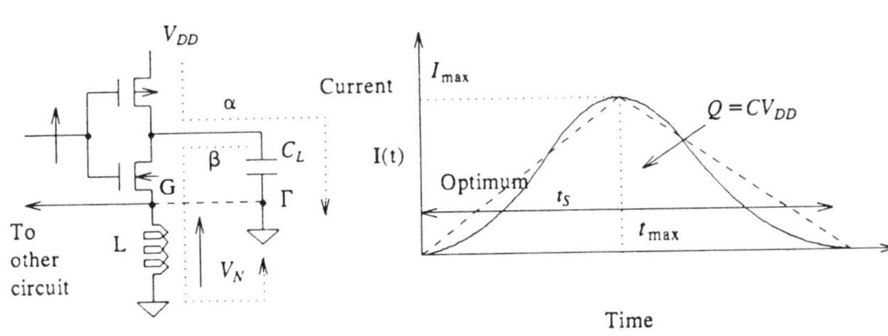

Figure 1 (Left)Equivalent circuit of power bus noise. (Right) Discharge current waveforms. The dashed curve shows the optimum waveform.

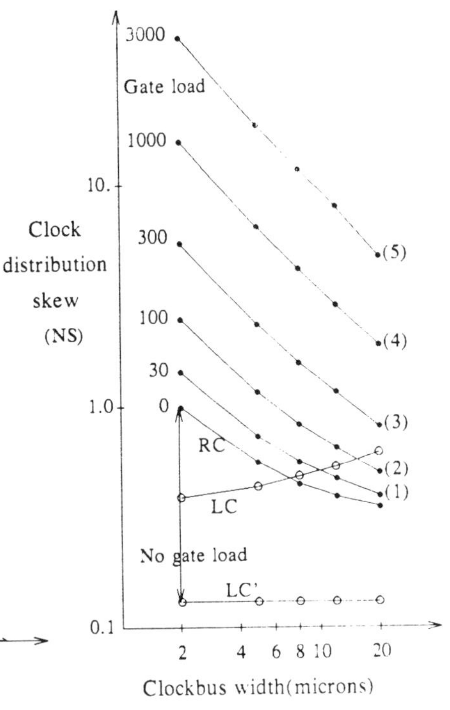

Figure 2 Clock distribution delay versus clockbus width.

[10] H.C.Lin and L.W.Linholm,"An optimized output stage for MOS integrated circuits",IEEE J. Solid-State Circuits,vol.SC-10,pp.106-109,1975
S.M.Kang,"A design of CMOS polycells for LSI circuits",IEEE Trans. on Circuits and Systems,vol.CAS-28,pp.838-843,Aug.,1981
A.Kanuma,"CMOS circuit optimization",Solid-State Electronics,vol.26,pp.47-58, January,1983
A.M.Moshen and C.A.Mead,"Delay-time optimization for driving and sensing of signals on high capacitance paths of VLSI systems",IEEE J. Solid-State Circuits,vol.SC-14,pp.462, April,1979
[11] R.H.Krambeck,D.E.Blahut,S.F.Law,B.W.Colbry,H.C.So and M.L.Harrison, "Top-down design of one-chip 32bit CPU",VLSI81,Edinbrough,Scotland, September,1982

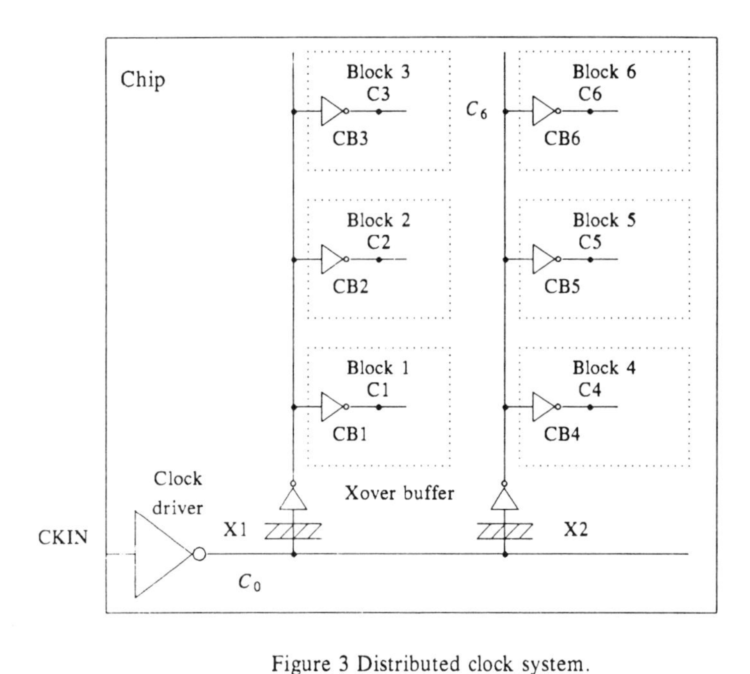

Figure 3 Distributed clock system.

Figure 5 Delay of node X cannot be reduced effectively by FET scaling, if number of FET's connected to the node, N, is large.

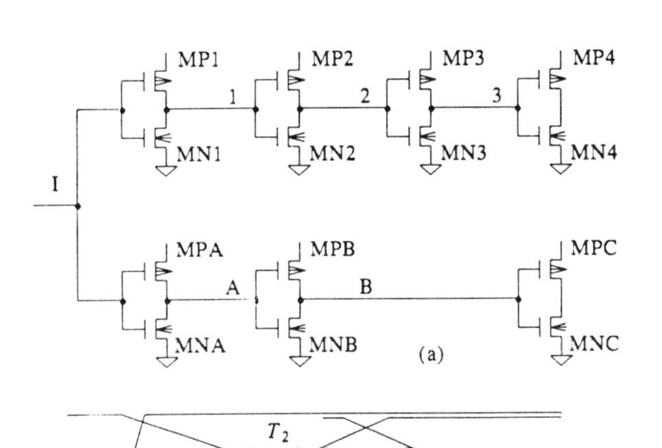

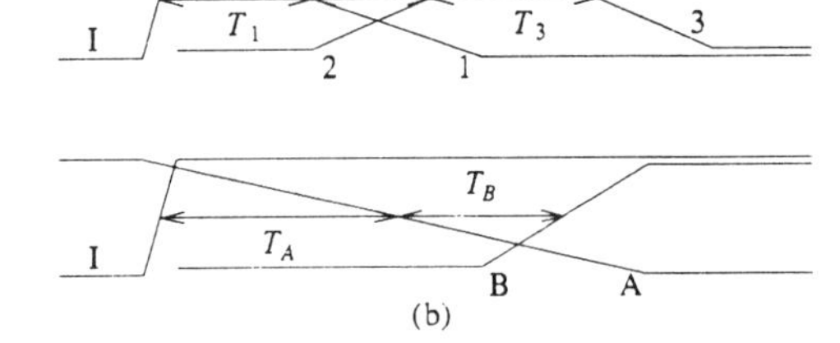

Figure 4 Elimination of process-dependent skew by FET scaling.

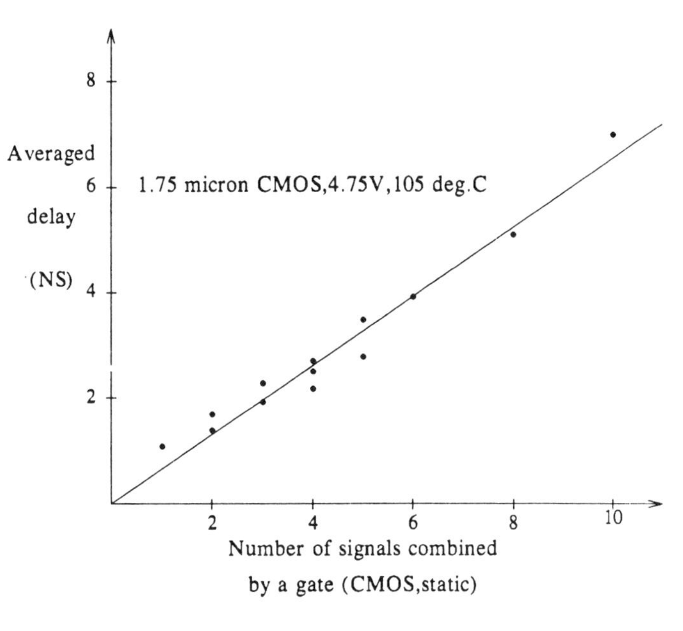

Figure 6 Gate delay versus the number of signals combined by the gate in a VLSI environment where the FET's are scaled to the optimum.

A 200-MHz 64-b Dual-Issue CMOS Microprocessor

Daniel W. Dobberpuhl, *Member, IEEE,* Richard T. Witek, Randy Allmon, Robert Anglin, David Bertucci, Sharon Britton, Linda Chao, *Member, IEEE,* Robert A. Conrad, Daniel E. Dever, Bruce Gieseke, Soha M. N. Hassoun, Gregory W. Hoeppner, *Member, IEEE,* Kathryn Kuchler, *Member, IEEE,* Maureen Ladd, *Member, IEEE,* Burton M. Leary, Liam Madden, Edward J. McLellan, Derrick R. Meyer, James Montanaro, Donald A. Priore, Vidya Rajagopalan, Sridhar Samudrala, and Sribalan Santhanam

***Abstract*—A 400-MIPS/200-MFLOPS (peak) custom 64-b VLSI CPU chip is described. The chip is fabricated in a 0.75-μm CMOS technology utilizing three levels of metalization and optimized for 3.3-V operation. The die size is 16.8 mm × 13.9 mm and contains 1.68M transistors. The chip includes separate 8-kilobyte instruction and data caches and a fully pipelined floating-point unit (FPU) that can handle both IEEE and VAX standard floating-point data types. It is designed to execute two instructions per cycle among scoreboarded integer, floating-point, address, and branch execution units. Power dissipation is 30 W at 200-MHz operation.**

I. Introduction

A RISC-style microprocessor has been designed and tested that operates up to 200 MHz. The chip implements a new 64-b architecture, designed to provide a huge linear address space and to be devoid of bottlenecks that would impede highly concurrent implementations. Fully pipelined and capable of issuing two instructions per clock cycle, this implementation can execute up to 400 million operations per second. The chip includes an 8-kilobyte I-cache, 8-kilobyte D-cache and two associated translation look-aside buffers, a four-entry 32-byte/entry write buffer, a pipelined 64-b integer execution unit with a 32-entry register file, and a pipelined floating-point unit (FPU) with an additional 32 registers. The pin interface includes integral support for an external secondary cache. The package is a 431-pin PGA with 140 pins dedicated to V_{DD}/V_{SS}. The chip is fabricated in 0.75-μm n-well CMOS with three layers of metalization. The die measures 16.8 mm × 13.9 mm and contains 1.68 million transistors. Power dissipation is 30 W from a 3.3-V supply at 200 MHz.

II. CMOS Process Technology

The chip is fabricated in a 0.75-μm 3.3-V n-well CMOS process optimized for high-performance microprocessor design. Process characteristics are shown in Table I. The thin gate oxide and short transistor lengths result in the fast transistors required to operate at 200 MHz. There are no explicit bipolar devices in the process as the incremental process complexity and cost were deemed too large in comparison to the benefits provided—principally more area-efficient large drivers such as clock and I/O.

The metal structure is designed to support the high operating frequency of the chip. Metal 3 is very thick and has a relatively large pitch. It is important at these speeds to have a low-resistance metal layer available for power and clock distribution. It is also used for a small set of special signal wires such as the data buses to the pins and the control wires for the two shifters. Metal 1 and metal 2 are maintained at close to their maximum thickness by planarization and by filling metal 1 and metal 2 contacts with tungsten plugs. This removes a potential weak spot in the electromigration characteristics of the process and allows more freedom in the design without compromising reliability.

III. Alpha Architecture

The computer architecture implemented is a 64-b load/store RISC architecture with 168 instructions, all 32 b wide [1]. Supported data types include 8-, 16-, 32-, and 64-b integers and 32- and 64-b floats of both DEC and IEEE formats. The two register files, integer and floating point, each contains 32 entries of 64 b with one entry in each being a hardwired zero. The program counter and virtual address are 64 b. Implementations can subset the virtual address size but are required to check the full 64-b address for sign extension. This insures that when later implementations choose to support a larger virtual address, programs will still run and not find addresses that have dirty bits in the previously ''unused'' bits.

Reprinted from *IEEE J. Solid-State Circuits,* vol. 27, no. 11, pp. 1555–1567, Nov. 1992.

TABLE I
PROCESS DESCRIPTION

Feature Size	0.75 μm
Channel Length	0.5 μm
Gate Oxide	10.5 nm
V_{TN}/V_{TP}	0.5 V/−0.5 V
Power Supply	3.3 V
Substrate	P-epi with n-well
Salicide	Cobalt disilicide in diffusions and gates
Buried Contact	Titanium nitride
Metal 1	0.75-m AlCu, 2.25-μm pitch (contacted)
Metal 2	0.75-m AlCu, 2.625-μm pitch (contacted)
Metal 3	2.0-m AlCu, 7.5-μm pitch (contacted)

The architecture is designed to support high-speed multi-issue implementations. To this end the architecture does not include condition codes, instructions with fixed source or destination registers, or byte writes of any kind (byte operations are supported by extract and merge instructions within the CPU itself). Also, there are no first-generation artifacts that are optimized around today's technology, which would represent a long-term liability to the architecture.

IV. Chip Microarchitecture

The block diagram (Fig. 1) shows the major functional blocks and their interconnecting buses, most of which are 64 b wide. The chip implements four functional units: the integer unit (IRF + EBOX), the floating-point unit (FRF + FBOX), the load/store unit (ABOX), and the branch unit (distributed). The bus interface unit (BIU), described in the next section, handles all communication between the chip and external components. The microphotograph (Fig. 2) shows the boundaries of the major functional units. The dual-issue rules are a direct consequence of the register-file ports, the functional units, and the I-cache interface. The integer register file (IRF) has two read ports and one write port dedicated to the integer unit, and two read and one write port shared between the branch unit and the load store unit. The floating-point register file (FRF) has two read ports and one write port dedicated to the floating unit, and one read and one write port shared between the branch unit and the load store unit. This leads to dual-issue rules that are quite general:

- any load/store in parallel with any operate,
- an integer op in parallel with a floating op,
- a floating op and a floating branch,
- an integer op and an integer branch,

except that integer store and floating operate and floating store and integer operate are disallowed as pairs.

As shown in Fig. 3(a), the integer pipeline is seven stages deep, where each stage is a 5-ns clock cycle. The first four stages are associated with instruction fetching, decoding, and scoreboard checking of operands. Pipeline stages 0 through 3 can be stalled. Beyond 3, however, all pipeline stages advance every cycle. Most ALU operations complete in cycle 4 allowing single-cycle latency, with the shifter being the exception. Primary cache accesses complete in cycle 6, so cache latency is three cycles. The chip will do hits under misses to the primary DCACHE.

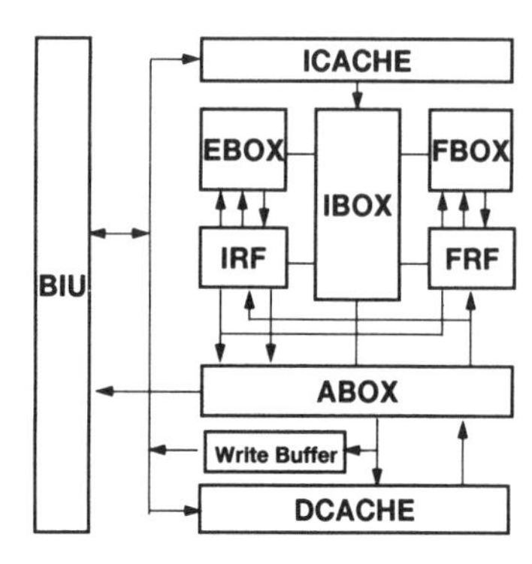

Fig. 1. CPU chip block diagram.

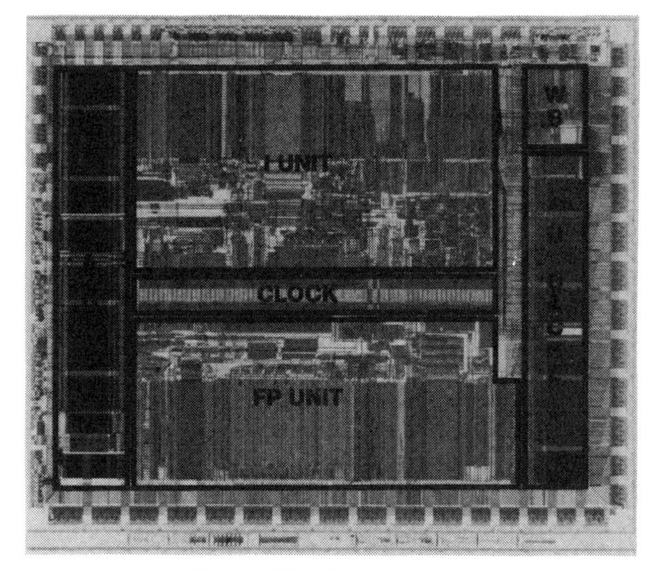

Fig. 2. Chip microphotograph.

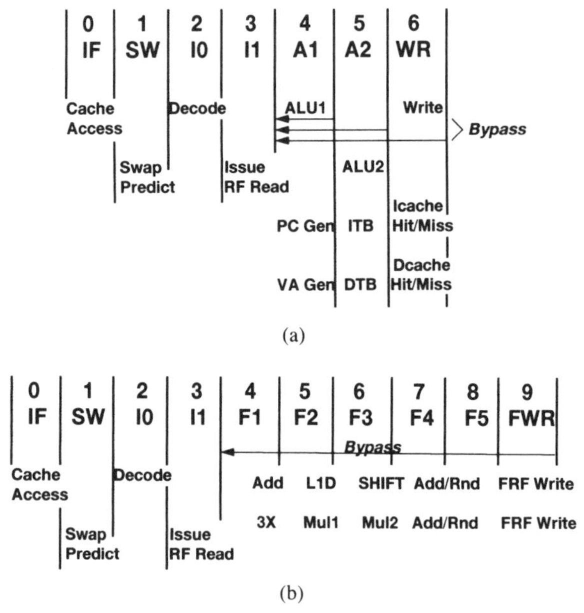

Fig. 3. (a) Integer unit pipeline timing. (b) Floating-point unit pipeline timing.

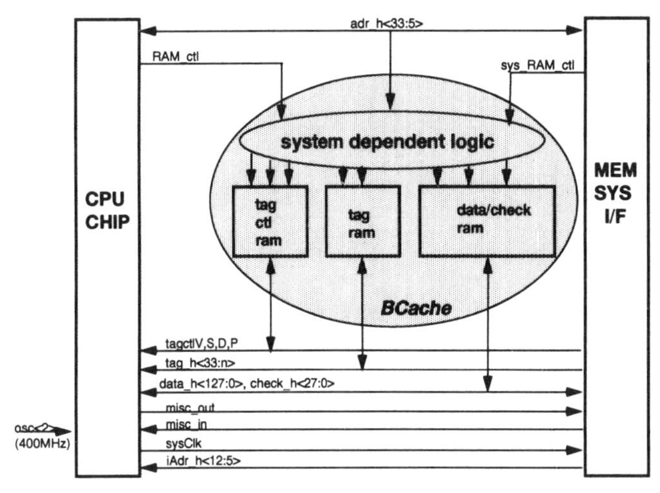

Fig. 4. CPU external interface.

The ISTREAM is based on autonomous prefetching in cycles 0 and 1 with the final resolution of ICACHE hit not occurring until cycle 5. The prefetcher includes a branch history table and a subroutine return stack. The architecture provides a convention for compilers to predict branch decisions and destination addresses including those for register indirect jumps. The penalty for branch mispredict is four cycles.

The floating-point unit is a fully pipelined 64-b floating-point processor that supports both VAX standard and IEEE standard data types and rounding modes. It can generate a 64-b result every cycle for all operations except division. As shown in Fig. 3(b), the floating-point pipeline is identical and mostly shared with the integer pipeline in stages 0 through 3, however, the execution phase is three cycles longer. All operations, 32 and 64 b, (except division) have the same timing. Division is handled by a nonpipelined, single bit per cycle, dedicated division unit.

In cycle 4, the register file data are formatted to fraction, exponent, and sign. In the first-stage adder, exponent difference is calculated and a $3\times$ multiplicand is generated for multiplies. In addition, a predictive leading 1 or 0 detector using the input operands is initiated for use in result normalization. In cycles 5 and 6, for add/subtract, alignment or normalization shift and sticky-bit calculation are performed. For both single- and double-precision multiplication, the multiply is done in a radix-8 pipelined array multiplier. In cycles 7 and 8, the final addition and rounding are performed in parallel and the final result is selected and driven back to the register file in cycle 9. With an allowed bypass of the register write data, floating-point latency is six cycles.

The CPU contains all the hardware necessary to support a demand paged virtual memory system. It includes two translation look-aside buffers to cache virtual to physical address translations. The instruction translation buffer contains 12 entries, eight that map 8-kilobyte pages and four that map 4-megabyte pages. The data translation buffer contains 32 entries that can map 8-kilobyte, 64-kilobyte, 512-kilobyte or 4-megabyte pages.

The CPU supports performance measurement with two counters that accumulate system events on the chip such as dual-issue cycles and cache misses or external events through two dedicated pins that are sampled at the selected system clock speed.

V. External Interface

The external interface (Fig. 4) is designed to directly support an off-chip backup cache that can range in size from 128 kilobytes to 16 megabytes and can be constructed from standard SRAM's. For most operations, the CPU chip accesses the cache directly in a combinatorial loop by presenting an address and waiting N CPU cycles for control, tag, and data to appear, where N is a mode-programmable number between 3 and 16 set at power-up time. For writes, both the total number of cycles and the duration and position of the write signal are programmable in units of CPU cycles. This allows the system designer to select the size and access time of the SRAM's to match the desired price/performance point.

The interface is designed to allow all cache policy decisions to be controlled by logic external to the CPU chip. There are three control bits associated with each back-up cache (BCACHE) line: valid, shared, and dirty. The chip completes a BCACHE read as long as valid is true. A write is processed by the CPU only if valid is true and shared is false. When a write is performed the dirty bit is set to true. In all other cases, the chip defers to an external state machine to complete the transaction. This state machine operates synchronously with the SYS_CLK output of the chip, which is a mode-controlled submultiple of the CPU clock rate ranging from divide by 2 to divide by 8. It is also possible to operate without a back-up cache.

As shown in the diagram, the external cache is connected between the CPU chip and the system memory interface. The combinatorial cache access begins with the desired address delivered on the *adr_h* lines and results in ctl, tag, data, and check bits appearing at the chip receivers within the prescribed access time. In 128-b mode, BCACHE accesses require two external data cycles to transfer the 32-byte cache line across the 16-byte pin bus. In 64-b mode, it is four cycles. This yields a maximum backup cache read bandwidth of 1.2 gigabyte/s and a write bandwidth of 711 megabyte/s. Internal cache lines can be invalidated at the rate of one line/cycle using the dedicated invalidate address pins, *iAdr_h* $\langle 12:5 \rangle$.

In the event external intervention is required, a request code is presented by the CPU chip to the external state machine in the time domain of the SYS_CLK as described previously. Fig. 5 shows the read miss timing where each cycle is a SYS_CLK cycle. The external transaction starts with the address, the quadword within block and instruction/data indication supplied on the *cWMask_h* pins, and READ_BLOCK function supplied on the *cReq_h* pins. The external logic returns the first 16 bytes of data on the *data_h* and ecc or parity on the *check_h* pins. The CPU latches the data based on receiving acknowledgment on

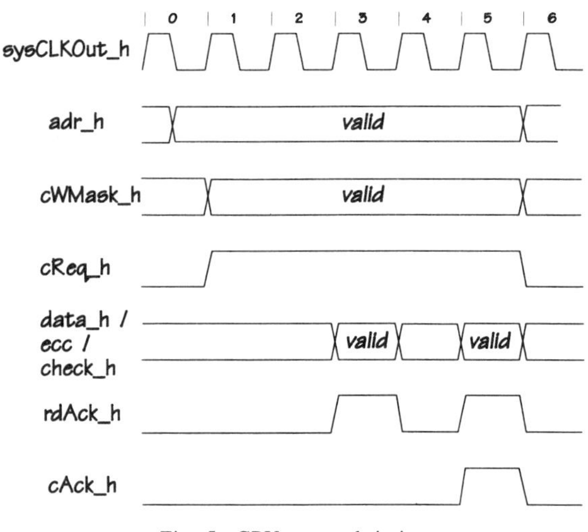

Fig. 5. CPU external timing.

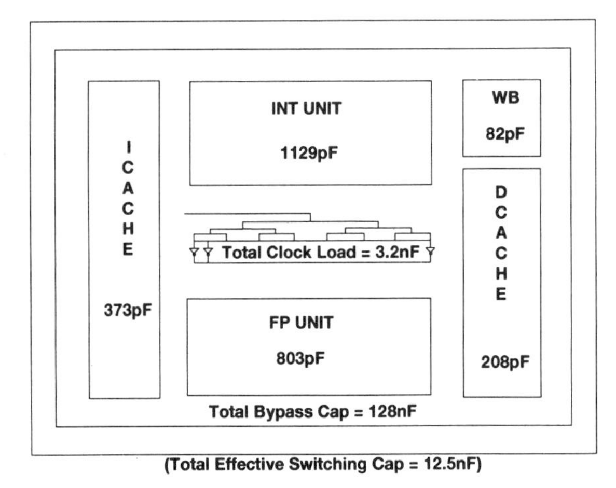

Fig. 6. Clock load distribution.

rdAck_H. The diagram shows a stall cycle (cycle 4) between the request and the return data; this depends on the external logic and could range from zero to many cycles. The second 16 bytes of data are returned in the same way with *rdAck_h* signaling the return of the data and *cAck_h* signaling the completion of the transaction. *cReq_h* returns to idle and a new transaction can start at this time.

The chip implements a novel set of optional features supporting chip and module test. When the chip is reset, the first attempted action is to read from a serial read-only memory (SROM) into the I-cache via a private three-wire port. The CPU is then enabled and the PC is forced to 0. Thus, with only three functional components (CPU chip, SROM, and clock input) a system is able to begin executing instructions. This initial set of instructions is used to write the bus control registers inside the CPU chip to set the cache timing and to test the chip and module from the CPU out. After the SROM loads the I-cache, the pins used for the SROM interface are enabled as serial-in and out ports. These ports can be used to load more data or to return status of testing and setup.

VI. Circuit Implementation

Many novel circuit structures and detailed analysis techniques were developed to support the clock rate in conjunction with the complexity demanded by the concurrence and wide data paths. The clocking method is single wire level sensitive. The bus interface unit operates from a buffered version of the main clock. Signals that cross this interface are deskewed to eliminate races. This clocking method eliminates dead time between phases and requires only a single clock signal to be routed throughout the chip.

One difficulty inherent in this clocking method is the substantial load on the clock node, 3.25 nF in our design. This load and the requirement for a fast clock edge led us to take particular care with clock routing and to do extensive analysis on the resulting grid. Fig. 6 shows the distribution of clock load among the major functional units. The clock drives into a grid of vertical metal 3 and horizontal metal 2. Most of the loading occurs in the integer and floating-point units that are fed from the most robust metal 3 lines. To ensure the integrity of the clock grid across the chip, the grid was extracted from the layout and the resulting network, which contained 63 000 *RC* elements, was simulated using a circuit simulation program based on the AWEsim simulator from Carnegie-Mellon University. Fig. 7 shows a three-dimensional representation of the output of this simulation and shows the clock delay from the driver to each of the 63 000 transistor gates connected to the clock grid.

The 200-MHz clock signal is fed to the driver through a binary fanning tree with five levels of buffering. There is a horizontal shorting bar at the input to the clock driver to help smooth out possible asymmetry in the incoming wavefront. The driver itself consists of 145 separate elements each of which contains four levels of prescaling into a final output stage that drives the clock grid.

The clock driver and predriver represent about 40% of the total effective switching capacitance determined by power measurement to be 12.5 nF (worst case including output pins). To manage the problem of di/dt on the chip power pins, explicit decoupling capacitance is provided on-chip. This consists of thin oxide capacitance that is distributed around the chip, primarily under the data buses. In addition, there are horizontal metal 2 power and clock shorting straps adjacent to the clock generator and the thin oxide decoupling cap under these lines supplies charge to the clock driver. di/dt for the driver alone is about 2×10^{11} A/s. The total decoupling capacitance as extracted from the layout measures 128 nF. Thus the ratio of decoupling capacitance to switching cap is about 10 : 1. With this capacitance ratio, the decoupling cap could supply all the charge associated with a complete CPU cycle with only a 10% reduction in the on-chip supply voltage.

A. Latches

As previously described, the chip employs a single-phase approach with nearly all latches in the core of the chip receiving the clock node, CLK, directly. A repre-

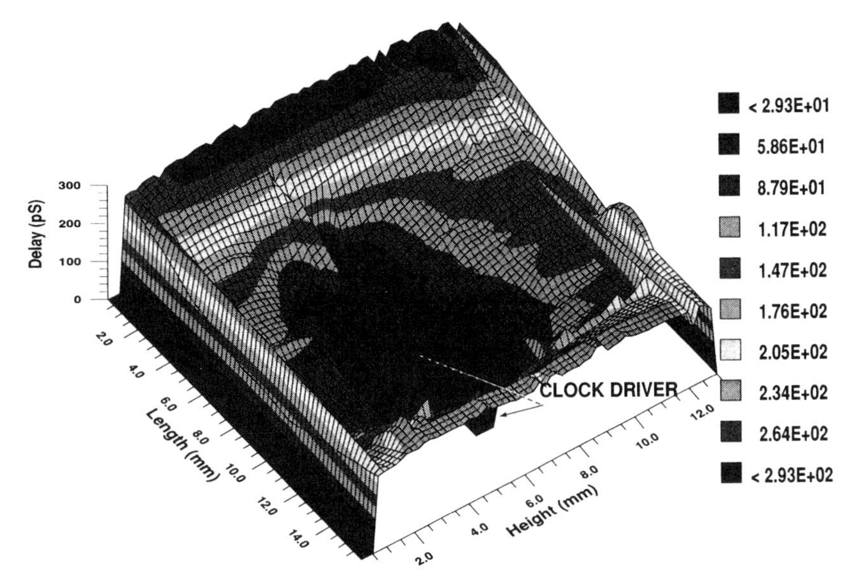

Fig. 7. CPU clock skew.

sentative example is illustrated in Fig. 8. Notice that $L1$ and $L2$ are transparent latches separated by random logic and are not simultaneously active: $L1$ is active when CLK is high and $L2$ is active when CLK is low. The minimum number of delays between latches is zero and the maximum number of delays is constrained only by the cycle time and the details of any relevant critical paths. The bus interface unit, many data-path structures, and some critical paths deviate from this approach and use buffered versions and/or conditionally buffered versions of CLK. The resulting clock skew is managed or eliminated with special latch structures.

The latches used in the chip can be classified into two categories: custom and standard. The custom latches were used to meet the unique needs of data-path structures and the special constraints of critical paths. The standard latches were used in the design of noncritical control and in some data-path applications. These latches were designed prior to the start of implementation and were included in the library of usable elements for logic synthesis. All synthesized logic used only this set of latches.

The standard latches are extensions of previously published work [2] and examples are shown in Figs. 9–11. To understood the operation of these latches refer to Fig. 9(a). When CLK is high, $P1$, $N1$, and $N3$ function as an inverter complementing IN1 to produce X. $P2$, $N2$, and $N4$ function as a second inverter and complement X to produce OUT. Therefore, the structure passes IN1 to OUT. Then CLK is "low," $N3$ and $N4$ are cut off. If IN1, X, and OUT are initially "high," "low," and "high," respectively, a transition of IN1 FALLING pulls X "high" through $P1$ causing $P2$ to cut off, which tristates OUT "high." If IN1, X, and OUT are initially "low," "high," and "low," respectively, a transition of IN1 RISING causing $P1$ to cut off, which tristates X "high" leaving OUT tristated "low." In both cases, additional transitions of IN1 leave X tristated or driven "high" with OUT tristated to its initial value. Therefore,

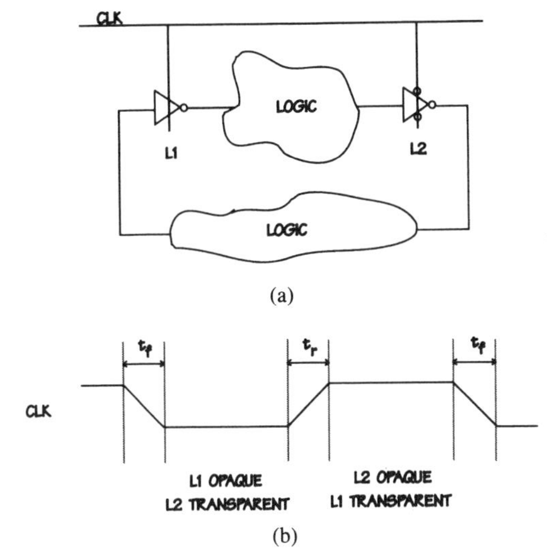

Fig. 8. (a) Latching schema. (b) Latch timing.

the structure implements a latch that is transparent when CLK is "high" and opaque when CLK is "low." Fig. 9(c) shows the dual of the latch just discussed; this structure implements a latch that is transparent when CLK is "low" and opaque when CLK is "high." Fig. 9(b) and (d) depicts latches with an output buffer used to protect the sometimes dynamic node OUT and to drive large loads.

The design of the standard latches stressed three primary goals: flexibility, immunity to noise, and immunity to race-through. To achieve the desired flexibility, a variety of latches like those in Figs. 9–11 in a variety of sizes were characterized for the implementors. Thus, the designer could select a latch with an optional output buffer and an embedded logic function that was sized appropriately to drive various loads. Furthermore, it was decided to allow zero delay between latches, completely freeing the designer from race-through considerations when designing static logic with these latches.

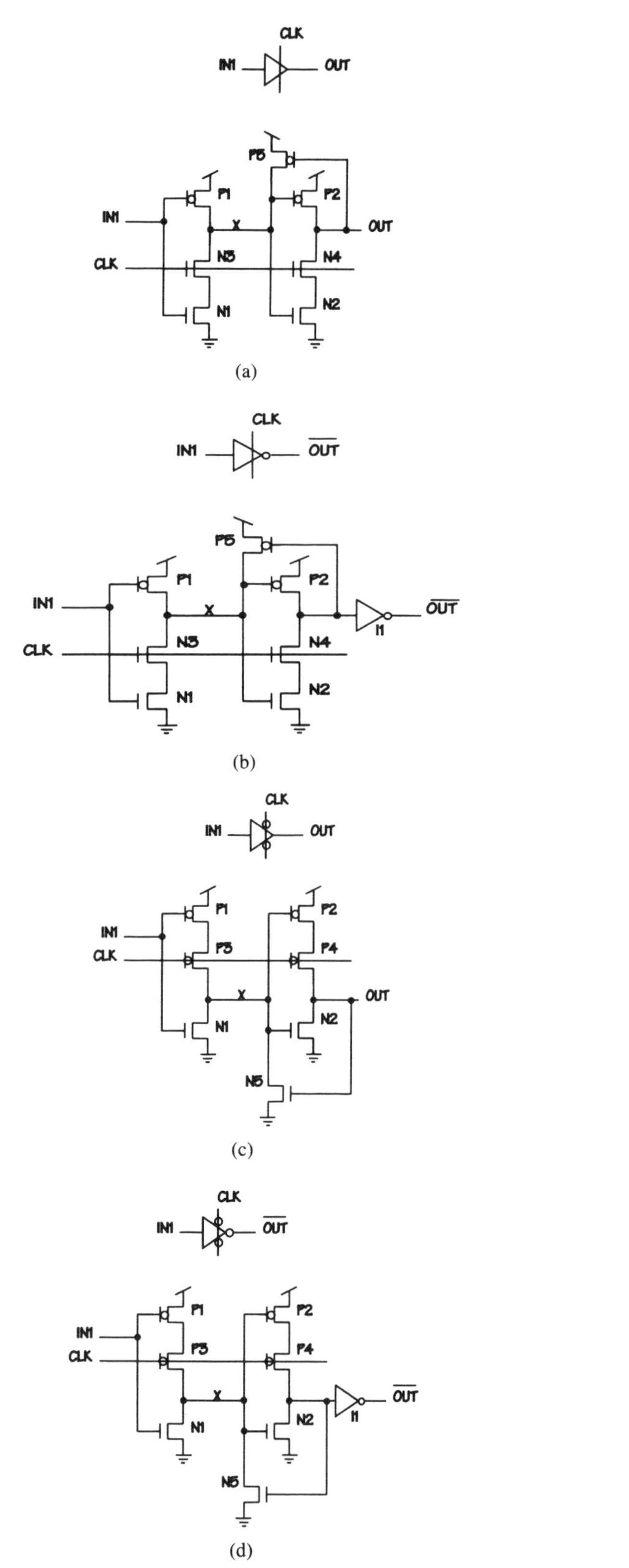

Fig. 9. (a) Noninverting active-high latch. (b) Inverting active-high latch. (c) Noninverting active-low latch. (d) Inverting active-low latch.

In the circuit methodology adopted for the implementation, only one node, *X* (Fig. 9(a)), poses inordinate noise margin risk. As noted above, *X* may be tristated "high" with OUT tristated "low" when the latch is opaque. This

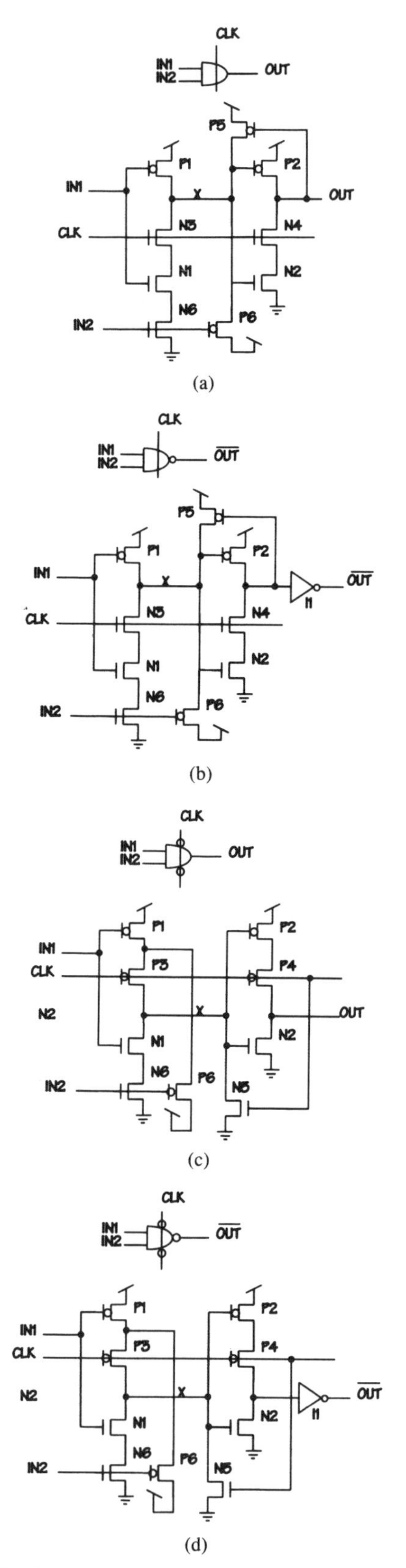

Fig. 10. (a) Two input AND active-high latch. (b) Two-input NAND active-high latch. (c) Two-input AND active-low latch. (d) Two-input NAND active-low latch.

maps into a dynamic node driving into a dynamic gate that is very sensitive to noise that reduces the voltage on *X*, causing leakage through $P2$, thereby destroying OUT.

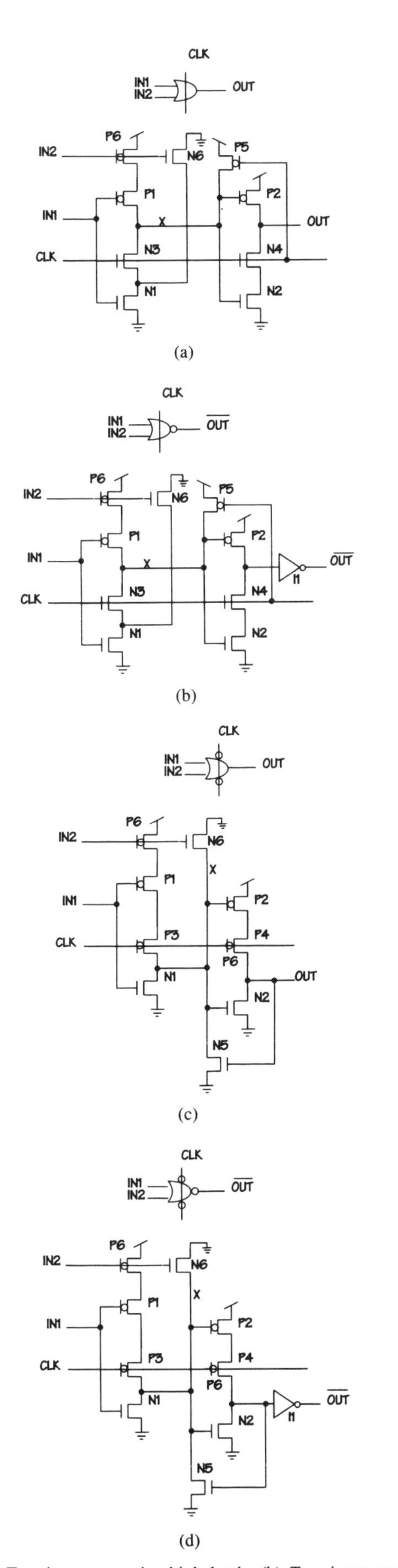

Fig. 11. (a) Two-input OR active-high latch. (b) Two-input NOR active-high latch. (c) Two-input OR active-low latch. (d) Two-input NOR active-low latch.

This problem was addressed by the addition of $P5$. This weak feedback device is sized to source enough current to counter reasonable noise and hold $P2$ in cutoff. $N5$ plays an analogous role in Fig. 9(c).

Race-through was the major functional concern with the latch design. It is aggravated by clock skew, the variety of available latches and the zero delay goal between latches. The clock skew concern was actually the easiest to address. If data propagate in a direction that opposes the propagation of the clock wavefront, clock skew is functionally harmless and tends only to reduce the effective cycle time locally. Minimizing this effect is of concern when designing the clock generator. If data propagate in a direction similar to the propagation of the clock wavefront, clock skew is a functional concern. This was addressed by radially distributing the clock from the center of the chip. Since the clock wavefront moves out radially from the clock driver toward the periphery of the die, it is not possible for the data to overtake the clock if the clock network is properly designed.

To verify the remaining race-through concerns, a mix and match approach was taken. All reasonable combinations of latches were cascaded together and simulated. The simulations were stressed by eliminating all interconnect and diffusion capacitance and by pushing each device into a corner of the process that emphasized race-through. Then many simulations with varying CLK rise and fall times, temperatures, and power supply voltages were performed. The results showed no appreciable evidence of race-through for CLK rise and fall times at or below 0.8 ns. With 1.0-ns rise and fall times, the latches showed signs of failure. To guarantee functionality, CLK was specified and designed to have an edge rate of less than 0.5 ns. This was not a serious constraint since other circuits in the chip required similar edge rates of the clock.

A last design issue worth noting is the feedback devices, $N5$ and $P5$, in Figs. 10(c), 10(d), 11(a), and 11(b). Notice that these devices have their gates tied to CLK instead of OUT like the other latches. This difference is required to account for an effect not present in the other latches. In these latches a stack of devices is connected to node X without passing through the clocked transistors $P3$ or $N3$. Referring to Fig. 11(a), assume CLK is "low," X is "high," and OUT is "low." If multiple random transitions are allowed by IN1 with IN2 "high," then coupling through $P1$ can drive X down by more than a threshold even with weak feedback, thereby destroying OUT. To counter this phenomenon, $P5$ cannot be a weak feedback device and therefore cannot be tied to OUT if the latch is to function properly when CLK is "high." Note that taller stacks aggravate this problem because the devices become larger and there are more devices to participate in coupling. For this reason stacks in these latches were limited to three high. Also, note that clocking $P5$ introduces another race-through path since X will unconditionally go "high" with CLK falling and OUT must be able to retain a stored ONE. So there is a two-sided constraint: $P5$ must be large enough to counter coupling and

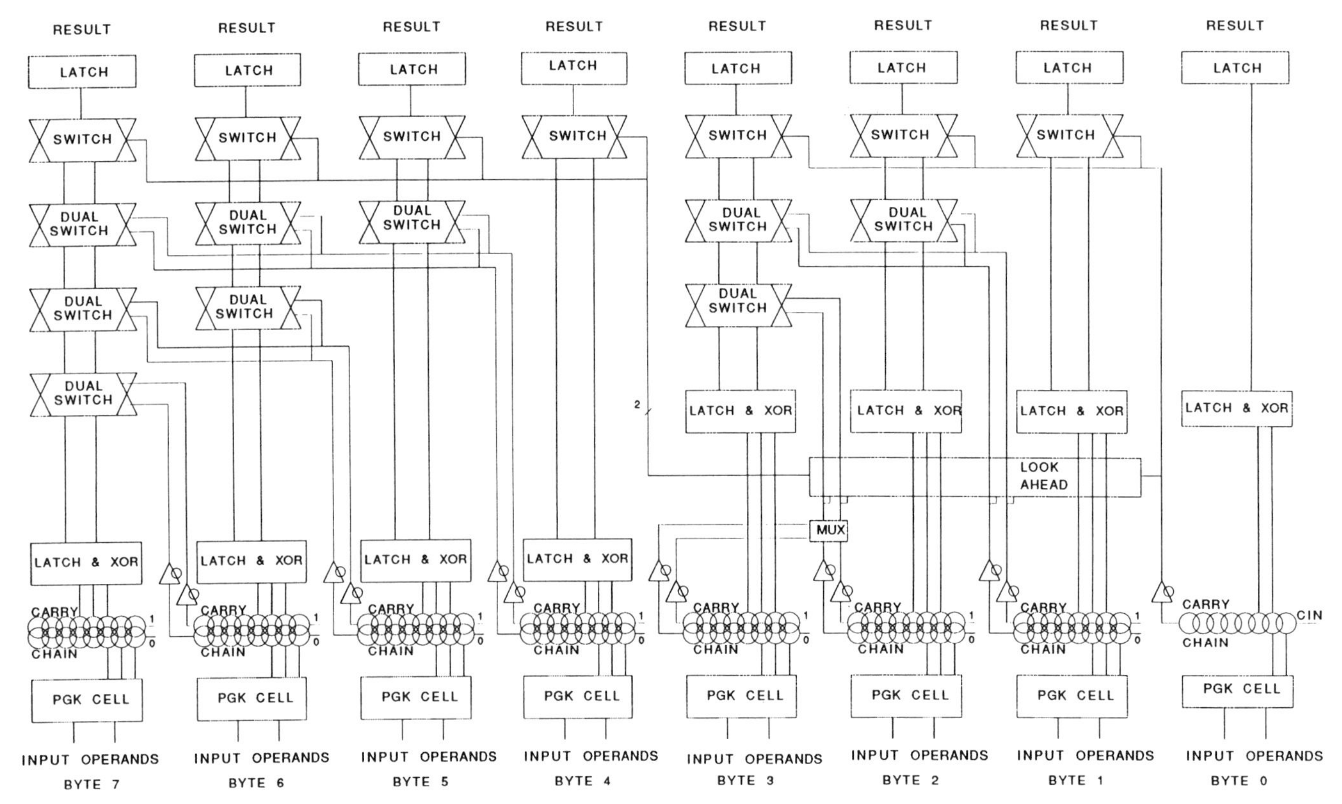

Fig. 12. 64-b adder block diagram.

small enough not to cause race-through. These trade-offs were analyzed by simulation in manner similar to the one outlined above.

B. 64-b Adder

A difficult circuit problem was the 64-b adder portion of the integer and floating-point ALU's. Unlike a previous high-speed design [3], we set a goal to achieve single-cycle latency in this unit. Fig. 12 has an organizational diagram of its structure. Every path through the adder includes two latches, allowing fully pipelined operation. The result latches are shown explicitly in the diagram, however, the input latches are somewhat implicit, taking advantage of the predischarge characteristics of the carry chains. The complete adder is a combination of three methods for producing a binary add: a byte-long carry chain, a long-word (32 b) carry select, and local logarithmic carry select [4]. The carry select is built as a set of nMOS switches that direct the data from byte carry chains. The 32-b long-word lookahead is implemented as a distributed differential circuit controlling the final stage of the upper longword switches. The carry chains are organized in groups of 8 b.

Carry chain width was chosen to implement a byte compare function specified by the architecture. The carry chain implemented with nMOS transistors is shown in Fig. 13(a). Operation begins with the chain predischarged to V_{SS}, with the controlling signal an OR of CLK and the kill function. Evaluation begins along the chain length without the delay associated with the V_{gs}–V_t threshold found in a chain precharged to V_{DD}. An alternative to a predischarged state was to precharge to V_{DD}–V_t, but the resulting low noise margins were deemed unacceptable. From the LSB to MSB, the width of the nMOS gates for each carry chain stage is tapered down, reducing the loading presented by the remainder of the chain. The local carry nodes are received by ratioed inverters. Each set of propagate, kill, and generate signals controls two carry chains, one that assumes a carry-in and one that assumes no carry-in. The results feed the bit-wise data switches as well as the carry selects.

The long-word carry select is built as a distributed cascode structure used to combine the byte generate, kill, and propagate signals across the lower 32-b long word. It controls the final data selection into the upper long-word output latch and is out of the critical path.

The nMOS byte carry select switches are controlled by a cascade of closest neighbor byte carry-outs. Data in the most significant byte of the upper long word are switched first by the carry-out data of the next lower byte, byte 6, then by byte 5, and finally byte 4. The switches direct the sum data from either the carry-in channel or the no-carry channel (Fig. 13(b)). Sign extension is accomplished by disabling the upper long-word switch controls on long-word operations and forcing the sign of the result into both data channels.

C. I/O Circuitry

To provide maximum flexibility in applications, the external interface allows for several different modes of operation all using common on-chip circuitry. This includes

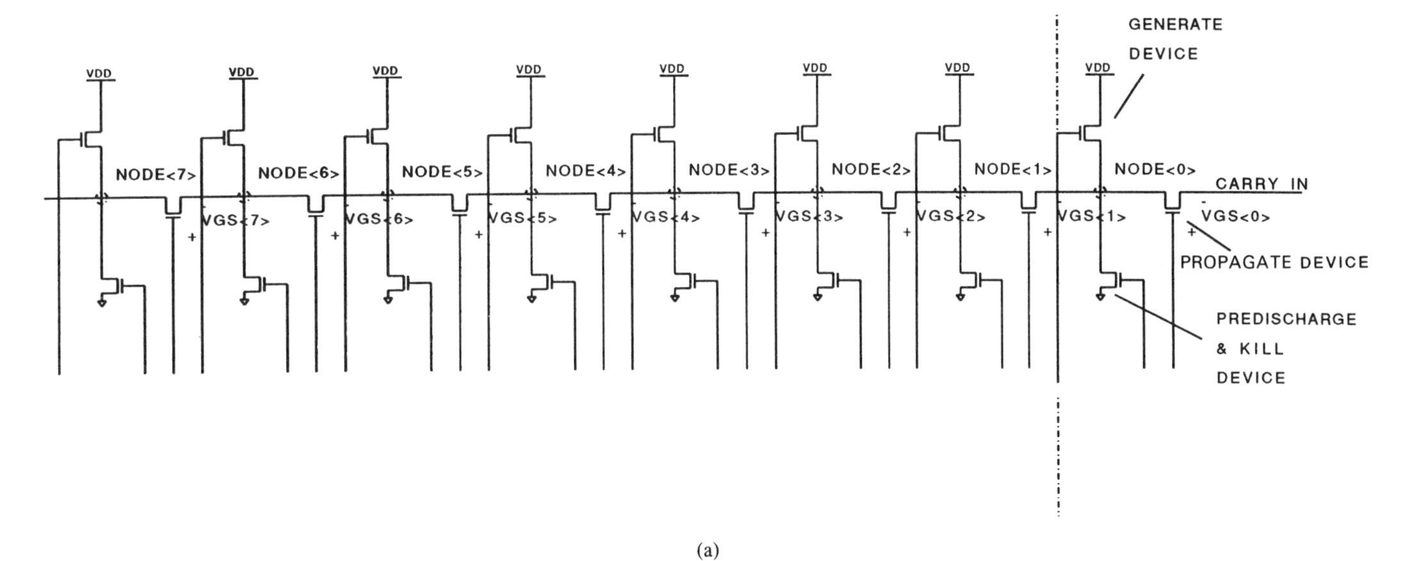

(a)

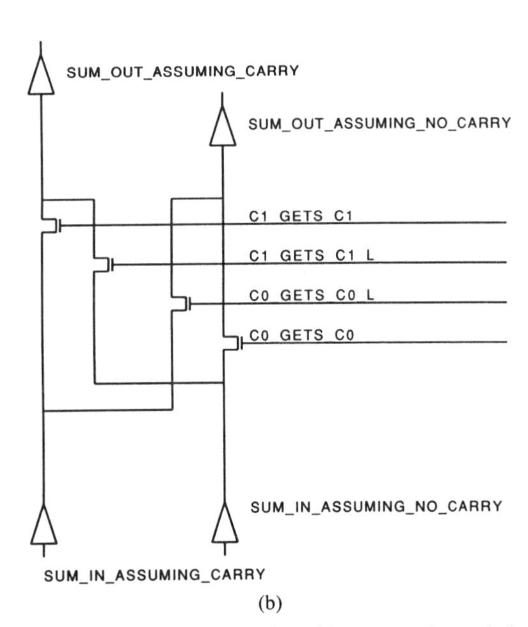

(b)

Fig. 13. (a) Adder carry chain. (b) Adder carry-select switches.

choice of logic family (CMOS/TTL or ECL) as well as bus width (64/128 b), external cache size and access time, and BIU clock rate. These parameters are set into mode registers during chip power-up. The logic family choice provided an interesting circuit challenge. The input receivers are differential amplifiers that utilize an external reference level which is set to the switching midpoint of the external logic family. To maintain signal integrity of this reference voltage, it is resistively isolated and *RC* filtered at each receiver.

The output driver presented a more difficult problem due to the 3.3-V V_{DD} chip power supply. To provide a good interface to ECL, it is important that the output driver pull to the V_{DD} rail (for ECL operation $V_{DD} = 0$ V, $V_{SS} = -3.3$ V). This precludes using NMOS pull-ups. PMOS pull-ups have the problem of well-junction forward bias and PMOS turn-on when bidirectional outputs are connected to 5-V logic in CMOS/TTL mode. The solution, as shown in Fig. 14, is a unique floating-well driver circuit that avoids the cost of series PMOS pull-ups in the final stage [5], while providing direct interface to 5-V CMOS/TTL as well as ECL.

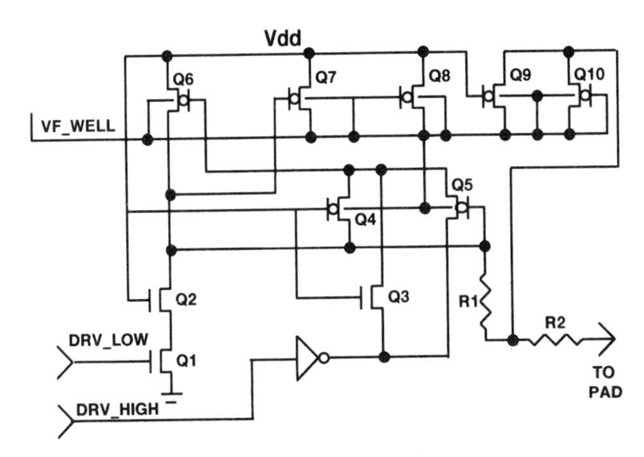

Fig. 14. Floating-well driver.

Transistors $Q1$, $Q2$, and $Q6$ are the actual output devices. $Q1$ and $Q2$ are NMOS devices arranged in cascode fashion to limit the voltages across a single transistor to no more than 4 V. $Q6$ is a PMOS pull-up device that shares a common n-well with $Q7$–$Q10$, which have responsibility for supplying the well with a positive bias voltage of either V_{DD} or the I/O pin potential, whichever is higher. $Q3$–$Q5$ control the source of voltage for the gate of $Q6$—either the output of the inverter or the I/O pad if it moves above V_{DD}. $R1$ and $R2$ provide 50-Ω series termination in either operating mode.

D. Caches

The two internal caches are almost identical in construction. Each stores up to 8 kilobytes of data (DCACHE) or instruction (ICACHE) with a cache block size of 32 bytes. The caches are direct mapped to realize a single cycle access, and can be accessed using untranslated bits of the virtual address since the page size is also 8 kilobytes. For a read, the address stored in the tag and a 64-b quadword of data are accessed from the caches and sent to either the memory management unit for the DCACHE or the instruction unit for the ICACHE. A write-through protocol is used for the DCACHE.

The DCACHE incorporates a pending fill latch that accumulates fill data for a cache block while the DCACHE services other load/store requests. Once the pending fill latch is full, an entire cache block can be written into the cache on the next available cycle. The ICACHE has a similar facility called the stream buffer. On an ICACHE miss, the IBOX fetches the required cache block from memory and loads it into the ICACHE. In addition, the IBOX will prefetch the next sequential cache block and place it in the stream buffer. The data are held in the stream buffer and are written into the ICACHE only if the data are requested by the IBOX.

Each cache is organized into four banks to reduce power consumption and current transients during precharge. Each array is approximately 1024 cells wide by 66 cells tall with the top two rows used as redundant elements. A six-transistor 98-μm^2 static RAM cell is used. The cell utilizes a local interconnect layer that connects between polysilicon and active area, resulting in a 20% reduction in cell area compared to a conventional six-transistor cell. A segmented word line is used to accommodate the banked design, with a global word line implemented in third-level metal and a local word line implemented in first-metal layer. The global word line feeds into local decoders that decode the lower 2 b of the address to generate the local word lines. As shown in Fig. 15, the word lines are enabled while the clock is high and the sense amplifiers are fired on the falling edge of the clock.

VII. Summary

A single-chip microprocessor that implements a new 64-b high-performance architecture has been described. By using a highly optimized design style in conjunction with a high-performance 0.75-μm technology, operating speeds up to 200 MHz have been achieved.

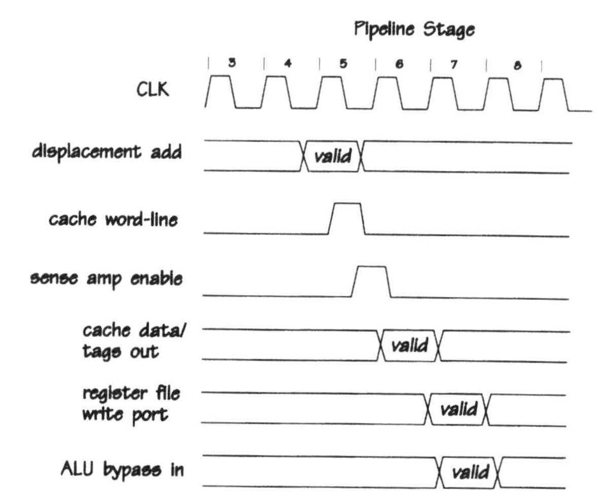

Fig. 15. DCACHE timing diagram.

The chip is superscalar degree 2 and has seven and ten stage pipelines for integer and floating-point instructions. The chip includes primary instruction and data caches, each 8 kilobytes in size. In each 5-ns cycle, the chip can issue two instructions to two of four units yielding a peak execution rate of 400 MIPS and 200 MFLOPS.

The chip is designed with a flexible external interface providing integral support for a secondary cache constructed of standard SRAM's. The interface is fully compatible with virtually any multiprocessor write cache coherence scheme, and can accommodate a wide range of timing parameters. It can interface directly to standard TTL and CMOS as well as ECL technology.

References

[1] *Alpha Architecture Handbook*, EC-H1689-10, Digital Equipment Corp., 1992.

[2] J. Yuan and C. Svensson, "High-speed CMOS circuit techniques," *IEEE J. Solid-State Circuits*, vol. SC-24, no. 1, pp. 62–70, Feb. 1989.

[3] R. Conrad *et al.*, "A 50 MIPS (peak) 32/64-b microprocessor," in *ISSCC Dig. Tech. Papers*, Feb. 1989, pp. 76–77.

[4] J. Sklansky, "Conditional-sum addition logic," *IRE Trans. Electron. Comput.*, vol. EC-9, pp. 226–231, 1960.

[5] H. Lee *et al.*, "An experimental 1 Mb CMOS SRAM with configurable organization and operation," in *ISSCC Dig. Tech. Papers*, Feb. 1988, pp. 180–181.

Clocking Considerations for a Pentium™-based CPU Module with 512K Byte Secondary Cache

Robert M. Reinschmidt and Dale H. Leuthold

MicroModule Systems
10500-A Ridgeview Court,
Cupertino, CA 95014-0736

Abstract

A method of distributing clock to and on a Pentium™-based CPU module was designed and produced. Multiple clock driver outputs are connected in parallel and drive a PC board trace through a series termination resistor to the input of the module. On the module, the clock inputs to 20 dice are connected together and brought to a single PGA pin. The effect of varying termination resistance, PWB foil characteristic impedance, and foil length on the clock waveform characteristics including skew, edge rate, overshoot, and time of flight were determined.

1. Overview of CPU Module

Figure 1 shows a block diagram of a typical microprocessor system using the CPU Module. The Multi-chip Module (MCM) contains 512K Bytes of secondary cache memory and consists of a Pentium™ CPU chip, a cache controller chip, and 18 SRAM chips. The high performance of the system is achieved by isolating the critical address, data, and control signals associated with the second level cache so that they remain in the superior signal environment of the MCM.

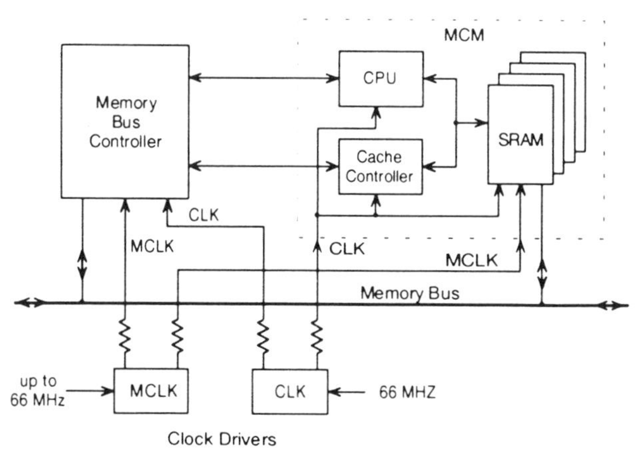

Figure 1. System block diagram.

The critical internal signals, as well as certain external signals to the Memory Bus Controller (MBC) are synchronous with the 66 MHz system clock, CLK. The interface from the MCM to the Memory Bus is clocked by MCLK, which can be up to 66 MHz in frequency, and may or may not be synchronous with CLK.

The most critical clock signal, and the subject of this paper, is the 66 MHz CLK, which is distributed to all 20 chips in the MCM. It is critical because: (1) signal quality requirements must be met at the destination chips, (2) skew between chips in the MCM must be minimized, and (3) absolute delay must be controlled in order to minimize clock skew between the MBC and the MCM.

2. MCM Physical Characteristics

The interconnect substrate for the MCM is fabricated on an aluminum wafer which serves as a ground plane. Layers of polyimide and copper are deposited on the substrate to make one power plane, two interconnect layers, and a surface bonding and die attach layer. The interconnect layers have a minimum routing pitch of 62 μm, but special widths are used to optimize the CLK distribution tree. Electrical parameters for traces on the high density interconnect (HDI) layers are given in Table 1. Note that these have significant resistance which results in "lossy" transmission characteristics [1].

Table 1. HDI Electrical Parameters.

Route Layer	Width μm	R Ω/cm	L nH/cm	C pF/cm	Z_0 Ω
X	23	1.96	2.70	1.70	39.9
Y	19	2.37	2.67	1.64	40.4
X	40	1.13	2.13	2.21	31.1
Y	40	1.13	1.78	2.53	26.5
X	80	0.56	1.40	3.57	19.8
Y	80	0.56	1.11	4.03	16.6

Note: X=Metal 3, Y=Metal 2

Reprinted from *Proc. IEEE Multi-Chip Module Conf.*, pp. 26–31, March 1994.

The HDI substrate is mounted in a 349-pin Pin Grid Array (PGA) package having 0.1 inch pin pitch. Figure 2 shows the layout of the HDI, the PGA package, and the CLK distribution net.

3. CLK Signal Requirements

The critical requirements on the 66 MHz CLK signal come from Reference [2] and are listed below. These requirements apply at the CLK input pads on the receiving chips.

Clock Skew, at 1.5V reference level:
 200 psec (max) Pentium to cache controller
 700 psec (max) Pentium or cache controller to SRAM
Rise time, 0.8 V to 2.0 V level:
 0.15 nsec to 1.5 nsec.
Pulse width (above 2.0 V or below 0.8V):
 4.0 nsec (min)
Overshoot (beyond VCC supply):
 1.6V (max) with clamp diodes absent from simulation.

The CLK driver and distribution scheme described in this paper easily exceeds these requirements. In particular, skew between chips was simulated to be less than 80 psec.

4. MCM Clock Distribution Approach

The important features of the CLK distribution network are the controlled impedance driving source, the PWB trace to the PGA pin, and the CLK distribution tree on the MCM substrate. Each of these was optimized as described below.

Module CLK network CLK starts at the PGA pin and branches in a tree structure to all 20 driven chips, as shown in Fig. 2. The trace is 80 μm wide at the entry point and narrows to 40 μm and eventually 19 μm width after branching. The resulting network has very low skew between the 20 driven chips. However, an unusual feature is that the highly capacitive network presents a low effective characteristic impedance (about 9 Ohms) at the PGA pin.

The ideal way to drive such a network is to match the source impedance and the PWB trace with the MCM input impedance. This would produce very predictable propagation delay and as well as high quality waveform characteristics. However, since a 9 Ohm PWB trace is impractical to route, the use of higher impedance traces was investigated.

An alternative to the single low impedance distribution tree would be to use multiple entry points to drive the different chips. This approach would require separate PWB traces and PGA pins, and was eliminated since it is much more difficult to achieve the low CLK skew requirement between chips in the MCM.

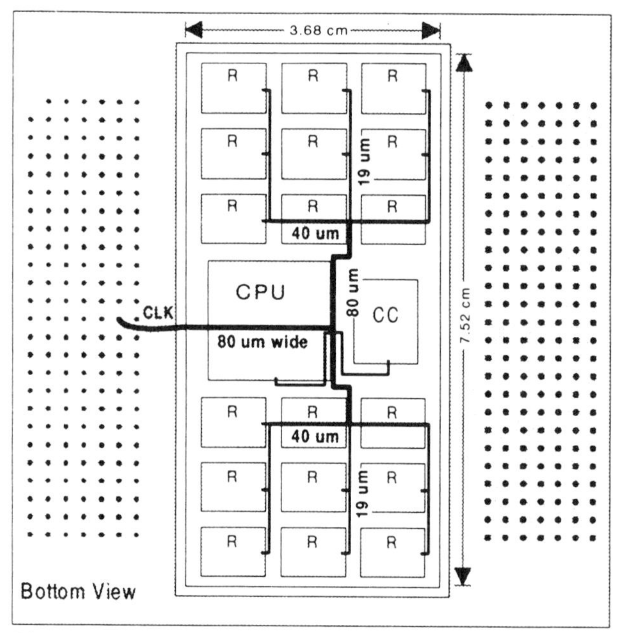

Figure 2. Layout of MCM CLK network

Driver There are a number of Bipolar and CMOS multi-output clock driver chips available. They are subject to large variations in output impedance resulting from semiconductor processing. To reduce the contribution of driver output tolerance, the approach taken was to tie 8 identical driver outputs in parallel. A series discrete resistor was added to bring the total source impedance to the desired value. Simulations were done to examine the effect of driver impedance on performance (see simulation results). The node that ties the drivers together is also a convenient source for the CLK signal to the MBC.

PWB Trace While typical signal routing traces have a characteristic impedance of 50 Ohms, the CLK trace to the MCM pin performs better if the impedance is 25 Ohms or less. The reason is that the load presented by the MCM pin is quite capacitive, and the impedance of the trace limits the charge and discharge currents. With a lower impedance trace, the load can be charged and discharged faster, resulting in a faster edge rate.

A 25 Ohm trace requires a conductor width about 3 times as wide as a 50 Ohm trace. The PWB line is heavily loaded, and not terminated in its characteristic impedance, so SPICE simulations were used to determine propagation delay and waveform characteristics.

Driving the low impedance CLK input of the MCM through a higher impedance PWB trace requires careful analysis. For example, the propagation delay through the PWB trace increases as compared to an ideally terminated line. The simple method of matching routing lengths is inadequate for matching delays to minimize clock skew between the MCM and other chips.

5. CLK Distribution Network Modeling

The entire CLK distribution network was modeled and SPICE simulated over a range of conditions to show the effect of circuit parameters on performance. The models used are described below.

Drivers Drivers were represented by a simplified model consisting of a linear ramp voltage source and a series resistance. An ideal voltage source with a ramp time of 3 ns from 0 to 3.3 Volts was used to represent a TTL driver. Series resistance values from 9.0 to 23.0 Ohms were used. The CLK Period was 15 ns (66.7 MHz), with 50% duty cycle.

PWB trace The PWB trace was broken into 1.0 cm sections, with each section represented by a lumped R-L-C equivalent. The assumed dielectric constant was 4.5, and both 25 Ohm and 50 Ohm characteristic impedance values were modeled. PWB length values were 4, 8, 12, 16, and 20 cm (1.6 inch to 7.9 inch).

CLK network in MCM The lossy foil segments were represented as R-L-C lumped elements, representing lengths of 1.16 cm or less. Lumped equivalent circuits were included to represent the PGA pin, the PGA package trace, the PGA to HDI bond wire, the HDI to chip bond wire, and the chip input capacitance.

6. Time of Flight Analysis

Time of flight analysis partitions the total propagation delay of a signal path into a chip related portion and a load related portion. The chip related portion of delay can be determined by simulating with no load on the output. The load related portion can be determined by simulating the driver and the interconnect foil with loading applied. Time of flight is the added propagation delay caused by the combination of interconnect foil loading the source and propagation delay along the foil. It is defined as the time difference between a reference level on the unloaded source waveform and the same reference level on the input waveform at the destination.

When the driver is represented as an ideal voltage source in series with a resistor, the source reference time is the 1.5 Volt level on the rising edge of the voltage source waveform. In this case, the voltage waveform on the driver output pin (after the resistor) need not be known. Because of transmission line effects, this waveform has discontinuities that make time measurements meaningless.

7. CLK Network Simulation Results

The full CLK network was simulated to determine time of flight, rise time, overshoot, and chip to chip skew. Simulation results were obtained for various conditions of PWB foil length, foil impedance, and driver series resistance. Time of flight was measured from the 1.5 Volt level of the voltage source to the 1.5 Volt level on the destination waveform. Rise time was measured from the 0.8 Volt level to the 2.0 Volt level at the destination. Overshoot is the peak voltage beyond the 3.3 Volt and 0 volt driver levels. A negative value of overshoot means that the signal did not reach full amplitude.

Figure 3 shows the simulation results as a function of PWB foil length for a foil characteristic impedance of 25 Ohms. Figure 4 shows the same results for a 50 Ohm PWB foil. The 25 Ohm case is preferred, because there is a wide range of resistor values that meet the requirements for rise time and still give acceptable overshoot. For the 50 Ohm case, a resistor value that is high enough to control overshoot may cause rise time to exceed 1.5 ns. An optimum source resistor value for the 25 Ohm case appears to be 11 to 13 Ohms.

Figure 5 shows the simulated waveforms for a case of 8 cm of 25 Ohm PWB foil and a driver source resistance of 11 Ohms. Waveforms were evaluated at the second clock cycle. Waveform descriptions, in order of appearance are:

1. Voltage source: 0 to 3.3 Volts, 3.0 ns. rise time (0 to 100%)
2. Driver output, after the series resistor. Note that full amplitude is never reached at this frequency.
3. MCM input pin. Waveform at the far end of the PWB trace. Some discontinuities are apparent.
4. A composite of the CLK pins of the driven chips. The signal is very clean, and skew between chips is less than 80 psec.

Figure 6 shows the oscilloscope waveform measured by probing the CLK pad of an SRAM die. Note that the waveform qualities closely match the simulated waveform in Figure 5. In this case, the driver consisted of a National 100390 TTL driver with 4 outputs tied in parallel, and the series resistance was 8.3 Ohms. Three 8 cm 50 Ohm co-axial cables were connected in parallel to form a 16.7 Ohm transmission line to the MCM pin.

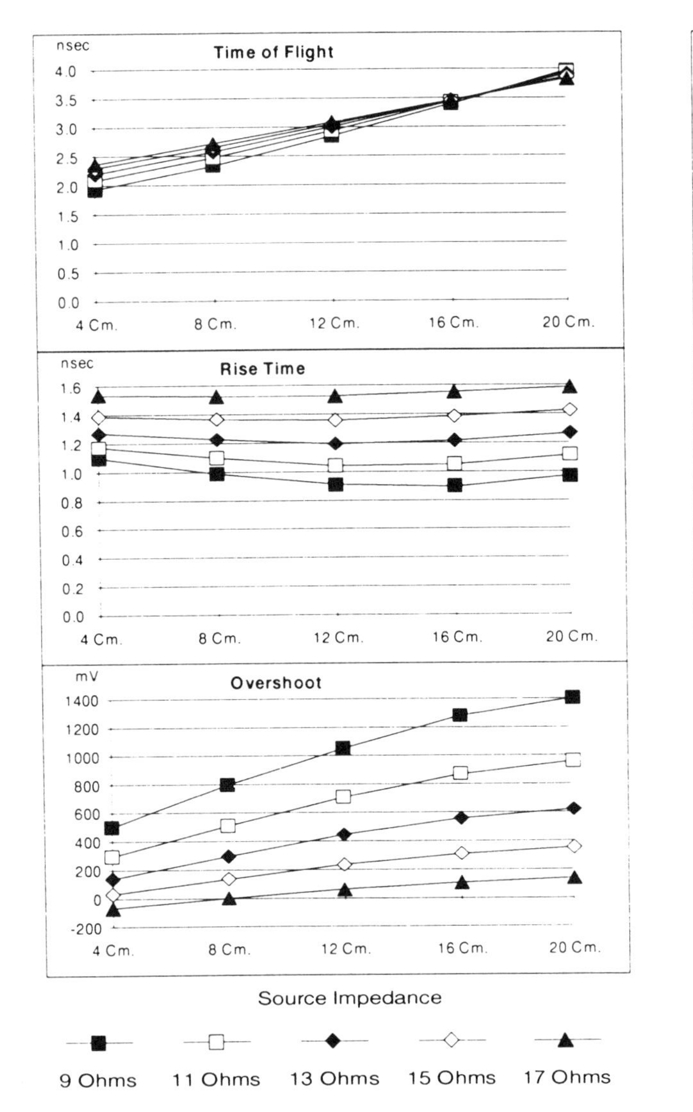

Figure 3. Simulated CLK Characteristics for 25 Ohm PWB Trace

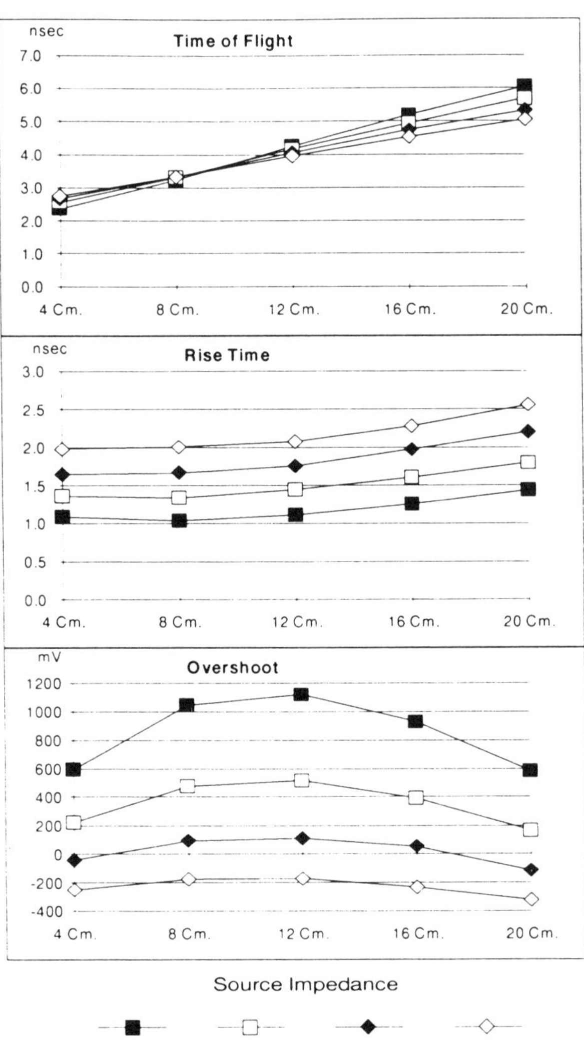

Figure 4. Simulated CLK Characteristics for 50 Ohm PWB Trace

8. External Interface Synchronization

CLK is distributed to the CPU Module and to three chips in the MBC. It is necessary to equalize the propagation delay of these paths in order to minimize clock skew. As noted earlier, SPICE simulation of these paths is necessary, since propagation delay of a heavily loaded line does not follow usual "rules of thumb" of delay vs. length.

Figure 7 shows a CLK distribution network that was tuned for equal delays to the MCM and the ASIC chips in the MBC. A 25 Ohm PWB trace of 9.0 cm length goes to the MCM and three 50 Ohm PWB traces of 15 cm length go to each of the ASIC chips in the MBC.

The clock driver used was a National 74CT2527 CMOS device with 5.0 Volt swing. A SPICE model of this device was included in the simulations. All 8 outputs of the driver were connected in parallel.

For these simulations, a simplified lumped equivalent circuit was created to represent the load that the MCM pin presents to the driver and foil. The simplified model is shown in Figure 7 and consists of a shunt capacitor of 128 pF (equal to the total MCM net capacitance), a series resistance of 1.8 Ohms, and a series inductance of 9.8 nH.

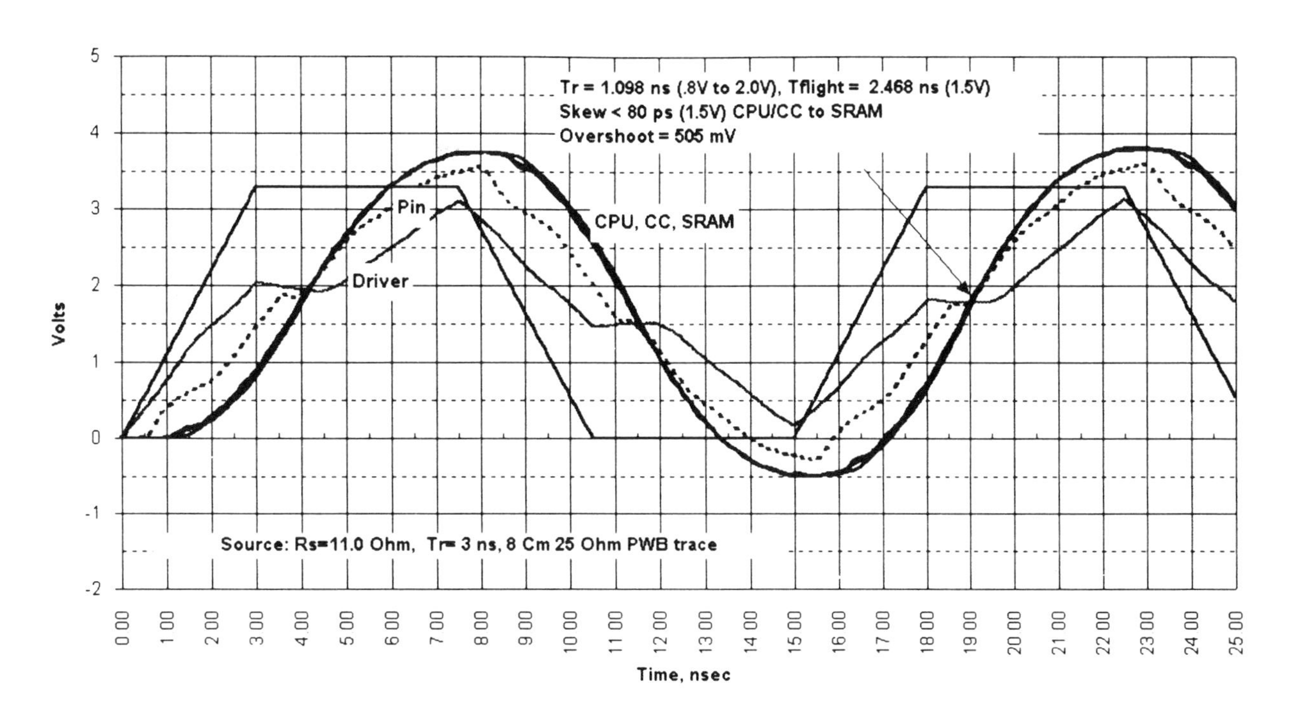

Figure 5. Simulated CLK Waveform

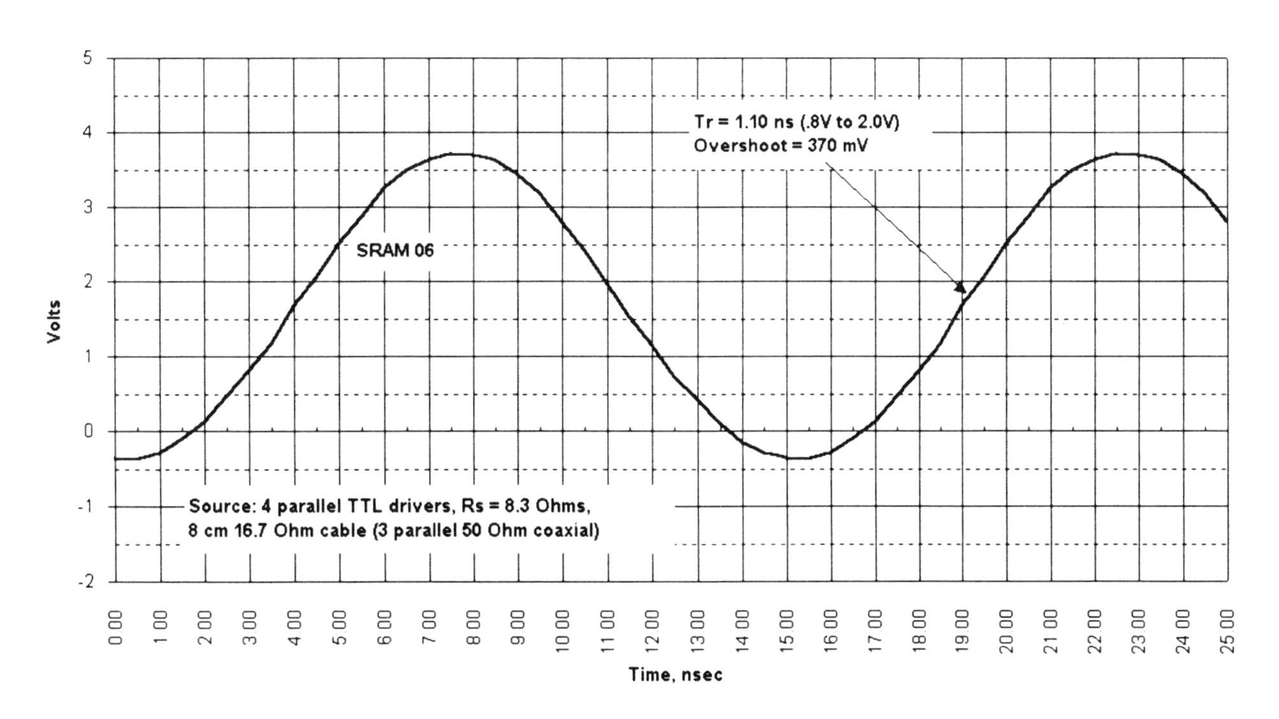

Figure 6. Measured CLK Waveform

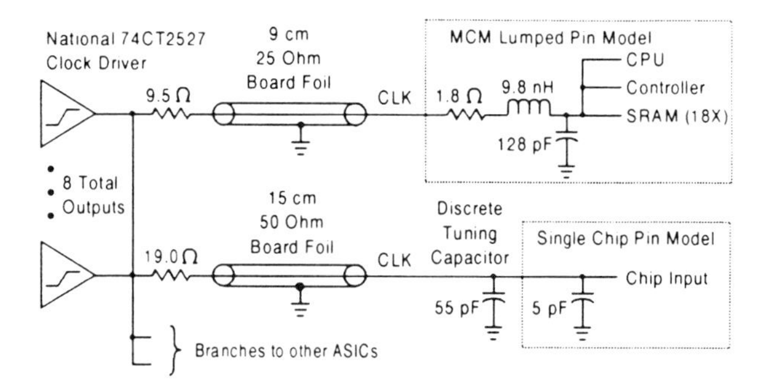

Figure 7. CLK Driver and Distribution

The values of resistance and inductance in the lumped model were empirically arrived at to create a SPICE result that looks the same as the full model with regard to source loading, propagation delay, and signal quality.

Delay tuning was accomplished by adding 6 cm foil length to the MBC paths and by adding 55 pF shunt capacitors at the ASIC pins. With the 2:1 component ratios chosen, the paths to the ASICs have time constants similar to the path to the MCM, resulting in closely matched edge rates and delays on the nets. The simulated waveforms at the MCM destination and the ASIC destinations (Single Chip Loads) are shown in Figure 8.

9. Conclusions

Low clock skew between chips in a MCM can be achieved by connecting the clock input pads in a single centrally driven tree. The resulting low impedance network can be successfully driven using a driver with controlled source termination and a practical width of PWB trace. Matching clock delay between the MCM chips and the other clock destinations on the board requires careful simulation and tuning of the paths involved, since the heavily loaded path to the MCM has a slower propagation delay than an ideal terminated line.

References

1. A. Deutsch and C.W. Ho, "Triplate structure design for thin film lossy unterminated transmission lines," *1981 International Symposium on Circuits and Systems*, Chicago, April 27-29, 1981.
2. Pentium™ Processor User's Manual Volume 2: 82496 Cache Controller and 82491 Cache SRAM Data Book, Intel Corp., 1993.

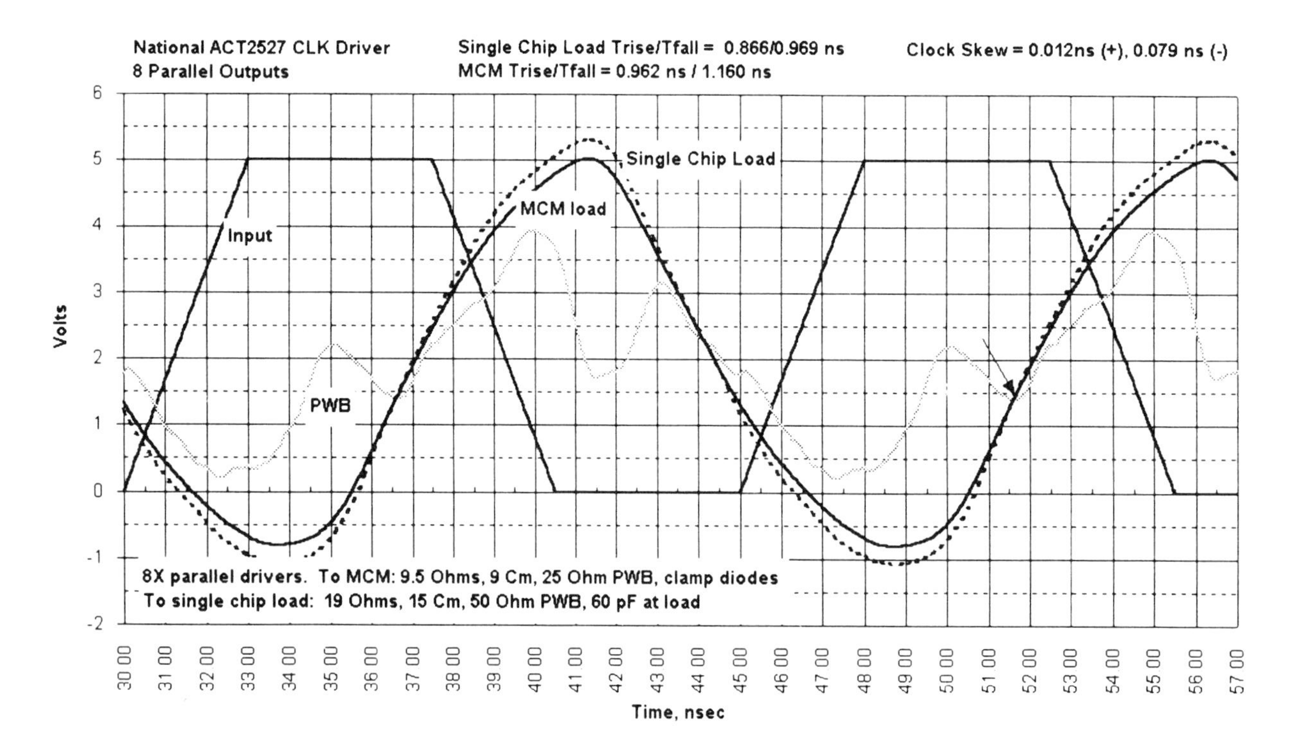

Figure 8. Simulated CLK Waveforms for MCM and Single Chip Load

Author Index

Subject Index

D

E

R

S

T

Editor's Biography

Eby G. Friedman was born in Jersey City, New Jersey, in 1957. He received the BS degree in electrical engineering from Lafayette College, Easton, Pennsylvania, in 1979 and the MS and PhD degrees in electrical engineering from the University of California, Irvine, in 1981 and 1989, respectively.

He was previously employed by Philips Gloeilampen Fabrieken, Eindhoven, The Netherlands, in 1978, where he worked on the design of bipolar differential amplifiers. From 1979 to 1983 he was with Hughes Aircraft Company, Newport Beach, California, working in the areas of custom IC design, software compatible gate array design, one- and two-dimensional device modeling, circuit modeling, and double-level metal process development. From 1983 to 1991 he was with Hughes Aircraft Company, Carlsbad, California, rising to the position of manager of the Signal Processing Design and Test Department, responsible for the design and test of high performance VLSI/VHSIC CMOS and BIMOS digital and analog ICs, the development of supporting design and test methodologies and CAD tools, functional and parametric test, and the development of high performance and high resolution DSP and oversampled systems. Dr. Friedman has been with the Department of Electrical Engineering at the University of Rochester, Rochester, New York, since 1991, where he is an associate professor and director of the High Performance VLSI/IC Design and Analysis Laboratory. His current research and teaching interests are in the areas of high speed and ultra-low power VLSI-based CMOS circuit design and analysis and related systems applications.

He has authored a book chapter and many papers in the fields of CMOS design techniques and CAD tools, pipelining and retiming, and the theory and application of synchronous clock distribution networks to high performance VLSI-based digital systems. Dr. Friedman was a recipient of the Howard Hughes Masters and Doctoral Fellowships, an NSF Research Initiation Award, a Department of Defense Augmentation Award for Science and Engineering Research Training, and a University of Rochester College of Engineering Teaching Excellence Award. He is a senior member of the IEEE, a member of the editorial board of the *IEEE Transactions on Circuits and Systems II: Analog and Digital Signal Processing,* chair of the VLSI Systems and Applications CAS Technical Committee, a member of the technical program committee of several conferences (ISCAS, APAW, GLSVLSI, MWSCAS, ASIC), and an officer of the Electron Devices chapter of the IEEE Rochester Section.